NEUROLOGICAL COMPLICATIONS OF SYSTEMIC CANCER AND ANTINEOPLASTIC THERAPY

NEUROLOGICAL COMPLICATIONS OF SYSTEMIC CANCER AND ANTINEOPLASTIC THERAPY

SECOND EDITION

Edited by

HERBERT B. NEWTON, MD, FAAN

Director, Neuro-Oncology Center, Orlando, FL, United States

Medical Director, CNS Oncology Program, Advent Health Cancer Institute, Advent Health Orlando Campus & Advent Health Medical Group, Orlando, FL, United States

Professor of Neurology, UCF School of Medicine, Orlando, FL, United States

Professor of Neurology & Neurosurgery (Retired), Division of Neuro-Oncology, Esther Dardinger Endowed Chair in Neuro-Oncology, James Cancer Hospital & Solove Research Institute, Wexner Medical Center at the Ohio State University, Columbus, OH, United States

MARK G. MALKIN, MD, FRCPC, FAAN

Director of the Neuro-Oncology Program at Massey Cancer Center, Virginia Commonwealth University School of Medicine, Richmond, VA, United States

Director of the Neuro-Oncology Division, Department of Neurology, Virginia Commonwealth University School of Medicine, Richmond, VA, United States

Professor of Neurology and Neurosurgery, William G. Reynolds, Jr. Chair in Neuro-Oncology, Virginia Commonwealth University School of Medicine, Richmond, VA, United States

ACADEMIC PRESS

An imprint of Elsevier

Academic Press is an imprint of Elsevier
125 London Wall, London EC2Y 5AS, United Kingdom
525 B Street, Suite 1650, San Diego, CA 92101, United States
50 Hampshire Street, 5th Floor, Cambridge, MA 02139, United States
The Boulevard, Langford Lane, Kidlington, Oxford OX5 1GB, United Kingdom

ISBN: 978-0-12-821976-8

For information on all Academic Press publications
visit our website at https://www.elsevier.com/books-and-journals

Publisher: Nikki P. Levy
Acquisitions Editor: Natalie Farra
Editorial Project Manager: Kristi Anderson
Production Project Manager: Swapna Srinivasan
Cover Designer: Mark Rogers

Typeset by STRAIVE, India

Working together
to grow libraries in
developing countries

www.elsevier.com • www.bookaid.org

Dedication

I would like to thank my wife, Cindy, and all of the children in my wonderful mixed family—Alex and Ashley Newton, and Sammi, Skylar, and Cameron Burrell, for their love, patience, and support while I worked on this book project.

I would also like to thank my Neuro-Oncology patients and their families for their amazing courage and strength in the face of adversity, which has been a constant inspiration.

Herbert B. Newton

To my peers and support staff for their confidence in me.

To my patients and their families for their trust in me.

To the memory of my parents, Drs. Dina Gordon Malkin and Aaron Malkin, who instilled in me a love of learning and respect for others—particularly those less fortunate.

To my son Adam whose intellectual curiosity knows no bounds.

And to my wife, Joy, whose unconditional love and support sustains me.

Mark G. Malkin

Contents

I

Metastatic neurological complications

1. Common symptoms at presentation of nervous system metastases

TONI CAO, ANTHONY ROSENBERG, PRIYA KUMTHEKAR, AND KARAN S. DIXIT

2. Neuroimaging of systemic metastatic disease

JOHN VINCENT MURRAY, Jr., RICHARD DOUGLAS BEEGLE, AND SEAN DODSON

3. Nonimaging evaluation of patients with nervous system metastases

KAITLYN MELNICK, VARALAKSHMI BALLUR NARAYANA REDDY, DAVID SHIN, AND ASHLEY GHIASEDDIN

4. Biology and pathophysiology of central nervous system metastases

MOHINI SINGH, ASHISH DAHAL, MAGALI DE SAUVAGE, JULIANA LARSON, AND PRISCILLA K. BRASTIANOS

5. Intracranial metastases

HERBERT B. NEWTON, SEEMA SHROFF, AND MARK G. MALKIN

6. Neurosurgical approaches to the treatment of intracranial metastases

MARK A. DAMANTE, JOSHUA L. WANG, AND J. BRADLEY ELDER

II

Nonmetastatic neurological complications

32. Neurologic complications of immune modulatory therapy

BRIAN M. ANDERSEN AND DAVID A. REARDON

33. Neurological complications of steroids and of supportive care

SHANNON FORTIN ENSIGN AND ALYX B. PORTER

V
Psychiatric, pain, psychosocial, and supportive care issues

34. Psychiatric aspects of care in the cancer patient

WILLIAM S. BREITBART, YESNE ALICI, AND MARK KURZROK

35. Chronic cancer pain syndromes and their treatment

NATHAN CHERNEY, ALAN CARVER, AND HERBERT B. NEWTON

36. Psychosocial issues in cancer patients with neurological complications

ASHLEE R. LOUGHAN, KELCIE WILLIS, AUTUMN LANOYE, DEBORAH ALLEN, MORGAN REID, SCOTT RAVYTS, RACHEL BOUTTE, AND JULIA BRECHBEIL

37. Supportive care

ALICIA M. ZUKAS, MARK G. MALKIN, AND HERBERT B. NEWTON

Contributors

Manmeet S. Ahluwalia   Miami Cancer Institute, Baptist Health South Florida, Miami, FL, United States

Yesne Alici   Department of Psychiatry and Behavioral Sciences, Memorial Sloan-Kettering Cancer Center, New York, NY, United States

Deborah Allen   Nursing Research, Duke University, Durham, NC, United States

Brian M. Andersen   Center for Neuro-Oncology, Department of Medical Oncology, Dana-Farber Cancer Institute, Boston, MA, United States

Joachim M. Baehring   Departments of Neurology and Neurosurgery, Yale University School of Medicine, New Haven, CT, United States

Onyinye Balogun   Department of Radiation Oncology, Weill Cornell Medicine and New York Presbyterian Hospital, New York, NY, United States

Taylor Beal   Southern Methodist University, Dallas, TX, United States

Richard Douglas Beegle   Department of Neuroradiology, AdventHealth Medical Group Central Florida Division, Orlando, FL, United States

Ankush Bhatia   Department of Neurology, The University of Houston Health Science Center at Houston, McGovern Medical School, Houston, TX, United States

Rachel Boutte   Department of Psychology, Virginia Commonwealth University, Richmond, VA, United States

Priscilla K. Brastianos   Divisions of Neuro-Oncology and Medical Oncology, Departments of Medicine and Neurology, Massachusetts General Hospital, Boston, MA, United States

Julia Brechbeil   Department of Psychology, Virginia Commonwealth University, Richmond, VA, United States

William S. Breitbart   Department of Psychiatry, Weill Medical College of Cornell University; Department of Psychiatry and Behavioral Sciences, Memorial Sloan-Kettering Cancer Center; Department of Medicine, Pain and Palliative Care Service, Memorial Sloan-Kettering Cancer Center, New York, NY, United States

Toni Cao   Department of Neurology, Northwestern University, Chicago, IL, United States

Alan Carver   Department of Neuro-Oncology, Memorial Sloan-Kettering Cancer Center, New York, NY, United States

Marc C. Chamberlain   Department of Neurology and Neurological Surgery, University of Washington, Fred Hutchinson Cancer Research Center, Seattle Cancer Care Alliance, Seattle, WA, United States

Samuel T. Chao   Department of Radiation Oncology, Taussig Cancer Center, Cleveland Clinic, Cleveland, OH, United States

Eloise Chapman-Davis   Department of Obstetrics and Gynecology, Division of Gynecologic Oncology, Weill Cornell Medicine and New York Presbyterian Hospital, New York, NY, United States

Zhi-Jian Chen   Neuro-Oncology Program at Massey Cancer Center; Neuro-Oncology Division, Department of Neurology, Virginia Commonwealth University School of Medicine, Richmond, VA, United States

Nathan Cherney   Department of Medical Oncology, Shaare Zedek Medical Center, Jerusalem, Israel

Ashish Dahal   Divisions of Neuro-Oncology and Medical Oncology, Departments of Medicine and Neurology, Massachusetts General Hospital, Boston, MA, United States

Mark A. Damante   Department of Neurological Surgery, The Ohio State University Wexner Medical Center, Columbus, OH, United States

Annick Desjardins   Neurosurgery, The Preston Robert Tisch Brain Tumor Center at Duke, Durham, NC, United States

Karan S. Dixit   Department of Neurology; Lou & Jean Malnati Brain Tumor Institute, Northwestern University, Chicago, IL, United States

Sean Dodson   Department of Neuroradiology, AdventHealth Medical Group Central Florida Division, Orlando, FL, United States

J. Bradley Elder   Department of Neurological Surgery, The Ohio State University Wexner Medical Center, Columbus, OH, United States

Marc S. Ernstoff   Immuno-Oncology Branch, Developmental Therapeutics Program, Division of Cancer Therapy & Diagnosis, National Cancer Institute, Bethesda, MD, United States

Camilo E. Fadul   Department of Neurology, Division of Neuro-Oncology, University of Virginia, Charlottesville, VA, United States

Shannon Fortin Ensign   Department of Hematology and Oncology, Mayo Clinic, Phoenix, AZ, United States

Ashley Ghiaseddin   Department of Neurosurgery, University of Florida, Gainesville, FL, United States

Sarah Goldberg   Section of Medical Oncology, Department of Internal Medicine, Yale University School of Medicine, New Haven, CT, United States

David Gritsch Department of Neurology, Mayo Clinic and Mayo Clinic Cancer Center, Phoenix/Scottsdale, AZ, United States

Craig Horbinski Lou & Jean Malnati Brain Tumor Institute; Department of Neurological Surgery; Department of Pathology, Northwestern University, Chicago, IL, United States

Jana Ivanidze Department of Radiology, Divisions of Neuroradiology and Nuclear Medicine, Weill Cornell Medicine and New York Presbyterian Hospital, New York, NY, United States

Larry Junck Department of Neurology, University of Michigan, Ann Arbor, MI, United States

Jeffrey M. Katz Departments of Neurology and Radiology, Donald and Barbara Zucker School of Medicine at Hofstra/Northwell, Hempstead, NY, United States

Leon D. Kaulen Department of Neurology, Heidelberg University Hospital, Heidelberg, Germany

Moh'd Khushman Department of Hematology-Oncology, The University of Alabama at Birmingham, Birmingham, AL, United States

Cassie Kline Children's Hospital of Philadelphia, Philadelphia, PA, United States

Priya Kumthekar Department of Neurology; Lou & Jean Malnati Brain Tumor Institute, Northwestern University, Chicago, IL, United States

Mark Kurzrok Department of Psychiatry and Behavioral Sciences, Memorial Sloan-Kettering Cancer Center, New York, NY, United States

Autumn Lanoye Department of Health Behavior, Virginia Commonwealth University, Richmond, VA, United States

Juliana Larson Divisions of Neuro-Oncology and Medical Oncology, Departments of Medicine and Neurology, Massachusetts General Hospital, Boston, MA, United States

Eudocia Q. Lee Center for Neuro-Oncology, Department of Medical Oncology, Dana-Farber Cancer Institute; Division of Cancer Neurology; Department of Neurology, Brigham and Women's Hospital; Harvard Medical School, Boston, MA, United States

Denise Leung Department of Neurology, University of Michigan, Ann Arbor, MI, United States

Angela Liou Children's Hospital of Philadelphia, Philadelphia, PA, United States

Simon S. Lo Department of Radiation Oncology, University of Washington School of Medicine, Seattle, WA, United States

Ashlee R. Loughan Department of Neurology, Virginia Commonwealth University, Richmond, VA, United States

Benjamin Lu Section of Medical Oncology, Department of Internal Medicine, Yale University School of Medicine, New Haven, CT, United States

Rimas V. Lukas Department of Neurology; Lou & Jean Malnati Brain Tumor Institute, Northwestern University, Chicago, IL, United States

Mark G. Malkin Neuro-Oncology Program at Massey Cancer Center; Neuro-Oncology Division, Department of Neurology; Neurology and Neurosurgery, William G. Reynolds, Jr. Chair in Neuro-Oncology, Virginia Commonwealth University School of Medicine, Richmond, VA, United States

Jacob Mandel Department of Neurology, Baylor College of Medicine, Houston, TX, United States

Kaitlyn Melnick Department of Neurosurgery, University of Florida, Gainesville, FL, United States

Jennifer Moliterno Department of Neurosurgery, Yale School of Medicine, New Haven, CT, United States

Maciej M. Mrugala Department of Neurology, Mayo Clinic and Mayo Clinic Cancer Center, Phoenix/Scottsdale, AZ, United States

Sabine Mueller University of California, San Francisco, San Francisco, CA, United States; University Children's Hospital Zurich, Zurich, Switzerland

Erin S. Murphy Department of Radiation Oncology, Taussig Cancer Center, Cleveland Clinic, Cleveland, OH, United States

John Vincent Murray, Jr. Neuroradiology, Radiology, Mayo Clinic, Jacksonville, FL, United States

Herbert B. Newton Neuro-Oncology Center; CNS Oncology Program, Advent Health Cancer Institute, Advent Health Orlando Campus & Advent Health Medical Group; Neurology, UCF School of Medicine, Orlando, FL; Neurology & Neurosurgery (Retired), Division of Neuro-Oncology, Esther Dardinger Endowed Chair in Neuro-Oncology, James Cancer Hospital & Solove Research Institute, Wexner Medical Center at the Ohio State University, Columbus, OH, United States

Evan K. Noch Department of Neurology, Division of Neuro-Oncology, Weill Cornell Medicine and New York Presbyterian Hospital, New York, NY, United States

Barbara J. O'Brien Department of Neuro-Oncology, University of Texas MD Anderson Cancer Center, Houston, TX, United States

Patrick O'Shea Case Western Reserve Medical School, Cleveland, OH, United States

Eseosa Odigie Weill Cornell Medicine, New York, NY, United States

Alexander C. Ou Department of Neuro-Oncology, University of Texas MD Anderson Cancer Center, Houston, TX, United States

Nina A. Paleologos Department of Neurology, Advocate Medical Group, Advocate Healthcare, Rush University Medical School, Chicago, IL, United States

Susan C. Pannullo Department of Neurological Surgery, Weill Cornell Medicine and New York Presbyterian Hospital, New York; Department of Biomedical Engineering, Cornell University, Ithaca, NY, United States

Kester A. Phillips Department of Neurology, The Ben and Catherine Ivy Center for Advanced Brain Tumor Treatment at Swedish Neuroscience Institute, Seattle, WA, United States

Alberto Picca Service de Neurologie 2—Mazarin, Neurology Department, Pitié-Salpêtrière Hospital, APHP; OncoNeuroTox Group, Center for Patients with Neurological Complications of Oncologic Treatments, Pitié-Salpetrière Hospital, Paris, France

Alyx B. Porter Department of Neurology, Mayo Clinic, Phoenix, AZ, United States

Amy A. Pruitt Department of Neurology, University of Pennsylvania School of Medicine, Philadelphia, PA, United States

Dimitri Psimaras Service de Neurologie 2—Mazarin, Neurology Department, Pitié-Salpêtrière Hospital, APHP; OncoNeuroTox Group, Center for Patients with Neurological Complications of Oncologic Treatments, Pitié-Salpetrière Hospital, Paris, France

Yasmeen Rauf University of North Carolina, Chapel Hill, NC, United States

Scott Ravyts Department of Psychology, Virginia Commonwealth University, Richmond, VA, United States

David A. Reardon Center for Neuro-Oncology, Department of Medical Oncology, Dana-Farber Cancer Institute, Boston, MA, United States

Varalakshmi Ballur Narayana Reddy Department of Neurology, University of Florida, Gainesville, FL, United States

Morgan Reid Department of Psychology, Virginia Commonwealth University, Richmond, VA, United States

Maricruz Rivera Department of Neurological Surgery, Weill Cornell Medicine and New York Presbyterian Hospital, New York, NY, United States

Anthony Rosenberg Department of Neurology, Northwestern University, Chicago, IL, United States

Amber Nicole Ruiz Departments of Neurology, Neurological Surgery and Medicine, University of Washington, Seattle Cancer Care Alliance, Seattle, WA, United States

Magali de Sauvage Divisions of Neuro-Oncology and Medical Oncology, Departments of Medicine and Neurology, Massachusetts General Hospital, Boston, MA, United States

Shreya Saxena Miami Cancer Institute, Baptist Health South Florida, Miami, FL, United States

David Schiff Division of Neuro-Oncology, University of Virginia Health System, Charlottesville, VA, United States

David Shin Department of Neurosurgery, University of Florida, Gainesville, FL, United States

Seema Shroff Department of Pathology, Advent Health, Orlando, FL, United States

Karanvir Singh Miami Cancer Institute, Baptist Health South Florida, Miami, FL, United States

Mohini Singh Divisions of Neuro-Oncology and Medical Oncology, Departments of Medicine and

Neurology, Massachusetts General Hospital, Boston, MA, United States

Prathusan Subramaniam Office of Clinical Research, Feinstein Institutes for Medical Research, Manhasset, NY, United States

John H. Suh Department of Radiation Oncology, Taussig Cancer Center, Cleveland Clinic, Cleveland, OH, United States

Ashley L. Sumrall Department of Oncology, Levine Cancer Institute, Atrium Health, Charlotte, NC, United States

Lynne P. Taylor Departments of Neurology, Neurological Surgery and Medicine, University of Washington, Seattle Cancer Care Alliance, Seattle, WA, United States

Jigisha P. Thakkar Department of Neurology, Division of Neuro-Oncology, Loyola University Chicago, Stritch School of Medicine, Maywood, IL, United States

Joshua L. Wang Department of Neurological Surgery, The Ohio State University Wexner Medical Center, Columbus, OH, United States

Patrick Y. Wen Center for Neuro-Oncology, Department of Medical Oncology, Dana-Farber Cancer Institute; Division of Cancer Neurology, Brigham and Women's Hospital; Harvard Medical School, Boston, MA, United States

Timothy G. White Department of Neurological Surgery, North Shore University Hospital, Manhasset, NY, United States

Kelcie Willis Department of Psychology, Virginia Commonwealth University, Richmond, VA, United States

Jean-Paul Wolinsky Lou & Jean Malnati Brain Tumor Institute; Department of Neurology Surgery, Northwestern University, Chicago, IL, United States

Kailin Yang Department of Radiation Oncology, Taussig Cancer Center, Cleveland Clinic, Cleveland, OH, United States

Lalanthica V. Yogendran Department of Neurology, Division of Neuro-Oncology, University of Virginia, Charlottesville, VA, United States

Gilbert Youssef Center for Neuro-Oncology, Department of Medical Oncology, Dana-Farber Cancer Institute; Division of Cancer Neurology, Brigham and Women's Hospital; Harvard Medical School, Boston, MA, United States

Michael N. Youssef Department of Neurology, UT Southwestern Medical Center, Dallas, TX, United States

Zhen Ni Zhou Department of Obstetrics and Gynecology, Division of Gynecologic Oncology, Weill Cornell Medicine and New York Presbyterian Hospital, New York, NY, United States

Alicia M. Zukas Department of Neurosurgery, Division of Neuro-Oncology, Hollings Cancer Center, Medical University of South Carolina, Charleston, SC, United States

About the editors

Herbert B. Newton

Dr. Newton graduated from SUNY Buffalo with High Honors in psychology, before pursuing a master's degree in cancer chemotherapy and pharmacology from the Roswell Park Cancer Institute. He then entered the SUNY Buffalo School of Medicine, graduating in 1984. After an internship in internal medicine within the SUNY Buffalo Health System, he was trained in neurology at the University of Michigan Medical Center in Ann Arbor, MI, followed by a fellowship in neuro-oncology at the Memorial Sloan-Kettering Cancer Center (MSKCC). After the completion of his fellowship at MSKCC, Dr. Newton joined the faculty in neurology at the Ohio State University Medical Center and James Cancer Hospital, establishing a division of neuro-oncology within the Department of Neurology. Dr. Newton achieved the rank of Full Professor of Neurology with tenure and was bestowed the Esther Dardinger Endowed Chair in Neuro-Oncology in 2003. His interests include clinical neuro-oncology, chemotherapy, molecular biology, and clinical trials. After retiring from OSUMC in 2015, he has since become the Director of the Neuro-Oncology Center for the Advent Health Cancer Institute and Advent Health Orlando campus. He has published more than 220 scientific articles and book chapters. Dr. Newton is the editor or coeditor of 10 books in the fields of neurology and neuro-oncology.

Mark G. Malkin

Dr. Malkin is currently the Director of the Neuro-Oncology Program at Massey Cancer Center, and Director of the Neuro-Oncology Division, in the Department of Neurology at Virginia Commonwealth University School of Medicine. He holds the William G. Reynolds, Jr. Chair in Neuro-Oncology, and is Professor of Neurology. Before his appointment at VCU, he was the Director of Neuro-Oncology and Professor of Neurology at Froedtert and the Medical College of Wisconsin. Prior to his move to Milwaukee, he was an Associate Clinical Member at Memorial Sloan-Kettering Cancer Center and Associate Clinical Professor of Neurology at Weill Cornell Medical College. He also trained in neuro-oncology with Dr. Jerome Posner and Dr. William Shapiro at Memorial Sloan-Kettering Cancer Center, and since then has been in academic neuro-oncology for more than 33 years. He has published more than 100 peer-reviewed articles and book chapters and is Coeditor of several textbooks in the field of neuro-oncology.

Preface

Involvement of the nervous system is one of the most dreaded complications of systemic cancer, striking fear into the hearts of patients and their loved ones and often causing considerable trepidation in their physicians. This issue is still prevalent despite numerous advances in cancer therapy over the past 60 years. Neurologists, medical oncologists, radiation oncologists, neurosurgeons, and neuro-oncologists in practice need to be aware of the broad spectrum of neurological complications of cancer that can occur in their patients and be well versed in the newest approaches to diagnosis and treatment. This is critical since rapid diagnosis and institution of appropriate therapy may minimize neurological injury and preserve quality of life.

The field of neuro-oncology has made considerable strides since it was nurtured by the likes of Drs. Jerome Posner, Harry Greenberg, Victor Levin, and Nicholas Vick many years ago. In the 10 years since the first edition of this book, dramatic advances have been made in the molecular biology of cancer, molecular targeted therapies, immunotherapy and immune modulation treatment approaches, and neuro-imaging of the nervous system—with advanced MRI techniques that permit much earlier detection of metastatic involvement and nervous system injury. Radiation therapy protocols have also become much more varied, with the advent of stereotactic radiosurgery, IMRT, and TomoTherapy and the ability to execute conformal treatment plans. Neurosurgical approaches to neuro-oncological complications, such as image-guided surgical resections and intraoperative MRI, are now more widely accepted and considered important for maximizing survival and avoiding neurological deficits. Progress in molecular biology has improved our understanding of the neoplastic genotype and phenotype and has ushered in a new era of individualized oncology treatment—molecular therapeutics—with its own unique spectrum of toxicity and effects on the nervous system.

The goal of this second edition is to provide an up-to-date, comprehensive, single-volume reference work on the neurological complications of systemic cancer and its treatment. This book should be helpful to medical students, residents, and fellows as well as to more seasoned attending clinicians. The book is divided into five sections that cover a wide range of topics, including metastatic neurological complications, nonmetastatic neurological complications, complications related to specific neoplasms, complications of antineoplastic therapy, and supportive care issues. Each chapter is written by one or more experts in the field of neuro-oncology or a similar field of study and offers in-depth information on the oncological background, differential diagnosis, and diagnostic evaluation for each topic as well as more practical management strategies for patient care. We hope that the book will improve the ability of clinicians caring for cancer patients to expeditiously and accurately recognize and treat neuro-oncological complications, thereby optimizing neurological functions and quality of life.

Herbert B. Newton
Orlando, FL, United States

Mark G. Malkin
Richmond, VA, United States

Metastatic neurological complications

1

Common symptoms at presentation of nervous system metastases

Toni Cao, Anthony Rosenberg, Priya Kumthekar, and Karan S. Dixit

Department of Neurology, Northwestern University, Chicago, IL, United States

1 Introduction

Metastatic cancers can affect both the central nervous system (CNS) and peripheral nervous system (PNS). Nervous system metastases are being diagnosed more often due to a combination of more effective therapeutics leading to improved overall survival, as well as better availability and more frequent use of advanced imaging. Nervous system metastases are often a feature of late-stage disease, though they can occur at the initial presentation of cancer, with the majority of patients having neurologic signs or symptoms. However, nearly one-third of patients may be neurologically intact and asymptomatic.[1] Diagnosing nervous system metastases is important as they are associated with increased morbidity and mortality. The presence of any new neurologic symptom in a patient with existing cancer should raise suspicion for nervous system involvement and prompt further evaluation. Here, in this chapter, we will discuss the most common presenting symptoms of central and peripheral nervous system metastases.

2 Brain metastases

Brain metastases are the most common neurologic complication of cancer and are also the most common intracranial tumor in adults. They occur in nearly 10%–30% of cancer patients with differing incidences based on specific primary histology.[2] Patients with brain metastases can present with a diverse range of symptoms, similar to any intracranial space-occupying lesion, including headaches, seizures, altered mentation and neurocognitive decline, as well as focal motor and somatosensory deficits.

2.1 Headache

Headache occurs in approximately 50% of patients with brain tumors.[3] Patients at increased risk of headache include those who are younger and those with a prior history of headaches.[4] In patients with brain tumors, headaches rarely occur in isolation. In a study by Vazquez-Barquero et al., less than 10% of brain tumor patients presented with headache as an isolated symptom with most patients subsequently developing other localizable symptoms to the central nervous system within 3 months.[5]

The classic "brain tumor headache" is described as severe, occurring in the early morning or at night during sleep, and associated with nausea and vomiting. Oftentimes, they are positional and worsened with any maneuvers that increase intrathoracic pressure, such as coughing, sneezing, or the Valsalva maneuver. However, in practice, the stereotypical brain tumor headache is neither sensitive nor specific for the diagnosis of brain metastases. In a study by Forsyth and Posner, 77% of patients experienced tension-type headaches, 9% had headaches similar to migraines, and the remainder had other types of headaches. Nausea and vomiting are associated with 40% of brain tumor headaches.[3]

There may be a volume-dependent nature of brain tumor headaches with increasing risk with each cubic centimeter of tumor.[6] Headaches occur more often with rapidly growing tumors and are also more common with multiple metastases. Up to 90% of patients with intraventricular and midline tumors experience headaches, likely from elevated intracranial pressure. Infratentorial tumors are more likely to cause headaches, in up to 84% of patients, compared to supratentorial tumors (55%–60%). Supratentorial tumor location does not correlate well with headache location as patients may

TABLE 1   Headache red flag symptoms.

Red flags	Examples
Systemic symptoms	Fevers, chills, myalgias, night sweats, rash, unintentional weight loss, immunocompromised state (HIV), malignancy, pregnancy or postpartum state
Neurologic symptoms or signs	Altered mental status or level of consciousness, cranial neuropathies, pulsatile tinnitus, weakness, sensory changes, ataxia, seizures, loss of consciousness, papilledema
Sudden onset	Abrupt onset, severe, "worst headache of life," thunderclap headache
Older onset	After 50 years of age
Change in pattern	Change in frequency, intensity, or quality, positional component, aggravated by Valsalva maneuver

experience ipsilateral, contralateral, or bilateral headaches.[4,7] Of note, only 15%–25% of patients have signs of papilledema.[7]

Due to the variable characteristics of the brain tumor headaches, physicians should be aware of "red flag" headache symptoms, which should prompt further investigation with brain imaging (see Table 1).[8]

2.2 Seizures

Seizures as the primary presenting symptom of malignancy are more common with primary brain tumors; however, patients with brain metastases are at an increased risk for seizures. Older reports estimate a 20%–35% incidence of seizures in patients with brain metastases, though more contemporary reviews performed in the MRI era estimate a lower incidence of 15%.[9–11] Seizures from brain metastases often occur at presentation, estimated at 80%, while the rest of patients develop them during the course of their disease.[11] The incidence of seizures from brain metastases was highest in patients with melanoma (67%) and lung cancer (29%), compared to GI malignancies (21%) and breast cancer (16%).[9]

Tumor location is also associated with risk of seizure. Metastases in the temporal lobe and adjacent to the motor cortex are associated with a higher risk of seizure, compared to occipital, brainstem, or cerebellar tumors.[12,13] Intractable epilepsy can result from lesions that affect the mesiotemporal and insular regions.[14] The risk of seizure also increases with an increased number of metastatic lesions and with leptomeningeal involvement.[11] Despite the higher risk of seizures in brain metastases, recently published guidelines by the Congress of Neurologic Surgeons advise against the use of prophylactic antiepileptic drugs in routine cases of brain metastases, as well as postcraniotomy.[15]

2.3 Alerted mental status/neurocognitive symptoms

Up to 20%–25% of patients with brain metastases experience mental status changes. The most common symptoms include lethargy, irritability, memory loss, and personality changes.[16] Patients may also have a change in their mood and affect with depression and apathy. Neuropsychological testing may reveal deficits across multiple domains in 65% of patients with intracranial metastases.[17] Altered mental status, or encephalopathy, is more commonly seen with multiple metastases and/or increased intracranial pressure.[7] However, it should be noted that encephalopathy and cognitive dysfunction in patients with brain metastases are more often attributed to toxic-metabolic encephalopathy (61%) rather than structural abnormalities from intracranial metastases (15%).[18]

2.4 Focal neurologic deficits

Focal neurologic deficits may also occur in patients with metastatic disease. Symptoms are typically gradual in onset and vary by neuroanatomical location. Supratentorial lesions can cause motor or sensory deficits, visual field deficits, and/or language deficits, where metastases to the infratentorium can cause cranial nerve deficits and cerebellar dysfunction. Motor deficits, which can vary by anatomical location, include hemiparesis or hemiplegia, gait impairment, ataxia, and/or incoordination. More than half of patients with brain metastases experience hemiparesis caused by a contralateral hemispheric lesion.[16] Symptoms and anatomic localization are listed in Table 2.

3 Leptomeningeal disease

Leptomeningeal metastasis (LM), also known as carcinomatous meningitis or leptomeningeal carcinomatosis, has an incidence ranging from 5% to 8% in patients with solid tumors and up to 5%–15% in patients with

TABLE 2   Deficits and localization.

Focal neurologic deficit	Localization
Hemiparesis	Primary, premotor, or supplementary motor cortex, thalamus, internal capsule, brainstem, corticospinal tract
Incoordination	Cerebellum, basal ganglia, brainstem
Ataxia	Cerebellum, thalamus, parietal lobe
Gait instability	Primary, premotor, or supplementary motor cortex, thalamus, internal capsule, cerebellum, basal ganglia, brainstem

hematologic malignancies.[19] Unfortunately, LM typically carries a poor prognosis with an average survival of 2–4 months despite treatment. The most common solid tumors associated with LM include breast, lung, and melanoma.[20] Recognition of signs and symptoms of LM is critical for early diagnosis to potentially prevent more rapid neurologic decline.[21]

Symptoms of LM are broad and can be localized to any part of the nervous system since cerebrospinal fluid encases the brain and spinal cord. The broad array of clinical findings can be explained by cranial and spinal nerve dysfunction, elevated intracranial pressure, meningeal irritation, and diffuse cerebellar dysfunction.[20] Symptoms may include headache (39%), nausea and vomiting (25%), weakness (21%), ataxia (17%), encephalopathy (16%), diplopia (14%), and facial palsy (13%),[22] which tend to develop over days to weeks.

Cranial neuropathies, especially multiple cranial neuropathies, are a defining feature of leptomeningeal metastases. The most common cranial nerves affected are the abducens (CN VI), facial (CN VII), and vestibulocochlear (CN VIII) causing diplopia, upper and lower facial weakness, and hearing loss, respectively. Diplopia is the most common symptom of LM-associated cranial neuropathy due to multiple nerves responsible for ocular motility (oculomotor, trochlear, and abducens).[20] Trigeminal neuralgia, presenting as facial pain/paresthesias, has also been reported in about one in five patients and can predict disease progression.[23,24] Hearing loss is less common at first diagnosis of LM (< 5%), but can ultimately develop with disease progression. Lower cranial nerve dysfunction can also cause dysphagia, dysarthria, and dysphonia. Spinal nerve roots can also be affected by LM, leading to radiculopathy or cauda equina syndrome (to be discussed separately).

Diffuse cerebellar dysfunction (gait instability, dizziness, falls) occurs in nearly 20% of patients at diagnosis, with examination notable for either truncal ataxia or limb dysmetria.[22] Additionally, patients can develop elevated intracranial pressure and present with headache, nausea/vomiting, blurred vision, and/or encephalopathy.

Multifocal, concurrent, or progressive neurologic symptoms (encephalopathy, focal motor deficits, seizures, cranial neuropathies, cerebellar dysfunction) should be worked up promptly with MRI of the entire neuraxis and cerebrospinal fluid evaluation for malignant cells.[25]

4 Spinal cord metastases

Metastatic lesions to the spinal cord can result in intradural intramedullary, intradural extramedullary, or spinal epidural lesions. The vast majority (94%–98%) of cancer patients with spinal metastases develop either vertebral or epidural disease, whereas 5%–6% and 0.9%–2.1% of patients, respectively, have intradural extramedullary and intradural intramedullary involvement.[26,27] About half of all cases of intramedullary metastases arise from primary lung carcinoma, specifically small cell carcinoma, followed by breast cancer, melanoma, lymphoma, and renal cell carcinoma.[27]

Intramedullary spinal cord metastases are often solitary lesions, which preferentially affect the cervical and thoracic cord. Back pain is a common presenting sign, which is oftentimes located in the mid-back and may progress to radicular pain. Myelopathic symptoms are also common with patients presenting with asymmetric weakness and can be associated with spasticity and rapidly progress to paraplegia. Sensory disturbances, including paresthesias, burning, and tingling, can also be seen and are associated with a sensory level. Motor and sensory findings can follow a pattern similar to a Brown-Séquard's syndrome (ipsilateral paralysis and loss of position with contralateral loss of pain and temperature) and can progress to a complete transection syndrome. Autonomic dysfunction and bowel and bladder incontinence can occur later. The majority of patients with intramedullary spinal cord metastases experience a relatively rapid progression of symptoms within a month, in contrast to primary intramedullary tumors, such as gliomas, that oftentimes have a slower time course of symptoms.[28]

5 Epidural spinal cord metastases/epidural spinal cord compression

Metastatic epidural spinal cord compression is a medical emergency and is defined as symptomatic spinal cord or nerve root compression and mechanical instability of the spinal column.[29] This occurs in up to 10% of cancer patients and may be the first sign of cancer in up to 20% of cases.[30] Breast, lung, and prostate cancers are the most frequent solid tumors associated with epidural spinal cord compression, but they can occur with any malignancy.[31]

The most common presenting symptom is progressive back pain, occurring in up to 83%–95% of patients and often precedes other myelopathic signs and symptoms including motor and bladder dysfunction.[32] Frequently, metastatic epidural spinal cord compression can be confused with degenerative disc disease as the pain is worsened with movement, Valsalva maneuvers, and neck flexion. Pain most commonly arises in the thoracic spine (60%–70%), followed by lumbosacral (20%–30%) and cervical regions (10%).[33] Pain only on movement suggests spinal cord instability. Unstable cervical metastases cause neck and/or interscapular pain with flexion/extension/rotation of the neck, whereas unstable thoracic metastases typically cause pain while lying down. Lumbar mechanical instability is generally secondary to mechanical radiculopathy, a severe radicular pain elicited by axial loading while ambulating or standing.[34]

Weakness is the second most common symptom of epidural spinal cord compression, seen in 60%–80% of patients.[30] Patients may describe being increasingly clumsy; however, this may actually be confirmed as weakness on physical examination. Weakness typically follows an upper motor neuron pattern, with symmetric upper extremity extensor and/or lower extremity flexor weakness, but can also present with a lower motor neuron pattern and asymmetric (distal greater than proximal) weakness. Due to (often severe) weakness, 50%–68% of patients are nonambulatory at the time of their diagnosis,[29,35] while 50%–70% of patients have sensory deficits. Both motor and sensory deficits typically begin distally, moving proximally with disease progression. Patients can present with gait ataxia due to the involvement of the spinocerebellar tracts, and late findings can include bowel/bladder incontinence due to autonomic dysfunction.[29,35,36]

Patients with symptoms concerning for metastatic epidural spinal cord compression should undergo urgent imaging, preferably with MRI (w/wo contrast), of the neuraxis given the possibility of multiple spinal level involvement. If MRI is contraindicated, CT myelogram or CT spine should be performed.[33]

6 Cauda equina syndrome

Cauda equina syndrome is a medical emergency oftentimes caused by large, space-occupying lesions within the lumbosacral spinal canal.[37,38] Patient symptoms, caused by compression of the spinal nerves beginning at L1, include low back pain, sciatica, lower extremity sensorimotor loss, and bowel/bladder dysfunction. Urinary retention is a defining feature of a complete cauda equina syndrome, while patients without it may be considered to have an incomplete cauda equina syndrome.[39] The five components of complete cauda equina syndrome are urinary retention, saddle anesthesia, bilateral lower extremity pain, numbness, and weakness. Decreased rectal tone has been found as a late finding, but is not classically part of the syndrome.[38] Cauda equina syndrome can also occur in the setting of leptomeningeal metastases.[22]

7 Neoplastic and radiation plexopathy

Cancer may also affect the cervical, brachial, or lumbosacral plexus. Plexus involvement is usually due to either direct invasion from adjacent tumor, or indirect invasion after metastases to neighboring soft tissue, bone, or lymph nodes.[40,41]

Brachial plexopathies are commonly associated with breast and lung cancer, as well as lymphoma, and head and neck cancers.[42] Lumbosacral plexopathy is commonly seen in patients with colorectal cancer, sarcoma, gynecologic cancers, and lymphoma.[43] Sacral plexopathy is mostly associated with prostate, anorectal, and gynecological cancers.[44] Limb pain is a defining feature of neoplastic plexopathy, which often starts as intermittent and becomes constant with a radicular character. Focal weakness and sensory loss affecting multiple nerve distributions, decreased deep tendon reflexes, and limb edema are also common symptoms.[45] Lumbosacral plexopathy symptoms also include perineal and buttock pain, as well as bowel, bladder, and sexual dysfunction.[46]

Radiation therapy can also cause delayed plexopathies with a reported incidence of approximately 2%–5% and occur at a median of 1.5 years (range of 3 months to 14 years) after radiotherapy.[47,48] There is a relationship between dose per fraction and the risk of developing radiation plexopathy with an estimated incidence of 66% in breast cancer patients who received 60 Gy in 5 Gy fractions compared to less than 1% in those who received 50 Gy in 2 Gy fractions.[49]

Differentiating neoplastic from radiation plexopathy is important as it may impact cancer therapy, though can be challenging given significant clinical overlap. Compared to neoplastic plexopathies, radiation-induced plexopathies are more likely to be painless, though can develop pain later in the course. Radiation plexopathy is usually a slowly progressive process, which is more likely to involve the entire plexus rather than a specific trunk or division. The presence of a Horner's syndrome should alert the clinician for a neoplastic rather than radiation plexopathy.[45] Swelling, caused by lymphatic obstruction, is more commonly seen in radiation-induced plexopathy.[43]

In patients with suspected plexopathy, MRI of the affected area should be obtained given its superior ability to evaluate soft tissue. Nonuniform, focal, and nodular enhancement is more suggestive of neoplastic involvement, while diffuse, uniform edema, and mild enhancement corresponding to known radiation field suggests radiation plexopathy.[50,51] In cases where distinguishing between the two entities is difficult on MRI, PET/CT is another modality to consider where hypermetabolic activity is more consistent with a neoplastic etiology.[52,53] Finally, electromyography is another study, which can help differentiate between the two as up to 40%–60% of radiation plexopathies can have myokymia, while it is rarely seen in neoplastic involvement.[54,55]

8 Conclusion

Systemic cancers can affect any part of the nervous system at any point along the disease course. With modern therapeutics improving the overall survival of many cancers, the incidence of nervous system metastases will likely continue to increase. Malignancy of the nervous system is associated with significant morbidity and mortality, making it increasingly important to recognize the signs/symptoms in pursuit of earlier diagnoses and possible interventions.

References

1. Tosoni A, Ermani M, Brandes AA. The pathogenesis and treatment of brain metastases: a comprehensive review. *Crit Rev Oncol Hematol.* 2004;52(3):199–215. https://doi.org/10.1016/j.critrevonc.2004.08.006.
2. Cagney DN, Martin AM, Catalano PJ, et al. Incidence and prognosis of patients with brain metastases at diagnosis of systemic malignancy: a population-based study. *Neuro-Oncology.* 2017. https://doi.org/10.1093/neuonc/nox077.
3. Forsyth PA, Posner JB. Headaches in patients with brain tumors: a study of 111 patients. *Neurology.* 1993;43(9):1678–1683.
4. Kirby S, Purdy RA. Headaches and brain tumors. *Neurol Clin.* 2014;32(2):423–432. https://doi.org/10.1016/j.ncl.2013.11.006.
5. Vazquez-Barquero A, Ibanez FJ, Herrera S, Izquierdo JM, Berciano J, Pascual J. Isolated headache as the presenting clinical manifestation of intracranial tumors: a prospective study. *Cephalalgia.* 1994;14(4):270–272. https://doi.org/10.1046/j.1468-2982.1994.1404270.x.
6. Valentinis L, Tuniz F, Valent F, et al. Headache attributed to intracranial tumours: a prospective cohort study. *Cephalalgia.* 2010;30(4):389–398. https://doi.org/10.1111/j.1468-2982.2009.01970.x.
7. Soffietti R, Ruda R, Mutani R. Management of brain metastases. *J Neurol.* 2002;249(10):1357–1369. https://doi.org/10.1007/s00415-002-0870-6.
8. Dodick DW. Pearls: headache. *Semin Neurol.* 2010;30(1):74–81. https://doi.org/10.1055/s-0029-1245000.
9. Oberndorfer S, Schmal T, Lahrmann H, Urbanits S, Lindner K, Grisold W. The frequency of seizures in patients with primary brain tumors or cerebral metastases. An evaluation from the Ludwig Boltzmann Institute of Neuro-Oncology and the Department of Neurology, Kaiser Franz Josef Hospital, Vienna. *Wien Klin Wochenschr.* 2002;114(21–22):911–916. Haufigkeit von epileptischen Anfallen bei Patienten mit primaren Hirntumoren oder zerebralen Metastasen. Eine Untersuchung des Ludwig Boltzmann Institutes fur NeuroOnkologie und der Neurologischen Abteilung des Kaiser Franz Josef Spitals in Wien.
10. Lynam LM, Lyons MK, Drazkowski JF, et al. Frequency of seizures in patients with newly diagnosed brain tumors: a retrospective review. *Clin Neurol Neurosurg.* 2007;109(7):634–638. https://doi.org/10.1016/j.clineuro.2007.05.017.
11. Chan V, Sahgal A, Egeto P, Schweizer T, Das S. Incidence of seizure in adult patients with intracranial metastatic disease. *J Neuro-Oncol.* 2017;131(3):619–624. https://doi.org/10.1007/s11060-016-2335-2.
12. Wu A, Weingart JD, Gallia GL, et al. Risk factors for preoperative seizures and loss of seizure control in patients undergoing surgery for metastatic brain tumors. *World Neurosurg.* 2017;104:120–128. https://doi.org/10.1016/j.wneu.2017.05.028.
13. Cohen N, Strauss G, Lew R, Silver D, Recht L. Should prophylactic anticonvulsants be administered to patients with newly-diagnosed cerebral metastases? A retrospective analysis. *J Clin Oncol.* 1988;6(10):1621–1624. https://doi.org/10.1200/JCO.1988.6.10.1621.
14. Ruda R, Bello L, Duffau H, Soffietti R. Seizures in low-grade gliomas: natural history, pathogenesis, and outcome after treatments. *Neuro-Oncology.* 2012;14(Suppl. 4):iv55–iv64. https://doi.org/10.1093/neuonc/nos199.
15. Chang SM, Messersmith H, Ahluwalia M, et al. Anticonvulsant prophylaxis and steroid use in adults with metastatic brain tumors: ASCO and SNO endorsement of the congress of neurological surgeons guidelines. *J Clin Oncol Off J Am Soc Clin Oncol.* 2019;37(13):1130–1135. https://doi.org/10.1200/JCO.18.02085.
16. Newton HB. Neurologic complications of systemic cancer. *Am Fam Physician.* 1999;59(4):878–886.
17. Chang EL, Wefel JS, Maor MH, et al. A pilot study of neurocognitive function in patients with one to three new brain metastases initially treated with stereotactic radiosurgery alone. *Neurosurgery.* 2007;60(2):277–283. discussion 283–284 https://doi.org/10.1227/01.NEU.0000249272.64439.B1.
18. Clouston PD, DeAngelis LM, Posner JB. The spectrum of neurological disease in patients with systemic cancer. *Ann Neurol.* 1992;31(3):268–273. https://doi.org/10.1002/ana.410310307.
19. Beauchesne P. Intrathecal chemotherapy for treatment of leptomeningeal dissemination of metastatic tumours. *Lancet Oncol.* 2010;11(9):871–879. https://doi.org/10.1016/S1470-2045(10)70034-6.
20. Wang N, Bertalan MS, Brastianos PK. Leptomeningeal metastasis from systemic cancer: review and update on management. *Cancer.* 2018;124(1):21–35. https://doi.org/10.1002/cncr.30911.
21. Wasserstrom WR, Glass JP, Posner JB. Diagnosis and treatment of leptomeningeal metastases from solid tumors: experience with 90 patients. *Cancer.* 1982;49(4):759–772. https://doi.org/10.1002/1097-0142(19820215)49:4<759::aid-cncr2820490427>3.0.co;2-7.
22. Clarke JL, Perez HR, Jacks LM, Panageas KS, Deangelis LM. Leptomeningeal metastases in the MRI era. *Neurology.* 2010;74(18):1449–1454. https://doi.org/10.1212/WNL.0b013e3181dc1a69.
23. Shapiro WR, Posner JB, Ushio Y, Chemik NL, Young DF. Treatment of meningeal neoplasms. *Cancer Treat Rep.* 1977;61(4):733–743.
24. Lossos A, Siegal T. Numb chin syndrome in cancer patients: etiology, response to treatment, and prognostic significance. *Neurology.* 1992;42(6):1181–1184. https://doi.org/10.1212/wnl.42.6.1181.
25. Nayar G, Ejikeme T, Chongsathidkiet P, et al. Leptomeningeal disease: current diagnostic and therapeutic strategies. *Oncotarget.* 2017;8(42):73312–73328. https://doi.org/10.18632/oncotarget.20272.
26. Tubiana-Hulin M. Incidence, prevalence and distribution of bone metastases. *Bone.* 1991;12(Suppl. 1):S9–10. https://doi.org/10.1016/8756-3282(91)90059-r.
27. Goetz C. Direct metastatic disease. In: *Textbook of Clinical Neurology.* 2nd ed. Saunders; 2003:1042–1051.
28. Mut M, Schiff D, Shaffrey ME. Metastasis to nervous system: spinal epidural and intramedullary metastases. *J Neuro-Oncol.* 2005;75(1):43–56. https://doi.org/10.1007/s11060-004-8097-2.
29. Cole JS, Patchell RA. Metastatic epidural spinal cord compression. *Lancet Neurol.* 2008;7(5):459–466. https://doi.org/10.1016/S1474-4422(08)70089-9.
30. Grossman SA, Lossignol D. Diagnosis and treatment of epidural metastases. *Oncology (Williston Park).* 1990;4(4):47–54. discussion 55, 58.
31. Schaberg J, Gainor BJ. A profile of metastatic carcinoma of the spine. *Spine (Phila Pa 1976).* 1985;10(1):19–20. https://doi.org/10.1097/00007632-198501000-00003.
32. Gabriel K, Schiff D. Metastatic spinal cord compression by solid tumors. *Semin Neurol.* 2004;24(4):375–383. https://doi.org/10.1055/s-2004-861532.
33. Byrne TN. Spinal cord compression from epidural metastases. *N Engl J Med.* 1992;327(9):614–619. https://doi.org/10.1056/NEJM199208273270907.

34. Moliterno J, Veselis CA, Hershey MA, Lis E, Laufer I, Bilsky MH. Improvement in pain after lumbar surgery in cancer patients with mechanical radiculopathy. *Spine J.* 2014;14(10):2434–2439. https://doi.org/10.1016/j.spinee.2014.03.006.

35. Helweg-Larsen S, Sorensen PS. Symptoms and signs in metastatic spinal cord compression: a study of progression from first symptom until diagnosis in 153 patients. *Eur J Cancer.* 1994;30A(3):396–398. https://doi.org/10.1016/0959-8049(94)90263-1.

36. Gilbert RW, Kim JH, Posner JB. Epidural spinal cord compression from metastatic tumor: diagnosis and treatment. *Ann Neurol.* 1978;3(1):40–51. https://doi.org/10.1002/ana.410030107.

37. Brouwers E, van de Meent H, Curt A, Starremans B, Hosman A, Bartels R. Definitions of traumatic conus medullaris and cauda equina syndrome: a systematic literature review. *Spinal Cord.* 2017;55(10):886–890. https://doi.org/10.1038/sc.2017.54.

38. Spector LR, Madigan L, Rhyne A, Darden 2nd B, Kim D. Cauda equina syndrome. *J Am Acad Orthop Surg.* 2008;16(8):471–479. https://doi.org/10.5435/00124635-200808000-00006.

39. Gardner A, Gardner E, Morley T. Cauda equina syndrome: a review of the current clinical and medico-legal position. *Eur Spine J.* 2011;20(5):690–697. https://doi.org/10.1007/s00586-010-1668-3.

40. Kori SH, Foley KM, Posner JB. Brachial plexus lesions in patients with cancer: 100 cases. *Neurology.* 1981;31(1):45–50. https://doi.org/10.1212/wnl.31.1.45.

41. Gwathmey KG. Plexus and peripheral nerve metastasis. *Handb Clin Neurol.* 2018;149:257–279. https://doi.org/10.1016/B978-0-12-811161-1.00017-7.

42. Basso-Ricci S, Della Costa C, Viganotti G, Ventafridda V, Zanolla R. Report on 42 cases of postirradiation lesions of the brachial plexus and their treatment. *Tumori.* 1980;66(1):117–122.

43. Pettigrew LC, Glass JP, Maor M, Zornoza J. Diagnosis and treatment of lumbosacral plexopathies in patients with cancer. *Arch Neurol.* 1984;41(12):1282–1285. https://doi.org/10.1001/archneur.1984.04050230068022.

44. Ladha SS, Spinner RJ, Suarez GA, Amrami KK, Dyck PJ. Neoplastic lumbosacral radiculoplexopathy in prostate cancer by direct perineural spread: an unusual entity. *Muscle Nerve.* 2006;34(5):659–665. https://doi.org/10.1002/mus.20597.

45. Jaeckle KA. Neurologic manifestations of neoplastic and radiation-induced plexopathies. *Semin Neurol.* 2010;30(3):254–262. https://doi.org/10.1055/s-0030-1255219.

46. Brejt N, Berry J, Nisbet A, Bloomfield D, Burkill G. Pelvic radiculopathies, lumbosacral plexopathies, and neuropathies in oncologic disease: a multidisciplinary approach to a diagnostic challenge. *Cancer Imaging.* 2013;13(4):591–601. https://doi.org/10.1102/1470-7330.2013.0052.

47. Fathers E, Thrush D, Huson SM, Norman A. Radiation-induced brachial plexopathy in women treated for carcinoma of the breast. *Clin Rehabil.* 2002;16(2):160–165. https://doi.org/10.1191/0269215502cr470oa.

48. Killer HE, Hess K. Natural history of radiation-induced brachial plexopathy compared with surgically treated patients. *J Neurol.* 1990;237(4):247–250. https://doi.org/10.1007/BF00314628.

49. Delanian S, Lefaix JL, Pradat PF. Radiation-induced neuropathy in cancer survivors. *Radiother Oncol.* 2012;105(3):273–282. https://doi.org/10.1016/j.radonc.2012.10.012.

50. Iyer VR, Sanghvi DA, Merchant N. Malignant brachial plexopathy: a pictorial essay of MRI findings. *Indian J Radiol Imaging.* 2010;20(4):274–278. https://doi.org/10.4103/0971-3026.73543.

51. Castagno AA, Shuman WP. MR imaging in clinically suspected brachial plexus tumor. *AJR Am J Roentgenol.* 1987;149(6):1219–1222. https://doi.org/10.2214/ajr.149.6.1219.

52. Chandra P, Purandare N, Agrawal A, Shah S, Rangarajan V. Clinical utility of (18)F-FDG PET/CT in brachial plexopathy secondary to metastatic breast cancer. *Indian J Nucl Med.* 2016;31(2):123–127. https://doi.org/10.4103/0972-3919.178263.

53. Ahmad A, Barrington S, Maisey M, Rubens RD. Use of positron emission tomography in evaluation of brachial plexopathy in breast cancer patients. *Br J Cancer.* 1999;79(3–4):478–482. https://doi.org/10.1038/sj.bjc.6690074.

54. Ko K, Sung DH, Kang MJ, et al. Clinical, electrophysiological findings in adult patients with non-traumatic plexopathies. *Ann Rehabil Med.* 2011;35(6):807–815. https://doi.org/10.5535/arm.2011.35.6.807.

55. Harper Jr CM, Thomas JE, Cascino TL, Litchy WJ. Distinction between neoplastic and radiation-induced brachial plexopathy, with emphasis on the role of EMG. *Neurology.* 1989;39(4):502–506. https://doi.org/10.1212/wnl.39.4.502.

Neuroimaging of systemic metastatic disease

John Vincent Murray, Jr.[a], Richard Douglas Beegle[b], and Sean Dodson[b]

[a]Neuroradiology, Radiology, Mayo Clinic, Jacksonville, FL, United States, [b]Department of Neuroradiology, AdventHealth Medical Group Central Florida Division, Orlando, FL, United States

Metastatic disease is the most commonly encountered neoplasm involving the central nervous system (CNS) in adults. Advancements in treatment regimens have led to prolonged survival and therefore increased incidence of metastatic disease, occurring 10 times more likely than primary CNS malignancy.[1] Up to 40% of advanced systemic malignancy will eventually metastasize to the CNS.[1–3]

Metastatic disease can spread to the CNS via hematogenous dissemination, CNS dissemination, direct geographic spread, and perineural spread. Hematogenous spread is the most common mode of dissemination and will be the focus of this chapter. Predilection for CNS involvement varies depending on the age of the patient and primary site of malignancy. In adults, the most common primary tumors to involve the brain are lung, breast, skin, gastrointestinal, and renal cancer.[4] The most common primary neoplasms to spread to the spine and calvarium are breast, lung, and prostate cancer.[5] Prostate cancer rarely spreads to brain parenchyma, but often involves the calvarium and pachymeninges.[6] In children with hematologic malignancies, Ewing's sarcoma, neuroblastoma, and osteogenic sarcoma are the most common to metastasize to the CNS.[7,8] This chapter will focus on the technical considerations for optimal imaging, the imaging appearance of metastatic disease to the central nervous system organized by anatomic location, potential mimics, and treatment-related complications.

1 Technical considerations

Magnetic resonance imaging (MRI) with intravenous gadolinium-based contrast is the gold standard for diagnosing CNS metastatic disease and for monitoring response to treatment. Computed tomography (CT) has decreased sensitivity when compared to MRI for detection and characterization of CNS metastatic disease. Furthermore, noncontrast CT has no role in detection

of metastatic disease, treatment planning, or treatment follow-up. Contrast-enhanced CT should only be utilized when there is a true contraindication to MRI which has become much less common over the past decade. Performing a CT without and with IV contrast rarely adds any diagnostic value, does not justify the increased dose to the patient, and therefore should not be routinely obtained. Noncontrast head CT should be reserved for patients experiencing acute neurologic symptoms that could necessitate emergent intervention. However, in everyday practice, this is often the initial imaging obtained upon the patient presenting to the emergency department.

Assessment of follow-up imaging in the setting of metastatic disease can be challenging and often results in a high degree of inter-reader variability. Appropriate interpretation requires an understanding of the tumor subtype and a detailed history of the patient's treatment including date of surgical intervention, chemotherapeutic regimen, radiation field, and if the patient is receiving steroids or antiangiogenic therapy. In order to improve radiologic interpretation and prevent patients from unnecessary treatment change, standardized response assessment criteria have been established and include neurooncology (RANO-BM), immunotherapy RANO (iRANO), and RANO leptomeningeal metastases (RANO-LM). Standardized and optimized MRI protocols are essential in this endeavor. Early and accurate detection of intracranial metastatic disease leads to appropriate and timely intervention with improved quality of life. The number and size of intracranial metastases will determine if the patient receives targeted radiosurgery or whole brain radiation.[9]

The ability to detect postcontrast enhancement is paramount in imaging of intracranial metastatic disease as this is the primary biomarker for blood/brain barrier breakdown which is almost always present with CNS metastatic disease. When designing an optimal protocol for imaging CNS metastases, there must be an adequate delay between injection of gadolinium-based contrast and the postcontrast imaging. Suboptimal delay leads to decreased sensitivity in detecting metastases.[10–12] At our institution, postcontrast sequences are obtained following a 10-min delay after contrast injection. This can be accomplished, without disrupting workflow, by obtaining sequences such as T2 FSE, fluid attenuation inversion recovery (FLAIR), and perfusion imaging prior to obtaining the postcontrast T1-weighted sequence. Obtaining the FLAIR sequence after administration of contrast improves detection of leptomeningeal metastatic disease.[13–15]

There have been multiple studies examining the optimal postcontrast MRI sequence for detecting intracranial metastatic disease. If there are no scanner limitations and if obtained on a 3T magnet, 3D FSE T1-weighted techniques such as CUBE (GE), SPACE (Siemens), and VISTA (Philips) are the most sensitive sequences for detecting enhancement associated with intracranial metastatic disease. Magnetization-prepared 3D gradient echo pulse sequences such as MPRAGE (Siemens), BRAVO (GE), and TFE (Philips) are the standard sequences widely available in the community and academic settings at both 1.5T and 3T. Contrast enhancement is more conspicuous on the spin echo (SE)-based volumetric sequences in comparison to the gradient echo (GRE)-based sequences. This is of particular importance when metastatic lesions are ≤ 5 mm in diameter with studies showing that there is a significant number of smaller metastases detectable on the SE-based techniques which are missed on the GRE-based techniques.[16–21] Undetected small metastases could lead to suboptimal patient outcomes.

Fat suppression can also be applied on the SE-based techniques which significantly increases the ability to detect osseous metastases. Significant limitations to 3D TSE techniques are the sensitivity to motion and inability to be used with metallic frames for stereotactic radiosurgery planning due to artifact.

MR perfusion has also been shown to increase accuracy in differentiating posttreatment changes (pseudo-progression, radiation necrosis, and immunotherapy effects) from true progression. True progression demonstrates increased relative cerebral blood volume (rCBV) and permeability in comparison to posttreatment effects.

PET using both glucose and more recently amino acid-based radiolabels can also complement MRI to evaluate disease response and to differentiate true progression from pseudo-progression.[22]

The optimal MRI protocol for imaging intracranial metastatic disease ultimately depends on the MRI scanner. An updated consensus recommendation for a standardized brain tumor imaging protocol for clinical trials in brain metastases has been recently released. The "minimum standard" recommended pulse sequences include parameter-matched pre- and postcontrast inversion-recovery prepared, isotropic 3D T1-wieghted gradient echo sequences; axial 2D T2 TSE acquired after contrast injection, axial 2D or 3D T2-weighted FLAIR; axial diffusion-weighted imaging (DWI); and postcontrast 2D T1-weighted spin echo sequence. The "ideal" protocol recommendation would replace the 3D GRE T1-weighted technique with 3D TSE fat-saturated pre- and postgadolinium sequences and add dynamic susceptibility contrast (DSC) perfusion. For detecting osseous metastases, fat-saturated T2-weighted technique is helpful. If possible, it is recommended to obtain brain imaging on a 3T rather than a 1.5T magnet. It is strongly encouraged that the same imaging protocol and field strength be utilized for patient follow-up to aid in accurately evaluating change over time.[9] Postcontrast T2 FLAIR imaging has been shown to increase sensitivity for detection of leptomeningeal disease in comparison to standard postcontrast T1-weighted imaging and should be considered routinely.[13–15] At our institution, susceptibility-weighted imaging (SWI) is also performed to improve sensitivity for detection of ferromagnetic material, namely blood products and calcification. Microhemorrhages visualized utilizing SWI can serve as a biomarker for neovascularity. When not possible, given scanner limitations, a GRE technique is utilized.

2 Anatomic locations

2.1 Brain parenchyma

The brain parenchyma is the most common site of CNS metastatic disease in adults.[23] The majority occur in the cerebral hemisphere followed by the cerebellum and basal ganglia. Symptoms vary based on tumor site. Parenchymal tumors most often present with focal neurological deficit or seizure.

Hematogenous metastases tend to occur at the gray/white matter junction and arterial watershed zones (Fig. 1).[24] This propensity is based on the relatively small size of the end arterioles in these locations. However, establishment of parenchymal metastasis is much more complex than the simple delivery of tumor emboli to end arterioles. There is a metastatic cascade for a metastasis to establish itself which includes specific receptor-mediated attachment as well as promotion of matrix proteins, cytokines, and growth factors to establish an environment where the micrometastasis can flourish. This cascade also includes inactivation of tumor suppressor genes as well as activation of proto-oncogenes.[3]

Half of all parenchymal metastases will be solitary. Up to one-third of patients presenting with CNS metastatic disease will present without a known primary malignancy. The most sensitive imaging modality for detecting metastatic disease to the CNS is MRI without

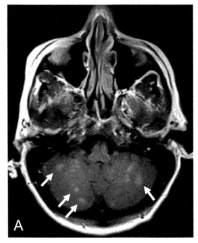
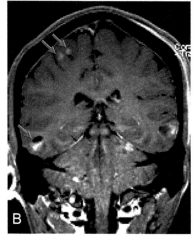

FIG. 1 Multiple parenchymal metastases. Axial (A) and coronal (B) fat-saturated postcontrast T1WI showing multiple enhancing lesions, mostly at the gray/white matter junction *(red arrows)*. Within the cerebellum, there is predominant involvement of the cerebellar watershed zones *(white arrows)*.

and with intravenous contrast.[25] CT is less sensitive, but is more widely available and is usually utilized first in the emergency department setting (Fig. 2). Noncontrast CT is relatively insensitive for detecting small metastatic lesions, as most metastases tend to be iso- to mildly hypodense relative to gray matter (Fig. 3). Hypercellularity, hemorrhage, and calcification increase sensitivity for detection of intracranial metastatic disease with noncontrast CT (Fig. 2). Often, the presence of vasogenic edema may be the only clue that underlying metastatic disease is present (Fig. 4).

Parenchymal metastases are typically round and well circumscribed. This imaging appearance is explained by understanding that a parenchymal metastasis is a small metastatic deposit of foreign cells that grow in a centrifugal pattern. They typically do not infiltrate along the

axonal scaffolding of the brain in the manner that a high-grade glioma does. Rarely, metastatic lesions are infiltrative and thus can be very difficult to differentiate from glioblastoma.

Imaging characteristics can vary depending on degree of intratumoral hemorrhage and necrosis but almost all will demonstrate enhancement after contrast administration secondary to breakdown of the blood brain barrier. Enhancement patterns are variable with lesions demonstrating solid, ring, and heterogeneous enhancement with areas of cystic change and necrosis. On MRI, most metastases are iso- to hypointense on T1-weighted imaging (T1WI). Melanocytic melanoma metastases, lesions with subacute hemorrhage, and mucinous metastases are the exception demonstrating intrinsic T1 hyperintense signal (Figs. 5 and 6). The appearance of

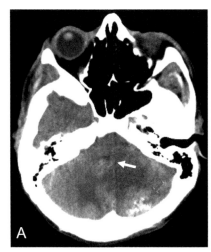
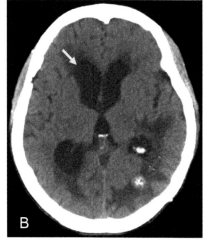

FIG. 2 History of breast and cervical cancer presenting with altered mental status. Axial noncontrast CT images through the level of the cerebellum (A) and mid-convexity (B) showing multiple calcified lesions within the brain parenchyma consistent with metastatic disease *(red arrows)*. Majority of metastatic lesions on CT will be iso- to hypodense relative to gray matter, but some can be hyperdense secondary to intratumoral hemorrhage or calcification as in this case. There is associated vasogenic edema surrounding these lesions resulting in partial effacement of the fourth ventricle *(white arrow)* and dilatation of the lateral ventricles *(yellow arrow)* consistent with obstructive hydrocephalus.

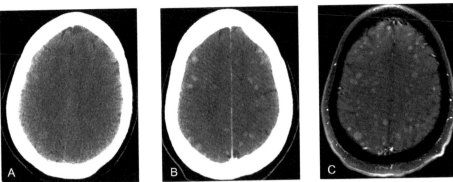

FIG. 3   Multiple metastases occult on CT. Noncontrast axial CT (A) demonstrates a fairly normal appearing CT. Postcontrast axial CT (B) and fat-saturated postcontrast axial T1WI (C) demonstrate multiple enhancing masses predominantly within the watershed zones and gray/white junction consistent with metastatic disease. This case demonstrates that a noncontrast CT can be insensitive for detection of metastases.

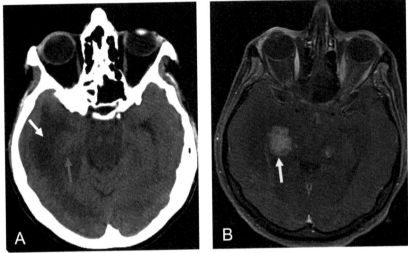

FIG. 4   Intraparenchymal metastasis on CT. Axial noncontrast CT of the brain (A) demonstrates vasogenic edema in the right temporal lobe *(white arrow)*. The underlying metastatic lesion is difficult to visualize and is isodense *(red arrow)*. Axial fat-saturated postcontrast T1WI (B) demonstrates the enhancing metastatic lesion within the right temporal lobe *(yellow arrow)*.

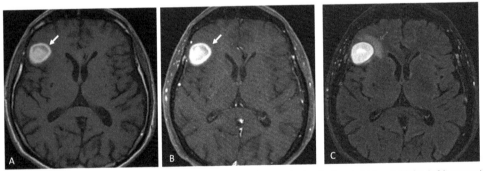

FIG. 5   Axial noncontrast T1WI (A) demonstrates an intrinsically T1 hyperintense lesion in the right frontal lobe *(white arrow)*. Axial fat-saturated postcontrast T1WI (B) demonstrates persistent hyperintense signal *(white arrow)*. It is difficult to assess for true underlying enhancement given the intrinsic T1 hyperintense signal. Axial FLAIR image (C) showing mild surrounding vasogenic edema *(red arrow)*.

diffusion-weighted imaging can also vary depending on tumor subtype with more highly cellular tumors demonstrating restricted diffusion (Fig. 7).

The differential diagnosis for parenchymal metastatic disease is broad. The pattern of surrounding edema is an important diagnostic consideration when presented with an abnormal CT or MRI. Many patients with metastatic disease can present with extremity weakness/numbness which can be confused with stroke. In the setting of infarction, cell death ultimately leads to cytotoxic edema which involves the cortex and subjacent white matter. Alternatively, vasogenic edema is associated with

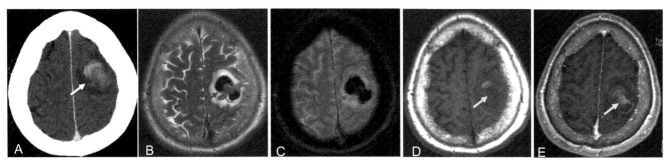

FIG. 6 Hemorrhagic testicular cancer metastasis. Noncontrast CT (A) demonstrates a hyperdense lesion in the left frontal high convexity secondary to underlying hemorrhagic metastasis *(white arrow)*. Axial T2WI (B) demonstrates heterogeneous but predominantly T2 hypointense signal secondary to the hemorrhage *(red arrow)*. Axial GRE (C) confirms the presence of hemorrhage with hypointense signal *(red arrow)*. When comparing the noncontrast T1WI (D) with the postcontrast T1WI (E), solid regions of internal enhancement are identified consistent with the underlying metastatic mass *(yellow arrows)*.

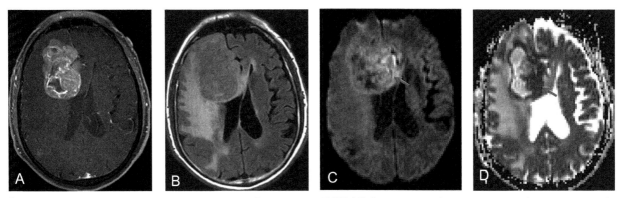

FIG. 7 Hypercellular melanoma metastasis. Fat-saturated postcontrast T1WI (A) demonstrates a large metastatic enhancing mass in the right frontal lobe. The FLAIR image (B) demonstrates prominent surrounding vasogenic edema. DWI (C) and ADC (D) demonstrate restricted diffusion *(red arrows)*. Hypercellular masses can demonstrate this finding. The tightly compacted cellular nature of the tumors results in restriction of random Brownian motion of water.

metastatic disease and results in surrounding edema involving the adjacent white matter with sparing of the cortex. Infarcts in the subacute phase can show enhancement secondary to breakdown of the blood/brain barrier, and therefore, small scattered subacute infarcts can appear very similar to metastatic disease. The enhancement in the setting of subacute infarcts tends to be localized to the cortex in a gyriform pattern whereas metastatic disease is more often localized to the gray/white junction and is mass like with ring or solid enhancement (Fig. 8). If there is suspicion that the lesions could be subacute infarcts, then a follow-up MRI without and with intravenous contrast in 4–8 weeks could be performed to confirm appropriate evolution.

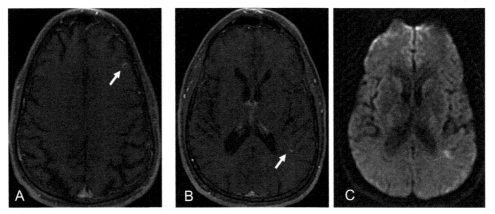

FIG. 8 Multiple subacute cortical infarcts. Axial fat-saturated postcontrast T1WI (A and B) demonstrates multiple small regions of cortical gyriform enhancement *(white arrows)*. This is a classic enhancing pattern of evolving subacute cortical infarcts and is an important differentiating feature from metastatic disease. Axial DWI (C) demonstrates restricted diffusion in the region of enhancement *(red arrow)*.

Intraparenchymal infections are on the differential diagnosis, especially for ring-enhancing lesions. Intraparenchymal abscesses can appear as a ring-enhancing mass and can mimic a ring-enhancing metastatic lesion with central necrosis. The enhancing wall of an abscess is classically thinner and less nodular than that of a metastasis. DWI is very important in these cases as the nonenhancing central purulent material within an abscess will show restricted diffusion; however, the central necrotic component will show facilitated diffusion in an untreated metastasis (Fig. 9). Treated lesions may demonstrate restricted diffusion in the nonenhancing component of the lesions which suggests posttreatment effect.[26] Some infections can be multifocal with more solid enhancement such as toxoplasmosis (Fig. 10). Clinical history and follow-up MRI after treatment for a presumed infection can be beneficial in these cases.

As mentioned earlier, metastatic lesions can be hemorrhagic. Solitary or multifocal intracranial hemorrhage can mimic metastatic disease. These can occur in the setting of uncontrolled hypertension, trauma, amyloid angiopathy, coagulopathy, and cavernous malformations (Fig. 11). In the era of coronavirus 2019 (COVID-19), it is important to understand that COVID-19 results in varying intracranial pathology including intracranial microhemorrhage. The pathophysiology of such is multifactorial and may be secondary to ischemic infarct, diffuse thrombotic microangiopathy, delayed hypoxic leukoencephalopathy with microhemorrhage, cytokine storm with resultant vascular endothelial damage, or iatrogenic secondary to anticoagulation and ECMO treatment (Fig. 12).[27]

Primary CNS malignancy is also a differential consideration for metastatic disease. Glioblastomas (GBM) tend to be solitary lesions located within the deep white matter with an infiltrating growth pattern, in contrast to metastatic disease which tends to be multiple, located at the gray/white junction, and demonstrate a

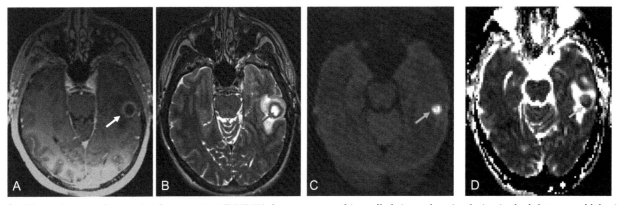

FIG. 9   Ring enhancing abscess. Axial postcontrast T1WI (A) demonstrates a thin-walled ring enhancing lesion in the left temporal lobe *(white arrow)*. There is central T2 hyperintense signal *(red arrow)* on the axial T2WI (B). DWI (C) and the corresponding ADC map (D) demonstrate internal restricted diffusion *(blue arrows)* consistent with purulent material which results in restricted diffusion. Central necrosis in the setting of an aggressive neoplasm or metastasis will not restrict.

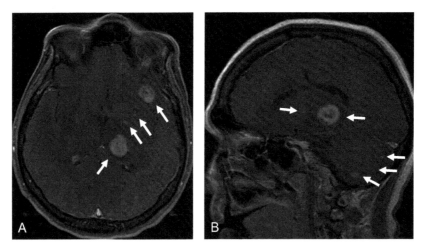

FIG. 10   Toxoplasmosis infection in HIV-positive patient. Axial (A) and sagittal (B) postcontrast T1WI demonstrates multiple enhancing lesions within the cerebral and cerebellar hemispheres *(white arrows)*. Biopsy revealed toxoplasmosis in this HIV-positive patient.

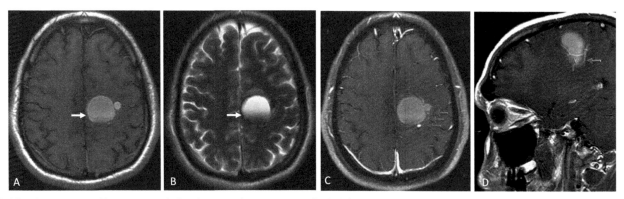

FIG. 11 Cavernous malformation with developmental venous anomaly. Axial noncontrast T1WI (A) and T2WI (B) demonstrate a lesion within the posterior left frontal lobe which demonstrates heterogenous internal T1 and T2 signal with a fluid-hematocrit level *(white arrows)*. There is a dark rim of T2 hypointense signal surrounding this lesion consistent with a hemosiderin ring. Axial (C) and sagittal (D) postcontrast T1WI demonstrates no significant enhancement of the mass and an adjacent developmental venous anomaly *(red arrows)*.

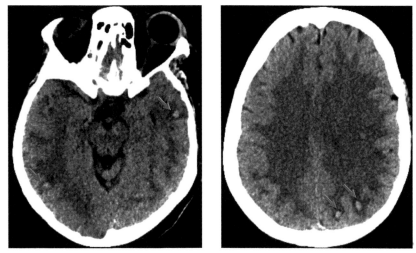

FIG. 12 Multiple intraparenchymal microhemorrhages in COVID-19. Axial noncontrast CT of the head demonstrates multiple subcentimeter hyperdense lesions with mild surrounding vasogenic edema *(red arrows)*. This appearance is similar to multiple hemorrhagic metastases. This case was secondary to multiple intraparenchymal hemorrhages related to COVID-19 infection.

circular growth pattern. However, a solitary ring-enhancing GBM in the periphery of the brain can appear very similar to a metastatic lesion (Fig. 13). The vast majority of solitary masses in the cerebellum in an adult are metastases which should always be at the top of the differential for an enhancing cerebellar mass in an adult. However, primary CNS neoplasms, such as hemangioblastoma, can still occur in the posterior fossa in an adult. Hemangioblastomas classically appear as a mass with both solid and cystic components, although metastatic lesions can have this same appearance (Fig. 14). A helpful differentiating feature is that the enhancing nodule abuts the pial surface in the setting of hemangioblastoma.

Lastly, tumefactive demyelinating lesions are on the differential diagnosis. Classically, a demyelinating lesion will demonstrate an incomplete ring of enhancement helping differentiate it from a metastatic lesion (Fig. 15).

2.2 Calvarium and skull base

The skull and the dura are the second most frequent sites of CNS metastatic disease. Osseous involvement is most frequent in breast, prostate, lung, and renal cell carcinomas.[5,28] Osseous metastatic disease can be confined to the bone marrow, but may also result in cortical breakthrough with extraosseous extension into the intracranial compartment or adjacent soft tissues (Fig. 16).

As with parenchymal metastases, the patient's presentation depends on the site of tumor involvement. Patients are most commonly asymptomatic, though often present with headache or palpable soft tissue mass. With intracranial extension, there may be mass effect on the brain resulting in focal neurological deficit or seizure. There can be compression or invasion of the dural venous sinuses resulting in thrombosis. Additionally, there can be involvement of the cranial nerve foramina resulting in cranial neuropathy (Fig. 17).

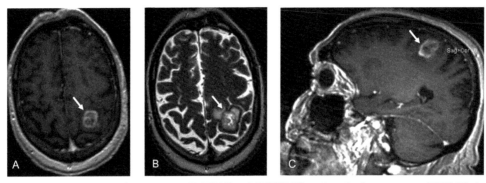

FIG. 13   GBM at gray/white junction. Axial postcontrast T1WI (A), axial T2WI (B), and sagittal postcontrast T1WI (C) demonstrate a solitary ring enhancing lesion at the gray/white junction in the left precentral gyrus *(white arrows)*. This demonstrates central T2 hyperintense signal suggestive of internal necrosis. There is mild surrounding vasogenic edema. This was originally thought to represent a metastatic lesion given its location at the gray/white junction and lack of infiltrating features. However, pathology showed glioblastoma.

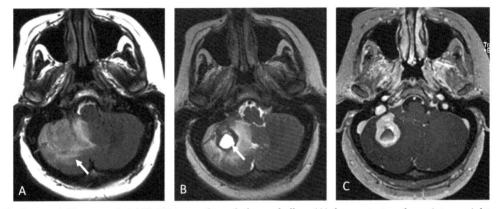

FIG. 14   Cerebellar hemangioblastoma. Axial FLAIR image through the cerebellum (A) demonstrates a hyperintense right cerebellar mass *(red arrow)* with surrounding vasogenic edema *(white arrow)*. Axial T2WI (B) demonstrates the mass *(red arrow)* to have both solid *(red arrow)* and cystic *(white arrow)* components. Postcontrast T1WI (C) shows enhancement of the solid components of the mass which abut the pia *(red arrow)*.

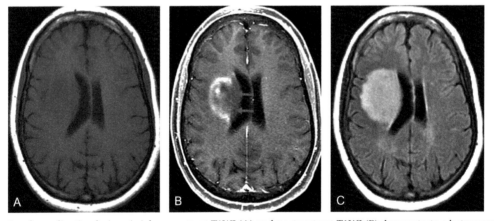

FIG. 15   Tumefactive demyelinating lesion. Axial noncontrast T1WI (A) and postcontrast T1WI (B) demonstrate a heterogeneously enhancing mass in the right frontal corona radiata with involvement of the periventricular white matter. There is a characteristic incomplete ring-type enhancement along the medial margin of the lesion *(red arrow)*. Axial FLAIR image (C) demonstrates hyperintense signal within this lesion.

Osseous metastatic disease can be seen on plain radiography, CT, and MRI with MRI being most sensitive. Depending on the type of malignancy, metastatic lesions can be lytic, blastic, or mixed on CT and radiography. These can be circumscribed or permeative in appearance. The majority are lytic (Fig. 18) with prostate and treated breast cancer the most frequent to result in blastic metastases. CT and MRI are complementary when evaluating bone lesions. CT is best for evaluating bone destruction and internal matrix. MRI is best for assessing marrow involvement, extraosseous soft tissue infiltration, and enhancement pattern. Although MRI is best for

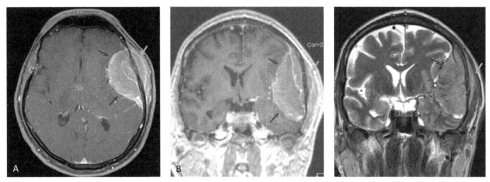

FIG. 16 Calvarial-based prostate metastasis with extraosseous extent. Axial (A) and coronal (B) postcontrast T1WI and coronal T2WI (C) demonstrate a prostate metastasis involving the left calvarium with extraosseous extent into the adjacent scalp *(blue arrows)*. There is also intracranial extension with invasion of the dura *(red arrows)* with associated mass effect on the brain parenchyma.

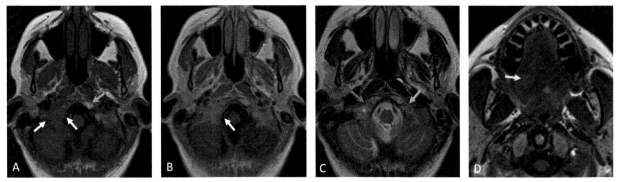

FIG. 17 Occipital skull base metastasis. Axial noncontrast T1WI (A) demonstrates a T1 hypointense lesion *(white arrow)* within the right occipital condyle. Metastases *(white arrows)* are T1 hypointense (A) and enhance (B) as they infiltrate the normal T1 hyperintense fatty bone marrow *(green arrow* in (A) showing normal bone marrow in the left clivus). Axial T2WI (C) demonstrates T2 mixed predominantly hyperintense signal. There is involvement of the right hypoglossal canal *(red arrow)* with the normal left hypoglossal canal shown for comparison *(blue arrow)*. Axial noncontrast T1WI (D) demonstrates mild T1 hyperintense fatty infiltration of the right tongue compared to the left *(yellow arrow)* consistent with denervation atrophy secondary to involvement of the right hypoglossal nerve.

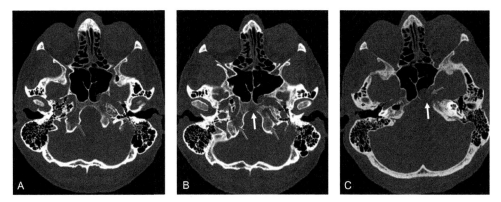

FIG. 18 Lytic calvarial metastases. Axial noncontrast CT images of the skull base demonstrate multiple lytic metastases involving the skull base *(red arrows)* with multiple cortical erosions *(white arrows)*.

evaluating soft tissue infiltration, CT can also detect extraosseous and intracranial extension; therefore, osseous lesions should be viewed on both bone and soft tissue windows. Nuclear medicine Tc-99 bone scan and FDG PET/CT have high negative predictive value for bone metastases.

The typical imaging appearance on MRI is replacement of the normal T1 hyperintense fatty bone marrow resulting in a T1 hypointense lesion which may have well-circumscribed or indistinct margins. Untreated metastases almost always show enhancement after contrast administration. They can show varying degrees of hyperintense signal on T2-weighted imaging (T2WI) and short tau inversion recovery (STIR) depending on the degree of associated sclerosis (Fig. 19).

The use of fat-saturated postcontrast T1WI can increase the sensitivity for detection of intraosseous metastases as some lesions may enhance enough to become isointense with the normal fatty marrow on non-fat-saturated T1WI.[3,5,24,28] The diffusion-weighted imaging (DWI) sequence is a useful screen for bony metastases, with some institutions using DWI of the entire body as a screening tool. Osseous metastases will be hyperintense compared to the normal, hypointense background of the calvarium (Fig. 19). The exception is for purely blastic metastases which are diffusely dark and difficult to detect on DWI. DWI is also useful for monitoring response to treatment. Hypercellular metastases will demonstrate restricted diffusion and enhancement. Subsequently with effective treatment, the ADC can normalize and the enhancement pattern can resolve or become more ring like although the lytic defects may persist particularly early in the treatment course.[29] Edema in the adjacent brain parenchyma suggests brain parenchymal invasion and/or compromised venous drainage.[3]

The differential diagnosis is broad and can include multiple myeloma as well as many other primary benign and malignant bone tumors. Differentiating lytic calvarial metastases from multiple myeloma by imaging is often not possible. Arachnoid granulations, venous channels/lakes, hemangiomas, and surgical defects can also be mistaken for metastatic disease although all often have characteristic imaging features.

2.3 Dura

Dural involvement is most often a result of direct tumor extension from adjacent calvarial metastases, and is often obvious on imaging. However, thin smooth reactive benign dural enhancement can occur in association with calvarial metastases. Nodular dural enhancement, > 5mm thick dural enhancement, loss of the hypointense line separating the osseous tumor from the dural enhancement, leptomeningeal enhancement, and parenchymal edema are all signs that there is dural invasion present. Dural metastasis without coexisting calvarial disease is less common, but can occur secondary to hematogenous spread and is most often seen with breast, prostate, lung, and stomach cancers.[3,28] The most common presenting symptoms include headache, other signs of increased intracranial pressure, and focal neurological deficits, though a portion may be asymptomatic.[28] Similar to calvarial metastatic disease, dural metastases can spread to involve cranial nerves leading to cranial neuropathy and involve the dural venous sinuses (Fig. 20).

Dural-based metastases will appear as an extra-axial mass with convex margins that may or may not cause mass effect on the adjacent brain parenchyma. Since they are located outside of the blood/brain barrier, they will demonstrate intense enhancement (Fig. 21).[30] A less common pattern is diffuse dural thickening and enhancement which may be smooth or nodular, and can be associated with diffuse calvarial tumor infiltration (Fig. 22). T2/FLAIR hyperintensity and enhancement in the adjacent sulci is concerning for leptomeningeal invasion. Edema in the adjacent brain parenchyma suggests brain parenchymal invasion or compromised venous drainage (Fig. 23).[3]

The primary differential consideration for focal dural metastatic disease is a meningioma. Up to 44% of dural-based metastases demonstrate a "dural tail," and therefore, this is not a useful finding in differentiating dural metastatic disease from meningioma.[30] Dural-based metastases from breast or prostate cancer can appear indistinguishable from meningiomas (Figs. 21 and 23).

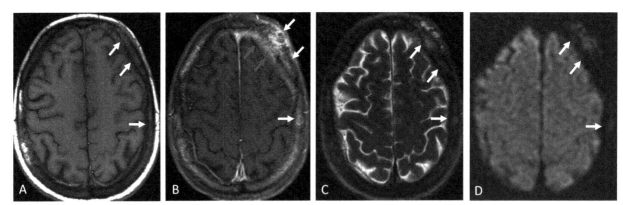

FIG. 19   Calvarial-based lung metastases with dural involvement. Axial noncontrast T1WI (A) demonstrates multifocal lesions within the calvarium replacing the normal fatty marrow *(white arrows)*. Axial fat-saturated postcontrast T1WI (B) demonstrates enhancement of the multifocal osseous metastatic lesions *(white arrows)*. There is intracranial extension with involvement of the dura which is thickened and enhances *(red arrows)*. Axial T2WI (C) demonstrates heterogenous T2 hyperintense signal within the lesions, which is typical for metastatic disease *(white arrows)*. DWI (D) shows subtle restricted diffusion *(white arrows)*.

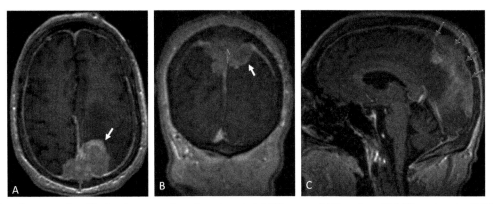

FIG. 20   Dural-based metastasis with dural venous sinus invasion. Axial (A), coronal (B), and sagittal (C) postcontrast T1WI demonstrates a large dural-based enhancing metastatic lesion *(white arrows)*. There is long segment invasion of the superior sagittal sinus *(red arrows)*.

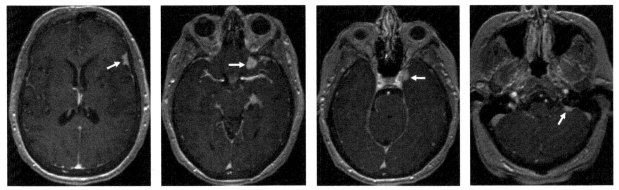

FIG. 21   Multifocal dural-based metastases mimicking meningiomas. Multiple axial postcontrast T1WI images demonstrating multiple enhancing extra-axial dural-based nodular lesions *(white arrows)*.

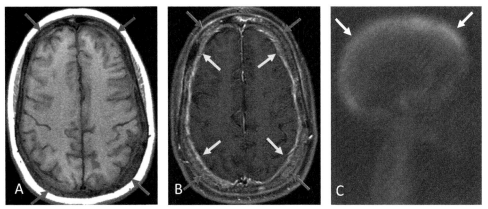

FIG. 22   Diffuse calvarial infiltration with dural involvement. Axial noncontrast T1WI (A) demonstrates diffuse patchy T1 hypointense signal throughout the bone marrow of the calvarium *(red arrows)*. Fat-saturated postcontrast T1WI (B) demonstrates diffuse patchy enhancement of the calvarium *(red arrows)*. There is also diffuse dural thickening and enhancement consistent with dural invasion *(yellow arrows)*. Nuclear medicine bone scan (C) demonstrates diffuse radiotracer uptake in the calvarium *(white arrows)*.

Aggressive features including parenchymal invasion, leptomeningeal spread, calvarial destruction, and invasion into the adjacent soft tissues, particularly in a patient with known primary malignancy, strongly favor metastatic disease. However, aggressive atypical meningiomas can demonstrate these features. Meningiomas may demonstrate hyperostosis of the adjacent calvarium, which can be a differentiating feature (Fig. 24). A challenging diagnosis is a collision tumor where two different tumor types can occur at the same anatomic site. For example, metastatic disease can spread to vascular tumors such as meningiomas (Fig. 25).

The less common diffuse pattern of dural metastatic disease has a separate differential diagnosis. IgG4-related

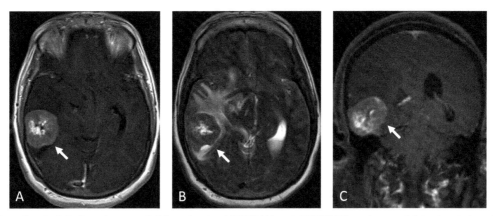

FIG. 23   Dural-based metastatic breast cancer. Axial postcontrast T1WI (A), axial T2WI (B), and coronal fat-saturated postcontrast T1WI (C) demonstrate a heterogeneous enhancing extra-axial dural-based mass *(white arrow)* along the right temporal convexity. This results in vasogenic edema in the adjacent brain parenchyma *(red arrow)* suggestive of brain invasion or compromise of venous drainage.

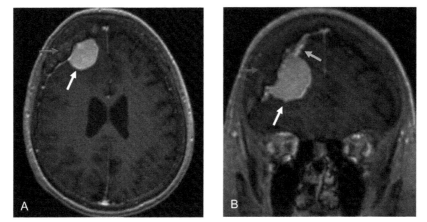

FIG. 24   Benign meningioma demonstrating hyperostosis of the adjacent calvarium. Axial (A) and coronal (B) postcontrast T1WI demonstrates an enhancing extra-axial dural-based mass *(white arrow)*. This case demonstrates hyperostosis of the adjacent calvarium *(red arrow)* which differentiates meningiomas from metastatic disease. There is a dural tail *(blue arrow)*, although this can also be seen in metastatic disease.

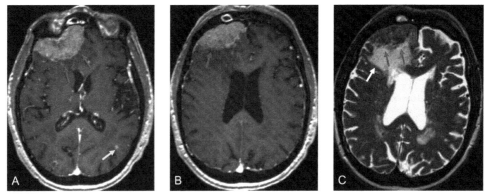

FIG. 25   Pathologically proven collision tumor. Axial postcontrast T1WI (A, B) demonstrates an enhancing right frontal extra-axial mass. This demonstrates atypical features for a benign meningioma. There are irregular margins with leptomeningeal and parenchymal invasion *(red arrows)*. There is another small intraparenchymal lesion *(yellow arrow)*. Axial T2WI (C) demonstrates ill-defined margins with respect to the brain parenchyma *(red arrows)* and severe surrounding vasogenic edema *(white arrows)* concerning for brain invasion. Pathology revealed a grade II meningioma with multiple internal subcentimeter nests of metastatic adenocarcinoma, consistent with a collision tumor.

pachymeningitis, chronic subdural hematoma, intracranial hypotension, and postprocedural reactive dural changes can appear as diffuse dural thickening and enhancement. These entities tend to be smooth in appearance, and dural-based metastatic disease tends to be more nodular. Also, marrow infiltration of the adjacent calvarium suggests diffuse dural metastatic disease.

2.4 Leptomeningeal

The subarachnoid space contains the cerebrospinal fluid and is contained by the leptomeninges. The leptomeninges are composed of the pia mater and arachnoid mater. There is both dura-arachnoid and pia arachnoid. However, the term "leptomeningeal metastasis" refers to the subarachnoid space and pia arachnoid.

Leptomeningeal spread of metastatic disease is less common than the other sites previously mentioned, but does portend a poor prognosis.[30] Leptomeningeal metastases tend to occur in widely disseminated and progressive systemic cancer, most often in breast cancer, leukemia, lymphoma, and small cell lung cancer.[24] The most common route of spread is hematogenous, which can be through both arterial and venous seeding. In high-grade primary CNS neoplasms, drop metastases can occur. Direct extension of tumor can occur through multiple routes including extension from the brain parenchyma through the pia matter, extension through the dura with invasion into the subarachnoid space, invasion of the ependyma by parenchymal or choroid plexus metastases, or spreading along perivascular spaces.[24] Once the metastatic cells enter the subarachnoid space, they can be transported along the surface of the brain and throughout the spinal canal.

Early diagnosis is essential for optimal therapy, and as such, there should be a low threshold for imaging the entire neural axis with contrasted MRI and CSF sampling. CSF sampling and imaging are complementary to each other as neither has optimal sensitivity. False-negative CSF cytology is common, and often repeated CSF sampling is needed to confirm diagnosis with up to 90% diagnostic accuracy after three lumbar punctures.[31]

Both communicating- and noncommunicating-type hydrocephalus can occur as a result of leptomeningeal disease.[24] There can be coating of the cisternal segments of the cranial nerves, which may or may not be symptomatic (Fig. 26).

CT imaging is of limited utility for detection of leptomeningeal metastatic disease and often appears normal. Rarely, a "dirty" CSF pattern can be seen where the CSF demonstrates subtle increased attenuation. Contrast-enhanced CT can show linear leptomeningeal enhancement. Involvement of the folia of the rostral cerebellum is a common pattern. In some cases, nodules along the surface of the brain may be seen and are typically difficult to differentiate from peripheral intraparenchymal metastases. Secondary effects suggesting the presence of leptomeningeal disease may be seen such as increased intracranial pressure or hydrocephalus.

Contrast-enhanced MRI is the imaging modality of choice for detection of leptomeningeal metastatic disease. MRI is considered complimentary to lumbar puncture with CSF cytological examination, which is regarded as the gold standard for diagnosis.[24,32,33] The advent of advanced MRI techniques, such as 3D T1-SPACE imaging, has greatly increased the sensitivity for detection of leptomeningeal disease and therefore has in some instances become the initial and sole diagnostic tool (Fig. 27).[14,34] Postcontrast T1WI demonstrates smooth or nodular enhancement involving the pia/subarachnoid space. Patterns of leptomeningeal spread include diffuse linear coating of the surface of the brain and/or spinal cord or multiple nodular lesions along the surface of the brain (Figs. 27 and 28). There can be intraventricular involvement with enhancement along the ependymal surface. Linear or nodular enhancement along the cranial nerves may also be seen (Fig. 26). FLAIR sequences are very sensitive for detection of leptomeningeal disease showing loss of the normal CSF suppression resulting in increased signal in the subarachnoid space (Fig. 29).[3,14] Obtaining the FLAIR sequence after the administration of contrast further increases sensitivity for detecting leptomeningeal disease.

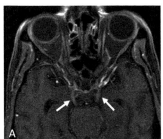 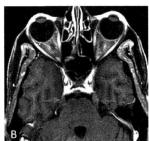 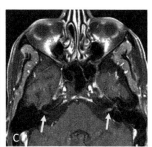 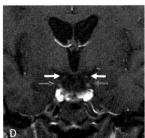

FIG. 26 Leptomeningeal carcinomatosis resulting in multiple cranial nerve palsies. Axial fat-saturated postcontrast T1WI through the brain (A–C) and coronal fat-saturated postcontrast T1WI (D) demonstrate thickening and enhancement of CN III (*white arrows*), CN V (*red arrows*), and CN VII (*yellow arrows*).

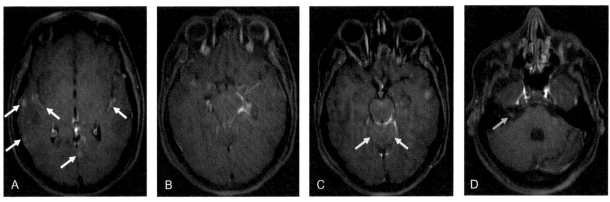

FIG. 27   Breast cancer with leptomeningeal carcinomatosis. Multiple fat-saturated postcontrast T1 SPACE images through the brain demonstrate multifocal linear and nodular enhancement along the pial surface of the brain consistent with leptomeningeal carcinomatosis. In (A), there is both linear and nodular enhancement seen in the sylvian fissures and multiple sulci *(white arrows)*. Seen in (B) and (C), there is coating of the brainstem with enhancement identified in the left ambient cistern and within the interpeduncular fossa *(red arrows)*. There is coating of the cerebellar folia in the superior cerebellum *(yellow arrows)*. There is involvement of the right cranial nerve VII/VIII complex within the right IAC *(blue arrow)* as seen in (D).

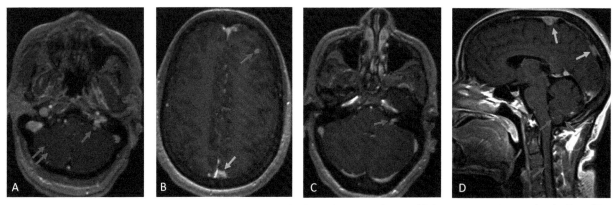

FIG. 28   Leptomeningeal and dural-based metastastic disease. Multiple axial (A–C) and sagittal (D) postcontrast T1-weighted MPRAGE images demonstrate multiple nodular foci of enhancement along the pial surface of the brain consistent with leptomeningeal disease *(red arrows)*. There is also multifocal dural-based metastatic disease *(blue arrows)*. Leptomeningeal metastases can involve the cranial nerves resulting in cranial neuropathies, as seen in (C) where there is involvement of the left cranial nerve VII/VIII complex.

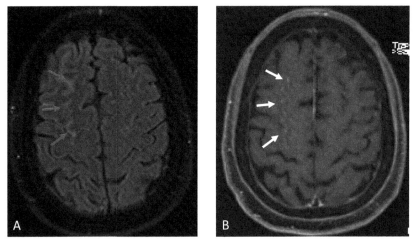

FIG. 29   Leptomeningeal metastatic disease. Axial FLAIR image (A) demonstrates hyperintense signal within the sulci of the right cerebral hemisphere *(red arrows)*. Subarachnoid FLAIR hyperintense signal can be associated with hemorrhage, proteinaceous fluid in the setting of infection/meningitis, or secondary to leptomeningeal carcinomatosis. Axial fat-saturated postcontrast T1WI (B) shows leptomeningeal enhancement *(white arrows)* corresponding to the areas of FLAIR hyperintense signal.

Differential diagnostic considerations include infectious meningitis, neurosarcoidosis, other inflammatory etiologies, and rarely the postictal state.

2.5 Miscellaneous intracranial metastases

Metastatic disease can spread to other less common intracranial locations such as the ventricles, choroid plexus, pituitary gland/infundibular stalk, and to the pineal gland. Intraventricular metastases account for 0.9% to 4.6% of cerebral metastases from extracranial malignancies, the most common being renal cell carcinoma and lung cancer.[35] The lateral ventricle is the most common site of metastases followed by the third ventricle. There are two patterns of intraventricular metastatic disease which include metastatic disease to the choroid plexus (most common) and to the ependymal surface (less common).[3]

Choroid plexus metastases should not be confused with xanthogranulomas (choroid plexus cysts) which are frequently observed, particularly in older patients, and are of no clinical significance. They are typically bilateral with a septated cystic appearance often with associated restricted diffusion.

Differential diagnosis for a choroid plexus mass also includes intraventricular meningioma and choroid plexus papilloma/carcinoma. The imaging appearance of metastatic disease can be indistinguishable from these entities. On imaging, there is mass-like enlargement of the choroid plexus, which usually is T2 and FLAIR hyperintense with heterogeneous enhancement (Fig. 30). These metastases tend to be hypervascular and can lead to intraventricular hemorrhage.[3,36] Less common are ependymal metastases which will demonstrate an enhancing mass along the ependymal margin of the ventricular system (Fig. 31).

Pituitary metastases are found in up to 1.9% of autopsy specimens with 71.5% of these arising from breast, lung, renal, and prostate carcinomas.[37] The most common presenting symptoms are headache, visual field defects, ophthalmoplegia, hypopituitarism, diabetes insipidus, and pituitary apoplexy.[37,38] MRI is the best

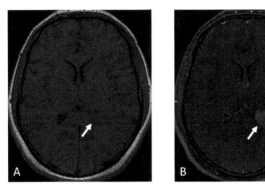 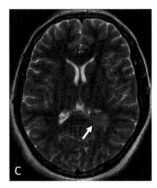

FIG. 30 Metastatic disease to the choroid plexus. Axial noncontrast T1WI (A) and fat-saturated postcontrast T1WI (B) show an avidly enhancing lesion involving the glomus of the left choroid plexus *(white arrows)*. Axial T2WI (C) confirms the intraventricular location and demonstrates mild surrounding vasogenic edema.

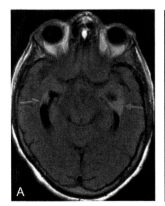 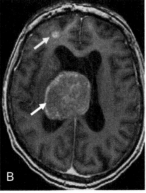 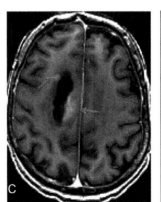 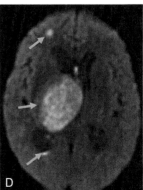

FIG. 31 High-grade neuroendocrine cancer with intraventricular and ependymal metastases. Axial FLAIR image (A) demonstrates nodular areas of FLAIR hyperintense signal along the ependymal surface of the lateral ventricles *(red arrows)*. Axial postcontrast T1WIs (B, C) demonstrate enhancement of a large intraventricular metastasis *(white arrow)*, multiple ependymal metastases *(red arrows)*, and a right frontal intraparenchymal metastatic lesion *(yellow arrow)*. Axial DWI (D) demonstrates restricted diffusion of the metastatic lesions secondary to high cellularity *(blue arrows)*.

imaging modality for evaluating the sella and pituitary gland. Imaging will show an infiltrating and enhancing mass involving the pituitary gland and/or pituitary infundibulum with loss of the posterior pituitary bright spot (Fig. 32). There may be associated bone erosion or cavernous sinus invasion.

Differential diagnostic considerations include pituitary macroadenoma, lymphocytic hypophysitis, and immune checkpoint inhibitor-induced hypophysitis. Pituitary metastases can look very similar to pituitary macroadenomas; however, pituitary macroadenomas rarely present with diabetes insipidus.[3] In addition, rapid growth of a pituitary mass on serial imaging suggests a metastasis.[37]

Metastatic disease can also spread to the pineal gland, although rare, occurring in only 0.3% of extracranial malignancies.[3] MRI is the best imaging modality and will demonstrate an enlarged heterogeneously enhancing pineal region mass (Fig. 33). Pineal region metastases will look very similar to other pineal region masses such as germ cell neoplasms, meningiomas, and primary pineal neoplasms.

2.6 Spine

Metastatic disease to the spine can involve the bone, the epidural space, the leptomeninges, and the spinal cord. Approximately 60%–70% of systemic cancer will have spinal metastases. The spine is the third most common site of metastatic disease behind lung and liver and is the most common site for osseous metastases.[31] Although osseous spinal metastases can be asymptomatic, they can also present with unrelenting back pain. Symptomatic cord compression occurs in 10%–20% of cases with spinal osseous metastatic disease.[31] With epidural tumor extension and cord compression, weakness can develop. Sensory levels may be 1–2 segments below the site of compression due to spinothalamic tract crossing pattern.[39] The thoracic spine is the most common site followed by the lumbar spine and cervical spine, which corresponds to distribution of red marrow.

Metastasis to the spine is most often a result of hematogenous spread. Less common modes of spread include direct extension and CSF dissemination. Hematogenous spread to the vertebral bodies with subsequent spread

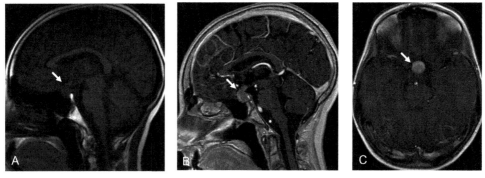

FIG. 32   Metastatic disease to the pituitary gland. Sagittal noncontrast T1WI (A), sagittal postcontrast T1WI (B), and axial postcontrast T1WI (C) demonstrate an enhancing mass involving the pituitary gland and pituitary infundibulum (*white arrows*).

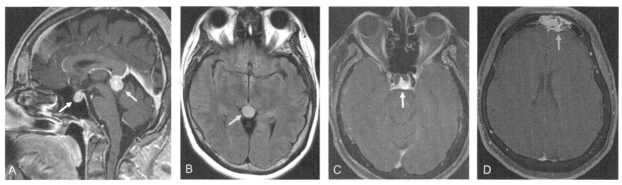

FIG. 33   Multiple sites of extra-axial metastases in this patient with thyroid cancer. Sagittal postcontrast T1WI (A) and axial fat-saturated postcontrast T1WI (B–D) demonstrate nodular enhancing mass lesions involving the pituitary gland (*white arrow*), pineal gland (*yellow arrow*), and left frontal sinus (*red arrow*) consistent with metastatic disease. In (D), the left frontal sinus metastasis (*red arrow*) extends through the inner table and involves the adjacent dura (*blue arrow*).

to the epidural space is common. Hematogenous spread can occur via arterial seeding or through the valveless Batson plexus which provides venous drainage of the vertebral bodies. Flow within Batson plexus is variable due to changes in the intrathoracic and intrabdominal pressures. The homing and receptive properties of the tumor cells are more important than the vascular route. Cancers with a high rate of metastases to bone include prostate, breast, and lung which account for most osseous spinal metastases.[39]

Extradural metastases account for up to 95% of spinal metastatic disease.[31] This includes osseous metastases, osseous metastases with epidural/paraspinal extension, and the rare isolated epidural metastasis. Spinal osseous metastases can be lytic, mixed, or blastic. Lytic metastases are most often seen in breast, lung, kidney, thyroid, oropharyngeal, melanoma, adrenal, and uterine primaries. Breast, lung, ovarian, testicular, and cervical cancer can be mixed lytic and blastic. Prostate, bladder, nasopharynx, medulloblastoma, neuroblastoma, and carcinoid primaries are most often blastic.[31]

Radiography should be avoided in evaluation for bony metastatic disease as it requires a 1 cm diameter mass and 50%–70% bone mineral loss for detection in the setting of a lytic metastasis.[39] Blastic metastases often present as multiple mottled areas of sclerosis.

Nuclear medicine bone scan is the standard imaging modality for screening of the entire skeleton. Most often, metastases will appear as focal areas of increased uptake. Sometimes, aggressive lytic metastases can appear as "cold" lesions. A nuclear medicine bone scan is not imaging the tumor itself but rather bone production in reaction to the tumor. Another pattern to be aware of is the "super scan" in the setting of diffuse osseous metastatic disease most often seen in the setting of prostate cancer (Fig. 34). False-negative scans are most common in multiple myeloma, leukemia, and anaplastic carcinomas. Any process which leads to bone turnover including degenerative disease, trauma, and infection can lead to increased uptake and a false positive. Comparing contemporaneous imaging and anatomical knowledge of where metastases occur and where degenerative disease occurs can help in preventing false positives. In patients with known primary malignancy, 50% of solitary foci of increased uptake will reflect metastatic disease. Multiple lesions of increased uptake are more likely to reflect osseous metastatic disease.[31]

FDG PET is useful in detecting osseous metastatic disease due to increased glucose metabolism of hypermetabolic malignancy (Fig. 35). Typically, FDG PET is fused with CT which has increased specificity in comparison with FDG PET alone.[39] Most studies indicate that FDG PET is more accurate than bone scan although this is predominantly due to improved specificity rather than sensitivity. Studies have suggested that FDG PET is more

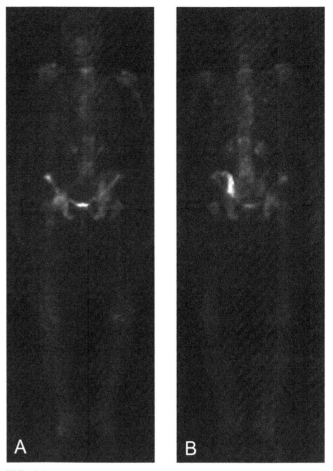

FIG. 34 Multifocal osseous metastases. Anterior (A) and posterior (B) whole-body bone scan demonstrates multiple focal and patchy areas of radiotracer uptake secondary to osseous metastatic disease to the axial and appendicular skeleton.

sensitive for lytic metastases and less sensitive for blastic metastases than bone scan.[40] Treated osteolytic metastases can be FDG negative, yet remain positive on bone scan.[39]

CT in the setting of osteoblastic disease will show sclerotic lesions often with coexisting areas of osteolysis. Lytic disease will show destruction of cancellous and cortical bone. Often lesions are round and have indistinct margins. Destruction of the posterior cortex and involvement of the pedicle is common (Fig. 36). Osteolytic metastases often become sclerotic with treatment. Both blastic and lytic metastases can be associated with paraspinal and epidural soft tissue masses which enhance and are therefore easier to visualize with the addition of iodinated contrast. CT myelography is indicated when there is concern for cord compression and the patient is unable to undergo MRI.

MRI is the most useful imaging modality in the setting of spinal metastatic disease. It can detect early bone marrow metastatic deposits which would not be detected by CT or nuclear medicine bone scan. It not

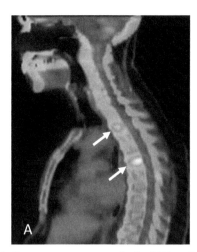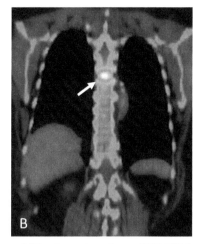

FIG. 35   Focal hypermetabolic osseous metastases. Sagittal (A) and coronal (B) reformations from a PET/CT demonstrate focal regions of increased radiotracer activity involving multiple vertebral bodies *(white arrows)* consistent with hypermetabolic metastases.

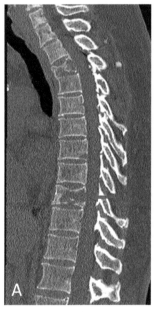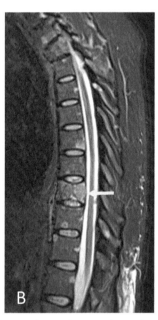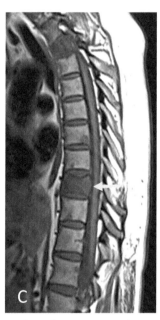

FIG. 36   Metastatic disease to the thoracic spine with pathological fracture. Sagittal CT (A) shows lytic metastatic disease with destruction of posterior cortex *(red arrow)*. STIR (B) and T1WI (C) in the same patient demonstrates bowing of the posterior cortical margin *(yellow arrows)* and complete fatty marrow replacement typical of a pathological fracture.

only answers the clinical question of is there metastatic disease present but localizes the levels of involvement and determines if there is associated cord or nerve root compression. Metastases will demonstrate decreased T1 signal relative to normal marrow signal, and therefore, a traditional T1WI is a basic but essential aspect of an MRI protocol evaluating spinal metastatic disease (Figs. 37 and 38). One of the most important sequences for detection of osseous metastatic disease is a fat-saturated T2-weighted sequence which typically demonstrates hyperintense signal although blastic metastases can be hypointense. Traditionally, this is accomplished with a STIR sequence. Alternatively, at our institution, Dixon

fat saturation is utilized when possible to provide homogeneous fat saturation and increase the signal-to-noise ratio in comparison with STIR.[41] The Dixon acquisition produces a fat-only sequence which is very useful when evaluating osseous metastases as they appear as "black holes" in the background of bright fatty marrow signal. The other essential sequence in evaluation of osseous metastatic disease is the T1 fat-saturated postcontrast sequence. Osseous metastases, with the rare exception of a completely blastic/sclerotic metastasis, will demonstrate enhancement which is made conspicuous with fat saturation. The Dixon postcontrast T1-weighted sequence is employed when possible to produce more uniform fat

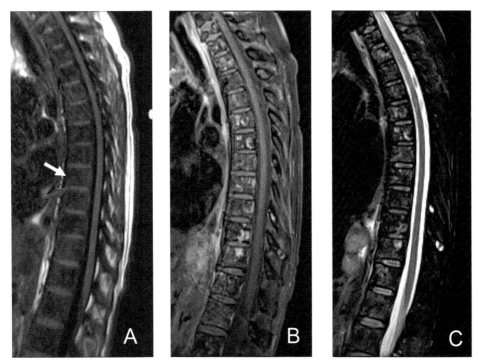

FIG. 37 Diffuse osseous infiltration. Sagittal noncontrast T1WI (A), sagittal fat-saturated postcontrast T1WI (B), and sagittal STIR image (C) of the thoracic spine demonstrate diffuse osseous metastatic disease. There is diffuse marrow infiltration as seen in (A) with loss of the normal T1 hyperintense fatty bone marrow signal. The bone marrow *(white arrow)* is darker than the adjacent intervertebral disc *(red arrow)*. There is diffuse heterogeneous enhancement seen in (B) and diffuse heterogeneous STIR hyperintense signal seen in (C) secondary to the underlying metastatic disease.

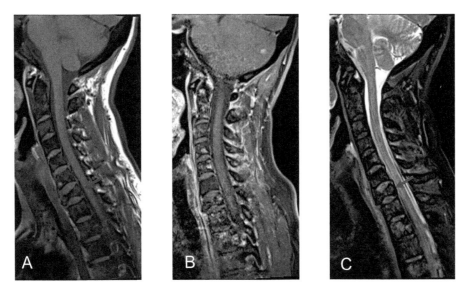

FIG. 38 Diffuse osseous metastastic disease with pathologic fracture. Sagittal noncontrast T1WI (A), sagittal fat-saturated postcontrast T1WI (B), and sagittal STIR image (C) of the cervical spine demonstrate diffuse osseous metastatic disease with diffuse T1 hypointense signal within the bone marrow (A), heterogenous enhancement (B), and heterogeneous STIR hyperintense signal (C). There is a chronic pathologic fracture at C7 *(red arrows)*. Note the rounded bowing of the posterior cortical margin, which is often seen with pathologic fractures caused by underlying metastatic disease.

saturation than traditional chemical fat saturation. Some institutions obtain a DWI sequence when imaging spinal metastatic disease although its utility in this setting is controversial.

Typically, spinal metastatic disease presents as multiple focal T1 hypointense enhancing lesions. A diffuse marrow infiltration pattern on imaging will appear as diffusely decreased T1 marrow signal with associated enhancement. If the T1 signal of the marrow is hypointense relative to the adjacent disc, this must be viewed with suspicion (Fig. 37). MRI provides exquisite soft tissue detail to evaluate for epidural disease, cord compression, and if there is associated cord edema (Fig. 39). The traversing nerve roots within the spinal canal and the exiting nerve roots can also be evaluated for compression resulting clinically in radiculopathy. Isolated epidural disease without adjacent osseous involvement is rare and most often seen in lymphoma. Epidural lesions are typically T1 hypointense and T2 hyperintense with avid enhancement.

Pathologic fractures are often encountered in the setting of spinal metastatic disease. Both benign and malignant compression fractures will demonstrate increased T2 signal and enhancement. Imaging findings which raise concern for malignant fracture include associated paraspinal or epidural soft tissue mass, other osseous lesions consistent with metastases, and rounded bowing of the posterior cortical margin (Fig. 36). Benign fractures tend to have angulated margins not rounded margins. Also, a low-intensity well-delineated line representing the fracture line or a circumscribed band of fluid signal is often seen in benign compression fractures.[31] Some report that pedicle involvement suggests malignant fracture although in practice it is common for a benign fracture to demonstrate edema within the pedicle.

Common mimics of spinal osseous metastases include benign intraosseous hemangiomas. Typically, hemangiomas demonstrate increase T1 signal on MRI due to their fatty content while spinal osseous metastatic disease demonstrates decreased T1 signal. Atypical lipid poor hemangiomas can occur which demonstrate decreased T1 signal although they will often demonstrate the characteristic coarsened vertical trabeculae with a "corduroy" appearance of a hemangioma. This pattern is typically better delineated on CT. Acute Schmorl's nodes can demonstrate signal intensity matching that of a metastasis and can enhance. However, they are identified by their characteristic circumscribed involvement of the vertebral body endplate. Discitis/osteomyelitis is distinguished by its characteristic edema and enhancement of the disc with extension to involve the adjacent endplates and marrow spaces. Metastatic disease typically does not cross the disc space. Patchy fatty marrow signal can be seen normally in older patients. T1 marrow signal remains hyperintense to the adjacent disc in the setting of heterogeneous marrow. Hematopoietic malignancy and nonmalignant marrow proliferative and marrow replacing processes can mimic diffuse spinal osseous metastatic disease. However, diffuse marrow involvement is much more common in these entities.

Other patterns to be aware of are marrow signal alterations related to treatment effect. Following radiotherapy, there will be increased fatty T1 and T2 hyperintense marrow signal in a geographic pattern which conforms to the treatment field (Fig. 40). Chemotherapeutic agents and/or anemia can result in conversion of yellow marrow to red marrow resulting in diffusely decreased T1 marrow signal which can mimic diffuse marrow infiltration or proliferation.

The incidence of leptomeningeal carcinomatosis is increasing due to improved systemic therapies resulting in patients living longer with systemic cancer. Contrast-enhanced CT may show enhancement of the

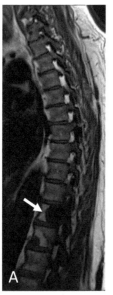  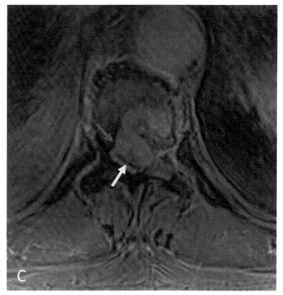

FIG. 39   Prostate osseous metastatic disease with epidural extent. Sagittal noncontrast T1WI (A) and fat-saturated postcontrast T1WI (B, C) demonstrate a T1 hypointense and enhancing lesion *(white arrows)*. There is epidural extension of disease into the ventral and left lateral epidural space *(red arrows)* resulting in severe stenosis of the thecal sac and compression of the spinal cord *(yellow arrow)*.

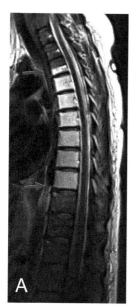
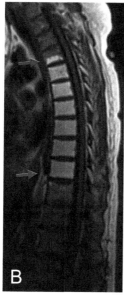

FIG. 40 Typical radiation induced bone marrow changes. Sagittal T2WI (A) and T1WI (B) of the thoracic spine demonstrate an abrupt transition of bone marrow signal within the upper-to-mid thoracic spine (outlined by the *red arrows*), which correlated with the radiation field. There is increased T2 and T1 signal within the vertebral bodies secondary to diffuse fatty replacement of the red marrow.

diagnostic considerations include pyogenic, granulomatous, and chemical meningitis. Recent lumbar puncture can also lead to leptomeningeal and dural enhancement.

Intramedullary metastases are rare and are often solitary. They are most often seen in small cell lung cancer.[31] On MRI, the appearance is of a small focal circumscribed enhancing intramedullary nodule with extensive surrounding edema (Fig. 42). There is often associated cord expansion and rarely syrinx formation. Intramedullary metastases can be hemorrhagic. Differential diagnostic considerations include demyelinating disease and inflammatory myelitis which typically present with less associated cord edema and expansion. Additional differential considerations include the more common primary cord neoplasms including ependymoma, astrocytoma, and hemangioblastoma.

2.7 Malignant and treatment-induced plexopathy

Patients with malignancy often present with neurologic symptoms. A high degree of clinical suspicion is necessary for accurate diagnosis, particularly in the setting of brachial and lumbosacral plexopathy. Clinically, differentiation between local tumor invasion, metastatic disease, and radiation-induced plexopathy can be challenging.

In the acute setting, radiation-induced plexitis often demonstrates diffuse T2 hyperintense signal and non-mass-like thickening with acute denervation edema of the innervated musculature (Fig. 43).

cauda equina or along the surface of cord, although it may also be normal. Contrasted MRI is a much more sensitive modality for detecting leptomeningeal disease. Multiple imaging patterns exist and include thin linear enhancement along the surface of the cord and nerve roots of the cauda equina, a solitary enhancing mass at the caudal terminus of the thecal sac or along the surface of the cord, diffuse thickening of the cauda equina, and a miliary appearance with numerous small enhancing leptomeningeal nodules (Fig. 41). Primary differential

More commonly, patients present with radiation-induced brachial and lumbosacral plexopathy. Patients receiving greater than 60Gy are at greatest risk. Symptoms develop after 5–6months and peak between 12 and 20months posttreatment.[42] Imaging demonstrates

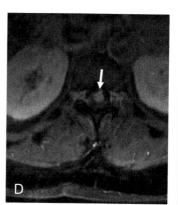
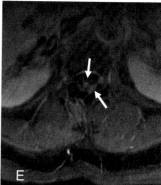

FIG. 41 Spinal leptomeningeal metastases. Sagittal fat-saturated postcontrast T1WIs through the cervical spine (A), thoracic spine (B), and lumbar spine (C), and axial fat-saturated postcontrast T1WI through the lumbar spine (D and E) demonstrate multiple nodular foci of enhancement along the pial surface of the spinal cord and along the nerve roots of the cauda equina (*white arrows* on D and E). There is a prominent conglomerate of metastatic deposits in the dependent portion of the thecal sac seen in the sacral spinal canal on (C) *(red arrows)*.

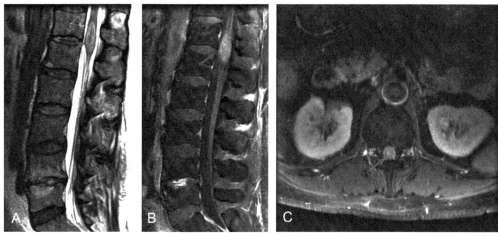

FIG. 42   Spinal intramedullary metastasis. Sagittal T2WI (A), sagittal (B), and axial (C) fat-saturated postcontrast T1WI demonstrate an intramedullary enhancing T2 hyperintense mass involving the distal spinal cord/conus *(red arrows)*.

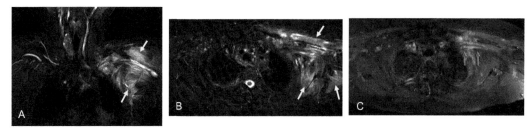

FIG. 43   Radiation-induced brachial plexopathy. Coronal STIR (A) and axial STIR (B) through the level of the axilla demonstrate increased signal along the left brachial plexus *(red arrows)* with associated enhancement on the axial fat-saturated postcontrast T1WI (C). There is also increased STIR hyperintense signal involving the muscles of the left shoulder which is secondary to acute denervation changes *(yellow arrows)*.

diffuse thickening and enhancement without focal mass. Diffuse T2 hypointense signal along the involved plexus has been shown to be the most useful differentiating imaging feature between radiation-induced plexopathy and tumor infiltration.[42–44]

Metastatic breast and lung cancers are the most common causes of malignant plexopathy. Characteristic imaging features of metastatic disease and focal tumor invasion demonstrate enhancing lesions with corresponding T1 hypointense and T2 hyperintense signal when compared to the adjacent muscles. Rarely, lesions may demonstrate T2 hypointense signal.[42–46] PET imaging may be useful as untreated tumor involvement can demonstrate increased FDG activity (Fig. 44).

3 Intracranial complications of systemic disease

Patients suffering from malignancy often experience neurologic complications, ranging from metastatic disease to paraneoplastic syndromes to treatment-related complications. In this section, we review the imaging findings associated with the complications of systemic cancer and paraneoplastic syndromes.

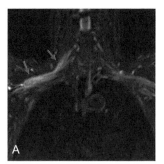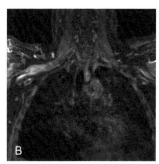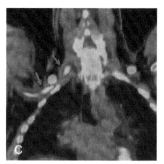

FIG. 44   Peripheral neurolymphomatosis of the brachial plexus. Coronal STIR (A) and fat-saturated postcontrast T1WI (B) demonstrate asymmetric hyperintense signal, thickening, and enhancement of the right brachial plexus. Coronal PET/CT image (C) through this region demonstrates corresponding increased FDG uptake *(red arrows)*.

3.1 Posterior reversible encephalopathy syndrome

Posterior reversible encephalopathy syndrome (PRES), initially described in 1996 by Hinchey et al., is a heterogeneous disorder characterized by sudden onset headache, altered mental status, seizure, and/or visual disturbance.[47,48] The underlying pathophysiology is poorly understood; however, the most commonly accepted theory is impaired cerebral autoregulation resulting in hyperperfusion and breakdown of the blood brain barrier. The posterior circulation is most susceptible due to the lack of sympathetic tone in the basilary artery.[49,50] Although the name implies a clear disease process, patients can present with anterior circulation involvement, irreversible ischemia, and/or hemorrhage. Furthermore, patients are not always encephalopathic.

The list of potential risk factors for development of PRES is long, with the most frequently described including hypertension, pregnancy, hypercoagulability, autoimmune disorders, use of immunosuppressive agents, and chemotherapeutic treatment. Numerous studies have shown a clear association between PRES and patients being treated for cancer. In 2015, Singer et al. conducted a retrospective review which demonstrated a 2:1 increased risk in women. Patients being treated for solid tumors (71%) were more commonly impacted than those with hematologic malignancies (26%). This study showed half of patients who developed PRES-received systemic therapy within the prior month.[51]

At the time of imaging, the majority of patients will demonstrate vasogenic edema within the parietal and/or occipital lobes, which is typically symmetric. Not infrequently, patients with PRES will have atypical imaging findings which include vasogenic edema outside of the parietal and occipital lobes, asymmetry, involvement of the cortex that may or may not progress to infarction, and hemorrhage. Atypical PRES frequently involves the watershed territories within the cerebral hemispheres, brainstem, and cerebellum (Fig. 45).

3.2 Methotrexate-induced leukoencephalopathy

Methotrexate (MTX), a commonly used chemotherapeutic agent, is known to have potential neurotoxicity. MTX inhibits dihydrofolate reductase, preventing the conversion of folic acid to tetrahydrofolic acid thereby inhibiting cell replication. Folate deficiency leads to a buildup of homocysteine which can induce a small-vessel vasculopathy.[52]

In a subset of patients, MTX has been shown to induce a toxic leukoencephalopathy. The reported incidence ranges from 0.8% to 4.5%, with rates increasing up to 40% in patients receiving intrathecal chemotherapy, high-dose systemic therapy, and concurrent radiation.[48] Three main imaging patterns have been described and include toxic leukoencephalopathy, disseminated necrotizing encephalopathy, and methotrexate-induced myelopathy.

Toxic leukoencephalopathy typically presents in the acute setting though can be seen years after drug administration. In the acute setting, brain MRI demonstrates a combination of asymmetric restricted diffusion and vasogenic edema throughout the centrum semiovale and corona radiata that crosses multiple vascular territories and spares the subcortical U fibers (Fig. 46).

Disseminated necrotizing encephalopathy is a rare, often fatal complication of combined intrathecal methotrexate and whole-brain radiation.[48] Clinically, patients present with rapidly progressive dementia, urinary incontinence, gait disturbance, and hemiparesis. Imaging demonstrates more extensive white matter involvement with multifocal regions of mass-like signal abnormality, characterized by central hemorrhage (T2 hypointensity and susceptibility artifact) with variable peripheral or solid enhancement.[48] Unfortunately, there are no definitive imaging features that allow for differentiation

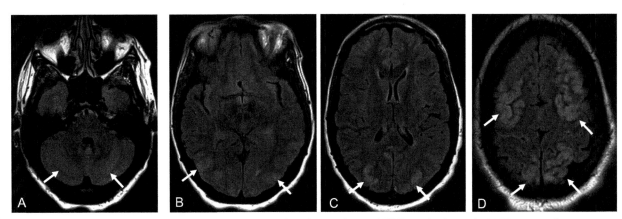

FIG. 45 Chemotherapy-induced posterior reversible encephalopathy syndrome (PRES). Multiple axial FLAIR images demonstrate patchy symmetric areas of cortical and subcortical hyperintense signal involving the parieto-occipital regions and watershed territories *(white arrows)*.

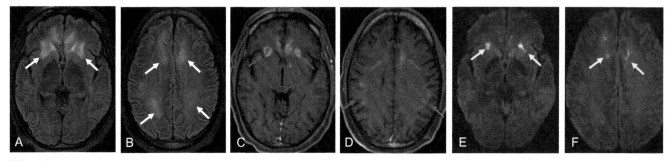

FIG. 46　Acute methotrexate-induced leukoencephalopathy. Axial FLAIR images of the brain (A and B) demonstrate patchy and confluent regions of hyperintense signal involving the supratentorial white matter *(white arrows)*. These areas demonstrate heterogeneous enhancement *(red arrows)* seen on the fat-saturated postcontrast T1WI (C and D) and patchy areas of restricted diffusion *(yellow arrows)* on DWI (E and F).

between disseminated necrotizing encephalopathy and recurrent disease.

Finally, methotrexate-induced myelopathy is an exceedingly rare complication that mimics subacute combined degeneration clinically and from an imaging standpoint.[53] Spine MRI demonstrates long segment signal abnormality involving the dorsal columns (Fig. 47). The only clear distinction between MTX-induced myelopathy and subacute combined degeneration is a normal B12 level.

Similar reversible and irreversible toxic leukoencephalopathies have been demonstrated in patients receiving other chemotherapeutic agents.

3.3　Cytarabine-induced vasospasm

High-dose IV cytarabine has been shown to have neurotoxic effects including seizures, encephalopathy, and acute cerebellar syndrome. Rarely, pediatric patients with ALL being treated with intrathecal cytarabine present with thunderclap headache secondary to underlying cerebral vasospasm with or without developing infarct.[54,55] Some reports suggest reversible cerebral

vasoconstriction syndrome or vasculitis in this setting though when further evaluated they demonstrate no appreciable vessel wall enhancement.

3.4　Immune checkpoint inhibitor-induced hypophysitis

Immune checkpoint inhibitor-induced hypophysitis was first described by Phan et al. in 2003 with ipilimumab therapy.[56] Since that time, hypophysitis has been encountered with multiple immune checkpoint inhibitors and is most frequently encountered with combination therapy. Barroso-Sousa et al. demonstrated incidence rates of immune checkpoint inhibitor-induced hypophysitis of 6.4% when treated with ipilimumab and PD-1 inhibitors, 3.2% with ipilimumab alone, 0.4% with PD-1 inhibitors alone, and < 0.1% with PD-L1 inhibitors alone.[57] Endocrine toxicity, requiring hormone replacement, is seen in up to 90% of patients with immune checkpoint inhibitor-induced hypophysitis and rarely can be fatal.[58]

Imaging features are often nonspecific. The most common finding is diffuse enlargement and enhancement of

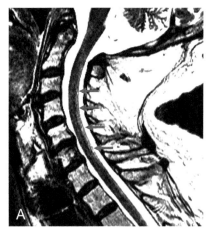
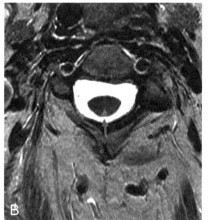

FIG. 47　Methotrexate-induced myelopathy. This patient was treated with intrathecal methotrexate for lymphoma. Sagittal (A) and axial (B) T2WI demonstrates longitudinal T2 hyperintense signal involving the dorsal columns of the spinal cord *(red arrows)* consistent with methotrexate-induced myelopathy which mimics subacute combined degeneration.

the pituitary infundibulum, with some reports demonstrating this finding in 100% of patients (Fig. 48).[58] Patients can also demonstrate variable enlargement of the pituitary gland, heterogeneous hypoenhancement throughout the adenohypophysis, and loss of the normal T1 hyperintensity throughout the neurohypophysis.[58,59] If left untreated, the gland will demonstrate progressive necrosis with corresponding volume loss.

3.5 Progressive multifocal leukoencephalopathy

Progressive multifocal leukoencephalopathy (PML) is a demyelinating condition resulting from reactivation of the JC virus within oligodendrocytes. PML most commonly impacts immunocompromised patients and has been seen in patients treated with immune modulatory therapy including many of the monoclonal antibody treatments. If left untreated, PML is usually fatal within 1 year.

Imaging requires a high degree of clinical suspicion for differential consideration. CT is nonspecific with asymmetric periventricular and subcortical hypoattenuation. MR imaging is more specific, demonstrating asymmetric periventricular and subcortical T2/FLAIR hyperintense signal with extension to involve the subcortical U fibers (Fig. 49). Often there is faint diffusion restriction along the leading edge of demyelination. Contrast enhancement is atypical though has been reported in the setting of immune reconstitution syndrome.

3.6 Autoimmune and paraneoplastic encephalitis

Autoimmune and paraneoplastic encephalitis is a heterogeneous group of disorders with widely variable clinical and imaging presentations. Although the limbic system is most frequently impacted, involvement of the neocortex, basal ganglia, brainstem, cerebellum, and spinal cord has been well documented.[60–62] As such, the imaging appearance is variable and requires heightened clinical suspicion.

There are two commonly accepted classification schemes which categorize the autoimmune

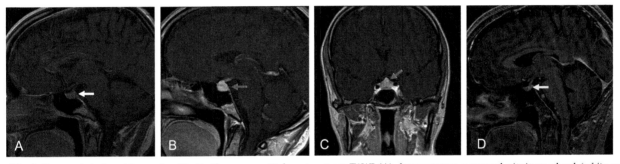

FIG. 48   Ipilimumab-induced hypophysitis. Pretherapy sagittal postcontrast T1WI (A) demonstrates a normal pituitary gland (*white arrow*). Sagittal (B) and coronal (C) postcontrast T1WI obtained 3 months after initiating treatment with Ipilimumab demonstrates enlargement of the pituitary gland and infundibulum secondary to immune-related hypophysitis (*red arrows*). Follow-up sagittal postcontrast T1WI (D) performed 5 months later after terminating Ipilimumab demonstrates resolution of the hypophysitis (*yellow arrow*).

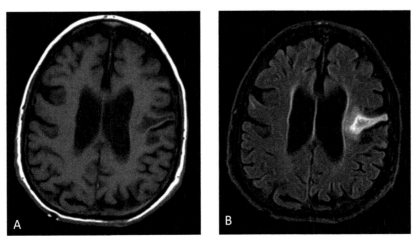

FIG. 49   Progressive multifocal leukoencephalopathy. Axial noncontrast T1WI (A) and axial FLAIR image (B) demonstrate a T1 hypointense and FLAIR hyperintense lesion involving the subcortical white matter including the subcortical U-fibers. For the size of the lesion, there is no significant mass effect. PML tends to be multifocal, asymmetric, and demonstrates little or no enhancement. These lesions are caused by the JC virus which infects the oligodendrocytes resulting in demyelination.

encephalitides based on the location of the neuronal antigens and whether or not the encephalitis is paraneoplastic or nonparaneoplastic. Distinction based on location of the neuronal antigen is said to be more clinically relevant as it factors in an association with underlying malignancy, treatment response, and prognosis.[62] Group I antibodies target intracellular neuronal antigens, are closely associated with an underlying malignancy, and are characterized by decreased response to therapy resulting in poor clinical outcomes. Given the heterogeneity, imaging features are often nonspecific (Table 1). Group II antibodies target cell-surface neuronal antigens, are less commonly associated with malignancy, and respond better to immune therapy. Furthermore, Group II antibodies have been shown to be more specific clinical markers of disease and are more closely linked with imaging findings (Table 2).

Classic limbic involvement will demonstrate T2/FLAIR hyperintense signal in the cortex and subcortical white matter throughout the mesial temporal lobe with or without restricted diffusion (Fig. 50). Enhancement is variable and can be mass-like mimicking metastatic disease. Ultimately, the imaging appearance necessitates a differential diagnosis including herpes encephalitis, postictal changes, and rarely metastatic disease. Symmetric involvement of the mesial temporal lobes strongly favors limbic encephalitis over herpes encephalitis.

TABLE 1   Group I antibodies.[62]

Antibody	Clinical association	Symptoms
Anti-Hu	Small cell lung cancer	Encephalomyelitis, cerebellar degeneration, epilepsia partialis continua, and sensory neuropathy
Anti-Ma/Ta	Testicular cancer in young males. Small cell lung and breast cancers in older adults	Combined limbic, diencephalic, or brainstem dysfunction often manifesting as ophthalmoplegia. Isolated limbic encephalitis is rare
Anti-glutamic acid decarboxylase	Rarely associated with malignancy	Classic limbic system involvement resulting in seizures. Stiff person syndrome
Anti-amphiphysin	Small cell lung and breast cancers	Encephalomyelitis, myoclonus, and stiff person syndrome
Anti-Ri (neuronal nuclear antibody 2)	Small cell lung and breast cancers	Brainstem encephalitis and opsoclonus-myoclonus syndrome
Anti-Yo	Ovarian and breast cancers	Cerebellar degeneration and less commonly encephalitis

TABLE 2   Group II antibodies.[62]

Antibody	Clinical association	Symptoms	Imaging
Anti-NMDA	Young women and children. Rare association with ovarian teratoma	Initial viral prodrome followed by psychiatric symptoms, temporal lobe dysfunction, and severe neurologic dysfunction	Often normal at the time of presentation. Progresses to cortical and subcortical edema with cortical enhancement and no restricted diffusion. PET imaging may show increased metabolic activity even when the brain MRI is normal
Anti-VGKC	No association with malignancy	Classic limbic encephalitis with development of medically refractory epilepsy. Extralimbic involvement is rare	In the acute setting, there is edema in the mesial temporal lobes. A small subset of patients demonstrate restricted diffusion and contrast enhancement. Commonly progress to mesial temporal sclerosis
Anti-VGCC	Rare subtype described in women and young children	Initial viral prodrome followed by psychiatric symptoms, temporal lobe dysfunction, and seizures	Mesial temporal edema with gyriform cortical enhancement progressing to cortical laminar necrosis. Extralimbic involvement is common
Anti-GABAr A receptor	No association with malignancy. Good prognosis		Mesial temporal and extensive extra-limbic T2/FLAIR hyperintense signal with or without enhancement
Anti-GABAr B receptor	Small cell lung carcinoma and pulmonary neuroendocrine tumors	Limbic encephalitis with frequent seizures often preceding the cancer diagnosis	Mesial temporal lobe edema with or without enhancement
Anti-AMPAr	Lung, breast, and thymic malignancy	Subacute onset of psychiatric symptoms	Edema isolated to the hippocampus

TABLE 2   Group II antibodies—cont'd

Antibody	Clinical association	Symptoms	Imaging
Anti-GluR3	Rasmussen encephalitis		
Anti-mGluR1	Lymphoma	Cerebellar ataxia	
Anti-mGluR5	Hodgkin lymphoma	Limbic encephalitis	
Anti-D2 dopamine receptor		Basal ganglia encephalitis	
Anti-GlyR1		Stiff leg syndrome, stiff person syndrome, or progressive encephalomyelitis with rigidity and myoclonus	

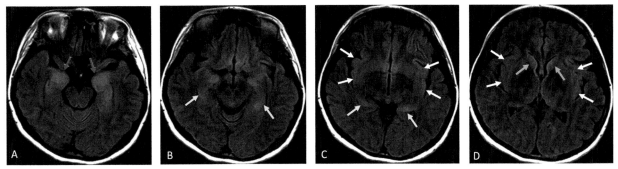

FIG. 50   Paraneoplastic encephalitis in a patient with small cell lung cancer. Contiguous FLAIR images demonstrate symmetric increased FLAIR hyperintense signal involving the bilateral amygdala *(red arrows)* and bilateral hippocampi *(yellow arrows)*, which is classic for limbic (paraneoplastic) encephalitis. Other areas that can also be involved include the bilateral insula *(white arrows)* and bilateral caudate nuclei *(blue arrows)*. This pattern can be similar to herpes encephalitis, but these entities can be differentiated based on clinical history and time course. Herpes is also typically hemorrhagic, bilateral, and asymmetric.

Paraneoplastic cerebellar degeneration is rare and most commonly seen with small cell lung cancer, Hodgkin lymphoma, breast cancer, and gynecologic cancer. It presents with vertigo, dizziness, and gait instability. It is most often associated with anti-Yo and ant-Ri antibodies. Acutely, MR is often normal but can show diffuse enlargement with associated increased T2/FLAIR signal. In the subacute to chronic period, cerebellar atrophy and hypometabolism on PET are characteristic (Fig. 51).

3.7 Radiation necrosis and pseudoprogression

Differentiation between treatment-related effects (radiation necrosis and pseudoprogression) and disease progression is often difficult as both can produce similar clinical and imaging features. Distinction is necessary to drive patient management. This is particularly true in the setting of targeted radiotherapy. A thorough understanding of the patient's treatment regimen and time course is essential when approaching imaging for this patient population.

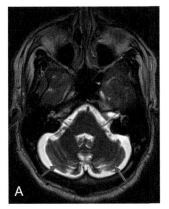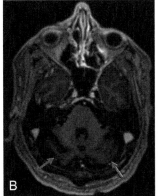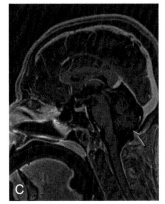

FIG. 51   Paraneoplastic cerebellar degeneration. Axial T2WI (A), axial postcontrast T1WI (B), and sagittal postcontrast T1WI (C) demonstrate diffuse cerebellar atrophy *(red arrows)*. This patient had a history of breast cancer and presented with progressive ataxia.

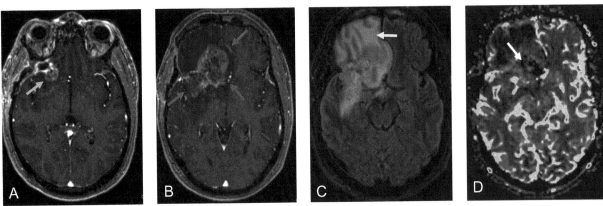

FIG. 52   Pseudoprogression. Axial postcontrast T1WI (A) demonstrates postoperative changes following partial resection of a metastatic lesion involving the inferior right frontal lobe with residual enhancement remaining (*blue arrow*). Axial postcontrast T1WI (B), axial FLAIR (C), and rCBV map (D) were obtained 4 months after radiation therapy. Findings show progression of the heterogeneous enhancement involving the right frontal lobe (*red arrows*) with progressive surrounding edema on the FLAIR image (*yellow arrows*). MR perfusion shows no corresponding elevation of the cerebral blood volume in the region of abnormal enhancement (*white arrow*).

Advanced imaging techniques such as MR perfusion increase specificity in differentiating true disease progression from posttreatment changes (Fig. 52). When compared to the contralateral normal white matter, disease progression will typically demonstrate rCBV ratios of greater than 2 to 1. Treatment effect typically demonstrates rCBV ratios of less than 1.8 to 1. When the rCBV ratio falls between 1.8 and 2, findings are indeterminate.[63,64] Numerous permeability parameters have been shown to improve diagnostic accuracy; however, technical variability between scanners and institutions limits widespread applicability without determination of institutional thresholds.

3.8 Radiation-induced arteriopathy

Intracranial radiation-induced arteriopathy is an underrecognized complication of whole-brain and stereotactic radiation therapy. With the advent of new imaging techniques, including vessel wall imaging and 3D T1 SPACE postcontrast sequences, imaging allows for more accurate diagnosis. Intracranial radiation-induced arteriopathy is most often thought of as an acute complication though more recent literature suggests that it can occur up to 19 years after radiation.[65] Vessel wall imaging and black blood postcontrast sequences demonstrate smooth concentric wall enhancement with luminal narrowing in the affected vessel (Fig. 53). Unfortunately, this complication will often first be encountered when the patient presents with stroke-like symptoms.

It is common to see confluent increased T2/FLAIR signal involving the periventricular and deep supratentorial white matter following chemoradiation (Fig. 54). Mineralizing microangiopathy is another pattern which can be seen following chemoradiation and appears as symmetric subcortical calcification on CT (Fig. 55).

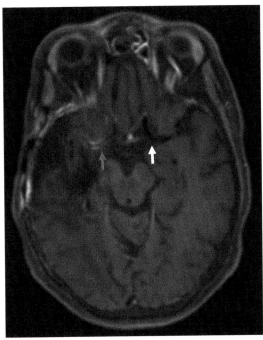

FIG. 53   Radiation-induced arteriopathy. Fat-saturated postcontrast T1 SPACE image demonstrates circumferential vessel wall enhancement of the supraclinoid right ICA and proximal right MCA (*red arrow*). In contrast, the left ICA and MCA are normal (*white arrow*).

3.9 Stroke-like migraine attacks after radiation therapy syndrome

Stroke-like migraine attacks after radiation therapy (SMART) syndrome is a well-documented delayed complication of whole-brain radiation. Patients typically present years after radiation with stroke-like symptoms with classic imaging findings. Historically, symptoms resolve completely within a couple of weeks. However, there is increasing evidence that symptoms and imaging findings are not always

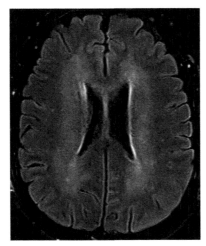 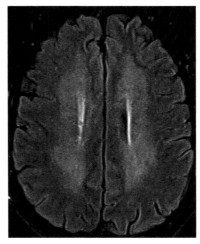

FIG. 54 Typical radiation-induced white matter changes. Axial FLAIR images through the brain demonstrate indistinct confluent hyperintense signal within the periventricular and deep white matter of the cerebral hemispheres with sparing of the subcortical white matter. This is the typical appearance of radiation-induced white matter changes, which mimics chronic microvascular ischemic change.

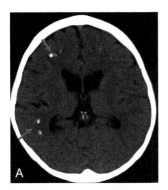 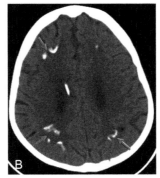 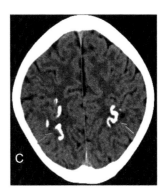

FIG. 55 Chemotherapy-induced mineralizing microangiopathy. Axial noncontrast CT images in this 12 year old who was treated with chemotherapy for acute lymphocytic leukemia demonstrates multiple scattered foci of subcortical calcification (red arrows). Chemotherapy-induced mineralizing microangiopathy typically occurs 2 or more years following treatment.

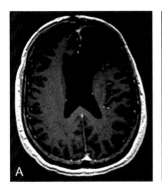 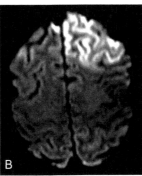 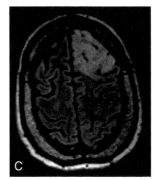 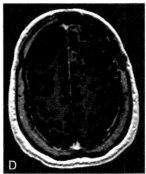

FIG. 56 Stroke-like migraine attacks after radiation therapy (SMART) syndrome. Postcontrast T1WI (A) shows a resection cavity involving the anterior left frontal lobe. DWI (B), FLAIR (C), and postcontrast T1WI (D) demonstrate restricted diffusion, increased FLAIR signal, and patchy enhancement with sulcal effacement. A 6-month follow-up without treatment showed resolution.

reversible.[66] MR imaging features are characteristic and include unilateral cortical thickening with T2/FLAIR hyperintense signal and gyral cortical enhancement that does not conform to a specific vascular territory (Fig. 56). Perfusion imaging can further complicate the clinical picture depending on when it is performed. Some studies have shown decreased perfusion mimicking infarction while others demonstrate increased perfusion.[67,68]

Acknowledgment

Figs. 3C, 16A, 41A, and 50A from Richard D. Beegle, MD on behalf of AdventHealth.

I. Metastatic neurological complications

References

1. Nathoo M, Chahlavi A, Barnett GH, Toms SA. Pathobiology of brain metastases. *J Clin Pathol.* 2005;58(3):237–242. https://doi.org/10.1136/jcp.2003.013623.

2. Wanleenuwat P, Iwanowski P. Metastases to the central nervous system: molecular basis and clinical considerations. *J Neurol Sci.* 2020;412(116755). https://doi.org/10.1016/j.jns.2020.116755.

3. Osborn A, Hedlund G, Salzman K. *Osborn's Brain: Imaging, Pathology, and Anatomy.* 2nd ed. Elsevier; 2017.

4. Schroeder T, Bittrich P, Kuhne JF, et al. Mapping distribution of brain metastases: does the primary tumor matter? *J Neuro-Oncol.* 2020;147(1):229–235. https://doi.org/10.1007/s11060-020-03419-6.

5. Mitsuya K, Nakasu Y, Horiguchi S, et al. Metastatic skull tumors: MRI features and a new conventional classification. *J Neuro-Oncol.* 2011;104(1):239–245. https://doi.org/10.1007/s11060-010-0465-5.

6. Barajas RF, Cha S. Metastasis in adult brain tumors. *Neuroimaging Clin N Am.* 2016;26(4):601–620. https://doi.org/10.1016/j.nic.2016.06.008.

7. Sánchez Fernández I, Loddenkemper T. Seizures caused by brain tumors in children. *Seizure.* 2017;44:98–107. https://doi.org/10.1016/j.seizure.2016.11.028.

8. Bouffet E, Doumi N, Thiesse P, et al. Brain metastases in children with solid tumors. *Cancer.* 1997;79(2):403–410.

9. Kaufmann TJ, Smits M, Boxerman J, et al. Consensus recommendations for a standardized brain tumor imaging protocol for clinical trials in brain metastases. *Neuro-Oncology.* 2020;22(6):757–772.

10. Fan B, Li M, Wang X, et al. Diagnostic value of gadobutrol versus gadopentetate dimeglumine in enhanced MRI of brain metastases. *J Magn Reson Imaging.* 2016;45(6):1827–1834.

11. Jeon J, Choi JW, Roh HG, Moon W. Effect of imaging time in the magnetic resonance detection of intracerebral metastases using single dose gadobutrol. *Korean J Radiol.* 2014;15(1):145–150.

12. Kim ES, Chang JH, Choi HS, Kim J, Lee S. Diagnostic yield of double-dose gadobutrol in the detection of brain metastasis: intraindividual comparison with double-dose gadopentetate. *Am J Neuroradiol.* 2010;31:1055–1058. https://doi.org/10.3174/ajnr.A2010.

13. Kremer S, Eid MA, Bierry G, et al. Accuracy of delayed post-contrast FLAIR MR imaging for the diagnosis of leptomeningeal infectious or tumoral diseases. *J Neuroradiol.* 2006;33(5):285–291.

14. Jeevanandham B, Kalyanpur T, Gupta P, Cherian M. Comparison of post-contrast 3D-T1-MPRAGE, 3D-T1-SPACE and 3D-T2-FLAIR MR images in evaluation of meningeal abnormalities at 3-T MRI. *Br J Radiol.* 2017;90(1074):1–10. https://doi.org/10.1259/bjr.20160834.

15. Ercan N, Gultekin S, Celik H, Tali TE, Oner YA, Erbas G. Diagnostic value of contrast-enhanced fluid-attenuated inversion recovery MR imaging of intracranial metastases. *Am J Neuroradiol.* 2004;25:761–765.

16. Oh J, Choi SH, Lee E, et al. Application of 3D fast spin-echo T1 black-blood imaging in the diagnosis and prognostic prediction of patients with leptomeningeal carcinomatosis. *Am J Neuroradiol.* 2018;1–7. Published online.

17. Suh CH, Jung SC, Kim KW, Pyo J. The detectability of brain metastases using contrast-enhanced spin-echo or gradient-echo images: a systematic review and meta-analysis. *J Neuro-Oncol.* 2016;129(2):363–371. https://doi.org/10.1007/s11060-016-2185-y.

18. Reichert M, Morelli JN, Runge VM, et al. Contrast-enhanced 3-dimensional SPACE versus MP-RAGE for the detection of brain metastases: considerations with a 32-channel head coil. *Investig Radiol.* 2013;48(1):55–60. https://doi.org/10.1097/RLI.0b013e318277b1aa.

19. Danieli L, Riccitelli GC, Distefano D, et al. Brain tumor-enhancement visualization and morphometric assessment: a comparison of MPRAGE, SPACE, and VIBE MRI techniques. *Am J Neuroradiol.* 2019;40:1140–1148.

20. Kato Y, Higano S, Tamura H, et al. Usefulness of contrast-enhanced T1-weighted sampling perfection with application-optimized contrasts by using different flip angle evolutions in detection of small brain metastasis at 3T MR imaging: comparison with magnetization-prepared rapid acquisition. *Am J Neuroradiol.* 2009;30:923–929.

21. Kim D, Heo YJ, Jeong HW, et al. Usefulness of the delay alternating with nutation for tailored excitation pulse with T1-weighted sampling perfection with application-optimized contrasts using different flip angle evolution in the detection of cerebral metastases: comparison with MPRAGE imaging. *AJNR Am J Neuroradiol.* 2019;40(9):1469–1475.

22. Chukwueke UN, Wen PY. Use of the response assessment in neuro-oncology (RANO) criteria in clinical trials and clinical practice. *CNS Oncol.* 2019;8(1):CNS28. https://doi.org/10.2217/cns-2018-0007.

23. Pekmezci M, Perry A. Neuropathology of brain metastases. *Surg Neurol Int.* 2013;4(Suppl. 4):245–255. https://doi.org/10.4103/2152-7806.111302.

24. Maroldi R, Ambrosi C, Farina D. Metastatic disease of the brain: extra-axial metastases (skull, dura, leptomeningeal) and tumour spread. *Eur Radiol.* 2005;15(3):617–626. https://doi.org/10.1007/s00330-004-2617-5.

25. Sze G, Shin J, Krol G, Johnson C, Liu D, Deck M. Intraparenchymal brain metastases: MR imaging versus contrast-enhanced CT. *Radiology.* 1988;168(1):187–194.

26. Alcaide-Leon P, Cluceru J, Lupo JM, et al. Centrally reduced diffusion sign for differentiation between treatment-related lesions and glioma progression: a validation study. *Am J Neuroradiol.* 2020;41(11):2049–2054. https://doi.org/10.3174/ajnr.A6843.

27. Gulko E, Oleksk ML, Gomes W, et al. MRI brain findings in 126 patients with COVID-19: initial observations from a descriptive literature review. *Am J Neuroradiol.* 2020;1–5. https://doi.org/10.3174/ajnr.a6805. Published online.

28. Harrison RA, Nam JY, Weathers SP, DeMonte F. *Intracranial Dural, Calvarial, and Skull Base Metastases.* vol. 149. 1st ed. Elsevier B.V.; 2018. https://doi.org/10.1016/B978-0-12-811161-1.00014-1.

29. Herneth AM, Friedrich K, Weidekamm C, et al. Diffusion weighted imaging of bone marrow pathologies. *Eur J Radiol.* 2005;55(1):74–83. https://doi.org/10.1016/j.ejrad.2005.03.031.

30. Nayak L, Abrey LE, Iwamoto FM. Intracranial dural metastases. *Cancer.* 2009;115(9):1947–1953. https://doi.org/10.1002/cncr.24203.

31. Shah LM, Salzman KL. Imaging of spinal metastatic disease. *Int J Surg Oncol.* 2011;2011(Figure 2):1–12. https://doi.org/10.1155/2011/769753.

32. Harris P, Diouf A, Guilbert F, et al. Diagnostic reliability of leptomeningeal disease using magnetic resonance imaging. *Cureus.* 2019;11(Lmd):9–15. https://doi.org/10.7759/cureus.4416.

33. Pan Z, Yang G, He H, et al. Leptomeningeal metastasis from solid tumors: clinical features and its diagnostic implication. *Sci Rep.* 2018;8(1):1–13. https://doi.org/10.1038/s41598-018-28662-w.

34. Clarke JL, Perez HR, Jacks LM, Panageas KS, Deangelis LM. Leptomeningeal metastases in the MRI era. *Neurology.* 2010;74(18):1449–1454. https://doi.org/10.1212/WNL.0b013e3181dc1a69.

35. Koeller KK, Sandberg GD. From the archives of the AFIP. *Radiographics.* 2002;22(6):1473–1505. https://doi.org/10.1148/rg.226025118.

36. Umehara T, Okita Y, Nonaka M, et al. Choroid plexus metastasis of follicular thyroid carcinoma diagnosed due to intraventricular hemorrhage. *Intern Med.* 2015;54(10):1297–1302. https://doi.org/10.2169/internalmedicine.54.3560.

37. Castle-Kirszbaum M, Goldschlager T, Ho B, Wang YY, King J. Twelve cases of pituitary metastasis: a case series and review of the literature. *Pituitary.* 2018;21(5):463–473. https://doi.org/10.1007/s11102-018-0899-x.

38. He W, Chen F, Dalm B, Kirby PA, Greenlee JDW. Metastatic involvement of the pituitary gland: a systematic review with pooled individual patient data analysis. *Pituitary*. 2015;18(1):159–168. https://doi.org/10.1007/s11102-014-0552-2.

39. Ross JS, Moore KR. *Diagnostic Imaging: Spine*. 3rd ed. Elsevier; 2015.

40. Cook GJR. PET and PET/CT imaging of skeletal metastases. *Cancer Imaging*. 2010;10(1):153–160. https://doi.org/10.1102/1470-7330.2010.0022.

41. Del Grande F, Santini F, Herzka D, et al. Fat-suppression techniques for 3-T MR imaging of the musculoskeletal system. *Radiographics*. 2014;34(1):217–233.

42. Wittenberg KH, Adkins MC. MR imaging of nontraumatic brachial plexopathies: frequency and spectrum of findings. *Radiographics*. 2000;20(4):1023–1032. https://doi.org/10.1148/radiographics.20.4.g00jl091023.

43. Glazer H, Lee J, Levitt R, et al. Radiation fibrosis: differentiation from recurrent tumor by MR imaging-work in progress. *Radiology*. 1985;156:721–726.

44. Ebner F, Kressel H, Mintz M, et al. Tumor recurrence versus fibrosis in the female pelvis: differentiation with MR imaging at 1.5 T. *Radiology*. 1988;166:333–340.

45. Castagno A, Shuman W. MR imaging in clinically suspected brachial plexus tumor. *AJR Am J Roentgenol*. 1987;149:1219–1222.

46. Iyer V, Sanghvi DA, Merchant N. Malignant brachial plexopathy: a pictorial essay of MRI findings. *Indian J Radiol Imaging*. 2010;20(4):274–278. https://doi.org/10.4103/0971-3026.73543.

47. López-García F, Amorós-Martínez F, Sempere AP. A reversible posterior leukoencephalopathy syndrome. *Rev Neurol*. 2004;38(3):261–266. https://doi.org/10.33588/rn.3803.2003342.

48. De Oliveira AM, Ana P, Mckinney AM, Leite C, Luis F, Lucato LT. Imaging patterns of toxic and metabolic brain disorders. *Radiographics*. 2019;1672–1695. Published online.

49. Schwartz RB, Feske SK, Polak JF, et al. Preeclampsia-eclampsia: clinical and neuroradiographic correlates and insights into the pathogenesis of hypertensive encephalopathy. *Radiology*. 2000;217(2):371–376. https://doi.org/10.1148/radiology.217.2.r00nv44371.

50. Cipolla MJ. Cerebrovascular function in pregnancy and eclampsia. *Hypertension*. 2007;50(1):14–24. https://doi.org/10.1161/HYPERTENSIONAHA.106.079442.

51. Singer S, Grommes C, Reiner AS, Rosenblum MK, Deangelis LM. Posterior reversible encephalopathy syndrome in patients with cancer. *Oncologist*. 2015;20:806–811.

52. Rollins N, Winick N, Bash R, Booth T. Acute methotrexate neurotoxicity: findings on diffusion-weighted imaging and correlation with clinical outcome. *Am J Neuroradiol*. 2004;25(10):1688–1695.

53. Pinnix CC, Chi L, Jabbour EJ, et al. Dorsal column myelopathy after intrathecal chemotherapy for leukemia. *Am J Hematol*. 2017;92(2):155–160. https://doi.org/10.1002/ajh.24611.

54. Tibussek D, Natesirinilkul R, Sun LR, Wasserman BA, Brandão LR, DeVeber G. Severe cerebral vasospasm and childhood arterial ischemic stroke after intrathecal cytarabine. *Pediatrics*. 2016;137(2). https://doi.org/10.1542/peds.2015-2143.

55. Yoon J, Yoon J, Park H, et al. Diffuse cerebral vasospasm with infarct after intrathecal cytarabine in childhood leukemia. *Pediatr Int*. 2014;56(6):921–924.

56. Phan GQ, Yang JC, Sherry RM, et al. Cancer regression and autoimmunity induced by cytotoxic T lymphocyte-associated antigen 4 blockade in patients with metastatic melanoma. *Proc Natl Acad Sci USA*. 2003;100(14):8372–8377. https://doi.org/10.1073/pnas.1533209100.

57. Barroso-Sousa R, Barry WT, Garrido-Castro AC, et al. Incidence of endocrine dysfunction following the use of different immune checkpoint inhibitor regimens a systematic review and meta-analysis. *JAMA Oncol*. 2018;4(2):173–182. https://doi.org/10.1001/jamaoncol.2017.3064.

58. Kurokawa R, Ota Y, Gonoi W, et al. MRI findings of immune checkpoint inhibitor-induced hypophysitis: possible association with fibrosis. *Am J Neuroradiol*. 2020;41(9):1683–1689. https://doi.org/10.3174/ajnr.A6692.

59. Carpenter KJ, Murtagh RD, Lilienfeld H, Weber J, Murtagh FR. Ipilimumab-induced hypophysitis: MR imaging findings. *Am J Neuroradiol*. 2009;30(9):1751–1753. https://doi.org/10.3174/ajnr.A1623.

60. Da Rocha AJ, Nunes RH, Maia ACM, Do Amaral LLF. Recognizing autoimmune-mediated encephalitis in the differential diagnosis of limbic disorders. *Am J Neuroradiol*. 2015;36(12):2196–2205. https://doi.org/10.3174/ajnr.A4408.

61. Demaerel P, Van Dessel W, Van Paesschen W, Vandenberghe R, Van Laere K, Linn J. Autoimmune-mediated encephalitis. *Neuroradiology*. 2011;53(11):837–851.

62. Kelley BP, Patel SC, Marin HL, Corrigan JJ, Mitsias PD, Griffith B. Autoimmune encephalitis: pathophysiology and imaging review of an overlooked diagnosis. *Am J Neuroradiol*. 2017. https://doi.org/10.3174/ajnr.A5086. Published online February 9.

63. Mitsuya K, Nakasu Y, Horiguchi S, et al. Perfusion weighted magnetic resonance imaging to distinguish the recurrence of metastatic brain tumors from radiation necrosis after stereotactic radiosurgery. *J Neuro-Oncol*. 2010;99(1):81–88. https://doi.org/10.1007/s11060-009-0106-z.

64. Huang J, Wang A-M, Shetty A, et al. Differentiation between intra-axial metastatic tumor progression and radiation injury following fractionated radiation therapy or stereotactic radiosurgery using MR spectroscopy, perfusion MR imaging or volume progression modeling. *Magn Reson Imaging*. 2011;29(7):993–1001. https://doi.org/10.1016/j.mri.2011.04.004.

65. Chen H, Li X, Zhang X, et al. Late delayed radiation-induced cerebral arteriopathy by high-resolution magnetic resonance imaging: a case report. *BMC Neurol*. 2019;19(1):1–5. https://doi.org/10.1186/s12883-019-1453-9.

66. Black DF, Morris JM, Lindell EP, et al. Stroke-like migraine attacks after radiation therapy (SMART) syndrome is not always completely reversible: a case series. *Am J Neuroradiol*. 2013;34(12):2298–2303. https://doi.org/10.3174/ajnr.A3602.

67. Olsen AL, Miller JJ, Bhattacharyya S, Voinescu PE, Klein JP. Cerebral perfusion in stroke-like migraine attacks after radiation therapy syndrome. *Neurology*. 2016;86(8). https://doi.org/10.1212/WNL.0000000000002400. 787 LP-789.

68. Wai K, Balabanski A, Chia N, Kleinig T. Reversible hemispheric hypoperfusion in two cases of SMART syndrome. *J Clin Neurosci*. 2017;43:146–148. https://doi.org/10.1016/j.jocn.2017.05.013.

3

Nonimaging evaluation of patients with nervous system metastases

Kaitlyn Melnick[a], Varalakshmi Ballur Narayana Reddy[b], David Shin[a], and Ashley Ghiaseddin[a]

[a]Department of Neurosurgery, University of Florida, Gainesville, FL, United States, [b]Department of Neurology, University of Florida, Gainesville, FL, United States

1 Introduction

Cancer remains one of the leading causes of death in the United States. As the ability to promptly diagnose and treat cancer improves, metastatic complications of the disease have become increasingly pervasive and problematic. Cancer can metastasize to the central and peripheral nervous systems, including the brain, spinal cord, leptomeninges, dura, cranial nerves, and peripheral nerves. In fact, the most common brain tumor is metastasis from a distant primary source. Cancer can also affect the nervous system by making a patient more susceptible to infections, thromboembolic events, paraneoplastic disorders, antineoplastic therapy side effects, including neurotoxicity, and metabolic derangements.[1] However, these disorders are outside the scope of this chapter and will be discussed elsewhere.

Regarding nervous system metastases, diagnosis can be evasive as symptoms may be attributed to other causes. For example, pain symptoms may be attributed to nearby musculoskeletal metastases and neuropathy or cognitive decline may be attributed to the neurotoxic effect of antineoplastic therapy. Neuroimaging is typically the mainstay of diagnosis for nervous system metastases. Conversely, this chapter will focus on the nonimaging evaluation of patients with nervous system metastases. Venerable methods, including electroencephalography, evoked potentials, electromyography, and nerve conduction studies, still have utility in patients where imaging is not possible or nondiagnostic. Analysis of cerebrospinal fluid is also particularly helpful in diagnosing and measuring treatment response, especially for leptomeningeal metastases. A novel technique in the analysis of cerebrospinal fluid is liquid biopsy. This minimally invasive method may potentially distinguish metastatic disease from primary brain cancer and can be helpful in identifying targeted therapy and monitoring for disease response, remission, and relapse.

2 Electroencephalography

Electroencephalography (EEG) is a technique that began in the 1920s in Germany with the Neuropsychiatrist Hans Berger.[2] At that time, there was a paucity of techniques to scientifically analyze the brain, and the popularization of this method allowed for a boom in the study of epilepsy, brain tumors, and sleep. Despite the interval development of computed tomography and magnetic resonance imaging, EEG remains an important tool in diagnosing and treating intracranial pathology.

2.1 Epilepsy

The incidence of epilepsy in patients with metastatic brain tumors is 15%–35%, and seizure control results in reduction of disease morbidity.[3,4] The epileptogenesis of metastatic tumors is thought to be related to their rapid rate of growth, tissue necrosis, gliosis, and hemosiderin deposition.[5] Importantly, treatment of metastatic brain tumors may still be associated with seizures despite local tumor control due to the effects of surgery, radiation, and systemic treatment. Seizure activity in the surgical setting may be related to tissue damage caused by hypoxia as well as postoperative vasogenic edema.[6] Radiation may lead to radiation necrosis or development of vascular malformations, such as cavernomas, increasing seizure risk. Furthermore, certain cytotoxic chemotherapies and metabolic encephalopathies may lead to seizure occurrence.[7] Predilection to the development of epilepsy is related to tumor location where cortical tumors, especially those of the temporal, frontal, and parietal lobes, portend the greatest risk for seizure development.[3] Tumor histology is also predictive of the development of seizures with melanoma being high risk and adenocarcinoma, prostate cancer, and hepatocellular carcinoma being low risk.[4,8] High seizure rates in melanoma, up to 67% in one retrospective series, may be related to the higher likelihood of intratumoral hemorrhage with these tumors.[9,10] EEG is an important tool to diagnose seizures and monitor the treatment for patients with brain metastases.

Patients with metastatic disease to the brain are at high risk of developing status epilepticus, which is defined as persistent or repeated seizure activity without a return to baseline between episodes, and has a mortality of nearly 20%.[11] Furthermore, brain tumor patients who develop status epilepticus have a significantly higher mortality compared to nonbrain tumor patients.[12] Most frequently, the development of status epilepticus is associated with a new, growing, or hemorrhagic metastatic lesion.[13] Although it is a clinical diagnosis, utilization of EEG can be helpful in managing a patient with status epilepticus. The capture rate of seizure activity while on EEG is highly variable depending on the study, ranging from 19% to 68%.[8,13] Classic EEG evidence of status epilepticus includes an initial flattening of background rhythms with building of low-voltage fast activity that progressively increases in amplitude and decreases in frequency, with eventual obstruction of encephalographic pattern due to muscle activity.[14] Interictal EEG findings in patients with confirmed status epilepticus include focal, hemispheric, and generalized slowing in addition to focal discharges.[13]

Altered mental status is a common presentation to the hospital with multiple etiologic factors. Cancer patients presenting with AMS have seizures as the major culprit in approximately 9% of cases.[15] Unlike convulsive seizures, nonconvulsive status epilepticus is an entity that is more rare and far more difficult to diagnose with only 2% of patients with brain tumors experiencing this problem.[16] However, this value may be a gross underestimation of the disorder, since Fox et al. demonstrated that nonconvulsive status epilepticus is more common than convulsive status epilepticus in patients with brain metastases.[13] It is important to consider that nonconvulsive status epilepticus may be the presenting symptom of metastatic disease, of hemorrhage within known metastatic disease, or as a new symptom of stable metastatic disease. Neuroimaging and EEG are both accepted diagnostic tests when nonconvulsive status is suspected. However, radiographic changes should be taken in context as one study has suggested that nonconvulsive status

epilepticus can result in transient cortical enhancement in the absence of tumor recurrence.[17] The current criteria for the diagnosis of nonconvulsive status epilepticus require epileptiform discharges > 2.5 Hz or epileptiform discharges < 2.5 Hz or delta or theta (< 0.5 Hz) with one of the following: clinical improvement with the administration of intravenous antiepileptic agent, subtle ictal phenomena during EEG changes, or spatiotemporal evolution.[18,19] Fig. 1 shows an example EEG for nonconvulsive status epilepticus. For patients with nonconvulsive status epilepticus in the setting of brain metastases, the epileptiform discharges can be variable in morphology and may not be localized to known metastatic disease.[20] In summary, patients with brain metastases are at risk of convulsive and nonconvulsive seizures, and an EEG should always be considered in a patient with cancer presenting with altered mental status.

2.2 Nonepileptiform changes

In the absence of convulsive or nonconvulsive seizures, EEG is still an effective tool in the management of patients with brain metastases. Numerous studies written prior to the invention of cranial imaging have demonstrated the efficacy of EEG in diagnosing brain tumors. Namely, this work was pioneered by Walter in the 1930s, who reported the method as a technique to localize primary brain tumors.[21,22] In 1958, Daly et al. demonstrated changes in the EEG readings over time in patients with brain tumors with most eventually developing focal or generalized "dysarrthymias."[23] In 1961, Strang and Marsan reported on using electroencephalography, specifically focal delta, to diagnose metastases. Although effective with tumors larger than two centimeters in diameter, the technique added little to localizing the lesions because most patients were already symptomatic.[24] However, in 1973, Hildebrand published a retrospective review of 300 patients with cancer, 50 of whom had confirmed brain metastases on pathologic examination.[25] Of those with brain metastases, 22 had localizing neurologic symptoms and an abnormal EEG. Of the 28 without localizing symptoms, 25 had an abnormal EEG. In 1974, a small retrospective review by Rowan et al. analyzed 20 EEGs of patients with systemic cancer with and without brain metastases. The review showed that patients with brain metastases were more likely to have delta activity that was persistent focal; intermittent focal; or local, persistent lateralized, or intermittent lateralized.[26] Of the patients with brain metastases, only one had a history of seizures and five had no lateralizing neurologic signs or symptoms. Then, in 1978, Matsouka et al. demonstrated that the administration of dexamethasone improved abnormal delta signals in patients with brain tumors.[27] Prior to the development and widespread adoption of intracranial imaging, EEG

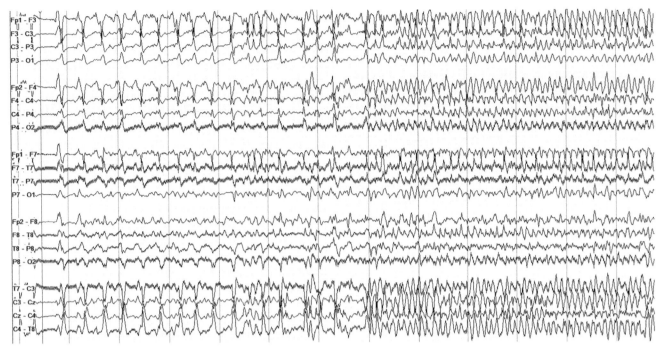

FIG. 1 Nonconvulsive status epilepticus: bifrontal predominant generalized periodic discharges (GPDs) evolving in morphology and frequency to faster frequency discharges. *Courtesy Dr. Jean Cibula.*

served as an important tool to identify and localize brain tumors. Since these early studies, it is now well established that focal delta activity is predictive of an intracranial lesion.[28] Additionally, focal theta activity is felt to be related to perilesional edema. Modern imaging techniques have made EEG-dependent diagnosis of brain metastases obsolete, but it is important to consider this option when imaging is not feasible as well as its use in a complimentary role to modern diagnostic techniques.

3 Evoked potentials

The technique of evoked potentials refers generically to recording the electrical activity from an area of the nervous system after a particular stimulus. Both sensory and motor evoked potentials can be measured, and sensory potentials can include visual, auditory, and somatosensory modalities. Evoked potentials have diagnostic utility in the management of brain metastases, but their primary utility is in intraoperative management.

3.1 Diagnostic utility

Brainstem auditory evoked potentials or responses (BAEPs or BAERs) are electrical responses to auditory stimuli recorded from the scalp and can identify lesions anywhere along the auditory pathway, including the auditory nerve, cochlear nucleus, superior olive, lateral

lemniscus, inferior colliculus, medial geniculate nucleus, and auditory cortex. Fig. 2 shows normal and abnormal BAERs. BAERs are helpful in the identification, localization, and lateralization of brainstem lesions especially in patients with metastatic cancer who can present with hearing loss from normal aging or effects of chemotherapy.[29] Moreover, BAERs can be assessed in the setting of large supratentorial metastases, which can result in prolonged ipsilateral wave V (lateral lemniscus) latency even in the absence of obvious radiographic brainstem compression.[30]

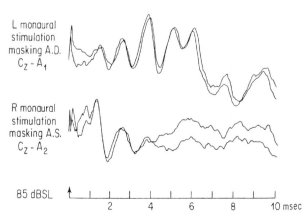

L monaural stimulation masking A.D. $C_Z - A_1$

R monaural stimulation masking A.S. $C_Z - A_2$

85 dBSL

2   4   6   8   10 msec

FIG. 2 Normal BAER from the left ear (upper). Abnormal BAER from the right ear (lower) with absent wave III to V due to caudal pontine lesion on right side. *Courtesy of the American Society of Electroneurodiagnostic Technologists, Inc.*

3.2 Intraoperative utility

Previously, the mainstay of treatment for cerebral metastases was considered to be whole brain radiation. However, concerns for cognitive effects have led to support of other modalities such as maximal safe surgical resection or radiosurgery, depending on tumor and patient characteristics.[31] Recently, targeted therapy and immunotherapy have entered the treatment discussion for patients with certain primary malignancies. If resection is undertaken on a tumor near an eloquent area, evoked potentials can be utilized for intraoperative neuromonitoring. Neuromonitoring is not often utilized in brain metastases as they are felt to be encapsulated or noninfiltrative (differing from gliomas). However, the discussion below will provide evidence for efficacy in these tumors.

3.2.1 SSEPs

Somatosensory evoked potentials (SSEPs) can be used to identify the central sulcus due to phase reversal of SSEPs in the region as demonstrated by Boughton.[32] The phase reversal is demonstrated in Fig. 3.[33] Intraoperatively, this technique is helpful in determining eloquent areas, identifying surface anatomy for deep tumors, and preventing postoperative neurological deficits.[34,35] However, the technique is not always reliable, and the advent of intraoperative image guidance has made this technique nearly obsolete in modern times.[33]

3.2.2 MEPs

Motor evoked potentials (MEPs) are recorded as electromyographic responses in muscles contralateral to the stimulated precentral gyrus. Normal MEPs imply that the entire corticospinal tract is intact, making this technique helpful for tumors that abut the white matter of this long tract. Although the goal of MEPs is to prevent a permanent neurologic deficit, critics fear any decline in amplitude would likely be precipitated by an irreversible step during tumor resection. Likewise, false positives may limit how aggressively a surgeon resects a tumor. Krieg et al. published an analysis of 53 patients who underwent craniotomy for the resection of metastasis with the use of intraoperative MEPs, and demonstrated that a 50% amplitude decline had a sensitivity of 60% and a specificity of 77% in detecting a postoperative neurologic deficit.[36] For an amplitude decline of 80%, MEPs had a sensitivity of 40% and a specificity of 89%. These values demonstrate that MEPs are useful but not infallible in predicting postoperative motor deficits. Of the patients with a false-positive change in MEPs, they were much more likely to have residual tumor on postoperative imaging. Obermueller et al. then compared the utility of MEPs in patients with brain metastases to patients with primary brain tumors.[37] These findings echoed those by Krieg et al., which found a high false-positive rate of MEP decline in brain metastases resection, and this false positivity was associated with an increased risk of subtotal resection. The authors recommended using a decrease in amplitude of > 80% for resection in metastases patients to circumvent this problem. Despite shortcomings, MEPs have definite utility in surgical resection of brain metastases.

4 Electromyography and nerve conduction studies

Electromyography (EMG) is an electrodiagnostic test that records spontaneous activity of muscles at rest and during conscious activation. The study can be used to identify radiculopathies, neuropathies, and myopathies. Nerve conduction study (NCS) is typically performed concomitantly with EMG and measures the conduction of motor and sensory nerves of the peripheral nervous system, which can be helpful in identifying the site of neuronal involvement.

Cancer can result in neuromuscular disease through a variety of mechanisms. Tumors can compress adjacent nerves or metastasize directly to nerves. Additionally, antineoplastic therapy is often neurotoxic, and cancers can result in paraneoplastic manifestations of neuromuscular disease. In these settings, EMG/NCS can be helpful in diagnosing, prognosticating, and assessing the treatment response. Historically, EMG was a primary tool for diagnosing tumors of the spinal cord and peripheral nervous system; however, the widespread accessibility of high-quality neuroimaging has made this technique nearly obsolete.[38] However, case reports have demonstrated abnormal EMG/NCS test results despite normal imaging in patients who were subsequently found to have tumors of the peripheral nervous system.[39]

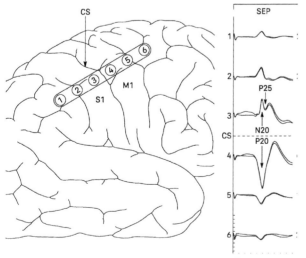

FIG. 3 Phase reversal of SSEPs over central sulcus.[33]

4.1 Spine tumors: Radiculopathy

Primary and metastatic tumors of the spine can compress both the central neural elements and the nerve roots. Common primary bone tumors of the spine include multiple myeloma, plasmacytoma, osteoid osteoma, osteoblastoma, osteosarcoma, Ewing's sarcoma, chondroma, chordoma, osteochondroma, chondrosarcoma, giant cell tumor, aneurysmal bone cyst, and others. Metastatic lesions to the spine are more common than primary bone tumors, and the most common metastatic lesions to the spine are from breast, lung, and prostate cancer. Large spine tumors can compress adjacent nerve roots, resulting in classic radiculopathy findings on EMG, including increased insertional activity (i.e., fibrillation and positive sharp waves), large polyphasic motor units, and reduced recruitment. Since the lesion is proximal to the dorsal root ganglion, sensory NCS are normal. Findings can be confirmed with imaging.

Polyradiculopathy in the setting of cancer is suggestive of leptomeningeal metastases. Leptomeningeal involvement is common in hematologic malignancies, including leukemia and lymphoma, as well as solid carcinomas such as breast, lung, and melanoma. Patients with polyradiculopathy present with appendicular pain, weakness, and numbness. A review by Kaplan et al. touted that EMG was the primary technique of diagnosing leptomeningeal disease in 10 patients presenting with these symptoms.[40] On EMG, patients had fibrillations in multiple radicular levels, most commonly of the lumbar roots, but cervical roots were involved as well. Fig. 4 demonstrates fibrillations. They also found abnormal motor nerve conduction studies with preserved sensory nerve conduction studies. All patients eventually had confirmed leptomeningeal disease by cytology. Similarly, a review of 25 patients by Argov and Siegal identified that F wave abnormalities (absent or prolonged F waves) as a sensitive finding of nerve root involvement in diagnosing radiculopathy due to leptomeningeal disease, and suggested repeating EMG to monitor response to therapy.[42] However, the F wave abnormalities are not specific to leptomeningeal disease.[43]

4.2 Peripheral nerve tumors: Plexopathies and neuropathies

4.2.1 Plexopathies

Compression of the brachial or lumbosacral plexuses can cause pain and neurologic deficits, which may be the presenting symptoms of primary tumors and cancer metastases. The most classic example of this is that of Pancoast tumors, which are typically nonsmall-cell lung cancers situated at the pulmonary apex. Among other symptoms, the associated compression of the lower brachial plexus can result in pain in the hand and medial arm as well as weakness in the hand.[44] Although simple to identify radiographically, the symptomatology of these tumors often results in EMG being the first diagnostic modality to suggest their presence. Other tumors that commonly involve the brachial plexus include carcinomas of the breast, head, and neck cancers, and lymphomas.[45] Involvement of the cervical plexus may present concomitantly with Horner's syndrome due to the proximity of the sympathetic chain.

Lumbosacral plexopathies occur in the setting of colon cancer, prostate cancer, gynecologic cancers, and lymphomas. These tumors may be difficult to visualize on neuroimaging so EMG can be helpful in confirming the diagnosis based on history and clinical symptoms. EMG is also vital in distinguishing tumor plexopathies from radiation plexopathies.[46,47] Specifically, in the setting of neoplastic plexopathies, myokymic discharges are identified far more commonly than in patients with radiation-induced plexopathies.

4.2.2 Neuropathies

Primary nerve tumors, including schwannomas, neurofibromas, and malignant peripheral nerve sheath tumors, can all present with mononeuropathies of their involved nerve. However, metastases to peripheral nerves from melanoma as well as primary melanoma presenting in a peripheral nerve have been described.[48] Other tumor types that can result in mononeuropathies include metastases from lymphomas and, in rare cases, solid tumors.[49,50] Similar to the evaluation of radiculopathy, electromyography can be helpful in identifying which nerve is involved and can even help localize the area of involvement. EMG can also be helpful intraoperatively to identify and preserve motor fascicles during tumor resection.[51]

5 Nerve biopsy

Primary lesions of peripheral nerves are far more common than metastases, even in the setting of known metastatic disease. These include lipoma, ganglion cyst, desmoid, neurofibroma, schwannoma, perineuroma, malignant peripheral nerve sheath tumor, and primary peripheral nerve lymphoma. Furthermore, in the setting of metastatic disease, damage to peripheral nerves from

FIG. 4   Fibrillations seen in recorded muscle in the absence of conscious effort.[41]

extraneural compression by adjacent tumors is much more common than malignant spread to the nerves themselves. The estimated prevalence of peripheral nerve metastases is less than 1% among cancer patients.[45] The very low incidence is likely related to relative impermeability of tumor cells to the nerve sheath despite ample blood supply. Spread is believed to be hematologic, but the possibility of spread through the cerebrospinal fluid has been postulated. Imaging and electromyography are often inconclusive, and surgical exploration is at times necessary.

Pathologically, carcinomas are the most likely to metastasize to peripheral nerves, followed by sarcomas and lymphomas.[52] Melanoma is another tumor type that can metastasize to peripheral nerves and can clinically and radiographically resemble a malignant peripheral nerve sheath tumor.[48] Regardless of the tumor type, patients often present with debilitating pain, which may progress to include numbness and weakness. The pain may be attributed to nearby skeletal metastases or medication side effects. However, due to a metastasis to the peripheral nervous system, the pain can be treated with surgery and radiation with good outcomes.[52]

6 Cerebrospinal fluid

Cerebrospinal fluid (CSF) is the clear acellular fluid that bathes the brain and spinal cord. The fluid can be collected from the lumbar cistern with a lumbar puncture and analyzed to assess for infections, hemorrhages, autoimmune disease, and cancer. Alternatively, CSF can be collected from the ventricular system via a ventricular puncture or rarely from the cisterna magna via a suboccipital puncture. Related to cancer, CSF can be utilized to diagnose and measure treatment response of leptomeningeal metastasis. In some cases, CSF can also be utilized to diagnose solid metastatic tumors in the brain and spinal cord. Simple techniques such as cell counts and chemistries may be informative, but more complex investigation of specific proteins and pathologic review of cytology and flow cytometry are imperative for diagnostic accuracy. Lastly, there is a growing body research on the utility of liquid biopsy, which is the examination of a specific fluid for nucleic acids, including DNA and RNA, that are characteristic of particular tumors.

The primary utility of CSF analysis in the setting of cancer is for leptomeningeal metastasis, also called neoplastic meningitis or carcinomatous meningitis. The diagnosis of leptomeningeal metastasis is currently done by a combination of a standardized neurologic examination, cytology or flow cytometry, and radiographic evaluation.[53] The incidence of leptomeningeal carcinomatosis is 4%–15% in solid tumors, most frequently breast cancer, small cell lung cancer, and melanoma.[54–56] Breast cancer and lung cancer are far more common than melanoma

and thus account for the majority of patients with leptomeningeal metastasis. However, although melanoma itself is a rare cancer, the incidence of leptomeningeal metastasis in these patients is 22%–46%.[56] Cancer cells can obtain access to the CNS through a variety of mechanisms related to mutations in proteins involved in cell trafficking and adhesion. Once within the CNS, cancer cells seed the subarachnoid space and preferentially settle along the skull base and cauda equina, making these areas typically the most symptomatic and radiographically abnormal.

Depending on the location of involvement, specific symptoms can include cranial nerve dysfunction or lumbar radiculopathy. Patients may also develop increased intracranial pressure due to communicating hydrocephalus because of protein and cells clogging the arachnoid granulations that drain CSF. In these patients, common symptoms include headaches, nausea, and vomiting. In severe cases, symptoms can progress to lethargy, coma, and death. It is important for physicians to consider this disease entity in patients with cancer as the symptoms can often be vague and may be attributed to the effects of antineoplastic therapy.

If imaging is obtained, there is often diffuse enhancement of the leptomeninges of the skull base and cauda equina. The primary differential diagnosis for patients with these imaging findings includes tuberculosis, fungal meningitis, and neurosarcoidosis. The most effective way to definitively diagnose the pathology is with CSF analysis. Timely and accurate diagnosis is important not only to initiate new treatment strategies but also in terms of prognostication. Unfortunately, despite significant recent advances in oncologic care, the mortality of leptomeningeal carcinomatosis remains very high and portends a poor prognosis.[53,57]

6.1 Opening pressure

When CSF is obtained, the opening pressure is recorded and should be less than 20 cm H_2O. Up to 50% of patients with leptomeningeal metastasis have an elevation in opening pressure.[56] Elevation in opening pressure is diagnostic of increased intracranial pressure, which can be due to mass effect from a large tumor, cerebral edema, venous outflow obstruction, or hydrocephalus among other pathologies. In the setting of leptomeningeal metastasis, hydrocephalus is often communicating and due to clogging of the arachnoid granulations with malignant cells and proteins. Radionucleotide flow studies demonstrate impairment in CSF flow in approximately 33% of patients with known leptomeningeal metastasis, with symptomatic hydrocephalus occurring in up to 8%–10%.[55,58] Development of hydrocephalus portends a very poor prognosis and poses a technical and ethical dilemma in management as surgical treatment with shunting is often futile in these patients.[59]

6.2 Laboratory analysis

6.2.1 Chemistries and cell counts

The basic review of cell counts and chemistries in cerebrospinal fluid is a long-standing practice for diagnosing central nervous system diseases. CSF should be relatively acellular although having up to five white blood cells and up to five red blood cells per µL is considered normal. More cells can be seen in the setting of a traumatic tap. These cells are not actually in the CSF, but are artifactual from traversing tissues superficial to the lumbar cistern. Glucose should be roughly two-third the value of serum glucose, and protein should be < 40mg/dL although specific reference ranges may vary by laboratory. Nonetheless, the characteristic findings of leptomeningeal carcinomatosis include elevated cell counts (predominantly white blood cells and neoplastic cells), elevated protein, and low glucose. These findings are nonspecific, and as such, the sensitivity of these findings is not high. The most sensitive finding is elevated protein followed by low glucose. However, even if these are all normal, there is still a 5% chance that the patient has leptomeningeal carcinomatosis.[60]

In part, false negatives may be due to regional variations in the CSF throughout the neuroaxis.[61] For example, CSF in the lumbar cistern may be normal if the primary area of involvement is the skull base. For this reason, CSF sampling from multiple locations may be necessary for accurate diagnosis. Correlation between radiographic findings and CSF sampling location is important because derangement in these basic studies may be suggestive of disease severity. When abnormal, they have been used to prognosticate survival such that low protein and high glucose portend a better prognosis.[62–64] In summary, although nonspecific, abnormalities in cell counts, proteins, and glucose in the CSF of patients with cancer can be suggestive of CNS metastasis and may be helpful for prognostication.

6.2.2 Other small molecules and proteins

Due to the low sensitivity and specificity of basic laboratory analysis of CSF for the diagnosis of leptomeningeal carcinomatosis, there has been extensive research into identifying sensitive and specific protein biomarkers for this disease. There are two theories for how these proteins arrive in the CSF. The first is that the cancer causes breakdown of the blood-brain barrier, allowing capillaries to leak proteins into the subarachnoid space, and the second is that the metastatic tumors actively secrete proteins into the CSF.[65] Due to the heterogeneity of cancer, there has been little success in identifying one particular protein which is diagnostic of all leptomeningeal carcinomatosis. Likewise, a particular marker may be suggestive of multiple types of cancer.

As far back as the 1940s, analysis of CSF in patients with known CNS metastases was undertaken. Barone identified that lysozyme was elevated in the CSF of patients with brain tumors.[65,66] Then, in the 1970s, Hildebrand demonstrated elevations in lactic dehydrogenase (LDH), glutamic transaminases, and phospholipids, including lysophosphatidylcholines, sphingomyelins, phosphatidylcholines, and phosphatidylethanolamines, as well as other lipids.[25] Next, Koch and Lichtenfeld demonstrated an elevation in CSF beta-2-microglobulin, which is a protein structurally similar to immunoglobin G in patients with CNS metastasis of leukemia, lymphoma, or small cell lung cancer.[67] These findings are all nonspecific and can also be found in infectious and inflammatory diseases; however, in the setting of known cancer, it can be highly suggestive of leptomeningeal involvement or other CNS metastasis.[68,69] Another group of proteins that are abnormal in leptomeningeal metastases are the proteases cathepsin B and H, which are elevated with reciprocal depressions in their inhibitor cystatin C.[70] These proteins are often signs of CNS penetration by tumor cells; so again, they are more suggestive of neurologic damage than of leptomeningeal carcinomatosis.[71] Identification of tumor markers, including CEA, CA125, CA15-3, CA19-9, CA72-4, CYFRA21-1, AFP, and NSE, is more suggestive of leptomeningeal carcinomatosis than infectious or inflammatory conditions, but not all of these markers have been validated in specific tumor models.[72]

Recently, increased focus has been placed on identifying markers that are highly specific to certain cancer types. Cancers that are common with a predilection of spread to the leptomeninges have been studied extensively. The results are summarized in Table 1, where in most cases cytology was used as the gold standard for calculating sensitivity and specificity. It is important to recognize that although studied in the setting of a particular form of cancer, biomarkers are not necessarily diagnostic of a particular primary cancer. The exception to this statement is in the measurement of proteins highly specific to only one cancer type, e.g., prostate-specific antigen (PSA) in prostate cancer and alpha-fetoprotein (AFP) or human chorionic gonadotrophin (HCG) in germ cell tumors.[65,78,79] Although the understanding and utility of biomarkers has grown in recent years, more advanced techniques, including cytology and liquid biopsy, can diagnose a particular cancer from a CSF sample.

6.3 Flow cytometry and cytology

Cytology is the qualitative visualization of cells, usually by a pathologist, although automated systems are increasingly used. Since the 1960s, CSF cytology has been performed to diagnose CNS malignancies.[80] Cytology from CSF is obtained from the lumbar cistern, cisterna magna, or ventricles although the former is by far the common due to lower risk profile. As with other

TABLE 1   Summary of the sensitivity and specificity of specific tumor markers in diagnosing leptomeningeal metastases.

Cancer	CSF marker	Sensitivity and specificity
Breast cancer	Beta glucuronidase[68]	Sensitivity = 87% Specificity = 93%
	Beta-2 microglobulin[68]	Sensitivity = 60% Specificity = 88%
	Carcinoembryonic antigen (CEA)[68]	Sensitivity = 60% Specificity = 93%
	Creatinine kinase BB isoenzyme (CK-BB)[73]	Sensitivity = 83% Specificity = 87%
	Lactate dehydrogenase[68]	Sensitivity = 93% Specificity = 93%
	Stromal-derived factor (SDF)[74]	Sensitivity = 67% Specificity = 90%
	Tissue polypeptide antigen (TPA)[73]	Sensitivity = 73%
	Vascular endothelial growth factor (VEGF)[74]	Sensitivity = 75% Specificity = 97%
Lung cancer	CEA[75]	Sensitivity = 91% Specificity = 91%
	CYFRA 21-1[75]	Sensitivity = 83% Specificity = 97%
	NSE[75]	Sensitivity = 51% Specificity = 91%
	Stromal-derived factor (SDF)[74]	Sensitivity = 38% Specificity = 88%
	Vascular endothelial growth factor (VEGF)[74]	Sensitivity = 70% Specificity = 95%
Lymphoma	Beta-2 microglobulin[76]	Sensitivity = 100% Specificity = 76%
	Chemokine (C-X-X) ligand 13 (CXCL13)[77]	Sensitivity = 71% Specificity = 95%
	Interleukin 10 (IL-10)[77]	Sensitivity = 64% Specificity = 94%
Melanoma	Stromal-derived factor (SDF)[74]	Sensitivity = 100% Specificity = 100%
	Vascular endothelial growth factor (VEGF)[74]	Sensitivity = 67% Specificity = 100%

CSF sampling, the yield of doing so is highest when the sample is obtained from a site close to the area with the highest radiographic or symptomatic burden. Although obtaining CSF from the cisterna magna carries significantly greater risk than lumbar puncture, it may be necessary to diagnose leptomeningeal metastasis in some patients.[81] Obtaining a sample from a particular location carries associated risks, and performing a lumbar puncture is generally contraindicated in the setting of obstructive hydrocephalus and very large supratentorial tumors due to concerns of inducing herniation.

Nonetheless, the examination of CSF for malignant cells is considered the most important tool for diagnosing leptomeningeal carcinomatosis. Unfortunately, the sensitivity of analyzing just one sample is very low, often cited as 50% or less.[55] The sensitivity of CSF cytological analysis increases with higher volumes (greater than 10 mL) and with a second sample; yet still, 10%–15% of patients will remain negative and diagnosis will be dependent on a combination of interpretation of CSF and radiographic findings. One study of over 500 patients with leptomeningeal metastases demonstrated that 18% of patients had positive CSF cytology with negative imaging and 29% of patients had negative CSF cytology but positive imaging.[64] This demonstrates the importance of a multifactorial assessment of patients with symptoms concerning for leptomeningeal carcinomatosis. In regard to prognostication, negative CSF cytology is not associated with improved survival.[82] However, in patients with positive cytology, conversion to negative cytology after treatment is associated with improved survival compared to patients who have persistently positive cytology.[53]

A large review of 5951 CSF cytological samples sent from patients with suspected infectious, inflammatory, or neoplastic CNS diseases by Prayson and Fishler demonstrated that the vast majority of samples analyzed were negative for any pathology.[83] Five percent of adult samples and 8% of pediatric samples did demonstrate neoplastic cells. Of the adult samples, 41.3% showed adenocarcinoma, 28.3% showed lymphoma, 12% showed leukemia, 2.9% showed melanoma, and 2.5% showed a primary brain tumor. Of the pediatric samples, 87% showed a primary brain tumor (mostly medulloblastoma), 6.5% showed leukemia, and 6.5% showed lymphoma. This demonstrates the utility of cytology as a tool for the diagnosis of leptomeningeal carcinomatosis. Examples of various cytological findings are shown in Figs. 5–8.

Flow cytometry is a technique that was first developed in the 1950s, although it took many years to gain

FIG. 5   Cytology from normal CSF showing lymphocytes and monocytes.[84]

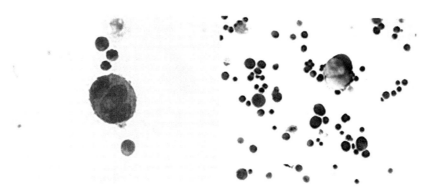

FIG. 6   Two cytological images showing abnormal large tumor cells.[84]

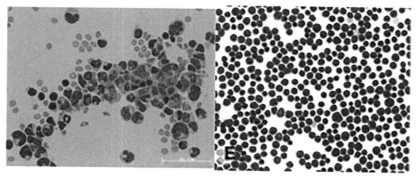

FIG. 7   Cytology showing abnormal lymphocytes indicative of leukemia or lymphoma.[84,85]

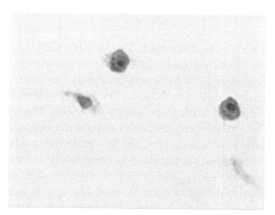

FIG. 8   Cytology showing melanoma cells.[86]

traction in clinical medicine. Initially, and presently, flow cytometry has been utilized predominantly to characterize populations of white blood cells, meaning that most of its clinical applications revolve around hematologic malignancies. The science behind flow cytometry is very complex but in simplest terms, after staining with fluorescent antibodies, the machine analyzes a single-cell suspension with a series of lasers to interpret cell size (scatter) and biomarkers (fluorescence). The biomarkers analyzed are often surface proteins, but intracellular proteins and nucleic acids can be labeled as well. Fig. 9 illustrates these basic principles. A more detailed explanation of the procedure has been described by Dux et al.[88] Flow cytometry gives both qualitative and quantitative data about a population of cells. A predominance of a particular type of cell is usually suggestive of malignancy (e.g., a clonal population).

The clinical application of flow cytometry is predominantly in hematologic malignancies, so CSF flow cytometry can be performed to assess for CNS involvement in leukemia and lymphoma, although the technique can be applied to metastasis of solid tumors as well.[89] Multiple studies have demonstrated that flow cytometry is far more sensitive for diagnosing CNS leukemia or lymphoma than convention cytology.[90] Even with the added sensitivity, the false-positive rate remains very low, although false positives can occur in inflammatory conditions.[91] The primary benefit of flow cytometry is that it is an objective, automated procedure and thus not prone to the biases of human interpretation. Although more sensitive, there are instances where cytology is positive and flow cytometry is negative, so the two modalities should always be used in combination.[53,91]

As with cytology, this modality is limited if a sample is very small. Additionally, flow cytometry requires live cells for accurate results. The longevity of cells in CSF after a sample has been obtained is very short so it is recommended that samples be analyzed immediately.[90] Techniques for preserving samples can marginally improve the life span of cells but are not very effective. In terms of prognosis, relapse after treatment resulting in positive flow cytometry, but negative cytology is associated with clinical relapse and poor prognosis.[92]

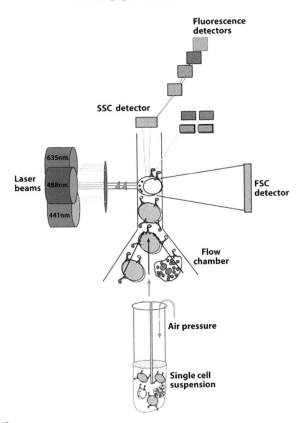

FIG. 9 Schematic of flow cytometry.[87]

6.4 Liquid biopsy

Liquid biopsy is an innovative technique in oncology that analyzes components of tumor cells obtained from bodily fluids. Components that can be analyzed include circulating tumor cells (CTCs), cell-free tumor DNA (ctDNA), RNAs (ctRNA), and exosomes. The technique was first applied to blood samples with the goal of identifying serologic biomarkers that are specific to certain cancers and can be utilized for establishing a diagnosis, tailoring antineoplastic therapy, assessing treatment response, and monitoring for recurrence.[93] More recently, the technique is being applied to different fluid samples, namely, to CSF in the setting of CNS metastasis, including both leptomeningeal and intraparenchymal tumors. Previously discussed techniques, including cytology and flow cytometry, are encompassed under the umbrella of liquid biopsy; however, the following section will focus on novel techniques in liquid biopsy that have not yet been discussed.

6.4.1 Circulating tumor cells

The Response Assessment in Neuro-Oncology (RANO) working groups for brain metastases and leptomeningeal metastasis published a consensus review of current literature on liquid biopsy in 2019.[94] In this publication, the Veridex CellSearch assay, which is an advanced technique for identifying circulating tumor cells in CSF, was discussed. The Veridex CellSearch assay was FDA-approved in 2004 and was designed to identify tumor cells from whole blood. The assay touts the ability to detect as little as one tumor cell in 7.5 mL of whole blood with a specificity of > 99%. The system works by immunomagnetically enhancing signals from cell surface markers specific to epithelial malignancies. The sample preparation is completely automated, but the processing and interpretation of the sample does require a highly trained user. The applicability of the Veridex CellSearch assay has been investigated in the context of leptomeningeal and intraparenchymal lung cancer, breast cancer, and melanoma metastasis. Although this technique appears promising, more research is needed to validate the findings of small studies. Additionally, a limitation of the technique is that the institution must use the Veridex CELLTRACKS system and reagents. Another limitation is that CellSearch can only be used to identify epithelial malignancies and can miss carcinomas that have undergone epithelial-mesenchymal transition, which is common in metastatic tumors. Other novel techniques for identifying circulating tumor cells, including methods utilizing polymerase chain reaction, microchips, nanoparticles, fluorescence in situ hybridization, and size-based filtration, have been investigated for blood samples but have not yet been applied to samples of CSF.[95]

6.4.2 Nucleic acids

Cell-free tumor DNA (ctDNA) can be found in CSF in the setting of leptomeningeal and intraparenchymal CNS metastasis, although the yield is much higher in leptomeningeal disease.[94,96] The acellular nature of CSF makes it an ideal fluid to analyze for ctDNA. In whole blood, a large portion of the cell-free DNA is normal genomic DNA. However, in the CSF, there is a much higher relative concentration of DNA from cancer cells. Another exciting feature of analyzing ctDNA in brain metastasis is that the metastatic tumors often have new actionable mutations not present in the primary tumor.[97] This unique feature may allow oncologists the opportunity to tailor changes in therapy. Additionally, ctDNA can be used to measure treatment response and diagnose relapse.[96] Lastly, analysis of ctDNA may eventually be used as a way to quantify the severity of leptomeningeal disease, as more DNA has been correlated with a higher disease burden.[94] As with studies on circulating tumor cells, this technique has been employed predominantly in the setting of metastatic lung cancer, breast cancer, and melanoma. The applicability to other cancer types still requires investigation.

RNAs, specifically micro RNAs (miRNAs), have also been investigated as targets for liquid biopsy. These RNAs can be cell free or found in exosomes, which are small extracellular vesicles. The advantage of searching for RNA is that it serves as a direct marker of protein expression unlike DNA, which may or may not be actively expressed. It is postulated that miRNAs, especially those in exosomes, are used by cancer cells as a means of cell signaling.[98] The majority of studies looking at miRNAs in brain tumors have focused on identifying miRNAs specific to primary brain tumors such as gliomas. Using these miRNAs, there have been a few small studies attempting to distinguish a primary brain tumor from metastatic disease with liquid biopsy alone.[99,100] Since the treatment for a primary brain tumor often differs from that of a metastatic tumor, miRNAs are an important discovery especially in tumors where direct tissue biopsy is high risk. Furthermore, a metaanalysis by Wei et al. has indicated the technique is extremely useful in the diagnosis of CNS lymphoma.[101] Specifically, miR-21, miR-19, and miR-92a levels together are highly sensitive and specific for CNS lymphoma compared to controls with inflammatory CNS disorders.[102] As previously discussed, the yield of cytology and flow cytometry for diagnosing CNS lymphoma is notoriously low, so this technique may eventually be preferred to diagnose CNS lymphoma, although further research is still needed.

7 Conclusion

Metastatic disease to the central nervous system is often diagnosed with advanced neuroimaging, followed by tissue examination. However, established and novel techniques, including EEG, EMG, CSF analysis, nerve biopsy, and liquid biopsy, have significant utility in diagnosing and treating CNS metastases. Novel techniques such as the Veridex CellSearch assay may allow us to more accurately diagnose patients with malignancy when biopsy of the primary site is note feasible. Lastly, the techniques described in this chapter may allow for a less invasive approach to monitoring for treatment response and/or recurrence.

Disclosures

Ashley Ghiaseddin has received personal fees from Monteris Medical and Novocure. Ashley Ghiaseddin has also received research funding support from Orbus Therapeutics.

Kaitlyn Melnick has no disclosures.

David Shin has no disclosures.

Lakshmi Reddy has no disclosures.

References

1. Khasraw M, Posner JB. Neurological complications of systemic cancer. *Lancet Neurol.* 2010;9(12):1214–1227.

2. Tudor M, Tudor L, Tudor KI. Hans Berger (1873-1941)—the history of electroencephalography. *Acta Med Croatica.* 2005;59(4):307–313.

3. van Breemen MS, Wilms EB, Vecht CJ. Epilepsy in patients with brain tumours: epidemiology, mechanisms, and management. *Lancet Neurol.* 2007;6(5):421–430.

4. Chan V, Sahgal A, Egeto P, Schweizer T, Das S. Incidence of seizure in adult patients with intracranial metastatic disease. *J Neuro-Oncol.* 2017;131(3):619–624.

5. Doria JW, Forgacs PB. Incidence, implications, and management of seizures following ischemic and hemorrhagic stroke. *Curr Neurol Neurosci Rep.* 2019;19(7):37.

6. Al-Dorzi HM, Alruwaita AA, Marae BO, et al. Incidence, risk factors and outcomes of seizures occurring after craniotomy for primary brain tumor resection. *Neurosciences.* 2017;22(2):107.

7. Singh G, Rees JH, Sander JW. Seizures and epilepsy in oncological practice: causes, course, mechanisms and treatment. *J Neurol Neurosurg Psychiatry.* 2007;78(4):342–349.

8 Ajinkya S, Fox J, Houston P, et al. Seizures in patients with metastatic brain tumors: prevalence, clinical characteristics, and features on EEG. *J Clin Neurophysiol.* 2021;38(2):143–148.

9. Englot DJ, Chang EF, Vecht CJ. Epilepsy and brain tumors. In: *Handbook of Clinical Neurology.* vol. 134. Elsevier; 2016:267–285.

10. Oberndorfer S, Schmal T, Lahrmann H, Urbanits S, Lindner K, Grisold W. The frequency of seizures in patients with primary brain tumors or cerebral metastases. An evaluation from the Ludwig Boltzmann Institute of Neuro-Oncology and the Department of Neurology, Kaiser Franz Josef Hospital, Vienna. *Wien Klin Wochenschr.* 2002;114(21–22):911–916.

11. Lowenstein DH, Alldredge BK. Status epilepticus. *N Engl J Med.* 1998;338(14):970–976.

12. Arik Y, Leijten FS, Seute T, Robe PA, Snijders TJ. Prognosis and therapy of tumor-related versus non-tumor-related status epilepticus: a systematic review and meta-analysis. *BMC Neurol.* 2014;14(1):1–5.

13. Fox J, Ajinkya S, Greenblatt A, et al. Clinical characteristics, EEG findings and implications of status epilepticus in patients with brain metastases. *J Neurol Sci.* 2019;407:116538.

14. Brenner RP. EEG in convulsive and nonconvulsive status epilepticus. *J Clin Neurophysiol.* 2004;21(5):319–331.

15. Tuma R, DeAngelis LM. Altered mental status in patients with cancer. *Arch Neurol.* 2000;57(12):1727–1731.

16. Marcuse LV, Lancman G, Demopoulos A, Fields M. Nonconvulsive status epilepticus in patients with brain tumors. *Seizure.* 2014;23(7):542–547.

17. Hormigo A, Liberato B, Lis E, DeAngelis LM. Nonconvulsive status epilepticus in patients with cancer: imaging abnormalities. *Arch Neurol.* 2004;61(3):362–365.

18. Beniczky S, Hirsch LJ, Kaplan PW, et al. Unified EEG terminology and criteria for nonconvulsive status epilepticus. *Epilepsia.* 2013;54:28–29.

19. Kaplan PW. EEG criteria for nonconvulsive status epilepticus. *Epilepsia.* 2007;48:39–41.

20. Blitshteyn S, Jaeckle KA. Nonconvulsive status epilepticus in metastatic CNS disease. *Neurology.* 2006;66(8):1261–1263.

21. Walter WG. The location of cerebral tumours by electroencephalography. *Lancet.* 1936;228(5893):305–308.

22. Walter WG. *The Electro-Encephalogram in Cases of Cerebral Tumour.* SAGE Publications; 1937.

23. Daly DD, Thomas JE. Sequential alterations in the electroencephalograms of patients with brain tumors. *Electroencephalogr Clin Neurophysiol.* 1958;10(3):395–404.

24. Strang R, Marsan CA. Brain metastases: pathological—electroencephalographic study. *Arch Neurol.* 1961;4(1):8–20.

25. Hildebrand J. Early diagnosis of brain metastases in an unselected population of cancerous patients. *Eur J Cancer.* 1973;9(9). 621-IN621.

26. Rowan A, Rudolf NDM, Scott D. EEG prediction of brain metastases: a controlled study with neuropathological confirmation. *J Neurol Neurosurg Psychiatry.* 1974;37(8):888–893.

27. Matsuoka S, Arakaki Y, Numaguchi K, Ueno S. The effect of dexamethasone on electroencephalograms in patients with brain tumors: with specific reference to topographic computer display of delta activity. *J Neurosurg.* 1978;48(4):601–608.

28. Fernández-Bouzas A, Harmony T, Bosch J, et al. Sources of abnormal EEG activity in the presence of brain lesions. *Clin Electroencephalogr.* 1999;30(2):46–52.

29. Oh SJ, Kuba T, Soyer A, Choi IS, Bonikowski FP, Vitek J. Lateralization of brainstem lesions by brainstem auditory evoked potentials. *Neurology.* 1981;31(1):14.

30. Jandolo B, Pietrangeli A, Pace A, Carapella C, Finocchiaro R, Morace E. Brain-stem auditory evoked potentials in supratentorial brain tumors. *Electromyogr Clin Neurophysiol.* 1992;32(6):307–309.

31. Lin X, DeAngelis LM. Treatment of brain metastases. *J Clin Oncol.* 2015;33(30):3475.

32. Broughton R, Rasmussen T, Branch C. Scalp and direct cortical recordings of somatosensory evoked potentials in man (circa 1967). *Can J Psychol Rev Can Psychol.* 1981;35(2):136.

33. MacDonald D, Dong C, Quatrale R, et al. Recommendations of the International Society of Intraoperative Neurophysiology for intraoperative somatosensory evoked potentials. *Clin Neurophysiol.* 2019;130(1):161–179.

34. Gregorie EM, Goldring S. Localization of function in the excision of lesions from the sensorimotor region. *J Neurosurg.* 1984;61(6):1047–1054.

35. Rowed DW, Houlden DA, Basavakumar DG. Somatosensory evoked potential identification of sensorimotor cortex in removal of intracranial neoplasms. *Can J Neurol Sci.* 1997;24(2):116–120.

36. Krieg SM, Schäffner M, Shiban E, et al. Reliability of intraoperative neurophysiological monitoring using motor evoked potentials during resection of metastases in motor-eloquent brain regions. *J Neurosurg.* 2013;118(6):1269–1278.

37. Obermueller T, Schaeffner M, Shiban E, et al. Intraoperative neuromonitoring for function-guided resection differs for supratentorial motor eloquent gliomas and metastases. *BMC Neurol.* 2015;15(1):211.

38. Hoefer PF, Cohen SM. Localization of cord tumors by electromyography. *J Neurosurg.* 1950;7(3):219–226.

39. Grisold W, Piza-Katzer H, Jahn R, Herczeg E. Intraneural nerve metastasis with multiple mononeuropathies. *J Peripher Nerv Syst.* 2000;5(3):163–167.

40. Kaplan JG, Portenoy RK, Pack DR, DeSouza T. Polyradiculopathy in leptomeningeal metastasis: the role of EMG and late response studies. *J Neuro-Oncol.* 1990;9(3):219–224.

41. Rubin DI. Normal and abnormal spontaneous activity. In: *Handbook of Clinical Neurology.* vol. 160. Elsevier; 2019:257–279.

42. Argov Z, Siegal T. Leptomeningeal metastases: peripheral nerve and root involvement—clinical and electrophysiological study. *Ann Neurol.* 1985;17(6):593–596.

43. Briemberg HR, Amato AA. Neuromuscular complications of cancer. *Neurol Clin.* 2003;21(1):141–165.

44. Pancoast HK. Importance of careful roentgen-ray investigations of apical chest tumors. *J Am Med Assoc.* 1924;83(18):1407–1411.

45. Jaeckle KA. Nerve plexus metastases. *Neurol Clin.* 1991;9(4):857–866.

46. Thomas JE, Cascino TL, Earle JD. Differential diagnosis between radiation and tumor plexopathy of the pelvis. *Neurology.* 1985;35(1):1.

47. Harper CM, Thomas JE, Cascino TL, Litchy WJ. Distinction between neo plastic and radiation-induced brachial plexopathy, with emphasis on the role of EMG. *Neurology.* 1989;39(4):502.

48. King R, Busam K, Rosai J. Metastatic malignant melanoma resembling malignant peripheral nerve sheath tumor: report of 16 cases. *Am J Surg Pathol.* 1999;23(12):1499.

49. Antoine J-C, Camdessanché J-P. Peripheral nervous system involvement in patients with cancer. *Lancet Neurol.* 2007;6(1):75–86.

50. Rogers L, Borkowski G, Albers J, Levin K, Barohn R, Mitsumoto H. Obturator mononeuropathy caused by pelvic cancer: six cases. *Neurology.* 1993;43(8):1489.

51. Yingling CD, Ojemann S, Dodson B, Harrington MJ, Berger MS. Identification of motor pathways during tumor surgery facilitated by multichannel electromyographic recording. *J Neurosurg.* 1999;91(6):922–927.

52. Metter I, Alkalay D, Mozes M, Geffen DB, Ferit T. Isolated metastases to peripheral nerves. Report of five cases involving the brachial plexus. *Cancer.* 1995;76(10):1829–1832.

53. Chamberlain M, Junck L, Brandsma D, et al. Leptomeningeal metastases: a RANO proposal for response criteria. *Neuro-Oncology.* 2017;19(4):484–492.

54. Chamberlain MC. Leptomeningeal metastasis. *Curr Opin Oncol.* 2010;22(6):627–635.

55. Gleissner B, Chamberlain MC. Neoplastic meningitis. *Lancet Neurol.* 2006;5(5):443–452.

56. Taillibert S, Laigle-Donadey F, Chodkiewicz C, Sanson M, Hoang-Xuan K, Delattre J-Y. Leptomeningeal metastases from solid malignancy: a review. *J Neuro-Oncol.* 2005;75(1):85–99.

57. El Shafie RA, Böhm K, Weber D, et al. Palliative radiotherapy for leptomeningeal carcinomatosis–analysis of outcome, prognostic factors, and symptom response. *Front Oncol.* 2019;8:641.

58. Chamberlain MC. Radioisotope CSF flow studies in leptomeningeal metastases. *J Neuro-Oncol.* 1998;38(2–3):135–140.

59. Lamba N, Fick T, Tewarie RN, Broekman ML. Management of hydrocephalus in patients with leptomeningeal metastases: an ethical approach to decision-making. *J Neuro-Oncol.* 2018;140(1):5–13.

60. Grossman SA, Krabak MJ. Leptomeningeal carcinomatosis. *Cancer Treat Rev.* 1999;25(2):103–119.

61. Murray JJ, Greco FA, Wolff SN, Hainsworth JD. Neoplastic meningitis: marked variations of cerebrospinal fluid composition in the absence of extradural block. *Am J Med.* 1983;75(2):289–294.

62. Palma J-A, Fernandez-Torron R, Esteve-Belloch P, et al. Leptomeningeal carcinomatosis: prognostic value of clinical, cerebrospinal fluid, and neuroimaging features. *Clin Neurol Neurosurg.* 2013;115(1):19–25.

63. Bruna J, González L, Miró J, et al. Leptomeningeal carcinomatosis: prognostic implications of clinical and cerebrospinal fluid features. *Cancer.* 2009;115(2):381–389.

64. Hyun J-W, Jeong IH, Joung A, Cho HJ, Kim S-H, Kim HJ. Leptomeningeal metastasis: clinical experience of 519 cases. *Eur J Cancer.* 2016;56:107–114.

65. Kaye S, Bagshawe K. Chemical markers in spinal fluid for tumours of the central nervous system (CNS). In: *CNS Complications of Malignant Disease.* Springer; 1979:306–323.

66. Barone A. L'attività lisozimica attuale del liquor nelle malatti nervose. *Acta Neurol Italia.* 1948;3:434.

67. Koch TR, Lichtenfeld KM. Detection of central nervous system metastasis with cerebrospinal fluid beta-2-microglobulin. *Cancer.* 1983;52(1):101–104.

68. Twijnstra A, Van Zanten A, Nooyen W, de Visser BO. Sensitivity and specificity of single and combined tumour markers in the diagnosis of leptomeningeal metastasis from breast cancer. *J Neurol Neurosurg Psychiatry.* 1986;49(11):1246–1250.

69. Van Zanten A, Twijnstra A, Hart A, De Visser BO. Cerebrospinal fluid lactate dehydrogenase activities in patients with central nervous system metastases. *Clin Chim Acta.* 1986;161(3):259–268.

70. Nagai A, Terashima M, Harada T, et al. Cathepsin B and H activities and cystatin C concentrations in cerebrospinal fluid from patients with leptomeningeal metastasis. *Clin Chim Acta.* 2003;329(1–2):53–60.

71. Nagai A, Murakawa Y, Terashima M, et al. Cystatin C and cathepsin B in CSF from patients with inflammatory neurologic diseases. *Neurology.* 2000;55(12):1828–1832.

72. Shi Q, Pu C, Wu W, et al. Value of tumor markers in the cerebrospinal fluid in the diagnosis of meningeal carcinomatosis. *Nan Fang Yi Ke Da Xue Xue Bao.* 2010;30(5):1192.

73. Bach F, Bach FW, Pedersen AG, Larsen PM, Dombernowsky P. Creatine kinase-BB in the cerebrospinal fluid as a marker of CNS metastases and leptomeningeal carcinomatosis in patients with breast cancer. *Eur J Cancer Clin Oncol.* 1989;25(12):1703–1709.

74. Groves MD, Hess KR, Puduvalli VK, et al. Biomarkers of disease: cerebrospinal fluid vascular endothelial growth factor (VEGF) and stromal cell derived factor (SDF)-1 levels in patients with neoplastic meningitis (NM) due to breast cancer, lung cancer and melanoma. *J Neuro-Oncol.* 2009;94(2):229–234.

75. Wang P, Piao Y, Zhang X, Li W, Hao X. The concentration of CYFRA 21-1, NSE and CEA in cerebro-spinal fluid can be useful indicators for diagnosis of meningeal carcinomatosis of lung cancer. *Cancer Biomark.* 2013;13(2):123–130.

76. Hansen P, Kjeldsen L, Dalhoff K, Olesen B. Cerebrospinal fluid beta-2-microglobulin in adult patients with acute leukemia or lymphoma: a useful marker in early diagnosis and monitoring of CNS-involvement. *Acta Neurol Scand.* 1992;85(3):224–227.

77. Rubenstein JL, Wong VS, Kadoch C, et al. CXCL13 plus interleukin 10 is highly specific for the diagnosis of CNS lymphoma. *Blood.* 2013;121(23):4740–4748.

78. Sundaresan N, Vugrin D, Nisselbaum J, Galicich JH, Cvitkovic E, Schwartz MK. Cerebrospinal fluid markers in central nervous system metastases from testicular carcinoma. *Neurosurgery.* 1979;4(4):292–295.

79. Cone LA, Koochek K, Henager HA, et al. Leptomeningeal carcinomatosis in a patient with metastatic prostate cancer: case report and literature review. *Surg Neurol.* 2006;65(4):372–375.

80. Bots GT, Went L, Schaberg A. Results of a sedimentation technique for cytology of cerebrospinal fluid. *Acta Cytol.* 1964;8(3):234.

81. Rogers L, Duchesneau P, Nunez C, et al. Comparison of cisternal and lumbar CSF examination in leptomeningeal metastasis. *Neurology.* 1992;42(6):1239.

82. Chamberlain MC, Johnston SK. Neoplastic meningitis: survival as a function of cerebrospinal fluid cytology. *Cancer.* 2009;115(9):1941–1946.

83. Prayson RA, Fischler DF. Cerebrospinal fluid cytology: an 11-year experience with 5951 specimens. *Arch Pathol Lab Med.* 1998;122(1):47.

84. Preusser M, Hainfellner JA. CSF and laboratory analysis (tumor markers). In: *Handbook of Clinical Neurology.* vol. 104. Elsevier; 2012:143–148.

85. Feng L, Chen D, Zhou H, et al. Spinal primary central nervous system lymphoma: case report and literature review. *J Clin Neurosci.* 2018;50:16–19.

86. Hironaka K, Tateyama K, Tsukiyama A, Adachi K, Morita A. Hydrocephalus secondary to intradural extramedullary malignant melanoma of spinal cord. *World Neurosurg.* 2019;130:222–226.

87. Pedreira CE, Costa ES, Lecrevisse Q, van Dongen JJ, Orfao A, Consortium E. Overview of clinical flow cytometry data analysis: recent advances and future challenges. *Trends Biotechnol.* 2013;31(7):415–425.

88. Dux R, Kindler-Röhrborn A, Annas M, Faustmann P, Lennartz K, Zimmermann C. A standardized protocol for flow cytometric analysis of cells isolated from cerebrospinal fluid. *J Neurol Sci.* 1994;121(1):74–78.

89. Cibas ES, Malkin MG, Posner JB, Melamed MR. Detection of DNA abnormalities by flow cytometry in cells from cerebrospinal fluid. *Am J Clin Pathol.* 1987;88(5):570–577.

90. de Graaf MT, de Jongste AH, Kraan J, Boonstra JG, Smitt PAS, Gratama JW. Flow cytometric characterization of cerebrospinal fluid cells. *Cytometry B Clin Cytom.* 2011;80(5):271–281.

91. Bromberg J, Breems D, Kraan J, et al. CSF flow cytometry greatly improves diagnostic accuracy in CNS hematologic malignancies. *Neurology.* 2007;68(20):1674–1679.

92. Hegde U, Filie A, Little RF, et al. High incidence of occult leptomeningeal disease detected by flow cytometry in newly diagnosed aggressive B-cell lymphomas at risk for central nervous system involvement: the role of flow cytometry versus cytology. *Blood.* 2005;105(2):496–502.

93. Bettegowda C, Sausen M, Leary RJ, et al. Detection of circulating tumor DNA in early-and late-stage human malignancies. *Sci Transl Med.* 2014;6(224):224ra224.

94. Boire A, Brandsma D, Brastianos PK, et al. Liquid biopsy in central nervous system metastases: a RANO review and proposals for clinical applications. *Neuro-Oncology.* 2019;21(5):571–584.

95. Alix-Panabières C, Pantel K. Circulating tumor cells: liquid biopsy of cancer. *Clin Chem.* 2013;59(1):110–118.

96. De Mattos-Arruda L, Mayor R, Ng CK, et al. Cerebrospinal fluid-derived circulating tumour DNA better represents the genomic alterations of brain tumours than plasma. *Nat Commun.* 2015;6(1):1–6.

97. Li Y, Pan W, Connolly ID, et al. Tumor DNA in cerebral spinal fluid reflects clinical course in a patient with melanoma leptomeningeal brain metastases. *J Neuro-Oncol.* 2016;128(1):93–100.

98. Valadi H, Ekström K, Bossios A, Sjöstrand M, Lee JJ, Lötvall JO. Exosome-mediated transfer of mRNAs and microRNAs is a novel mechanism of genetic exchange between cells. *Nat Cell Biol.* 2007;9(6):654–659.

99. Teplyuk NM, Mollenhauer B, Gabriely G, et al. MicroRNAs in cerebrospinal fluid identify glioblastoma and metastatic brain cancers and reflect disease activity. *Neuro-Oncology.* 2012;14(6):689–700.

100. Drusco A, Bottoni A, Lagana A, et al. A differentially expressed set of microRNAs in cerebro-spinal fluid (CSF) can diagnose CNS malignancies. *Oncotarget.* 2015;6(25):20829.

101. Wei D, Wan Q, Li L, et al. MicroRNAs as potential biomarkers for diagnosing cancers of central nervous system: a meta-analysis. *Mol Neurobiol.* 2015;51(3):1452–1461.

102. Baraniskin A, Kuhnhenn J, Schlegel U, et al. Identification of microRNAs in the cerebrospinal fluid as marker for primary diffuse large B-cell lymphoma of the central nervous system. *Blood.* 2011;117(11):3140–3146.

4

Biology and pathophysiology of central nervous system metastases

Mohini Singh, Ashish Dahal, Magali de Sauvage*, Juliana Larson*,
and Priscilla K. Brastianos*

Divisions of Neuro-Oncology and Medical Oncology, Departments of Medicine and Neurology, Massachusetts General
Hospital, Boston, MA, United States

1 Introduction

Cancer is a disease that is triggered by progressive accumulation of genetic alterations in regulatory systems that control normal cellular homeostasis. These alterations can be inherited, arise spontaneously (i.e., DNA damage, mutations from replication errors), occur through posttranscriptional modification by microRNAs, or regulated through epigenetic controls (i.e., DNA methylation, histone modification, altered chromatin).[1] Genetic aberrations can be subdivided into two gene groups: overexpression of tumor-promoting genes, or oncogenes, and inactivation or repression of tumor suppressor genes (TSGs). Extensive investigations into the molecular underpinnings of neoplastic transformation have revealed the process to be multifaceted and a result of the dysregulated interplay of oncogenes and TSGs.[1-4] Several oncogenes are involved in pathways that positively control internal signaling pathways that in turn regulate tumor cell proliferation and growth, such as platelet-derived growth factor and its receptor (PDGF, PDGFR), epidermal growth factor and its receptor (EGF, EGFR), fibroblast growth factor (FGF), insulin-like growth factor (IGF), Ras, Akt, myc, and mTOR.[5] Conversely, tumor suppressor genes are commonly negative regulators of the cell cycle, DNA repair, such as p53, Rb, p16, and p15 (i.e., INK4a, INK4b), cyclins (i.e., CCND1, CCNE1), BRCA2, and PTEN.[5,6]

The instigation of these tumor-inducing processes confers a selective survival advantage as a normal cell progressively evolves into tumorigenic and subsequent malignant states. With the identification of over 100 cancer types and organ-specific subtypes, innumerable studies have been conducted to determine the existence, if any, of universal similarities during neoplastic development. As such, there are ten hallmarks that have been proposed as defining biological traits of a tumorigenic cell: sustained proliferative signaling, evasion of growth suppressors, resistance of cell death, replicative immortality, induction of angiogenesis, activation of invasion and metastasis, avoidance of immune destruction, promotion of inflammation, genomic instability and mutation, and dysregulation of cellular energetics.[7]

While tumor initiation has been determined to be genetically or epigenetically regulated, progression in solid tumors requires an intricate interplay between the tumor cells and the surrounding microenvironment.[1] The extracellular matrix (ECM), vascularization, immune cells, and noncancerous cells (i.e., fibroblasts) all collaborate with the tumor cell through disrupted signaling pathways. Tumor evolution is accompanied by the evolution of the "ecosystem" and has been shown to mimic signaling seen in tissue development and repair.[1]

Further genetic and epigenetic alterations within a tumor cell can lead to the transition into a metastatic phenotype, whereby the cell gains the ability to travel throughout the body to initiate tumor growth at a secondary location, or "metastasize." The metastatic tumor cell cycles through several stages of a "metastatic cascade"; briefly, the stages involve invasion, dissemination into the circulation and avoidance of immune surveillance, arrest, and then extravasation at a secondary site.[8] This process is initiated and culminates in an evolutionary conserved developmental program that has been implicated in carcinogenesis, termed the epithelial-mesenchymal transition (EMT),

---

*These authors contributed equally as co-second authors.

and its reversal termed the mesenchymal-epithelial transition (MET). Essentially, the metastatic cell shifts from a sedentary, adherent, and proliferative phenotype into a detached cell with enhanced mobility, invasive, and apoptosis-resistant properties.[8–10] The blood–brain barrier (BBB) plays a significant role in the prevention, establishment, and treatment of brain metastases. While a healthy BBB functions to maintain the delicate environment of the brain by regulating the passage of solutes and proteins, once breached by metastatic cells the BBB becomes dysfunctional, and in studies has been seen to assist in metastasis development. Furthermore, the BBB retains enough of its barrier ability to restrict the passage of therapeutics, which limits treatment options.

This chapter will review the major pathway components that regulate neoplastic transformation in solid tumors, with a focus on how these pathways relate to brain metastases, the molecular mechanisms that regulate the transition into a metastasis, and the genetics underlying brain metastasis formation.

2 Molecular biology of neoplasms

Under physiological conditions, tissue homeostasis is maintained by fate-determining signals transduced by growth factors. Integration of these growth factors and their receptor-mediated signaling pathways play a critical role in driving oncogenesis in a number of cancers. The first evidence linking soluble growth factors to cancer was discovered in the 1950s with the isolation of nerve growth factor (NGF) and epidermal growth factor (EGF) by Cohen et al.,[11] followed by the isolation of two transforming growth factors, TGF-α and TGF-β.[12] Studies conducted by Waterfield et al. concerning the structure of platelet-derived growth factor (PDGF) supported evidence of growth factor secretion by virally transformed cells, chemically transformed cells, and tumor cells to promote self-stimulated (autocrine) growth.[13] Cloning of the EGF receptor improved upon the knowledge of the intracellular GF mechanisms[14]: The majority of growth factor receptors are single-pass transmembrane proteins that possess an intracellular tyrosine kinase (RTK) domain (serine/threonine for TGF-β receptors).[15] Upon binding their respective growth factor, RTKs undergo dimerization and autophosphorylation and induce activation of signaling pathway proteins along with their associated secondary messengers. These secondary messengers are ultimately translocated to the nucleus to activate gene expression and cell survival activities, such as proliferation and neoangiogenesis, and metastatic properties such as invasion and therapeutic resistance.[16] In addition to the autocrine effects, growth factors are essential mediators of paracrine loops, conduct signals between tumor cells, neighboring tumor cells, the extracellular matrix (ECM),

and stromal cells such as fibroblasts, macrophages, and endothelial cells of the vasculature.[17]

The key growth factors implicated in tumorigenesis include PDGF, EGF, FGF, and IGF, which are key constituents of the RAS/Raf MAP kinase pathway, PI3K/PTEN/Akt cascade, and the mTOR signaling pathways.[18–21] Due to their essential and pervasive roles in tumor cell growth, survival, angiogenesis, and metastasis as well as their elevated expression on cancer cell surfaces, growth factor receptors remain an attractive molecular target in the treatment of cancer.[22] Numerous monoclonal antibodies (mAb) have been engineered as "magic bullet" targeted therapies for several cancers; they are designed to bind antigens found on the cell surface and block or modulate the function of growth factor receptors. Subsequently, this mechanism can restore or even enhance the host's immune response to cancer cells.[22] Tyrosine kinase inhibitors (TKIs) target the active site of the kinase domain in RTKs and prevent phosphorylation of intracellular targets, essentially blocking the downstream activation cascade.[23,24] Nucleic acid-based therapies are a diverse class of plasmids, antisense oligonucleotides (ASO), small interfering RNA (siRNA) and microRNA, messenger RNA (mRNA), immunomodulatory DNA/RNA, and gene-editing guide RNA (gRNA) therapies that are promising therapies being investigated due to their versatile capabilities to alter target gene expression and modulate immune response. This is significant when developing immunotherapies; however, delivery challenges are faced with the intrinsic physiochemical properties such as hydrophilicity and susceptibility to enzymatic degradation, especially when attempting to deliver therapeutics across the blood–brain barrier to reach CNS neoplasms.[24] Multitargeting of tumors at several levels by combinations of chemotherapies is a promising option in the treatment of cancer, and some delivery methods are discussed later.

2.1 Growth factor signaling—PDGF and PDGFR

Platelet-derived growth factor (PDGF) is a multigene regulated protein with four peptide chains (A, B, C, and D), with each chain expressed by a different gene. To form the active configuration of PDGF, two chains dimerize to form one of five different dimers (AA, AB, BB, CC, or DD). The receptor of PDGF (PDGFR) is a single-chain transmembrane glycoprotein that has five immunoglobulin (Ig)-like extracellular domains and a tyrosine kinase domain. There are two receptor forms, α and β, whose expressions are controlled by two separate genes. The functional form of PDGFR is also a dimer (aa, ab, or bb). The ligand and the receptor sub-parts have varying affinities for each other, adding another level of specificity to the formation of the ligand-receptor complex and greater difficulty in developing targeted therapies.[19,25,26] Binding of PDGF homo−/heterodimers causes dimerization and conformational changes of the respective PDGFR, activating the receptors for downstream trans-phosphorylation of transduction molecules with Src Homology 2 (SH2) domains and relaying signals via mainly the RAS/Raf MAP kinase and PI3K/PTEN/Akt pathways to ultimately promote cell division, differentiation, and migration.[19,25,26] In addition to its autocrine role, PDGFB has a paracrine role after its release from neighboring stromal cells, which then activates PDGFR-β on tumor cells.[27]

Amplification and/or overexpression of *PDGFR-α* and *PDGFR-β* can cause aberrant PDGFR expression, which has been seen in high-grade gliomas,[28,29] prostate cancer,[30] pancreatic cancer[31] malignant melanoma,[32,33] and soft tissue sarcomas.[34] In preclinical models, mammary tumor cells expressing high PDGFB had a high propensity to metastasize to the brain when injected intravenously, shown to be promoted by stromal expression of PDGFRβ[D849V] Furthermore, high PDGFB protein expression was shown to be a prognostic indicator for brain metastases.[27]

Therapies to inhibit and decrease the activity of overexpressed PDGFR are clinically available, such as imatinib (leukemia and gastrointestinal stromal tumors), sunitinib (renal cell carcinoma (RCC) and imatinib-resistant gastrointestinal stromal tumor (GIST)), pazopanib (advanced renal cell carcinoma), and sorafenib (hepatocellular carcinoma).[32,33,35–37] Crenolanib, a benzimidazole that has potent and selective inhibition of PDGFR-α and PDGFR-β, was shown to inhibit intracranial growth of mammary tumor cells.[27] Unfortunately, these drugs are general TKIs that antagonize multiple tyrosine kinase signaling pathways. Therefore, the therapeutic success of these drugs is a result of blocking several tyrosine kinases and specific PDGFR inhibition cannot be evaluated.[38] For instance, imatinib is a proven inhibitor of PDGFR-α and PDGFR-β. In a preclinical model with oxygen-induced retinopathy (OIR) mice, angiogenesis and neovascularization were effectively reduced as well as the expression of PDGFR-α and PDGFR-β. However, the mice also showed lower levels of other growth factors, VEGF and FGF.[39]

2.2 Growth factor signaling—FGF and FGFR

Fibroblast growth factors (FGFs) are a family of 19 peptides that each bind to and activate one of four FGF closely related single-chain transmembrane receptors (FGFR 1–4), with each isoform undergoing differential splicing of the *FGFR* mRNA to result in altered ligand specificity.[22] Each FGFR possesses an extracellular domain with three immunoglobulin repeats (Ig I-III) to bind ligands and a transmembrane domain and one intracellular domain with kinase activity at the carboxy-terminus. FGFs are members of the tyrosine kinase receptor family, which dimerize, autophosphorylate, and undergo a conformational change

for activation and phosphorylation of internal transduction molecules. FGFRs function as a docking hub for several downstream effectors, subsequently organizing multiple context-dependent cellular functions such as cell proliferation, growth, migration, and wound healing.[22] In cancer, FGF-FGFR interactions are vital to tumorigenesis, activating pathways similar to those of PDGFR and EGFR, such as the RAS-MAPK and its converged PLC/PKC pathways as well as the PI3K-AKT pathway to induce tumor cell growth, inhibition of cell death, and angiogenesis.[40] Importantly, aberrant FGFR expression is implicated in acquired resistance to oncotherapies, especially those targeting other growth factor receptors via several mechanisms. One such mechanism is through the gene amplification of FGFR, which causes upregulated expression and constitutive activation of signaling pathways that are both ligand-dependent and ligand-independent processes. Another mechanism is producing an excess of FGF, which leads to a similar effect of overstimulation of FGFR and the ligand-dependent pathways. Increased FGF/FGFR signaling can also exaggerate angiogenesis to such a degree that it facilitates resistance to antiangiotherapy.[41]

Holistically, upregulation and genomic alterations in the FGFR exome play a role in oncogenesis. Each of the four FGF receptors has been shown to have varying expressions in different cancers. FGFR1 gene amplifications have been demonstrated in 22% of squamous cell lung cancer[42] and 10% of primary breast cancers.[43,44] FGFR1 amplifications were found to be enriched in brain metastases of lung adenocarcinomas.[45]

Targeting of FGFR is attractive to circumvent the associated resistance mechanisms, which if used in conjunction with other therapies could make the combined treatment more effective. Several drugs have been developed to target FGF-/FGFR-driven tumor cell proliferation. For example, dovitinib is a reversible inhibitor that targets FGFR1 and FGFR3 and has been used in clinical trials for renal cell carcinoma,[46] advanced breast cancer,[44] and prostate cancer.[47] Other reversible inhibitors of FGF include AZD4547, BGJ398, Debio-1347, JNJ-42756493, and LY2874455. BLU9931 and FIIN-2 bind irreversibly to FGFR by forming a covalent bond to each cysteine in the kinase domain.[48]

2.3 Growth factor signaling—IGF and IGFR

The insulin-like growth factor (IGF) family is comprised of two proteins, IGF1, also known as somatomedin, and IGF2, that binds the respective transmembrane receptors IGF1R and IGF2R, as well as several growth factor-binding proteins involved with growth factor transport. IGF1 binds to IGF1R with the higher affinity than to IGF2R, and vice versa with IGF2 and IGF2R. IGF1R is a polypeptide possessing two extracellular α chains and two transmembrane β chains, with each α and β subunit connected by a disulfide bond to for an αβ chain.[22,49] Upon IGF1 binding to IGF1R, the receptor's intracellular kinase domain autophosphorylates to activate downstream components of the Ras, PI3K, and Akt signal transduction pathways. Perturbation in IGFR1 activity, either by overexpression on constitutive activation, leads to enhanced tumor cell proliferation, transformation, and metastasis in multiple cancers.[50] IGF2R is a single polypeptide chain that behaves as a "scavenger receptor" for IGF2; IGF2R acts as an antagonist to IGF2 and breaks it down upon binding, leading to suppression of tumor growth, modified invasion, and suppressed angiogenesis.[51]

IGF1 and IGF2 also act as mitogens that accelerate mitotic division by prompting the expression of cyclin D1 and pushing the cell from G1 to S phase, and by upregulating Bcl and downregulating Bax proteins, effectively inhibiting apoptosis.[52] These properties support the overexpression of these mitogens and IGF2R commonly seen in cancer. In fact, preclinical data have demonstrated that IGF1R is overexpressed in lung cancer,[53] prostate cancer,[54] breast cancer,[55] glioma,[56] and gastrointestinal cancers.[57] Additionally, increased IGF1R expression has been associated with resistance to treatment,[53] metastases, and a shorter overall survival of the patient.[58] IGF1R notably plays a role in the development of breast cancer brain metastases. Saldana et al. showed that IGF1R was constitutively autophosphorylated and active in brain-seeking breast cancer cells and that knockdown of IGF1R reduced potential for brain metastasis generation by these cells in an experimental in vivo model.[59] Interestingly, brain pericytes drive brain metastasis formation by secreting high levels of IGF to act as a chemoattractant, enticing breast cancer cells to travel to the brain and promote their proliferation.[60] Ireland et al. showed that breast cancer-associated macrophages had high expression of IGFR1 and IGFR2 and that the addition of IGF blockers enhanced the activity of paclitaxel to reduce breast cancer metastasis.[61] Clinical trials of IGF1R targeted therapies have been moderately successful. For example, a phase I clinical trial treated four patients with advanced squamous cell carcinoma of the lung over the course of seven months with three IGF1R inhibitors (picropodophyllin, PPP, AXL1717).[62] Preclinical data suggest that picropodophyllin attenuates the malignant phenotype of brain-seeking breast cancer cells.[59] BMS-754807, a potent and reversible IGF1R inhibitor, is in clinical trials to treat advanced metastatic solid cancers (NCT00908024).

2.4 Growth factor signaling—EGF

The family of epidermal growth factor (EGF) is very large and comprised of proteins with highly similar structural and functional characteristics. Each family member contains one or more repeats of a conserved amino acid sequence comprised of six cysteine residues forming three disulfide bonds, which then forms three structural loops that are vital for bonding between the ligand and its receptor. In addition to the well-known EGF, other family members include heparin-binding EGF-like growth factors (HB-EGF), transforming growth factor-α (TGF-α), amphiregulin (AR), epiregulin (ER), epigen, betacellulin (BTC), and neuregulin 1 through 4 (NRG1–4).[63,64]

EGF receptor family encompasses 4 proteins, specifically EGFR 1 to 4, otherwise known as ErbB-1 to ErbB-4 or HER-1 to HER-4. Each receptor is a single-chain glycoprotein that contains an amino terminal extracellular ligand-binding domain, a hydrophobic membrane-spanning region, and a cytoplasmic region that contains the tyrosine kinase domain, critical tyrosine residues, and receptor regulatory motifs. Like the PDGF pathway, receptors will form either homo- or heterodimers upon ligand binding to become active, and subsequently activate downstream regulators of multiple signaling pathways such as the RAS/MAPK pathway, the PI3K/AKT pathway, and the phospholipase C/protein kinase C (PKC) pathways to promote tumor cell proliferation, survival, differentiation, and migration.[20,22] Each of the 11 ligands can bind combinations of EGFRs with varying affinities, thereby controlling the specificity of the cellular response and strength of the signaling output.[65]

EGF and EGFR are known to be important drivers in malignant cancer phenotypes in a number of cancers including glioblastoma, nonsmall cell lung cancer (NSCLC), colorectal cancer, and head and neck cancer.[66] HER2 (EGFR2) is overexpressed in about 25%–30% of breast cancer.[67] Alterations in EGFR are common and include gene amplification, protein overexpression, mutations, or in-frame deletions.[19] In NSCLC, two specific mutations account for over 90% of the EGFR activation: exon-19 deletions and/or exon-21 L858R substitution.[65,68] However, over 200 EGFR mutations have been discovered in NSCLC.[69]

Discoveries and characterization of mutations in this pathway have led to the development of targeted antibodies and small molecule inhibitors that have revolutionized patient outcomes through personalized treatment. Several types of drug inhibitors have been developed and are in clinical use to target these specific EGFR mutations. Gefitinib and erlotinib are type I reversible inhibitors. Gefitinib was the first targeted therapy approved for lung cancer with a known molecular target.[70] Cetuximab and panitumumab are monoclonal antibodies that target EGFR1 and are approved to treat metastatic colorectal cancers.[22,71,72] Trastuzumab is a monoclonal antibody that targets EGFR2 and is approved to treat metastatic breast and stomach cancers.[22,73]

Notably, brain metastases occur frequently in patients with EGFR mutant lung cancer and HER2 overexpressing breast cancer, thus, there is significant interest in the development of CNS penetrant EGFR inhibitors. Osimertinib is an irreversible type VI CNS penetrant inhibitor, which targets EGFR T790M mutations, common resistant mutations to EGFR inhibitors in lung cancer.[74] Drug resistance can develop to osimertinib with a C797S mutation, causing the drug to lose the ability to form the covalent bond to its target molecule, EGFR.[68] In a study looking at osimertinib compared to gefitinib/erlotinib, the overall response rate in brain metastases was much higher in patients receiving osimertinib.[75]

2.5 Growth factor signaling—MAPK pathway

The mitogen-activated protein kinase (MAPK, otherwise known as ERK) pathway consists of a chain of proteins that conveys a signal from the surface receptor to the cell nucleus to regulate cell growth, differentiation, cytoskeletal organization, membrane trafficking, and apoptosis. The pathway consists of several members: Ras, Raf, MEK, and MAPK. MAPK pathway activation begins with Ras. All Ras protein family members belong to the class of small GTPase proteins, comprised of H, K, M, N, and R types, Rap (types 1 and 2), and Ral. As a GTPase, RAS acts as a molecular switch that cycles between an inactive GDP-bound form and an active GTP-bound form. For activation to occur, RAS must first undergo a series of posttranslational modifications that increase the hydrophobicity of the C-terminus through farnesylation of the cysteine residue in the C-terminal of the CaaX motif located in the HVR.[33] This step is essential for the RAS protein to associate with the inner cell membrane to activate downstream signaling pathways.[76] The conversion between the stable, inactive RAS-GDP form to the active RAS-GTP form is stimulated by guanine nucleotide exchange factors (GEFs), while conversion to the inactive form is mediated by GTPase-activating proteins (GAPs).[32] When proximal to the membrane-bound RAS protein, GEFs and GAPs act as highly regulatory molecules that are crucial for regulating levels of active and inactive RAS. Upstream activation of RAS proteins begins with specific ligand binding to an RTK, such as EGFR, followed by autophosphorylation of the receptor site. An adapter protein with an SH2 domain, such as Grb2, then binds the receptor and recruits the GEF protein Sos-1. Sos-1 stimulates the RAS-GTP conformation allowing stimulation of several downstream effectors, including Raf-1, Rac and Rho, MEKK, PI2K, and phospholipase C.[32]

Ras activation then leads to a phosphorylation cascade. First comes the Raf family of serine/threonine-selective kinases, comprised of A-Raf, B-Raf, and C-Raf (or Raf-1); these are considered the "gatekeepers" of the MAPK pathway. All Rafs share the same domain architecture, regulation, and structure but differ in the length of their nonconserved N- and C-terminal ends. Raf will go on to phosphorylate mitogen-activated protein kinase (also known as MEK MAP2K, MAPKK) at tandem serine residues in an activation loop. The MEK family is limited to only two isoforms, MEK1 and MEK2, and shares ~85% sequence homology.[77] Each Raf isoform differs in its ability to phosphorylate either MEK1 or MEK2; B-Raf is the strongest MEK activator, A-Raf is the weakest activator with preference for MEK1, and C-Raf activates MEK1 and MEK2 with equal efficiency.[78] The conserved KDD motif in the kinase domain of MEK encourages ATP coordination and lends MEK1/2 its dual-specificity threonine/tyrosine protein kinase, allowing for downstream activation of MAPK.[79,80]

Defects at multiple levels of the MAPK pathway have been shown to promote tumorigenesis and have been a target of interest for more than 15 years.[78] Within the RAS gene family, HRAS, KRAS, and NRAS are the only members recognized as oncogenes, discovered in human tumors more than 30 years ago.[29,76] Approximately 19% of cancer patients harbor RAS mutations, which are most common in pancreatic, colorectal, and lung carcinomas.[34] An effective RAS inhibitor would prevent the interaction between RAS and its downstream binding agents by either decreasing the proportion of RAS in the GTP state, disrupting RAS-GTP-GEF interactions, or decreasing the concentration of RAS localization at the membrane.[35] RAS inhibitors are in active clinical development currently (NCT03785249; NCT04685135). The B-Raf gene is mutated in ~66% of melanomas and at lower frequencies in numerous other cancers.[81] BRAF inhibitors have revolutionized therapeutic paradigms in melanomas,[82–84] including in the setting of brain metastases.[85] MEK inhibitors are typically combined with BRAF inhibitors to delay the emergence of resistance to BRAF inhibition.[86]

2.6 PI3K/PTEN/AKT signaling pathway

The PI3K/Akt/mTOR pathway is another intracellular signaling pathway vital to the regulation of the cell cycle, thus directly related to controlling quiescence, longevity, and proliferation. The pathway consists of several integral components: RTKs and GPCRs, PI3K, PIP2, PIP3, and AKT/protein kinase B.

Activation of the PI3K pathway can begin with two different cell surface receptors, RTKs and G-protein coupled receptors (GPCRs). As mentioned previously, RTKs have three functional domains: an extracellular ligand-binding domain, a transmembrane domain, and an intracellular tyrosine kinase domain, and are stimulated by binding of growth factors, cytokines, and hormones, which causes dimerization and autophosphorylation.[87,88] GPCRs are a group of evolutionarily-related proteins possessing a domain coupled to G proteins and a transmembrane domain that passes through the cell membrane seven times.[89] Several ligands can active GPCR, including stromal cell-derived factor, sphingosine 1-phosphate, lyso-phosphatidic acid, carbachol, isoproterenol, prosta-glandin E2, thyroid-stimulating hormone (TSH), FSH, and luteinizing hormone (LH)/choriogonadotropin (CG). However, the activation of PI3K is predom-inantly tissue-specific.[90] Upon ligand binding, the GPCR undergoes a conformational change that allows it to act as a guanine nucleotide exchange factor (GEF), which essentially allows the GPCR to activate its as-sociated G protein by exchanging its GDP for a GTP. The G protein along with its bound GTP will then dis-sociate from the GPCR to activate further intracellular signaling proteins like PI3K.[91]

The phosphoinositide 3-kinases (PI3Ks, also called phosphatidylinositol 3-kinases) family of enzymes can be grouped into three classes (I-III), with each class playing a specific role in signal transduction. Generally, classes I and II of PI3Ks are important for cell signaling, whereas classes II and III are implicated in membrane trafficking.[37] Class I PI3Ks are divided into Class IA PI3Ks, which are activated by growth factor RTKs, and Class IB PI3Ks, which are activated by GPCRs.[92] Class IA contains regulatory subunit p85 and catalytic subunit p110, where p85 regulates the levels of p110 depending on if the cell is quiescent or activated by the growth fac-tor receptors or adapter proteins.[92]

The transfer of the γ-phosphate of ATP to the D3 posi-tion of the head group of phosphatidylinositols, a form of membrane lipid, sets into motion the PI3K signaling cascade.[92] GPCRs can activate Ras to directly bind p110, where PI3K activity is further stimulated and the sub-strate phosphatidylinositol-4,5-bisphosphate [PI(4,5)P2] is converted to phosphatidylinositol-3,4,5-trisphosphate [PI(3,4,5)P3].[87,88] In addition to PI(4,5)P2, other PI3K substrates include Rac, p70S6K, and certain isoforms of protein kinase C. After phosphorylation, PI(3,4,5)P3 ac-tivates membrane-associated kinases PDK1 and PDK2, while simultaneously binding a pleckstrin homology (pH) domain on Akt (or protein kinase B), an essential domain for recruiting Akt to the juxtamembrane re-gion.[93,94] PIP3 binding of Akt recruits Akt to the plasma membrane and into contact with PDK1, where Akt is then phosphorylated at the threonine 308 position by PDK1, and at the serine 473 position by PDK2. After phosphor-ylation at both sites, Akt is fully activated and detaches from the membrane to activate downstream targets that mediate cell survival and growth. As a serine/threonine kinase, Akt phosphorylates downstream effectors such as glycogen synthase kinase 3 a and b (GSK3a, GSK3b), BAD, 6-phosphofructo-2-kinase (PFK-2), GLUT-4, p70S6K, E2F, mTOR, and others.[95,96] Akt phosphoryla-tion of protein targets most often inhibits activity. For ex-ample, Akt phosphorylation of the proapoptotic protein

Bad inhibits Bad from heterodimerizing and binding the pro-survival proteins Bcl-2 and Bcl-XL.[92] Consequently, this leads to increased activity of Bcl-2 and Bcl-XL and promotion of cell survival. Other pro-survival targets of Akt inhibition include forkhead transcription factors (FKHR), caspase-9, and NF-kB, all proteins involved in pro-apoptotic pathways.[87,97]

The PTEN tumor suppressor gene is an important reg-ulator of the PI3K/Akt pathway as it can dephosphor-ylate tyrosine, serine, and threonine-phosphorylated peptides. More importantly, PTEN acts as an antago-nist by dephosphorylating PI(3,4,5)P3 with specificity for the phosphate group at the D3 position of the inosi-tol ring. Therefore, in PTEN deficient tumor cells, there are high basal levels of PI(3,4,5)P3 and phosphorylated Akt.

Aberrations in the pathway can lead to cancer dis-ease progression due to its involvement in inhibition of apoptosis, angiogenesis, tissue invasion, signal au-tonomy, and other hallmarks of cancer and have been implicated in several cancers including NSCLC, ovar-ian carcinoma, breast cancer, pancreatic cancer, and prostate cancer.[36,37,87,97] Gene mutations or amplifica-tions in the p110α isoform have been found across a range of cancers.[98] Hemizygous deletion of the wild-type PTEN allele is frequent in affected tumor cells, and the remaining copy of the gene is then inactivated by frameshift, nonsense, or missense mutations. Mutation and loss of PTEN function have a higher rate of inci-dence later in disease progression. In primary prostate tumors, there is a 12%–15% PTEN mutation rate, com-pared to at the time of metastasis when the phenotype is more aggressive and the PTEN mutation increases to 60%.[36,87] Rarely, activating mutations of p85 are found in ovarian and colon cancers, PIK3CA is amplified in up to 50% of breast, lung, and ovarian cancers, and PDK1 is infrequently mutated in colon cancers.[99] Up to 70% of breast-derived brain metastases show aberrant activa-tion of the PI3K pathway. PI3K pathway alterations in the context of brain metastases will be discussed later in the chapter.[100]

Small molecule drug inhibitors to inhibit the activ-ity of the PI3K/Akt signal transduction pathway are in clinical use or in active development, including in brain metastases. Several PI3K inhibitors have been studied. Alpelisib is FDA-approved for PIK3CA-mutated hormone receptor-positive metastatic breast cancer.[101] Buparlisib, a pan-class I PI3K inhibitor, is in phase II clinical trials in metastatic triple-negative breast cancer patients.[102] Idelalisib is the first FDA-approved PI3K inhibitor for the treatment of relapsed/refractory chronic lymphocytic leukemia/small lym-phocytic lymphoma and follicular lymphoma.[103] ARQ-092 is a potent pan-Akt inhibitor currently in phase I clinical studies.[104]

2.7 mTOR signaling pathway

The mechanistic target of rapamycin (mTOR, otherwise known as mammalian target of rapamycin and FRAP1) pathway converges with the PI3K pathway and maintains cellular homeostasis; it is responsible for monitoring the availability of nutrients, mitogenic signals and cellular energy and oxygen levels and consequently is significant in the regulation of cell growth and proliferation.[105] mTOR is another serine/threonine protein kinase that belongs to the PI3K-related kinase (PIKK) family. It is the catalytic subunit of two structurally and functionally distinct proteins, mTOR complex 1 (mTORC1) and complex 2 (mTORC2). mTORC1 is characterized by the core components mTOR, regulatory protein associated with mTOR (Raptor), and mammalian lethal with Sec13 protein 8 (mLST8, also known as GβL). Raptor facilitates the recruitment of substrates to mTORC1 and its subcellular localization, while mLST8 associates with the mTORC1 catalytic domain to stabilize the kinase activation loop. mTORC1 also possesses two inhibitory subunits: proline-rich Akt substrate of 40 kDa (PRAS40) and DEP domain containing mTOR-interacting protein (DEPTOR).[106] mTORC2 is also comprised of mTOR and mLST8, but contains rapamycin-insensitive companion of mTOR (Rictor) instead of Raptor. mTORC2 also contains the regulatory subunits DEPTOR, mSin1, and Protor1/2.[106] Both complexes localize to different subcellular compartments, which influences their activation and function[107]; mTORC1 is a master controller of cell growth and metabolism, while mTORC2 controls cell survival and proliferation.[108]

mTOR is activated by a chain of upstream components of the PI3K pathway. Akt phosphorylates the protein tuberous sclerosis complex (TSC), which is a heterotrimeric complex comprised of TSC1, TSC2, and TBC1D7 that functions as a GTPase-activator protein (GAP) for small G-protein Rheb (Ras-homolog enriched in brain).[109,110] Phosphorylation of TSC inhibits its dissociation from the lysosomal membrane and keeps it close to localized portions of Rheb for binding.[111] Rheb then directly binds and activates mTOR to stimulate its kinase activity, though the exact process is still unknown.[106]

There are several substrates of mTORC1 that ultimately regulate the production of anabolic-promoting proteins, lipids, and nucleotides while also suppressing catabolic pathways such as autophagy. mTORC1 phosphorylates p70S6 Kinase 1 (S6K1) on Thr389, enabling its subsequent phosphorylation and activation by PDK1, thus inducing phosphorylation and activation of several substrates that promote mRNA translation.[112] mTORC1 also phosphorylates eIF4E binding protein (4EBP) at multiple sites, which under normal circumstances binds and sequesters eIF4E to prevent the assembly of the eIF4F complex. However, this phosphorylation triggers 4EBP dissociation from eIF4E to permit the eIF4F complex to form.[113] mTORC1 activation of sterol-responsive element-binding protein (SREBP) promotes de novo lipid synthesis, which provides lipids for new membrane formation and expansion of growing cells.[114] Other substrates of mTORC1 include HIF1-α, MTHFD2, TFEB, and ULK1.[106] mTORC2 activation regulates cell proliferation and survival through activation of members of the AGC (PKA/PKG/PKC) family of protein kinases. mTORC2 substrates include protein kinase C (PKC) to regulate the actin cytoskeleton,[115] Akt to regulate the insulin/PI3K signaling,[116] and glucocorticoid-inducible kinase (SGK) to regulate ion transport as well as cell survival.[117]

Innumerable clinical studies have shown that mTOR is deregulated in several cancers and implicated as tumorigenic. The antifungal Rapamycin and its analogues are first-generation mTOR inhibitors, showing immunosuppressive and antiproliferative properties. Rapamycin has poor solubility and pharmacokinetics.[118] Sirolimus, a macrocyclic triene antibiotic, and its analogues Everolimus and Temsirolimus are three commercially available inhibitors approved by the FDA for the treatment of metastatic renal cell carcinoma, pancreatic neuroendocrine tumors, and postmenopausal HR⁺ advanced breast cancer. However, their use as single agents in other solid cancers is limited.[119] Unfortunately, the efficacy of rapalogs is rather modest in clinic to date, and the response variable. Good biomarkers are being investigated that are associated with response or resistance to mTOR inhibition. Resistance mutations to mTOR inhibitors include BRAF, KRAS, and TSC mutations, whereas PIK3CA mutations are markers associated with sensitivity.[119,120]

Second-generation ATP-competitive mTOR kinase inhibitors (TORKinibs) have been developed to more fully inhibit mTORC1 and mTORC2, specifically in tumors that are addicted to mTOR signaling pathways. These treatments inhibit cell growth and induce apoptosis.[121] Vistusertib and its analogue AZD8055 have shown significant efficacy in treating ER⁺ breast cancer and can even suppress growth of breast cancers that have acquired resistance to endocrine therapy, rapalogs, and paclitaxel.[122,123] Third-generation bivalent mTOR inhibitors that specifically target mTOR resistance mutations have been in development as potential cancer therapeutic agents.[119,120,124] Preclinical studies have shown GDC-0068, a highly selected AKT inhibitor, and GDC-0084, a CNS penetrant inhibitor of PI3K and mTOR, to be promising treatments for breast-derived brain metastases.[125,126]

2.8 Angiogenesis signaling pathways—VEGF

Angiogenesis is the physiological process of generating new blood vessels and vasculature from preexisting networks. In normal conditions, angiogenesis is a product of ischemic and hypoxic signaling and is vital to development, muscle hypertrophy, menstruation, pregnancy, tissue regeneration, and wound healing.[127] The process is complex and tightly regulated, requiring a delicate balance between pro-angiogenic and antiangiogenic factors (see Table 1).[128–131] In cancer, angiogenesis is induced by the hypoxic microenvironment and is crucial to growth, progression, invasion, and metastasis.[127] Tumor cells will tip this delicate equilibrium of factors during an "angiogenic switch," which initiates vessel sprouting, upregulating the production of angiogenic factors to bind with and activate endothelial cells within the extracellular matrix (ECM). These endothelial cells will migrate toward the source of these angiogenic factors, secreting enzymes that break down the basement membrane to create tiny pores they can migrate through. As the endothelial cells migrate, they proliferate and undergo a tubule formation phase, where additional mural cells (vascular smooth muscle cells, pericytes), fibroblasts, and immune cells are recruited for structural support.[132] Compared to normal vasculature, tumor vasculature is characterized by uncontrolled upregulation of these pro-angiogenic factors, atypical morphology, high permeability, and poor blood flow.[133] Angiogenesis is particularly important for the growth of brain tumors. Hypoxia and acidosis within the brain tumor environment lead to the upregulation of vascular endothelial growth factor, which in turn promotes angiogenesis.[134]

In addition to classical angiogenesis, tumor cells have been shown to activate an additional mechanism to acquire microcirculation, termed "vasculogenic mimicry" (VM). The distinction from typical angiogenesis is the lack of dependence on endothelial cells—tumor cells can undergo tubular VM, creating tubular channels that are lined with tumor cells as opposed to endothelial cells and are encased by secretory glycoproteins,[135] and a patterned matrix VM, where tumor cells and tissue are encased in periodic acid-Schiff (PAS)-positive matrix proteins (e.g., laminins, heparan sulfate proteoglycans, collagens IV VI).[136] Without vessels to provide resources, tumor growth is limited to approximately 2–3 mm^3, obtains nutrients through diffusion, but can become susceptible to the hypoxic environment and undergo necrosis.[137] Achievement of neoangiogenesis enables the tumor to rapidly grow, and the imperfection of the vascular matrix allows easy penetration by the tumors cells to permit dissemination throughout the body.[135]

2.9 Apoptosis and other pathways

Apoptosis is a highly regulated mechanism by which the body eliminates unnecessary or unwanted cells and plays a vital role in development and homeostasis. Apoptosis can be induced via two different pathways, an intrinsic mitochondrial pathway that is signaled by DNA damage, growth factor deprivation, and cytokine deprivation, and an extrinsic death receptor path that is signaled by cytotoxic T cells from the immune system responding to cells that are damaged or infected.[138] Both pathways converge at the execution of apoptosis by caspases. Cysteine aspartyl-specific proteases (caspases) are a class of cysteine proteins that cleave target proteins and are divided into two groups: initiator (caspase-2, -8, -9, 10) and executioner (caspase-3, -6, -7).[138] The intrinsic pathway is regulated by B-cell lymphoma-2 (BCL-2) proteins, which include proapoptotic effector proteins, proapoptotic BH3-only proteins, and antiapoptotic BCL-2 proteins.[139] A hallmark of cancer is the ability to

TABLE 1   Pro- and anti-angiogenic factors and their corresponding receptors.

Proangiogenic factors		Antiangiogenic factors	
Factor	Receptor	Factor	Receptor
VEGF	Tyrosine kinase receptors (VEGFR1, VEGFR2, and VEGFR3)	Thrombospondins (TSP)	CD36, CD47, integrins
PDGF	Tyrosine kinase receptors (PDGFR-α and PDGFR-β)	Tissue inhibitor of matrix metalloproteinases (TIMPs)	Metalloproteinases (MMPs)
FGF	Tyrosine kinase receptors (FGFR1, FGFR2, FGFR3, and FGFR4)	Angiostatins	Angiomotin, integrin alpha-v beta-3 (αvβ3), tissue-type plasminogen activator, annexin II
EGF	Tyrosine kinase receptors: EGFR (ErbB1, HER1), ErbB2 (HER2), ErbB3 (HER3), and ErbB4 (HER4)	Endostatin	Biglycan, lipoproteins
TGF	Serine/threonine kinase receptors (type I and type II)		
MMP	Low-density lipoprotein receptor-related protein (LRP)		
TNF	Tyrosine kinase receptors (TNFR-I and TNFR-II)		
Angiopoietin	Tyrosine kinase receptors (Tie-1 and Tie-2)		

avoid this mechanism, allowing the cancer cell to survive longer, which in turn also allows the cell to amass mutations that increase its tumorigenicity and metastatic potential.[140] Tumor cells can evade apoptosis using several methods: inhibition of caspase function or disabling the triggers for inducing apoptosis, but upregulation of antiapoptotic BCL-2 proteins and loss of BAX and/or BAK have been shown to be the principal methods of invasion.[139] While the vast majority of limited FDA-approved apoptosis-inducing antineoplastic therapies rely on the BCL-2/BAX-dependent mechanisms to kill cancer cells, unfortunately if this pathway is disrupted, therapies may fail.[141]

3 Mechanisms of brain metastasis

The metastatic spread of tumors from a primary site to secondary sites throughout the body is responsible for much of the mortality and morbidity of cancer.[2] Cancer metastasis proves especially fatal when neoplastic cells migrate into the central nervous system, after which prognosis remains poor with a 5-year survival rate at an approximated 10%.[142] The onset of metastatic disease occurs in a series of sequential steps[9]: initial separation from primary tumor, avoiding anoikis, dysregulation of extracellular matrix (ECM) dynamics, increased cell motility, intravasation, immune escape, adhesion to the metastatic site, extravasation, and angiogenesis. Apoptosis is carried out by cysteine aspartyl-specific proteases (caspases), which cleave target proteins.

3.1 Epithelial-mesenchymal transition

The epithelial-mesenchymal transition (EMT) is a process that denotes a cellular change from an epithelial phenotype to a mesenchymal phenotype.[8] Normal epithelial cells exhibit 3 main characteristics—polarity (basal and apical), cell–cell interactions, and basal membrane—and possess a flat and polygonal shape. A mesenchymal cell possesses characteristics that enable its migration and invasion: front-to-back polarity, no cellular adhesion, and are morphologically irregular with a more spindle-like cell body.[8] In normal cellular processes, EMT is involved in gastrulation and organogenesis during early development[143] and in re-epithelialization during would healing.[144] EMT in normal processes is subtle and controlled. The induction of the EMT in metastasis initiation is contrarily aggressive and uncontrolled[145] and lends tumor cells their specialized abilities to complete the metastatic cascade. Though the precise controlling events underlying EMT are still under investigation, particularly in brain metastases, the transition is regulated by several transcription factors.

3.2 SNAIL proteins

The SNAIL family is comprised of Snail1 (SNAIL or SNAI1), Snail2 (Slug), and Snail3 (Smuc) that encode zinc finger transcriptional repressors. All members possess a highly conserved C-terminal domain, which incorporates 4–6 C_2H_2-type zinc fingers and binds to a E-box motif in target gene promoters, as well as an evolutionarily conserved SNAG domain at the N-terminus that is essential for binding of transcriptional co-repressors.[146] In EMT, SNAIL1 and SNAIL2 have been shown to recruit factors like polycomb repressive complex 2 (PRC2), which harmonizes histone hypermethylation and deacetylation. This ultimately causes repression of epithelial gene expression (E-cadherin, occludins, cadherins, cytokeratins) to promote cellular detachment, while in turn increasing expression of mesenchymal factors (N-cadherin and ZEB1, fibronectin) and proteases (MMP2 and MMP9) to enhance invasion.[147] Several signaling pathways within the tumor microenvironment converge to regulate SNAIL1 and SNAIL2, such as the canonical TGF-β,[148] MAPK,[149,150] GSK3β-mediated Wnt,[151] receptor-tyrosine kinases (RTKs),[152] and Notch.[153]

3.3 Zinc finger E-box binding (ZEB) proteins

The ZEB family is comprised of two structurally conserved multidomain proteins, ZEB1 (originally called ZFHX1A, TCF8, or δEF1) and ZEB2 (originally called ZFHX1B or SIP1)[154] that encode zinc finger homeobox proteins. ZEB1 and ZEB2 are overexpressed in brain metastases compared to primary tumors.[155] They contain an N-terminal (NZF) and C-terminal zinc finger cluster (CZF) that allow them to bind E2-box sequences.[156] ZEBs repress target gene transcription through the recruitment of the C-terminal binding protein (CtBP) corepressor complex via their CtBP interaction domain (CID).[157] ZEBs can inhibit expression of cadherins, polarity proteins (i.e., CRB2, HUGL2), and gap junction proteins (i.e., connexions), while simultaneously upregulating the expression of mesenchymal genes (i.e., vimentin, N-cadherin, fibronectin).[158,159] ZEBs possess additional conserved domains to recruit other transcriptional complexes. For example, a SMAD-binding domain (SBD) resides in between the NZF and central homeobox domain in both ZEBs but has different downstream results. In ZEB1, there is synergistic binding with SMAD to activate SMAD-mediated transcription whereas binding with ZEB2 inhibits this activity.[160]

3.4 TWIST proteins

TWIST1 and TWIST2 (Dermo1) belong to a family of β helix–loop–helix (βHLH) proteins that form homo- and heterodimers with helix–loop–helix factors E12 and E47 (TCF3), to regulate DNA binding that leads to downregulated expression of epithelial genes and upregulated expression on mesenchymal genes.[161] They recruit the methyltransferase SET8 to mediate H4k20 mono-methylation and subsequently repress E-cadherin promoters and activate N-cadherin promoters,[162] which indirectly increases the expression of α-smooth muscle actin (α-SMA) and MMP-p to promote invasion.[147,163] The downregulation of E-cadherin upregulates TWIST in a feed-forward loop, which in turn helps to activate SNAIL and maintain EMT.[164] TWIST will bind p53 to induce its degradation in order to abrogate anoikis, an apoptotic death that is incurred during cellular detachment, and overcome senescence.[165] Similar to the SNAIL TFs, several pathways collaborate to activate TWIST, such as transcription factor hypoxia-inducible factor 1α (HIF-1α), Akt, signal transducer and activator of transcription 3 (STAT3), mitogen-activated protein kinase, Ras, and Wnt signaling.[147,166] In an immunohistochemical analysis of brain metastases and nonmetastatic tissue, brain metastases showed expression of TWIST1, again pointing toward EMT playing a role in brain metastases.[167,168]

3.5 FOX proteins

Forkhead box (FOX) proteins are a subgroup of helix-turn-helix transcription factors that are defined by the forkhead box (aka winged helix), a sequence of 80–100 amino acids that form a motif to bind DNA.[169] Several FOX proteins have been identified with important functions in multiple biological processes, such as cell cycle control, cell differentiation, proliferation, and development. Within EMT, FOXC1, FOXC2, and FOXQ1 downregulate E-cadherin and upregulate fibronectin, vimentin, and N-cadherin. Several FOX proteins repress the expression of polarity complexes and cell–cell junction proteins; other FOX proteins conversely relieve repression of other factors to ultimately promote EMT.[170–172] FOXO has been implicated in the upregulation of gluconeogenesis in breast cancer cells, suggesting increased energy production can induce brain metastases.[173]

3.6 KLF proteins

Kruppel-like factors (KLF) zinc finger proteins have the ability to bind CACCC or GT box DNA elements to regulate cell metabolism, proliferation, differentiation, development, and programmed cell death.[174] Several KLFs have been implicated as potent EMT role players. For instance, KLF8 represses E-cadherin expression by binding the GT box within its promoter.[175,176] KLF6 behaves as a tumor suppressor by mediating E-cadherin expression through transactivation of its promoter,

which leads to subcellular localization of β-catenin and c-Myc.[177] KLF4 has been shown to antagonize EMT activity by provoking epithelial differentiation through the targeting of numerous adhesion and cytoskeletal genes and negative regulation of β-catenin-mediated E-cadherin gene expression.[178–180] KLF4 has been shown to impair EMT through repression of SNAIL1 and SNAIL2,[181] whereas in active EMT, SNAIL will repress KLF4.[182] KLF4-specific binding sites have been detected in the promoter sequences of vimentin, VEGF-A, endothelin-1, and JNK-1 (MAPK8), next to E-cadherin, N-cadherin, and CTNNB1.[179] KLF4 has been found to be upregulated in a preclinical model of breast-derived brain metastases.[183]

3.7 Other EMT factors

Several other transcription factors play vital and shared roles in EMT and the metastatic phenotype. Paired-related homeobox (PRRX) factors belong to the family of homeobox proteins that function with the Wnt pathway and localize to the nucleus to act as transcription co-activators. PRRXX1 and PRRX2 are relatively newly discovered EMT inducers, where their overexpression regulates expression of E-cadherin, N-cadherin, and vimentin.[184] Grainyhead-like (GRHL) transcription factors regulate EMT in wound healing and embryonic development.[185] Within metastasis, GRHL2 shares a reciprocal feedback loop with ZEB1 and TGF-β to regulate expression of genes responsible for cell junctions and differentiation but has a tumor-type specific role of either EMT suppression or activation.[186] Brachyury is a transcription factor within the T-box gene family vital to the formation and differentiation of posterior mesoderm and has recently been implicated as a novel tumor antigen and EMT driver through induction of SNAIL1, SNAIL2 and downstream signaling.[187] Yes-associated protein (YAP) and PDZ-binding domain (TAZ) are key transcriptional co-activators within the Hippo signaling pathway, which likely plays a role in brain metastases, as will be discussed below.[188] Increased expression of TEA domain transcription factor 2 (TEAD) results in increased nuclear localization of YAP/TAZ and subsequent association with TEAD, which regulates expression of genes involved in cell–cell adhesion and actin cytoskeletal remodeling to promote EMT.[189] Nuclear factor kappaB (NFκB) family consists of five subunits: p50 (NF-κB1), p52 (NF-κB2), p65 (RelA), c-Rel (Rel), and RelB, that associate to form functional homo- and heterodimers. NF-κB is inactive when bound to IκB; upon IκB phosphorylation and subsequent ubiquitination, NF-κB translocates to the nucleus to bind promoter regions of SNAIL2, TWIST1, and SIP1 to ultimately induce and maintain EMT.[190,191]

4 Biological mechanisms of the metastatic cascade

The metastatic cascade is comprised of several intricate processes divided into specific stages (see Fig. 1). It begins with metastatic tumor cells detaching from their primary mass.[9] The primary tumor is anchored together by numerous transmembrane adhesion molecules such as cadherins, integrins, and immunoglobulins.[192] In order to leave the primary tumor site, a tumor cell must interrupt intercellular adhesion interactions. The downregulation or dysfunction of E-cadherin, a glycoprotein involved in homophilic cellular adhesion, has been associated with metastasis development in a number of cancers.[193] Similarly, the loss of immunoglobulin function such as ICAM-1 and ICAM-2 is correlated with cancer progression and the metastatic phenotype.[194] Interestingly, the role of a particular cell adhesion molecule can change drastically across different cancer subtypes. High E-cadherin levels are required for metastasis in certain models of ductal carcinoma,[195] while high ICAM-1 levels are correlated with cancer progression in certain lung carcinomas.[196] These observations suggest that cell adhesion molecules are also involved in additional signaling pathways, making it difficult to identify disruption of any one adhesion molecule as a hallmark feature of all metastases.

Upon leaving the primary tumor, the migrating tumor cells must avoid anoikis, the process by which cells undergo apoptosis upon losing contact with the ECM.[197] A cell typically mediates its connection with the ECM via transmembrane integrin receptors, which activate focal adhesion kinase (FAK) and prevent apoptosis by triggering the PI3k/Akt survival pathway.[198] One strategy that detached tumor cells utilize to avoid anoikis is through integrin switching, a mechanism in which the cells alter their integrin repertoire to upregulate integrins that provide high resistance to apoptosis.[197] These cells also undergo autoactivation of FAK and the PI3k/Akt survival pathway, allowing detached cancer cells to survive without being anchored to the ECM.

After surviving the initial separation from the primary mass, migrating tumor cells continue their journey through intravasation of the surrounding tissue toward the vasculature.[9] In order to make space for movement, tumor cells secrete numerous proteases such as matrix metalloproteinases (MMP) that degrade collagen, elastin, laminin, proteoglycans, and fibronectin in the ECM.[199] The tumor cells also hyperactivate the urokinase-type plasminogen activator (u-PA) system, a series of extracellular proteolytic enzymes that further facilitate ECM degradation.[200] Furthermore, cancer cells bypass many regulatory elements of the proteolytic system by downregulating levels of proteolytic inhibitors in the ECM.[199]

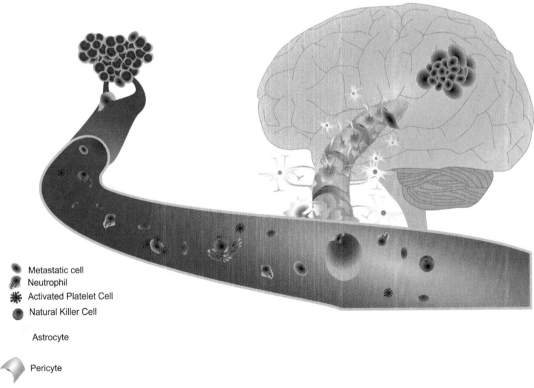

● Metastatic cell
🦠 Neutrophil
✳ Activated Platelet Cell
● Natural Killer Cell
° Astrocyte
◣ Pericyte

FIG. 1 Simplified schematic representing the metastatic cascade, depicting the stages of tumor intravasation, survival and circulation, arrest, extravasation, and brain metastasis initiation.

The process of ECM degradation during tumor metastasis releases many distinctive peptide fragments into the bloodstream, and these peptides may serve as potential molecular markers for assessing cancer progression in future diagnostic tools.[199]

Migrating tumor cells further augment their ability to invade surrounding tissue by undergoing phenotypic change that ultimately increase their motility. Tumor cells upregulate the release of cytokines such as autocrine motility factors (AMF) that work to increase both cancer cell proliferation and migration through autocrine and paracrine mechanisms.[201] The migrating cancers also release additional growth factors such as PDGF, EGF, FGF, and IGF that further drive cancer progression.[17] Notably, in an in vivo brain metastasis model of brain metastasis, inhibition of IGF-IR signaling attenuates the development of brain metastases, likely through decreased migration and invasion.[59]

Another key feature during this stage of the metastatic cascade is the upregulation of TWIST, a transcription factor that mediates the epithelial-mesenchymal transition that promotes cancerous motility and therapeutic resistance.[202] TWIST upregulation is implicated in the metastasis of cancers such as basal-like breast,[203] head and neck,[204] and gastric cancer.[205] The motile cancer cells eventually reach blood vessels and lymph nodes, secrete angiogenic factors to loosen basal membrane, and intravasate into systemic circulation.[9]

Once tumor cells enter the systemic circulation, they must avoid immune detection or risk elimination by cells from both the innate and adaptive immune systems.[9] One strategy that circulating cancer cells employ to remain invisible is interfering with the antigen presentation machinery used by immune surveillance cells. Cancer cells accomplish this feat through various means, from physically shielding their cell surface neoantigens with blood platelets to downregulating proteins involved in the antigen presentation process.[206] The cancer cells also secrete immunosuppressive cytokines such as TGF-β that actively prevent immune mobilization.[207] Another strategy involves hijacking existing immune mechanisms that are designed to limit inflammation and prevent excessive immune responses. The protein PDL-1 is expressed in a myriad of cancers and binds to the PD1 immunosuppressive receptor found in T cells, B cells, natural killer cells, dendritic cells, and monocytes.[208] PD1/PDL-1 interaction effectively inactivates both innate and adaptive immune cells, allowing metastasizing cancer cells to circulate uninterrupted. The PD1/PDL-1 pathway is associated with very aggressive cancers such as prostate,[209] gastric, lung, renal, and pancreatic cancer.[210] Targeting of this pathway has revolutionized the management of multiple cancers, including in the setting of brain metastases.[211,212]

Upon reaching capillaries near their final metastatic site, cancer cells roll along the surface of endothelial cells and adhere using various integrins, selectins, cadherins, and immunoglobulins such as SPARC, APP, and ANGPLT4 in a process that resembles wound healing.[206] Many of these same proteins also increase permeability between neighboring endothelial cells by disrupting inter-endothelial adhesion molecules, allowing migrating cancer cells to extravasate into the basement membrane of the organ. Some cancer cells may become dormant in the basement membrane upon extravasation and only continue the metastatic cascade years later—a mechanism that is hypothesized to account for metastasis recurrence.[213] The remaining cancer cells release proteases to degrade the ECM and travel into the parenchymal tissue.[9] Finally, the metastasized cancer cells anchor onto their new site, utilize nutrients from the surrounding tissue, and begin growing a secondary tumor site. The cancer cells secrete angiogenic growth factors such VEGF, FGF, IGF-1, PDGF, and EGF to increase penetrance of blood vessels and nutrient supplies at the secondary site.[9] The metastasized cancer cells are thus able to establish a new colony far from the original tumor.

The metastatic cascade explains how metastasis occurs but fails to explain why particular cancers regularly metastasize to the same few organs. For example, prostate cancer most commonly metastasizes to the bone and rarely to the brain, whereas approximately 50% of melanoma patients have metastasis to the nervous system.[214] Two accepted theories in the field attempt to address this question of where cancers metastasize: the mechanical hypothesis and the seed-soil hypothesis. The mechanical hypothesis indicates that metastasis is probably nonspecific and is likely influenced by circulatory and lymphatic drainage routes throughout the body.[215] Migrating tumors nonspecifically colonize the first organs they encounter during circulation that contain an adequate environment for growth. Anatomical evidence also supports the theory. Liver metastasis is very common in gastric cancers,[216] highlighting the mechanical hypothesis because most blood from the abdominal region passes through the hepatic portal vein. Similarly, some colon cancers can metastasize to the lung because they bypass the liver and encounter the lung when passing through the pulmonary artery during circulation.[217]

The seed-soil hypothesis postulates that metastasizing tumors cells are akin to "seeds" that require a nurturing environment at target sites in order to colonize a secondary tumor.[218] Accordingly, the spread of metastatic disease is not random but rather follows specific patterns that are highly contingent on the molecular profiles of both the cancer cell and the host site. For instance, cancers such as lung, breast, renal, and melanoma may frequently metastasize to the brain due to their high neurotrophin receptor expression levels, making these cancers more suited to handle the brain's unique chemical environment.[219] Prevailing opinion in the field gives credence to both theories of metastasis, suggesting that the spread of cancer is dependent on both mechanical and molecular influences.

5 Blood–brain barrier

In the development of brain metastases, the metastatic cells have an additional barrier to overcome in the blood–brain barrier (BBB). The BBB is a network of specialized cells that coordinate to form a tightly regulated neurovascular unit to regulate the homeostasis of the CNS by controlling the molecular and cellular transport into and out of the CNS.[220] The structure of the BBB consists of the endothelial cells that comprise the capillary vasculature of the neural parenchyma, surrounded by a basal lamina shared between pericytes and the end-feet of astrocytes, and interconnected intermittently with neural cell ends and microglia.[221] This network defines the physical properties of the BBB, possessing transport systems that allow the influx of circulating molecules essential to CNS function while allowing the efflux of toxic cellular by-products back into the circulation. Immune cells have also been seen to pass across the BBB in situations of neuroinflammation or reparation of damaged nervous tissue.[222,223] The BBB is supposed to restrict the flow of solutes as well as cells; however, some cancer types give rise to metastases preferentially to the CNS, implicating a role for the BBB in supporting metastatic cell transmigration, penetration, and survival within the neural parenchyma.[224] While the process of trans-endothelial migration has been studied with immune cells, this process is similar but remains less understood with metastatic cells.[225] The extravasation of metastatic cells into the brain has been shown to take longer than into other organs and is even primary cancer-type-dependent. It appears that, while the diapedesis process is swift, the survival of the metastatic cell in the brain parenchyma takes much longer than extravasation into other organs and is even suggested to be the rate-limiting step in brain metastasis formation.[226,227] During brain metastasis development, the BBB is recruited to aid the survival of the metastatic cells; the vascular basement membrane supports metastatic growth prior to the metastasis developing its own vasculature, termed vessel cooption, and offers protection from antitumoral immune surveillance.[228,229] The tumor cooption compromises the integrity of the BBB and is dubbed as the blood-tumor barrier (BTB). This BTB is characterized by increased "leakiness," heterogeneous permeability, aberrant pericyte distribution, and reduced astrocytic end-feet and neural connections, but retains the critical BBB functional aspects including retaining the expression of efflux transporters.[230] The BBB is also a critical hindrance to the permeability of cancer treatments. Several physiochemical traits are required for a drug to pass the BBB and remain in the neural parenchyma: small molecules, liposolubility, charge, interactions with plasma proteins, and interactions with efflux pumps and transporters.[231] Even with the disrupted BTB, targeted therapies are faced with limited penetration and rapid efflux from the brain into the blood resulting in very low pharmacologically irrelevant levels of the therapies within the brain.[232] Fortunately, therapies that can circumvent this barrier are under examination, with several positive outcomes. Investigations into the mechanical disruption of the BBB are ongoing, which involve exerting pressure variations to manipulate the conformation of the endothelial cells and increase BBB permeability, achieved by either osmotic or ultrasound disruption.[233,234] Nanoparticles are natural or artificial particles ranging from 10 to 1000 nm in size, and it is a term applied to several drug delivery systems including vehicles, including dendrimers, micelles, liposomes, nanoscale ceramics, metallic and polymer nanoparticles. Nanoparticles are efficacious drug delivery systems in that they are taken up by the BBB and protect the drug during delivery.[235–237]

6 Genetics of metastases

The process of metastasis is quite inefficient: Only ~0.01% of cancer cells can survive and complete the entire metastatic cascade; many cells that leave the primary tumor will die during intermediate steps.[238] The genetic profiles of cancer cells at the primary tumor site differ substantially from those at the site of metastasis[239] including in brain metastases,[100] suggesting that cancer cells continue evolving and undergo positive selection for survival traits during the migratory process.[240] Since the genetic and regulatory factors that promote primary tumor growth have been shown to differ substantially from metastatic factors, identifying any genetic or epigenetic changes may offer clues on why metastasis occurs. A common method utilized in research to identify brain metastasis genes is in vitro genetic screens, allowing researchers to interrogate pools of genes and determine their roles in organ-specific metastasis, and comparative genome studies, which allow researchers to compare the genomic profiles of primary tumors to their metastatic counterpart and identify tumor-specific genetic profiles. Another method utilizes in vivo "brain-trained" cell lines, where cells are inoculated into a specific route, harvested from the brain, and re-inoculated. This process is repeated several times, where each time the frequency of brain metastases increases while the latency to brain metastasis development decreases and can also aid in the identification of genes that drive brain metastasis formation.

6.1 Suppressors of metastasis

Metastasis suppressor genes (MSGs) work to inhibit the processes that occur during each step of the metastatic cascade from transcriptional changes and cellular migration to actin remodeling and growth signal secretion. Interestingly, unlike oncogenes, MSGs are not mutated but rather downregulated,[240] highlighting this gene family as a target of high interest for potential therapeutics. MSGs are a relatively new discovery in oncology and have yet to be extensively characterized, with less than 30 MSGs identified across various cancer histologies at the present time.[240]

One of the first MSGs discovered is Nm23, a protein kinase involved in inhibiting signals that activate the ERK–MAPK signaling pathway.[241] ERK–MAPK is involved in allowing initial tumor breakage and increasing cell motility and tissue invasion in certain breast cancer models. Cancer cells often downregulate Nm23 activity by releasing TGF-Beta, a transforming growth factor that binds the protein and switches it to an inactive state.[242] Nm23 is also implicated in suppressing the lysophosphatidic acid receptor EDG2.[242] Induction of Nm23 activity has been shown to suppress metastasis by 58% in experimental models of melanomas, and low levels of NM23 are associated with aggressive melanomas[243] and ductal breast carcinomas[244] as well as brain metastasis.[245]

Another important MSG is RhoGDI2, a key modulator of Rho GTPases.[246] An important function of Rho GTPases is remodeling cell cytoskeleton components, and recent evidence points to Rho GTPases as a key component of signaling cascades that allow for epithelial-mesenchymal transition. RhoGDI2 therefore suppresses cancer metastasis by inhibiting signaling pathways that lead to EMT, as evidenced in several lung cancer models.[247] RhoGDI2 also prevented lung cancers from invading surrounding tissue in both in vitro and in vivo models.[248] In addition, RhoGDI2 is associated with metastasis suppression in bladder cancers.[249]

KAI1 is another MSG that has been implicated in prostate, breast, bladder, and lung cancers.[250] As a member of the tetraspanin protein family, KAI1 functions by stabilizing complexes of large transmembrane receptor such as RTKs, integrins, and cell adhesion molecules.[251] KAI1 downregulation has been identified in metastases of endometrial[252] and prostate cancers, with evidence from prostate suggesting that KAI1 works to prevent activation of the metastasis-promoting transcription factor ATF3.[253] Similarly, KAI1 expression is decreased in brain metastases from breast cancer when compared to primary tumors.[254] Furthermore, KAI1 is also involved in regulating EGFR-mediated complexes of breast cancer cells, with increased KAI1 attenuating EGFR signaling and internalizing the receptor.[255] Both pathways elucidate possible mechanisms through which KAI1 works to suppress the spread of cancer cells.

KISS1 is a MSG that is found in models of both melanoma and breast cancer.[256] Expression of KISS1 yields kisspeptin, which interacts with GPR54 and regulates phospholipase C-β activation.[257] Kisspeptin is also secreted in high quantities from the cell to modulate GPR54 activity in surrounding cells. The primary role of KISS1 is dormancy activation—the high KISS1 activity causes the secretion of dormancy-inducing factors by stromal cells as well as the endocrine system.[258] Therefore, even if a cancer cell successfully enters circulation and extravasates into new tissue, KISS1 prevents the cell from colonizing the new site.[258] This intriguing mechanism highlights the gene as a desirable target for treating advanced metastasis. Moreover, when comparing brain metastases and primary breast cancer tumors, KISS1 expression was significantly higher in primary breast cancer.[254,259]

BRMS-1 is a MSG that has been studied in breast, melanoma, ovarian, and bladder cancers.[260] BRMS-1 functions by blocking signaling cascades related to extracellular growth signals.[261] Additionally, it prevents activation and translocation of NF-κB, a fundamental protein transcription factor related to growth. Knockdown of BRMS-1 has been shown to increase aggressive cancer phenotypes such as increased high growth rates, rapid migration, and invasion of surrounding tissue.[262] Furthermore, in mouse models, intracardiac inoculation of cells expressing BRMS-1 showed fewer metastases to multiple organs including brain compared to parental cells.[261] Expression of BRMS1 is positively correlated with improved patient outcomes.[263]

The Gas1 gene has been identified as a MSG through models of melanoma, thyroid, colorectal cancer.[264] Gas1 increases the activity of caspase 3, a protease involved in apoptosis, and also inhibits the GDNF-mediated survival signaling pathway.[265] In thyroid carcinoma, MiR-34a specifically targets Gas1 to promote cell proliferation and prevent apoptosis.[266] Colorectal cancer models also show that Gas1 negatively regulates the AMPK/mTOR/p70S6K signaling cascade, a well-established cancer malignancy pathway that allows for metastasis.[267]

DRG1 is a MSG that has been studied in a number of cancers such as pancreatic and breast cancer.[268,269] DRG1 regulates cellular microtubule dynamics[270] and prevents angiogenesis in cancers through VEGF protein inhibition.[271] Without angiogenesis, migrating cancer cells are unable to establish large colonies in distant sites.[269] Low expression of DRG1 was found in several patients who died of breast metastasis.[272] Furthermore, knockdown of DRG1 significantly increases the spread and motility of breast cancer cells in functional assays.[268] Many of DRG1's binding partners are still unknown, and DRG1 may actually function as an oncogene in lung adenocarcinoma.[273]

The LSD1 gene has been characterized as a MSG in breast cancer models and produces a histone-modifying enzyme that regulates gene expression. The LSD1 protein positively regulates GATA3, a key survival transcription factor, to target genes highly related to breast cancer. The LSD1/GATA3 interaction upregulates genes involving E-cadherin stabilization, cell adhesion, and the integrin pathway.[274] In functional experiments, LSD1 was shown to also suppress invasive luminal breast cancer cells by preventing EMT. Furthermore, LSD1 appears to actively repress expression of TRIM37, a breast epithelial oncogene. High TRIM37 is associated with reduced survival of patients with luminal breast cancer.[274]

The gene CD44 encodes for a nonkinase transmembrane glycoprotein involved in cell adhesion and has been implicated as a MSG in prostate cancer.[272] Downregulation of CD44 is implicated in increased metastatic potential of prostate cancers, and transfection of CD44 into highly metastatic rat prostatic cells significantly reduced the cells' metastatic ability.[272] CD44 has also been identified as a MSG involved in brain metastasis.[245]

The aforementioned genes encompass only a small subset of MSGs, many of which have only been discovered in recent years.[240] Since many of their molecular mechanisms remain to be uncovered, especially in brain metastases, future work with MSGs might provide additional insight on how cancer cells avoid checkpoints to progress to each step of the metastatic cascade.

6.2 Clinically actionable mutations in brain metastases

With the advent of next-generation sequencing technologies, our understanding of the genomic evolution specifically in brain metastases has substantially increased. In the largest comprehensive genomic sequencing study of brain metastases across histologies to date, approximately 100 brain metastases paired with primary tumors underwent whole exome sequencing.[100] The study demonstrated that brain metastases demonstrate branched evolution, whereby brain metastases and primary tumors share a common genetic ancestor, yet there is divergent evolution such that the brain metastases harbor novel mutations not detected in the primary tumor; many of these mutations are "clinically actionable," in other words have therapeutic targets. These data have significant clinical implications given that the primary tumor is often used to make clinical decisions for patients with brain metastases. A national genomically guided brain metastasis trial is now ongoing to investigate the efficacy of targeted therapies in patients with alterations in their brain metastases (Alliance A071701).

6.3 CDK pathway

More than 50% of brain metastases harbor mutations in the cyclin-dependent kinase pathway,[100] including loss of CDKN2A (cyclin-dependent kinase inhibitor 2A) and CDK4/6 (cyclin-dependent kinase 4/6). Inhibitors of CDK4/6 are FDA-approved in breast cancer, including palbociclib, ribociclib, and abemaciclib. There are ongoing trials to explore CDK inhibition in brain metastases. A recent genomically driven trial evaluated palbociclib in patients with brain metastases that harbored alterations in the CDK pathway in their tumors. At the prespecified interim analysis, the trial has met its primary endpoint with more than 50% of patients having intracranial benefit.[275]

6.4 PI3K/AKT pathway

The PI3K/AKT signaling pathway is also commonly mutated in brain metastases with approximately 40% of brain metastases from breast and lung cancers harboring clinically actionable mutations in this pathway.[100] In a study comparing hotspot mutations, copy number variations, mRNA and protein expression in matched brain metastases and extracranial metastases from melanoma, brain metastases had increased expression in the PI3K/AKT pathway compared to extracranial metastases.[276] Preclinical work in patient-derived xenograft models of breast cancer brain metastases has demonstrated that inhibition of the PI3K/AKT signaling pathway inhibits the growth of *PI3KCA* mutant, but not wild-type, brain metastases.[125,126] The efficacy of PI3K inhibition with a CNS penetrant inhibitor is now being examined in clinical trials of patients with *PIK3CA*-mutant brain metastases (NCT04192981; NCT03994796).

6.5 Oxidative phosphorylation pathway

The oxidative phosphorylation metabolic pathway (OXPHOS) is heightened in brain metastases and may play an important role in cancer progression.[277] The genes and metabolites of OXPHOS are upregulated in brain metastases compared to patient-matched extracranial metastases. Furthermore, inhibition of OXPHOS increased survival and prevented formation of brain metastases in xenograft mouse models of melanoma.[277] Since increased OXPHOS can mediate cancer resistance to BRAFi- and MEKi-targeted therapies,[278] the enhancement of OXPHOS in brain metastasis may explain the difficulty of treating cancers that have progressed to the brain.

6.6 YAP1

YAP1 is a transcriptional mediator of the Hippo signaling pathway, which is involved in a number of cellular processes including EMT.[279] In a genomic study of 73 brain metastases from lung adenocarcinoma compared to 503 primary lung cancers, YAP1 was amplified at a significantly higher rate in brain metastases.[188] Overexpression of YAP1 in a patient-derived cell line that underwent intracardiac inoculation into a mouse model of metastasis led to an increased incidence of brain metastases.[188] Consistent with these data, inhibition of YAP with shRNA decreased the burden of brain metastases.[280] YAP inhibitors are in clinical development.

6.7 FGFR

As mentioned earlier in the chapter, amplifications in FGFR are common in squamous cell carcinoma. Data are emerging that alterations in FGFR are associated with brain metastases. A study of 30 breast cancer brain metastases compared to 165 extracranial metastases showed that aberrations in FGFR increase the risk of brain metastases and are associated with a poor prognosis.[281] Similarly, another study in 175 NSCLC brain metastases showed that FGFR1 amplifications are enriched in brain metastases from lung adenocarcinomas, up to fivefold compared to reports in primary tumors.[45] The exact mechanisms by which they can exert a pro-brain metastatic phenotype still need to be elucidated.

6.8 MMP13

The matrix metalloproteinase (MMP) family of proteins likely plays a role in brain metastases. MMPs assist the transmigration and extravasation of metastatic cells across the brain capillaries and BBB by proteolyzing tight junction and adherence junction proteins.[226] A genomic study of lung adenocarcinoma brain metastases showed that amplifications in MMP13 were enriched compared to primary lung cancers. Overexpression of MMP13 led to an increased incidence of brain metastasis in a preclinical mouse model of metastasis.[188] In another preclinical study, silencing of MMP-1 led to reduced formation of brain metastases by breast cancer cells.[282] An in vivo study showed that increased MMP-1 was associated with the promotion of breast-derived brain metastasis initiation and growth.[282]

6.9 SPOCK1

SPOCK1 is an extracellular plasma proteoglycan that belongs to the secreted protein acidic and rich in cysteine (SPARC) family. It possesses a multidomain core and glycosaminoglycan side chain, and several studies implicate SPOCK1 as an inducer of EMT and driver of metastasis in cancer.[283–286] Through in vitro and in vivo pooled shRNA screens performed with primary patient lung-derived brain metastasis cells, Singh et al. identified SPOCK1 to be a novel putative brain metastasis

regulator, showing abrogation of tumor growth and metastasis in a novel intrathoracic model.[168] Similarly, the synaptic cell adhesion molecule CADM2 was found to induce EMT and potentially drive brain metastasis development from NSCLC in in vitro studies.[287]

6.10 Caveolin-1

Cav-1 is a scaffolding protein component of caveolae plasma membrane, functioning to link integrin subunits in the Ras–ERK pathway.[288] In cancer, Cav-1 was found to be overexpressed in several cancers, including liver, colon, breast, kidney, and lung,[289] but behaves as a tumor suppressor of promoter dependent on the tumor type.[290] Cav-1 is a critical downstream target of Stat3, where in breast cancer cells Cav-1 upregulation mediates suppression of tumor growth and brain metastasis.[291] Conversely, Cav-1 expression is predictive of poor prognosis and radioresistance in lung-derived brain metastasis patient samples[292] and was found to enhance brain metastases potentially through SNAIL.[293]

6.11 Other genes

In vitro and in vivo studies have implicated several other genes to have roles as brain metastasis drivers, leading to the identification of several genetic signatures. In a genome-wide comparative study, Massague et al. implicated several genes to be up- and downregulated in breast-derived brain metastatic populations. However, they identified α2,6-sialyltransferase ST6GALNAC5 as an important factor in the transmigration of breast cancer cells across the BBB to form brain metastases.[294] Another study conducted with frozen patient NSCLC samples identified N-cadherin to play a vital role in the interaction of metastatic cells with brain endothelial cells of the BBB. In fact, N-cadherin is a true hallmark of EMT, where its expression increases in cells undergoing metastasis in conjunction with the decreased expression of E-cadherin.[295] Klein et al. identified 51 genes to have significantly higher expression in brain metastases as compared to bone metastases from breast cancer cells, where several of the genes are associated with translation, metabolism, signal transduction and transport, and cell adhesion.[296] In another gene expression study, Palmieri et al. identified a signature of six downregulated genes and of two upregulated genes, hexokinase and laminin-γ3, in breast-derived brain metastases.[297] The one drawback to such studies is that there is little overlap between the different gene signatures and gene sets identified thus far; more work in this area is needed. The recent study of tumor exosomes through proteomic analysis has led to the discovery of cell migration-inducing and hyaluronan-binding protein (CEMIP) to be elevated in exosomes from brain metastases but not the corresponding primary tumor exosomes (bone and lung). Where depletion of CEMIP impeded brain metastasis development, uptake of CEMIP[+] exosomes by brain endothelial and microglial cells promoted induction of endothelial cell branching and inflammation in the perivascular niche to enhance brain vascular remodeling and metastasis. Taken together, this study shows one intriguing driver of brain metastases, though much more work is needed to determine if CEMIP will be a useful therapeutic target.[298]

7 Conclusion

Understanding the molecular dynamics of intracranial metastasis is crucial for developing new biomarkers for diagnosing disease progression and developing novel therapeutics. While there is great promise for the treatment of brain metastases, future discoveries will aim to better understand the differences of metastatic cells from the remainder of the tumor bulk, better elucidate the intricacies of the metastatic process including crossing the BBB, and organ selection and specificity for a neural environment over other organs. It is anticipated that the ultimate goal with such knowledge will be to produce additional targeted therapies for treating patients with metastatic disease.

Disclosure

Unrelated to this work, P.K.B. has consulted for Angiochem, Genentech-Roche, Lilly, Tesaro, ElevateBio, Pfizer (Array), Dantari, SK Life Sciences, Advise Connect Inspire (ICI), Voyager Therapeutics, and Sintetica, and has received grant/research support to MGH from Merck, BMS, Mirati and Lilli and honoraria from Merck, Genentech-Roche, Pfizer, and Lilly. P.K.B. receives funding from Damon Runyon Cancer Research Foundation, Ben and Catherine Ivy Foundation, Demetra Fund from the Hellenic Women's Club, Breast Cancer Research Foundation, MGH Research Scholar Program and the NCI (1R01CA227156-01, 5R21CA220253-02, and 1R01CA244975-01).

References

1. Sever R, Brugge JS. Signal transduction in cancer. *Cold Spring Harb Perspect Med*. 2015;5(4):2–17.
2. Martin GS. Cell signaling and cancer. *Cancer Cell*. 2003;4(3):167–174.
3. Matsui WH. Cancer stem cell signaling pathways. *Medicine (Baltimore)*. 2016;95(1 Suppl 1):S8–S19.
4. da Silva HB, Amaral EP, Nolasco EL, et al. Dissecting major signaling pathways throughout the development of prostate cancer. *Prostate Cancer*. 2013;2013, 920612.
5. Cooper GM. *The Cell: A Molecular Approach*. 2nd ed. Sunderland, MA: Sinauer Associates; 2000.
6. Wang LH, Wu CF, Rajasekaran N, Shin YK. Loss of tumor suppressor gene function in human cancer: an overview. *Cell Physiol Biochem*. 2018;51(6):2647–2693.
7. Hanahan D, Weinberg RA. Hallmarks of cancer: the next generation. *Cell*. 2011;144(5):646–674.
8. Singh M, Yelle N, Venugopal C, Singh SK. EMT: mechanisms and therapeutic implications. *Pharmacol Ther*. 2018;182:80–94.
9. Valastyan S, Weinberg RA. Tumor metastasis: molecular insights and evolving paradigms. *Cell*. 2011;147(2):275–292.
10. Heerboth S, Housman G, Leary M, et al. EMT and tumor metastasis. *Clin Transl Med*. 2015;4:6.
11. Cohen S, Levi-Montalcini R, Hamburger V. A nerve growth-stimulating factor isolated from sarcom as 37 and 180. *Proc Natl Acad Sci U S A*. 1954;40(10):1014–1018.
12. Roberts AB, Anzano MA, Lamb LC, et al. Isolation from murine sarcoma cells of novel transforming growth factors potentiated by EGF. *Nature*. 1982;295(5848):417–419.
13. Waterfield MD, Scrace GT, Whittle N, et al. Platelet-derived growth factor is structurally related to the putative transforming protein p28sis of simian sarcoma virus. *Nature*. 1983;304(5921):35–39.
14. Ullrich A, Coussens L, Hayflick JS, et al. Human epidermal growth factor receptor cDNA sequence and aberrant expression of the amplified gene in A431 epidermoid carcinoma cells. *Nature*. 1984;309(5967):418–425.
15. Yarden Y, Ullrich A. Growth factor receptor tyrosine kinases. *Annu Rev Biochem*. 1988;57:443–478.
16. Manning G, Whyte DB, Martinez R, Hunter T, Sudarsanam S. The protein kinase complement of the human genome. *Science*. 2002;298(5600):1912–1934.
17. Witsch E, Sela M, Yarden Y. Roles for growth factors in cancer progression. *Physiology (Bethesda)*. 2010;25(2):85–101.
18. Roskoski Jr R. Small molecule inhibitors targeting the EGFR/ErbB family of protein-tyrosine kinases in human cancers. *Pharmacol Res*. 2019;139:395–411.
19. Sigismund S, Avanzato D, Lanzetti L. Emerging functions of the EGFR in cancer. *Mol Oncol*. 2018;12(1):3–20.
20. Lemmon MA, Schlessinger J. Cell signaling by receptor tyrosine kinases. *Cell*. 2010;141(7):1117–1134.
21. Singh SB, Lingham RB. Current progress on farnesyl protein transferase inhibitors. *Curr Opin Drug Discov Devel*. 2002;5(2):225–244.
22. Tiash S, Chowdhury EH. Growth factor receptors: promising drug targets in cancer. *J Cancer Metastasis Treat*. 2015;1:190–200.
23. Pottier C, Fresnais M, Gilon M, Jerusalem G, Longuespee R, Sounni NE. Tyrosine kinase inhibitors in cancer: breakthrough and challenges of targeted therapy. *Cancers (Basel)*. 2020;12(3):3.
24. Chen J, Tang Y, Liu Y, Dou Y. Nucleic acid-based therapeutics for pulmonary diseases. *AAPS PharmSciTech*. 2018;19(8):3670–3680.
25. Santos E, Nebreda AR. Structural and functional properties of ras proteins. *FASEB J*. 1989;3(10):2151–2163.
26. Simanshu DK, Nissley DV, McCormick F. RAS proteins and their regulators in human disease. *Cell*. 2017;170(1):17–33.
27. Thies KA, Hammer AM, Hildreth 3rd BE, et al. Stromal platelet-derived growth factor receptor-beta signaling promotes breast cancer metastasis in the brain. *Cancer Res*. 2021;81(3):606–618.
28. Hermanson M, Funa K, Hartman M, et al. Platelet-derived growth factor and its receptors in human glioma tissue: expression of messenger RNA and protein suggests the presence of autocrine and paracrine loops. *Cancer Res*. 1992;52(11):3213–3219.
29. Mo SP, Coulson JM, Prior IA. RAS variant signalling. *Biochem Soc Trans*. 2018;46(5):1325–1332.
30. Sitaras NM, Sariban E, Bravo M, Pantazis P, Antoniades HN. Constitutive production of platelet-derived growth factor-like proteins by human prostate carcinoma cell lines. *Cancer Res*. 1988;48(7):1930–1935.

31. McCarty MF, Somcio RJ, Stoeltzing O, et al. Overexpression of PDGF-BB decreases colorectal and pancreatic cancer growth by increasing tumor pericyte content. *J Clin Invest.* 2007;117(8):2114–2122.

32. Cherfils J, Zeghouf M. Regulation of small GTPases by GEFs, GAPs, and GDIs. *Physiol Rev.* 2013;93(1):269–309.

33. Apolloni A, Prior IA, Lindsay M, Parton RG, Hancock JF. H-ras but not K-ras traffics to the plasma membrane through the exocytic pathway. *Mol Cell Biol.* 2000;20(7):2475–2487.

34. Prior IA, Hood FE, Hartley JL. The frequency of ras mutations in cancer. *Cancer Res.* 2020;80(14):2969–2974.

35. Ostrem JM, Shokat KM. Direct small-molecule inhibitors of KRAS: from structural insights to mechanism-based design. *Nat Rev Drug Discov.* 2016;15(11):771–785.

36. Chen H, Zhou L, Wu X, et al. The PI3K/AKT pathway in the pathogenesis of prostate cancer. *Front Biosci (Landmark Ed).* 2016;21:1084–1091.

37. Tan AC. Targeting the PI3K/Akt/mTOR pathway in non-small cell lung cancer (NSCLC). *Thorac Cancer.* 2020;11(3):511–518.

38. Cao Y. Multifarious functions of PDGFs and PDGFRs in tumor growth and metastasis. *Trends Mol Med.* 2013;19(8):460–473.

39. Zhou L, Sun X, Huang Z, et al. Imatinib ameliorated retinal neovascularization by suppressing PDGFR-alpha and PDGFR-beta. *Cell Physiol Biochem.* 2018;48(1):263–273.

40. Katoh M, Nakagama H. FGF receptors: cancer biology and therapeutics. *Med Res Rev.* 2014;34(2):280–300.

41. Zhou Y, Wu C, Lu G, Hu Z, Chen Q, Du X. FGF/FGFR signaling pathway involved resistance in various cancer types. *J Cancer.* 2020;11(8):2000–2007.

42. Weiss J, Sos ML, Seidel D, et al. Frequent and focal FGFR1 amplification associates with therapeutically tractable FGFR1 dependency in squamous cell lung cancer. *Sci Transl Med.* 2010;2(62):62ra93.

43. Courjal F, Cuny M, Simony-Lafontaine J, et al. Mapping of DNA amplifications at 15 chromosomal localizations in 1875 breast tumors: definition of phenotypic groups. *Cancer Res.* 1997;57(19):4360–4367.

44. Musolino A, Campone M, Neven P, et al. Phase II, randomized, placebo-controlled study of dovitinib in combination with fulvestrant in postmenopausal patients with HR(+), HER2(−) breast cancer that had progressed during or after prior endocrine therapy. *Breast Cancer Res.* 2017;19(1):18.

45. Preusser M, Berghoff AS, Berger W, et al. High rate of FGFR1 amplifications in brain metastases of squamous and non-squamous lung cancer. *Lung Cancer.* 2014;83(1):83–89.

46. Motzer RJ, Porta C, Vogelzang NJ, et al. Dovitinib versus sorafenib for third-line targeted treatment of patients with metastatic renal cell carcinoma: an open-label, randomised phase 3 trial. *Lancet Oncol.* 2014;15(3):286–296.

47. Choi YJ, Kim HS, Park SH, et al. Phase II study of Dovitinib in patients with castration-resistant prostate cancer (KCSG-GU11-05). *Cancer Res Treat.* 2018;50(4):1252–1259.

48. Katoh M. FGFR inhibitors: effects on cancer cells, tumor microenvironment and whole-body homeostasis (review). *Int J Mol Med.* 2016;38(1):3–15.

49. Brahmkhatri VP, Prasanna C, Atreya HS. Insulin-like growth factor system in cancer: novel targeted therapies. *Biomed Res Int.* 2015;2015:538019.

50. Kaleko M, Rutter WJ, Miller AD. Overexpression of the human insulinlike growth factor I receptor promotes ligand-dependent neoplastic transformation. *Mol Cell Biol.* 1990;10(2):464–473.

51. Forbes BE, Blyth AJ, Wit JM. Disorders of IGFs and IGF-1R signaling pathways. *Mol Cell Endocrinol.* 2020;518:111035.

52. Cullen KJ, Yee D, Rosen N. Insulinlike growth factors in human malignancy. *Cancer Invest.* 1991;9(4):443–454.

53. Yeo CD, Park KH, Park CK, et al. Expression of insulin-like growth factor 1 receptor (IGF-1R) predicts poor responses to epidermal growth factor receptor (EGFR) tyrosine kinase inhibitors in non-small cell lung cancer patients harboring activating EGFR mutations. *Lung Cancer.* 2015;87(3):311–317.

54. Pollak M, Beamer W, Zhang JC. Insulin-like growth factors and prostate cancer. *Cancer Metastasis Rev.* 1998;17(4):383–390.

55. Sun WY, Yun HY, Song YJ, et al. Insulin-like growth factor 1 receptor expression in breast cancer tissue and mammographic density. *Mol Clin Oncol.* 2015;3(3):572–580.

56. Trojan J, Cloix JF, Ardourel MY, Chatel M, Anthony DD. Insulin-like growth factor type I biology and targeting in malignant gliomas. *Neuroscience.* 2007;145(3):795–811.

57. Nakajima N, Kozu K, Kobayashi S, et al. The expression of IGF-1R in *Helicobacter pylori*-infected intestinal metaplasia and gastric cancer. *J Clin Biochem Nutr.* 2016;59(1):53–57.

58. Heskamp S, Boerman OC, Molkenboer-Kuenen JD, et al. Upregulation of IGF-1R expression during neoadjuvant therapy predicts poor outcome in breast cancer patients. *PLoS One.* 2015;10(2), e0117745.

59. Saldana SM, Lee HH, Lowery FJ, et al. Inhibition of type I insulin-like growth factor receptor signaling attenuates the development of breast cancer brain metastasis. *PLoS One.* 2013;8(9), e73406.

60. Molnar K, Meszaros A, Fazakas C, et al. Pericyte-secreted IGF2 promotes breast cancer brain metastasis formation. *Mol Oncol.* 2020;14(9):2040–2057.

61. Ireland L, Santos A, Campbell F, et al. Blockade of insulin-like growth factors increases efficacy of paclitaxel in metastatic breast cancer. *Oncogene.* 2018;37(15):2022–2036.

62. Ekman S, Frodin JE, Harmenberg J, et al. Clinical phase I study with an insulin-like growth factor-1 receptor inhibitor: experiences in patients with squamous non-small cell lung carcinoma. *Acta Oncol.* 2011;50(3):441–447.

63. Dreux AC, Lamb DJ, Modjtahedi H, Ferns GA. The epidermal growth factor receptors and their family of ligands: their putative role in atherogenesis. *Atherosclerosis.* 2006;186(1):38–53.

64. Harris RC, Chung E, Coffey RJ. EGF receptor ligands. *Exp Cell Res.* 2003;284(1):2–13.

65. Yang YP, Ma H, Starchenko A, et al. A chimeric Egfr protein reporter mouse reveals Egfr localization and trafficking in vivo. *Cell Rep.* 2017;19(6):1257–1267.

66. Hubbard SR, Till JH. Protein tyrosine kinase structure and function. *Annu Rev Biochem.* 2000;69:373–398.

67. Baselga J, Tripathy D, Mendelsohn J, et al. Phase II study of weekly intravenous trastuzumab (Herceptin) in patients with HER2/neu-overexpressing metastatic breast cancer. *Semin Oncol.* 1999;26(4 Suppl 12):78–83.

68. Pao W, Miller V, Zakowski M, et al. EGF receptor gene mutations are common in lung cancers from "never smokers" and are associated with sensitivity of tumors to gefitinib and erlotinib. *Proc Natl Acad Sci U S A.* 2004;101(36):13306–13311.

69. Massarelli E, Johnson FM, Erickson HS, Wistuba II, Papadimitrakopoulou V. Uncommon epidermal growth factor receptor mutations in non-small cell lung cancer and their mechanisms of EGFR tyrosine kinase inhibitors sensitivity and resistance. *Lung Cancer.* 2013;80(3):235–241.

70. Lynch TJ, Bell DW, Sordella R, et al. Activating mutations in the epidermal growth factor receptor underlying responsiveness of non-small-cell lung cancer to gefitinib. *N Engl J Med.* 2004;350(21):2129–2139.

71. Messersmith WA, Ahnen DJ. Targeting EGFR in colorectal cancer. *N Engl J Med.* 2008;359(17):1834–1836.

72. Douillard JY, Oliner KS, Siena S, et al. Panitumumab-FOLFOX4 treatment and RAS mutations in colorectal cancer. *N Engl J Med.* 2013;369(11):1023–1034.

73. Drebin JA, Stern DF, Link VC, Weinberg RA, Greene MI. Monoclonal antibodies identify a cell-surface antigen associated with an activated cellular oncogene. *Nature.* 1984;312(5994):545–548.

74. Ballard P, Yates JW, Yang Z, et al. Preclinical comparison of osimertinib with other EGFR-TKIs in EGFR-mutant NSCLC brain metastases models, and early evidence of clinical brain metastases activity. *Clin Cancer Res.* 2016;22(20):5130–5140.

75. Reungwetwattana T, Nakagawa K, Cho BC, et al. CNS response to osimertinib versus standard epidermal growth factor receptor tyrosine kinase inhibitors in patients with untreated EGFR-mutated advanced non-small-cell lung cancer. *J Clin Oncol.* 2018;, JCO2018783118.

76. Fernandez-Medarde A, Santos E. Ras in cancer and developmental diseases. *Genes Cancer.* 2011;2(3):344–358.

77. Zhou L, Tan X, Kamohara H, et al. MEK1 and MEK2 isoforms regulate distinct functions in pancreatic cancer cells. *Oncol Rep.* 2010;24(1):251–255.

78. Dhillon AS, Hagan S, Rath O, Kolch W. MAP kinase signalling pathways in cancer. *Oncogene.* 2007;26(22):3279–3290.

79. Gentry L, Samatar AA, Der CJ. Inhibitors of the ERK mitogen-activated protein kinase cascade for targeting RAS mutant cancers. *Enzyme.* 2013;34 Pt. B:67–106.

80. Avruch J, Khokhlatchev A, Kyriakis JM, et al. Ras activation of the Raf kinase: tyrosine kinase recruitment of the MAP kinase cascade. *Recent Prog Horm Res.* 2001;56:127–155.

81. Davies H, Bignell GR, Cox C, et al. Mutations of the BRAF gene in human cancer. *Nature.* 2002;417(6892):949–954.

82. Chapman PB, Robert C, Larkin J, et al. Vemurafenib in patients with BRAFV600 mutation-positive metastatic melanoma: final overall survival results of the randomized BRIM-3 study. *Ann Oncol.* 2017;28(10):2581–2587.

83. Dummer R, Ascierto PA, Gogas HJ, et al. Encorafenib plus binimetinib versus vemurafenib or encorafenib in patients with BRAF-mutant melanoma (COLUMBUS): a multicentre, open-label, randomised phase 3 trial. *Lancet Oncol.* 2018;19(5):603–615.

84. Bollag G, Hirth P, Tsai J, et al. Clinical efficacy of a RAF inhibitor needs broad target blockade in BRAF-mutant melanoma. *Nature.* 2010;467(7315):596–599.

85. Davies MA, Saiag P, Robert C, et al. Dabrafenib plus trametinib in patients with BRAF(V600)-mutant melanoma brain metastases (COMBI-MB): a multicentre, multicohort, open-label, phase 2 trial. *Lancet Oncol.* 2017;18(7):863–873.

86. Long GV, Stroyakovskiy D, Gogas H, et al. Combined BRAF and MEK inhibition versus BRAF inhibition alone in melanoma. *N Engl J Med.* 2014;371(20):1877–1888.

87. Vivanco I, Sawyers CL. The phosphatidylinositol 3-kinase AKT pathway in human cancer. *Nat Rev Cancer.* 2002;2(7):489–501.

88. Wymann MP, Pirola L. Structure and function of phosphoinositide 3-kinases. *Biochim Biophys Acta.* 1998;1436(1–2):127–150.

89. Trzaskowski B, Latek D, Yuan S, Ghoshdastider U, Debinski A, Filipek S. Action of molecular switches in GPCRs--theoretical and experimental studies. *Curr Med Chem.* 2012;19(8):1090–1109.

90. Vanhaesebroeck B, Guillermet-Guibert J, Graupera M, Bilanges B. The emerging mechanisms of isoform-specific PI3K signalling. *Nat Rev Mol Cell Biol.* 2010;11(5):329–341.

91. Gilman AG. G proteins: transducers of receptor-generated signals. *Annu Rev Biochem.* 1987;56:615–649.

92. Cantley LC. The phosphoinositide 3-kinase pathway. *Science.* 2002;296(5573):1655–1657.

93. Klippel A, Kavanaugh WM, Pot D, Williams LT. A specific product of phosphatidylinositol 3-kinase directly activates the protein kinase Akt through its pleckstrin homology domain. *Mol Cell Biol.* 1997;17(1):338–344.

94. Andjelkovic M, Alessi DR, Meier R, et al. Role of translocation in the activation and function of protein kinase B. *J Biol Chem.* 1997;272(50):31515–31524.

95. Datta SR, Brunet A, Greenberg ME. Cellular survival: a play in three Akts. *Genes Dev.* 1999;13(22):2905–2927.

96. Datta SR, Dudek H, Tao X, et al. Akt phosphorylation of BAD couples survival signals to the cell-intrinsic death machinery. *Cell.* 1997;91(2):231–241.

97. Scheid MP, Woodgett JR. PKB/AKT: functional insights from genetic models. *Nat Rev Mol Cell Biol.* 2001;2(10):760–768.

98. Samuels Y, Wang Z, Bardelli A, et al. High frequency of mutations of the PIK3CA gene in human cancers. *Science.* 2004;304(5670):554.

99. Fruman DA, Chiu H, Hopkins BD, Bagrodia S, Cantley LC, Abraham RT. The PI3K pathway in human disease. *Cell.* 2017;170(4):605–635.

100. Brastianos PK, Carter SL, Santagata S, et al. Genomic characterization of brain metastases reveals branched evolution and potential therapeutic targets. *Cancer Discov.* 2015;5(11):1164–1177.

101. Andre F, Mills D, Taran T. Alpelisib for PIK3CA-mutated advanced breast cancer. Reply. *N Engl J Med.* 2019;381(7):687.

102. Garrido-Castro AC, Saura C, Barroso-Sousa R, et al. Phase 2 study of buparlisib (BKM120), a pan-class I PI3K inhibitor, in patients with metastatic triple-negative breast cancer. *Breast Cancer Res.* 2020;22(1):120.

103. Greenwell IB, Ip A, Cohen JB. PI3K inhibitors: understanding toxicity mechanisms and management. *Oncology (Williston Park).* 2017;31(11):821–828.

104. Huck BR, Mochalkin I. Recent progress towards clinically relevant ATP-competitive Akt inhibitors. *Bioorg Med Chem Lett.* 2017;27(13):2838–2848.

105. Cheng JX, Liu BL, Zhang X. How powerful is CD133 as a cancer stem cell marker in brain tumors? *Cancer Treat Rev.* 2009;35(5):403–408.

106. Saxton RA, Sabatini DM. mTOR signaling in growth, metabolism, and disease. *Cell.* 2017;168(6):960–976.

107. Betz C, Hall MN. Where is mTOR and what is it doing there? *J Cell Biol.* 2013;203(4):563–574.

108. Unni N, Arteaga CL. Is dual mTORC1 and mTORC2 therapeutic blockade clinically feasible in cancer? *JAMA Oncologia.* 2019;5(11):E1–E2.

109. Manning BD, Cantley LC. Rheb fills a GAP between TSC and TOR. *Trends Biochem Sci.* 2003;28(11):573–576.

110. Dibble CC, Elis W, Menon S, et al. TBC1D7 is a third subunit of the TSC1-TSC2 complex upstream of mTORC1. *Mol Cell.* 2012;47(4):535–546.

111. Menon S, Dibble CC, Talbott G, et al. Spatial control of the TSC complex integrates insulin and nutrient regulation of mTORC1 at the lysosome. *Cell.* 2014;156(4):771–785.

112. Holz MK, Ballif BA, Gygi SP, Blenis J. mTOR and S6K1 mediate assembly of the translation preinitiation complex through dynamic protein interchange and ordered phosphorylation events. *Cell.* 2005;123(4):569–580.

113. Gingras AC, Gygi SP, Raught B, et al. Regulation of 4E-BP1 phosphorylation: a novel two-step mechanism. *Genes Dev.* 1999;13(11):1422–1437.

114. Porstmann T, Santos CR, Griffiths B, et al. SREBP activity is regulated by mTORC1 and contributes to Akt-dependent cell growth. *Cell Metab.* 2008;8(3):224–236.

115. Jacinto E, Loewith R, Schmidt A, et al. Mammalian TOR complex 2 controls the actin cytoskeleton and is rapamycin insensitive. *Nat Cell Biol.* 2004;6(11):1122–1128.

116. Sarbassov DD, Guertin DA, Ali SM, Sabatini DM. Phosphorylation and regulation of Akt/PKB by the rictor-mTOR complex. *Science.* 2005;307(5712):1098–1101.

117. Garcia-Martinez JM, Alessi DR. mTOR complex 2 (mTORC2) controls hydrophobic motif phosphorylation and activation of serum- and glucocorticoid-induced protein kinase 1 (SGK1). *Biochem J.* 2008;416(3):375–385.

118. Brown EJ, Albers MW, Shin TB, et al. A mammalian protein targeted by G1-arresting rapamycin-receptor complex. *Nature.* 1994;369(6483):756–758.

119. Sun SY. mTOR kinase inhibitors as potential cancer therapeutic drugs. *Cancer Lett.* 2013;340(1):1–8.

120. Rodrik-Outmezguine VS, Okaniwa M, Yao Z, et al. Overcoming mTOR resistance mutations with a new-generation mTOR inhibitor. *Nature.* 2016;534(7606):272–276.

121. Hua H, Kong Q, Zhang H, Wang J, Luo T, Jiang Y. Targeting mTOR for cancer therapy. *J Hematol Oncol.* 2019;12(1):71.

122. Guichard SM, Curwen J, Bihani T, et al. AZD2014, an inhibitor of mTORC1 and mTORC2, is highly effective in ER+ breast cancer when administered using intermittent or continuous schedules. *Mol Cancer Ther.* 2015;14(11):2508–2518.

123. Jordan NJ, Dutkowski CM, Barrow D, et al. Impact of dual mTORC1/2 mTOR kinase inhibitor AZD8055 on acquired endocrine resistance in breast cancer in vitro. *Breast Cancer Res.* 2014;16(1):R12.

124. Zhang YJ, Duan Y, Zheng XF. Targeting the mTOR kinase domain: the second generation of mTOR inhibitors. *Drug Discov Today.* 2011;16(7–8):325–331.

125. Ippen FM, Grosch JK, Subramanian M, et al. Targeting the PI3K/Akt/mTOR pathway with the pan-Akt inhibitor GDC-0068 in PIK3CA-mutant breast cancer brain metastases. *Neuro Oncol.* 2019;21(11):1401–1411.

126. Ippen FM, Alvarez-Breckenridge CA, Kuter BM, et al. The dual PI3K/mTOR pathway inhibitor GDC-0084 achieves antitumor activity in PIK3CA-mutant breast cancer brain metastases. *Clin Cancer Res.* 2019;25(11):3374–3383.

127. Dimova I, Popivanov G, Djonov V. Angiogenesis in cancer—general pathways and their therapeutic implications. *J BUON.* 2014;19(1):15–21.

128. Cavallaro U, Christofori G. Molecular mechanisms of tumor angiogenesis and tumor progression. *J Neurooncol.* 2000;50(1–2):63–70.

129. Loizzi V, Del Vecchio V, Gargano G, et al. Biological pathways involved in tumor angiogenesis and bevacizumab based anti-Angiogenic therapy with special references to ovarian cancer. *Int J Mol Sci.* 2017;18(9):1–5.

130. Yadav L, Puri N, Rastogi V, Satpute P, Sharma V. Tumour angiogenesis and angiogenic inhibitors: a review. *J Clin Diagn Res.* 2015;9(6):XE01–XE05.

131. Huang Z, Bao SD. Roles of main pro- and anti-angiogenic factors in tumor angiogenesis. *World J Gastroenterol.* 2004;10(4):463–470.

132. Pandya NM, Dhalla NS, Santani DD. Angiogenesis—a new target for future therapy. *Vascul Pharmacol.* 2006;44(5):265–274.

133. Gavalas NG, Liontos M, Trachana SP, et al. Angiogenesis-related pathways in the pathogenesis of ovarian cancer. *Int J Mol Sci.* 2013;14(8):15885–15909.

134. Fukumura D, Xu L, Chen Y, Gohongi T, Seed B, Jain RK. Hypoxia and acidosis independently up-regulate vascular endothelial growth factor transcription in brain tumors in vivo. *Cancer Res.* 2001;61(16):6020–6024.

135. Luo Q, Wang J, Zhao W, et al. Vasculogenic mimicry in carcinogenesis and clinical applications. *J Hematol Oncol.* 2020;13(1):19.

136. Ayala-Dominguez L, Olmedo-Nieva L, Munoz-Bello JO, et al. Mechanisms of Vasculogenic mimicry in ovarian cancer. *Front Oncol.* 2019;9:998.

137. Folkman J. Tumor angiogenesis: therapeutic implications. *N Engl J Med.* 1971;285(21):1182–1186.

138. Zaman S, Wang R, Gandhi V. Targeting the apoptosis pathway in hematologic malignancies. *Leuk Lymphoma.* 2014;55(9):1980–1992.

139. Lopez J, Tait SW. Mitochondrial apoptosis: killing cancer using the enemy within. *Br J Cancer.* 2015;112(6):957–962.

140. Hassan M, Watari H, AbuAlmaaty A, Ohba Y, Sakuragi N. Apoptosis and molecular targeting therapy in cancer. *Biomed Res Int.* 2014;2014:150845.

141. Yip KW, Reed JC. Bcl-2 family proteins and cancer. *Oncogene.* 2008;27(50):6398–6406.

142. Rastogi K, Bhaskar S, Gupta S, Jain S, Singh D, Kumar P. Palliation of brain metastases: analysis of prognostic factors affecting overall survival. *Indian J Palliat Care.* 2018;24(3):308–312.

143. Thiery JP, Acloque H, Huang RY, Nieto MA. Epithelial-mesenchymal transitions in development and disease. *Cell.* 2009;139(5):871–890.

144. Kothari AN, Mi Z, Zapf M, Kuo PC. Novel clinical therapeutics targeting the epithelial to mesenchymal transition. *Clin Transl Med.* 2014;3:35.

145. Kalluri R, Weinberg RA. The basics of epithelial-mesenchymal transition. *J Clin Invest.* 2009;119(6):1420–1428.

146. Nieto MA. The snail superfamily of zinc-finger transcription factors. *Nat Rev Mol Cell Biol.* 2002;3(3):155–166.

147. Lamouille S, Xu J, Derynck R. Molecular mechanisms of epithelial-mesenchymal transition. *Nat Rev Mol Cell Biol.* 2014;15(3):178–196.

148. Zhang J, Tian XJ, Xing J. Signal transduction pathways of EMT induced by TGF-beta, SHH, and WNT and their Crosstalks. *J Clin Med.* 2016;5(4):1–2.

149. Kim J, Kong J, Chang H, Kim H, Kim A. EGF induces epithelial-mesenchymal transition through phospho-Smad2/3-snail signaling pathway in breast cancer cells. *Oncotarget.* 2016;7(51):85021–85032.

150. Han Y, Luo Y, Wang Y, Chen Y, Li M, Jiang Y. Hepatocyte growth factor increases the invasive potential of PC-3 human prostate cancer cells via an ERK/MAPK and Zeb-1 signaling pathway. *Oncol Lett.* 2016;11(1):753–759.

151. Komiya Y, Habas R. Wnt signal transduction pathways. *Organogenesis.* 2008;4(2):68–75.

152. Wang P, Gao Q, Suo Z, et al. Identification and characterization of cells with cancer stem cell properties in human primary lung cancer cell lines. *PLoS One.* 2013;8(3), e57020.

153. Leong KG, Niessen K, Kulic I, et al. Jagged1-mediated notch activation induces epithelial-to-mesenchymal transition through slug-induced repression of E-cadherin. *J Exp Med.* 2007;204(12):2935–2948.

154. Fortini ME, Lai ZC, Rubin GM. The Drosophila zfh-1 and zfh-2 genes encode novel proteins containing both zinc-finger and homeodomain motifs. *Mech Dev.* 1991;34(2–3):113–122.

155. Nagaishi M, Nakata S, Ono Y, et al. Tumoral and stromal expression of slug, ZEB1, and ZEB2 in brain metastasis. *J Clin Neurosci.* 2017;46:124–128.

156. Comijn J, Berx G, Vermassen P, et al. The two-handed E box binding zinc finger protein SIP1 downregulates E-cadherin and induces invasion. *Mol Cell.* 2001;7(6):1267–1278.

157. Wang J, Lee S, Teh CE, Bunting K, Ma L, Shannon MF. The transcription repressor, ZEB1, cooperates with CtBP2 and HDAC1 to suppress IL-2 gene activation in T cells. *Int Immunol.* 2009;21(3):227–235.

158. Sanchez-Tillo E, Siles L, de Barrios O, et al. Expanding roles of ZEB factors in tumorigenesis and tumor progression. *Am J Cancer Res.* 2011;1(7):897–912.

159. Vandewalle C, Comijn J, De Craene B, et al. SIP1/ZEB2 induces EMT by repressing genes of different epithelial cell-cell junctions. *Nucleic Acids Res.* 2005;33(20):6566–6578.

160. Postigo AA. Opposing functions of ZEB proteins in the regulation of the TGFbeta/BMP signaling pathway. *EMBO J.* 2003;22(10):2443–2452.

161. Peinado H, Olmeda D, Cano A. Snail, Zeb and bHLH factors in tumour progression: an alliance against the epithelial phenotype? *Nat Rev Cancer.* 2007;7(6):415–428.

162. Yang F, Sun L, Li Q, et al. SET8 promotes epithelial-mesenchymal transition and confers TWIST dual transcriptional activities. *EMBO J.* 2012;31(1):110–123.

163. Margetts PJ. Twist: a new player in the epithelial-mesenchymal transition of the peritoneal mesothelial cells. *Nephrol Dial Transplant.* 2012;27(11):3978–3981.

164. Yang J, Mani SA, Donaher JL, et al. Twist, a master regulator of morphogenesis, plays an essential role in tumor metastasis. *Cell.* 2004;117(7):927–939.

165. Puisieux A, Valsesia-Wittmann S, Ansieau S. A twist for survival and cancer progression. *Br J Cancer.* 2006;94(1):13–17.

166. Yang J, Weinberg RA. Epithelial-mesenchymal transition: at the crossroads of development and tumor metastasis. *Dev Cell.* 2008;14(6):818–829.

167. Jeevan DS, Cooper JB, Braun A, Murali R, Jhanwar-Uniyal M. Molecular pathways mediating metastases to the brain via epithelial-to-mesenchymal transition: genes, proteins, and functional analysis. *Anticancer Res.* 2016;36(2):523–532.

168. Singh M, Venugopal C, Tokar T, et al. RNAi screen identifies essential regulators of human brain metastasis-initiating cells. *Acta Neuropathol.* 2017;134(6):923–940.

169. Lehmann OJ, Sowden JC, Carlsson P, Jordan T, Bhattacharya SS. Fox's in development and disease. *Trends Genet.* 2003;19(6):339–344.

170. Han B, Qu Y, Jin Y, et al. FOXC1 activates smoothened-independent hedgehog signaling in basal-like breast cancer. *Cell Rep.* 2015;13(5):1046–1058.

171. Mani R, St Onge RP, JLt H, Giaever G, Roth FP. Defining genetic interaction. *Proc Natl Acad Sci U S A.* 2008;105(9):3461–3466.

172. Ou-Yang L, Xiao SJ, Liu P, et al. Forkhead box C1 induces epithelialmesenchymal transition and is a potential therapeutic target in nasopharyngeal carcinoma. *Mol Med Rep.* 2015;12(6):8003–8009.

173. Chen J, Lee HJ, Wu X, et al. Gain of glucose-independent growth upon metastasis of breast cancer cells to the brain. *Cancer Res.* 2015;75(3):554–565.

174. McConnell BB, Yang VW. Mammalian Kruppel-like factors in health and diseases. *Physiol Rev.* 2010;90(4):1337–1381.

175. Lahiri SK, Zhao J. Kruppel-like factor 8 emerges as an important regulator of cancer. *Am J Transl Res.* 2012;4(3):357–363.

176. Wang F, Sloss C, Zhang X, Lee SW, Cusack JC. Membrane-bound heparin-binding epidermal growth factor like growth factor regulates E-cadherin expression in pancreatic carcinoma cells. *Cancer Res.* 2007;67(18):8486–8493.

177. DiFeo A, Narla G, Camacho-Vanegas O, et al. E-cadherin is a novel transcriptional target of the KLF6 tumor suppressor. *Oncogene.* 2006;25(44):6026–6031.

178. Sellak H, Wu S, Lincoln TM. KLF4 and SOX9 transcription factors antagonize beta-catenin and inhibit TCF-activity in cancer cells. *Biochim Biophys Acta.* 2012;1823(10):1666–1675.

179. Tiwari N, Meyer-Schaller N, Arnold P, et al. Klf4 is a transcriptional regulator of genes critical for EMT, including Jnk1 (Mapk8). *PLoS One.* 2013;8(2), e57329.

180. Limame R, Op de Beeck K, Lardon F, De Wever O, Pauwels P. Kruppel-like factors in cancer progression: three fingers on the steering wheel. *Oncotarget.* 2014;5(1):29–48.

181. Yori JL, Seachrist DD, Johnson E, et al. Kruppel-like factor 4 inhibits tumorigenic progression and metastasis in a mouse model of breast cancer. *Neoplasia.* 2011;13(7):601–610.

182. De Craene B, Gilbert B, Stove C, Bruyneel E, van Roy F, Berx G. The transcription factor snail induces tumor cell invasion through modulation of the epithelial cell differentiation program. *Cancer Res.* 2005;65(14):6237–6244.

183. Okuda H, Xing F, Pandey PR, et al. miR-7 suppresses brain metastasis of breast cancer stem-like cells by modulating KLF4. *Cancer Res.* 2013;73(4):1434–1444.

184. Guo J, Fu Z, Wei J, Lu W, Feng J, Zhang S. PRRX1 promotes epithelial-mesenchymal transition through the Wnt/beta-catenin pathway in gastric cancer. *Med Oncol.* 2015;32(1):393.

185. Wang S, Samakovlis C. Grainy head and its target genes in epithelial morphogenesis and wound healing. *Curr Top Dev Biol.* 2012;98:35–63.

186. Cieply B, Farris J, Denvir J, Ford HL, Frisch SM. Epithelial-mesenchymal transition and tumor suppression are controlled by a reciprocal feedback loop between ZEB1 and Grainyhead-like-2. *Cancer Res.* 2013;73(20):6299–6309.

187. Fernando RI, Litzinger M, Trono P, Hamilton DH, Schlom J, Palena C. The T-box transcription factor Brachyury promotes epithelial-mesenchymal transition in human tumor cells. *J Clin Invest.* 2010;120(2):533–544.

188. Shih DJH, Nayyar N, Bihun I, et al. Genomic characterization of human brain metastases identifies drivers of metastatic lung adenocarcinoma. *Nat Genet.* 2020;52(4):371–377.

189. Diepenbruck M, Waldmeier L, Ivanek R, et al. Tead2 expression levels control the subcellular distribution of yap and Taz, zyxin expression and epithelial-mesenchymal transition. *J Cell Sci.* 2014;127(Pt 7):1523–1536.

190. Gilmore TD, Garbati MR. Inhibition of NF-kappaB signaling as a strategy in disease therapy. *Curr Top Microbiol Immunol.* 2011;349:245–263.

191. Pires BR, Mencalha AL, Ferreira GM, et al. NF-kappaB is involved in the regulation of EMT genes in breast cancer cells. *PLoS One.* 2017;12(1), e0169622.

192. Cavallaro U, Christofori G. Cell adhesion and signalling by cadherins and Ig-CAMs in cancer. *Nat Rev Cancer.* 2004;4(2):118–132.

193. Makrilia N, Kollias A, Manolopoulos L, Syrigos K. Cell adhesion molecules: role and clinical significance in cancer. *Cancer Invest.* 2009;27(10):1023–1037.

194. Wai Wong C, Dye DE, Coombe DR. The role of immunoglobulin superfamily cell adhesion molecules in cancer metastasis. *Int J Cell Biol.* 2012;2012:340296.

195. Padmanaban V, Krol I, Suhail Y, et al. E-cadherin is required for metastasis in multiple models of breast cancer. *Nature.* 2019;573(7774):439–444.

196. Kotteas EA, Boulas P, Gkiozos I, Tsagkouli S, Tsoukalas G, Syrigos KN. The intercellular cell adhesion molecule-1 (icam-1) in lung cancer: implications for disease progression and prognosis. *Anticancer Res.* 2014;34(9):4665–4672.

197. Paoli P, Giannoni E, Chiarugi P. Anoikis molecular pathways and its role in cancer progression. *Biochim Biophys Acta.* 2013;1833(12):3481–3498.

198. Moreno-Layseca P, Streuli CH. Signalling pathways linking integrins with cell cycle progression. *Matrix Biol.* 2014;34:144–153.

199. Brassart-Pasco S, Brezillon S, Brassart B, Ramont L, Oudart JB, Monboisse JC. Tumor microenvironment: extracellular matrix alterations influence tumor progression. *Front Oncol.* 2020;10:397.

200. Andreasen PA, Kjoller L, Christensen L, Duffy MJ. The urokinase-type plasminogen activator system in cancer metastasis: a review. *Int J Cancer.* 1997;72(1):1–22.

201. Funasaka T, Raz A. The role of autocrine motility factor in tumor and tumor microenvironment. *Cancer Metastasis Rev.* 2007;26(3–4):725–735.

202. Wang Y, Liu J, Ying X, Lin PC, Zhou BP. Twist-mediated epithelial-mesenchymal transition promotes breast tumor cell invasion via inhibition of hippo pathway. *Sci Rep.* 2016;6:24606.

203. Cao J, Wang X, Dai T, et al. Twist promotes tumor metastasis in basal-like breast cancer by transcriptionally upregulating ROR1. *Theranostics.* 2018;8(10):2739–2751.

204. Zhuo X, Luo H, Chang A, Li D, Zhao H, Zhou Q. Is overexpression of TWIST, a transcriptional factor, a prognostic biomarker of head and neck carcinoma? Evidence from fifteen studies. *Sci Rep.* 2015;5:18073.

205. Hsu KW, Hsieh RH, Huang KH, et al. Activation of the Notch1/STAT3/twist signaling axis promotes gastric cancer progression. *Carcinogenesis.* 2012;33(8):1459–1467.

206. Strilic B, Offermanns S. Intravascular survival and extravasation of tumor cells. *Cancer Cell.* 2017;32(3):282–293.

207. Dahmani A, Delisle JS. TGF-beta in T cell biology: implications for cancer immunotherapy. *Cancers (Basel).* 2018;10(6):1–21.

208. Han Y, Liu D, Li L. PD-1/PD-L1 pathway: current researches in cancer. *Am J Cancer Res.* 2020;10(3):727–742.

209. Gevensleben H, Dietrich D, Golletz C, et al. The immune checkpoint regulator PD-L1 is highly expressed in aggressive primary prostate cancer. *Clin Cancer Res.* 2016;22(8):1969–1977.

210. Wang X, Teng F, Kong L, Yu J. PD-L1 expression in human cancers and its association with clinical outcomes. *Onco Targets Ther.* 2016;9:5023–5039.

211. Tawbi HA, Forsyth PA, Algazi A, et al. Combined Nivolumab and Ipilimumab in melanoma metastatic to the brain. *N Engl J Med.* 2018;379(8):722–730.

212. Goldberg SB, Schalper KA, Gettinger SN, et al. Pembrolizumab for management of patients with NSCLC and brain metastases: long-term results and biomarker analysis from a non-randomised, open-label, phase 2 trial. *Lancet Oncol.* 2020;21(5):655–663.

213. Giancotti FG. Mechanisms governing metastatic dormancy and reactivation. *Cell.* 2013;155(4):750–764.

214. Riihimaki M, Thomsen H, Sundquist K, Sundquist J, Hemminki K. Clinical landscape of cancer metastases. *Cancer Med.* 2018;7(11):5534–5542.

215. Arvelo F, Sojo F, Cotte C. Cancer and the metastatic substrate. *Ecancer Medical Science.* 2016;10:701.

216. Luo Z, Rong Z, Huang C. Surgery strategies for gastric cancer with liver metastasis. *Front Oncol.* 2019;9:1353.

217. Li WH, Peng JJ, Xiang JQ, Chen W, Cai SJ, Zhang W. Oncological outcome of unresectable lung metastases without extrapulmonary metastases in colorectal cancer. *World J Gastroenterol.* 2010;16(26):3318–3324.

218. Langley RR, Fidler IJ. The biology of brain metastasis. *Clin Chem.* 2013;59(1):180–189.

219. Choy CAA, Duenas MJ, et al. Microenvironmental landscape of brain metastases. In: *Metastatic Cancer: Clinical and Biological Perspectives.* Austin, TX: Landes Bioscience; 2013.

220. Hobbs SK, Monsky WL, Yuan F, et al. Regulation of transport pathways in tumor vessels: role of tumor type and microenvironment. *Proc Natl Acad Sci U S A.* 1998;95(8):4607–4612.

221. Daneman R, Prat A. The blood-brain barrier. *Cold Spring Harb Perspect Biol.* 2015;7(1), a020412.

222. Prinz M, Priller J. The role of peripheral immune cells in the CNS in steady state and disease. *Nat Neurosci.* 2017;20(2):136–144.

223. Arvanitis CD, Ferraro GB, Jain RK. The blood-brain barrier and blood-tumour barrier in brain tumours and metastases. *Nat Rev Cancer.* 2020;20(1):26–41.

224. Wilhelm I, Molnar J, Fazakas C, Hasko J, Krizbai IA. Role of the blood-brain barrier in the formation of brain metastases. *Int J Mol Sci.* 2013;14(1):1383–1411.

225. Strell C, Entschladen F. Extravasation of leukocytes in comparison to tumor cells. *Cell Commun Signal.* 2008;6:10.

226. Lorger M, Felding-Habermann B. Capturing changes in the brain microenvironment during initial steps of breast cancer brain metastasis. *Am J Pathol.* 2010;176(6):2958–2971.

227. Paku S, Dome B, Toth R, Timar J. Organ-specificity of the extravasation process: an ultrastructural study. *Clin Exp Metastasis.* 2000;18(6):481–492.

228. Fidler IJ. The role of the organ microenvironment in brain metastasis. *Semin Cancer Biol.* 2011;21(2):107–112.

229. Carbonell WS, Ansorge O, Sibson N, Muschel R. The vascular basement membrane as "soil" in brain metastasis. *PLoS One.* 2009;4(6), e5857.

230. Watkins S, Robel S, Kimbrough IF, Robert SM, Ellis-Davies G, Sontheimer H. Disruption of astrocyte-vascular coupling and the blood-brain barrier by invading glioma cells. *Nat Commun.* 2014;5:4196.

231. Liebner S, Fischmann A, Rascher G, et al. Claudin-1 and claudin-5 expression and tight junction morphology are altered in blood vessels of human glioblastoma multiforme. *Acta Neuropathol.* 2000;100(3):323–331.

232. Angeli E, Nguyen TT, Janin A, Bousquet G. How to make anticancer drugs cross the blood-brain barrier to treat brain metastases. *Int J Mol Sci.* 2019;21(1):22.

233. Arvanitis CD, Askoxylakis V, Guo Y, et al. Mechanisms of enhanced drug delivery in brain metastases with focused ultrasound-induced blood-tumor barrier disruption. *Proc Natl Acad Sci U S A.* 2018;115(37):E8717–E8726.

234. Knuutinen O, Kuitunen H, Alahuhta S, et al. Case report: chemotherapy in conjunction with blood-brain barrier disruption for a patient with germ cell tumor with multiple brain metastases. *Clin Genitourin Cancer.* 2018;16(5):e993–e996.

235. Patel T, Zhou J, Piepmeier JM, Saltzman WM. Polymeric nanoparticles for drug delivery to the central nervous system. *Adv Drug Deliv Rev.* 2012;64(7):701–705.

236. Khaitan D, Reddy PL, Ningaraj N. Targeting brain tumors with nanomedicines: overcoming blood brain barrier challenges. *Curr Clin Pharmacol.* 2018;13(2):110–119.

237. Kreuter J. Nanoparticulate systems for brain delivery of drugs. *Adv Drug Deliv Rev.* 2001;47(1):65–81.

238. Luzzi KJ, MacDonald IC, Schmidt EE, et al. Multistep nature of metastatic inefficiency: dormancy of solitary cells after successful extravasation and limited survival of early micrometastases. *Am J Pathol.* 1998;153(3):865–873.

239. Chen R, Goodison S, Sun Y. Molecular profiles of matched primary and metastatic tumor samples support a linear evolutionary model of breast cancer. *Cancer Res.* 2020;80(2):170–174.

240. Yan J, Yang Q, Huang Q. Metastasis suppressor genes. *Histol Histopathol.* 2013;28(3):285–292.

241. Guo YJ, Pan WW, Liu SB, Shen ZF, Xu Y, Hu LL. ERK/MAPK signalling pathway and tumorigenesis. *Exp Ther Med.* 2020;19(3):1997–2007.

242. Steeg PS, Horak CE, Miller KD. Clinical-translational approaches to the Nm23-H1 metastasis suppressor. *Clin Cancer Res.* 2008;14(16):5006–5012.

243. Caligo MA, Grammatico P, Cipollini G, Varesco L, Del Porto G, Bevilacqua G. A low NM23.H1 gene expression identifying high malignancy human melanomas. *Melanoma Res.* 1994;4(3):179–184.

244. Barnes R, Masood S, Barker E, et al. Low nm23 protein expression in infiltrating ductal breast carcinomas correlates with reduced patient survival. *Am J Pathol.* 1991;139(2):245–250.

245. Caffo M, Barresi V, Caruso G, et al. Innovative therapeutic strategies in the treatment of brain metastases. *Int J Mol Sci.* 2013;14(1):2135–2174.

246. Bozza WP, Zhang Y, Hallett K, Rivera Rosado LA, Zhang B. RhoGDI deficiency induces constitutive activation of rho GTPases and COX-2 pathways in association with breast cancer progression. *Oncotarget.* 2015;6(32):32723–32736.

247. Niu H, Wu B, Jiang H, et al. Mechanisms of RhoGDI2 mediated lung cancer epithelial-mesenchymal transition suppression. *Cell Physiol Biochem.* 2014;34(6):2007–2016.

248. Gildea JJ, Seraj MJ, Oxford G, et al. RhoGDI2 is an invasion and metastasis suppressor gene in human cancer. *Cancer Res.* 2002;62(22):6418–6423.

249. Stone L. Bladder cancer: rho-sensitive pathway mediates metastasis. *Nat Rev Urol.* 2016;13(11):630.

250. Mashimo T, Watabe M, Hirota S, et al. The expression of the KAI1 gene, a tumor metastasis suppressor, is directly activated by p53. *Proc Natl Acad Sci U S A.* 1998;95(19):11307–11311.

251. Miller J, Dreyer TF, Bacher AS, et al. Differential tumor biological role of the tumor suppressor KAI1 and its splice variant in human breast cancer cells. *Oncotarget.* 2018;9(5):6369–6390.

252. Liu FS, Dong JT, Chen JT, et al. KAI1 metastasis suppressor protein is down-regulated during the progression of human endometrial cancer. *Clin Cancer Res.* 2003;9(4):1393–1398.

253. Liu W, Iiizumi-Gairani M, Okuda H, et al. KAI1 gene is engaged in NDRG1 gene-mediated metastasis suppression through the ATF3-NFkappaB complex in human prostate cancer. *J Biol Chem.* 2011;286(21):18949–18959.

254. Stark AM, Tongers K, Maass N, Mehdorn HM, Held-Feindt J. Reduced metastasis-suppressor gene mRNA-expression in breast cancer brain metastases. *J Cancer Res Clin Oncol.* 2005;131(3):191–198.

255. Odintsova E, Berditchevski F. Role of the metastasis suppressor tetraspanin CD82/KAI 1 in regulation of signalling in breast cancer cells. *Breast Cancer Res.* 2006;8(2):P21.

256. Harms JF, Welch DR, Miele ME. KISS1 metastasis suppression and emergent pathways. *Clin Exp Metastasis.* 2003;20(1):11–18.

257. Nash KT, Welch DR. The KISS1 metastasis suppressor: mechanistic insights and clinical utility. *Front Biosci.* 2006;11:647–659.

258. Beck BH, Welch DR. The KISS1 metastasis suppressor: a good night KISS for disseminated cancer cells. *Eur J Cancer.* 2010;46(7):1283–1289.

259. Ulasov IV, Kaverina NV, Pytel P, et al. Clinical significance of KISS1 protein expression for brain invasion and metastasis. *Cancer.* 2012;118(8):2096–2105.

260. Slipicevic A, Holm R, Emilsen E, et al. Cytoplasmic BRMS1 expression in malignant melanoma is associated with increased disease-free survival. *BMC Cancer.* 2012;12:73.

261. Phadke PA, Vaidya KS, Nash KT, Hurst DR, Welch DR. BRMS1 suppresses breast cancer experimental metastasis to multiple organs by inhibiting several steps of the metastatic process. *Am J Pathol.* 2008;172(3):809–817.

262. Zhang Y, Ye L, Tan Y, Sun P, Ji K, Jiang WG. Expression of breast cancer metastasis suppressor-1, BRMS-1, in human breast cancer and the biological impact of BRMS-1 on the migration of breast cancer cells. *Anticancer Res.* 2014;34(3):1417–1426.

263. Zimmermann RC, Welch DR. BRMS1: a multifunctional signaling molecule in metastasis. *Cancer Metastasis Rev.* 2020;39(3):755–768.

264. Gobeil S, Zhu X, Doillon CJ, Green MR. A genome-wide shRNA screen identifies GAS1 as a novel melanoma metastasis suppressor gene. *Genes Dev.* 2008;22(21):2932–2940.

265. Zarco N, Gonzalez-Ramirez R, Gonzalez RO, Segovia J. GAS1 induces cell death through an intrinsic apoptotic pathway. *Apoptosis.* 2012;17(6):627–635.

266. Ma Y, Qin H, Cui Y. MiR-34a targets GAS1 to promote cell proliferation and inhibit apoptosis in papillary thyroid carcinoma via PI3K/Akt/Bad pathway. *Biochem Biophys Res Commun.* 2013;441(4):958–963.

267. Li Q, Qin Y, Wei P, et al. Gas1 inhibits metastatic and metabolic phenotypes in colorectal carcinoma. *Mol Cancer Res.* 2016;14(9):830–840.

268. Baig RM, Sanders AJ, Kayani MA, Jiang WG. Association of differentiation-related gene-1 (DRG1) with breast cancer survival and in vitro impact of DRG1 suppression. *Cancers (Basel).* 2012;4(3):658–672.

269. Maruyama Y, Ono M, Kawahara A, et al. Tumor growth suppression in pancreatic cancer by a putative metastasis suppressor gene Cap43/NDRG1/Drg-1 through modulation of angiogenesis. *Cancer Res.* 2006;66(12):6233–6242.

270. Schellhaus AK, Moreno-Andres D, Chugh M, et al. Developmentally regulated GTP binding protein 1 (DRG1) controls microtubule dynamics. *Sci Rep.* 2017;7(1):9996.

271. Bandyopadhyay S, Pai SK, Gross SC, et al. The Drg-1 gene suppresses tumor metastasis in prostate cancer. *Cancer Res.* 2003;63(8):1731–1736.

272. Gao AC, Lou W, Dong JT, Isaacs JT. CD44 is a metastasis suppressor gene for prostatic cancer located on human chromosome 11p13. *Cancer Res.* 1997;57(5):846–849.

273. Lu L, Lv Y, Dong J, Hu S, Peng R. DRG1 is a potential oncogene in lung adenocarcinoma and promotes tumor progression via spindle checkpoint signaling regulation. *Oncotarget.* 2016;7(45):72795–72806.

274. Hu X, Xiang D, Xie Y, et al. LSD1 suppresses invasion, migration and metastasis of luminal breast cancer cells via activation of GATA3 and repression of TRIM37 expression. *Oncogene.* 2019;38(44):7017–7034.

275. Brastianos PK, Kim KE, Wang N. Palbociclib demonstrates intracranial activity in progressive brain metastases harboring cyclin-dependent kinase pathway alterations. *Nat Cancer.* 2021;2:498–502.

276. Chen G, Chakravarti N, Aardalen K, et al. Molecular profiling of patient-matched brain and extracranial melanoma metastases implicates the PI3K pathway as a therapeutic target. *Clin Cancer Res.* 2014;20(21):5537–5546.

277. Fischer GM, Jalali A, Kircher DA, et al. Molecular profiling reveals unique immune and metabolic features of melanoma brain metastases. *Cancer Discov.* 2019;9(5):628–645.

278. Gopal YN, Rizos H, Chen G, et al. Inhibition of mTORC1/2 overcomes resistance to MAPK pathway inhibitors mediated by PGC1alpha and oxidative phosphorylation in melanoma. *Cancer Res.* 2014;74(23):7037–7047.

279. Overholtzer M, Zhang J, Smolen GA, et al. Transforming properties of YAP, a candidate oncogene on the chromosome 11q22 amplicon. *Proc Natl Acad Sci U S A.* 2006;103(33):12405–12410.

280. Hsu PC, Miao J, Huang Z, et al. Inhibition of yes-associated protein suppresses brain metastasis of human lung adenocarcinoma in a murine model. *J Cell Mol Med.* 2018;22(6):3073–3085.

281. Xie N, Tian C, Wu H, et al. FGFR aberrations increase the risk of brain metastases and predict poor prognosis in metastatic breast cancer patients. *Ther Adv Med Oncol.* 2020;12, 1758835920915305.

282. Liu H, Kato Y, Erzinger SA, et al. The role of MMP-1 in breast cancer growth and metastasis to the brain in a xenograft model. *BMC Cancer.* 2012;12:583.

283. Fan LC, Jeng YM, Lu YT, Lien HC. SPOCK1 is a novel transforming growth factor-beta-induced myoepithelial marker that enhances invasion and correlates with poor prognosis in breast Cancer. *PLoS One.* 2016;11(9), e0162933.

284. Miao L, Wang Y, Xia H, Yao C, Cai H, Song Y. SPOCK1 is a novel transforming growth factor-beta target gene that regulates lung cancer cell epithelial-mesenchymal transition. *Biochem Biophys Res Commun.* 2013;440(4):792–797.

285. Sun LR, Li SY, Guo QS, Zhou W, Zhang HM. SPOCK1 involvement in epithelial-to-mesenchymal transition: a new target in cancer therapy? *Cancer Manag Res.* 2020;12:3561–3569.

286. Chen D, Zhou H, Liu G, Zhao Y, Cao G, Liu Q. SPOCK1 promotes the invasion and metastasis of gastric cancer through slug-induced epithelial-mesenchymal transition. *J Cell Mol Med.* 2018;22(2):797–807.

287. Dai L, Zhao J, Yin J, Fu W, Chen G. Cell adhesion molecule 2 (CADM2) promotes brain metastasis by inducing epithelial-mesenchymal transition (EMT) in human non-small cell lung cancer. *Ann Transl Med.* 2020;8(7):465.

288. Liu P, Rudick M, Anderson RG. Multiple functions of caveolin-1. *J Biol Chem.* 2002;277(44):41295–41298.

289. Burgermeister E, Liscovitch M, Rocken C, Schmid RM, Ebert MP. Caveats of caveolin-1 in cancer progression. *Cancer Lett.* 2008;268(2):187–201.

290. Gupta R, Toufaily C, Annabi B. Caveolin and cavin family members: dual roles in cancer. *Biochimie.* 2014;107 Pt B:188–202.

291. Chiu WT, Lee HT, Huang FJ, et al. Caveolin-1 upregulation mediates suppression of primary breast tumor growth and brain metastases by stat3 inhibition. *Cancer Res.* 2011;71(14):4932–4943.

292. Duregon E, Senetta R, Pittaro A, et al. CAVEOLIN-1 expression in brain metastasis from lung cancer predicts worse outcome and radioresistance, irrespective of tumor histotype. *Oncotarget.* 2015;6(30):29626–29636.

293. Kim YJ, Kim JH, Kim O, et al. Caveolin-1 enhances brain metastasis of non-small cell lung cancer, potentially in association with the epithelial-mesenchymal transition marker SNAIL. *Cancer Cell Int.* 2019;19:171.

294. Bos PD, Nguyen DX, Massague J. Modeling metastasis in the mouse. *Curr Opin Pharmacol.* 2010;10(5):571–577.

295. Loh CY, Chai JY, Tang TF, et al. The E-cadherin and N-cadherin switch in epithelial-to-mesenchymal transition: signaling, therapeutic implications, and challenges. *Cells.* 2019;8(10):1118.

296. Klein A, Olendrowitz C, Schmutzler R, et al. Identification of brain- and bone-specific breast cancer metastasis genes. *Cancer Lett.* 2009;276(2):212–220.

297. Palmieri D, Fitzgerald D, Shreeve SM, et al. Analyses of resected human brain metastases of breast cancer reveal the association between up-regulation of hexokinase 2 and poor prognosis. *Mol Cancer Res.* 2009;7(9):1438–1445.

298. Rodrigues G, Hoshino A, Kenific CM, et al. Tumour exosomal CEMIP protein promotes cancer cell colonization in brain metastasis. *Nat Cell Biol.* 2019;21(11):1403–1412.

5

Intracranial metastases

Herbert B. Newton[a,b,c,d], *Seema Shroff*[e], *and Mark G. Malkin*[f,g,h]

[a]Neuro-Oncology Center, Orlando, FL, United States, [b]CNS Oncology Program, Advent Health Cancer Institute, Advent Health Orlando Campus & Advent Health Medical Group, Orlando, FL, United States, [c]Neurology, UCF School of Medicine, Orlando, FL, United States, [d]Neurology & Neurosurgery (Retired), Division of Neuro-Oncology, Esther Dardinger Endowed Chair in Neuro-Oncology, James Cancer Hospital & Solove Research Institute, Wexner Medical Center at the Ohio State University, Columbus, OH, United States, [e]Department of Pathology, Advent Health, Orlando, FL, United States, [f]Neuro-Oncology Program at Massey Cancer Center, Virginia Commonwealth University School of Medicine, Richmond, VA, United States, [g]Neuro-Oncology Division, Department of Neurology, Virginia Commonwealth University School of Medicine, Richmond, VA, United States, [h]Neurology and Neurosurgery, William G. Reynolds, Jr. Chair in Neuro-Oncology, Virginia Commonwealth University School of Medicine, Richmond, VA, United States

1 Introduction and epidemiology

Brain metastases (MBT) are the most common complication of systemic cancer, with estimated incidence rates of 8.3–11 cases per 100,000 population.[1–6] Hospital and autopsy-based studies estimate that these tumors develop in 20%–40% of all adult cancer patients, which corresponds to approximately 150,000–170,000 new cases per year in the United States. More recent data using population-based estimates would suggest a lower incidence of MBT, in the range of 10%.[7] The presence of MBT does not always correlate with clinical sequelae; it is estimated that only 60%–75% of patients with MBT will become symptomatic. The frequency of MBT appears to be rising due to more successful systemic treatment and longer patient survival, earlier detection and implementation of therapy, and improved imaging techniques. MBT most often arise from primary tumors of the lung (50%–60%), breast (15%–20%), melanoma (5%–10%), and gastrointestinal tract (4%–6%).[1–6] Empiric screening of patients with newly diagnosed nonsmall cell lung cancer identifies MBT in 3%–10% of cases.[6] However, MBT can develop from virtually any systemic malignancy, including primary tumors of the prostate, ovary and female reproductive system, kidney, esophagus, soft tissue sarcoma, bladder, and thyroid.[8–17] In addition, between 10% and 15% of patients will develop MBT from an unknown primary.[1,18] Autopsy studies in adults would suggest that melanoma (20%–45% of patients) has the most neurotropism of all primary tumors;

however, small cell lung carcinoma, renal carcinoma, breast, and testicular carcinoma also have a strong propensity for spread to the brain.[1] Tumors with a low degree of neurotropism include prostate, gastrointestinal tract, ovarian, and thyroid malignancies. In children and young adults, MBT arise most often from sarcomas (e.g., osteogenic, Ewing's), germ cell tumors, and neuroblastomas.[1–5,19] In 65%–75% of patients, two or more metastatic tumors will develop simultaneously and be present at the time of cancer diagnosis. Single brain metastases are less common and are most often noted in patients with breast, colon, and renal cell carcinoma. Patients with malignant melanoma and lung carcinoma are more likely to have multiple metastatic lesions.

The prognosis for patients with MBT is quite poor and is dependent on the histological tumor type, number and size of the metastatic lesions, neurological status, and degree of systemic involvement. Overall, the presence of MBT is associated with high morbidity and mortality, with approximately one-third of all patients dying from the brain tumor.[1,4] The natural history is such that, left untreated, patients with MBT will usually die of neurological deterioration within 4 weeks. The addition of steroids will typically extend survival to 8 weeks. External beam radiotherapy, the most common modality of treatment, can further extend survival to 12–20 weeks in many patients.[1–5] However, survival is also dependent on the type of primary malignancy, as shown in a report by Hall and colleagues.[20] In their study, the overall 2-year survival rate for patients with MBT was 8.1%, with a range from 1.7% in patients with small cell lung carcinoma, up to 23.9% for those with ovarian cancer. Several studies have assessed how various prognostic factors relate to MBT patients at the time of diagnosis. A recursive partitioning analysis (RPA) of three RTOG trials evaluated a wide range of prognostic factors and their impact on patient survival.[21] The most important favorable factors were younger age (younger vs older than 65 years; $P < .0001$), higher Karnofsky Performance Status (KPS) score (greater or less than 70; $P < 0.0001$), and limited extent of systemic disease (controlled vs widespread disease; $P < .0001$). Using these criteria, patients could be grouped into three distinct classes. Class 1 included patients who were less than 65 years of age, had KPS scores greater than 70, and had well-controlled systemic disease; Class 3 consisted of all patients with KPS scores less than 70, while Class 2 included all other patients who did not fit into Class 1 or Class 3. The median overall survival varied significantly between groups: 28.4 weeks for patients in Class 1, 16.8 weeks for those in Class 2, and 9.2 weeks for Class 3 patients. In addition, by univariate analysis, patients with multiple MBT had a significantly reduced survival compared to that of those with solitary lesions ($P = .021$).

In a similar study by Nussbaum and colleagues, the number of metastatic lesions present at diagnosis was found to correlate with overall survival.[22] They noted a significant difference ($P = .0001$) in median survival between patients with solitary brain metastases and those with multifocal disease: 5 months versus 3 months, respectively.

2 Pathology

As noted above, in an autopsy series, brain metastases were found in up to a quarter of cancer patients, the most common being lung, breast, gastrointestinal tract, and melanoma (Table 1).[6] While a known clinical history of a primary systemic malignancy is most helpful, sometimes a new or unknown patient will present with an acute neurologic event and emergent neurosurgical intervention. In these cases, the surgical specimen is the first tissue and the pathologic evaluation is critical not only in making the diagnosis, but also in directing subsequent appropriate clinical care.

Clinicopathologic correlation: When the primary malignancy is known and well-characterized, the morphologic features of the metastatic tumor and a targeted immunostain panel are sufficient for a diagnosis and the remaining tissue can be saved for potential further molecular studies, if needed. However, if the history is unknown or no primary source is apparent, a more elaborate workup and correlation with clinical findings are essential. Metastases can be solitary (25%–35%) or multiple

TABLE 1   Primary sites of metastatic brain tumors.

Primary tumor	Percentage (%)
Lung	50–60
Squamous cell	25–30
Adenocarcinoma	12–15
Small cell	10–13
Large cell	2
Breast	15–20
Melanoma	5–10
Gastrointestinal	4–6
Genitourinary	3–5
Unknown	4–8
Other	3–5

Data compiled from Valiente M, Ahluwalia MS, Boire A, Brastianos PK, et al. The evolving landscape of brain metastasis. Trends Cancer 2018;4:176–196; Takei H, Rouah E, Ishida Y. Brain metastasis: clinical characteristics, pathological findings and molecular subtyping for therapeutic implications. Brain Tumor Pathol 2016;33:1–12; Langer CJ, Mehta MP. Current management of brain metastases, with a focus on systemic options. J Clin Oncol 2005;23:6207–6219; Bajaj GK, Kleinberg L, Terezakis S. Current concepts and controversies in the treatment of parenchymal brain metastases: improved outcomes with aggressive management. Cancer Invest 2005;23:363–376; Lassman AB, DeAngelis LM. Brain metastases Neurol Clin N Am 2003;21:1–23.

(65%–75%), or may even present as a miliary pattern of numerous tiny masses. Primary tumors most likely to cause multifocal metastatic deposits include small cell and adenocarcinoma of the lung, melanoma, and choriocarcinoma. Single metastatic deposits are more likely to arise from renal cell, gastrointestinal, breast, prostatic, and uterine carcinomas. The location of the lesion may be helpful, since different primary tumors metastasize to different parts of the CNS with varying frequencies. Parenchymal metastases are the most common and often occur in the arterial watershed zones and at the gray-white junction and are mostly from the frequent primaries mentioned above. Four to 15% of tumors can cause leptomeningeal spread and, less frequently, dural metastases can be seen. In the spine, epidural metastases far outnumber leptomeningeal or parenchymal lesions; common sources include prostate cancer, lymphomas, myeloma, and renal cell carcinoma.[23] Rarely, entirely intravascular carcinomatoses have been reported, leading to infarction.[24] In children, the diagnostic considerations for primary tumors are different from adults and include hematopoietic malignancies, germ cell tumors, osteosarcoma, and "small round blue cell tumors" like Ewing sarcoma, neuroblastoma, and rhabdomyosarcoma.

On microscopic examination, the histological features of the metastasis are usually similar, if not identical, to those of the primary neoplasm (Fig. 1). In some cases, there may be a vigorous angiogenic response, with more prominent vascular proliferation and the formation of

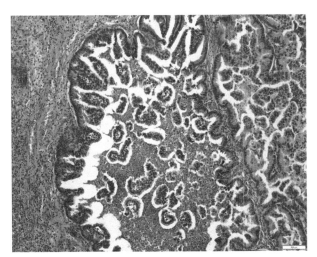

FIG. 1   The morphologic features of most brain metastases are similar, if not identical, to the primary tumor; e.g., this metastatic lung adenocarcinoma retains the papillary morphology reminiscent of the primary tumor.

glomeruloid structures. In other tumors, there may be extensive necrosis, with only small regions of recognizable neoplastic tissue at the periphery of the lesion or adjacent to blood vessels. However, unlike glioblastoma multiforme, pseudopalisading of tumor nuclei around necrotic foci is very uncommon. The tumor mass will usually have well-defined borders, tending to displace adjacent brain parenchyma without significant infiltration. Areas of hemorrhage and gliosis are often noted.

A particular challenge in the absence of clinical history is when the morphology of the metastasis does not recapitulate the primary tumor or organ, or is poorly differentiated; in these cases, a detailed immunohistochemical workup is probably the most helpful tool in the Neuropathologist's kit (Fig. 2, Table 2). A basic panel of stains for a morphologically undifferentiated epithelioid tumor includes cytokeratins (to prove epithelial origin—e.g., cytokeratin [CK], including AE1/3, CK7, CK20, EMA) and lineage-suggestive or lineage-specific markers (TTF-1 and Napsin A: lung; GATA-3: breast and urothelial; p40 and p63: squamous; CDX2: lower gastrointestinal tract; Pax8: gynecologic and renal; PSA and PAP: prostate). With increasing survival times in recent years, more exotic metastases also have to be considered and investigated on occasion, e.g., pancreas (Smad-4) and thyroid (thyroglobulin) (Fig. 3). Immunohistochemical staining is also useful to determine the origin of the metastases when the patient has more than one primary tumor. In morphologically undifferentiated cases, a melanoma panel is often included (S100, HBM45, MelanA, Sox10, PRAME), since melanomas can be nonpigmented and very heterogeneous in their morphologic appearance. Melanomas are often very hemorrhagic, a clue to their nature on imaging. Other differential diagnostic considerations for a hemorrhagic mass would include

renal cell cancer, choriocarcinoma, and even nonmetastatic conditions such as amyloid angiopathy. With or without hemorrhage, the tumor is usually surrounded by an extensive amount of vasogenic edema, which often seems out of proportion to the size of the mass, and contributes to regional mass effect. Lastly, in very rare cases, the pathologist may also have to investigate the possibility of transdifferentiation in a morphologically distinct primary. Fig. 4 demonstrates esophageal adenocarcinoma with choriocarcinomatous differentiation that presented with brain and liver metastasis with choriocarcinomatous morphology. Thus, tissue sampling and integration of clinicopathologic information are paramount in subsequent therapy and management.

Pathogenesis: The formation of brain metastases is increasingly believed to be a targeted/orchestrated process rather than chance colonization by disseminated tumor cells. Many studies have begun to recognize a series of steps/events in the formation of brain metastases, and each step involves many unanswered questions and possible theoretical scenarios. The "seed and soil" hypothesis portends a specific affinity of circulating tumor cells (seed) to the microenvironment (soil) of specific organs. The origin of the circulating tumor cells is also debated; many theories have been proposed regarding the heterogeneity of primary tumor cells, the routes they take to reach the systemic circulation (direct versus via initial secondary sites such as lymph nodes), and the timing of formation of brain macrometastases due to possible dormancy of tumor cells in perivascular niches. On a microscopic scale, this entire process is highly inefficient, and more than 90% of the time, a tumor cell arrested in the brain circulation will not form a macrometastasis.[25]

Many studies have begun to recognize a series of steps/events in the formation of brain metastases, the most fundamental of which include circulatory arrest, invasion (to reach the brain), penetration of the blood–brain barrier (BBB), survival, proliferation and growth in the brain environment, interaction with resident brain cells, recruitment of blood vessels, and various growth patterns.[26,27] A brief discussion of these steps is outlined in the next few paragraphs.

Reaching the brain: Due to the absence of lymphatics, brain metastases are exclusively thought to arise from hematogenous dissemination, mainly from the arterial tree, settling at branch points, watershed zones, and the gray-white junction. For abdominal malignancies, a propensity to form posterior fossa metastases was noted, which led some to believe that retrograde spread through Batson's venous plexus may be responsible for metastases from tumors of the kidney, bladder, and uterus.

BBB invasion and penetration: Once arrested in the cerebral circulation, tumor cells encounter the BBB—and whether viewed in its more traditional structural form as endothelial tight junctions and astrocytic end-feet,

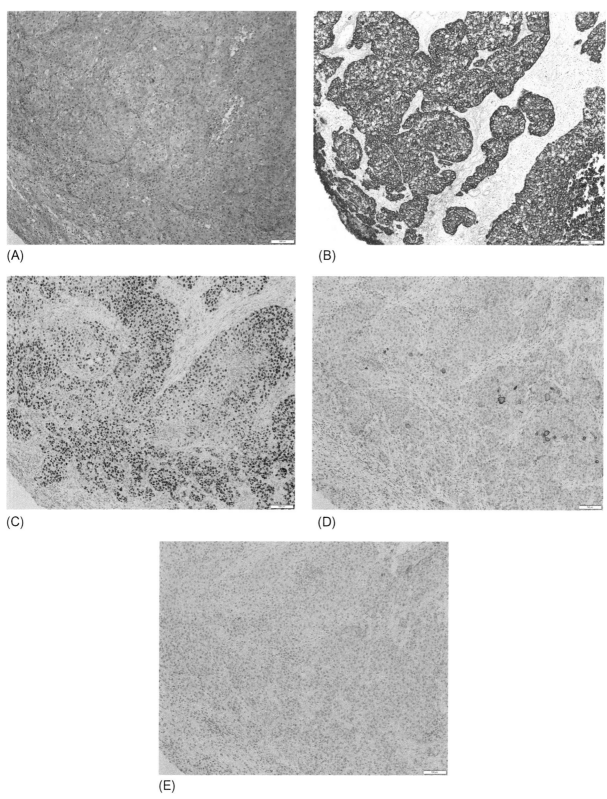

FIG. 2   Basic immunohistochemical workup of a metastatic adenocarcinoma: (A) H&E stain; (B) the tumor cells are positive for CK7; (C) the tumor cells are also positive for TTF-1; (D) rare cells are positive for CK20; and (E) the cells are negative for CDX2. This immunohistochemical profile is suggestive of metastatic pulmonary adenocarcinoma.

TABLE 2 Immunocytochemical staining techniques used in the diagnosis of metastatic brain tumors.

Initial screening panel:

Cytokeratins: Cytokeratin cocktail (AE1/3) or CK7 and CK20, epithelial membrane antigen (EMA)

Glial: GFAP

Helpful lineage markers:

Lung cancer: TTF-1, Napsin A (adenocarcinoma); p40 (squamous carcinoma); TTF-1 and synaptophysin (small cell carcinoma)

Breast cancer: cytokeratin 7, GATA-3, estrogen and progesterone receptors

Gastrointestinal cancer: cytokeratin 20, CDX2

Gynecologic cancer: Pax8, estrogen receptors

Neuroendocrine: cytokeratin, synaptophysin, chromogranin, INSM1, CD56

Thyroid cancer: thyroglobulin, TTF-1, Pax8

Prostate cancer: prostate-specific antigen, prostatic acid phosphatase

Germ cell tumors:

Placental alkaline phosphatase (PLAP), SALL4, HCG (choriocarcinoma)

Sarcomas:

Desmin, smooth-muscle actin

Malignant melanoma:

S-100, HMB45, MelanA, PRAME, MART-1

Lymphoma:

CD45

CD3

CD20

Data derived from Kienast Y, von Baumgarten L, Fuhrmann M, Klinkert WE, et al. Real-time imaging reveals the single steps of brain metastasis formation. Nat Med 2010;16:116–122; Preusser M, Capper D, Ilhan-Mutlu A, Berghoff AS, et al. Brain metastases: pathobiology and emerging targeted therapies. Acta Neuropathol 2012;123:205–222; Winkler F. The brain metastatic niche. J Mol Med 2015;93:1213–1220; Aravantis CD, Gino GB, Jain RK. The blood-brain barrier and blood tumour barrier in brain tumours and metastases. Nat Rev Cancer 2020;20:26–41.

or the more dynamic concept of the neurovascular unit (NVU), tumor cells acquire an arsenal to successfully penetrate this barrier.[28,29] Mentioned briefly are some key molecules in this milieu that facilitate the above process. Selectins (P—on platelets and endothelial cells; L—on leukocytes; and E—on endothelial cells) provide anchors to selectin-ligand expressing tumor cells (leukocyte mimicry) or their natural interactions in above-mentioned cells, acting as bridges for transmigration of tumor cells through the BBB. Binding to platelets via P-selectin may also help tumor cells evade immune detection. Integrins are another class of molecules facilitating cell–cell interaction, and certain types of integrins are expressed at higher levels in brain metastases over

bone metastases of nonsmall cell lung cancer. They can cause powerful downstream signaling, such as upregulation of VEGF, which decreases endothelial integrity and facilitates transendothelial migration. Other implicated molecules include chemokines (CXCR4 and its ligand CXCL12), COX2, HBEGF, and ST6GALNAC5.[30]

Survival in brain microenvironment and growth: After crossing the BBB, the tumor cells reside close to the blood vessels, in the "brain metastatic niche." In an elegant study by Winkler et al., in vivo imaging showed that tumor cells not in direct contact with the blood vessels failed to survive and proliferate.[27] This perivascular niche is a fascinating haven where the tumor cells can adapt to survive the alien environment, remain dormant, evade immune detection, and begin to proliferate. ECM-modifying molecules, heparinase and MMPs, are expressed by tumor cells to create a less hostile environment. In vivo imaging has shown that both melanoma and lung cancer cells can remain dormant after crossing the BBB and, in parallel to evidence from gliomas, acquire "stemness" due to a complex interplay of many factors in this area including VEGF, NO, CXCL12, and NOTCH signaling.

Role of resident CNS cells: Although astrocytes play a macroscopic "protective" role by surrounding tumor foci with reactive gliosis and some paracrine secretion that is detrimental to tumor cells, they are often a double-edged sword. Heparinase production by astrocytes promotes degradation of the ECM, in addition to secreted factors that increase MMP activity. Paracrine secretions from astrocytes can increase stemness and proliferation. Lin et al. demonstrated that reactive astrocytes protect melanoma cells from chemotherapy.[31] Microglia, when activated, can be tumoricidal. Melanoma cells express neurotrophin receptors, signaling from which promotes migration and growth.

Growth pattern: Most brain metastases grow to form a distinct mass with a well-demarcated border (Fig. 5). Some tumors, like melanoma, demonstrate prominent vascular co-option and are also frequently hemorrhagic (Fig. 6). Leptomeningeal metastasis often continues along the perivascular spaces of cerebral vasculature (Fig. 7). Rarely, small cell lung cancer can diffusely invade the brain in a pseudo-gliomatous pattern. And more recently, an entirely intravascular spread has been described.[24]

Angiogenesis: Sustained growth of tumor foci requires an adequate blood supply. Melanoma cells can achieve this by using preexisting brain vessels (i.e., vascular co-option) mediated via L1CAM. In contrast, the formation of new vessels is an early and requisite event in nonsmall cell lung cancer metastases (i.e., neoangiogenesis). Tumor angiogenesis is predominantly powered by the activation of the VEGF pathway.

Genetics and molecular pathology: In the era of personalized medicine, molecular characterization of the patient's tumor guides the selection of targeted therapies, and knowledge of the genetic landscape of metastases can be

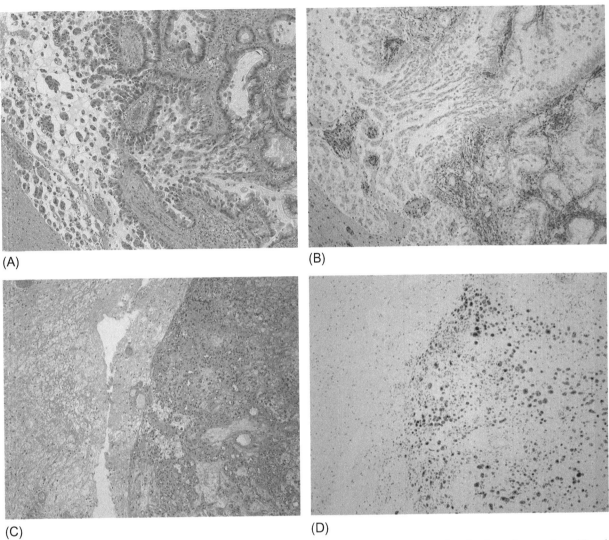

FIG. 3   Lineage-specific markers in metastases from more exotic primary cancers: (A, B) pancreas, showing loss of expression of Smad4; and (C, D) thyroid, positive for thyroglobulin.

very helpful. For the most part, the genetic alterations, especially driver mutations, are similar in the primary and metastatic tumors, and hence, an adequate sampling of the metastatic focus to perform molecular characterization is imperative if the primary tumor material is unavailable or cannot be biopsied. Depending on the availability of resources and amount of tumor tissue, a comprehensive tumor sequencing assay (NGS panel) or a targeted tumor-specific test (BRAF v600e mutation for melanoma) can be performed. Metastatic tumors are also commonly tested for PD-L1 and microsatellite instability, which guides the use of immune checkpoint inhibitors in the treatment plan.

As detailed molecular analyses of pathologic material from various cancers become the norm, trends of certain types with predilection for brain metastasis are revealed,

e.g., Her2-amplified or triple-negative breast cancer, or Alk-rearranged/Alk-positive nonsmall cell lung carcinoma. Seminal work by Bos et al. analyzing genes altered in breast cancer metastasis to the brain identified COX2, EGFR ligand, HBEGF, and ST6GALNAC5 as those that potentially mediate formation of metastatic deposits.[30] Of these, the first two are associated with lung metastases, while the latter are thought to be a site-specific adaptation of malignant cells to their environment. ST6GALNAC5, which is normally restricted to the brain, is expressed by breast cancer cells to increase invasiveness or transmigration through the BBB. A recent multicenter study comparing profiles of various primary and metastatic brain tumors highlighted enrichment of several markers, including TOP2A (DNA transcription and regulation) in primary tumors and RRM1, ERCC1, and TS (DNA

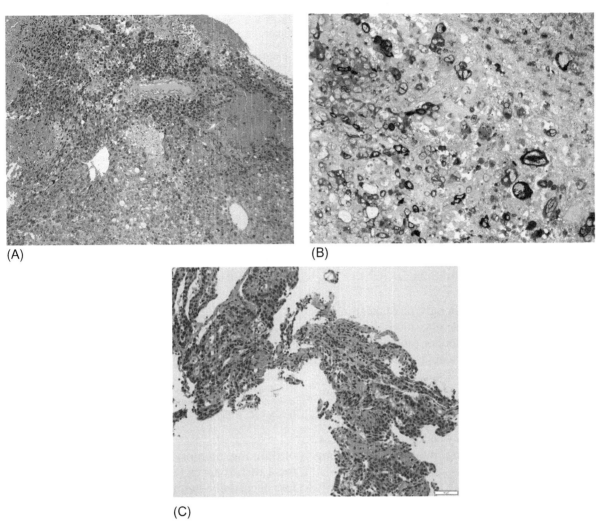

FIG. 4 Focal transdifferentiation in primary cancers may lead to a different phenotype in metastases: (A) hemorrhagic brain metastasis with morphologic features consistent with choriocarcinoma; (B) positive for HCG; (C) biopsy of esophageal mass from the same patient showing adenocarcinoma.

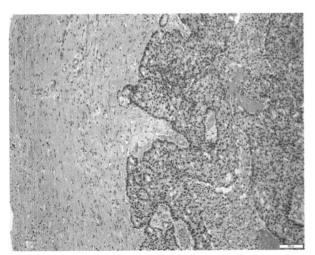

FIG. 5 Metastatic tumors usually have a sharp demarcation to surrounding brain parenchyma, in contrast to infiltrating primary brain tumors.

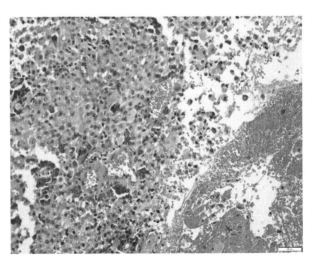

FIG. 6 Melanoma with hemorrhage and pigment production. Prominent nucleoli and nuclear pseudoinclusions as seen in this image are additional helpful features, especially during intraoperative evaluation.

I. Metastatic neurological complications

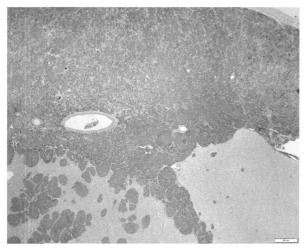

FIG. 7   Leptomeningeal disease from breast cancer, with spread into the brain parenchyma along perivascular spaces.

replication, repair, and chemotherapy resistance) in metastases.[32] In addition to genetic alterations (mainly involving DNA repair, cell adhesion, or canonical pathway drivers), cancer cells also invoke epigenetic changes and microRNAs to increase their metastatic potential.[33]

Treatment implications: The efforts to understand each step in the process of metastatic tumor formation, and the complex interactions at each step, ultimately offer avenues to inhibit various key components in the cascade, in the hopes of preventing, retarding, and killing metastatic tumor cells.

3 Clinical presentation

In more than two-thirds of patients with metastatic brain lesions, the tumors will produce a variety of neurological symptoms or signs that can be focal or generalized in character (see Table 3).[1,5] The most common symptoms are headache, alterations of mental status, and focal weakness. The headaches are usually generalized, often occur during sleep or in the morning, and become progressively more severe. If the patient has a history of a certain type of headache (e.g., migraine, tension), the tumor-related headache will be different in quality and intensity. Alterations of mental status are variable and include lethargy, loss of interest in activities, irritability, confusion, and memory loss. Changes in cognition and personality are most often noted by family and friends and may not be apparent to the patient. Weakness will vary depending on the location of the tumor; however, a hemiparetic pattern is most common. Seizures can be generalized, focal (e.g., hemimotor, arm, leg, or face), or both. On neurological examination, hemiparesis and impaired cognition are most common, each noted in over 50% of patients (see Table 3). Alterations of cognition include impairment

TABLE 3   Symptoms and signs in patients with metastatic brain tumors.

Symptom	%	Sign	%
Headache	25–40	Hemiparesis	55–60
Altered mental status	20–25	Impaired cognition	55–60
Focal weakness	20–30	Sensory loss	20
Seizure activity	15–20	Papilledema	20
Gait disturbance	10–20	Gait abnormality	15–20
Speech difficulty	5–10	Aphasia	15–20
Visual disturbance	5–8	Hemianopsia	5–7
Sensory disturbance	5	Limb ataxia	5–7
Nausea/vomiting	5	Somnolence	5
None	5–10		

Data derived from Valiente M, Ahluwalia MS, Boire A, Brastianos PK, et al. The evolving landscape of brain metastasis. Trends Cancer 2018;4:176–196; Takei H, Rouah E, Ishida Y. Brain metastasis: clinical characteristics, pathological findings and molecular subtyping for therapeutic implications. Brain Tumor Pathol 2016;33:1–12; Langer CJ, Mehta MP. Current management of brain metastases, with a focus on systemic options. J Clin Oncol 2005;23:6207–6219; Bajaj GK, Kleinberg L, Terezakis S. Current concepts and controversies in the treatment of parenchymal brain metastases: improved outcomes with aggressive management. Cancer Invest 2005;23:363–376; Lassman AB, DeAngelis LM. Brain metastases Neurol Clin N Am 2003;21:1–23.

of thinking, memory loss, poor judgment, and various focal disturbances (e.g., dyscalculia, apraxia). Sensory loss is usually hemifocal, but may involve only one limb. Gait abnormalities are typically hemiparetic in patients with cerebral tumors or, less often, ataxic in patients with lesions of the cerebellum or brainstem. Aphasia may present as an expressive disturbance (i.e., Broca's type) with impaired, nonfluent speech output and retained comprehension, a receptive disturbance (i.e., Wernicke's type) with fluent, nonsensical speech output and poor comprehension, or a more global syndrome with both components.

4 Neuroimaging

The diagnosis of MBT can be confirmed with an enhanced computerized tomography (CT) or magnetic resonance imaging (MRI) scan (see Figs. 8 and 9).[1–5,34] On both CT and MRI, metastatic tumors are typically rounded, well-circumscribed, noninfiltrative, and surrounded by a large amount of edema. With administration of contrast, enhancement (homogeneous or ringlike) is almost always present. The enhancing nodules can be quite variable in size, ranging from punctate lesions a few mm in diameter to large masses several cm across. In some patients, a miliary pattern may occur, with numerous small enhancing nodules

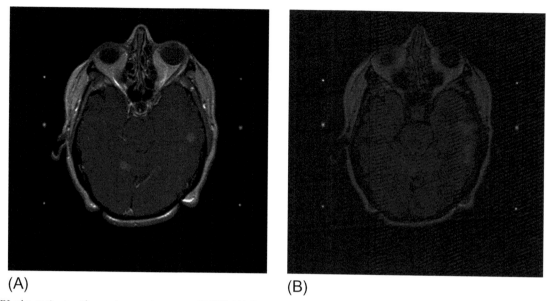

FIG. 8   MRI of a patient with ovarian carcinoma and MBT. (A) T1-weighted, gadolinium-enhanced axial image demonstrating two diffusely enhancing nodules of tumor within the left temporal lobe and right superior cerebellum. (B) Axial FLAIR image demonstrating high signal abnormality surrounding the lesions.

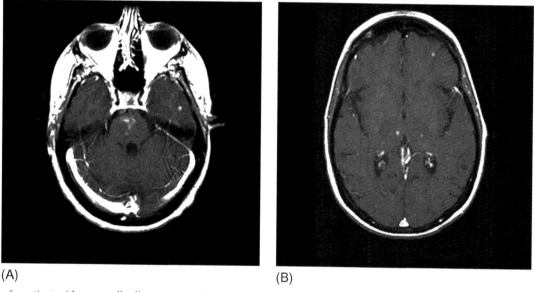

FIG. 9   MRI of a patient with nonsmall cell carcinoma of the lung and MBT. (A) T1-weighted, gadolinium-enhanced axial image demonstrating enhancing nodules of tumor within the pons and the left temporal lobe. (B) T1-weighted, gadolinium-enhanced axial image revealing small nodules of tumor affecting the left frontal lobe and both thalami.

scattered throughout the brain. Although CT remains an excellent screening tool, MRI is more sensitive to multifocal and small tumors, as well as to lesions in the cerebellum and brainstem. The neuroimaging differential diagnosis of brain metastases, especially solitary lesions, includes primary brain tumors, abscesses, infarcts, and hemorrhage.[1,5,34] In certain patients, a surgical biopsy may be necessary to definitively establish the diagnosis.

5 Surgical therapy

In the modern era of neurosurgery, there is now an important role for surgical resection of MBT, in carefully selected patients.[35–38] Surgical removal should be considered in all patients with an MRI-documented solitary metastasis. Unfortunately, this constitutes only 25%–35% of all patients. Among those patients with solitary lesions, only half will be appropriate for surgery because of factors such as inaccessibility of the tumor (e.g., brainstem, eloquent cortex), extensive systemic tumor burden, or other medical problems (e.g., cardiac ischemia, pulmonary insufficiency). Using second-generation image-guided, neuronavigation systems with frameless stereotaxy, patients with MBT can undergo aggressive surgical resection with significantly less risk for neurological injury.[39] In a review of 49 patients by Tan and Black, the use of image-guided craniotomy allowed for a gross total resection of the tumor and complete resolution of symptoms in 96% and 70% of the cohort, respectively. Neurological deterioration was only noted in 2 patients (3.6%), in whom significant deficits were present prior to surgery. The median survival for the entire group was 16.2 months, with a local recurrence rate of 16%. When neuronavigation and image guidance are integrated with intraoperative magnetic resonance imaging (iMRI), the extent of surgical resection can be monitored and maximized in the operating room.[40,41] This often allows for a more complete resection of tumor and the potential for improved local control.

Class I evidence from two phase III trials is available to support the use of surgical resection in MBT patients.[35–38,42,43] In the seminal study by Patchell and colleagues, 48 patients with solitary MBT were randomly assigned to receive surgical resection plus irradiation versus irradiation alone.[42] Local recurrence at the site of the original metastasis was significantly less frequent in the surgical cohort in comparison with the irradiation alone cohort (20% versus 52%; $P < .02$). Overall survival was significantly longer in the surgical group (median 40 weeks versus 15 weeks; $P < .01$). In addition, functional independence was maintained longer in the surgical cohort (median 38 weeks versus 8 weeks; $P < .005$). In a similar European phase III trial, 63 evaluable patients with solitary MBT were randomized to receive hyperfractionated irradiation (200 cGy × 2 per day; total of 4000 cGy) with or without surgical resection.[43] The overall survival was significantly longer in the surgical cohort (median 10 months versus 6 months; $P = .04$). A survival advantage was also noted for the surgical group in the 12-month (41% versus 23%) and 24-month (19% versus 10%) overall survival rates. The effect of the surgical procedure on survival was most pronounced in the patient cohort with stable systemic disease, with significant differences in overall survival (median 12 months versus 7 months; $P = .02$), 12-month survival rate (50% versus 24%), and 24-month survival rate (27% versus 10%). For patients with active systemic disease, the surgical resection and irradiation alone cohorts had the same median overall survival (5 months). One negative phase III trial has been reported by Mintz and co-workers, in their review of 84 patients randomized to receive irradiation with or without surgical resection.[44] The overall survival was similar between the surgical and irradiation alone groups

(median 5.6 months versus 6.3 months; $P = .24$). There was also no difference between treatment cohorts in the ability of patients to maintain Karnofsky Performance Status equal to or above 70%. However, it should be mentioned that this study had several methodological shortcomings, including the fact that 73% of all patients had poorly controlled systemic disease, there was an unequal distribution of primary pathologies between treatment cohorts (i.e., more radioresistant colorectal cancer in the surgical group and more radiosensitive breast cancer in the irradiation alone group), and nonuniform calculation of survival times.[38]

There is also Class II and Class III evidence to support the use of surgical resection for selected patients with a solitary MBT, mainly reflecting individual institutional experience.[34–38,45–49] This has been demonstrated in patients with solitary MBT from various types of primary tumors, including those from lung, breast, colon and rectum, melanoma, renal cell, and others. In general, these studies also demonstrate improved local control rates and longer survival in patients with solitary, accessible MBT that receive surgical resection followed by external beam irradiation.

For patients with multiple MBT, the use of surgical treatment is more controversial and remains unclear.[34–38] Some authors advocate the removal of all metastatic tumors, if the lesions are accessible and not located in eloquent regions of brain.[50] Using this approach with carefully selected patients, the survival can be similar to that of patients undergoing surgery for solitary metastases. Other authors suggest limiting the use of surgical resection for the "dominant or symptomatic" lesion, if it is accessible.[37,49] The smaller and less symptomatic tumors can then be controlled by postoperative irradiation.

6 Radiation therapy

Whole brain external beam irradiation (WBRT) remains the primary form of therapy for the majority of patients with brain metastases.[1–5,51–53] It is still the treatment of choice for tumors that are located in eloquent cortex or are too large or too numerous for surgical resection or radiosurgical approaches. Early randomized trials in the 1970s and 1980s by the Radiation Therapy Oncology Group (RTOG) and others evaluated variable dosing (10–54.4 Gy) and fractionation (1–34 fractions) schemes, in an attempt to determine the optimal therapeutic regimen.[52,53] The median survival across all studies ranged from 2.4 to 4.8 months, thereby proving that differences in dosing, timing, and fractionation schedules did not significantly influence the results in MBT patients. Objective tumor responses (i.e., CR, PR, MR) were noted in approximately 60% of patients in the randomized RTOG trials. The most widely used WBRT regimen delivers a total of 30 Gy in ten 3 Gy fractions over two weeks. Although this dose has limited potential for long-term tumor control, it is well-tolerated and designed to minimize the neurotoxicity associated with WBRT. An analysis of RTOG clinical trial data suggests that this regimen can provide control of disease in roughly 35% of patients at 6 months. After receiving WBRT, most MBT patients note an improvement or stabilization of neurologic symptoms, including headache, seizures, impaired mentation, cerebellar dysfunction, and motor deficits.[52]

A randomized trial has also evaluated the utility of WBRT in the context of patients with a solitary MBT that have undergone surgical resection.[54] In this study, 95 patients with solitary MBT were treated with complete surgical resection and then randomized into a postoperative radiotherapy group or an observation group. The overall recurrence rate of MBT anywhere in the brain was significantly reduced in the radiotherapy group (18% versus 70%; $P < .001$). Postoperative WBRT was able to reduce the rate of MBT recurrence at the site of the original metastasis (10% versus 46%; $P < .001$) and at distant sites in the brain (14% versus 37%; $P < .01$). In addition, patients in the radiotherapy cohort were less likely to die of neurological causes than patients in the observation group (14% versus 44%; $P = .003$). However, there was no significant difference between groups in terms of the overall length of survival or the length of time that patients were able to maintain functional independence. This is not surprising since one would not expect WBRT to have any effect on the course of the systemic cancer.

Prophylactic cranial irradiation (PCI) is an "up-front" application of WBRT that is only appropriate for consideration in selected patients with lung cancer. The efficacy of PCI was first demonstrated in patients with small cell lung cancer (SCLC), especially those with well-controlled systemic disease.[55,56] Initial reports demonstrated a survival benefit of 5.4% at 3 years, with a 25.3% reduction in the cumulative incidence of MBT in the cohort of patients achieving a complete systemic remission with chemotherapy.[55] A subsequent analysis of 505 patients that had participated in randomized trials has further characterized the benefit of PCI in SCLC patients.[56] The 5-year cumulative incidence of MBT as an isolated first site of relapse was 20% in the PCI cohort and 37% in control patients ($P < .001$). The overall 5-year incidence of MBT for the PCI and control groups was 43% and 59%, respectively (relative risk [RR] 0.50; $P < .001$). However, the effect on overall survival was modest, with 5-year rates for the PCI and control groups of 18% and 15%, respectively (RR 0.84; $P = .06$). Presumably, this is because the majority of SCLC patients ultimately die of systemic metastases, an issue not addressed by PCI. Prophylactic cranial irradiation has also been investigated in patients with nonsmall cell lung cancer (NSCLC), but with less compelling evidence of benefit.[57,58] Although there does appear to be a reduction in the incidence of MBT in the PCI cohorts, no survival benefit has been observed. This view is consistent with a Cochrane Review of the use of PCI in NSCLC patients.[59] The authors concluded that there was insufficient evidence at this time to recommend the use of PCI in clinical practice and that it should only be offered in the context of a clinical trial.

Stereotactic radiosurgery (SRS) is a method of delivering focused irradiation to the boundaries of a tumor (i.e., conformal dosing), in a single or few fractions, using great precision.[51–53,60–64] SRS has become an important therapeutic option for brain metastases for several reasons, including the fact that most MBT are spherical and small at the time of diagnosis, the degree of infiltration into surrounding brain is usually quite limited, the gray-white matter junction is considered a relatively "noneloquent" area of the brain, and improved local control in the brain may extend patient survival. The treatment is most often administered using a Gamma Knife (i.e., Co[60] sources); however, linear accelerator (e.g., CyberKnife) and proton beam units are also used and demonstrate comparable local control and complication rates. SRS is most effective for tumors less than or equal to 3 cm in diameter. However, some authors recommend treatment of tumors up to 4 cm in diameter. Typical doses are in the range of 15 to 20 Gy to the margins of the tumor, with higher doses administered at the center of the mass. Optimal dosing will depend on the size of the tumor, previous exposure to irradiation, and proximity to delicate neural structures (e.g., optic chiasm).

There are two reports that provide Class I evidence for the efficacy of SRS in the context of a boost to WBRT.[65,66] In the first study from the University of Pittsburgh, 27 patients with two to four MBT were randomized to receive WBRT (30 Gy over 12 fractions) plus SRS (tumor margin dose of 16 Gy) or WBRT alone.[65] Local control was improved by the use of the SRS boost, with local failure rates at 1 year of 8% for the combined treatment group and 100% for the WBRT alone group. The median time to local failure was 36 months for the WBRT plus SRS cohort and 6 months for the WBRT alone group ($P = .0005$). In addition, median time to overall brain failure (local or distant) was longer for the combined treatment cohort in comparison with the WBRT alone group (34 months versus 5 months; $P = .002$). However, the addition of the SRS boost did not significantly influence overall survival between the two groups (11 months versus 7.5 months, respectively; $P = .22$). Again, this lack of effect on overall survival could simply reflect the effect of systemic metastases in these patients. In a similar study by the RTOG (RTOG 9508), 333 patients with one to three MBT were randomized to receive either WBRT (37.5 Gy over 15 fractions) or WBRT plus a SRS boost of 15 to 24 Gy, depending on tumor size.[66] Local control at 1 year was significantly better for the SRS group in comparison with the WBRT alone group (82% versus 71%; $P = .01$). In addition, time to local progression was extended in the combined treatment cohort ($P = .0132$). Overall median survival was similar between groups; however, for patients with a single MBT, median survival was longer in the WBRT plus SRS cohort (6.5 months versus 4.9 months; $P = .0393$). The Karnofsky Performance Status (KPS) was more likely to be stable or improved at 6 months follow-up in the WBRT plus SRS group (43% versus 27%; $P = .03$). This is consistent with the multivariate analysis, which demonstrated improved survival in patients with RPA Class 1 disease ($P < .0001$).

In a variation on the studies cited above, another randomized, controlled trial administered up-front stereotactic radiosurgery to a series of 132 patients with one to four MBT (up to 3 cm in diameter), followed by WBRT in half of the cohort.[67] The median survival time and 1-year actuarial survival rate were similar between the SRS alone and SRS + WBRT groups (8.0 months and 28.4% versus 7.5 months and 38.5%, respectively; $P = .42$). However, the 12-month MBT recurrence rates were significantly different between the two cohorts (SRS alone—76.4% versus SRS + WBRT—46.8%; $P < .001$). In addition, salvage brain treatment was less frequently required in the SRS + WBRT group (10 versus 29; $P < .001$). No difference was noted between groups in terms of systemic and neurological functional preservation or toxicity from radiation treatments. The authors concluded that the addition of WBRT to SRS did not improve survival in this group of patients, but did significantly reduce the local and distant relapse rate.

There are numerous reports in the literature describing Class II and Class III evidence supporting the use of SRS for the treatment of MBT.[51–53,60–64] A review of the larger trials (i.e., 100 or more patients) would suggest that SRS is as effective as, if not more effective than, WBRT.[68–79] In most of the studies, the median survival ranged between 5.5 and 13.5 months, with overall local control rates of 85%–95%. The increase in local control rates did not translate into an improvement in survival, with most patients dying of systemic disease progression. Several factors have been found to influence the degree of local control, including primary tumor histology (e.g., melanoma versus lung carcinoma), tumor volume, tumor location, presentation (e.g., new versus recurrent), and pattern of MRI enhancement (e.g., homogeneous versus heterogeneous versus ring). Some authors are recommending the use of SRS as the primary, "up-front" mode of irradiation in high-performance patients with well-controlled systemic disease, instead of WBRT.[68–79] However, this view is not supported by the conclusions of an ASTRO meta-analysis of SRS treatment of MBT.[80] The ASTRO recommendations are to advise an SRS boost to WBRT in selected patients with one to four newly diagnosed MBT. The omission of WBRT results in significantly lower rates of local and distant brain control.

7 Chemotherapy

Chemotherapy has become a more viable option for the treatment of MBT in recent years, especially for recurrent disease.[81–87] The prior reluctance to use chemotherapy stemmed from concerns about the ability of chemotherapy drugs to cross the blood–brain barrier (BBB) and penetrate tumor cells, intrinsic chemoresistance of metastatic disease, and the high probability of early death from systemic progression. However, recent animal data suggest that metastatic tumors that strongly enhance on CT or MRI have an impaired BBB and will allow entry of chemotherapeutic drugs.[81,83] In addition, systemic resistance to a given drug does not always preclude sensitivity of the metastasis within the brain.[81] Several types of metastatic brain tumors are relatively chemosensitive and may respond, including breast cancer, small cell lung cancer, nonsmall cell lung cancer, germ cell tumors, and ovarian carcinoma.

The most common older approach to chemotherapy for brain metastases was to administer it "up-front," before or during conventional WBRT or SRS.[88–97] Several authors have demonstrated that combination regimens given intravenously can be active in this context. The most frequently used agents included cisplatin (CDDP), etoposide (VP16), and cyclophosphamide (CTX). In a series of 19 patients with small cell lung cancer and brain metastases, Twelves and co-workers used intravenous (IV) CTX, vincristine, and VP16 every three weeks before any form of irradiation.[88] Ten of the 19 patients (53%) had a radiological or clinical response. In 9 patients, there was CT evidence of tumor shrinkage, while in 1 patient, there was neurological improvement, without neuroimaging follow-up. The mean time to progression (TTP) was 22 weeks, with a median overall survival of 28 weeks. Cocconi and colleagues used up-front IV cisplatin and etoposide every three weeks for 22 evaluable patients with MBT from breast carcinoma.[89] There were 12 objective responses, for an overall objective response rate of 55%. The median TTP was 25 weeks overall and 40 weeks in the objective response cohort. Overall median survival was 58 weeks. The same authors have expanded their series to include 89 patients with MBT from breast, nonsmall lung carcinoma, and malignant melanoma.[90] Objective responses were noted in the breast and lung cohorts. None of the patients with melanoma had objective responses. The overall objective response rate was 30% (34/89). Median TTP was 15 weeks, with a median survival for the cohort of 27 weeks. Similar responses have been noted in a series of patients with MBT from lung and breast carcinoma.[91–97] However, although objective responses were noted in many of these studies, they did not translate into improvements in patient survival.

Topotecan is a semisynthetic camptothecin derivative that selectively inhibits topoisomerase I in the S phase of the cell cycle.[98] It demonstrates excellent penetration of the BBB in primate animal models and humans. Summating the data of more than 60 patients in several European studies of single-agent topotecan, the objective response rates have been encouraging, with 30%–60% of patients demonstrating a complete response (CR) or partial response (PR).[99–102] Topotecan is also being investigated in combination with radiotherapy and other cytotoxic chemotherapy agents, such as temozolomide. A phase I trial has evaluated the tolerability of temozolomide ($50–200\,mg/m^2$) and topotecan ($1–1.5\,mg/m^2$), given daily for five days every 28 days.[103] Twenty-five patients with systemic solid tumors were treated. Toxicity was mainly hematological, with frequent neutropenia and thrombocytopenia. Three patients were noted to have a PR.

Temozolomide is an imidazotetrazine derivative of the alkylating agent dacarbazine with activity against systemic and CNS malignancies.[83,103–105] The drug undergoes chemical conversion at physiological pH to the active species 5-(3-methyl-1-triazeno)imidazole-4-carboxamide (MTIC). Temozolomide exhibits schedule-dependent antineoplastic activity by interfering with DNA replication through the methylation of DNA at the following sites: N^7-guanine (70%), N^3-adenine (9.2%), and O^6-guanine (5%). Several reports have suggested activity of single-agent temozolomide against MBT, with occasional objective responses.[106,107] Temozolomide is also under investigation as a radiation sensitizer, including a randomized phase II trial by Antonadou and associates.[108] In this study, 52 newly diagnosed MBT patients (lung and breast) were treated with either WBRT alone (40 Gy) or WBRT plus conventional temozolomide. The addition of temozolomide improved the objective response rate when compared to WBRT alone (CR 38%, PR 58% versus CR 33%, PR 33%). In addition, neurologic improvement during treatment was more pronounced in the cohort of patients receiving chemotherapy. A similar randomized phase II trial by Verger and colleagues treated 82 patients with MBT (mostly lung and breast) using combined WBRT (30 Gy) and temozolomide ($75\,mg/m^2/day$ during irradiation, plus two cycles of conventional adjuvant dosing) versus WBRT alone.[109] The objective response rate and overall survival were similar between treatment groups. However, there was a significantly higher rate of progression-free survival at 90 days in the combined treatment cohort (72% versus 54%, $P = .03$). In addition, the percentage of patients dying from the MBT was lower in the chemotherapy arm (41% versus 69%; $P = .03$). Temozolomide has also been shown to have activity, as a single agent and in combination with other drugs (e.g., cisplatin, docetaxel, thalidomide), against MBT from malignant melanoma.[110–113]

In an effort to improve dose intensity to MBT, some authors have given some or all of the chemotherapy drugs by the intra-arterial (IA) route.[83,114–118] There are several advantages to administering chemotherapy IA instead of by the conventional IV route, including augmentation of the peak concentration of drug in the region of the tumor and an increase in the local area under the concentration-time curve.[114] Pathologically, metastatic brain tumors are excellent candidates for IA approaches, because they tend to be well-circumscribed and noninfiltrative.[1] In addition, MBT almost always enhance on MRI imaging, indicating excellent arterial vascularization and impairment of the BBB. Pharmacologic studies using animal models of IA and IV drug infusion have shown that the IA route can increase the intra-tumoral concentration of a given agent by at least a factor of threefold to fivefold.[119,120] For chemosensitive tumors, improving the intra-tumoral concentrations of drug should augment tumor cell kill and the ability to achieve objective responses.[114] Initial applications of IA chemotherapy to MBT involved the use of BCNU and cisplatin.[115–118] Although objective responses were noted in patients with lung and breast tumors, significant neurotoxicity occurred (e.g., seizures, confusion). More recent reports have used carboplatin as the primary IA agent and have resulted in similar objective response rates, with significantly less neurotoxicity.[121–123]

The recent expansion of knowledge regarding the molecular biology of neoplasia and the metastatic phenotype has led to intense development of therapeutic strategies designed to exploit this new information.[1,124] Numerous targets of therapeutic intervention have been developed, including growth factor receptors and their tyrosine kinase activity, disruption of aberrant internal signal transduction pathways, inhibition of excessive matrix metalloproteinase activity, downregulation of cell cycle pathways, and manipulation of the apoptosis pathways.[124] The most promising approach thus far has been the development of small-molecule drugs or monoclonal antibodies to the major growth factor receptors (e.g., PDGFR, EGFR, Her2, CD20).[125–129] Monoclonal antibody agents such as rituximab (i.e., Rituxan) and trastuzumab (i.e., Herceptin) have proven to be clinically active against non-Hodgkin's lymphoma and breast cancer, respectively. Several first-generation small-molecule inhibitors of the tyrosine kinase activity of the EGFR (e.g., gefitinib, erlotinib) have been evaluated in clinical trials of patients with solid tumors.[1,126–128] Similar efforts have been made to develop agents that can target the tyrosine kinase activity of PDGFR and the ras signaling pathway.[1,128,129] An early report using imatinib, a tyrosine kinase inhibitor with activity against C-KIT and PDGFR, describes a 75-year-old male with a C-KIT positive GI stromal tumor that developed neurological deterioration and gait difficulty.[130] An MRI demonstrated leptomeningeal disease with brain infiltration and edema. After treatment with imatinib mesylate (400 mg bid) for 2 months, his neurological function and gait improved. A follow-up MRI scan revealed complete resolution of the meningeal and intra-parenchymal abnormalities. Several authors have described case reports of the use of gefitinib, an oral tyrosine kinase inhibitor of EGFR, in patients with MBT from NSCLC.[130–134] A few of these initial patients had objective responses, including CR, that were quite durable. These early reports lead Ceresoli and colleagues to perform a prospective phase II trial of gefitinib in patients with MBT from NSCLC.[135] Forty-one consecutive patients were treated with gefitinib (250 mg/day); 37 had received prior chemotherapy and 18 had undergone WBRT. There were four patients with a PR and seven with SD. The overall progression-free survival was only 3 months. However, the median duration of responses in the patients with a PR was an encouraging 13.5 months.

More recent reports of targeted treatment of brain metastases have shown increasing efficacy and durability when using this approach (for additional discussion of this topic, see Chapter 4).[1,124,136–138] This includes patients with brain metastases from nonsmall cell lung cancer, breast cancer, melanoma, and others. For example, in nonsmall cell lung cancer, there are frequent mutations noted in EGFR, as well as fusions that activate ALK, in the primary tumors.[139] Brain metastases from EGFR mutant primary tumors will typically maintain these main driver mutations, as well as additional mutations and alterations in some cases. Treatment with first- and second-generation EGFR tyrosine kinase inhibitors (TKIs) such as erlotinib, gefitinib, and afatinib has been shown to have modest activity in this setting.[1,124,136–139] Third-generation TKIs (e.g., osimertinib) have also shown direct activity against brain metastases from EGFR mutant primary lung tumors.[137,139] In patients with ALK-activated lung cancer and brain metastases, newer generation TKIs such as alectinib have demonstrated excellent CNS activity, with intracranial response rates of 70% or higher in some studies.[124,136–139] In fact, some authors are now questioning the use of WBRT in this group of patients, since the brain metastases are so responsive to anti-ALK TKIs.[140] In patients with HER2-positive breast cancer and brain metastases, lapatinib (dual HER2 and EGFR TKI) has shown some activity in the CNS.[136–138,141] In monotherapy studies, lapatinib has demonstrated a CNS objective response rate of 6%, with increased efficacy when used in combination with other drugs (e.g., capecitabine). Melanoma patients often have BRAF-mutant disease, which can respond well to BRAF inhibitors such as vemurafenib and dabrafenib.[136–138] In a prospective multicenter phase II trial, 172 patients with BRAF-mutant melanoma and brain metastases were treated with dabrafenib and had intracranial response rates of 39.2% in the treatment-naïve cohort and 30.8% in the previously treated cohort.[142]

Some authors are also endorsing the concept of performing a biopsy of brain metastases whenever possible, in order to perform advanced molecular phenotyping of the tumor and then a direct comparison to the primary neoplasm (see Chapter 4).[138] In metastatic cases with new driver mutations, more accurate targeted therapy can then be delivered.

References

1. Valiente M, Ahluwalia MS, Boire A, et al. The evolving landscape of brain metastasis. *Trends Cancer.* 2018;4:176–196.
2. Takei H, Rouah E, Ishida Y. Brain metastasis: clinical characteristics, pathological findings and molecular subtyping for therapeutic implications. *Brain Tumor Pathol.* 2016;33:1–12.
3. Langer CJ, Mehta MP. Current management of brain metastases, with a focus on systemic options. *J Clin Oncol.* 2005;23:6207–6219.
4. Bajaj GK, Kleinberg L, Terezakis S. Current concepts and controversies in the treatment of parenchymal brain metastases: improved outcomes with aggressive management. *Cancer Invest.* 2005;23:363–376.
5. Lassman AB, DeAngelis LM. Brain metastases. *Neurol Clin N Am.* 2003;21:1–23.
6. Gavrilovic IT, Posner JB. Brain metastases: epidemiology and pathophysiology. *J Neurooncol.* 2005;75:5–14.
7. Barnhholtz-Sloan JS, Sloan AE, Davis FG, Vigneau FD, Lai P, Sawaya RE. Incidence proportions of brain metastases in patients diagnosed (1973 to 2001) in the metropolitan detroit cancer surveillance system. *J Clin Oncol.* 2004;22:2865–2872.
8. Leroux PD, Berger MS, Elliott JP, Tamimi HK. Cerebral metastases from ovarian carcinoma. *Cancer.* 1991;67:2194–2199.
9. Martinez-Manas RM, Brell M, Rumia J, Ferrer E. Case report. Brain metastases in endometrial carcinoma. *Gynecol Oncol.* 1998;70:282–284.
10. Mccutcheon IE, Eng DY, Logothetis CJ. Brain metastasis from prostate carcinoma. Antemortem recognition and outcome after treatment. *Cancer.* 1999;86:2301–2311.
11. Lowis SP, Foot A, Gerrard MP, et al. Central nervous system metastasis in Wilms' tumor. A review of three consecutive United Kingdom trials. *Cancer.* 1998;83:2023–2029.
12. Culine S, Bekradda M, Kramar A, et al. Prognostic factors for survival in patients with brain metastases from renal cell carcinoma. *Cancer.* 1998;83:2548–2553.
13. Qasho R, Tommaso V, Rocchi G, et al. Choroid plexus metastasis from carcinoma of the bladder: case report and review of the literature. *J Neurooncol.* 1999;45:237–240.
14. Salvati M, Frati A, Rocchi G, et al. Single brain metastasis from thyroid cancer: report of twelve cases and review of the literature. *J Neurooncol.* 2001;51:33–40.
15. Ogawa K, Toita T, Sueyama H, et al. Brain metastases from esophageal carcinoma. Natural history, prognostic factors, and outcome. *Cancer.* 2002;94:759–764.
16. Espat NJ, Bilsky M, Lewis JJ, et al. Soft tissue sarcoma brain metastases. Prevalence in a cohort of 33829 patients. *Cancer.* 2002;94:2706–2711.
17. Schouten LJ, Rutten J, HAM H, Twijnstra A. Incidence of brain metastases in a cohort of patients with carcinoma of the breast, colon, kidney, and lung and melanoma. *Cancer.* 2002;94:2698–2705.
18. Ruda R, Borgognone M, Benech F, Vasario E, Soffietti R. Brain metastases from unknown primary tumour. A prospective study. *J Neurol.* 2001;248:394–398.

19. Kebudi R, Ayan I, Görgün O, Agaoglu FY, Vural S, Darendeliler E. Brain metastasis in pediatric extracranial solid tumors: survey and literature review. *J Neurooncol.* 2005;71:43–48.

20. Hall WA, Djalilian HR, Nussbaum ES, et al. Long-term survival with metastatic cancer to the brain. *Med Oncol.* 2000;17:279–286.

21. Gaspar L, Scott C, Rotman M, et al. Recursive partitioning analysis (RPA) of prognostic factors in three radiation therapy oncology group (RTOG) brain metastases trials. *Int J Radiat Oncol Biol Phys.* 1997;37:745–751.

22. Nussbaum ES, Djalilian HR, Cho KH, Hall WA. Brain metastases: histology, multiplicity, surgery, and survival. *Cancer.* 1996;78:1781–1788.

23. Mut M, Schiff D, Shaffrey ME. Metastasis to the nervous system: spinal epidural and intramedullary metastases. *J Neurooncol.* 2005;75:43–55.

24. Chan J, Magaki S, Zhang XR, et al. Intravascular carcinomatosis of the brain: a report of two cases. *Brain Tumor Pathol.* 2020;37:118–125.

25. Kienast Y, von Baumgarten L, Fuhrmann M, et al. Real-time imaging reveals the single steps of brain metastasis formation. *Nat Med.* 2010;16:116–122.

26. Preusser M, Capper D, Ilhan-Mutlu A, et al. Brain metastases: pathobiology and emerging targeted therapies. *Acta Neuropathol.* 2012;123:205–222.

27. Winkler F. The brain metastatic niche. *J Mol Med.* 2015;93:1213–1220.

28. Aravantis CD, Gino GB, Jain RK. The blood-brain barrier and blood tumour barrier in brain tumours and metastases. *Nat Rev Cancer.* 2020;20:26–41.

29. Hasko J, Fazakas C, Molnar K, et al. Response of the neurovascular unit to brain metastatic breast cancer cells. *Acta Neuropathol Commun.* 2019;7(1):133. https://doi.org/10.1186/s40478-07688-1.

30. Bos PD, Zhang XH, Nadal C, et al. Genes that mediate breast cancer metastasis to the brain. *Nature.* 2009;459:1005–1009.

31. Lin Q, Balasubramanian K, Fan D, et al. Reactive astrocytes protect melanoma cells from chemotherapy by sequestering intracellular calcium through gap junction communication channels. *Neoplasia.* 2010;12:748–754.

32. Ferguson SD, Zheng S, Xiu J, et al. Profiles of brain metastes: prioritization of therapeutic targets. *Int J Cancer.* 2018;143:3019–3026.

33. Custodio-Santos T, Videira M, Brito MA. Brain metastasization of breast cancer. *Biochim Biophys Acta Rev Cancer.* 2017;1868:132–147.

34. Vossough A, Henson JW. Intracranial metastases. In: Newton HB, ed. *Handbook of Neuro-Oncology Neuroimaging.* vol. 52. 2nd ed. Amsterdam: Academic Press/Elsevier; 2016:643–652.

35. Carapella CM, Gorgoglione N, Oppido PA. The role of surgical resection in patients with brain metastases. *Curr Opin Oncol.* 2018;30:390–395.

36. Phang I, Leach J, Leggate JRS, et al. Minimally invasive resection of brain metastases. *World Neurosurg.* 2019;130:e362–e367.

37. Chua TH, See AAQ, Ang BT, King NKK. Awake craniotomy for resection of brain metastases: a systematic review. *World Neurosurg.* 2018;120:e1128–e1135.

38. Vogelbaum MA, Suh JH. Resectable brain metastases. *J Clin Oncol.* 2006;24:1289–1294.

39. Tan TC, Black PM. Image-guided craniotomy for cerebral metastases: techniques and outcomes. *Neurosurgery.* 2003;53:82–90.

40. Albayrak B, Samdani AF, Black PM. Intra-operative magnetic resonance imaging in neurosurgery. *Acta Neurochir.* 2004;146:543–556.

41. Nimsky C, Ganslandt O, von Keller B, Romstöck J, Fahlbusch R. Intraoperative high-field-strength MR imaging: implementaton and experience in 200 patients. *Radiology.* 2004;233:67–78.

42. Patchell RA, Tibbs PA, Walsh JW, et al. A randomized trial of surgery in the treatment of single metastases to the brain. *N Engl J Med.* 1990;322:494–500.

43. Vecht CJ, Haaxma-Reiche H, Noordijk EM, et al. Treatment of single brain metastasis: radiotherapy alone or combined with neurosurgery? *Ann Neurol.* 1993;33:583–590.

44. Mintz AH, Kestle J, Rathbone MP, et al. A randomized trial to assess the efficacy of surgery in addition to radiotherapy in patients with a single cerebral metastasis. *Cancer.* 1996;78:1470–1476.

45. Arbit E, Wronski M, Burt M, Galicich JH. The treatment of patients with recurrent brain metastases. A retrospective analysis of 109 patients with nonsmall cell lung cancer. *Cancer.* 1995;76:765–773.

46. Wronski M, Arbit E, McCormick B. Surgical treatment of 70 patients with brain metastases from breast cancer. *Cancer.* 1997;80:1746–1754.

47. Wronski M, Arbit E. Resection of brain metastases from colorectal carcinoma in 73 patients. *Cancer.* 1999;85:1677–1685.

48. O'Neill BP, Iturria NJ, Link MJ, Pollock BE, Ballman KV, O'Fallon JR. A comparison of surgical resection and stereotactic radiosurgery in the treatment of solitary brain metastases. *Int J Radiat Oncol Biol Phys.* 2003;55:1169–1176.

49. Paek SH, Audu PB, Sperling MR, Cho J, Andrews DW. Reevaluation of surgery for the treatment of brain metastases: review of 208 patients with single or multiple brain metastases treated at one institution with modern neurosurgical techniques. *Neurosurgery.* 2005;56:1021–1034.

50. Bindal RK, Sawaya R, Leavens ME, Lee JJ. Surgical treatment of multiple brain metastases. *J Neurosurg.* 1993;79:210–216.

51. Thiagarajan A, Yamada Y. Radiobiology and radiotherapy of brain metastases. *Clin Exp Metastasis.* 2017;34:411–419.

52. Lam TC, Sahgal A, Lo SS, Chang EL. An update on radiation therapy for brain metastases. *Chin Clin Oncol.* 2017;6(4):35. https://doi.org/10.21037/cco.2017.06.02.

53. Wang TJC, Brown PD. Brain metastases: fractionated whole-brain radiotherapy. *Handb Clin Neurol.* 2018;149:123–127.

54. Patchell RA, Tibbs PA, Regine WF, et al. Postoperative radiotherapy in the treatment of single metastases to the brain: a randomized trial. *JAMA.* 1998;280:1485–1489.

55. Aupérin A, Arriagada R, Pignon JP, et al. Prophylactic cranial irradiation for patients with small-cell lung cancer in complete remission. Prophylactic cranial irradiation overview collaborative group. *N Engl J Med.* 1999;341:476–484.

56. Arriagada R, Le Chevalier T, Rivière A, et al. Patterns of failure after prophylactic cranial irradiation in small-cell lung cancer: analysis of 505 randomized patients. *Ann Oncol.* 2002;13:748–754.

57. Laskin JJ, Sandler AB. The role of prophylactic cranial radiation in the treatment of non-small-cell lung cancer. *Clin Adv Hematol Oncol.* 2003;1:731–740.

58. Gore EM. Prophylactic cranial irradiation for patients with locally advanced non-small-cell lung cancer. *Oncologia.* 2003;17:775–779.

59. Lester JF, MacBeth FR, Coles B. Prophylactic cranial irradiation for preventing brain metastases in patients undergoing radical treatment for non-small-cell lung cancer: a cochrane review. *Int J Radiat Oncol Biol Phys.* 2005;63:690–694.

60. Specht HM, Combs SE. Stereotactic radiosurgery of brain metastases. *J Neurosurg Sci.* 2016;60:357–366.

61. Sahgal A, Ruschin M, Ma L, et al. Stereotactic radiosurgery alone for multiple brain metastases? A review of clinical and technical issues. *Neuro Oncol.* 2017;19(Suppl 2):ii2–ii15.

62. Hatiboglu MA, Tuzgen S, Akdur K, Chang EL. Treatment of high numbers of brain metastases with gamma knife radiosurgery: a review. *Acta Neurochir.* 2016;158:625–634.

63. McDermott MW, Sneed PK. Radiosurgery in metastatic brain cancer. *Neurosurgery.* 2005;57. S4-45–S4-53.

64. Bhatnagar AK, Flickinger JC, Kondziolka D, Lunsford LD. Stereotactic radiosurgery for four or more intracranial metastases. *Int J Radiat Oncol Biol Phys.* 2006;64:898–903.

65. Kondziolka D, Patel A, Lunsford LD, Kassam A, Flickinger JC. Stereotactic radiosurgery plus whole brain radiotherapy versus

radiotherapy alone for patients with multiple brain metastases. *Int J Radiat Oncol Biol Phys.* 1999;45:427–434.

66. Andrews DW, Scott CB, Sperduto PW, et al. Whole brain radiation therapy with or without stereotactic radiosurgery boost for patients with one to three brain metastases: phase III results of the RTOG 9508 randomised trial. *Lancet.* 2004;363:1665–1672.

67. Aoyama H, Shirato H, Tago M, et al. Stereotactic radiosurgery plus whole-brain radiation therapy vs stereotactic radiosurgery alone for treatment of brain metastases. A randomized controlled trial. *JAMA.* 2006;295:2483–2491.

68. Flickinger JC, Kondziolka D, Lunsford LD, et al. A multi-institutional experience with stereotactic radiosurgery for solitary brain metastases. *Int J Radiat Oncol Biol Phys.* 1994;28:797–802.

69. Alexander E, Moriarty TM, Davis RB, et al. Stereotactic radiosurgery for the definitive, noninvasive treatment of brain metastases. *J Natl Cancer Inst.* 1995;87:34–40.

70. Gerosa M, Nicolato A, Severi F, et al. Gamma knife radiosurgery for intracranial metastases: from local control to increased survival. *Stereotact Funct Neurosurg.* 1996;66:184–192.

71. Joseph J, Adler JR, Cox RS, Hancock SL. Linear accelerator-based stereotactic radiosurgery for brain metastases: the influence of number of lesions on survival. *J Clin Oncol.* 1996;14:1085–1092.

72. Pirzkall A, Debus J, Lohr F, et al. Radiosurgery alone or in combination with whole-brain radiotherapy for brain metastases. *J Clin Oncol.* 1998;16:3563–3569.

73. Chen JC, Petrovich Z, O'Day S, et al. Stereotactic radiosurgery in the treatment of metastatic disease to the brain. *Neurosurgery.* 2000;47:268–279.

74. Hoffman R, Sneed PK, McDermott MW, et al. Radiosurgery for brain metastases from primary lung carcinoma. *Cancer J.* 2001;7:121–131.

75. Gerosa M, Nicolato A, Foroni R, et al. Gamma knife radiosurgery for brain metastases: a primary therapeutic option. *J Neurosurg.* 2002;97:515–524.

76. Petrovich Z, Yu C, Giannotta SL, O'Day S, Apuzzo MLJ. Survival and pattern of failure in brain metastases treated with stereotactic gamma knife radiosurgery. *J Neurosurg.* 2002;97:499–506.

77. Hasegawa T, Kondziolka D, Flickinger JC, Germanwala A, Lunsford LD. Brain metastases treated with radiosurgery alone: an alternative to whole brain radiotherapy? *Neurosurgery.* 2003;52:1318–1326.

78. Lutterbach J, Cyron D, Henne K, Ostertag CB. Radiosurgery followed by planned observation in patients with one to three brain metastases. *Neurosurgery.* 2003;52:1066–1073.

79. Muacevic A, Kreth FW, Tonn JC, Wowra B. Stereotactic radiosurgery for multiple brain metastases from breast carcinoma. *Cancer.* 2004;100:1705–1711.

80. Mehta MP, Tsao MN, Whelan TJ, et al. The American Society for Therapeutic Radiology and Oncology (ASTRO) evidence-based review of the role of radiosurgery for brain metastases. *Int J Radiat Oncol Biol Phys.* 2005;63:37–46.

81. Newton HB. Chemotherapy of brain metastases. In: Newton HB, ed. *Handbook of Brain Tumor Chemotherapy, Molecular Therapeutics, and Immunotherapy.* vol. 41. 2nd ed. London: Elsevier Medical Publishers/Academic Press; 2018:527–546.

82. Lesser GJ. Chemotherapy of cerebral metastases from solid tumors. *Neurosurg Clin N Am.* 1996;7:527–536.

83. Newton HB. Chemotherapy for the treatment of metastatic brain tumors. *Expert Rev Anticancer Ther.* 2002;2:495–506.

84. Tosoni A, Lumachi F, Brandes AA. Treatment of brain metastases in uncommon tumors. *Expert Rev Anticancer Ther.* 2004;4:783–793.

85. van den Bent MJ. The role of chemotherapy in brain metastases. *Eur J Cancer.* 2003;39:2114–2120.

86. Schuette W. Treatment of brain metastases from lung cancer: chemotherapy. *Lung Cancer.* 2004;45(suppl 2):S253–S257.

87. Bafaloukos D, Gogas H. The treatment of brain metastases in melanoma patients. *Cancer Treat Rev.* 2004;30:515–520.

88. Twelves CJ, Souhami RL, Harper PG, et al. The response of cerebral metastases in small cell lung cancer to systemic chemotherapy. *Br J Cancer.* 1990;61:147–150.

89. Cocconi G, Lottici R, Bisagni G, et al. Combination therapy with platinum and etoposide of brain metastases from breast carcinoma. *Cancer Invest.* 1990;8:327–334.

90. Franciosi V, Cocconi G, Michiarava M, et al. Front-line chemotherapy with cisplatin and etoposide for patients with brain metastases from breast carcinoma, nonsmall lung carcinoma, or malignant melanoma. A prospective study. *Cancer.* 1999;85:1599–1605.

91. Bernardo G, Cuzzoni Q, Strada MR, et al. First-line chemotherapy with vinorelbine, gemcitabine, and carboplatin in the treatment of brain metastases from non-small-cell lung cancer: a phase II study. *Cancer Invest.* 2002;20:293–302.

92. Rosner D, Nemoto T, Lane WW. Chemotherapy induces regression of brain metastases in breast carcinoma. *Cancer.* 1986;58:832–839.

93. Boogerd W, Dalesio O, Bais EM, Van Der Sande JJ. Response of brain metastases from breast cancer to systemic chemotherapy. *Cancer.* 1992;69:972–980.

94. Robinet G, Thomas P, Breton JL, et al. Results of a phase III study of early versus delayed whole brain radiotherapy with concurrent cisplatin and vinorelbine combination in inoperable brain metastasis of non-small-cell lung cancer: Groupe Français de Pneumo-Cancérologie (GFPC) protocol 95-1. *Ann Oncol.* 2001;12:59–67.

95. Postmus PE, Haaxma-Reiche H, Smit EF, et al. Treatment of brain metastases of small-cell lung cancer: comparing teniposide and teniposide with whole-brain radiotherapy – a phase III study of the European organization for the research and treatment of cancer lung cancer cooperative group. *J Clin Oncol.* 2000;18:3400–3408.

96. Ushio Y, Arita N, Hayakawa T, et al. Chemotherapy of brain metastases from lung carcinoma: a controlled randomized study. *Neurosurgery.* 1991;28:201–205.

97. Guerrieri M, Wong K, Ryan G, Millward M, Quong G, Ball DL. A randomized phase III study of palliative radiation with concomitant carboplatin for brain metastases from non-small cell carcinoma of the lung. *Lung Cancer.* 2004;46:107–111.

98. Slichenmyer WJ, Rowinsky EK, Donehower RC, Kaufmann SH. The current status of camptothecin analogues as antitumor agents. *J Natl Cancer Inst.* 1993;85:271–291.

99. Ardizzoni A, Hansen H, Dombernowsky P, et al. Topotecan, a new active drug in the second-line treatment of small-cell lung cancer: a phase II study in patients with refractory and sensitive disease. *J Clin Oncol.* 1997;15:2090–2096.

100. Korfel A, Oehm C, von Pawel J, et al. Response to topotecan of symptomatic brain metastases of small-cell lung cancer also after whole-brain irradiation: a multicentre phase II study. *Eur J Cancer.* 2002;38:1724–1729.

101. Oberhoff C, Kieback DG, Würstlein R, et al. Topotecan chemotherapy in patients with breast and brain metastases: results of a pilot study. *Onkologie.* 2001;24:256–260.

102. Wong ET, Berkenblit A. The role of topotecan in the treatment of brain metastases. *Oncologist.* 2004;9:68–79.

103. Stupp RK, Gander M, Leyvraz S, Newlands E. Current and future developments in the use of temozolomide for the treatment of brain tumors. *Lancet Oncol.* 2001;2:552–560.

104. Newlands ES, Stevens MFG, Wedge SR, et al. Temozolomide: a review of its discovery, chemical properties, pre-clinical development and clinical trials. *Cancer Treat Rev.* 1997;23:35–61.

105. Zhu W, Zhou L, Qian JQ, et al. Temozolomide for treatment of brain metastases: a review of 21 clinical trials. *World J Clin Oncol.* 2014;5:19–27.

106. Abrey LE, Olson JD, Raizer JJ, et al. A phase II trial of temozolomide for patients with recurrent or progressive brain metastases. *J Neurooncol.* 2001;53:259–265.

107. Christodoulou C, Bafaloukos D, Kosmidos P, et al. Phase II study of temozolomide in heavily pretreated cancer patients with brain metastases. *Ann Oncol.* 2001;12:249–254.

108. Antonadou D, Paraskaveidis M, Sarris N, et al. Phase II randomized trial of temozolomide and concurrent radiotherapy in patients with brain metastases. *J Clin Oncol.* 2002;20:3644–3650.

109. Verger E, Gil M, Yaya R, et al. Temozolomide and concomitant whole brain radiotherapy in patients with brain metastases: a phase II randomized trial. *Int J Radiat Oncol Biol Phys.* 2005;61:185–191.

110. Biasco G, Pantaleo MA, Casadei S. Treatment of brain metastases of malignant melanoma with temozolomide. *N Engl J Med.* 2001;345:621–622.

111. Agarwala SS, Kirdwood JM, Gore M, et al. Temozolomide for the treatment of brain metastases associated with metastatic melanoma: a phase II study. *J Clin Oncol.* 2004;22:2101–2107.

112. Bafaloukos D, Tsoutsos D, Fountzilas G, et al. The effect of temozolomide-based chemotherapy in patients with cerebral metastases from melanoma. *Melanoma Res.* 2004;14:289–294.

113. Hwu WJ, Raizer JJ, Panageas KS, Lis E. Treatment of metastatic melanoma in the brain with temozolomide and thalidomide. *Lancet Oncol.* 2001;2:634–635.

114. Stewart DJ. Pros and cons of intra-arterial chemotherapy. *Oncologia.* 1989;3:20–26.

115. Yamada K, Bremer AM, West CR, et al. Intra-arterial BCNU therapy in the treatment of metastatic brain tumor from lung carcinoma. A preliminary report. *Cancer.* 1979;44:2000–2007.

116. Madajewicz S, West CR, Park HC, et al. Phase II study – intra-arterial BCNU therapy for metastatic brain tumors. *Cancer.* 1981;47:653–657.

117. Cascino TL, Byrne TN, Deck MDF, Posner JB. Intra-arterial BCNU in the treatment of metastatic tumors. *J Neurooncol.* 1983;1:211–218.

118. Madajewicz S, Chowhan N, Iliya A, et al. Intracarotid chemotherapy with etoposide and cisplatin for malignant brain tumors. *Cancer.* 1991;67:2844–2849.

119. Barth RF, Yang W, Rotaru JH, et al. Boron neutron capture therapy of brain tumors: enhanced survival following intracarotid injection of either sodium borocaptate or boronophenylalanine with or without blood-brain barrier disruption. *Cancer Res.* 1997;57:1129–1136.

120. Kroll RA, Neuwelt EA. Outwitting the blood-brain barrier for therapeutic purposes: osmotic opening and other means. *Neurosurgery.* 1998;42:1083–1100.

121. Gelman M, Chakares D, Newton HB. Brain tumors: complications of cerebral angiography accompanied by intra-arterial chemotherapy. *Radiology.* 1999;213:135–140.

122. Newton HB, Stevens C, Santi M. Brain metastases from fallopian tube carcinoma responsive to intra-arterial carboplatin and intravenous etoposide: a case report. *J Neurooncol.* 2001;55:179–184.

123. Newton HB, Snyder MA, Stevens C, et al. Intra-arterial carboplatin and intravenous etoposide for the treatment of brain metastases. *J Neurooncol.* 2003;61:35–44.

124. Niranjan A, Lunsford LD, Ahluwalia MS. Targeted therapies for brain metastases. *Prog Neurol Surg.* 2019;34:125–137.

125. Livitzki A, Gazit A. Tyrosine kinase inhibition: an approach to drug development. *Science.* 1995;267:1782–1788.

126. Gibbs JB. Anticancer drug targets: growth factors and growth factor signaling. *J Clin Investig.* 2000;105:9–13.

127. Dillman RO. Monoclonal antibodies in the treatment of malignancy: basic concepts and recent developments. *Cancer Invest.* 2001;19:833–841.

128. Hao D, Rowinsky EK. Inhibiting signal transduction: recent advances in the development of receptor tyrosine kinase and ras inhibitors. *Cancer Invest.* 2002;20:387–404.

129. Newton HB. Molecular neuro-oncology and the development of "targeted" therapeutic strategies for brain tumors. Part 1 – growth factor and ras signaling pathways. *Expert Rev Anticancer Ther.* 2003;3:595–614.

130. Poon ANY, Ho SSM, Yeo W, Mok TSK. Brain metastases responding to gefitinib alone. *Oncology.* 2004;67:174–178.

131. Cappuzzo F, Ardizzoni A, Soto-Parra H, et al. Epidermal growth factor receptor targeted therapy by ZD 1839 (Iressa) in patients with brain metastases from non-small cell lung cancer (NSCLC). *Lung Cancer.* 2003;41:227–231.

132. Cappuzzo F, Calandri C, Bartolini S, Crinò L. ZD 1839 in patients with brain metastases from non-small cell lung cancer (NSCLC): report of four cases. *Br J Cancer.* 2003;89:246–247.

133. Ishida A, Kanoh K, Nishisaka T, et al. Gefitinib as a first line of therapy in non-small cell lung cancer with brain metastases. *Intern Med.* 2004;43:718–720.

134. Katz A, Zalewski P. Quality-of-life benefits and evidence of antitumor activity for patients with brain metastases treated with gefitinib. *Br J Cancer.* 2003;89:S15–S18.

135. Ceresoli GL, Cappuzzo F, Gregorc V, Bartolini S, Crinò L, Villa E. Gefitinib in patients with brain metastases from non-small cell lung cancer: a prospective trial. *Ann Oncol.* 2004;15:1042–1047.

136. Bohn JP, Pall G, Stockhammer G, Steurer M. Targeted therapies for the treatment of brain metastasis in solid tumors. *Target Oncol.* 2016;11:263–275.

137. Lazaro T, Prastianos PK. Immunotherapy and targeted therapy in brain metastases: emerging options in precision medicine. *CNS Oncol.* 2017;6(2):139–151.

138. Chukwueke UN, Brastianos PK. Sequencing brain metastases and opportunities for targeted therapies. *Pharmacogenomics.* 2017;18:585–594.

139. Yousefi M, Bahrami T, Salmaninejad A, et al. Lung cancer – associated brain metastasis: molecular mechanisms and therapeutic options. *Cell Oncol.* 2017;40:419–441.

140. Martinez P, Mak RH, Oxnard GR. Targeted therapy as an alternative to whole-brain radiotherapy in EGFR-mutant or ALK-positive non-small-cell lung cancer with brain metastases. *JAMA Oncol.* 2017;3:1274–1275.

141. Venur VA, Leone JP. Targeted therapies for brain metastases from breast cancer. *Int J Mol Sci.* 2016;17:1543. https://doi.org/10.3390/ijms 17091543.

142. Long GV, Trefzer U, Davies MA, et al. Dabrafinib in patients with Val600Glu or Val600Lys BRAF-mutant melanoma metastatic to the brain (BREAK-MB): a multicentre, open-label, phase 2 trial. *Lancet Oncol.* 2012;13:1087–1095.

Further reading

143. Eckardt JR, Martin KA, Schmidt AM, White LA, Greco AO, Needles BM. A phase I trial of IV topotecan in combination with temozolomide daily time 5 every 28 days. *Proc ASCO.* 2002;21:83b.

144. Brooks BJ, Bani JC, Fletcher CDM, Demeteri GD. Response of metastatic gastrointestinal stromal tumor including CNS involvement to imatinib mesylate (STI-571). *J Clin Oncol.* 2002;20:870–872.

6

Neurosurgical approaches to the treatment of intracranial metastases

Mark A. Damante, Joshua L. Wang, and J. Bradley Elder

Department of Neurological Surgery, The Ohio State University Wexner Medical Center, Columbus, OH, United States

1 Background

Brain metastases are suggested to occur in roughly 25% of all cancer patients at autopsy.[1] With the ever improving management and prolonged survival of cancer patients without brain metastases, it is likely that this percentage will increase.[2,3] Intracranial lesions are most often found supra- or infratentorial, at the gray matter-white matter junction, though skull base, dural-based lesions, and/or leptomeningeal disease (LMD) can occur. Early in the 20th century, surgical resection was avoided due to the associated high morbidity and mortality.[4] Advancements in neurosurgical technique and neuro-anesthesia have effectively circumvented the perioperative complications that once plagued neurosurgical care of brain metastases. Today, surgical resection of brain metastases is often a critical component of multimodal treatment for single or dominant metastases, particularly for primary cancers that are inherently radiation-intermediate or radiation-resistant.

In the 1990s, two randomized controlled trials for solitary brain metastases, resection versus either needle biopsy (Patchell et al.) or no surgery (Vecht et al.) prior to whole brain radiotherapy (WBRT) demonstrated improvement in rates of local recurrence (20% vs. 52%), overall survival (OS, 40–42 vs. 15–24 weeks), and sustained functional independence (38 vs. 8 weeks).[5,6] Stable extracranial disease was an important positive prognostic factor in both the Patchell and Vecht studies. Interestingly, a subsequent study by Mintz and colleagues showed no significant benefit in terms of OS of surgical resection plus WBRT compared to WBRT alone. However, nearly half of these patients had progressive extracranial disease, and the study included patients with poor performance status (KPS ≥50).[7] In a 2005 Cochrane meta-analysis, surgical resection with adjuvant WBRT provided the best outcome for patients with single brain metastases, good performance statuses (KPS ≥70), and controlled systemic disease, suggesting that the appropriately selected patient would benefit from surgical resection of intracranial metastatic disease.[8]

Since the Patchell, Vecht, and Mintz studies, efforts have been made to further define which patients specifically would benefit from surgical resection. Several groups have reported that, among patients with as many as three brain metastases, resection of all metastases significantly increases survival when compared to resection of only some, but not all, brain metastases. Additionally, resection of all intracranial disease provided similar survival benefits compared to patients undergoing resection of a solitary brain metastasis.[9,10] Reoperation upon first recurrence and second recurrence of intracranial disease has also been associated with prolonged OS and improved quality of life, regardless of whether the recurrence was local or distant.[11] Systemic disease control and KPS >70 remain predictors of improved OS, in addition to prolonged time to recurrence (>4 months), younger age, and primary malignancy.

Over the years since these early studies, neurosurgeons have developed and refined surgical techniques and operative adjuncts to optimize surgical outcomes. Certain techniques limit potential neurologic morbidity associated with surgery and often allow for less invasive and/or more efficient approaches for metastasis resection. This chapter discusses the selection of the appropriate patient for surgical intervention, as well as the most common surgical strategies, intraoperative localization and verification techniques, and mapping and monitoring of eloquent function used in the management of brain metastases.

2 Initial approach

2.1 Presentation and tissue diagnosis

For patients with newly diagnosed brain metastases, tissue diagnosis is critical in determining subsequent medical, radiation, and surgical management. Thus, one of the first surgical decisions is whether a patient with a newly diagnosed brain tumor consistent with metastasis needs biopsy. Although most patients with brain metastases will already have a cancer diagnosis, some will present without a known cancer histology. In a large analysis of over two million patients from the SEER database diagnosed between 2011 and 2015, the incidence of brain metastases synchronously diagnosed (42,047 patients) with the primary malignancy was estimated to be 7.3 per 100,000 US persons, with the overwhelming majority originating from a lung primary (80%), followed by melanoma (3.8%), breast (3.7%), and renal cell carcinoma (3.0%).[12] Another review of 26,430 patients with brain metastases from the SEER database between 2010 and 2013 estimated that 12.1% of brain metastases were synchronously diagnosed with extracranial metastases.[13] Melanoma, non-small cell and small cell lung cancer, and renal cell carcinoma had the highest reported incidence of synchronously diagnosed intra- and extra-cranial metastases.[13] Given the advanced nature of the disease at the time of presentation, synchronously diagnosed patients tend to have a poorer prognosis than those diagnosed with brain metastases metachronously.[12] In these patients, if biopsy of systemic disease is not possible, brain biopsy may be indicated to establish diagnosis. Brain biopsy should also be considered in the setting of intracranial lesions with radiographic findings inconsistent with metastasis.

2.2 Stereotactic biopsy

Tissue acquisition allows for histologic diagnosis, as well as for molecular and genetic analysis, which may further fine-tune clinical decision-making. There are several surgical methods for brain biopsy, the most common being frameless stereotactic biopsy. Other methods include frame-based stereotactic biopsy, MR-guided stereotactic biopsy, newer robotic systems, and open biopsy.

Stereotactic intracranial biopsy is accomplished with the use of cross-sectional MRI or CT scans combined with neuronavigation biopsy planning software. The images should be obtained close to the day of surgery to minimize any change in tumor-associated anatomy due to progression or treatment effects or brain shift associated with decreased edema due to initiation of steroids. The trajectory through the brain to the biopsy target is planned prior to surgery and avoids traversing eloquent anatomic areas, blood vessels, or the ventricular system.

All preoperative trajectory planning and intraoperative feedback regarding location of the biopsy needle is provided by the neuronavigation system. For example, the stereotactic biopsy needle is attached to a neuronavigation wand, which has sensors allowing live tracking of its location within the brain. Accuracy of these procedures, estimated to be within 2 mm, is dependent on accuracy of the process of registering the patient's facial landmarks, or previously placed fiducials, with the preoperative 3D imaging. Current frameless stereotactic biopsy systems allow for planning of multiple depths and trajectories to ensure adequate tissue harvest and minimize the risk of sampling error (Fig. 1). For deep-seated or brainstem lesions, for which navigation inaccuracies would result in devastating neurologic deficits, MR-guided biopsy provides potentially increased accuracy, though this procedure typically requires longer anesthesia times (Fig. 2).

In addition to surgical complications related to brain biopsy, missed diagnosis is possible, though diagnostic success rates are quite high, reaching 96.5%–99.7% in some studies.[14–16] With improved surgical technology and increased experience, survival outcomes and diagnostic yield have similarly improved. Dammers et al. initially published data in 2008 from their institution among patients who underwent framed ($n = 227$) or frameless ($n = 164$) stereotactic brain biopsy between 1996 and 2006, in which successful diagnosis was made in 89.4%.[15] In a subsequent study on the frameless biopsy system, the Dammers group evaluated additional patients ($n = 160$) between 2006 and 2010 and compared them with patients ($n = 164$) of the previous study.[14] The associated mortality from biopsy trended down (0.6% vs. 3.7%, $P = .121$), and successful diagnosis was significantly higher (98.2% vs. 89.0%, $P = .001$) in the follow-up study.[14,15] Given the importance of operative technique, it is not surprising that the number of biopsies performed annually by a single institution was the only significant predictor of diagnostic success rate reported by the Kickingereder et al. meta-analysis.[16]

One limitation of current stereotactic biopsy strategies is the reliance on facial landmark registration and utilization of preoperative imaging. For deep-seated or small lesions, for which relatively small targeting inaccuracies would result in significant neurologic deficits, MR-guided biopsies provide potentially increased accuracy by allowing for real-time imaging of the trajectory and depth of the biopsy needle. Real-time MR-guided percutaneous intracranial biopsy is an adaptation of similar procedures used for deep brain stimulator (DBS) lead placement, laser diode applicators for laser interstitial thermal therapy, and convection-enhanced delivery infusion.[17] Preoperative imaging is used for planning entry point, trajectory, and biopsy target. Under general anesthesia, the patient's skull is rigidly fixed in a head frame and the MRI-compatible targeting system is attached

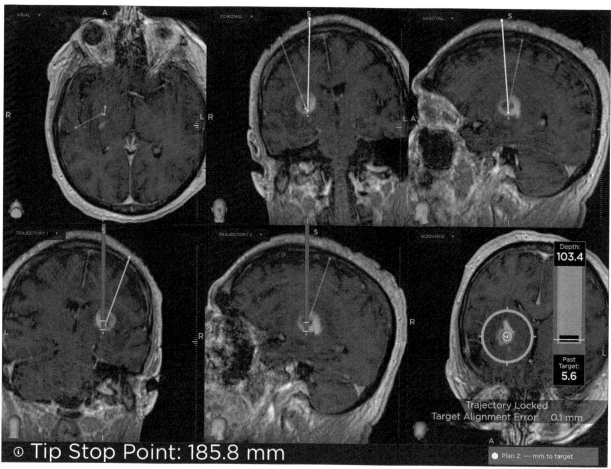

FIG. 1   Intraoperative neuronavigation images demonstrating two separate trajectories and biopsy depths for a single lesion within the right basal ganglia. Clockwise from top left: axial, coronal, and sagittal views of the preoperative MRI, probe's eye view, and two views orthogonal to the trajectory. Thin *blue* and *yellow lines*: planned trajectories; *thick blue line* in bottom row: live view of location of biopsy needle.

percutaneously to the skull with screws. Correct positioning and trajectory of the guide cannula and biopsy target are then confirmed in real time with a T1-weighted MRI. The entry point is made with a percutaneous twist drill through which the biopsy needle is placed to target. As with stereotactic biopsy procedures, multiple biopsy cores may be taken to improve sample size and accuracy. Real-time MRI allows the neurosurgeon to account for any changes in the location or configuration of the target lesion and allows for intraoperative adjustments if necessary. Additionally, the ability to observe the exact location of the biopsy core precludes the need for confirming lesional tissue with frozen section (Fig. 1). The accuracy of this system utilized for intracranial biopsy has been reported to be similar to its accuracy when used for DBS lead implantation as measured by both mean radial error (1.3-mm vs. 1.2-mm, respectively) and mean absolute tip error (1.5-mm vs. 2.2-mm, respectively).[17,18]

The overall complication rate of stereotactic intracranial biopsy varies by study (1%–10.5%), although the risks of stereotactic biopsies are generally outweighed by the potential diagnostic benefits.[14,19–21] Despite a lower risk than craniotomy for open biopsy, there are associated complications with potentially devastating neurologic injury. In a meta-analysis by Kickingereder and colleagues of 38 studies and 1480 brainstem tumor biopsies, the overall rate of morbidity was 7.8%; permanent morbidity was 1.7% and mortality rate was 0.9%.[16] Intracranial hemorrhage was the most common operative complication.[16] Perioperative bleeds are likely due to vessel injury as the biopsy needle traverses the parenchyma, despite careful planning to avoid radiographically apparent vasculature and sulci. While reported hemorrhage rates on postoperative imaging can be strikingly high (up to 59.8% in one study), the rate of clinically significant hemorrhage in the literature is lower.[20,22] One study of 296 patients identified 26 (8.8%) cases that sustained hemorrhage noted on postoperative CT scan, though only 3 (1%) were clinically significant.[23] A similar evaluation of 622 stereotactic biopsies described 4.8% of patients experiencing symptomatic hemorrhage.[24]

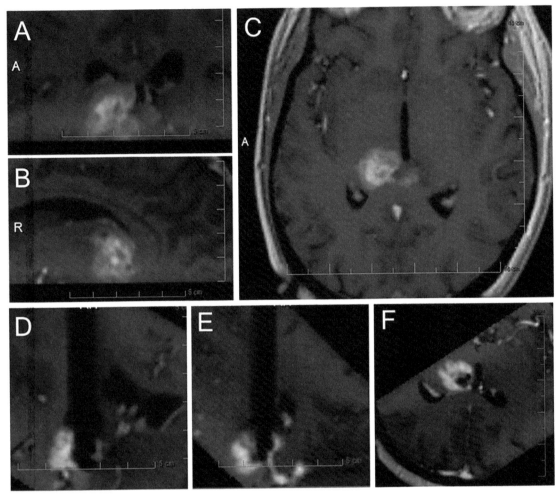

FIG. 2   Intraoperative MR images demonstrating a homogeneously enhancing mass located in the right thalamus (A: coronal, B: sagittal, C: axial planes) and the biopsy needle placed under MR guidance (D and E: orthogonal views of the biopsy tract, F: probe's eye view) into the target lesion.

2.3 Craniotomy

Surgical resection improves survival outcomes compared to those patients receiving biopsy. Patchell et al. prospectively analyzed 48 patients, 25 treated with surgery followed by radiation and 23 who received a brain biopsy followed by radiation. This study described both increased time to recurrence (> 59 vs. 21 weeks, $P < .0001$, respectively) and improved survival (40 vs. 15 weeks, $P < .01$) among patients who underwent a resection and radiation compared to those who underwent brain biopsy and radiation.[5]

Although metastases may be found anywhere in the brain, they most commonly arise at the junction of the gray and white matter, where the tapering caliber of the capillaries no longer permits tumor cell passage, resulting in tumor deposition and growth. A number of standard neurosurgical approaches provide access to the tumor, whether the tumor sits near the cortical surface, along the skull base, or deep in the parenchyma. For superficial tumors, the craniotomy must encompass the entire area of the tumor. However, deep tumors can potentially be accessed through smaller craniotomies since the intracranial operative field widens with increasing distance from the skull. For deep-seated tumors or those in eloquent locations, surgical and neurophysiologic techniques minimize damage to normal brain during resection and are discussed further below.

Deep-seated tumors can often be accessed through minimally invasive (or "keyhole") craniotomies, which are refinements of conventional craniotomies.[25] These keyhole craniotomies can be smaller than the lesion by subtending the full angles of approach at the extent of the craniotomy; further exposure of the tumor can be achieved using endoscopes, though these are not required for all minimally invasive craniotomies. Multiple case series have demonstrated safety and success of keyhole craniotomies to achieve gross total resection for intrinsic and extrinsic lesions, including brain metastases.[26–30] In multiple studies evaluating keyhole craniotomies for resection brain metastases, gross total resection was achieved in 74%–87% of patients, including in patients undergoing simultaneous multiple minimally invasive craniotomies.[26–30] Complications were reported in 2%–9% of patients, and the majority of patients demonstrated performance status improvement.[28–30] The postoperative complications experienced by patients undergoing minimally invasive craniotomies largely match those of conventional craniotomies: infection and cerebrospinal fluid leak, though rare, were most commonly reported. In a study of 38 metastases receiving minimally invasive resection, median WHO performance status improved from 2 to 1 and all of these patients had performance status 2 or better postoperatively (preoperative performance status ranged from 0 to 4).[30]

One such minimally invasive approach is the supraorbital "eyebrow" craniotomy, which provides access to the frontal pole, and subfrontal, suprasellar, and retrosellar regions, for lesions less than 5 cm in diameter.[31–33] An incision is made in or just above the eyebrow, following the curve of the eyebrow with the option of extending further laterally, through which a low frontal craniotomy is made with or without removal of the orbital rim. In a series of 418 patients who underwent eyebrow craniotomies, high patient satisfaction was reported with cosmetic results (93% reporting "very pleasant" or "quite pleasant," the two best options on a five-point scale) and pain (89% reported no or "minimal" postoperative scar pain and headache on a five-point scale).[34] Approach-associated complications were rare, with only 2% of patients reporting difficulty chewing (all temporary), 5.6% reporting frontalis muscle palsy (2.1% of total patients had permanent palsy), and 8.3% reporting frontal hypesthesia (3.4% of all patients had permanent hypesthesia).

2.4 Tubular retractor

For deep-seated lesions, neurosurgeons have long utilized retractor systems to maintain adequate visualization of the operative field.[35] In contrast to bladed retractors, cylindrical, or tubular retractor systems, (colloquially, "ports") apply even pressure to the walls of the operative corridor, thereby minimizing retractor-induced injury but maintaining the operative corridor.[35–38] Further refinement of the technique uses preoperative diffusion tensor imaging (DTI) to identify the projecting white matter fibers and select a trajectory that minimally disrupts white matter tracts, which may not be the shortest intraparenchymal route.[37] Ports began as metal cylinders and were attached to stereotactic head frames to allow accurate placement during surgery.[35] The development of frameless stereotaxy and improvement in microscope optics over time have improved the ease of repositioning the port during surgery and overall visualization of the whole tumor. Endoscopes have also been used to give better views at the end of the port, particularly using angled lenses to view the proximal walls of the resection cavity, but these also limit the working space within the port.[39] Even newer iterations use exoscopes, which can further improve mobility of the port and microscopic perspective without occupying space inside the port.

3 Tumor resection

Brain metastases are composed of solid and/or cystic tumor without intervening brain tissue.[40] Metastatic deposits tend to be well-circumscribed, and any infiltrative tumor cells into the surrounding brain tend to spread no further than 5 mm deep.[41] Typically, a rim of gliotic tissue separates the tumor from the surrounding brain, and dissection within this plane generally allows for safe resection.

3.1 En bloc

En bloc resection (or removal of the entire tumor as a single specimen) of brain metastases is associated with decreased likelihood of developing LMD. In a retrospective single-center study of 542 patients with supratentorial brain metastases, the rate of developing LMD was significantly lower with en bloc resection (111 of 351 patients, 3%) compared to piecemeal resection (17 of 191 patients, 9%).[42] Similarly, in another retrospective study of 260 patients with posterior fossa metastases, 7 of 123 patients (6%) receiving en bloc resection developed LMD compared to 19 of 137 patients (14%) receiving piecemeal resection.[43] In both studies, these findings remained significant after controlling for tumor size, location, and patient characteristics. For highly vascular tumors, such as renal cell carcinoma, en bloc resection can reduce operative blood loss and thus operative time and complications. In a study of 1033 patients, en bloc resection was associated with decreased postoperative complications compared to piecemeal removal, including tumors in eloquent regions and larger tumors.[44] Additionally, en bloc resection of tumors with volume $< 10\,cm^3$ was associated with significantly decreased rates of local recurrence in a retrospective study of 570 patients with single brain metastasis.[45] However, en bloc resection is not always technically feasible or safe, particularly in cases of large tumors or those located in eloquent areas.

3.2 Supratotal

Theoretically, "supramarginal" or "microscopic total" resection could improve local recurrence rates of brain metastases by resecting the infiltrative malignant cells in otherwise normal appearing parenchyma.[41,46–48] In four retrospective studies encompassing 165 patients, an additional 5 mm of surrounding parenchyma was resected in patients following standard of care gross microsurgical resection. In one study, "microscopic total" resection (or "clean margins") was verified by obtaining intraoperative frozen biopsy samples from the additional resection margin until no tumor cells were seen.[41] In all four studies, supramarginal resection was well-tolerated and could be safely achieved in eloquent areas using awake craniotomies and intraoperative monitoring.[46] Additionally, survival outcomes suggested improved local tumor control with supramarginal resection compared to in-study and historical controls.[41,47] In the study presented by Yoo and colleagues, 43 patients with tumors located in noneloquent areas received microscopic total resection, with intraoperatively confirmed negative margins of brain parenchyma beyond the tumor border, compared with 51 patients with tumors in eloquent areas, who received standard of care gross total resection. Local recurrence at 1 and 2 years was noted in only 29% and 29% of patients in the microscopic total resection group, compared to 59% and 63% of gross total resection patients, respectively ($P = .04$). Though there was no difference in median overall survival (10.3 vs. 11.0 months, respectively) between microscopic and gross total resection, there was significantly increased 2-year survival rates in the microscopic total resection group (27% vs. 4%, $P = .0001$). Prospective data are needed to help delineate the role of supramarginal resection in the management of brain metastases.

3.3 Piecemeal

For tumors in eloquent areas or very large tumors (>3 cm), aggressive brain dissection around the tumor may result in retraction-induced ischemia and unacceptable neurologic deficits, and thus, the risks of en bloc resection through a longitudinal corticotomy may outweigh the potential benefits of increased local control and decreased rates of LMD. In such cases, the tumor can be resected from the inside out in a piecemeal fashion. Ultrasonic aspirators disrupt tumor architecture at the tool tip while aspirating the resultant debris and thus may assist in tumor debulking while minimizing retraction and manipulation of normal brain tissue. In a study of 1028 intracranial tumor resections (encompassing metastases as well as primary CNS neoplasms and extra-axial tumors), use of ultrasonic aspirators demonstrated relatively low rates of surgical morbidity, with 10% of patients experiencing a significant decrease in KPS at discharge (defined as a decrease of 20 points or greater if baseline KPS $\geq$ 80 or of 10 points or greater if baseline KPS <70).[49] For deep-seated tumors approached through a tubular retractor, the small working diameter of the port limits the feasibility of achieving en bloc removal. Furthermore, these lesions are often located in

deep eloquent nuclei, including the basal ganglia and thalamus, where excessive retraction during circumferential dissection would lead to significant deficits.

3.4 Patients with multiple brain metastases

With modern imaging techniques, as many as 80% of patients are found to have more than one brain metastasis simultaneously diagnosed on MRI, and approximately 50% will have three or more.[50] Traditionally, the presence of multiple brain metastases was considered a relative contraindication to surgical resection given the presumed poor prognosis. These patients were often treated either with continuation of systemic therapies and/or local intervention with WBRT. WBRT may still be considered a first-line therapy for patients with innumerable metastases or in patients with multiple radiosensitive lesions. There is a higher risk for radiation-induced cognitive decline associated with WBRT compared with SRS, although newer methods that include concurrent memantine or hippocampal-sparing WBRT may diminish this concern.[51,52] The role of surgical resection in patients with multiple brain metastases has not been well-established.[9,53–55] The goal of surgical resection is to relieve mass effect and/or associated peritumoral edema to reduce neurologic deficit, relieve an epileptogenic focus of disease, and potentially improve survival. Strategies for surgical management include resection of a single dominant or symptomatic lesion among multiple lesions or resection of multiple lesions simultaneously through one or more craniotomies. The data regarding the resection of a single dominant or symptomatic lesion in a patient with multiple intracranial lesions are inconsistent. Paek et al. retrospectively reviewed 208 patients with brain metastases, in which 191 underwent resection of a single brain metastasis (132 with solitary lesions and 59 with multiple lesions) and 17 patients had multiple brain metastases resected.[54] Among patients where a single metastasis was removed, postoperative Karnofsky performance status (KPS) was improved in 30% of patients with solitary lesions and 37% of patients with multiple metastases.[54] Only 9% of patients with multiple metastases worsened compared to 4% of those with solitary lesions resected.[54] Though patients with multiple intracranial lesions are presumed to have poorer prognoses, surgical resection in this cohort equalized median survival among patients with up to three lesions compared to those with solitary lesions.[54] However, it should be noted that tumors located in noneloquent or near-eloquent areas had a significantly greater likelihood of gross total resection ($P < .001$) and improvement of KPS ($P = .0018$) after surgery compared to tumors located in eloquent areas. Ultimately, only recursive partitioning analysis (RPA) class I patients who underwent resection of one or more lesions demonstrated reversed or stabilized neurologic symptoms and improved OS, without increased morbidity and mortality.[54] Most patients received postoperative radiation therapy and systemic therapy, which were associated with improved OS.[54]

In a related study, Wroński et al. studied 70 patients with brain metastases from breast cancer and demonstrated no survival benefit after surgery comparing patients with multiple metastases ($n = 16$, OS 14.8 months) to those with single metastases ($n = 54$, OS 13.9 months, $P = .28$), although they did not specify how many lesions were resected among patients with multiple lesions.[56] It should be noted that additional salvage WBRT, at some point after surgery, was utilized in the majority of patients in the studies reviewed above and significantly improved OS compared to the few patients that did not receive WBRT.[54,56] Unfortunately, the data regarding resection of a single dominant or symptomatic lesion in the setting of multiple intracranial metastases are inconclusive. The decision remains highly selective based on patient performance status, tumor-associated anatomy, and primary pathology and thus should be made following multidisciplinary discussion with neurosurgery and medical and radiation oncology.

In certain settings, simultaneous resection of multiple (2–4) intracranial lesions through single or multiple craniotomies may benefit appropriately selected patients.[9,53,56] However, the additional benefit following resection of multiple tumors is unclear, particularly regarding whether all lesions will be resected or only some of the lesions while leaving others behind. Bindal et al. evaluated a cohort of patients with multiple brain metastases (from 2 to 4 lesions) for which multiple, but not all, brain metastases were removed (group A, $n = 30$) or all brain metastases were removed (group B, $n = 26$) and compared them to 26 patients, matched with group B, in which a solitary brain metastasis was resected (group C).[9] Groups B (14 months, $P = .003$) and C (14 months, $P = .012$) both demonstrated significantly improved OS compared to group A (6 months). OS of groups B and C did not differ significantly ($P > .05$). Additionally, the complication and mortality rates were similar across all three groups and the rate of recurrence between groups B and C was similar. In a more recently published study by Salvati et al., 32 patients with multiple brain metastases (25 with 2 lesions, 7 with 3 lesions) underwent resection of all lesions and demonstrated a similar prognosis to 30 matched patients with a solitary lesion that was resected (OS 14.6 vs. 17.4 months, respectively, $P = .2$).[55] However, patients with multiple lesions only demonstrated a survival benefit from surgery if intracranial disease was limited to ≤ 3 lesions, KPS > 60, systemic disease was well-controlled, and prognosis > 3 months.[55] Based on the current literature, the benefit of surgical resection of multiple intracranial lesions may only benefit the appropriately selected

patient if all lesions can be resected. This may limit the appropriate patient to someone with up to four lesions that may all be safely and completely resected. Otherwise, it is likely still the case that other modalities including systemic therapies and radiation therapies should be prioritized over surgical resection.

4 Surgical adjuncts

Since surgery for brain metastases was first shown to improve outcomes, further refinements of techniques and technologies have improved operative efficiency and clinical outcomes. Use of these technologies for brain metastases ranges from routine standard of care, such as frameless stereotactic neuronavigation, to uncommon, such as 3D ultrasound.

4.1 Fluorescence

Two fluorescent agents used in neurosurgery to guide tumor resection are 5-ALA and fluorescein. Although the use of 5-aminolevulinic acid (5-ALA) improves extent of resection and progression-free survival in glioma, brain metastases do not display consistent 5-ALA fluorescence (only 41% in a series of 84 metastases), and there was no correlation with primary site of the metastasis or histopathological subtype.[57] In that study by Kamp and colleagues, use of 5-ALA to guide surgical resection did not correlate with improved survival outcomes.[57] Median OS among patients receiving 5-ALA for resection of solitary brain metastases was 15 months. Furthermore, 5-ALA fluorescence did not predict local recurrence rates in this patient series.

In a study of 95 patients with brain metastases, intraoperative fluorescein was associated with 83% gross total resection (83%), which the authors noted was improved compared to historical control gross total resection rates of 54%–76%, and no adverse events were registered during the postoperative course.[58–60] No survival outcomes were published with intraoperative fluorescein use.

4.2 Neuronavigation

Frame-based stereotactic systems that map intraoperative cross-sectional imaging to a reference grid associated with the stereotactic frame have largely been replaced by frameless navigation systems that register preoperative (or intraoperative) cross-sectional imaging reconstructed into a three-dimensional model with facial landmarks or surface fiducials. These frameless systems provide registration accuracy of 2–4 mm, and the lack of need for a reference frame allows for nearly instant localization using a navigation "wand."[61] These "neuronavigation" systems are particularly useful for planning skin incision, craniotomy, and trajectory to the lesion. However, changes in anatomy intraoperatively due to brain shift (which can be caused by loss of cerebrospinal fluid, osmotic agents, manipulation of normal brain tissue, and tumor resection) will not be accounted for by the navigation system, since the images used are obtained preoperatively. The navigation systems can be updated intraoperatively with further cross-sectional imaging, whether CT, MRI, or 3D ultrasound. Although no outcomes data exist regarding neuronavigation solely for brain metastases, early data in glioblastoma resection demonstrated significantly improved extent of resection without significantly prolonging operative time.[62]

4.3 Intraoperative imaging

Intraoperative CT, ultrasound, and MRI can be used to identify tumor in real time and update the navigation system. Intraoperative ultrasound is widely used and is the least expensive and obtrusive option with rapid and repeatable utilization. Brain metastases are often hyperechoic on ultrasound and easily distinguishable from the surrounding normal brain given the high density of tumor cells. Tumor cysts usually demonstrate hyperechoic walls with central hypoechoic fluid. Ultrasound can be easily brought to the field repeatedly to evaluate extent of resection (Fig. 3). Navigated three-dimensional ultrasound is a further refinement of this technology and can update neuronavigation systems to account for brain shift.[63,64] In multiple trials, use of intraoperative ultrasound increased extent of resection and postoperative KPS.[65]

Although still costly, intraoperative MRI (iMRI) systems are becoming more widespread in their use. These iMRI systems can update navigation systems, assess extent of resection, and identify surgical complications such as hematoma formation. The increased imaging resolution is offset by additional workflow and increased time under general anesthesia that makes iMRI significantly less convenient than intraoperative ultrasound.[66,67] Since most brain metastases are well-circumscribed and easily distinguished from normal brain, the use of iMRI in neurosurgical oncology has usually focused on glioma surgery. In Livne and colleagues' study of 163 patients, iMRI was associated with increased extent of resection in all enhancing lesions, including gross total resection in 73% of metastases.[67] In their cohort of 163 patients, iMRI use resulted in extending resection in 69 patients, including in a 28 of 53 patients with tumors located in eloquent brain regions, for whom incomplete resection was initially planned. However, patients with metastases were half as likely as low-grade gliomas to have iMRI findings requiring further resection to achieve maximal extent of resection.

Intraoperative CT systems, whether portable or on a sliding gantry, have workflow and speed advantages over iMRI, but are limited by their soft tissue definition. As such, intraoperative CT is most useful for tumors invading the skull base and for CT angiography of vascular lesions.[68]

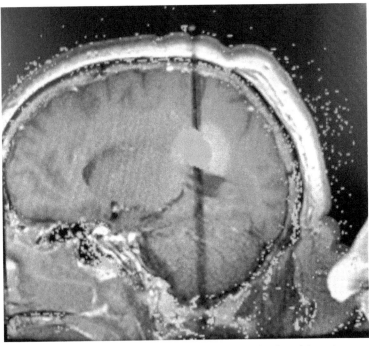

FIG. 3   Intraoperative MR image orthogonal to the LITT fiber demonstrating live monitoring of tissue temperature.

4.4 Intraoperative brain mapping

For tumors in eloquent brain, mapping of critical functions such as motor or language can be performed both preoperatively and intraoperatively. Functional MRI and diffusion tensor imaging are used to identify important cortical foci and white matter tracts, respectively.[69,70] However, as with neuronavigation, these imaging modalities are limited by their susceptibility to brain shift intraoperatively.

Intraoperative neurophysiologic monitoring utilizes both cortical responses to peripheral inputs and direct stimulation of the cortex with recording of the motor response peripherally. Somatosensory evoked potentials (SSEPs) are recorded from a strip electrode placed on the cortical surface as the contralateral median, ulnar, or posterior tibial nerves are stimulated. Because sensory evoked potentials are negative and motor evoked potentials are positive, the central sulcus can be identified between two adjacent leads that demonstrate "phase reversal," thereby localizing the adjoining primary motor and primary somatosensory gyri. Intraoperative identification of the motor and sensory cortices allows planning of the surgical corridor to minimize the risk of postoperative motor or sensory deficits.

Conversely, electrical stimulation can be administered cortically or subcortically and recorded peripherally. Motor evoked potentials (MEPs) are measured following cortical stimulation using peripherally placed electrodes or direct visualization of movement in an extremity, depending on the type of anesthesia.[71] Similarly, subcortical motor white matter pathways can be stimulated to elicit motor responses, but tend to be less reliable than cortical stimulation.[72] For deep-seated lesions, such "white matter mapping" can be used to confirm preoperative DTI and the planned surgical trajectory.

For motor mapping, awake craniotomy may increase the sensitivity and specificity of direct cortical stimulation. Conversely, for tumors arising from the motor strip, several groups support intraoperative monitoring and mapping using SSEPs and MEPs under general anesthesia (with or without preoperative mapping) as adequately safe and associated with minimal neurologic complications,[73–75] despite the fact that MEPs under general anesthesia do have an associated false negative rate.[76] In Sanmillan and colleagues' series of 33 patients with brain metastasis undergoing surgical resection, 4 had worsening motor function and 2 had new sensory disturbances postoperatively, but all 6 patients had total recovery of these postoperative changes by the third month of follow-up. However, in Krieg and colleagues' series of 56 supratentorial metastases resected using intraoperative neuromonitoring, 7 patients (12.5%) had new permanent motor deficit, 5 (9%) had temporary motor deficit, and 12 patients (21%) had improvement in

strength following surgery. In contrast, the 135 patients (across eight studies) with brain metastases located in motor or speech areas underwent awake craniotomy with 76% of patients demonstrating improvement or stability of neurologic function immediately postoperatively.[77,78] Of the 24% with worse postoperative neurologic symptoms, 96% experienced long-term improvement in neurologic function to their preoperative baseline or better. In a retrospective study of 49 patients with intrinsic brain tumors located in the primary motor cortex, Magill and colleagues noted no difference in long-term functional outcome or extent of resection in patients undergoing awake versus asleep craniotomies.[79] Without high-quality prospective data available to provide further guidance, selection of awake versus asleep craniotomies often comes down to a case-by-case basis and depends largely on surgeon preference, anesthesia familiarity, quality of preoperative and intraoperative mapping and monitoring, and individual patient selection.

5 Other neurosurgical techniques

5.1 Stereotactic radiosurgery

Historically, WBRT was standard of care radiation therapy modality for brain metastases. With the evolution of stereotactic radiation techniques such as Gamma Knife, CyberKnife, and LINAC-based technologies, WBRT has become largely reserved for patients with large numbers of metastases, radiation-sensitive histologies, and as a salvage technique. This is primarily due to the improved side effect profile and equivalent local control offered by stereotactic radiosurgery (SRS).[52] SRS allows for the delivery of higher radiation doses to a precise region of the brain with minimal exposure of healthy brain tissue to significant radiation doses. The lesion targeted for SRS is mapped via cross-sectional scans to identify the volume to be radiated and the point of rapid dose falloff to avoid damaging surrounding healthy parenchyma. SRS has been modified to include a multitude of dosages, dosing schedules, fractionation of radiation delivery and timing of delivery in relation to surgical resection and/or systemic therapies. The shorter course of SRS, typically 1–5 treatment sessions, is also an advantage.

The utility of SRS alone is determined based on the size of and number of lesions to be irradiated. SRS alone has been shown effective in patients with a single brain metastasis of ≤3cm, but when >3cm,[80] the survival benefit is lost and the risk of side effects, such as radiation necrosis, increases. There is no universally agreed maximum number of metastases that can be safely and effectively irradiated with SRS. While SRS alone was historically considered for a single lesion, Chang et al. suggested that patients with up to 3 brain metastases benefited from SRS,[81] and Suh et al. provided evidence in favor of up to 4 lesions treated with SRS.[82] However more recently, the Japanese Leksell Gamma Knife Society suggested that patient with 2–4 and 5–10 brain metastases had similar OS. While OS for patients with just 1 lesion was significantly better than the 2–4 and 5–10 lesion groups, the OS of the 2–4 and 5–10 lesion groups was not significantly different when treated with SRS alone.[83] The rates of salvage WBRT, SRS, or surgical resection did not differ between the 2–4 and 5–10 brain metastasis groups. This study suggests the noninferiority of SRS in patients with 2–10 intracranial lesions compared to WBRT in terms of OS, though WBRT continues to be recommended in patients with high intracranial disease burden.[84,85]

Bindal and colleagues reported that surgical resection alone (n = 62) provided longer OS (16.4 vs. 7.5 months, respectively, P = .0018) and decreased neurologic mortality (P = .0001) compared to patients treated with SRS alone (n = 31).[86] However, this observation has not been consistently supported by subsequent studies. For example, Auchter et al. retrospectively compared 122 patients treated with SRS plus WBRT to the historical Patchell et al. surgery plus WBRT cohort, using the same inclusion criteria, and observed longer OS with SRS plus WBRT (56 vs. 40 weeks, respectively).[5,87] Additionally, Muacevic et al. compared 33 patients treated with surgery plus WBRT to 31 receiving SRS alone and determined that there was no difference in local control between groups, suggesting that patients with solitary brain metastases may be amenable to minimally invasive approach with SRS.[88] Given the inconsistent data, it is difficult to characterize the benefit of OS and local control in a head-to-head comparison of surgery to SRS. The combination of surgical resection and radiation may provide the best local control and survival benefit in the appropriate patient.

With respect to its role as an adjuvant therapy to treat potential microscopic disease following surgical resection, SRS has become standard of care. Although WBRT was originally the radiation strategy of choice, postoperative adjuvant SRS has gained popularity given many retrospective and prospective studies demonstrating its similar rates of local control and fewer cognitive side effects. One of the earliest studies examining SRS for treatment of brain metastases, by Choi et al., retrospectively evaluated 120 resection cavities treated with postoperative SRS with an additional 2-mm margin surrounding the resection cavity, which resulted in an improved 12-month local failure rate of 3% compared to 16% without a margin (P = .042) and could be done with no increased risk of toxicity (3% vs. 8%, P = .27).[89] The authors found that WBRT could be avoided in 72% of patients,

though they also highlight a need for cavity margin expansion to improve the efficacy of postoperative SRS.[89] These findings were confirmed by a prospective evaluation of 39 patients, and 40 lesions treated with postoperative SRS had significantly lower incidence of local failure compared to the 10 patients with unirradiated cavities (15% vs. 50%, HR 0.24, P = .008).[89,90] Risk of local failure was increased when tumor size was ≥ 3 cm.[90]

A multiinstitutional analysis retrospectively studied 223 large brain metastases ($\geq 4 \text{ cm}^3$), in which 66 were treated with SRS alone and 157 underwent gross total resection with either pre- (n = 63) or postoperative (n = 94) SRS. OS was significantly improved with surgery plus SRS (2-year OS 38.9% compared to 19.8%, P = .01) with SRS alone.[91] This study is one of few that suggests OS benefit following postoperative SRS compared to SRS alone. However, as discussed above, SRS is not typically used as a primary treatment for tumors larger than 3 cm, and therefore, the survival benefit may not be directly attributable to the addition of SRS, but the addition of surgical resection.

The SRS method also decreases the risk of neurocognitive decline associated with WBRT. In a phase III randomized clinical trial comparing postoperative WBRT to postoperative SRS, SRS demonstrated a longer cognitive-decline free interval (3.7 vs. 3.0 months, respectively, P < .0001), although there was not a significant effect on OS (12.2 vs. 11.6 months, respectively, P = .70).[92] A large meta-analysis containing eight retrospective studies with 646 patients (238 treated with postoperative SRS and 408 treated with postoperative WBRT) confirmed that SRS offers comparable effect on OS and similar local control compared to postoperative WBRT, but with lower risk of cognitive decline.[93] One important addition to the literature was the authors' description of a higher associated risk of leptomeningeal disease development with postoperative SRS compared to postoperative WBRT (relative risk 2.99, 95%CI 1.55–5.76), though associated OS was unchanged.[93]

A newer treatment strategy for combining radiation and surgery employs preoperative or neoadjuvant SRS. Among the multiple theoretical benefits of neoadjuvant SRS being studied are improved conformation of the treatment target (resulting in decreased normal brain irradiated and thus decreased rates of symptomatic radiation necrosis), sterilization of tumor cells (decreasing rates of tumor dissemination and LMD), and increased tumor antigen exposure (increasing effectiveness of checkpoint inhibition therapy). Prior to surgery, there is a better-delineated target compared to the ill-defined margins of a resection cavity with a potential decreased need for an expanded cavity margin of radiation.[94,95] Furthermore, variability in the patient's postoperative clinical course may lead to unpredictable delay or cancelation in postoperative SRS, which is mitigated by neoadjuvant SRS.[90,95] Additionally, neoadjuvant SRS may reduce intraoperative tumor cell dissemination during piecemeal resection, thereby reducing the risk of LMD development.[94,95]

In a small prospective trial, 47 patients with 51 lesions were treated with neoadjuvant SRS at a median of 1 day prior to operation.[94] These patients demonstrated a 1- and 2-year local control rate of 85.6% and 71.8%, respectively, both of which are higher than the reported local control observed in the SRS studies.[94] A multiinstitutional retrospective analysis of 180 patients with 189 resected brain metastases compared those receiving neoadjuvant SRS within 48 hours of resection ($n = 66$) to the postoperative SRS ($n = 114$).[96] There was no associated difference in OS (HR 0.74, 95%CI 0.52–1.06, $P = .1$), local recurrence at 1 year (HR 1.55, 95%CI 0.75–3.2, $P = .24$), or distant intracranial recurrence at 1 year (HR 1.08, 95%CI 0.68–1.7, $P = .75$) between the two treatment groups.[96] Most importantly, patients in the neoadjuvant SRS group had significantly lower rates of leptomeningeal disease (3.2% vs. 16.6%, HR 4.03, 95%CI 1.2–13.6, $P = .02$) and rates of symptomatic radiation necrosis (4.9% vs. 16.4%, HR 8.14, 95%CI 2.6–30.74, $P = .002$) at 2 years out from treatment.[96] The lower incidence of radiation necrosis may be explained by the ability to use lower peripheral radiation doses in the neoadjuvant SRS group (14.5 Gy vs. 18 Gy, $P < .001$), a smaller median margin (0-mm vs. 2-mm, $P < .001$), and lower prescription isodose line (80% vs. 92%, $P < .001$).[96]

There are several potential drawbacks to neoadjuvant SRS. As suggested in postoperative SRS data, lesion size is a predictor of local recurrence, with smaller tumors showing a lower rate of recurrence than larger tumors, particularly when > 3 cm, and resection cavities are often smaller than the original volume of the metastasis. Secondly, neoadjuvant SRS does not allow for CNS tissue confirmation prior to radiation delivery. However, despite these drawbacks, the data seem favorable for neoadjuvant SRS over the postoperative SRS method specifically with respect to lower rates of symptomatic RN and LMD.[94,96]

5.2 Laser interstitial thermal therapy

Laser interstitial thermal therapy (LITT) is a minimally invasive surgical technique in which a small burr hole is created for the passage of an optical fiber to deliver thermal energy to a targeted lesion. This technique has recently emerged as a treatment option for patients with epilepsy, neurodegenerative disorders, and brain tumors. After placement of the LITT optic fiber, intraoperative MRI is used to monitor tissue thermodynamics to ensure effective thermal ablation without compromising surrounding healthy brain parenchyma (Fig. 4), along with adjuvant therapies such as stereotactic radiosurgery and systemic chemotherapies. Similar to SRS, LITT allows for access to deep-seated tumors that may be surgically inaccessible.[97] LITT is limited to a linear trajectory that does not encounter vital structures such as vasculature, the ventricular system, or eloquent anatomical regions.[97,98] Lesions that are hypervascular and/or diffuse in nature may not be amenable to LITT.[97] Lesion size is a critical factor in LITT, as the degree of ablation is strongly associated with effectiveness, with one study suggesting significantly prolonged PFS when ablation of at least 97% of the tumor volume was achieved.[98–100] Very large lesions may be considered if treatment is staged to avoid complications.[98] In addition to the treatment of metastases, LITT can be used for the treatment of radiation necrosis from previous radiation treatment, of which 2%–10% of SRS patients develop symptomatic radiation necrosis.[101]

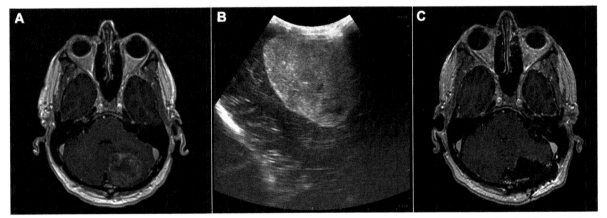

FIG. 4   Intraoperative ultrasound for resection of gastrointestinal adenocarcinoma metastasis. (A) Axial T1-weighted with gadolinium MR image demonstrating large, heterogeneously enhancing mass located in the left cerebellar hemisphere, with associated vasogenic edema and mass effect on the fourth ventricle. (B) Representative intraoperative B-mode ultrasound image demonstrating the hyperechoic mass in the near field and the hyperechoic tentorium in the far field along the leading edge (left side of image). (C) Postoperative axial T1-weighted with gadolinium MR image demonstrating gross total resection.

Evidence for LITT is increasing and showing to be useful in the treatment of both brain metastasis recurrence and radiation necrosis. Its application may be beneficial in symptom relief, as well as reducing steroid dependence.[100,102–104] The Laser Ablation After Stereotactic Radiosurgery (LAASR) trial, the first multicenter prospective trial of LITT related to brain tumors (42 patients, 19 with radiation necrosis, 20 with recurrent tumor, and 3 unknown pathology), showed 30% of patients were able to stop or taper steroid dosage 12 weeks after LITT.[100] The LAASR trial demonstrated the importance of total versus subtotal ablation of metastases as it relates to risk of progression. In the LAASR study, just 25% of totally ablated tumors progressed compared to 62.5% progression among subtotally ablated tumors.[100] A similar multicenter retrospective study evaluated 30 patients with previously irradiated brain metastases and demonstrated 73.3% of patients were able to discontinue steroids at a median of 5 weeks and nearly half saw improvement (32% complete and 16% partial resolution) in symptoms.[104] While it seems that LITT may be useful for both recurrent tumors and radiation necrosis, it is not clear if LITT offers a particular advantage over craniotomy.

The role of LITT in the treatment of brain metastases compared to open resection is not well-understood. The largest retrospective review in the literature consisted of 75 patients with previously irradiated tumors (42 with recurrence and 33 with radiation necrosis) that were treated with LITT ($n = 16$ for recurrence and $n = 18$ for radiation necrosis) or craniotomy ($n = 26$ for tumor recurrence and $n = 15$ for radiation necrosis).[102] In this study, both PFS (1-year PFS 72.2% vs. 61.1%, 2-years: 60.0% vs. 61.1%, respectively, $P = .72$) and OS (1-year OS, 69.0% vs. 69.3%, 2-years: 56.6% vs. 49.5%, respectively, $P = .90$) between groups were not significantly different. Clinical outcomes for LITT compared to craniotomy, including perioperative complication rate (35.3% vs. 24.4%, $P = .32$) and steroid cessation (34.8% vs. 47.4%, $P = .53$), were not significantly different. While overall (complete or partial) symptom improvement (87% vs. 90%) was similar between groups, the rate of complete symptom resolution was significantly higher following craniotomy compared to LITT (72.4% vs. 26.1%, respectively, $P < .01$). The need for salvage radiation therapy was approximately 40% following both craniotomy and LITT ($P > .99$), and specifically, among patients treated for radiation necrosis, the rate of postoperative bevacizumab was not significantly different (38.9% vs. 20%, respectively, $P = .28$). While the largest of its kind, the retrospective nature of this study limits the conclusiveness of LITT for brain metastases, as patient selection bias, particularly for those with recurrent brain metastases, is a concern.

The current literature demonstrates rather consistent benefit from LITT in the treatment of brain metastasis recurrence or radiation necrosis in previously irradiated tumors. However, in comparison with traditional craniotomy for the management of these pathologies, its utility is not as clear. In the retrospective study by Hong et al. above, there were no significant differences between craniotomy and LITT for tumor recurrence and radiation necrosis. However, time under anesthesia was significantly long for LITT (7.6 hours) compared to craniotomy (4.5 hours, $P < .0001$).[102] While there are no data on the topic currently, the cost-benefit of LITT compared to craniotomy, as well as allocation of hospital resources, such as OR time and anesthesia, for a LITT procedure has yet to be determined. Additional prospective and randomized clinical trials are needed to compare LITT versus craniotomy for the management of recurrent brain metastasis or radiation necrosis.

6 Management of recurrence—When to repeat resection

Brain metastases recur in 20%–40% of patients following surgical resection alone.[5,9,11,105] When treated with SRS alone, distant intracranial recurrence was noted in up to 45% of patients.[106] Comparatively, in a randomized clinical trial, local recurrence among patients with a single brain metastasis has been shown to be lower following resection (5/25, 20%) compared to radiation alone (12/23, 52%, $P < .02$).[5] Risk of recurrence is dependent on number and size of brain metastases at the time of initial treatment and extracranial disease burden.[107] Recurrence is defined as either local recurrence, due to subtotal resection or residual microscopic disease, or distant recurrence, which reflects a new focus of metastasis to the brain. Particularly in the setting of previously irradiated metastases or patients receiving systemic checkpoint inhibitor agents, suspected recurrence can be confused with treatment effect (pseudoprogression, radiation necrosis). Treatment options available for the management of recurrent brain metastases include surgical resection of the recurrent lesion, LITT, or irradiation with SRS or WBRT.

The clinical decision to resect recurrent metastases depends on factors including anatomic location, tumor histology, and patient clinical presentation and performance status. Sundaresan and colleagues prospectively identified patients with stable extracranial disease who presented with recurrent symptomatic brain metastases for re-resection.[105] The result of their 21 patient analysis suggested re-operation was an important therapeutic option for recovery of neurologic baseline and resulted in improved KPS to 80 or greater in two-thirds of patients.[105] Additionally, Bindal and colleagues retrospectively reviewed 48 patients who recurred following initial resection. Following reoperation, median survival was 11.5 months, which was concordant with median survival of 12–14 months among patients undergoing an initial craniotomy in previous studies.[6,9] Following a second craniotomy, 26 patients recurred, of whom 17 underwent a third craniotomy. Repeat craniotomy once again significantly improved median survival (8.6 months vs. 2.8 months, $P \leq .0001$).[11] Among patients who underwent a third craniotomy after re-recurrence, all but four had just one or two lesions.

Patient selection bias is critical when considering the results of these studies. The decision to re-resect is highly variable and should be determined on a case-by-case basis. These studies, although limited to retrospective review, reflected increased survival and improved neurological outcomes following repeat resection for recurrent brain metastasis among those who are good surgical candidates with tumors that are symptomatic and surgically accessible.[11,105]

Stereotactic radiosurgery may also be pursued for local or distant recurrence of brain metastases. Minniti and colleagues demonstrated the safety and efficacy of re-irradiating recurrent or progressive brain metastases after initial SRS.[108] Forty-seven lesions were treated with three fractions of 7–8 Gy and achieved 1- and 2-year local control rates of 70% and 60%, respectively, and 1- and 2-year survival rates of 37% and 20%. Local control was significantly influenced by primary malignancy, achieving 1-year local control rates for breast (78%) and non-small cell lung cancer (73%) compared to just 38% for melanoma metastases ($P = .01$).[108] As mentioned above for resection following recurrence, patient performance status (KPS, $P = .03$) and control of extracranial disease ($P = .01$) were critical in predicting survival among these patients.[108] Other positive prognostic indicators for response to re-irradiation, as suggested by Kurtz et al., include extended progression-free survival periods (> 365 days from initial radiation) and younger age.[109]

7 Management of CNS metastases in uncommon locations

7.1 Skull base metastases

The incidence of skull base metastasis is highly variable depending on mechanism of study, though ranging from 9% to 14% in a series of early autopsy studies.[110–112] Similar to parenchymal metastases, these lesions evolve via hematogenous dissemination of systemic disease, though direct invasion of certain regional malignancies may occur.[113] Most common malignancies to metastasize to the skull base include breast, lung, and prostate cancers.[114–116] Malignancies involving the head and neck such as squamous cell carcinomas (e.g., nasopharyngeal), nasal sinus malignancies, such as adenocystic carcinoma, as well as other malignancies

including esthesioneuroblastoma can directly invade the skull base. In end-stage disease, these tumors may progress to involve the dura and/or brain parenchyma, and intradural metastases directly spreading from calvarial metastases were noted in 61% of patients in one series.[116]

Metastatic invasion of the skull base often occurs insidiously and therefore is typically found incidentally (between 11% and 50% of all cases)[114–117] or presents with headache, signs of increased intracranial pressure, or focal neurologic deficits such as weakness or cranial neuropathies.[113] Upon discovery, evaluation for a primary malignancy remains crucial to determining treatment course. In the setting of a solitary lesion without a known primary cancer, biopsy may be indicated. Often, management of the systemic disease is most effective for prognosis, particularly in the setting of asymptomatic skull base or dural lesions, as skull base involvement is usually a sign of end-stage disease. However, with symptomatic lesions or extensive lesions causing mass effect, particularly in patients with good prognosis, local management may be necessary. Treatment can mitigate symptoms and improve morbidity, but ultimately, few patients will die due to complications related to local skull base metastases, but rather due to progression of extracranial disease.[118]

Targeted radiotherapy (SRS, Gamma Knife) is typically preferred for the management of skull base metastases, as it has been shown to both alleviate cranial nerve deficits caused by tumor mass effect and improve overall survival.[118,119] In a small prospective study, 11 patients with skull base metastases were treated with SRS, and though the follow-up period was just 9–36 months, all 11 patients demonstrated local control (7 died from progression of extracranial disease) and 10 of the 11 (91%) had complete or partial resolution of their presenting cranial neuropathies.[119] Surgical resection is reserved for those with limited disease and good prognosis. Additionally, the adjacent anatomy associated with these lesions may make surgical resection technically challenging as the skull base neurovascular structures may become encased by tumor or adhered to the metastasis, thus increasing risk of morbidity or mortality. Appropriate radiographic evaluation of the tumor is critical to determining surgical candidates and for operative planning. However, surgical resection management of skull base lesions has not generally been suggested to improve either overall survival or local progression-free survival.[120] The role of surgery for this type of metastasis is likely for palliation of associated neurologic symptoms, such as vision loss or other cranial nerve deficits, caused by mass effect.[120]

7.2 Dural metastases

As with skull base metastases, dural metastases are clinically uncommon, but are diagnosed in as many as 20% of patients at the time of autopsy. The most common malignancies to seed the dura are breast, prostate, lung, and melanoma. Presentation of dural metastases is typically insidious, often first developing nonspecific symptoms, such as headache. However, as the tumor progresses and/or invades parenchyma, focal neurologic deficits and seizures may occur.

There is no consensus on the best treatment of dural metastases. Clinical decision-making depends on the primary malignancy, as well as the burden of dural disease. Since these tumors are within the dura and therefore not protected by the blood-brain barrier, they should receive adequate dosing of systemic therapies appropriate for the primary malignancy. In patients with a solitary, or dominant, lesion, or symptomatic lesion, surgical resection is safe and effective. The outcomes data regarding dural metastases are sparse. In a retrospective review of 122 patients with isolated dural metastases, resection was shown to improve survival (HR 0.57, 95% CI 0.33–0.98, $P = .04$) and PFS (HR 0.50, 95% CI 0.40–2.29, $P = .006$).[114] Patients that undergo resection often receive postoperative radiation.[117] It is possible to treat with radiation alone if a patient is not a good surgical candidate due to disseminated dural involvement, medical comorbidities, or a primary cancer that is considered radiosensitive. However, radiation therapy in the form of WBRT, focal external beam radiation, or SRS did not significantly improve survival, though it may be the only option among patients with dural involvement that is not amenable to surgical resection (HR 0.78, 95% CI 0.52–1.15, $P = .21$).[114]

7.3 Leptomeningeal disease

LMD is most common in lymphoma, breast, lung, and melanoma histologies. Overall survival among solid tumor LMD is measured in weeks to months.[121] Surgical intervention for LMD has not been shown to benefit patient survival due to diffuse involvement of the central nervous system. Systemic therapy to treat the primary malignancy can be continued if the drug readily crosses the blood-brain barrier at an effective dose. Otherwise, intrathecal chemotherapy with either methotrexate or cytarabine can be used for pharmacologic treatment directed at the entire neuroaxis. Intrathecal chemotherapy administration via an Ommaya reservoir into the central nervous systemic ventricular system or by infusion into the thecal sac via lumbar puncture can be achieved. Intraventricular delivery systems are favored, as it allows for more uniform distribution of adequate doses throughout the central nervous system.[122–124] A trend toward improved overall survival, though statistically insignificant, with intraventricular delivery compared to lumbar puncture delivery was observed in a randomized clinical trial (ORR 65% vs. 48%, $P > .1$, respectively).[125]

Some clinical studies describe benefits of radiation therapy for LMD, including improved symptom relief from radiation therapy targeted at regions of symptomatic LMD compared to intrathecal chemotherapy alone.[126] Radiation therapy to the entirety of the neuroaxis should be avoided due to significant risk for neurotoxicity and significant myelosuppression.[121] Despite symptomatic benefit from radiation therapy, there has been no conclusive evidence for significant improvement in overall survival following radiation therapy in patients with LMD.[127,128] Overall survival among these patients is dependent on clinical performance status (Karnofsky performance score), younger age, concurrent administration of intrathecal chemotherapy, and primary malignancy.[128] In a systematic review of 7 breast cancer and 10 non-small cell lung cancer studies, patients with such demographics showed improved survival following WBRT (breast: 17 weeks vs. 11.9 weeks, $P = .015$; NSCLC: 17.6 weeks vs. 12.2 weeks, $P = .041$).[128]

From a neurosurgical perspective, surgical management of LMD is directed at palliation of clinical manifestations related to involvement of the leptomeninges. For example, the development of obstructive/noncommunicating hydrocephalus in a patient may warrant some form of CSF diversion therapy, such as ventriculoperitoneal shunting. As such, surgical intervention will have no effect on LMD progression, but can prevent morbidity and mortality due to complications associated with LMD.

8 Conclusion

Brain metastases remain a leading cause of morbidity and mortality for cancer patients, but options for surgical management continue to expand. Advances in techniques and technologies have allowed the modern neurosurgeon to resect tumors that were once deemed unresectable, either due to their depth or due to their proximity to eloquent brain. Keyhole approaches and tubular retractors decrease approach morbidity and improve patient satisfaction. Preoperative and intraoperative mapping and monitoring allow for maximal resection with minimal neurologic compromise. LITT and SRS are minimally invasive modalities that allow access to surgically inaccessible and/or radiosensitive malignancies. Until systemic therapies become viable options, continued advances in surgical strategies are needed to optimize outcomes for patients with brain metastases.

Disclosures

The authors have nothing to disclose.

References

1. Ostrom QT, Wright CH, Barnholtz-Sloan JS. Brain metastases: epidemiology. *Handb Clin Neurol.* 2018;149:27–42.
2. Johnson JD, Young B. Demographics of brain metastasis. *Neurosurg Clin N Am.* 1996;7(3):337–344.
3. Wen PY, Loeffler JS. Management of brain metastases. *Oncology (Williston Park).* 1999;13(7):941–954. 957–961; discussion 961–942, 949.
4. Grant FC. Concerning intracranial malignant metastases: their frequency and the value of surgery in their treatment. *Ann Surg.* 1926;84(5):635–646.
5. Patchell RA, Tibbs PA, Walsh JW, et al. A randomized trial of surgery in the treatment of single metastases to the brain. *N Engl J Med.* 1990;322(8):494–500.
6. Vecht CJ, Haaxma-Reiche H, Noordijk EM, et al. Treatment of single brain metastasis: radiotherapy alone or combined with neurosurgery? *Ann Neurol.* 1993;33(6):583–590.
7. Mintz AH, Kestle J, Rathbone MP, et al. A randomized trial to assess the efficacy of surgery in addition to radiotherapy in patients with a single cerebral metastasis. *Cancer.* 1996;78(7):1470–1476.
8. Hart MG, Grant R, Walker M, Dickinson H. Surgical resection and whole brain radiation therapy versus whole brain radiation therapy alone for single brain metastases. *Cochrane Database Syst Rev.* 2005;(1), CD003292.
9. Bindal RK, Sawaya R, Leavens ME, Lee JJ. Surgical treatment of multiple brain metastases. *J Neurosurg.* 1993;79(2):210–216.
10. Iwadate Y, Namba H, Yamaura A. Significance of surgical resection for the treatment of multiple brain metastases. *Anticancer Res.* 2000;20(1B):573–577.
11. Bindal RK, Sawaya R, Leavens ME, et al. Reoperation for recurrent metastatic brain tumors. *J Neurosurg.* 1995;83(4):600–604.
12. Singh R, Stoltzfus KC, Chen H, et al. Epidemiology of synchronous brain metastases. *Neurooncol Adv.* 2020;2(1), vdaa041.
13. Cagney DN, Martin AM, Catalano PJ, et al. Incidence and prognosis of patients with brain metastases at diagnosis of systemic malignancy: a population-based study. *Neuro Oncol.* 2017;19(11):1511–1521.
14. Dammers R, Schouten JW, Haitsma IK, et al. Towards improving the safety and diagnostic yield of stereotactic biopsy in a single centre. *Acta Neurochir.* 2010;152(11):1915–1921.
15. Dammers R, Haitsma IK, Schouten JW, et al. Safety and efficacy of frameless and frame-based intracranial biopsy techniques. *Acta Neurochir.* 2008;150(1):23–29.
16. Kickingereder P, Willeit P, Simon T, Ruge MI. Diagnostic value and safety of stereotactic biopsy for brainstem tumors: a systematic review and meta-analysis of 1480 cases. *Neurosurgery.* 2013;72(6):873–881. discussion 882; quiz 882.
17. Mohyeldin A, Lonser RR, Elder JB. Real-time magnetic resonance imaging-guided frameless stereotactic brain biopsy: technical note. *J Neurosurg.* 2016;124(4):1039–1046.
18. Starr PA, Martin AJ, Ostrem JL, et al. Subthalamic nucleus deep brain stimulator placement using high-field interventional magnetic resonance imaging and a skull-mounted aiming device: technique and application accuracy. *J Neurosurg.* 2010;112(3):479–490.
19. Hall WA. The safety and efficacy of stereotactic biopsy for intracranial lesions. *Cancer.* 1998;82(9):1749–1755.
20. Akshulakov SK, Kerimbayev TT, Biryuchkov MY, et al. Current trends for improving safety of stereotactic brain biopsies: advanced optical methods for vessel avoidance and tumor detection. *Front Oncol.* 2019;9:947.
21. Teixeira MJ, Fonoff ET, Mandel M, et al. Stereotactic biopsies of brain lesions. *Arq Neuropsiquiatr.* 2009;67(1):74–77.
22. Kulkarni AV, Guha A, Lozano A, Bernstein M. Incidence of silent hemorrhage and delayed deterioration after stereotactic brain biopsy. *J Neurosurg.* 1998;89(1):31–35.
23. Frati A, Pichierri A, Bastianello S, et al. Frameless stereotactic cerebral biopsy: our experience in 296 cases. *Stereotact Funct Neurosurg.* 2011;89(4):234–245.
24. Kongkham PN, Knifed E, Tamber MS, Bernstein M. Complications in 622 cases of frame-based stereotactic biopsy, a decreasing procedure. *Can J Neurol Sci.* 2008;35(1):79–84.
25. Garrett M, Consiglieri G, Nakaji P. Transcranial minimally invasive neurosurgery for tumors. *Neurosurg Clin N Am.* 2010;21(4):595–605. v.
26. Raza SM, Garzon-Muvdi T, Boaehene K, et al. The supraorbital craniotomy for access to the skull base and intraaxial lesions: a technique in evolution. *Minim Invasive Neurosurg.* 2010;53(1):1–8.
27. Gazzeri R, Nishiyama Y, Teo C. Endoscopic supraorbital eyebrow approach for the surgical treatment of extraaxial and intraaxial tumors. *Neurosurg Focus.* 2014;37(4):E20.
28. Baker CM, Glenn CA, Briggs RG, et al. Simultaneous resection of multiple metastatic brain tumors with multiple keyhole craniotomies. *World Neurosurg.* 2017;106:359–367.
29. Eroglu U, Shah K, Bozkurt M, et al. Supraorbital keyhole approach: lessons learned from 106 operative cases. *World Neurosurg.* 2019;124:e667–e674.
30. Phang I, Leach J, Leggate JRS, et al. Minimally invasive resection of brain metastases. *World Neurosurg.* 2019;130:e362–e367.
31. Reisch R, Perneczky A. Ten-year experience with the supraorbital subfrontal approach through an eyebrow skin incision. *Neurosurgery.* 2005;57(4 suppl):242–255. discussion 242–255.
32. Ormond DR, Hadjipanayis CG. The supraorbital keyhole craniotomy through an eyebrow incision: its origins and evolution. *Minim Invasive Surg.* 2013;2013, 296469.
33. Ditzel Filho LF, McLaughlin N, Bresson D, et al. Supraorbital eyebrow craniotomy for removal of intraaxial frontal brain tumors: a technical note. *World Neurosurg.* 2014;81(2):348–356.
34. Reisch R, Marcus HJ, Hugelshofer M, et al. Patients' cosmetic satisfaction, pain, and functional outcomes after supraorbital craniotomy through an eyebrow incision. *J Neurosurg.* 2014;121(3):730–734.
35. Kelly PJ, Goerss SJ, Kall BA. The stereotaxic retractor in computer-assisted stereotaxic microsurgery. Technical note. *J Neurosurg.* 1988;69(2):301–306.
36. Plaha P, Livermore LJ, Voets N, et al. Minimally invasive endoscopic resection of intraparenchymal brain tumors. *World Neurosurg.* 2014;82(6):1198–1208.
37. Eliyas JK, Glynn R, Kulwin CG, et al. Minimally invasive transsulcal resection of intraventricular and periventricular lesions through a tubular retractor system: multicentric experience and results. *World Neurosurg.* 2016;90:556–564.
38. Newman WC, Engh JA. Stereotactic-guided dilatable endoscopic port surgery for deep-seated brain tumors: technical report with comparative case series analysis. *World Neurosurg.* 2019;125:e812–e819.
39. Hong CS, Prevedello DM, Elder JB. Comparison of endoscope-versus microscope-assisted resection of deep-seated intracranial lesions using a minimally invasive port retractor system. *J Neurosurg.* 2016;124(3):799–810.
40. Sawaya R, Bindal RK, Lang FF, Suki D. Metastatic brain tumors. In: Kaye AH, Laws RR, eds. *Brain Tumors: An Encyclopedic Approach.* 3rd ed. London: Elsevier Saunders; 2011:864–892.
41. Yoo H, Kim YZ, Nam BH, et al. Reduced local recurrence of a single brain metastasis through microscopic total resection. *J Neurosurg.* 2009;110(4):730–736.
42. Suki D, Hatiboglu MA, Patel AJ, et al. Comparative risk of leptomeningeal dissemination of cancer after surgery or stereotactic radiosurgery for a single supratentorial solid tumor metastasis. *Neurosurgery.* 2009;64(4):664–674. discussion 674–666.

43. Suki D, Abouassi H, Patel AJ, et al. Comparative risk of lep-tomeningeal disease after resection or stereotactic radiosurgery for solid tumor metastasis to the posterior fossa. *J Neurosurg.* 2008;108(2):248–257.

44. Patel AJ, Suki D, Hatiboglu MA, et al. Impact of surgical methodology on the complication rate and functional outcome of patients with a single brain metastasis. *J Neurosurg.* 2015;122(5):1132–1143.

45. Patel AJ, Suki D, Hatiboglu MA, et al. Factors influencing the risk of local recurrence after resection of a single brain metastasis. *J Neurosurg.* 2010;113(2):181–189.

46. Kamp MA, Dibue M, Niemann L, et al. Proof of principle: supra-marginal resection of cerebral metastases in eloquent brain areas. *Acta Neurochir.* 2012;154(11):1981–1986.

47. Kamp MA, Rapp M, Slotty PJ, et al. Incidence of local in-brain progression after supramarginal resection of cerebral metastases. *Acta Neurochir.* 2015;157(6):905–910. discussion 910–901.

48. Pessina F, Navarria P, Cozzi L, et al. Role of surgical resection in patients with single large brain metastases: feasibility, morbidity, and local control evaluation. *World Neurosurg.* 2016;94:6–12.

49. Henzi S, Krayenbuhl N, Bozinov O, et al. Ultrasonic aspiration in neurosurgery: comparative analysis of complications and outcome for three commonly used models. *Acta Neurochir.* 2019;161(10):2073–2082.

50. Khuntia D, Brown P, Li J, Mehta MP. Whole-brain radiotherapy in the management of brain metastasis. *J Clin Oncol.* 2006;24(8):1295–1304.

51. Brown PD, Pugh S, Laack NN, et al. Memantine for the prevention of cognitive dysfunction in patients receiving whole-brain radiotherapy: a randomized, double-blind, placebo-controlled trial. *Neuro Oncol.* 2013;15(10):1429–1437.

52. Brown PD, Gondi V, Pugh S, et al. Hippocampal avoidance during whole-brain radiotherapy plus memantine for patients with brain metastases: Phase III Trial NRG Oncology CC001. *J Clin Oncol.* 2020;38(10):1019–1029.

53. Pollock BE, Brown PD, Foote RL, et al. Properly selected patients with multiple brain metastases may benefit from aggressive treatment of their intracranial disease. *J Neurooncol.* 2003;61(1):73–80.

54. Paek SH, Audu PB, Sperling MR, et al. Reevaluation of surgery for the treatment of brain metastases: review of 208 patients with single or multiple brain metastases treated at one institution with modern neurosurgical techniques. *Neurosurgery.* 2005;56(5):1021–1034. discussion 1021–1034.

55. Salvati M, Tropeano MP, Maiola V, et al. Multiple brain metastases: a surgical series and neurosurgical perspective. *Neurol Sci.* 2018;39(4):671–677.

56. Wronski M, Arbit E, McCormick B, Wronski M. Surgical treatment of 70 patients with brain metastases from breast carcinoma. *Cancer.* 1997;80(9):1746–1754.

57. Kamp MA, Fischer I, Buhner J, et al. 5-ALA fluorescence of cerebral metastases and its impact for the local-in-brain progression. *Oncotarget.* 2016;7(41):66776–66789.

58. Hohne J, Hohenberger C, Proescholdt M, et al. Fluorescein sodium-guided resection of cerebral metastases-an update. *Acta Neurochir.* 2017;159(2):363–367.

59. Patchell RA, Tibbs PA, Regine WF, et al. Postoperative radiotherapy in the treatment of single metastases to the brain: a randomized trial. *JAMA.* 1998;280(17):1485–1489.

60. Schackert G, Steinmetz A, Meier U, Sobottka SB. Surgical management of single and multiple brain metastases: results of a retrospective study. *Onkologie.* 2001;24(3):246–255.

61. Pfisterer WK, Papadopoulos S, Drumm DA, et al. Fiducial versus nonfiducial neuronavigation registration assessment and considerations of accuracy. *Neurosurgery.* 2008;62(3 suppl 1):201–207. discussion 207–208.

62. Wirtz CR, Albert FK, Schwaderer M, et al. The benefit of neuro-navigation for neurosurgery analyzed by its impact on glioblastoma surgery. *Neurol Res.* 2000;22(4):354–360.

63. Hammoud MA, Ligon BL, elSouki R, et al. Use of intraoperative ultrasound for localizing tumors and determining the extent of resection: a comparative study with magnetic resonance imaging. *J Neurosurg.* 1996;84(5):737–741.

64. Lindner D, Trantakis C, Renner C, et al. Application of intraoperative 3D ultrasound during navigated tumor resection. *Minim Invasive Neurosurg.* 2006;49(4):197–202.

65. de Lima Oliveira M, Picarelli H, Menezes MR, et al. Ultrasonography during surgery to approach cerebral metastases: effect on Karnofsky index scores and tumor volume. *World Neurosurg.* 2017;103:557–565.

66. Garcia-Baizan A, Tomas-Biosca A, Bartolome Leal P, et al. Intraoperative 3 tesla magnetic resonance imaging: our experience in tumors. *Radiologia.* 2018;60(2):136–142.

67. Livne O, Harel R, Hadani M, et al. Intraoperative magnetic resonance imaging for resection of intra-axial brain lesions: a decade of experience using low-field magnetic resonance imaging, Polestar N-10, 20, 30 systems. *World Neurosurg.* 2014;82(5):770–776.

68. Schichor C, Terpolilli N, Thorsteinsdottir J, Tonn JC. Intraoperative computed tomography in cranial neurosurgery. *Neurosurg Clin N Am.* 2017;28(4):595–602.

69. Heilbrun MP, Lee JN, Alvord L. Practical application of fMRI for surgical planning. *Stereotact Funct Neurosurg.* 2001;76(3–4):168–174.

70. Witwer BP, Moftakhar R, Hasan KM, et al. Diffusion-tensor imaging of white matter tracts in patients with cerebral neoplasm. *J Neurosurg.* 2002;97(3):568–575.

71. Berger MS, Ojemann GA. Intraoperative brain mapping techniques in neuro-oncology. *Stereotact Funct Neurosurg.* 1992;58(1–4):153–161.

72. Skirboll SS, Ojemann GA, Berger MS, et al. Functional cortex and subcortical white matter located within gliomas. *Neurosurgery.* 1996;38(4):678–684. discussion 684–675.

73. Krieg SM, Schaffner M, Shiban E, et al. Reliability of intraoperative neurophysiological monitoring using motor evoked potentials during resection of metastases in motor-eloquent brain regions: clinical article. *J Neurosurg.* 2013;118(6):1269–1278.

74. Krieg SM, Picht T, Sollmann N, et al. Resection of motor eloquent metastases aided by preoperative nTMS-based motor maps-comparison of two observational cohorts. *Front Oncol.* 2016;6:261.

75. Sanmillan JL, Fernandez-Coello A, Fernandez-Conejero I, et al. Functional approach using intraoperative brain mapping and neurophysiological monitoring for the surgical treatment of brain metastases in the central region. *J Neurosurg.* 2017;126(3):698–707.

76. Obermueller T, Schaeffner M, Shiban E, et al. Intraoperative neuromonitoring for function-guided resection differs for supratentorial motor eloquent gliomas and metastases. *BMC Neurol.* 2015;15:211.

77. Chua TH, See AAQ, Ang BT, King NKK. Awake craniotomy for resection of brain metastases: a systematic review. *World Neurosurg.* 2018;120:e1128–e1135.

78. Groshev A, Padalia D, Patel S, et al. Clinical outcomes from maximum-safe resection of primary and metastatic brain tumors using awake craniotomy. *Clin Neurol Neurosurg.* 2017;157:25–30.

79. Magill ST, Han SJ, Li J, Berger MS. Resection of primary motor cortex tumors: feasibility and surgical outcomes. *J Neurosurg.* 2018;129(4):961–972.

80. Lehrer EJ, Peterson JL, Zaorsky NG, et al. Single versus multifraction stereotactic radiosurgery for large brain metastases: an international meta-analysis of 24 trials. *Int J Radiat Oncol Biol Phys.* 2019;103(3):618–630.

81. Chang EL, Wefel JS, Hess KR, et al. Neurocognition in patients with brain metastases treated with radiosurgery or radiosurgery plus whole-brain irradiation: a randomised controlled trial. *Lancet Oncol.* 2009;10(11):1037–1044.

82. Suh JH, Chao ST, Angelov L, et al. Role of stereotactic radiosurgery for multiple (>4) brain metastases. *J Radiosurg SBRT.* 2011;1(1):31–40.

83. Yamamoto M, Serizawa T, Shuto T, et al. Stereotactic radiosurgery for patients with multiple brain metastases (JLGK0901): a multi-institutional prospective observational study. *Lancet Oncol.* 2014;15(4):387–395.

84. Linskey ME, Andrews DW, Asher AL, et al. The role of stereotactic radiosurgery in the management of patients with newly diagnosed brain metastases: a systematic review and evidence-based clinical practice guideline. *J Neurooncol.* 2010;96(1):45–68.

85. Tsao MN, Rades D, Wirth A, et al. Radiotherapeutic and surgical management for newly diagnosed brain metastasis(es): an American Society for Radiation Oncology evidence-based guideline. *Pract Radiat Oncol.* 2012;2(3):210–225.

86. Bindal AK, Bindal RK, Hess KR, et al. Surgery versus radiosurgery in the treatment of brain metastasis. *J Neurosurg.* 1996;84(5):748–754.

87. Auchter RM, Lamond JP, Alexander E, et al. A multiinstitutional outcome and prognostic factor analysis of radiosurgery for resectable single brain metastasis. *Int J Radiat Oncol Biol Phys.* 1996;35(1):27–35.

88. Muacevic A, Wowra B, Siefert A, et al. Microsurgery plus whole brain irradiation versus gamma knife surgery alone for treatment of single metastases to the brain: a randomized controlled multicentre phase III trial. *J Neurooncol.* 2008;87(3):299–307.

89. Choi CY, Chang SD, Gibbs IC, et al. Stereotactic radiosurgery of the postoperative resection cavity for brain metastases: prospective evaluation of target margin on tumor control. *Int J Radiat Oncol Biol Phys.* 2012;84(2):336–342.

90. Brennan C, Yang TJ, Hilden P, et al. A phase 2 trial of stereotactic radiosurgery boost after surgical resection for brain metastases. *Int J Radiat Oncol Biol Phys.* 2014;88(1):130–136.

91. Prabhu RS, Press RH, Patel KR, et al. Single-fraction stereotactic radiosurgery (SRS) alone versus surgical resection and SRS for large brain metastases: a multi-institutional analysis. *Int J Radiat Oncol Biol Phys.* 2017;99(2):459–467.

92. Brown PD, Ballman KV, Cerhan JH, et al. Postoperative stereotactic radiosurgery compared with whole brain radiotherapy for resected metastatic brain disease (NCCTG N107C/CEC.3): a multicentre, randomised, controlled, phase 3 trial. *Lancet Oncol.* 2017;18(8):1049–1060.

93. Lamba N, Muskens IS, DiRisio AC, et al. Stereotactic radiosurgery versus whole-brain radiotherapy after intracranial metastasis resection: a systematic review and meta-analysis. *Radiat Oncol.* 2017;12(1):106.

94. Asher AL, Burri SH, Wiggins WF, et al. A new treatment paradigm: neoadjuvant radiosurgery before surgical resection of brain metastases with analysis of local tumor recurrence. *Int J Radiat Oncol Biol Phys.* 2014;88(4):899–906.

95. Prabhu RS, Patel KR, Press RH, et al. Preoperative vs postoperative radiosurgery for resected brain metastases: a review. *Neurosurgery.* 2019;84(1):19–29.

96. Patel KR, Burri SH, Asher AL, et al. Comparing preoperative with postoperative stereotactic radiosurgery for resectable brain metastases: a multi-institutional analysis. *Neurosurgery.* 2016;79(2):279–285.

97. Kamath AA, Friedman DD, Akbari SHA, et al. Glioblastoma treated with magnetic resonance imaging-guided laser interstitial thermal therapy: safety, efficacy, and outcomes. *Neurosurgery.* 2019;84(4):836–843.

98. Holste KG, Orringer DA. Laser interstitial thermal therapy. *Neuro-Oncology Adv.* 2019;1–6.

99. Salehi A, Kamath AA, Leuthardt EC, Kim AH. Management of intracranial metastatic disease with laser interstitial thermal therapy. *Front Oncol.* 2018;8:499.

100. Ahluwalia M, Barnett GH, Deng D, et al. Laser ablation after stereotactic radiosurgery: a multicenter prospective study in patients with metastatic brain tumors and radiation necrosis. *J Neurosurg.* 2018;130(3):804–811.

101. Telera S, Fabi A, Pace A, et al. Radionecrosis induced by stereotactic radiosurgery of brain metastases: results of surgery and outcome of disease. *J Neurooncol.* 2013;113(2):313–325.

102. Hong CS, Deng D, Vera A, Chiang VL. Laser-interstitial thermal therapy compared to craniotomy for treatment of radiation necrosis or recurrent tumor in brain metastases failing radiosurgery. *J Neurooncol.* 2019;142(2):309–317.

103. Rao MS, Hargreaves EL, Khan AJ, et al. Magnetic resonance-guided laser ablation improves local control for postradiosurgery recurrence and/or radiation necrosis. *Neurosurgery.* 2014;74(6):658–667. discussion 667.

104. Chaunzwa TL, Deng D, Leuthardt EC, et al. Laser thermal ablation for metastases failing radiosurgery: a multicentered retrospective study. *Neurosurgery.* 2018;82(1):56–63.

105. Sundaresan N, Sachdev VP, DiGiacinto GV, Hughes JE. Reoperation for brain metastases. *J Clin Oncol.* 1988;6(10):1625–1629.

106. Stockham AL, Suh JH, Chao ST, Barnett GH. Management of recurrent brain metastasis after radiosurgery. *Prog Neurol Surg.* 2012;25:273–286.

107. McTyre E, Ayala-Peacock D, Contessa J, et al. Multi-institutional competing risks analysis of distant brain failure and salvage patterns after upfront radiosurgery without whole brain radiotherapy for brain metastasis. *Ann Oncol.* 2018;29(2):497–503.

108. Minniti G, Scaringi C, Paolini S, et al. Repeated stereotactic radiosurgery for patients with progressive brain metastases. *J Neurooncol.* 2016;126(1):91–97.

109. Kurtz G, Zadeh G, Gingras-Hill G, et al. Salvage radiosurgery for brain metastases: prognostic factors to consider in patient selection. *Int J Radiat Oncol Biol Phys.* 2014;88(1):137–142.

110. Posner JB, Chernik NL. Intracranial metastases from systemic cancer. *Adv Neurol.* 1978;19:579–592.

111. Lesse S, Netsky MG. Metastasis of neoplasms to the central nervous system and meninges. *AMA Arch Neurol Psychiatry.* 1954;72(2):133–153.

112. Meyer PC, Reah TG. Secondary neoplasms of the central nervous system and meninges. *Br J Cancer.* 1953;7(4):438–448.

113. Harrison RA, Nam JY, Weathers SP, DeMonte F. Intracranial dural, calvarial, and skull base metastases. *Handb Clin Neurol.* 2018;149:205–225.

114. Nayak L, Abrey LE, Iwamoto FM. Intracranial dural metastases. *Cancer.* 2009;115(9):1947–1953.

115. Da Silva AN, Schiff D. Dural and skull base metastases. *Cancer Treat Res.* 2007;136:117–141.

116. Laigle-Donadey F, Taillibert S, Mokhtari K, et al. Dural metastases. *J Neurooncol.* 2005;75(1):57–61.

117. Newton HB. Skull and dural metastases. In: Schiff D, Kesari S, Wen PY, eds. *Cancer Neurology in Clinical Practice: Neurologic Complications of Cancer and its Treatment.* Totowa, NJ: Humana Press; 2008:145–161.

118. Pan J, Liu AL, Wang ZC. Gamma knife radiosurgery for skull base malignancies. *Clin Neurol Neurosurg.* 2013;115(1):44–48.

119. Mori Y, Hashizume C, Kobayashi T, et al. Stereotactic radiotherapy using Novalis for skull base metastases developing with cranial nerve symptoms. *J Neurooncol.* 2010;98(2):213–219.

120. Chaichana KL, Flores M, Acharya S, et al. Survival and recurrence for patients undergoing surgery of skull base intracranial metastases. *J Neurol Surg Part B, Skull Base.* 2013;74(4):228–235.

121. Nayar G, Ejikeme T, Chongsathidkiet P, et al. Leptomeningeal disease: current diagnostic and therapeutic strategies. *Oncotarget.* 2017;8(42):73312–73328.

122. Shapiro WR, Young DF, Mehta BM. Methotrexate: distribution in cerebrospinal fluid after intravenous, ventricular and lumbar injections. *N Engl J Med.* 1975;293(4):161–166.

123. Kesari S, Batchelor TT. Leptomeningeal metastases. *Neurol Clin.* 2003;21(1):25–66.

124. Mack F, Baumert BG, Schäfer N, et al. Therapy of leptomeningeal metastasis in solid tumors. *Cancer Treat Rev.* 2016;43:83–91.

125. Hitchins RN, Bell DR, Woods RL, Levi JA. A prospective randomized trial of single-agent versus combination chemotherapy in meningeal carcinomatosis. *J Clin Oncol.* 1987;5(10):1655–1662.

126. El Shafie RA, Böhm K, Weber D, et al. Palliative radiotherapy for leptomeningeal carcinomatosis-analysis of outcome, prognostic factors, and symptom response. *Front Oncol.* 2018;8:641.

127. Yan W, Liu Y, Li J, et al. Whole brain radiation therapy does not improve the overall survival of EGFR-mutant NSCLC patients with leptomeningeal metastasis. *Radiat Oncol.* 2019;14(1):168.

128. Kim J. Examining the survival benefit of radiation therapy on leptomeningeal carcinomatosis and identifying factors associated with survival benefit of whole-brain radiation therapy. *Precis Radiat Oncol.* 2018;2(2):44–51.

7

Epidural metastasis and spinal cord compression

Kester A. Phillips[a] and David Schiff[b]

[a]Department of Neurology, The Ben and Catherine Ivy Center for Advanced Brain Tumor Treatment at Swedish Neuroscience Institute, Seattle, WA, United States, [b]Division of Neuro-Oncology, University of Virginia Health System, Charlottesville, VA, United States

1 Introduction

Therapeutic advances in medical oncology in the 21st century have dramatically changed the treatment landscape in cancer care, leading to improved survival rates for patients harboring malignancy and, consequently, an increase in the incidence of spinal metastasis. Approximately 40% of all cancer patients develop metastatic extension into the spinal column, and in that group, roughly 90%–95% have either vertebral or spinal epidural metastasis (SEM).[1–3] This condition represents an ominous clinical event in patients with malignancy and confers significant suffering and a decreased remaining survival time and quality of remaining life. SEM can occur at any time during a patient's disease course but disproportionally arise in patients with advanced-stage disease. However, in some patients, it may also be the initial sign of malignancy.[4] Among the frequent offenders, solid neoplasms, especially those arising from the breast, lung, and prostate, account for most cases. At the same time, hematologic malignancies, such as non-Hodgkin's lymphoma and multiple myeloma, are relatively common. Tumor seeding can develop anywhere along the spinal axis, but display a high tropism for the thoracic spine (69%), followed by the lumbosacral (29%) and cervical regions (10%).[5] About one-third of patients have multilevel involvement; therefore, diagnostic imaging of the entire spine is crucial to exclude metastases at other locations.[6–8] The clinical manifestations of SEM, such as sensory changes and motor weakness, may not be readily discernible, especially in the early stages. However, they can evolve to a fulminant course with calamitous clinical consequences if the problem is undiscovered and not treated. SEM progression can trigger severe and disabling bone pain, pathological fractures, and vertebral body collapse. Moreover, the tumor itself or these catastrophic events may lead to metastatic spinal cord compression (MSCC) or nerve root impingement that can precipitate para- or quadriplegia and a host of subsequent medical complications. For this reason, MSCC is an oncological emergency that requires urgent assessment and aggressive treatment. Accordingly, clinicians caring for patients with malignancy should maintain a heightened degree of suspicion for the possibility of this problem.

In short, the goal of treatment for symptomatic SEM and MSCC remains palliative. It includes the maintenance or restoration of the patient's functional independence, pain relief, reestablishment of biomechanical stability, and local tumor control. A multidisciplinary approach, including corticosteroid administration, decompressive surgery, various radiation delivery techniques, and physical rehabilitation, collectively has a beneficial impact on both functional and survival outcomes.[1,9–12] The role of systemic therapies in the management of SEM and MSCC is relatively limited. Nevertheless, in select cases, treatment with chemotherapy and biologic agents has yielded favorable results.[13–15] Fortunately, as therapeutic approaches for neoplastic disease continue to gain momentum, other potential treatment strategies such as laser interstitial thermal therapy (LITT) and immunotherapy have emerged with promising results.[16,17] Ultimately, early diagnosis and judicious treatment, preferably before the onset of symptoms, remain the mainstay of management. Optimal care requires a multidisciplinary team effort, including medical oncologists, neuro-oncologists, radiation oncologists, radiologists, neurosurgical oncologists, physiatrists, pain specialists, and palliative medicine physicians.

2 Epidemiology

The skeletal system bears the brunt of metastatic deposits after the lung and the liver, with the spinal column being the most common site of tumor burden.[18] Evidence from cadaveric studies has established that upward of 70% of patients with cancer will have spine metastases at death.[19] Although the true incidence of SEM is unknown, about 5%–10% of patients with terminal cancer suffer from epidural metastases.[20–22] Putting this into perspective, on January 1, 2019, cancer survivorship in America was approximately 17 million; that number is projected to climb to more than 22.1 million by 2030.[23] Sadly, this means that about 2.2 million cancer patients would suffer from SEM by that time. Even more sobering, according to reports by Loblaw et al., this number will more than double for patients with MSCC.[24] This projected increase in spinal metastases may create economic pressure on the healthcare system and a substantial financial burden for patients and their families for medical and nonmedical end-of-life care. An earlier report estimated that 2.5% of patients dying of cancer had at least one hospitalization for MSCC within the last 5 years of life.[24] However, recent evidence suggests that the average annual incidence of hospitalization related to MSCC among patients with terminal metastatic cancer in the United States was 3.4%. Of over 75,000 hospitalizations from 1998 to 2006 in the study period, lung cancer (24.9%), prostate cancer (16.2%), and multiple myeloma (11.1%) were the most prevalent tumor histologies.[25] Moreover, the study authors also noted a significant increase in MSCC-related hospitalization costs accentuating the need for updated treatment guidelines for tailored care. To this end, multiple studies have illustrated evidence-based treatment algorithms for cost-effective therapy.[1,26–28]

Spinal metastases affect all age-groups, but the incidence is highest between the ages 40 and 70.[29] The median age at the time of the first episode of SEM is 62 years with a slight male predominance, probably owing to a higher incidence of prostate cancer relative to breast cancer.[24,29] Virtually any systemic neoplasm can disseminate to the spine, but the most frequent cancers associated with SEM are breast cancer (29%), lung cancer (17%), and prostate cancer (14%).[30] Other common primary tumors accounting for the remainder include renal cell carcinoma, colorectal carcinoma, melanoma, and tumors of unknown primary.[8,24,31] SEM also presents a challenge in pediatric oncology. In a series of 2259 solid malignant tumors in children, 5% developed SEM with cord compression during treatment. The most common cause was Ewing's sarcoma and neuroblastoma, followed by osteogenic sarcoma, rhabdomyosarcoma, Hodgkin's disease, soft tissue sarcoma, germ cell tumor, Wilms' tumor, and hepatoma.[32] In a recent study, MSCC was the presenting symptom of new cancer in 75% of pediatric cases.[33]

3 Pathophysiology

Several pathways facilitate tumor embolism from the primary site to the spine. Metastatic cells may exploit the venous, arterial, or lymphatic vasculature or extend directly from paraspinal disease. Moreover, from the viewpoint of locoregional recurrence, surgical insult itself can potentially precipitate new metastatic foci and the acceleration of locoregional tumor growth.[34] Experimental studies have shown that neurosurgical techniques to address vertebral tumors may pose a risk for tumor cell contamination and contribute to local recurrence.[35]

Hematogenous arterial spread to the marrow in the vertebral body is the principal conduit for cancer cell dissemination. This process involves a multistep cascade of events commencing with local invasion of tumor cells into the neighboring stroma, angiogenesis and survival in the systemic circulation, extravasation to secondary sites, and culminates with the formation of ectopic metastatic niches.[36] Because of the highly vascular vertebrae and the marrow within, the spinal canal's ventral aspect is more vulnerable to metastatic seeding than the posterior elements such as the pedicles, lamina, and spinous processes.

Less commonly, direct invasion of the tumor via transforaminal extension from paravertebral sources can occur. This infiltration route is observed more commonly in patients with MSCC due to lymphoma (75%) and, to a lesser extent, in patients with solid tumors (15%).[37–39] Transforaminal extension is also frequent in childhood cancer. For example, approximately 5% of all newly diagnosed patients with neuroblastoma will present with symptomatic MSCC due to transforaminal extension, representing one of the few true emergencies in pediatric oncology.[40]

Lastly, preclinical evidence suggests that SEM may also occur by retrograde venous spread from pelvic tumors via a valveless paravertebral venous plexus (Batson's plexus).[41] Experimentally, the injection of tumor cells into the femoral vein of mice when intrathoracic pressure was normal led to the development of metastases in the lungs. On the contrary, injections in the setting of increased intrathoracic pressure resulted in epidural metastases without vertebral involvement. These findings led to the notion that fluxes in blood flow direction due to intrathoracic and abdominal pressure changes may provide an additional opportunity for metastatic seeding.[42] Other murine animal models have validated this concept.[43,44]

Once tumor cells arrest and colonize the bone marrow, the interplay of several systemic hormones, growth factors, proteases, and locally produced cytokines within the tumor microenvironment (e.g., IL-6, IL-1, TGF-beta, nuclear factor-κB ligand, proteinases, parathyroid hormone, 1,25-dihydroxyvitamin D3, prostaglandins, and osteoprotegerin) ignites homeostatic imbalances favoring the production of osteoclasts that are responsible for bone resorption.[45] This cycle supports tumorigenesis and permits bone matrix degradation, collapse, and compression of the dural sac, root sleeve, and their contents, leading to early vascular compromise, vasogenic edema, and demyelination.[46,47] Prolonged mechanical compression will push cord injury beyond a repairable set point and precipitate irreversible neurodegeneration.

4 Clinical presentation

4.1 Pain

Symptomatic SEM and MSCC often herald devastating clinical consequences. Patients present with a wide array of problems, but the most common initial complaint is pain. Roughly 20% of adult patients with SEM will complain of back pain as the initial manifestation of their disease.[4] Additionally, about 96% of patients with MSCC complain of prodromal pain.[48] In a prospective observational study, pain was present for about 3 months before the diagnosis of MSCC.[49] In a review of 70 children with MSCC, pain was the most common symptom (94%).[50] On account of this, pain in patients with malignancy should not be discounted—it warrants a comprehensive assessment to avoid a missed opportunity for detecting early-stage disease. Treating physicians should also remember that pain symptoms cannot reliably discriminate cancer patients at risk for SEM. Therefore, magnetic resonance (MR) imaging of the entire spine is pivotal during standard workup.[51] It is also essential that clinicians recognize that patients may present with pain in the absence of overlapping neurological deficits.

Pain may be localized, mechanical, or radicular. Local pain is precipitated by periosteal stretching as the tumor expands within the vertebral body, or by epidural venous congestion. Patients often describe this pain as an unremitting deep aching at the site of the involved vertebral body. It is often aggravated by coughing, bending, and sneezing.[49] Valsalva maneuver may also trigger local pain, but it is typically not provoked by movement unless there is spinal instability. Characteristically, local pain is more severe when the patient is recumbent. It is relieved by arising and walking—a key distinction from discogenic pain, which improves with lying down. Local pain occurs at night but resolves during the day. The widely accepted explanation is that local pain is attenuated shortly after awakening when endogenous cortisol production reaches peak levels and worsens when cortisol levels nadir at night. History and physical examination may provide clues about localization. For example, neck pain exacerbated by neck flexion may be a telltale sign of cervical, upper, or midthoracic vertebral body involvement. Moreover, straight-leg raising maneuver will elicit pain of lower thoracic and lumbar spinal metastatic disease. However, on rare occasions, patients may experience pain distant from the metastatic focus posing a diagnostic conundrum. For instance, thoracic and cervical cord lesions can produce pain in the lower extremities simulating lumbar spine pathology.[52,53] Conversely, lumbosacral compression may also precipitate thoracic pain.[49] Clinicians should keep these false localizing signs in mind to avoid an unfruitful diagnostic workup and treatment delays. A broad range of nociceptive pathways mediate bone pain, but inflammatory mediators produced in response to mechanical loading appear to play a key role in bone hyperalgesia.[54] Thus, bone pain responds to traditional analgesics such as nonsteroidal antiinflammatory drugs, nonopiates, and opiates. Fortunately, in situations of severe MSCC-related pain, corticosteroids ease symptoms within hours following administration.[55] For pharmacoresistant bone pain, palliative spinal irradiation yields adequate pain control.

Biomechanical instability produces pain primarily from pathological fracture or vertebral body compression (with or without collapse). Additionally, insults to the ligamentous apparatus, particularly the highly pain-sensitive posterior longitudinal ligament, are also provocative. Mechanical pain is triggered by movement and with the loading of the spine during sitting and standing. In most cases, lying down is the only position where patients find relief. This pain is debilitating and usually requires narcotics, spinal instrumentation, or external orthosis. Moreover, minimally invasive procedures such as vertebroplasty or kyphoplasty, radiofrequency ablation, and cryoablation may provide pain relief in patients without neurological sequela. These techniques aim to overcome the high complication rates associated with conventional surgery and allow faster initiation of adjuvant therapy. Yet still, they are not without limitations.[56] For example, vertebroplasty and kyphoplasty involve percutaneous injection of bone cement into the fractured vertebral body. Vertebroplasty-associated cement leakage into the vertebral canal can induce iatrogenic cord compression. For this reason, patients with blatant spinal instability with symptomatic MSCC are ineligible for both procedures. Despite the appeal of vertebral augmentation, these techniques may not be a logical approach for patients with a poor functional status and a reduced survival time. Finally, external beam radiotherapy has no impact on spinal stability or mechanical pain.

Radicular pain is triggered when tumors become intimately involved with the spinal nerve roots or cauda equina. Radicular pain is present in 80% of cervical, 55% of thoracic, and 90% of patients with lumbosacral MSCC.[48] Patients often report hyperalgesia or a perceived burning or shock-like sensation radiating to the extremities with or without weakness. Pain intensity is a function of tumor location and is most severe with tumor spread to the anterolateral/lateral portion of the vertebral body.[57] Cervical or lumbar radiculopathy may be unilateral or bilateral and instigate pain or weakness in the upper or lower extremities, respectively. Thoracic radiculopathy is usually bilateral and manifests as a band-like pain around the torso. Patients may also develop mechanical radiculopathy due to instability and neuroforaminal compression by the tumor in the lumbar spine. In this scenario, increasing the axial load (i.e., sitting or standing) exacerbates pain. Generally, radicular pain is managed conservatively with anticonvulsants, tricyclic antidepressants, and physical rehabilitation. However, clinical evidence suggests that transforaminal endoscopic nerve root decompression may be a viable treatment option for patients suffering from radicular pain due to nerve compression from spinal metastasis disease.[58–60]

4.2 Motor dysfunction

Motor dysfunction is the second most common complaint in patients with spinal metastases.[48,55] Eighty percent of patients with MSCC present with weakness of varying degrees.[48] Roughly 50% of patients are ambulatory, 35% are paraparetic, and 15% are paraplegic at the time of diagnosis.[48] Motor dysfunction results from disruption of the descending motor fibers within the spinal cord, functionally divided into the pyramidal (responsible for voluntary control) and the extrapyramidal tracts (responsible for involuntary and autonomic control). Regardless of the etiology, insults to either tract will precipitate upper motor neuron (UMN) weakness, whereas compression of the ventral (anterior) or motor root axons within the ventral horn of the spinal cord as well as nerve roots of the cauda equina will produce lower motor neuron (LMN) weakness. UMN signs include hyperreflexia, clonus, spasticity, flexor and extensor spasms, synkinesis, spastic dystonia, and extensor plantar reflex on examination. Weakness usually presents bilaterally and symmetric, but although rare, a Brown-Séquard's syndrome producing ipsilateral motor paralysis can occur.[48,61] UMN weakness in the lower extremities is more pronounced proximally than distally in the early phase of the disease course. Patients often report difficulty climbing stairs or rising from a seated position. As the disease progresses, patients may report unsteady gait, leg heaviness, or knee buckling and falls even when walking on a flat surface. A thorough assessment should include documentation of baseline manual muscle test (MMT) grades, which helps with predicting the chance of regaining ambulatory ability.

In general, the development of motor deficits hinges upon the tumor's growth rate.[62] Rapid onset and quick progression confer a poor prognosis for functional recovery.[63] Moreover, the degree of motor weakness at presentation is a key prognostic variable for functional recovery.[48] Approximately 80% of patients with MSCC who were ambulatory at presentation maintain that status after treatment. Between 30% and 45% of nonambulatory patients with antigravity proximal leg function will regain ambulatory activity, while only 5% of patients without proximal antigravity function will walk again.[48] Spinal cord compression anterolaterally and circumferentially is associated with rapid paralysis.[64] In cases with abrupt paraplegia, the underlying mechanism may be spinal cord infarction due to anterior spinal artery compression. Sadly, once paraplegia develops, it is generally irreversible despite aggressive therapy. It is for this reason that making the diagnosis while patients are ambulatory is strongly advocated. In a recent pediatric study, motor deficit was the initial symptom of MSCC in all patients. Approximately 35% exhibited grade 1 weakness, 43% grade 2 weakness, and 43% grade 3. Severe motor deficit was more frequent in infants (age ≤ 2 years) and was associated with worst motor outcomes.[33]

LMN dysfunction creates hypotonia, flaccid paralysis, muscle atrophy, fasciculations, and areflexia. Weakness predominates distally rather than proximally. Patients typically present with foot drop and give a history of tripping, falls, hitting or scraping their toes on the floor while walking. They generally adopt a high steppage gait to compensate for the weakness in the foot muscles. On examination, dorsiflexion will be reduced.

Finally, patients with advanced-stage systemic cancer are at an increased risk for venous thromboembolism (VTE), and motor weakness and immobility confer a high additional risk. Germane to the optimal VTE management in patients with cancer is the identification of (1) patients who are most likely to benefit from prophylactic anticoagulation and (2) agents that reduce the risk of VTE relapse and mortality. In a series of over 230 consecutive patients requiring surgical treatment of spinal metastases, screening Doppler ultrasonography identified the highest rates of preoperative deep vein thrombosis (DVT) in nonambulatory patients.[65] Generally, in the absence of active bleeding or other contraindications, pharmacologic thromboprophylaxis is administered during periods that pose the most significant risk for VTE (e.g., hospitalization, immobility, cancer-related surgery). In many institutional protocols, short-term use of low-molecular-weight heparin (LMWH), unfractionated heparin, or fondaparinux is used for VTE prophylaxis in patients without contraindications for anticoagulation. Preoperatively, anticoagulation is withheld, and patients are commenced on provisional mechanical prophylaxis, such as pneumatic venous compression devices or graduated compression stockings. In the postoperative setting, pharmacologic thromboprophylaxis for patients undergoing major surgery for cancer should be continued for at least 7–10 days.[66] Over the past decade, LMWH has been the staple of standard treatment for cancer-related thrombosis and its prevention. However, in its most current update, the American Society of Clinical Oncology has added rivaroxaban and edoxaban (selective inhibitors of factor X) as options for both thromboprophylaxis and the treatment of cancer-associated VTE.[66] In the absence of high-quality clinical trial evidence, apixaban and the oral direct thrombin inhibitor dabigatran are not recommended in the management of cancer-related VTE.

4.3 Sensory

Sensory disturbance is a frequent precursor to the diagnosis of MSCC. Sixty-eight percent of patients experience sensory changes 4–41 days before the diagnosis of MSCC.[49] Alterations such as tingling and numbness in its corresponding dermatome usually occur in parallel to severity of motor dysfunction. However, sensory level does not accurately predict the site of MSCC and may vary by up to 3 dermatomes below or above the compressive lesion.[49] On rare occasions, Lhermitte's phenomenon, an electric shock-like sensation radiating down the back into the extremities produced by neck flexion, can indicate thoracic spine metastasis involving the posterior columns.[67,68] This may be clinically relevant for patients with adenocarcinoma of the prostate, papillary and follicular carcinoma of thyroid, seminoma, transitional cell carcinoma of bladder, and clear cell carcinoma of kidney, which may have an affinity for the dorsal spine.[62] Saddle region sensory loss is common in cauda equina lesions.

4.4 Dysautonomia

Spinal cord or cauda equina compression can also evoke dysregulation of the autonomic neural circuitry producing a spectrum of definable clinical symptoms. The consequences can be dramatic and often interfere with daily self-care routines. These dreadful repercussions directly result from disruption of the sympathetic preganglionic neurons in the thoracolumbar gray matter (T1-L2) and parasympathetic preganglionic neurons in the S2-4 nerve roots. Consequently, the gastrointestinal and genitourinary systems are often compromised. Approximately 57% of patients experience bladder and bowel symptoms at the time of diagnosis.[48] Patients may report urinary incontinence (15%), frequency (6%), urgency (3%), and hesitancy (14%).[49] Initial management typically involves a bladder ultrasound, anticholinergic medications, and intermittent catheterization. Regardless of the etiology, a neurogenic bladder introduces a host of medical complications such as recurrent urinary tract infections and sepsis making further care difficult. Bladder issues often accompany sexual dysfunction marked by erectile, orgasmic, and ejaculatory

dysfunction in men, while a decrease in vaginal lubrication is common in women.[69] Moreover, about 74% of patients report bowel disturbances, with the most common complaint being constipation (66%), granted that most patients were medicated with opioids.[49] Depending on the tumor location, neurogenic bowel dysfunction may result in constipation with fecal retention (UMN). However, about 5% of patients report fecal incontinence (LMN).[49] Uniquely, compression of the conus medullaris will produce early autonomic dysfunction, contrasting it from cauda equina-associated dysautonomia, which typically presents late. Overall, these challenges create immense psychosocial disturbances for patients due to the loss of personal independence and dignity. Many patients and caregivers facing these consequences benefit from onco-psychosocial rehabilitation, underscoring the need for a comprehensive patient care approach. Autonomic disturbances can also lead to life-threatening issues. In rare cases, compression in the midcervical region can lead to diaphragmatic paralysis and respiratory failure requiring intubation and mechanical ventilation.[33,70] In an autopsy study of patients who succumbed to thyroid cancer, necrosis in the cervical spinal cord due to MSCC was the immediate cause of death in one case.[71] Finally, Horner's syndrome, an exceedingly rare complication of spinal metastasis, has been reported.[72]

5 Diagnosis

5.1 Magnetic resonance imaging

MR imaging is the current modality of choice for the diagnosis of SEM and MSCC. By far, the capabilities of this technique are unmatched when compared to other imaging modalities. MR imaging provides outstanding soft-tissue characterization, can produce images in several anatomical planes, does not require ionizing radiation to generate images, is less time-consuming, and may serve to guide biopsy of areas of abnormal signal intensity. Furthermore, MR imaging of the spine can detect different pathological conditions without intravenous contrast administration. Unenhanced T(1)-weighted sagittal MRI shows hypointense marrow signal (Fig. 1A). Specialized images such as spin-echo T(1)-weighted and T(2)-weighted pulse sequences (Fig. 1C and D) as well as short tau inversion recovery (STIR) imaging in the sagittal plane (Fig. 1B) permit excellent visualization of spinal bone marrow pathology and can distinguish benign from malignant marrow changes. Images in the axial plane, on the other hand, have proved beneficial for detecting the presence of extramedullary, neural, and paraspinal neoplastic involvement. The standardized MR protocol includes contrast-enhanced sequences, which allows for a detailed topographical assessment of tumors (Fig. 2).

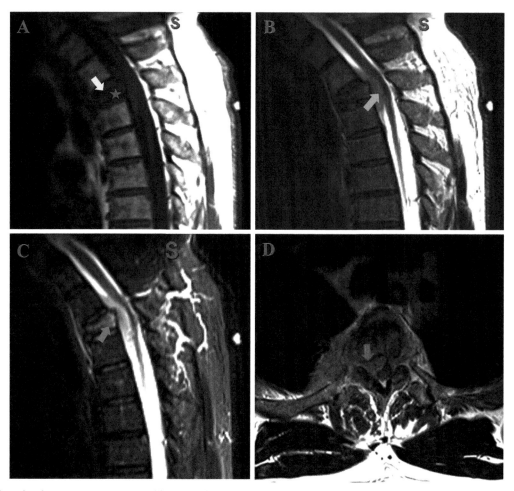

FIG. 1   Spinal epidural metastasis in 61-year-old man with multiple myeloma. Sagittal T1-weighted MRI (A) with T3 hypointense marrow signal (*red star*) and pathological compression fracture (*yellow arrow*), sagittal T2-weighted (B) illustrated a T3 hypointense lesion within the epidural space (*green arrow*) with spinal cord displacement and compression, STIR image (C) shows T3 hyperintense marrow signal (*blue arrow*), axial T2-weighted (D) with right ventral and lateral epidural soft tissue component centered at the T3 vertebral body resulting in moderate to severe central canal stenosis and leftward displacement and mild flattening of the cord (*red arrow*).

MR physics also provides for the suppression of normal adipose tissue to reliably assess contrast-enhanced images. MR imaging can delineate intramedullary, intradural extramedullary, and extradural lesions, particularly in the epidural compartment, resulting in compression of the spinal cord. The technique has a reported diagnostic accuracy of 98.7% for detecting vertebral metastasis[73] and a 93% sensitivity and 97% specificity for detecting MSCC.[74] T(1)-weighted sagittal images alone have a specificity of 89% for detecting epidural metastasis.[75]

Advanced MR imaging techniques offer promising advantages to conventional modalities for the assessment of response to treatment. Dynamic contrast-enhanced MRI can be used to predict local tumor recurrence in patients with spinal metastases undergoing high-dose RT.[76] Furthermore, decreased signal intensity on diffusion-weighted MR imaging following successful radiotherapy correlated with favorable treatment response.[77] Despite its strengths, MR imaging does not provide morphological data about the extent of bony destruction and therefore is not adequate for preoperative evaluation for spinal stabilization surgery.

5.2 Computed tomography

Computed tomography (CT) provides excellent visualization of tumor-induced bone destruction. CT provides excellent bone surface delineation and enables the detection of cortical damage. An epidural metastatic lesion may appear as an amorphous soft tissue displacing the thecal sac of filling the neural foramen. CT also provides superb visualization of vertebral body fractures and has the advantage of rapid image acquisition, making it very convenient in the emergency setting. Moreover, CT findings are instrumental for preoperative planning in complex cases. Although this modality is beneficial, it has several limitations. Metastatic lesions without bone destruction may go undetected. Additionally, CT does

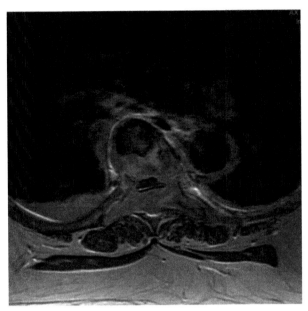

FIG. 2 Spinal epidural metastasis in 68-year-old woman with T6 plasmacytoma. Axial postcontrast T1-weighted image shows the involvement of the vertebral body with expansile changes in the posterior elements with small extraosseous posterior epidural component resulting in severe central stenosis.

not exclude cortical bone destruction due to intrinsic degenerative diseases such as osteoporosis. Finally, the health risks associated with radiation exposure are public health concerns, particularly in the pediatric patient population.

5.3 CT myelography

Before the advent of MRI, myelography was the procedure of choice to diagnose spinal metastasis and associated MSCC. Nowadays, CT myelography is a useful

procedure for evaluating patients with contraindications to undergoing MRI (e.g., electronic implantable device, severe claustrophobia, morbid obesity). The technique not only permits assessment of the thecal sac contents, but it also allows for concurrent cerebrospinal fluid (CSF) sampling. Postoperative CT myelogram (Fig. 3B) is the preferred modality for guiding adjuvant radiotherapy and for delineation of the CSF space and spinal cord in the setting of spinal instrumentation. However, the procedure carries inherent shortcomings and risks. The risk of neurological deterioration after removing CSF below the level of a complete spinal subarachnoid block is well known.[78] Moreover, in patients with complete myelographic block, CT myelogram may not detect widely separated metastatic lesions. Other postmyelographic complications include nerve root avulsion, hematomas, seizures, paraplegia, and contrast-related encephalopathy.[79–81]

5.4 Radionucleotide bone scan

Bone scintigraphy utilizing the radiotracer technetium-99m-labeled diphosphonates is frequently performed as a screening tool for skeletal metastases. Radionucleotide bone scans are not specific, but more sensitive than plain films for detecting bone metastases.[82] The tracers accumulate rapidly in bone, but the extent of uptake relies on the rate of new bone formation and blood flow. As such, the scan plays a vital role in disease staging and management, particularly for carcinoma of the breast, lung, and prostate. However, the modality has limited utility in those tumors with minuscule osteoblastic response to bone destruction (e.g., multiple myeloma). More importantly, bone scintigraphy does not detect adjacent soft tissue abnormalities.

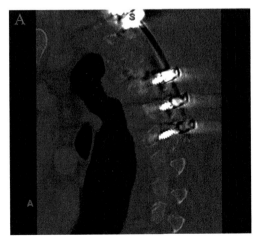
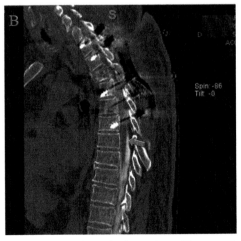

FIG. 3 Sagittal, CT spine without contrast (A) illustrating T3 corpectomy and right lateral vertebral resection with posterolateral fusion with transpedicular screws at T1, T2, T4, T5, and T6. Postoperative CT myelogram (B) demonstrating myelographic block at the T6 level (*red arrow*).

5.5 Radiography

Plain radiographs are widely available and inexpensive and provide a rapid assessment of bone integrity. Radiographs can reveal vertebral pathology, including pedicle erosion, widening of the neural foramina, altered bone density, and vertebral compression fractures. Although this modality was historically useful, it has limited diagnostic utility at present. Radiography is insensitive for detecting early or small metastatic deposits and has a false negative rate of up to 17% for the diagnosis of SEM.[20] Most importantly, plain radiographs cannot identify MSCC. A prospective study showed that in patients with suspected MSCC, MR imaging led to a change in the radiotherapy treatment plans in 53% of cases that were originally based on plain radiographs, underscoring the pitfall of plain films.[31]

6 Prognosis

Survival estimates following active treatment of SEM and MSCC have proven to be inexact and unreliable.[83–85] As a result, many groups have recommended integrating more accurate models into clinical practice. These scoring systems vary but incorporate elements from the medical history, physical examination, laboratory findings (e.g., serum albumin and lactate dehydrogenase levels), imaging, and pathology to help forecast patient outcomes. Most predictive models combine primary tumor histology, the patient's functional status, metastatic disease burden (e.g., extraspinal bone metastases, number of vertebral body metastases, visceral metastases), and spinal cord palsy into scoring systems.[86–90] Furthermore, the interval between tumor diagnosis and MSCC, duration of motor deficits, and tumor growth rate all influence survival.[28,62] In addition, presurgical pain also portends poor postsurgical survival.[91,92] Given the complexity of these systems, other groups have encouraged a more personalized treatment approach.[93] Not to mention, in the era of molecular oncology, more modern prognostication scoring systems have taken into account the survival gains attributed to the use of molecularly targeted drugs.[93–97] To illustrate this point, in an updated scoring system for patients with skeletal metastases, primary cancer was categorized into three groups: tumors that exhibited rapid growth, moderate growth, or slow growth. Cancer patients with a median survival time of > 20 months were classified into the slow-growth group, which included hormone-dependent prostate cancer and hormone-dependent breast cancer. On the other hand, patients with a median survival time from 10 to 20 months were classified into the moderate-growth category, including nonsmall cell lung cancer with molecularly targeted therapy (e.g., gefitinib and/or erlotinib), hormone-independent breast cancer, and hormone-independent prostate cancer. However, patients who harbored lung cancer without molecularly targeted drugs were regarded as a rapid-growth tumor.[97] Although these algorithms help guide appropriate clinical decision-making, they are less than perfect; each has its inherent limitations.[94,98]

Prognostic scoring systems are largely employed to avoid neurosurgical intervention in patients with poor survival. Generally, an expected remaining survival time of 3–6 months is the accepted threshold for excisional procedures.[87,88,99] Perhaps the most widely recognized prognostic scoring systems are the Tomita score[90,100] and the revised Tokuhashi system[88] (rTs). The Tomita score was introduced in 2001 and comprised three parameters based on the tumor growth rate, visceral metastases, and the number of bone metastasis lesions.[90] Tokuhashi and colleagues initially published their scoring system in 1989 but later published a revised version in 2005 to include life expectancy predictions for patients receiving conservative treatment (radiation, chemotherapy, hormonal therapy, or analgesics only). This semiprospective study evaluated a revised scoring system's accuracy for predicting metastatic spinal tumor prognosis and selected management strategies (palliative surgery, excisional surgery, or conservative treatment) for the predicted prognosis. The study group incorporated several parameters, including performance status, number of extraspinal bone metastases, number of metastases in the vertebral body, presence or absence of metastases to major internal organs, site of the primary lesion, and severity of palsy.[88] Although the rTs is widely used, several authors have challenged its usability, reporting a total predictive ability of about 60%.[101–103] On the grounds of this, a recent single-institution retrospective study proposed adjusting the Tokuhashi category scores to make allowances for modern spinal surgery techniques.[104] Furthermore, in light of primary tumor heterogeneity and the low number of patients with lung cancer metastases included in the rTs, Cai et al. also suggested modifying rTs to improve predictive accuracy in patients with lung cancer harboring spinal metastases.[105] The study found that targeted therapy and tumor marker levels were additional independent predictors of survival in this patient population. Based on the data analysis, the prognosis of patients who did not receive targeted therapy was poorer than that of patients who underwent targeted therapy. Equally, patients with abnormal tumor markers fared much worse than patients with normal tumor markers.[105] Additionally, a retrospective study also noted an increase in the mean posttreatment survival period from 5.1 to 9.3 months in patients with metastatic spine tumor from lung primary treated with molecularly targeted drugs.[106] Finally, the Skeletal Oncology Research Group (SORG) recently compared the ability of several well-known scoring systems to provide estimates for both overall survival at various time points and tumor-specific survival for patients undergoing surgery for the treatment of metastatic spine disease. Algorithms were accurate if the area under the curve (AUC) was > 0.70. Of the nine scoring systems evaluated, the SORG Nomogram demonstrated the highest accuracy at predicting 30-day (area under the curve [AUC] 0.81) and 90-day (AUC 0.70) survival after surgery.[107] Other current systems such as the SORG machine learning algorithms that provide validation of clinical prediction models are accessible through an open-access Web application for smartphones, tablets, or desktop computers and currently output probabilities and explanations for predicted survival at 90 days and 1 year after surgery.[108]

There is general agreement that pretreatment ambulatory status is an overriding prognostic factor for guiding therapy.[109] However, the speed of progression of motor deficits also influences the postoperative outcome. In a retrospective study, the speed of progression of motor deficits independently and strongly predicted the chance of postoperative ambulatory recovery.[62] The authors discovered that a threshold of 6 days from the onset of neurological symptoms (e.g., numbness or motor weakness) to gait inability was an independent risk factor for failing to regain ambulatory status. Additionally, preoperative weakness in hip flexion (MMT < 3) also foretells irrecoverable gait dysfunction.[62]

The radiosensitivity profile of tumors is one of the most important prognostic factors in the management of SEM.[48] Lymphoma, myeloma, breast cancer, and prostate cancer are associated with improved posttreatment ambulatory status and survival when compared to radioresistant tumors (e.g., nonsmall cell lung, renal cell, esophagus, and melanoma). Fortunately, patients harboring these less radio-responsive tumors may still derive benefit from palliative radiotherapy. In one study, 75% of nonambulatory patients harboring radiosensitive tumors regained ambulatory status after radiotherapy, compared to only 34% of comparable patients with radioresistant tumors.[48] Following treatment, the probability of ambulatory patients surviving 1 year is 0.73, while that of nonambulatory patients is 0.09.[110] Survival estimates among radiation oncologists tend to be inaccurate and propitious.[83] Accordingly, Rades et al. proposed several scoring systems based on data obtained from patients who underwent radiation therapy for MSCC. The first was based on the analysis of 1852 patients. Tumor type, the interval between tumor diagnosis and MSCC, the presence of other bone or visceral metastases at the time of RT, ambulatory status, and duration of motor deficits were significant prognostic factors.[28] Patients who developed motor weakness > 14 days before radiotherapy had better survival than patients who developed weakness within 1–7 days of treatment.[28] Other tumor-specific survival scoring systems for patients with MSCC from prostate cancer,[111] breast cancer,[112] and in cancer of unknown primary[113] were later developed. In general, patients with cancer of unknown primary bear a worse fate, with a median survival of 5–11 months.[114]

7 Treatment

7.1 Corticosteroids

Corticosteroids are an effective temporizing agent for symptomatic management of MSCC. Steroids rapidly alleviate pain and improve neurological function.[115,116] After randomly allocating 57 patients with MSCC to treatment with either high-dose corticosteroid (dexamethasone 96 mg bolus intravenously, followed by 96 mg orally for 3 days then tapered in 10 days) or no corticosteroids during radiation therapy, Sorenson and colleagues found significant improvement in gait function in patients treated with dexamethasone (81% vs. 63%, $P = .046$).[116] Six months after treatment, 59% of the dexamethasone group patients were still ambulatory compared to 33% in the control arm.[116] Historically, controversy existed between the use of high-dose dexamethasone (100 mg loading, then 96 mg daily) and moderate dose (10 mg loading, then 16 mg daily) for the treatment of MSCC. Nevertheless, a retrospective study provided compelling evidence against the use of dexamethasone in high doses for patients with MSCC.[115] When a dexamethasone loading dose of either 10 mg or 100 mg was administered and both followed by dexamethasone 16 mg orally per day in patients with MSCC, there was a significant decrease in pain rating (from an average of 5.2 before treatment to an average of 1.4 one week posttreatment). Nevertheless, a comparison of initial doses of 10 mg IV bolus versus 100 mg IV bolus showed no outcome differences concerning pain, ambulation, or bladder function.[115] There are no guidelines for corticosteroid dose and schedule for MSCC. Typically, a bolus of 10 mg dexamethasone (or equivalent) is given, followed by 16 mg/day (usually twice-daily to four-time-daily). A significant drawback of corticosteroid use is the high rates of treatment-related systemic side effects. The most frequent unwanted neurologic and systemic complications include weight gain, hyperglycemia, peripheral edema, infection, myopathy, and psychiatric disorders.[117] Asymptomatic patients with radiographic evidence of MSCC do not require steroids.[118]

7.2 Surgery

Surgical stabilization and decompression of SEM provide rapid pain relief and restoration of neurological function, which improves the quality of remaining life.

A landmark prospective randomized trial provided the primary evidence for surgical decompression. Patchell et al. illustrated improved ambulation outcomes following direct surgical decompression when compared to radiotherapy alone for patients with SEM.[119] Significant improvements were observed in overall ambulation, maintenance of ambulation, recovery of ambulation, bowel and bladder continence, narcotic requirement, and survival. Notably, 57% of patients in the radiation arm maintained ambulatory status, but the duration was only 13 days compared to ambulation until death (122 days) in the surgical arm. Surgery also offers an opportunity to establish a diagnosis in patients with unknown primaries. The most commonly used decompression techniques include corpectomy (Fig. 3A),[120,121] laminectomies,[122–124] and transpedicular decompression.[125]

Conventional open surgery has inherent drawbacks, including its invasive nature. It carries the risk of neurological deterioration. It is costly, and it may interfere with or delay the administration of adjuvant therapies. Moreover, although spinal stabilization is an indication for surgery, laminectomy is associated with iatrogenic spinal destabilization.[126] As with any surgical procedure, wound infection is also of great concern, especially in patients who have had prior radiotherapy.[127] Given that SBRT has expanded armamentarium to treat spinal tumors, there has been a shift from conventional open surgery to less invasive techniques with an emphasis on local control and spinal stabilization.

Separation surgery and minimally invasive surgery are among the few recent advancements in spinal tumor decompression and stabilization techniques. Since its inception in 2000, separation surgery has garnered much attention, with data validating its safe use and efficacy. The procedure creates a small margin of 2–3 mm between the tumor (particularly radioresistant tumors) and spinal cord, thereby permitting optimal radiation dosimetry to the gross tumor volume, while minimizing collateral damage to the neighboring spinal cord. At the same time, the technique circumferentially decompresses and stabilizes the spinal cord column.[128] Surgeons typically utilize a single-stage posterolateral transpedicular approach to achieve this aim. Adequate separation between the tumor and the spinal cord followed by SBRT achieves > 90% local control at 1 year.[129] However, high-grade epidural tumor extension following separation surgery predicts a greater chance for posttreatment progression.[130] In addition to spinal cord decompression, the technique provided pain relief in 86% of patients with spinal metastases.[131] Additional benefits of separation surgery include a shortened hospital stay, decreased blood loss, reduced perioperative complications, and earlier initiation of adjuvant radiation.[132]

As cutting-edge advancements in surgical technology continue to improve, less invasive operative corridors for decompression and corpectomy are emerging.[133,134] Other promising novel modalities for the treatment of MSCC, such as the use of LITT [17,135] and radiofrequency ablation,[136] seem promising, but strong data for their use are limited. Still, minimally invasive spine surgery techniques such as the intraoperative stereotactic navigation system and percutaneous pedicle screw instrumentation are being incorporated into spinal metastasis surgery.[137]

7.3 Conventional radiotherapy

Conventional external beam radiotherapy (cEBRT) has played a vital role in both upfront and adjuvant therapy for spinal metastasis for decades. The technique delivers high-dose conformal radiation in ablative doses and improved local control rates from 30% at 3 months to over 90% when used as definitive therapy or in the postoperative setting.[138–142] cEBRT also has proven effectiveness for partial pain relief; however, the full analgesic effect rates are unsatisfactory, ranging from 0% to 20%.[143,144] Typical doses include 8 Gy in 1 fraction, 20 Gy in 5 fractions, and 30 Gy in 10 fractions.[143] As discussed earlier, the radiobiology of tumors dictates treatment response.[138,140] In general, most hematologic malignancies (e.g., lymphoma, multiple myeloma, and plasmacytoma) are radiosensitive and respond favorably to cEBRT. Similarly, a few solid tumors such as breast, prostate, ovarian, and seminoma are also quite susceptible to radiotherapy.[145,146] In contrast, other solid tumors such as renal cell carcinoma, colon cancer, nonsmall cell lung cancer, thyroid carcinoma, hepatocellular carcinoma, melanoma, and sarcoma are relatively radioresistant.[138,147] In one study, patients with spinal metastases from breast and prostate primaries were more likely to achieve early pain palliation following cEBRT than lung primary.[148] In patients with recalcitrant pain, a meta-analysis determined that re-irradiation of painful bone metastases is practical and achieves a 58% response rate.[149] A significant pitfall of cEBRT is the lack of precision required to spare adjacent critical structures.

7.4 Stereotactic body radiation therapy

Palliative stereotactic body radiation therapy (SBRT) has been a game-changer in the management of spinal metastases. This technique seeks to overcome tumor radioresistance by delivering high ablative doses in 1–5 fractions while limiting exposure to the neighboring spinal cord.[150,151] Typical doses range from 16 to 24 Gy in 1 fraction, 24–27 Gy in 2–3 fractions, 30 Gy in 4 fractions, and 30–50 Gy in 5 fractions. The proposed radiobiologic effects of SBRT differ from those of cEBRT. Notably, SBRT's tumoricidal effects may involve the induction of tumor necrosis, desmoplasia, radiation-induced tumor-antigen specific immune response, tumor apoptosis, and decreased tumor vessel density in addition to DNA damage.[152–154] The safety and efficacy of SBRT are well-established. In a large single-institution study, Yamada et al. showed that SBRT provides local control rates of up to 98% over 4 years in 811 targeted spine lesions, irrespective of underlying radiation resistance.[155] Similar rates of local control were corroborated in studies evaluating single-fraction and hypofractionated SBRT.[155,156]

The clinical benefit of SBRT extends beyond achieving tumor control; successful treatment also provides adequate pain relief and neurological recovery. Anand et al. observed complete pain relief in 90% of patients with MSCC following hypofractionated SBRT and noted neurological improvement in 60%.[157] In a cohort of 500 cases of spinal metastases (73 cervical, 212 thoracic, 112 lumbar, and 103 sacral), SBRT provided lasting pain relief in 86% and achieved durable tumor control in 90% when used as the primary treatment modality, and in 88% of lesions treated for radiographic tumor progression.[45] Moreover, pretreatment neurological deficits improved in 84% following treatment.[45] With regard to pain control, recent data suggest that SBRT is superior to cEBRT in providing complete pain control.[158] The phase 3 component of the Radiation Therapy Oncology Group study [NCT00922974] will also aim to answer this lingering question.

Historically, cEBRT has been the standard postoperative therapy; however, local failure rates upward of 69.3% at 1 year prompted the adoption of SBRT in the postoperative setting.[159] Long-term outcomes of postoperative spine SBRT are excellent. In a retrospective study of 80 patients treated with surgery followed by postoperative SBRT to a median dose of 24 Gy in 2 fractions, the 1-year local control and overall survival rates were 84% and 64%, respectively.[130] The authors noted the highest failure rates among tumors confined to the epidural space (71%). In another report on 186 patients with MSCC who underwent epidural spinal cord decompression followed by adjuvant spine SBRT, high-dose hypofractionated SBRT was superior to low-dose hypofractionated SBRT and demonstrated 1-year local progression rates of less than 5%.[129] Similarly, in a review of 57 patients with 69 lesions treated with spine SBRT for spinal metastases (48 lesions received radiotherapy alone, while 21 lesions were treated with surgery before radiation), the local control rates were outstanding at 94.2% for all patients and 90.5% for those who underwent epidural spinal cord decompression followed by adjuvant spine SBRT.[160] A recent prospective phase 2 study examined the efficacy of postoperative SBRT in patients who underwent surgical intervention for spine metastases.[161] Radiographic and symptomatic local control at 1 year were 90%.

Spine SBRT is generally well tolerated. Still, transient pain following treatment,[162] vertebral compression fractures,[163] and radiation myelopathy[164] are well-known adverse events. Nevertheless, sufficient clinical experience and guidelines are available to address these concerns.[165–168] Re-irradiation of spinal metastasis using SBRT achieves effective pain relief and neurologic improvement, with minimal toxicity.[169–171]

7.5 Systemic therapy

The role of systemic therapy in the management of SEM remains undefined. Chemotherapy may play a role in managing asymptomatic SEM or in patients with disease at other sites; however, in patients with calamitous neurological events, chemotherapy by itself has little value. Systemic therapy is typically administered as an adjunct to or following more rapidly effective therapies such as radiotherapy or surgery, but seldom on a stand-alone basis. In appropriately selected patients with chemosensitive epidural tumors, systemic therapy is highly effective, as illustrated in a few cases. For example, epidural cord compression due to lymphoma responds dramatically to systemic chemotherapy.[172,173] Similarly, in exceptional cases of SEM from breast primary, treatment with chemotherapy and/or hormonal therapy can provide durable response rates.[14] Wilson et al. reported complete radiographic resolution and recovery from symptomatic MSCC from breast cancer following weekly administration of docetaxel.[174] Additionally, Sinoff and Blumsohn reported a dramatic response in five patients with symptomatic MSCC due to multiple myeloma treated solely with melphalan and prednisone.[175] Furthermore, cisplatin-based chemotherapy on a stand-alone basis can effectively manage germ cell tumors with epidural involvement.[176] Patients with symptomatic MSCC due to prostatic cancer may benefit from hormonal therapy.[15]

In the era of molecular and immuno-oncology, tumor profiling for therapeutic targets has reshaped cancer therapeutic strategies. Although the clinical utility of targeted therapy for spinal metastasis seems promising, an established role is still in question. Yet still, the advent of new treatments offers extended survival among subgroups of nonsmall cell lung carcinoma (NSCLC). An analysis of published literature revealed an increase in patients' overall survival with NSCLC with spine metastases from 3.6 to 9 months to a median reported survival of 18 months in patients with targetable epidermal growth factor receptor (EGFR) mutations.[177]

As mentioned earlier, once tumor cells arrest and colonize the bone marrow, the interplay of several systemic hormones, growth factors, proteases, and locally produced cytokines within the tumor microenvironment influences tumorigenesis. Bearing this in mind, bone-targeted therapies in a prophylactic setting are a growing research interest area. Bone loss and bone degradation are commonly observed in metastatic breast cancer.[178] Bone resorption requires osteoclasts, which are responsible for bone resorption.[45] Osteoclast activation is triggered by the receptor activator of nuclear factor-kappa B (RANK)/RANK-ligand pathway, rendering inhibition of the RANK-RANKL axis an attractive target for exploitation to retard bone loss.[179] Preclinical evidence suggests that this pathway may be implicated in breast and prostate tumorigenesis, thus opening a new potential avenue for breast cancer prophylaxis and therapy.[180,181] Additionally, RANK expression on prostate cancer cells may augment the metastatic behavior of tumor cells, with RANKL serving as a potential homing signal to bone marrow.[182] Denosumab is a human IgG2 monoclonal antibody that blocks the binding of RANKL to its receptor RANK expressed on osteoclasts, resulting in the subsequent decline of the formation, function, and survival of these osteoclasts, thus reducing bone resorption.[181] Denosumab is currently registered for the treatment of cancer patients at high risk for bone fractures and other skeletal-related events. Antitumor effects of RANKL inhibition were observed in several experimental models.[182–188] Evidence from several animal studies suggests that RANKL inhibition can prevent bone metastases.[189–191] A large randomized phase 3 study demonstrated that targeting the bone microenvironment with denosumab prevented bone metastasis in men with prostate cancer.[192] Patients were randomly assigned to either denosumab 120 mg every 4 weeks or placebo. Denosumab significantly increased bone-metastasis-free survival by a median of 4.2 months compared with placebo. Denosumab also significantly delayed time to first bone metastasis.[192] A recent meta-analysis showed that denosumab is equally as effective as the bisphosphonate, zoledronic acid, in reducing the likelihood of spinal cord compression in patients with bone metastases from advanced cancer.[193]

In addition to regulating cell growth and survival, the mammalian target of rapamycin (mTOR) signaling pathway plays a role in bone tissue signaling and bone cancer formation.[194] Tumor angiogenesis, orchestrated by vascular endothelial growth factor receptors (VEGFR), is also an essential component of the metastatic pathway. Neoangiogenesis provides the principal route by which tumor cells exit the primary tumor site and enter the circulation.[195] Preliminary data showed the potential of targeted therapies against mTOR and VEGFR to prevent and retard the formation of symptomatic MSCC.[196] Future studies are needed to evaluate the clinical effectiveness of this approach to treatment.

8 Conclusion

The management and treatment of SEM and associated MSCC remains a challenging problem. The goal of therapy is to relieve pain, stabilize the spinal structures,

maintain neurologic function, and improve quality of life. Cutting-edge advancements in surgical technology, radiotherapy, and pharmacotherapy have expanded the treatment approaches to tackle this complex issue. Ultimately, timely diagnosis and appropriate treatment selection are crucial in optimizing patient outcomes. Although different decision-making frameworks are proposed, more studies are needed to evaluate their clinical utility.

References

1. Barzilai O, Laufer I, Yamada Y, et al. Integrating evidence-based medicine for treatment of spinal metastases into a decision framework: neurologic, oncologic, mechanicals stability, and systemic disease. J Clin Oncol Off J Am Soc Clin Oncol. 2017;35(21):2419–2427.
2. Perrin RG, Livingston KE, Aarabi B. Intradural extramedullary spinal metastasis. A report of 10 cases. J Neurosurg. 1982;56(6):835–837.
3. Schick U, Marquardt G, Lorenz R. Intradural and extradural spinal metastases. Neurosurg Rev. 2001;24(1):1–5. Discussion 6–7.
4. Schiff D, O'Neill BP, Suman VJ. Spinal epidural metastasis as the initial manifestation of malignancy: clinical features and diagnostic approach. Neurology. 1997;49(2):452–456.
5. Lu C, Gonzalez RG, Jolesz FA, Wen PY, Talcott JA. Suspected spinal cord compression in cancer patients: a multidisciplinary risk assessment. J Support Oncol. 2005;3(4):305–312.
6. Hammack JE. Spinal cord disease in patients with cancer. Continuum (Minneapolis, Minn). 2012;18(2):312–327.
7. van der Sande JJ, Kröger R, Boogerd W. Multiple spinal epidural metastases; an unexpectedly frequent finding. J Neurol Neurosurg Psychiatry. 1990;53(11):1001–1003.
8. Schiff D, O'Neill BP, Wang CH, O'Fallon JR. Neuroimaging and treatment implications of patients with multiple epidural spinal metastases. Cancer. 1998;83(8):1593–1601.
9. Fehlings MG, Nater A, Tetreault L, et al. Survival and clinical outcomes in surgically treated patients with metastatic epidural spinal cord compression: results of the prospective multicenter AOSpine study. J Clin Oncol Off J Am Soc Clin Oncol. 2016;34(3):268–276.
10. Barzilai O, Versteeg AL, Goodwin CR, et al. Association of neurologic deficits with surgical outcomes and health-related quality of life after treatment for metastatic epidural spinal cord compression. Cancer. 2019;125(23):4224–4231.
11. Raj VS, Lofton L. Rehabilitation and treatment of spinal cord tumors. J Spinal Cord Med. 2013;36(1):4–11.
12. Fortin CD, Voth J, Jaglal SB, Craven BC. Inpatient rehabilitation outcomes in patients with malignant spinal cord compression compared to other non-traumatic spinal cord injury: a population based study. J Spinal Cord Med. 2015;38(6):754–764.
13. Burch PA, Grossman SA. Treatment of epidural cord compressions from Hodgkin's disease with chemotherapy. A report of two cases and a review of the literature. Am J Med. 1988;84:555–558. ed1988.
14. Boogerd W, van der Sande JJ, Kröger R, Bruning PF, Somers R. Effective systemic therapy for spinal epidural metastases from breast carcinoma. Eur J Cancer Clin Oncol. 1989;25(1):149–153.
15. Sasagawa I, Gotoh H, Miyabayashi H, Yamaguchi O, Shiraiwa Y. Hormonal treatment of symptomatic spinal cord compression in advanced prostatic cancer. Int Urol Nephrol. 1991;23(4):351–356.
16. Fareed MM, Pike LRG, Bang A, et al. Palliative radiation therapy for vertebral metastases and metastatic cord compression in patients treated with anti-PD-1 therapy. Front Oncol. 2019;9:199.

17. Tatsui CE, Stafford RJ, Li J, et al. Utilization of laser interstitial thermotherapy guided by real-time thermal MRI as an alternative to separation surgery in the management of spinal metastasis. *J Neurosurg Spine*. 2015;23(4):400–411.

18. Black P. Spinal metastasis: current status and recommended guidelines for management. *Neurosurgery*. 1979;5(6):726–746.

19. Abrams HL, Spiro R, Goldstein N. Metastases in carcinoma; analysis of 1000 autopsied cases. *Cancer*. 1950;3(1):74–85.

20. Bach F, Larsen BH, Rohde K, et al. Metastatic spinal cord compression. Occurrence, symptoms, clinical presentations and prognosis in 398 patients with spinal cord compression. *Acta Neurochir*. 1990;107(1–2):37–43.

21. Barron KD, Hirano A, Araki S, Terry RD. Experiences with metastatic neoplasms involving the spinal cord. *Neurology*. 1959;9(2):91–106.

22. Schiff D. Spinal cord compression. *Neurol Clin*. 2003;21(1):67–86. viii.

23. Miller KD, Nogueira L, Mariotto AB, et al. Cancer treatment and survivorship statistics, 2019. *CA Cancer J Clin*. 2019;69(5):363–385.

24. Loblaw DA, Laperriere NJ, Mackillop WJ. A population-based study of malignant spinal cord compression in Ontario. *Clin Oncol*. 2003;15(4):211–217.

25. Mak KS, Lee LK, Mak RH, et al. Incidence and treatment patterns in hospitalizations for malignant spinal cord compression in the United States, 1998-2006. *Int J Radiat Oncol Biol Phys*. 2011;80(3):824–831.

26. Abrahm JL, Banffy MB, Harris MB. Spinal cord compression in patients with advanced metastatic cancer: "all I care about is walking and living my life". *JAMA*. 2008;299(8):937–946.

27. Jennelle RL, Vijayakumar V, Vijayakumar S. A systematic and evidence-based approach to the management of vertebral metastasis. *ISRN Surg*. 2011;2011:719715.

28. Rades D, Dunst J, Schild SE. The first score predicting overall survival in patients with metastatic spinal cord compression. *Cancer*. 2008;112(1):157–161.

29. Janjan NA. Radiation for bone metastases: conventional techniques and the role of systemic radiopharmaceuticals. *Cancer*. 1997;80(8 suppl):1628–1645.

30. Fuller BG, Heiss JD, Oldfield EH. Spinal cord compression. In: Devita VT, Hellman S, Rosenberg SA, eds. Philadelphia, PA: Lippincott Williams and Wilkins; 2001:2617–2633. Cancer: Principles and Practice of Oncology; vol. 1.

31. Husband DJ, Grant KA, Romaniuk CS. MRI in the diagnosis and treatment of suspected malignant spinal cord compression. *Br J Radiol*. 2001;74(877):15–23.

32. Klein SL, Sanford RA, Muhlbauer MS. Pediatric spinal epidural metastases. *J Neurosurg*. 1991;74(1):70–75.

33. De Martino L, Spennato P, Vetrella S, et al. Symptomatic malignant spinal cord compression in children: a single-center experience. *Ital J Pediatr*. 2019;45(1):80.

34. Tohme S, Simmons RL, Tsung A. Surgery for cancer: a trigger for metastases. *Cancer Res*. 2017;77(7):1548–1552.

35. Abdel-Wanis Mel S, Tsuchiya H, Kawahara N, Tomita K. Tumor growth potential after tumoral and instrumental contamination: an in-vivo comparative study of T-saw, Gigli saw, and scalpel. *J Orthop Sci*. 2001;6(5):424–429.

36. Bendas G, Borsig L. Cancer cell adhesion and metastasis: selectins, integrins, and the inhibitory potential of heparins. *Int J Cell Biol*. 2012;2012:676731.

37. Haddad P, Thaell JF, Kiely JM, Harrison EG, Miller RH. Lymphoma of the spinal extradural space. *Cancer*. 1976;38(4):1862–1866.

38. Torma T. Malignant tumours of the spine and the spinal extradural space; a study based on 250 histologically verified cases. *Acta Chir Scand Suppl*. 1957;225:1–176.

39. Wright RL. Malignant tumors in the spinal extradural space: results of surgical treatment. *Ann Surg*. 1963;157(2):227–231.

40. De Bernardi B, Balwierz W, Bejent J, et al. Epidural compression in neuroblastoma: diagnostic and therapeutic aspects. *Cancer Lett*. 2005;228(1–2):283–299.

41. Batson OV. The function of the vertebral veins and their role in the spread of metastases. *Ann Surg*. 1940;112(1):138–149.

42. Coman DR, de LR. The role of the vertebral venous system in the metastasis of cancer to the spinal column; experiments with tumor-cell suspensions in rats and rabbits. *Cancer*. 1951;4(3):610–618.

43. Nishijima Y, Uchida K, Koiso K, Nemoto R. Clinical significance of the vertebral vein in prostate cancer metastasis. *Adv Exp Med Biol*. 1992;324:93–100.

44. Tatsui CE, Lang FF, Gumin J, Suki D, Shinojima N, Rhines LD. An orthotopic murine model of human spinal metastasis: histological and functional correlations. *J Neurosurg Spine*. 2009;10(6):501–512.

45. Gerszten PC, Burton SA, Ozhasoglu C, Welch WC. Radiosurgery for spinal metastases: clinical experience in 500 cases from a single institution. *Spine*. 2007;32(2):193–199.

46. Altman DG. Systematic reviews of evaluations of prognostic variables. *BMJ*. 2001;323(7306):224–228.

47. Perrin RG, Laxton AW. Metastatic spine disease: epidemiology, pathophysiology, and evaluation of patients. *Neurosurg Clin N Am*. 2004;15(4):365–373.

48. Gilbert RW, Kim JH, Posner JB. Epidural spinal cord compression from metastatic tumor: diagnosis and treatment. *Ann Neurol*. 1978;3(1):40–51.

49. Levack P, Graham J, Collie D, et al. Don't wait for a sensory level—listen to the symptoms: a prospective audit of the delays in diagnosis of malignant cord compression. *Clin Oncol*. 2002;14(6):472–480.

50. Pollono D, Tomarchia S, Drut R, Ibañez O, Ferreyra M, Cédola J. Spinal cord compression: a review of 70 pediatric patients. *Pediatr Hematol Oncol*. 2003;20(6):457–466.

51. Kienstra GE, Terwee CB, Dekker FW, et al. Prediction of spinal epidural metastases. *Arch Neurol*. 2000;57(5):690–695.

52. Rousseff RT, Tzvetanov P. False localising levels in spinal cord compression. *NeuroRehabilitation*. 2006;21(3):219–222.

53. Scott M. Lower extremity pain simulating sciatica; tumors of the high thoracic and cervical cord as causes. *J Am Med Assoc*. 1956;160(7):528–534.

54. Haegerstam GA. Pathophysiology of bone pain: a review. *Acta Orthop Scand*. 2001;72(3):308–317.

55. Greenberg HS, Kim JH, Posner JB. Epidural spinal cord compression from metastatic tumor: results with a new treatment protocol. *Ann Neurol*. 1980;8(4):361–366.

56. Zairi F, Vieillard MH, Assaker R. Spine metastases: are minimally invasive surgical techniques living up to the hype? *CNS Oncol*. 2015;4(4):257–264.

57. Hsu HC, Liao TY, Ro LS, Juan YH, Liaw CC. Differences in pain intensity of tumors spread to the anterior versus anterolateral/lateral portions of the vertebral body based on CT scans. *Pain Res Manag*. 2019;2019:9387941.

58. Telfeian AE, Oyelese A, Fridley J, Doberstein C, Gokaslan ZL. Endoscopic surgical treatment for symptomatic spinal metastases in long-term cancer survivors. *J Spine Surg*. 2020;6(2):372–382.

59. Gao Z, Wu Z, Lin Y, Zhang P. Percutaneous transforaminal endoscopic decompression in the treatment of spinal metastases: a case report. *Medicine*. 2019;98(11), e14819.

60. Tsai SH, Wu HH, Cheng CY, Chen CM. Full endoscopic interlaminar approach for nerve root decompression of sacral metastatic tumor. *World Neurosurg*. 2018;112:57–63.

61. Stark RJ, Henson RA, Evans SJ. Spinal metastases. A retrospective survey from a general hospital. *Brain J Neurol*. 1982;105(Pt. 1):189–213.

62. Ohashi M, Hirano T, Watanabe K, et al. Preoperative prediction for regaining ambulatory ability in paretic non-ambulatory patients with metastatic spinal cord compression. *Spinal Cord*. 2017;55(5):447–453.

63. Brice J, McKissock W. Surgical treatment of malignant extradural spinal tumours. *Br Med J.* 1965;1(5446):1341–1344.

64. Uei H, Tokuhashi Y, Maseda M. Analysis of the relationship between the epidural spinal cord compression (ESCC) scale and paralysis caused by metastatic spine tumors. *Spine.* 2018;43(8):E448–e455.

65. Zacharia BE, Kahn S, Bander ED, et al. Incidence and risk factors for preoperative deep venous thrombosis in 314 consecutive patients undergoing surgery for spinal metastasis. *J Neurosurg Spine.* 2017;27(2):189–197.

66. Key NS, Khorana AA, Kuderer NM, et al. Venous thromboembolism prophylaxis and treatment in patients with cancer: ASCO clinical practice guideline update. *J Clin Oncol Off J Am Soc Clin Oncol.* 2020;38(5):496–520.

67. Ventafridda V, Caraceni A, Martini C, Sbanotto A, De Conno F. On the significance of Lhermitte's sign in oncology. *J Neuro-Oncol.* 1991;10(2):133–137.

68. Broager B. Lhermitte's sign in thoracic spinal tumour. Personal observation. *Acta Neurochir.* 1978;41(1–3):127–135.

69. Podnar S, Oblak C, Vodusek DB. Sexual function in men with cauda equina lesions: a clinical and electromyographic study. *J Neurol Neurosurg Psychiatry.* 2002;73(6):715–720.

70. Kara M, Isik M, Ozcakar L, et al. Unilateral diaphragm paralysis possibly due to cervical spine involvement in multiple myeloma. *Med Princ Pract.* 2006;15(3):242–244.

71. Kitamura Y, Shimizu K, Nagahama M, et al. Immediate causes of death in thyroid carcinoma: clinicopathological analysis of 161 fatal cases. *J Clin Endocrinol Metab.* 1999;84(11):4043–4049.

72. Zhou D, Ibrahim M, Malach D, Tomsak RL. Unusual cause of horner syndrome 13 years after in situ ductal carcinoma. *Neuro-Ophthalmology.* 2016;40(3):130–132.

73. Buhmann Kirchhoff S, Becker C, Duerr HR, Reiser M, Baur-Melnyk A. Detection of osseous metastases of the spine: comparison of high resolution multi-detector-CT with MRI. *Eur J Radiol.* 2009;69(3):567–573.

74. Li KC, Poon PY. Sensitivity and specificity of MRI in detecting malignant spinal cord compression and in distinguishing malignant from benign compression fractures of vertebrae. *Magn Reson Imaging.* 1988;6(5):547–556.

75. Kim JK, Learch TJ, Colletti PM, Lee JW, Tran SD, Terk MR. Diagnosis of vertebral metastasis, epidural metastasis, and malignant spinal cord compression: are T(1)-weighted sagittal images sufficient? *Magn Reson Imaging.* 2000;18(7):819–824.

76. Kumar KA, Peck KK, Karimi S, et al. A pilot study evaluating the use of dynamic contrast-enhanced perfusion mri to predict local recurrence after radiosurgery on spinal metastases. *Technol Cancer Res Treat.* 2017;16(6):857–865.

77. Byun WM, Shin SO, Chang Y, Lee SJ, Finsterbusch J, Frahm J. Diffusion-weighted MR imaging of metastatic disease of the spine: assessment of response to therapy. *AJNR Am J Neuroradiol.* 2002;23(6):906–912.

78. Hollis PH, Malis LI, Zappulla RA. Neurological deterioration after lumbar puncture below complete spinal subarachnoid block. *J Neurosurg.* 1986;64(2):253–256.

79. Smoker WR, Godersky JC, Knutzon RK, Keyes WD, Norman D, Bergman W. The role of MR imaging in evaluating metastatic spinal disease. *AJR Am J Roentgenol.* 1987;149(6):1241–1248.

80. Kieffer SA, Binet EF, Esquerra JV, Hantman RP, Gross CE. Contrast agents for myelography: clinical and radiological evaluation of Amipaque and Pantopaque. *Radiology.* 1978;129(3):695–705.

81. Bain PG, Colchester AC, Nadarajah D. Paraplegia after iopamidol myelography. *Lancet (London, England).* 1991;338(8761):252–253.

82. Algra PR, Bloem JL, Tissing H, Falke TH, Arndt JW, Verboom LJ. Detection of vertebral metastases: comparison between MR imaging and bone scintigraphy. *Radiographics.* 1991;11(2):219–232.

83. Chow E, Davis L, Panzarella T, et al. Accuracy of survival prediction by palliative radiation oncologists. *Int J Radiat Oncol Biol Phys.* 2005;61(3):870–873.

84. Chow E, Harth T, Hruby G, Finkelstein J, Wu J, Danjoux C. How accurate are physicians' clinical predictions of survival and the available prognostic tools in estimating survival times in terminally ill cancer patients? A systematic review. *Clin Oncol.* 2001;13(3):209–218.

85. Parkes CM. Accuracy of predictions of survival in later stages of cancer. *Br Med J.* 1972;2(5804):29–31.

86. van der Linden YM, Dijkstra SP, Vonk EJ, Marijnen CA, Leer JW. Prediction of survival in patients with metastases in the spinal column: results based on a randomized trial of radiotherapy. *Cancer.* 2005;103(2):320–328.

87. Tokuhashi Y, Matsuzaki H, Toriyama S, Kawano H, Ohsaka S. Scoring system for the preoperative evaluation of metastatic spine tumor prognosis. *Spine.* 1990;15(11):1110–1113.

88. Tokuhashi Y, Matsuzaki H, Oda H, Oshima M, Ryu J. A revised scoring system for preoperative evaluation of metastatic spine tumor prognosis. *Spine.* 2005;30(19):2186–2191.

89. Enkaoua EA, Doursounian L, Chatellier G, Mabesoone F, Aimard T, Saillant G. Vertebral metastases: a critical appreciation of the preoperative prognostic tokuhashi score in a series of 71 cases. *Spine.* 1997;22(19):2293–2298.

90. Tomita K, Kawahara N, Kobayashi T, Yoshida A, Murakami H, Akamaru T. Surgical strategy for spinal metastases. *Spine.* 2001;26(3):298–306.

91. Pointillart V, Vital JM, Salmi R, Diallo A, Quan GM. Survival prognostic factors and clinical outcomes in patients with spinal metastases. *J Cancer Res Clin Oncol.* 2011;137(5):849–856.

92. Hosono N, Ueda T, Tamura D, Aoki Y, Yoshikawa H. Prognostic relevance of clinical symptoms in patients with spinal metastases. *Clin Orthop Relat Res.* 2005;436:196–201.

93. Laufer I, Rubin DG, Lis E, et al. The NOMS framework: approach to the treatment of spinal metastatic tumors. *Oncologist.* 2013;18(6):744–751.

94. Lei M, Li J, Liu Y, Jiang W, Liu S, Zhou S. Who are the best candidates for decompressive surgery and spine stabilization in patients with metastatic spinal cord compression?: a new scoring system. *Spine.* 2016;41(18):1469–1476.

95. Yu W, Tang L, Lin F, Yao Y, Shen Z. Accuracy of Tokuhashi score system in predicting survival of lung cancer patients with vertebral metastasis. *J Neuro-Oncol.* 2015;125(2):427–433.

96. Gregory TM, Coriat R, Mir O. Prognostic scoring systems for spinal metastases in the era of anti-VEGF therapies. *Spine.* 2013;38(11):965–966.

97. Katagiri H, Okada R, Takagi T, et al. New prognostic factors and scoring system for patients with skeletal metastasis. *Cancer Med.* 2014;3(5):1359–1367.

98. Chang SY, Mok S, Park SC, Kim H, Chang BS. Treatment strategy for metastatic spinal tumors: a narrative review. *Asian Spine J.* 2020;14(4):513–525.

99. L'Espérance S, Vincent F, Gaudreault M, et al. Treatment of metastatic spinal cord compression: cepo review and clinical recommendations. *Curr Oncol.* 2012;19(6):e478–e490.

100. Kawahara N, Tomita K, Murakami H, Demura S. Total en bloc spondylectomy for spinal tumors: surgical techniques and related basic background. *Orthop Clin North Am.* 2009;40(1):47–63. vi.

101. Luksanapruksa P, Buchowski JM, Hotchkiss W, Tongsai S, Wilartratsami S, Chotivichit A. Prognostic factors in patients with spinal metastasis: a systematic review and meta-analysis. *Spine J.* 2017;17(5):689–708.

102. Quraishi NA, Manoharan SR, Arealis G, et al. Accuracy of the revised Tokuhashi score in predicting survival in patients with metastatic spinal cord compression (MSCC). *Eur Spine J.* 2013;22(suppl 1):S21–S26.

103. Bollen L, Wibmer C, Van der Linden YM, et al. Predictive value of six prognostic scoring systems for spinal bone metastases: an analysis based on 1379 patients. *Spine*. 2016;41(3):E155–E162.

104. Mezei T, Horváth A, Pollner P, Czigléczki G, Banczerowski P. Research on the predicting power of the revised Tokuhashi system: how much time can surgery give to patients with short life expectancy? *Int J Clin Oncol*. 2020;25(4):755–764.

105. Cai Z, Tang X, Yang R, Yan T, Guo W. Modified score based on revised Tokuhashi score is needed for the determination of surgical intervention in patients with lung cancer metastases to the spine. *World J Surg Oncol*. 2019;17(1):194.

106. Uei H, Tokuhashi Y, Maseda M. Treatment outcome of metastatic spine tumor in lung cancer patients: did the treatments improve their outcomes? *Spine*. 2017;42(24):E1446–e1451.

107. Ahmed AK, Goodwin CR, Heravi A, et al. Predicting survival for metastatic spine disease: a comparison of nine scoring systems. *Spine J*. 2018;18(10):1804–1814.

108. Karhade AV, Ahmed AK, Pennington Z, et al. External validation of the SORG 90-day and 1-year machine learning algorithms for survival in spinal metastatic disease. *Spine J*. 2020;20(1):14–21.

109. Grant R, Papadopoulos SM, Sandler HM, Greenberg HS. Metastatic epidural spinal cord compression: current concepts and treatment. *J Neuro-Oncol*. 1994;19(1):79–92.

110. Ruff RL, Lanska DJ. Epidural metastases in prospectively evaluated veterans with cancer and back pain. *Cancer*. 1989;63(11):2234–2241.

111. Rades D, Douglas S, Veninga T, et al. A survival score for patients with metastatic spinal cord compression from prostate cancer. *Strahlenther Onkol*. 2012;188(9):802–806.

112. Rades D, Douglas S, Schild SE. A validated survival score for breast cancer patients with metastatic spinal cord compression. *Strahlenther Onkol*. 2013;189(1):41–46.

113. Douglas S, Schild SE, Rades D. Metastatic spinal cord compression in patients with cancer of unknown primary. Estimating the survival prognosis with a validated score. *Strahlenther Onkol*. 2012;188(11):1048–1051.

114. Fehri R, Rifi H, Alboueiri A, et al. Carcinoma of unknown primary: retrospective study of 437 patients treated at Salah Azaiez Institute. *Tunis Med*. 2013;91(3):205–208.

115. Vecht CJ, Haaxma-Reiche H, van Putten WL, de Visser M, Vries EP, Twijnstra A. Initial bolus of conventional versus high-dose dexamethasone in metastatic spinal cord compression. *Neurology*. 1989;39(9):1255–1257.

116. Sørensen S, Helweg-Larsen S, Mouridsen H, Hansen HH. Effect of high-dose dexamethasone in carcinomatous metastatic spinal cord compression treated with radiotherapy: a randomised trial. *Eur J Cancer*. 1994;30a(1):22–27.

117. Phillips KA, Fadul CE, Schiff D. Neurologic and medical management of brain tumors. *Neurol Clin*. 2018;36(3):449–466.

118. Maranzano E, Latini P, Beneventi S, et al. Radiotherapy without steroids in selected metastatic spinal cord compression patients. A phase II trial. *Am J Clin Oncol*. 1996;19(2):179–183.

119. Patchell RA, Tibbs PA, Regine WF, et al. Direct decompressive surgical resection in the treatment of spinal cord compression caused by metastatic cancer: a randomised trial. *Lancet (London, England)*. 2005;366(9486):643–648.

120. Azad TD, Varshneya K, Ho AL, Veeravagu A, Sciubba DM, Ratliff JK. Laminectomy versus corpectomy for spinal metastatic disease-complications, costs, and quality outcomes. *World Neurosurg*. 2019;131:e468–e473.

121. Zhou X, Cui H, He Y, Qiu G, Zhou D, Liu Y. Treatment of spinal metastases with epidural cord compression through corpectomy and reconstruction via the traditional open approach versus the mini-open approach: a multicenter retrospective study. *J Oncol*. 2019;2019:7904740.

122. Sherman RM, Waddell JP. Laminectomy for metastatic epidural spinal cord tumors. Posterior stabilization, radiotherapy, and preoperative assessment. *Clin Orthop Relat Res*. 1986;207:55–63.

123. Younsi A, Riemann L, Scherer M, Unterberg A, Zweckberger K. Impact of decompressive laminectomy on the functional outcome of patients with metastatic spinal cord compression and neurological impairment. *Clin Exp Metastasis*. 2020;37(2):377–390.

124. Schoeggl A, Reddy M, Matula C. Neurological outcome following laminectomy in spinal metastases. *Spinal Cord*. 2002;40(7):363–366.

125. Molina C, Goodwin CR, Abu-Bonsrah N, Elder BD, De la Garza Ramos R, Sciubba DM. Posterior approaches for symptomatic metastatic spinal cord compression. *Neurosurg Focus*. 2016;41(2), E11.

126. Loblaw DA, Laperriere NJ. Emergency treatment of malignant extradural spinal cord compression: an evidence-based guideline. *J Clin Oncol Off J Am Soc Clin Oncol*. 1998;16(4):1613–1624.

127. Sundaresan N, Steinberger AA, Moore F, et al. Indications and results of combined anterior-posterior approaches for spine tumor surgery. *J Neurosurg*. 1996;85(3):438–446.

128. Barzilai O, Laufer I, Robin A, Xu R, Yamada Y, Bilsky MH. Hybrid therapy for metastatic epidural spinal cord compression: technique for separation surgery and spine radiosurgery. *Oper Neurosurg*. 2019;16(3):310–318.

129. Laufer I, Iorgulescu JB, Chapman T, et al. Local disease control for spinal metastases following "separation surgery" and adjuvant hypofractionated or high-dose single-fraction stereotactic radiosurgery: outcome analysis in 186 patients. *J Neurosurg Spine*. 2013;18(3):207–214.

130. Al-Omair A, Masucci L, Masson-Cote L, et al. Surgical resection of epidural disease improves local control following postoperative spine stereotactic body radiotherapy. *Neuro-Oncology*. 2013;15(10):1413–1419.

131. Xiaozhou L, Xing Z, Xin S, et al. Efficacy analysis of separation surgery combined with SBRT for spinal metastases-a long-term follow-up study based on patients with spinal metastatic tumor in a single-center. *Orthop Surg*. 2020;12(2):404–420.

132. Turel MK, Kerolus MG, O'Toole JE. Minimally invasive "separation surgery" plus adjuvant stereotactic radiotherapy in the management of spinal epidural metastases. *J Craniovertebr Junction Spine*. 2017;8(2):119–126.

133. Donnelly DJ, Abd-El-Barr MM, Lu Y. Minimally invasive muscle sparing posterior-only approach for lumbar circumferential decompression and stabilization to treat spine metastasis—technical report. *World Neurosurg*. 2015;84(5):1484–1490.

134. Saigal R, Wadhwa R, Mummaneni PV, Chou D. Minimally invasive extracavitary transpedicular corpectomy for the management of spinal tumors. *Neurosurg Clin N Am*. 2014;25(2):305–315.

135. Thomas JG, Al-Holou WN, de Almeida Bastos DC, et al. A novel use of the intraoperative mri for metastatic spine tumors: laser interstitial thermal therapy for percutaneous treatment of epidural metastatic spine disease. *Neurosurg Clin N Am*. 2017;28(4):513–524.

136. Gevargez A, Groenemeyer DH. Image-guided radiofrequency ablation (RFA) of spinal tumors. *Eur J Radiol*. 2008;65(2):246–252.

137. Park J, Ham DW, Kwon BT, Park SM, Kim HJ, Yeom JS. Minimally invasive spine surgery: techniques, technologies, and indications. *Asian Spine J*. 2020;14(5):694–701.

138. Maranzano E, Latini P. Effectiveness of radiation therapy without surgery in metastatic spinal cord compression: final results from a prospective trial. *Int J Radiat Oncol Biol Phys*. 1995;32(4):959–967.

139. Yamada Y, Lovelock DM, Yenice KM, et al. Multifractionated image-guided and stereotactic intensity-modulated radiotherapy of paraspinal tumors: a preliminary report. *Int J Radiat Oncol Biol Phys*. 2005;62(1):53–61.

140. Gerszten PC, Mendel E, Yamada Y. Radiotherapy and radiosurgery for metastatic spine disease: what are the options, indications, and outcomes? *Spine*. 2009;34(22 suppl):S78–S92.

141. Garg AK, Shiu AS, Yang J, et al. Phase 1/2 trial of single-session stereotactic body radiotherapy for previously unirradiated spinal metastases. *Cancer.* 2012;118(20):5069–5077.

142. Joaquim AF, Ghizoni E, Tedeschi H, Pereira EB, Giacomini LA. Stereotactic radiosurgery for spinal metastases: a literature review. *Einstein (Sao Paulo, Brazil).* 2013;11(2):247–255.

143. Chow E, Harris K, Fan G, Tsao M, Sze WM. Palliative radiotherapy trials for bone metastases: a systematic review. *J Clin Oncol Off J Am Soc Clin Oncol.* 2007;25(11):1423–1436.

144. Howell DD, James JL, Hartsell WF, et al. Single-fraction radiotherapy versus multifraction radiotherapy for palliation of painful vertebral bone metastases-equivalent efficacy, less toxicity, more convenient: a subset analysis of Radiation Therapy Oncology Group trial 97-14. *Cancer.* 2013;119(4):888–896.

145. Rades D, Fehlauer F, Stalpers LJ, et al. A prospective evaluation of two radiotherapy schedules with 10 versus 20 fractions for the treatment of metastatic spinal cord compression: final results of a multicenter study. *Cancer.* 2004;101(11):2687–2692.

146. Rades D, Fehlauer F, Schulte R, et al. Prognostic factors for local control and survival after radiotherapy of metastatic spinal cord compression. *J Clin Oncol Off J Am Soc Clin Oncol.* 2006;24(21):3388–3393.

147. Mizumoto M, Harada H, Asakura H, et al. Radiotherapy for patients with metastases to the spinal column: a review of 603 patients at Shizuoka Cancer Center Hospital. *Int J Radiat Oncol Biol Phys.* 2011;79(1):208–213.

148. Nguyen J, Chow E, Zeng L, et al. Palliative response and functional interference outcomes using the Brief Pain Inventory for spinal bony metastases treated with conventional radiotherapy. *Clin Oncol.* 2011;23(7):485–491.

149. Huisman M, van den Bosch MA, Wijlemans JW, van Vulpen M, van der Linden YM, Verkooijen HM. Effectiveness of reirradiation for painful bone metastases: a systematic review and meta-analysis. *Int J Radiat Oncol Biol Phys.* 2012;84(1):8–14.

150. Bilsky MH, Yamada Y, Yenice KM, et al. Intensity-modulated stereotactic radiotherapy of paraspinal tumors: a preliminary report. *Neurosurgery.* 2004;54(4):823–830. Discussion 830–831.

151. Yamada Y, Bilsky MH, Lovelock DM, et al. High-dose, single-fraction image-guided intensity-modulated radiotherapy for metastatic spinal lesions. *Int J Radiat Oncol Biol Phys.* 2008;71(2):484–490.

152. Steverink JG, Willems SM, Philippens MEP, et al. Early tissue effects of stereotactic body radiation therapy for spinal metastases. *Int J Radiat Oncol Biol Phys.* 2018;100(5):1254–1258.

153. Dewan MZ, Galloway AE, Kawashima N, et al. Fractionated but not single-dose radiotherapy induces an immune-mediated abscopal effect when combined with anti-CTLA-4 antibody. *Clin Cancer Res.* 2009;15(17):5379–5388.

154. Lugade AA, Moran JP, Gerber SA, Rose RC, Frelinger JG, Lord EM. Local radiation therapy of B16 melanoma tumors increases the generation of tumor antigen-specific effector cells that traffic to the tumor. *J Immunol.* 2005;174(12):7516–7523.

155. Yamada Y, Katsoulakis E, Laufer I, et al. The impact of histology and delivered dose on local control of spinal metastases treated with stereotactic radiosurgery. *Neurosurg Focus.* 2017;42(1), E6.

156. Guckenberger M, Mantel F, Gerszten PC, et al. Safety and efficacy of stereotactic body radiotherapy as primary treatment for vertebral metastases: a multi-institutional analysis. *Radiat Oncol.* 2014;9:226.

157. Anand AK, Venkadamanickam G, Punnakal AU, et al. Hypofractionated stereotactic body radiotherapy in spinal metastasis—with or without epidural extension. *Clin Oncol.* 2015;27(6):345–352.

158. Sahgal A, Myrehaug S, Siva S, et al. CCTG SC.24/TROG 17.06: A randomized phase II/III study comparing 24Gy in 2 stereotactic body radiotherapy (SBRT) fractions versus 20Gy in 5 conventional palliative radiotherapy (CRT) fractions for patients with painful spinal metastases. In: *Presented at the American Society for Radiation Oncology (ASTRO) Annual Meeting (Abstract LBA-2);* 2020.

159. Klekamp J, Samii H. Surgical results for spinal metastases. *Acta Neurochir.* 1998;140(9):957–967.

160. Bate BG, Khan NR, Kimball BY, Gabrick K, Weaver J. Stereotactic radiosurgery for spinal metastases with or without separation surgery. *J Neurosurg Spine.* 2015;22(4):409–415.

161. Redmond KJ, Sciubba D, Khan M, et al. A phase 2 study of post-operative stereotactic body radiation therapy (SBRT) for solid tumor spine metastases. *Int J Radiat Oncol Biol Phys.* 2020;106(2):261–268.

162. Chow E, Meyer RM, Ding K, et al. Dexamethasone in the prophylaxis of radiation-induced pain flare after palliative radiotherapy for bone metastases: a double-blind, randomised placebo-controlled, phase 3 trial. *Lancet Oncol.* 2015;16(15):1463–1472.

163. Redmond KJ, Sahgal A, Foote M, et al. Single versus multiple session stereotactic body radiotherapy for spinal metastasis: the risk-benefit ratio. *Future Oncol.* 2015;11(17):2405–2415.

164. Hall WA, Stapleford LJ, Hadjipanayis CG, Curran WJ, Crocker I, Shu HK. Stereotactic body radiosurgery for spinal metastatic disease: an evidence-based review. *Int J Surg Oncol.* 2011;2011:979214.

165. Sahgal A, Roberge D, Schellenberg D, et al. The Canadian Association of Radiation Oncology scope of practice guidelines for lung, liver and spine stereotactic body radiotherapy. *Clin Oncol.* 2012;24(9):629–639.

166. Sahgal A, Bilsky M, Chang EL, et al. Stereotactic body radiotherapy for spinal metastases: current status, with a focus on its application in the postoperative patient. *J Neurosurg Spine.* 2011;14(2):151–166.

167. Husain ZA, Thibault I, Letourneau D, et al. Stereotactic body radiotherapy: a new paradigm in the management of spinal metastases. *CNS Oncol.* 2013;2(3):259–270.

168. Lo SS, Lutz ST, Chang EL, et al. ACR Appropriateness Criteria ® spinal bone metastases. *J Palliat Med.* 2013;16(1):9–19.

169. Sasamura K, Suzuki R, Kozuka T, Yoshimura R, Yoshioka Y, Oguchi M. Outcomes after reirradiation of spinal metastasis with stereotactic body radiation therapy (SBRT): a retrospective single institutional study. *J Radiat Res.* 2020;61(6):929–934.

170. Myrehaug S, Sahgal A, Hayashi M, et al. Reirradiation spine stereotactic body radiation therapy for spinal metastases: systematic review. *J Neurosurg Spine.* 2017;27(4):428–435.

171. Chiu N, Chiu L, Popovic M, et al. Re-irradiation for painful bone metastases: evidence-based approach. *Ann Palliat Med.* 2015;4(4):214–219.

172. Toprak A, Kodalli N, Alpdogan TB, et al. Stage IV Hodgkin's disease presenting with spinal epidural involvement and cauda equina compression as the initial manifestation: case report. *Spinal Cord.* 1997;35(10):704–707.

173. Wong ET, Portlock CS, O'Brien JP, DeAngelis LM. Chemosensitive epidural spinal cord disease in non-Hodgkins lymphoma. *Neurology.* 1996;46(6):1543–1547.

174. Wilson B, Sapp C, Abdeen G, Kamona A, Massarweh S. Resolution of extensive leptomeningeal metastasis and clinical spinal cord compression from breast cancer using weekly docetaxel chemotherapy. *Breast Cancer Res Treat.* 2012;131(1):343–346.

175. Sinoff CL, Blumsohn A. Spinal cord compression in myelomatosis: response to chemotherapy alone. *Eur J Cancer Clin Oncol.* 1989;25(2):197–200.

176. Grommes C, Bosl GJ, DeAngelis LM. Treatment of epidural spinal cord involvement from germ cell tumors with chemotherapy. *Cancer.* 2011;117(9):1911–1916.

177. Batista N, Tee J, Sciubba D, et al. Emerging and established clinical, histopathological and molecular parametric prognostic factors for metastatic spine disease secondary to lung cancer: Helping surgeons make decisions. *J Clin Neurosci.* 2016;34:15–22.

178. Coleman RE. Metastatic bone disease: clinical features, pathophysiology and treatment strategies. *Cancer Treat Rev.* 2001;27(3):165–176.

179. Boyce BF, Xing L. Biology of RANK, RANKL, and osteoprotegerin. *Arthritis Res Ther.* 2007;9(suppl 1):S1.

180. Beleut M, Rajaram RD, Caikovski M, et al. Two distinct mechanisms underlie progesterone-induced proliferation in the mammary gland. *Proc Natl Acad Sci U S A.* 2010;107(7):2989–2994.

181. Schramek D, Leibbrandt A, Sigl V, et al. Osteoclast differentiation factor RANKL controls development of progestin-driven mammary cancer. *Nature.* 2010;468(7320):98–102.

182. Jones DH, Nakashima T, Sanchez OH, et al. Regulation of cancer cell migration and bone metastasis by RANKL. *Nature.* 2006;440(7084):692–696.

183. Morony S, Capparelli C, Sarosi I, Lacey DL, Dunstan CR, Kostenuik PJ. Osteoprotegerin inhibits osteolysis and decreases skeletal tumor burden in syngeneic and nude mouse models of experimental bone metastasis. *Cancer Res.* 2001;61(11):4432–4436.

184. Kiefer JA, Vessella RL, Quinn JE, et al. The effect of osteoprotegerin administration on the intra-tibial growth of the osteoblastic LuCaP 23.1 prostate cancer xenograft. *Clin Exp Metastasis.* 2004;21(5):381–387.

185. Yonou H, Kanomata N, Goya M, et al. Osteoprotegerin/osteoclastogenesis inhibitory factor decreases human prostate cancer burden in human adult bone implanted into nonobese diabetic/severe combined immunodeficient mice. *Cancer Res.* 2003;63(9):2096–2102.

186. Zhang J, Dai J, Yao Z, Lu Y, Dougall W, Keller ET. Soluble receptor activator of nuclear factor kappaB Fc diminishes prostate cancer progression in bone. *Cancer Res.* 2003;63(22):7883–7890.

187. Tannehill-Gregg SH, Levine AL, Nadella MV, Iguchi H, Rosol TJ. The effect of zoledronic acid and osteoprotegerin on growth of human lung cancer in the tibias of nude mice. *Clin Exp Metastasis.* 2006;23(1):19–31.

188. Feeley BT, Liu NQ, Conduah AH, et al. Mixed metastatic lung cancer lesions in bone are inhibited by noggin overexpression and Rank:Fc administration. *J Bone Miner Res Off J Am Soc Bone Miner Res.* 2006;21(10):1571–1580.

189. Canon J, Bryant R, Roudier M, et al. Inhibition of RANKL increases the anti-tumor effect of the EGFR inhibitor panitumumab in a murine model of bone metastasis. *Bone.* 2010;46(6):1613–1619.

190. Holland PM, Miller R, Jones J, et al. Combined therapy with the RANKL inhibitor RANK-Fc and rhApo2L/TRAIL/dulanermin reduces bone lesions and skeletal tumor burden in a model of breast cancer skeletal metastasis. *Cancer Biol Ther.* 2010;9(7):539–550.

191. Canon J, Bryant R, Roudier M, Branstetter DG, Dougall WC. RANKL inhibition combined with tamoxifen treatment increases anti-tumor efficacy and prevents tumor-induced bone destruction in an estrogen receptor-positive breast cancer bone metastasis model. *Breast Cancer Res Treat.* 2012;135(3):771–780.

192. Smith MR, Saad F, Coleman R, et al. Denosumab and bone-metastasis-free survival in men with castration-resistant prostate cancer: results of a phase 3, randomised, placebo-controlled trial. *Lancet (London, England).* 2012;379(9810):39–46.

193. Al Farii H, Frazer A, Farahdel L, Alfayez S, Weber M. Zoledronic acid versus denosumab for prevention of spinal cord compression in advanced cancers with spine metastasis: a meta-analysis of randomized controlled trials. *Global Spine J.* 2020;10(6):784–789.

194. Bertoldo F, Silvestris F, Ibrahim T, et al. Targeting bone metastatic cancer: role of the mTOR pathway. *Biochim Biophys Acta.* 2014;1845(2):248–254.

195. Zetter BR. Angiogenesis and tumor metastasis. *Annu Rev Med.* 1998;49:407–424.

196. Kratzsch T, Piffko A, Broggini T, Czabanka M, Vajkoczy P. Role of MTOR and VEGFR inhibition in prevention of metastatic tumor growth in the spine. *Front Oncol.* 2020;10:174.

8

Leptomeningeal metastasis

Jigisha P. Thakkar[a] and Marc C. Chamberlain[b]

[a]Department of Neurology, Division of Neuro-Oncology, Loyola University Chicago, Stritch School of Medicine, Maywood, IL, United States, [b]Department of Neurology and Neurological Surgery, University of Washington, Fred Hutchinson Cancer Research Center, Seattle Cancer Care Alliance, Seattle, WA, United States

1 Introduction

Leptomeningeal metastasis (LM) is a secondary complication of cancer wherein malignant cells infiltrate the cerebrospinal fluid (CSF), the pia mater, and arachnoid membranes. In patients with solid tumors, LM is referred to as carcinomatous meningitis. In patients with leukemia or lymphoma, it is termed leukemic or lymphomatous meningitis, respectively. LM is also known as neoplastic meningitis.

The incidence of solid tumor-related LM is difficult to estimate and cannot clearly be defined as this information is not tracked at the population level. However, autopsy studies including single-institution series lend some insight.[1] Autopsy studies in patients with solid tumors have shown that 19% of cancer patients with neurologic signs and symptoms have evidence of LM.[1] LM is the third most common site of metastases affecting the central nervous system following parenchymal brain metastases and epidural spinal cord compression.

LM is most common in leukemia and non-Hodgkin's lymphoma but is being observed with increased frequency in patients with solid tumors. Among solid tumors, the highest incidence of LM appears to be in melanoma (23%) and lung cancer (9%–25%) followed by breast cancer (5%).[2–4] The median time from systemic cancer diagnosis to the diagnosis of LM is approximately 1–2 years.[5,6] Patients with hormone receptor-positive breast cancers have the longest interval between initial cancer diagnosis and development of LM.[7] Also, 30%–60% of patients with LM have, in addition, associated parenchymal brain metastases.[5,8]

2 Prognosis

A diagnosis of LM portends a poor prognosis with an estimated median survival of only 2–4 months with therapy and only 4–6 weeks if untreated.[9–15] Favorable prognostic factors include a good performance status, minimal neurologic deficits, low CSF protein (< 50 mg/dL), minimal or well-controlled systemic disease, and treatment-responsive histology with remaining systemic treatment options.[7,16–18] The presence or absence of CSF cytology does not influence survival.[19] Poor prognostic factors include an impaired performance status, moderate-to-severe neurologic deficits, extensive systemic disease with few treatment options, bulky radiographic CNS disease, obstruction to CSF circulation, and LM-related encephalopathy.[2,20,21]

3 Anatomy

The meninges are a three-layer membrane composed of the dura mater found immediately beneath the skull, the arachnoid, and the pia mater, the latter of which covers the brain and spinal cord. The leptomeninges consist of two membranes, the arachnoid and pia mater. CSF is contained within the subarachnoid space, which separates these two membranes. In adults, 800 mL of CSF is produced daily wherein the entire volume of CSF turns over three to four times per day; approximately 140 mL of CSF is contained in the combined cerebral and spinal subarachnoid space.[22] By 3–4 years of age, the CSF volume of the child and the adult is equivalent.

CSF is produced by the choroid plexus, located primarily in the lateral ventricles. It flows out of the lateral ventricles through the foramen of Monro into the third ventricle through the aqueduct of Sylvius into the fourth ventricle and through the foramen of Magendie and Luschka to the base of the brain. From there, the CSF travels inferiorly to the base of the spine (lumbar cistern) and superiorly over the cerebral convexities exiting through the arachnoid granulations in the superior sagittal sinus. Because the subarachnoid space is partitioned by irregular trabeculae throughout its length and delicate blood vessels of the pia traverse it, tumor cells have a significant opportunity to accumulate and interrupt CSF flow at any point along its circulatory path.[22,23]

4 Pathogenesis

Tumor deposits can be either a thin/nonbulky layer (Fig. 1B) coating the leptomeninges versus plaque-like or nodular/bulky deposits (Fig. 1A). Malignant deposits of either type may be local or diffuse.

As with any metastatic process, tumor cells exit the primary tumor, invade the circulation as circulating tumor cells, and then enter the CSF through one of the several mechanisms mentioned below. Within the CSF, tumor cells are relatively protected from immune surveillance and an intact BCSFB (blood-CSF barrier).[24] This likely contributes to the facilitation of their malignant cell survival when treated with non-CNS-penetrating systemic therapy.

Malignant cells can gain access to the CSF through several routes. Tumor cells can extend from the brain or spinal cord parenchymal metastases that abut the subarachnoid space, the ventricular surface, or the choroid plexus and directly seed the CSF. Tumor cells can gain access to the CSF through hematogenous spread via the arterial circulation to the choroid plexus and then into the cerebral ventricles. Tumor cells may also enter the CSF through leptomeningeal veins. LM may occur when tumor cells infiltrate the bone marrow of the vertebrae or the skull, grow along the veins exiting the marrow to reach the dura, and ultimately invade perivenous adventitial tissue connecting the dura mater with the subarachnoid space. Tumor cells may grow along the cranial or peripheral nerves and enter the subarachnoid

space along the nerve, so-called centrifugal spread. LM can develop after the surgical resection of parenchymal brain metastases as a result of inadvertent spillage of malignant cells into the CSF. Patients who undergo a piecemeal resection, particularly of posterior fossa metastases, are at a higher risk of developing LM than those who undergo *en bloc* resection.[25,26]

5 Clinical presentation

Clinical manifestations of LM arise from the involvement of three regions in the nervous system: (1) the cerebral hemispheres, (2) the cranial nerves, and (3) the spinal cord and associated exiting nerve roots.

Infiltration of leptomeninges with cancer cells and/or the associated inflammation may cause an impedance of the normal CSF circulation at any point within the subarachnoid space.[27] If the outflow of CSF from the ventricles or arachnoid granulations is obstructed, increased intracranial pressure (ICP) and hydrocephalus may result (either obstructive or communicating/nonobstructive hydrocephalus).[28] Consequently, a patient with intracranial CSF flow disturbance may exhibit signs of increased ICP and hydrocephalus. Common manifestations of increased ICP include headaches which are more pronounced when lying down and upon first awakening, gait ataxia, cognitive slowing, urinary incontinence, vision decline, papilledema, nausea, vomiting, horizontal diplopia (resulting from abducens nerve compression at the base of the skull), and somnolence.[29] Plateau waves of sustained elevated CSF pressure may occur spontaneously or be precipitated by changes in the position.[30] A finding of communicating hydrocephalus without a radiologic correlate in a patient with cancer is most often a consequence of LM.

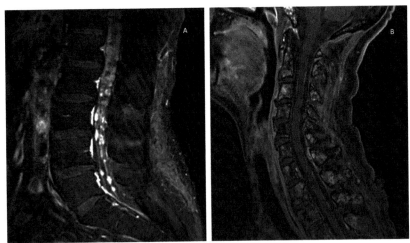

FIG. 1   Magnetic resonance imaging (MRI) spine with gadolinium contrast demonstrating enhancement of the leptomeningeal space. (A) MRI L-spine T1 postcontrast sagittal view demonstrating multiple bulky nodular foci of enhancement along the lumbar cord and cauda equina leptomeninges. (B) MRI C-spine T1 postcontrast sagittal view demonstrating multiple linear foci of enhancement along the cervical cord leptomeninges.

Common symptoms of LM involving the cerebral hemispheres are headache, encephalopathy, and seizures. Invasion, compression, or spasm of blood vessels located on the brain convexity or in the Virchow-Robin spaces may cause ischemia, leading to transient ischemic attacks, strokes, and encephalopathy secondary to a global decrease in cerebral blood flow or diffuse infiltration of the cortex by malignant cells..[31]

Cranial nerve involvement by LM may result in diplopia, ptosis, facial pain or numbness, facial palsy, tinnitus, hearing loss, vertigo, dysarthria, and dysphagia. Involvement of the cranial nerves that subserve eye movement (cranial nerves III, IV, and VI) are the most common manifestation of LM. An isolated VIth nerve palsy in a patient with leukemia is most often a result of leukemic meningitis.

Infiltration of the spine or exiting nerve roots causes motor and/or sensory symptoms. These manifest as flaccid weakness, hypoesthesia, paresthesia, and radicular pain. Involvement of the most caudal spinal nerve roots can result in a cauda equina (asymmetric radicular sensory loss in the lower extremities, paraparesis, and delayed incontinence) or conus medullaris syndrome (saddle anesthesia, urinary incontinence, fecal incontinence, and constipation from parasympathetic denervation of the rectum and sigmoid and anal sphincters).[5,32] A finding of a cauda equina syndrome in a patient with breast cancer and without a radiographic correlate is most often a consequence of LM.

6 Diagnostic studies/staging

Diagnosis of LM is formally established based on a positive (malignant) CSF cytology, radiologic findings (CT or MRI alterations as enumerated below) with corresponding clinical findings, and/or signs and symptoms suggestive of CSF involvement in a patient with known malignancy irrespective of CSF cytology and radiology.[20,33] Workup of LM includes CNS imaging with contrast MRI brain and entire spine (brain CT and CT myelogram if MRI is not feasible), CSF analysis, and systemic staging/restaging with contrast CT and/or CT-PET.

6.1 CSF examination

Cytological evaluation of the CSF is the pathologic standard for diagnosing LM and is used as a component of response criteria.[28,34] A CSF volume greater than 10 mL is recommended for CSF cytology as with a higher volume of CSF the likelihood of capturing malignant cells increases.[35] Delays in CSF processing results in a decrease of tumor cell viability and diagnostic yield, thus immediate transport to and processing by the cytology lab is important.[36] Indirect and nonspecific indicators of CSF

involvement by LM include an elevated opening pressure (> 20 cm H_2O), elevated protein (> 45 mg/dL), decreased glucose (< 50–60 mg/dL), and elevated white blood cell count (> 4/mm^3, most often lymphocytes).[5,37] The sensitivity of CSF cytology for detecting malignant cells is 45% with a single lumbar puncture (LP) and increases to 80% with subsequent LPs.[5] Little further sensitivity is gained following two CSF analyses unless performed at a different site. CSF flow cytometry is two to three times more sensitive than cytology for the detection of hematologic malignancy in the CSF and is commonly combined with CSF cytology in assessing LM or response to treatment.[38,39] In patients with lymphoma, CSF polymerase chain reaction (PCR) analysis looking for clonal immunoglobin gene rearrangement may be useful.

Epithelial cell adhesion molecule (EpCAM) is expressed by solid tumors of epithelial origin, such as non-small-cell lung cancer, breast cancer, and ovarian cancer. EpCAM-based flow cytometry is superior to CSF cytology for the diagnosis of LM in patients with epithelial tumors but is limited in applicability due to laboratory availability.[40]

A number of CSF biomarkers have been investigated both for the diagnosis as well as monitoring the response to the treatment of LM. These include beta-glucuronidase, carcinoembryonic antigen (CEA), alpha-fetoprotein (AFP), beta-human chorionic gonadotropin (β-HCG), CA 15-3, and vascular endothelial growth factor (VEGF), among others.[41–43] While many of these have demonstrated promise in small studies, none have been established as a component of the standard evaluation of LM or suspected LM.

6.2 Neuroimaging

Gadolinium-enhanced MRI of the brain and spine is the standard imaging modality to diagnose LM. In patients with supportive clinical findings, MRI has a sensitivity of 76% and specificity of 77%.[44–46] MRI should precede LP to avoid dural enhancement caused by intracranial hypotension following an LP. MRI helps assess nodular versus linear enhancement, focal versus diffuse disease, and the presence of hydrocephalus, parenchymal brain metastases, dural metastases, and bony metastases. The nodular disease is defined as enhancing nodules greater than 5×10 mm in orthogonal diameter.[34] The most frequent MRI findings include focal or diffuse enhancement of leptomeninges along the sulci, cerebellar folia, cranial nerves, and spinal nerve roots, linear or nodular enhancement of the spinal cord, and thickening of the lumbosacral nerve roots (Figs. 1 and 2). MRI abnormalities are seen more often in LM from solid tumors as compared to hematologic malignancies.[47] If MRI is contraindicated, then contrast CT brain and CT myelography can be used although it is viewed as a less

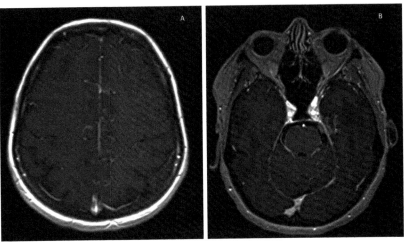

FIG. 2   Magnetic resonance imaging (MRI) brain with gadolinium contrast demonstrating enhancement of the cerebral sulci/leptomeningeal space. (A) MRI brain T1 postcontrast axial view demonstrating diffuse leptomeningeal enhancement in the cerebral sulci. (B) MRI brain T1 postcontrast axial view demonstrating leptomeningeal enhancement of the cerebellar folia.

sensitive imaging modality.[29] CT myelography, although invasive, has a similar sensitivity to MRI for the evaluation of the spine.[48]

CSF flow studies (radionuclide cisternograms) are performed to evaluate the patency of the CSF pathway in cases where intra-CSF (also called intrathecal or intraventricular) chemotherapy is considered.[49–51] Abnormal CSF flow interferes with the uniform distribution of intra-CSF chemotherapy and can lead to toxicity due to excess local accumulation of chemotherapy.[52] CSF flow studies involve injection of a radionuclide tracer such as indium-111 DTPA or technetium-99m into the CSF which can demonstrate areas of obstruction that may be either partial or complete. A radionuclide cisternogram before intra-CSF therapy is initiated is recommended to ensure uniform distribution of the drug and should be performed by the same route as the planned route of administration of intra-CSF drug (intrathecal or intraventricular). Hence, in the instance intraventricular chemotherapy is to be administered, a ventricular access device such as an Ommaya reservoir should be placed before examining CSF flow dynamics. Of note, bulky disease in LM seen on imaging is frequently associated with impaired CSF flow. CSF flow studies can help guide therapeutic decisions by determining whether a patient is a candidate for intra-CSF treatment in instances where no CSF flow obstruction is demonstrated or whether involved-field radiotherapy to the site of CSF flow obstruction is required before proceeding with intra-CSF chemotherapy.[50,53]

7 Treatment

Management goals of LM include improving a patient's neurologic function, quality of life, preventing further neurologic deterioration, and prolonging survival.[54]

The treatment of LM should be individualized and is based on an assessment of prognostic factors including tumor histology, presence of hydrocephalus, bulky radiographic CNS disease and CSF flow obstruction, status of the systemic disease, availability of systemic therapy, severity of neurologic deficits including LM-related encephalopathy, and performance status.[55–57]

Symptomatic treatment of LM entails managing the myriad of symptoms related to elevated ICP and infiltration of various parts of the neuraxis.[2,58] Symptom management utilizes involved-field radiation to the sites of symptomatic disease, e.g., treatment of the lumbosacral spine in patients with a cauda equina syndrome and to the sites of CSF flow obstruction. Corticosteroids are used to manage cerebral edema related to the coexisting parenchymal brain metastases but in general, have little utility in managing symptoms of LM. Additional symptom-directed therapies include placement of a ventriculoperitoneal shunt for hydrocephalus and use of analgesics for pain, anticonvulsant medications in the instance seizures occur, antiemetics for the alleviation of nausea and vomiting, and stimulants such as modafinil and amphetamine derivatives to treat attention problems, fatigue, and somnolence caused by the disease or due to radiation therapy to the brain.[59]

Cancer-directed treatment of LM is multimodal and includes the use of radiation, intra-CSF therapy as well as systemic therapy.

7.1 Radiation therapy

Radiation is utilized primarily for symptom palliation, e.g., skull-based radiotherapy for cranial nerve involvement, site of CSF flow obstruction, and treatment of bulky disease, e.g., coexistent parenchymal brain metastases.[60] Radiotherapy will transiently improve penetration of coadministered systemic treatments by mechanical disruption of the blood–brain barrier (BBB) and blood-CSF barrier.[61,62]

A variety of radiation regimens are used including involved-field radiation, whole-brain radiation therapy (WBRT), and craniospinal irradiation (CSI).[60] CSI is rarely used as it is not curative, is time- and resource-intensive, and is associated with multiple toxicities, such as myelosuppression, neurocognitive dysfunction, fatigue, headache, alopecia, odynophagia, and diarrhea.

Simultaneous or close sequential coadministration of radiation and systemic or intra-CSF chemotherapy may augment radiation-associated side effects. Severe leukoencephalopathy may occur with concomitant WBRT or CSI and chemotherapy, in particular methotrexate. WBRT or CSI followed by chemotherapy is more likely to cause delayed leukoencephalopathy than administering chemotherapy first.

A strategy for potentially minimizing radiation-associated cognitive toxicity would be WBRT with hippocampal avoidance and the addition of memantine.[63]

7.2 Chemotherapy and targeted therapies

7.2.1 Intra-CSF therapy

Intra-CSF therapy is a method of delivering a chemotherapeutic agent directly into the CSF which circumvents the BCSFB. However, regionally (intra-CSF) administered chemotherapy only penetrates the tumor several millimeters adjacent to the CSF as transport into the tumor is diffusion-dependent. Therefore, in patients with bulky LM disease, intra-CSF chemotherapy has limited effectiveness and usually requires involved-field radiotherapy or coadministered systemic therapy.[64–66] The addition of intra-CSF therapy to systemic treatment and radiation may lead to a survival benefit though combined modality therapy has not been studied in an adequately powered randomized clinical trial and may be associated with an increased risk of neurotoxicity.[67]

Intra-CSF treatment can be administered via lumbar puncture (intrathecal; IT) or an intraventricular route. Intraventricular treatment is often preferred as it allows for a comparatively straightforward and relatively rapid procedure for drug administration, assures drug delivery into the CSF, and achieves a more uniform distribution in the CSF compared to intrathecal drug administration.[68] Furthermore, superior survival has been demonstrated with this route of delivery.[69] The most frequently utilized intraventricular access device is an Ommaya reservoir which requires a neurosurgical procedure for placement. A noncontrast CT should be done prior to the initial administration of any drugs in order to verify the correct placement of the intraventricular catheter. Complications due to the Ommaya reservoir

placement include malposition rates between 3% and 12%, bleeding, obstruction, and infection, usually from skin flora.[70,71]

The technique of administering intra-CSF chemotherapy is of utmost importance. It is critical that the CSF fluid volume not be greater after chemotherapy than before administration.[22] Patients with LM may be precariously poised on the edge of the ventricular compliance ("pressure-volume") curve. Even small amounts of fluid added to the total CSF volume can cause symptoms of raised ICP. Hence, techniques to facilitate isovolumetric administration should be followed (removal of CSF prior to chemotherapy administration should be equivalent or greater to the volume of drug and diluent injected).[22]

Potential complications of intra-CSF treatment include infection (bacterial meningitis), encephalopathy, seizure, myelopathy, aseptic or chemical meningitis/arachnoiditis, and delayed leukoencephalopathy.[67,72] Preservative-free medications should be utilized for intra-CSF administration to avoid the risk of preservative-induced anaphylaxis and neurotoxicity. Erroneous administration of a higher dose of intra-CSF therapy or incorrect therapy can be fatal.

The five intra-CSF chemotherapy agents principally used include methotrexate, cytarabine (including liposomal cytarabine), thioTEPA, topotecan, and etoposide. Methotrexate is the drug most commonly used to treat solid tumors, whereas cytarabine with or without methotrexate is most commonly used to treat lymphomatous and leukemic meningitis.[22] Combination chemotherapy may be useful in hematologic malignancies, but for patients with LM from solid tumors, there is no proven benefit of combination treatment.[73,74]

7.2.2 Methotrexate

Methotrexate, a cell-cycle-specific chemotherapy, is an inhibitor of dihydrofolate reductase, the enzyme that generates reduced folates for purine synthesis, thereby inhibiting DNA synthesis. CSF half-life is 4.5–8 h, and cytotoxic concentrations are maintained for 24 h after each dose. A common dosing schedule is 10–15 mg twice weekly for 4 weeks, then 10–15 mg once weekly for 4 weeks, then 10–15 mg once monthly. CSF half-life is 4.5–8 h.[75]

An alternative regimen of intra-CSF methotrexate that provides improved pharmacokinetics is administered as 2 mg per day for five consecutive days (defined as a cycle) every other week.[76] Induction is administered over 8 weeks followed by maintenance therapy given as a single cycle administered once monthly. Progression-free survival is significantly better when methotrexate is administered via an intraventricular route versus the intrathecal route.[77]

In the rare instance of an accidental overdose of methotrexate, intra-CSF carboxypeptidase G2 can be used.[78]

7.2.3 Cytarabine (Ara-C)

Ara-C, a cell-cycle-specific chemotherapy, is a cytidine analog antimetabolite that inhibits DNA synthesis. The dosing schedule is 25–100 mg twice weekly for 4 weeks, 25–100 mg once weekly for 4 weeks, then 25–100 mg once monthly. CSF half-life is 6 h, and cytotoxic concentrations are maintained for 24 h after each dose. Use is most often reserved for patients with either leukemic or lymphomatous LM.[79]

A sustained-release multivesicular liposome-encapsulated form of Ara-C (DepoCyt) has been approved for the treatment of patients with lymphomatous meningitis.[80] A major advantage of this preparation is its long half-life within the CSF (141 h), which permits a reduced dosing frequency (initially 50 mg once every 2 weeks, followed by once monthly) and the ability to achieve cytotoxic drug levels within the ventricles following intrathecal instillation, limitations to short half-life chemotherapy agents, such as methotrexate and Ara-C.[80–83]

7.2.4 ThioTEPA

ThioTEPA is a DNA-alkylating agent that damages cancer cells in a cell-cycle-independent manner. It is highly lipophilic and therefore penetrates the CNS after systemic administration. The rationale for administering intra-CSF thioTEPA is to mitigate the systemic side effects, primarily cytopenia. ThioTEPA has been shown to be equally effective as methotrexate in the management of LM from solid tumors and has a different toxicity profile (predominantly cytopenias).[84]

ThioTEPA is administered 10 mg twice weekly for 4 weeks, then 10 mg once weekly for 4 weeks, then 10 mg once a month.[85] Although thioTEPA has a half-life within the CSF of only a few minutes and is more myelosuppressive than methotrexate, it has a wide spectrum of activity against many solid tumors. ThioTEPA may be considered for patients in whom prior methotrexate failed, in those with methotrexate-induced leukoencephalopathy or in those for whom concurrent WBRT is required.[22]

7.2.5 Topotecan

Topotecan, a cell-cycle-specific chemotherapy, is a camptothecin analog. Topotecan's active lactone form intercalates between DNA bases in the topoisomerase I cleavage complex. The binding of topotecan in the cleavage complex prevents topoisomerase I from religating the nicked DNA strand after relieving the strain. It has a CSF half-life of 2.6 h and is administered as 0.4 mg twice weekly for 4–6 weeks followed by once weekly for 4–6 weeks, then twice monthly as maintenance therapy.[86,87]

7.2.6 Etoposide

Etoposide, another cell-cycle-specific chemotherapy, is a topoisomerase II inhibitor and has a CSF half-life of 9.6 h. Etoposide forms a ternary complex with DNA and the topoisomerase II enzyme (which aids in relaxing negative or positive supercoils in DNA), prevents religation of the DNA strands, and by doing so causes DNA strands to break. Etoposide is administered at a dose of 0.5 mg once daily for five consecutive days (defined as a single cycle) given every other week for 8 weeks followed by a single cycle once monthly as maintenance therapy.[88,89]

7.2.7 Miscellaneous intra-CSF agents

Additionally, monoclonal antibodies like trastuzumab (an anti-HER2 antibody) and rituximab (an anti-CD20 antibody) may be given as intra-CSF therapies.[90,91] Rituximab has a half-life of 34.9 h and has been successfully used in patients with primary CNS lymphoma. Similarly, trastuzumab has been used for patients with HER2-positive breast cancer LM. Other agents like intra-CSF alpha-interferon have shown modest activity against LM, but toxicity limits its use.[92]

7.3 Systemic chemotherapy and targeted agents

The superiority of regional versus systemic therapy in the treatment of LM has not been established. An advantage of systemic therapy is its ability to treat nodular or bulky CNS disease and the availability of a wide variety of agents with known cancer-specific activity. However, most systemic agents fail to penetrate the CSF in sufficient concentrations due to the BCSFB. Higher doses of chemotherapy and utilization of small lipid-soluble molecules help achieve therapeutic concentrations of these agents in the CSF.[93]

7.3.1 Chemotherapeutic agents

Systemically administered high-dose methotrexate (HD-MTX) achieves CSF concentrations comparable to those achieved with intra-CSF therapy and allows for wider tissue distribution in the CNS.[71] HD-MTX given as 3.5–7 g/m^2 with leucovorin rescue has been successfully used in CNS malignancies such as primary CNS lymphoma involving the brain parenchyma and CSF.[71] Nonetheless, HD-MTX requires hospitalization for several days, alkalinization of the urine, and hyperhydration to mitigate potential renal toxicity.

Systemic administration of high-dose Ara-C (HDAC) achieves a CSF/serum ratio of approximately 20% as compared to 1%–3% for HD-MTX. There are several different drug schedules for administering HDAC that achieve cytotoxic levels in the CSF. These include 2–3 g/m^2 given every 12 h or a 72-h continuous IV infusion of doses of more than 4 g/m^2. HDAC has considerable toxicity including myelosuppression, cerebellar toxicity, encephalopathy, nausea, vomiting, and mucositis.[22]

Temozolomide is an orally bioavailable alkylating agent that achieves CSF levels of approximately 20% of those achievable in serum but has never been formally evaluated in a trial in patients with LM.[94,95]

7.3.2 Molecularly targeted therapies

Various targeted therapies are available for HER2-overexpressing breast cancer, NSCLC with epidermal growth factor receptor (EGFR) mutation or anaplastic lymphoma kinase (ALK) positivity as well as melanoma with BRAF (V600E or V600K) mutation.

For patients with epidermal growth factor (EGFR)-mutated NSCLC, various oral tyrosine kinase inhibitors (TKIs) have good BCSFB penetration. Gefitinib (250 mg daily) and erlotinib (2000 mg weekly) are first-generation TKIs which modestly penetrate the CSF.[96–98] Afatinib is a second-generation TKI which is reported to improve LM lesions at a higher dose (40 mg daily) in combination with cetuximab (250 mg/m^2 biweekly).[99] Osimertinib is a third-generation TKI which demonstrated encouraging results in the phase I BLOOM study in patients with LM at a dose of 160 mg/day.[100,101]

For ALK-translocated NSCLC, newer ALK inhibitors alectinib and ceritinib have higher CSF penetration rates and have been reported to have activity in LM.[102–106] The next-generation ALK inhibitor brigatinib (90–240 mg) has shown promising results in a phase I/II study of patients with parenchymal brain metastases including patients with LM.[107]

Targeted treatments for melanoma include BRAF and MEK inhibitors as well as immune therapy with checkpoint inhibition. Of note, dabrafenib has higher CNS penetration as compared to vemurafenib.[108–115]

7.4 Hormone therapy

LM from hormone-dependent cancers, such as breast and prostate cancer, may be responsive to hormonal manipulation. LM from prostate cancer and hormone-positive breast cancer has been reported to have responded to hormonal therapy but only in isolated case reports.[116–118]

7.5 Immunotherapy

Immune effectors recognize and destroy tumor cells and provide long-term immune surveillance. A variety of immunotherapies, including cellular therapies, have been incorporated into cancer treatment. Treatment with immune checkpoint blockade (ICB) anti-programmed death 1 (PD1), anti-programmed death ligand 1 (PD-L1), or anti-cytotoxic T lymphocyte-associated protein 4 (CTLA4) blocks inhibitory signals for T cells and allows the adaptive immune system to target tumor cells.

For solid tumors like melanoma and NSCLC, immune checkpoint inhibitors have safety and activity in parenchymal brain metastases and are now also being evaluated in LM. In a retrospective series of melanoma patients with LM, improved survival was seen with BRAF inhibitors and anti-PD-1 antibodies.[119]

8 Summary

Treatment of LM is rarely curative and essentially never in instances of carcinomatous meningitis wherein the median survival after diagnosis is only 2–3 months. However, palliation of LM-related neurologic signs and symptoms may temporarily improve a patient's quality of life.

Novel treatments of LM include proton radiation to the entire craniospinal axis to circumvent toxicities associated with photon craniospinal irradiation (NCT03520504). Immunotherapy, as well as chemotherapy with better BCSFB penetration, holds promise for future improvement in the prognosis of LM. Chemotherapy with paclitaxel trevatide (designed to cross the BCSFB) for HER2-negative breast with newly diagnosed LM (NCT03613181) is under investigation as is the novel tyrosine kinase inhibitor, tucatinib, for HER2-positive breast cancer-related LM (NCT03501979).[120] Multiple immune therapy trials aim to investigate immune checkpoint inhibitors as well as chimeric antigen receptor (CAR)-engineered T cell therapy cells for LM.[121–123] Novel approaches for drug delivery using ultrasound technologies to increase the permeability of the BBB and BCSF are under investigation. Increasingly targeted therapies with small molecule inhibitors are utilized for tumors with defined oncogenic drivers. Late-generation-targeted therapies have been developed that are CNS penetrant and are increasingly utilized for parenchymal brain metastases and will be explored for the treatment of LM. Improvements in diagnostic tools to aid in rendering a diagnosis of LM continue to be explored including the use of circulating tumor cells by cell sorting, tumor-free DNA, and epithelial cell markers such as epithelial adhesion molecules. An ongoing challenge in treating LM has been evaluating response to treatment for which both the Response Assessment in Neuro-Oncology (RANO) working group and the European Association of Neuro-Oncology (EANO) have suggested guidelines that are increasingly being incorporated into ongoing trials in LM.[34]

References

1. Glass JP, Melamed M, Chernik NL, Posner JB. Malignant cells in cerebrospinal fluid (CSF): the meaning of a positive CSF cytology. *Neurology.* 1979;29(10):1369–1375. https://doi.org/10.1212/wnl.29.10.1369.
2. Taillibert S, Chamberlain MC. Leptomeningeal metastasis. *Handb Clin Neurol.* 2018;149:169–204. https://doi.org/10.1016/B978-0-12-811161-1.00013-X.
3. Rosen ST, Aisner J, Makuch RW, et al. Carcinomatous leptomeningitis in small cell lung cancer: a clinicopathologic review of the National Cancer Institute experience. *Medicine (Baltimore).* 1982;61(1):45–53.

4. Aroney RS, Dalley DN, Chan WK, Bell DR, Levi JA. Meningeal carcinomatosis in small cell carcinoma of the lung. *Am J Med.* 1981;71(1):26–32.

5. Wasserstrom WR, Glass JP, Posner JB. Diagnosis and treatment of leptomeningeal metastases from solid tumors: experience with 90 patients. *Cancer.* 1982;49(4):759–772.

6. Waki F, Ando M, Takashima A, et al. Prognostic factors and clinical outcomes in patients with leptomeningeal metastasis from solid tumors. *J Neurooncol.* 2009;93(2):205–212. https://doi.org/10.1007/s11060-008-9758-3.

7. Yust-Katz S, Garciarena P, Liu D, et al. Breast cancer and leptomeningeal disease (LMD): hormone receptor status influences time to development of LMD and survival from LMD diagnosis. *J Neurooncol.* 2013;114(2):229–235. https://doi.org/10.1007/s11060-013-1175-6.

8. Clarke JL, Perez HR, Jacks LM, Panageas KS, Deangelis LM. Leptomeningeal metastases in the MRI era. *Neurology.* 2010;74(18):1449–1454. https://doi.org/10.1212/WNL.0b013e3181dc1a69.

9. Clatot F, Philippin-Lauridant G, Ouvrier MJ, et al. Clinical improvement and survival in breast cancer leptomeningeal metastasis correlate with the cytologic response to intrathecal chemotherapy. *J Neurooncol.* 2009;95(3):421–426. https://doi.org/10.1007/s11060-009-9940-2.

10. de Azevedo CR, Cruz MR, Chinen LT, et al. Meningeal carcinomatosis in breast cancer: prognostic factors and outcome. *J Neurooncol.* 2011;104(2):565–572. https://doi.org/10.1007/s11060-010-0524-y.

11. Gauthier H, Guilhaume MN, Bidard FC, et al. Survival of breast cancer patients with meningeal carcinomatosis. *Ann Oncol.* 2010;21(11):2183–2187. https://doi.org/10.1093/annonc/mdq232.

12. Harstad L, Hess KR, Groves MD. Prognostic factors and outcomes in patients with leptomeningeal melanomatosis. *Neuro Oncol.* 2008;10(6):1010–1018. https://doi.org/10.1215/15228517-2008-062.

13. Morris PG, Reiner AS, Szenberg OR, et al. Leptomeningeal metastasis from non-small cell lung cancer: survival and the impact of whole brain radiotherapy. *J Thorac Oncol.* 2012;7(2):382–385. https://doi.org/10.1097/JTO.0b013e3182398e4f.

14. Park JH, Kim YJ, Lee JO, et al. Clinical outcomes of leptomeningeal metastasis in patients with non-small cell lung cancer in the modern chemotherapy era. *Lung Cancer.* 2012;76(3):387–392. https://doi.org/10.1016/j.lungcan.2011.11.022.

15. Rudnicka H, Niwinska A, Murawska M. Breast cancer leptomeningeal metastasis—the role of multimodality treatment. *J Neurooncol.* 2007;84(1):57–62. https://doi.org/10.1007/s11060-007-9340-4.

16. Hyun JW, Jeong IH, Joung A, Cho HJ, Kim SH, Kim HJ. Leptomeningeal metastasis: clinical experience of 519 cases. *Eur J Cancer.* 2016;56:107–114. https://doi.org/10.1016/j.ejca.2015.12.021.

17. Kingston B, Kayhanian H, Brooks C, et al. Treatment and prognosis of leptomeningeal disease secondary to metastatic breast cancer: a single-centre experience. *Breast.* 2017;36:54–59. https://doi.org/10.1016/j.breast.2017.07.015.

18. Abouharb S, Ensor J, Loghin ME, et al. Leptomeningeal disease and breast cancer: the importance of tumor subtype. *Breast Cancer Res Treat.* 2014;146(3):477–486. https://doi.org/10.1007/s10549-014-3054-z.

19. Chamberlain MC, Johnston SK. Neoplastic meningitis: survival as a function of cerebrospinal fluid cytology. *Cancer.* 2009;115(9):1941–1946. https://doi.org/10.1002/cncr.24210.

20. Guidelines N. *National Comprehensive Cancer Network*; 2019.

21. Chamberlain MC, Kormanik PA. Prognostic significance of coexistent bulky metastatic central nervous system disease in patients with leptomeningeal metastases. *Arch Neurol.* 1997;54(11):1364–1368. https://doi.org/10.1001/archneur.1997.00550230037013.

22. Chowdhary S, Chamberlain M. Leptomeningeal metastases: current concepts and management guidelines. *J Natl Compr Canc Netw.* 2005;3(5):693–703. https://doi.org/10.6004/jnccn.2005.0039.

23. Shafique S, Rayi A. *Anatomy, Head and Neck, Subarachnoid Space.* Treasure Island (FL): StatPearls; 2020.

24. Leal T, Chang JE, Mehta M, Robins HI. Leptomeningeal metastasis: challenges in diagnosis and treatment. *Curr Cancer Ther Rev.* 2011;7(4):319–327. https://doi.org/10.2174/157339411797642597.

25. Suki D, Abouassi H, Patel AJ, Sawaya R, Weinberg JS, Groves MD. Comparative risk of leptomeningeal disease after resection or stereotactic radiosurgery for solid tumor metastasis to the posterior fossa. *J Neurosurg.* 2008;108(2):248–257. https://doi.org/10.3171/JNS/2008/108/2/0248.

26. Ahn JH, Lee SH, Kim S, et al. Risk for leptomeningeal seeding after resection for brain metastases: implication of tumor location with mode of resection. *J Neurosurg.* 2012;116(5):984–993. https://doi.org/10.3171/2012.1.JNS111560.

27. Wang N, Bertalan MS, Brastianos PK. Leptomeningeal metastasis from systemic cancer: review and update on management. *Cancer.* 2018;124(1):21–35. https://doi.org/10.1002/cncr.30911.

28. Chang EL, Lo S. Diagnosis and management of central nervous system metastases from breast cancer. *Oncologist.* 2003;8(5):398–410.

29. Thakkar JP, Kumthekar P, Dixit KS, Stupp R, Lukas RV. Leptomeningeal metastasis from solid tumors. *J Neurol Sci.* 2020;411. https://doi.org/10.1016/j.jns.2020.116706, 116706.

30. Sagar SM. Carcinomatous meningitis: it does not have to be a death sentence. *Oncology (Williston Park).* 2002;16(2):237–243. discussion 44, 49–50.

31. Siegal T, Mildworf B, Stein D, Melamed E. Leptomeningeal metastases: reduction in regional cerebral blood flow and cognitive impairment. *Ann Neurol.* 1985;17(1):100–102. https://doi.org/10.1002/ana.410170121.

32. Winge K, Rasmussen D, Werdelin LM. Constipation in neurological diseases. *J Neurol Neurosurg Psychiatry.* 2003;74(1):13–19. https://doi.org/10.1136/jnnp.74.1.13.

33. Chamberlain MC, Glantz M, Groves MD, Wilson WH. Diagnostic tools for neoplastic meningitis: detecting disease, identifying patient risk, and determining benefit of treatment. *Semin Oncol.* 2009;36(4 Suppl. 2):S35–S45. https://doi.org/10.1053/j.seminoncol.2009.05.005.

34. Chamberlain M, Junck L, Brandsma D, et al. Leptomeningeal metastases: a RANO proposal for response criteria. *Neuro Oncol.* 2017;19(4):484–492. https://doi.org/10.1093/neuonc/now183.

35. Glantz MJ, Cole BF, Glantz LK, et al. Cerebrospinal fluid cytology in patients with cancer: minimizing false-negative results. *Cancer.* 1998;82(4):733–739.

36. Dux R, Kindler-Rohrborn A, Annas M, Faustmann P, Lennartz K, Zimmermann CW. A standardized protocol for flow cytometric analysis of cells isolated from cerebrospinal fluid. *J Neurol Sci.* 1994;121(1):74–78.

37. Le Rhun E, Weller M, Brandsma D, et al. EANO-ESMO Clinical Practice Guidelines for diagnosis, treatment and follow-up of patients with leptomeningeal metastasis from solid tumours. *Ann Oncol.* 2017;28(suppl_4):iv84–iv99. https://doi.org/10.1093/annonc/mdx221.

38. Hegde U, Filie A, Little RF, et al. High incidence of occult leptomeningeal disease detected by flow cytometry in newly diagnosed aggressive B-cell lymphomas at risk for central nervous system involvement: the role of flow cytometry versus cytology. *Blood.* 2005;105(2):496–502. https://doi.org/10.1182/blood-2004-05-1982.

39. French CA, Dorfman DM, Shaheen G, Cibas ES. Diagnosing lymphoproliferative disorders involving the cerebrospinal fluid: increased sensitivity using flow cytometric analysis. *Diagn Cytopathol.* 2000;23(6):369–374.

40. Milojkovic Kerklaan B, Pluim D, Bol M, et al. EpCAM-based flow cytometry in cerebrospinal fluid greatly improves diagnostic accuracy of leptomeningeal metastases from epithelial tumors. *Neuro Oncol.* 2016;18(6):855–862. https://doi.org/10.1093/neuonc/nov273.

41. Nakagawa H, Kubo S, Murasawa A, et al. Measurements of CSF biochemical tumor markers in patients with meningeal carcinomatosis and brain tumors. *J Neurooncol.* 1992;12(2):111–120. https://doi.org/10.1007/bf00172659.

42. Le Rhun E, Kramar A, Salingue S, et al. CSF CA 15-3 in breast cancer-related leptomeningeal metastases. *J Neurooncol.* 2014;117(1):117–124. https://doi.org/10.1007/s11060-014-1361-1.

43. Chamberlain MC. Cytologically negative carcinomatous meningitis: usefulness of CSF biochemical markers. *Neurology.* 1998;50(4):1173–1175. https://doi.org/10.1212/wnl.50.4.1173.

44. Straathof CS, de Bruin HG, Dippel DW, Vecht CJ. The diagnostic accuracy of magnetic resonance imaging and cerebrospinal fluid cytology in leptomeningeal metastasis. *J Neurol.* 1999;246(9):810–814.

45. Chamberlain MC, Sandy AD, Press GA. Leptomeningeal metastasis: a comparison of gadolinium-enhanced MR and contrast-enhanced CT of the brain. *Neurology.* 1990;40(3 Pt 1):435–438. https://doi.org/10.1212/wnl.40.3_part_1.435.

46. Chamberlain MC. Comparative spine imaging in leptomeningeal metastases. *J Neurooncol.* 1995;23(3):233–238. https://doi.org/10.1007/BF01059954.

47. Chamberlain MC. Comprehensive neuraxis imaging in leptomeningeal metastasis: a retrospective case series. *CNS Oncol.* 2013;2(2):121–128. https://doi.org/10.2217/cns.12.45.

48. Schuknecht B, Huber P, Buller B, Nadjmi M. Spinal leptomeningeal neoplastic disease. Evaluation by MR, myelography and CT myelography. *Eur Neurol.* 1992;32(1):11–16. https://doi.org/10.1159/000116780.

49. Chamberlain MC. Spinal 111Indium-DTPA CSF flow studies in leptomeningeal metastasis. *J Neurooncol.* 1995;25(2):135–141. https://doi.org/10.1007/BF01057757.

50. Chamberlain MC, Kormanik PA. Prognostic significance of 111indium-DTPA CSF flow studies in leptomeningeal metastases. *Neurology.* 1996;46(6):1674–1677. https://doi.org/10.1212/wnl.46.6.1674.

51. Chamberlain MC. Radioisotope CSF flow studies in leptomeningeal metastases. *J Neurooncol.* 1998;38(2–3):135–140. https://doi.org/10.1023/a:1005982826121.

52. Siegal T. Toxicity of treatment for neoplastic meningitis. *Curr Oncol Rep.* 2003;5(1):41–49.

53. Chamberlain MC, Corey-Bloom J. Leptomeningeal metastases: 111indium-DTPA CSF flow studies. *Neurology.* 1991;41(11):1765–1769. https://doi.org/10.1212/wnl.41.11.1765.

54. Chamberlain MC. Leptomeningeal metastasis. *Semin Neurol.* 2010;30(3):236–244. https://doi.org/10.1055/s-0030-1255220.

55. Chamberlain MC, Tsao-Wei D, Groshen S. Neoplastic meningitis-related encephalopathy: prognostic significance. *Neurology.* 2004;63(11):2159–2161.

56. Chamberlain MC, Johnston SK, Glantz MJ. Neoplastic meningitis-related prognostic significance of the Karnofsky performance status. *Arch Neurol.* 2009;66(1):74–78. https://doi.org/10.1001/archneurol.2008.506.

57. Glantz MJ, Walters BC. Diagnosis and outcome measures in trials for neoplastic meningitis: a review of the literature and clinical experience. *Neurosurg Focus.* 1998;4(6). https://doi.org/10.3171/foc.1998.4.6.7, e4.

58. Le Rhun E, Taillibert S, Chamberlain MC. Carcinomatous meningitis: leptomeningeal metastases in solid tumors. *Surg Neurol Int.* 2013;4(Suppl 4):S265–S288. https://doi.org/10.4103/2152-7806.111304.

59. Chamberlain MC. Carcinomatous meningitis. *Arch Neurol.* 1997;54(1):16–17. https://doi.org/10.1001/archneur.1997.00550130008003.

60. El Shafie RA, Bohm K, Weber D, et al. Palliative radiotherapy for leptomeningeal carcinomatosis-analysis of outcome, prognostic factors, and symptom response. *Front Oncol.* 2018;8:641. https://doi.org/10.3389/fonc.2018.00641.

61. Miot E, Hoffschir D, Pontvert D, et al. Quantitative magnetic resonance and isotopic imaging: early evaluation of radiation injury to the brain. *Int J Radiat Oncol Biol Phys.* 1995;32(1):121–128. https://doi.org/10.1016/0360-3016(94)00413-F.

62. Lumniczky K, Szatmari T, Safrany G. Ionizing radiation-induced immune and inflammatory reactions in the brain. *Front Immunol.* 2017;8:517. https://doi.org/10.3389/fimmu.2017.00517.

63. Vinai Gondi SD, Brown PD, Wefel JS, et al. NRG oncology CC001: a phase III trial of hippocampal avoidance (HA) in addition to whole-brain radiotherapy (WBRT) plus memantine to preserve neurocognitive function (NCF) in patients with brain metastases (BM). *J Clin Oncol.* 2019;37(15_suppl):2009.

64. Burch PA, Grossman SA, Reinhard CS. Spinal cord penetration of intrathecally administered cytarabine and methotrexate: a quantitative autoradiographic study. *J Natl Cancer Inst.* 1988;80(15):1211–1216.

65. Benjamin JC, Moss T, Moseley RP, Maxwell R, Coakham HB. Cerebral distribution of immunoconjugate after treatment for neoplastic meningitis using an intrathecal radiolabeled monoclonal antibody. *Neurosurgery.* 1989;25(2):253–258.

66. Blasberg RG, Patlak C, Fenstermacher JD. Intrathecal chemotherapy: brain tissue profiles after ventriculocisternal perfusion. *J Pharmacol Exp Ther.* 1975;195(1):73–83.

67. Boogerd W, van den Bent MJ, Koehler PJ, et al. The relevance of intraventricular chemotherapy for leptomeningeal metastasis in breast cancer: a randomised study. *Eur J Cancer.* 2004;40(18):2726–2733. https://doi.org/10.1016/j.ejca.2004.08.012.

68. Shapiro WR, Young DF, Mehta BM. Methotrexate: distribution in cerebrospinal fluid after intravenous, ventricular and lumbar injections. *N Engl J Med.* 1975;293(4):161–166. https://doi.org/10.1056/NEJM197507242930402.

69. de Oca M, Delgado M, Cacho Diaz B, et al. The comparative treatment of intraventricular chemotherapy by Ommaya reservoir vs. lumbar puncture in patients with leptomeningeal carcinomatosis. *Front Oncol.* 2018;8:509. https://doi.org/10.3389/fonc.2018.00509.

70. Chamberlain MC, Kormanik PA, Barba D. Complications associated with intraventricular chemotherapy in patients with leptomeningeal metastases. *J Neurosurg.* 1997;87(5):694–699. https://doi.org/10.3171/jns.1997.87.5.0694.

71. Zairi F, Le Rhun E, Bertrand N, et al. Complications related to the use of an intraventricular access device for the treatment of leptomeningeal metastases from solid tumor: a single centre experience in 112 patients. *J Neurooncol.* 2015;124(2):317–323. https://doi.org/10.1007/s11060-015-1842-x.

72. Byrnes DM, Vargas F, Dermarkarian C, et al. Complications of intrathecal chemotherapy in adults: single-institution experience in 109 consecutive patients. *J Oncol.* 2019;2019:4047617. https://doi.org/10.1155/2019/4047617.

73. Pullen J, Boyett J, Shuster J, et al. Extended triple intrathecal chemotherapy trial for prevention of CNS relapse in good-risk and poor-risk patients with B-progenitor acute lymphoblastic leukemia: a Pediatric Oncology Group study. *J Clin Oncol.* 1993;11(5):839–849. https://doi.org/10.1200/JCO.1993.11.5.839.

74. Pui CH, Mahmoud HH, Rivera GK, et al. Early intensification of intrathecal chemotherapy virtually eliminates central nervous system relapse in children with acute lymphoblastic leukemia. *Blood.* 1998;92(2):411–415.

75. Bleyer WA, Dedrick RL. Clinical pharmacology of intrathecal methotrexate. I. Pharmacokinetics in nontoxic patients after lumbar injection. *Cancer Treat Rep.* 1977;61(4):703–708.

76. Chamberlain MC, Kormanik P. Carcinoma meningitis secondary to non-small cell lung cancer: combined modality therapy. *Arch Neurol.* 1998;55(4):506–512. https://doi.org/10.1001/archneur.55.4.506.

77. Glantz MJ, Van Horn A, Fisher R, Chamberlain MC. Route of intracerebrospinal fluid chemotherapy administration and efficacy of therapy in neoplastic meningitis. *Cancer.* 2010;116(8):1947–1952. https://doi.org/10.1002/cncr.24921.

78. Widemann BC, Balis FM, Shalabi A, et al. Treatment of accidental intrathecal methotrexate overdose with intrathecal carboxypeptidase G2. *J Natl Cancer Inst.* 2004;96(20):1557–1559. https://doi.org/10.1093/jnci/djh270.

79. Fulton DS, Levin VA, Gutin PH, et al. Intrathecal cytosine arabinoside for the treatment of meningeal metastases from malignant brain tumors and systemic tumors. *Cancer Chemother Pharmacol.* 1982;8(3):285–291. https://doi.org/10.1007/BF00254052.

80. Chamberlain MC, Khatibi S, Kim JC, Howell SB, Chatelut E, Kim S. Treatment of leptomeningeal metastasis with intraventricular administration of depot cytarabine (DTC 101). A phase I study. *Arch Neurol.* 1993;50(3):261–264. https://doi.org/10.1001/archneur.1993.00540030027009.

81. Kim S, Chatelut E, Kim JC, et al. Extended CSF cytarabine exposure following intrathecal administration of DTC 101. *J Clin Oncol.* 1993;11(11):2186–2193. https://doi.org/10.1200/JCO.1993.11.11.2186.

82. Glantz MJ, Jaeckle KA, Chamberlain MC, et al. A randomized controlled trial comparing intrathecal sustained-release cytarabine (DepoCyt) to intrathecal methotrexate in patients with neoplastic meningitis from solid tumors. *Clin Cancer Res.* 1999;5(11):3394–3402.

83. Jaeckle KA, Phuphanich S, Bent MJ, et al. Intrathecal treatment of neoplastic meningitis due to breast cancer with a slow-release formulation of cytarabine. *Br J Cancer.* 2001;84(2):157–163. https://doi.org/10.1054/bjoc.2000.1574.

84. Grossman SA, Finkelstein DM, Ruckdeschel JC, Trump DL, Moynihan T, Ettinger DS. Randomized prospective comparison of intraventricular methotrexate and thiotepa in patients with previously untreated neoplastic meningitis. Eastern Cooperative Oncology Group. *J Clin Oncol.* 1993;11(3):561–569. https://doi.org/10.1200/JCO.1993.11.3.561.

85. Le Rhun E, Taillibert S, Chamberlain MC. Neoplastic meningitis due to lung, breast, and melanoma metastases. *Cancer Control.* 2017;24(1):22–32. https://doi.org/10.1177/107327481702400104.

86. Groves MD, Glantz MJ, Chamberlain MC, et al. A multicenter phase II trial of intrathecal topotecan in patients with meningeal malignancies. *Neuro Oncol.* 2008;10(2):208–215. https://doi.org/10.1215/15228517-2007-059.

87. Blaney SM, Heideman R, Berg S, et al. Phase I clinical trial of intrathecal topotecan in patients with neoplastic meningitis. *J Clin Oncol.* 2003;21(1):143–147. https://doi.org/10.1200/JCO.2003.04.053.

88. van der Gaast A, Sonneveld P, Mans DR, Splinter TA. Intrathecal administration of etoposide in the treatment of malignant meningitis: feasibility and pharmacokinetic data. *Cancer Chemother Pharmacol.* 1992;29(4):335–337.

89. Chamberlain MC, Tsao-Wei DD, Groshen S. Phase II trial of intracerebrospinal fluid etoposide in the treatment of neoplastic meningitis. *Cancer.* 2006;106(9):2021–2027. https://doi.org/10.1002/cncr.21828.

90. Gabay MP, Thakkar JP, Stachnik JM, Woelich SK, Villano JL. Intra-CSF administration of chemotherapy medications. *Cancer Chemother Pharmacol.* 2012;70(1):1–15. https://doi.org/10.1007/s00280-012-1893-z.

91. Rubenstein JL, Fridlyand J, Abrey L, et al. Phase I study of intraventricular administration of rituximab in patients with recurrent CNS and intraocular lymphoma. *J Clin Oncol.* 2007;25(11):1350–1356. https://doi.org/10.1200/JCO.2006.09.7311.

92. Chamberlain MC. A phase II trial of intra-cerebrospinal fluid alpha interferon in the treatment of neoplastic meningitis. *Cancer.* 2002;94(10):2675–2680. https://doi.org/10.1002/cncr.10547.

93. Banks WA. Characteristics of compounds that cross the blood-brain barrier. *BMC Neurol.* 2009;9(Suppl. 1):S3. https://doi.org/10.1186/1471-2377-9-S1-S3.

94. Addeo R, De Rosa C, Faiola V, et al. Phase 2 trial of temozolomide using protracted low-dose and whole-brain radiotherapy for nonsmall cell lung cancer and breast cancer patients with brain metastases. *Cancer.* 2008;113(9):2524–2531. https://doi.org/10.1002/cncr.23859.

95. Ostermann S, Csajka C, Buclin T, et al. Plasma and cerebrospinal fluid population pharmacokinetics of temozolomide in malignant glioma patients. *Clin Cancer Res.* 2004;10(11):3728–3736. https://doi.org/10.1158/1078-0432.CCR-03-0807.

96. Togashi Y, Masago K, Masuda S, et al. Cerebrospinal fluid concentration of gefitinib and erlotinib in patients with non-small cell lung cancer. *Cancer Chemother Pharmacol.* 2012;70(3):399–405. https://doi.org/10.1007/s00280-012-1929-4.

97. Milton DT, Azzoli CG, Heelan RT, et al. A phase I/II study of weekly high-dose erlotinib in previously treated patients with nonsmall cell lung cancer. *Cancer.* 2006;107(5):1034–1041. https://doi.org/10.1002/cncr.22088.

98. Clarke JL, Pao W, Wu N, Miller VA, Lassman AB. High dose weekly erlotinib achieves therapeutic concentrations in CSF and is effective in leptomeningeal metastases from epidermal growth factor receptor mutant lung cancer. *J Neurooncol.* 2010;99(2):283–286. https://doi.org/10.1007/s11060-010-0128-6.

99. Lin CH, Lin MT, Kuo YW, Ho CC. Afatinib combined with cetuximab for lung adenocarcinoma with leptomeningeal carcinomatosis. *Lung Cancer.* 2014;85(3):479–480. https://doi.org/10.1016/j.lungcan.2014.06.002.

100. Yang JCH, Cho BC, Kim D-W, et al., eds. Osimertinib for patients (pts) with leptomeningeal metastases (LM) from EGFR-mutant non-small cell lung cancer (NSCLC): updated results from the BLOOM study. *Chicago: J Clin Oncol.* 2017;38:538–547.

101. Yang JC-H, Cho BC, Kim D-W, et al. Osimertinib for patients (pts) with leptomeningeal metastases (LM) from EGFR-mutant non-small cell lung cancer (NSCLC): updated results from the BLOOM study. *J Clin Oncol.* 2017;35(15_suppl):2020. https://doi.org/10.1200/JCO.2017.35.15_suppl.2020.

102. Gadgeel SM, Gandhi L, Riely GJ, et al. Safety and activity of alectinib against systemic disease and brain metastases in patients with crizotinib-resistant ALK-rearranged non-small-cell lung cancer (AF-002JG): results from the dose-finding portion of a phase 1/2 study. *Lancet Oncol.* 2014;15(10):1119–1128. https://doi.org/10.1016/S1470-2045(14)70362-6.

103. Bui N, Woodward B, Johnson A, Husain H. Novel treatment strategies for brain metastases in non-small-cell lung cancer. *Curr Treat Options Oncol.* 2016;17(5):25. https://doi.org/10.1007/s11864-016-0400-x.

104. Arrondeau J, Ammari S, Besse B, Soria JC. LDK378 compassionate use for treating carcinomatous meningitis in an ALK translocated non-small-cell lung cancer. *J Thorac Oncol.* 2014;9(8):e62–e63. https://doi.org/10.1097/JTO.0000000000000174.

105. Gainor JF, Sherman CA, Willoughby K, et al. Alectinib salvages CNS relapses in ALK-positive lung cancer patients previously treated with crizotinib and ceritinib. *J Thorac Oncol.* 2015;10(2):232–236. https://doi.org/10.1097/JTO.0000000000000455.

106. Crino L, Ahn MJ, De Marinis F, et al. Multicenter phase II study of whole-body and intracranial activity with ceritinib in patients with ALK-rearranged non-small-cell lung cancer previously treated with chemotherapy and crizotinib: results from ASCEND-2. *J Clin Oncol.* 2016;34(24):2866–2873. https://doi.org/10.1200/JCO.2015.65.5936.

107. Camidge DR, Kim D-W, Tiseo M, et al. Exploratory analysis of brigatinib activity in patients with anaplastic lymphoma kinase-positive non–small-cell lung cancer and brain metastases in two clinical trials. *J Clin Oncol.* 2018;36(26):2693–2701. https://doi.org/10.1200/jco.2017.77.5841.

108. Mittapalli RK, Vaidhyanathan S, Dudek AZ, Elmquist WF. Mechanisms limiting distribution of the threonine-protein kinase B-RaF(V600E) inhibitor dabrafenib to the brain: implications for the treatment of melanoma brain metastases. *J Pharmacol Exp Ther.* 2013;344(3):655–664. https://doi.org/10.1124/jpet.112.201475.

109. Sakji-Dupre L, Le Rhun E, Templier C, Desmedt E, Blanchet B, Mortier L. Cerebrospinal fluid concentrations of vemurafenib in patients treated for brain metastatic BRAF-V600 mutated melanoma. *Melanoma Res.* 2015;25(4):302–305. https://doi.org/10.1097/CMR.0000000000000162.

110. McArthur GA, Maio M, Arance A, et al. Vemurafenib in metastatic melanoma patients with brain metastases: an open-label, single-arm, phase 2, multicentre study. *Ann Oncol.* 2017;28(3):634–641. https://doi.org/10.1093/annonc/mdw641.

111. Kim DW, Barcena E, Mehta UN, et al. Prolonged survival of a patient with metastatic leptomeningeal melanoma treated with BRAF inhibition-based therapy: a case report. *BMC Cancer.* 2015;15:400. https://doi.org/10.1186/s12885-015-1391-x.

112. Floudas CS, Chandra AB, Xu Y. Vemurafenib in leptomeningeal carcinomatosis from melanoma: a case report of near-complete response and prolonged survival. *Melanoma Res.* 2016;26(3):312–315. https://doi.org/10.1097/CMR.0000000000000257.

113. Schafer N, Scheffler B, Stuplich M, et al. Vemurafenib for leptomeningeal melanomatosis. *J Clin Oncol.* 2013;31(11):e173–e174. https://doi.org/10.1200/JCO.2012.46.5773.

114. Lee JM, Mehta UN, Dsouza LH, Guadagnolo BA, Sanders DL, Kim KB. Long-term stabilization of leptomeningeal disease with whole-brain radiation therapy in a patient with metastatic melanoma treated with vemurafenib: a case report. *Melanoma Res.* 2013;23(2):175–178. https://doi.org/10.1097/CMR.0b013e32835e589c.

115. Wilgenhof S, Neyns B. Complete cytologic remission of V600E BRAF-mutant melanoma-associated leptomeningeal carcinomatosis upon treatment with dabrafenib. *J Clin Oncol.* 2015;33(28):e109–e111. https://doi.org/10.1200/JCO.2013.48.7298.

116. Boogerd W, Dorresteijn LD, van Der Sande JJ, de Gast GC, Bruning PF. Response of leptomeningeal metastases from breast cancer to hormonal therapy. *Neurology.* 2000;55(1):117–119. https://doi.org/10.1212/wnl.55.1.117.

117. Ozdogan M, Samur M, Bozcuk HS, et al. Durable remission of leptomeningeal metastasis of breast cancer with letrozole: a case report and implications of biomarkers on treatment selection. *Jpn J Clin Oncol.* 2003;33(5):229–231. https://doi.org/10.1093/jjco/hyg046.

118. Mencel PJ, DeAngelis LM, Motzer RJ. Hormonal ablation as effective therapy for carcinomatous meningitis from prostatic carcinoma. *Cancer.* 1994;73(7):1892–1894. https://doi.org/10.1002/1097-0142(19940401)73:7<1892::aid-cncr2820730720>3.0.co;2-c.

119. Arasaratnam M, Hong A, Shivalingam B, et al. Leptomeningeal melanoma-A case series in the era of modern systemic therapy. *Pigment Cell Melanoma Res.* 2018;31(1):120–124. https://doi.org/10.1111/pcmr.12652.

120. Kumthekar P, Tang S-C, Brenner AJ, et al. ANG1005, a novel brain-penetrant taxane derivative, for the treatment of recurrent brain metastases and leptomeningeal carcinomatosis from breast cancer. *J Clin Oncol.* 2016;34(15_suppl):2004. https://doi.org/10.1200/JCO.2016.34.15_suppl.2004.

121. Barrett DM, Singh N, Porter DL, Grupp SA, June CH. Chimeric antigen receptor therapy for cancer. *Annu Rev Med.* 2014;65:333–347. https://doi.org/10.1146/annurev-med-060512-150254.

122. Chen X, Han J, Chu J, et al. A combinational therapy of EGFR-CAR NK cells and oncolytic herpes simplex virus 1 for breast cancer brain metastases. *Oncotarget.* 2016;7(19):27764–27777. https://doi.org/10.18632/oncotarget.8526.

123. Priceman SJ, Tilakawardane D, Jeang B, et al. Regional delivery of chimeric antigen receptor-engineered T cells effectively targets HER2(+) breast cancer metastasis to the brain. *Clin Cancer Res.* 2018;24(1):95–105. https://doi.org/10.1158/1078-0432.CCR-17-2041.

9

Cranial nerve involvement by metastatic cancer

Ashley L. Sumrall

Department of Oncology, Levine Cancer Institute, Atrium Health, Charlotte, NC, United States

1 Introduction

In this chapter, we will review malignancies that directly spread or metastasize such that they involve the cranial nerves. In addition to the sites of involvement, we will review imaging characteristics and clinical presentation. We will not cover leptomeningeal involvement or carcinomatous meningitis. Given that they are covered elsewhere, we will not include primary brain tumors.

As we consider malignancies that affect the central nervous system, they may initially spread to structures outside of the brain, such as the dura, the base of the skull, soft tissues, sinuses or cavities, and bone.[1] This may provide a conduit to the cranial nerve involvement. Cranial nerve involvement may occur via multiple mechanisms including direct spread from bony metastases, direct perineural spread from locally occurring malignancies, and spread via soft tissue and/or nodal metastases. As prognosis from various malignancies improves, the opportunity for metastases also increases.

Cranial nerve involvement by cancer is not rare, but most of the medical literature discussing this topic also includes leptomeningeal involvement. There are few resources devoted to isolated cranial nerve involvement, and many are case reports or small case series. In a prospective study of 242 patients with cancers, 20 patients exhibited isolated cranial nerve deficits. Depending on the population of patients studied (and the responsible primary malignancy), the incidence of cranial nerve involvement varies. In 16 of 20 patients, metastatic disease was the etiology for the cranial nerve deficit. Hematologic cancers were the most common cause, affecting 9 of 20 patients.[2] In a more focused population of patients with cancer known to spread to the bone, there may be a higher incidence. For example, a team at Memorial Sloan Kettering Cancer Center (MSKCC) reviewed cases of 851 patients seen in a neurologic consultation over a six-month period. Although they did not classify cranial nerve involvement, they reported findings of 5.1% with leptomeningeal metastases and 2.7% with skull base metastases.[3] These categories are closely linked to cranial nerve involvement and may provide a starting point to estimate the true incidence of cranial neuropathies.[3]

Cancer can spread along the cranial nerves by various mechanisms, as noted above. In particular, tumors can spread along the nerves using the same or nearby cranial nerves as a bridge. They can also crawl along the distribution of one cranial nerve into another via anastomosis. These sites can facilitate a rapid spread of malignancy, especially in areas that are filled with anastomosed regions (such as the caudal cranial nerves). In addition to these mechanisms of cancer spreading to the cranial nerves, additional environmental factors may promote this activity. Factors such as nerve growth factor, neural cell adhesion molecule, p75, and other immune regulators of the nervous system have been shown to enhance or inhibit nerve growth. One can hypothesize that these environmental factors may help or hinder the spread of cancer in the central nervous system. These mechanisms may become important in developing systemic treatments for cancers affecting the nervous system.[1]

2 Clinical presentation of cranial nerve involvement

Depending on the cranial nerve and/or adjacent structures affected, patients will present with various findings. Patients may exhibit deficits related to one or multiple cranial nerves affected. It is also feasible to have asymptomatic cranial nerve involvement, such as that which may occur when cranial nerves are victims of perineural spread.[1] For this reason and more, it is important to understand the pathophysiology of individual cancers, as well as the neuroanatomy being affected. Imaging can play an important role in detection.

In the case of perineural invasion, the trigeminal (V2) and facial nerves (VII) seem to be the most frequently affected, although any cranial nerve can be affected. One example of a conduit between the affected nerves involves studying the path of the auriculotemporal nerve—a nerve that connects cranial nerve V to cranial nerve VII. A typical presentation of a patient experiencing this cranial nerve involvement may include pain, sensory changes, or focal deficits of cranial nerves V or VII. Furthermore, this may allow additional perineural spread along the pathways of the affected nerves.[1]

2.1 Bony metastases

Bony metastases are common, and those affecting the skull are most often observed with solid tumors. We will distinguish between calvarial metastases and skull base metastases. Given the wide availability of advanced imaging, there appears to be an increased incidence of skull metastases. Calvarial metastases will typically be identified on imaging prior to clinical presentation, and patients with calvarial involvement may often describe nonspecific symptoms. Conversely, individuals with skull base metastases often describe distinct clinical syndromes depending on the involvement of adjacent vascular structures and nerves.[1]

2.2 Calvarial metastases

Individuals with calvarial metastases may present with nonspecific symptoms such as headaches or may remain asymptomatic.[4] Metastases to this part of the skull rarely cause a direct nerve injury by infiltration or compression. Interestingly, if metastases involve the torcula, they may produce a significant clinical presentation. Occlusion of part of the dural venous sinus system could lead to increased intracranial pressure. This may be evident by a patient exhibiting papilledema and bilateral cranial nerve VI palsies. The most common causes are breast and prostate cancer and neuroblastoma.[5]

2.3 Skull base metastases

The skull base can be conceptually divided into three regions: the anterior or parasellar region, the posterior region comprising the lower clivus with the jugular foramen and hypoglossal canal, and the lateral skull base formed by the temporal bone.[6] Different tumor types may spread to different portions of this region. Tumors that originate in the head and neck regions obviously spread to this location. Also, other cancers such as lung, breast, renal, and prostate may metastasize to the base of the skull and result in single or multiple cranial nerve palsies. Expected overall survival following a diagnosis of metastatic skull tumor resulting from cancer varies significantly with the following estimates: 2 years for

prostate cancer, 1.5 years for breast cancer, 6 months for lymphoma, and 5 months for lung cancer. The incidence is difficult to estimate, as there are limited reports in the medical literature.[1] Breast cancer often metastasizes to the bone and can affect any part of the skull. In a small series of 10 patients with breast cancer who developed cranial nerve involvement, 8 of the 10 patients studied had skull base metastases. The remaining 2 patients had soft tissue metastases adjacent to the skull base.[7]

Tumors affecting the anterior or parasellar region may induce headaches. If these malignancies extend into the adjacent cavernous sinus, a variety of cranial nerves may be affected including III, IV, the ophthalmic (V1) and maxillary (V2) divisions of V, and VI. This is typically unilateral and referred to as cavernous sinus syndrome (CSS).[5] Diplopia is the most common symptom after a headache, with cranial nerves III and VI being the most affected nerves. Facial pain, dysesthesia, and numbness may occur.[8] The most common cancers affecting the cavernous sinus are caused by direct invasion from pituitary tumors, perineural spread of head and neck cancers, or hematogenous spread from distant lesions.[9] In a review covering 26 years of patients with CSS, 151 patients were included. Tumors (45 patients, 30%) were the most frequent cause of cavernous sinus syndrome. The leading cancers were nasopharyngeal malignancies, metastases, and lymphoma.[10] CSS related to nasopharyngeal carcinoma may begin with cranial nerve V2 and VI deficits. This may translate to a patient reporting midfacial numbness while demonstrating a cranial nerve VI palsy.[5]

Tumors affecting the posterior portion of the skull base often report facial numbness since the V2 and V3 branches pass through this region. Symptoms may mimic trigeminal neuralgia, but patients will often have concurrent sensory deficits. Unilateral masseter and pterygoid weakness may also occur.[5]

Tumors involving the temporal bone occur often, but neurologic sequelae from this affected region are rare. An autopsy-based study of the temporal bones of 212 patients with cancer showed microscopic metastases in 47 (22%) of them. Breast cancer was the most common culprit, followed by lung cancer, prostate cancer, melanoma, and head and neck cancers. The authors determined that the temporal bones from 14 patients showed facial nerve involvement. Of those 14 patients, only 6 patients (42.8%) had reported facial paralysis. This is consistent with prior reports that invasion of the facial canal does not necessarily produce facial paralysis. No other examples of direct intracranial cranial nerve infiltration were reported. Notably, 36% of the patients studied had no neurologic symptoms.[11]

As cranial nerves exit the skull, there is ample opportunity for the tumors to spread to the nerves, especially near the jugular foramen. Cranial nerves IX, X, and XI reside in this space, and cranial nerve XII is nearby. Lesions affecting these four nerves commonly produce what is known as the "jugular foramen syndrome.[12,13]" The most common presenting symptom of jugular foramen syndrome is voice hoarseness or dysphagia, due to an affected vagal nerve. Pain may also be reported (periauricular pain or headache). With further involvement, patients may demonstrate unilateral paralysis of the soft palate with a deviation of the uvula to the unaffected side. If the glossopharyngeal nerve is affected, the patient may report decreased sensation to the posterior ipsilateral tongue and loss of the ipsilateral gag reflex. With the involvement of the accessory nerve, one would expect the ipsilateral shoulder to droop, as well as the inability to abduct the arm. Significant headaches and papilledema can also result if the venous sinuses and/ or veins are obstructed, leading to increased intracranial edema.[14]

2.4 Mandibular metastases

One of the easier presentations of nerve involvement to detect is related to mandibular involvement by metastases. Originating from the trigeminal nerve, the inferior alveolar nerve or its terminal branch, the mental nerve, is compressed by bone metastases, leading to "numb chin syndrome.[15,16]" In a retrospective analysis of 42 patients with cancer and numb chin syndrome, 50% of the patients had mandibular metastases. Breast cancer was the most common etiology, represented by nearly 65% of the cases. Median overall survival after patients received this diagnosis was 5 months due to bone metastases.[16]

2.5 Orbital metastases

Although rare, cancers may affect the orbital space. This may occur through direct extension, hematogenous spread, or perineural spread. A direct extension is suspected to be the most likely mechanism of spread. The most common cancers to affect this area include breast, lung, and prostate cancer, as well as melanoma.[17,18] A small series describes 5 patients with partial or complete ophthalmoplegia following fifth and seventh cranial nerve involvement by the perineural spread of cutaneous squamous cell carcinoma.[18]

Diplopia, ocular motility limitation, and mass effect with displacement, proptosis, or palpable mass are reported signs and symptoms. Diplopia is the most common presenting symptom. The mechanism for diplopia appears to be related to globe positioning or muscle weakness, rather than the involvement of oculomotor or optic nerves. Of course, these nerves could be affected via infiltration or compression, leading to deficits.[19]

Visual loss from optic neuropathy and ophthalmoplegia involving multiple cranial nerves are the presenting deficits for orbital apex syndrome (OAS). There may be overlap with the cavernous sinus syndrome[20] or the superior orbital fissure syndrome,[21] as the anatomy is closely related. Typically, patients present with a quickly evolving, painful ophthalmoplegia. This may be caused by a tumor or an inflammatory condition. Patients with superior orbital fissure syndrome report retro-orbital pain and have ophthalmoplegia as well as impairment of the first trigeminal branches. The optic nerve is frequently involved. Cases of superior orbital fissure syndrome that include visual loss are typically not tumor-related, but there are reports where tumors cause this condition.[21]

2.6 Facial nerve involvement

Although facial nerve palsy is nearly always due to nonmalignant causes, it can be seen as a presenting sign of cancer. The authors performed a retrospective review of patients who presented to medical attention with acute facial nerve paralysis that mimicked Bell's palsy over 13 years. All patients were subsequently found to harbor an occult skull base cancer. All patients experienced a delay in diagnosis, and seven patients died of their disease. Eight patients reported acute onset of symptoms over 48 h or less. Five of the patients had parotid tumors.[22]

In a large, retrospective study of the Dutch Head and Neck cooperative group (NWHHT), 324 patients with parotid carcinoma were studied. Prior to treatment, 77% of the patients had an intact function of cranial nerve VII. Perineural invasion noted in the surgical specimen was the only independent factor in a multivariate analysis that correlated with facial nerve function at presentation. This phenomenon was observed in 86% of those individuals with complete cranial nerve VII paralysis. Paralysis was related to tumor location and was most commonly seen with medial lobe tumors.[23]

2.7 Cranial nerve involvement arising from cancer of the neck or chest

Now, we will turn our attention to more distal cancers that can affect cranial nerves. In the neck, squamous cell carcinoma is the most common malignancy to affect the cranial nerves. Other cancers include thyroid cancer, metastatic lung cancer, and lymphoma.

Squamous cell carcinomas originating in the head and neck regions such as the hypopharynx and larynx can give rise to perineural invasion. Extracapsular extension of nodal metastases from squamous cell carcinoma can result in cranial nerve involvement. In a study of resection specimens from 30 patients undergoing radical neck dissections for clinically positive nodes, an extracapsular extension was identified in 25 patients (83%), but a perineural invasion of the cervical plexus was seen in only one (4%) of these.[24] Similarly, in the extensive studies of perineural invasion by Carter, an incidence of 8% was found.[25]

When the vagus nerve is affected, syncope can occur. In a series of 17 patients with head and neck cancer who experienced recurrent syncope, 16 of those individuals had recurrent cancer. Autopsy in 2 of these patients showed a tumor involving the glossopharyngeal and vagus nerves.[26]

When the recurrent laryngeal nerve is affected, vocal cord paralysis occurs. Voice fatigue, hoarseness, and stridor may be noted. The vagus nerve travels along with the carotid artery and internal jugular vein, and the left recurrent laryngeal nerve arises just beneath the arch of the aorta. Both recurrent laryngeal nerves ascend along the lateral trachea, then pass along the thyroid gland to enter the larynx. In a series of 100 patients with vocal cord paralysis, the most likely cause was cancer. Lung cancer was the cause in 15 patients, and the left vocal cord was affected in almost 90%.[1] Thyroid carcinoma, breast cancer, and Hodgkin's disease were the next most common causes.[27] The relationship between thyroid cancers and recurrent laryngeal nerve palsy is noteworthy. In women, thyroid cancer has been shown to top lung cancer as the most common tumor associated with recurrent laryngeal nerve palsy. If the recurrent laryngeal nerve palsy is part of the patient's initial symptoms prior to surgery, the likelihood of a malignant thyroid cancer is very high (> 90%).[28]

3 Typical imaging findings of cranial nerve involvement

The most popular noninvasive diagnostic approaches to identify cranial nerve involvement by cancer include computed tomography (CT), magnetic resonance imaging (MRI), and positron emission tomography (PET).[1,29] A contrast-enhanced MRI is often obtained first when looking for cranial nerve involvement. Expected changes on MRI of the cranial nerves affected by perineural tumor spread include the breakdown of the blood-nerve barrier. Due to this breakdown, contrast media is expected to leak and become visible before any thickening of the affected nerves is apparent. As the nerve's thickness grows, the perineurial fat tissue will decrease.[29]

An MRI of nerves affected by perineural tumor spread usually shows a breakdown of the blood-nerve barrier and leakage of contrast media, which may become apparent even before any thickening of the affected nerves. Once the diameter of the nerves has increased, the perineurial fat tissue (particularly that at the foraminal openings) tends to decrease.[29]

When motor branches from cranial nerves are affected, signal changes within the muscle indicating a cranial nerve lesion can be detected via MRI. As time passes, fatty infiltration of the muscle occurs. Eventually, muscle atrophy may be visible on T1 images or fast spin-echo sequences.[30] Additional MRI findings of perineural involvement include smooth thickening and enhancement of the nerve, concentric expansion of the skull base foramina, enlargement of the cavernous sinus, and muscular atrophy.[31]

Ultrasound can also be a useful modality, especially in approachable areas of the skull and neck.[32] Additional more invasive methods of diagnosis include fine needle and open biopsies, surgical exploration, and CSF examination. When considering cancer as part of the differential diagnosis, it is also prudent to search for other etiologies such as infections, immune-mediated diseases, and other neurologic conditions such as neurofibromatosis or nonmalignant tumors such as schwannoma or meningioma.[1]

4 Cancers of special interest

4.1 Cancers of the head and neck

Primary neoplasms of the head and neck are most commonly squamous cell carcinomas, but other histologic subtypes of tumors also occur in this area. In fact, cranial nerve involvement is a key element in staging cancers, such as nasopharyngeal carcinoma, malignant salivary gland neoplasms, and adenoid cystic carcinoma. These tumor subtypes move along the nerve branches, making resection difficult. Other cancers may occur in this region as well, including lymphoma, soft tissue sarcomas, melanoma, and primary bone cancer. Nerves may become affected due to these cancers as a result of perineural invasion or compression.[5]

4.2 Squamous cell carcinoma

Squamous cell carcinoma is a common cancer of the head and neck region, spreading via local invasion and regional spread along lymph nodes. These cancers are staged regarding such features as nerve involvement and tissue invasion. Treatment-related damage to cranial nerves and other nerves is unfortunately common, given the potential for injury from surgery and/or radiation therapy.

4.3 Nasopharyngeal carcinoma

Nasopharyngeal carcinoma represents a unique form of poorly differentiated or undifferentiated squamous cell carcinoma. It is closely related to an infection with an Epstein-Barr virus, and it is endemic in certain parts of the Eastern world.[33,34]

Combining the tumor location with the capacity for quick growth and spread, nasopharyngeal carcinoma can induce neurological damage. Direct extension to the skull base or perineural spread along the cranial nerves is both associated with a poor prognosis and results in T4 staging.[34] A prospective imaging evaluation of 262 patients with nasopharyngeal cancer by a CT scan was completed. Eighty-four of the patients had T4 lesions at presentation, and 82 patients had skull base erosion. Thirty-four patients had cranial nerve palsies. Eighty-eight percent of the cranial nerve palsies involved cranial nerves III, IV, V, and/or VI. A meager 12% involved cranial nerves IX–XII. Only two patients had cranial nerve involvement without a skull base erosion.[34] In searching for prognostic factors, only cranial nerve involvement had independent prognostic values for T4 patients. The median survival for patients with cranial nerve involvement was 17 months.[34]

One of the few areas where treatment may improve cranial nerve function is seen with nasopharyngeal carcinoma. After definitive radiation therapy for this condition, recovery of cranial nerve function is observed in 50%–80% of patients. For many patients, recovery may be complete. If a patient has a cranial nerve deficit for more than 3 months prior to treatment, the chance of improvement decreases. And, 25%–50% of patients do not experience recovery of cranial nerve function. The overall survival is higher in patients whose cranial nerve deficits are improved (47% vs 26%).[35]

As mentioned above, a treatment-related injury is seen in this group of patients. Given the location of nasopharyngeal carcinoma, at least a small portion of patients develops progressive cranial neuropathies after radiation. The incidence is estimated to be 1% when patients are treated with daily fractions near 180–200 cGy, with cumulative doses of 7000–13,000 cGy to the nasopharynx. Higher dose fractions appear to be associated with an increased risk. Based on a series by Lin et al., the hypoglossal nerve is most frequently affected and may cause dysphagia, dysarthria, and trouble swallowing. Also affected were the vagus nerve, recurrent laryngeal nerve, and accessory nerve. An important point to note is that the interval between the completion of radiation and onset of cranial nerve palsy was variable, occurring between 12 and 240 months; the median was 61 months. Given the delayed onset, the authors hypothesize that neck fibrosis due to radiation may be a significant contributing risk factor. For these reasons, these patients should continue to have long-term follow-up with careful attention to the neuro exam.[36]

4.4 Adenoid cystic carcinoma

Adenoid cystic carcinoma is a salivary gland malignancy with a propensity for perineural invasion. Given its behavior, it carries a high risk of local recurrence and skull base involvement, leading to poor long-term outcomes.[5] Both surgery and radiation serve an important role in initial management.[37] Adenoid cystic carcinoma is a good model to study the mechanisms of perineural spread. Local control rates for adenoid cystic carcinoma with surgery and radiation are about 85% at 10 years, and perineural invasion is an adverse prognostic factor when a major (named) nerve is involved. In a series by Garden et al., it was noted that when both a positive margin and involvement of a named nerve were observed, control rates dropped. Still, the control rate was estimated to be 70% at 10 years.[38]

4.5 Primary cutaneous neoplasms of the head and neck region

Another example of perineural spread to consider is that of skin cancers of the head and neck region, as these may spread by perineural invasion. Via clinical presentation or pathologic confirmation on surgical specimens, an estimated 2%–6% of basal cell and cutaneous squamous cell cancers exhibits perineural invasion. This phenomenon is associated with midface location, recurrence of disease, high histologic grade, and larger tumor size. The cranial nerves most involved are the 5th and 7th nerves.[39]

Initial neurologic symptoms of these cancers tend to be sensory, including paresthesias (53%) or pain (27%). The reason for this is that the area is innervated by the trigeminal nerve. The facial nerve is the second most common nerve involved.[40] With careful attention when viewing the cross-sectional imaging, visualization of perineural spread may be noted in 50%–75% of symptomatic patients.[41] These findings may appear as thickening of the nerve, abnormal enhancement, loss of fat planes in adjacent soft tissues, and/or foraminal enlargement.[42]

4.6 Other solid tumors of importance

In addition to the types of cancer mentioned above, there are multiple examples of other cancers that may spread to affect the cranial nerves. Cancers that spread to the bones may be more likely to affect cranial nerves, given examples listed above. Prostate and breast cancer will be the two examples covered below.

4.7 Prostate cancer

Given prostate cancer's propensity to spread to the bone, skull involvement is a common finding. In a report of 11 patients with metastatic prostate cancer and cranial nerve involvement, patients were found to have skull base involvement via imaging in 8 cases. Two cases did not show imaging abnormalities despite compelling clinical presentations. Of the 11 patients, diplopia, speech changes, tongue deviation, and headache were the common presenting symptoms. Most patients responded positively to treatment.[43] In another series of 11 patients with advanced prostate cancer and skull metastases, cranial nerve involvement correlated to a dismal prognosis. Ten of the 11 patients studied passed away within a median of 5 months (range 1–16 months) after presentation with cranial nerve deficits. This was despite some impressive responses to radiation therapy. In this group of patients, the seventh cranial nerve was the most frequently affected nerve, as an isolated finding or in combination with other nerves.[44]

An additional series of patients with hormone-refractory metastatic prostate cancer and skull base lesions confirm the dismal prognosis associated with cranial nerve involvement. In this series, patients were pretreated as 12 patients had received prior chemotherapy, and 13 patients had received previous radiation to the areas of bony pain. Presenting symptoms varied from recognized clinical syndromes affecting multiple cranial nerves to isolated cranial nerve lesions. After undergoing palliative radiation, 14 of the 15 patients had a clinical (either partial or complete) response. Unfortunately, these individuals died of prostate cancer despite their responses. In fact, 10 of the 15 patients (67%) died within 3 months of developing symptoms.[45]

4.8 Breast cancer

Like prostate cancer, breast cancer has an affinity for metastasizing to the bone, including the skull. Often, neurologic involvement by breast cancer is discussed in terms of brain metastases or leptomeningeal disease. Breast cancer is one of the most common tumors to metastasize to the head and neck. In two large series, breast cancer constitutes approximately 15%–20% of all metastases to this region[46,47] and accounts for the majority of the orbital soft tissue and temporal bone metastases.[46–49] Given that metastases to the head and neck are uncommon, to begin with, breast carcinoma metastases are still relatively rare in clinical practice.[50]

In a retrospective review of cases of patients seen at two academic institutions with breast cancer with head and neck metastases, 25 patients were identified. Clinical presentation varied considerably, but 2 patients presented with cranial nerve involvement due to skull base involvement. Also noted from this review was the fact that 5 patients did not have a preexisting diagnosis of breast cancer. For the 20 patients that did already have documented breast cancer, 16 (80%) presented with their head and neck metastatic disease for 5 or more years after their primary diagnosis, with 9 of those patients developing metastases after 10 or more years. The mean interval was 10.9 years, with the longest being 33 years.[50]

There have been several case reports of breast cancer spreading to the clivus or nearby areas, resulting in cranial nerve VI palsies.[51–54] Due to the long course of this nerve through the brain, the abducens nerve is especially vulnerable to damage from trauma or metastatic disease.[51] Reyes and team reported abducens nerve palsy due to an isolated brainstem metastasis from breast cancer.[53] Another case report described a metastatic mass in the lower pons involving the abducens nucleus, resulting in gaze palsy.[54] In another case, MRI showed an enhancing mass in the clivus extending to the left cavernous sinus, which also demonstrated contrast enhancement.[51]

4.9 Hematologic involvement

4.9.1 Lymphoma

Neurolymphomatosis is a term that traditionally was used to describe the involvement of the peripheral nervous system by lymphoma. This definition was updated in an important paper by Baehring and team in 2003, and cranial neuropathies were included. In a review of patient encounters at Massachusetts General Hospital (MGH) spanning over 25 years, 25 patients with neurolymphomatosis were identified. The authors identified 47 patients with similar characteristics in the medical literature. Fifteen patients of the combined series (7 patients from MGH and 8 patients from the medical literature) presented with involvement of a single cranial nerve. Reported findings included 3 patients with Bell's palsy, 4 patients with lateral rectus muscle palsies, 4 patients with oculomotor neuropathy, 2 patients with trigeminal neuropathy, and 1 patient each with hearing loss and vocal cord paralysis. Only 4 patients had pain associated with cranial neuropathy. An impressive 33 patients from the combined series were not correctly diagnosed until autopsy. As with primary central nervous system (CNS) lymphoma, most of these lymphomas were diffuse large B-cell lymphomas.[55]

In an Italian series of patients with hematologic malignancies, 12 patients who presented with cranial nerve deficits were studied. They used contrasted MR imaging as well as analysis of cerebrospinal fluid (CSF), including flow cytometry. They found that patients developed cranial neuropathies at variable times during their courses of illness. Eleven patients had isolated cranial nerve deficits, and 1 patient had multiple cranial nerves affected. In 5 of the reported cases, the patients were in complete remission after hematopoietic stem cell transplantation.[56] In a small series from China, 5 adult patients were examined who presented with systemic lymphoma with multiple cranial nerve deficits. Presenting symptoms included blurred vision, diplopia, ptosis, and facial palsy. The 3rd and 6th cranial nerves were the most affected. All these patients were diagnosed with non-Hodgkin's lymphoma (4 with B-cell lymphoma and 1 with T-cell lymphoma).[57]

Lymphoma can present in many ways. For example, a group in Korea reported an immunocompetent patient who presented with facial palsy and then progressively developed other cranial nerve palsies over several months. He was noted to have diffuse large B-cell lymphoma originating from the frontal sinus.[58] The medical literature has multiple interesting case reports and small case series by which lymphoma has produced cranial neuropathies. Even the exceptionally rare primary meningeal lymphoma has presented with cranial nerve involvement.[59]

4.9.2 Leukemia

Leukemia can affect any part of the nervous system. Involvement of the nervous system is often discovered via cerebrospinal fluid examination but is also not identified until postmortem examination. Leukemic involvement of the CNS in adults with systemic leukemia has been reported at autopsy in approximately 25% of patients. Imaging may disclose cranial nerve involvement instead. For example, in a series of patients with leukemia and documented CSF involvement, MRI showed abnormal findings in 25 patients (74%). Of those patients, 9 had cranial nerve involvement.[60]

In a retrospective analysis of cases of acute leukemia with neuro-ophthalmic manifestations that were seen over a 6-year period by ophthalmology, 12 adult patients with acute leukemia had cranial nerve findings, including isolated sixth and fourth nerve palsies, and some patients had multiple cranial nerve palsies. These findings announced new CNS involvement for 3 of the patients with known systemic disease and helped to diagnose 3 patients with relapse. In nearly 60% of the patients, the identification of neuro-ophthalmic leukemic involvement led to a change in leukemia treatment.[61] Cases of cranial nerve involvement by chronic leukemia have also been described.[62,63]

5 Summary

Involvement of cranial nerves by metastatic cancer may herald a new diagnosis, signal a recurrence of disease, or be part of an ongoing progression of cancer. Cranial nerves may become affected by various mechanisms such as direct invasion, perineural spread, or other less obvious means. This involvement may bring pain, sensory loss, motor weakness, or other neurologic challenges, which may range from subtle to advanced. In fact, some patients may even be asymptomatic despite cranial nerve involvement. It may be challenging to detect the involvement despite examining a cross-sectional imaging or cerebrospinal fluid. Clinicians should use knowledge of neuroanatomy, cranial nerve pathways, and the pathophysiology of cancer to determine how to

pursue these findings. With all cancers, the finding of an involved cranial nerve should prompt careful evaluation and consideration of the morbidity and mortality risks posed to the affected individual.

References

1. Grisold W, Grisold A. Cancer around the brain. *Neurooncol Pract.* 2014;1:13–21.
2. Gokce M. Analysis of isolated cranial nerve manifestations in patients with cancer. *J Clin Neurosci.* 2005;12:882–885.
3. Clouston PD, De Angelis LM, Posner JB. The spectrum of neurological disease in patients with systemic cancer. *Ann Neurol.* 1992;31:268–273.
4. Harrison RA, Nam JY, Weathers SP, Demonte F. Intracranial dural, calvarial, and skull base metastases. *Handb Clin Neurol.* 2018;149:205–225.
5. Moots P. Cranial nerve involvement by metastatic cancer. In: Newton HB, Malkin MG, eds. *Neurological Complications of Systemic Cancer and Antineoplastic Therapy.* 1st ed. Florida: CRC Press; 2010.
6. Laigle-Donadey F, Taillibert S, Martin-Duverneuil N, et al. Skull-base metastases. *J Neurooncol.* 2005;75:63–69.
7. Hall SM, Buzdar AU, Blumenschien GR. Cranial nerve palsies in metastatic breast cancer due to osseous metastasis without intracranial involvement. *Cancer.* 1983;52:180–184.
8. Johnston J. Parasellar syndromes. *Curr Neurol Neurosci Rep.* 2002;2:423–431.
9. Lee JH, Lee HK, Park JK, et al. Cavernous sinus syndrome: clinical features and differential diagnosis with MR imaging. *Am J Roentgenol.* 2003;181:583–590.
10. Keane JR. Cavernous sinus syndrome: analysis of 151 cases. *Arch Neurol.* 1996;53:967–971.
11. Gloria-Gray TI, Schachern PA, Paparella MM, et al. Metastases to the temporal bones from primary nonsystemic malignant neoplasms. *Arch Otolaryngol Head Neck Surg.* 2000;126:209–214.
12. Hadley W, Johnson DH. Jugular foramen syndrome as a complication of metastatic cancer of the prostate. *South Med J.* 1984;77:92–93.
13. Robbins KT, Fenton RS. Jugular foramen syndrome. *J Otolaryngol.* 1980;9:505–516.
14. Das JM, Khalili YA. Jugular Foramen Syndrome. In: *StatPearls.* Florida: StatPearls Publishing; 2021 [Internet] https://www.ncbi.nlm.nih.gov/books/NBK549871/. [Accessed 11 September 2021].
15. Bruyn RP, Boogerd W. The numb chin. *Clin Neurol Neurosurg.* 1991;93:187–193.
16. Lossos A, Siegal T. Numb chin syndrome in cancer patients: etiology, response to treatment, and prognostic significance. *Neurology.* 1992;42:1181–1184.
17. Goldberg RA, Rootman J, Cline RA. Tumors metastatic to the orbit: a changing picture. *Surv Ophthalmol.* 1990;35:1–24.
18. Clouston PD, Sharpe DM, Corbett AJ, et al. Perineural spread of cutaneous head and neck cancer. Its orbital and central neurologic complications. *Arch Neurol.* 1990;47:73–77.
19. Golberg RA, Rootman J. Clinical characteristics of metastatic orbital tumors. *Ophthalmology.* 1990;97:620–624.
20. Yeh S, Foroozan R. Orbital apex syndrome. *Curr Opin Ophthalmol.* 2004;15:490–498.
21. Lenzi GL, Fieschi C. Superior orbital fissure syndrome. Review of 130 cases. *Eur Neurol.* 1977;16:23–30.
22. Marzo SJ, Leonetti JP, Petruzzelli G. Facial paralysis caused by malignant skull base neoplasms. *Ear Nose Throat J.* 2002;81:845–849.
23. Terhaard C, Lubsen H, Tan B, et al. Facial nerve function in carcinoma of the parotid gland. *Eur J Cancer.* 2006;42:2744–2750.
24. Esclamado RM, Carroll WR. Extracapsular spread and the perineural extension of squamous cell cancer in the cervical plexus. *Arch Otolaryngol Head Neck Surg.* 1992;118:1157–1158.

25. Carter RL, Pittam MR. Squamous carcinomas of the head and neck: some patterns of spread. *J R Soc Med.* 1980;73:420–427.

26. Macdonald DR, Strong E, Nielsen S, Posner JB. Syncope from head and neck cancer. *J Neurooncol.* 1983;1:257–267.

27. Panell FW, Brandenburg JH. Vocal cord paralysis. A review of 100 cases. *Laryngoscope.* 1970;80:1036–1045.

28. Chiang FY, Lin JC, Lee KW, et al. Thyroid tumors with preoperative recurrent laryngeal nerve palsy: clinicopathologic features and treatment outcome. *Surgery.* 2006;140:413–417.

29. Saremi F, Helmy M, Farzin S, et al. MRI of cranial nerve enhancement. *Am J Roentgenol.* 2005;185:1487–1497.

30. Russo CP, Smoker WR, Weissman JL. MR appearance of trigeminal and hypoglossal motor denervation. *Am J Neuroradiol.* 1997;18:1375–1383.

31. Ginsberg LE, De Monte F, Gillenwater AM. Greater superficial petrosal nerve: anatomy and MR findings in perineural tumor spread. *Am J Neuroradiol.* 1996;17:389–393.

32. Nemzek WRSH, Gandour-Edwards R, Donald P, et al. Perineural spread of head and neck tumors: how accurate is MR imaging? *Am J Neuroradiol.* 1998;19:701–706.

33. Wei WI, Sham JS. Nasopharyngeal carcinoma. *Lancet.* 2005;365:2041–2065.

34. Sham JST, Cheung YK, Choy D, et al. Cranial nerve involvement and base of skull erosion in nasopharyngeal carcinoma. *Cancer.* 1991;68:422–426.

35. Chang JT, Lin CY, Chen TM, et al. Nasopharyngeal carcinoma with cranial nerve palsy: the importance of MRI for radiotherapy. *Int J Radiat Oncol Biol Phys.* 2005;1(63):1354–1360.

36. Lin YS, Jen YM, Lin JC. Radiation-related cranial nerve palsy in patients with nasopharyngeal carcinoma. *Cancer.* 2002;95:404–409.

37. Chen AM, Bucci MK, Weinberg V, et al. Adenoid cystic carcinoma of the head and neck treated by surgery with or without postoperative radiation therapy: prognostic features of recurrence. *Int J Radiat Oncol Biol Phys.* 2006;66:152–159.

38. Garden AS, Weber RS, Morrison WH, et al. The influence of positive margins and nerve invasion in adenoid cystic carcinoma of the head and neck treated with surgery and radiation. *Int J Radiat Oncol Biol Phys.* 1995;32:619–626.

39. Mendenhall WM, Amdur RJ, Hinerman RW, et al. Skin cancer of the head and neck with perineural invasion. *Am J Clin Oncol.* 2007;30:93–96.

40. McCord MW, Mendenhall WM, Parsons JT, et al. Skin cancer of the head and neck with clinical perineural invasion. *Int J Radiat Oncol Biol Phys.* 2000;47:89–93.

41. Williams LS, Mancusso AA, Mendenhall WM. Perineural spread of cutaneous squamous and basal cell carcinoma: CT and MR detection and its impact on patient management and prognosis. *Int J Radiat Oncol Biol Phys.* 2001;49:1061–1069.

42. Galloway TJ, Morris CG, Mancuso AA, et al. Impact of radiographic findings on prognosis for skin carcinoma with clinical perineural invasion. *Cancer.* 2005;103:1254–1257.

43. Svare A, Fosså SD, Heier MS. Cranial nerve dysfunction in metastatic cancer of the prostate. *Br J Urol.* 1988;61:441–444.

44. Seymore CH, Peeples WJ. Cranial nerve involvement with carcinoma of prostate. *Urology.* 1988;31:211–213.

45. McDermott RS, Anderson PR, Greenberg RE, et al. Cranial nerve deficits in patients with metastatic prostate carcinoma: clinical features and treatment outcomes. *Cancer.* 2004;101:1639–1643.

46. Barnes L. Metastases to the head and neck: an overview. *Head Neck Pathol.* 2009;3:217–224.

47. Lee YT. Patterns of metastasis and natural courses of breast carcinoma. *Cancer Metastasis Rev.* 1985;4:153–172.

48. Raap M, Antonopoulos W, Dammrich M, et al. High frequency of lobular breast cancer in distant metastases to the orbit. *Cancer Med.* 2015;4:104–111.

49. Nelson EG, Hinojosa R. Histopathology of metastatic temporal bone tumors. *Arch Otolaryngol Head Neck Surg.* 1991;117:189–193.

50. Dikson DG, Chernock R, El-Mofty S, et al. The great mimicker: metastatic breast carcinoma to the head and neck with emphasis on unusual clinical and pathologic features. *Head Neck Pathol.* 2017;11:306–313.

51. Kapoor A, Beniwal V, Beniwal S, et al. Isolated clival metastasis as the cause of abducens nerve palsy in a patient of breast carcinoma: a rare case report. *Indian J Ophthalmol.* 2015;63:354–357.

52. Amouzgarhashemi F, Vakilha M, Sardari M. An unusual metastatic breast cancer presentation: report of a case. *Iran J Radiat Res.* 2005;3:43–45.

53. Reyes KB, Lee HY, Ng I, Goh KY. Abducens (sixth) nerve palsy presenting as a rare case of isolated brainstem metastasis from a primary breast carcinoma. *Singapore Med J.* 2011;52:e220–e222.

54. Han SB, Kim JH, Hwang JM. Presumed metastasis of breast cancer to the abducens nucleus presenting as gaze palsy. *Korean J Ophthalmol.* 2010;24:186–188.

55. Baehring JM, Damek D, Martin EC, et al. Neurolymphomatosis. *Neuro Oncol.* 2003;5:104–115.

56. Diamanti L, Berzero G, Franciotta D, et al. Cranial nerve palsies in patients with hematological malignancies: a case series. *Int J Neurosci.* 2020;130:777–780.

57. Li JJ, Qiu BS, Chen JX, et al. Multiple cranial nerve deficits as preceding symptoms of systemic non-Hodgkin's lymphoma. *CNS Neurosci Ther.* 2019;25:409–411.

58. Kim K, Kim MJ, Ahn S, et al. Frontal sinus lymphoma presenting as progressive multiple cranial nerve palsy. *Yonsei Med J.* 2011;52:1044–1047.

59. Tortosa A, Rubio F, Reñé R, et al. Progressive infiltration of cranial nerves as first manifestation of primary meningeal lymphoma. *Med Clin (Barc).* 1993;100:137–139.

60. Guenette JP, Tirumani SH, Keraliya AR, et al. MRI findings in patients with leukemia and positive CSF cytology: a single-institution 5-year experience. *AJR Am J Roentgenol.* 2016;207:1278–1282.

61. Alrobaian MA, Henderson AD. Neuro-ophthalmic manifestations of acute leukemia. *J Neuroophthalmol.* 2021;41(4):E584–E590.

62. Kim JH, Kang HG, Noh SM. Chronic myeloid leukemia presenting as multiple cranial nerve palsy. *Geriatr Gerontol Int.* 2017;17:1331–1333.

63. Cramer SC, Glaspy JA, Efird JT, Louis DN. Chronic lymphocytic leukemia and the central nervous system: a clinical and pathological study. *Neurology.* 1996;46:19–25.

10

Cancer-related plexopathies

Ashley L. Sumrall

Department of Oncology, Levine Cancer Institute, Atrium Health, Charlotte, NC, United States

1 Introduction

Cancer-related plexopathy may occur during the course of a malignancy or as a result of treatment received for a cancer. Any of the nerve plexuses may be affected, and clinical presentations vary depending on the location. Typically, the cervical, brachial, or lumbosacral plexuses are involved. Cancer-related plexopathy is estimated to occur in approximately 1 per 100 patients with cancer.[1] The two most common plexopathies, brachial and lumbosacral, occur at rates of 0.43% and 0.71%, respectively.[2,3] The incidence of plexopathy based on tumor histology varies considerably. For example, the estimated incidence of plexopathy in breast cancer patients ranges between 1.8% and 4.9%.[4–6]

When a plexopathy is suspected, the clinician should try to localize the neurologic lesion. If a peripheral nerve is affected, there is a greater possibility of recovery than observed with damage to nerves of the central nervous system. Preservation of function is optimal when plexopathy is diagnosed and treated early. Treatment varies, depending on the location of the affected plexus and the cause of the plexopathy. A lymphoma, for example, may respond quickly to systemic therapy. A superior sulcus tumor may require surgical resection.

An optimal treatment strategy will also include aggressive symptom management. Pain is often a presenting feature and may become chronic and disabling. Supportive multidisciplinary care is also indicated, utilizing resources such as physical and occupational therapists, to maintain or improve function and quality of life.[1]

2 Review of plexus anatomy

The anatomy of the major plexuses is complex, and a review of the neuroanatomy can be helpful in considering the etiologies of plexopathy. There are four basic plexuses, but they are considered three for clinical purposes, as lumbosacral is considered one entity. The ability to localize the suspected lesion can assist with determining diagnostic studies needed, such as magnetic resonance imaging (MRI), electromyography (EMG), and more. Even a basic knowledge of neuroanatomy can help a clinician to distinguish tumor plexopathy from neoplastic meningitis and/or spinal cord compression, conditions that can have very similar presentations.[1]

2.1 Cervical plexus

The cervical plexus forms from the ventral branches of C2-C4. For clinical purposes, this provides sensory input from the skin and soft tissues of the anterosuperior and lateral neck, submandibular area, and the mandibular angle. There are motor components in this plexus that control the diaphragm (phrenic nerve), deep cervical and hyoid muscles, as well as the sternocleidomastoid and trapezius muscles (spinal accessory nerve).[1]

2.2 Brachial plexus

The brachial plexus is complex and is comprised of fibers from the lower four cervical and the first thoracic roots (C5-T1) (see Fig. 1). From the five roots, three trunks are formed: the superior (C5-C6), medial (C7), and inferior (C8-T1) trunks. These trunks further divide and then reorganize into three cords (posterior, lateral, and medial). The posterior cord gives rise to the thoracodorsal nerve (latissimus dorsi), subscapular nerve (subscapularis), axillary nerve (deltoid), and radial nerve (triceps, brachioradialis, and wrist and finger extensors). The posterior cord also provides sensory input to the posterior arm and forearm. The lateral cord divides into two branches which supply the hand and arm. One branch forms the musculocutaneous nerve (C5-C7) which

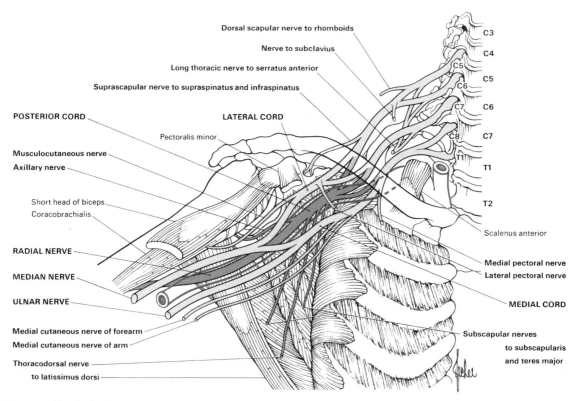

FIG. 1 Diagram of brachial plexus. *From Aids to Examination of the Peripheral Nervous System, Fourth Edition, Copyright 2000. Used with permission from Elsevier.*

provides motor input to biceps, brachialis, coracobrachialis, as well as sensory input to the radial forearm. The second branch joins a branch from the medial cord to form the median nerve (C8-T1) which supplies the forearm flexors and pronators, radial hand flexor, and lumbrical muscles. The median nerve also supplies sensation to the radial palm, first three fingers, and radial one-half of the fourth finger. The remainder of the medial cord becomes the medial brachial cutaneous nerve (T1) which provides sensory input to the medial arm, the medial antebrachial cutaneous nerve (C8-T1) which provides sensory input to the medial forearm, and the ulnar nerve (C8-T1) which innervates the ulnar forearm flexor muscles, interossei, and lumbricals while supplying sensation to the ulnar aspect of the hand and fifth finger.[1]

2.3 Lumbosacral plexus

The lumbosacral plexus is technically two plexuses, but for clinical consideration, they are grouped together (see Fig. 2). The lumbar plexus (L1-L4) and sacral plexus (L5-S5) are connected by the lumbosacral trunk (L4-L5). The lumbar plexus forms within the iliacus muscle, and the nerve roots join to form anterior and posterior divisions. The anterior division gives rise to the iliohypogastric, ilioinguinal, and genitofemoral nerves (L1-L2), which provide sensory fibers from the lower abdominal skin, lateral genitalia, and upper thigh. The anterior division also forms the obturator nerve (L2-L4), which provides motor supply to the adductors and gracilis muscles and sensation to the medial thigh. The posterior division divides into the iliohypogastric and lateral femoral cutaneous nerves (L2-L3), which provide sensation to the lateral hip and thigh, and the femoral nerve (L2-L3), which brings motor fibers to the iliopsoas and quadriceps muscles as well as sensory fibers from the anterior thigh and medial upper foreleg.[1]

The sacral plexus forms from ventral rami of S1-S3. The anterior division of the sacral plexus provides motor fibers to the gemelli, quadratus femoris, obturator internus, and hamstrings. The remainder of the anterior division continues as the tibial nerve (L4-S3), which provides motor supply to the foot plantar flexors (gastrocnemius, soleus, posterior tibialis, and intrinsic cutaneous nerve [S1-S3]). Also, the pudendal nerve (S2-S4), along with many smaller nerve bundles, supplies the pelvic floor, genital musculature, and external anal and urethral sphincters, and carries afferent sensory fibers from the perineum.[1]

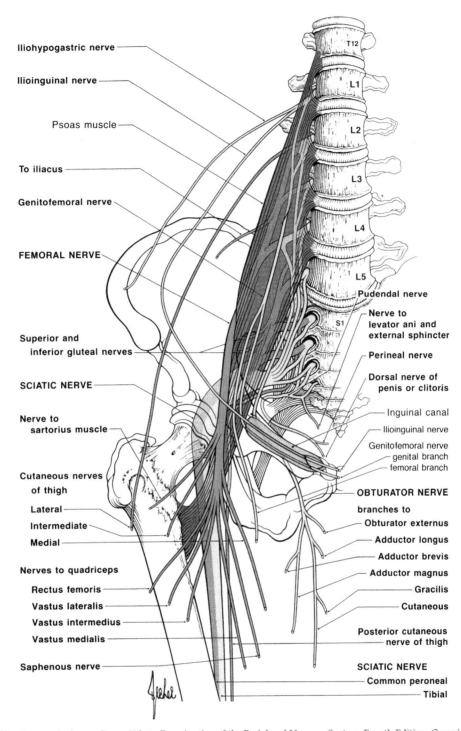

Iliohypogastric nerve
Ilioinguinal nerve
Psoas muscle
To iliacus
Genitofemoral nerve
FEMORAL NERVE
Superior and inferior gluteal nerves
SCIATIC NERVE
Nerve to sartorius muscle
Cutaneous nerves of thigh
Lateral
Intermediate
Medial
Nerves to quadriceps
Rectus femoris
Vastus lateralis
Vastus intermedius
Vastus medialis
Saphenous nerve

T12
L1
L2
L3
L4
L5
S1
Pudendal nerve
Nerve to levator ani and external sphincter
Perineal nerve
Dorsal nerve of penis or clitoris
Inguinal canal
Ilioinguinal nerve
Genitofemoral nerve
genital branch
femoral branch
OBTURATOR NERVE
branches to
Obturator externus
Adductor longus
Adductor brevis
Adductor magnus
Gracilis
Cutaneous
Posterior cutaneous nerve of thigh
SCIATIC NERVE
Common peroneal
Tibial

FIG. 2   Diagram of lumbosacral plexus. *From Aids to Examination of the Peripheral Nervous System, Fourth Edition, Copyright 2000. Used with permission from Elsevier.*

3 Plexopathy

3.1 Cervical plexopathy

The most common cause of a cervical plexopathy is direct tumor invasion, as seen in Fig. 3. The tumor may be local or maybe a metastatic deposit from a distant source. The most commonly associated tumors include squamous cell carcinoma of the head and neck, lymphoma, and adenocarcinomas of the lung and breast. Clinical presentation is typically pain, often described as a deep pain in the neck or shoulder. Patients may describe exacerbating factors such as

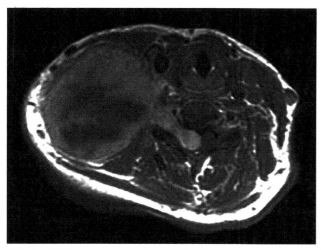

FIG. 3   Malignant peripheral nerve sheath tumor causing cervical plexopathy. Axial T1-weighted postcontrast MRI of the neck of a patient with neurofibromatosis type 1. A large, heterogeneously enhancing malignant peripheral nerve sheath tumor is seen extending through the neural foramen, resulting in compression of the cervical spinal cord.

TABLE 1   Tumor types commonly causing cancer-related plexopathy, classified by plexus.

Cervical plexus	Brachial plexus	Lumbosacral plexus
Lymphoma	Lungs	Colorectal
Head and neck	Breasts	Sarcoma
Lungs	Lymphoma	Breasts
Breasts	Sarcoma	Lymphoma
	Other cancers	Cervix
		Other cancers

cough, neck movement, and swallowing. Sensory loss is often patchy and difficult to demonstrate, but patients may use terms such as "pressure" or "burning" to describe it. If the presenting patient has recently had a neck dissection, then numbness of the upper anterior neck and the submandibular area would not be an unexpected finding for a typical patient. If the patients have involvement of the deep cervical and hyoid muscles, there may be no symptoms. If the spinal accessory cranial nerve is affected, shoulder weakness is likely. If the tumor involves the phrenic nerve, then a paralyzed hemidiaphragm may be identified. A patient with this affected nerve typically reports dyspnea, especially when lying down.[1] Perhaps the most worrisome part of cervical plexopathy is the proximity to the cervical spine. If sharp pain occurs with movement or examination of the neck or spine, it is important to consider epidural tumor as an extension of the tumor.[1]

3.2 Brachial plexopathy

As cancer affects the brachial plexus, patients typically report local pain for weeks to months with the development of progressive neurological deficits. Pain was the most common presenting symptom (75%) in a large study of neoplastic brachial plexopathy, and the axilla and shoulder were commonly affected.[2] Given the anatomy of the brachial plexus, it is unusual to encounter pure motor or sensory abnormalities when there is tumor involvement. Most tumors affecting the brachial plexus (Table 1) originate from the lung or breast and invade the lower plexus. This is referred to as Pancoast syndrome. Primary squamous cell carcinoma of the head and neck can also invade this space and directly affect the plexus.[1,2]

As tumor nears the first thoracic vertebrae, patients are at risk of developing Horner's syndrome. Nearly one-fourth of the patients in the study by Kori et al. developed Horner's syndrome. For clinicians, this should prompt additional imaging to assess for epidural involvement.[1,2]

Breast and lung cancers comprise over 70% of metastatic brachial plexopathies in the Kori series.[2] Based on the anatomy and typical tumor behavior, both breast and lung cancers tend to affect the medial cord of the plexus and its continuation as the ulnar nerve. The upper plexus is more often involved by lymphoma or primary head and neck cancers.

3.3 Lumbosacral plexopathy

Lumbosacral plexopathy is preceded by pain for weeks or months, similar to the presentation seen with brachial plexopathy. As with brachial plexopathy, the pain is described as constant, dull, and aching. Patients may report some sharp pain superimposed on a background of aching, as well as some muscle cramping. They may report exacerbation of pain when supine, during bearing down, and after prolonged walking.[4] Often, these patients will experience immobility and conditions related to that, such as venous thrombosis.[3]

Based on a series of 85 patients with cancer and lumbosacral plexopathy at Memorial Sloan Kettering Cancer Center (MSKCC) studied by Jaeckle et al., pain was present in all but one patient. Pain was of three types: local (72 of 85; 85%), radicular (72 of 85; 85%), or referred (37 of 85; 44%).[3] Following pain, patients will often develop asymmetric weakness and sensory loss. With time, the affected muscles will atrophy.[1] In the above series, weakness became symptomatic in nearly two-thirds, and over half of the patients developed sensory symptoms.[3] Given that this is a large plexus that derives input from multiple sources, the whole plexus is rarely affected. The sacral portion is affected most often as rectal cancers will locally invade. Enlarged retroperitoneal nodes may also directly invade this area, and bony metastases may extend to affect the plexus. With deep, midline tumors, bilateral plexopathy is possible due to nerve root involvement as nerves exit the sacral foramen. When this occurs, epidural extension is common. For this reason, advanced neuroimaging is recommended in the setting of bilateral involvement. Based on the series by Jaeckle et al., 45% of patients had epidural extension.[3]

Some interesting clinical presentations may be seen with lumbosacral plexopathy. For example, patients with L3-4 involvement or a femoral nerve lesion may exhibit rigid knee extension. With an L2-3 lesion, an individual may have trouble standing from a sitting position. A foot drop may indicate the presence of a lumbosacral trunk lesion.[3] One unique clinical presentation is the "hot and dry foot" which occurs due to sympathetic deafferentation by the tumor. This may happen without other physical signs.[7]

4 Diagnosis of plexopathy in patients with cancer

4.1 Neurologic differential diagnosis

It is prudent to always consider a nononcologic reason for plexopathy. Localization of these neurologic lesions can be very difficult, especially given that plexopathies may have incomplete nerve involvement. It is important to also keep in mind that patients can have more than one process happening. Imaging is always a good first step after examination to differentiate this. If a plexopathy is not identified, the clinician should keep searching for peripheral nerve involvement. For example, a patient with a foot drop could have a lumbosacral plexopathy or a peroneal nerve compression.

The metastatic disease affecting the peripheral nervous system occurs most often as a solitary, regional process with plexopathies being the most common. Multifocal peripheral nervous system injury due to metastatic disease is rare, but two notable exceptions occur. The more common scenario is the patient with carcinomatous meningitis with extensive spinal nerve root involvement that clinically mimics diffuse peripheral neuropathy. This presentation of carcinomatous meningitis typically involves the lumbar roots. The other example is that of a diffuse lymphoma, often called neurolymphomatosis.[8]

Additional nonmalignant causes should be considered as part of the differential diagnosis. In a series of 851 patients with systemic cancer seen for neurologic consultation at MSKCC, multiple noncancer diagnoses were made based on careful neurologic examination and evaluation. Metastatic disease involving the nervous system was identified in less than half of the patients. The most common nonmetastatic manifestation of cancer was

metabolic encephalopathy (10.2%, 87 patients). There were a significant number of diagnoses that were felt to be uncertain or unrelated. Headache was the most common of these diagnoses (8.5%; 72 patients). Other diagnoses included cerebrovascular disease (6.1%), degenerative disease of the spine (4.6%), syncope (1.8%), epilepsy (1.5%), peripheral neuropathy (1.2%), psychiatric disorder (1.2%), dementia (0.9%), and movement disorder (0.8%).[9]

4.2 Oncologic differential diagnosis

4.2.1 Cancer-related plexopathy

The two most common causes of plexopathy in the population of patients with cancer are direct invasion by tumor and radiation-induced plexopathy. Distinguishing radiation-induced plexopathy from cancer-related plexopathy based on clinical features can be challenging, and Table 2 lists key similarities and differences between the two presentations.[2]

As reported by Kori et al., tumor-related brachial plexopathy outnumbered radiation-induced plexopathy by 43 to 2 in a consecutive series of new patients seen over 12 months.[2] Although metastatic injury to distal peripheral nerves is rare, cancer patients are at increased risk of developing peripheral nervous system complications in the form of multifocal or diffuse neuropathies. Individuals undergoing therapy for cancer are at risk of chemotherapy-related neuropathy, neuropathy due

TABLE 2 Distinguishing radiation-induced plexopathy from cancer-related plexopathy by clinical features.

Clinical features	Cancer-related plexopathy	Radiation-induced plexopathy
Patient's presenting symptom	Pain	Paresthesias, weakness
Pain characteristics	Early in patient's course, severe	Later in patient's course
Edema	Occasional	Common
Plexus involvement:		
• Brachial plexopathy	Often involves lower plexus	Often involves the entire plexus
• Lumbosacral plexopathy	Often involves lower plexus, typically unilateral	Typically bilateral
Horner's syndrome	Commonly seen	Unusual
Myokymia on EMG	Unusual	Present
Nerve enhancement on contrasted MRI	Present	Absent
PET scan	Positive	Usually negative

to nutritional deficiencies, inflammatory or immune-mediated neuropathies, as well as multiple other causes of neuropathy. Paraneoplastic plexopathies, although rare, have also been described. As cancers progress and patients experience malnutrition, muscle atrophy, and immobility, the risk for compression-related neuropathies increases.[2]

4.2.2 Treatment-related plexopathy

Treatment for cancer may also produce plexopathy as an undesired effect. Radiation therapy induces nerve injuries by direct damage to the axons as well as directly affecting the vasa nervorum, with secondary microinfarction of the nerve.[1] In the report by Kori et al., authors performed a retrospective review across a 12-year period which showed that cancer-related plexopathies outnumbered radiation-induced plexopathies by a ratio of 3.5 to 1. Radiation plexopathy was rare with doses of less than 60 Gy and often occurred after a considerable latency, with a mean interval of 6 years posttreatment.[2]

In a series from Denmark, 161 patients with breast cancer that were definitively treated were examined for radiation-induced brachial plexopathy (RIBP) after a median follow-up period of 50 months (13–99 months). One hundred twenty-eight patients were treated with postoperative radiotherapy with 50 Gy, and 82 of these patients received cytotoxic therapy. Mild, symptomatic RIBP occurred in 9% of the patients, and 5% experienced disabling RIBP. Individuals who received cytotoxic therapy were more likely to develop RIBP, but this could have been due to the presence of more advanced diseases.[10]

In a case report by Harris and Tugwell, a patient developed RIBP over 20 years following definitive treatment for a primary breast cancer. She developed motor weakness as well as sensory complaints. Cross-sectional imaging did not reveal the recurrence of cancer. After neurologic evaluation including careful examination, it was suspected that her decreased grip and pinch strength as well as diminished deep tendon reflexes were due to plexopathy. These suspicions were confirmed with nerve conduction testing demonstrating markedly reduced amplitudes of sensory and motor nerves on the affected side. Motor nerve conduction testing also revealed decreased amplitudes in the affected hand. On needle electromyography, the right biceps demonstrated both chronic neurogenic changes and myokymic discharges, the latter is indicative of RIBP.[11]

The reported frequency of radiation-induced lumbosacral plexopathy (RILSP) ranges from 1.3% to 6.67%.[12,13] The clinical presentation varies and is associated with pain, numbness, weakness, and rarely, urinary or fecal incontinence. In a series of 50 patients with cervical cancer who received intensity-modulated radiation therapy

and high-dose-rate brachytherapy, RILSP was defined as "occurrence of paresthesias, numbness, dysesthesias, pain, or lower extremity weakness confirmed on T2-weighted MRI (diffuse marrow and perineural foraminal hyperintensity)." In that series, with 60 months of follow-up, four patients (8%) were found to have symptomatic RILSP. There were no patients identified with disabling RILSP in this series. The mean maximal lumbosacral plexus delineation dose in patients with RILSP was 59.6 Gy compared with 53.9 Gy in patients without RILSP (control; $P = .04$).[13]

5 Diagnostic studies

As mentioned above, it is prudent to have objective data when attempting to make a diagnosis of cancer-related plexopathy. Cross-sectional imaging, nerve conduction studies, and electromyography are common tools that are employed. If investigating other causes of neuropathy, lab work may also be appropriate.

5.1 Imaging

The clinical diagnosis of metastatic plexopathy may be confirmed by cross-sectional imaging via magnetic resonance imaging (MRI) or computed tomography (CT). MRI is the preferred choice as it provides more details and is more sensitive in tumor plexopathy (see Fig. 4). Increased T2 intensity within nerve trunks which may or may not be contrast-enhancing can be visualized.[14] If imaging is not revealing, it is recommended to repeat imaging in 4–6 weeks.[1] When a regional tumor recurrence is identified, the diagnosis of tumor plexopathy is more likely. Positron-emission tomography (PET) scans can be helpful in detecting active cancer in the vicinity of the plexus. Although the specificity and sensitivity of PET for tumor plexopathy are not fully identified, there are small series revealing the worth of this test. In a series of 19 patients with breast cancer and symptoms of brachial plexopathy, 14 patients exhibited abnormal fluorodeoxyglucose uptake by PET in the involved plexus.[15]

In a retrospective study, 31 patients seen with lumbosacral plexopathy at Mayo Clinic Rochester between 1987 and 1993 were studied. Tumor types included prostate, colorectal, bladder, cervical, and other pelvic cancers. Eighteen patients had received pelvic radiotherapy before the diagnosis of lumbosacral plexopathy. Twenty-seven patients had MRI, and lumbosacral involvement was evident on 23 of those patients. Six other patients had diffuse, metastatic disease evident on MRI. In 13 patients who had CT imaging, direct involvement of the lumbosacral plexus by tumor was noted. In four patients, MRI findings were abnormal and CT findings

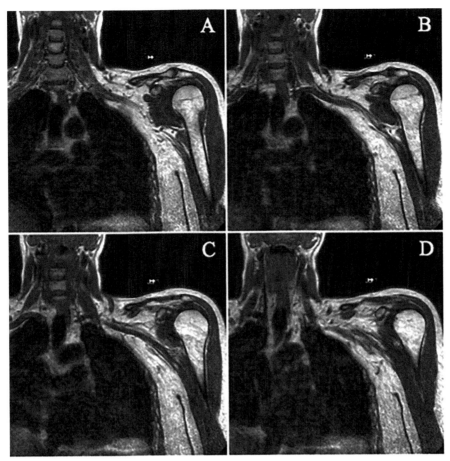

FIG. 4   MRI of normal brachial plexus. Coronal T1-weighted MRI of a normal brachial plexus. (A)–(D) are adjacent sections taken at 3 mm slice thickness from posterior to anterior. The nerves of the left brachial plexus can be seen exiting near the vertebral bodies and traveling in close proximity to the subclavian artery but differentiating components of the brachial plexus is difficult.

were normal. No patient had abnormal CT findings and normal MRI findings. The authors determined that MRI was more sensitive than CT for diagnosing cancer-induced lumbosacral plexopathy.[16]

Most MRI cross-sections of the affected plexus will include some views of the adjacent spinal canal (see Fig. 5). Given the high likelihood of concurrent epidural extension near an involved plexus, it is prudent to examine regional spinal canal images as well.

5.2 Electromyography

Electromyography (EMG) can provide important diagnostic information in the study of plexopathies in cancer patients. As motor nerve fibers are damaged, evidence of denervation in the affected muscles may be detected. The pattern of denervated muscles demonstrated by EMG can assist the clinician in determining the anatomic level at which the nerve or nerves are injured.

Another useful finding from EMG is the finding of myokymia, an unusual form of spontaneous muscle activity. Myokymia is strongly associated with radiation-induced plexopathies, where it is typically seen in many of the involved muscles. Plexopathy due to tumor invasion rarely causes myokymia. When imaging findings are nondiagnostic, the presence of myokymia can be a helpful tool.[17]

5.3 Treatment of cancer-related plexopathy

Treatment of metastatic plexopathy is required to prevent progressive and irreversible nerve damage and may improve outcomes. Early detection is helpful but can be challenging.

Corticosteroids may be used to reduce pain and manage pain and associated symptoms. Improvement in muscle strength and sensation is unlikely. Surgical management is seldom employed as a management tool unless debulking for relief of symptoms is needed. For example, surgery may be used to manage Pancoast syndrome.[18]

Radiation therapy is often employed to assist with local control of cancer or to manage pain. The utility for plexopathy management remains unclear. In a series of 49 patients with cancer-associated brachial plexopathy

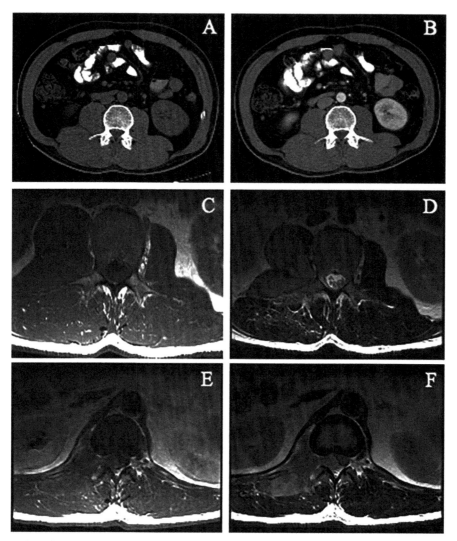

FIG. 5   Image of abnormal lumbosacral plexus. (A) and (B) are pre- and postcontrasted axial CT images, respectively. Slight enlargement of the right psoas muscle can be appreciated, though this was not read as abnormal on initial interpretation. (C) and (D) are T1- and T2-weighted axial MRI lumbar spine images, respectively, at the same level as that in (A) and (B). The T1-weighted image (C) is precontrasted and does not reveal changes in signal intensity. The T2-weighted image (D) clearly reveals abnormal hyperintensity in the right psoas. (E) and (F) are T1- and T2-weighted axial MRI images of the thoracic spine, respectively, in the same patient and reveal epidural extension via the neural foramen, resulting in cord compression.

from MSKCC, just over half of the patients who received radiation continued to have progressive symptoms. Pain control was reported in 46%, but neurological impairment did not improve. In another small series of patients with cancer-associated brachial plexopathy, 77% of 23 patients achieved pain control and 46% had objective responses. Radiation doses and fields varied, but higher doses were the most effective.[19]

Unfortunately, the effects of radiation therapy for lumbosacral plexopathy are not significantly better. Of 65 patients with cancer-associated lumbosacral plexopathy who received radiation, 15% improved clinically, 20% stabilized, and 65% progressed. Only 10% had definite improvement in strength. The median survival in this group was only 5.5 months.[3]

In the setting of plexopathy due to a tumor subtype that is sensitive to systemic therapy, a quick response to systemic therapy may be used. Lymphoma is a good example of a disease that may respond quickly to cytotoxic chemotherapy, offering neurologic improvement. Chemotherapy may also be a treatment option for individuals with a disease that is refractory to radiation or those who are not eligible for radiation therapy.

6 Cancers of special interest

6.1 Lung cancer

Individuals with nonsmall cell lung cancer (NSCLC) who present with superior sulcus tumors are at high risk of brachial plexopathy. This rare subtype of NSCLC only represents about 5% of NSCLC. Superior sulcus tumors are usually locally advanced and may be invading adjacent structures. If surgical cure is not feasible, then chemotherapy and/or radiation therapy are employed. Surgical planning uses cross-sectional imaging, and yet plexus imaging rarely provides conclusive evidence for the extent of the nerve root and plexus involvement that is definitively established during surgery.[20–22] Surgeons can excise the T1 and T2 nerve roots, but C8 root or lower trunk involvement precludes surgical resection given the risk of hand weakness. Resection of involved vertebral bodies and epidural disease has also become an important part of the operative goal. Some patients may receive induction chemotherapy prior to a surgical attempt. Complete resection, where feasible, significantly improves median overall survival.[22]

Actual plexus involvement leading to the classical Pancoast syndrome occurs in 15%–50% of patients with superior sulcus tumors, and one-fourth of the affected patients will eventually have an epidural extension. The presence of Horner's syndrome and involvement of other neural structures has been associated with a worse prognosis. In a retrospective review by Attar et al., Horner's syndrome was present in 20 of 105 patients and in 14 of the 67 individuals who underwent surgical resection. The median overall survival of the 84 patients without Horner's syndrome was 9.9 months compared with 6.4 months in patients who exhibited Horner's syndrome ($P < .05$). In the group of patients who experienced surgical resection, the 53 patients without Horner's syndrome had a median survival of 27.5 months compared with a median survival of 9.1 months in the 14 patients with Horner's syndrome ($P < .01$). Horner's syndrome is associated with a high likelihood of epidural disease and appears to signal a greater difficulty in achieving a complete resection.[23]

Long-term outcome data specific to patients with plexopathy are difficult to determine from the published trials. Current clinical trials typically exclude patients with plexus involvement that extends beyond the lower trunk. Furthermore, those patients with plexus involvement may not be reported on as a specific subgroup.

6.2 Breast cancer

As mentioned above, breast cancer is one of the most common causes of brachial plexopathy. From the series by Kori et al., pain was the most common presenting feature of tumor-associated plexopathy.[2] In another series from Downstate Medical College in New York, 28 patients with breast cancer and brachial plexopathy were studied. Supraclavicular, axillary, or chest wall metastases developed concurrently with brachial plexopathy in 26 of those patients. Of those 26 patients, 21 patients

were diagnosed with recurrent breast cancer. This diagnosis occurred 6–94 months (median 34 months) after primary therapy for breast cancer. The remaining five patients experienced brachial plexopathy in the setting of progressive, metastatic breast cancer. In 22 affected patients, treatment for brachial plexopathy was initiated: 8 patients received radiation and 14 patients had systemic therapy. Nineteen patients (86%) had partial or complete remission of pain and neurologic deficits, with a median duration of response of 8 months.[24]

In a retrospective review of 44 patients with breast cancer and cancer-related brachial plexopathy, it was noted that one-fourth of the affected individuals had been diagnosed with stage IV breast cancer at initial diagnosis. The mean age of affected patients was 51.9 years ± 9.3 years. The most common presenting symptom was muscle weakness. Most patients reported pain and decreased range of motion of the shoulders. Two-thirds of patients exhibited malignant lymphedema. Nineteen patients underwent brachial plexus MRI, and supraclavicular metastatic disease was the most common finding (57.9%).[25]

Radiation-induced brachial plexopathy (RIBP) as mentioned above may occur in patients with breast cancer. It may start with dysesthesias and numbness and then eventually include sensory changes and lymphedema. Pain generally develops later in the course than is seen with cancer-related plexopathy.[1]

7 Supportive treatment of cancer-related plexopathy

Surgical resection, systemic therapy, and radiation therapy are the three tools available for the treatment of cancer-related plexopathies. Durable responses from these options are not commonly seen. Given the fact that these conditions may result in debilitating consequences for patients, any supportive therapies should be used early.

In a report describing experiences of a "Plexus Clinic" that treated patients with brachial plexopathy of various etiologies, patients reported satisfaction as well as an improved sense of well-being and independence. The multidisciplinary team included a neurologist, rehabilitation physician, as well as a physical and occupational therapist. Patients were assessed and given personalized treatment plans. Interventions included physical and occupational therapy, medications, orthotics, or functional aids. The goals of the treatment plan included education, support in problem solving and decision-making, and empowering patients to be more independent.[26]

Physical therapy for muscle weakness, immobility, and other neurologic deficits related to plexopathy should be initiated promptly. Therapists can conduct a full physical assessment, advise on any recommendations for safety, as well as make recommendations for any helpful home equipment. Goals of therapy should include improvement of muscle strength, prevention of muscular contractures related to decreased muscle usage, and prevention of joint issues related to decreased mobility.

Occupational therapy can complement physical therapy and may be useful for individuals who have decreased grip strength, decreased ability to perform routine activities of daily living, and patients with foot drop. Additional benefits of participating in occupational therapy include a better sense of life balance and utilizing self-management strategies known to reduce fatigue and improve quality of life. In a small pilot study performed by the clinic noted above, results showed significant and clinically important differences in participation (performance and satisfaction) scores on the Canadian Occupational Performance Measure as well as significant improvement in self-reported shoulder function, pain, and performance of activities of daily living.[26]

Additional palliative needs for patients with plexopathy should address pain control. As mentioned above, patients may experience acute and/or chronic pain that can be difficult to control. Typical pain management may start with antiinflammatory agents and physical relief with hot or cold compresses. Even with the administration of antitumor therapy, additional pain management is required by a high percentage of patients. In addition to narcotics for pain relief, these individuals may benefit from agents designed to treat neuropathic pain such as gabapentin or pregabalin.[27] Other pharmacologic choices include antidepressants, antiepileptic medications, topical pharmacologic agents, or muscle relaxants. Nerve blocks and regional sympathetic blockades are used occasionally. Approximately 5% of patients may develop severe, intractable pain that necessitates a more aggressive approach, is uncontrolled by these medical approaches, and will undergo surgical procedures. Dorsal root entry zone (DREZ) operations, CT-guided ablative procedures, epidural electrode stimulators, transcutaneous electrical nerve stimulation (TENS) units, and other tools may be used.[28] These types of pain problems are difficult and often disabling and require sophisticated multimodality therapy.[27,29]

8 Summary

Cancer-related plexopathy typically presents with pain and/or functional loss affecting the brachial or lumbosacral plexus. It may occur in patients undergoing active therapy for cancer or 20 years after completing definitive cancer therapy in individuals who have been deemed "cancer-free." Early identification through examination and imaging is promoted. Care provided by a multidisciplinary team is encouraged, as patients will need pain management, supportive care, and even psychological support. For those individuals whose plexopathy heralds a new or recurrent cancer diagnosis, a treatment plan should be promptly initiated.

Brachial plexopathy is often attributed to breast and lung cancers and may be related to direct local extension of disease. Brachial plexopathy may also be caused by therapy, as in the case of radiation-induced brachial plexopathy (RIBP). Lumbosacral plexopathy is less common and more tumor types may contribute to this condition. Plexus involvement by the tumor is painful and usually foreshadows motor weakness and sensory loss. In distinction, radiation-induced plexopathy is much less often painful and is predominantly a sensory disturbance in onset that slowly evolves. Early diagnosis is a key to functional preservation. With a clear diagnosis, treatment plans can be developed including radiation therapy, chemotherapy or alternate systemic therapy, or surgical resection. Long-term management should include efforts at physical rehabilitation and often require aggressive pain control measures to deal with both somatic and neuropathic pain.

References

1. Jaeckle KA. Neurological manifestations of neoplastic and radiation-induced plexopathies. *Semin Neurol.* 2010;30(3):254–262.
2. Kori SH, Foley KM, Posner JB. Brachial plexus lesions in patients with cancer: 100 cases. *Neurology.* 1981;31:45–50.
3. Jaeckle KA, Young DF, Foley KM. The natural history of lumbosacral plexopathy in cancer. *Neurology.* 1985;35:8–15.
4. Pierce SM, Recht A, Lingos TI, et al. Long-term radiation complications following conservative surgery (CS) and radiation therapy (RT) in patients with early-stage breast cancer. *Int J Radiat Oncol Biol Phys.* 1992;23:915–923.
5. Powell S, Cooke J, Parsons C. Radiation-induced brachial plexus injury: follow-up of two different fractionation schedules. *Radiother Oncol.* 1990;18:213–220.
6. Sheldon T, Hayes DF, Cady B, et al. Primary radiation therapy for locally advanced breast cancer. *Cancer.* 1987;60:1219–1225.
7. Evans RJ, Watson CP. The hot foot syndrome: Evans' sign and the old way. *Pain Res Manag.* 2012;17(1):31–34.
8. Moots P, Edgeworth M. Plexopathies. In: Newton HB, Malkin MG, eds. *Neurological Complications of Systemic Cancer and Antineoplastic Therapy.* 1st ed. Florida: CRC Press; 2010.
9. Clouston PD, DeAngelis LM, Posner JB. The spectrum of neurological disease in patients with systemic cancer. *Ann Neurol.* 1992;31(3):268–273.
10. Olsen NK, Pfeiffer P, Johannsen L, et al. Radiation-induced brachial plexopathy: neurological follow-up in 161 recurrence-free breast cancer patients. *Int J Radiat Oncol Biol Phys.* 1993;26(1):43–49.
11. Harris SR, Tugwell KE. Neurological and dexterity assessments in a woman with radiation-induced brachial plexopathy after breast cancer. *Oncologist.* 2020;25(10):e1583–e1585.
12. Dahele M, Davey P, Reingold S, Shun WC. Radiation-induced lumbo-sacral plexopathy (RILSP): an important enigma. *Clin Oncol (R Coll Radiol).* 2006;18(5):427–428.

13. Tunio M, Al Asiri M, Bayoumi Y, et al. Lumbosacral plexus delineation, dose distribution, and its correlation with radiation-induced lumbosacral plexopathy in cervical cancer patients. *Onco Targets Ther.* 2014;8:21–27.

14. Thyagarajan D, Cascino T, Harms G. Magnetic resonance imaging in brachial plexopathy of cancer. *Neurology.* 1995;45:421–427.

15. Ahmad A, et al. Use of positron emission tomography in evaluation of brachial plexopathy in breast cancer patients. *Br J Cancer.* 1999;79:478–482.

16. Taylor BV, Kimmel DW, Krecke KN, Cascino TL. Magnetic resonance imaging in cancer-related lumbosacral plexopathy. *Mayo Clin Proc.* 1997;72(9):823–829.

17. Harper Jr CM, et al. Distinction between neoplastic and radiation-induced brachial plexopathy, with emphasis on the role of EMG. *Neurology.* 1989;39:502–506.

18. Rusch VW. Management of Pancoast tumours. *Lancet Oncol.* 2006;7:997–1005.

19. Ampil FL. Radiotherapy for carcinomatous brachial plexopathy. A clinical study of 23 cases. *Cancer.* 1985;56(9):2185–2188.

20. Bilsky MH, et al. Surgical treatment of superior sulcus tumors with spinal and brachial plexus involvement. *J Neurosurg.* 2002;97:301–309.

21. Kent MS, Bilsky MH, Rusch VW. Resection of superior sulcus tumors (posterior approach). *Thorac Surg Clin.* 2004;14:217–228.

22. Sundaresan N, Hilaris BS, Martini N. The combined neurosurgical-thoracic management of superior sulcus tumors. *J Clin Oncol.* 1987;5:1739–1745.

23. Attar S, Krasna J, Sonett JR, et al. Superior sulcus (Pancoast) tumor: experience with 105 patients. *Ann Thorac Surg.* 1998;66(1):193–198.

24. Kamenova B, Braverman AS, Schwartz M, et al. Effective treatment of the brachial plexus syndrome in breast cancer patients by early detection and control of loco-regional metastases with radiation or systemic therapy. *Int J Clin Oncol.* 2009;14(3):219–224.

25. Kim J, Jeon JY, Choi YJ, et al. Characteristics of metastatic brachial plexopathy in patients with breast cancer. *Support Care Cancer.* 2020;28(4):1913–1918.

26. Janssen RMJ, Satink T, Ijspeert J, et al. Reflections of patients and therapists on a multidisciplinary rehabilitation programme for persons with brachial plexus injuries. *Disabil Rehabil.* 2019;41(12):1427–1434.

27. Namaka M, Leong C, Grossberndt A, et al. A treatment algorithm for neuropathic pain: an update. *Consult Pharm.* 2009;24(12):885–902.

28. Sundaresan N, DiGiacinto GV, Hughes JE. Neurosurgery in the treatment of cancer pain. *Cancer.* 1989;63:2365–2377.

29. Vecht CJ. Cancer pain: a neurological perspective. *Curr Opin Neurol.* 2000;13:649–653.

Nonmetastatic neurological complications

Cerebrovascular complications of malignancy

Jeffrey M. Katz[a], Prathusan Subramaniam[b], and Timothy G. White[c]

[a]Departments of Neurology and Radiology, Donald and Barbara Zucker School of Medicine at Hofstra/Northwell, Hempstead, NY, United States, [b]Office of Clinical Research, Feinstein Institutes for Medical Research, Manhasset, NY, United States, [c]Department of Neurological Surgery, North Shore University Hospital, Manhasset, NY, United States

Cerebrovascular disease commonly occurs in cancer patients, often from mechanisms and processes specific to malignancy. Stroke may arise by direct tumor effects, through restriction of cerebral blood flow from tumor-associated mass effect and edema, direct blood vessel infiltration, or intratumoral hemorrhage. Embolization of tumor debris and of infectious and noninfectious cardiac vegetations, related to the immunocompromised and hypercoagulable states found in many cancer patients, commonly targets cerebral blood vessels. The neoplastic disease may additionally incite a coagulopathic state, leading to either cerebral thrombosis, hemorrhage, or both in settings such as disseminated intravascular coagulation (DIC) or microangiopathic hemolytic anemia. Cancer treatment itself may be complicated by cerebrovascular disease, as both chemotherapy and radiation therapy may damage the blood vessels or disrupt the coagulation cascade. Herein, we explore the fairly common and often complicated relationship between cerebrovascular disease and malignancy.

1 Epidemiology of cerebrovascular disease in malignancy

1.1 Ischemic stroke

Stroke is the most common intracranial complication of cancer after metastases, with about 15% of cancer patients having cerebrovascular pathology at autopsy and only about half of these patients displaying clinical symptoms. The most common causes of ischemic stroke in cancer patients found at autopsy include nonbacterial thrombotic endocarditis, cerebral intravascular coagulation, and septic emboli.[1] Vascular risk factors do not significantly differ in cancer patients compared to noncancer patients;

however, the underlying mechanisms are typically different, and the proportion of symptomatic intracranial hemorrhages is much greater (57% in malignancy versus 15% in noncancer patients).[1,2] Studies have indicated variable contributions of atherosclerotic disease to stroke etiology in cancer patients, ranging from 15% to 33%.[3] The most common cancer type associated with ischemic stroke is lung cancer, followed by primary brain tumors and prostate cancer.[4] Ischemic stroke is much more likely in the setting of metastatic cancer than with solitary tumors. Patients with active cancer tend to have higher stroke severity on admission and a higher rate of in-hospital mortality.[5] Patients presenting with a large vessel occlusion in the setting of malignancy should be screened for mechanical thrombectomy. Previous series have demonstrated that outcomes of patients with cancer-related strokes who undergo mechanical thrombectomy are similar to those with a nonmalignancy-related stroke.[6] Furthermore, the rates of recanalization are similar to other etiologies of stroke.[7,8]

1.2 Intracranial hemorrhage

Cancer-associated intracranial hemorrhage is typically associated with an underlying coagulopathy or hemorrhage within a hypervascular tumor. The most common location is intracerebral, although spontaneous subdural hematoma and subarachnoid hemorrhage (SAH) may also develop. Only about half of all cerebrovascular lesions that occur in the setting of malignancy are symptomatic, with hemorrhages resulting in symptoms slightly more than half of the time and ischemic strokes slightly less than half.[1] The rate of symptomatic intracranial hemorrhage is highest in leukemia, being more prevalent in acute myelogenous leukemia (AML, 84% symptomatic) than in acute lymphocytic leukemia

(ALL, 73% symptomatic).[9,10] Similar to ischemic stroke, hemorrhagic stroke is much more likely in the setting of metastatic disease compared to localized tumors.[11]

Intracranial hemorrhage has been reported in 20%–50% of patients with metastatic brain tumors (Fig. 1).[12] While any metastasis may bleed, melanoma, renal cell carcinoma, and choriocarcinoma have a higher tendency to be hemorrhagic.[13] In addition, as lung cancer is the most common intracranial metastasis, a fair number of hemorrhagic intracranial neoplasms are found pathologically to be metastatic lung cancer. Despite the prevalence of intracranial metastasis from metastatic lung cancer, patients with melanoma or renal cell carcinoma were found to have a fourfold higher risk of intracranial hemorrhage.[12] Among primary brain tumors, glioblastoma multiforme (GBM) is the most likely to bleed and oligodendrogliomas are more likely than lower grade astrocytomas to be associated with hemorrhage. Vascular tumors like hemangioblastoma (Fig. 2) infrequently bleed into the parenchyma or adjacent subarachnoid space, and the hemorrhage risk is greater with larger tumors.[14] Hypervascular primary dural-based tumors, such as hemangiopericytomas and angioblastic meningiomas, usually hemorrhage into the subarachnoid or subdural spaces because of their anatomical proximity to these compartments.[9,15]

2 Direct tumor effects and intracranial hemorrhage

Often, a symptomatic hemorrhage into a tumor is the initial presentation of an intracerebral malignancy.[15] A new sudden neurological deterioration, in the case of a known cerebral tumor, is another classic presentation of intratumoral hemorrhage. Clinical sequelae of cerebral

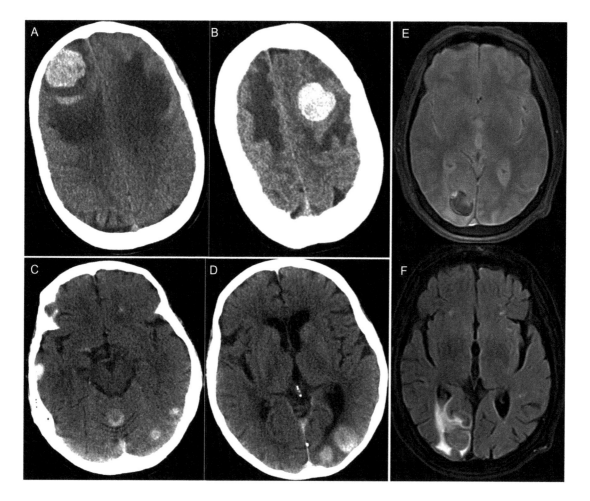

FIG. 1   Hemorrhagic brain metastases. (A) and (B) Noncontrast CT head showing right (A) and left frontal (B) hemorrhagic brain metastases with extensive surrounding vasogenic edema in a patient with metastatic melanoma. (C) and (D) Noncontrast CT head showing multifocal infra- and supratentorial hemorrhagic brain metastases in a patient subsequently diagnosed with metastatic breast cancer. (E) and (F) Brain MRI, gradient echo (E) and fluid attenuation inversion recovery (F) sequences showing a hemorrhagic brain metastasis with surrounding vasogenic edema in a patient with known renal cell carcinoma. Histopathological analysis of the resected specimen determined that the lesion was actually metastatic melanoma.

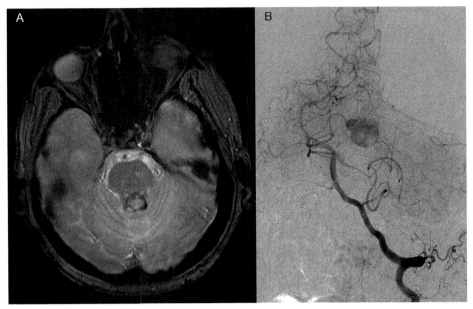

FIG. 2 Hemangioblastoma. (A) Brain MRI, gradient echo sequence, showing a small round tumor in the superior medullary velum with peripheral calcification. (B) Left vertebral artery angiogram, frontal oblique view, showing the hypervascular tumor blush typical of hemangioblastoma.

hemorrhage include headache, seizure, change in mental status or level of consciousness, and focal neurological deficits.[16] When the hemorrhage is large, signs of cerebral herniation may ensue. Noncontrast cranial CT scan is the quickest and simplest way to identify a new intracranial hemorrhage. Other causes of intracerebral hemorrhage, including hypertension, cerebral amyloid angiopathy, trauma, and rupture of a cerebral aneurysm or arteriovenous malformation may be difficult to differentiate from an intratumoral hemorrhage by noncontrast CT scan, especially as tumor bleeds account for only 4.4% of all intracerebral hemorrhages.[17] Brain MRI, on the other hand, may be very useful in defining the underlying etiology as a tumor-related hemorrhage. MRI findings that suggest intratumoral hemorrhage include enhancement around the hematoma and marked perihematomal edema (Fig. 3).[18] MRI specificity for a tumor-based hemorrhage reaches 70% if the diameter of the lesion, including the vasogenic edema, is twice the diameter of the hematoma alone.[19] Tumor bleeds also tend to evolve slower and deposit less hemosiderin than other spontaneous hemorrhages.[20] Also, a brain MRI may reveal other nonhemorrhagic mass lesions suggesting cerebral metastasis. Single-photon-emission computed tomography (SPECT) scanning may also be helpful in this regard by distinguishing the elevated blood flow demands of a neoplasm.[21] MR spectroscopy allows for the visualization of the neoplastic tissue while also quantifying its metabolic rate. This may be utilized to monitor a potential response to treatment or to identify a relapse.[22] Amino acid PET has also been shown to be useful for differentiating neoplastic versus nonneoplastic tissue and may be an effective method for diagnosis, tumor grading, and

prognostication.[23,24] While noninvasive examinations are illuminating, the only definitive diagnostic tool remains the pathological examination of a surgical specimen.

Of all cancer-associated hemorrhages in each brain compartment, intraparenchymal hemorrhage is the most frequent, followed by subdural hemorrhage, subarachnoid hemorrhage, and epidural hemorrhage, respectively.[15] Tumor localization determines the compartment of intracranial hemorrhage. For example, leukemic leptomeningeal infiltration will typically hemorrhage into the subdural space, whereas GBMs generally cause an intracerebral hemorrhage. The mechanism of hemorrhage involves several tumor-specific factors, including intratumoral necrosis, neovascularization, and local blood vessel infiltration.[10] Tumor histology is the major determinant of its vascularity, and tumor expression of vascular endothelial growth factor (VEGF) has been correlated with the propensity of a neoplasm to bleed.[25,26]

Neoplastic aneurysm formation is a rare cause of tumor-associated hemorrhage resulting from a tumor embolic occlusion of the vasa vasorum. The consequent loss of arterial wall integrity culminates in segmental dilatation of the vessel into a pseudoaneurysm. If a neoplastic aneurysm ruptures, the patient generally presents with a SAH. The location and mechanism of neoplastic aneurysm formation are analogous to the development of infectious mycotic aneurysms. Cardiac myxoma, lung cancer, and choriocarcinoma are most commonly associated with a neoplastic aneurysm formation.[27–29] Prognosis for patients with neoplastic aneurysm formation from cardiac myxoma is better than for those with choriocarcinoma and other tumors, with mortality rates of 11.4%, 60.9%, and 92.3%, respectively.[30]

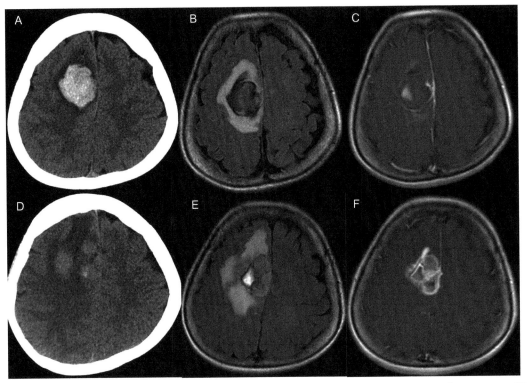

FIG. 3   Glioblastoma multiforme presenting initially as a right frontal cortical hemorrhage. (A)–(C) The initial noncontrast CT head (A), and brain MRI, fluid attenuation inversion recovery (FLAIR, B), and postcontrast T1-weighted sequence (C) showing an acute right frontal hemorrhage with minimal contrast enhancement and only mildly more edema than what might be expected. (D)–(F) The same imaging modalities and sequences 3 months later show resolving hemorrhage, but significantly increased edema and FLAIR signal as well as contrast enhancement of the central mass. Excisional biopsy confirmed the lesion to be a glioblastoma multiforme.

Intraventricular hemorrhage and SAH are rare complications of cancer associated with leptomeningeal metastases,[1] AML-associated hyperleukocytosis,[1,31] and meningiomas.[9,32] Spinal hemangioblastoma,[33] oligodendroglioma,[34] clival chordoma,[35] and central neurocytoma[36] have also been reported to present with SAH. Epidural hematoma is a rare initial presentation of malignant disease, particularly arising from skull metastases, and has been reported in hepatocellular carcinoma.[37,38] Tumor invasion with obstruction of dural veins results in venous hemorrhage into the subdural space. Subdural hematomas occasionally develop with systemic cancers, such as leukemia, prostate cancer, and breast cancer. Gliomas were also found to predispose patients to SDH.[39]

3 Direct tumor effects and cerebral infarction

Ischemic cerebrovascular disease is less frequently a direct tumor effect than intracranial hemorrhage. Most commonly, tumors directly cause cerebral infarction by invasion or compression of neighboring cerebral arteries, pial veins, and dural venous sinuses, and by tumor embolism. Invasion of perivascular (Virchow-Robin) spaces by leptomeningeal metastases resulting in small vessel

thrombosis is another direct mechanism.[40] The extracranial carotid arteries are occasionally invaded or compressed by nasopharyngeal carcinomas, squamous cell carcinoma, and carotid body paragangliomas (Fig. 4A and B). Carotid blowout is an uncommon but highly fatal complication of salvage reirradiation for recurrent head and neck cancer (Fig. 4C). A carotid blowout can develop when a damaged arterial wall cannot sustain its integrity against the patient's blood pressure.[41] This can be caused by pathological changes to the vascular structure caused by radiation therapy.[42] Some studies have shown incidence rates of carotid blowout postreirradiation for recurrent head and neck cancer ranging from 0% to 17%,[43] with one study showing a mortality rate as high as 76%.[41] The intracranial carotid arteries may be strangled or encased by tumors in the cavernous sinuses or along the supraclinoid segment, particularly by pituitary tumors and parasellar meningiomas. Clival chordomas and base of skull metastases may involve the vertebral and basilar arteries. B-cell lymphoma may rarely produce intravascular lymphoproliferation causing vessel occlusion and infarction. Similarly, acute leukemia-associated hyperleukocytosis may obstruct the cerebral vessels resulting in cerebral infarction or hemorrhage.[1]

Tumor cerebral emboli are rare and occur in the setting of advanced metastatic disease, usually of sarcoma

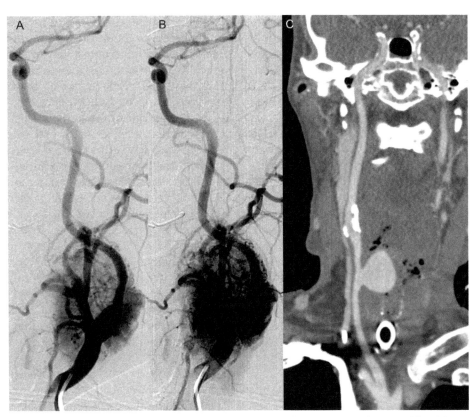

FIG. 4 Local tumor effects on the carotid artery. (A) and (B) Left common carotid angiogram, early (A) and late (B) arterial phases showing a hypervascular carotid body tumor encasing the left distal common and proximal internal and external carotid arteries. Pretreatment embolization was performed prior to tumor resection. (C) CT angiogram, neck, coronal reconstruction shows a carotid blow-out (*blue arrow*) in an elderly woman with squamous cell carcinoma of the larynx status post chemotherapy and radiation.

or carcinoma, especially if the heart or lung is involved. Tumor emboli occur either spontaneously, with direct invasion of the pulmonary veins or the left atrium, paradoxically through a venous-to-arterial (cardiac septal or pulmonary) shunt, or iatrogenically, during surgical tumor manipulation at the time of pneumonectomy.[44] The latter may be prevented by minimizing tumor manipulation, ligating the pulmonary vein early in the operation, or utilizing cardiopulmonary bypass.[45] Through the invasion of the vessel, a tumor embolus can also lead to vessel dilation, causing a cerebral aneurysm.[27] Symptoms of tumor embolization include focal neurological deficits and encephalopathy. Evidence of systemic embolization on clinical examination and imaging is supportive of this uncommon stroke mechanism.

Dural venous sinus thrombosis classically presents with headache, seizures, papilledema, and symptoms and signs that are not atypical for patients with brain metastases or primary cerebral neoplasms, making this clinical diagnosis particularly challenging in these patients. Dural venous sinus thrombosis is typically associated with lymphoma, neuroblastoma, or lung cancer metastases to the calvarium or dura. Sinus thrombosis developing during a hematological malignancy is more likely to be complicated by secondary hemorrhage.

Other precipitating factors include cancer-associated hypercoagulability and the effects of chemotherapy. Sinus thrombosis can be diagnosed with almost 100% sensitivity using MRI and MR venography.[46] Contrast-enhanced MR venography is preferred. Treatment generally involves anticoagulation, even in the setting of venous infarction with secondary hemorrhage.

4 Coagulation disorders

4.1 Hemorrhagic diatheses and disseminated intravascular coagulation

Cancer-associated coagulopathy accounts for about half of all symptomatic strokes in cancer patients and ranges from hemorrhagic diatheses to prothrombotic states. Intracerebral hemorrhage due to an underlying hemorrhagic diathesis is most commonly seen in leukemia, especially AML, and the majority of these bleeds are symptomatic.[1] There are several potential mechanisms by which cancer patients acquire a bleeding diathesis and, typically, multiple etiologies play a role in an individual patient. Thrombocytopenia and clotting factor deficiencies are commonly found during the course of many malignancies. Bone marrow infiltration, chemotherapy, and radiation therapy may each cause thrombocytopenia, and hepatic failure, from liver metastases or chemotherapy, may lower serum clotting factor concentrations. Disseminated intravascular coagulation (DIC), associated in various degrees with many cancers, results in the consumption of both platelets and clotting factors and may cause hemorrhage, infarctions, or both. Tissue factor, released by promyelocytes in AML, and expressed by certain tumor cells, particularly adenocarcinoma, is the major precipitant for DIC. This results in the production of excessive thrombin, which in turn develops a thrombotic cascade. Depletion of platelets, fibrinogen, and prothrombin leads to a consumptive coagulopathy which in severe cases can lead to bleeding.[40] Tumor lysis syndrome and sepsis may exacerbate DIC, possibly due to additional tissue factor release. Unfortunately, DIC treatment usually hinges on treating the underlying cause, in this case, the cancer. Heparin may be useful in chronic DIC, especially in leukemia, where patients on heparin may have fewer DIC-related thromboembolic complications.[47]

4.2 Microangiopathic hemolytic anemia

Microangiopathic hemolytic anemia (MAHA) is an uncommon coagulopathy, usually associated with mucin-secreting adenocarcinomas, such as lung, breast, and gastric adenocarcinoma. MAHA is a clinical constellation of DIC, hemolytic anemia with schistocytes, renal failure, severely low platelets, and thrombotic or hemorrhagic cerebrovascular events. In contrast to the microvascular thrombi of DIC, composed primarily of fibrin, clots in MAHA have a significant platelet component. This principle consumption of platelets is one of the primary causes of intracerebral hemorrhage in MAHA.[48]

MAHA is incited by tumor cell expression of integrins like glycoprotein IIb/IIIa and Gp1b. These adhesion proteins allow tumor cells to stick to the extracellular matrix. A consequence of this expression is the induction of an inflammatory response that includes the formation of immune complexes that attack the vascular endothelium. MAHA is pathophysiologically similar to hemolytic-uremic syndrome (HUS) and thrombotic thrombocytopenic purpura (TTP), which respond to plasmapheresis. Unfortunately, plasma exchange is less effective in MAHA, although platelet transfusions, considered harmful in HUS and TTP, are not contraindicated in the management of MAHA.[48]

4.3 Nonbacterial thrombotic endocarditis

Nonbacterial thrombotic endocarditis (NBTE) is the most common cause of stroke in advanced cancers (Fig. 5).[49] In one study, NBTE was reported in 19% of patients with metastatic adenocarcinoma.[50] Noninfectious platelet-fibrin vegetations fragment off cardiac valves and embolize to the intracerebral arteries causing multifocal cerebral infarctions. Showered emboli travel systemically and numerous infarctions are found in other organ systems. While many of these systemic infarctions are asymptomatic, the kidneys, liver, and spleen are the organs most frequently affected, followed by the myocardium and gastrointestinal tract. Systemic complications include central retinal artery occlusion, deep venous thrombosis, extremity arterial thrombosis, myocardial infarction, and internal hemorrhage. NBTE has a 1.3% prevalence in end-stage cancer patients. Half of these patients will have an embolic stroke at autopsy, of which three-quarters will have been symptomatic. Fifty percent of these patients also have a concurrent intracerebral hemorrhage, either primary or secondary hemorrhagic transformation of a cerebral infarction. NBTE arises most frequently in patients with adenocarcinoma. More specifically, NBTE arises in mucin-producing carcinomas of the lungs, gastrointestinal tract, lymphoma, as well as in carcinomas of the pancreas, ovaries, and biliary system.[1,51,52]

NBTE presents with acute focal neurological deficits or encephalopathy and may be the final act of advanced cancer or the initial manifestation of an incipient malignancy. The most common NBTE-related neurological deficit is aphasia.[51] MRI diffusion-weighted imaging reveals multiple areas of infarction of different ages and sizes.[53] Transesophageal echocardiography is usually necessary to diagnose NBTE, as the vegetations are generally too small to be visible by transthoracic studies.[4] In addition, multiple sterile blood cultures are required to eliminate infectious endocarditis from the differential diagnosis.[53] A postmortem analysis is needed to make a definite diagnosis.[50]

Small case series allude that heparin is the preferred anticoagulant in the management of NBTE due to its ability to reduce thromboembolic complications, as well as the rate of ischemic stroke in NBTE associated with cancer while minimizing the risk of anticoagulation-associated intracerebral hemorrhage.[51,54,55] In those who fail heparin therapy, warfarin may be effective.[56] Lifelong anticoagulation, usually with low-molecular-weight heparin, is necessary, as the recurrence rate of thrombosis after discontinuing heparin is very high. No data are available to support the use of oral direct thrombin or factor Xa inhibitors in these patients. Valvular repair or replacement may be considered for patients with severe valvular dysfunction.[50]

4.4 Cerebral intravascular coagulation

Cerebral intravascular coagulation (CIVC) is a postmortem diagnosis characterized pathologically by in situ occlusion of numerous small blood vessels by

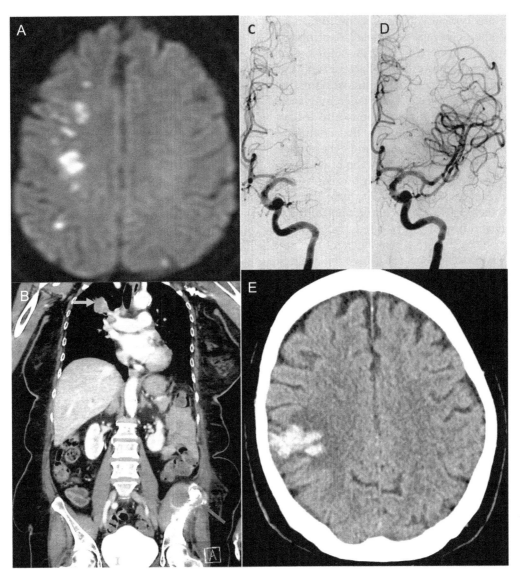

FIG. 5 Nonbacterial thrombotic endocarditis. (A) Brain MRI, diffusion-weighted imaging, showing multiple infarctions in a watershed distribution between the right middle cerebral artery (MCA) and anterior cerebral artery territories in a patient with metastatic lung cancer. (B) Coronal reconstruction of a contrast-enhanced CT scan of the chest, abdomen, and pelvis showing the same patient's lung cancer (*green arrow*) and a bone metastasis to the left pelvis (*blue arrow*). (C) Preembolectomy cerebral angiogram, AP view, showing a left MCA M1 segment occlusion in a patient with metastatic adenocarcinoma of unknown primary. (D) On the postembolectomy angiogram, the left MCA is completely recanalized. (E) Noncontrast CT head, axial slice, taken one day postintervention, shows no infarction in the recanalized left MCA territory and hemorrhagic conversion of a separate cortical infarction in the right MCA territory.

fibrin-laden microthrombi without an identified embolic source. Symptoms include focal neurological deficits and seizures, but typically patients are encephalopathic and may deteriorate to coma.[16] Pathologically, only NBTE causes more symptomatic cerebral infarctions. However, because there are numerous etiologies for encephalopathy in end-stage cancer patients and no noninvasive test is diagnostic, this condition is very difficult to confirm in vivo. CIVC is most frequently associated with leukemia, lymphoma, gastrointestinal adenocarcinoma, and breast cancer. Evidence of concurrent systemic thrombosis, commonly involving the spleen, kidneys, adrenal glands, and lungs, was found in almost three-quarters

of patients with CIVC examined at autopsy. In addition, multifocal systemic hemorrhage and sepsis are also commonly identified in these patients, and death typically occurs within 3 weeks of symptom onset.[1]

5 Cancer-related infectious disease and stroke: Septic cerebral infarction

Multifocal ischemic and hemorrhagic cerebrovascular lesions characterize septic cerebral infarction, which often complicates advanced leukemia, lymphoma, and carcinoma. Patients with septic infarctions of fungal or bacterial origin are typically immunocompromised, a complication of chemotherapy-associated neutropenia, radiation therapy-induced bone marrow suppression, corticosteroid use, or the cancer itself. Bacterial endocarditis may present in the absence of immunosuppression. Septic emboli to the distal cerebral arteries may incite a local arteritis, with resulting infectious and inflammatory destruction of the arterial wall that culminates in the formation of a mycotic aneurysm.[57] These aneurysms have a tendency to rupture into the brain parenchyma or the subarachnoid space. Unruptured mycotic aneurysms may respond to a prolonged course of appropriate antibiotic therapy. Endovascular embolization or surgical ablation is indicated for ruptured mycotic aneurysms and may be the preferred management for large or solitary lesions, especially prior to valve replacement surgery.

Fungal infections are the second most common in cancer patients, after bacterial infections. *Aspergillus* is a frequent microbe causing septic cerebral infarction. Cerebral *Aspergillus* infections are typically secondary to lung infections.[27,58] *Aspergillus* infections infiltrate the vessel wall, causing mycotic arteritis or aneurysms.[57] These infections also spread along the blood vessels into the brain, leading to in situ arterial occlusion or tears through the arterial wall causing large intracerebral hemorrhages. Similarly, the *Zygomycetes* Mucor and Rhizopus, fungi that are more commonly invasive in poorly controlled diabetic patients, may spread intracranially from paranasal sinus infections and cause clinical and radiographic manifestations that are indistinguishable from *Aspergillus*. These infections present with seizures or acute focal deficits. *Candida* species invade along central venous catheters or through the gastrointestinal or genitourinary tract and present similarly, particularly with embolic disease from *Candida*-associated endocarditis. *Candida* septic infarction, in contrast, is typified by diffuse encephalopathy.[1]

Fungal sepsis is difficult to diagnose, as blood, sputum, and CSF cultures are usually negative. *Aspergillus*- and *Zygomycetes*-infected patients usually have pulmonary infiltrates by the time they develop septic cerebral infarction. Brain biopsy is the most definitive diagnostic test and polymerase chain reaction and assaying for galactomannan antigen may be useful diagnostic adjuncts in the detection of *Aspergillus* species. Treatment is often difficult and incomplete, particularly with amphotericin B, leading to a high mortality rate from these infections. Several antifungals have efficacy against both *Candida* and *Aspergillus* species, where voriconazole is considered first-line therapy for *Aspergillus* infection and echinocandins are commonly used for a *Candida* infection.[59–61] Posaconazole, in conjunction with surgical debridement, is the treatment of choice for *Zygomycetes* infections.[62]

6 Infectious vasculitis

Immunocompromised cancer patients have an elevated risk of mycobacterial and fungal basal meningitis, which may be complicated by infectious invasion and inflammation of the large arteries at the base of the brain. Cerebral arterial thrombosis may ensue, resulting in large multifocal infarctions. Patients with malignancy also have a higher rate of Varicella Zoster virus reactivation, which may incite a particular granulomatous vasculitis of the small- and medium-sized cerebral vessels, first described in patients with lymphoma. The virus is transported along the axons of affected trigeminal or cervical sensory nerves and causes vasculitis of the innervated blood vessels. Patients present with fever, headache, altered sensorium, and focal deficits with or without a dermal zoster eruption.[1,63] MRI shows multifocal ischemic and hemorrhagic infarctions in the cerebral cortex and in the subcortical white matter. Similar clinical and radiographic findings are seen with cerebral cytomegalovirus infection that plagues severely immunocompromised cancer patients.[1]

7 Complications of cancer treatment

7.1 Chemotherapy

Certain antineoplastic agents have effects on blood vessels or the clotting cascade and may be associated with ischemic strokes and intracranial hemorrhages. L-Asparaginase (L-asp), a component of the regimen used to treat ALL is also associated with ischemic and hemorrhagic stroke. L-Asp-linked cerebral hemorrhage may be cortical or subcortical and thrombosis affects the cerebral arteries or the dural venous sinuses.[64] L-Asp affects the clotting and fibrinolytic cascades by decreasing hepatic production of antithrombin III, and to a more limited degree, the output of fibrinogen, plasminogen, and protein C.[2,65] Fresh frozen plasma does not effectively reverse the L-asp-associated coagulopathy, and prothrombin complex concentrate has not been studied. Instead, infusion of AT III concentrate has been shown to have greater efficacy.[66] High-dose methotrexate and cisplatin rarely cause a characteristic stroke-like syndrome that includes aphasia, alternating hemiparesis, and encephalopathy. Cisplatin has also been implicated with thrombosis causing cortical blindness.[67] Patients with doxorubicin-associated cardiomyopathy may develop intracardiac thrombus, elevating the potential for embolic cerebral infarction. Mitomycin C, cytarabine (ara-C), bleomycin, and methyl-CCNU (lomustine) may trigger microangiopathic hemolytic anemia and cause multifocal hemorrhagic and ischemic strokes.[48] BiCNU (carmustine), when injected intraarterially, may cause a rare necrotizing encephalopathy with hypodensity on a CT scan relevant to the injected vascular territory.[68] Antiestrogenic agents may increase the risk of cerebrovascular disease. Raloxifene carries a risk of stroke and other thromboembolic events.[69] While tamoxifen therapy raises the risk of venous thromboembolic disease, it does not seem to increase the likelihood of ischemic stroke. One study demonstrated that tamoxifen therapy actually tempered the general elevated stroke risk associated with breast cancer chemotherapy.[70] Androgen-deprivation therapy (ADT), utilized to treat prostate cancer, was associated with an increased risk of ischemic stroke in three large-scale studies.[71–73] Other studies have challenged the link between ADT and ischemic stroke while accepting the association between ADT and thromboembolic events.[74]

Bevacizumab (Avastin), a humanized monoclonal antibody directed against vascular endothelial growth factor and functioning as an antiangiogenic agent, is used to treat multiple malignancies, including recurrent glioblastoma multiforme, renal cell carcinoma, colorectal cancer, nonsquamous nonsmall-cell lung cancer, and metastatic breast cancer, and has been reported to cause intracerebral hemorrhage secondary to endothelial dysfunction.[64,68,75] Central nervous system bleeding in patients receiving bevacizumab may be spontaneous or occurs in patients with primary or secondary brain tumors. In a retrospective review of 99 bevacizumab-treated patients with intracranial hemorrhage from the Food and Drug Administration MedWatch database, 71% did not have documented intracranial metastases or primary brain tumors, 30% had hypertension, and 31% had concomitant use of antithrombotic medications or NSAIDs.[76] Bevacizumab-related hemorrhage seems to be rare, and in a meta-analysis, bevacizumab did not increase the risk of intracranial hemorrhage when used in patients with brain metastases despite these patients being excluded from many clinical trials.[75]

7.2 Radiation therapy

Radiation injures the vascular endothelium and causes occlusion of the vasa vasorum, resulting in arterial wall ischemia and subendothelial fibrosis.[77,78] Arteries within previous radiation portals are thus damaged and develop accelerated atherosclerotic disease. Radiation is linked to the formation of inflammatory plaques, which have a greater chance of rupturing and causing an atheroembolic stroke.[79] The vascular effects of radiation may be seen years after treatment. Extracranial and intracranial carotid and vertebral arteries may be affected. Typically, radiation-induced carotid atherosclerotic disease is not limited to the carotid bifurcation. Plaques are usually long and may extend distally or involve the common carotid artery (Fig. 6).[68,70] Classic risk factors for

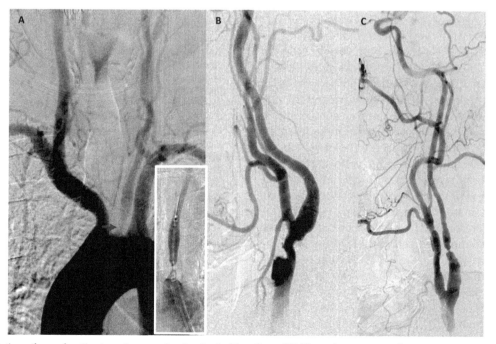

FIG. 6 Postradiation atherosclerotic stenosis occurring in atypical locations. (A) Thoracic aortogram showing a severe concentric stenosis of the left common carotid artery origin (*red arrow*) in a middle-aged woman with a history of radiation therapy to the chest for breast cancer. Inset: Left common carotid angiogram during angioplasty and stenting of the symptomatic lesion shows the tight stenosis more clearly. (B) Left common carotid angiogram, lateral view, showing a symptomatic irregular stenosis of the distal left common carotid artery, proximal to the carotid bifurcation, that was discovered 15 years post radiation therapy for laryngeal cancer. (C) Right common carotid angiogram, lateral view, showing a symptomatic severe stenosis of an irregular atherosclerotic plaque distal to the carotid bulb that was found 15 years post radiation therapy for oropharyngeal cancer.

atherosclerotic diseases such as hypertension, hyperlipidemia, tobacco smoking, and diabetes mellitus elevate the stroke risk associated with radiation-induced carotid atherosclerosis and need to be strictly managed.[77] Angioplasty with stenting is preferred to carotid endarterectomy in the surgical treatment of these lesions.

Intracranial arteries also develop delayed radiation-induced stenoses. When these stenoses are multifocal and of long duration, neovascularization by small friable collateral channels develops to alleviate cerebral ischemia. The angiographic appearance of this neovascularization is analogous to the "puff of smoke" seen in Moyamoya disease. These patients present with transient ischemic attacks, cerebral infarctions, progressive vascular dementia, seizures, and intracerebral hemorrhage. Other delayed vascular consequences of radiation therapy include fusiform or saccular aneurysms in atypical locations, telangiectasias, and cavernomas which may rupture and hemorrhage into the cerebrospinal parenchyma or subarachnoid space.[68,70]

7.3 Hematopoietic stem-cell transplantation

Hematopoietic stem-cell transplantation (HSCT) is a component of some treatment protocols for breast cancer, neuroblastoma, multiple myeloma, leukemia, and lymphoma. Cerebrovascular complications of HSCT are frequent and often have a high mortality rate, with one study showing a 5-year survival rate of 17.9%[80] and another study showing an 8-month median survival rate for post-HSCT patients with cerebrovascular complications.[81] Etiologies are numerous and involve influences of HSCT on coagulation, consequences of graft-versus-host-disease (GVHD) and its treatment, and infections related to a severely immunocompromised state. Precipitants of hemorrhagic stroke include an association of HSCT with thrombosis of the dural venous sinuses, endothelial damage from preconditioning irradiation or conditioning therapy with alkylating agents, infection with *Aspergillus* species, and postlumbar puncture subdural hematoma. Toxicity from cyclosporine used for the prevention of GVHD is associated with both intraparenchymal and subarachnoid bleeding. HSCT is an independent risk factor for NBTE, which results in multifocal ischemic infarcts. Other causes of ischemic cerebrovascular disease associated with HSCT include induction of a hypercoagulable state via deficiency of protein C, antithrombin III, factor VII, and factor XII, and the development of anticardiolipin antibodies. Chronic GVHD is associated with cerebral granulomatous angiitis and patients treated with cyclosporine and methylprednisolone, used to prevent GVHD, may develop MAHA secondary to infection or reactivation of viruses of the herpes family.[82–84]

8 Conclusion

Cancer patients have multiple malignancy-specific risk factors for intracranial thrombotic and hemorrhagic disease. Direct tumor effects, coagulopathies, and opportunistic infections can manifest at any time during the cancer course. A cerebrovascular event may even be the presenting symptom of an occult malignancy. Stroke treatment and secondary prevention include standard cerebrovascular therapy and treating the underlying malignancy. Unfortunately, cancer treatments themselves may cause vessel damage or coagulopathy, both during treatment and as long-term sequelae in cancer survivors. Cancer patients also have a significantly higher mortality rate following an ischemic or hemorrhagic stroke relative to the general stroke population. Only with enhanced stroke prevention and safer and more effective cancer treatment, will the morbidity and high mortality of cerebrovascular events in cancer patients be mitigated.

References

1. Graus F, Rogers LR, Posner JB. Cerebrovascular complications in patients with cancer. *Medicine*. 1985;64:16–35.
2. Zhang YY, Chan DK, Cordato D, Shen Q, Sheng AZ. Stroke risk factor, pattern and outcome in patients with cancer. *Acta Neurol Scand*. 2006;114(6):378–383.
3. Chaturvedi S, Ansell J, Recht L. Should cerebral ischemic events in cancer patients be considered a manifestation of hypercoagulability? *Stroke*. 1994;25(6):1215–1218.
4. Cestari DM, Weine DM, Panageas KS, et al. Stroke in patients with cancer: incidence and etiology. *Neurology*. 2004;62:2025–2030.
5. Kneihsl M, Enzinger C, Wünsch G, et al. Poor short-term outcome in patients with ischaemic stroke and active cancer. *J Neurol*. 2016;263(1):150–156.
6. Jung S, Jung C, Hyoung Kim J, et al. Procedural and clinical outcomes of endovascular recanalization therapy in patients with cancer-related stroke. *Interv Neuroradiol*. 2018;24(5):520–528.
7. Cho BH, Yoon W, Kim JT, et al. Outcomes of endovascular treatment in acute ischemic stroke patients with current malignancy. *Neurol Sci*. 2020;41(2):379–385.
8. Jeon Y, Baik SH, Jung C, et al. Mechanical thrombectomy in patients with acute cancer-related stroke: is the stent retriever alone effective? *J Neurointerv Surg*. 2020. neurintsurg-2020-016144.
9. Little JR, Dial B, Belanger G, Carpenter S. Brain hemorrhage from intracranial tumor. *Stroke*. 1979;10:283–288.
10. Lieu AS, Hwang SL, Howng SL, Chai CY. Brain tumors with hemorrhage. *J Formos Med Assoc*. 1999;98:365–367.
11. Zöller B, Ji J, Sundquist J, Sundquist K. Risk of haemorrhagic and ischaemic stroke in patients with cancer: a nationwide follow-up study from Sweden. *Eur J Cancer*. 2012;48(12):1875–1883.
12. Donato J, Campigotto F, Uhlmann EJ, et al. Intracranial hemorrhage in patients with brain metastases treated with therapeutic enoxaparin: a matched cohort study. *Blood*. 2015;126(4):494–499.
13. Dearborn JL, Urrutia VC, Zeiler SR. Stroke and cancer—a complicated relationship. *J Neurol Transl Neurosci*. 2014;2(1):1039.
14. Gläsker S, Van Velthoven V. Risk of hemorrhage in hemangioblastomas of the central nervous system. *Neurosurgery*. 2005;57(1):71–76.
15. Velander AJ, DeAngelis LM, Navi BB. Intracranial hemorrhage in patients with cancer. *Curr Atheroscler Rep*. 2012;14(4):373–381.

16. Rogers LR. Cerebrovascular complications in patients with cancer. *Semin Neurol.* 2004;24:453–460.

17. Licata B, Turazzi S. Bleeding cerebral neoplasms with symptomatic hematoma. *J Neurosurg Sci.* 2003;47:201–210.

18. Wong AA, Henderson RD, O'Sullivan JD, et al. Ring enhancement after hemorrhagic stroke. *Arch Neurol.* 2004;61:1790.

19. Tung GA, Julius BD, Rogg JM. MRI of intracerebral hematoma: value of vasogenic edema ratio for predicting the cause. *Neuroradiology.* 2003;45:357–362.

20. Atlas SW, Grossman RI, Gomori JM, et al. Hemorrhagic intracranial malignant neoplasms: spin-echo MR imaging. *Radiology.* 1987;164:71–77.

21. Minutoli F, Angileri FF, Cosentino S, et al. 99mTc-MIBI SPECT in distinguishing neoplastic from nonneoplastic intracerebral hematoma. *J Nucl Med.* 2003;44:1566–1573.

22. Reni M, Mazza E, Zanon S, Gatta G, Vecht CJ. Central nervous system gliomas. *Crit Rev Oncol Hematol.* 2017;113:213–234.

23. Albert NL, Weller M, Suchorska B, et al. Response Assessment in Neuro-Oncology working group and European Association for Neuro-Oncology recommendations for the clinical use of PET imaging in gliomas. *Neuro Oncol.* 2016;18(9):1199–1208.

24. Najjar AM, Johnson JM, Schellingerhout D. The emerging role of amino acid PET in neuro-oncology. *Bioengineering (Basel).* 2018;5(4):104.

25. Cheng SY, Nagane M, Huang HS, Cavenee WK. Intracerebral tumor-associated hemorrhage caused by overexpression of the vascular endothelial growth factor isoforms VEGF121 and VEGF165 but not VEGF189. *Proc Natl Acad Sci U S A.* 1997;94:12081–12087.

26. Rubenstein J, Fischbein N, Aldape K, et al. Hemorrhage and VEGF expression in a case of primary CNS lymphoma. *J Neurooncol.* 2002;58:53–56.

27. Grisold W, Oberndorfer S, Struhal W. Stroke and cancer: a review. *Acta Neurol Scand.* 2009;119(1):1–16.

28. Gliemroth J, Nowak G, Kehler U, et al. Neoplastic cerebral aneurysm from metastatic lung adenocarcinoma associated with cerebral thrombosis and recurrent subarachnoid haemorrhage. *J Neurol Neurosurg Psychiatry.* 1999;66:246–247.

29. Ho KL. Neoplastic aneurysm and intracranial hemorrhage. *Cancer.* 1982;50:2935–2940.

30. Zheng J, Zhang J. Neoplastic cerebral aneurysm from metastatic tumor: a systematic review of clinical and treatment characteristics. *Clin Neurol Neurosurg.* 2015;128:107–111.

31. Yamauchi K, Umeda Y. Symptomatic intracranial haemorrhage in acute nonlymphoblastic leukaemia: analysis of CT and autopsy findings. *J Neurol.* 1997;244:94–100.

32. Bruno MC, Santangelo M, Panagiotopoulos K, et al. Bilateral chronic subdural hematoma associated with meningioma: case report and review of the literature. *J Neurosurg Sci.* 2003;47:215–227.

33. Berlis A, Schumacher M, Spreer J, et al. Subarachnoid haemorrhage due to cervical spinal cord haemangioblastomas in a patient with von Hippel-Lindau disease. *Acta Neurochir.* 2003;145:1009–1013.

34. Hentschel S, Toyota B. Intracranial malignant glioma presenting as subarachnoid hemorrhage. *Can J Neurol Sci.* 2003;30:63–66.

35. Nakau R, Kamiyama H, Kazumata K, Andou M. Subarachnoid hemorrhage associated with clival chordoma—case report. *Neurol Med Chir (Tokyo).* 2003;43:605–607.

36. Vates GE, Arthur KA, Ojemann SG, et al. A neurocytoma and an associated lenticulostriate artery aneurysm presenting with intraventricular hemorrhage: case report. *Neurosurgery.* 2001;49:721–725.

37. McIver JI, Scheithauer BW, Rydberg CH, Atkinson JL. Metastatic hepatocellular carcinoma presenting as epidural hematoma: case report. *Neurosurgery.* 2001;49:447–449.

38. Hayashi K, Matsuo T, Kurihara M, et al. Skull metastasis of hepatocellular carcinoma associated with acute epidural hematoma: a case report. *Surg Neurol.* 2000;53:379–382.

39. Reichman J, Singer S, Navi B, et al. Subdural hematoma in patients with cancer. *Neurosurgery.* 2012;71(1):74–79.

40. Katz JM, Segal AZ. Incidence and etiology of cerebrovascular disease in patients with malignancy. *Curr Atheroscler Rep.* 2005;7(4):280–288.

41. McDonald MW, Moore MG, Johnstone PA. Risk of carotid blowout after reirradiation of the head and neck: a systematic review. *Int J Radiat Oncol Biol Phys.* 2012;82(3):1083–1089.

42. Dorresteijn LD, Kappelle AC, Scholz NM, et al. Increased carotid wall thickening after radiotherapy on the neck. *Eur J Cancer.* 2005;41(7):1026–1030.

43. Suárez C, Fernández-Alvarez V, Hamoir M, et al. Carotid blowout syndrome: modern trends in management. *Cancer Manag Res.* 2018;10:5617–5628.

44. Lefkovitz NW, Roessmann U, Kori SH. Major cerebral infarction from tumor embolus. *Stroke.* 1986;17:555–557.

45. Brown DV, Faber LP, Tuman K. Perioperative stroke caused by arterial tumor embolism. *Anesth Analg.* 2004;98:806–809.

46. Raizer JJ, DeAngelis LM. Cerebral sinus thrombosis diagnosed by MRI and MR venography in cancer patients. *Neurology.* 2000;54:1222–1226.

47. Carey MJ, Rodgers GM. Disseminated intravascular coagulation: clinical and laboratory aspects. *Am J Hematol.* 1998;59(1):65–73.

48. Kwaan HC, Gordon LI. Thrombotic microangiopathy in the cancer patient. *Acta Haematol.* 2001;106:52–56.

49. Taccone FS, Jeangette SM, Blecic SA. First-ever stroke as initial presentation of systemic cancer. *J Stroke Cerebrovasc Dis.* 2008;17(4):169–174.

50. Liu J, Frishman WH. Nonbacterial thrombotic endocarditis: pathogenesis, diagnosis, and management. *Cardiol Rev.* 2016;24:244–247.

51. Rogers LR, Cho ES, Kempin S, Posner JB. Cerebral infarction from non-bacterial thrombotic endocarditis: clinical and pathological study including the effects of anticoagulation. *Am J Med.* 1987;83:746–756.

52. Royter V, Cohen SN. Recurrent embolic strokes and cardiac valvular disease in a patient with non-small cell adenocarcinoma of lung. *J Neurol Sci.* 2006;241(1–2):99–101.

53. Singhal AB, Topcuoglu MA, Buonanno FS. Acute ischemic stroke patterns in infective and nonbacterial thrombotic endocarditis: a diffusion-weighted magnetic resonance imaging study. *Stroke.* 2002;33:1267–1273.

54. Lopez JA, Ross RS, Fishbein MC, Siegel RJ. Nonbacterial thrombotic endocarditis: a review. *Am Heart J.* 1987;113(3):773–784.

55. Salem DN, Stein PD, Al-Ahmad A, et al. Antithrombotic therapy in valvular heart disease–native and prosthetic: the Seventh ACCP Conference on Antithrombotic and Thrombolytic Therapy. *Chest.* 2004;126(3 Suppl):457S–482S.

56. Sack Jr GH, Levin J, Bell WR. Trousseau's syndrome and other manifestations of chronic disseminated coagulopathy in patients with neoplasms: clinical, pathophysiologic, and therapeutic features. *Medicine (Baltimore).* 1977;56:1–37.

57. Fugate JE, Lyons JL, Thakur KT, Smith BR, Hedley-Whyte ET, Mateen FJ. Infectious causes of stroke. *Lancet Infect Dis.* 2014;14(9):869–880.

58. Zembower TR. Epidemiology of infections in cancer patients. *Cancer Treat Res.* 2014;161:43–89.

59. Marr KA, Schlamm HT, Herbrecht R, et al. Combination antifungal therapy for invasive aspergillosis: a randomized trial [published correction appears in Ann Intern Med. 2015 Mar 17;162(6):463] [published correction appears in Ann Intern Med. 2019 Feb 5;170(3):220]. *Ann Intern Med.* 2015;162(2):81–89.

60. Gafter-Gvili A, Vidal L, Goldberg E, Leibovici L, Paul M. Treatment of invasive candidal infections: systematic review and meta-analysis. *Mayo Clin Proc.* 2008;83(9):1011–1021.

61. Karthaus M, Cornely OA. Recent developments in the management of invasive fungal infections in patients with hematological malignancies. *Ann Hematol.* 2005;84:207–216.

62. Greenberg RN, Anstead G, Herbrecht R, et al. Posaconazole (POS) experience in the treatment of zygomycosis. In: *43rd ICAAC abstracts*. American Society for Microbiology; September 2003:476. M-1757.

63. Gilden DH, Cohrs RJ, Mahalingam R. VZV vasculopathy and postherpetic neuralgia: progress and perspective on antiviral therapy. *Neurology*. 2005;64:21–25.

64. Saynak M, Cosar-Alas R, Yurut-Caloglu V, Caloglu M, Kocak Z, Uzal C. Chemotherapy and cerebrovascular disease. *J BUON*. 2008;13(1):31–36.

65. Bushman JE, Palmieri D, Whinna HC, Church FC. Insight into the mechanism of asparaginase-induced depletion of antithrombin III in treatment of childhood acute lymphoblastic leukemia. *Leuk Res*. 2000;24:559–565.

66. Hongo T, Okada S, Ohzeki T, et al. Low plasma levels of hemostatic proteins during the induction phase in children with acute lymphoblastic leukemia: a retrospective study by the JACLS. Japan Association of Childhood Leukemia Study. *Pediatr Int*. 2002;44:293–299.

67. Keime-Guibert F, Napolitano M, Delattre JY. Neurological complications of radiotherapy and chemotherapy. *J Neurol*. 1998;245:695–708.

68. Nguyen TD, Abrey LE. Intracranial hemorrhage in patients treated with bevacizumab and low-molecular weight heparin. *Clin Adv Hematol Oncol*. 2007;5(5):375–376.

69. Mosca L, Grady D, Barrett-Connor E, et al. Effect of raloxifene on stroke and venous thromboembolism according to subgroups in postmenopausal women at increased risk of coronary heart disease. *Stroke*. 2009;40(1):147–155.

70. Geiger AM, Fischberg GM, Chen W, Bernstein L. Stroke risk and tamoxifen therapy for breast cancer. *J Natl Cancer Inst*. 2004;96:1528–1536.

71. Robinson D, Garmo H, Lindahl B, et al. Ischemic heart disease and stroke before and during endocrine treatment for prostate cancer in PCBaSe Sweden. *Int J Cancer*. 2012;130(2):478–487.

72. Azoulay L, Yin H, Benayoun S, Renoux C, Boivin JF, Suissa S. Androgen-deprivation therapy and the risk of stroke in patients with prostate cancer. *Eur Urol*. 2011;60(6):1244–1250.

73. Jespersen CG, Nørgaard M, Borre M. Androgen-deprivation therapy in treatment of prostate cancer and risk of myocardial infarction and stroke: a nationwide Danish population-based cohort study. *Eur Urol*. 2014;65(4):704–709.

74. Liao KM, Huang YB, Chen CY, Kuo CC. Risk of ischemic stroke in patients with prostate cancer receiving androgen deprivation therapy in Taiwan. *BMC Cancer*. 2019;19(1):1263.

75. Yang L, Chen CJ, Guo XL, et al. Bevacizumab and risk of intracranial hemorrhage in patients with brain metastases: a meta-analysis. *J Neurooncol*. 2018;137(1):49–56.

76. Letarte N, Bressler LR, Villano JL. Bevacizumab and central nervous system (CNS) hemorrhage. *Cancer Chemother Pharmacol*. 2013;71(6):1561–1565.

77. Anderson NE. Late complications in childhood central nervous system tumour survivors. *Curr Opin Neurol*. 2003;16:677–683.

78. Dorresteijn LD, Kappelle AC, Boogerd W, et al. Increased risk of ischemic stroke after radiotherapy on the neck in patients younger than 60 years. *J Clin Oncol*. 2001;20:282–288.

79. Stewart FA, Hoving S, Russell NS. Vascular damage as an underlying mechanism of cardiac and cerebral toxicity in irradiated cancer patients. *Radiat Res*. 2010;174(6):865–869.

80. Najima Y, Ohashi K, Miyazawa M, et al. Intracranial hemorrhage following allogeneic hematopoietic stem cell transplantation. *Am J Hematol*. 2009;84(5):298–301.

81. Lin TA, Gau JP, Liu YC, et al. Cerebrovascular disease after allogeneic hematopoietic stem cell transplantation: incidence, risk, and clinical outcome. *Int J Hematol*. 2019;109(5):584–592.

82. Krouwer HG, Wijdicks EF. Neurologic complications of bone marrow transplantation. *Neurol Clin North Am*. 2003;21:319–352.

83. Ma M, Barnes G, Pulliam J, et al. CNS angiitis in graft vs host disease. *Neurology*. 2002;59:1994–1997.

84. Belford A, Myles O, Magill A, Wang J, Myhand RC, Waselenko JK. Thrombotic microangiopathy (TMA) and stroke due to human herpesvirus-6 (HHV-6) reactivation in an adult receiving high-dose melphalan with autologous peripheral stem cell transplantation. *Am J Hematol*. 2004;76:156–162.

12

Metabolic and nutritional nervous system dysfunction in cancer patients[*]

Michael N. Youssef[a], Taylor Beal[b], and Jacob Mandel[c]

[a]Department of Neurology, UT Southwestern Medical Center, Dallas, TX, United States, [b]Southern Methodist University, Dallas, TX, United States, [c]Department of Neurology, Baylor College of Medicine, Houston, TX, United States

1 Introduction

Metabolic and nutritional disorders are common in cancer patients and can result from organ damage secondary to tumor metastases (e.g., liver, kidney, and bone), chemotherapy, drugs used for cancer treatment and symptom control, secretion of tumor-derived substances that impair systemic organ function, malnutrition, and infection.[1-4] In a seminal study from Memorial Sloan-Kettering Cancer Center (MSKCC), metabolic and nutritional disorders accounted for over 10% of all diagnoses in a series of 851 patients seen by the neurology consult service.[2] The spectrum of disease that can occur in this setting is very broad and may involve dysfunction of the brain, spinal cord, peripheral nerves, muscle, or various combinations (see Table 1). The signs and symptoms often develop insidiously and are typically not accompanied by focal signs. For example, patients with metabolic encephalopathy are usually symptomatic for a day or two before they become confused and agitated enough to warrant consultation. Therefore, an extensive workup is required in most patients to clarify the etiology of the underlying metabolic and/or nutritional disorder.

2 Metabolic and toxic encephalopathy

Metabolic and toxic encephalopathy is diagnosed in cancer patients with delirium when the workup shows an absence of structural disease within the central nervous system (CNS) (see Table 2).[3,4] Structural diseases that would need to be ruled out include brain metastases, leptomeningeal metastases, hydrocephalus, intracranial hemorrhage, stroke, and other forms of ischemia. Metabolic encephalopathy is usually noted in the context of "mental status changes" or a "confusional syndrome" and is very common in cancer patients.[2-4] In the MSKCC study, it was the most common cause of altered mental status, present in 61% (80 of 132) of the cohort.[2] However, it should be noted that in 5.3% of patients in the MSKCC study, metabolic encephalopathy and structural disease of the CNS were present concomitantly when it was determined that reversible metabolic factors (e.g., hyponatremia) had contributed to a change in the level of alertness. In a similar study of 162 inpatients at the Johns Hopkins Oncology Center, a change in mental status was noted in 16% of the cohort.[5] Among this group of patients, the underlying cause was determined to be metabolic in 6%.

The metabolic disturbance is due to a single cause in 30%–35% of patients with metabolic and toxic encephalopathy.[6,7] In another study from MSKCC, a single cause of delirium was present in approximately one-third of patients (see Table 3). Among this group of 140 patients, the most common metabolic and toxic etiologies included poor oxygenation, drug intoxications, and organ failure. In a similar study of 229 elderly inpatients in a general hospital, it was noted that 22% were delirious.[7] A definite single cause could be identified in 36% of this group, while another 20% had a single cause that was considered to be probable. In the two-thirds of patients that have multifactorial causes of encephalopathy, the most common factors were metabolic in origin, including drug intoxications, organ failure, electrolyte disturbances, and infections (see Table 4).[6,7]

Metabolic and toxic encephalopathy is the result of abnormal brain function that is due to interference with

---

[*] Adapted from Version 1 by Herbert Newton, MD, Advent Health, Orlando, FL.

TABLE 1 Spectrum of nervous system complications in cancer patients related to metabolic and nutritional disorders.

Brain

    Diffuse encephalopathy

    Mild (e.g., mainly lethargy and/or agitation)

    Moderate to severe

    Focal encephalopathy

Spinal cord

    Nutritional myelopathy

Peripheral nerves

    Polyneuropathy

    Focal neuropathy

Muscle and/or neuromuscular junction

    Nutritional myopathy

Data derived from Spinazze S, Schrijvers D. Metabolic emergencies. Crit Rev Oncol Hematol. 2006;58(1):79–89; Clouston PD, DeAngelis LM, Posner JB. The spectrum of neurological disease in patients with systemic cancer. Ann Neurol. 1992;31(3):268–273; Posner JB. Metabolic and nutritional complications of cancer. In: Posner J, ed. Neurologic Complications of Cancer. Philadelphia, PA: F.A. Davis Company; 1995:264–281; Boerman RH, Padberg GW. Metabolicnervous system dysfunction in cancer. In: Vecht CJ, ed. Handbook of Clinical Neurology. Amsterdam: Elsevier Medical Publishers; 1997:395–412.

brain metabolism by an extracerebral factor.[2–4] It is important to make the proper diagnosis and identify all of the contributing factors since many of them may be reversible and could lead to normalization or improvement of neurological function. Cancer patients with disseminated disease or who are older than 60 years of age are more likely to develop a metabolic and toxic encephalopathy.[6] The occurrence of encephalopathy in a young cancer patient is a poor prognostic sign for survival since it is presumed that more severe systemic disease is necessary to induce delirium in a younger brain.

Patients with metabolic and toxic encephalopathy demonstrate cognitive and behavioral changes that are quite variable in their features and severity (see Table 5).[1–6] The severity of the metabolic encephalopathy correlates with the metabolic disorder that caused it.[8] In most patients, the initial symptoms are very mild and slowly progressive and remain undetected by family and medical staff. At this stage, patients exhibit mild deficits in concentration and the ability to sustain attention. They often appear apathetic or disinterested in their medical care, with poor eye contact, and may not understand explanations of medical procedures or test results. As the encephalopathy worsens, patients often become more apathetic, with pronounced lethargy, drowsiness, and inability to attend to external stimuli. Restlessness, insomnia, other disturbances of the sleep/wake cycle, and nightmares are fairly common. A crucial step in diagnosing delirium is to determine the patient's baseline mental status; this often

TABLE 2 Metabolic and toxic encephalopathies in cancer patients.

Organ failure	Respiratory dysfunction
	Hypoxemia
	Hypercapnia
	Liver
	Hepatic failure (hyperammonemia)
	Kidneys
	Uremia
Electrolytes	Hypercalcemia
	Hyponatremia
	Other electrolyte disorders
Hypoglycemia	Diabetes
	Terminal cancer
	Paraneoplastic
	Increased consumption by neoplasm
	Drugs
Hyperglycemia	Diabetes
	Corticosteroids
Vitamin deficiency	Vitamin B1
	Vitamin B12
Endocrine disorders	Adrenal
	Thyroid
Septic encephalopathy	
Drug reactions	Chemotherapy
Intoxication	Anticonvulsants
	Morphine
	Sedatives
	Neuroleptics
	Antidepressants
	Manganese intoxication

Data derived from Spinazze S, Schrijvers D. Metabolic emergencies. Crit Rev Oncol Hematol. 2006;58(1):79–89; Clouston PD, DeAngelis LM, Posner JB. The spectrum of neurological disease in patients with systemic cancer. Ann Neurol. 1992;31(3):268–273; Posner JB. Metabolic and nutritional complications of cancer. In: Posner J, ed. Neurologic Complications of Cancer. Philadelphia, PA: F.A. Davis Company; 1995:264–281; Boerman RH, Padberg GW. Metabolicnervous system dysfunction in cancer. In: Vecht CJ, ed. Handbook of Clinical Neurology. Amsterdam: Elsevier Medical Publishers; 1997:395–412.

involves obtaining a history from a knowledgeable informant (such as the primary caregiver of the patient).[9] The Confusion Assessment Method (CAM) is a useful screening instrument to verify the prescience of delirium.[10] CAM was designed to be used along with the Mini-Mental State Examination (MMSE).[10,11] The MMSE can be administered by nonmedical personnel

TABLE 3 Single causes of delirium in 140 patients with cancer.

Poor oxygenation	11
Hypoxia	2
Hypoperfusion (shock)	6
DIC	3
Organ failure	3
Liver	2
Kidneys	1
Hyperosmolarity	2
Brain metastases	19
Other focal lesions (e.g., meningitis, infarction)	2
Drugs intoxication (e.g., steroids, opioids)	6
Total	43

Data adapted from Tuma R, DeAngelis LM. Acute encephalopathy in patients with systemic cancer (abstract). Ann Neurol. 1992;32:288.

TABLE 4 Etiologies involved in multifactorial delirium in 140 patients with cancer.

Etiology	% of patients
Drugs	59
Organ failure	51
Fluid electrolyte imbalance	45
Infection	45
Hypoxia	35
Brain lesions	21
Environment	21

Data adapted from Tuma R, DeAngelis LM. Acute encephalopathy in patients with systemic cancer (abstract). Ann Neurol. 1992;32:288.

TABLE 5 Clinical features in 140 patients with cancer and altered mental status.

Clinical features	% of patients
Consciousness	
Lethargic	52
Alert	46
Comatose	2
Cognition	
Inattention	92
Dyscalculia	92
Memory impairment	91
Disorientation	83
Language impairment	35
Behavior	
Agitation	44
Delusions and/or hallucinations	28
Motor behavior	
Focal findings	40
Asterixis	36
Seizure activity	9

Data adapted from Tuma R, DeAngelis LM. Acute encephalopathy in patients with systemic cancer (abstract). Ann Neurol. 1992;32:288.

and has a sensitivity and specificity (0.90 and 0.80, respectively) that compares well to the judgment of a neurologist or psychiatrist. In some patients, encephalopathy is severe enough to result in varying degrees of reduced consciousness, such as obtundation or coma. In the previously noted study from MSKCC by Tuma and DeAngelis, more than half of the cohort was lethargic; agitation and delusions and/or hallucinations were also noted in 44% and 28%, respectively (see Table 5).[6] Coma was an uncommon finding present in only 2% of patients. Cognitive features such as impaired memory, inattention, and dyscalculia were very frequent and noted in greater than 90% of patients. In addition, the disorientation was also quite common and present in 83% of the cohort, while disturbances of language were only noted in approximately one-third. Although generally considered to be uncommon in patients with encephalopathy, focal motor abnormalities can occur. In fact, focal motor signs were present in 40% of the cohort in the MSKCC study.[6] Motor signs that have been described include flaccidity, tremor, decerebrate posturing, hemiparesis, asterixis (unilateral or bilateral), myoclonus, flaccid paralysis, and seizure activity (focal, generalized, or nonconvulsive status).[3,4,6] In most cases of encephalopathy or delirium, the pupils remain small but reactive, with intact oculovestibular reflexes. Focal neurological symptoms can originate either from hemispheres or the brain stem.[8] The hemispheric symptoms consist of vision disorders, apraxia, aphasia, hemispasticity, hemiataxia, and hemisensory syndrome.[8] Symptoms of brain stem lesions include cranial nerve signs, pathological brain stem reflexes, dysarthria, dysphagia, ataxia, hemi- or quadriparesis, and various sensitive end-respiratory disorders.[8]

The workup for patients with metabolic and toxic encephalopathy is extensive and will include imaging and various laboratory tests (see Table 6).[1,3,4] The most important initial test is a magnetic resonance imaging (MRI) scan of the brain, with and without contrast, to screen for brain metastases, meningeal enhancement, abscesses, and other forms of infection. The basal ganglia, thalamus, cerebral cortex, and hemispheric

TABLE 6   Laboratory evaluation for metabolic and toxic encephalopathy.

MRI scan (with and without contrast)

   Metastases

   Abscess

   Infection

Lumbar puncture

   Leptomeningeal tumor

   Infection

Electroencephalography

Blood cultures

   Sepsis

   Septic emboli

Blood gases—pO_2, pCO_2, pH

Electrolytes—Na, K, Ca, Mg, PO_4

Complete blood count

Renal panel

Liver panel, including NH_3

Lactate

Coagulation profile

Endocrine tests—T4, cortisol

Glucose

Vitamin B12, folic acid, thiamine

Drug levels—anticonvulsants, theophylline, digoxin, etc.

Data derived from Spinazze S, Schrijvers D. Metabolic emergencies. Crit Rev Oncol Hematol. 2006;58(1):79–89; Clouston PD, DeAngelis LM, Posner JB. The spectrum of neurological disease in patients with systemic cancer. Ann Neurol. 1992;31(3):268–273; Posner JB. Metabolic and nutritional complications of cancer. In: Posner J, ed. Neurologic Complications of Cancer. Philadelphia, PA: F.A. Davis Company; 1995:264–281; Boerman RH, Padberg GW. Metabolic nervous system dysfunction in cancer. In: Vecht CJ, ed. Handbook of Clinical Neurology. Amsterdam: Elsevier Medical Publishers; 1997:395–412.

white matter are typically the target of toxic or metabolic encephalopathy.[8] A lumbar puncture will also be necessary for many patients to further rule out infection and to screen for the presence of leptomeningeal tumor spread. An electroencephalogram (EEG) is helpful to evaluate patients for seizures and nonconvulsive status epilepticus. Other common blood tests that should be ordered include complete blood count, blood cultures and gases, electrolytes, renal and liver panels, ammonia, lactate, coagulation profile, endocrine testing, glucose, vitamin levels (B12, folate, and thiamine), and appropriate drug levels (e.g., digoxin, anticonvulsants).

The following sections will review the most common causes for metabolic and toxic encephalopathy in cancer patients (see Table 2).

2.1 Organ failure

Metastases or primary tumors in the lungs and liver, as well as metastatic obstruction of the urinary drainage system and other forms of acute renal failure, can lead to metabolic encephalopathy.[1,3,4] Lung metastases are a frequent complication of systemic cancer and can occasionally lead to respiratory failure with hypoxemia and hypercapnia. They are most commonly noted in sarcomas and various carcinomas and are the sole site of metastases in 80% and 5%–10% of these patients, respectively.[12] Most lung metastases are asymptomatic or cause regional pain in the chest wall, with or without hemoptysis. However, patients with extensive disease may have chronic hypoxia and hypercapnia, resulting in drowsiness, lethargy, confusion, morning headache, tremor, and myoclonus. An underlying pneumonia may also contribute to hypoxemia.

Liver metastases are most often noted in patients with colorectal cancer, with 20% present at diagnosis and another 40% that arise later in the course of diseases.[3,13] Other solid tumors that have a predilection for liver metastases include melanoma (24%), lung cancer (15%), and breast cancer (4%). The liver maintains normal function until there is widespread, diffuse parenchymal damage from metastatic deposits—usually in the context of a terminal disease. Hepatic encephalopathy occurs due to a combination of pathophysiological mechanisms such as inflammation, oxidative stress, impaired blood-brain barrier permeability, neurotoxins, and impaired energy metabolism of the brain.[14] As the liver begins to fail, the patient begins to develop the symptoms of mild hepatic encephalopathy, with irritability, restlessness, confusion, and clouding of the sensorium.[13,15] Other common symptoms include asterixis, tremor, and myoclonus. In some patients, dysconjugate gaze, skew deviation, ocular bobbing, and other neuro-ophthalmological abnormalities can be noted. As the encephalopathy deepens, some patients may become obtunded or comatose. Rarely, the liver failure can be acute or subacute and be associated with hyperammonemia, cerebral edema, and increased intracranial pressure.[16]

Bilateral obstruction of the urinary system leading to uremia is rare from primary tumors of the kidney and ureters, but can be noted more often in pelvic tumors arising from the ovaries, uterus, bladder, prostate, and retroperitoneum (e.g., sarcomas, metastases).[4,17] Other causes for acute renal failure and severe uremia in cancer patients include nephrotoxic chemotherapy, tumor lysis syndrome, elevated uric acid, sepsis, and contrast agent nephropathy.[18] Failure to excrete metabolic products leads to their accumulation which may cause severe intoxication.[19] Symptoms of metabolic encephalopathy worsen as the uremia progresses from mild to severe. Severe uremia is uncommon as the sole cause of metabolic encephalopathy and is more often only one component in a patient with multiorgan failure. Impaired kidney blood flow can deplete levels of intracellular ATP which disrupts intracellular calcium homeostasis.[20]

2.2 Electrolyte abnormalities

Imbalance of the electrolytes is a relatively common cause of metabolic encephalopathy in cancer patients, especially disorders of Ca^{2+} and Na^{+}.[1,3,4] Arterial blood gas analysis is used to evaluate the electrolyte status of patients; based on this information, clinicians can focus on the detection of the underlying etiology of encephalopathy.[8] Hypercalcemia is a frequent complication, occurring in 8%–10% of all patients with a malignant disease.[1,21,22] It is caused by one or more of the following mechanisms: mobilization from bone via increased absorption or osteoclastic activity, increased renal reabsorption, increased intestinal absorption, decreased renal excretion, or production of vitamin D, cytokines, or parathyroid-like hormones by the tumor. Hypercalcemia is most often associated with bone metastases, typically from breast cancer, lung cancer, multiple myeloma, leukemia, and lymphoma. Calcium is released from the bone through the activity of osteoclasts stimulated by humoral factors secreted by the tumor tissue. On occasion, occult tumors may present with hypercalcemia. The calcium level is graded as mild (< 3.0 mg/dL), moderate (3.0–3.5 mg/dL), or severe (> 3.5 mg/dL). Levels in the severe range are considered a medical emergency and require urgent treatment.[1] The symptoms of hypercalcemia are similar to other metabolic encephalopathies; there is no linear relationship between serum calcium levels and the severity of the mental status changes. The most common symptoms are apathy, confusion, disorientation, and proximal muscle weakness; rarely, seizures can be noted. Management of hypercalcemia includes parenteral hydration with saline and treatment with bisphosphonate or calcitonin.[10] Calcitonin is a hormone that reduces blood calcium levels. Hypocalcemia is uncommon in cancer patients, but can cause encephalopathy and proximal muscle weakness.[1,3] Hyponatremia is very common in cancer patients and is usually dilutional and of the hypoosmolar type.[1,23,24] It typically occurs in the setting of the syndrome of inappropriate secretion of antidiuretic hormone (SIADH), but can have other underlying causes (e.g., renal dysfunction, drugs, glucocorticoid deficiency). The reduction in sodium occurs slowly in most patients (i.e., chronic), but can be acute or subacute. Rapid or excessive hyponatremia will cause water to shift from the body into the brain, with resultant brain edema. For example, rapid reduction of serum sodium to less than 130 mEq/L will result in encephalopathy, headaches, nausea, and emesis. If the serum sodium is further reduced to 120 mEq/L or less, obtundation, coma, and seizures can occur. Initial treatment of mild-to-moderate hyponatremia is a fluid restriction; the principal drawback is that patients with hyponatremia find it difficult to restrict fluids as thirst is inappropriately high from a downward resetting of the osmotic thirst threshold.[25] Chronic hyponatremia, with slow lowering of the serum sodium, even to levels below 115 mEq/L, is usually associated with mild symptoms and an uneventful recovery with treatment. In cancer patients, SIADH is common and can be induced by excessive secretion of antidiuretic hormone or similar substances, as well as from fluid overload during chemotherapy, pneumonia, and various CNS lesions. It is most often noted in patients with small cell lung cancer and carcinoid. Certain patient-specific risk factors are predictive of a poor outcome and are important to identify; gender (premenopausal and postmenopausal females), age (children), and the presence of hypoxia are the three main clinical risk factors and are more predictive of a poor outcome than the rate of development of hyponatremia.[26] Treatment of severe hyponatremia should be performed carefully (i.e., slowly; at rates of 1–2 mEq/L/h or less) since overly rapid correction can result in CNS demyelination, especially in the pons (i.e., central pontine myelinolysis).[1,4,27] Vaptans, which block vasopressin type 2 receptors, have been shown by several clinical trials to be effective in increasing serum sodium levels in patients with SIADH.[28] In the MSKCC study of acute delirium in cancer patients, hyponatremia was not the primary cause of mental status changes in any of the cohorts.[6] However, it was considered to be a contributing factor in a significant percentage of patients with multifactorial encephalopathy. The severity of delirium is often more pronounced when hyponatremia occurs in patients with concomitant structural brain disease. Hypernatremia is an occasional cause of encephalopathy in cancer patients.[1,3,4] It can be seen in patients with altered thirst mechanisms from brain disease (i.e., metastases), prolonged fever, poor tolerance for swallowing liquids, and inadequately replaced fluid losses (e.g., chemotherapy-induced diarrhea). Diabetes insipidus, secondary to pituitary gland damage from metastases, has been noted in patients with breast cancer. Hypomagnesemia is rare in cancer patients and most often occurs in the context of chemotherapy with cis-platinum.[1,3,4] It is usually asymptomatic, although seizures have been reported. Magnesium is necessary for axon stabilization, and magnesium deficiency lowers the threshold for axon stimulation causing increased nerve conduction velocity with hypomagnesemia.[29] Other potential causes of hypomagnesemia include decreased oral intake, impaired intestinal absorption, and intravenous hyperalimentation. Hypophosphatemia, when it is mild, is relatively common in cancer patients and can arise from decreased dietary intake, reduced intestinal absorption, or intracellular shifts of phosphorus from respiratory alkalosis.[1,3,4] Symptoms of hypophosphatemia can be confused with an underlying iron deficiency.[30] More severe hypophosphatemia is infrequent but has been reported in the setting of rapidly proliferating tumor burden (e.g., leukemia and lymphoma), oncogenic osteomalacia, and intravenous hyperalimentation. A mild reduction in phosphorus is usually not implicated in delirium unless the patient has other electrolyte and metabolic disturbances. *Hypocalcemia* and *hyperkalemia* are not specifically related to cancer patients.[1,4]

2.3 Hypoglycemia

Hypoglycemia is an uncommon cause of metabolic encephalopathy in cancer patients.[1,3,4,31–33] However, since the symptoms can be quite variable, they should be included in the differential diagnosis and metabolic evaluation of these patients. The symptoms can range from mild confusion to frank coma, depending on the glucose level and how rapidly it has fallen. Other possible signs include agitation, alterations of affect, dizziness, tremor, seizure activity, and focal deficits such as hemiparesis. The underlying cause of hypoglycemia in cancer patients is quite broad and often is related to common comorbidities like diabetes mellitus.[31–33] Diabetic cancer patients are still at risk for hypoglycemia caused by diabetic therapy. For example, hypoglycemia may develop if the insulin dose has not been properly adjusted in patients with inadequate food intake. Hypoglycemia can also be noted in patients with primary or metastatic tumors of the liver, with secondarily reduced glucose synthesis. Metastatic tumors to the adrenal glands or the pituitary gland may result in adrenal or growth hormone deficiency, which can interfere with the stimulation of hepatic glucose production. Insulinomas are tumors of the pancreatic beta cells that secrete insulin and typically cause severe hypoglycemia, even during fasting conditions.[34,35] These tumors are often small and may be difficult to diagnose. Hypoglycemia will not improve until the tumor has been resected. Other tumors can produce insulin-like growth factors (e.g., hepatocellular carcinoma) and antiinsulin antibodies (e.g., multiple myeloma), which can also induce severe hypoglycemia.[1,4] On occasions, large tumors in the retroperitoneum and abdomen can utilize large amounts of serum glucose, resulting in intermittent hypoglycemia and symptoms such as confusion, hunger, disorientation, and occasional obtundation or coma. Hypoglycemia can also arise in some terminal cancer patients, either from poor appetite and reduced food intake or because of alterations in carbohydrate metabolism.[31] Certain drugs are also known to occasionally cause hypoglycemia in cancer patients.[4,31] The most common drugs to cause this complication include pentamidine, β-adrenergic blocking agents, sulfonylureas, aspirin, and alcohol.

2.4 Hyperglycemia

Hyperglycemia is an infrequent cause of encephalopathy in cancer patients unless there is uncontrolled primary or secondary diabetes mellitus.[1,3,4] A diffuse encephalopathy can occur if the elevated glucose level becomes excessive (600–800 mg/dL or higher), with some patients progressing into *nonketotic hyperosmolar coma*. The mechanism of hyperosmolar coma appears to be cellular dehydration of the brain, due to the severe diuresis from elevated glucose levels. Secondary diabetes can arise in patients placed on long-term corticosteroids (e.g., dexamethasone for the control of edema from brain metastases). Other drugs (i.e., phenytoin) can also induce hyperglycemia in rare patients.

2.5 Vitamin deficiency

Vitamin deficiencies can occur in cancer patients as a result of prolonged emesis, malnutrition, and conditions affecting proper intestinal absorption (e.g., irradiation).[3,4] The most well-characterized conditions are Wernicke-Korsakoff syndrome, from vitamin B1 deficiency, and vitamin B12 deficiency. *Wernicke's encephalopathy* is a neurological emergency with features of acute mental status changes and confusion, ophthalmoplegia, and ataxia.[36] Other possible signs include nystagmus, autonomic insufficiency, and postural hypotension. Occasionally, the symptoms will not include the entire "classic triad" or may have a more protracted onset. Without rapid treatment, patients continue to worsen and develop an irreversible amnestic syndrome—Korsakoff's psychosis. Neuroimaging of the brain with MRI may demonstrate T2 and FLAIR high signal abnormalities in the medial thalami and periaqueductal gray matter but can be normal.[36,37] Cancer patients are at high risk for Wernicke's encephalopathy due to chronic malnutrition, persistent emesis from chemotherapy, consumption of thiamine by rapidly growing tumors, frequent diarrhea, and intestinal malabsorption from chemotherapy and radiation-induced intestinal damage. In addition, the onset of symptoms can be precipitated in patients placed on parenteral nutrition or glucose infusions without adequate replacement of thiamine.[3,4,37] A high index of suspicion is necessary for any malnourished cancer patient with encephalopathy and nystagmus since the disorder is potentially reversible with early treatment. Therefore, until the workup is completed and levels of vitamin B1 are available, empiric treatment with parenteral vitamin B1 is recommended. *Vitamin B12 deficiency* has been reported as an occasional complication in cancer patients.[3,4] Like vitamin B1 deficiency, it is most likely in patients with intestinal dysfunction and malabsorption syndromes induced by cancer and/or its treatment. For example, vitamin B12 deficiency has been reported at a high rate in patients with bladder cancer that have undergone pelvic radiotherapy.[38] It is theorized that the irradiation caused damage to the terminal ileum, thereby interfering with vitamin B12 absorption. The cardinal clinical features of vitamin B12 deficiency include megaloblastic anemia and various neuropsychiatric syndromes, which can occur in up to 40% of patients.[39,40] The most prominent neurological signs are peripheral neuropathy and corticospinal tract damage; encephalopathy and confusional states are uncommon but can occur. A sense of weariness and mild muscular weakness are also reported.

2.6 Endocrine disorders

Endocrine disorders are an uncommon cause of metabolic encephalopathy in cancer patients and usually involve dysfunction of the thyroid or adrenal glands.[1,4] Hypothyroidism can occur after irradiation to the head and neck for regional cancers and for patients with Hodgkin's disease. The most common symptoms include encephalopathy, ataxia, and peripheral neuropathy. The symptoms can arise at any time following radiotherapy and may not be accompanied by other systemic signs of the disease. Thyroid function studies should be evaluated in any patient with encephalopathy who has received prior irradiation to the neck. Adrenal insufficiency can be noted in patients with metastases to the adrenal gland or hypothalamus or after withdrawal from prolonged corticosteroid treatment.[41] Metastases to the adrenal glands result in a loss of glucocorticoids and mineralocorticoids, which presents as a mild but progressive delirium followed by cardiovascular collapse.

2.7 Septic encephalopathy

Although the underlying mechanisms remain incompletely understood, it is well known that sepsis is a frequent cause of metabolic encephalopathy.[1,3,4,42] The most likely cause for CNS dysfunction is an imbalance in the "inflammatory network," with excessive production of pro-inflammatory cytokines, including tumor necrosis factor-α (TNF-α), interleukin-1β, interleukin-6, bradykinin, and various leukotrienes.[43–45] Another separate or contributing mechanism is the alteration of the blood-brain barrier via sepsis-induced inflammation, which might allow entry of normally excluded neurotoxic substances.[46] Mental status changes and encephalopathy are noted in up to 70% of patients with sepsis and can occur as the sole cause or, more frequently, as a contributing cofactor in a multifactorial delirium.[42] The symptoms of encephalopathy may precede the onset of fever or begin simultaneously. In rare cases of severe sepsis, delirium may be noted in patients with hypothermia. Most patients become confused and agitated, with decreased alertness and attention. Other signs such as

tremor, asterixis, myoclonus, generalized and focal seizures, and paratonic rigidity can be noted. Rarely, focal neurologic signs such as hemiparesis and oculomotor palsies can occur. The electroencephalographic (EEG) record is typically abnormal, demonstrating slowing of the background and prominent delta wave activity, with excessive theta activity and occasional triphasic waves. Septic encephalopathy is most common with infections from gram-negative organisms but can occur during sepsis from any type of bacteria. The diagnosis requires exclusion of other common causes of delirium, including brain metastases, leptomeningeal metastases, bacterial meningitis, and encephalitis. Following treatment with appropriate antibiotics, symptoms of septic encephalopathy will typically resolve. Sepsis-related seizures will usually not require long-term treatment with anticonvulsants.

2.8 Drug reactions

In the previously cited study by Tuma and DeAngelis from MSKCC, it was noted that drugs of different varieties were major factors in the causation of delirium in 59% of their cohort.[6] Many of these patients were intoxicated from opioid pain medications, anticonvulsants, neuroleptics, and similar medications (see next Section). Others were receiving single-drug or multiagent chemotherapy and developing encephalopathy as part of a reaction to the therapeutic regimen. Encephalopathy and confusional syndromes are very common complications of chemotherapy agents and have been reported for numerous drugs (see Table 7).[47–49] Most often, the delirium occurs in the setting of high-dose intravenous (IV) therapy but can also be noted in patients after intrathecal (IT) and intra-arterial (IA) drug administration. The drugs that are most likely to cause encephalopathy include methotrexate, cisplatin (IA>IV), 5-fluorouracil, capecitabine, ifosfamide, L-asparaginase, BCNU (IA≫IV), and biological agents such as the interferons and interleukins.

2.9 Drug intoxications

The drugs that most commonly cause toxic encephalopathy in cancer patients are listed in Table 2 and include opioid analgesics, anticonvulsants, sedatives, neuroleptics, and antidepressants.[3,4] The delirium can be caused by an overdose or a withdrawal reaction from some of these drugs (e.g., opioids, sedative/hypnotics). Mental status changes from drug overdosage are characterized by a passive delirium, with drowsiness, poor attention, disorientation, and somnolence. During drug withdrawal, the delirium has more prominent tremulousness, hallucinations, and possible seizures. In some cases, the symptoms of drug overdosage and withdrawal can overlap and appear very similar. Furthermore, some

TABLE 7   Chemotherapy drugs associated with encephalopathy.

Methotrexate

5-Fluorouracil

Capecitabine

Cisplatin (IA>IV)

BCNU (IA≫IV)

Vinca alkaloids

Mechlorethamine

Procarbazine

Levamisole

Cytosine arabinoside (Ara-C)

Fludarabine

Gemcitabine

Hydroxyurea

Pentostatin

Chorambucil

Thiotepa (high dose)

Ifosfamide

Hexamethylmelamine

Cyclophosphamide

Etoposide

Paclitaxel

Docetaxel

Pyrazoloactidine (PZA)

Doxorubicin (IT administration)

Nitroimidazole

Plicamycin

Thalidomide

Interferon

Interleukin-2

Interleukin-3

Interleukin-6

Interleukin-11

Tumor necrosis factor-α (TNF-α)

Tamoxifen

Mitotane

L-Asparaginase

Abbreviations: *IA*, intra-arterial; *IV*, intravenous; *IT*, intrathecal.
Data derived from Cavaliere R, Schiff D. Neurologic toxicities of cancer therapies. Curr Neurol Neurosci Rep. 2006;6(3):218–226; Hammack JE. Neurologiccomplications of chemotherapy and biologic therapies. In: Schiff D, O'Neill BP, eds. Principles of Neuro-Oncology. New York: McGraw-Hill Medical Publishing Division; 2005:679–709; Newton HB. Intra-arterial chemotherapy. In: Newton HB, ed. Handbook of Brain Tumor Chemotherapy. Amsterdam: Elsevier/Academic Press; 2006:247–261.

patients can be sensitive to the effects of certain drugs because of advanced age or the presence of other metabolic abnormalities and become delirious at low or "nontoxic" doses. Cancer-related pain is a very common problem in oncology and many patients remain undertreated.[50,51] Opioid analgesics are the mainstay of treatment for moderate-to-severe cancer pain and are the class of drugs most often associated with delirium in cancer patients.[51,52] Opioid intoxication usually induces a "quiet," diffuse encephalopathy that may progress to a deeper state of obtundation or coma. More severely affected patients may also exhibit seizures, symmetric pinpoint reactive pupils, myoclonus, hypothermia, and hypoventilation. Any cancer patient with severe lethargy or coma that has been receiving opioids should be treated with an opioid antagonist, such as naloxone (0.4 mg in 10 mL of saline over several minutes). The resultant improvement in alertness may be brief since naloxone is a short-acting drug. More persistent results can be achieved by repeated injections of naloxone or using a slow naloxone drip. Opioid withdrawal can also cause encephalopathy, along with anxiety, restlessness, chills, insomnia, nausea, anorexia, abdominal cramping, emesis, dilated pupils, tachycardia, and mild hyperthermia.[3,52] Meperidine should not be used for cancer pain since repeated usage will result in the accumulation of a toxic metabolite, normeperidine, and the onset of delirium, multifocal myoclonus, and seizures. Other drugs, such as benzodiazepines, neuroleptics, antidepressants, and anticonvulsants are rarely the primary cause of toxic encephalopathy in cancer patients.[3,4] More often, they are contributors to a multifactorial delirium induced by multiple drugs and other metabolic derangements. The measurement of drug levels can often be helpful in this situation. Rarely, the use of herbal medicine by cancer patients can lead to a toxic encephalopathy.[4,53] For example, some Chinese herbal medicines can result in manganese intoxication, with encephalopathy and extrapyramidal motor signs.

3 Tumor lysis syndrome

Tumor lysis syndrome (TLS) is an oncological emergency that results from massive cytolysis of malignant cells after treatment, with sudden release of intracellular contents into the bloodstream at high concentrations.[1,54,55] It occurs most often in patients with large tumor burdens, usually with hematological malignancies such as high-grade lymphoma and leukemia. TLS can also occur in solid tumors but is much less common.[54] The cardinal features include hyperuricemia, hyperkalemia, hyperphosphatemia, and hypocalcemia; less often, hyperammonemia can also be noted. Hyperuricemia does not have any direct effects on the

CNS but can induce acute renal failure and uremia (i.e., uric acid nephropathy), which can lead to encephalopathy as noted above. Similarly, hyperphosphatemia can also lead to acute renal failure, uremia, and mental status changes. Hypocalcemia, which occurs as a result of the hyperphosphatemia, can cause muscle cramping, muscle tetany, and cardiac arrhythmias. Severe and life-threatening cardiac arrhythmias are the most common complication of hyperkalemia. Hyperammonemia can lead to delirium and seizure activity (similar to hepatic encephalopathy), especially if the increase in ammonia level occurs rapidly. Hemodialysis should be implemented early for any patient with TLS that exhibits excessive uric acid, phosphate, potassium, or acute renal failure.[55] In patients with low-to-intermediate risk of developing TLS, prophylaxis with monitoring, hydration, and allopurinol is recommended.[56] Specifically, in low-risk patients, allopurinol is given only when there are signs of metabolic changes, bulky and/or advanced disease, or highly proliferative disease. In patients with a high risk of developing TLS, prophylaxis with monitoring, hydration, and rasburicase is recommended.[56] Here, rasburicase is given once and given again only if clinically indicated and is contraindicated in patients with a glucose-6-phosphate dehydrogenase deficiency. In these patients, allopurinol is used instead of rasburicase.[56]

4 Cancer anorexia-cachexia syndrome

The cancer anorexia-cachexia syndrome (CACS) is very common and generally occurs at the terminal phases of the disease.[3,4,57–60] The syndrome is complicated and affects protein, carbohydrate, and fat metabolism, resulting in anorexia, weight loss, negative nitrogen balance, and skeletal muscle wasting. It is estimated that 60%–65% of late-stage cancer patients will have anorexia and some degree of cachexia. For most patients, the causes of the CACS are complex and multifactorial, including reduced intake, metabolic dysfunction, and increased energy requirement.[61] In some cases, it occurs because of direct effects of the neoplasm, including primary or metastatic tumors of the gastrointestinal (GI) tract, including the oropharynx, esophagus and stomach, small bowel, and large intestine. Similar effects can develop as a result of radical surgical resection of the tumor in these regions of the GI tract. Treatment with external beam radiotherapy and chemotherapy can also contribute to the development of the CACS and are often the instigating factors when it develops early in the course of the disease. Adding to the negative effects of irradiation and chemotherapy is the process of "food aversion learning".[59,60] In this setting, patients learn to avoid many types of foods after associating them with nausea, emesis, and other unpleasant symptoms of chemoradiation. Ultimately,

food aversion learning leads to abnormal food selection, appetite dysregulation, and reduced oral intake. Another important group of mediators of the CACS is the numerous cytokines, neuropeptides (e.g., neuropeptide Y), and hormones (e.g., leptin) secreted by primary and metastatic tumors.[57,58] Of these compounds, cytokines such as TNF-α, interferon-δ, and the interleukins (− 1 and − 6) appear to play the most prominent role. TNF-α is a cytokine related to cachexia and infusion has long been known to cause loss of skeletal muscle in mice, and TNF-α-blocking immunoglobulin reduced muscle loss in mice with tumors.[61–63] In a study of mice with elevated IL-6, mice were noted to have decreased skeletal muscle and fat mass, as well as increased tumor burden, lending credence to the theory of the role of increased acute phase reactants through the presence of IL-6.[61,64] It is theorized that the expression of cytokines prevents the arcuate nucleus of the hypothalamus from responding appropriately to peripheral signals (i.e., adiposity signals versus energy signals), via the stimulation of anorexigenic systems and by inhibiting prophagic pathways.[60] In very advanced stages of the disease, chronic pain, depression, and the effects of intense treatment contribute to the persistence and escalation of the CACS.

During the course of the CACS, with its severe malnutrition, weight loss, and wasting, the CNS and peripheral neuromuscular systems are often affected.[3,4] The neurological manifestations can include encephalopathy and neurocognitive symptoms (see above), myelopathy, progressive myopathy, and peripheral mono- and polyneuropathies.

Etanercept, an anti-TNF-α agent, has improved chemotherapy adherence and fatigue during cancer treatment.[61,65] It has also been shown to stop the TNF-α-mediated activation of other cytokines, mitigating further effects of anorexia.

5 Nutritional myelopathy

As noted above, cancer patients with malnutrition or the CACS are at risk of developing vitamin B12 deficiency, with its resultant neuropsychiatric manifestations.[3,4,39,40,66] In many cases, signs and symptoms referable to the spinal cord are prominent and demonstrate a myelopathy, with variable lower extremity weakness and spasticity, upper motor neuron dysfunction, hyperactive reflexes, abnormal gait, and Babinski's signs. A sensory level may be present in some patients. Pathologically, spongiform changes and foci of demyelination and axonal destruction are noted in the spinal cord white matter in the dorsal and lateral columns, typically in the cervical and upper thoracic regions. Neuroimaging with MRI will sometimes show increased signals on T2 and FLAIR images in the appropriate level of the cord, especially within the dorsal columns.

6 Peripheral neuropathy

Cancer patients are at high risk for peripheral nervous system dysfunction from malnutrition, nutritional deficiencies, and exposure to toxic effects of chemotherapy.[2,3,48,66,67] Patients with malnutrition and CACS have been documented to develop a high percentage of nerve compression neuropathies. For example, a study by Hawley and coworkers noted a rate of 13% in a series of patients with small-cell lung cancer.[68] Each of the patients continued to progress and eventually develop polyneuropathy. Other investigators have noted abnormalities of the neuromuscular junction in cancer patients with significant weight loss and cachexia.[69] Biopsies demonstrated axonal degeneration of intramuscular nerve fibers, atrophy of type-I and type-II muscle fibers, and the sprouting of nerve endings. Cancer patients with neurological diseases related to nutritional deficiencies are also known to develop peripheral neuropathy.[3,66] In patients with vitamin B12 deficiency, a sensorimotor polyneuropathy can be noted clinically, with EMG evidence for axonal degeneration with or without associated demyelination. Somatosensory-evoked potentials are also typically abnormal. Vitamin E deficiency, which is rare in cancer patients but can occur in cases with extensive small bowel resection, may exhibit an axonal neuropathy, mainly of centrally directed fibers of large myelinated neurons.[66] Vitamin B1 deficiency can present as dry beriberi, which is characterized by a sensorimotor, distal, axonal peripheral neuropathy, often associated with burning feet, muscular tenderness, and calf cramping.[66,70] An autonomic neuropathy with orthostatic symptoms has also been described in dry beriberi.

Peripheral neuropathy is a common complication of systemic chemotherapy in cancer patients.[2,47,48,67,71] Approximately 30%–40% of patients treated with neurotoxic chemotherapy will develop chemotherapy-induced peripheral neuropathy.[71] Numerous chemotherapy drugs are known to cause neuropathy, which can be quite variable in terms of clinical severity and relevance (see Table 8). The agents most likely to cause a clinically significant neuropathy are vincristine (and other vinca alkaloids), cisplatin, paclitaxel, docetaxel, thalidomide, suramin, etoposide, and cytosine arabinoside.[71] In general, patients develop a dose-dependent, generalized symmetric axonal sensorimotor polyneuropathy, with numbness and paresthesia of the feet and hands, distal sensory loss, variable reduction of reflexes, and mild weakness. Each different chemotherapeutic agent has a different mechanism of chemotherapy-induced peripheral neuropathy, with platinum agents causing nuclear and mitochondrial damage, vinca alkaloids causing destabilization of microtubule polymers, and taxols causing stabilization of microtubule polymers.[71]

TABLE 8   Chemotherapy drugs associated with toxic peripheral neuropathy.

Vincristine and other vinca alkaloids
Cisplatin
Carboplatin
Paclitaxel (Taxol)
Docetaxel (Taxotere)
Suramin
Cladribine
Fludarabine
Etoposide
Ifosfamide
Cytosine arabinoside
Hexamethylmelamine
Thalidomide
Procarbazine

Data derived from Cavaliere R, Schiff D. Neurologic toxicities of cancer therapies. Curr Neurol Neurosci Rep. 2006;6(3):218–226; Hammack JE. Neurologiccomplications of chemotherapy and biologic therapies. In: Schiff D, O'Neill BP, eds. Principles of Neuro-Oncology. New York: McGraw-Hill Medical Publishing Division; 2005:679–709; Briemberg HR, Amato AA. Neuromuscular complications of cancer. Neurol Clin. 2003;21(1):141–165.

7 Nutritional and toxic myopathy

Cancer patients with malnutrition or CACS often complain of a generalized lack of strength, especially those with significant weight loss and cachexia.[3,67,72] However, on formal neurological testing, many of these patients will have normal strength but may lack endurance and stamina. When weakness is present, it is usually mild (even in patients with atrophy) and affects the proximal lower extremities. The creatine phosphokinase (CPK) level is typically normal. EMG is also often normal but may show myopathic-appearing motor units. Cancer patients are also at risk for a toxic myopathy from chronic corticosteroid use and several chemotherapy drugs.[68] Vincristine can occasionally cause a myopathy with an acute or insidious onset, characterized by proximal muscle weakness and myalgias. Similarly, hydroxyurea and imatinib are known to induce an inflammatory myopathy in rare patients, characterized by elevated CPK and myopathic changes on EMG.

8 Immunotherapy side effects

Immune checkpoints are small molecules expressed by immune cells that play a critical role in maintaining immune homeostasis.[73] Cancer patients are increasingly being treated with immune checkpoint inhibitors. Ipilimumab was approved in 2011 by the FDA for the treatment of metastatic melanoma, and since that time a myriad of other agents have been approved for a variety of other solid tumors. Multiple potential side effects of treatment with ICPi (immune checkpoint inhibitors) exist, as detailed in Table 9.

TABLE 9 Immune checkpoint inhibitor side effects.

Condition	Diagnostic symptoms	Lab abnormalities	Treatment
Autonomic neuropathy	• Constipation • Nausea • Urinary problems • Sexual difficulties • Sweating abnormalities • Sluggish pupil reaction • Orthostatic hypertension	• Abnormal morning orthostatic vitals • Check for antineutrophil cytoplasmic antibodies • Check for ganglionic acetylcholine receptor antibody testing	• Grade I: Hold immunotherapy and monitor symptoms for a week; if symptoms continue, then monitor very closely for symptom progression • Grade II: Hold immunotherapy and resume once returned to grade I, initiate prednisone, ad neurologic consultation • Grade III–IV: Permanently discontinue immunotherapy, admit patient, initiate methylprednisolone 1 g daily for 3 days followed by oral corticosteroid taper, and neurologic consultation
Peripheral neuropathy	• Asymmetric or symmetric sensory, motor, or sensorimotor deficit • Focal mononeuropathies, including cranial neuropathies • Numbness and paresthesias may be painful or painless • Hypo- or areflexia or sensory ataxia may be present	• Screen for reversible neuropathy causes • Check for MRI abnormalities • Lumbar puncture abnormalities checking for cell count and protein	• Grade I: Low threshold to hold immunotherapy and monitor symptoms for a week; if symptoms continue, then monitor very closely for any symptom progressions • Grade II: Hold immunotherapy and resume once returned to Grade I, initiate prednisone 0.5–1 mg/kg; gabapentin, pregabalin, or duloxetine for pain • Grade III–IV: Permanently discontinue immunotherapy, admit patient, neurologic consultation, and initiate IV methylprednisolone 2–4 mg/kg
Aseptic meningitis	• Headache • Photophobia • Neck stiffness • Nausea • Vomiting	• Abnormal brain MRI • Lumbar puncture abnormalities checking for cell count and protein glucose • May see elevated white blood cell count with normal glucose • May see reactive lymphocytes or histiocytes on cytology	• All grades: Low threshold to hold immunotherapy and discuss resumption with patients after taking into account risks and benefits • All grades: Consider empirical antiviral and antibacterial therapy until CSF results • Grades II–IV: Once bacterial and viral infections are negative, closely monitor off corticosteroids or consider oral prednisone or IV methylprednisolone
Myasthenia gravis	• Fatigable or fluctuating muscle weakness (generally more proximal than distal) • Ptosis • Extraocular movement abnormalities • Neck and/or respiratory muscle weakness	• Acetylcholine receptor and antistriated muscle antibodies in the blood • Pulmonary function assessment abnormalities • Brain MRI abnormalities • Abnormal CPK (creatine phosphokinase), aldolase, ESR (erythrocyte sedimentation rate), CRP (C-reactive protein) levels	• All grades: Permanently discontinue immunotherapy, administer methylprednisolone 2 mg/kg, strongly consider intravenous immunoglobulin
Hypophysitis	• Headache • Visual field impairment • Endocrine blood test to check for hormone abnormalities	• ACTH deficiency • Growth hormone deficiency • Follicle-stimulating hormone deficiency • Antibodies against TSH-secreting cells • Pituitary imaging with MRI • FDG PET scan evidence of immune-mediated adverse events	• Grade I: Continue with immunotherapy and hormone replacement in physiological doses • Grade II: Delay immunotherapy, hormone replacement as needed, consider the use of high-dose steroids • Grade III–IV: Stop immunotherapy, high-dose steroids, close follow-up

Continued

II. Nonmetastatic neurological complications

TABLE 9   Immune checkpoint inhibitor side effects—cont'd

Condition	Diagnostic symptoms	Lab abnormalities	Treatment
Thyroid dysfunction	• Fatigue • Weakness • Anorexia • Headache • Visual field defects • Nausea • Fever • Lethargy • Impotence	• High levels of TSH with low levels of T4	• Grade I: Continue with immunotherapy and frequently monitor thyroid function and hormone levels • Grade II: Continue with immunotherapy while treating thyroid disorder, administer thyroid hormone and/or steroid replacement therapy, frequently monitor thyroid function and hormone levels, consider consultation with endocrinologist • Grade III–IV: Discontinue immunotherapy, rule out infection and sepsis, treat with IV methylprednisolone followed by oral prednisone, replacement of appropriate hormones may be required • For symptomatic hyperthyroidism, prescribe beta-blockers
Type 1 diabetes	• Weight loss • Nausea • Vomiting • Abdominal pain • Hyperventilation • Lethargy • Seizure • Coma	• Fasting C-peptide • Insulin autoantibody • Glutamic acid decarboxylase antibody • Insulinoma-associated protein 2 antibody • Zinc transporter antibody	• Insulin therapy • Monitor glucose levels with each dose of immunotherapy
Primary adrenal insufficiency	• Fatigue • Dizziness • Orthostatic hypotension • Anorexia • Weight low • Altered mental status	• Electrolyte imbalance • Low cortisol or inappropriate cortisol stimulation test • High ACTH • Dehydration	• Grade I: Continue with immunotherapy and treat electrolyte imbalance • Grade II: Delay immunotherapy and administer glucocorticoid therapy • Grade III–IV: Discontinue immunotherapy and administer IV corticosteroids, patient should be hospitalized immediately
Pneumonitis	• Cough • Shortness of breath • Dyspnea • Fever • Asymptomatic radiographic changes	• Chest X-ray abnormalities • CT scan of the thorax to check for abnormalities	• Grade I: Continue with immunotherapy and monitor • Grade II: Delay immunotherapy, administer corticosteroids and upon improvement to grade I, initiate corticosteroid taper • Grade III and IV: Discontinue immunotherapy, consider pulmonary function tests, bronchoscopy with biopsy, treat with steroids until symptoms improve to grade I
Hepatoxicity	• Fever	• Elevated serum levels of hepatic enzymes • Elevated aspartate aminotransferase • Elevated alanine aminotransferase • CT scan may show mild abnormalities	• Grade I: Continue immunotherapy and monitor • Grade II: Delay immunotherapy, and administer corticosteroids • Grade III or IV: Immunotherapy should be permanently discontinued, and administer corticosteroids

Data derived from Chang LS, et al. Endocrine toxicity of cancer immunotherapy targeting immune checkpoints. Endocr Rev. 2019;40(1):17–65; Doughty CT, Amato AA. Toxic myopathies. Continuum (Minneap Minn). 2019;25(6):1712–1731; Nadeau BA, et al. Liver toxicity with cancer checkpoint inhibitor therapy. Semin Liver Dis. 2018;38(4):366–378; Weber JS, Kahler KC, Hauschild A. Management of immune-related adverse events and kinetics of response with ipilimumab. J Clin Oncol. 2012;30(21):2691–2697; Su Q, et al. Risk of pneumonitis and pneumonia associated with immune checkpoint inhibitors for solid tumors: a systematic review and meta-analysis. Front Immunol. 2019;10:108; Suresh K, et al. Immune checkpoint immunotherapy for non-small cell lung cancer: benefits and pulmonary toxicities. Chest. 2018;154(6):1416–1423; Linardou H, Gogas H. Toxicity management of immunotherapy for patients with metastatic melanoma. Ann Transl Med. 2016;4(14):272; Joshi MN, et al. Immune checkpoint inhibitor-related hypophysitis and endocrine dysfunction: clinical review. Clin Endocrinol (Oxf). 2016;85(3):331–339.

8.1 Myositis

Myositis can occur with immune checkpoint inhibitors targeting PD-1, PD-L1, and CTLA-4, with patients typically developing weakness and myalgia early after initiation, with a median onset of symptoms 25 days after initiation of therapy. Weakness most commonly involves the axial muscles and proximal extremities. Also commonly seen and often co-occurring with myositis is myasthenia gravis (MG), which is commonly seen with bulbar and oculomotor weakness. Creatine kinase is typically elevated in myositis, and a decremental response on repetitive nerve stimulation is helpful in diagnosing MG.[74] Current grading for MG includes moderate (G2) and severe (G3–4). Moderate MG is defined as some symptoms interfering with daily activities, Myasthenia Gravis Foundation of America (MGFA) severity class 1 (ocular symptoms and findings only) and MGFA severity class 2 (mild generalized weakness). Management of moderate MG includes holding ICPi therapy which may resume if the symptoms resolve, pyridostigmine starting at 30 mg orally three times a day, or administering corticosteroids and weaning based on symptom improvement. Severe MG is defined as limiting self-care and aids warranted, weakness limiting walking, dysphagia, facial weakness, respiratory muscle weakness, rapidly progressive symptoms, or MGFA severity class 3–4 (moderate-to-severe generalized weakness to a myasthenic crisis). Management for severe MG is to permanently discontinue ICPi therapy, admit the patient for monitoring, continue corticosteroids, and initiate intravenous immunoglobulin (IVIG) 2 g over 5 days or plasmapheresis for 5 days, frequent pulmonary function assessment, and daily neurologic review.

8.2 Hepatotoxicity

Hepatotoxicity refers to a state of toxic damage to the liver. Hepatotoxicity from checkpoint inhibitors is a less common toxicity from ICPi therapy and is often mild.[75] Symptoms of hepatotoxicity include nausea, vomiting, abdominal pain, anorexia, diarrhea, fatigue, weakness, and jaundice. Most episodes of hepatotoxicity are asymptomatic laboratory abnormalities.[76] One clinical manifestation of hepatotoxicity is elevations in serum levels of the hepatic enzymes, aspartate aminotransferase (AST) and alanine aminotransferase (ALT).[76] Grade 2 hepatotoxicity is defined as AST or ALT > 2.5 times the upper limit of normal (ULN) but < 5 times the ULN or total bilirubin > 1.5 times the ULN but < 3 times the ULN.[76] Grade 3 or greater hepatotoxicity is defined as AST or ALT > 5 times the ULN or total bilirubin > 3 times the ULN and immunotherapy should be permanently discontinued.[76]

8.3 Pneumonitis

When combating some solid tumors, ICPi therapy alters the balance of immune cells in the body, which in turn can damage certain organ systems.[77] Pneumonitis refers to inflammation of the lung tissue due to noninfectious causes. Symptoms of pneumonitis include cough, dyspnea, tachypnea, and hypoxia.[77] Immune-mediated lung injury from ICPi use, termed checkpoint inhibitor pneumonitis (CIP), occurs in about 3%–5% of patients receiving ICPi therapy.[78] Most cases of CIP require treatment with high doses of oral or parenteral steroids.[78] It is typically difficult to diagnose CIP because of the lack of specific clinical or radiographic markers.[78] The clinical symptoms are often nonspecific, and the radiographic appearance of CIP is varied and may mimic tumor progression.[78] Pneumonitis can occur in isolation without toxicity in other organ systems; therefore, absence of other immune-related toxicities does not help in the determination of CIP.[78] CIP is responsible for a spectrum of lung injuries, from the acute phase (acute interstitial pneumonia) to the organizing (organizing pneumonia) and fibrotic phases (nonspecific interstitial pneumonia).[78] The primary method of CIP management is corticosteroid therapy. High doses of prednisone (or equivalent corticosteroids) are initiated once a formal diagnosis has been made.[78] Patients typically require high-dose corticosteroids for several weeks, followed by a slow taper, along with cessation of the ICPi. Grade 1 pneumonitis involves asymptomatic radiographic changes. Grade 2 pneumonitis involves dyspnea and exertion; the patient should delay ICPi therapy and receive steroids until symptoms improve. Grade 3–4 pneumonitis is shortness of breath while at rest and the patient requires oxygen or assisted ventilation; ICPi therapy should be discontinued.[79] Because the onset of symptoms for pneumonitis is often vague, clinicians should consider diagnostic radiology or bronchoscopy to rule out other causes.[79]

8.4 Endocrinopathies

Endocrinopathies have emerged as one of the most common immune-related adverse effects of immune checkpoint inhibitors (ICPi) for the treatments of several malignancies.[73] An endocrinopathy is a disease of an endocrine gland, caused by a disorder of the endocrine system. Hypophysitis, thyroid dysfunction, type 1 diabetes, and primary adrenal insufficiency have been reported as endocrinopathies due to ICPi therapy. ICPi therapy is increasingly being used in oncologic practice so it is critical for endocrinologists and oncologists to be aware of the clinical manifestations, diagnosis, and management and ICPi-related endocrinopathies.[73] Hypophysitis is associated with anti-CTLA-4 therapy while thyroid dysfunction is associated with anti-PD-1 therapy.[73] Hypophysitis refers to inflammation of the pituitary gland and can result in pituitary dysfunction, particularly in the anterior pituitary, leading to deficiencies in hormones produced by the anterior pituitary.[73] Symptoms of ICPi-related hypophysitis are typically nonspecific, with common features including headache and fatigue. Other symptoms may include nausea, decreased appetite, dizziness, decreased libido, cold intolerance, hot flashes, and weight loss.[73] Mild (grade 1) hypophysitis involves no symptoms with clinical or diagnostic observations only. Moderate hypophysitis (grade 2) involves symptoms that interfere with the patient's daily life. Severe hypophysitis (grade 3–4) involves severe symptoms and care is needed. Thyroid dysfunction is one of the most common endocrinopathies associated with ICPi therapy and is described in the literature as hyperthyroidism, hypothyroidism, and/or thyroiditis.[73] Several case series have shown that ICPi-related thyroid dysfunction appears to be due to destructive thyroiditis.[73] Symptoms of ICPi-related hyperthyroidism include fatigue, weight loss, and palpitations. Physical examination may reveal tachycardia, warm smooth skin, lid lag, and brisk deep tendon reflexes.[73] Symptoms of ICPi-related hypothyroidism include fatigue, weight gain, cold intolerance, constipation, and dry skin.[73] The symptoms of thyroid dysfunction are nonspecific and have an overlap with symptoms experienced in patients with malignancy so it is recommended that providers check TSH and free T4 in any patient who is receiving or has received ICPi therapy.[73] Since ICPi-related thyroid dysfunction is a manageable condition with monitoring and treatment, ICPi therapy need not be discontinued in most mild-to-moderate cases of thyroid dysfunction.[73] Grade 1 thyroid dysfunction is asymptomatic with no intervention needed, but hormone levels should be frequently monitored. Grade 2 thyroid dysfunction includes symptoms that limit the daily activities of the patient. Grade 3–4 involves severe symptoms, limiting self-care, and hospitalization is indicated.[80] ICPi-related diabetes mellitus (DM) is characterized by rapid-onset hyperglycemia, swift progression of insulin deficiency, and high risk of diabetic ketoacidosis if not promptly treated with insulin therapy.[73] ICPi-related DM has been reported to occur as early as after a single dose of ICPi therapy to as late as after 17 doses of ICPi therapy.[73] Symptoms of ICPi-related DM include weight loss, nausea, vomiting, abdominal pain, hyperventilation, lethargy, seizure, or coma.[73] ICPi-related DM typically results in long-term need for insulin. Once the insulin therapy has begun, ICPi therapy can continue. Glucose levels should be monitored with each dose of immunotherapy. Primary adrenal insufficiency (PAI) is a rare condition in which the adrenal glands do not produce adequate amounts of steroid hormones, particularly cortisol. PAI requires mineralocorticoid replacement in addition to glucocorticoid replacement.[73]

Symptoms of adrenal insufficiency include fatigue, dizziness, orthostatic hypotension, anorexia, weight loss, abdominal discomfort, altered mental status, and delirium.[73] Hypernatremia and hyperkalemia are common in PAI due to the presence of both glucocorticoid and mineralocorticoid deficiency. Corticosteroids should be administered to any patient with adrenal insufficiency who presents with an acute adrenal crisis. Grade 1 adrenal insufficiency involves no symptoms and clinical observation. Grade 2 adrenal insufficiency involves moderate symptoms and medical intervention is indicated. Grade 3–4 involves severe symptoms, hospitalization indicated, and life-threatening consequences, and urgent intervention is indicated.[80] It has been theorized that the development of endocrinopathies during ICPi therapy may be a positive indicator that the treatment is working, as reduction of immune self-tolerance triggering endocrinopathies may be associated with an enhanced ability of the immune system to recognize and destroy cancer cells.[73]

8.5 Neuropathies

Immune checkpoint inhibitor (ICPi) therapy can also result in either autonomic or peripheral neuropathy. Autonomic neuropathy includes nerve damage to those nerves that control involuntary bodily functions. Mild autonomic neuropathy is defined as having no interference with function. Treatment for this grade includes a low threshold to hold ICPi therapy and continue monitoring symptoms. Moderate (grade 2) autonomic neuropathy involves some interference with daily life activities, and the symptoms concern the patient. Treatment includes holding ICPi therapy and resuming once the patient returns to mild grade, observation or initiating prednisone, and a neurologic consultation. Severe (grade 3–4) autonomic neuropathy involves limiting self-care and aids warranted. Treatment includes permanently discontinuing ICPi therapy, admitting the patient for a neurologic consultation, and initiating methylprednisolone.

Grading for peripheral neuropathy varies from mild (G1) to severe (G3–4). Mild peripheral neuropathy is classified as having no interference with function and the symptoms are not concerning to the patient. Treatment for mild is to hold ICPi therapy and monitor symptoms for a week; if they continue, monitor closely for symptom progression. Moderate peripheral neuropathy involves some interference with daily life activities, the symptoms are concerning to the patient, and the patient experiences pain but no weakness or gait limitation. Treatment for moderate neuropathy is to hold ICPi and resume when the patient returns to mild grade neuropathy, observation or initiating prednisone, and pain management. Severe peripheral neuropathy is defined as limiting self-care with aids warranted, weakness limiting the ability to walk, or respiratory problems. A severe complication also includes symptoms consistent with Guillain-Barré syndrome and should be managed as such. Treatment involves permanently discontinuing ICPi therapy, admitting the patient for a neurologic consultation, and initiating IV methylprednisolone.

8.6 Aseptic meningitis

Aseptic meningitis ranges from mild to severe but the treatment is the same for each grade. Treatment involves holding ICPi therapy, considering empirical antiviral and antibacterial therapy until CSF results, and possibly administering prednisone or IV methylprednisolone for moderate/severe symptoms.

9 CAR-T cell complications

Chimeric antigen receptor (CAR)-T cell therapy is a type of cancer treatment that uses cells from your own immune system. The T-cells used in CAR-T cell therapy are taken from the patient, modified in the lab to spot specific cancer cells, and put back into the patient. CAR-T cell therapies are revolutionizing the management of B-cell leukemias and lymphomas and are being extended to numerous other malignancies.[81] Cytokine release syndrome (CRS) is a systemic inflammatory response that can be triggered by a variety of factors including infections and drug reactions.[82] CRS is caused by the large and rapid release of cytokines into the blood from immune cells affected by immunotherapy. CRS symptoms include fever, fatigue, headache, rash, and myalgia; more severe symptoms include hypotension, high fever, and progression to multiorgan system failure.[82] Some patients develop neurotoxicity after the administration of T-cell-engaging therapies. In the case of CAR-T cell therapy, neurotoxicity is the second most common serious adverse event and therefore referred to as "CAR-T cell-related encephalopathy syndrome" (CRES).[82] The neurotoxicity of CAR-T cell therapy does not seem to be directly related to CRS since the neurologic symptoms do not always coincide with CRS onset and neurotoxicity can occur prior to CRS or after CRS has resolved.[82] Many CRS-inducing agents display a "first-dose effect": the most severe symptoms only occur after the first administered dose and do not recur after subsequent administrations.[82] The strength of T-cell activation and the degree of T-cell expansion seem to correlate with the severity of CRS.[82] Low-grade CRS is treated symptomatically with antihistamines, antipyretics, and fluids.[82] Since T-cell therapies are relatively recent, effective management of severe CRS requires close collaboration between the physicians of different specialties.[82]

References

1. Spinazze S, Schrijvers D. Metabolic emergencies. *Crit Rev Oncol Hematol.* 2006;58(1):79–89.
2. Clouston PD, DeAngelis LM, Posner JB. The spectrum of neurological disease in patients with systemic cancer. *Ann Neurol.* 1992;31(3):268–273.
3. Posner JB. Metabolic and nutritional complications of cancer. In: Posner J, ed. *Neurologic Complications of Cancer.* Philadelphia, PA: F.A. Davis Company; 1995:264–281.
4. Boerman RH, Padberg GW. Metabolic nervous system dysfunction in cancer. In: Vecht CJ, ed. *Handbook of Clinical Neurology.* Amsterdam: Elsevier Medical Publishers; 1997:395–412.
5. Gilbert MR, Grossman SA. Incidence and nature of neurologic problems in patients with solid tumors. *Am J Med.* 1986;81(6):951–954.
6. Tuma R, DeAngelis LM. Acute encephalopathy in patients with systemic cancer (abstract). *Ann Neurol.* 1992;32:288.
7. Francis J, Martin D, Kapoor WN. A prospective study of delirium in hospitalized elderly. *JAMA.* 1990;263(8):1097–1101.
8. Berisavac I, et al. How to recognize and treat metabolic encephalopathy in neurology intensive care unit. *Neurol India.* 2017;65(1):123–128.
9. Oh ES, et al. Delirium in older persons: advances in diagnosis and treatment. *JAMA.* 2017;318(12):1161–1174.
10. Bush SH, Tierney S, Lawlor PG. Clinical assessment and management of delirium in the palliative care setting. *Drugs.* 2017;77(15):1623–1643.
11. Albert MS, et al. The delirium symptom interview: an interview for the detection of delirium symptoms in hospitalized patients. *J Geriatr Psychiatry Neurol.* 1992;5(1):14–21.
12. Pass HI, Donington JS. Metastatic cancer to the lung. In: DeVita Jr VT, Hellman S, Rosenberg SA, eds. *Cancer. Principles & Practice of Oncology.* vol. 2. 5th ed. Philadelphia, PA: Lippincott-Raven Publishers; 1997:2536–2551.
13. Daly JM, Kemeny NE. Metastatic cancer to the liver. In: DeVita Jr VT, Hellman S, Rosenberg SA, eds. *Cancer. Principles & Practice of Oncology.* vol 2. 5th ed. Philadelphia, PA: Lippincott-Raven Publishers; 1997:2551–2570.
14. Hadjihambi A, et al. Hepatic encephalopathy: a critical current review. *Hepatol Int.* 2018;12(suppl 1):135–147.
15. Munoz SJ. Hepatic encephalopathy. *Med Clin North Am.* 2008;92(4):795–812. viii.
16. Wendon J, Lee W. Encephalopathy and cerebral edema in the setting of acute liver failure: pathogenesis and management. *Neurocrit Care.* 2008;9(1):97–102.
17. Kouba E, Wallen EM, Pruthi RS. Management of ureteral obstruction due to advanced malignancy: optimizing therapeutic and palliative outcomes. *J Urol.* 2008;180(2):444–450.
18. Lameire NH, et al. Acute renal failure in cancer patients. *Ann Med.* 2005;37(1):13–25.
19. Baluarte JH. Neurological complications of renal disease. *Semin Pediatr Neurol.* 2017;24(1):25–32.
20. Koza Y. Acute kidney injury: current concepts and new insights. *J Inj Violence Res.* 2016;8(1):58–62.
21. Lumachi F, et al. Medical treatment of malignancy-associated hypercalcemia. *Curr Med Chem.* 2008;15(4):415–421.
22. Shepard MM, Smith III JW. Hypercalcemia. *Am J Med Sci.* 2007;334(5):381–385.
23. Gross P. Treatment of hyponatremia. *Intern Med.* 2008;47(10):885–891.
24. Hoorn EJ, Zietse R. Hyponatremia revisited: translating physiology to practice. *Nephron Physiol.* 2008;108(3):46–59.
25. Dineen R, Thompson CJ, Sherlock M. Hyponatraemia—presentations and management. *Clin Med (Lond).* 2017;17(3):263–269.
26. Achinger SG, Ayus JC. Treatment of hyponatremic encephalopathy in the critically ill. *Crit Care Med.* 2017;45(10):1762–1771.

27. Sterns RH, et al. Current perspectives in the management of hyponatremia: prevention of CPM. *Expert Rev Neurother.* 2007;7(12):1791–1797.

28. Hoorn EJ, Zietse R. Diagnosis and treatment of hyponatremia: compilation of the guidelines. *J Am Soc Nephrol.* 2017;28(5):1340–1349.

29. Ahmed F, Mohammed A. Magnesium: the forgotten electrolyte-a review on hypomagnesemia. *Med Sci (Basel).* 2019;7(4).

30. Fang W, et al. Symptomatic severe hypophosphatemia after intravenous ferric carboxymaltose. *JGH Open.* 2019;3(5):438–440.

31. Malouf R, Brust JC. Hypoglycemia: causes, neurological manifestations, and outcome. *Ann Neurol.* 1985;17(5):421–430.

32. Mendoza A, Kim YN, Chernoff A. Hypoglycemia in hospitalized adult patients without diabetes. *Endocr Pract.* 2005;11(2):91–96.

33. Pourmotabbed G, Kitabchi AE. Hypoglycemia. *Obstet Gynecol Clin North Am.* 2001;28(2):383–400.

34. Alexakis N, Neoptolemos JP. Pancreatic neuroendocrine tumours. *Best Pract Res Clin Gastroenterol.* 2008;22(1):183–205.

35. Tucker ON, Crotty PL, Conlon KC. The management of insulinoma. *Br J Surg.* 2006;93(3):264–275.

36. Kuo SH, et al. Wernicke's encephalopathy: an underrecognized and reversible cause of confusional state in cancer patients. *Oncology.* 2009;76(1):10–18.

37. Vortmeyer AO, Hagel C, Laas R. Haemorrhagic thiamine deficient encephalopathy following prolonged parenteral nutrition. *J Neurol Neurosurg Psychiatry.* 1992;55(9):826–829.

38. Kinn AC, Lantz B. Vitamin B12 deficiency after irradiation for bladder carcinoma. *J Urol.* 1984;131(5):888–890.

39. Hvas AM, Nexo E. Diagnosis and treatment of vitamin B12 deficiency—an update. *Haematologica.* 2006;91(11):1506–1512.

40. Savage DG, Lindenbaum J. Neurological complications of acquired cobalamin deficiency: clinical aspects. *Baillieres Clin Haematol.* 1995;8(3):657–678.

41. Torrey SP. Recognition and management of adrenal emergencies. *Emerg Med Clin North Am.* 2005;23(3):687–702. viii.

42. Bolton CF, Young GB, Zochodne DW. The neurological complications of sepsis. *Ann Neurol.* 1993;33(1):94–100.

43. Hotchkiss RS, Karl IE. The pathophysiology and treatment of sepsis. *N Engl J Med.* 2003;348(2):138–150.

44. Rittirsch D, Flierl MA, Ward PA. Harmful molecular mechanisms in sepsis. *Nat Rev Immunol.* 2008;8(10):776–787.

45. Russell JA. Management of sepsis. *N Engl J Med.* 2006;355(16):1699–1713.

46. Jeppsson B, et al. Blood-brain barrier derangement in sepsis: cause of septic encephalopathy? *Am J Surg.* 1981;141(1):136–142.

47. Cavaliere R, Schiff D. Neurologic toxicities of cancer therapies. *Curr Neurol Neurosci Rep.* 2006;6(3):218–226.

48. Hammack JE. Neurologic complications of chemotherapy and biologic therapies. In: Schiff D, O'Neill BP, eds. *Principles of Neuro-Oncology.* New York: McGraw-Hill Medical Publishing Division; 2005:679–709.

49. Newton HB. Intra-arterial chemotherapy. In: Newton HB, ed. *Handbook of Brain Tumor Chemotherapy.* Amsterdam: Elsevier/Academic Press; 2006;17:247–261.

50. Deandrea S, et al. Prevalence of undertreatment in cancer pain. A review of published literature. *Ann Oncol.* 2008;19(12):1985–1991.

51. Cherny NI. The pharmacologic management of cancer pain. *Oncology (Williston Park).* 2004;18(12):1499–1515. discussion 1516, 1520–1, 1522, 1524.

52. Benyamin R, et al. Opioid complications and side effects. *Pain Physician.* 2008;11(2 suppl):S105–S120.

53. De Smet PA. Herbal remedies. *N Engl J Med.* 2002;347(25):2046–2056.

54. Gemici C. Tumour lysis syndrome in solid tumours. *Clin Oncol (R Coll Radiol).* 2006;18(10):773–780.

55. Rampello E, Fricia T, Malaguarnera M. The management of tumor lysis syndrome. *Nat Clin Pract Oncol.* 2006;3(8):438–447.

56. Cairo MS, et al. Recommendations for the evaluation of risk and prophylaxis of tumour lysis syndrome (TLS) in adults and children with malignant diseases: an expert TLS panel consensus. *Br J Haematol.* 2010;149(4):578–586.

57. Nelson KA, Walsh D, Sheehan FA. The cancer anorexia-cachexia syndrome. *J Clin Oncol.* 1994;12(1):213–225.

58. Tisdale MJ. Cachexia in cancer patients. *Nat Rev Cancer.* 2002;2(11):862–871.

59. Laviano A, Meguid MM, Rossi-Fanelli F. Cancer anorexia: clinical implications, pathogenesis, and therapeutic strategies. *Lancet Oncol.* 2003;4(11):686–694.

60. Laviano A, et al. Neurochemical mechanisms for cancer anorexia. *Nutrition.* 2002;18(1):100–105.

61. Kim DH. Nutritional issues in patients with cancer. *Intest Res.* 2019;17(4):455–462.

62. Costelli P, et al. Tumor necrosis factor-alpha mediates changes in tissue protein turnover in a rat cancer cachexia model. *J Clin Invest.* 1993;92(6):2783–2789.

63. Fong Y, et al. Cachectin/TNF or IL-1 alpha induces cachexia with redistribution of body proteins. *Am J Physiol.* 1989;256(3 Pt 2):R659–R665.

64. Baltgalvis KA, et al. Interleukin-6 and cachexia in ApcMin/+ mice. *Am J Physiol Regul Integr Comp Physiol.* 2008;294(2):R393–R401.

65. Monk JP, et al. Assessment of tumor necrosis factor alpha blockade as an intervention to improve tolerability of dose-intensive chemotherapy in cancer patients. *J Clin Oncol.* 2006;24(12):1852–1859.

66. Kumar N. Nutritional neuropathies. *Neurol Clin.* 2007;25(1):209–255.

67. Briemberg HR, Amato AA. Neuromuscular complications of cancer. *Neurol Clin.* 2003;21(1):141–165.

68. Hawley RJ, et al. The carcinomatous neuromyopathy of oat cell lung cancer. *Ann Neurol.* 1980;7(1):65–72.

69. Hildebrand J, Coers C. The neuromuscular function in patients with malignant tumours. Electromyographic and histological study. *Brain.* 1967;90(1):67–82.

70. Koike H, et al. Postgastrectomy polyneuropathy with thiamine deficiency is identical to beriberi neuropathy. *Nutrition.* 2004;20(11–12):961–966.

71. Staff NP, et al. Chemotherapy-induced peripheral neuropathy: a current review. *Ann Neurol.* 2017;81(6):772–781.

72. Walsh RJ, Amato AA. Toxic myopathies. *Neurol Clin.* 2005;23(2):397–428.

73. Chang LS, et al. Endocrine toxicity of cancer immunotherapy targeting immune checkpoints. *Endocr Rev.* 2019;40(1):17–65.

74. Doughty CT, Amato AA. Toxic myopathies. *Continuum (Minneap Minn).* 2019;25(6):1712–1731.

75. Nadeau BA, et al. Liver toxicity with cancer checkpoint inhibitor therapy. *Semin Liver Dis.* 2018;38(4):366–378.

76. Weber JS, Kahler KC, Hauschild A. Management of immune-related adverse events and kinetics of response with ipilimumab. *J Clin Oncol.* 2012;30(21):2691–2697.

77. Su Q, et al. Risk of pneumonitis and pneumonia associated with immune checkpoint inhibitors for solid tumors: a systematic review and meta-analysis. *Front Immunol.* 2019;10:108.

78. Suresh K, et al. Immune checkpoint immunotherapy for non-small cell lung cancer: benefits and pulmonary toxicities. *Chest.* 2018;154(6):1416–1423.

79. Linardou H, Gogas H. Toxicity management of immunotherapy for patients with metastatic melanoma. *Ann Transl Med.* 2016;4(14):272.

80. Joshi MN, et al. Immune checkpoint inhibitor-related hypophysitis and endocrine dysfunction: clinical review. *Clin Endocrinol (Oxf).* 2016;85(3):331–339.

81. Lee DW, et al. ASTCT consensus grading for cytokine release syndrome and neurologic toxicity associated with immune effector cells. *Biol Blood Marrow Transplant.* 2019;25(4):625–638.

82. Shimabukuro-Vornhagen A, et al. Cytokine release syndrome. *J Immunother Cancer.* 2018;6(1):56.

13

Central nervous system infections in cancer patients

Amy A. Pruitt

Department of Neurology, University of Pennsylvania School of Medicine, Philadelphia, PA, United States

1 Introduction

Central nervous system (CNS) infections remain a source of significant morbidity and mortality among cancer patients, despite the evolution of effective prophylactic regimens and better antimicrobials for active infection. With the past two decades' introduction of more intensive immunosuppressive regimens, including solid organ and hematopoietic cell transplantation, and novel biologic response modifiers, as well as refinement of hematopoietic growth factor support of chemotherapy, longer survival has been achieved along with a change in the spectrum of infections in increasing types of vulnerable populations.[1] For example, since the 1980s, the numbers of patients at risk for CNS infection after nontransplant hematologic malignancy treatment have almost reached the levels encountered in recipients of hematopoietic stem cell transplantation (HSCT), and nosocomial invasive fungal disease has doubled, while better control of bacteria and viruses has reduced the incidence of infections from these organisms.[2] Rapid diagnosis is essential if patients are to experience meaningful survival without serious neurologic sequelae, a problem in at least one-third of long-term pediatric cancer survivors of CNS infections.[3]

Neurologic infectious disease consultants face formidable challenges. The presentation and course of infections in cancer patients often differ from those of patients without cancer and new syndromes related to drug combinations, particularly with the advent of immune checkpoint inhibitor therapies, surface regularly. The precise list of infections varies not only with geographic location but also with local medical practices and nosocomial infection patterns. The special dilemmas of cancer patients during the COVID-19 pandemic and concerns particular to this vulnerable population illustrate the constant challenge of new and still-evolving pathogens. The clinician also must remain sensitive to a host of conditions that mimic CNS infections, including adverse effects of drug treatments, vascular lesions, radiation effects, and tumor recurrence.

Nevertheless, it is possible to approach each patient with an organized diagnostic strategy based on patient neoplastic disease and neurologic presentation. The two most commonly affected groups are neurosurgical patients and those receiving versions of hematopoietic stem cell transplantation. The two most common clinical syndromes are meningoencephalitic syndromes and those due to focal brain lesions. The range of pathogens can be narrowed by considering the type(s) of immune deficits pertinent to the patient under evaluation.

This chapter presents a diagnostic approach to CNS infections beginning with a summary of the epidemiology of CNS infections in cancer patients, and then discusses neuroimaging and cerebrospinal fluid (CSF) diagnostic testing. The two major groups of at-risk cancer patients described earlier are then covered in detail. Additionally, some recently recognized infectious syndromes and potentially confusing treatment complications are covered including reversible posterior leukoencephalopathy syndrome (RPLS, also known as posterior reversible encephalopathy syndrome or PRES), immune reconstitution inflammatory syndrome (IRIS), and Epstein-Barr virus (EBV) reactivation with posttransplantation lymphoproliferative disorder (PTLD), and complications of immune checkpoint inhibitor therapy that may mimic infection. Finally, the chapter discusses general recommendations for steroid use in acute infection, choice of antiepileptic drugs, and clinical, pathophysiologic, treatment information about organism-specific clinical syndromes including bacterial meningitis, endocarditis, *Aspergillus* and cryptococcal infections, varicella-zoster virus (VZV) and its complications, and progressive multifocal leukoencephalopathy (PML).

---

2 Approach to cancer patients with suspected CNS infection

2.1 Five clinical challenges

The timely diagnosis of CNS infection is hindered by a number of obstacles to the prompt institution of therapy:

1. **Diversity of potential pathogens:** The long list of potential pathogens includes many organisms of low pathogenicity in the immunocompetent host. The list is dependent on conditioning regimens and chemotherapeutic agents, is geographically diverse, and must be differentiated from antibiotic and other drug toxicity, engraftment syndromes, graft-vs-host disease (GVHD), central nervous system (CNS) vasculitis, and a host of acronymic problems including nonbacterial thrombotic endocarditis (NBTE), reversible posterior leukoencephalopathy syndrome (RPLS), immune reconstitution inflammatory syndrome (IRIS), and posttransplantation lymphoproliferative disorder (PTLD).

2. **Multiple simultaneous infections:** Infection with more than one agent or sequential infection is common and is often complicated by prior antibiotic use and coexisting metabolic or treatment-related encephalopathy.

3. **Altered inflammatory response:** The cancer patient's diminished inflammatory response may make clinical clues mild or, conversely, may mimic recurrent infection as the host's immune system reconstitutes after effective treatment.

4. **Ambiguous neuroimaging results:** Neuroimaging of CNS infections may be nonspecific and may mimic treatment-related changes such as radiation necrosis or drug-induced leukoencephalopathy. Knowledge of when to use more specialized tests such as magnetic resonance spectroscopy (MRS) or positron emission tomography (PET) is important to expedite diagnosis and minimize invasive biopsy procedures.

5. **Changing cancer treatment and infection prophylaxis regimens:** The clinician's fifth and major challenge is to remain up-to-date with cancer treatment regimens and the expected neurological consequences of new drug combinations. Major changes in infection risks, patterns, and syndromes that have occurred in the past 10 years in non-HIV-infected cancer patients include the following:

    (a) With increased-dose intense regimens including potent immunosuppressive purine analogues such as fludarabine, pentostatin, and cladribine, as well as anti-T- and anti-B-cell antibodies such as alemtuzumab and rituximab, the number of at-risk patients with hematologic malignancies who did not have transplants nearly equals those receiving allogeneic stem cell transplants. *BOX of pathogens specifically to suspect with different drugs.*

    (b) Nonmyeloablative or minitransplant allogeneic transplant patients are also at risk for serious infections. Selective CD34 + depletion to reduce tumor recurrence has been associated with particular risk for cerebral toxoplasmosis.[4] Similarly, conditioning regimens with imatinib, rituximab, and alemtuzumab change the spectrum and timing of opportunistic fungal and viral infections.

    (c) Extensive antibiotic use has been accompanied by the selection of resistant organisms. Methicillin-resistant *Staphylococcus aureus* (MRSA) acquired both in the hospital and, increasingly, in the community where over 60% of the acquired infections are MRSA, assumes a larger role and can be associated with both CNS infections and lethal systemic complications such as necrotizing fasciitis.[5] So-called community-onset MRSA infections are associated with recent hospitalization, an invasive medical device, prior colonization, dialysis, or residence in a long-term care facility within 12 months of culture.[6]

    (d) Opportunistic fungi have become the most frequent and lethal pathogens in the past 25 years as nosocomial fungal infection rates have doubled. There is an increased incidence of fungal and other opportunistic diseases among patients who are not terminally ill and the timing with respect to organ transplantation has changed so that, for example, *Aspergillus* infections occur later posttransplantation than they did a decade ago and patients with chronic myelogenous leukemia (CML) now experience increased VZV after treatment with imatinib.[7]

    (e) Nearly 200 million intravascular devices are sold in the USA each year and their most common complications are bloodstream infection, ranging from local colonization to bacteremia or candidemia.[8]

    (f) There has been a decided change in the spectrum of organisms causing bacterial and fungal meningitis in cancer patients, with marked decreased incidence of *Listeria* and dominance of the risk pool by patients who have had neurosurgery. The lower incidence of meningitis in transplant patients likely results from trimethoprim/sulfamethoxazole prophylaxis for *Toxoplasma gondii* and *Pneumocystis jirovecii*. However, the increasing use of alemtuzumab (Campath) results in long-term T-cell depletion which reactivates cytomegalovirus (CMB), also increasing the risk of *Listeria*. Acyclovir prophylaxis has reduced the incidence of herpes viruses and cytomegalovirus (CMV). However, PML is encountered in a broader group of patients, including those who have received rituximab.[9]

(g) Reactivation of viruses such as EBV has led to a higher incidence of PTLD, the diagnosis and management of which remain controversial.[10]

(h) Changes in immunosuppression from standard regimens of corticosteroids, azathioprine, and calcineurin inhibitors to approaches using sirolimus, mycophenolate mofetil, T-cell and B-cell, and costimulatory blockade have reduced *pneumocystis* infection while increasing activation of CMV, EBV, and HIV. Late infections due to cellular depletion favor CMV and John Cunningham JC virus (JCV) as well as fungal conditions, with even later development of secondary malignant conditions.

(i) In brain tumor patients, the increasing use of concurrent chemotherapy and radiation regimens has been associated with ambiguous MRI abnormalities including chemoradiation necrosis that produces a ring-enhancing lesion termed "pseudoprogression."[11]

2.2 Diagnostic approach to potential CNS infection

The neurologic consultant evaluating a patient for possible CNS infection adheres to a rigid algorithm at his own (and his patients') peril, but the responsible pathogens can be narrowed systematically by the following guidelines involving four steps of data acquisition.

2.2.1 Use epidemiologic clues

CNS infections occur in a relatively small subset of cancer patients. Patients with hematopoietic stem cell transplantation (HSCT) are a particularly high-risk group, while patients with leukemia or lymphoma represent more than a quarter of those with CNS infections, and 16% of the CNS infections in cancer patients occur among those who have primary CNS tumors.

Barrier disruption: Barrier disruption by shunts, monitoring devices, ventricular reservoirs, cranial surgery, central lines or ports, gastrointestinal surgery, urinary catheters, and loss of cutaneous or mucosal integrity, often treatment induced, leads to both bacterial and fungal infections. Two organisms, with potential to cause bacterial meningitis, particularly associated with gastrointestinal procedures, are *Streptococcus bovis* and *Listeria monocytogenes*. Another gastrointestinal pathogen of growing importance is *Strongyloides stercoralis*, a nematode that typically colonizes the gut without symptoms, but whose larvae carrying enteric pathogens may circulate causing Gram-negative bacillary meningitis.

Neutropenia: Bone marrow infiltration by leukemia, lymphoma, or solid tumors or treatment-induced

marrow failure predispose to all types of bacterial organisms, as well as *Aspergillus*, and special consideration must be given to blood transfusion-related infections such as adenovirus or, most recently, West Nile virus.

B-lymphocyte/immunoglobulin deficiency: Predisposing disorders include the leukemias, IgA deficiency, and multiple myeloma. Profound B-cell depletion occurs with the increasing use of the monoclonal antibodies alemtuzumab, rituximab, and related therapies. Patients with profound B-cell depletion are at risk for unusually severe forms of illness due to pathogens such as *Borrelia miyamotoi. Babesiosis,* and *West Nile virus.*[12]

T-lymphocyte depletion: A brief list of drugs introduced in the 20 years suffices to illustrate this diverse and numerically largest group of patients at risk for T-lymphocyte/macrophage-deficiency-related infections. The drugs include *alemtuzumab, cyclosporine, tacrolimus, sirolimus, azathioprine, bortezomib, fludarabine mycophenolate,* and *temozolomide.* Disorders producing T-cell deficiency as part of the disease process or requiring the aforementioned drugs include HIV/AIDS, lymphoreticular neoplasms, organ transplantation, and chronic corticosteroid use. It is in this group of patients that the various herpes viruses become most important as well as a spectrum of fungi including *Cryptococcus neoformans, Mucoraceae, Pseudallescheria boydii, Aspergillus* species, and *Candida* species. Toxoplasmosis is relevant in particular epidemiologic subgroups with high seroprevalence of this parasite.

2.2.2 Recognize clinical syndromes on physical examination

Meningitis/meningoencephalitis: The second step in diagnosis is to consider the broad category of neuroanatomical presentation. The clinician should categorize the patient's problem generally as either a meningitis/meningoencephalitis pattern versus one with focal signs suggesting a brain parenchymal process. This latter group can be further subdivided to suggest brain abscess, leukoencephalopathy, stroke-like vascular distribution (endocarditis, VZV, *C. neoformans*), and more clinically and radiographically restricted processes such as viral tropisms producing limbic encephalitis (HHV-6), movement disorders (West Nile), and brainstem syndromes (*Listeria*). The focal syndromes are outlined in Table 1 that categories common infections by CNS presenting site and magnetic resonance imaging (MRI).

2.2.3 Exclude noninfectious conditions mimicking CNS infection

The third diagnostic step is to consider the possibility that a noninfectious process is mimicking an infectious one. Focal deficit mimes are summarized in Table 1. However, this problem arises most commonly in the evaluation of lymphocyte-predominant meningitis

TABLE 1 Differential diagnosis of CNS infection by predominant focal clinical syndrome and MRI appearance: infections and infectious mimes.

Leukoencephalopathy	Stroke (s)	Limbic encephalitis	Mass lesion(s)	Brainstem	Spinal cord
Infections					
Varicella-zoster virus (VZV)	VZV	Herpes simplex types 1 and 2	Aspergillus. Fumigatus	Listeria monocytogenes	VZV
Progressive multifocal leukoencephalopathy(PML)	Emboli due to endocarditis	Human herpesvirus 1, 6	Bacteria (S. aureus/ bacteroides, P. acnes)	Cryptococcus neoformans	HTLV-1[a]
	Aspergillus		Nocardia asteroides	Varicella-zoster virus	Aspergillus
			Toxoplasma gondii	PML	PML
			EBV-virus-associated CNS lymphoma		
Noninfectious conditions					
Immune reconstitution inflammatory syndrome (IRIS)	Radiation-related arteriopathy Vasculitis	Hashimoto's encephalopathy	Immune reconstitution inflammatory syndrome (IRIS)	Wernicke's encephalopathy	
Reversible posterior leukoencephalopathy syndrome (RPLS) methotrexate, cyclosporine, cisplatin, L-asparaginase, tacrolimus, DMSO-treated stem cells, metronidazole, ifosfamide, cytosine arabinoside, gemcitabine	Nonbacterial thrombotic endocarditis	Paraneoplastic syndromes: Anti-Hu, Ma1, Ma2, voltage-gated potassium channel antibodies Anti-NMDA, AMPA, LGI1	Secondary tumor: Lymphoma Astrocytic tumor Metastases	Osmotic demyelination syndrome	
Acute disseminated (toxic) leukoencephalopathy	CNS vasculitis (graft-vs-host disease, granulomatous arteritis)	Repetitive seizures	Radiation necrosis	Graft-vs-host disease	
Osmotic demyelination syndrome (pontine and extrapontine)	Chemotherapy				
Amphotericin (mainly frontal and post-XRT)	Cytomegalovirus[b]				
Rituximab	DMSO preservative stem cell infusion				
Valproate					
Acyclovir					

[a] HTLV-1 = human T-lymphotropic virus, type 1, acquired from contaminated blood.
[b] Variable manifestations: diffuse encephalitis, mass lesions, myelitis, polyradiculitis.
ADEM, acute disseminated encephalomyelitis; CIDP, chronic inflammatory demyelinating polyneuropathy; GBS, Guillain-Barré syndrome; PML, progressive multifocal leukoencephalopathy; PNET, primitive neuroectodermal tumor.

whose infectious and noninfectious etiologies are summarized in Table 2. Of particular importance are drug-related adverse effects that should be familiar to the consultant. Amphotericin B and valproic acid have been associated with a parkinsonian-like state that is reversible.[13] Cefetaxime and cefepime have been associated with encephalopathy and nonconvulsive status epilepticus, even in the absence of renal failure.[14,15] Particularly important to recognize is ifosfamide encephalopathy, a potentially fatal development that can occur in 10%–30% of the patients during or after the intravenous infusion, which is seen more commonly in hypoalbuminemic patients or following cisplatin therapy, and whose specific antidote is methylene blue.[16] Limbic encephalitis can be caused by several types of herpes viruses but also can be due to paraneoplastic processes associated with many different tumors. Infection-triggered autoimmunity is a recently recognized phenomenon, leading to descriptions of NMDA receptor encephalitis following herpes virus infection.[17,18] Immune checkpoint inhibitor (ICI)-related encephalitis can lead to clinical and neuroradiographic findings similar to those of infections.[19,20] Autoimmunity in the form of symptomatic emergence of paraneoplastic antibody syndromes represents another potential confounder in the exclusion of infectious processes.[21]

TABLE 2 Major causes of lymphocyte-predominant meningitis syndrome in cancer patients.

Infectious etiologies		Noninfectious Etiologies
Viruses	Diagnostic test (on CSF, unless indicated)	
Enteroviruses	Polymerase chain reaction (PCR)	Adverse drug reactions
Herpes simplex Types 1,2,6	PCR[a]	NSAIDS
Varicella-zoster	PCR, virus-specific antibody	Cyclooxygenase-2 (Cox2)
Epstein-Barr	PCR	Inhibitors
HIV	Blood	OKT3
West Nile	PCR, virus-specific antibody[a]	Valacyclovir
		Azathioprine
		Isoniazid
Bacteria		IVIG
		Intrathecal chemotherapy
Partially-treated bacterial meningitis	Culture, low CSF glucose	methotrexate, ARA-c cephalosporins
Endocarditis	TEE, blood cultures	
Parameningeal infection	MRI/CT of suspected region(s)	ADEM
Ventriculitis postoperatively		CNS vasculitis
Neurosurgical procedure	Shunt/reservoir tap and removal	Arachnoiditis
M. pneumoniae	Chest X ray, serology	Prolonged status epilepticus
M. tuberculosis	PCR, culture	PTLD
B. burgdoferi	Blood: ELISA, Western blot	
T. pallidum	VDRL/RPR,blood and CSF	
	MHATP, FTA-ABS, blood	
Fungi		
Cryptococcus neoformans	India ink, cryptococcal antigen	
Histoplasma capsulatum[b]	serum: CSF antibody ratio	
Coccidioides immitis[b]	serum: CSF antibody ratio	

[a] For viruses such as HHV-6, PCR does not establish active infection and causation for clinical syndrome; acute and convalescent sera for virus-specific IgG to viruses may be obtained to confirm pathogenicity of specific infectious agent. PCR may be negative in WNV neuroinvasive disease-see text for the discussion of serum:CSF antibody ratios.
[b] If patient lives in or has been in appropriate geographic regions.
ADEM, acute disseminated encephalomyelitis; ADEM, acute disseminated leukoencephalopathy; ELISA, enzyme-linked immunosorbent assay; HIV, human immunodeficiency virus; IVIG, intravenous immunoglobulin; NSAIDS, nonsteroidal anti-inflammatory agents; PTLD, posttransplantation lymphoproliferative disorder; TEE, transesophageal echocardiogram.

II. Nonmetastatic neurological complications

2.2.4 Order cost-effective laboratory studies

Based on the first three sets of data, a cost-effective laboratory workup is designed that can include serologic tests, lumbar puncture, CT, MRI, and, at times, brain or meningeal biopsy. Spinal fluid interpretation can be very confusing since a positive polymerase chain reaction (PCR) test does not necessarily indicate pathogenicity and an initially negative test may need to be repeated to confirm the diagnosis. Similarly, neuroimaging poses special issues: contrast enhancement on both CT and MRI may be reduced in cancer patients taking corticosteroids. Renal function must be considered in patients with glomerular filtration rate (GFR) under 30 cc/min, which precludes administration of gadolinium contrast due to the possibility of nephrogenic systemic fibrosis syndrome.[22] Diffuse meningeal enhancement and hyperintensity of CSF on fluid-attenuated inversion recovery (FLAIR) MRI sequences can be seen not only during infections but also after lumbar puncture or after blood-brain barrier disruption due to neoplastic, chemical, or ictal processes. The use of MRI sequences including diffusion-weighted imaging (DWI), apparent diffusion coefficient (ADC) maps, and MR spectroscopy have improved the ability to distinguish between tumor-, infection-, and radiation-related tissue injuries. For example, using ADC cutoffs, it is possible to differentiate lymphoma from toxoplasmosis with good reliability.[23] Magnetic resonance angiography (MRA) or conventional intraarterial arteriography can be helpful in the evaluation of venous sinus thrombosis; can reveal arteritis associated with VZV, mycobacteria, and mucormycosis; and can exclude large infectious aneurysms in endocarditis.

Lumbar puncture is a major diagnostic procedure. In a cancer patient with a known solid tumor, a screening CT or MRI is recommended prior to spinal tap to exclude metastatic disease or other mass lesions. Platelet counts of less than 50,000 should be corrected with platelet transfusion prior to the lumbar puncture. The interpretation of the resulting CSF analysis depends on the immunosuppressed patient's ability to mount an inflammatory response. All cerebrospinal fluid should be sent for a cell count and differential, glucose, and protein concentration, routine bacterial cultures and, in appropriate situations, cytology and flow cytometry. A predominantly polymorphonuclear leukocyte (PMN) pleocytosis and CSF cell count of greater than 200 is suggestive of bacterial meningitis, though this range can be seen in West Nile meningoencephalitis as well as with some fungi.[24,25] CSF glucose less than 50% of the concomitant blood glucose supports a bacterial, fungal, or neoplastic process. Diagnostic tests include PCR and antibody studies. The detection of viral or bacterial specific IgM is definitive evidence of CNS infection, while IgG in CSF is not definitive as it can be passively transferred from serum to CSF. The *antibody index* is the ratio of CSF fluid/serum quotients for specific antibodies and total IgG. This index discriminates between a blood-derived and a CNS-derived specific antibody. For example, in VZV encephalitis, myelitis, or vasculopathy, VZV DNA by PCR and VZV IgM are detected early but may then become negative at which time anti-VZV-IgG is detectable. More than 1 week after symptom-onset, viral-specific IgG and the antibody index are helpful in determining causation of a CNS infection.

Meningitis/encephalitis panels provide a broad screen for multiple organs. Metagenomic next-generation sequencing (mNGS), available at an increasing number of centers, is a rapid unbiased screen for a broad range of human pathogens when selective serology and CSF microbiological testing is unrevealing. The Wilson UCSF Lab identified 35 infections in 151 patients, more than one-third of which were not diagnosed by conventional serologic and CSF testing.[26] This method eventually may supplant many of the current single-agent laboratory tests.

Brain or meningeal biopsy remains the definitive procedure in the relatively few situations in which the etiology of a process remains unclear after MRI, MRS, and CSF studies.

3 High-risk patient groups

3.1 Transplant recipients (hematopoietic stem cell and solid organ)

More than 30,000 autologous transplantations are performed annually worldwide, two-thirds for multiple myeloma or non-Hodgkin lymphoma. Acute myelocytic leukemia (AML), neuroblastoma, ovarian cancer, and germ cell tumors are other neoplasms for which the procedure is used. Allogeneic transplantation is performed for more than 15,000 patients worldwide annually, nearly half for acute leukemias. The majority of the others are for CML, myelodysplastic syndromes, non-Hodgkin lymphoma, chronic lymphocytic leukemia (CLL), multiple myeloma, and Hodgkin's disease.[27]

Hematopoietic stem cell transplantation may be from a closely matched sibling or unrelated donor (allogeneic transplantation) or from the patient himself (autologous transplantation). HSCT refers both to collection of peripheral blood after stimulation by hematopoietic growth factors and to harvesting of cells from bone marrow. Peripheral blood stem cells, which contain more T-cells than marrow, increase the incidence and duration of graft-versus-host disease (GVHD). The cell surface marker CD34 is used to estimate peripheral blood stem cells mobilized from the marrow. Leukemia patients have the highest rate of neurological complications among HSCT recipients.[28] Many developments in transplantation technology have occurred in the last few years, including increasing use of unrelated donors and peripheral blood stem cells and changes in both conditioning regimens and antibacterial, antifungal, and antiviral prophylactic regimens.

Risk factors for infection in HSCT and solid transplant recipients include exposures to infections in the recent or distant past that may affect either the organ donor or the recipient. What has been called the "net state of immunosuppression" is a concept encompassing all factors that may contribute to a host's inability to combat infection, including exogenous immunosuppression, complications of surgery, underlying immune deficits, drugs, and viral coinfections.[29,30] Table 3 summarizes the likely infectious and noninfectious syndromes based on time after transplantation. HSCT recipients are vulnerable to many organisms presenting with meningoencephalitis and to a more restricted number of pathogens producing mass lesions.

The "top ten" organisms producing mass lesions include four fungal pathogens, *Aspergillus*, *Zygomyces*, *C. neoformans*, and *Candida* species; three viruses, JCV, VZV, and Epstein-Barr virus (EBV); two bacterial organisms,

TABLE 3   Time course of CNS infections after hematopoietic stem cell transplantation.

	<1 Month	1–6 Months	>6 Months
Mechanisms/risk factors	**Donor-derived** Nosocomial Barrier disruption[a] Neutropenia	Activation of latent or new opportunistic infection Type of immunosuppressive regimen/steroids/calcineurin inhibitors	Intensity of B-/T-cell depletion GVHD Viral reactivation
Community-acquired organisms			
Bacteria	MRSA, VREC Coagulase-negative staphylococci		**MRSA** *S. pneumoniae* Nocardia Listeria
Fungi	*Aspergillus, Candida, Cryptococcus*	*Cryptococcus* *Aspergillus*	*Aspergillus* *Mucoraceae*
Viruses	**LCMV, HIV WNV, Rabies, CMV, HHV6** Adenovirus, Coxsackie B4 *Toxoplasma gondii* *Strongyloides stercoralis*	VZV, HHV6, HSV 1,2	**VZV, WNV** CMV PML EBV (PTLD or lymphoma)
Parasite	***Trypanosoma cruzi***	*Toxoplasma gondii*	*Toxoplasma gondii*
Noninfectious processes	Metabolic encephalopathy (including CPM) Drug-related encephalopathy: chemotherapy, antiepileptic therapy, antibiotics—see text) Parkinsonism (valproate, amphotericin B) DMSO-related stroke RPLS Calcineurin inhibitors sirolimus: Seizures (cefepime, imipenem) Intracranial hypotension after LP Engraftment syndrome Delirium due to organ failure SDH due to coagulopathy Intraparenchymal brain hemorrhages (AML)	Metabolic, including, reactivation of HBV, HCV Engraftment syndrome ADEM Steroid: psychosis/brain atrophy Wernicke's encephalopathy	GVHD (polymyositis, myasthenia GBS, or CIDP) Secondary malignancy, including brain tumors[b] Disease relapse

[a] *Mucositis, bowel surgery, skin (dialysis, catheters/ports), craniotomy +/– hardware.*
[b] *Almost all patients had received cranial or craniospinal radiation- tumors include astrocytoma, primitive neuroectodermal tumors (PNETs), and meningioma).*
Organisms associated with donor- and community-acquired infections are indicated in boldface type in their respective columns.
ADEM, acute disseminated encephalomyelitis; *AML*, acute myelocytic leukemia; *CIDP*: chronic inflammatory demyelinating polyneuropathy; *CMV*, cytomegalovirus; *DMSO*, dimethyl sulfoxide (preservative for stem cell infusion); *EBV*, Epstein-Barr virus; *GBS*, Guillain-Barré syndrome; *GVHD*, graft-vs-host disease; *HHV 6*, human herpesvirus 6; *HSV*, herpes simplex virus; *LCMV*, lymphocytic choriomeningitis virus; *LP*, lumbar puncture; *PTLD*, posttransplantation lymphoproliferative disorder; *RPLS*, reversible posterior leukoencephalopathy syndrome; *VZV*, varicella-zoster virus; *WNV*, West Nile virus.

Nocardia asteroides, Mycoplasma tuberculosis; and one parasite, *Toxoplasma gondii*. Infections occur within the first 4 months in 87% of the patients with a high mortality rate of 47%.[20] A recent French series emphasizes cerebral toxoplasmosis and CMV in CD 34+ selected autologous HSCT recipients.[4] The emphasis on local clinical patterns of disease is noteworthy, as seroprevalence for toxoplasmosis is higher in Europe than in the United States. Invasive fungal infections are more common in the first month after transplantation in the USA.

3.1.1 Infections according to time from transplantation

Early posttransplant period (0–30 days)

Nosocomial infections. The period of neutropenia before engraftment is a period of infection risk from hospital-acquired organisms and from infections acquired from the host tissue. Candida fungemia occurs during this period as a complication of line sepsis and may be difficult to diagnose because the picture is one of nonspecific meningoencephalitis.

Reactivation of infection in recipient. More than 350 million people worldwide have hepatitis B virus (HBV). In the United States, the serologic prevalence of hepatitis B surface antigen (HBsAg) is less than 1%, but up to 5%–15% in immigrants from Asia, Africa, the Middle East, and Eastern Europe. Preventive therapy with lamivudine for patients who test positive for HBsAg and are undergoing chemotherapy may reduce the risk for HBV reactivation and HBV-associated morbidity and mortality.[31,32] Activation of HBV with attendant hepatic dysfunction can produce encephalopathy, making recognition of CNS infection more difficult, and may also make patients more susceptible to chemotherapeutic or infectious toxicity. Hepatitis C virus (HCV) rarely affects the CNS but can increase susceptibility to and morbidity from other pathogens.

Donor-derived infections. In the past several years, the evolving spectrum of tissue-acquired infections has been highlighted by several dramatic examples of West Nile virus (WNV), *Trypanosoma cruzi*, lymphocytic choriomeningitis virus, *Balamuthia mandrillaris*, *C. neoformans*, lymphocytic choriomeningitis virus, and rabies from donors whose disease was not recognized before the organ donation, and whose mortality rate was close to 100% in these heavily immunosuppressed patients.[33,34] The listed cause of donor death included stroke, anoxia, and meningoencephalitis, emphasizing that careful donor assessment is necessary to exclude potentially fatal transmission of infection to vulnerable recipients.[35]

West Nile virus. Transmission of WNV by transplantation yields particularly virulent neuroinvasive disease in immunocompromised hosts.[36] WNV, unlike HSV, can produce either diffuse or focal encephalitis. Clinical manifestations range from meningoencephalitis to acute focal, flaccid paralysis. The presence of a febrile illness with poliomyelitis-like anterior horn cell syndrome should raise suspicion of this virus.[37] In a minority of patients, movement disorders including parkinsonism, involuntary movements such as tremor, or myoclonus develop during or after the acute encephalitis.[38] Seizures are less frequent in WNV (5%) than HSV (40%) reflecting the subcortical WNV viral injury to thalamus, globus pallidus, substantia nigra, dentate nuclei, mesencephalon, and spinal cord.[39] Early on MRI may be normal but later will show progressive involvement of deep gray matter structures.

WNV antibodies may take more than 1 week to become positive. Viral DNA detection by PCR has a low yield since the viremia may have cleared by the time of clinical presentation. Positive serum IgG and IgM are evidence of exposure to WNV and should not be considered to confirm the diagnosis of CNS parenchymal infection. IgM in serum may persist longer than 500 days after infection.[40] The definitive test is CSF IgM antibody. A defining characteristic of WNV meningitis is a persistent neutrophil-predominant pleocytosis with normal glucose. No drugs are effective for WNV, but high WNV-antibody titer immunoglobulin, antisense oligonucleotides, and several vaccines are under study.[24]

Endemic infections. The differential diagnosis of reactivation of preexisting recipient infections must be considered in light of the patient's recent and remote geographic location and antimicrobial prophylactic regimen. Routine prophylaxis has eliminated most early infections due to *Listeria*, toxoplasmosis, *Nocardia*, and herpes viruses. Reactivation of tuberculosis in demographic groups at high risk, including AIDS patients and those of Asian or African origin, must be considered. Endemic mycoses such as *C. neoformans*, *Aspergillus* species, and parasites are currently the most common early host-derived pathogens. Risk factors for invasive fungal infection include HCV and HBV infections, hepatic and renal dysfunction, and antibacterial therapy.

Engraftment syndrome or human herpesvirus 6. Rash, fever, and headache mimicking CNS infection may occur in HSCT recipients just as the absolute neutrophil count exceeds 500 mg/m^3. This presentation, usually at 2–4 weeks after transplantation, is known as the *engraftment syndrome* due to upregulation of cytokines by neutrophils, after administration of colony-stimulating factors and must be distinguished from HHV-6 encephalitis.[41] The rather specific syndrome of posttransplant acute limbic encephalitis (PALE) with confusion, short-term memory problems, sleep disturbance, and seizures has a specific MRI appearance, with hyperintense signal abnormality in the amygdala uncus, entorhinal area, and hippocampus on T2-weighted and FLAIR MRI (see Fig. 1).[42,43] Memory impairment dominates the clinical picture, whose differential diagnosis includes

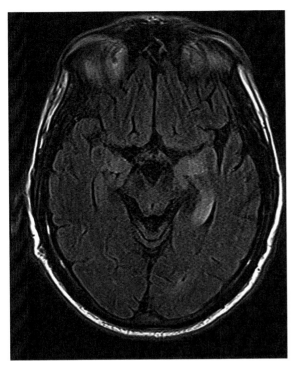

FIG. 1 MRI FLAIR image demonstrates abnormal signal in hippocampi bilaterally in HSCT patient with confusion and memory loss at the time of engraftment. HHV-6 PCR was positive in CSF.

infections in organ transplant recipients, with nearly one-third involving the CNS. Because of antifungal prophylaxis with voriconazole at the onset of transplantation, the infections tend to occur later than they did in earlier cohorts of patients.[53] It is also during this interval that *C. neoformans*, the most common infection in this patient group, becomes important, though it can occur earlier.

More than six months posttransplantation. Infectious complications remain a major problem long after HSCT. Patients requiring continued high-dose immunosuppression continue to be at risk for CNS infection throughout their course and confusion with disease recurrence becomes part of the difficult differential diagnosis. Graft-vs-host disease (GVHD) should be viewed as a marker for degree of immunosuppression, though it infrequently involves the CNS. However, autopsy series reveal lymphocytic infiltration consistent with possible viral infection.[54] Regardless of mechanism, GVHD is an independent risk factor for late infection as is CMV infection.

Toxoplasma gondii remains an important pathogen during this period, producing multiple mass lesions, predominantly in the basal ganglia (see Fig. 2).[55] Routine use of trimethoprim-sulfamethoxazole reduces the risk. Low CD4+ count (<200 cells/μL) and GVHD are risk

paraneoplastic limbic encephalitis associated with a variety of neoplasms,[44-46] voltage-gated potassium channel autoimmune limbic encephalitis,[47] herpes simplex virus encephalitis, status epilepticus,[48] Wernicke's encephalopathy, and immunosuppressive treatment-related toxicity. Additional clinical features include inappropriate antidiuretic hormone secretion and temporal EEG abnormalities. Cognitive recovery is often incomplete despite appropriate foscarnet or ganciclovir. Autopsy in a recent series showed profound neuronal loss in the amygdala and hippocampus.[49] Human herpesvirus 7 also has been isolated from HSCT recipients whose clinical syndromes included transverse myelitis and optic neuritis, as well as meningitis.[50-52] See further discussion of diagnosis and treatment of HHV-6 in agent-specific management section at the end of this chapter.

One to six months posttransplantation. This is the period of greatest CNS infection risk. As neutrophil counts rise, the risk of bacterial infection decreases, but opportunistic fungi and parasites, herpes viruses, and CMV emerge. Ganciclovir use during the first 3 posttransplant months has reduced herpes virus encephalitis and CMV infections. *Aspergillus*, PML, and VZV become important during this period. Other unusual pathogens are increasingly recognized in the HSCT population. *Scedosporium apiospermum*, a form of *P. boydii*, is a mold that is widespread in the environment and can produce a disseminated infection, including the CNS. These fungi now account for one quarter of non-*Aspergillus* mold

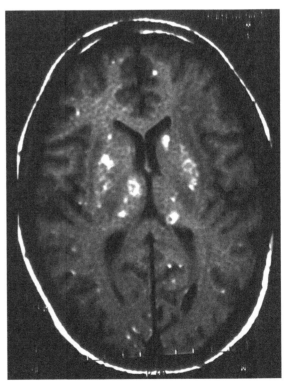

FIG. 2 Two months after receiving allogeneic stem cell transplant patient developed confusion. MRI FLAIR shows multiple deep bilateral lesions. Serology was positive for *Toxoplasma gondii* and lesions resolved with appropriate antibiotic treatment.

factors. PCR is the primary diagnostic tool and can be performed on blood, ocular samples, and bronchoalveolar lavage fluid.[56]

Patients on high-dose maintenance immunosuppression remain at risk for bacterial infections such as *Listeria* and *Nocardia* (see Fig. 3). Immunocompromised hosts tend to have a more diffuse cerebritis picture, as compared to the rhombencephalitis of immunocompetent individuals. *Nocardia* may be hard to distinguish from other abscess-producing organisms. PML due to JCV, a papovavirus, becomes a concern at this stage posttransplantation and can appear after many years, while in a recent series, all fungal infections occurred within the first year after HSCT.[57]

3.1.2 Six special situations

The changing spectrum of conditioning regimens and immunosuppressive agents has led to several newly recognized diseases and risk patterns, and to an expansion of agents associated with the syndromes. Six special clinical situations occur predictably in HSCT patients, due both to the types of diseases treated and to the transplantation treatment:

(1) *Posttransplantation lymphoproliferative disorder (PTLD)*. PTLD is a heterogeneous group of disorders with polyclonal-B-cell infiltration of multiple organ systems, including the allograft, in up to 10% of the solid organ transplant recipients. In up to 35% of these patients, the CNS is involved and is the only site of abnormality in 85% of these patients or at least 3% of all PTLD patients.[58] PTLD

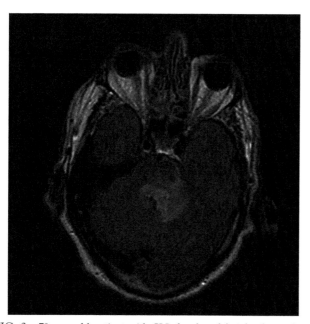

FIG. 3   79-year-old patient with CLL developed facial palsy and ear pain. Examination did not suggest VZV. MRI shows extensive FLAIR abnormality in brainstem that enhanced after administration of intravenous gadolinium contrast (not shown). *Listeria* was cultured from CSF.

may occur months to years after transplantation. Clinically, PTLD can vary from a mononucleosis-like syndrome to malignant lymphoma. Primary EBV infection in seronegative allograft recipients from seropositive donors, T-cell depletion by anti-thymocyte globulin, nonmyeloablative stem cell transplants, and CMV coinfection are all risk factors. More than 90% of these lesions are EBV-associated lymphocytic proliferation of B-cell origin and thus on brain biopsy can be distinguished from even rarer cases of CNS GVHD.

The introduction of PCR to monitor EBV reactivation is a potential tool to predict the risk of developing PTLD. B-cell depletion of the graft or restoration of EBV-specific T-cell immunity by infusion of donor lymphocytes and preemptive treatment with B-cell depletion by the CD 20 antibody rituximab have been used, but an optimal treatment strategy remains unclear.[59]

(2) *Rituximab*. Rituximab is a chimeric mouse-human monoclonal antibody against CD20 on B-lymphocytes. While it does not affect T-cells and therefore had been deemed safe with respect to opportunistic infection risk,[60] it produces profound B-cell depletion that persists for several months. Delayed-onset neutropenia has been reported between 1 and 5 months after infusion. Reactivation of hepatitis B virus has been reported and cases of PML and CMV are reported unusually early in patients treated with rituximab.[61] A recent advisory from the American College of Rheumatology underscores the labeling change for rituximab that now includes information about patients with non-Hodgkin lymphoma who developed serious viral infections with CMV, HSV, VZV, and PML after rituximab. Multiple cases of PML have been reported following rituximab therapy in patients with non-Hodgkin's lymphoma and other hematologic malignancies, mostly in combination with chemotherapy or stem cell transplantation.[62] A case of BK papovavirus-related leukoencephalopathy after rituximab has been described as well.[63] While rituximab may improve the immediate outcome for such lymphoma patients when added to the CHOP chemotherapy regimen (cyclophosphamide, doxorubicin, vincristine, and prednisone), the increasing incidence of infections may outweigh the benefits.[64] At the author's institution, we have seen a case of PML after only three cycles of CHOP with rituximab (see Fig. 4).[65]

Perhaps not surprisingly, the use of other monoclonal antibodies, including imatinib, has become the focus of attention with respect to CNS infections. Imatinib mesylate, used for CML and Ph + acute

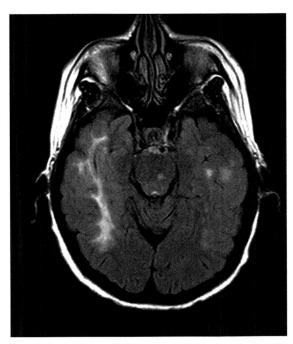

FIG. 4 Patient with recent diagnosis follicular lymphoma who had received only three courses of CHOP with rituximab. White matter lesions progressed despite cessation of chemotherapy and were rapidly fatal. CSF was PCR-positive for JCV confirming diagnosis of progressive multifocal leukoencephalopathy.

lymphoblastic leukemia (ALL), has led to reports of nondisseminated varicella-zoster infection without stem cell transplantation, a previously extremely rare complication.[7]

(3) *Alemtuzumab (Campath, Lemtrada).* Alemtuzumab is a monoclonal antibody directed against the CD52 glycoprotein expressed on both B- and T-lymphocytes, monocytes, and natural killer cells. The drug is used for B-cell lymphocytic leukemia and in stem cell and organ transplantation. CMV reactivation with attendant *Listeria* risk is an important problem. In one series, 10% of the patients developed infections including CMV, BK virus, PTLD, HHV-6, *Nocardia,* and fungal infections.[66,67] Patients receiving alemtuzumab for rejection therapy are at higher risk for infection than are those who receive the drug for induction therapy, with a median time to infection after alemtuzumab of 84 days.

(4) *Immune reconstitution inflammatory syndrome (IRIS).* IRIS is a series of infection reactivations recognized initially in the HIV/AIDS population, with the advent of effective antiretroviral therapy. However, the scope of IRIS must be expanded to include many non-HIV clinical situations. For example, administration of certain drugs (especially anticonvulsants) for 2–6 weeks triggers immunosuppression and cessation of such therapy

provokes the reactivation of various viruses, including Epstein-Barr virus (EBV), human herpesvirus (HHV) 6, and cytomegalovirus (CMV) whose inflammatory manifestations can resemble IRIS, and which most properly can be referred to as drug-induced hypersensitivity syndrome (DIHS).[68] In recent years, many novel drugs with immunological actions have been developed; these complement "conventional" immunosuppressants. The new drugs principally target malignant tumors and autoimmune diseases. Various immune-related adverse events (irAEs) have been described, many of which are common with sequelae of DIHS and events following antiretroviral therapy (ART); thus, these adverse events can be regarded as IRIS. If immunosuppressive agents are stopped or reduced abruptly, immune reconstitution will be accelerated, possibly exacerbating adverse events. In the neurologist's purview, the primary infections are cryptococcal meningitis, VZV, PML, and tuberculosis, which are increasingly seen in cancer patients whose immune systems are rebounding from intensive immunosuppression.[69,70] An immune reconstitution syndrome can be associated with cryptococcal meningitis infection in organ transplant recipients, as the degree of immune suppression is reduced to help fight infection. The optimal safe schedule for tapering of immune suppression remains unclear.[71] The host's inflammatory response may be vigorous, leading to exuberant leptomeningeal enhancement and new parenchymal enhancement and/or edema in the areas of prior CNS involvement, possibly causing confusion with recurrent infection or tumor (see Fig. 5).[72] The role of corticosteroids to blunt the host's inflammatory response remains uncertain, though the author has found them to be useful in patients with markedly raised intracranial pressure due to IRIS.

(5) *Reversible posterior leukoencephalopathy syndrome (RPLS).* The term RPLS describes headaches, seizures, cortical visual disturbances, and delirium associated with transient cerebral lesions on FLAIR MRI (see Fig. 6). First described in 1996; it is a form of vasogenic edema occurring in association with renal failure, hypertensive encephalopathy, preeclampsia, arteriography, and a growing list of immunosuppressive agents— including many common chemotherapy drugs[73] and angiogenesis modifiers, such as the vascular endothelial growth factor inhibitor bevacizumab, a drug indicated for colon and renal cancers. The drug increasingly is used for primary brain tumors and cerebral metastases to control radiation-related edema.[74] The syndrome usually involves

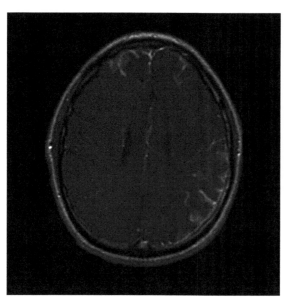
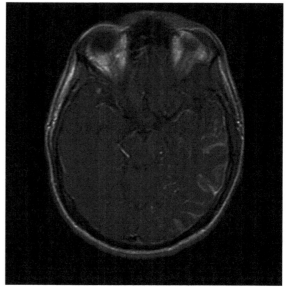

FIG. 5   Patient with cryptococcal meningitis returned 1 month after successful treatment with prolonged left hemisphere focal seizures and markedly elevated pressure. Gadolinium-enhanced MRI shows prominent leptomeningeal enhancement. Cryptococcal antigen was negative and CD4 + count had trebled since his prior admission. Patient was treated for IRIS with corticosteroids with resolution of symptoms and MRI abnormalities.

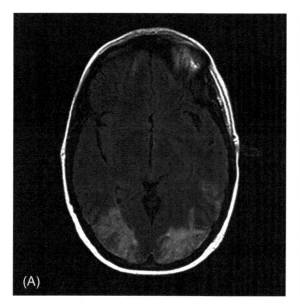
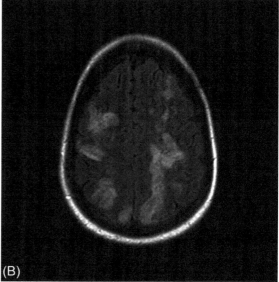

FIG. 6   Patient treated for a widespread systemic relapse of breast cancer with one dose of gemcitabine and presented 9 days later with confusion and visual changes. MRI FLAIR (A) and gadolinium (B) showed multiple areas of largely white matter pathology consistent with RPLS (PRES) lesions partly resolved over the next several weeks.

the posterior cerebrum but may involve the brain asymmetrically, as well as producing isolated cerebellar or brainstem syndromes with variable enhancement.[75,76] Because the radiographic findings are so variable, the syndrome can be confused with many other processes such as acute disseminated encephalomyelitis (ADEM), arteritis, CNS lymphoma, PML, and, in the appropriate settings, radiation therapy, brain metastases, and stroke.

(6) *Graft-**versus**-host disease (GVHD)*. HSCT recipients are more heavily immunosuppressed than their counterparts with solid organ transplants, and the immunosuppression required chronically to prevent GVHD renders them susceptible to viral, parasitic, and fungal infections. Episodes of acute GVHD are treated with increased immunosuppression and can lead to CMV or *Aspergillus* infection. Most of the manifestations of GVHD involve skin, bowel, and liver and when

the nervous system is affected polymyositis, peripheral neuropathy (consistent with chronic inflammatory demyelinating polyneuropathy [CIDP]), and myasthenia gravis are the principal nervous system syndromes. The CNS is involved infrequently by a recurring cerebellar syndrome, associated with progressive motor weakness, that parallels GVHD activity and CNS angiitis; stroke-like onset also has been described.[77,78] Brain biopsy has revealed a granulomatous encephalitis with profound perivascular lymphocytic infiltration, composed predominantly of T-lymphocytes (CD3) of donor origin.[79,80] MRI changes in chronic GVHD include atrophy and white matter lesions, but most of these patients have also been treated with corticosteroids and calcineurin inhibitors, so that the relative role of each factor is not identifiable.

3.2 Neurosurgical patients

Patients with brain tumors, including both primary CNS malignancies and systemic metastases, who have had neurosurgical procedures account for at least one quarter of CNS infections among cancer patients. Risk factors include barrier disruption, often with poor wound healing after multiple procedures, prolonged corticosteroid use, and radiation therapy. Recent trends in long-term chemotherapy with attendant low CD4+ counts in patients with low-grade tumors add to the potential cancer population at risk for infection. For diagnostic purposes, it is useful to think of neurosurgical patients in three distinct groups: recent surgical, nonsurgical, and patients with indwelling CNS "hardware." Patients should be considered in the surgical group if they have undergone a procedure, other than hardware placement, within 30 days, while nonsurgical patients have had either no procedure or one more remote than a month before infection. The hardware group is patients with CNS infection at any time afterward device placement.

3.2.1 Recent neurosurgery group

Bacterial meningitis: Bacterial meningitis usually arises from hematogenous dissemination after blood stream infection or from contiguous spread from an otic or sinus source. Illustrating the change in likely bacterial meningitis pathogens in cancer patients is a Memorial Sloan Kettering 2008 report updating a 1977 retrospective review. The earlier study showed an overall decrease in the incidence of CNS infections with an increase in meningitis caused by *L. monocytogenes.*[81] The 2008 report revealed that 78% of the patients had had prior neurosurgery. Over 60 percent had indwelling ventricular catheters, includ-

ing Ommaya reservoirs and 15 ventriculoperitoneal shunts. In contrast to the earlier report, few patients had Gram-negative bacilli or fungi. Organisms included 68% Gram-positive cocci, a distribution these and other authors have attributed to the significant increase in neurosurgical procedures in cancer patients, use of prophylactic antibiotics, and shortening of neutropenia with colony-stimulating factors.[82,83] Only 2 of 77 patients had *Listeria.* Only 5% of the presentations included the classic triad of fever, nuchal rigidity, and mental status changes. Among neutropenic patients, 2 had completely normal CSF. VP shunt infection presented with mental status changes and 30-day mortality was 13%.[84] In this same study, 91% of the staphylococcal infections occurred in the surgical and hardware groups.

Patients with "hardware": In the hardware group, infections were roughly equally divided between coagulase-negative staphylococci and *S. aureus* infections. Patients with VP shunts were more likely to present with altered mental status and less likely to have headache or seizure as part of their symptoms. Access of an Ommaya reservoir led to 31 positive CSF cultures, only 29% of which were felt to be indicative of active infection.[72]

Nonsurgical brain tumor patients: With increasing use of chronic chemotherapy in patients with fairly indolent neoplasms, infectious complications are likely to increase. Preliminary data on 60 patients chronically receiving temozolomide showed that median CD4+ count decreased from 640 to 250 after 3 months of treatment. Over half of the patients (57%) had CD4 counts <200 and 13 patients (22%) had CD4<100 at some point during the study period.[85]

The nonsurgical group of brain tumor patients also remains at risk for meningitis, but *S. aureus* becomes a less likely pathogen more than 1 month from the surgical procedure. In the Memorial Sloan Kettering study, all six cryptococcal infections occurred in the nonsurgical group. Both cases of *Streptococcus pneumoniae* meningitis occurred in the nonsurgical group. Cranial irradiation can exacerbate wound-healing problems and has been reported to reactivate HSV, causing encephalitis.[86]

Brain tumor patients are exposed chronically to some of the highest doses of corticosteroids used clinically and are therefore at risk for *P. jirovecii* pneumonia. They should receive trimethoprim/sulfamethoxazole prophylaxis three times weekly. Such therapy also decreases the risk of listeriosis, toxoplasmosis, nocardiosis, and urinary tract infections.

At the present time, there are no specific investigations of the efficacy of VZV vaccination to patients under age 60 about to undertake a prolonged corticosteroid course, but current data suggest that vaccination for everyone over age 60 may reduce the frequency of symptomatic VZV reactivation. Neurosurgical patients on corticosteroids who develop zoster should receive intravenous acyclovir therapy (see Table 4).

Reactivation of other viruses such as hepatitis B and C complicates the course, and the ensuing metabolic encephalopathy or liver function abnormalities erroneously may be attributed to either brain tumor or chemotherapy, respectively. Rituximab has been associated with early or delayed hepatitis B when given alone or in combination with R-CHOP therapy. An example of hepatitis B reactivation in a glioblastoma patient treated with temozolomide that was successfully treated with lamivudine illustrates the increasing scope of the problem, and screening for HBV prior to chemotherapy is suggested by the authors.[87]

4 Management of common CNS infections

The concluding sections of this chapter provide management recommendations for pathologically verified infections. Consultants should be attuned to specific antibiotic resistance patterns and nosocomial infections in their own institutions, as well as to renal and hepatic impairment in specific patients and adjust recommendations accordingly. Table 4 summarizes specific antimicrobial recommendations.

4.1 General medical management issues

Corticosteroid supplementation. Corticosteroid supplementation is necessary in every cancer patient with a CNS infection who has been treated with systemic corticosteroids in the recent past and, as a result, is at risk for adrenal insufficiency. Adrenal insufficiency presents as hypotension unresponsive to volume repletion and requiring urgent intravenous hydrocortisone.[88] The role of corticosteroids in improving survival from shock during sepsis in patients not recently on corticosteroids remains a source of some clinical controversy. Two meta-analyses and an international consensus guideline suggest stress-dose steroid therapy only after blood pressure is poorly responsive to fluid and vasopressor therapy. Recombinant activated protein C can be helpful in patient stabilization as well.[89–91] Another recent report, however, showed no survival advantage.[92] Beyond providing steroid support, the clinician should be aware of other strategies to modulate the inflammatory cascade during severe sepsis. A recent meta-analysis of intravenous immunoglobulin therapy concluded that a survival benefit was observed for patients who received polyclonal intravenous immunoglobulin.[93]

Seizures. Treatment of seizures in cancer patients with CNS infections is complex. Most seizures are iso-

TABLE 4　Treatment of common CNS infections in cancer patients.

Organism	Antibiotic regimen and alternatives intravenous route except as noted		
Bacteria			
Staphylococcus aureus	Methicillin-sensitive: Nafcillin 2g q4h plus Cefotaxime 2g q6h		
	Methicillin-resistant: Vancomycin 500 mg q6h +/− intraventricular Vancomycin 20 mg/d		
Streptococcus pneumoniae			
Penicillin intermediate resistance	MIC[a] < 0.1–1 µg/mL	Cefepime 2g q12h or Ceftriaxone 2g q12h or Cefotaxime 2g q4h	
Penicillin resistant	MIC > 1 µg/mL	One of the cephalosporins mentioned previously plus plus Vancomycin 500 mg q6h +/− intraventricular Vancomycin 20 mg/d	
Listeria monocytogenes	Ampicillin 2–3g q4h plus Gentamicin 2mg/kg q8h		
Gram negatives (except *Pseudomonas*)	Ceftriaxone plus Gentamicin 1.5 mg/kg q8h granulocyte transfusions for neutropenia[82]		
Pseudomonas aeruginosa	Ceftazidime 2g q8h or cefepime 2g q12h or Meropenem 2g q8h		
Nocardia asteroides	Sulfadiazine 8–12 g/d		
Viruses			
Herpes simplex (encephalitis)	Acyclovir 10-12 mg/kg q8h[b] (? chronic po Valacyclovir [b])		
Varicella zoster (encephalitis) (dermatomal)	Acyclovir 10–12 mg/kg q8h		
	Valacyclovir 1000 mg po twice daily for 10 days or		
	Famciclovir 500 mg po three times daily for 10 days or		
	Acyclovir 200 mg po five times daily for 10 days		
HHV 6, types A and B	Foscarnet 60 mg/kg q8h		
Cytomegalovirus	Ganciclovir 5 mg/kg q12h plus Foscarnet		
Epstein-Barr virus (PTLD)	Acyclovir 10 mg/kg q8h		
Enteroviruses	Pleconaril 200 mg po 3 times daily for 7 days		
Fungic			
Cryptococcus neoformans	Amphotericin B 0.7 mg/kg per day followed by		
	Fluconazole 400-800 mg/d po or		

TABLE 4 Treatment of common CNS infections in cancer patients—cont'd

Organism	Antibiotic regimen and alternatives intravenous route except as noted
	AmBisome/ABLC[d] 5 mg/kg per day
	for 2 weeks plus
	Flucytosine 150 mg/kg/d po for 6 weeks
	Itraconazole 400 mg/d may substitute for Fluconazole
	Aspergillus species[e] Amphotericin B 0.8–1.25 mg/kg per day or
	AmBisome or
	ABLC plus (for all three drugs mentioned already)
	Itraconazole 600–800 mg/day po X 4d or
Mucoraceae	ABLC 5 m/kg per day or
	AmBisome 5 mg/kg per day plus surgical debridement
Candida species[f]	Amphotericin 0.7–1.0 mg/kg per day plus
	Flucytosine 25 mg/kg four times daily
Histoplasma capsulatum	Amphotericin 0.7–1.0 mg/kg per day plus
	Itraconazole 400 mg/d suppressive treatment
Coccidioides immitis	Fluconazole 800 mg/d or Itraconazole 400–600 mg/d or
	Voriconazole 400–600 mg/d
Parasites	
Toxoplasma gondii	Sulfadiazine 1.5–2 g po 4 times daily plus
	Pyrimethamine 100–200 mg po load then 75–100 mg po qd plus
	Folinic acid 10–50 mg po qd

[a] MIC, minimal inhibitory concentration.
[b] Adjustment of dose required in renal failure/dialysis patients Chronic po prophylaxis currently under study.
[c] Voriconazole under study for many of the mycoses.
[d] AmBisome and ABLC (amphotericin B lipid complex) are liposomal formulations of amphotericin B.
[e] Aspergillus species treatment with caspofungin and voriconazole or posaconazole under study.
[f] New class of antifungal agents anidulafungin under study.

lated events due to toxic drug reactions, metabolic abnormalities, or electrolyte imbalance. Seizures occurring during the course of an episode of PRES usually do not recur and AEDs are discontinued within a month of resolution of the episode and correction of the inciting drug or condition. Drugs associated with seizures in the cancer transplantation population include tacrolimus and cyclosporine (as part of the RPLS/PRES syndrome), muromonab-CD3 (OKT3), busulfan, quinolone antibiotics, beta lactams, penicillin, cephalosporins, ifosfamide, and bupropion. None of these syndromes likely requires long-term antiseizure medicine, but some clinicians treat with antiepileptic drugs (AEDS) for at least 1 month after a recognized seizure.[94] For those conditions such as intracerebral hemorrhage, infarction, venous sinus thrombosis, or abscess that may cause longer term seizure risk, the choice of seizure medicine should reflect interactions between drug and HSCT regimen. The older AEDS such as phenytoin and carbamazepine may reduce immunosuppressant blood levels and accelerate metabolism of many cytotoxic and biologic agents used in cancer chemotherapy. Drugs such as phenytoin, carbamazepine, and valproate are heavily protein-bound and in hypoproteinemic transplant patients, free-level AED drug measurement is indicated. Hypersensitivity reactions to phenytoin, carbamazepine, oxcarbazepine, and lamotrigine present with fever rash, and, at times, a Stevens-Johnson syndrome along with pancytopenia. This may be particularly common in the setting of patients undergoing a steroid taper and radiation therapy. Because of these considerations, agents such as levetiracetam and lacosamide, available in intravenous form, are appealing due to their rapid onset of activity, oral availability, and absence of significant protein binding or hepatic enzyme induction. Renal failure may reduce the clearance of levetiracetam, and hemodialysis removes small nonprotein-bound drugs requiring dosage adjustment and supplementation as suggested in manufacturers' recommendations.

4.2 Bacterial meningitis

The spectrum of bloodstream infections in patients with malignancies is an important clue to bacterial meningitis etiology. Hospital prophylaxis patterns have altered the distribution and sensitivity of potential meningitis pathogens as illustrated by the fact that 85% of the patients with hematologic malignancies and HSCT were on quinolone (ciprofloxacin or levofloxacin) prophylaxis when they developed their infection. In this series from MD Anderson, clinicians looked at Gram-positive bloodstream infections and found coagulase-negative staphylococci (33%), *S. aureus* (15%), viridans group streptococci (10%), and enterococci 8%. The majority of the patients were neutropenic when they developed their infection and the new-generation quinolones, moxifloxacin and gatifloxacin, had severalfold more potency against these infections than ofloxacin and ciprofloxacin.[95] Multiresistant coagulase-negative staphylococci may cause systemic and CNS infections in patients undergoing bone marrow transplantation. Daptomycin, a lipopeptide, and tigecycline, a new glycylcycline, have excellent activity against methicillin-resistant staphylococci, though most isolates still remain susceptible to vancomycin.[96]

General medical support. In experimental studies, outcomes of acute bacterial meningitis (BM), including hearing loss and survival rates, have been correlated with the severity of the inflammatory process in the subarachnoid space. This response can be reduced with corticosteroids. In the original Dutch studies, a dose of dexamethasone of 10 mg intravenously 15 min before or simultaneous with the first dose of antibiotics was continued every 6 h for 4 days.[97] Details of this regimen remain under investigation, but the principle that steroid treatment before antibiotics reduces mortality and neurological sequelae in adults with bacterial meningitis has been confirmed in a Cochrane meta-analysis.[98] However, according to a recent multicenter observational study, the same benefit does not accrue to pediatric patients.[99] While the clinician caring for an immunocompromised patient with acute BM can be advised to follow these current guidelines, the question arises as to whether the immunocompromised host's impaired inflammatory response may be less dangerous than that of a normal host and thus the benefits of steroids may be offset by the risk of reducing blood-brain barrier permeability and antibiotic access in the special situation of the cancer population. There are no Class I guidelines to help with this management issue.

Potentially, the most immediately lethal of cancer patient CNS infections, BM in the cancer population, is disproportionately due to Gram-negative organisms and *S. aureus*. Additional causal pathogens in the neurosurgical population are *S. bovis* and, in decreasing numbers, *Listeria*.[100,101] *Listeria* does not present in the immunocompromised patients with rhombencephalitis which can be confused with leptomeningeal metastases or, in recent years, with West Nile virus.[102,103] In a study from Barcelona conducted over the 30 year period 1982 to 2012, 15% of the patients with BM had active cancer and were in general older than those without cancer. The triad of fever, neck stiffness, and headache was less frequent and there was a longer interval between admission and antibiotic therapy. Listeria meningitis was more common in cancer than noncancer patients. Mortality among patients with cancer was double that of those without cancer.[104] A Dutch study from the interval 2006–14 recorded a similar preponderance of Listeria in cancer patients, but noted that cancer patients had a similar clinical presentation of BM compared to those without cancer. Those with active cancer had lower leukocyte counts in blood and CSF and had higher mortality than noncancer patients.[105]

Empiric treatment for meningoencephalitis in a HSCT recipient should include coverage for likely bacterial and viral pathogens. This includes combination of a third-generation cephalosporin (cefotaxime or ceftriaxone) or a fourth-generation cephalosporin (cefepime) plus vancomycin plus ampicillin plus acyclovir for community-acquired infections. Nosocomially acquired infections require ceftazidime or cefepime for *P. aeruginosa* coverage (see Table 4). Amphotericin B is not usually introduced as part of the empiric regimen. Vancomycin should never be given alone with dexamethasone in the treatment of bacterial meningitis due to reduced access to CSF. Some authors recommend the addition of rifampin when vancomycin and dexamethasone are coadministered.[106] Dexamethasone is not currently recommended for the adjunctive treatment of Gram-negative bacillary meningitis.

4.3 Endocarditis

Epidemiology. Endocarditis is a life-threatening, largely nosocomial complication in cancer patients, many of whom spend long periods of time in the hospital and have intravenous catheters. In one recent study, 13% of the patients with catheter-associated *S. aureus* blood stream infections had complications, including endocarditis.[107] A large series from M.D. Anderson Cancer Center reviewed transthoracic (TTE) and transesophageal (TEE) echocardiograms in 654 patients between 1994 and 2004 and found that 7% met the modified Duke criteria for endocarditis, the diagnostic TEE discovering vegetations in 42% of the patients with initially nondiagnostic TTE.[108] Among the 58% who had positive cultures, *S. aureus* was the most common organism isolated, followed by coagulase-negative *Staphylococcus* species. Culture-positive endocarditis (CPE) patients more often had central venous catheters and larger vegetations, but the incidence of cerebrovascular embolic complications was higher in the culture-negative endocarditis group. Thirty-one percent of the CPE cases were nosocomially acquired, a figure more than three times higher than that of nonimmunosuppressed patients with native heart valve infection.[109] *Candida* endocarditis is a rare complication even with prolonged *Candida* bloodstream infection. The presence of structural heart disease, particularly with a prosthetic valve, increases the risk.[110] Caspofungin (see later) appears to be a promising treatment of candidial infection, so that some patients can be treated without replacement of the infected heart valve.[111] The culture-negative endocarditis patients in this series may have been pretreated with antibiotics or could have had nonbacterial thrombotic endocarditis (NBTE), a condition now diagnosed with increasing frequency premortem in cancer patients. In a study from Weill Cornell Medical College, 18% of the endocarditis suspects with cancer had NBTE vegetations and about half of these experienced cerebral embolism.[112] These recent studies support an important role for TEE in the evaluation of cancer patients with cerebral ischemia, provided that esophageal pathology and coagulopathy can be excluded or reversed.

The most common neurological complication of endocarditis is cerebral embolism, with a resulting pathologic spectrum from stroke to cerebritis to abscess formation. In the setting of cancer and suspected endocarditis, the differential diagnosis of a ring-enhancing cerebral lesion is large and initial neuroimaging is nonspecific as to causative organism. Fig. 7A shows a bacterial abscess indistinguishable from that of Fig. 7B, which is due to *Aspergillus*.

4.4 Fungal infections

Invasive fungal infections occur primarily in heavily immunocompromised patients receiving HSCT or intensive chemotherapy for acute leukemia. In a multi-center study, the most important pathogen in children and adolescents was *Aspergillus spp*, but there was also an increasing proportion of molds such as Fusarium, Scedosporium, and the Mucorales.[113] Management is

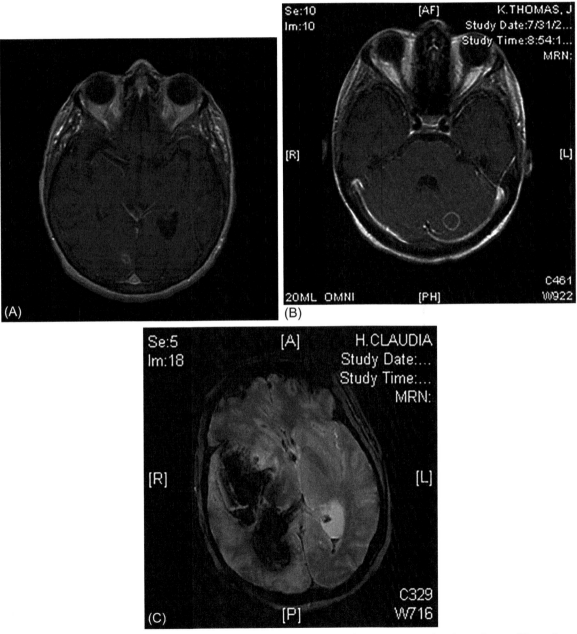

FIG. 7 (A) Endocarditis due to *S. aureus* in patient with prolonged hospitalization for neutropenia after chemotherapy. Ring-enhancing lesion not radiographically distinguishable from the ring of an *Aspergillus* abscess in a heavily immunosuppressed patient with non-Hodgkin's lymphoma in relapse. (B). Patient with fungal abscess deteriorated rapidly with massive hemorrhage (C).

usually multidisciplinary as there is a need for vessel imaging and sometimes neurosurgical debridement for optimal outcome.

Aspergillus fumigatus. *Aspergillus* remains the most common fungal infection in HSCT and solid organ transplant recipients and increasingly is a problem in other cancer patients on prolonged antibiotics.[114–116] *A. fumigatus* and *A. flavus* account for most invasive aspergillosis whose entry portal is usually the lung, though infection can be disseminated through skin, ear, or corneal lesions. Most patients have pulmonary disease or sinusitis, and neutropenia is a major risk factor. Up to 20% of the *Aspergillus* cases have cerebral involvement.[117] *Aspergillus* accounts for 50% of the brain abscesses in bone marrow transplant recipients.[118] A particularly serious clinical feature is its tendency to cause vascular thrombosis with stroke and rapid deterioration due to hemorrhage from septic aneurysm or vasculitis (see Fig. 8C). Solitary or multiple cerebral abscesses are the most common clinical manifestation, but infectious aneurysms and carotid artery invasion have been reported and extensively reviewed.[119] Invasive disease of the spine has been reported as well (see Fig. 8).[120]

Diagnosis. CSF cultures are almost always negative and biopsy is the method for definitive diagnosis. CT and MRI showing multiple embolic lesions with ring enhancement in the clinical setting of a neutropenic patient with abnormal chest X-ray can be helpful. A promising test for early diagnosis of fungal infection is ELISA for galactomannan, a cell wall component that has shown 90% sensitivity and 98% specificity. This abnormality can be detected up to a week before the appearance of chest radiography abnormalities. Another antigenic test is 1,3-β-D-glycan, a cell wall component that can be detected on either blood or CSF.[121]

Treatment. Although amphotericin B remains the recommendation for first-line treatment of CNS aspergillosis (see Table 4), the efficacy of voriconazole has been reported in some studies to be superior to amphotericin.[122] The echinocandins are a new class of drugs, represented by caspofungin and micafungin, that have a more favorable adverse event profile, particularly for patients with renal failure. Although few patients with CNS aspergillosis have been studied, caspofungin monotherapy recently achieved a 45% response rate in refractory invasive aspergillosis.[123] Combination therapy with voriconazole plus caspofungin may improve results.[124,125] Anidulafungin, a novel echinocandin with excellent in vitro activity against *Aspergillus* and many *Candida* species, effectively treats esophageal candidiasis, including azole-refractory disease and, unlike caspofungin, its levels are not affected by concomitant treatment with cytochrome P450 modifiers. It also does not alter cyclosporine levels. This agent may be used safely with liposomal amphotericin B and further trials for combination efficacy in invasive aspergillosis are in progress.[126]

Prophylaxis of Aspergillus infection. Since the prognosis of invasive disease is so poor, effective prophylaxis is important. Fluconazole does not protect against *Aspergillus* species, though posaconazole has activity

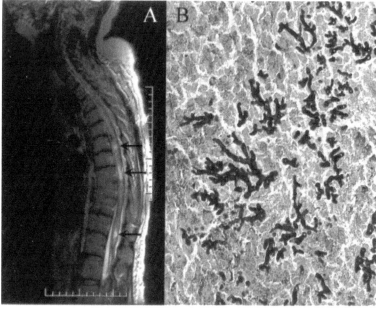

FIG. 8   Paraplegia caused by invasive spinal aspergillosis in patient with infliximab for Crohn's disease and also a history of multiple sclerosis. MRI showed edematous cord at all levels below C5 (A) and biopsy showed caseating granulomata with fungal hyphae on Gomori methenamine silver stain (B). Patient responded to posaconazole after failing voriconazole and caspofungin therapy. Granulomatous infections including invasive fungi are a risk after tumor necrosis factor antagonist therapy. *From Karakousis PC, Magill SS, Gupta A. Paraplegia caused by invasive spinal aspergillosis. Neurology. 2007;68(2):158. https://doi.org/10.1212/01.wnl.0000238981.65444.f2.*

against a wide range of yeasts (including *Candida* species) and molds (including *Aspergillus* and *Zygomycetes*). In the first of two recent controlled trials, patients undergoing chemotherapy for acute leukemia had lower rates of invasive fungal disease than those receiving fluconazole or itraconazole, though there were more adverse events in the posaconazole group.[127] In the second study, posaconazole had a favorable profile for fungal prophylaxis in HSCT recipients immunosuppressed for GVHD when compared to fluconazole.[128] Editorial opinion accompanying these studies emphasizes that prophylactic regimens remain problematic. Voriconazole continues to be the preferred treatment for proven or probable aspergillosis, while caspofungin and liposomal amphotericin B are options for empiric therapy in patients with persistent fever and neutropenia.[129,130] A noteworthy drug interaction is the 90% reduction in sirolimus required when patients are cotreated with voriconazole.[131]

Candida species. Central venous catheters, antibiotics, corticosteroids, diabetes mellitus, prolonged neutropenia, and HSCT are all risk factors for Candida infection. *C albicans* is no longer the predominant species and represents only about 45% of the isolates. *Candida glabrata, parapsilosis,* and *tropicalis,* which, unfortunately, are less susceptible to traditional antifungal agents, are an increasing problem.[132] CSF may be negative, though neutrophilic meningitis may be present. A possible improvement for what has been a difficult antemortem diagnosis is an assay in serum and CSF for mannan, a surface antigen on the cell wall of *C. albicans*.[133] Voriconazole has adequate CSF penetration and has been described as effective against invasive candidiasis.[134] The recommended treatment for CNS candidiasis remains amphotericin B plus 5-flucytosine for at least 4 weeks.[135]

Zygomycetes. The order *Mucorales* has become a more frequent pathogen in immunosuppressed patients, perhaps because of the excessive use of some antifungal agents. Vascular invasion and tissue necrosis are the hallmarks of this poor prognostic disease (see Fig. 9).[136] Risk factors include acute leukemia or lymphoma, HSCT, prolonged neutropenia, diabetes mellitus, and renal failure. The CNS invasive condition presents as a rhinocerebral process with facial pain and frontal headache. Spread to the eyes is frequent and, since the fungus invades blood vessels, infarction of areas within the orbit is common. Internal carotid artery thrombosis and cranial nerve dysfunction are also common. Transfusion of colony-stimulating factors or granulocytes may be helpful in controlling infection, as are reduction of immunosuppression and control of hyperglycemia, along with surgical debridement. Fluconazole and 5-flucytosine are ineffective, and caspofungin and voriconazole are also of little utility.[137]

Endemic fungal pathogens. Endemic or primary pathogens acquired in restricted geographic areas have

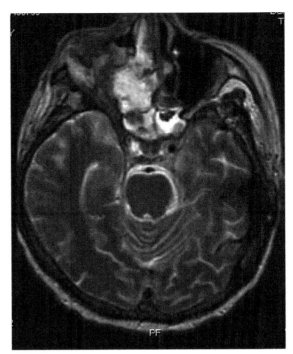

FIG. 9 Patient with AML who received allogeneic stem cell transplant and had persistent GVHD as well as diabetes mellitus while on prednisone developed eye pain and diplopia. FLAIR MRI shows mass involving orbit that was positive for mucormycosis.

increased in the immunocompromised population and more frequent travel has increased the range of these organisms. The disease can remain dormant only to be reactivated by corticosteroids or chemotherapy. *Histoplasma capsulatum,* whose geographic range includes South America as well as the Ohio and Mississippi River valleys, can manifest as chronic meningitis or a mass lesion that can be mistaken for a malignancy or abscess. CSF glucose is usually low, but cell counts may be normal. Organisms can be identified in blood, tissue, or sputum specimens.[138]

Areas endemic for coccidioidomycosis include Central America and the southwestern United States. The most common sites of disseminated disease are the skin, joints, and skeletal bone. CNS disease usually comes from pulmonary infection and can include basilar meningitis, vasculitis, encephalitis, and space-occupying lesions. Almost all patients have pulmonary involvement.[139] A report from Arizona noted that 3% of the renal transplant patients contracted coccidioidomycosis, a third of whom had disseminated infection. No high-risk patient who received targeted antifungal prophylaxis had a reactivation after transplantation.[140]

C. neoformans. Worldwide, the most common fungal infection of the CNS is cryptococcal meningoencephalitis. Cryptococcal infection occurs in up to 3% of the organ recipients with a mortality rate exceeding 40%.[141] Reactivation of infection in the recipient and acquisition of the infection from donors means that the disease can occur fairly early on in the transplant period, though the period from 1 to 6 months is the most common time frame. Routine use of posttransplantation prophylaxis with trimethoprim/sulfamethoxazole and valacyclovir or ganciclovir prevents many meningitides due to listeriosis, toxoplasmosis, nocardiosis, and herpes viruses, leaving *Cryptococcus* among the most common of early meningitis etiologies.[142] Among patients with malignancies, those with lymphoma, CLL, and AML are at highest risk, while among solid organ transplant recipients liver recipients are at highest risk. HCV coinfection and alcoholism further increase the risk.

The disease may present either as a focal mass (more common with *Cryptococcus gattii*) or as a more diffuse meningoencephalitis, characteristically with greatly elevated intracranial pressure and variable inflammatory response correlated with the host's degree of immunosuppression.[143] Small vessel strokes can occur early in the illness course.[12] Headache, fever, and mental status changes are the most common clinical manifestations. Cutaneous ulcers can be seen and any suspicious skin lesions should be biopsied. CSF diagnosis is straightforward, with detection of cryptococcal polysaccharide antigen. One of the major early risks in cryptococcal meningitis is visual loss either from sustained elevation of intracranial pressure or from direct invasion of the optic nerves.[144] Therapy is triple drug 5-flucytosine, amphotericin B (possibly the liposomal form), and fluconazole, though posaconazole is being studied. The echinocandins have poor activity against cryptococcus, as indicated in Table 4. Morbidity includes obstructive hydrocephalus requiring ventriculostomy or shunting. The mortality rate is nearly 50% in one large series.[145] Immune reconstitution poses a problem for these patients as discussed previously under IRIS (see Fig. 5).[60]

4.5 Viral infections

4.5.1 Varicella-zoster virus

As many as 15% of the patients with leukemia and lymphoma, and those with HSCT, will develop symptomatic dermatomal VZV infection, but prophylactic regimens appear to reduce the incidence of fatal dissemination. Reactivation can occur at any time posttransplantation, but particular risk is during reconstitution of the immune system following bone marrow engraftment.[146] Most VZV-related fatalities occur from disseminated disease and the use of intravenous acyclovir is indicated for all patients with allogeneic HSCTS and HSCT with moderate or severe GVHD.[147]

Postherpetic neuralgia (PHN) may occur as much as three times more frequently in cancer patients as in the general population. Early use of antiviral agents may help to reduce the frequency of PHN. The American Academy of Neurology Practice Parameter suggests that gabapentin, pregabalin, tricyclic antidepressants,

and topical lidocaine patches may be helpful.[148] A possibly promising strategy for postherpetic pain is continued treatment with oral valacyclovir after intravenous acyclovir.[149] There is no evidence that systemic corticosteroids are helpful in reducing the incidence of PHN. VZV vaccine, known as Shingrix, should be given in two doses 2–6 months apart, if possible, to all adults about to undertake a regimen that will cause significant T-cell immunosuppression. Recent studies suggest that long-term acyclovir (800 mg bid × 12 months) will reduce the risk of zoster reactivation in transplant recipients.[150]

While rash confined to one or two dermatomes or disseminated skin lesions are the most common manifestations of VZV reactivation, focal segmental weakness, peripheral facial palsy, and hearing loss from VZV oticus with segmental pontine myelitis have been reported.[151] Dissemination of VZV to the brain takes the form of either an acute necrotizing encephalitis[152] or stroke syndrome, including large and small cell arteritis or carotid occlusion, which may precede the rash.[153] Acute retinal necrosis may be early or delayed, and all patients with ophthalmic zoster should be referred for ophthalmologic care.[154]

Spinal dissemination takes the form of acute or progressive myelitis. Spinal cord involvement, which can occur weeks after the dermatomal involvement, is often extensive and suggests ischemia. Recent use of diffusion-weighted MRI (DWI) confirms the ischemic nature of the acute disease process.[154] Patients presenting with more subacute myelopathy, which may also be inflammatory or demyelinating by conventional spinal cord imaging, have reduced serum/CSF VZV IgG ratio consistent with intrathecal VZV IgG synthesis; this appears to be the most sensitive test for VZV rather than VZV DNA by PCR.[156]

Nagel and colleagues updated information on patients with virologically verified VZV vasculopathy. An important point emerging from this 30 patient series was that nearly 40% of the patients had no history of rash. Thus, VZV must be considered as a cause of transient ischemic attack (TIA) or stroke without rash or with a long delay from rash to symptoms (average was 4.1 months). There was no CSF pleocytosis in one-third of the vasculopathy patients. Only 70% of the patients who had MRA or angiography had evidence of focal narrowing because the disease involves small arteries frequently. The most important tests were MRI, which showed ischemic lesions in 97% of the patients, and among immunocompromised patients, there was no statistical difference between rash, CSF pleocytosis, or anti-VZV IgG antibody compared with immunocompetent patients. The best diagnostic test for VZV encephalitis is the detection of VZV IgM antibodies in CSF. However, VZV DNA detection by PCR was more common in the immunocompromised patients. In general, the diagnostic value of anti-VZV IgG antibody was greater than that of VZV DNA (93% positive vs 30%), probably because of the protracted course. To determine the intrathecal synthesis of VZV IgG, the ratio of [anti-VZV IgG in CSF/anti-VZV IgG in serum] to [total IgG in CSF/total IgG in serum] is measured. Values greater than or equal to 1.5 demonstrate CSF antibody synthesis. Regimens including acyclovir and/or steroids were not directly comparable. Thus, no recommendations can be made about the combination.[157]

4.5.2 Human herpesvirus 6

Most HHV-6 infections are considered to be reactivation of host infection, though transmission of the virus from donor with the allograft has been demonstrated.[158] HHV-6 predominantly type B has been documented in 38% of the HSCT recipients, and in at least that number of solid organ transplant recipients,[159] while human herpesvirus 6 variant A has been described in a single case of acute encephalitis syndrome with paraparesis.[160] Risk factors in the transplant population include conditioning regimens with OKT3 monoclonal antibodies or antithymocyte globulin, and seroconversion may be more frequent in patients who received immunosuppressive regimens containing sirolimus and IL-12 receptor antibodies as induction.[161] There is also some evidence that hypersensitivity reactions to drugs such as phenytoin or carbamazepine increase the risk of HHV-6 reactivation.[162,163] The group at greatest risk is HSCT patients receiving grafts from unrelated donors or patients with severe GVHD. Increased levels of HHV-6 DNA, an immunosuppressive virus, are associated with several medical complications including an increased requirement for platelet transfusions due to delayed platelet engraftment and reactivation of hepatitis C disease, as well as facilitation of superinfections with other opportunistic pathogens like CMV and fungi.[164] The indirect medical effects of HHV-6 appear to be more common in solid organ transplant recipients, while the encephalitic complications dominate the picture for HSCT patients.[165,166] Mortality rates are reported between 25% and 40%, but prognosis is not clear because many patients received antiviral therapy only rather late in their course.[167]

Early suspicion of HHV-6 must be guided by the concept that the detection of HHV-6 nucleic acid in CSF is not definitive proof of HHV-6 as the etiology of encephalitis.[168] Following primary infection in early childhood (manifested as exanthema subitum), virus can be integrated into host chromosomes and such patients have a high concentration of viral DNA in peripheral blood leukocytes, as well as HHV-6 DNA in the CSF. They can also acquire chromosomally integrated HHV-6 from donors and such patients may not need to be treated with antiviral therapy.[169] The number of copies of HHV-6 DNA in serum and/or whole blood and CSF, as well as hair

follicles, helps make the distinction between viral chromosomal integration and active infection, as can be a fourfold or greater rise in IgG between acute and convalescent sera. Conversely, study of patients with encephalitis who were HHV-6-negative by CSF PCR has found postmortem evidence of high HHV-6 DNA and mRNA in hippocampi. Further studies are needed to assess the role of PCR for HHV-6 in determining the etiology of limbic encephalitis.[170]

Given the hippocampal localization of MRI abnormalities in both HSV-1, whose incidence is not increased in immunocompromised patients, and HHV-6, the clinical context must be considered in the interpretation of initial radiographic results (see Fig. 1). The CSF PCR for HSV-1 DNA may be negative in the first 72h of symptoms. Herpes simplex virus antibodies can be detected within 2 weeks of symptom onset so that a serum:CSF antibody ratio of <20:1 is evidence of HSV-1 infection.[171] The distinction becomes important since HHV-6 responds better to ganciclovir or foscarnet. Foscarnet may be the preferred agent in heavily hematologically suppressed patients, though nephrotoxicity is a concern with both antivirals.

4.5.3 Progressive multifocal leukoencephalopathy

Fifty years ago, Richardson and colleagues described two patients with CLL and one with Hodgkin's disease who had multiple demyelinating lesions in the brain, with inclusion bodies in oligodendrocytes that ultimately were shown to be papovavirus particles.[172] In the HIV era, PML has become a major opportunistic infection. Since the advent of highly active antiretroviral therapy (HAART), the percentage of PML patients with AIDS has decreased from 80% to around 50%, while an increasing percentage have hematologic malignancies (13%) and 5% are transplant recipients.[61,173] Among HIV-negative cases, rheumatologic disease and multiple sclerosis treated with immune-modifying agents are emerging as populations at risk.[174] Among HIV-negative cases at the Mayo Clinic, 55% have hematologic malignancies. Among HIV-negative patients with lymphoproliferative disorders, those treated with purine analogues such as fludarabine appear to be at the greatest risk.[175] Other drugs used in cancer and transplantation and associated with PML risk are belatacept, brentuximab, cyclosporine, ibrutinib, mycophenolate, obinutuzumab, ofatumumab, ruxolitinib, sirolimus, and tacrolimus.

Roughly 86% of the healthy adults are JCV seropositive. In immunocompromised hosts, JCV is caused by a reactivation of latent infection. The majority of the cases lack anti-JC IgM antibodies, in the presence of preexisting anti-JC IgG antibodies whose levels remain stable.[176] Once within the CNS, JCV infects oligodendrocytes and astrocytes. PML predominantly involves the subcortical white matter. Unlike HIV patients, HIV-negative patients and HIV patients on HAART may have enhancing white matter lesions correlated with CD3+ lymphocyte infiltration pathologically and with rising peripheral CD4+ counts, as part of the IRIS (see Fig. 10).[177] JCV may also cause the JCV granule cell neuronopathy, a cerebellar syndrome with ataxia and atrophy in the internal granule cell cerebellar layer, without classic PML demyelinating lesions.[178]

Suspicion of PML is raised by the MRI lesions in the white matter. Clinical and radiographic characteristics of PML are quite variable, but a recently described notable feature is a rim of diffusion restriction in the advancing border of the white matter lesions, and also susceptibility- weighted abnormalities developing just before clinical symptoms become evident.[179] Spinal cord lesions have been reported rarely in PML as well, further raising confusion over the issue of possible multiple sclerosis.[180] Definitive diagnosis of PML is by brain biopsy, though until recently CSF JCV PCR had a sensitivity of close to 92% and a specificity of nearly 100%. However, with the advent of HAART, partial recovery of immune function has been associated with clearance of JCV from CSF with a reduction in PCR sensitivity to about 58%.[181] There is no specific treatment for PML. In most immunocompromised patients, reduction of immunosuppressives is attempted. Cytosine, arabinoside, cidofovir, topotecan, interferon β-1a,[182] and interferon α-2b have been ineffective.[183,184] Clinical improvement was noted in one natalizumab- treated multiple sclerosis patient with PML at the time of massive IRIS inflammatory reaction 3 months after discontinuation of natalizumab and concomitant with treatment with

cytarabine.[185] Discovery of JCV penetration into glial cells via the cellular 5-hydroxytryptamine 2a (serotonin) receptor has led to attempts to block this receptor with agents such as mirtazapine.[186] Cortese and colleagues reported successful stabilization of disease in five of eight patients treated with pembrolizumab.[187] While optimal therapy has not been determined, it is clear that both the MRI and clinical presentation of PML and its outcome are more varied than previously described. Immunotherapeutic interventions, such as the use of checkpoint inhibitors and adoptive T-cell transfer, have shown some success in small numbers of patients, but these therapies run the risk of inducing immune reconstitution inflammatory syndrome, at which time an exuberant immune response can cause morbidity and death. Cortese and colleagues point out that many people who survive PML are left with neurological deficits and some with persistent, low-level viral replication in the CNS.[61] Koralnik has suggested that the name PML is a misnomer since the disease is neither invariably progressive nor always multifocal or even exclusively in white matter.[163]

5 Vaccine-preventable infections: The COVID-19 pandemic and cancer patients

While it appears that direct CNS invasion by SARS-CoV-2 is uncommon, a host of CNS sequelae of the virus include stroke, persistent anosmia, late movement disorders, Guillain-Barré syndrome, transverse myelitis, and both white matter and microvascular damage

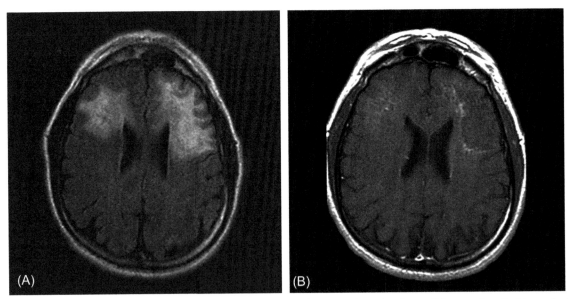

FIG. 10 Progressive multifocal leukoencephalopathy develops 2 years after transplantation in a solid organ (cardiac) transplant recipient. Multifocal white matter process consistent with PML is seen on FLAIR images (A). Unlike the patient in Fig. 4, this less heavily immunosuppressed patient had mild contrast enhancement on gadolinium sequence (B) consistent with inflammatory reaction to the viral pathogen.

with long-term encephalopathy. Further, the mortality of COVID-19 is higher in cancer patients than in the general population, with hospitalized lung cancer patients reaching 25% mortality in a recent study from Madrid.[188] The reasons for cancer patients' susceptibility to COVID and the implications for prognosis are being elucidated.

In a cohort of patients with hematologic cancers hospitalized at the University of Pennsylvania combined with a cohort from Memorial Sloan Kettering, it was found that those with hematologic malignancies had higher mortality relative to patients with solid tumors or patients without cancer. Patients with hematologic cancers had significant impairment of B-cells and SARS-CoV-2-specific antibody responses. However, patients with hematologic cancers and preserved CD8 T-cells fared better, while an immune phenotype characterized by CD8 T-cell depletion was associated with a high viral load and the highest mortality of 71%, among all cancer patients. Further, depletion of B-cells with antiCD20 therapy such as rituximab resulted in almost complete abrogation of SARS-CoV-2-specific IgG and IgM 65 antibodies, but was not associated with increased mortality compared to other hematologic cancers, when adequate CD8 T-cells were present. Thus, CD8 T-cells likely compensate for deficient humoral immunity and influence clinical recovery from COVID-10.[189]

Therefore, knowledge of vaccine efficacy in this vulnerable cancer patient group is critical. Revision of vaccination guidelines is an evolving process and clinicians should be well informed about specific vaccination requirements before prescribing immune system-altering medications. See Table 5[190] for vaccine safety guidelines and recommendations. Based on the author's assessment of current theoretical knowledge about mRNA-based COVID-19 vaccines, most immunocompromised patients should be advised to accept COVID-19 vaccination. However, a robust antibody response to the SARS-CoV-2 spike protein may not be achieved in these patients as reported in the transplantation population, and continued masking and social distancing remain cautionary recommendations.[191] Ongoing studies have begun to elucidate appropriate intervals for maximum vaccine response in immunocompromised patients on B-cell-depleting therapies.[192]

6 Conclusions

The diagnosis and management of CNS infections remain ever-challenging as the conditions, presentations, therapies, and neuroimaging features evolve. Despite efforts to stratify patients by epidemiologic risk factors, clinical syndromes, and appropriate diagnostic tests, uncertainties are common and morbidity and mortality

TABLE 5 Vaccine-Preventable infections: safety recommendations in immunocompromised patients.

Inactivated or recombinant vaccine (generally safe on any therapy)	Live/attenuated vaccines (do not use while on specific therapies)	mRNA vaccines (theoretically safe)a
Inactivated influenza vaccineb	Live nasal spray influenza vaccine	Pfizer SARS-CoV-2-vaccine
Hepatitis A	Yellow fever	Moderna SARS-CoV-2-vaccine
Hepatitis B	Measles, mumps, and rubella (MMR)	
Human papilloma virus (HPV)	Cholera	
DTaP		
TD		
Inactivated polio vaccine		
Meningococcal vaccine		
Pneumococcal vaccine		
Rabies		
Varicella-zoster (Shingrix)c		
Viral vector: Johnson and Johnson SARS-CoV-2		

a Timing of vaccination with respect to B-cell-depleting infusions should be optimized to afford the best change of a robust immune response, current recommendations being second vaccination at least 3 weeks before next B-cell-depleting infusion.
b Antibody response may be attenuated for patients on some disease-modifying therapies: high-dose (quadrivalent) influenza vaccination may be recommended.
c Recommended, if possible, before treatment with alemtuzumab and B-cell-depleting regimens. From Grebenciucova E, Pruitt A. Infections in patients receiving multiple sclerosis disease-modifying therapies. Curr Neurol Neurosci Rep. 2017;17(11):88. https://doi.org/10.1007/s11910-017-0800-8.

from CNS infections remain high in the cancer population. The clinician should remain aware of emerging infections, transfusion safety issues, changing microbial susceptibilities, synergistic coinfections, and novel cancer therapies that will continue to impact the nervous system in new ways. Finally, the clinician should not forget that two infections may coexist and that the specter of recurrence of the original neoplasm or a secondary neoplasm must remain in the differential diagnosis.[193]

References

1. Pruitt AA. Central nervous system infections complicating immunosuppression and transplantation. *Continuum.* 2018;24(5):1370–1396.
2. Blijlevens NM, Donnelly JP, dePauw BE. Microbiologic consequences of new approaches to managing hematologic malignancies. *Rev Clin Exp Hematol.* 2005;9(2):E2.

3. Schmidt K, Schulz AS, Debatin KM, et al. CNS complications in children receiving chemotherapy or hematopoietic stem cell transplantation: retrospective analysis and clinical study of survivors. *Pediatr Blood Cancer*. 2008;50(2):331–336.

4. Denier C, Bourhis J-H, Lacroix C, et al. Spectrum and prognosis of neurologic complications after hematopoietic transplantation. *Neurology*. 2006;67:1990–1997.

5. King MD, Humphrey BJ, Wang YF, et al. Emergence of community-acquired methicillin-resistant *Staphylococcus aureus* USA 300 clone as the predominant cause of skin and soft-tissue infections. *Ann Intern Med*. 2006;144:309–317.

6. Klevens RM, Morrison MA, Nadle J, et al. Invasive methicillin-resistant Staphylococcus aureus infections in the United States. *JAMA*. 2007;298(15):1763–1771.

7. Mattiuzzi GN, Cortes JE, Talpaz M, et al. Development of varicella-zoster virus infection in patients with chronic myelogenous leukemia treated with imatinib mesylate. *Clin Cancer Res*. 2003;9:976–980.

8. Safdar N, Fine JP, Maki DG. Meta-analysis: methods for diagnosing intravascular device-related bloodstream infection. *Ann Intern Med*. 2005;142:451–466.

9. Goldberg SLS, Pecora AL, Aler RS, et al. Unusual viral infections after high-dose chemotherapy with autologous blood stem cell rescue and peritransplantation rituximab. *Blood*. 2002;99:1486–1488.

10. Annels NE, Kalpoe JS, Brodius RGM, et al. Management of Epstein Barr virus reactivation after allogeneic stem cell transplantation by simultaneous analysis of EBV DNA and EBV-specific T cell reconstitution. *Clin Infect Dis*. 2006;42:1743–1748.

11. Chamberlain MC, Glantz MJ, Chalmers L, et al. Early necrosis following concurrent Temodar and radiotherapy in patients with glioblastoma. *J Neurooncol*. 2007;82:81–83.

12. Mukerji SS, Ard K, Schaefer PW, Brana JA. Case 32-2020: a 63-year- old man with confusion, fatigue and garbled speech. *N Engl J Med*. 2020;383:1578–1586.

13. Antonini G, Morino S, Fiorelli M, et al. Reversal of encephalopathy during treatment with amphotericin B. *J Neurol Sci*. 1996;144:212–213.

14. Capparelli FJ, Diaz MF, Hlavnika A, et al. Cefepime and cefixime-induced encephalopathy in a patient with normal renal function. *Neurology*. 2005;65:1840.

15. Fernandez-Torre JL, Martinez-Martinez M, Gonzalez-Rato J, et al. Cephalosorin-induced nonconvulsive status epilepticus: clinical and electroencephalographic features. *Epilepsia*. 2005;46:1550–1552.

16. Ajithkumar T, Parkinson C, Shamshad F, et al. Ifosfamide encephalopathy. *Clin Oncol*. 2007;19:108–114.

17. Armangue T, Leypoldt F, Malaga I, et al. Herpes simplex virus encephalitis is a trigger of brain autoimmunity. *Ann Neurol*. 2014;75(2):317–323.

18. Dale RC, Nosadini M. Infection triggered autoimmunity: the case of herpes simplex virus type 1 and NMDA receptor encephalitis. *Neurol Neuroimmunol Neuroinflamm*. 2018;5. https://doi.org/10.1212/NXI.0000000000000471, e471.

19. Stuby J, Herren T, Guidoc SN, et al. Immune checkpoint inhibitor therapy-associated encephalitis: a case series and review of the literature. *Swiss Med Wkly*. 2020;150, w20377. Epub 23 November 2020 https://doi.org/10.4414/smw.2020.20377.

20. Larkin J, Chmielowski B, Lado CD, et al. Neurologic serious adverse events associated with nivolumab plus ipilimumab or nivolumab alone in advanced melanoma, including a case series of encephalitis. *Oncologist*. 2017;22:709–718.

21. Sechi E, Markovic SN, McKeon A, et al. Neurologic autoimmunity and immune checkpoint inhibitors: autoantibody profiles and outcomes. *Neurology*. 2020;95:e2442–e2452.

22. Kay J, Bazri H, Avery LL, et al. Case 6-2008 A 46-year-old woman with renal failure and stiffness of the joints and skin. *N Engl J Med*. 2008;358:827–838.

23. Camacho DL, Smith JK, Castaillo M. Differentiation of toxoplasmosis and lymphoma in AIDS patients by using apparent diffusion coefficients. *Am J Neuroradiol*. 2003;24:633–637.

24. Davis LE, DeBiasi R, Goade DE, et al. West Nile virus neuroinvasive disease. *Ann Neurol*. 2006;60:286–300.

25. Singh N, Alexander BC, Lortholary O, et al. *Cryptococcus neoformans* in organ transplant recipients: impact of calcineurin inhibitor agents on mortality. *J Infect Dis*. 2007;195:756–774.

26. Wilson MR, Sample HA, Zorn KC, et al. Clinical metagenomic sequencing for diagnosis of meningitis and encephalitis. *N Engl J Med*. 2019;380:2327–2340.

27. Copelan EA. Hematopoietic stem-cell transplantation. *N Engl J Med*. 2006;354:1813–1826.

28. Pruitt A, Graus F, Rosenfeld MR. Neurological complications of transplantation. Part I: hematopoietic cell transplantation. *Neurohospitalist*. 2013;3:24–38.

29. Fishman JA, Gonzalez RG, Branda JA. Case 11-2008: a 45 year old man with changes in mental status after liver transplantation. *N Engl J Med*. 2008;358:1604–1613.

30. Fishman JA. Infection in solid-organ transplant recipients. *N Engl J Med*. 2007;357:2601–2614.

31. Hoofnagle JH, Doo E, Liang TJ, et al. Management of hepatitis B: summary of a clinical research workshop. *Hepatology*. 2007;45:1056–1075.

32. Loomba R, Rowley A, Wesley R, et al. Systematic review: the effect of preventive lamivudine on hepatitis B reactivation during chemotherapy. *Ann Intern Med*. 2008;148(7):519–528.

33. Srinivasan A, Burton EC, Kuehnert MJ, et al. Transmission of rabies virus from an organ donor to four transplant recipients. *N Engl J Med*. 2005;352:1103–1111.

34. Fischer SA, Graham MB, Kuehnert MJ, et al. Transmission of lymphocytic choriomeningitis virus by organ transplantation. *N Engl J Med*. 2006;354:2235–2249.

35. Kaul DR, Covington S, Taranato S, et al. Solid organ transplant donors with central nervous system infection. *Transplantation*. 2014;98:666–670.

36. Iwamoto M, Jrnigan DB, Guasch A, et al. Transmission of West Nile virus from an organ donor to four transplant recipients. *N Engl J Med*. 2003;348:2196–2203.

37. Hollander H, Schaefer PW, Hedley-Whyte TE. Case records of the Massachusetts General Hospital. Case 22-2005. An 81-year-old man with cough, fever and altered mental status. *N Engl J Med*. 2005;353:287–295.

38. Solomon T, Fisher AF, Beasley DW, et al. Natural and nosocomial infection in a patient with West Nile encephalitis and extrapyramidal movement disorders. *Clin Infect Dis*. 2003;36:E140–E145.

39. Tyler KL, Aksamit AJ, Keegan BM, et al. An 85 year old man with chronic lymphocytic leukemia and altered mental status. *Neurology*. 2007;68:460–466.

40. Prince HE, Hogrefe WR. Detection of West Nile virus (WNV)-specific immunoglobulin M in a reference laboratory setting during the 2002 WNV season in the United States. *Clin Diagn Lab Immunol*. 2003;10:764–768.

41. Spitzer TR. Engraftment syndrome following hematopoietic stem cell transplantation. *Bone Marrow Transplant*. 2001;27:893–898.

42. Wainwright MS, Martin PL, Morse RP, et al. Human herpes virus-6 limbic encephalitis after stem cell transplantation. *Ann Neurol*. 2001;50:612–619.

43. MacLean HJ, Douen AG. Severe amnesia associated with human herpsvirus-6 encephalitis after bone marrow transplantation. *Transplantation*. 2002;73:1086–1089.

44. Vitaliani R, Mason W, Ances B, et al. Paraneoplastic encephalitis, psychiatric symptoms, and hypoventilation in ovarian teratoma. *Ann Neurol.* 2005;58:594–604.

45. Bataller L, Kleopa KA, Wu GF, et al. Autoimmune limbic encephalitis in 39 patients: immunophenotypes and outcomes. *J Neurol Neurosurg Psychiatry.* 2007;78:381–385.

46. Dalmau J, Tuzunn E, Wu HY, et al. Paraneoplastic anti N-methyl-D-aspartate receptor encephalitis associated with ovarian teratoma. *Ann Neruol.* 2007;61:25–36.

47. Vincent A, Buckley C, Schott JM, et al. Potassium channel antibody-associated encephalopathy: a potential immunotherapy-responsive form of limbic encephalitis. *Brain.* 2004;127:701–712.

48. Szabo K, Poepel A, Pohlmann-Eden B, et al. Diffusion-weighted and perfusion MRI demonstrates parenchymal changes in complex partial status epilepticus. *Brain.* 2005;128(Pt 6):1369–1376.

49. Seeley WW, Marty FM, Holmes TM, et al. Post-transplant acute limbic encephalitis: clinical features and relationship to HHV6. *Neurology.* 2007;69:156–165.

50. Yoshikawa T, Yoshida J, Hamaguchi M, et al. Human herpes virus-7-associated meningitis and optic neuritis in a patient after allogeneic stem cell transplantation. *J Med Virol.* 2003;70:4400–4443.

51. Ward KN, White RP, Mackinnon S, et al. Human herpes virus-7 infection of the CNS with acute myelitis in an adult bone marrow recipient. *Bone Marrow Transplant.* 2002;30:983–985.

52. Dewhurst S. Human herpesvirus type 6 and human herpesvirus type 7 infections of the central nervous system. *Herpes.* 2004;11(suppl 2):105A–111A.

53. Husain S, Munoz P, Forrest G, et al. Infections due to *Scedosporium apiospermum* and *Scedosporium prolificans* in transplant recipients: clinical characteristics and impact of antifungal agent therapy on outcome. *Clin Infect Dis.* 2005;40:89–999.

54. Ma M, Barnes G, Oulliam J, et al. CNS angiitis in graft-vs.-host disease. *Neurology.* 2002;59:1884–1997.

55. Mueller-Mang C, Mang TG, Kalhs P, et al. Imaging characteristics of toxoplasmosis encephalitis after bone marrow transplantation: report of two cases and review of the literature. *Neuroradiology.* 2006;48:84–89.

56. Edvinsson B, Lundquist J, Ljungman P, et al. A prospective study of diagnosis of *Toxoplasma gondii* infection after bone marrow transplantation. *APMIS.* 2008;116:345–351.

57. Bjorklund A, Aschan J, Labopin M, et al. Risk factors for fatal infectious complications developing late after allogeneic stem cell transplantation. *Bone Marrow Transplant.* 2007;40:1055–1062.

58. Neill TA, Lineberry K, Nabors LB. Incidence of post-transplant lymphoproliferative disorder isolated to the central nervous system in renal transplant patients. *Neurology.* 2004;62(Suppl 5):A479.

59. Annels NE, Kalpoe JS, Bredius RGM, et al. Management of EBV reactivation after allogeneic stem cell transplantation by simultaneous analysis of EBV DNA load and EBV-specific T cell reconstitution. *Clin Infect Dis.* 2006;42:1743–1748.

60. Kimby E. Tolerability and safety of rituximab (MabThera). *Cancer Treat Rev.* 2005;31:456–473.

61. Siakantaris MP, Argyropoulos KV, Ioannou S, et al. Cytomegalovirus meningoencephalitis after rituximab treatment for primary central nervous system lymphoma. *Neurologist.* 2015;19(2):35–37.

62. Cortese I, Reich DS, Nath A. Progressive multifocal leukoencephalopathy and the spectrum of JC virus-related disease. *Nat Rev Neurol.* 2020;17:37–51.

63. Matteucci P, Magni M, DiNIcola M, et al. Leukoencephalopathy and papovavirus infection after treatment with chemotherapy and anti-CD20 antibody. *Blood.* 2002;100(3):1104–1105.

64. Kaplan LD, Lee JY, Ambinder RF, et al. Rituximab does not improve clinical outcome in a randomized phase 3 trial of CHOP with or without rituximab in patients with HIV-associated non-Hodgkin lymphoma: AIDS malignancies consortium trial 010. *Blood.* 2005;106:1538–1543.

65. Kranick SM, Mowry EM, Rosenfeld MR. Progressive multifocal leukoencephalopathy after rituximab in a patient with non-Hodgkin lymphoma. *Neurology.* 2007;69(7):704–706.

66. Peleg AY, Husain S, Kwak EJ, et al. Opportunistic infections in 547 organ transplant recipients receiving alemtuzumab, a humanized monoclonal CD-52 antibody. *Clin Infect Dis.* 2007;44:202–212.

67. Nath DS, Kandaswaym R, Gruessner R, et al. Fungal infections in transplant recipients receiving alemtuzumab. *Transplant Proc.* 2005;37:934–936.

68. Sueki H, Mizukawa Y, Aoyama Y. Immune reconstitution inflammatory syndrome in non-HIV immunosuppressed patients. *J Dermatol.* 2018;45:3–9.

69. Shelburne III SA, Hamill RJ. The immune reconstitution inflammatory syndrome. *AIDS Rev.* 2003;5:67–79.

70. Venkataramana A, Pardo CA, McArthur JC, et al. Immune reconstitution inflammatory syndrome in the CNS of HIV-infected patients. *Neurology.* 2006;67:383–388.

71. Singh N, Lortholary O, Alexander BD, et al. An immune reconstitution syndrome-like illness associated with *Cryptococcus neoformans* infection in organ transplant recipients. *Clin Infect Dis.* 2005;40:1756–1761.

72. Powles T, Thirlwell C, Nelson M, et al. Immune reconstitution inflammatory syndrome mimicking relapse of AIDS related lymphoma in patients with HIV-1 infection. *Leuk Lymphoma.* 2003;44:1417–1419.

73. Fugate JE, Rabinstein AA. Posterior reversible encephalopathy syndrome: clinical and radiological manifestations, pathophysiology and outstanding questions. *Lancet Neurol.* 2015;14(9):914–925.

74. Allen JA, Aadlakha A, Bergethon PR. Reversible posterior leukoencephalopathy syndrome after bevacizumab/FOLFIRI regimen for metastatic colon cancer. *Arch Neurol.* 2006;63:1475–1478.

75. Soysal DD, Caliskan M, Aydin K, et al. Isolated cerebellar involvement in a case of posterior reversible leukoencephalopathy. *Clin Radiol.* 2006;61:983–986.

76. Kitaguchi H, Tomimoto H, Miki Y, et al. A brainstem variant of reversible posterior leukoencephalopathy syndrome. *Neuroradiology.* 2005;47:652–656.

77. Campbell JN, Morris PP. Cerebral vasculitis in graft-vs.-host disease: a case report. *Am J Neuroradiol.* 2005;26:654–656.

78. Padovan CS, Bise K, Hahn J, et al. Angiitis of the central nervous system after allogeneic bone marrow transplantation. *Stroke.* 1999;30:1651–1656.

79. Kamble RT, Chan CC, Sanchez S, et al. Central nervous system graft-versus-host disease: report of two cases and literature review. *Bone Marrow Transplant.* 2007;39:49–52.

80. Kew AK, Macaulay R, Burrell S, et al. Central nervous system graft-versus-host disease presenting with granulomatous encephalitis. *Bone Marrow Transplant.* 2007;40:183–184.

81. Chernik NL, Armstrong D, Posner JB. Central nervous system infections in patients with cancer: changing patterns. *Cancer.* 1977;40:268–274.

82. Wang KW, Chang WN, Hunag CR, et al. Post-neurosurgical nosocomial bacterial meningitis in adults: microbiology, clinical features, and outcomes. *J Clin Neurosci.* 2005;12:647–650.

83. Aapro MS, Cameron DA, Pettengell R, et al. EORTC guidelines for the use of granulocyte-colony stimulating factor to reduce the incidence of chemotherapy-induced febrile neutropenia in adult patients with lymphomas and solid tumours. *Eur J Cancer.* 2006;42:2433–2453.

84. Safdieh JE, Mead PA, Sepkowitz KA, et al. Bacterial and fungal meningitis in patients with cancer. *Neurology.* 2008;70:943–947.

85. Grossman SA, Desideri S, Ye X, et al. Iatrogenic immunosuppression in patients with newly diagnosed high-grade gliomas. *J Clin Oncol.* 2007;25(18 Suppl):2012.

II. Nonmetastatic neurological complications

86. Riel-Romero RM, Baumann RJ. Herpes simplex encephalitis and radiotherapy. *Pediatr Neurol*. 2003;29:69–71.

87. Chheda MG, Drappatz J, Greenberger NJ, et al. Hepatitis B reactivation during glioblastoma treatment with temozolomide: a cautionary note. *Neurology*. 2007;68:955–956.

88. Coursin D, Wood K. Corticosteroid supplementation for adrenal insufficiency. *JAMA*. 2002;287:236–240.

89. Annane D, Bellissant E, Bollaert PE, et al. Corticosteroids for severe sepsis and septic shock: a systematic review and meta-analysis. *BMJ*. 2004;329:480–488.

90. Minneci PC, Deans KJ, Banks SM, et al. Meta-analysis: the effect of steroids on survival and shock during sepsis depends on the dose. *Ann Intern Med*. 2004;141:47–56.

91. Dellinger RP, Levy MM, Carlet JM, et al. Surviving sepsis campaign: internal guidelines for the management of severe sepsis and septic shock: 2008. *Crit Care Med*. 2008;36(1):296–327.

92. Sprung CL, Annane D, Keh D, et al. Hydrocortisone therapy for patients with septic shock. *N Engl J Med*. 2008;358:111–124.

93. Turgeon AF, Hutton B, Fergusson DA, et al. Meta-analysis: intravenous immunoglobulin in critically ill adult patients with sepsis. *Ann Intern Med*. 2007;146:193–203.

94. Chabolla DR, Wszolek ZK. Pharmacologic management of seizures in organ transplant. *Neurology*. 2006;67(Suppl. 4):S34–S38.

95. Rolston KV, Yadegarynia D, Knotoyiannis DP, et al. The spectrum of Gram-positive bloodstream infections in patients with hematologic malignancies, and the in vitro activity of various quinolones against Gram-positive bacteria isolated from cancer patients. *Int J Infect Dis*. 2006;10(3):223–230.

96. Kratzer C, Rabitsch W, Hirschl AM, et al. In vitro activity of daptomycin and tigecycline against coagulase-negative staphylococcus blood isolates from bone marrow transplant recipients. *Eur J Haematol*. 2007;79(5):405–409.

97. Van de Beek D, de Gans J, McINtyre P, et al. Steroids in adults with acute bacterial meningitis: a systematic review. *Lancet Infect Dis*. 2004;4:139–143.

98. Van de Beek D, De Gans JK, McIntyre P, et al. Corticosteroids for acute bacterial meningitis. *Cochrane Database Syst Rev*. 2007;(1):CD004405.

99. Mongelluzzo J, Mohamad Z, Ten Have TR, et al. Corticosteroids and mortality in children with bacterial meningitis. *JAMA*. 2008;299(17):2048–2055.

100. Cohen LF, Dunbar SA. *Streptococcus bovis* infection of the central nervous system: report of two cases and review. *Clin Infect Dis*. 1997;25:819–823.

101. Bartt R. Listeria and atypical presentations of *Listeria* in the central nervous system. *Semin Neurol*. 2000;20:361–373.

102. Mileshkin L, Michael M. CNS listeriosis confused with leptomeningeal carcinomatosis in a patient with a malignant insulinoma. *Am J Clin Oncol*. 2002;25:576–579.

103. Cunha BA, Filozov A, Reme P. *Listeria monocytogenes* encephalitis mimicking West Nile encephalitis. *Heart Lung*. 2004;33:61–64.

104. Pomar V, Benito N, Lopez-Contreras J, et al. Characteristics and outcome of spontaneous bacterial meningitis in patients with cancer compared to patients without cancer. *Medicine*. 2017;96, e6899.

105. Costerus JM, Brouwer MC, Vanderende A, Van de Beek D. Community-acquired bacterial meningitis in adults with cancer or a history of cancer. *Neurology*. 2016;86:860–866.

106. Mace SE. Acute bacterial meningitis. *Emerg Med Clin N Am*. 2008;38:281–317.

107. Fowler Jr VG, Justice A, Moore C, et al. Risk factors for hematogenous complications of intravascular catheter-associated *Staphylococcus aureus* bacteremia. *Clin Infect Dis*. 2005;40:695–703.

108. Yusuf SW, Ali SS, Swafford J, et al. Culture-positive and culture-negative endocarditis in patients with cancer. *Medicine*. 2006;85(2):86–94.

109. Martin-Davila P, Fortun J, Navas E, et al. Nosocomial endocarditis in a tertiary hospital: an increasing trend in native valve cases. *Chest*. 2005;128:772–779.

110. Antoniadou A, Torres HA, Lewis RE, et al. Candidemia in a tertiary care cancer center: in vitro susceptibility and its association with outcome of initial antifungal therapy. *Medicine*. 2003;82:309–321.

111. Rajendram R, Alp NJ, Mitchell AR. Candida prosthetic valve endocarditis cured by caspofungin therapy without valve replacement. *Clin Infect Dis*. 2005;40:e72–e74.

112. Dutta T, Karas MG, Segal A, et al. Yield of transesophageal echocardiography for nonbacterial thrombotic endocarditis and other cardiac sources of embolism in cancer patients with cerebral ischemia. *Am J Cardiol*. 2006;97:894–898.

113. Lauten M, Attarbaschi A, Cario G, et al. Invasive mold disease of the central nervous system in children and adolescents with cancer or undergoing hematopoietic stem cell transplantation: analysis of 29 contemporary patients. *Pediatr Blood Cancer*. 2019;66, e27806.

114. Gavalda J, Len O, San Juan R, et al. Risk factors for invasive aspergillosis in solid–organ transplant recipients: a case-control study. *Clin Infect Dis*. 2005;41:52–59.

115. Guermazi A, Gluckman E, Tabti B, et al. Invasive central nervous system aspergillosis in bone marrow transplantation recipients: an overview. *Eur Radiol*. 2003;13:377–388.

116. Pfeiffer CD, Fine JP, Safdar N, et al. Diagnosis of invasive aspergillosis using a galactomannan assay: a meta-analysis. *Clin Infect Dis*. 2006;42:1417–1427.

117. Denning DW. Invasive aspergillosis. *Clin Infect Dis*. 1998;26:781–805.

118. Mathisen GE, Johnson JP. Brain abscess. *Clin Infect Dis*. 1997;25:763–781.

119. Singh N, Paterson D. Aspergillus infections in transplant recipients. *Clin Microbiol Rev*. 2005;18:44–69.

120. Karakousis PC, Magill SS, Gupta A. Paraplegia caused by invasive spinal aspergillosis. *Neurology*. 2007;68(2):158. https://doi.org/10.1212/01.wnl.0000238981.65444.f2.

121. Kami M, Ogawa S, Kanda Y, et al. Early diagnosis of central nervous system aspergillosis using polymerase chain reaction, latex agglutination test, and enzyme-linked immunosorbent assay. *Br J Haematol*. 1999;106:536–537.

122. Herbrecht R, Denning DW, Pattersron TG, et al. Voriconazole versus amphotericin B for primary therapy of invasive aspergillosis. *N Engl J Med*. 2002;347:408–415.

123. Kartsonis NA, Saah AJ, Joy Lipka C, et al. Salvage therapy with caspofungin for invasive aspergillosis: results from the caspofungin compassionate use study. *J Infect*. 2005;50:196–205.

124. Singh N, Limaye AP, Forrest G, et al. Combination of voriconazole and caspofungin as primary therapy for invasive aspergillosis in solid organ transplant recipients: a prospective, multicenter, observational study. *Transplantation*. 2006;81:3320–3326.

125. Munoz P, Singh N, Bouza E. Treatment of solid organ transplant patients with invasive fungal infections: should a combination of antifungal drugs be used? *Curr Opin Infect Dis*. 2006;19:365–370.

126. Vazquez JA, Sobel JD. Anidulafungin: a novel echinocandin. *Clin Infect Dis*. 2006;43:215–222.

127. Cornely OA, Maertens J, Winston DJ, et al. Posaconazole vs. fluconazole or itraconazole prophylaxis in patients with neutropenia. *N Engl J Med*. 2007;356:348–359.

128. Ullmann AJ, Lipton JH, Vesole DH, et al. Posaconazole or fluconazole for prophylaxis in severe graft-versus-host disease. *N Engl J Med*. 2007;356:335–347.

129. DePAuw BE, Donnelly JP. Prophylaxis and aspergillosis: has the principle been proven? *N Engl J Med*. 2007;356:409–411.

130. Walsh TJ, Teppler H, Donowitz GR, et al. Caspofungin versus liposomal amphotericin B for empirical antifungal therapy in patients with persistent fever and neutropenia. *N Engl J Med*. 2004;351:1391–1402.

131. Marth FM, Lowry CM, Cutler CS, et al. Voriconazole and sirolimus coadministration after allogeneic hematopoietic stem cell transplantation. *Biol Blood Marrow Transplant.* 2006;12:552–559.

132. Hajjeh RA, Sofair AN, Harrison LH, et al. Incidence of bloodstream infections due to *Candida* species and in vitro susceptibilities of isolates collected from 1998 to 2000 in a population-based active surveillance program. *J Clin Microbiol.* 2004;42:1519–1527.

133. Verduyn-Lunel FM, Voss A, Kuijper EJ. Detection of the *Candida* antigen mannan in cerebrospinal fluid specimens from patients suspected of having *Candida meningitis. J Clin Microbiol.* 2004;42:867–870.

134. Perfect JR, Marr KA, Walsh TJ, et al. Voriconazole treatment for less-common, emerging, or refractory fungal infections. *Clin Infect Dis.* 2003;36:1112–1131.

135. Mattiuzzi G, Giles FJ. Management of intracranial fungal infections in patients with haematological malignancies. *Br J Haematol.* 2005;131:287–300.

136. Kauffman CA. Zygomycosis: reemergence of an old pathogen. *Clin Infect Dis.* 2004;39:588–590.

137. Greenberg RN, Scott LH, Vaughn HH, et al. Zygomycosis (mucormycosis): emerging clinical importance and new treatments. *Curr Opin Infect Dis.* 2004;17:517–525.

138. Kauffman CA. Endemic mycoses in patients with hematologic malignancies. *Semin Respir Infect.* 2002;117:106–112.

139. Blair JE, Smilack JD, Caples SM. Coccidioidomycosis in patients with hematologic malignancies. *Arch Intern Med.* 2005;165:113–117.

140. Braddy CM, Heilman RL, Blair JE. Coccidioidomycosis after renal transplantation in an endemic area. *Am J Transplant.* 2006;6(2):340–345.

141. Husain S, Wagener MM, Singh N. *Cryptococcus neoformans* infection in organ transplant recipients: variables influencing clinical characteristics and outcome. *Emerg Infect Dis.* 2001;7:375–381.

142. Horn DL, Fishman JA, Steinbach WJ, et al. Presentation of the PATH alliance registry of prospective data collection and analysis of the epidemiology, therapy and outcomes of invasive fungal infections. *Diagn Microbiol Infect Dis.* 2007;59:407–414.

143. Schwartz S, Kontoyiannis DP, Harrison T, Ruhnke M. Advances in the diagnosis and treatment of fungal infections of the CNS. *Lancet Neurol.* 2018;17:362–372.

144. Rex JH, Larsen RA, Dismukes WE, et al. Catastrophic visual loss due to *Cryptococcus neoformans meningitis. Medicine (Baltimore).* 1993;72:207–224.

145. Wu G, Vilchez RA, Eidelman B, et al. Cryptococcal meningitis: an analysis among 5,521 consecutive organ transplant recipients. *Transpl Infect Dis.* 2002;4:183–188.

146. Rosenfeld MR, Pruitt AA. Neurologic complications of bone marrow, stem cell, and organ transplantation in patients with cancer. *Semin Oncol.* 2006;33:352–361.

147. Dworkin RH, Johnson RW, Breuer J, et al. Recommendations for the management of herpes zoster. *Clin Infect Dis.* 2007;44:S1–S26.

148. Dubinsky RM, Kabbani AH, El-Chami Z, et al. Practice parameter: treatment of post-herpetic neuralgia: an evidence-based report of the Quality Standards Subcommittee of the American Academy of Neurology. *Neurology.* 2004;63:959–965.

149. Quan D, Hammack BN, Kittelson J, et al. Improvement of post-herpetic neuralgia after treatment with intravenous acyclovir followed by oral valacyclovir. *Arch Neurol.* 2006;63:940–942.

150. Boeckh M, Kim HW, Flowers ME, et al. Long-term acyclovir for prevention of varicella zoster virus disease after allogeneic hematopoietic cell transplantation- a randomized double-blind placebo-controlled study. *Blood.* 2006;107:1800–1805.

151. Baldwin KJ, Cummings CL. Herpesvirus infections of the nervous system. *Continuum.* 2018;24(5):1349–1369.

152. Weaver S, Rosenblum MK, DeAngelis LM. Herpes varicella-zoster encephalitis in immunocompromised patients. *Neurology.* 1999;52:192–195.

153. Gilden DH, Lipton H, Wolf J, et al. Two patients with unusual forms of varicella-zoster virus vasculopathy. *N Engl J Med.* 2002;347:1500–1503.

154. Pawlitzki M, Teube J, Campe C, et al. VZV-associated acute retinal necrosis in a patient with MS treated with natalizumab. *Neurol Neuroimmunol Neuroinflamm.* 2018;5. https://doi.org/10.1212/NXI.0000000000000475, e475.

155. Orme HT, Smith AG, Nagel MA, et al. VZV spinal cord infarction identified by diffusion-weighted MRI (DWI). *Neurology.* 2007;69:398–400.

156. Nagel MA, Forghani B, Mahalingam R, et al. The value of detecting anti-VZV IGG antibody in CSF to diagnosis VZV vasculopathy. *Neurology.* 2007;68:1069–1073.

157. Nagel MA, Cohrs RJ, Mahalingam R, et al. The varicella zoster virus vasculopathies. Clinical, CSF, imaging and virologic features. *Neurology.* 2008;70:853–860.

158. Lau YL, Peiris M, Chan GC, et al. Primary human herpes virus-6 infection transmitted from donor to recipient through bone marrow infusion. *Bone Marrow Transplant.* 1998;21:1063–1066.

159. Ogata M, Kikuchi H, Satou T, et al. Human herpes virus-6 DNA in plasma after allogeneic stem cell transplantation: incidence and clinical significance. *J Infect Dis.* 2006;193:69–79.

160. Pot C, Burkhard PR, Villard J, et al. Humanherpesvirus-6 variant A encephalomyelitis. *Neurology.* 2008;70:974–975.

161. Deborska D, Durlik M, Sadowska A, et al. Human herpes virus-6 in renal transplant recipients: potential risk factors for the development of human herpes virus-6 seroconversion. *Transplant Proc.* 2003;35:2199–2201.

162. Hashimoto K, Yasukawa M, Tohyama M. Human herpes virus-6 and drug allergy. *Curr Opin Allergy Clin Immunol.* 2003;3:255–260.

163. Fujino Y, Nakajima M, Inoue H, et al. Human herpes virus-6 encephalitis associated with hypersensitivity syndrome. *Ann Neurol.* 2002;51:771–774.

164. Ljungman P, Singh N. Human herpesvirus-6 infection in solid organ and stem cell transplant recipients. *J Clin Virol.* 2006;37(Suppl. 1):S87–S91.

165. Zerr DM, Corey L, Kin HW, et al. Clinical outcomes of human herpes virus-6 reactivation after hematopoietic cell transplantation. *Clin Infect Dis.* 2005;40:932–940.

166. Bhanushali MJ, Kranick SM, Freeman AF, et al. Human herpes 6 virus encephalitis complicating allogeneic hematopoietic stem cell transplantation. *Neurology.* 2013;80(16):1494–1500.

167. Zerr DM. Human herpes virus-6 and central nervous system disease in hematopoietic cell transplantation. *J Clin Virol.* 2006;37(Suppl. 1):S52–S56.

168. Ward KN, Leong HM, Thiruchelvam AD, et al. Human herpes virus-6 DNA levels in cerebrospinal fluid due to primary infection differ from those due to chromosomal viral integration and have implications for diagnosis of encephalitis. *J Clin Microbiol.* 2007;45:1298–1304.

169. Kamble RT, Clark DA, Leong HN, et al. Transmission of integrated human herpesvirus-6 in allogeneic hematopoietic stem cell transplantation. *Bone Marrow Transplant.* 2007;40:563–566.

170. Fotheringham J, Akhyani N, Vortmeyer A, et al. Detection of active human herpesvirus-6 infection in the brain: correlation with polymerase chain reaction detection in cerebrospinal fluid. *J Infect Dis.* 2007;195:450–454.

171. Weil AA, Glaser CA, Amad Z, Forghani B. Patients with suspected herpes simplex encephalitis: rethinking an initial negative polymerase chain reaction result. *Clin Infect Dis.* 2002;34:1154–1157.

172. Astrom KE, Mancall EL, Richardson EP. Progressive multifocal leukoencephalopathy. *Brain.* 1958;81:93–127.

173. Koralnik IJ, Schellingerhout D, Frosch MP. Case records of the Massachusetts General Hospital. Weekly clinicopathological exercises Case 14-2004. *N Engl J Med*. 2004;350:1882–1893.

174. Boren EJ, Cheema GS, Naguwa SM, et al. The emergence of progressive multifocal leukoencephalpathy (PML) in rheumatic diseases. *J Autoimmun*. 2008;30:90–98.

175. Garcia-Suarez J, de Miguel D, Krasnik I, et al. Changes in the natural history of progressive multifocal leukoencephalopathy in HIV-negative lymphoproliferative disorders; impact of novel therapies. *Am J Hematol*. 2005;80:271–281.

176. Kharfan-Dabaja MA, Ayala E, Greene J, et al. Two case of progressive multifocal leukoencephalopathy after allogeneic hematopoietic cell transplantation and a review of the literature. *Bone Marrow Transplant*. 2007;39:101–107.

177. Huang D, Cossoy M, Li M, et al. Inflammatory progressive multifocal leukoencephalopathy in human immunodeficiency virus-negative patients. *Ann Neurol*. 2007;62:34–39.

178. Koralnik IJ, Wuthrich C, Dang X, et al. JC virus granule cell neuronopathy: a novel clinical syndrome distinct from progressive multifocal leukoencephalopathy. *Ann Neurol*. 2005;57:576–580.

179. Ueno H, Kikumto M, Takebayashi Y, et al. Pomalidomide-associated PML in multiple myeloma: cortical susceptibility weight imaging hypointense findings prior to clinical deterioration. *J Neurovirol*. 2020;26(3):452–455.

180. Bernal-Cano F, Joseph JT, Koralnik IJ. Spinal cord lesions of progressive multifocal leukoencephalopathy in an acquired immunodeficiency syndrome patients. *J Neurovirol*. 2007;13:474–476.

181. Koralnik IJ. Progressive multifocal leukoencephalopathy revisited: has the disease outgrown its name? *Ann Neurol*. 2006;60:162–173.

182. Nath A, Venkataramana A, Reich DS, et al. Progression of PML despite treatment with beta-interferon. *Neurology*. 2006;66(1):149–150.

183. Marra CM, Rajicic N, Barker DE, et al. A pilot study of cidofovir for progressive multifocal leukoencephalopathy in AIDS. *AIDS*. 2002;16:1791–1797.

184. Viallard JF, Lazaro E, Lafon ME, et al. Successful cidofovir therapy of PML preceding angioimmunoblastic T-cell lymphoma. *Leuk Lymphoma*. 2005;46:1659–1662.

185. Langer-Gould A, Atlas SW, Green AJ, et al. Progressive multifocal leukoencephalopathy in a patient treated with natalizumab. *N Engl J Med*. 2005;353:375–381.

186. Vulliemoz S, Lurati-Ruiz F, Borruat FX, et al. Favourable outcome of progressive multifocal leukoencephalopathy in two patients with dermatomyositis. *J Neurol Neurosurg Psychiatry*. 2006;77:1079–1082.

187. Cortese I, Muranski P, Enose-Akahata Y, et al. Pembrolizumab treatment for progressive multifocal leukoencephalopathy. *N Engl J Med*. 2019;380:1597–1605.

188. Calvo V, Fernandez-Cruz A, Nunez B. Cancer and the SARS-CoV-2 infection: a third level hospital experience. *Clin Epidemiol*. 2021;13:317–324.

189. Bange EM, Han NA, Wileyto P, et al. CD8 T cells compensate for impaired humoral immunity in COVID-19 patients with 2 hematologic cancer. *Nat Med*. 2021;27:1280–1289. epub May 20 34017137.

190. Grebenciucova E, Pruitt A. Infections in patients receiving multiple sclerosis disease-modifying therapies. *Curr Neurol Neurosci Rep*. 2017;17(11):88. https://doi.org/10.1007/s11910-017-0800-8.

191. Boyarski BM, Werbel WA, Avery RK, et al. Antibody response to 2-dose SARS-CoV-2 mRNA vaccine series in solid organ transplant recipients. *JAMA*. 2021;325:2204–2206 [epub 5 May 2021].

192. Bar-Or A, Calkwood JC, Chognot C, et al. Effect of ocrelizumab on vaccine responses I patients with multiple sclerosis: the VELOCE study. *Neurology*. 2020;95(14):1999–2008.

193. Hijiya N, Hudson NM, Lensing S, et al. Cumulative incidence of secondary neoplasms as a first event after childhood acute lymphoblastic leukemia. *JAMA*. 2007;296:1207–1215.

14

Diagnosis and treatment of paraneoplastic neurological disorders

Annick Desjardins

Neurosurgery, The Preston Robert Tisch Brain Tumor Center at Duke, Durham, NC, United States

1 Introduction

The term paraneoplastic neurological disorder (PND) refers to an extensive and heterogeneous group of neurological disorders caused by mechanisms other than metastasis and any of the following complications of cancer: (i) metabolic and nutritional deficits, (ii) infections, (iii) cerebrovascular disease, (iv) coagulopathy, and (v) neurotoxicity from chemotherapy and radiation therapy.[1]

Despite an incidence that varies with each type of tumor, overall less than 1% of all patients with cancer will develop symptoms of PND. About 3% of patients with small-cell lung cancer (SCLC) develop PND, 30% of patients with thymoma, and 5%–15% of patients with plasma cell dyscrasias associated with malignant monoclonal gammopathies.[2] As such, even neurologists considered specialists for PND will diagnose only a few patients each year.[3]

Thus, in 2004, an international panel of neurologists with an interest in PND met to publish proposed diagnostic criteria (Table 1).[3] The proposed diagnostic criteria not only allow neurologists to diagnose PND, but also provide a universal language allowing data pooling for clinical trials and the rapid recognition of new diagnosis and tumor association.[3] Two levels of evidence to determine a neurological syndrome as paraneoplastic—"definite" and "possible"—were proposed based on a number of criteria established by the presence or absence of cancer and the definitions of "classical" syndrome and "well-characterized" onconeural antibody.[3] Since then, the list of onconeural antibodies and syndromes has been increasing steadily warranting the need for an update of the criteria.[4]

The rapid recognition of a PND is crucial, as early diagnosis and prompt aggressive treatment of the malignancy in conjunction with immunotherapy may stabilize or improve the neurological injury.[5–7] Patients already known for a primary malignancy can develop symptoms of PND at the time of recurrence. In those patients, the diagnosis of PND is a diagnosis of exclusion, as neurological complications of cancer itself or its treatment are significantly more frequent. The detection of antineuronal antibodies in the patient's serum or cerebrospinal fluid (CSF) might help in some instances. However, more than 40% of patients do not present any antibodies and, in some cases, antibodies are detected in cancer patients without PND.[3]

A significant advance in the treatment of cancer in recent years has been the approval of many checkpoint inhibitors for the treatment of different neoplasias. As PND is known to derive from an autoimmune reaction, an increase in the incidence/frequency of PND has been observed; the use of immune checkpoint inhibitors targets T-cell responses directed against the tumor, and in turn the nervous system.[8–10]

2 Pathogenesis

It was first suggested more than 60 years ago that PND could have an autoimmune origin,[11,12] where specific antibodies recognize antigens expressed both in the tumor and in the nervous system.[1,5,6,13,14] It has now been demonstrated that two patterns of immune response may be associated with PND[15]: (i) an antibody-mediated immune response directed against neuronal receptors or other cell membrane antigens[4,16–18] and (ii) an immune response directed against intracellular neuronal proteins.[4,15] Tumor antigens released after tumor-cell death are presented to T cells by antigen-presenting cells, which in turn lead to an onconeural antigen-specific antibody

TABLE 1 Diagnostic criteria for paraneoplastic disorder per the 2004 international panel.

Diagnostic criteria for definite PND

1. A classical syndrome and cancer that develops within 5 years of the diagnosis of the neurological disorder

2. A nonclassical syndrome that resolves or significantly improves after cancer treatment without concomitant immunotherapy, provided that the syndrome is not susceptible to spontaneous remission

3. A nonclassical syndrome with onconeural antibodies (well characterized or not) and cancer that develops within 5 years of the diagnosis of the neurological disorder

4. A neurological syndrome (classical or not) with well-characterized onconeural antibodies (anti-Hu, Yo, CV2, Ri, Ma2, or amphiphysin), and no cancer

Diagnostic criteria for possible PND

1. A classical syndrome, no onconeural antibodies, no cancer but at high risk to have an underlying tumor

2. A neurological syndrome (classical or not) with partially characterized onconeural antibodies and no cancer

3. A nonclassical syndrome, no onconeural antibodies, and cancer present within 2 years of diagnosis

---

or cell-mediated autoimmune response.[4,6] When the immune response is directed against neuronal receptors or other cell membrane antigens, a neuronal dysfunction is observed and a cancer diagnosis might or might not be present. Fortunately, this type of immune response normally responds well to therapy.[15] Conversely, when the immune response is directed against intracellular neuronal proteins, a neoplasia is almost always present. Neuronal death occurs as a consequence of the immune response, a cytotoxic T-lymphocyte-mediated process,[5,16,18] leading to an irreversible neurological deficit.[15] Quick recognition of the PND with rapid initiation of empiric treatment is warranted to limit as much as possible the extent of permanent neurological damage or even death.[15]

This autoimmune origin for PND supports the hypothesis that the tumors of patients with PND are able to circumvent immune tolerance and trigger a strong B- and T-cell responses,[15] explaining why almost two-thirds of patients develop symptoms of neurological dysfunction secondary to a PND before a diagnosis of malignancy is known[6,14] or why the extent of disease is limited at the time of PND diagnosis.[4,19] In addition, it has been demonstrated that cancer patients who have onconeural autoantibodies have a better prognosis than patients with the same tumor diagnosis who do not.[4,20]

Also, it has been demonstrated that the type of immunological stimulation against a tumor can determine the course of the neurological disorder and the tumor. For example, patients with SCLC and CV2-Ab have a

median survival time that is 2.5 times greater than that of patients with the same cancer, similar neurological symptoms, and Hu-Ab.[21]

3 Clinical diagnosis

Detailed descriptions of the main PNDs are available in later sections, but some generalities apply to most PNDs. First, they are difficult to diagnose for two main reasons: (i) frequently, the neurological symptoms present before a diagnosis of a primary malignant process is made and (ii) symptoms normally develop rapidly in days or a few weeks, followed by a stabilization.[3] The rapid development of symptoms suggests that by the time most PNDs are recognized, the damage affecting the nervous system is irreversible, making PNDs difficult to treat. Severe disability can occur, with patients rapidly becoming wheelchair-bound or bedridden. Most authors agree that treatment of the primary malignant process is the main requisite to stabilize the neurological disorder or, in rare cases, control it. Exceptions do occur, and some spontaneous improvements in the neurological disorder have occasionally been reported.[22,23]

3.1 Clinical symptoms

Given different patterns of injury, seizures improve more rapidly after initiation of treatment, while for cognitive deficits, paresis, and sensory deficits, improvement is more protracted.[4,19,24]

3.2 Cerebrospinal fluid

CSF studies are important to rule out disorders that may mimic PND, mainly infectious or neoplastic meningitis.[7] Findings consistent with PND include mild lymphocytic pleocytosis, elevated proteins, and elevated immunoglobulin (IgG) index, sometimes with oligoclonal bands or paraneoplastic antibodies.[25] However, similar CSF abnormalities can be encountered in any inflammatory or immune-mediated disorders of the CNS, and some patients with PND may have normal CSF studies.[26]

3.3 Radiographic assessment

Radiological evaluations are necessary to eliminate metastatic lesions or other complications of cancer.[27,28] Magnetic resonance imaging (MRI) is the study of choice. More frequently than not, affected brain regions do not enhance with contrast, given that the blood-brain barrier function is preserved. T2-weighted and fluid-attenuated inversion recovery (FLAIR) sequences are normally the first to demonstrate any anomaly. One study reported brain T2-weighted abnormalities in 71% of patients with paraneoplastic limbic encephalitis and SCLC.[29] Another series of 24 patients with paraneoplastic limbic encephalitis demonstrated that all patients had FLAIR abnormalities, involving one or both temporal lobes.[30] In a series of 29 patients with anti-Ma2-associated encephalitis, brain MRI abnormalities were present in 71%; some of these patients had abnormalities that appeared as nodular enhancing lesions, suggesting tumor metastasis.[31] All patients who underwent repeated brain MRI had new abnormalities in subsequent studies, including patients whose initial MRI was normal.

There is also increasing evidence that brain fluorodeoxyglucose-positron emission tomography (FDG-PET) is of diagnostic utility in PND.[32] In early stages of PND, FDG-PET may show hypermetabolism in abnormal brain regions identified by MRI, and even in some patients whose MRI is normal.[27]

3.4 Tissue sampling

Brain biopsy is rarely required for diagnosis of PND. Biopsy of an abnormal area identified by MRI or FDG-PET may be considered if a neoplastic process is suspected or if the clinical, CSF, and MRI findings are unusual. Abnormalities supporting, but not specific to PND, include infiltrates of mononuclear cells, neuronophagic nodules, neuronal degeneration, microglial proliferation, and gliosis.[33]

4 Antineuronal antibodies

Patients suspected to have a PND should be examined for antineuronal antibodies in serum and CSF. As the clinical presentation of PND varies significantly, since new onconeural antibodies are being regularly identified, and because rapid treatment is warranted, commercially available antibody panels should be obtained rather than single-antibody testing.[15,34] These antibodies preferentially associate with restricted histological types of tumors (Table 2). Therefore, in addition to supporting the diagnosis of PND, the presence of antineuronal antibodies helps narrow the search for the primary malignancy.

Some antibodies, for example, P/Q-type voltage-gated calcium channels (VGCCs), voltage-gated potassium channels (VGKCs), and nicotinic or ganglionic acetylcholine receptor (AchR) antibodies, associate with specific disorders, but do not differentiate between paraneoplastic and nonparaneoplastic cases.[35–37] Although some of these disorders (i.e., Lambert-Eaton Myasthenia Syndrome (LEMS) and myasthenia gravis (MG)) can be diagnosed on clinical and electrophysiological grounds, analysis for specific antibodies is useful in several settings. For example, the recognition of overlapping syndromes, such as LEMS and paraneoplastic cerebellar degeneration (PCD), is improved by antibody analysis.[38] For other syndromes such as dysautonomia, the detection of antibodies to the neuronal acetylcholine

TABLE 2   Most common autoantibodies with associated paraneoplastic disorders and cancers.

Antibody	Most common paraneoplastic disorders	Most common associated cancers
Autoantibodies directed against intracellular neuronal proteins		
Anti-Hu (ANNA1)	Limbic encephalitis, paraneoplastic cerebellar degeneration, encephalomyelitis, sensory neuronopathy, gastroparesis	SCLC, NSCLC, extrathoracic cancers, neuroendocrine tumors, retinoblastoma (infants)
Anti-CV2/CRMP5	Limbic encephalitis, encephalomyelitis, paraneoplastic cerebellar degeneration, mixed axonal-demyelinating sensorimotor neuropathy, sensory neuronopathy, optic neuritis, chronic gastrointestinal pseudo-obstruction, chorea	SCLC, NSCLC, breast, thymoma
Antiamphiphysin	Limbic encephalitis, paraneoplastic cerebellar degeneration, encephalomyelitis, stiff-person syndrome, sensory or sensorimotor neuropathy	Breast, SCLC
Anti-Yo (PCA1)	Paraneoplastic cerebellar degeneration, brainstem encephalitis, myelitis	Gynecological, breast
Anti-GAD65	Stiff-person syndrome, paraneoplastic cerebellar degeneration, epilepsy, limbic encephalitis	Thymoma, breast
Anti-Ri (ANNA-2)	Opsoclonus-myoclonus, brainstem encephalitis	Neuroblastoma, breast, SCLC
Anti-Zic4	Encephalomyelitis, paraneoplastic cerebellar degeneration	SCLC, ovarian adenocarcinoma
Anti-Ma and/or anti-Ma2 (Ta)	Limbic encephalitis, upper brainstem encephalitis, opsoclonus-myoclonus, paraneoplastic cerebellar degeneration	Testicular, NSCLC

Continued

TABLE 2   Most common autoantibodies with associated paraneoplastic disorders and cancers—cont'd

Antibody	Most common paraneoplastic disorders	Most common associated cancers
Anti-MaP1B (PCA2)	Limbic encephalitis, paraneoplastic cerebellar degeneration, encephalomyelitis, polyradiculoneuropathy	SCLC, NSCLC
Anti-ANNA3	Encephalomyelitis, limbic encephalitis, paraneoplastic cerebellar degeneration, brainstem encephalitis, neuropathies	SCLC, NSCLC, bronchogenic cancers
Anti-SOX1 (AGNA)	Lambert–Eaton Myasthenia Syndrome, paraneoplastic cerebellar degeneration, limbic encephalitis, sensory neuronopathy	SCLC, NSCLC
Anti-NIF	Paraneoplastic cerebellar degeneration, encephalitis, myelitis	Neuroendocrine neoplasms, SCLC, hepatocellular carcinoma, Merkel cell carcinoma
Anti-ITPR1	Sensory-motor and dysautonomic neuropathy, paraneoplastic cerebellar degeneration	Breast, endometrial, lung
Anti-GFAP	Meningoencephalitis, myelitis, epilepsy, movement disorder, paraneoplastic cerebellar degeneration	Gastrointestinal, gynecological, thymoma
Autoantibodies directed against neuronal receptors or other cell membrane antigens		
Anti-P/Q-type VGCC	Lambert–Eaton Myasthenia Syndrome, paraneoplastic cerebellar degeneration	SCLC
Anti-VGKC	Limbic encephalitis	SCLC, thymoma
Anti-AChR	Myasthenia gravis	Thymoma
Anti-gAChR	Autoimmune autonomic ganglionopathy	Thymoma, SCLC, adenocarcinoma
Anti-DNER (PCA-Tr)	Paraneoplastic cerebellar degeneration	Hodgkin's lymphoma
Anti-mGlurR1	Paraneoplastic cerebellar degeneration	Hodgkin's lymphoma
Anti-mGlur2	Paraneoplastic cerebellar degeneration	SCLC, alveolar rhabdomyosarcoma
Anti-mGlurR5	Limbic encephalitis (Ophelia's syndrome), extralimbic encephalitis	Hodgkin's lymphoma, SCLC
Anti-NMDAR	Encephalitis (initial psychiatric presentation)	Ovarian teratoma
Anti-Lgi1	Limbic encephalitis, faciobrachial dystonic seizures, sleep disorder	Thymoma, SCLC, breast
Anti-Caspr2	Neuromyotonia, Morvan's syndrome, limbic encephalitis, paraneoplastic cerebellar degeneration, autoimmune epilepsy	Thymoma, SCLC, melanoma
Anti-AMPAR	Limbic encephalitis	Thymoma, SCLC, breast
Anti-GABA$_A$R	Encephalitis	Hodgkin lymphoma
Anti-GABA$_B$R	Limbic encephalitis, isolated status epilepticus, paraneoplastic cerebellar degeneration, opsoclonus-myoclonus	SCLC, thymoma
Anti-DPPX	CNS hyperexcitability, encephalitis, GI dysmotility	Lymphoma
Anti-aquaporin-4	Neuromyelitis Optica Spectrum Disorders (optic neuritis, longitudinally extensive transverse myelitis, area postrema syndrome)	Thymoma, breast, lung
Anti-GluD2	Opsoclonus-myoclonus ataxia syndrome	Neuroblastoma
Anti-GlyR	Encephalomyelitis (with rigidity and myoclonus), stiff-person syndrome	Thymoma

receptor identifies a subset of patients that benefit from immunosuppression.[37] Similar neurological disorders can associate with different immune response, which suggests clinical-immunological heterogeneity (Table 2). For instance, antibodies to glutamic acid decarboxylase (GAD), amphiphysin, or gephyrin have all been reported to be associated with stiff-person syndrome, but the majority of patients with antibodies to GAD do not have cancer.[39–41] Furthermore, several antineuronal antibodies may co-occur in the same patient, particularly if the underlying tumor is SCLC. Concurrent Hu, CRMP5, and Zic4 antibodies were identified in 27% of patients with SCLC and paraneoplastic encephalomyelitis (PEM).[42]

Cancer patients without PND may have antineuronal antibodies,[43] but usually the antibody titers are lower than those in patients with PND. It was reported that antibodies to Hu, CRMP5, and Zic4 were encountered in 19%, 9%, and 16% of SCLC patients without PND, respectively.[42] A practical implication is that the detection of any of these antibodies in a patient with neurological symptoms of unknown cause suggests a PND and predicts an underlying cancer, usually SCLC.

Paraneoplastic antibodies are classified into two categories based on their clinical relevance: well-characterized paraneoplastic antibodies and partially characterized antibodies.[3] Well-characterized paraneoplastic antibodies—anti-Hu, Yo, Ma2, Ri, CV2/CRMP5, and amphiphysin—have been characterized by different laboratories and reported in large series of patients with PND. If any of these antibodies are identified, the diagnosis of PND is strongly supported, even if no tumor is found at the time of the initial workup. Some antibodies are more specific for a syndrome, for example, anti-Yo antibodies for cerebellar degeneration and anti-Ma2 for limbic or upper brain stem dysfunction. However, anti-Hu or anti-CV2/CRMP5 antibodies are associated with a wider spectrum of symptoms. The second category, partially characterized antibodies, for which limited clinical experience is available or the target antigens are unknown, is normally identified by clinical laboratories examining sera of thousands of patients suspected to have PND.[44–48]

5 Treatment

As stated earlier, the goal of PND treatment is to prevent permanent neurological deficit or death. As syndromes can progress rapidly and antibody panels require time to be completed, empiric initiation of treatment is recommended.[15]

5.1 Antibody-mediated to cell membrane antigens

When a PND secondary to antibody-mediated immune response directed against neuronal receptors or other cell membrane antigens is suspected, first-line therapy should involve intravenous methylprednisolone, 1000 mg daily for 3–5 days.[15,49] Plasma exchange and/or IVIG are the next treatment strategies. Plasma exchanges consist of every-other-day exchanges for five to seven sessions,[50] while IVIG is normally dosed at 2 g/kg over 3–5 days.[51] As a reminder, plasmapheresis should be trialed before IVIGs, as the process of plasma exchange would remove the IgG previously administered.

In the event that a patient fails to respond, additional treatment options include rituximab (1000 mg intravenously for two doses 14 days apart followed by

every 6 months dosing)[4,52] and cyclophosphamide (500–1000 mg/m^2 every month).[53] Finally, identification and treatment of the underlying malignancy are the best options for long-term control.

5.2 Autoantibodies directed against intracellular neuronal proteins (cytotoxic T-cell-mediated process)

In PND where it is suspected that a cytotoxic T-cell-mediated process is most probably the cause due to the presence of onconeural antibodies targeting intracellular antigens or based on pathological confirmation, treatment is more difficult. Corticosteroids, plasma exchange, IVIG, rituximab, and cyclophosphamide normally have limited efficacy.[54] Thus, therapies targeting lymphocytic lineage (B and T cells) are rationale first approaches, including cyclophosphamide, mycophenolate mofetil, and azathioprine.[4] Intravenous or oral cyclophosphamide should be considered when the presentation of the PND is more rapid and severe.

5.3 Other possible options

Other possible options in the management of PND are tacrolimus, sirolimus, and human chorionic gonadotropin (hCG).[4]

5.4 Additional considerations

Most patients suffering from PND will remain on corticosteroids for a prolonged period of time, and thus calcium and vitamin D supplementation should be initiated. Bone densitometry should be monitored and, as needed, bisphosphonate initiated. Other important measures are the prevention of gastric ulcerative disease[55] and *Pneumocystis jiroveci* pneumonia.

As it is postulated that the immune response responsible for PND is also limiting or delaying progression of the responsible malignancy, it should be kept in mind that immunosuppressive therapy directed against the PND might result in tumor growth. Furthermore, drug-drug interaction or similar side effect profile might potentiate the side effects of chemotherapy used to treat the tumor.[4,56]

6 Underlying malignancy

The discovery of an occult neoplasm that is associated with a particular neurological syndrome remains the gold standard to diagnose PND. However, diagnosis of the malignancy may be difficult at first because of its location or small size.[57] In addition to the clinical history and evaluation of risk factors for cancer and serological cancer markers such as carcinoembryonic antigen, Ca-125, CA-15.3, or prostate-specific antigen, most patients will need chest, abdomen, and pelvis-computed tomography (CT) as part of the initial evaluation or restaging of cancer. The need for other tests will vary with the patient's gender and type of syndrome or antineuronal antibody. These may include pelvic, testicular, or vaginal ultrasound and mammograms. Total-body FDG-PET is useful in identifying tumors not visible with other studies.[32,58–60] Despite the sensitivity of FDG-PET scan, there are instances where antineuronal antibodies can lead to the discovery of neoplasms that escape FDG-PET detection.[61] The presence of a second cancer should be suspected if the tumor identified is different from the cancer usually associated with a specific syndrome or antineuronal antibody, or if the tumor does not express the neuronal antigen.[20,62]

Close cancer surveillance is needed in patients with classic PND, or with nonclassic PND but positive antineuronal antibodies, whose tumor is not found. A common practice is to repeat evaluations every 6 months (i.e., body CT or FDG-PET). The cancer usually manifests within the first 4 years of the PND, and in 90% of cases, it manifests in the first year after presentation of the neurological symptoms. There are rare instances where the expected type of tumor has been demonstrated 10 years later.[63] If no malignancy is demonstrated with the initial workup and a gynecological tumor is suspected in view of the antibodies, surgical exploration and removal of pelvic organs should be considered.[64] Patients presenting with a neuropathy of unclear etiology should have a search for monoclonal gammopathy in the serum and urine. If positive, a bone series and bone marrow biopsy should be performed.[2] An exception to this is a diagnosis of LEMS, where if after 2 years an underlying malignancy has not been identified, additional investigation can be discontinued.[4,56,65]

7 Paraneoplastic disorders and immune checkpoint inhibitors

New in cancer therapy is the use of immune checkpoint inhibitors and their propensity to trigger flares in patients known to suffer from an autoimmune disorder. Case series have reported a flare-up of the underlying autoimmune disorder in 29%–55% of patients.[9,66–69] Similarly, cases of patients already diagnosed with a PND have been shown to worsen following initiation of immune checkpoint inhibitors,[69,70] which lead to death in half of the cases.[9,71–76] Given this risk, a paraneoplastic antibody screen could be considered prior to initiation of immune checkpoint inhibitor in patients diagnosed with a neoplasia with a higher frequency of PND, for example, SCLC and thymoma.[71,75] Given the widespread use

of immune checkpoint inhibitors, one might be ready to observe an increased incidence of PND, especially in patients with SCLC, gynecological cancer, and thymoma.[9] In a case report of a fatal autoimmune encephalitis following treatment with immune checkpoint inhibitors, autopsy demonstrated infiltration of the brain by CD4 and CD8 T-cells, with minimal CD20 infiltration.[77]

8 Paraneoplastic neurological disorders

8.1 Paraneoplastic cerebellar degeneration

Representing 37% of all antibody-associated PNDs, PCD is one of the most common and characteristic PNDs.[33,78] The pancerebellar dysfunction normally presents acutely with nausea, vomiting, dizziness, and mild gait ataxia. It then progresses rapidly over a few weeks or months to appendicular and truncal ataxia, dysarthria, dysphagia, downbeat nystagmus, and diplopia (examination does not usually allow an objective determination of any oculomotor abnormality). In most cases, the process stabilizes after a few months of evolution. However, by that time most patients are severely debilitated. The evolution is usually symmetrical, although one side might be more affected than the other. If a careful neurological examination is performed, about half of the patients will present with additional mild neurological abnormalities, including sensorineural hearing loss, pyramidal and/or extrapyramidal signs, and mental status abnormalities.[64,78,79] The Paraneoplastic Neurological Syndrome Euronetwork established that for a paraneoplastic cerebellar syndrome to be labeled as classical, a pancerebellar syndrome with no imaging evidence of cerebellar atrophy outside of the expected changes from age needs to develop in less than 12 weeks.[3]

This disorder has been associated with any type of malignant tumor or antibodies, but the malignancies more frequently involved are lung cancer (more than half of the cases and in majority SCLC cases),[38] breast carcinoma, ovarian or uterine cancer,[80] and Hodgkin's lymphoma.[81] In most instances (60%–70%), the cerebellar syndrome develops before any primary cancer has been identified.[78,82,83]

A pure or predominant cerebellar dysfunction is more likely to be associated with anti-Yo/PCA-1 (breast or gynecological malignancies),[84,85] PCA-2 (lung cancer),[48,86,87] anti-Hu (ANNA-1), anti-Ri (breast or lung cancer), and then anti-Tr[88] (Hodgkins' disease).[15,64,80,82,89–91] Less frequently, mGluR1 (Hodgkin's disease) and mGlur2[47] are responsible. Anti-Tr is associated with Hodgkin's disease and produces a characteristic punctate pattern involving the molecular layer of the cerebellum secondary to its reaction with the cytoplasm of Purkinje cells.[88] Since the cerebellum is a common target of most paraneoplastic immune injuries, other antibodies have been associated with cerebellar degeneration, but normally are part of a more diffuse neurological process.[38,92–94]

The characteristic pathological changes in PCD are a degeneration and loss of Purkinje cells, with a variable degree of involvement of other cerebellar neurons. On occasion, inflammatory infiltrates can be found in the deep cerebellar nuclei.

Anderson et al. performed FDG-PET in a few patients and observed hypometabolism in all areas of the neuraxis (cerebral cortex, cerebellum, and brainstem).[95] In most instances, the MRI is normal at the time symptoms appear, diffuse cerebellar atrophy appearing after months to a few years.[38] T2-weighted image abnormalities involving the cerebral and cerebellar white matter have been described occasionally, as well as transient contrast enhancement of the cerebellar folia, suggesting leptomeningeal disease. Early on, the CSF demonstrates mild lymphocytic pleocytosis, slight proteinorachia, and elevated IgG concentrations. Oligoclonal bands may also be present. The pleocytosis normally resolves with time.

Prompt diagnosis and treatment of the disorder before Purkinje cells are destroyed can, on occasion, allow a stabilization of the disorder and protect the remnants of the nervous system. Once the deficits have reached their maximum, most patients remain stable despite treatment of the underlying malignancy. Case reports of patients benefiting from treatment of the underlying tumor, plasma exchange, intravenous immunoglobulin (IVIg), immune suppression with cyclophosphamide, steroids, or rituximab have been published, but are the exception rather than the rule.[96–99] In anti-Yo-positive PCD, breast cancer patients do better than gynecological cancer patients.[83] However, as a general rule, patients presenting with a paraneoplastic disorder associated with anti-Yo or anti-Hu are particularly resistant to any therapy. Patients with Hodgkin's disease anti-Tr or anti-mGlurR1 positive have a much better prognosis. Symptoms can be reversible and the antibodies can disappear with successful treatment of the tumor and/or immunosuppression.[47,82] Symptomatic improvement of ataxia with clonazepam 0.5–1.5 mg daily is possible. Buspirone has also been described as giving modest relief.[100]

8.2 PEM and focal encephalitis

PEM describes cancer patients presenting with symptoms of multifocal involvement of the nervous system, including the temporal lobes and limbic system (limbic encephalitis), brainstem (brainstem encephalitis), cerebellum (subacute cerebellar degeneration), spinal cord (myelitis), dorsal root ganglia (subacute sensory neuronopathy), and autonomous nervous system (autonomic neuropathy).[33,101] Signs of inflammation can be demonstrated by multiple techniques: CSF studies,

biopsy, or gadolinium-positive lesions on MRI. Patients are said to present the isolated clinical syndrome when they present with predominant involvement of one area and only mild involvement of other areas. On pathological evaluation, a multifocal process is demonstrated, even in patients presenting with symptoms of focal encephalitis.[23,101,102]

8.3 Paraneoplastic limbic encephalopathy

Paraneoplastic limbic encephalopathy (PLE) normally presents with confusion, irritability, sleep disorders, depression, agitation, anxiety, hallucinations, short-term memory deficits, dementia, and partial complex seizures, which develop in a subacute fashion in days to 12 weeks.[3,23] On occasion, hypothalamic dysfunctions are associated and present with somnolence, hyperthermia, and endocrine abnormalities. Three groups of patients with PLE have been described.[23] In the first group, patients are usually older, with a median age of 62, frequently smoke, are more often female, and present with anti-Hu antibodies and lung cancer (usually SCLC).[23,29] In those patients, other areas of the nervous system are involved, which makes it a form of PEM. The second group consists of young males, with a median age of 34, who have testicular cancer with anti-Ma2 antibodies.[103–105] Usually, the symptomatology is limited to the limbic system, hypothalamus, and brainstem. The third group represents about 40% of patients with PLE. The median age is around 57, with lung cancer in 50%–60% of cases (40%–55% being SCLC), and testicular germ cell tumor in 20% of patients,[23,29,30] and no antineuronal antibodies can be identified.[23,30] The majority of patients present with a pure limbic encephalopathy.[23,29] Breast cancer (anti-AMPAR), thymoma (anti-AMPAR),[106] Hodgkin's disease (anti-mGluR5),[90] and immature teratoma have also been described.[23,30] Anti-GABAbR has also been described with SCLC.[107]

In limbic encephalitis, the absence of specific clinical markers and symptoms, as well as the neurological syndrome preceding the diagnosis of cancer, makes the recognition of the disorder difficult.[23] Increased signal on T2-weighted and FLAIR images of one or both medial temporal lobes, hypothalamus, and brainstem is present in 65%–80% of patients on MRI[23,30]; more rarely, contrast enhancement in those areas is observed.[108] Repeat imaging and coregistration with FDG-PET may be necessary to improve the sensitivity.[109] Many confused patients or those with a depressed level of consciousness are, in fact, in nonconvulsive status epilepticus.[110] The electroencephalographic study often demonstrates unilateral or bilateral temporal lobe epileptic discharges or slow background activity. In 80% of patients, CSF examination shows transient mild lymphocytic pleocytosis with elevated protein, IgG, or oligoclonal bands.[23,30]

The detection of paraneoplastic antibodies helps in the diagnosis. In the absence of paraneoplastic antibodies, a search for lung, breast, or testicular cancer is necessary. About 60% of PLE patients have paraneoplastic antibodies. SCLC is usually associated with anti-Hu or CV2/CRMP5 antibodies. Anti-CV2/CRMP5 can also be associated with thymoma.[111] Anti-Ma2 (or anti-Ta) antibodies suggest testicular cancer in young men and non-SCLC in older men or women,[29,30,103] and antiamphiphysin antibodies can also be associated.[23,103,111] Anti-Ma1 antibodies are normally associated with neoplasia other than testicular; a cerebellar dysfunction is usually observed. Anti-VGKC antibodies can be observed in paraneoplastic and nonparaneoplastic limbic encephalitis; thymoma or SCLC is normally the cause.[36,112] On pathological examination, changes are usually observed in the limbic and insular cortex for patients known to have PLE. However, other deep gray matter and white matter regions can be affected at the same time. Neuronal loss with reactive gliosis, perivascular lymphocytic infiltration, and microglial proliferation are also normally observed. Spontaneous remissions[29,113] or improvement following prompt treatment of the neoplastic tumor have been described, as well as following treatment with corticosteroids and IVIg.[23,29,31] However, no treatment has been demonstrated to be consistently beneficial.[114,115] Gultekin et al. observed that irrespective of treatment, 38% of patients with anti-Hu antibodies, 30% of patients with anti-Ma2 antibodies, and 64% of patients without antibodies showed a partial neurological recovery.[23]

In about 15% of patients, symptoms of cerebellar dysfunction predominate, mostly gait ataxia. However, most patients progress to a pancerebellar syndrome.

8.4 Brainstem encephalitis

One-third of patients with PEM develop symptoms suggestive of brainstem encephalopathy. Symptoms are dependent on the level of the brainstem involved: oscillopsia, diplopia, vertigo, hypoventilation, dysarthria, dysphagia, gaze abnormalities (nuclear, internuclear, or supranuclear), subacute hearing loss, facial weakness, and facial numbness. Also, movement disorders can present with chorea, dystonia, bradykinesia, myoclonus, and typical parkinsonian syndrome. It is unusual to obtain a pure paraneoplastic brainstem syndrome.[116] On examination, the pathological findings normally predominate in the lower brainstem, involving usually the medulla and inferior olivary nuclei.[101] Patients with Ma2-Ab generally develop limbic and brainstem encephalitis with testis tumors, but patients with Ma2-Ab and additional antibodies directed against other members of the Ma proteins have developed additional cerebellar symptoms and had tumors other than testis neoplasm.[31] Anti-Hu antibodies due to SCLC should also be considered.[15]

8.5 Myelitis

Approximately 20% of patients with PEM develop myelitis and demonstrate a progressive weakness with lower motor neuron signs: fasciculations, sensory loss, and autonomic dysfunction (e.g., incontinence and postural hypotension). A predominant involvement of posterior column function and cervical cord segments can occur. Respiratory failure and death are possible. Imaging is necessary to exclude compressive or intrinsic spinal cord masses, inflammatory or infectious myelopathies, and radiation injury in previously treated patients. On occasion, an enlargement or hyperintensity of the spinal cord can be identified, as well as contrast enhancement, but most of the time the spinal cord appears normal. Inflammatory changes are normally seen with CSF analysis. Pathological examination demonstrates neuronal loss with intense inflammatory reaction, involving the anterior and posterior horns. Secondary nerve root degeneration and neurogenic muscle atrophy can also develop. A degeneration of the white matter tracts is also a possibility. No treatment has been demonstrated effective.

About 25% of patients develop autonomic nervous system dysfunction secondary to involvement of the dorsal root ganglia and sympathetic or parasympathetic peripheral nerves and ganglia. Orthostatic hypotension, gastroparesis, and intestinal dysmotility, cardiac arrhythmias, hyperhidrosis, pupillary abnormalities to light stimulation, sweating abnormalities, neurogenic bladder, and impotence can be observed. Patients known to have a malignancy, as well as patients with an idiopathic pandysautonomia, can present antibodies to the ganglionic acetylcholine receptors.[37] Death can occur secondary to respiratory and autonomic failures.[101] SCLC is the cancer most often associated, and those patients normally harbor anti-Hu or anti-CVZ/CRMP5 antibodies or both.[62,111,117] Thymoma and other cancers are also associated with anti-CV2/CRMP5. Breast neoplasms with anti-Ri antibodies should also be considered.[118] Antibodies to Ma proteins can also be present in older patients with similar neurological complaints and other cancers.

PEM is associated with several autoantibodies. The major antibody is anti-Hu.[20,62,101,119] Other antibodies include anti-Ta,[105] anti-Ma,[120] anti-CV2/CRMP5,[93,111] anti-Zic, and, less frequently, antiamphiphysin,[42,111,121,122] ANNA-3,[45] and Purkinje cell antibody PCA-2.[48] SCLC is the neoplasia presented by a majority of patients affected by PEM.[20,33,62,101,119]

CSF analysis is quasi-consistently abnormal with at least mild lymphocytic pleocytosis, proteinorachia, oligoclonal bands, and increased IgG.[62] On brain imaging, FLAIR and/or T2-weighted signal abnormalities can be observed in symptomatic areas and also, on occasion, in asymptomatic regions. No abnormal enhancement is normally observed on postgadolinium sequences, except for limbic-diencephalic encephalitis associated with anti-Ma2 antibodies.[29,103]

As a general rule, encephalomyelitis and paraneoplastic focal encephalitis are poorly responsive to any type of treatment. One exception to this rule is limbic encephalitis, which normally improves with the treatment of the underlying tumor; steroids also have a minor role in the control of the condition.[23] Alamowitch et al. demonstrated that SCLC patients with anti-Hu antibodies are less likely to improve than SCLC patients without these antibodies.[29] Patients with anti-Ma antibodies are the exception; they can demonstrate significant neurological improvement with rapid treatment of the tumor and immunomodulation.[31]

Keime-Guibert et al.[123] reported that patients experiencing a complete tumor response were the ones more likely to demonstrate a neurological stabilization. Rapid diagnosis of PEM, along with rapid demonstration and treatment of the neoplasia, is the goal. Although immunotherapy has not been demonstrated to modify the outcome of patients, one or two immunosuppressive treatments should be initiated, since some cases of neurological improvement have been described.[62,119,124] It has been suggested that since plasma exchange, IVIg, and corticosteroids have at the maximum a modest effect, a more aggressive immunosuppression with cyclophosphamide, tacrolimus, or cyclosporine can be considered. Long-term treatment, for example, cyclophosphamide with corticosteroids, may be envisioned in the rare responder.[26,125] However, as a general rule, patients become severely debilitated and bedridden, and only half of the patients are still alive 1 year after diagnosis.[62,119,124] Negative prognostic factors are poor functional status at the time of diagnosis, age above 60, involvement of multiple areas of the nervous system, and absence of treatment.[62]

8.6 Opsoclonus-myoclonus

Opsoclonus is a disorder of saccadic stability, consisting of involuntary, arrhythmic, multidirectional, high-amplitude conjugate saccades. It can occur intermittently or, in more severe cases, in a more constant fashion. Darkness or eye closing does not help. Diffuse or focal myoclonus and truncal titubation, with or without other cerebellar and brainstem signs, are often associated with opsoclonus.[7,126,127] In most instances, it represents a self-limited viral infection of the brainstem affecting children. Half of the paraneoplastic cases affecting children are a manifestation of a neuroblastoma and usually occur in children younger than 4 years of age. Of children with neuroblastoma, about 2%–3% will develop the syndrome.[128–130] The incidence is at its peak at 18 months of age and more girls are affected than boys.

Half of the time, the neurological picture appears before the identification of the tumor, making the recognition of the neurological signs important. Ataxia, irritability, hypotonia, vomiting, and dementia may accompany opsoclonus-myoclonus.[131] Some patients present with an excessive startle response.[132] When opsoclonus-myoclonus accompanies neuroblastoma, the incidence of intrathoracic tumors and tumors with a benign histology is higher. The prognosis of neuroblastoma is better if it is accompanied by opsoclonus-myoclonus. Anti-Hu is positive in about 10% of children with paraneoplastic opsoclonus-myoclonus and neuroblastoma.[133,134] However, 4%–15% of children with neuroblastoma are anti-Hu positive without opsoclonus-myoclonus.[133,135] The neurological disorder responds to ACTH and intravenous immunoglobulin, but not prednisone.[136,137] It also frequently responds to treatment of the tumor.[131,137–139] Each patient has different responses to different treatments. More than two-thirds of patients are left with behavioral or psychomotor retardation.[131,140]

Adult-onset opsoclonus-myoclonus is paraneoplastic in 20% of cases.[127] Myoclonus, dysarthria, truncal ataxia, vertigo, and encephalopathy often affect adults experiencing opsoclonus. Also, patients appear to present a PCD.[141] On occasion, ophthalmoplegia can also be present.[142] Different tumors have been associated with adult-onset paraneoplastic opsoclonus-myoclonus. However, anti-Hu antibody-positive SCLC is the most frequent malignancy.[126] Most patients do not harbor antineuronal antibodies. When anti-Hu is present, it is usually part of a more widespread PEM. Anti-Ri antibodies (ANNA-2) may be present in patients with breast and gynecological cancers and occasionally in male patients with bladder carcinoma or SCLC.[7,22,141,143,144] Some patients with lung cancer harbor anti-Ma antibodies.[31] Other antineuronal antibodies described are CRMP5/CV2, Zic2, amphiphysin, and Yo.[44,64,111,122]

MRI is usually normal, but may show abnormalities involving the brainstem or cerebellum in T2-weighted images.[145] CSF examination demonstrates mild pleocytosis and slight protein elevation. In children, a CT of the chest, abdomen, and pelvis; urine catecholamine measurements; and metaiodobenzylguanidine scan should be obtained in search of an occult neuroblastoma.[146] If negative, those tests should be repeated a few months later.[147]

While some patients have no neuropathological abnormalities, in others loss of olivary neurons, loss of Purkinje cells, inflammatory infiltrates, gliosis of the Bergmann astrocytes, and loss of cells in the granular layer of the cerebellum have been observed.[148]

The degree of improvement in adults is mostly correlated with the rapidity of treatment of the underlying malignancy.[22,149] Some patients do improve with IVIg or other immunological treatments, including depletion of serum IgG using protein-A columns, but most of the time, the degree of improvement is limited. Patients in whom the underlying malignancy is not treated often develop severe encephalopathy and die. Symptomatic improvement can occur with clonazepam, piracetam, valproic acid, and thiamine.[149–151] The course of opsoclonus-myoclonus may on occasion be remitting and relapsing, which is in contrast to most PNDs.[126] Remissions can be spontaneous or following treatment directed at the tumor or immunosuppression.[150]

8.7 Stiff-person syndrome

Stiff-person syndrome (SPS) is characterized by a fluctuating rigidity of the axial musculature, representing a cocontraction of the agonist and antagonist muscles. It affects primarily the lower trunk and legs but can extend to the shoulders, upper limbs, and neck. Emotions and auditory or somesthetic stimuli can trigger painful spasms. Sleep and local or general anesthesia relieve the rigidity. When electrophysiological studies are performed, continuous activity of the motor units in the stiffened muscles is typically observed, which improves with the administration of diazepam.

Evidence suggests that SPS might result from an autoimmunity directed toward the spinal cord interneurons that control motor neuron activity and that cosecrete g-aminobutyric acid (GABA) and glycine. The syndrome is nonparaneoplastic in 70%–80% of cases and develops in conjunction with diabetes, polyendocrinopathy, and often antibodies to GAD.[41,152] Breast cancer is the neoplasia most commonly associated with the paraneoplastic form, with associated antibodies to amphiphysin.[40,153] Hodgkin's disease and lung cancers have also been associated.[40] One patient presenting a mediastinal mass and antibodies to gephyrin also presented symptoms suggestive of SPS.[39] Pathological findings in paraneoplastic SPS are not as impressive as those seen in other PNDs.

Immunosuppression has been described as possibly effective for both paraneoplastic and nonparaneoplastic SPS. Treatment of the underlying tumor has also been described as useful in improving the symptomatology.[40,154] Symptomatic improvement can occur with the use of drugs that enhance GABA neurotransmissions such as diazepam, gabapentin, and baclofen.

8.8 Lambert-Eaton myasthenic syndrome

Lambert-Eaton myasthenic syndrome (LEMS) normally presents with a fatigable proximal weakness involving mostly the lower extremities. Other common features are myalgia and paresthesias. The neurological picture usually develops before the tumor is diagnosed, and symptoms gradually progress over weeks or months; in some patients, LEMS presents acutely.[40] Some patients experience transitory bulbar symptoms—diplopia, ptosis, or dysphagia—but these are normally milder than with Myasthenia Gravis (MG).[155] On occasion, patients can experience respiratory weakness. On neurological examination, a proximal weakness predominantly involving the legs and an absence or diminution of tendon reflexes is characteristically observed. However, brief exercise can improve both the weakness and the quality of tendon reflexes. On the contrary, continued exercise worsens the picture. About 95% of patients will eventually develop autonomic symptoms; the most frequent ones are xerophthalmia, xerostomia, impotence, orthostatic hypotension, blurry vision due to abnormal pupillary responses, and mild-to-moderate ptosis.[155–157] LEMS may be part of a more generalized paraneoplastic disorder, and some patients experience features of PCD and encephalomyelitis.[38] At tumor recurrence, symptoms of LEMS can reappear.

More than half of patients have an underlying cancer; SCLC is almost always the culprit.[157–159] However, only about 3% of all SCLC patients present symptoms of LEMS,[156,160] and those patients normally present a more progressive course.[155] Other tumors have been described: small-cell carcinoma of the prostate and cervix, lymphomas, and adenocarcinoma. Wirtz et al. demonstrated that in patients presenting an absence of the HLAB8 genotype, showing symptoms of LEMS, and known for a history of smoking, an underlying SCLC is strongly predicted.[161] Also, patients with SCLC and LEMS survive significantly longer than SCLC patients who do not have a paraneoplastic disorder.[162,163] All patients diagnosed with LEMS should have a CT of the chest and abdomen and, if negative, a whole body PET scan. Patients presenting with LEMS who do not have cancer normally have evidence of an autoimmune disease, such as thyroiditis and insulin-dependent diabetes mellitus.[164]

The hallmark of LEMS is the typical electrophysiological pattern. At rest, small amplitude compound-muscle action potentials are observed. A decremental response of greater than 10% is obtained with low rates of repetitive nerve stimulation (2–5 Hz). However, facilitation and an incremental response of at least 100% are observed at high rates of repetitive nerve stimulation (20 Hz or greater) or 15–30 s of maximal voluntary muscle contraction.[35]

P/Q-type VGCC antibodies are the antibodies most often associated with LEMS. Those antibodies are located presynaptically at the neuromuscular junction.[35] At the presynaptic neuromuscular junction, quantal release of acetylcholine occurs, allowing neuromuscular transmission. In LEMS, P/Q-type VGCC antibodies interfere with this quantal release, which results in a reduced release of acetylcholine at the presynaptic nerve terminals, blocking neuromuscular transmission. The presence of P/Q-type VGCC antibodies is not specific for a paraneoplastic process.[165] The richest concentration of P/Q-type VGCC is in the cerebellum, supporting the presence of a PCD in association with LEMS.[166] The presence of anti-Hu antibodies suggests an association of PEM with LEMS. Anti-MysB antibodies reactive with the B subunit of neuronal calcium channels are present in about 20% of patients with LEMS.[167]

Removal of the IgG antibodies to the VGCC allows an improvement of the neurological function, underscoring the fact that LEMS is a classic autoimmune disorder.[165] Paraneoplastic LEMS can respond to either immunosuppression or treatment of the underlying neoplasia.[168,169] Symptomatic treatment with drugs facilitating the release of acetylcholine from the motor nerve terminals, such as 3,4-diaminopyridine (DAP), is possible.[170] Sanders et al. completed a placebo-controlled randomized trial with DAP and demonstrated that at a dose of 5–20 mg three to four times daily, DAP was effective as monotherapy for long-term management, as well as in combination with other treatments.[171] Eighty percent of patients showed at least a moderate improvement with DAP. The maximum recommended daily dose of DAP is 80 mg; at higher doses, seizures can occur. Cholinesterase inhibitors like pyridostigmine, 30–60 mg every 6 h, are useful mostly for xerostomia; however, this therapy does not normally improve weakness. A combination of pyridostigmine with guanidine may be an effective replacement for DAP if it is unavailable.[172] If those treatments do not control the symptomatology appropriately, immunosuppressive treatments with steroids, azathioprine, or cyclosporine may be necessary.[173] Plasma exchange and IVIg are useful for quick improvement of a severe weakness; however, the results are transient[157,174–176] and immunotherapy is not as effective in LEMS as it is in MG.

8.9 Paraneoplastic sensory neuronopathy

Sensory neuronopathy occurs in a variety of conditions, in previously healthy individuals and individuals with a variety of underlying autoimmune conditions, such as Sjögren's syndrome, heavy metal intoxication, and the chemotherapeutic agent cisplatin. Paraneoplastic sensory neuronopathy is a rare syndrome, representing about 20% of all subacute sensory neuronopathy. It is characterized by a progressive sensory loss, at first asymmetric or multifocal, and affects vibration and proprioception more than it does nociception.[177] It is commonly associated with painful dysesthesias; the limbs, chest, abdomen, and face can be affected. The upper extremities are almost invariably involved and are the site of presentation in most cases. Some patients also present sensorineural hearing loss and autonomic neuropathy, with gastrointestinal pseudo-obstruction. With time, ataxia, gait difficulty, and involuntary pseudoathetoid movements, mostly of the hands, develop secondary to the sensory deficits.[177] Depressed or absent tendon reflexes are also observed on neurological examination, but motor function is normal. In most cases, the disease leaves the patient severely disabled after a few weeks or months. About 75% of patients with PEM are affected by a subacute sensory neuronopathy, but 25% of all paraneoplastic sensory neuronopathy cases are clinically pure.[62,119]

In 70%–80% of cases, a pulmonary malignancy is the cause, most commonly SCLC, and anti-Hu antibodies are present.[20,62,119,178,179] Breast cancer, ovarian cancer, sarcoma, and Hodgkin's lymphoma are other malignancies, on occasion.[177,180] In most instances, the sensory neuronopathy develops 3.5–4.5 months before the diagnosis of cancer.[62,119]

Prominent inflammatory infiltration and neuronal loss involving the dorsal root ganglia are the main pathological findings.[101] Those changes, in combination with demyelination, can also be observed in the posterior and anterior spinal cord nerve roots and peripheral nerves.[181] In about half the cases, the nervous system is otherwise normal. The CSF normally shows inflammatory abnormalities, with mild pleocytosis, elevated IgG level, and oligoclonal bands.[119,177,180] When the dorsal root ganglia are primarily affected, small or absent sensory nerve action potentials are observed. Motor nerve conduction studies and F-wave studies are usually normal or sometimes reveal mildly reduced motor nerve conduction. Electromyographic (EMG) examination does not show evidence of denervation. Sural nerve biopsy is rarely required for the diagnosis, but helps differentiate the disorder from vasculitic neuropathy.[181,182]

The different treatment modalities used—plasma exchange, steroids, and IVIg—do not alter the course of the disease in most patients.[99,123] However, some reports of response to immunotherapy do exist.[62,183] Keime-Guibert et al.[124] reported a study of 10 patients treated with a combination of steroids, cyclophosphamide, and IVIg, with two patients responding. In those cases, a stabilization or slight improvement of the dorsal root ganglia dysfunction, with electrophysiological improvement, has been reported.[183] Some patients present a mild or indolent case of neuropathy. In those patients, the best chance of stabilizing the neurological syndrome relies on the early detection and treatment of the underlying malignancy.[119,123] Antitumor treatment may also be considered for anti-Hu-positive patients older than 50 and smokers, even if the primary malignancy is not demonstrated. Symptomatic treatment of the neuropathic pain and dysautonomia is warranted.

8.10 Paraneoplastic neuropathies

Some patients do present sensorimotor neuropathies that are debilitating and usually present before or at the time the responsible neoplasia is discovered. Frequently, those sensorimotor neuropathies are progressive, and both axonal and demyelinating features are present. Very rarely, a paraneoplastic relapsing-remitting neuropathy can occur.[184] Tumors normally associated with this disorder are lung and breast cancers. The frequency of neuropathies increases as more advanced stages of cancer are reached and an extensive number of other etiologies must be considered before a diagnosis of paraneoplastic neuropathy is made.[185]

Several malignancies of plasma cells and lymphocytes are associated with sensorimotor neuropathies, and they often show a similarity to chronic inflammatory demyelinating neuropathy. Malignancies involved are multiple myeloma, sclerotic myeloma, Waldenström's macroglobulinemia, B-cell lymphoma, and Castleman's disease.[2] Sclerotic myeloma can trigger a sensorimotor neuropathy similar to chronic inflammatory demyelinating neuropathy. Neurological improvement is normally experienced with the treatment of the sclerotic myeloma. The opposite is observed with multiple myeloma, where the sensorimotor or sensory axonal neuropathy rarely improves with treatment. When amyloidosis is associated with myeloma, autonomic dysfunction often accompanies the neuropathic symptoms, as well as lancinating and burning dysesthesias.[186] Five to ten percent of patients with Waldenström's macroglobulinemia present at some point with a symmetric sensorimotor polyneuropathy; large sensory fibers are predominantly involved.[186] Immunoglobulin M antibodies to myelin-associated glycoprotein can be observed. Several treatments can improve the neuropathy, including treatment of the Waldenström's macroglobulinemia, plasma exchange, IVIg, chlorambucil, cyclophosphamide, fludarabine, or rituximab.

Pathological examination normally demonstrates axonal degeneration with inflammatory infiltrates and demyelination. As a general rule, antibodies are negative in paraneoplastic neuropathies. Exceptions are anti-Hu antibodies, which suggest involvement of the dorsal root ganglia, and anti-CV2/CRMP5 antibodies, which are associated with a mixed axonal demyelinating sensorimotor neuropathy.[58] Patients with anti-CV2/CRMP5 antibodies usually present with a sensory or sensorimotor neuropathy, with less frequent involvement of the arms, but often associated with cerebellar ataxia.[58,111,117] SCLC, neuroendocrine tumors, and thymoma are some types of neoplasia associated with anti-CV2/CRMP5 antibodies. Patients with antiamphiphysin antibodies can also present with a sensory or sensorimotor neuropathy, and SCLC, breast cancer, and melanoma are forms of neoplasia that are possibly associated with these cases.[121,122,187]

When a predominance of demyelinating features exists, immunotherapies are useful: corticosteroids, IVIg, and plasma exchange. Although axonal neuropathies are refractory to most therapies, sclerotic myeloma, other plasma cell dyscrasias, and lymphoma can respond to treatment of the tumor or immunosuppression.[184]

8.11 Vasculitis of the nerve and muscle

SCLC and lymphoma are the tumors more frequently responsible for a painful symmetric or asymmetric subacute sensorimotor polyneuropathy affecting mostly older men; less frequently, multiple mononeuropathy occurs.[183] On electrophysiological examination, axonal degeneration involving the motor and sensory nerves can be observed. An elevated erythrocyte sedimentation rate and proteinorachia are also observed. Pathological examination of nerve and muscle biopsies shows intramural and perivascular inflammatory infiltrates composed of CD8 + T cells. Most patients do not demonstrate any serological markers of paraneoplasia, but some patients with SCLC present anti-Hu antibodies. If recognized in a timely fashion, this disorder can respond to steroids, cyclophosphamide, or both.[183]

8.12 Myasthenia gravis

Antibodies against the AchR at the postsynaptic neuromuscular junction are usually responsible for MG, a postsynaptic disorder of neuromuscular transmission. Some patients express a different kind of antibody against muscle-specific kinase (MuSK). Weakness and muscle fatigability, which improve with rest and worsen with activity, are the characteristic complaints. A hallmark of MG is the early and prominent ocular paresis and preservation of reflexes.[188] The majority of patients experience ptosis and diplopia, and in 15% of these patients, symptoms remain localized to the extraocular and eyelid muscles. Other patients experience a generalized weakness that progresses to the point that mechanical ventilation may be necessary to overcome respiratory muscle weakness/insufficiency. In MG, tendon reflexes and sensation remain normal.

Ten percent of patients with MG have a thymic epithelial tumor (thymoma or thymic carcinoma). One-third of patients with thymoma develop MG.[189] A few cases of MG in association with other tumors have been described: thyroid gland tumors, SCLC, breast cancer, and lymphoma. Patients with thymoma almost invariably present AchR antibodies and also have additional antibodies against skeletal muscle proteins such as titin, but not MuSK antibodies. Eighty to ninety percent of patients with generalized MG, as well as 70% of patients with ocular MG, present acetylcholine receptor antibodies. Antistriated and antititin antibodies are also found in patients with MG. The detection of antititin antibodies is not a predictor of the presence of a tumor.[190,191] In generalized MG, antibodies to the acetylcholine receptor at the neuromuscular junction are detected in 80%–90% of the patients with or without thymoma.[192] Subsets of patients without these antibodies harbor antibodies to MuSK.[193] These patients do not have thymoma and preferentially develop cranial and bulbar weakness with a higher frequency of respiratory crisis. Detection of MuSK antibodies is associated with poor response to anticholinesterase treatment, but good response to plasma exchange and cyclosporine.[194]

Resection of the underlying tumor is the first priority, at which time symptomatic treatments, as in any case of MG, are initiated. These include cholinesterase inhibitors, immunomodulation with plasma exchange and/or IVIg, and immunosuppression with steroids, azathioprine, and others.[173]

8.13 Dermatomyositis

Dermatomyositis exists as a paraneoplastic syndrome in adults, but not polymyositis.[195] Paraneoplastic dermatomyositis presents as dermatomyositis without cancer. The hallmark purplish discoloration of the eyelids (heliotrope rash) with edema and erythematous lesions over the knuckles is present. It is suggested that the presence of an underlying cancer is more probable if necrotic skin ulcerations and pruritus are present.[196] It typically presents with the subacute onset of proximal muscle weakness and myalgia. Aspiration and hypoventilation can occur secondary to neck flexor, pharyngeal, and respiratory muscle weakness. Arthralgia, myocarditis, and congestive heart failure are other possible manifestations. Tendon reflexes and sensation are normal. An elevation of serum creatine kinase concentrations is often observed, although normal levels can be occasionally found, even in patients with profound muscle weakness. EMG examination demonstrates increased spontaneous activity, fibrillation, positive sharp waves, and complex repetitive discharges, as well as low-amplitude, short-duration, polyphasic motor units. Pathologically, dermatomyositis presents the same, regardless of whether a malignancy is present. Under microscopy, muscle biopsy shows predominantly CD4 + T-cell inflammatory infiltrates and muscle necrosis; the presence of perifascicular atrophy is characteristic of dermatomyositis.[197] Muscle imaging (CT or MRI) can be useful to confirm the diagnosis, suggest a type of inflammatory myopathy, and select a biopsy site.

The tumors most commonly associated with dermatomyositis are breast, lung, ovarian, pancreas, stomach, and colorectum; Hodgkin's lymphoma can also be responsible.[198] Other possible tumors are thymoma, germ cell tumors, melanoma, nasopharyngeal cancer, and lymphoma. No specific serological tests have been identified. However, all adult patients, particularly those older than 50, should undergo cancer screening, since dermatomyositis is frequently associated with cancer in adults.

About 50% of patients presenting with interstitial lung disease present an antibody directed against histidyl-tRNA synthetase (anti-Jo-1). About 35% of cases of dermatomyositis present high titers of antibodies to the Mi-2 protein complex.[199]

Some patients do present an improvement of the muscle and dermatological symptoms with treatment of the tumor. Otherwise, therapies similar to those used in nonparaneoplastic dermatomyositis can show some efficacy: steroids, azathioprine, or IVIg.[200]

8.14 Paraneoplastic visual syndromes

In these syndromes, the retina and, less frequently, the uvea and optic nerves show a paraneoplastic involvement.[201] A diagnosis of paraneoplastic visual syndrome can be made only after ruling out a metastatic infiltration of the optic nerve, toxic effects of chemotherapy or radiation therapy, and severe anemia.

Three well-characterized syndromes represent the paraneoplastic visual syndromes.[26] Cancer-associated retinopathy usually presents initially bilaterally and reflects simultaneous dysfunction of cones and rods: photosensitivity, progressive loss of vision and color perception, central and ring scotomas, night blindness, and prolonged dark adaptation.[202] If the disorder is initially unilateral, the other eye becomes symptomatic in days or weeks. SCLC is almost always the malignancy-associated.[203] On ophthalmoscopic examination, nonspecific arteriolar narrowing is seen. Electroretinogram records flat or severely attenuated photopic and scotopic responses. Imaging studies and CSF evaluation are negative. The antirecoverin antibody, a retinal-specific calcium-binding protein, is the antibody most frequently associated. Antibodies to the tubby-like protein 1 and the photoreceptor-specific nuclear receptor have been reported.[203,204]

Patients with metastatic cutaneous melanoma can present a melanoma-associated retinopathy (MAR).[205] The normal mode of presentation of the MAR syndrome is near normal visual acuity and color vision, but with acute apparition of shimmering, flickering, or pulsating photopsias; night blindness; and mild peripheral visual field loss.[202] It often progresses to complete visual loss. A dysfunction of the rods with preservation of the cones explains the symptomatology.[206] An electroretinogram usually demonstrates a markedly reduced or absent dark-adapted B-wave, and a slightly attenuated A-wave response to scotopic stimulus. Rod-bipolar-cell antibodies are usually present in patients with MAR.[207] A paraneoplastic optic neuritis can occur, mostly in patients presenting with PEM, but is very uncommon. It normally presents as a subacute painless bilateral visual loss. On ophthalmoscopic evaluation, papilledema may be present. Several antibodies, principally anti-CV2/CRMP5 and anti-Hu, have been described with paraneoplastic optic neuritis, and SCLC is the malignancy most often associated with paraneoplastic optic neuritis.[117,208,209]

As with most paraneoplastic disorders, the primary treatment for paraneoplastic visual syndromes is aimed at control of the tumor. Immunotherapy can then be initiated, although it has known limitations; however, some cases of a response to steroids, plasma exchange, and IVIg have been reported. Nevertheless, the majority of patients do not improve.

9 Conclusion

In conclusion, the diagnosis of PND is difficult and often missed. In most cases, the neurological symptoms develop before the neoplasia has been diagnosed. Paraneoplastic antibodies seem to be helpful diagnostic markers, but their role in the pathogenesis of the disorders is still not clear.

References

1. Dalmau JO, Posner JB. Paraneoplastic syndromes affecting the nervous system. *Semin Oncol.* 1997;24(3):318–328.
2. Rudnicki SA, Dalmau J. Paraneoplastic syndromes of the spinal cord, nerve, and muscle. *Muscle Nerve.* 2000;23(12):1800–1818.
3. Graus F, et al. Recommended diagnostic criteria for paraneoplastic neurological syndromes. *J Neurol Neurosurg Psychiatry.* 2004;75(8):1135–1140.
4. Devine MF, et al. Paraneoplastic neurological syndromes: clinical presentations and management. *Ther Adv Neurol Disord.* 2021;14, 1756286420985323.
5. Darnell RB. Onconeural antigens and the paraneoplastic neurologic disorders: at the intersection of cancer, immunity, and the brain. *Proc Natl Acad Sci U S A.* 1996;93(10):4529–4536.
6. Darnell RB, Posner JB. Paraneoplastic syndromes involving the nervous system. *N Engl J Med.* 2003;349(16):1543–1554.
7. Posner JB. Paraneoplastic syndromes. *Neurologic Complications of Cancer.* Phiadelphia, PA: FA Davis; 1995:353–384.
8. Pignolet BS, Gebauer CM, Liblau RS. Immunopathogenesis of paraneoplastic neurological syndromes associated with anti-Hu antibodies: a beneficial antitumor immune response going awry. *Oncoimmunology.* 2013;2(12), e27384.
9. Valencia-Sanchez C, Zekeridou A. Paraneoplastic neurological syndromes and beyond emerging with the introduction of immune checkpoint inhibitor cancer immunotherapy. *Front Neurol.* 2021;12:642800.
10. Yshii LM, Hohlfeld R, Liblau RS. Inflammatory CNS disease caused by immune checkpoint inhibitors: status and perspectives. *Nat Rev Neurol.* 2017;13(12):755–763.
11. Trotter JL, Hendin BA, Osterland CK. Cerebellar degeneration with Hodgkin disease. An immunological study. *Arch Neurol.* 1976;33(9):660–661.
12. Wilkinson PC, Zeromski J. Immunofluorescent detection of antibodies against neurones in sensory carcinomatous neuropathy. *Brain.* 1965;88(3):529–583.
13. Dalmau J, Rosenfeld MR. Paraneoplastic neurologic syndromes. In: Kasper DL, Braunwald E, Fauci AS, eds. *Harrison's Principles of Internal Medicine.* New York: McGraw-Hill; 2005:571–575.
14. Vianello M, et al. The spectrum of antineuronal autoantibodies in a series of neurological patients. *J Neurol Sci.* 2004;220(1–2):29–36.
15. Galli J, Greenlee J. Paraneoplastic diseases of the central nervous system. *F1000Res.* 2020;9:1–11.
16. Albert ML, et al. Tumor-specific killer cells in paraneoplastic cerebellar degeneration. *Nat Med.* 1998;4(11):1321–1324.
17. Binks SNM, et al. LGI1, CASPR2 and related antibodies: a molecular evolution of the phenotypes. *J Neurol Neurosurg Psychiatry.* 2018;89(5):526–534.
18. Zekeridou A, Lennon VA. Neurologic autoimmunity in the era of checkpoint inhibitor cancer immunotherapy. *Mayo Clin Proc.* 2019;94(9):1865–1878.
19. Dubey D, et al. Autoimmune CRMP5 neuropathy phenotype and outcome defined from 105 cases. *Neurology.* 2018;90(2):e103–e110.
20. Lucchinetti CF, Kimmel DW, Lennon VA. Paraneoplastic and oncologic profiles of patients seropositive for type 1 antineuronal nuclear autoantibodies. *Neurology.* 1998;50(3):652–657.
21. Honnorat J, Cartalat-Carel S. Advances in paraneoplastic neurological syndromes. *Curr Opin Oncol.* 2004;16(6):614–620.
22. Bataller L, et al. Clinical outcome in adult onset idiopathic or paraneoplastic opsoclonus-myoclonus. *Brain.* 2001;124(Pt 2): 437–443.
23. Gultekin SH, et al. Paraneoplastic limbic encephalitis: neurological symptoms, immunological findings and tumour association in 50 patients. *Brain.* 2000;123(Pt 7):1481–1494.
24. Husari KS, Dubey D. Autoimmune epilepsy. *Neurotherapeutics.* 2019;16(3):685–702.
25. Stich O, et al. Qualitative evidence of anti-Yo-specific intrathecal antibody synthesis in patients with paraneoplastic cerebellar degeneration. *J Neuroimmunol.* 2003;141(1–2):165–169.
26. Bataller L, Dalmau JO. Paraneoplastic disorders of the central nervous system: update on diagnostic criteria and treatment. *Semin Neurol.* 2004;24(4):461–471.
27. Dadparvar S, et al. Paraneoplastic encephalitis associated with cystic teratoma is detected by fluorodeoxyglucose positron emission tomography with negative magnetic resonance image findings. *Clin Nucl Med.* 2003;28(11):893–896.
28. Fakhoury T, Abou-Khalil B, Kessler RM. Limbic encephalitis and hyperactive foci on PET scan. *Seizure.* 1999;8(7):427–431.
29. Alamowitch S, et al. Limbic encephalitis and small cell lung cancer. Clinical and immunological features. *Brain.* 1997;120(Pt 6):923–928.
30. Lawn ND, et al. Clinical, magnetic resonance imaging, and electroencephalographic findings in paraneoplastic limbic encephalitis. *Mayo Clin Proc.* 2003;78(11):1363–1368.
31. Rosenfeld MR, et al. Molecular and clinical diversity in paraneoplastic immunity to Ma proteins. *Ann Neurol.* 2001;50(3):339–348.
32. Crotty E, Patz Jr EF. FDG-PET imaging in patients with paraneoplastic syndromes and suspected small cell lung cancer. *J Thorac Imaging.* 2001;16(2):89–93.
33. Henson RA, Urich H. Cancer and the nervous system. In: Henson RA, Urich H, eds. *The Neurological Maninfestations of Systemic Malignant Disease.* Oxford: Blackwell Scientific; 1982:657.
34. Dalmau J, Geis C, Graus F. Autoantibodies to synaptic receptors and neuronal cell surface proteins in autoimmune diseases of the central nervous system. *Physiol Rev.* 2017;97(2):839–887.
35. Motomura M, et al. An improved diagnostic assay for Lambert-Eaton myasthenic syndrome. *J Neurol Neurosurg Psychiatry.* 1995;58(1):85–87.
36. Pozo-Rosich P, et al. Voltage-gated potassium channel antibodies in limbic encephalitis. *Ann Neurol.* 2003;54(4):530–533.

37. Vernino S, et al. Autoantibodies to ganglionic acetylcholine receptors in autoimmune autonomic neuropathies. *N Engl J Med.* 2000;343(12):847–855.

38. Mason WP, et al. Small-cell lung cancer, paraneoplastic cerebellar degeneration and the Lambert-Eaton myasthenic syndrome. *Brain.* 1997;120(Pt 8):1279–1300.

39. Butler MH, et al. Autoimmunity to gephyrin in Stiff-Man syndrome. *Neuron.* 2000;26(2):307–312.

40. Folli F, et al. Autoantibodies to a 128-kd synaptic protein in three women with the stiff-man syndrome and breast cancer. *N Engl J Med.* 1993;328(8):546–551.

41. Solimena M, et al. Autoantibodies to GABA-ergic neurons and pancreatic beta cells in stiff-man syndrome. *N Engl J Med.* 1990;322(22):1555–1560.

42. Bataller L, et al. Antibodies to Zic4 in paraneoplastic neurologic disorders and small-cell lung cancer. *Neurology.* 2004;62(5):778–782.

43. Drlicek M, et al. Antibodies of the anti-Yo and anti-Ri type in the absence of paraneoplastic neurological syndromes: a long-term survey of ovarian cancer patients. *J Neurol.* 1997;244(2):85–89.

44. Bataller L, et al. Autoantigen diversity in the opsoclonus-myoclonus syndrome. *Ann Neurol.* 2003;53(3):347–353.

45. Chan KH, Vernino S, Lennon VA. ANNA-3 anti-neuronal nuclear antibody: marker of lung cancer-related autoimmunity. *Ann Neurol.* 2001;50(3):301–311.

46. Scheid R, et al. A new anti-neuronal antibody in a case of paraneoplastic limbic encephalitis associated with breast cancer. *J Neurol Neurosurg Psychiatry.* 2004;75(2):338–340.

47. Sillevis Smitt P, et al. Paraneoplastic cerebellar ataxia due to autoantibodies against a glutamate receptor. *N Engl J Med.* 2000;342(1):21–27.

48. Vernino S, Lennon VA. New Purkinje cell antibody (PCA-2): marker of lung cancer-related neurological autoimmunity. *Ann Neurol.* 2000;47(3):297–305.

49. Mann AP, Grebenciucova E, Lukas RV. Anti-N-methyl-D-aspartate-receptor encephalitis: diagnosis, optimal management, and challenges. *Ther Clin Risk Manag.* 2014;10:517–525.

50. Cortese I, Cornblath DR. Therapeutic plasma exchange in neurology: 2012. *J Clin Apher.* 2013;28(1):16–19.

51. Widdess-Walsh P, et al. Response to intravenous immunoglobulin in anti-Yo associated paraneoplastic cerebellar degeneration: case report and review of the literature. *J Neurooncol.* 2003;63(2):187–190.

52. Lee WJ, et al. Rituximab treatment for autoimmune limbic encephalitis in an institutional cohort. *Neurology.* 2016;86(18):1683–1691.

53. Thone J, et al. Effective immunosuppressant therapy with cyclophosphamide and corticosteroids in paraneoplastic cerebellar degeneration. *J Neurol Sci.* 2008;272(1–2):171–173.

54. Greenlee JE. Treatment of paraneoplastic neurologic disorders. *Curr Treat Options Neurol.* 2010;12(3):212–230.

55. McKeon A. Autoimmune encephalopathies and dementias. *Continuum (Minneap Minn).* 2016;22(2 Dementia):538–558.

56. Rosenfeld MR, Dalmau JO. Paraneoplastic disorders of the CNS and autoimmune synaptic encephalitis. *Continuum (Minneap Minn).* 2012;18(2):366–383.

57. Chartrand-Lefebvre C, et al. Association of small cell lung cancer and the anti-Hu paraneoplastic syndrome: radiographic and CT findings. *AJR Am J Roentgenol.* 1998;170(6):1513–1517.

58. Antoine JC, et al. Paraneoplastic anti-CV2 antibodies react with peripheral nerve and are associated with a mixed axonal and demyelinating peripheral neuropathy. *Ann Neurol.* 2001;49(2):214–221.

59. Rees JH, et al. The role of [18F]fluoro-2-deoxyglucose-PET scanning in the diagnosis of paraneoplastic neurological disorders. *Brain.* 2001;124(Pt 11):2223–2231.

60. Younes-Mhenni S, et al. FDG-PET improves tumour detection in patients with paraneoplastic neurological syndromes. *Brain.* 2004;127(Pt 10):2331–2338.

61. Debourdeau P, Gligorov J, Zammit C. A serologic marker of paraneoplastic limbic and brain-stem encephalitis in patients with testicular cancer. *N Engl J Med.* 1999;341(19):1475–1476.

62. Graus F, et al. Anti-Hu-associated paraneoplastic encephalomyelitis: analysis of 200 patients. *Brain.* 2001;124(Pt 6):1138–1148.

63. Rojas-Marcos I, et al. Spectrum of paraneoplastic neurologic disorders in women with breast and gynecologic cancer. *Medicine (Baltimore).* 2003;82(3):216–223.

64. Peterson K, et al. Paraneoplastic cerebellar degeneration. I. A clinical analysis of 55 anti-Yo antibody-positive patients. *Neurology.* 1992;42(10):1931–1937.

65. Titulaer MJ, et al. Screening for small-cell lung cancer: a follow-up study of patients with Lambert-Eaton myasthenic syndrome. *J Clin Oncol.* 2008;26(26):4276–4281.

66. Abdel-Wahab N, et al. Use of immune checkpoint inhibitors in the treatment of patients with cancer and preexisting autoimmune disease: a systematic review. *Ann Intern Med.* 2018;168(2):121–130.

67. Johnson DB, et al. Ipilimumab therapy in patients with advanced melanoma and preexisting autoimmune disorders. *JAMA Oncol.* 2016;2(2):234–240.

68. Menzies AM, et al. Anti-PD-1 therapy in patients with advanced melanoma and preexisting autoimmune disorders or major toxicity with ipilimumab. *Ann Oncol.* 2017;28(2):368–376.

69. Sechi E, et al. Neurologic autoimmunity and immune checkpoint inhibitors: autoantibody profiles and outcomes. *Neurology.* 2020;95(17):e2442–e2452.

70. Manson G, et al. Worsening and newly diagnosed paraneoplastic syndromes following anti-PD-1 or anti-PD-L1 immunotherapies, a descriptive study. *J Immunother Cancer.* 2019;7(1):337.

71. Gill A, et al. A case series of PD-1 inhibitor-associated paraneoplastic neurologic syndromes. *J Neuroimmunol.* 2019;334:576980.

72. Hottinger AF, et al. Natalizumab may control immune checkpoint inhibitor-induced limbic encephalitis. *Neurol Neuroimmunol Neuroinflamm.* 2018;5(2):e439.

73. Matsuoka H, et al. Nivolumab-induced limbic encephalitis with anti-Hu antibody in a patient with advanced pleomorphic carcinoma of the lung. *Clin Lung Cancer.* 2018;19(5):e597–e599.

74. Papadopoulos KP, et al. Anti-Hu-associated autoimmune limbic encephalitis in a patient with PD-1 inhibitor-responsive myxoid chondrosarcoma. *Oncologist.* 2018;23(1):118–120.

75. Raibagkar P, et al. Worsening of anti-Hu paraneoplastic neurological syndrome related to anti-PD-1 treatment: case report and review of literature. *J Neuroimmunol.* 2020;341:577184.

76. Raskin J, et al. Recurrent dysphasia due to nivolumab-induced encephalopathy with presence of Hu autoantibody. *Lung Cancer.* 2017;109:74–77.

77. Johnson DB, et al. A case report of clonal EBV-like memory CD4(+) T cell activation in fatal checkpoint inhibitor-induced encephalitis. *Nat Med.* 2019;25(8):1243–1250.

78. Shams'ili S, et al. Paraneoplastic cerebellar degeneration associated with antineuronal antibodies: analysis of 50 patients. *Brain.* 2003;126(Pt 6):1409–1418.

79. Hammack JE, et al. Paraneoplastic cerebellar degeneration: a clinical comparison of patients with and without Purkinje cell cytoplasmic antibodies. *Mayo Clin Proc.* 1990;65(11):1423–1431.

80. Cao Y, et al. Anti-Yo positive paraneoplastic cerebellar degeneration associated with ovarian carcinoma: case report and review of the literature. *Gynecol Oncol.* 1999;75(1):178–183.

81. Peltola J, et al. A reversible neuronal antibody (anti-Tr) associated paraneoplastic cerebellar degeneration in Hodgkin's disease. *Acta Neurol Scand.* 1998;98(5):360–363.

82. Bernal F, et al. Anti-Tr antibodies as markers of paraneoplastic cerebellar degeneration and Hodgkin's disease. *Neurology.* 2003;60(2):230–234.

83. Rojas I, et al. Long-term clinical outcome of paraneoplastic cerebellar degeneration and anti-Yo antibodies. *Neurology.* 2000;55(5):713–715.

84. Greenlee JE, et al. Association of anti-Yo (type I) antibody with paraneoplastic cerebellar degeneration in the setting of transitional cell carcinoma of the bladder: detection of Yo antigen in tumor tissue and fall in antibody titers following tumor removal. *Ann Neurol.* 1999;45(6):805–809.

85. Krakauer J, et al. Anti-Yo-associated paraneoplastic cerebellar degeneration in a man with adenocarcinoma of unknown origin. *Neurology.* 1996;46(5):1486–1487.

86. Pittock SJ, Lucchinetti CF, Lennon VA. Anti-neuronal nuclear autoantibody type 2: paraneoplastic accompaniments. *Ann Neurol.* 2003;53(5):580–587.

87. Venkatraman A, Opal P. Paraneoplastic cerebellar degeneration with anti-Yo antibodies – a review. *Ann Clin Transl Neurol.* 2016;3(8):655–663.

88. Graus F, et al. Localization of the neuronal antigen recognized by anti-Tr antibodies from patients with paraneoplastic cerebellar degeneration and Hodgkin's disease in the rat nervous system. *Acta Neuropathol.* 1998;96(1):1–7.

89. Graus F, et al. Immunological characterization of a neuronal antibody (anti-Tr) associated with paraneoplastic cerebellar degeneration and Hodgkin's disease. *J Neuroimmunol.* 1997;74(1–2):55–61.

90. Lancaster E, et al. Antibodies to metabotropic glutamate receptor 5 in the Ophelia syndrome. *Neurology.* 2011;77(18):1698–1701.

91. Vernino S. Paraneoplastic cerebellar degeneration. *Handb Clin Neurol.* 2012;103:215–223.

92. Degenhardt A, et al. Absence of antibodies to non-NMDA glutamate-receptor subunits in paraneoplastic cerebellar degeneration. *Neurology.* 1998;50(5):1392–1397.

93. Honnorat J, et al. Ulip/CRMP proteins are recognized by autoantibodies in paraneoplastic neurological syndromes. *Eur J Neurosci.* 1999;11(12):4226–4232.

94. Mason WP. Paraneoplastic cerebellar degeneration (PCD) in small-cell lung cancer: impact of Q8 anti-hu antibody (HuAb) on clinical presentation and survival. *Neurology.* 1996;46:127.

95. Anderson NE, et al. The metabolic anatomy of paraneoplastic cerebellar degeneration. *Ann Neurol.* 1988;23(6):533–540.

96. Blaes F, et al. Intravenous immunoglobulins in the therapy of paraneoplastic neurological disorders. *J Neurol.* 1999;246(4):299–303.

97. David YB, et al. Autoimmune paraneoplastic cerebellar degeneration in ovarian carcinoma patients treated with plasmapheresis and immunoglobulin. A case report. *Cancer.* 1996;78(10):2153–2156.

98. Shams'ili S, et al. An uncontrolled trial of rituximab for antibody associated paraneoplastic neurological syndromes. *J Neurol.* 2006;253(1):16–20.

99. Uchuya M, et al. Intravenous immunoglobulin treatment in paraneoplastic neurological syndromes with antineuronal autoantibodies. *J Neurol Neurosurg Psychiatry.* 1996;60(4):388–392.

100. Trouillas P, et al. Buspirone, a 5-hydroxytryptamine1A agonist, is active in cerebellar ataxia. Results of a double-blind drug placebo study in patients with cerebellar cortical atrophy. *Arch Neurol.* 1997;54(6):749–752.

101. Dalmau J, et al. Anti-Hu—associated paraneoplastic encephalomyelitis/sensory neuronopathy. A clinical study of 71 patients. *Medicine (Baltimore).* 1992;71(2):59–72.

102. Henson RA, Hoffman HL, Urich H. Encephalomyelitis with carcinoma. *Brain.* 1965;88(3):449–464.

103. Dalmau J, et al. Clinical analysis of anti-Ma2-associated encephalitis. *Brain.* 2004;127(Pt 8):1831–1844.

104. Greenlee JE, et al. Antibody types and IgG subclasses in paraneoplastic neurological syndromes. *J Neurol Sci.* 2001;184(2):131–137.

105. Voltz R, et al. A serologic marker of paraneoplastic limbic and brain-stem encephalitis in patients with testicular cancer. *N Engl J Med.* 1999;340(23):1788–1795.

106. Hoftberger R, et al. Encephalitis and AMPA receptor antibodies: novel findings in a case series of 22 patients. *Neurology.* 2015;84(24):2403–2412.

107. van Coevorden-Hameete MH, et al. The expanded clinical spectrum of anti-GABABR encephalitis and added value of KCTD16 autoantibodies. *Brain.* 2019;142(6):1631–1643.

108. Provenzale JM, Barboriak DP, Coleman RE. Limbic encephalitis: comparison of FDG PET and MR imaging findings. *AJR Am J Roentgenol.* 1998;170(6):1659–1660.

109. Kassubek J, et al. Limbic encephalitis investigated by 18FDG-PET and 3D MRI. *J Neuroimaging.* 2001;11(1):55–59.

110. Bataller L, Dalmau J. Paraneoplastic disorders of the nervous system. *Continuum.* 2005;11(5):69–92.

111. Yu Z, et al. CRMP-5 neuronal autoantibody: marker of lung cancer and thymoma-related autoimmunity. *Ann Neurol.* 2001;49(2):146–154.

112. Vincent A, et al. Potassium channel antibody-associated encephalopathy: a potentially immunotherapy-responsive form of limbic encephalitis. *Brain.* 2004;127(Pt 3):701–712.

113. Taylor RB, et al. Reversible paraneoplastic encephalomyelitis associated with a benign ovarian teratoma. *Can J Neurol Sci.* 1999;26(4):317–320.

114. Brennan LV, Craddock PR. Limbic encephalopathy as a nonmetastatic complication of oat cell lung cancer. Its reversal after treatment of the primary lung lesion. *Am J Med.* 1983;75(3):518–520.

115. Nokura K, et al. Reversible limbic encephalitis caused by ovarian teratoma. *Acta Neurol Scand.* 1997;95(6):367–373.

116. Baloh RW, et al. Novel brainstem syndrome associated with prostate carcinoma. *Neurology.* 1993;43(12):2591–2596.

117. Honnorat J, et al. Antibodies to a subpopulation of glial cells and a 66 kDa developmental protein in patients with paraneoplastic neurological syndromes. *J Neurol Neurosurg Psychiatry.* 1996;61(3):270–278.

118. Leypoldt F, et al. Successful immunosuppressive treatment and long-term follow-up of anti-Ri-associated paraneoplastic myelitis. *J Neurol Neurosurg Psychiatry.* 2006;77(10):1199–1200.

119. Sillevis Smitt P, et al. Survival and outcome in 73 anti-Hu positive patients with paraneoplastic encephalomyelitis/sensory neuronopathy. *J Neurol.* 2002;249(6):745–753.

120. Dalmau J, et al. Ma1, a novel neuron- and testis-specific protein, is recognized by the serum of patients with paraneoplastic neurological disorders. *Brain.* 1999;122(Pt 1):27–39.

121. Dropcho EJ. Antiamphiphysin antibodies with small-cell lung carcinoma and paraneoplastic encephalomyelitis. *Ann Neurol.* 1996;39(5):659–667.

122. Saiz A, et al. Anti-amphiphysin I antibodies in patients with paraneoplastic neurological disorders associated with small cell lung carcinoma. *J Neurol Neurosurg Psychiatry.* 1999;66(2):214–217.

123. Keime-Guibert F, et al. Clinical outcome of patients with anti-Hu-associated encephalomyelitis after treatment of the tumor. *Neurology.* 1999;53(8):1719–1723.

124. Keime-Guibert F, et al. Treatment of paraneoplastic neurological syndromes with antineuronal antibodies (anti-Hu, anti-Yo) with a combination of immunoglobulins, cyclophosphamide, and methylprednisolone. *J Neurol Neurosurg Psychiatry.* 2000;68(4):479–482.

125. Vernino S, et al. Immunomodulatory treatment trial for paraneoplastic neurological disorders. *Neuro Oncol.* 2004;6(1):55–62.

126. Anderson NE, et al. Opsoclonus, myoclonus, ataxia, and encephalopathy in adults with cancer: a distinct paraneoplastic syndrome. *Medicine (Baltimore).* 1988;67(2):100–109.

127. Digre KB. Opsoclonus in adults. Report of three cases and review of the literature. *Arch Neurol.* 1986;43(11):1165–1175.

128. Altman AJ, Baehner RL. Favorable prognosis for survival in children with coincident opso-myoclonus and neuroblastoma. *Cancer.* 1976;37(2):846–852.

129. Blaes F, Pike MG, Lang B. Autoantibodies in childhood opsoclonus-myoclonus syndrome. *J Neuroimmunol.* 2008;201–202:221–226.

130. Rudnick E, et al. Opsoclonus-myoclonus-ataxia syndrome in neuroblastoma: clinical outcome and antineuronal antibodies-a report from the Children's Cancer Group Study. *Med Pediatr Oncol.* 2001;36(6):612–622.

131. Russo C, et al. Long-term neurologic outcome in children with opsoclonus-myoclonus associated with neuroblastoma: a report from the Pediatric Oncology Group. *Med Pediatr Oncol.* 1997;28(4):284–288.

132. Wirtz PW, et al. Anti-Ri antibody positive opsoclonus-myoclonus in a male patient with breast carcinoma. *J Neurol.* 2002;249(12):1710–1712.

133. Dalmau J, et al. Major histocompatibility proteins, anti-Hu antibodies, and paraneoplastic encephalomyelitis in neuroblastoma and small cell lung cancer. *Cancer.* 1995;75(1):99–109.

134. Pranzatelli MR, et al. Screening for autoantibodies in children with opsoclonus-myoclonus-ataxia. *Pediatr Neurol.* 2002;27(5):384–387.

135. Antunes NL, et al. Antineuronal antibodies in patients with neuroblastoma and paraneoplastic opsoclonus-myoclonus. *J Pediatr Hematol Oncol.* 2000;22(4):315–320.

136. Borgna-Pignatti C, et al. Treatment with intravenously administered immunoglobulins of the neuroblastoma-associated opsoclonus-myoclonus. *J Pediatr.* 1996;129(1):179–180.

137. Hammer MS, Larsen MB, Stack CV. Outcome of children with opsoclonus-myoclonus regardless of etiology. *Pediatr Neurol.* 1995;13(1):21–24.

138. Mitchell WG, Snodgrass SR. Opsoclonus-ataxia due to childhood neural crest tumors: a chronic neurologic syndrome. *J Child Neurol.* 1990;5(2):153–158.

139. Yiu VW, et al. Plasmapheresis as an effective treatment for opsoclonus-myoclonus syndrome. *Pediatr Neurol.* 2001;24(1):72–74.

140. Koh PS, et al. Long-term outcome in children with opsoclonus-myoclonus and ataxia and coincident neuroblastoma. *J Pediatr.* 1994;125(5 Pt 1):712–716.

141. Luque FA, et al. Anti-Ri: an antibody associated with paraneoplastic opsoclonus and breast cancer. *Ann Neurol.* 1991;29(3):241–251.

142. Ohmer R, et al. Ophthalmoplegia associated with the anti-Ri antibody. *J Neuroophthalmol.* 1999;19(4):246–248.

143. Armangue T, et al. Clinical and immunological features of opsoclonus-myoclonus syndrome in the era of neuronal cell surface antibodies. *JAMA Neurol.* 2016;73(4):417–424.

144. Prestigiacomo CJ, Balmaceda C, Dalmau J. Anti-Ri-associated paraneoplastic opsoclonus-ataxia syndrome in a man with transitional cell carcinoma. *Cancer.* 2001;91(8):1423–1428.

145. Hormigo A, et al. Immunological and pathological study of anti-Ri-associated encephalopathy. *Ann Neurol.* 1994;36(6):896–902.

146. Swart JF, de Kraker J, van der Lely N. Metaiodobenzylguanidine total-body scintigraphy required for revealing occult neuroblastoma in opsoclonus-myoclonus syndrome. *Eur J Pediatr.* 2002;161(5):255–258.

147. Hayward K, et al. Long-term neurobehavioral outcomes in children with neuroblastoma and opsoclonus-myoclonus-ataxia syndrome: relationship to MRI findings and anti-neuronal antibodies. *J Pediatr.* 2001;139(4):552–559.

148. Scaravilli F, et al. The neuropathology of paraneoplastic syndromes. *Brain Pathol.* 1999;9(2):251–260.

149. Dropcho EJ, Kline LB, Riser J. Antineuronal (anti-Ri) antibodies in a patient with steroid-responsive opsoclonus-myoclonus. *Neurology.* 1993;43(1):207–211.

150. Batchelor TT, Platten M, Hochberg FH. Immunoadsorption therapy for paraneoplastic syndromes. *J Neurooncol.* 1998;40(2):131–136.

151. Cher LM, et al. Therapy for paraneoplastic neurologic syndromes in six patients with protein A column immunoadsorption. *Cancer.* 1995;75(7):1678–1683.

152. Brown P, Marsden CD. The stiff man and stiff man plus syndromes. *J Neurol.* 1999;246(8):648–652.

153. De Camilli P, et al. The synaptic vesicle-associated protein amphiphysin is the 128-kD autoantigen of Stiff-Man syndrome with breast cancer. *J Exp Med.* 1993;178(6):2219–2223.

154. Schmierer K, et al. Atypical stiff-person syndrome with spinal MRI findings, amphiphysin autoantibodies, and immunosuppression. *Neurology.* 1998;51(1):250–252.

155. Wirtz PW, Wintzen AR, Verschuuren JJ. Lambert-Eaton myasthenic syndrome has a more progressive course in patients with lung cancer. *Muscle Nerve.* 2005;32(2):226–229.

156. Elrington GM, et al. Neurological paraneoplastic syndromes in patients with small cell lung cancer. A prospective survey of 150 patients. *J Neurol Neurosurg Psychiatry.* 1991;54(9):764–767.

157. O'Neill JH, Murray NM, Newsom-Davis J. The Lambert-Eaton myasthenic syndrome. A review of 50 cases. *Brain.* 1988;111(Pt 3): 577–596.

158. Tim RW, Massey JM, Sanders DB. Lambert-Eaton myasthenic syndrome: electrodiagnostic findings and response to treatment. *Neurology.* 2000;54(11):2176–2178.

159. Wirtz PW, et al. The epidemiology of myasthenia gravis, Lambert-Eaton myasthenic syndrome and their associated tumours in the northern part of the province of South Holland. *J Neurol.* 2003;250(6):698–701.

160. Hawley RJ, et al. The carcinomatous neuromyopathy of oat cell lung cancer. *Ann Neurol.* 1980;7(1):65–72.

161. Wirtz PW, et al. HLA and smoking in prediction and prognosis of small cell lung cancer in autoimmune Lambert-Eaton myasthenic syndrome. *J Neuroimmunol.* 2005;159(1–2):230–237.

162. Maddison P, et al. Favourable prognosis in Lambert-Eaton myasthenic syndrome and small-cell lung carcinoma. *Lancet.* 1999;353(9147):117–118.

163. Wirtz PW, et al. P/Q-type calcium channel antibodies, Lambert-Eaton myasthenic syndrome and survival in small cell lung cancer. *J Neuroimmunol.* 2005;164(1–2):161–165.

164. Wirtz PW, et al. Differences in clinical features between the Lambert-Eaton myasthenic syndrome with and without cancer: an analysis of 227 published cases. *Clin Neurol Neurosurg.* 2002;104(4):359–363.

165. Vincent A. Antibodies to ion channels in paraneoplastic disorders. *Brain Pathol.* 1999;9(2):285–291.

166. Voltz R, et al. P/Q-type voltage-gated calcium channel antibodies in paraneoplastic disorders of the central nervous system. *Muscle Nerve.* 1999;22(1):119–122.

167. Rosenfeld MR, et al. Cloning and characterization of a Lambert-Eaton myasthenic syndrome antigen. *Ann Neurol.* 1993;33(1):113–120.

168. Chalk CH, et al. Response of the Lambert-Eaton myasthenic syndrome to treatment of associated small-cell lung carcinoma. *Neurology.* 1990;40(10):1552–1556.

169. Newsom-Davis J. A treatment algorithm for Lambert-Eaton myasthenic syndrome. *Ann N Y Acad Sci.* 1998;841:817–822.

170. McEvoy KM, et al. 3,4-Diaminopyridine in the treatment of Lambert-Eaton myasthenic syndrome. *N Engl J Med.* 1989;321(23):1567–1571.

171. Sanders DB, et al. A randomized trial of 3,4-diaminopyridine in Lambert-Eaton myasthenic syndrome. *Neurology.* 2000;54(3):603–607.

172. Oh SJ, et al. Low-dose guanidine and pyridostigmine: relatively safe and effective long-term symptomatic therapy in Lambert-Eaton myasthenic syndrome. *Muscle Nerve.* 1997;20(9):1146–1152.

173. Newsom-Davis J. Therapy in myasthenia gravis and Lambert-Eaton myasthenic syndrome. *Semin Neurol.* 2003;23(2):191–198.

174. Bain PG, et al. Effects of intravenous immunoglobulin on muscle weakness and calcium-channel autoantibodies in the Lambert-Eaton myasthenic syndrome. *Neurology.* 1996;47(3):678–683.

175. Bird SJ. Clinical and electrophysiologic improvement in Lambert-Eaton syndrome with intravenous immunoglobulin therapy. *Neurology.* 1992;42(7):1422–1423.

176. Newsom-Davis J, Murray NM. Plasma exchange and immunosuppressive drug treatment in the Lambert-Eaton myasthenic syndrome. *Neurology.* 1984;34(4):480–485.

177. Chalk CH, et al. The distinctive clinical features of paraneoplastic sensory neuronopathy. *Can J Neurol Sci.* 1992;19(3):346–351.

178. Gwathmey KG. Sensory neuronopathies. *Muscle Nerve.* 2016;53(1):8–19.

179. Molinuevo JL, et al. Utility of anti-Hu antibodies in the diagnosis of paraneoplastic sensory neuropathy. *Ann Neurol.* 1998;44(6):976–980.

180. Horwich MS, et al. Subacute sensory neuropathy: a remote effect of carcinoma. *Ann Neurol.* 1977;2(1):7–19.

181. Camdessanche JP, et al. Paraneoplastic peripheral neuropathy associated with anti-Hu antibodies. A clinical and electrophysiological study of 20 patients. *Brain.* 2002;125(Pt 1):166–175.

182. Lauria G, Pareyson D, Sghirlanzoni A. Neurophysiological diagnosis of acquired sensory ganglionopathies. *Eur Neurol.* 2003;50(3):146–152.

183. Oh SJ, Dropcho EJ, Claussen GC. Anti-Hu-associated paraneoplastic sensory neuropathy responding to early aggressive immunotherapy: report of two cases and review of literature. *Muscle Nerve.* 1997;20(12):1576–1582.

184. Antoine JC, et al. Carcinoma associated paraneoplastic peripheral neuropathies in patients with and without anti-onconeural antibodies. *J Neurol Neurosurg Psychiatry.* 1999;67(1):7–14.

185. Dropcho EJ. Neurotoxicity of cancer chemotherapy. *Semin Neurol.* 2004;24(4):419–426.

186. Ropper AH, Gorson KC. Neuropathies associated with paraproteinemia. *N Engl J Med.* 1998;338(22):1601–1607.

187. Pittock SJ, et al. Amphiphysin autoimmunity: paraneoplastic accompaniments. *Ann Neurol.* 2005;58(1):96–107.

188. Wirtz PW, et al. Difference in distribution of muscle weakness between myasthenia gravis and the Lambert-Eaton myasthenic syndrome. *J Neurol Neurosurg Psychiatry.* 2002;73(6):766–768.

189. Hart IK, et al. Phenotypic variants of autoimmune peripheral nerve hyperexcitability. *Brain.* 2002;125(Pt 8):1887–1895.

190. Buckley C, et al. Do titin and cytokine antibodies in MG patients predict thymoma or thymoma recurrence? *Neurology.* 2001;57(9):1579–1582.

191. Somnier FE, Engel PJ. The occurrence of anti-titin antibodies and thymomas: a population survey of MG 1970–1999. *Neurology.* 2002;59(1):92–98.

192. Vincent A, Newsom-Davis J. Acetylcholine receptor antibody as a diagnostic test for myasthenia gravis: results in 153 validated cases and 2967 diagnostic assays. *J Neurol Neurosurg Psychiatry.* 1985;48(12):1246–1252.

193. Hoch W, et al. Auto-antibodies to the receptor tyrosine kinase MuSK in patients with myasthenia gravis without acetylcholine receptor antibodies. *Nat Med.* 2001;7(3):365–368.

194. Evoli A, et al. Clinical correlates with anti-MuSK antibodies in generalized seronegative myasthenia gravis. *Brain.* 2003;126(Pt 10):2304–2311.

195. Leow YH, Goh CL. Malignancy in adult dermatomyositis. *Int J Dermatol.* 1997;36(12):904–907.

196. Gallais V, Crickx B, Belaich S. Prognostic factors and predictive signs of malignancy in adult dermatomyositis. *Ann Dermatol Venereol.* 1996;123(11):722–726.

197. Dalakas MC. Polymyositis, dermatomyositis and inclusion-body myositis. *N Engl J Med.* 1991;325(21):1487–1498.

198. Hill CL, et al. Frequency of specific cancer types in dermatomyositis and polymyositis: a population-based study. *Lancet.* 2001;357(9250):96–100.

199. Targoff IN. Update on myositis-specific and myositis-associated autoantibodies. *Curr Opin Rheumatol.* 2000;12(6):475–481.

200. Amato AA, Barohn RJ. Idiopathic inflammatory myopathies. *Neurol Clin.* 1997;15(3):615–648.

201. Jacobson DM. Paraneoplastic diseases of neuroophthalmologic interest. In: Miller NR, Newman NJ, eds. *Walsh and Hoyt's Clinical Neuro-Ophthalmology.* Philadelphia, PA: Lippincott Williams & Wilkins; 1998:2497–2551.

202. Keltner JL, Thirkill CE, Yip PT. Clinical and immunologic characteristics of melanoma-associated retinopathy syndrome: eleven new cases and a review of 51 previously published cases. *J Neuroophthalmol.* 2001;21(3):173–187.

203. Shiraga S, Adamus G. Mechanism of CAR syndrome: anti-recoverin antibodies are the inducers of retinal cell apoptotic death via the caspase 9- and caspase 3-dependent pathway. *J Neuroimmunol.* 2002;132(1–2):72–82.

204. Ohguro H, Nakazawa M. Pathological roles of recoverin in cancer-associated retinopathy. *Adv Exp Med Biol.* 2002;514:109–124.

205. Boeck K, et al. Melanoma-associated paraneoplastic retinopathy: case report and review of the literature. *Br J Dermatol.* 1997;137(3):457–460.

206. Alexander KR, et al. Contrast-processing deficits in melanoma-associated retinopathy. *Invest Ophthalmol Vis Sci.* 2004;45(1):305–310.

207. Milam AH, et al. Autoantibodies against retinal bipolar cells in cutaneous melanoma-associated retinopathy. *Invest Ophthalmol Vis Sci.* 1993;34(1):91–100.

208. Cross SA, et al. Paraneoplastic autoimmune optic neuritis with retinitis defined by CRMP-5-IgG. *Ann Neurol.* 2003;54(1):38–50.

209. de la Sayette V, et al. Paraneoplastic cerebellar syndrome and optic neuritis with anti-CV2 antibodies: clinical response to excision of the primary tumor. *Arch Neurol.* 1998;55(3):405–408.

Neurological complications of specific neoplasms

15

Neurological complications of lung cancer

Leon D. Kaulen[a],*, Benjamin Lu[b],*, Sarah Goldberg[b], and Joachim M. Baehring[c]

[a]Department of Neurology, Heidelberg University Hospital, Heidelberg, Germany, [b]Section of Medical Oncology, Department of Internal Medicine, Yale University School of Medicine, New Haven, CT, United States, [c]Departments of Neurology and Neurosurgery, Yale University School of Medicine, New Haven, CT, United States

1 Introduction

1.1 Epidemiology

Lung cancer is the most common type of cancer and the leading cause of cancer-related mortality in the world, with over 2.1 million new cases diagnosed and 1.8 million deaths in 2018.[1,2] In the United States, it is estimated that 230,000 new cases will be diagnosed in 2020 and will be responsible for 135,000–160,000 deaths.[3] While lung cancer is the second most commonly diagnosed type of cancer in men and women behind prostate cancer and breast cancer, respectively, lung cancer remains the leading cause of cancer-related death for either gender.[3] Neurologic complications arise from metastatic spread, paraneoplastic mechanisms, or adverse effects of lung cancer therapy (Table 1).

Based on Surveillance Epidemiology and End Results (SEER) data from 2013 to 2017, the median age of diagnosis is 71 years.[4] The incidence of lung cancer is higher in men (61.7 per 100,000) than women (48.6 per 100,000). Among men in the United States, the incidence is highest among black men (71.2 per 100,000) and lowest in Hispanic men (35.1 per 100,000). Among women, the incidence is highest in white women and lowest in Asian/Pacific Islanders.

The strongest risk factor for developing lung cancer is smoking tobacco, and global trends in lung cancer incidence and mortality are invariably associated with tobacco smoking. However, lung cancer may occur in a significant portion of patients with no or light smoking histories, which is estimated to occur in 15%–20% of men and 50% of women with lung cancer globally.[5] The underlying risk factor is frequently unclear in patients with light smoking histories, though radiation exposure and environmental toxins, such as radon, asbestos, heavy metals, and air pollution, have also been associated with an increased risk of lung cancer.[6] Alternative tobacco products, such as e-cigarettes, are associated with a relatively lower risk of pulmonary toxicity compared to tobacco smoking, though the association with lung cancer remains to be defined.[7]

In the United States, smoking cessation efforts implemented in the 1960s have led to a gradual decline in lung cancer incidence in men over the past three decades.[8] Among men, the annual rate of decline increased from 1.9% between 2001 and 2008 to 3.1% between 2008 and 2016.[9] Among women, a decline in incidence by 1.7% was observed from 2006 to 2016. Incidence-adjusted mortality has also improved for patients with nonsmall cell lung cancer, which is likely a reflection of the significant improvements in systemic treatment options, as discussed later.[9] No major improvements in incidence-adjusted mortality for patients with small cell lung cancer have been observed.

TABLE 1 Neurological complications of lung cancer.

Metastatic complications	Paraneoplastic syndromes	Treatment complications
Parenchymal brain metastases	Encephalomyelitis	Radiation necrosis
Leptomeningeal dissemination	Cerebellar syndrome	Neurocognitive decline
Epidural or intramedullary metastases	Sensory or autonomic neuropathy	Neurological syndromes associated with immune checkpoint inhibition
Peripheral nerve/plexus infiltration	Lambert-Eaton myasthenic syndrome	

---

* Both authors contributed equally to this chapter.

1.2 Tumor classification

Lung cancers encompass cancers arising from the trachea, bronchi, bronchioles, and alveoli. Histologically, lung cancers are broadly classified as nonsmall cell lung cancers (NSCLCs), which encompasses ~ 85% of all lung cancers,[10] and small cell lung cancers (SCLCs). In the 2015 World Health Organization (WHO) Classification System, NSCLC are further subdivided into seven groups (i) adenocarcinoma (40%), (ii) squamous cell (30%), (iii) adenosquamous, (iv) large cell, (v) sarcomatoid, (vi) neuroendocrine, and (vii) diffuse idiopathic pulmonary neuroendocrine cell hyperplasia.[11] These subgroups are often combined under the heading of NSCLC because they frequently coexist in a single tumor and are categorized based on the most prevalent histology.

The most commonly used staging system for lung cancer is the American Joint Committee on Cancer (AJCC) Tumor, Node, Metastasis (TNM) staging system, which is currently in its 8th edition.[12] Stages are grouped from I to IV, depending on the tumor size, nodal involvement, and distant metastatic involvement. For NSCLC, patients are clinically treated as localized, locally advanced, or advanced disease. For SCLC, management is based on limited- or extensive-stage disease. This chapter will primarily focus on neurological complications arising from locally advanced-stage NSCLC, advanced-stage NSCLC, and extensive-stage SCLC.

Molecular testing has also been incorporated into the standard of care for patients with advanced NSCLC, with an emphasis on targetable alterations. Standard testing includes assessing for alterations in *EGFR*, *ALK*, ROS1, *BRAF*, *NTRK*, *MET*, and *RET*. The histological quantification of programmed death-ligand 1 (PD-L1) expression also serves as an important predictive marker for response to immune checkpoint inhibitor therapy.

1.3 General oncologic management

1.3.1 Nonsmall cell lung cancer

The management of NSCLC has changed substantially over the last two decades as a result of an improved understanding of oncogenic drivers of cancer, the development of targeted therapies, and the emergence of immunotherapies.[13] As neurologic complications generally occur in association with advanced stages of disease and with the use of systemic treatments, we will briefly review the management of early-stage NSCLC and primarily focus on the management of advanced disease.

Localized disease

For patients with localized NSCLC (i.e., stage I or II in the AJCC 8th edition), the 5-year overall survival (OS) based on clinical staging ranges from 25% to 50%.[14] When possible, complete surgical resection is performed for all patients with localized disease. Limited resection (i.e., wedge resection) or lobectomy is preferred over pneumonectomy to preserve lung function if complete resections are attainable. Minimally invasive approaches, such as video-assisted thoracoscopic surgery, decrease morbidity and reduce the time to systemic treatment, when indicated.[15]

Despite complete resection, recurrence occurs in 36% of the patients with initial localized disease by 5 years, including local recurrences in 19% of the patients with stage I disease.[16,17] The addition of adjuvant chemotherapy is not indicated for patients with stage IA disease given a trend toward worse survival in the Lung Adjuvant Cisplatin Evaluation (LACE) metaanalysis; however, risk stratification for patients for stages IB remains an active area of investigation.[18–22] An unplanned subset analysis of the Cancer and Leukemia Group B (CALGB) 9633 study, which included 344 patients with node-negative disease, showed an OS benefit for 4 cycles of paclitaxel and carboplatin chemotherapy given every 3 weeks for patients with stage IB NSCLC with tumors greater than or equal to 4 cm in diameter. However, a subsequent metaanalysis using the National Cancer Database of 50,814 patients demonstrated no improvement in survival with adjuvant chemotherapy for patients with early-stage, node-negative NSCLC based solely on tumor size.[19] Adjuvant, postoperative chemotherapy for node-negative, stage II and IIIA disease with a cisplatin-based regimen is an accepted standard of care based on several, large randomized clinical trials that demonstrated an improvement in OS with adjuvant chemotherapy.[20,21,23–27]

For patients with localized disease who are unable to undergo surgical resection, definitive radiation therapy is an acceptable alternative.[24] Stereotactic ablative radiation therapy (SBRT) may result in local control rates up to ~ 90%. SBRT is the preferred modality for peripheral lesions, though may also have a role for centrally located lesions.[28] Trials are ongoing to define the role of SBRT for resectable stage I NSCLC. Postoperative radiation therapy may be considered if surgical margins are positive.

Locally advanced disease

Locally advanced, or stage III, NSCLC encompasses a heterogeneous group of patients. Surgical resection may be offered for patients with negative mediastinal lymph nodes (N0-1) if technically feasible followed by adjuvant chemotherapy, as discussed later. For patients with ipsilateral mediastinal lymph node involvement (N2), definitive radiation therapy with concurrent platinum-based chemotherapy is the most common treatment approach. A select group of patients may benefit from induction therapy based on possible improvements in local disease control and progression-free survival.[29–32] No OS benefit has been shown with induction therapy to date.

For patients with locally advanced disease who do not undergo surgical resection, definitive radiation with concurrent platinum-based chemotherapy regimens is considered standard. In the United States, two commonly used chemotherapy combinations include cisplatin 50 mg/m^2 (days 1, 8, 29, and 36) plus etoposide 50 mg/m^2 (days 1–5, 29–33) or weekly carboplatin AUC 2 plus paclitaxel 45 mg/m^2.[33] For patients with nonsquamous NSCLC, cisplatin plus pemetrexed is also an option.[34]

For patients who do not have disease progression following definitive chemoradiation therapy, consolidation therapy with the anti-PD-L1 monoclonal antibody, durvalumab is given. In the landmark, randomized phase III PACIFIC trial, patients with unresectable, stage III NSCLC who received durvalumab for up to 12 months had significantly longer progression-free survival (17.2 months vs 5.6 months) and OS (hazard ratio 0.68) when compared to placebo.[35–37]

Metastatic disease

The approach toward treating patients with advanced NSCLC has rapidly evolved over the last decade. From January to September 2020 alone, there were 10 new FDA drug approvals for treating advanced NSCLC. While cytotoxic chemotherapy remains an important cornerstone of therapy for most patients, the emergence of targeted therapies and immune checkpoint inhibitors have revolutionized treatment paradigms for a subset of patients. Therefore, the general approach toward treating patients with advanced NSCLC is to first evaluate the presence of molecular alterations to determine candidacy of targeted therapies. If absent, the expression of PD-L1 is the best predictive biomarker to date of benefit to immune checkpoint therapy. If PD-L1 is not expressed, combination immune checkpoint blockade or cytotoxic chemotherapy with or without immune checkpoint inhibitors is indicated.

Targeted therapies

For patients diagnosed with metastatic NSCLC, assessment for targetable gene alterations is an essential component for initial treatment selection. Potential oncogenic driver mutations may be detected in up to 64% of the patients with lung adenocarcinoma, and this proportion is only expected to increase as new targets are discovered.[38] To date, there have been 18 oral kinase inhibitors approved by the FDA in NSCLC targeting alterations in seven genes, with several numerous additional agents in development.[39]

The first tyrosine kinase inhibitor (TKI) to demonstrate significant clinical activity was the first-generation EGFR TKI, gefitinib.[40,41] Gefitinib and erlotinib, another first-generation TKIs, are oral agents that reversibly inhibit EGFR. Subsequent clinical trials demonstrated improvements in objective response rates (ORR) and

progression-free survival (PFS) compared to cytotoxic chemotherapy, including in the frontline setting.[42–47] The second-generation agent afatinib, which irreversibly inhibits EGFR, also improves ORR and PFS compared to cytotoxic chemotherapy.[48–51] When compared to gefitinib, the second-generation agents such as afatinib result in improved PFS and trends toward improved OS, though at the expense of greater toxicity.[52–54] While all EGFR TKIs may result in a characteristic acneiform rash, pulmonary toxicity, and diarrhea, serious treatment-related adverse events are more commonly observed in those treated with second-generation agents.[55]

Patients who initially respond to EGFR TKIs invariably develop resistance, and the most common mechanism of resistance to first- and second-generation EGFR TKIs is a threonine-to-methionine substitution in codon 790 (T790M).[56,57] The third-generation EGFR TKI, osimertinib, demonstrated activity against acquired T790M mutations and has become the preferred frontline agent for all patients with EGFR-mutant NSCLC after demonstrating improvement in PFS, OS, and tolerability compared with first-generation EGFR TKIs.[58–60] Importantly, osimertinib does penetrate the blood-brain barrier and results in improved outcomes in patients with brain metastases.[61]

In ~ 20% of the patients with acquired EGFR resistance, comutations amplifying the tyrosine kinase MET may be found.[62] Molecular alterations in MET may be found independent of EGFR mutations and include MET exon 14 skip mutations and MET gene amplifications. The FDA recently approved the MET inhibitor capmatinib for patients with MET exon 14 skip mutations based on a multicenter, nonrandomized phase II study including 28 newly diagnosed patients and 69 patients who had received at least one prior treatment.[63] ORR was 68% in the first-line setting and 41% for previously treated patients. The median duration of response was 12.6 months and 9.7 months, respectively. The multitarget TKI, crizotinib, is also active in MET-mutated disease.[64] Multiple additional agents are currently under investigation.

As in EGFR-positive NSCLC, TKIs targeting rearrangements in the anaplastic lymphoma kinase (ALK) gene have been developed and improve outcomes. The first ALK inhibitor approved by the FDA was crizotinib, which is a multitargeted TKI that demonstrated improved ORR and PFS compared to cytotoxic chemotherapy both in the first- and second-line settings.[65,66] Differences in OS did not meet statistical significance, though an improvement was observed after statistical adjustments for crossover.[67] The second-generation agents, alectinib and brigatinib, are now the preferred first-line agents after demonstrating improvements in PFS, better tolerability, and trend toward longer survival compared with crizotinib, though OS results have yet to mature.[68–73] Importantly, both alectinib and brigatinib have better CNS penetration compared with crizotinib resulting in longer time to CNS progression, as discussed in greater detail later. Ceritinib is another second-generation TKI with demonstrated efficacy, though was compared to platinum-based chemotherapy rather than crizotinib.[74] Unfortunately, as with EGFR TKIs, resistance inevitably develops, including L1196M and G1202R mutations in ALK, and the third-generation ALK/ROS1 inhibitor lorlatinib has been approved for the third-line setting based on results from a phase II trial that included a subgroup of 215 patients with ALK-positive disease, which showed a 48% ORR and median response duration of 19.5 months.[75] Trials are currently underway to evaluate the role of lorlatinib in the frontline setting. The most commonly observed side effect with ALK inhibitors include gastrointestinal side effects, including constipation in over one-third of patients on alectinib, and neurologic toxicity has been reported in up to 40% of the patients on lorlatinib.[75]

Given the high degree of homology between the kinase domains of ROS1 and ALK, many TKIs active in ALK rearranged disease are also active in patients with ROS1 rearrangements, including crizotinib, ceritinib, and lorlatinib.[76–79] Crizotinib obtained FDA approval for ROS1-rearranged disease after phase I/II studies showed a 72% ORR, median PFS of 19.3 months, and median OS 51 months.[77,78] Lorlatinib has also demonstrated activity in patients with ROS-1-positive disease who progressed on crizotinib.[79] Similarly, the kinase domain of ROS1 and tropomyosin receptor kinase (TRK) share a high degree of homology, and the TRK inhibitor, entrectinib, has shown clinical activity in ROS1-rearranged disease. Entrectinib obtained FDA approval in ROS1-rearranged disease after showing a 77% ORR and median PFS of 19 months.[80] Entrectinib has better CNS penetration compared to crizotinib and is therefore preferred for patients with brain metastases.

In addition to entrectinib, patients with a neurotrophic receptor tyrosine kinase (NTRK) gene fusion may also be treated with larotrectinib. Both agents obtained FDA approval in the second-line setting based on early-phase trial of patients with TRK-fusion solid tumors, including a total of 22 patients with lung tumors, demonstrating 70%–80% ORR.[80–84]

Selpercatinib is a RET inhibitor that was recently approved for NSCLC with RET fusion gene alterations. In a phase I/II trial, selpercatinib was found to have an ORR 85% among 39 treatment-naïve patients and an ORR of 64% in patients previously treated with platinum-based chemotherapy regimen.[85] Grade 3–4 toxicities were observed in 58% of the patients, including hypertension (14%), liver function derangements (10%), and hyponatremia (6%). Selpercatinib did show activity in patients with brain metastases, as discussed later.

For patients with BRAF V600E mutations, the combination of BRAF kinase inhibitor, dabrafenib, and MEK kinase inhibitor, trametinib, has been approved by the FDA. While not as commonly observed in NSCLC compared with melanoma, BRAF V600E-positive NSCLC is responsive to dabrafenib/trametinib in either frontline or subsequent line setting.[86–88]

Immunotherapy

In patients with tumors that do not harbor targetable molecular alterations, the use of immune checkpoint inhibitors (ICIs) with or without chemotherapy has become the frontline standard of care.[24] The emergence of ICIs over the last decade has revolutionized the treatment of multiple cancer types, including NSCLC. These agents target coinhibitory receptors essential for generating T-cell-mediated antitumor immune responses.[89,90] ICIs with current approvals target either the programmed death 1 (PD-1) and its ligand, PD-L1 or the cytotoxic T-lymphocyte antigen-4 (CTLA-4), though several additional checkpoints are currently under investigation.

Currently, the most predictive biomarker for response to ICIs is the level of PD-L1 expression on tumor cells, as assessed by immunohistochemistry.[91,92] In patients whose tumors are found to have high levels of PD-L1 expression ($\geq$ 50% membranous staining on tumor cells) with wild-type EGFR/ALK, both pembrolizumab and atezolizumab prolong survival as single agents in the frontline setting compared with platinum-based doublet chemotherapy.[93–95] Pembrolizumab is a highly selective, humanized IgG4 monoclonal antibody against PD-1, which prevents engagement with PD-L1 and PD-L2. In the KEYNOTE-024 randomized phase III study of 305 patients with previously untreated NSCLC, the median OS for patients receiving pembrolizumab was 30.2 months compared with 14.2 months in patients receiving platinum-based chemotherapy. Atezolizumab is a humanized IgG1 antibody against PD-L1. In the case of atezolizumab, patients with PD-L1 expression $\geq$ 10% on immune cells also derived benefit from single-agent therapy. In the IMPOWER-110 study of 572 patients with previously untreated NSCLC, the subgroup of patients ($n=205$) with high PD-L1 (tumor expression $\geq$ 50% or immune cell expression $\geq$ 10%), OS was 20 months compared with 13 months in patients treated with platinum-based chemotherapy. Pembrolizumab in combination with platinum-pemetrexed chemotherapy is also an accepted treatment regimen for patients with nonsquamous NSCLC and PD-L1 $\geq$ 50%, as supported by the KEYNOTE-189 study.[96,97] No study to date has directly compared ICI monotherapy with chemoimmunotherapy in patients with high PD-L1 expression.

For patients with PD-L1 expression < 50%, multiple ICI-based combination regimens have been approved. In nonsquamous NSCLC, pembrolizumab in combination with platinum-pemetrexed chemotherapy was shown to improve OS in all subgroups, including PD-L1 tumor proportion score (TPS) 1%–49% and < 1%. The median OS was 16.9 months and 12.6 months, respectively, compared with 9.1 months and 8.9 months for the chemotherapy-only groups. Several atezolizumab-based combinations have also been shown to be effective in nonsquamous NSCLC compared with chemotherapy alone, including atezolizumab in combination with carboplatin/paclitaxel with or without bevacizumab, carboplatin/nab-paclitaxel, and pemetrexed.[98–101] For patients with PD-L1 TPS < 50% and squamous histology, pembrolizumab in combination with platinum-based doublet chemotherapy is approved based on the results of the KEYNOTE-407 trial.[102] For patients with TPS > 1%, pembrolizumab is FDA-approved as a single agent based on findings from KEYNOTE-042, although the benefit was likely driven by the patients with PD-L1 TPS > 50%, and therefore, single-agent immunotherapy is not typically used in those with PD-L1 TPS 1%–49%.[103]

Another FDA-approved regimen for patients with metastatic NSCLC irrespective of PD-L1 expression is the anti-PD-1 nivolumab in combination with the anti-CTLA4 ipilimumab in the frontline setting based on the results from the CheckMate-227 trial.[104] While PD-L1-expressing tumors derived benefit from the combination ICIs compared with chemotherapy, perhaps most striking was the higher rates of complete response, longer duration of response, and the improved median OS (17.2 months vs 12.2 months) for patients with tumor PD-L1 < 1% (n = 373). Nivolumab plus ipilimumab was also superior to both nivolumab monotherapy and nivolumab plus chemotherapy in this study. Overall, these findings are consistent with the hypothesis that the immune-modulatory effects of CTLA-4 may be of particular importance in PD-L1-negative tumors.

In addition to efficacy in the frontline setting, pembrolizumab, atezolizumab, and nivolumab have demonstrated superiority to single-agent chemotherapy and are approved for use in the second-line setting following progression on first-line chemotherapy.[91,105–109] Pembrolizumab may be used for tumors with PD-L1 TPS ≥ 1%, whereas atezolizumab and nivolumab may be used regardless of PD-L1 expression.

As a class, ICIs are remarkable for the potentially durable responses conveyed by immune surveillance and overall tolerability relative to cytotoxic chemotherapies. However, toxicities observed with ICIs are distinct and primarily involved immune-related phenomenon. Of immune-related adverse events (irAEs), hypothyroidism, pneumonitis, and colitis are among the most commonly seen toxicities.[110,111] Patients treated with combination ICIs are at greater risk for developing irAEs.[112] Neurological irAEs may also be observed and encompass a wide spectrum of manifestations.[113]

While headache and sensory neuropathy are the most commonly observed neurological irAEs, more serious complications such as myasthenia gravis, Guillain-Barre syndrome, transverse myelitis, autoimmune encephalitis, and posterior reversible encephalopathy syndrome and among other conditions have all been reported.[114–116] Depending on the severity of the irAEs, glucocorticoids or other immunosuppressive therapies may be required and ICIs may need to be held or permanently discontinued.[117]

Cytotoxic chemotherapy

Palliative chemotherapy remains an important component in the treatment of advanced NSCLC, either given in combination with immunotherapy, or alone for patients who are not candidates for a checkpoint inhibitor. Chemotherapy can both improve quality of life and increase survival. Patients who have a good performance status should be treated with a platinum-based (cisplatin, carboplatin) regimen and a second active cytotoxic agent for a total of four to six cycles.[118–120] Carboplatin has a more favorable toxicity profile compared to cisplatin and is therefore commonly used in practice. Agents commonly used in conjunction with platinum include pemetrexed, taxanes (docetaxel, paclitaxel, nab-paclitaxel), and gemcitabine. Pemetrexed is not used for patients with squamous NSCLC due to inferior efficacy compared with gemcitabine in combination with cisplatin, whereas cisplatin/pemetrexed was shown to have similar efficacy in nonsquamous disease.[121,122] For patients who do not progress following four to six cycles of the platinum-doublet, maintenance therapy may be considered with single-agent pemetrexed or bevacizumab (if used as part of the first-line regimen).[123,124] Patients who progress either during the initial or maintenance phase of treatment may be treated with an alternative agent mentioned earlier. Ramucirumab, a monoclonal antibody targeting vascular endothelial growth factor receptor 2 (VEGF2), has also shown modest improvements in OS (10.5 months vs 9.1 months) in combination with docetaxel in the second-line setting.[125,126]

1.3.2 Small cell lung cancer

Small cell lung cancers (SCLCs) are aggressive, poorly differentiated, neuroendocrine tumors that most commonly present with advanced, metastatic disease. Disease confined to ipsilateral hemithorax and regional lymph nodes that can be encompassed by a single radiation field is termed "limited-stage" disease. If no clinical evidence of nodal involvement is present, mediastinal staging is recommended, and, if negative, patients may be considered for surgical resection. In most patients with limited-stage disease, there will be clinical evidence of lymph node involvement, and the standard approach is chemoradiotherapy (typically with concurrent therapy).[127] The most commonly used chemotherapy regimen is four cycles of etoposide/cisplatin.[128,129] Patients who are able to complete initial chemoradiotherapy with partial or complete responses may also be considered for prophylactic cranial irradiation, which decreases the rate of intracranial relapse and has a survival benefit (15.3 vs 20.7% survival at 3 years).[130,131] Overall, complete responses to initial therapy may be seen in 70%–90% of the patients with limited-stage disease, though these patients frequently relapse, as reflected by the 25%–30% 5-year survival rates.[131,132]

For patients with extensive-stage (ES) SCLC, ICIs have been incorporated into frontline treatment regimens. There are currently two ICIs approved for ES-SCLC in the frontline setting: the anti-PD-L1 antibody atezolizumab and the anti-PD-1 antibody durvalumab. Both were shown to improve OS when combined with four to six cycles of platinum-based chemotherapy in large, randomized phase III studies.[133,134] Patients who respond to initial chemoimmunotherapy are continued on single-agent immunotherapy maintenance therapy until disease progression. Prophylactic cranial irradiation and thoracic radiation for residual disease can be considered in patients who respond to systemic therapy. The median OS for both chemoimmunotherapy combinations was ~12 months compared with 10 months for chemotherapy alone. Unlike in NSCLC, the addition of an anti-CTLA-4 agent (in this case tremelimumab) to anti-PD-1 did not increase ORR or OS and resulted in greater toxicity.[135]

Patients who have primary progression or relapse within 6 months of completing induction platinum-based therapy are generally treated with single-agent therapy. Lurbinectedin is an alkylating agent that has been approved for progression after platinum-based chemotherapy after showing a 35% ORR with a 5.1-month median duration of response.[136] The camptothecin topotecan, which has both intravenous and oral formulations, is also approved for ES-SCLC that relapsed beyond 45 days based on improved tolerability compared to combination regimens.[137]

Patients who relapse beyond 6 months of completion of therapy may be reinitiated on the initial induction chemotherapy regimen. Single-agent pembrolizumab or nivolumab with or without ipilimumab may also be used in the third-line setting for patients who did not receive immunotherapy previously.[138,139]

2 Direct complications of lung cancer

2.1 Parenchymal brain metastases

2.1.1 Epidemiology

Lung cancer represents the most common tumor that spreads to the central nervous system (CNS)[140,141] accounting for 39%–56% of the brain metastases.[142,143] Reflecting differences in overall prevalence, approximately 80%–85% of all lung cancer metastases occur in the setting of NSCLC, the remaining 15%–20% in small cell lung cancer (SCLC).[142,143] At initial staging, over 20% of the patients with NSCLC and SCLC are diagnosed with brain metastases; 34% and 13%, respectively, are asymptomatic.[144,145] In SCLC, the risk for developing brain metastases within 2 years after initial diagnosis is equal or greater than 50%.[146] The utility of routine MRI scans in asymptomatic lung cancer patients remains subject to controversy. Known risk factors that predispose for CNS involvement of lung cancer include large size of the primary tumor, lymphovascular space invasion, hilar lymph node involvement, and younger age at initial diagnosis.[147] The incidence of brain metastases in patients with epidermal growth factor receptor (EGFR) mutations and those with anaplastic lymphoma kinase (ALK) rearrangement appears higher than in unselected cohorts suggesting the genetic landscape of primary tumors may be relevant for metastatic tropism to the brain.[148] Recent studies support the notion that targeted NSCLC therapy in tumors harboring somatic EGFR mutations may protect from CNS dissemination.[149] In SCLC, prophylactic whole-brain radiotherapy (WBRT) was shown to lower the incidence of brain metastases.[150]

2.1.2 Molecular pathogenesis

Recent technological advances including whole-exome, whole-genome, and RNA sequencing have led to a better understanding of the genetic landscape of brain tumors and pathways relevant to CNS dissemination. Findings supported the notion that while brain metastases and primary cancers shared a common ancestor, clinically relevant genetic alterations not found in primary tumor tissue were detected in 53% of the CNS metastasis samples.[151] The genetic signature of spatially and temporally separated CNS lesions from the same patient, however, was similar.[151] In a large whole-exome sequencing study of lung cancer brain metastases, single-nucleotide variants were commonly found in TP53 (64%), KRAS (40%), KEAP1 (29%), and EGFR (16%) genes.[152] Deletions of the tumor suppressor gene, CDKN2A/B, and amplification of oncogenes MYC, VAP1, and MMP13 were significantly more frequent in brain metastases compared with matched primary lung cancer tissue samples.[152] Independent overexpression of all three genes in patient-derived xenograft mouse models increased the incidence of metastatic lung cancer lesions, suggesting a crucial role during CNS dissemination of lung cancer. Additional single-cell sequencing studies are pending and will likely offer further insights with regard to heterogeneity and mechanisms of drug resistance in lung cancer brain metastases.

2.1.3 Clinical manifestations

Neurological symptoms of brain metastases from lung cancer reflect the localization of the cancer deposit, increased intracranial pressure, or seizures.[153] Symptoms resulting from increased intracranial pressure include headaches, nausea and vomiting, depressed sensorium, or neuropsychological deficits (impaired memory, attention, or reasoning, personality change, depression). Headaches are among the most common symptoms and are found in up to 50% of the cases.[153] They are typically worse upon awakening in the morning and aggravated by Valsalva maneuver. Formal neuropsychological testing yields abnormal results in the vast majority (75%) of the patients.[153] Common focal neurological deficits include weakness, sensory loss, gait disturbance, ataxia, dysphasia, or apraxia.[153,154] Focal seizures with or without secondary generalization are the presenting complaint in approximately 20% of the patients and occur at some point during the disease course in 30%–40% of the patients.[153,154]

2.1.4 Diagnosis

Brain metastases are commonly located at the junction of gray and white matter and of major arterial vascular territories.[155,156] The vast majority (80%) of lesions is found in the cerebral hemispheres, followed by cerebellum (11%) and basal ganglia (3%).[155] Solitary lesions are observed in approximately 25%–50% of the cases.[156,157]

Brain computed tomography (CT) is an emergency imaging modality for the rapid diagnosis of increased intracranial pressure, brain herniation, and hemorrhage.

Contrast administration is unnecessary unless there is a contraindication to MRI. Sensitivity is inferior to MRI, especially with respect to detection of small or infratentorial lesions.[158] The often necrotic masses appear hypodense in the center and hyperdense at the viable and cellular margin. Intralesional hemorrhage is seen as an area of markedly increased density.

Gadolinium-enhanced MRI represents the imaging modality of choice for the detection of brain metastases and ought to be obtained in lung cancer patients with a new neurological deficit, seizures, or in asymptomatic high-risk individuals.[155] Lung cancer metastases typically display ring-like enhancement (Fig. 1). The enhancing viable margin appears isointense to normal brain on T1 and iso- to hypointense on T2, whereas the largely necrotic center is T1-hypo- and T2-hyperintense.[157] One-third of the lesions are hemorrhagic in a susceptibility artifact on susceptibility-weighted imaging (SWI). Appearance on other sequences reflects the stage of hematoma evolution[157] Diffusion-weighted imaging (DWI) demonstrates relative restriction of water diffusion within the enhancing margin (DWI iso- to hyperintense; apparent diffusion coefficient map (ADC) iso- to hypointense) and increased diffusivity within the necrotic center. DWI findings of brain metastases from lung cancer may correlate with histological subtype and genetic landscape of the primary tumor.[159] Patients with small cell or large cell neuroendocrine carcinomas may display higher signal intensity on DWI.[159] However, despite observing a trend toward lower ADC values in SCLC, another study found no significant correlation between histology and DWI parameters.[159] Minimum ADC was

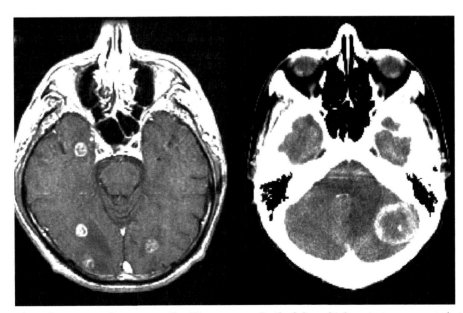

FIG. 1 Cerebral parenchymal metastases from nonsmall cell lung cancer. On the left, multiple metastases are noted at the gray-white matter junction of both hemispheres. Hypointense signal surrounding some of the lesion is consistent with vasogenic edema (T1-weighted MRI with gadolinium). On the right, a large rim-enhancing mass lesion is demonstrated within the left cerebellar hemisphere consistent with a cerebellar metastasis. The hypodensity in the pons and left cerebellar peduncle represents vasogenic edema.

found to be significantly lower in EGFR-mutant NSCLC and may hence provide a helpful diagnostic and prognostic tool.[159] On perfusion MR imaging, brain metastases stand out as areas of increased relative cerebral blood volume. Proton H^1 MR spectroscopy is most useful in the distinction of radiation necrosis and an insufficiently treated metastasis. An increase in choline/creatinine and choline to N-acetyl aspartate peak ratio is the result of higher cell membrane turnover and compromised neuronal integrity.[160] Lactate, a surrogate for anaerobic glycolysis, and lipid, a surrogate for necrosis, may be elevated, especially when tumor lesions are necrotic.

A tissue-based diagnosis of brain lesions with typical radiological appearance is not mandated in patients with known metastatic lung cancer. However, genetic studies of lung cancer brain metastases suggest that they may differ from the genetic landscape of the primary lung tumor.[152] Thus, a biopsy and molecular analysis may help to guide personalized care. A biopsy is required when no systemic cancer is known or history thereof is remote. Of note, one study reported diagnoses other than metastases in 11% of the patients with a known systemic cancer and solitary CNS lesions.[161]

Liquid biopsy represents a minimally invasive method in which several biomarkers (e.g., DNA fragments from tumor cells, exosomes) released from tumor cells into body fluids such as the bloodstream or the cerebrospinal fluid (CSF) are analyzed.[162,163] The main objectives are to identify actionable genetic alterations in order to tailor personalized treatment as well as to monitor treatment responses. Liquid biopsies were successfully implemented in clinical trials of patients with CNS dissemination of lung cancer. As part of the BLOOM trial evaluating osimertinib for leptomeningeal metastases from EGFR-mutant NSCLC, analysis of cell-free DNA identified EGFR-sensitizing mutations in all CSF specimen from the first 21 patients treated. Correlating with clinical responses, complete CSF clearance was noted in 28% of the cases.[164]

2.1.5 Treatment

Treatment options for lung cancer brain metastases include surgery, radiation therapy (stereotactic or whole-brain radiation therapy), systemic therapy (targeted therapy, immunotherapy, or cytotoxic chemotherapy), and multimodality approaches. Individual selection of therapy is guided by multiple factors including location and number of brain lesions, extent of extracranial disease, past treatment responses, performance status, patient comorbidities, age at diagnosis, and patient preference. CNS dissemination of lung cancer used to be considered a terminal complication and physicians resorted to palliative whole-brain radiation therapy (WBRT) or comfort measures. However, the recent introduction of therapies specifically targeting molecular vulnerabilities of lung

cancer cells or unleashing the cellular immune response has improved CNS disease control and patient outcome. Stereotactic approaches, further promoted by technological advances facilitating single-session therapy of multiple brain metastases, administered in combination with these new therapeutic modalities have substituted for WBRT resulting in not only improved survival but also cognitive outcome. Despite therapeutic advances, brain metastases remain a serious complication for lung cancer patients. Important prognostic factors include performance status, age, histological subtype of the primary tumor, the interval from diagnosis of the primary lung cancer to CNS dissemination, as well as location and extent of intra- and extracranial disease.[143,165] The recursive partitioning analysis (RPA) classification, which was developed based on the data from three RTOG trials including 1200 patients from 1979 to 2003, remains a useful prognostic scoring system for lung cancer brain metastases.[165]

Glucocorticoid therapy reduces vasogenic edema related to brain metastases, alleviates neurological symptoms in approximately 70% of the patients, and improves both response to further therapy and survival.[143,166] Dexamethasone has long been the steroid of choice in brain tumor patients, mainly due to its long half-life, low likelihood of inducing psychosis and its limited mineralocorticoid activity.[167] In the presence of increased intracranial pressure or herniation, it is customary to administer 10 mg intravenously. Subsequent maintenance treatment consists of daily dosages of 4–24 mg (intravenously or oral) typically divided into two to four doses. Further dose escalation only increases the risk of adverse reactions such as peptic ulcer disease. It is important to taper corticosteroids upon initiation of definitive therapy for metastatic disease in order to avoid the risks of long-term exposure (diabetes mellitus, myopathy, opportunistic infections).[168]

Antiepileptic treatment is indicated in patients with brain metastases who experience seizures. The prophylactic use of anticonvulsive medications is not generally recommended but may benefit selected patients at high risk in the pre- and early postoperative setting.[169]

Surgery

In limited intracranial metastatic disease, neurosurgical resection is considered a key treatment modality. Resection may be indicated to establish a diagnosis, to eradicate CNS metastasis or for palliation in patients with symptomatic, dominant intracranial lesions. Based on multiple clinical trials, neurosurgical resection and SRS are considered equivalent treatment options for patients with limited intracranial CNS metastases from lung cancer. Neurosurgical therapy, however, is usually preferred for lesions larger than about 1–1.5 in. in diameter or solitary cerebellar lesions to prevent hydrocephalus from brain edema following radiosurgery. Median OS following resection was 18 months in a prospective randomized trial.[170] Local disease-free recurrence rate was 43% at 12 months.[170] Surgical resection of two or three cerebral metastases was shown to prolong survival and improve quality of life to a similar extent as the removal of solitary lesions.[171] Supporting the use of postoperative radiation therapy, local disease-free recurrence rates improved with the addition of postoperative SRS (72%) or WBRT (87%), although the latter was accompanied by substantial decline in neurocognitive function.[170,172] Positive prognostic factors for local disease control and OS following surgical removal included extent of resection, postoperative radiation therapy, smaller size of the intracranial metastases, controlled systemic disease, KPS > 70, and age younger than 65 years.[161,170,172–174] Brain metastases were long considered well demarcated from surrounding brain tissue. This notion was recently challenged as a demarcated growth pattern was only found in 50% of the brain metastases.[175] Infiltrative growth of surrounding brain tissue was found in 32%, whereas distant perivascular extension was noted in 18% of the cases. In another study, infiltration of normal brain tissue was noted in 75% of the brain metastases from NSCLC beyond more than 2 mm of the boundaries of the resection cavity.[176] These observations may explain the relatively high local recurrence rates following surgical removal and the benefit from postoperative radiation therapy. Resection of a safety margin of at least 5 mm when feasible resulted in improved local disease control.[174] Whether use of 5-aminolevulinic acid (ALA), a fluorescence marker for intraoperative visualization of tumor tissue, improves local control rate remains uncertain.[177]

Re-resection of recurrent, surgically accessible brain metastases is considered when intracranial disease is limited (one to three lesions) or for palliation of symptomatic lesions. A second resection of brain metastases was previously shown to improve OS in NSCLC. Survival after re-resection was 15 months compared to 10 months for patients who did not undergo a second surgical intervention in a retrospective case series.[178]

Radiation therapy

Stereotactic radiation. Stereotactic radiosurgery (SRS) denotes treatment of a discrete target volume with a large dose of radiation (usually 12–24 Gy) in a single fraction. When treatment is delivered in several fractions, the term stereotactic radiation therapy (SRT) is used. Gamma Knife, CyberKnife, and linear accelerator (LINAC) represent commonly used delivery systems. Early and late adverse effects are locoregional and distinct from those following WBRT.[179] Early adverse effects include headaches typically within 24 h after treatment, stereotactic frame pin site infections and seizures within 48 h after SRS.[180] Radiation necrosis constitutes a key long-term adverse effect (6–18 months after SRS) with an estimated incidence between 5% and 25%. It may mandate steroid treatment, bevacizumab, or local interventions (surgery or laser interstitial thermal therapy).[181,182] The risk for radiation necrosis increases with tumor volume, volume of normal brain irradiated, dose, fraction size, and previous or concomitant therapies administered (specifically some targeted therapies or prior radiation therapy).[181] The RTOG 90-05 trial investigated the maximum safe radiation dose tolerated for single-fraction SRS and recommended selecting dosages based on the diameter of the tumor lesion.[183] Lesions ≤ 20 mm, 21–30 mm, and 31–40 mm were safely treated with 24 Gy, 18 Gy, and 15 Gy, respectively. However, it has been suggested that doses greater than 20 Gy cause increased toxicity without improving local disease control.[184] Prior radiation exposure is associated with an increased incidence of radiation necrosis following SRS. The risk of radiation necrosis at 1 year with SRS following prior SRS to the same location, after WBRT, with concomitant WBRT, or without prior radiotherapy was 20%, 4%, 8%, and 3%, respectively.[185] Prior administration of targeted (e.g., TKIs) or chemotherapy may additionally increase the risk for radiation necrosis.[186] However, another retrospective study found no significant correlation between radiation necrosis and targeted therapy. Instead, genetic alterations including EGFR and ALK alterations were identified as potential risk factors for development of radiation necrosis.[187] Extrapolating data from studies investigating radiation necrosis following irradiation of arteriovenous malformations, the risk may also depend on brain location with the frontal cortex carrying the highest and the brainstem carrying the lowest risk.[188]

In limited metastatic disease, SRS is equivalent to surgical removal of metastatic lesions from lung cancer regarding local disease control and outcome.[189] Local recurrence following SRS and surgery is similar. The risk for early (≤ 3 months) local disease recurrence may be higher following surgical removal (hazard ratio (HR) 5.94; 95% CI 1.72–20.45), whereas there may be a lower risk for late recurrence (≥ 9 months, HR 0.36, 95% CI 0.14–0.93).[189] SRS remains more relevant in the treatment of limited metastatic disease in NSCLC than SCLC given the latter histology's responsiveness to WBRT. However, the FIRE-SCLC trial recently demonstrated that SRS is both feasible and effective in SCLC.[190] Median OS (6.5 months for SRS vs 5.2 months for WBRT) and CNS progression-free survival (4 months for SRS vs 3.8 months for WBRT) were similar in both treatment arms. However, WBRT was associated with a longer time to CNS progression (HR 0.38, 95% CI 0.26–0.55). The utility of SRS in the setting of brain metastases from SCLC hence warrants further investigation, possibly including neurocognitive outcome as an additional endpoint.[190] SRS is also feasible in cases with up to 10 lesions.[191,192]

Recent trials have demonstrated a benefit for combination of surgical resection with postoperative radiation therapy administered to the resection cavity establishing it as a standard treatment for limited metastatic disease from lung cancer.[170,172] Mahajan et al. compared observation versus SRS following surgical resection of brain metastases.[170] Local tumor-free recurrence rates at 12 months significantly improved with postoperative SRS (72%, 95% CI 60%–87% for SRS versus 43%, 95% CI 31%–59% for observation). Median time to local recurrence was 7.6 months following observation only versus not reached in the SRS treatment arm. In addition to postoperative SRS, the size of metastatic lesions was identified as a significant predictor of improved local disease control. Local disease-free recurrence rates for patients with a maximum diameter of brain lesions ≤ 2.5 cm compared with lesions between 2.5 and 3.5 cm were 91% (95% CI 48%–76%) versus 40% (95% CI 27%–60%), respectively.[170] In another prospective, randomized phase III trial, Brown et al. compared postoperative SRS (single fractions of 12–20 Gy depending on the size of the lesions) with WBRT (30 Gy in 10 daily fractions or 37.5 Gy in 15 daily fractions).[172] Median OS was similar in both treatment arms (SRS: 12.2 months, 95% CI 9.7–16 versus WBRT: 11.6 months, 95% CI 9.9–18). Local disease control was slightly worse for postoperative SRS (local disease-free recurrence rate at 12 months: 61.8% with SRS versus 87.1% with WBRT). Median cognitive deterioration-free survival was significantly prolonged following SRS (3.7 months, 95% CI 3.45–5.06) when compared to postoperative WBRT (3 months, 95% CI 2.86–3.25). Cognitive deterioration at 6 months following SRS or WBRT was found in 52% and 85%, respectively.

The combination of SRS and WBRT has been explored in several prospective trials.[193–195] Local disease control was worse with SRS alone compared with SRS plus WBRT (HR 3.6, 95% CI 2.2–5.9) without an impact on median OS.[193] Cognitive deterioration at 3 months was noted in 63.5% after SRS alone and in 91.7% of the patients treated with WBRT following SRS. Cognitive decline also translated into worsening life quality (SRS plus WBRT: − 10.9 points versus SRS alone: − 1.3 points). Similarly, in another randomized controlled trial, Aoyama et al. found an improved local tumor recurrence rate at 12 months in the SRS plus WBRT treatment arm (46.8% versus 76.4%).[194] OS between both treatment arms was comparable. However, in a later published subgroup analysis SRS plus WBRT significantly prolonged OS (SRS plus WBRT: 16.7 months, 95% CI 7.5–72.9; SRS alone 10.6 months, 95% CI 7.7–15.5) in patients with a favorable prognosis (DS-GPA score 2.5–4.0), whereas no difference was observed for patients with an unfavorable prognosis (DS-GPA 0.5–2.0).[195]

Whole-brain radiotherapy. WBRT continues to be used for patients with multiple metastatic lesions (> 3 lesions). Using external beam, radiation of 30–37.5 Gy is administered over the course of 10–15 daily fractions.[196] Hyperfractionated schedules (e.g., two daily fractions) were found to have no benefit over conventional regimens. As mentioned earlier, the status of WBRT as standard of care for patients with brain metastases from SCLC has been challenged by the FIRE-SCLC trial.[190]

Common acute side effects of WBRT include nausea, vomiting, dermatitis, fatigue, temporary hair loss, headaches, and middle ear effusion. Worsening neurologic deficits including cognitive dysfunction may reflect a treatment-related increase in peritumoral edema, which can be effectively treated with glucocorticoids.[197] Late side effects include permanent hair loss, chronic middle ear effusion, fatigue, and most importantly, neurocognitive deterioration. Dementia manifests in only a small portion (< 5%) of patients. However, this finding has to be interpreted on the background of poor OS in recipients of WBRT. Radiation-induced dementia is often accompanied by urinary incontinence and gait disturbance—the classical triad of normal pressure hydrocephalus.[198]

WBRT continues to be an appropriate tool for palliation. Partial responses were found in more than 60% of the brain metastases following WBRT, and symptoms were improved in greater than 50% of the patients in prospective clinical trials.[199] The QUARTZ trial investigated the utility of WBRT in the setting of brain metastases with a poor prognosis.[200] Patients were randomized to receive either WBRT plus supportive care or best supportive care alone. Overall, no difference in quality of life (WBRT 46.4 QALYs; supportive care 41.7 QALYs), median OS (WBRT 9.2 weeks 95% CI 7.2–11.1; supportive care 8.5 weeks 95% CI 7.1–9.9), or steroid use was found between both treatment arms. In a subgroup with a favorable prognosis, specifically patients younger than 60 years, with a KPS > 70 and controlled primary tumor, WBRT treatment, however, prolonged OS. Although the QUARTZ trial challenged the use of palliative WBRT in patients with poor prognosis, WBRT remains a valid option in patients with a more favorable prognosis (NSCLC-GPA score ≥ 1.5).

Hippocampal sparing and addition of memantine may improve cognitive outcomes after WBRT.[201–203] Memantine is an N-methyl-D-aspartate receptor antagonist that blocks excessive, pathologic stimulation of the receptor and was shown to be neuroprotective. In the RTOG 0614 trial, patients were randomized to receive either placebo or memantine with WBRT.[201] Time to cognitive decline was significantly longer in the memantine arm (HR 0.78). RTOG 0933 trial found that hippocampal avoidance during WBRT was associated with a significantly lower cognitive decline at 4 months after treatment (− 7%) when compared to conventional WBRT (− 30%).[202] In a randomized controlled phase III trial WBRT with hippocampal sparing plus memantine was associated with a significantly lower risk for cognitive failure (HR 0.74, 95% CI 0.58–0.95) than conventional WBRT plus memantine.[203] Hippocampal avoidance significantly improved executive functions at 4 months, as well as learning and memory at 6 months in comparison with conventional WBRT.

Prophylactic cranial irradiation (PCI). Given the high incidence of brain metastases in NSCLC, especially in individuals with locally advanced tumor, young age at diagnosis, female gender, and adenocarcinoma, the role of *PCI* has been assessed in prospective clinical trials. With improved systemic disease control related to the discovery of first-generation TKIs with limited penetrance through the blood-brain barrier, the CNS importance as a sanctuary site in NSCLC was recognized. The design of TKIs with enhanced CNS bioavailability has mitigated this problem. In the RTOG 0214 trial, patients with stage III NSCLC were assigned to PCI or observation. PCI reduced the incidence of brain metastases (PCI 7.7% versus observation 18%) without prolonging OS. Mini mental state examination scores or quality of life was not affected by PCI. However, memory was impaired at 1 year following PCI.[204] Based on these data, PCI is not recommended for NSCLC at this time as toxicity outweighs the clinical benefit.

For limited-stage SCLC, Auperin et al. demonstrated in their metaanalysis that PCI lowered the incidence of brain metastases and improved OS.[205] Rate of survival at 3 years in the PCI treatment arm and the control group was 20.7% and 15.3%, respectively. Radiation doses correlated inversely with a reduced risk for brain metastases. Earlier administration of PCI after induction chemotherapy was also associated with a lower risk for CNS dissemination of SCLC. Since then, two prospective trials have assessed PCI in extensive-stage SCLC.[150,206] Slotman et al. showed that PCI reduced the risk for brain metastases in extensive-stage SCLC and prolonged OS.[150] However, patients in this study did not undergo routine follow-up MRI. In another randomized controlled trial, PCI was evaluated in extensive-stage SCLC patients who underwent routine follow-up MRIs.[206] The latter study found no survival differences between PCI (11.6 months, 95% CI 9.5–13.3) and observation only (13.7, 95% CI 10.2–16.4). Takahashi et al. therefore concluded that PCI is not essential in patients with extensive-stage disease SCLC when patients undergo periodic follow-up MRI examination. While PCI is recommended in limited-stage SCLC (category 1), recommendation level for extensive-stage disease is therefore lower (category 2A) and still subject to controversy.

Radiosensitizers. Radiosensitizers constitute pharmacological or chemical agents that sensitize cancer cells for the lethal effects of radiation therapy. However, in clinical studies thus far no radiosensitizer has been identified that prolonged OS in brain metastases from lung cancer without causing severe toxicity. Radiosensitizers that were previously assessed for the treatment of brain metastases include lonidamine, metronidazole, misonidazole, temozolomide, motexafin gadolinium, bromodeoxyuridine, and efaproxiral.[207–209]

Several studies have studied immune checkpoint inhibitors (ICIs) and TKIs as potential radiosensitizers for brain metastases. In preclinical models, SRS increased programmed death ligand 1 (PDL1) expression allowing the successful use of ICIs as radiosensitizers.[210] In a retrospective study, Ahmed et al. found that a combination of SRS and ICIs to treat NSCLC brain metastases was safe and well tolerated.[211] Distant brain control following SRS plus ICI at 6 months was 57% compared with 48% after SRS alone. Results for SRS treatment before and concurrently to ICI were superior to SRS performed after immunotherapy. Of note, the combination of immunotherapy and radiotherapy often causes pseudoprogression and may require multimodal imaging to evaluate their efficacy. EGFR-mutant and ALK-rearrangement-positive NSCLCs carry a higher risk for brain metastasis. In addition, signal transduction pathways downstream of EGFR and ALK are known to cause resistance to radiation therapy.[140] TKIs therefore could represent promising radiosensitizers. At the same time, disruption of the blood-brain barrier following radiation therapy may enhance CNS penetration of TKIs. In a retrospective study, SRS followed by EGFR-TKI resulted in the longest median OS (SRS plus TKI 46 months, WBRT plus TKI 30 months, TKI followed by SRS/WBRT 25 months) while avoiding neurocognitive side effects of WBRT.[212] Similar to findings regarding the most beneficial sequence of immuno- and radiotherapy, this study supports the administration of TKIs after and not before radiation. In another retrospective series, TKI monotherapy was compared with a combination of TKI and radiation therapy. Median progression-free survival (11.5 versus 16 months) and OS (15 versus 22 months) were significantly longer when TKIs were administered along with radiation therapy.[213] Additional prospective randomized clinical trials are warranted and pending to establish ICIs and TKIs as radiosensitizers for lung cancer brain metastases.

Systemic therapy

Given advancements in the CNS penetration of systemic therapies, their role in treating parenchymal brain metastases has increased. In this section, we highlight agents with high intracranial activity effective for treating brain metastases where forgoing radiation therapy with close monitoring may be considered for select patients.

Nonsmall cell lung cancer
Targeted therapy Many TKIs used for NSCLC have excellent CNS penetration and can be considered for primary management of brain metastases rather than

local therapy, particularly as a means to avoid WBRT. Because it is common for patients with advanced oncogene-driven lung cancer to live many years, avoiding WBRT is a priority in these patients whenever possible to avoid long-term toxicity.

For patients with *EGFR* mutant NSCLC, the third-generation TKI osimertinib was shown to have greater CNS penetration compared to earlier generation agents (gefitinib and erlotinib) in preclinical models.[214] When evaluated in large, randomized phase III studies, the ORR in the CNS was 90% in the frontline setting and 70% in subsequent lines.[215,216] PFS at 18 months among patients with CNS metastases was 58% compared with 40% in patients treated with first-generation TKIs. It is therefore reasonable to initiate systemic therapy with osimertinib for patients with asymptomatic or minimally symptomatic brain metastases at the time of diagnosis instead of radiation therapy. Observational studies and retrospective analyses suggest radiotherapy (SRS or WBRT) may improve outcomes over EGFR TKIs; however, these studies focused on earlier generation TKIs that have lower levels of CNS penetrance compared to osimertinib.[212,217–219] Other studies suggest that an intracranial response may be obtained with first- and second-generation EGFR TKIs in select patients.[220–222] EGFR therapy during WBRT should be managed carefully and may require dose reductions due to increased toxicity.[223]

In *ALK*-positive NSCLC, both alectinib and brigatinib have excellent CNS activity and may be used in the treatment-naïve and crizotinib-resistant settings. Ceritinib has also demonstrated intracranial activity, though with less durable responses than those observed with alectinib and brigatinib.[224] Alectinib was found to have a significantly higher intracranial response rate (81% vs 50%) and duration of response (17.3 months vs 5.5 months) compared with crizotinib in the first-line setting.[69] Brigatinib also demonstrated higher intracranial activity compared with crizotinib (78% vs 26%) and lower rates of intracranial disease progression (9% vs 19%).[225] In relapsed *ALK*-positive disease, the third-generation agent lorlatinib is also effective, demonstrating a 68% intracranial response rate and 14.5 month median duration of response.[226] The likelihood of response diminishes to 48% if previously treated with at least two prior *ALK* inhibitors.[226]

For patients with brain metastases and either a *ROS1* rearrangement or NTRK fusion, entrectinib may be used. In patients with *ROS1*-rearranged disease and brain metastases, entrectinib is preferred over crizotinib. While crizotinib does have some CNS penetrance, entrectinib showed an intracranial response in five of seven patients (71%).[80] As in *ALK*-positive disease, lorlatinib is active for patients with relapsed, *ROS1*-rearranged disease.

While data are limited for the *MET* inhibitor capmatinib and the *RET* inhibitor selpercatinib, both agents are active in the CNS and have generated promising responses. Seven of thirteen patients with brain metastases treated with capmatinib derived intracranial response, including four CRs.[227] Of 11 patients with brain metastases treated with selpercatinib, 10 achieved intracranial responses exceeding 6 months in duration.[85]

Immunotherapy Early trials investigating immunotherapy largely excluded patients with brain metastases due to concern for poor CNS penetrance and the frequent use of corticosteroids to relieve cerebral edema, which may counteract the antitumor immune effect. However, several studies have since demonstrated significant and durable intracranial responses for patients treated with ICIs.[228–233] In an open-label, phase II trial of patients with NSCLC with untreated brain metastases or intracranial progression following radiotherapy, intracranial response was observed in 29.7% of the patients with PD-L1 ≥ 1% with a median duration of response of 5.7 months. No responses were observed in five patients with PD-L1 expression < 1%. Pooled analyses of patients from four large clinical trials of patients with PD-L1-positive (TPS ≥ 1%) tumors comparing pembrolizumab to chemotherapy identified 293 patients with stable brain metastases at baseline. Regardless of whether baseline brain metastases were observed, similar improvements in PFS and OS were observed with pembrolizumab compared with chemotherapy.[234] Similar responses have also been described with nivolumab and atezolizumab.[230,232]

While ICIs have shown effectiveness for brain metastases, the prolonged time to initial response often necessitates the addition of local therapy in treating large or symptomatic lesions. Literature reporting on the concurrent use of ICIs and radiation therapy for NSCLC brain metastases is currently lacking. However, several studies on the concurrent use of ICIs and stereotactic radiosurgery (SRS) to treat melanoma brain metastases suggest similar if not enhanced local control without increasing the overall rate of adverse events.[235–237] The concurrent use of SRS and ICI does increase the risk of radiation necrosis.[238–240] To mitigate this risk, we recommend frequent radiographic surveillance following SRS and to minimize radiation doses, as able.[241]

Cytotoxic chemotherapy Given advancements in radiation techniques and the emergence of intracranially active targeted and immunotherapies, we generally do not use cytotoxic chemotherapy alone to control intracranial metastases for patients with NSCLC. Various single-agent and multidrug regimens have been tested in small series of patients with brain metastases from NSCLC. Platinum-based chemotherapy regimens have been associated with ORR in 30%–40% of the patients in the frontline setting.[242–245] Pemetrexed demonstrated intracranial response in 38.4% of the patients as a single agent.[246] Topotecan, a topoisomerase I inhibitor, freely crosses the blood-brain barrier and presents and alternative agent.[247] Several studies have also investigated the role of temozolomide as a single agent without or in combination with radiation therapy with overall mixed results.[223,248–255]

Small cell lung cancer

Unlike NSCLC, SCLC is exquisitely sensitive to cytotoxic chemotherapies. As SCLC also frequently metastasizes to the brain, standard systemic treatment regimens include agents with CNS activity regardless of whether patients have established brain metastases. The utility of chemotherapy alone (teniposide) or in combination with WBRT was previously evaluated in a phase III trial.[256] The radiographic response rate (57%: 30% CR and 27% PR) was significantly higher in the combined-therapy group than in the chemotherapy group (21%: 8% CR and 13% PR). Patients receiving teniposide alone had a significantly higher risk of failing in the brain, and median time to brain progression was longer in the combined modality group (11 vs 7 weeks). The median survival in the two groups was identical: 3.2 versus 3.5 months. While ICIs have been shown to have similar efficacy for patients with NSCLC and brain metastases, the benefit for patients with ES-SCLC and brain metastases remains to be defined. Initial subgroup analyses from the IMpower133, KEYNOTE-604, and CASPIAN phase III studies did not seem to confer the same survival benefit for patients with known brain metastases at diagnosis compared with those without.[133,134,257]

2.1.6 Prognosis

Median OS in patients with brain metastases from lung cancer remains poor. It ranges from weeks when palliative care is pursued to 17 months in the setting of clinical trials when lesions were resected followed by postoperative SRS (Table 2).[170,200]

KPS represents the main prognostic factor in patients with brain metastases from lung cancer.[143] Tumor histology does not seem to correlate with OS although some series have noted shorter OS for adenocarcinomas.[143] Patients younger than 70 years have a more favorable prognosis. Progressive systemic disease translates into worse OS. Of note, response to steroids was previously identified as a prognostic factor.[143] Median OS in patients who responded to steroids was 4.3 months in contrast to 1.6 months for nonresponders. A prognostic scoring system based on three radiation therapy oncology group (RTOG) brain metastases trials is available (Table 3).

2.2 Spinal cord metastases

Systemic cancer including lung cancer only rarely spreads to the spinal cord (8.5% of all CNS metastases). In autopsy studies on patients with systemic cancer, intramedullary spinal cord metastases (ISCM) are found in 2% of the cases.[258] However, metastases from lung cancer account for 50% of all cases, of which the majority (60%) occur in SCLC.[259] One-third of patients with ISCM has synchronous brain metastases.[259]

Upon presentation, symptoms from ISCM are usually mild but readily localizable while relatively

TABLE 2 Overall survival with various treatment options for brain metastases.

	OS (months)	Local recurrence-free rate at 12 months	Reference
Limited metastatic disease (1–3 metastases)			
Surgery alone	18	43%	Mahajan et al.[170]
Surgery + SRS	17	72%	Mahajan et al.[170]
	12.2	61.8%	Brown et al.[172]
Surgery + WBRT	11.6	87.1%	Brown et al.[172]
SRS alone	10.4	72.8%	Brown et al.[193]
SRS + WBRT	7.4	90.1%	Brown et al.[193]
Multiple metastases (≥ 3 metastases)			
WBRT	7		Sperduto et al.[223]
SRS	10.8		Yamamoto et al.[191]
Targeted therapy (TKI)	11		Zhu et al.[213]
Steroids	2		Mulvenna et al.[200]

Abbreviations: *SRS*, stereotactic radiosurgery; *WBRT*, whole-brain radiation therapy.

TABLE 3 Prognostic scoring system in brain metastases patients treated with radiation.

Recursive partitioning analysis class	Patient characteristics	Median survival (months)
I	KPS ≥ 70, age < 65 years, controlled primary cancer with no extracranial metastases	7.1
II	All others	4.2
III	KPS < 70	2.3

nonspecific.[260] Common complaints include para- or quadriparesis (93%), a truncal sensory level (78%), and neurogenic bladder dysfunction (62%). ISCM may be associated with funicular pain but radicular pain is much more common in extramedullary disease.

MRI is considered the imaging modality of choice to diagnose ISCM. Lesions typically display gadolinium enhancement and are surrounded by vasogenic edema.[261]

Given its rarity, no standard treatment regimen exists for ICSM. However, retrospective series found early initiation of focal radiotherapy was an effective treatment.[260] Radiotherapy can be combined with systemic targeted or chemotherapy to improve outcomes.[260]

Median OS in patients with ISCM is poor, particularly as ISCM mainly occurs late in the disease course when extensive systemic metastases are found, and ranges from 3 to 6 months.[259,260]

2.3 Leptomeningeal metastases

2.3.1 Incidence by tumor type

Leptomeningeal metastases (LMs) are defined as spread of tumor cells to the arachnoid and pia mater, subarachnoid space and other cerebrospinal fluid (CSF) compartments.[262] Similar to the brain parenchyma, the leptomeningeal space constitutes a sanctuary site for cancer cells as many systemic therapy agents only insufficiently penetrate the intact blood-brain barrier. Tumor cells can reach the leptomeningeal space via direct extension through the dura, hematogenously, via lymphatic spread or via perineurial or endoneurial dissemination.

In advanced-stage NSCLC, LMs are found in 3%–5% of the patients.[263] Retrospective studies suggest that LM is significantly more frequent in EGFR-mutated than in EGFR-wild-type NSCLC (9.4% versus 1.7%), which may simply reflect improved outcome following targeted therapy of the primary cancer.[264] The relative incidence of LM in ALK-rearrangement-positive NSCLC is about 5%.[262] In patients with SCLC, LMs are found in 25% of the patients after 3 years and approximately

half the patients with CNS dissemination will present with LMs.[265]

Risk factors for LMs include presence of other CNS metastases, male gender, limited response to therapy, and extensive systemic disease.[265]

2.3.2 Manifestations

Patients with LM present with diverse symptoms and signs that often reflect involvement of different levels of the neuraxis including spinal nerves, spinal cord, cranial nerves, and brain.[262]

Occurrence of headaches with nausea and vomiting may be indicative of increased cranial pressure. Signs of focal parenchymal brain dysfunction (motor or sensory deficits, aphasia, neglect, etc.) are uncommon in LM and coexistence of parenchymal brain metastases ought to be considered in those cases. Global neurocognitive deficits, however, can be found in approximately 50% of the patients. Seizures constitute the presenting complaint in 3%–12% of the patients.

Symptoms reflecting cranial nerve (CN) involvement include decreased visual acuity (CN II), diplopia (CN III, IV, VII), facial numbness (CN V), facial paresis (CN VII), hearing loss (CN VIII), dysphagia with decreased gag (CN IX, X), or tongue deviation (CN XII).[266] Spinal cord or radicular symptoms are found in approximately half of all LM cases.[266] These include radicular neck or back pain as well as segmental motor or sensory deficits. Large meningeal metastases give rise to symptoms related to spinal cord compression (see later).[266] As meningeal deposits follow gravity cauda equina syndrome with foot drop, perineal numbness, and neurogenic bladder dysfunction is commonly encountered. The initial presentation of LM may be entirely nonspecific, and thus, high clinical suspicion and vigilance are required for timely diagnosis.[262]

2.3.3 Diagnosis

Radiologic findings

Gadolinium-enhanced MRI of brain and spine constitutes modality of choice for the evaluation of LM. Enhancement of leptomeninges of brain or spinal cord, nerve roots, or cranial nerves may indicate LM on MRI and can be easily distinguished from the dural enhancement pattern accompanying metastatic involvement of the overlying bone or transient intracranial hypotension after a lumbar puncture. Enhancement may be focal or diffuse and appear nodular or linear (Fig. 2).[267] However, sensitivity and specificity of MRI with gadolinium for detection of LM in lung cancer is only 70%–85% and 75%–90%, respectively.[263,268] High-dose (0.3 mmol/kg) gadolinium-enhanced MRI may improve the diagnostic yield over standard-dose (0.1 mmol/kg) protocol.[269] Communicating hydrocephalus is another radiologic correlation of LM. In patients with typical clinical pre-

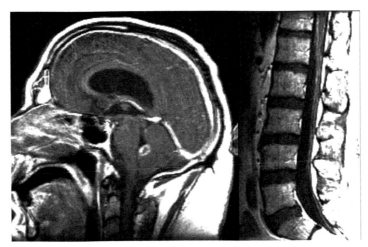

FIG. 2 Meningeal carcinomatosis from nonsmall cell lung cancer. This patient presented with excruciating headache related to obstructive hydrocephalus caused by a metastasis within the outflow of the fourth ventricle (left; axial T1-weighted MRI brain with gadolinium). Linear enhancement is noted along the nerve roots of the cauda equine indicative of coating by cancer cells.

sentation and known metastatic lung cancer, characteristic MRI findings may be sufficient to diagnose LM. In a large retrospective analysis, in which the majority of patients had lung cancer, diagnosis of LM was established with MRI alone in 35% of the cases.[268]

Lumbar puncture

Lumbar puncture with cytopathologic CSF analysis remains the gold standard for securing a diagnosis of LM. Repeated CSF sampling is often required as the sensitivity of the initial lumbar puncture is low (50%) and can be improved by obtaining a second (75%–85%) or third CSF sample.[270] To maximize diagnostic yield of the lumbar puncture, 10 mL of CSF should be obtained and processed expeditiously before cellular degradation can occur.[262] As a result of reduced CSF absorption, opening pressures are increased in 60% of the LM cases. CSF is further characterized by mild lymphocytic pleocytosis, elevated protein concentration, and hypoglycorrhachia (25%–68%), likely the results of impaired glucose transport across the blood-CSF barrier.[265] In a large retrospective series of 519 patients with LM, diagnosis was established using CSF analysis alone in 22%. In 42% of the patients, a combination of CSF and MRI findings was necessary to secure a diagnosis of LM.

Novel approaches to improve the diagnostic yield of CSF analysis are currently under investigation. Tumor marker immunostaining fluorescence in situ hybridization may be superior to conventional cytopathology regarding the detection of tumor cells in the CSF.[271] Immunomagnetic identification of circulating tumor cells within the CSF was also shown to enhance the detection of tumor cells compared to conventional cytopathology.[272] "Liquid biopsies" using CSF samples present one of the most promising new techniques. Detection of genetic alterations within the CSF may not only facilitate diagnosis but also help tailor individualized targeted therapy and detect minimal residual disease. Liquid bi-

opsies have hence already been implemented in recent clinical trials. In the BLOOM trial, which investigated osimertinib treatment for leptomeningeal metastases from EGFR-mutated NSCLC, EGFR-sensitizing mutations were found in CSF and CSF clearance correlated with clinical remission.[164]

2.3.4 Management

LM constitutes a serious complication of lung cancer and treatment aims at palliation including improvement of life quality and neurological symptoms and prolonging survival.

Treatment of LM in recent years has seen a gradual shift from radiotherapy or intrathecal chemotherapy to novel targeted, systemic agents with improved CNS penetration including TKIs and ICIs (Fig. 3). A retrospective series found that patients with LM who were treated with novel systemic therapies including TKIs, bevacizumab, pemetrexed had a significantly lower risk of death (HR 0.24) compared to patients who did not receive modern systemic agents.[273]

Based on the National Comprehensive Cancer Network (NCCN) guidelines, patients with LM from systemic cancer are divided into good (acceptable KPS, limited systemic disease, no major neurological deficits, additional systemic treatment options available) and poor risk categories (reduced KPS, extensive systemic disease, major neurological deficits, few additional treatment options, and bulky CNS disease). Aggressive systemic therapy is recommended for the good risk group, whereas best supportive care should be considered when patients fall into the poor risk category.

Systemic therapy

Targeted therapy While several retrospective studies provided good evidence for successful TKI treatment of LM, prospective randomized controlled studies

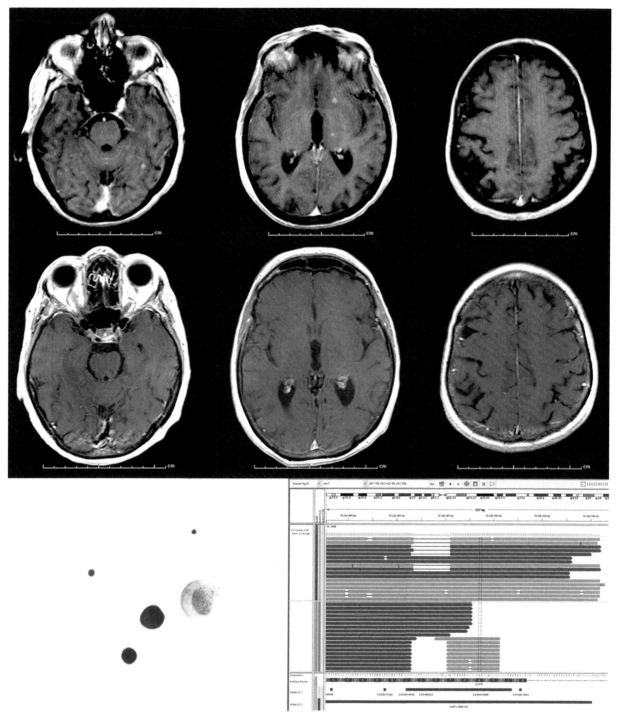

FIG. 3   Postcontrast MRI findings in a patient with leptomeningeal carcinomatosis from EGFR-mutant NSCLC before (upper row) and after (middle row) erlotinib treatment are shown together with cytopathologic and genetic CSF findings (bottom row). Linear enhancement in the foliage of the cerebellum and micronodular enhancement within Virchow-Robin spaces surrounding deep and superficial perforating arteries are observed (upper panel). Following erlotinib therapy, complete resolution of linear and nodular contrast enhancement is noted (middle panel). Cytopathological evaluation of the CSF at diagnosis of meningeal carcinomatosis revealed large atypical tumor cells next to normal lymphocytes (lower panel, left image). Exome sequencing of DNA isolated from the CSF was performed and the sequencing coverage across the EGFR locus is shown. Seven reads revealed EGFR exon 19 deletion (E746-A750; 2235-2249del15).

evaluating TKIs and other targeted therapies to treat LM from lung cancer are pending.

EGFR mutations   Li et al. found that TKI treatment of patients with LM prolonged median OS (10 months with TKIs versus 3.3 months without TKIs).[264] Combination of targeted systemic therapy with WBRT in this series did not improve OS. Performance status was identified as the key prognostic factor.

Among first-generation EGFR TKIs, erlotinib is superior to gefitinib with respect to CSF concentration (erlotinib 66.9 nM versus gefitinib 8.2 nM) and penetration (erlotinib 2.8% versus gefitinib 1.13%).[274] A retrospective study found erlotinib to be more effective in treating LM than gefitinib.[275] CSF cytological conversion rate in cases treated with erlotinib or gefitinib was 64.3% and 9.1%, respectively.[275] Several approaches have been evaluated to overcome pharmacological limitations of EGFR TKIs and treatment resistance including TKI dose escalation strategies, combination with other chemotherapy agents and development of novel EGFR-TKIs with improved CNS penetration. "Pulsed"-dose erlotinib treatment (200 or 300 mg every 2 days, 300 or 450 mg every 3 days, or 600 mg every 4 days) was compared in a retrospective study with conventional doses of erlotinib in EGFR-mutant NSCLC that had developed erlotinib resistance.[276] Although dose escalation was associated with radiographic (30%) and clinical responses (neurological improvements in 50%), median OS was not significantly different between the two groups (conventional dosing 3.1 months, pulsed dosing 2.4 months). The second-generation TKI afatinib is effective in LM after prior erlotinib or gefitinib therapy.[277] Clinical responses were noted in 27.3% with a median OS of 3.8 months. Median CSF concentration of is 2.9 nM and CSF penetration rate is 1.65%. The third-generation TKI osimertinib was found to be well tolerated and especially effective for refractory LM from EGFR-Thr790Met-mutant NSCLC.[278] Median CSF concentration and penetration rate were 8.1 nM and 2.0%, respectively.[278] In the prospective phase I BLOOM trial, patients with cytologically confirmed EGFR mutation-positive LM were treated with daily osimertinib doses of 160 mg.[164] Neuroradiologic blinded central independent review using response assessment in neuro-oncology criteria resulted in LM objective response rates of 62% (95% CI 45%–78%) and a median duration of response of 15.2 months (95% CI 7.5–17.5). Neurologic symptoms were improved in more than half (57%) of all patients treated with osimertinib. Median PFS and OS were 8.6 (95% CI 5.4–13.7 months) and 11 months (95% CI 8–18 months), respectively.[164]

AZD3759, a novel, selective EGFR TKI specifically designed to fully cross the blood-brain barrier with equal free concentrations in blood, brain tissue, and CSF (penetration rate 1:1), showed promising efficacy in a preclinical LM mouse model.[279] In a phase 1, prospective dose escalation and expansion study, the safety and efficacy of AZD3759 were assessed for the treatment of EGFR-mutant NSCLC with brain or leptomeningeal metastases.[280] In the expansion cohorts of the trial, AZD3759 was given at doses of 200 mg or 300 mg twice daily. Among 18 patients with LMs that were previously treated with other EGFR-TKIs, five patients had radiographic disease response (28%), whereas stable disease was noted in 50% of the patients.[280] Due to its excellent penetration rate and promising clinical activity, additional prospective clinical trials investigating AZD3759 for LM are warranted.

Overall several EGFR-TKIs have shown promising clinical activity against LM from lung cancer with limited toxicity. CNS-specific toxicities have not been identified for EGFR TKIs including novel agents with excellent CSF penetration rates. Prospective clinical trials are warranted to determine the optimal agent, dose, and schedule for treatment of LM.

ALK rearrangement Only very few studies have investigated the clinical activity of ALK inhibitors in LM.[262] Optimal agents, doses, and schedules remain to be determined in prospective clinical trials. Although CNS penetration of crizotinib is low, studies have shown better CNS disease control compared with conventional chemotherapy in ALK-positive NSCLC.[66,281] In two cases of LM from ALK-positive lung cancer, concurrent systemic crizotinib and intrathecal methotrexate achieved CSF clearance of tumor cells.[282] Ceritinib was found to be effective against brain and leptomeningeal metastases in a case of ALK-positive NSCLC that had failed cytotoxic chemotherapy and crizotinib.[283] Preliminary results of the phase II ASCEND-7 trial, which investigated ceritinib against brain or leptomeningeal metastases from ALK-positive lung cancer, are promising.[284] Overall intracranial response rates ranged from 27.6% (95% CI 12.7–42.2) in patients with prior ALK-inhibitor treatment to 51.5% (95% CI 33.5–69.2) in patients without prior radiotherapy or ALK-inhibitor treatment.[284] Disease control rates above 75% were achieved and median duration of response in patients without prior ALK inhibitor or radiation therapy was 7.5 months (95% CI 5.6–11.2). Gainor et al. describes four patients with LM from ALK-positive NSCLC who responded to alectinib (600 mg, twice daily; three clinical and radiographic responses, one stable disease).[285] All patients had been treated with both crizotinib and ceritinib prior to receiving alectinib. In a follow-up study, Gainor et al. showed that alectinib dose escalation (900 mg, twice daily) was able to reinduce CNS response in two ALK-positive patients with who progressed on lower alectinib doses (600 mg twice daily).[286] In two cases of LM from ALK-positive NSCLC, long-lasting clinical response greater than 6 months was achieved with brigatinib.[287] Activity of lorlatinib in LM remains to be determined.

Cytotoxic chemotherapy Systemic or intrathecal chemotherapy is appropriate for selected patients with LM from lung cancer devoid of actionable targets[263] So far, no standard of care regimen for LM has been established and the impact of novel agents such as pemetrexed and bevacizumab is subject to ongoing research.[262]

Intrathecal chemotherapy refers to the direct application of chemotherapy agents into the CSF allowing high drug concentration as the blood-brain barrier is bypassed. Chemotherapy is typically administered via an Ommaya reservoir, a dome-shaped collapsible silicone reservoir attached to an indwelling ventricular catheter, or lumbar puncture. The optimal drug and dosing schedule are subject to ongoing research. One of the major obstacles to the success of intrathecal chemotherapy is its distribution throughout the subarachnoid and ventricular compartment that is hampered by meningeal adhesions and CSF flow abnormalities commonly occurring in carcinomatosis from solid cancers. Distribution of chemotherapy is less reliable when administered through lumbar puncture. Chemotherapy agents that have been assessed for intrathecal therapy include methotrexate, thiotepa, and cytarabine.[263,288] None of the regimens appears more effective than administration of intrathecal methotrexate alone.[262,289] Single-agent intrathecal chemotherapy is preferred as combinations of multiple agents caused increased toxicity with similar efficacy.[289] When intrathecal chemotherapy is initiated, methotrexate is typically given once or twice weekly at a dose of 12 mg until neurological improvements or tumor cell clearance is achieved. The frequency of intrathecal treatment is then gradually decreased to every 2–4 weeks typically over a course of 3–6 months.[289]

In a pooled analysis of four prospective and five retrospective studies, Wu et al. found cytological, clinical, and radiographic response to intrathecal chemotherapy in 55%, 64%, and 53% of the patients, respectively.[290] Median OS following initiation of intrathecal chemotherapy was 6 months.

Immunotherapy Few studies have investigated the activity of ICIs against LM. Gion et al. reported clinical and radiographic response after nivolumab treatment of a patient with symptomatic LM.[291] The role of ICIs for treatment of LM, however, is yet to be determined in prospective clinical trials.

Radiation therapy

The goal of radiation therapy for LM in lung cancer is palliation by alleviating neurological symptoms, reducing bulky or nodular disease and correcting CSF flow.[262] Overall, there is no consensus with respect to benefit from WBRT in LM from lung cancer. One retrospective series was able to show prolonged OS for WBRT treatment LM (10.9 months for WBRT versus 2.4 months without WBRT).[292] Another retrospective study found no survival benefit.[293] There is no role for craniospinal radiation.[262]

Ventriculoperitoneal shunting

Hydrocephalus constitutes a well-known complication of LM as a result of reduced CSF absorption or obstruction of CSF flow. Normal CSF circulation can be restored with successful systemic therapy (TKIs for tu-

mors with targetable mutations) or radiation therapy. Ventriculoperitoneal shunting (VPS) may be considered in selected patients in the good risk NCCN category, for example, when symptoms of impaired CSF reabsorption persist in spite of serial lumbar punctures and response to therapy.[294] Alternatively, endoscopic third ventriculostomy may be appropriate for the rare patient with third ventricular outflow obstruction that is not reversed by otherwise successful therapy.

2.3.5 Prognosis

LM represents a serious neurological complication of advanced lung cancer and median OS remains poor. However, novel systemic/targeted therapies have translated into moderately improved outcomes. Prior to introduction of contemporary systemic therapies, median OS ranged from 1 to 3 months.[262,263] With recently approved systemic drugs, median OS ranging from 3 to 11 months can be achieved.[262,263] Performance status remains one of the best-established prognostic factors for LM.[264]

2.4 Dural-based metastases

2.4.1 Intracranial epidural space

Skull base brain metastases give rise to distinctive syndromes as a result of invasion of surrounding bones and compression or infiltration of cranial nerves or blood vessels in the middle cranial fossa, the parasellar regions, orbits, occipital condyle and jugular foramen. The detailed description of these syndromes that are by no means specific to cancer is beyond the scope of this chapter. Comprehensive reviews are available.[295]

CT scans with bone windows provide a helpful diagnostic tool to demonstrate bone erosion.[295] Disappearance of the fat pad surrounding cranial nerves within the skull base identified on precontrast T1-weighted MR images facilitates discovery of metastases.

Radiation therapy provides symptomatic relief to patients with skull base metastases.

2.4.2 Spinal epidural space

Epidural spinal cord compression

Incidence Metastatic epidural spinal cord compression (ESCC) occurs when systemic cancer spreads to spine or epidural space and as a result compresses the spinal cord.[296] It constitutes a neuro-oncologic emergency as any delay in its recognition or therapy may result in irreversible neurologic injury. The incidence of ESCC ranges from 3% in population-based studies in cancer patients to 5% in autopsy studies.[297,298] Prostrate, breast, and lung cancer each account for 15%–20% of all ESCC cases reflecting overall incidence of the primary cancer and its propensity to spread to bone.[296] ESCC in lung cancer typically results from hematogenous metastasis to the vertebral body. As a result of greatest relative blood flow and relative bone mass, ESCC most commonly occurs in the thoracic spine (60%) followed by lumbosacral spine (25%) and cervical spine (15%).[296,299] Multiple sites of ESCC are encountered in approximately one-third of all cases.[299] Paraspinal masses can invade the epidural space through the intervertebral neural foramen or by destroying anatomical barriers. Masses located in the posterior mediastinum may exploit the perivertbral venous plexus of Batson for metastatic spread into the epidural space. ESCC may be part of the initial presentation of cancer in up to 20% of the cases.[300] It typically occurs early in lung cancer patients. This is particularly true for SCLC, where 87% of ESCC manifestations occur within the first 3 months after initial diagnosis.[300]

Signs and symptoms

Pain Back pain typically represents the first and most frequent symptom caused by ESCC, occurring in 83%–95% of the cases.[296,301] ESCC is diagnosed after a median of 2 months following initial pain onset.[296] ESCC-related back pain can be either localized, radicular, or mechanical pain. Localized back pain, confined to the site of spinal metastasis, is often described as steady and aching pain, which localizes to the midline. Pathophysiologically, it results from tumor extension from bone marrow into the exquisitely pain-sensitive periosteum or surrounding soft tissue.[296]

Radicular pain is caused by compression or invasion of spinal nerve roots either by epidural tumor extension or by collapse of vertebral bodies. When cervical or lumbosacral spine are affected, radicular pain is typically unilateral, whereas bilateral pain can be found with thoracic spine involvement. Radicular pain is exacerbated by Valsalva maneuver or movement.[296] Localized and radicular pain is typically worse at night, maybe as a result of lengthening of the spine in the supine position or venous pooling within the tumor. Mechanical back pain results from pathological fractures leading to vertebral body collapse and spinal instability. It is also aggravated by movement but ameliorated by resting.[296]

Weakness Weakness represents the second most common symptom resulting from ESCC and is found in 35%–65% of the cases at diagnosis.[299,301] It often emerges weeks after pain onset. It can progress rapidly within a matter of days. Weakness manifests as progressive upper motor neuron syndromes, or segmental weakness. Upper motor neuron weakness results from invasion or compression of the corticospinal tracts. On neurologic examination, bilateral, proximal leg weakness or quadriparesis with hyperreflexia and positive pyramidal signs is a common initial finding. Lower motor neuron weakness results from cauda equina or nerve root com-

pression. Typical symptoms include asymmetrical weakness restricted to the affected myotome accompanied by hypo- or areflexia and decreased muscle tone. On occasion, in case of acute tumor hemorrhage or collapse of the infiltrated vertebral body, spinal shock with sudden onset of flaccid paraplegia and areflexia may occur with little prospect of full neurological recovery.[296]

Other symptoms Other less common symptoms caused by ESCC include sensory deficits, autonomic dysfunction, or ataxia. Sensory deficits typically occur following back pain and motor deficits. At diagnosis of ESCC, 50%–70% of the patients will present with decreased sensation or paresthesias.[296,301] Sensory deficits are found below the level of ESCC and initially are characterized by a distal distribution that ascends during the course of disease. Autonomic deficits including bowel and bladder dysfunction or impotence occur rather late during the course of ESCC and are considered a negative prognostic factor. Patients with ESCC involving the conus medullaris (T11-T12 vertebral body metastases) may present an exception with autonomic deficits often found prior to weakness. In a cohort of lung cancer patients with ESCC, 59% had bladder dysfunction usually characterized by urinary retention requiring catheterization, whereas urge incontinence was found in 12% of the cases.[299] Bowel dysfunction includes incontinence and more commonly retention. On occasion, ataxia is the presenting manifestation of ESCC when the spinocerebellar tract is affected. However, concurrent symptoms including weakness and sensory deficits typically obscure ataxia, which may be more prominent after successful treatment of ESCC.

Radiologic findings and diagnosis Back pain and weakness, the two most common symptoms found in ESCC, should raise clinical suspicion in a patient with known systemic cancer and prompt radiological evaluation to rule out ESCC.

MRI Contrast-enhanced MRI of the entire spine is the imaging modality of choice to diagnose ESCC. It allows for evaluation of the integrity of spinal column and cord as well as surrounding soft tissues. Sensitivity and specificity establishing a diagnosis of ESCC are 93% and 97%, respectively.[302] Information gathered by MRI compared with conventional CT was shown to alter radiation fields in 40% of the cases and can hence influence treatment in addition to securing the diagnosis.[303] Fat-suppressed T1-weighted images with gadolinium identify areas of epidural involvement by cancer, whereas T2-weighted sequences highlight secondary changes within the cord.[304]

Other imaging modalities Plain X-rays lack sufficient specificity and sensitivity for the diagnosis of ESCC. X-rays have a false-positive rate of 17% and erosion is only detected when more than 50% of the vertebral bone is affected.[296] Myelography with or without CT is an option when there are contraindications to MRI (e.g., metallic implants) or the patient does not tolerate MRI (e.g., due to claustrophobia or pain).[305]

Management A considerable number of patients suffers from ESCC late in the course of their disease and is treated with palliative intent. Patients with treatment options for the systemic illness and good quality of life are considered for decompressive surgery with adjuvant radiotherapy or radiotherapy alone. Whenever feasible, stabilization surgery is performed in patients with unstable metastatic disease of the spine. Otherwise, selection of surgery or radiotherapy is dependent on considerations of extent of systemic disease, performance status including ability to ambulate, prior therapies, multifocal disease, and radiosensitivity of the primary tumor. Early initiation of corticosteroid treatment reduces the risk of irreversible damage to the cord by maintaining intrinsic cord pressure below the critical level of perfusion pressure. A useful summary of treatment principles can be shown in the International Spine Oncology Consortium report.[306]

Glucocorticoids Steroids are among the first-line treatment for ESCC as they successfully reduce metastasis-associated cord edema and improve symptoms, particularly pain. In a randomized trial, Sorenson et al. showed that dexamethasone treatment improved 3- (81% versus 63% without steroids) and 6-month ambulatory rates (59% versus 33% without steroids) in patients with ESCC from solid tumors.[307] In their study, intravenous dexamethasone loading with 96 mg was followed by 3 days of oral dexamethasone (daily dose 96 mg) and a 10-day oral taper prior to initiation of radiotherapy. Other studies found a low loading dose of dexamethasone (10 mg) equally effective to a high dose (100 mg).[308] A commonly used dosing regimen is the initial administration of 10 mg intravenously followed by 4 mg four times daily. Oral and intravenous administration of the maintenance doses is considered equivalent.

Surgery Surgery for the treatment of ESCC had been controversial until clear patient selection criteria were established in a prospective, randomized controlled.[309]

Before radiotherapy became available, laminectomy, the surgical removal of the posterior portion of the spinal canal, presented the only treatment option for ESCC. Several clinical trials found no benefit for laminectomy when compared to radiotherapy, and surgical interventions were largely deferred for radiation therapy.[310,311] As ESCC is typically caused by vertebral body metastases, tumor removal and subsequent circumferential decompression via an anterior approach were then developed.

In contrast to laminectomy, which could cause destabilization of the spine, decompressive surgery allowed immediate reconstruction of the spine and provided instant stabilization.[296] In a randomized, controlled prospective trial, Patchell et al. compared decompressive surgery and postoperative radiotherapy with radiotherapy alone in patients with ESCC.[309] Compared to radiotherapy alone, decompressive surgery followed by radiation therapy resulted in a significantly higher percentage of patients who were able to walk after treatment (84% versus 57%); the ability to ambulate was retained longer (122 days versus 13 days); and median OS was prolonged (126 versus 100 days). The inclusion criteria for this trial now serve as selection criteria for surgical candidates (radioinsensitive primary tumor, displacement of spinal cord on MRI, a single site of ESCC, no paraplegia > 48 h). In addition, patients with unknown primary tumor, relapse after or progression on radiotherapy, unstable spine fractures, limited systemic disease and good KPS are considered for surgery.[296,309]

Radiation therapy Radiotherapy should be administered immediately in ESCC patients who are not deemed surgical candidates. If surgery is performed, postoperative radiation therapy is a key component of the treatment algorithm and typically started 2 weeks after surgery. Radiation therapy can be administered via conventional external beam radiation therapy (EBRT) or stereotactic body radiation techniques (stereotactic body radiation therapy (SBRT), stereotactic radiosurgery (SRS)).

Conventional EBRT still constitutes a first-line treatment for ESCC, especially in radiosensitive histologies such as SCLC, when no unstable spine metastases are found.[306] Optimum dose and schedule are subject to ongoing controversy. In the United States, 30 Gy are typically administered in 10 fractions, whereas in Europe, shorter courses or EBRT are administered. In a retrospective study, Rades et al. compared five different radiotherapy schedules: 1×8 Gy, 5×4 Gy in 1 week, 10×3 Gy in 2 weeks, 15×2.5 Gy in 3 weeks, and 20×2 Gy in 4 weeks.[312] Ambulatory rates (63%–74%) and motor improvement (26%–31%) were similar regardless of the schedule. However, local recurrence rates at 2 years were significantly lower for protracted schedules (e.g., 20×2 Gy in 4 weeks: 7% versus 1×8 Gy: 24%).[312] Short courses of EBRT are therefore currently recommended in the palliative setting (poor KPS, extensive systemic disease), whereas protracted schedules are preferred in patients with a favorable prognosis (good KPS, controlled, oligometastatic systemic disease).

SBRT and SRS are novel, focal radiotherapy modalities, which allow delivery of a more concentrated radiation dose to the tumor field (doses up to three times higher than with EBRT). They may therefore improve local disease control and expand the indication for radiotherapy to histologies with low sensitivity to conventional radiation, such as NSCLC.[306] Focal radiation also minimizes toxicity on surrounding tissue. As with EBRT, the optimum dose and schedule are yet to be determined. SBRT is most commonly administered in two to five fractions, whereas SRS constitutes a single fraction.[306]

Overall, radiotherapy may be preferred in patients with poor KPS and extensive systemic disease, radiosensitive primary tumors such as SCLC, multiple sites of ESCC, stable metastatic disease of the spine and patients who are able to ambulate at the time of diagnosis.[306]

Chemotherapy Cytotoxic chemotherapy plays a secondary role in the treatment of ESCC as responses are too slow and unreliable for management of this medical emergency. However, chemotherapy or novel targeted therapies administered in conjunction with radiotherapy may improve outcome in ESCC.[296]

Approach to recurrent disease Recurrence rates after successful treatment of ESCC range between 7% and 14% with approximately half of the recurrences occurring within the original site of ESCC.[313,314] If recurrence of ESCC is found at a different site, treatment options are identical (surgery or radiotherapy) as those for the original episode.[296] If the ESCC site of recurrence is within a previously irradiated field, options may be reduced as there is concern for radiation-induced myelopathy following reirradiation.[296] However, several retrospective series demonstrated a negligible risk of reirradiation-induced adverse effects, which in part may be owed to poor overall prognosis at recurrence of ESCC.[315]

Prognosis Both decompressive surgery and radiotherapy accomplish durable pain control in more than two-thirds of all patients with ESCC.[309,312] However, with respect to walking ability, surgery is superior as a single modality. With careful patient selection, decompressive surgery followed by radiotherapy preserves ambulatory function to a larger extent (84% versus 57%) and for a longer period of time (median 122 days versus 13 days) than radiotherapy alone. When patients were unable to walk prior to therapy, the chances of regaining this ability were significantly higher following surgery (surgery: 10/16 patients, radiotherapy 3/16 patients).[309] Of note, the vast majority of patients who were ambulatory prior to ESCC treatment retained this ability following therapy regardless of the chosen treatment modality. As ESCC often occurs in the context of extensive systemic metastatic disease, median OS following a diagnosis of ESCC remains poor and ranges from 3 to 6 months.[309,316] The most important predictor of survival is the ability to walk before and after therapy. Other prognostic factors indicating longer survival include good KPS, radiosen-

sitive tumor histology, limited systemic disease without brain or visceral metastases, single site of ESCC, and cervical location of ESCC.[296,312,316]

2.5 Plexus and peripheral nerve metastases

2.5.1 Brachial plexus

Lung cancer, particularly NSCLC in the apex of the lung, can infiltrate the brachial plexus and peripheral nerve fibers through local extension.[317] The entity of tumors invading the apical chest wall is termed superior sulcus tumors or Pancoast tumors, the latter referring to the first describer.[318] As the tumor infiltrates the brachial plexus from below, C8-T1 nerve fibers, which are the origin of the ulnar nerve, are first affected. As the tumor progresses, the medial and radial nerves follow.[317]

Manifestations

Shoulder pain is the initial symptom in most cases. Pain rapidly progresses and involves the upper arm, elbow, and forearm. As C8-T1 nerve fibers are typically infiltrated first, hypo- or paresthesias in the ulnar nerve distribution (e.g., numbness in the fourth and fifth finger) are among the first neurological deficits.[317] Following sensory deficits, weakness of muscles innervated by the ulnar nerve can be found on examination (e.g., intrinsic hand muscles) and patients may complain about difficulty grasping small items with their hand. As the tumor extends locally, it may infiltrate fibers giving rise to the medial and radial nerve.[317] Weakness in the flexors and extensors of hand and wrist as well as the extensors of the elbow may be found on neurologic examination as a result. Biceps and brachioradialis muscles are affected late during the course of superior sulcus tumors because they are innervated by nerve fibers originating from the upper brachial plexus. If tumors extend medially, invasion of the sympathetic trunk may cause a partial Horner's syndrome (miosis, ptosis, enophthalmos, and anhidrosis).[317]

Radiologic findings

MRI allows for precise assessment of the tumors relationship to nerve roots, brachial plexus, and subclavian blood vessels.[317] CT lacks the anatomical accuracy of MRI but is often sufficient for treatment planning.[317] Chest X-rays may show a mass in the lung apex possibly with invasion or destruction of surrounding bones. However, chest X-rays are not recommended for screening purposes as tumors may easily be missed.

Diagnosis

Given the location of lesions, diagnosis of superior sulcus tumor is usually secured by transthoracic needle biopsy.[317] Brachial plexus and peripheral nerve infiltration remains a radiological diagnosis, typically established by MRI. Electromyography is useful in the distinction of radiation plexopathy from locally recurrent tumor but is rarely indicated for diagnosis of the primary tumor.

Treatment

Standard of care for superior sulcus tumors includes induction radio-chemotherapy followed by radical surgical resection, which achieves 5-year survival rates of 56%.[319] However, when distant metastases, mediastinal nodal disease, brachial plexus involvement above T1, or invasion of esophagus or trachea is found, surgery is contraindicated.[317]

Infiltration of brachial plexus and peripheral nerves can effectively be treated with radiotherapy. Full neurological recovery can occur when superior sulcus tumors are diagnosed early and becomes less likely when symptoms indicative of advanced disease (e.g., weakness in the medial and radial nerve distribution on examination) are found on examination. In the context of induction radio-chemotherapy, 45 Gy are typically administered in 1.8 Gy daily fractions. If surgery is not feasible, tumors can be irradiated with 60 Gy or higher although this increases the risk of radiation injury to the brachial plexus.[317]

Prognosis

Neurological outcome in superior sulcus tumors depends on the severity of symptoms. If tumors are diagnosed early during the disease course, neurological deficits may fully disappear with successful treatment. However, with extensive brachial plexus involvement alleviation of neurological symptoms but not full recovery can be achieved in most cases.[317]

Extensive infiltration of surrounding tissues including brachial plexus, vertebral bodies or blood vessels and presence of a Horner's syndrome on examination correlate with early tumor recurrence and poor survival in superior sulcus tumors.[317,319] Additional established prognostic factors include degree of resection, lymph node status, as well as KPS.[319]

3 Indirect complications of lung cancer

Given the improved outcome of lung cancer patients following the introduction of novel treatments, indirect neurological complications of lung cancer and their management have become more relevant.

Important indirect complications include cerebrovascular accidents related to hypercoagulable state or marantic endocarditis, seizures in setting of brain metastases, metabolic disorders, nutritional deficiencies, treatment related complications such as neurocognitive decline or brain necrosis following radiotherapy or ICI-induced adverse events affecting the central or peripheral nervous system, and the rare paraneoplastic syndromes. Most of these complications are addressed in detail in other chapters of this book. We limit our discussion of indirect complications to paraneoplastic neurologic syndromes encountered in lung cancer patients.

3.1 Paraneoplastic neurologic syndromes

Paraneoplastic neurologic syndromes (PSs) are disorders that are caused by an immune attack on onconeural antigens—antigens shared by cancer cells and nervous system structures.[320–322] The immunologic response is either serologically mediated (antibodies targeting onconeural antigens) or through cytotoxic T-cells although the exact mechanisms remain unclear.[322] PSs are rare with an estimated incidence of 1:10,000 in patients with systemic cancer. However, serological screening of 60,000 cases with possible PSs detected antibodies with known association to paraneoplastic conditions in 0.9% of the cases.[323] SCLC constitutes the primary cancer most commonly found in patients with PSs. Based on retrospective series, incidence of PSs ranged between 3% and 5% in patients with SCLC.[324]

PSs may involve any level of the CNS or PNS and diagnosis often precedes that of the primary cancer. At diagnosis of PSs, primary tumors—if detectable at all—are found in their early stages. Syndromes are typically rapidly progressive and, in many cases, cause severe neurological deficits within weeks to months.[320–322] Symptoms are often irreversible due to early destruction of neuronal structures following the inflammatory process. In 60% and less than 20% of the patients with PSs affecting CNS and PNS, respectively, well-characterized antibodies can be detected.[325] Two broad categories of antibodies can be distinguished: onconeural antibodies targeting surface antigens and those targeting intracellular antigens. The detection of antibodies is useful diagnostically but does not prove their pathogenic relevance.[326] PSs with antibodies targeting intracellular antigens are presumed to be mediated by cytotoxic T-cells and commonly result in irreversible neurologic injury. PSs with neuronal cell surface antibodies are likely serologically mediated and reversible when readily diagnosed.[326]

The diagnostic approach to CNS PSs includes integration of clinical symptoms, serological findings in serum and CSF, and a cancer screening using whole-body CT or 18-flourodeoxyglucose positron emission tomography.[320] A diagnosis of definite PSs requires a classic neurologic syndrome in combination with cancer known to give rise to PSs or detection of classical paraneoplastic antibodies.[320]

3.1.1 Manifestations of paraneoplastic neurologic syndromes in lung cancer patients

Limbic encephalitis

Symptoms of paraneoplastic limbic encephalitis include subacute cognitive declines, hallucinations, seizures, and personality changes. MRI typically shows uni- or bilateral mesial temporal lobe hyperintensities on FLAIR or T2-weighted sequences (Fig. 4). Electroencephalography commonly reveals cortical hyperexcitability or epileptic activity in one or both temporal lobes.

In a retrospective series of 50 cases with paraneoplastic limbic encephalitis, the condition was most commonly associated with lung cancer (50% of the cases).[327] Limbic encephalitis was diagnosed a median of 3.5 months before cancer in 60% of the cases. Antibodies were detected in 60% of the cases with anti-Hu antibodies found in the majority of cases (60%). Hu-antibodies were almost invariably associated with SCLC (97%) and predictive of poor clinical outcome (clinical improvement in 38% versus 64% with anti-Hu-negative disease). In another retrospective study, Graus et al. assessed 200 cases with Hu-antibody-positive PSs, which were again associated with SCLC in the majority (74%) of cases.[328] Upon PS diagnosis, extensive disease SCLC was rare with extrathoracic tumors found in only 0.5% of the cases, which underlines early occurrence of PSs. Anti-Hu antibodies were found in a variety of PSs but were most frequently associated with paraneoplastic sensory neuropathy (54%), followed by multifocal paraneoplastic syndromes (11%), cerebellar ataxia (10%), and limbic encephalitis (9%). Median OS in Hu-antibody-positive PSs was 11.8 with age > 60 years, Rankin score at diagnosis > 3, multifocal paraneoplastic disease, and no treatment identified as prognostic factors in multivariate analysis.

Other antibodies found in limbic encephalitis that is frequently associated with lung cancer include onconeuronal CRMP5-antibodies and neuronal cell surface antibodies (e.g., AMPAR, GABA(B) antibodies). Patients with CRMP5-antibody-positive PSs, which are typically associated with either SCLC or thymoma, can present with sensorimotor neuropathy, uveitis, optic neuritis, cerebellar ataxia, or chorea in addition to encephalomyelitis.[326] Anti-AMPAR limbic encephalitis is characterized by prominent psychosis and two-thirds of cases are paraneoplastic with common primary tumors including SCLC, thymoma, and breast cancer.[329] Anti-GABA(B)R limbic encephalitis often presents with severe and refractory seizures and paraneoplastic etiology is found in half of all cases, which are typically associated with SCLC.[330]

Prognosis in patients with limbic encephalitis associated with antibodies targeting intracellular antigens (e.g., anti-Hu, anti-CRMP5) remains poor with immunotherapy and tumor-directed therapy rarely resulting in meaningful responses. In contrast limbic encephalitis associated with neuronal cell surface antibodies (e.g., anti-AMPAR, anti-GABA(B)R) typically respond to treatment. For example, over 50% of the patients with GABA(B)R-antibody-positive limbic encephalitis associated with SCLC respond to treatment and full recovery can be accomplished.[330]

Paraneoplastic cerebellar degeneration

Paraneoplastic cerebellar degeneration typically presents with subacute onset of limb and truncal ataxia, which may initially be asymmetrical but almost always becomes symmetric as the disease progresses.[331] Pancerebellar dysfunction results as the syndrome progresses with symptoms including dysarthria, vertigo, and nystagmus.[331] MRI on initial presentation in most

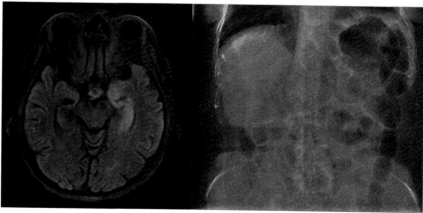

FIG. 4 Possible treatment complication and paraneoplastic syndromes in small cell lung cancer. The left image is a FLAIR sequence of an MRI brain demonstrating hyperintense signal in the left mesial temporal lobe consistent with limbic encephalitis in a patient with anti-Hu antibody-positive limbic encephalitis 4 weeks after initiation of chemoimmunotherapy including atezolizumab. The right image is an X-ray of the abdomen showing markedly dilated loops of bowel in a patient with anti-Hu-positive paraneoplastic gastrointestinal paresis.

cases is normal but occasionally shows transient cerebellar enlargement or corticomeningeal enhancement. Over time, cerebellar atrophy is found on MRI.[321]

SCLC is the most commonly found primary tumor in patients with paraneoplastic cerebellar degeneration and commonly associated with anti-Hu, anti-Ri, anti-ZIC4, anti-SOX1, or anti-VGCC antibodies.[326,332] Anti-Ri antibodies may additionally cause opsoclonus myoclonus syndrome and VGCC antibodies may be associated with concomitant LEMS.[332] The additional detection of SOX1 antibodies in cases of VGCC-positive cerebellar degeneration may be a powerful predictor of an underlying occult SCLC and hence help to differentiate between idiopathic and paraneoplastic etiology.[333] Of note, NSCLC represents the most common primary cancer in patients with paraneoplastic cerebellar degeneration without identifiable immune response.[321]

Response of paraneoplastic cerebellar degeneration to both tumor treatment and immunotherapy remains poor reflecting early irreversible cytotoxic T-cell-mediated destruction of cerebellar Purkinje cells.[326]

Opsoclonus-myoclonus

Opsoclonus refers to involuntary, multidirectional saccades with torsional, horizontal, or vertical components and may be accompanied by myoclonus of limbs or trunk and by additional cerebellar symptoms.[321] Paraneoplastic opsoclonus-myoclonus occurs in the setting of small cell lung cancer and is often associated with onconeuronal Ri-antibodies.

Subacute sensory neuropathy and dorsal root ganglionopathy

Subacute sensory neuropathy is characterized by a subacute onset of asymmetrical, multifocal burning paresthesias and pain involving lower and upper extremities as well as cranial nerves. Severe sensory ataxia mimicking cerebellar dysfunction results from an immune attack on dorsal root ganglion cells. The condition is most commonly associated with SCLC. Onconeuronal antibodies, particularly anti-Hu and anti-CRMP5, are found in 80% of the cases. Of note, clinical presentation of CRMP5-antibody-positive sensory neuronopathy differs from Hu-antibody-positive disease as predominant limb and motor involvement, and a mixed axonal and demyelinating neuropathy are more commonly found.[334]

Autonomic neuropathy

Symptoms of the rare entity of paraneoplastic autonomic neuropathy include orthostatic hypotension, neurogenic bladder, gastrointestinal dysmotility (Fig. 4), or pupillary asymmetries. The syndrome is typically associated with SCLC and Hu-antibodies.

Lambert-Eaton myasthenic syndrome (LEMS)

Patients with LEMS, a disorder of the neuromuscular junction, present with proximal muscle weakness and fatigability. In contrast to myasthenia gravis, oculobulbar muscles are less commonly affected and muscle strength will increase temporarily after maximum voluntary muscle contraction.[335]

Using electromyography, repetitive nerve stimulation at low frequencies (1–5 Hz) will reveal decrement > 10%, whereas increment > 100% is found after voluntary muscle contraction or at high frequency (50 Hz). Approximately half of LEMS cases are paraneoplastic and associated with a primary cancer, almost invariably SCLC.[335] VGCC antibodies are detected in the serum of more than 90% of the patients with LEMS and do not help to distinguish between autoimmune and paraneoplastic etiology.[335] However, additional onconeuronal SOX1 antibodies are invariably found in paraneoplastic VGCC-positive LEMS and may be a powerful predictor of an underlying occult SCLC.[333] Treatment in addition to targeted cancer and immunotherapy includes administration of 3,4-diaminopyridine, which blocks voltage-gated potassium channels, prolongs the action potential, and thereby lengthens the opening time of VGCC.

Polymyositis/dermatomyositis

Polymyositis and dermatomyositis constitute inflammatory myopathies that are characterized by limb girdle muscle weakness. In dermatomyositis, this is accompanied by the typical heliotrope rash. Paraneoplastic etiology is rare but especially encountered in dermatomyositis (20%) with SCLC among the common associated tumors. Antibodies found in (dermato-)myositis that is linked to systemic cancer include TIF1-gamma and NXP2.[336]

3.1.2 Management

Management of PSs includes cancer-directed and immunosuppressive therapy.[321,326,337] The latter may be especially relevant in cases where no systemic cancer is detected following radiological screening. Options include administration of high-dose glucocorticoids, intravenous immunoglobulins, or plasmapheresis.[321,326,337] Long-term immunosuppression can be accomplished with steroid-sparing agents such as cyclophosphamide or azathioprine.[321,337]

3.2 Treatment-related complications

3.2.1 Immunotherapy

In recent years, ICIs, especially PD-1/PD-L1 antibodies (e.g., nivolumab, pembrolizumab), have been introduced as first-line therapy for lung cancer based on the data from several prospective clinical trials.[338] As ICIs target inhibitory T-cell receptors and enhance immune responses against cancer, this has given rise to several immune-related adverse events (irAEs), which can involve any organ including the nervous system.[339]

In a metaanalysis including 9208 patients enrolled on 59 clinical trials, the incidence of neurological irAEs with anti-CTLA4 antibodies, anti-PD1 antibodies, or both was 3.8%, 6.1%, and 12%, respectively.[339] However, grade 3–4 neurological irAEs were rare (incidence < 1%) and included encephalopathy (19%), meningitis (15%), Guillain-Barré-like syndrome (7%), peripheral neuropathy (6%), and myasthenic syndrome (2%). We and others have diagnosed paraneoplastic neurologic syndromes after initiation of ICI therapy suggesting that there may be a pathogenetic link. Various ICIs were found to have similar neurological irAEs and increased drug doses did not correlate with an increased incidence. A meta-analysis of 27 case reports found a median interval of 6 weeks between administration of ICIs and onset of neurological irAEs.[339] Tumor response to ICIs was found in the majority (69%) of the cases with irAEs. Adverse events included encephalitis ($n=6$), meningoradiculitis ($n=6$), anti-AChR-positive myasthenic syndromes ($n=5$), Guillain-Barré syndrome ($n=4$), peripheral neuropathy ($n=3$), myelitis ($n=2$), and meningitis ($n=1$).[339] The majority of patients (73%) had a partial or full neurological recovery after discontinuation of ICIs and steroid treatment in most cases. Intravenous immunoglobulin therapy (IVIGs) or plasmapheresis were successfully used in 8 patients in addition to the aforementioned treatment.[339] Although discontinuation of ICIs appears to be required for recovery, a retrospective series of seven patients with checkpoint inhibition-related meningitis suggests that reintroduction of ICIs may be possible and well tolerated in selected patients.[340]

Overall, neurological irAEs occur early after initiation of ICIs. A high level of clinical suspicion for early diagnosis and rapid discontinuation of ICIs are required. Together with subsequent immunosuppressive treatment outcomes are generally favorable and sometimes even allow reintroduction of ICIs.

References

1. *World Cancer Report: Cancer Research for Cancer Prevention.* Lyon: International Agency for Research on Cancer; 2020. Available from: http://publications.iarc.fr/586.
2. Bray F, Ferlay J, Soerjomataram I, Siegel RL, Torre LA, Jemal A. Global cancer statistics 2018: GLOBOCAN estimates of incidence and mortality worldwide for 36 cancers in 185 countries. *CA Cancer J Clin.* 2018;68(6):394–424.
3. Siegel RL, Miller KD, Jemal A. Cancer statistics, 2020. *CA Cancer J Clin.* 2020;70(1):7–30.
4. NCI SEER Database. Available from: https://seer.cancer.gov/statfacts/html/lungb.html.
5. Parkin DM, Bray F, Ferlay J, Pisani P. Global cancer statistics, 2002. *CA Cancer J Clin.* 2005;55(2):74–108.
6. Alberg AJ, Samet JM. Epidemiology of lung cancer. *Chest.* 2003;123(1 Suppl):21S–49S.
7. Shields PG, Berman M, Brasky TM, et al. A review of pulmonary toxicity of electronic cigarettes in the context of smoking: a focus on inflammation. *Cancer Epidemiol Biomarkers Prev.* 2017;26(8):1175–1191.
8. Jeon J, Holford TR, Levy DT, et al. Smoking and lung cancer mortality in the United States from 2015 to 2065: a comparative modeling approach. *Ann Intern Med.* 2018;169(10):684–693.
9. Howlader N, Forjaz G, Mooradian MJ, et al. The effect of advances in lung-cancer treatment on population mortality. *N Engl J Med.* 2020;383(7):640–649.
10. Molina JR, Yang P, Cassivi SD, Schild SE, Adjei AA. Non-small cell lung cancer: epidemiology, risk factors, treatment, and survivorship. *Mayo Clin Proc.* 2008;83(5):584–594.
11. Travis WD, Brambilla E, Burke AP, Marx A, Nicholson AG. *WHO Classification of Tumours of the Lung, Pleura, Thymus, and Heart.* 7th ed. Lyon: IARC Press; 2014.
12. Detterbeck FC. The eighth edition TNM stage classification for lung cancer: what does it mean on main street? *J Thorac Cardiovasc Surg.* 2018;155(1):356–359.
13. Herbst RS, Morgensztern D, Boshoff C. The biology and management of non-small cell lung cancer. *Nature.* 2018;553(7689):446–454.
14. Goldstraw P, Chansky K, Crowley J, et al. The IASLC lung cancer staging project: proposals for revision of the TNM stage groupings in the forthcoming (eighth) edition of the TNM classification for lung cancer. *J Thorac Oncol.* 2016;11(1):39–51.
15. Petersen RP, Pham D, Burfeind WR, et al. Thoracoscopic lobectomy facilitates the delivery of chemotherapy after resection for lung cancer. *Ann Thorac Surg.* 2007;83(4):1245–1249. discussion 50.
16. Kelsey CR, Marks LB, Hollis D, et al. Local recurrence after surgery for early stage lung cancer: an 11-year experience with 975 patients. *Cancer.* 2009;115(22):5218–5227.
17. Fedor D, Johnson WR, Singhal S. Local recurrence following lung cancer surgery: incidence, risk factors, and outcomes. *Surg Oncol.* 2013;22(3):156–161.
18. Pignon JP, Tribodet H, Scagliotti GV, et al. Lung adjuvant cisplatin evaluation: a pooled analysis by the LACE Collaborative Group. *J Clin Oncol.* 2008;26(21):3552–3559.
19. Pathak R, Goldberg SB, Canavan M, et al. Association of survival with adjuvant chemotherapy among patients with early-stage non-small cell lung cancer with vs without high-risk clinicopathologic features. *JAMA Oncol.* 2020;6(11):1741–1750.
20. Waller D, Peake MD, Stephens RJ, et al. Chemotherapy for patients with non-small cell lung cancer: the surgical setting of the Big Lung Trial. *Eur J Cardiothorac Surg.* 2004;26(1):173–182.
21. Scagliotti GV, Novello S. Adjuvant therapy in completely resected non-small-cell lung cancer. *Curr Oncol Rep.* 2003;5(4):318–325.
22. Strauss GM, Herndon 2nd JE, Maddaus MA, et al. Adjuvant paclitaxel plus carboplatin compared with observation in stage IB non-small-cell lung cancer: CALGB 9633 with the Cancer and Leukemia Group B, Radiation Therapy Oncology Group, and North Central cancer Treatment Group Study Groups. *J Clin Oncol.* 2008;26(31):5043–5051.
23. Pisters KM, Evans WK, Azzoli CG, et al. Cancer Care Ontario and American Society of Clinical Oncology adjuvant chemotherapy and adjuvant radiation therapy for stages I-IIIA resectable non small-cell lung cancer guideline. *J Clin Oncol.* 2007;25(34):5506–5518.
24. *Non-Small Cell Lung Cancer;* September 15, 2020.
25. Arriagada R, Bergman B, Dunant A, et al. Cisplatin-based adjuvant chemotherapy in patients with completely resected non-small-cell lung cancer. *N Engl J Med.* 2004;350(4):351–360.

26. Winton T, Livingston R, Johnson D, et al. Vinorelbine plus cisplatin vs. observation in resected non-small-cell lung cancer. *N Engl J Med.* 2005;352(25):2589–2597.

27. Douillard JY, Rosell R, De Lena M, et al. Adjuvant vinorelbine plus cisplatin versus observation in patients with completely resected stage IB-IIIA non-small-cell lung cancer (Adjuvant Navelbine International Trialist Association [ANITA]): a randomised controlled trial. *Lancet Oncol.* 2006;7(9):719–727.

28. Senthi S, Lagerwaard FJ, Haasbeek CJ, Slotman BJ, Senan S. Patterns of disease recurrence after stereotactic ablative radiotherapy for early stage non-small-cell lung cancer: a retrospective analysis. *Lancet Oncol.* 2012;13(8):802–809.

29. Thomas M, Rube C, Hoffknecht P, et al. Effect of preoperative chemoradiation in addition to preoperative chemotherapy: a randomised trial in stage III non-small-cell lung cancer. *Lancet Oncol.* 2008;9(7):636–648.

30. Albain KS, Swann RS, Rusch VW, et al. Radiotherapy plus chemotherapy with or without surgical resection for stage III non-small-cell lung cancer: a phase III randomised controlled trial. *Lancet.* 2009;374(9687):379–386.

31. Eberhardt WE, Pottgen C, Gauler TC, et al. Phase III study of surgery versus definitive concurrent chemoradiotherapy boost in patients with resectable stage IIIA(N2) and selected IIIB non-small-cell lung cancer after induction chemotherapy and concurrent chemoradiotherapy (ESPATUE). *J Clin Oncol.* 2015;33(35):4194–4201.

32. van Meerbeeck JP, Kramer GW, Van Schil PE, et al. Randomized controlled trial of resection versus radiotherapy after induction chemotherapy in stage IIIA-N2 non-small-cell lung cancer. *J Natl Cancer Inst.* 2007;99(6):442–450.

33. Liang J, Bi N, Wu S, et al. Etoposide and cisplatin versus paclitaxel and carboplatin with concurrent thoracic radiotherapy in unresectable stage III non-small cell lung cancer: a multicenter randomized phase III trial. *Ann Oncol.* 2017;28(4):777–783.

34. Senan S, Brade A, Wang LH, et al. PROCLAIM: randomized phase III trial of pemetrexed-cisplatin or etoposide-cisplatin plus thoracic radiation therapy followed by consolidation chemotherapy in locally advanced nonsquamous non-small-cell lung cancer. *J Clin Oncol.* 2016;34(9):953–962.

35. Antonia SJ, Villegas A, Daniel D, et al. Durvalumab after chemoradiotherapy in stage III non-small-cell lung cancer. *N Engl J Med.* 2017;377(20):1919–1929.

36. Antonia SJ, Ozguroglu M. Durvalumab in stage III non-small-cell lung cancer. *N Engl J Med.* 2018;378(9):869–870.

37. Gray JE, Villegas A, Daniel D, et al. Three-year OS with durvalumab after chemoradiotherapy in stage III NSCLC-update from PACIFIC. *J Thorac Oncol.* 2020;15(2):288–293.

38. Kris MG, Johnson BE, Berry LD, et al. Using multiplexed assays of oncogenic drivers in lung cancers to select targeted drugs. *JAMA.* 2014;311(19):1998–2006.

39. Targeted Therapy: Lungevity. Available from: https://lungevity.org/for-patients-caregivers/lung-cancer-101/treatment-options/targeted-therapy.

40. Lynch TJ, Bell DW, Sordella R, et al. Activating mutations in the epidermal growth factor receptor underlying responsiveness of non-small-cell lung cancer to gefitinib. *N Engl J Med.* 2004;350(21):2129–2139.

41. Paez JG, Janne PA, Lee JC, et al. EGFR mutations in lung cancer: correlation with clinical response to gefitinib therapy. *Science.* 2004;304(5676):1497–1500.

42. Mok TS, Wu YL, Thongprasert S, et al. Gefitinib or carboplatin-paclitaxel in pulmonary adenocarcinoma. *N Engl J Med.* 2009;361(10):947–957.

43. Maemondo M, Inoue A, Kobayashi K, et al. Gefitinib or chemotherapy for non-small-cell lung cancer with mutated EGFR. *N Engl J Med.* 2010;362(25):2380–2388.

44. Mitsudomi T, Morita S, Yatabe Y, et al. Gefitinib versus cisplatin plus docetaxel in patients with non-small-cell lung cancer harbouring mutations of the epidermal growth factor receptor (WJTOG3405): an open label, randomised phase 3 trial. *Lancet Oncol.* 2010;11(2):121–128.

45. Zhou C, Wu YL, Chen G, et al. Erlotinib versus chemotherapy as first-line treatment for patients with advanced EGFR mutation-positive non-small-cell lung cancer (OPTIMAL, CTONG-0802): a multicentre, open-label, randomised, phase 3 study. *Lancet Oncol.* 2011;12(8):735–742.

46. Rosell R, Carcereny E, Gervais R, et al. Erlotinib versus standard chemotherapy as first-line treatment for European patients with advanced EGFR mutation-positive non-small-cell lung cancer (EURTAC): a multicentre, open-label, randomised phase 3 trial. *Lancet Oncol.* 2012;13(3):239–246.

47. Wu YL, Zhou C, Liam CK, et al. First-line erlotinib versus gemcitabine/cisplatin in patients with advanced EGFR mutation-positive non-small-cell lung cancer: analyses from the phase III, randomized, open-label, ENSURE study. *Ann Oncol.* 2015;26(9):1883–1889.

48. Sequist LV, Yang JC, Yamamoto N, et al. Phase III study of afatinib or cisplatin plus pemetrexed in patients with metastatic lung adenocarcinoma with EGFR mutations. *J Clin Oncol.* 2013;31(27):3327–3334.

49. Yang JC, Hirsh V, Schuler M, et al. Symptom control and quality of life in LUX-Lung 3: a phase III study of afatinib or cisplatin/pemetrexed in patients with advanced lung adenocarcinoma with EGFR mutations. *J Clin Oncol.* 2013;31(27):3342–3350.

50. Wu YL, Zhou C, Hu CP, et al. Afatinib versus cisplatin plus gemcitabine for first-line treatment of Asian patients with advanced non-small-cell lung cancer harbouring EGFR mutations (LUX-Lung 6): an open-label, randomised phase 3 trial. *Lancet Oncol.* 2014;15(2):213–222.

51. Yang JC, Sequist LV, Geater SL, et al. Clinical activity of afatinib in patients with advanced non-small-cell lung cancer harbouring uncommon EGFR mutations: a combined post-hoc analysis of LUX-Lung 2, LUX-Lung 3, and LUX-Lung 6. *Lancet Oncol.* 2015;16(7):830–838.

52. Paz-Ares L, Tan EH, O'Byrne K, et al. Afatinib versus gefitinib in patients with EGFR mutation-positive advanced non-small-cell lung cancer: OS data from the phase IIb LUX-Lung 7 trial. *Ann Oncol.* 2017;28(2):270–277.

53. Mok TS, Cheng Y, Zhou X, et al. Improvement in OS in a randomized study that compared dacomitinib with gefitinib in patients with advanced non-small-cell lung cancer and EGFR-activating mutations. *J Clin Oncol.* 2018;36(22):2244–2250.

54. Wu YL, Mok TS. Dacomitinib in NSCLC: a positive trial with little clinical impact—authors' reply. *Lancet Oncol.* 2018;19(1):e5.

55. Park K, Tan EH, O'Byrne K, et al. Afatinib versus gefitinib as first-line treatment of patients with EGFR mutation-positive non-small-cell lung cancer (LUX-Lung 7): a phase 2B, open-label, randomised controlled trial. *Lancet Oncol.* 2016;17(5):577–589.

56. Kobayashi S, Boggon TJ, Dayaram T, et al. EGFR mutation and resistance of non-small-cell lung cancer to gefitinib. *N Engl J Med.* 2005;352(8):786–792.

57. Sequist LV, Waltman BA, Dias-Santagata D, et al. Genotypic and histological evolution of lung cancers acquiring resistance to EGFR inhibitors. *Sci Transl Med.* 2011;3(75):75ra26.

58. Mok TS, Wu YL, Ahn MJ, et al. Osimertinib or platinum-pemetrexed in EGFR T790M-positive lung cancer. *N Engl J Med.* 2017;376(7):629–640.

59. Mok TS, Wu YL, Papadimitrakopoulou VA. Osimertinib in EGFR T790M-positive lung cancer. *N Engl J Med.* 2017;376(20):1993–1994.

60. Soria JC, Ohe Y, Vansteenkiste J, et al. Osimertinib in untreated EGFR-mutated advanced non-small-cell lung cancer. *N Engl J Med.* 2018;378(2):113–125.

61. Ramalingam SS, Vansteenkiste J, Planchard D, et al. OS with osimertinib in untreated, EGFR-mutated advanced NSCLC. *N Engl J Med.* 2020;382(1):41–50.

62. Bean J, Brennan C, Shih JY, et al. MET amplification occurs with or without T790M mutations in EGFR mutant lung tumors with acquired resistance to gefitinib or erlotinib. *Proc Natl Acad Sci U S A.* 2007;104(52):20932–20937.

63. Wolf J, Seto T, Han JY, et al. Capmatinib in MET exon 14-mutated or MET-amplified non-small-cell lung cancer. *N Engl J Med.* 2020;383(10):944–957.

64. Drilon A, Clark JW, Weiss J, et al. Antitumor activity of crizotinib in lung cancers harboring a MET exon 14 alteration. *Nat Med.* 2020;26(1):47–51.

65. Shaw AT, Kim DW, Nakagawa K, et al. Crizotinib versus chemotherapy in advanced ALK-positive lung cancer. *N Engl J Med.* 2013;368(25):2385–2394.

66. Solomon BJ, Mok T, Kim DW, et al. First-line crizotinib versus chemotherapy in ALK-positive lung cancer. *N Engl J Med.* 2014;371(23):2167–2177.

67. Solomon BJ, Kim DW, Wu YL, et al. Final OS analysis from a study comparing first-line crizotinib versus chemotherapy in ALK-mutation-positive non-small-cell lung cancer. *J Clin Oncol.* 2018;36(22):2251–2258.

68. Hida T, Nokihara H, Kondo M, et al. Alectinib versus crizotinib in patients with ALK-positive non-small-cell lung cancer (J-ALEX): an open-label, randomised phase 3 trial. *Lancet.* 2017;390(10089):29–39.

69. Peters S, Camidge DR, Shaw AT, et al. Alectinib versus crizotinib in untreated ALK-positive non-small-cell lung cancer. *N Engl J Med.* 2017;377(9):829–838.

70. Camidge DR, Kim HR, Ahn MJ, et al. Brigatinib versus crizotinib in ALK-positive non-small-cell lung cancer. *N Engl J Med.* 2018;379(21):2027–2039.

71. Camidge DR, Dziadziuszko R, Peters S, et al. Updated efficacy and safety data and impact of the EML4-ALK fusion variant on the efficacy of alectinib in untreated ALK-positive advanced non-small cell lung cancer in the global phase III ALEX study. *J Thorac Oncol.* 2019;14(7):1233–1243.

72. Camidge DR, Kim HR, Ahn MJ, et al. Brigatinib versus crizotinib in advanced ALK inhibitor-naive ALK-positive non-small cell lung cancer: second interim analysis of the phase III ALTA-1L trial. *J Clin Oncol.* 2020;JCO2000505.

73. Mok T, Camidge DR, Gadgeel SM, et al. Updated OS and final progression-free survival data for patients with treatment-naive advanced ALK-positive non-small-cell lung cancer in the ALEX study. *Ann Oncol.* 2020;31(8):1056–1064.

74. Soria JC, Tan DSW, Chiari R, et al. First-line ceritinib versus platinum-based chemotherapy in advanced ALK-rearranged non-small-cell lung cancer (ASCEND-4): a randomised, open-label, phase 3 study. *Lancet.* 2017;389(10072):917–929.

75. Solomon BJ, Besse B, Bauer TM, et al. Lorlatinib in patients with ALK-positive non-small-cell lung cancer: results from a global phase 2 study. *Lancet Oncol.* 2018;19(12):1654–1667.

76. Lim SM, Kim HR, Lee JS, et al. Open-label, multicenter, phase II study of ceritinib in patients with non-small-cell lung cancer harboring ROS1 rearrangement. *J Clin Oncol.* 2017;35(23):2613–2618.

77. Shaw AT, Ou SH, Bang YJ, et al. Crizotinib in ROS1-rearranged non-small-cell lung cancer. *N Engl J Med.* 2014;371(21):1963–1971.

78. Shaw AT, Riely GJ, Bang YJ, et al. Crizotinib in ROS1-rearranged advanced non-small-cell lung cancer (NSCLC): updated results, including OS, from PROFILE 1001. *Ann Oncol.* 2019;30(7):1121–1126.

79. Shaw AT, Solomon BJ, Chiari R, et al. Lorlatinib in advanced ROS1-positive non-small-cell lung cancer: a multicentre, open-label, single-arm, phase 1-2 trial. *Lancet Oncol.* 2019;20(12):1691–1701.

80. Drilon A, Siena S, Dziadziuszko R, et al. Entrectinib in ROS1 fusion-positive non-small-cell lung cancer: integrated analysis of three phase 1-2 trials. *Lancet Oncol.* 2020;21(2):261–270.

81. Doebele RC, Drilon A, Paz-Ares L, et al. Entrectinib in patients with advanced or metastatic NTRK fusion-positive solid tumours: integrated analysis of three phase 1-2 trials. *Lancet Oncol.* 2020;21(2):271–282.

82. Drilon A, Laetsch TW, Kummar S, et al. Efficacy of larotrectinib in TRK fusion-positive cancers in adults and children. *N Engl J Med.* 2018;378(8):731–739.

83. Hong DS, Bauer TM, Lee JJ, et al. Larotrectinib in adult patients with solid tumours: a multi-centre, open-label, phase I dose-escalation study. *Ann Oncol.* 2019;30(2):325–331.

84. Hong DS, DuBois SG, Kummar S, et al. Larotrectinib in patients with TRK fusion-positive solid tumours: a pooled analysis of three phase 1/2 clinical trials. *Lancet Oncol.* 2020;21(4):531–540.

85. Drilon A, Oxnard GR, Tan DSW, et al. Efficacy of selpercatinib in RET fusion-positive non-small-cell lung cancer. *N Engl J Med.* 2020;383(9):813–824.

86. Planchard D, Besse B, Groen HJM, et al. Dabrafenib plus trametinib in patients with previously treated BRAF(V600E)-mutant metastatic non-small cell lung cancer: an open-label, multicentre phase 2 trial. *Lancet Oncol.* 2016;17(7):984–993.

87. Planchard D, Kim TM, Mazieres J, et al. Dabrafenib in patients with BRAF(V600E)-positive advanced non-small-cell lung cancer: a single-arm, multicentre, open-label, phase 2 trial. *Lancet Oncol.* 2016;17(5):642–650.

88. Planchard D, Smit EF, Groen HJM, et al. Dabrafenib plus trametinib in patients with previously untreated BRAF(V600E)-mutant metastatic non-small-cell lung cancer: an open-label, phase 2 trial. *Lancet Oncol.* 2017;18(10):1307–1316.

89. Ishida Y, Agata Y, Shibahara K, Honjo T. Induced expression of PD-1, a novel member of the immunoglobulin gene superfamily, upon programmed cell death. *EMBO J.* 1992;11(11):3887–3895.

90. Leach DR, Krummel MF, Allison JP. Enhancement of antitumor immunity by CTLA-4 blockade. *Science.* 1996;271(5256):1734–1736.

91. Garon EB, Rizvi NA, Hui R, et al. Pembrolizumab for the treatment of non-small-cell lung cancer. *N Engl J Med.* 2015;372(21):2018–2028.

92. Herbst RS, Baas P, Kim DW, et al. Pembrolizumab versus docetaxel for previously treated, PD-L1-positive, advanced non-small-cell lung cancer (KEYNOTE-010): a randomised controlled trial. *Lancet.* 2016;387(10027):1540–1550.

93. Reck M, Rodriguez-Abreu D, Robinson AG, et al. Pembrolizumab versus chemotherapy for PD-L1-positive non-small-cell lung cancer. *N Engl J Med.* 2016;375(19):1823–1833.

94. Reck M, Rodriguez-Abreu D, Robinson AG, et al. Updated analysis of KEYNOTE-024: pembrolizumab versus platinum-based chemotherapy for advanced non-small-cell lung cancer with PD-L1 tumor proportion score of 50% or greater. *J Clin Oncol.* 2019;37(7):537–546.

95. Herbst RS, Giaccone G, de Marinis F, et al. Atezolizumab for first-line treatment of PD-L1-selected patients with NSCLC. *N Engl J Med.* 2020;383(14):1328–1339.

96. Gadgeel S, Rodriguez-Abreu D, Speranza G, et al. Updated analysis from KEYNOTE-189: pembrolizumab or placebo plus pemetrexed and platinum for previously untreated metastatic nonsquamous non-small-cell lung cancer. *J Clin Oncol.* 2020;38(14):1505–1517.

97. Gandhi L, Rodriguez-Abreu D, Gadgeel S, et al. Pembrolizumab plus chemotherapy in metastatic non-small-cell lung cancer. *N Engl J Med.* 2018;378(22):2078–2092.

98. West H, McCleod M, Hussein M, et al. Atezolizumab in combination with carboplatin plus nab-paclitaxel chemotherapy compared with chemotherapy alone as first-line treatment for metastatic non-squamous non-small-cell lung cancer (IMpower130): a multicentre, randomised, open-label, phase 3 trial. *Lancet Oncol.* 2019;20(7):924–937.

99. Reck M, Mok TSK, Nishio M, et al. Atezolizumab plus bevacizumab and chemotherapy in non-small-cell lung cancer (IMpower150): key subgroup analyses of patients with EGFR mutations or baseline liver metastases in a randomised, open-label phase 3 trial. *Lancet Respir Med.* 2019;7(5):387–401.

100. Socinski MA, Jotte RM, Cappuzzo F, et al. Atezolizumab for first-line treatment of metastatic nonsquamous NSCLC. *N Engl J Med.* 2018;378(24):2288–2301.

101. Papadimitrakopoulou V, Cobo M, Bordoni R, et al. OA05.07 IMpower132: PFS and safety results with 1L atezolizumab + carboplatin/cisplatin + pemetrexed in stage IV non-squamous NSCLC. *J Thorac Oncol.* 2018;13(10):S332–S333.

102. Paz-Ares L, Luft A, Vicente D, et al. Pembrolizumab plus chemotherapy for squamous non-small-cell lung cancer. *N Engl J Med.* 2018;379(21):2040–2051.

103. Mok TSK, Wu YL, Kudaba I, et al. Pembrolizumab versus chemotherapy for previously untreated, PD-L1-expressing, locally advanced or metastatic non-small-cell lung cancer (KEYNOTE-042): a randomised, open-label, controlled, phase 3 trial. *Lancet.* 2019;393(10183):1819–1830.

104. Hellmann MD, Paz-Ares L, Bernabe Caro R, et al. Nivolumab plus Ipilimumab in advanced non-small-cell lung cancer. *N Engl J Med.* 2019;381(21):2020–2031.

105. Borghaei H, Paz-Ares L, Horn L, et al. Nivolumab versus docetaxel in advanced nonsquamous non-small-cell lung cancer. *N Engl J Med.* 2015;373(17):1627–1639.

106. Brahmer J, Reckamp KL, Baas P, et al. Nivolumab versus docetaxel in advanced squamous-cell non-small-cell lung cancer. *N Engl J Med.* 2015;373(2):123–135.

107. Horn L, Spigel DR, Vokes EE, et al. Nivolumab versus docetaxel in previously treated patients with advanced non-small-cell lung cancer: two-year outcomes from two randomized, open-label, phase III trials (CheckMate 017 and CheckMate 057). *J Clin Oncol.* 2017;35(35):3924–3933.

108. Rittmeyer A, Barlesi F, Waterkamp D, et al. Atezolizumab versus docetaxel in patients with previously treated non-small-cell lung cancer (OAK): a phase 3, open-label, multicentre randomised controlled trial. *Lancet.* 2017;389(10066):255–265.

109. Herbst RS, Garon EB, Kim DW, et al. Long-term outcomes and retreatment among patients with previously treated, programmed death-ligand 1-positive, advanced non–small-cell lung cancer in the KEYNOTE-010 study. *J Clin Oncol.* 2020;38(14):1580–1590.

110. Naidoo J, Page DB, Li BT, et al. Toxicities of the anti-PD-1 and anti-PD-L1 immune checkpoint antibodies. *Ann Oncol.* 2015;26(12):2375–2391.

111. Puzanov I, Diab A, Abdallah K, et al. Managing toxicities associated with immune checkpoint inhibitors: consensus recommendations from the Society for Immunotherapy of Cancer (SITC) Toxicity Management Working Group. *J Immunother Cancer.* 2017;5(1):95.

112. Barroso-Sousa R, Barry WT, Garrido-Castro AC, et al. Incidence of endocrine dysfunction following the use of different immune checkpoint inhibitor regimens: a systematic review and meta-analysis. *JAMA Oncol.* 2018;4(2):173–182.

113. Dubey D, David WS, Reynolds KL, et al. Severe neurological toxicity of immune checkpoint inhibitors: growing spectrum. *Ann Neurol.* 2020;87(5):659–669.

114. Maur M, Tomasello C, Frassoldati A, Dieci MV, Barbieri E, Conte P. Posterior reversible encephalopathy syndrome during ipilimumab therapy for malignant melanoma. *J Clin Oncol.* 2012;30(6):e76–e78.

115. Liao B, Shroff S, Kamiya-Matsuoka C, Tummala S. Atypical neurological complications of ipilimumab therapy in patients with metastatic melanoma. *Neuro Oncol.* 2014;16(4):589–593.

116. Safa H, Johnson DH, Trinh VA, et al. Immune checkpoint inhibitor related myasthenia gravis: single center experience and systematic review of the literature. *J Immunother Cancer.* 2019;7(1):319.

117. Brahmer JR, Lacchetti C, Schneider BJ, et al. Management of immune-related adverse events in patients treated with immune checkpoint inhibitor therapy: American Society of Clinical Oncology clinical practice guideline. *J Clin Oncol.* 2018;36(17):1714–1768.

118. Rossi A, Chiodini P, Sun JM, et al. Six versus fewer planned cycles of first-line platinum-based chemotherapy for non-small-cell lung cancer: a systematic review and meta-analysis of individual patient data. *Lancet Oncol.* 2014;15(11):1254–1262.

119. Soon YY, Stockler MR, Askie LM, Boyer MJ. Duration of chemotherapy for advanced non-small-cell lung cancer: a systematic review and meta-analysis of randomized trials. *J Clin Oncol.* 2009;27(20):3277–3283.

120. Rajeswaran A, Trojan A, Burnand B, Giannelli M. Efficacy and side effects of cisplatin- and carboplatin-based doublet chemotherapeutic regimens versus non-platinum-based doublet chemotherapeutic regimens as first line treatment of metastatic non-small cell lung carcinoma: a systematic review of randomized controlled trials. *Lung Cancer.* 2008;59(1):1–11.

121. Syrigos KN, Vansteenkiste J, Parikh P, et al. Prognostic and predictive factors in a randomized phase III trial comparing cisplatin-pemetrexed versus cisplatin-gemcitabine in advanced non-small-cell lung cancer. *Ann Oncol.* 2010;21(3):556–561.

122. Scagliotti GV, Parikh P, von Pawel J, et al. Phase III study comparing cisplatin plus gemcitabine with cisplatin plus pemetrexed in chemotherapy-naive patients with advanced-stage non-small-cell lung cancer. *J Clin Oncol.* 2008;26(21):3543–3551.

123. Ciuleanu T, Brodowicz T, Zielinski C, et al. Maintenance pemetrexed plus best supportive care versus placebo plus best supportive care for non-small-cell lung cancer: a randomised, double-blind, phase 3 study. *Lancet.* 2009;374(9699):1432–1440.

124. Paz-Ares L, de Marinis F, Dediu M, et al. Maintenance therapy with pemetrexed plus best supportive care versus placebo plus best supportive care after induction therapy with pemetrexed plus cisplatin for advanced non-squamous non-small-cell lung cancer (PARAMOUNT): a double-blind, phase 3, randomised controlled trial. *Lancet Oncol.* 2012;13(3):247–255.

125. Garon EB, Ciuleanu TE, Arrieta O, et al. Ramucirumab plus docetaxel versus placebo plus docetaxel for second-line treatment of stage IV non-small-cell lung cancer after disease progression on platinum-based therapy (REVEL): a multicentre, double-blind, randomised phase 3 trial. *Lancet.* 2014;384(9944):665–673.

126. Paz-Ares LG, Perol M, Ciuleanu TE, et al. Treatment outcomes by histology in REVEL: a randomized phase III trial of Ramucirumab plus docetaxel for advanced non-small cell lung cancer. *Lung Cancer.* 2017;112:126–133.

127. *Small Cell Lung Cancer*; September 15, 2020.

128. Fukuoka M, Furuse K, Saijo N, et al. Randomized trial of cyclophosphamide, doxorubicin, and vincristine versus cisplatin and etoposide versus alternation of these regimens in small-cell lung cancer. *J Natl Cancer Inst.* 1991;83(12):855–861.

129. Evans WK, Shepherd FA, Feld R, Osoba D, Dang P, Deboer G. VP-16 and cisplatin as first-line therapy for small-cell lung cancer. *J Clin Oncol.* 1985;3(11):1471–1477.

130. Meert AP, Paesmans M, Berghmans T, et al. Prophylactic cranial irradiation in small cell lung cancer: a systematic review of the literature with meta-analysis. *BMC Cancer.* 2001;1:5.

131. Faivre-Finn C, Snee M, Ashcroft L, et al. Concurrent once-daily versus twice-daily chemoradiotherapy in patients with limited-stage small-cell lung cancer (CONVERT): an open-label, phase 3, randomised, superiority trial. *Lancet Oncol.* 2017;18(8):1116–1125.

132. Auperin A, Arriagada R, Pignon JP, et al. Prophylactic cranial irradiation for patients with small-cell lung cancer in complete remission. Prophylactic Cranial Irradiation Overview Collaborative Group. *N Engl J Med.* 1999;341(7):476–484.

133. Horn L, Mansfield AS, Szczesna A, et al. First-line atezolizumab plus chemotherapy in extensive-stage small-cell lung cancer. *N Engl J Med.* 2018;379(23):2220–2229.

134. Paz-Ares L, Dvorkin M, Chen Y, et al. Durvalumab plus platinum-etoposide versus platinum-etoposide in first-line treatment of extensive-stage small-cell lung cancer (CASPIAN): a randomised, controlled, open-label, phase 3 trial. *Lancet.* 2019;394(10212):1929–1939.

135. Paz-Ares LG, Dvorkin M, Chen Y, et al. Durvalumab ± tremelimumab + platinum-etoposide in first-line extensive-stage SCLC (ES-SCLC): updated results from the phase III CASPIAN study. *J Clin Oncol.* 2020;38(15_suppl):9002.

136. Trigo J, Subbiah V, Besse B, et al. Lurbinectedin as second-line treatment for patients with small-cell lung cancer: a single-arm, open-label, phase 2 basket trial. *Lancet Oncol.* 2020;21(5):645–654.

137. von Pawel J, Schiller JH, Shepherd FA, et al. Topotecan versus cyclophosphamide, doxorubicin, and vincristine for the treatment of recurrent small-cell lung cancer. *J Clin Oncol.* 1999;17(2):658–667.

138. Ott PA, Elez E, Hiret S, et al. Pembrolizumab in patients with extensive-stage small-cell lung cancer: results from the phase Ib KEYNOTE-028 study. *J Clin Oncol.* 2017;35(34):3823–3829.

139. Ready NE, Ott PA, Hellmann MD, et al. Nivolumab monotherapy and nivolumab plus Ipilimumab in recurrent small cell lung cancer: results from the CheckMate 032 randomized cohort. *J Thorac Oncol.* 2020;15(3):426–435.

140. Preusser M, Winkler F, Valiente M, et al. Recent advances in the biology and treatment of brain metastases of non-small cell lung cancer: summary of a multidisciplinary roundtable discussion. *ESMO Open.* 2018;3(1):e000262.

141. Nayak L, Lee EQ, Wen PY. Epidemiology of brain metastases. *Curr Oncol Rep.* 2012;14(1):48–54.

142. Nussbaum ES, Djalilian HR, Cho KH, Hall WA. Brain metastases. Histology, multiplicity, surgery, and survival. *Cancer.* 1996;78(8):1781–1788.

143. Lagerwaard FJ, Levendag PC, Nowak PJ, Eijkenboom WM, Hanssens PE, Schmitz PI. Identification of prognostic factors in patients with brain metastases: a review of 1292 patients. *Int J Radiat Oncol Biol Phys.* 1999;43(4):795–803.

144. Shi AA, Digumarthy SR, Temel JS, Halpern EF, Kuester LB, Aquino SL. Does initial staging or tumor histology better identify asymptomatic brain metastases in patients with non-small cell lung cancer? *J Thorac Oncol.* 2006;1(3):205–210.

145. Seute T, Leffers P, Wilmink JT, ten Velde GP, Twijnstra A. Response of asymptomatic brain metastases from small-cell lung cancer to systemic first-line chemotherapy. *J Clin Oncol.* 2006;24(13):2079–2083.

146. Komaki R, Cox JD, Whitson W. Risk of brain metastasis from small cell carcinoma of the lung related to length of survival and prophylactic irradiation. *Cancer Treat Rep.* 1981;65(9–10):811–814.

147. Hubbs JL, Boyd JA, Hollis D, Chino JP, Saynak M, Kelsey CR. Factors associated with the development of brain metastases: analysis of 975 patients with early stage nonsmall cell lung cancer. *Cancer.* 2010;116(21):5038–5046.

148. Doebele RC, Lu X, Sumey C, et al. Oncogene status predicts patterns of metastatic spread in treatment-naive nonsmall cell lung cancer. *Cancer.* 2012;118(18):4502–4511.

149. Heon S, Yeap BY, Britt GJ, et al. Development of central nervous system metastases in patients with advanced non-small cell lung cancer and somatic EGFR mutations treated with gefitinib or erlotinib. *Clin Cancer Res.* 2010;16(23):5873–5882.

150. Slotman B, Faivre-Finn C, Kramer G, et al. Prophylactic cranial irradiation in extensive small-cell lung cancer. *N Engl J Med.* 2007;357(7):664–672.

151. Brastianos PK, Carter SL, Santagata S, et al. Genomic characterization of brain metastases reveals branched evolution and potential therapeutic targets. *Cancer Discov.* 2015;5(11):1164.

152. Shih DJH, Nayyar N, Bihun I, et al. Genomic characterization of human brain metastases identifies drivers of metastatic lung adenocarcinoma. *Nat Genet.* 2020;52(4):371–377.

153. Posner JB. Neurological complications of systemic cancer. *Med Clin North Am.* 1979;63(4):783–800.

154. Das A, Hochberg FH. Clinical presentation of intracranial metastases. *Neurosurg Clin N Am.* 1996;7(3):377–391.

155. Fink KR, Fink JR. Imaging of brain metastases. *Surg Neurol Int.* 2013;4(Suppl. 4):S209–S219.

156. Delattre JY, Krol G, Thaler HT, Posner JB. Distribution of brain metastases. *Arch Neurol.* 1988;45(7):741–744.

157. Jena A, Taneja S, Talwar V, Sharma JB. Magnetic resonance (MR) patterns of brain metastasis in lung cancer patients: correlation of imaging findings with symptom. *J Thorac Oncol.* 2008;3(2):140–144.

158. Yokoi K, Kamiya N, Matsuguma H, et al. Detection of brain metastasis in potentially operable non-small cell lung cancer: a comparison of CT and MRI. *Chest.* 1999;115(3):714–719.

159. Jung WS, Park CH, Hong CK, Suh SH, Ahn SJ. Diffusion-weighted imaging of brain metastasis from lung cancer: correlation of MRI parameters with the histologic type and gene mutation status. *AJNR Am J Neuroradiol.* 2018;39(2):273–279 [1936-959X (Electronic)].

160. Chiang IC, Kuo YT, Lu CY, et al. Distinction between high-grade gliomas and solitary metastases using peritumoral 3-T magnetic resonance spectroscopy, diffusion, and perfusion imagings. *Neuroradiology.* 2004;46(8):619–627.

161. Patchell RA, Tibbs PA, Walsh JW, et al. A randomized trial of surgery in the treatment of single metastases to the brain. *N Engl J Med.* 1990;322(8):494–500.

162. Revelo AE, Martin A, Velasquez R, et al. Liquid biopsy for lung cancers: an update on recent developments. *Ann Transl Med.* 2019;7(15):349.

163. Pentsova EI, Shah RH, Tang J, et al. Evaluating cancer of the central nervous system through next-generation sequencing of cerebrospinal fluid. *J Clin Oncol.* 2016;34(20):2404–2415.

164. Yang JCH, Kim SW, Kim DW, et al. Osimertinib in patients with epidermal growth factor receptor mutation-positive non-small-cell lung cancer and leptomeningeal metastases: the BLOOM study. *J Clin Oncol.* 2020;38(6):538–547.

165. Gaspar LE, Scott C, Murray K, Curran W. Validation of the RTOG recursive partitioning analysis (RPA) classification for brain metastases. *Int J Radiat Oncol Biol Phys.* 2000;47(4):1001–1006.

166. Dietrich J, Rao K, Pastorino S, Kesari S. Corticosteroids in brain cancer patients: benefits and pitfalls. *Expert Rev Clin Pharmacol.* 2011;4(2):233–242.

167. Galicich JH, French LA, Melby JC. Use of dexamethasone in treatment of cerebral edema associated with brain tumors. *J Lancet.* 1961;81:46–53.

168. Vecht CJ, Hovestadt A, Verbiest HB, van Vliet JJ, van Putten WL. Dose-effect relationship of dexamethasone on Karnofsky performance in metastatic brain tumors: a randomized study of doses of 4, 8, and 16 mg per day. *Neurology.* 1994;44(4):675–680.

169. Skardelly M, Brendle E, Noell S, et al. Predictors of preoperative and early postoperative seizures in patients with intra-axial primary and metastatic brain tumors: a retrospective observational single center study. *Ann Neurol.* 2015;78(6):917–928.

170. Mahajan A, Ahmed S, McAleer MF, et al. Post-operative stereotactic radiosurgery versus observation for completely resected brain metastases: a single-centre, randomised, controlled, phase 3 trial. *Lancet Oncol.* 2017;18(8):1040–1048.

171. Bindal RK, Sawaya R, Leavens ME, Lee JJ. Surgical treatment of multiple brain metastases. *J Neurosurg.* 1993;79(2):210–216.

172. Brown PD, Ballman KV, Cerhan JH, et al. Postoperative stereotactic radiosurgery compared with whole brain radiotherapy for resected metastatic brain disease (NCCTG N107C/CEC·3): a multicentre, randomised, controlled, phase 3 trial. *Lancet Oncol.* 2017;18(8):1049–1060.

III. Neurological complications of specific neoplasms

173. Enders F, Geisenberger C, Jungk C, et al. Prognostic factors and long-term survival in surgically treated brain metastases from non-small cell lung cancer. *Clin Neurol Neurosurg*. 2016;142:72–80.

174. Yoo H, Kim YZ, Nam BH, et al. Reduced local recurrence of a single brain metastasis through microscopic total resection. *J Neurosurg*. 2009;110(4):730–736.

175. Berghoff AS, Rajky O, Winkler F, et al. Invasion patterns in brain metastases of solid cancers. *Neuro Oncol*. 2013;15(12):1664–1672.

176. Siam L, Bleckmann A, Chaung H-N, et al. The metastatic infiltration at the metastasis/brain parenchyma-interface is very heterogeneous and has a significant impact on survival in a prospective study. *Oncotarget*. 2015;6(30).

177. Kamp MA, Fischer I, Bühner J, et al. 5-ALA fluorescence of cerebral metastases and its impact for the local-in-brain progression. *Oncotarget*. 2016;7(41):66776–66789.

178. Arbit E, Wroński M, Burt M, Galicich JH. The treatment of patients with recurrent brain metastases. A retrospective analysis of 109 patients with nonsmall cell lung cancer. *Cancer*. 1995;76(5):765–773.

179. Nieder C, Grosu AL, Gaspar LE. Stereotactic radiosurgery (SRS) for brain metastases: a systematic review. *Radiat Oncol*. 2014;9:155.

180. Werner-Wasik M, Rudoler S, Preston PE, et al. Immediate side effects of stereotactic radiotherapy and radiosurgery. *Int J Radiat Oncol Biol Phys*. 1999;43(2):299–304.

181. Vellayappan B, Tan CL, Yong C, et al. Diagnosis and management of radiation necrosis in patients with brain metastases. *Front Oncol*. 2018;8:395.

182. Boothe D, Young R, Yamada Y, Prager A, Chan T, Beal K. Bevacizumab as a treatment for radiation necrosis of brain metastases post stereotactic radiosurgery. *Neuro Oncol*. 2013;15(9):1257–1263.

183. Shaw E, Scott C, Souhami L, et al. Single dose radiosurgical treatment of recurrent previously irradiated primary brain tumors and brain metastases: final report of RTOG protocol 90-05. *Int J Radiat Oncol Biol Phys*. 2000;47(2):291–298.

184. Shehata MK, Young B, Reid B, et al. Stereotatic radiosurgery of 468 brain metastases ≤2 cm: implications for SRS dose and whole brain radiation therapy. *Int J Radiat Oncol Biol Phys*. 2004;59(1):87–93.

185. Penny KS, Joe M, GMV-vdH J, et al. Adverse radiation effect after stereotactic radiosurgery for brain metastases: incidence, time course, and risk factors. *J Neurosurg*. 2015;123(2):373–386.

186. Kim JM, Miller JA, Kotecha R, et al. The risk of radiation necrosis following stereotactic radiosurgery with concurrent systemic therapies. *J Neurooncol*. 2017;133(2):357–368.

187. Miller JA, Bennett EE, Xiao R, et al. Association between radiation necrosis and tumor biology after stereotactic radiosurgery for brain metastasis. *Int J Radiat Oncol Biol Phys*. 2016;96(5):1060–1069.

188. Flickinger JC, Kondziolka D, Lunsford LD, et al. Development of a model to predict permanent symptomatic postradiosurgery injury for arteriovenous malformation patients. Arteriovenous Malformation Radiosurgery Study Group. *Int J Radiat Oncol Biol Phys*. 2000;46(5):1143–1148.

189. Churilla TM, Chowdhury IH, Handorf E, et al. Comparison of local control of brain metastases with stereotactic radiosurgery vs surgical resection: a secondary analysis of a randomized clinical trial. *JAMA Oncol*. 2019;5(2):243–247.

190. Rusthoven CG, Yamamoto M, Bernhardt D, et al. Evaluation of first-line radiosurgery vs whole-brain radiotherapy for small cell lung cancer brain metastases: the FIRE-SCLC cohort study. *JAMA Oncol*. 2020;6(7):1028–1037.

191. Yamamoto M, Serizawa T, Shuto T, et al. Stereotactic radiosurgery for patients with multiple brain metastases (JLGK0901): a multi-institutional prospective observational study. *Lancet Oncol*. 2014;15(4):387–395.

192. Yamamoto M, Serizawa T, Higuchi Y, et al. A multi-institutional prospective observational study of stereotactic radiosurgery for patients with multiple brain metastases (JLGK0901 study update): irradiation-related complications and long-term maintenance of mini-mental state examination scores. *Int J Radiat Oncol Biol Phys*. 2017;99(1):31–40.

193. Brown PD, Jaeckle K, Ballman KV, et al. Effect of radiosurgery alone vs radiosurgery with whole brain radiation therapy on cognitive function in patients with 1 to 3 brain metastases: a randomized clinical trial. *JAMA*. 2016;316(4):401–409.

194. Aoyama H, Shirato H, Tago M, et al. Stereotactic radiosurgery plus whole-brain radiation therapy vs stereotactic radiosurgery alone for treatment of brain metastases: a randomized controlled trial. *JAMA*. 2006;295(21):2483–2491.

195. Aoyama H, Tago M, Shirato H. Stereotactic radiosurgery with or without whole-brain radiotherapy for brain metastases: secondary analysis of the JROSG 99-1 randomized clinical trial. *JAMA Oncol*. 2015;1(4):457–464.

196. Loganadane G, Hendriks L, Le Péchoux C, Levy A. The current role of whole brain radiation therapy in non–small cell lung cancer patients. *J Thorac Oncol*. 2017;12(10):1467–1477.

197. Brown PD, Ahluwalia MS, Khan OH, Asher AL, Wefel JS, Gondi V. Whole-brain radiotherapy for brain metastases: evolution or revolution? *J Clin Oncol*. 2018;36(5):483–491.

198. DeAngelis LM, Delattre JY, Posner JB. Radiation-induced dementia in patients cured of brain metastases. *Neurology*. 1989;39(6):789–796.

199. Khuntia D, Brown P, Li J, Mehta MP. Whole-brain radiotherapy in the management of brain metastasis. *J Clin Oncol*. 2006;24(8):1295–1304.

200. Mulvenna P, Nankivell M, Barton R, et al. Dexamethasone and supportive care with or without whole brain radiotherapy in treating patients with non-small cell lung cancer with brain metastases unsuitable for resection or stereotactic radiotherapy (QUARTZ): results from a phase 3, non-inferiority, randomised trial. *Lancet*. 2016;388(10055):2004–2014.

201. Brown PD, Pugh S, Laack NN, et al. Memantine for the prevention of cognitive dysfunction in patients receiving whole-brain radiotherapy: a randomized, double-blind, placebo-controlled trial. *Neuro Oncol*. 2013;15(10):1429–1437.

202. Gondi V, Pugh SL, Tome WA, et al. Preservation of memory with conformal avoidance of the hippocampal neural stem-cell compartment during whole-brain radiotherapy for brain metastases (RTOG 0933): a phase II multi-institutional trial. *J Clin Oncol*. 2014;32(34):3810–3816.

203. Brown PD, Gondi V, Pugh S, et al. Hippocampal avoidance during whole-brain radiotherapy plus memantine for patients with brain metastases: phase III trial NRG oncology CC001. *J Clin Oncol*. 2020;38(10):1019–1029.

204. Sun A, Hu C, Wong SJ, et al. Prophylactic cranial irradiation vs observation in patients with locally advanced non-small cell lung cancer: a long-term update of the NRG oncology/RTOG 0214 phase 3 randomized clinical trial. *JAMA Oncol*. 2019;5(6):847–855.

205. Aupérin A, Arriagada R, Pignon J-P, et al. Prophylactic cranial irradiation for patients with small-cell lung cancer in complete remission. *N Engl J Med*. 1999;341(7):476–484.

206. Takahashi T, Yamanaka T, Seto T, et al. Prophylactic cranial irradiation versus observation in patients with extensive-disease small-cell lung cancer: a multicentre, randomised, open-label, phase 3 trial. *Lancet Oncol*. 2017;18(5):663–671.

207. Mehta MP, Rodrigus P, Terhaard CH, et al. Survival and neurologic outcomes in a randomized trial of motexafin gadolinium and whole-brain radiation therapy in brain metastases. *J Clin Oncol*. 2003;21(13):2529–2536.

208. Suh JH, Stea B, Nabid A, et al. Phase III study of efaproxiral as an adjunct to whole-brain radiation therapy for brain metastases. *J Clin Oncol.* 2006;24(1):106–114.

209. Antonadou D, Paraskevaidis M, Sarris G, et al. Phase II randomized trial of temozolomide and concurrent radiotherapy in patients with brain metastases. *J Clin Oncol.* 2002;20(17):3644–3650.

210. Deng L, Liang H, Burnette B, et al. Irradiation and anti-PD-L1 treatment synergistically promote antitumor immunity in mice. *J Clin Invest.* 2014;124(2):687–695.

211. Ahmed KA, Kim S, Arrington J, et al. Outcomes targeting the PD-1/PD-L1 axis in conjunction with stereotactic radiation for patients with non-small cell lung cancer brain metastases. *J Neurooncol.* 2017;133(2):331–338.

212. Magnuson WJ, Lester-Coll NH, Wu AJ, et al. Management of brain metastases in tyrosine kinase inhibitor-naïve epidermal growth factor receptor-mutant non-small-cell lung cancer: a retrospective multi-institutional analysis. *J Clin Oncol.* 2017;35(10):1070–1077.

213. Zhu Q, Sun Y, Cui Y, et al. Clinical outcome of tyrosine kinase inhibitors alone or combined with radiotherapy for brain metastases from epidermal growth factor receptor (EGFR) mutant non small cell lung cancer (NSCLC). *Oncotarget.* 2017;8(8):13304–13311.

214. Ballard P, Yates JW, Yang Z, et al. Preclinical comparison of osimertinib with other EGFR-TKIs in EGFR-mutant NSCLC brain metastases models, and early evidence of clinical brain metastases activity. *Clin Cancer Res.* 2016;22(20):5130–5140.

215. Wu YL, Ahn MJ, Garassino MC, et al. CNS efficacy of osimertinib in patients with T790M-positive advanced non-small-cell lung cancer: data from a randomized phase III trial (AURA3). *J Clin Oncol.* 2018;36(26):2702–2709.

216. Reungwetwattana T, Nakagawa K, Cho BC, et al. CNS response to osimertinib versus standard epidermal growth factor receptor tyrosine kinase inhibitors in patients with untreated EGFR-mutated advanced non-small-cell lung cancer. *J Clin Oncol.* 2018. JCO2018783118.

217. Gerber NK, Yamada Y, Rimner A, et al. Erlotinib versus radiation therapy for brain metastases in patients with EGFR-mutant lung adenocarcinoma. *Int J Radiat Oncol Biol Phys.* 2014;89(2):322–329.

218. Magnuson WJ, Yeung JT, Guillod PD, Gettinger SN, Yu JB, Chiang VL. Impact of deferring radiation therapy in patients with epidermal growth factor receptor-mutant non-small cell lung cancer who develop brain metastases. *Int J Radiat Oncol Biol Phys.* 2016;95(2):673–679.

219. Soon YY, Leong CN, Koh WY, Tham IW. EGFR tyrosine kinase inhibitors versus cranial radiation therapy for EGFR mutant non-small cell lung cancer with brain metastases: a systematic review and meta-analysis. *Radiother Oncol.* 2015;114(2):167–172.

220. Porta R, Sanchez-Torres JM, Paz-Ares L, et al. Brain metastases from lung cancer responding to erlotinib: the importance of EGFR mutation. *Eur Respir J.* 2011;37(3):624–631.

221. Iuchi T, Shingyoji M, Sakaida T, et al. Phase II trial of gefitinib alone without radiation therapy for Japanese patients with brain metastases from EGFR-mutant lung adenocarcinoma. *Lung Cancer.* 2013;82(2):282–287.

222. Schuler M, Wu YL, Hirsh V, et al. First-line afatinib versus chemotherapy in patients with non-small cell lung cancer and common epidermal growth factor receptor gene mutations and brain metastases. *J Thorac Oncol.* 2016;11(3):380–390.

223. Sperduto PW, Wang M, Robins HI, et al. A phase 3 trial of whole brain radiation therapy and stereotactic radiosurgery alone versus WBRT and SRS with temozolomide or erlotinib for non-small cell lung cancer and 1 to 3 brain metastases: Radiation Therapy Oncology Group 0320. *Int J Radiat Oncol Biol Phys.* 2013;85(5):1312–1318.

224. Kim DW, Mehra R, Tan DSW, et al. Activity and safety of ceritinib in patients with ALK-rearranged non-small-cell lung cancer (ASCEND-1): updated results from the multicentre, open-label, phase 1 trial. *Lancet Oncol.* 2016;17(4):452–463.

225. Kim DW, Tiseo M, Ahn MJ, et al. Brigatinib in patients with crizotinib-refractory anaplastic lymphoma kinase-positive non-small-cell lung cancer: a randomized, multicenter phase II trial. *J Clin Oncol.* 2017;35(22):2490–2498.

226. Shaw AT, Felip E, Bauer TM, et al. Lorlatinib in non-small-cell lung cancer with ALK or ROS1 rearrangement: an international, multicentre, open-label, single-arm first-in-man phase 1 trial. *Lancet Oncol.* 2017;18(12):1590–1599.

227. Garon EB, Heist RS, Seto T, et al. Abstract CT082: Capmatinib in METex14-mutated (mut) advanced non-small cell lung cancer (NSCLC): results from the phase II GEOMETRY *mono-1* study, including efficacy in patients (pts) with brain metastases (BM). *Cancer Res.* 2020;80(16 Suppl):CT082-CT.

228. Goldberg SB, Gettinger SN, Mahajan A, et al. Pembrolizumab for patients with melanoma or non-small-cell lung cancer and untreated brain metastases: early analysis of a non-randomised, open-label, phase 2 trial. *Lancet Oncol.* 2016;17(7):976–983.

229. Goldberg SB, Schalper KA, Gettinger SN, et al. Pembrolizumab for management of patients with NSCLC and brain metastases: long-term results and biomarker analysis from a non-randomised, open-label, phase 2 trial. *Lancet Oncol.* 2020;21(5):655–663.

230. Crino L, Bronte G, Bidoli P, et al. Nivolumab and brain metastases in patients with advanced non-squamous non-small cell lung cancer. *Lung Cancer.* 2019;129:35–40.

231. Dudnik E, Yust-Katz S, Nechushtan H, et al. Intracranial response to nivolumab in NSCLC patients with untreated or progressing CNS metastases. *Lung Cancer.* 2016;98:114–117.

232. Lukas RV, Gandhi M, O'Hear C, Hu S, Lai C, Patel JD. Safety and efficacy analyses of atezolizumab in advanced non-small cell lung cancer (NSCLC) patients with or without baseline brain metastases. *Ann Oncol.* 2017;28:ii28.

233. Hendriks LEL, Henon C, Auclin E, et al. Outcome of patients with non-small cell lung cancer and brain metastases treated with checkpoint inhibitors. *J Thorac Oncol.* 2019;14(7):1244–1254.

234. Mansfield AS, Herbst RS, Castro Jr G, et al. Outcomes with pembrolizumab (pembro) monotherapy in patients (pts) with PD-L1–positive NSCLC with brain metastases: pooled analysis of KEYNOTE-001, -010, -024, and -042. *Ann Oncol.* 2019;30:v602–v606.

235. Qian JM, Yu JB, Kluger HM, Chiang VL. Timing and type of immune checkpoint therapy affect the early radiographic response of melanoma brain metastases to stereotactic radiosurgery. *Cancer.* 2016;122(19):3051–3058.

236. Acharya S, Mahmood M, Mullen D, et al. Distant intracranial failure in melanoma brain metastases treated with stereotactic radiosurgery in the era of immunotherapy and targeted agents. *Adv Radiat Oncol.* 2017;2(4):572–580.

237. Yusuf MB, Amsbaugh MJ, Burton E, Chesney J, Woo S. Peri-SRS Administration of immune checkpoint therapy for melanoma metastatic to the brain: investigating efficacy and the effects of relative treatment timing on lesion response. *World Neurosurg.* 2017;100:632–640.e4.

238. Colaco RJ, Martin P, Kluger HM, Yu JB, Chiang VL. Does immunotherapy increase the rate of radiation necrosis after radiosurgical treatment of brain metastases? *J Neurosurg.* 2016;125(1):17–23.

239. Hubbeling HG, Schapira EF, Horick NK, et al. Safety of combined PD-1 pathway inhibition and intracranial radiation therapy in non-small cell lung cancer. *J Thorac Oncol.* 2018;13(4):550–558.

240. Martin AM, Cagney DN, Catalano PJ, et al. Immunotherapy and symptomatic radiation necrosis in patients with brain metastases treated with stereotactic radiation. *JAMA Oncol.* 2018;4(8):1123–1124.

241. Chiang V, Cheok S. Combining radiosurgery and systemic therapies for treatment of brain metastases. In: Ahluwalia M, Metellus P, Soffietti R, eds. *Central Nervous System Metastases.* Cham: Springer International Publishing; 2020:247–258.

III. Neurological complications of specific neoplasms

242. Franciosi V, Cocconi G, Michiara M, et al. Front-line chemotherapy with cisplatin and etoposide for patients with brain metastases from breast carcinoma, nonsmall cell lung carcinoma, or malignant melanoma: a prospective study. *Cancer.* 1999;85(7):1599–1605.

243. Fujita A, Fukuoka S, Takabatake H, Tagaki S, Sekine K. Combination chemotherapy of cisplatin, ifosfamide, and irinotecan with rhG-CSF support in patients with brain metastases from non-small cell lung cancer. *Oncology.* 2000;59(4):291–295.

244. Cortes J, Rodriguez J, Aramendia JM, et al. Front-line paclitaxel/cisplatin-based chemotherapy in brain metastases from non-small-cell lung cancer. *Oncology.* 2003;64(1):28–35.

245. Bailon O, Chouahnia K, Augier A, et al. Upfront association of carboplatin plus pemetrexed in patients with brain metastases of lung adenocarcinoma. *Neuro Oncol.* 2012;14(4):491–495.

246. Bearz A, Garassino I, Tiseo M, et al. Activity of pemetrexed on brain metastases from non-small cell lung cancer. *Lung Cancer.* 2010;68(2):264–268.

247. Wong ET, Berkenblit A. The role of topotecan in the treatment of brain metastases. *Oncologist.* 2004;9(1):68–79.

248. Dziadziuszko R, Ardizzoni A, Postmus PE, et al. Temozolomide in patients with advanced non-small cell lung cancer with and without brain metastases. A phase II study of the EORTC Lung Cancer Group (08965). *Eur J Cancer.* 2003;39(9):1271–1276.

249. Siena S, Crino L, Danova M, et al. Dose-dense temozolomide regimen for the treatment of brain metastases from melanoma, breast cancer, or lung cancer not amenable to surgery or radiosurgery: a multicenter phase II study. *Ann Oncol.* 2010;21(3):655–661.

250. Ebert BL, Niemierko E, Shaffer K, Salgia R. Use of temozolomide with other cytotoxic chemotherapy in the treatment of patients with recurrent brain metastases from lung cancer. *Oncologist.* 2003;8(1):69–75.

251. Abrey LE, Olson JD, Raizer JJ, et al. A phase II trial of temozolomide for patients with recurrent or progressive brain metastases. *J Neurooncol.* 2001;53(3):259–265.

252. Christodoulou C, Bafaloukos D, Kosmidis P, et al. Phase II study of temozolomide in heavily pretreated cancer patients with brain metastases. *Ann Oncol.* 2001;12(2):249–254.

253. Giorgio CG, Giuffrida D, Pappalardo A, et al. Oral temozolomide in heavily pre-treated brain metastases from non-small cell lung cancer: phase II study. *Lung Cancer.* 2005;50(2):247–254.

254. Athanassiou H, Synodinou M, Maragoudakis E, et al. Randomized phase II study of temozolomide and radiotherapy compared with radiotherapy alone in newly diagnosed glioblastoma multiforme. *J Clin Oncol.* 2005;23(10):2372–2377.

255. Verger E, Gil M, Yaya R, et al. Temozolomide and concomitant whole brain radiotherapy in patients with brain metastases: a phase II randomized trial. *Int J Radiat Oncol Biol Phys.* 2005;61(1):185–191.

256. Postmus PE, Haaxma-Reiche H, Smit EF, et al. Treatment of brain metastases of small-cell lung cancer: comparing teniposide and teniposide with whole-brain radiotherapy—a phase III study of the European Organization for the Research and Treatment of Cancer Lung Cancer Cooperative Group. *J Clin Oncol.* 2000;18(19):3400–3408.

257. Rudin CM, Awad MM, Navarro A, et al. Pembrolizumab or placebo plus etoposide and platinum as first-line Therapy for extensive-stage small-cell lung cancer: randomized, double-blind, phase III KEYNOTE-604 study. *J Clin Oncol.* 2020;38(21):2369–2379.

258. Costigan DA, Winkelman MD. Intramedullary spinal cord metastasis. A clinicopathological study of 13 cases. *J Neurosurg.* 1985;62(2):227–233.

259. Schiff D, O'Neill BP. Intramedullary spinal cord metastases: clinical features and treatment outcome. *Neurology.* 1996;47(4):906–912.

260. Potti A, Abdel-Raheem M, Levitt R, Schell DA, Mehdi SA. Intramedullary spinal cord metastases (ISCM) and non-small cell lung carcinoma (NSCLC): clinical patterns, diagnosis and therapeutic considerations. *Lung Cancer.* 2001;31(2):319–323.

261. Fredericks RK, Elster A, Walker FO. Gadolinium-enhanced MRI: a superior technique for the diagnosis of intraspinal metastases. *Neurology.* 1989;39(5):734–736.

262. Cheng H, Perez-Soler R. Leptomeningeal metastases in non-small-cell lung cancer. *Lancet Oncol.* 2018;19(1):e43–e55.

263. Remon J, Le Rhun E, Besse B. Leptomeningeal carcinomatosis in non-small cell lung cancer patients: a continuing challenge in the personalized treatment era. *Cancer Treat Rev.* 2017;53:128–137.

264. Li Y-S, Jiang B-Y, Yang J-J, et al. Leptomeningeal metastases in patients with NSCLC with EGFR mutations. *J Thorac Oncol.* 2016;11(11):1962–1969.

265. Rosen ST, Aisner J, Makuch RW, et al. Carcinomatous leptomeningitis in small cell lung cancer: a clinicopathologic review of the National Cancer Institute experience. *Medicine.* 1982;61(1):45–53.

266. Chamberlain MC. Leptomeningeal metastases: a review of evaluation and treatment. *J Neurooncol.* 1998;37(3):271–284.

267. Chamberlain M, Junck L, Brandsma D, et al. Leptomeningeal metastases: a RANO proposal for response criteria. *Neuro Oncol.* 2017;19(4):484–492.

268. Hyun JW, Jeong IH, Joung A, Cho HJ, Kim SH, Kim HJ. Leptomeningeal metastasis: clinical experience of 519 cases. *Eur J Cancer.* 2016;56:107–114.

269. Kallmes DF, Gray L, Glass JP. High-dose gadolinium-enhanced MRI for diagnosis of meningeal metastases. *Neuroradiology.* 1998;40(1):23–26.

270. Grossman SA, Krabak MJ. Leptomeningeal carcinomatosis. *Cancer Treat Rev.* 1999;25(2):103–119.

271. Ma C, Lv Y, Jiang R, Li J, Wang B, Sun L. Novel method for the detection and quantification of malignant cells in the CSF of patients with leptomeningeal metastasis of lung cancer. *Oncol Lett.* 2016;11(1):619–623.

272. Nayak L, Fleisher M, Gonzalez-Espinoza R, et al. Rare cell capture technology for the diagnosis of leptomeningeal metastasis in solid tumors. *Neurology.* 2013;80(17):1598–1605. discussion 603.

273. Riess JW, Nagpal S, Iv M, et al. Prolonged survival of patients with non-small-cell lung cancer with leptomeningeal carcinomatosis in the modern treatment era. *Clin Lung Cancer.* 2014;15(3):202–206.

274. Togashi Y, Masago K, Masuda S, et al. Cerebrospinal fluid concentration of gefitinib and erlotinib in patients with non-small cell lung cancer. *Cancer Chemother Pharmacol.* 2012;70(3):399–405.

275. Lee E, Keam B, Kim DW, et al. Erlotinib versus gefitinib for control of leptomeningeal carcinomatosis in non-small-cell lung cancer. *J Thorac Oncol.* 2013;8(8):1069–1074.

276. Kawamura T, Hata A, Takeshita J, et al. High-dose erlotinib for refractory leptomeningeal metastases after failure of standard-dose EGFR-TKIs. *Cancer Chemother Pharmacol.* 2015;75(6):1261–1266.

277. Tamiya A, Tamiya M, Nishihara T, et al. Cerebrospinal fluid penetration rate and efficacy of afatinib in patients with EGFR mutation-positive non-small cell lung cancer with leptomeningeal carcinomatosis: a multicenter prospective study. *Anticancer Res.* 2017;37(8):4177–4182.

278. Nanjo S, Hata A, Okuda C, et al. Standard-dose osimertinib for refractory leptomeningeal metastases in T790M-positive EGFR-mutant non-small cell lung cancer. *Br J Cancer.* 2018;118(1):32–37.

279. Yang Z, Guo Q, Wang Y, et al. AZD3759, a BBB-penetrating EGFR inhibitor for the treatment of EGFR mutant NSCLC with CNS metastases. *Sci Transl Med.* 2016;8(368):368ra172.

280. Ahn MJ, Kim DW, Cho BC, et al. Activity and safety of AZD3759 in EGFR-mutant non-small-cell lung cancer with CNS metastases (BLOOM): a phase 1, open-label, dose-escalation and dose-expansion study. *Lancet Respir Med.* 2017;5(11):891–902.

281. Costa DB, Shaw AT, Ou SH, et al. Clinical experience with crizotinib in patients with advanced ALK-rearranged non-small-cell lung cancer and brain metastases. *J Clin Oncol.* 2015;33(17):1881–1888.

282. Ahn HK, Han B, Lee SJ, et al. ALK inhibitor crizotinib combined with intrathecal methotrexate treatment for non-small cell lung cancer with leptomeningeal carcinomatosis. *Lung Cancer.* 2012;76(2):253–254.

283. Arrondeau J, Ammari S, Besse B, Soria JC. LDK378 compassionate use for treating carcinomatous meningitis in an ALK translocated non-small-cell lung cancer. *J Thorac Oncol.* 2014;9(8):e62–e63.

284. Chow LQ, Barlesi F, Bertino EM, et al. Results of the ASCEND-7 phase II study evaluating ALK inhibitor (ALKi) ceritinib in patients (pts) with ALK+ non-small cell lung cancer (NSCLC) metastatic to the brain. *Ann Oncol.* 2019;30:v602–v603.

285. Gainor JF, Sherman CA, Willoughby K, et al. Alectinib salvages CNS relapses in ALK-positive lung cancer patients previously treated with crizotinib and ceritinib. *J Thorac Oncol.* 2015;10(2):232–236.

286. Gainor JF, Chi AS, Logan J, et al. Alectinib dose escalation reinduces central nervous system responses in patients with anaplastic lymphoma kinase-positive non-small cell lung cancer relapsing on standard dose alectinib. *J Thorac Oncol.* 2016;11(2):256–260.

287. Geraud A, Mezquita L, Bigot F, et al. Prolonged leptomeningeal responses with brigatinib in two heavily pretreated *ALK*-rearranged non-small cell lung cancer patients. *J Thorac Oncol.* 2018;13(11):e215–e217.

288. Beauchesne P. Intrathecal chemotherapy for treatment of leptomeningeal dissemination of metastatic tumours. *Lancet Oncol.* 2010;11(9):871–879.

289. Chamberlain MC. Leptomeningeal metastasis. *Curr Opin Oncol.* 2010;22(6):627–635.

290. Wu YL, Zhou L, Lu Y. Intrathecal chemotherapy as a treatment for leptomeningeal metastasis of non-small cell lung cancer: a pooled analysis. *Oncol Lett.* 2016;12(2):1301–1314.

291. Gion M, Remon J, Caramella C, Soria JC, Besse B. Symptomatic leptomeningeal metastasis improvement with nivolumab in advanced non-small cell lung cancer patient. *Lung Cancer.* 2017;108:72–74.

292. Liao BC, Lee JH, Lin CC, et al. Epidermal growth factor receptor tyrosine kinase inhibitors for non-small-cell lung cancer patients with leptomeningeal carcinomatosis. *J Thorac Oncol.* 2015;10(12):1754–1761.

293. Morris PG, Reiner AS, Szenberg OR, et al. Leptomeningeal metastasis from non-small cell lung cancer: survival and the impact of whole brain radiotherapy. *J Thorac Oncol.* 2012;7(2):382–385.

294. Lamba N, Fick T, Nandoe Tewarie R, Broekman ML. Management of hydrocephalus in patients with leptomeningeal metastases: an ethical approach to decision-making. *J Neurooncol.* 2018;140(1):5–13.

295. Laigle-Donadey F, Taillibert S, Martin-Duverneuil N, Hildebrand J, Delattre JY. Skull-base metastases. *J Neurooncol.* 2005;75(1):63–69.

296. Cole JS, Patchell RA. Metastatic epidural spinal cord compression. *Lancet Neurol.* 2008;7(5):459–466.

297. Loblaw DA, Laperriere NJ, Mackillop WJ. A population-based study of malignant spinal cord compression in Ontario. *Clin Oncol (R Coll Radiol).* 2003;15(4):211–217.

298. Barron KD, Hirano A, Araki S, Terry RD. Experiences with metastatic neoplasms involving the spinal cord. *Neurology.* 1959;9(2):91–106.

299. Bach F, Agerlin N, Sørensen JB, et al. Metastatic spinal cord compression secondary to lung cancer. *J Clin Oncol.* 1992;10(11):1781–1787.

300. Schiff D, O'Neill BP, Suman VJ. Spinal epidural metastasis as the initial manifestation of malignancy: clinical features and diagnostic approach. *Neurology.* 1997;49(2):452–456.

301. Helweg-Larsen S, Sørensen PS. Symptoms and signs in metastatic spinal cord compression: a study of progression from first symptom until diagnosis in 153 patients. *Eur J Cancer.* 1994;30(3):396–398.

302. Li KC, Poon PY. Sensitivity and specificity of MRI in detecting malignant spinal cord compression and in distinguishing malignant from benign compression fractures of vertebrae. *Magn Reson Imaging.* 1988;6(5):547–556.

303. Colletti PM, Siegel HJ, Woo MY, Young HY, Terk MR. The impact on treatment planning of MRI of the spine in patients suspected of vertebral metastasis: an efficacy study. *Comput Med Imaging Graph.* 1996;20(3):159–162.

304. Mehta RC, Marks MP, Hinks RS, Glover GH, Enzmann DR. MR evaluation of vertebral metastases: T1-weighted, short-inversion-time inversion recovery, fast spin-echo, and inversion-recovery fast spin-echo sequences. *AJNR Am J Neuroradiol.* 1995;16(2):281–288.

305. Carmody RF, Yang PJ, Seeley GW, Seeger JF, Unger EC, Johnson JE. Spinal cord compression due to metastatic disease: diagnosis with MR imaging versus myelography. *Radiology.* 1989;173(1):225–229.

306. Spratt DE, Beeler WH, de Moraes FY, et al. An integrated multidisciplinary algorithm for the management of spinal metastases: an International Spine Oncology Consortium report. *Lancet Oncol.* 2017;18(12):e720–e730.

307. Sørensen S, Helweg-Larsen S, Mouridsen H, Hansen HH. Effect of high-dose dexamethasone in carcinomatous metastatic spinal cord compression treated with radiotherapy: a randomised trial. *Eur J Cancer.* 1994;30A(1):22–27.

308. Vecht CJ, Haaxma-Reiche H, van Putten WL, de Visser M, Vries EP, Twijnstra A. Initial bolus of conventional versus high-dose dexamethasone in metastatic spinal cord compression. *Neurology.* 1989;39(9):1255–1257.

309. Patchell RA, Tibbs PA, Regine WF, et al. Direct decompressive surgical resection in the treatment of spinal cord compression caused by metastatic cancer: a randomised trial. *Lancet.* 2005;366(9486):643–648.

310. Young RF, Post EM, King GA. Treatment of spinal epidural metastases. Randomized prospective comparison of laminectomy and radiotherapy. *J Neurosurg.* 1980;53(6):741–748.

311. Greenberg HS, Kim JH, Posner JB. Epidural spinal cord compression from metastatic tumor: results with a new treatment protocol. *Ann Neurol.* 1980;8(4):361–366.

312. Rades D, Stalpers LJ, Veninga T, et al. Evaluation of five radiation schedules and prognostic factors for metastatic spinal cord compression. *J Clin Oncol.* 2005;23(15):3366–3375.

313. Chamberlain MC, Kormanik PA. Epidural spinal cord compression: a single institution's retrospective experience. *Neuro Oncol.* 1999;1(2):120–123.

314. van der Sande JJ, Boogerd W, Kröger R, Kappelle AC. Recurrent spinal epidural metastases: a prospective study with a complete follow up. *J Neurol Neurosurg Psychiatry.* 1999;66(5):623–627.

315. Schiff D, Shaw EG, Cascino TL. Outcome after spinal reirradiation for malignant epidural spinal cord compression. *Ann Neurol.* 1995;37(5):583–589.

316. Rades D, Fehlauer F, Schulte R, et al. Prognostic factors for local control and survival after radiotherapy of metastatic spinal cord compression. *J Clin Oncol.* 2006;24(21):3388–3393.

317. Marulli G, Battistella L, Mammana M, Calabrese F, Rea F. Superior sulcus tumors (Pancoast tumors). *Ann Transl Med.* 2016;4(12):239.

318. Pancoast HK. Superior pulmonary sulcus tumor: tumor characterized by pain, Horner's syndrome, destruction of bone and atrophy of hand muscles chairman's address. *JAMA.* 1932;99(17):1391–1396.

319. Kunitoh H, Kato H, Tsuboi M, et al. Phase II trial of preoperative chemoradiotherapy followed by surgical resection in patients with superior sulcus non-small-cell lung cancers: report of Japan Clinical Oncology Group trial 9806. *J Clin Oncol.* 2008;26(4):644–649.

320. Graus F, Delattre JY, Antoine JC, et al. Recommended diagnostic criteria for paraneoplastic neurological syndromes. *J Neurol Neurosurg Psychiatry*. 2004;75(8):1135–1140.

321. Dalmau J, Rosenfeld MR. Paraneoplastic syndromes of the CNS. *Lancet Neurol*. 2008;7(4):327–340.

322. Graus F, Dalmau J. Paraneoplastic neurological syndromes in the era of immune-checkpoint inhibitors. *Nat Rev Clin Oncol*. 2019;16(9):535–548.

323. Pittock SJ, Kryzer TJ, Lennon VA. Paraneoplastic antibodies coexist and predict cancer, not neurological syndrome. *Ann Neurol*. 2004;56(5):715–719.

324. Elrington GM, Murray NM, Spiro SG, Newsom-Davis J. Neurological paraneoplastic syndromes in patients with small cell lung cancer. A prospective survey of 150 patients. *J Neurol Neurosurg Psychiatry*. 1991;54(9):764–767.

325. Giometto B, Grisold W, Vitaliani R, Graus F, Honnorat J, Bertolini G. Paraneoplastic neurologic syndrome in the PNS Euronetwork database: a European study from 20 centers. *Arch Neurol*. 2010;67(3):330–335.

326. Höftberger R, Rosenfeld MR, Dalmau J. Update on neurological paraneoplastic syndromes. *Curr Opin Oncol*. 2015;27(6):489–495.

327. Gultekin SH, Rosenfeld MR, Voltz R, Eichen J, Posner JB, Dalmau J. Paraneoplastic limbic encephalitis: neurological symptoms, immunological findings and tumour association in 50 patients. *Brain*. 2000;123(Pt 7):1481–1494.

328. Graus F, Keime-Guibert F, Reñe R, et al. Anti-Hu-associated paraneoplastic encephalomyelitis: analysis of 200 patients. *Brain*. 2001;124(Pt 6):1138–1148.

329. Höftberger R, van Sonderen A, Leypoldt F, et al. Encephalitis and AMPA receptor antibodies: novel findings in a case series of 22 patients. *Neurology*. 2015;84(24):2403–2412.

330. Lancaster E, Lai M, Peng X, et al. Antibodies to the GABA(B) receptor in limbic encephalitis with seizures: case series and characterisation of the antigen. *Lancet Neurol*. 2010;9(1):67–76.

331. Peterson K, Rosenblum MK, Kotanides H, Posner JB. Paraneoplastic cerebellar degeneration. I. A clinical analysis of 55 anti-Yo antibody-positive patients. *Neurology*. 1992;42(10):1931–1937.

332. Mason WP, Graus F, Lang B, et al. Small-cell lung cancer, paraneoplastic cerebellar degeneration and the Lambert-Eaton myasthenic syndrome. *Brain*. 1997;120(Pt 8):1279–1300.

333. Sabater L, Titulaer M, Saiz A, Verschuuren J, Güre AO, Graus F. SOX1 antibodies are markers of paraneoplastic Lambert-Eaton myasthenic syndrome. *Neurology*. 2008;70(12):924–928.

334. Antoine JC, Honnorat J, Camdessanché JP, et al. Paraneoplastic anti-CV2 antibodies react with peripheral nerve and are associated with a mixed axonal and demyelinating peripheral neuropathy. *Ann Neurol*. 2001;49(2):214–221.

335. Titulaer MJ, Maddison P, Sont JK, et al. Clinical Dutch-English Lambert-Eaton Myasthenic syndrome (LEMS) tumor association prediction score accurately predicts small-cell lung cancer in the LEMS. *J Clin Oncol*. 2011;29(7):902–908.

336. Yang H, Peng Q, Yin L, et al. Identification of multiple cancer-associated myositis-specific autoantibodies in idiopathic inflammatory myopathies: a large longitudinal cohort study. *Arthritis Res Ther*. 2017;19(1):259.

337. Graus F, Dalmau J. Paraneoplastic neurological syndromes. *Curr Opin Neurol*. 2012;25(6):795–801.

338. Carbone DP, Reck M, Paz-Ares L, et al. First-line nivolumab in stage IV or recurrent non-small-cell lung cancer. *N Engl J Med*. 2017;376(25):2415–2426.

339. Cuzzubbo S, Javeri F, Tissier M, et al. Neurological adverse events associated with immune checkpoint inhibitors: review of the literature. *Eur J Cancer*. 2017;73:1–8.

340. Cuzzubbo S, Tetu P, Guegan S, et al. Reintroduction of immune-checkpoint inhibitors after immune-related meningitis: a case series of melanoma patients. *J Immunother Cancer*. 2020;8(2).

16

Neurological complications of breast cancer[*]

Alexander C. Ou and Barbara J. O'Brien

Department of Neuro-Oncology, University of Texas MD Anderson Cancer Center, Houston, TX, United States

1 Introduction

Breast cancer is the second most common malignancy in the world, the most common malignancy in women, and an important cause of global cancer-related morbidity and mortality.[1] Nearly one in eight women living in the United States will be diagnosed with breast cancer in their lifetime. Neurological complications are quite common in this population and can arise from the cancer itself or the various treatment modalities, i.e., surgery, radiation, or chemotherapy. Ironically, with the continued refinement of anticancer therapies and consequently improved survival afforded patients, the incidence of central nervous system (CNS) metastasis has increased, likely reflecting the sanctuary site that the central nervous system provides cancer cells. CNS metastases typically occur in patients with disseminated systemic malignancy, but the CNS may also be the first or isolated site of relapse after systemic disease control is achieved. Metastases in the central and peripheral nervous systems usually carry a poor prognosis, but when breast cancer is the primary cancer, treatment is more likely to afford local control.

As in all medicine, the approach to diagnosing and managing the neurologic complications of breast cancer begins with a thorough understanding of the patient's pertinent medical and oncologic history, followed by a thoughtful and comprehensive neurologic examination. Broadly, the neurologic complications of breast cancer may be conceptualized as those attributable to: (1) metastatic involvement of the central or peripheral nervous system; (2) nonmetastatic etiologies from systemic cancer or cancer therapies such as toxic, metabolic, or hematologic derangements; (3) treatment-related complications; and (4) paraneoplastic syndromes. It is the goal of the forthcoming discussion to provide a framework to assist clinicians in caring for these patients.

2 Metastatic complications

2.1 Skull

Skull metastases are usually associated with metastases in other bones.[2] Most calvarial metastases are asymptomatic, but may cause headache, localized tenderness, and swelling of the scalp. Uncommonly, they produce neurological symptoms by invading the dura and leptomeninges or compressing the superior sagittal sinus.[3] Symptomatic lesions are treated with radiotherapy, chemotherapy, or hormonal therapy, but asymptomatic lesions do not require specific treatment.

Metastases to the skull base may cause symptoms by compressing cranial nerves as they exit the basal foramina.[4,5] The clinical manifestations depend on the location of the metastasis (see Table 1).[2,6] Magnetic resonance imaging (MRI) of the brain with thin slices through the skull base is the most useful investigation to establish the diagnosis (see Fig. 1), but plain films, CT with bone windows, radionuclide bone scans, and single-photon emission CT may be sensitive for bony lesions when MRI is unrevealing.[7,8] Radiotherapy typically alleviates pain, and cranial nerve palsies may improve with early treatment.[6] Chemotherapy and hormonal therapy may also produce improvement.[2]

Mandibular metastases may affect the mental nerve (i.e., mental neuropathy), producing numbness or paresthesias in the lower lip and chin, or pain and swelling in the mandible.[9,10] Skull base, dural, and leptomeningeal metastases may also cause facial numbness or paresthesias, but the altered sensation in these cases may affect a greater area of the face and other neurological signs are often present. Mandibular metastases are usually visible on plain radiographs and CT, though contrast-enhanced MRI of the mandible may be indicated if the aforementioned studies are uninformative and a high index of suspicion exists.[11] Radiotherapy is the treatment of choice.

[*] Originally authored by Neil E. Anderson

TABLE 1   Clinical syndromes associated with skull base metastases.

Site	Symptoms	Signs
Orbit	Supraorbital headache	Proptosis
	Diplopia	Ophthalmoplegia
		± Reduced visual acuity
		± Periorbital swelling
Parasellar (sella turcica, petrous apex)	Frontal headache	Ophthalmoplegia
	Diplopia	Facial numbness (V_1)
		Periorbital swelling
Gasserian ganglion	Facial numbness	Facial numbness (V_1, V_2)
	Facial pain	Abducens palsy (anterior ridge)
		Facial palsy (posterior ridge)
Jugular foramen	Occipital pain	Cranial nerve IX, X, XI palsies
	Hoarseness	
	Dysphagia	
Occipital condyle	Occipital pain	Cranial nerve XII palsy
	Dysarthria	

2.2 Dural metastasis

Dural metastases occur in 7%–18% of the patients with breast cancer.[12–14] In about half of these patients, the dura is the solitary site of metastasis in the CNS. Dural metastasis may be associated with a contiguous calvarial metastasis, while leptomeningeal or intraparenchymal metastases more often arise by hematogenous spread. Dural metastases may be solitary and nodular or cause diffuse dural thickening (see Fig. 1B and C). Microscopic extension into the leptomeninges and the cerebral cortex can occur, but macroscopic cortical invasion is uncommon.[15] Dural metastases are often asymptomatic, but can cause symptoms by compression or invasion of the brain, cranial nerves, pituitary, or venous sinuses. A subdural hematoma may coexist with a dural metastasis.[16,17] There is an increased incidence of meningioma with breast cancer and care must be taken to distinguish a dural metastasis from a meningioma (see Fig. 2).[18,19] Metastases may even lodge within a meningioma.[20] Symptomatic dural metastases may be treated with corticosteroids, radiotherapy, chemotherapy, and/or hormonal therapy. Surgical resection occasionally is indicated for a large, symptomatic dural metastasis.

2.3 Leptomeningeal metastasis

Leptomeningeal metastasis—historically known by other names including carcinomatous meningitis, leptomeningeal carcinomatosis, or simply leptomeningeal dissemination or leptomeningeal disease—is one of the most devastating complications of metastatic breast cancer, which untreated is associated with a mean survival of 4–8 weeks.[21,22] Leptomeningeal metastatic disease (LMD) fundamentally differs from other CNS metastases—i.e., those involving the brain parenchyma or dura—in that it involves the CSF space, normally kept anatomically separate from the bloodstream by the blood-brain barrier. If patients with hematological malignancies are excluded, breast cancer accounts for 50% of the cases of leptomeningeal metastasis.[20] Leptomeningeal metastasis occurs in 2%–5% of the patients with metastatic breast cancer, though in recent years the incidence of leptomeningeal metastasis has increased both as a result of earlier detection with modern imaging techniques as well as improvements in systemic therapy that allows for microscopic tumor cells to gain sanctuary behind the blood-brain barrier despite systemic control.[13,23–25] Leptomeningeal metastasis historically is associated with advanced disseminated systemic disease, but the leptomeninges may also be the first site of progression in 8% of the patients who have responded to chemotherapy.[26] Lobular breast carcinoma and triple-negative breast carcinoma have a propensity to metastasize to the leptomeninges.[25,27–29]

Leptomeningeal metastasis should be suspected with the subacute onset of focal or multifocal brain, cranial nerve, cauda equina, radicular, or obstructive hydrocephalus-like symptoms (e.g., vomiting; headache worsening with recumbency).[21,30] Cranial nerve palsies—such as complete facial weakness, trigeminal neuropathy, or ocular dysmotility—are present in up to 80% of the patients. Spinal leptomeningeal metastases may present with back pain, radicular pain, paresthesias, sensory loss, patchy weakness, loss of tendon reflexes, and bowel or bladder dysfunction. Headache, neck stiffness, confusion, nausea, vomiting, seizures, dysarthria, and gait abnormalities may also occur in patients with leptomeningeal metastasis.

Gadolinium-contrasted MRIs of the entire neuroaxis—i.e., brain, cervical, thoracic, and lumbar spine—are essential to the workup of leptomeningeal metastasis and may show enhancement of the cerebellar folia, cranial and spinal leptomeninges, cranial nerves, and subependyma; focal or diffuse dural enhancement; enhancing superficial cortical nodules; or hydrocephalus (see Fig. 3).[31,32] These MRI abnormalities are not specific for leptomeningeal metastasis, but in the context of the aforementioned typical clinical features and a known diagnosis of cancer, they are adequate to make a diagnosis of leptomeningeal metastasis, even when

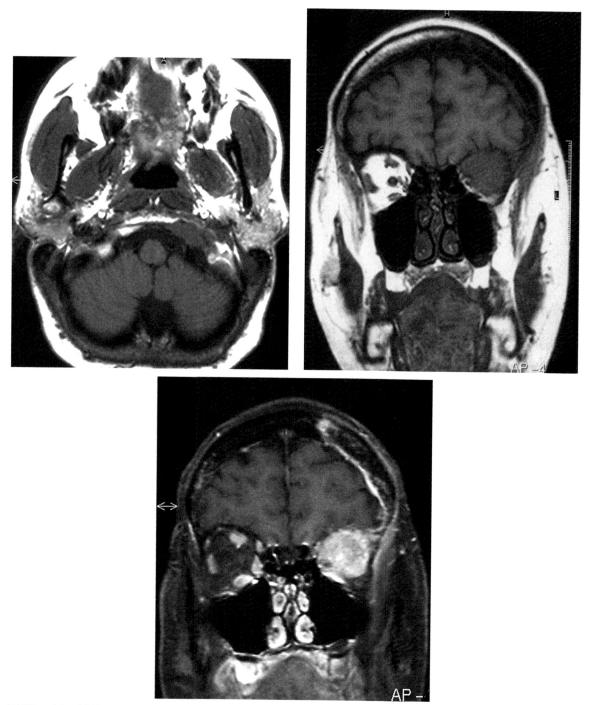

FIG. 1 (A) T1-weighted MR axial section that shows a metastasis in the clivus in a 44-year-old woman with breast cancer. (B and C) MR coronal sections showing a metastasis in the left orbit and diffuse infiltration of the dura overlying the left hemisphere in a 65-year-old woman with metastatic breast cancer.

the cerebrospinal fluid (CSF) cytology is negative.[32] CSF classically demonstrates a pleocytosis, elevated protein, and hypoglycorrhachia (low glucose concentration), but these abnormalities may occur in any combination in a patient with leptomeningeal metastasis, or be absent.[24] Cytopathologic detection of malignant cells in the CSF is the gold standard for definitive diagnosis of leptome-

ningeal metastasis, and several CSF samples of sufficient volume—i.e., at least 10 mL—may be required before a positive result is obtained. In fact, it is typical at many institutions to perform as many as three lumbar punctures in an attempt to confirm the diagnosis.[21] Although specific CSF biomarkers for leptomeningeal metastasis have not been heretofore characterized, recent advances in

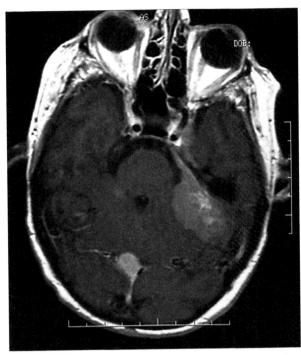

FIG. 2   MRI with gadolinium shows an enhancing, lobulated mass arising from the petrous temporal bone in a 53-year-old woman with no history of malignancy. Another enhancing lesion lies adjacent to the torcula. These lesions were thought to be meningiomas, but histology showed metastatic adenocarcinoma. Further investigation identified a primary breast tumor.

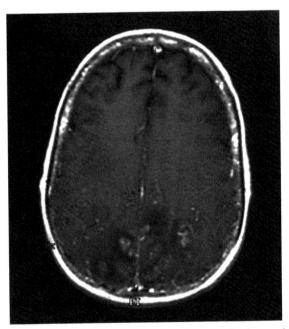

FIG. 3   Enhancement of the leptomeninges and edema of the underlying brain in a 53-year-old woman with a history of a previous mastectomy, but no evidence of metastatic disease is present on MRI. CSF examination confirmed a diagnosis of leptomeningeal metastases.

liquid biopsy techniques for detection of cell-free DNA or circulating tumor cells in the CSF hold promise.[33–36]

Leptomeningeal metastasis when originated from primary breast cancer carries a more favorable prognosis relative to that of other primaries. Nevertheless, the decision to treat is an individualized one that integrates factors such as the patient's functional status, prognosis, systemic disease control, personal values, and goals of care. Patients should be counseled that the intent of intrathecal chemotherapy is to forestall or stabilize neurologic decline rather than reverse it. Historically, the most often considered treatment for leptomeningeal metastasis was intraventricular or intrathecal methotrexate, with consideration of radiotherapy to symptomatic sites and areas of particularly bulky disease.[30,37] Hypofractionated courses of craniospinal irradiation with protons were shown to be safe in a recent phase I study, and a phase II trial is ongoing.[38]

Sites of disruption of CSF flow may be irradiated prior to administration of intraventricular chemotherapy.[30] Ventriculoperitoneal shunting may be necessary to relieve hydrocephalus.[39] In earlier studies, 60%–80% of the patients improved or stabilized after intraventricular methotrexate, with a median survival of 6–8 months.[21,22,40,41] Treated patients died from progressive leptomeningeal disease, systemic disease, or complications related to the latter or its treatment.[22,41] Yet other retrospective studies have questioned the efficacy of intrathecal methotrexate. These studies showed that while 50%–75% improved or stabilized initially, the median survival was only 7–14 weeks.[24,42,43] The duration of survival did not differ significantly from the survival of patients who did not receive intrathecal chemotherapy, and more than 50% of the patients who survived more than 4 months developed a leukoencephalopathy.[24]

Systemic methotrexate has been considered as an alternative to intrathecal methotrexate, avoiding the serious neurological complications of intraventricular treatment. Cytotoxic concentrations of methotrexate in the CSF are maintained for longer after high-dose intravenous methotrexate than when methotrexate is administered intrathecally.[44] In some retrospective studies, survival was longer when intravenous chemotherapy was used in addition to intrathecal treatment.[42,43] In a randomized trial of 35 patients with breast cancer and leptomeningeal metastases, the addition of intraventricular methotrexate to systemic chemotherapy and involved-field radiotherapy did not improve survival or the neurological response when compared to systemic chemotherapy and radiotherapy alone.[45]

Despite its potential for benefit, methotrexate carries a significant risk of neurotoxicity particularly with prolonged therapy, and alternative intrathecal chemotherapeutic agents have been investigated, including cytarabine, topotecan, and trastuzumab for patients with human epidermal growth factor receptor 2 (HER2)-positive cancers.[46–49] Additionally, a single-arm, phase II trial of the immune checkpoint inhibitor pembrolizumab recently demonstrated promising results in a cohort of 20 patients, 17 of whom had leptomeningeal metastasis from a breast primary.[50] Intraventricular chemotherapy with thiotepa, cytarabine, liposomal cytarabine, and combinations of these agents with IT methotrexate does not have a clear advantage over methotrexate monotherapy.[41,51–54] A study of systemic tucatinib, capecitabine, and trastuzumab for patients with HER2 + leptomeningeal metastasis is ongoing.[55]

2.4 Brain

Breast cancer is the most common source of parenchymal brain metastasis in women, and after lung cancer, it is the second most common cause overall.[12,56] The incidence of brain metastases in autopsy studies of patients with breast cancer ranges from 10% to 40%,[14,57] while the incidence in population-based studies is 5%.[58,59] For reasons similar to leptomeningeal metastasis (i.e., improved monitoring, more effective systemic therapies), the incidence of brain metastases among patients with breast cancer is increasing.[58,60]

Brain metastases typically occur late in the course of breast cancer and only occur as the first manifestation of breast cancer in less than 1% of the patients.[58,61–63] The interval between diagnosis of the primary tumor and presentation with brain metastases exceeds 5 years in one-quarter of patients, and moreover, there is no evidence of systemic disease in 30%–40% of the patients when they present with a brain metastasis.[63] In up to 50% of the patients with brain metastases, the brain is the first site of relapse after treatment of the primary tumor,[63–67] and the brain and leptomeninges may be the first sites of relapse after systemic disease has responded to adjuvant chemotherapy.[31,68–70] Failure of systemically administered drugs to penetrate an intact blood-brain barrier may allow micrometastases to develop in the brain while the systemic disease is controlled. The risk of developing brain metastases is greater in younger,[14,57–59,67,71] premenopausal,[64] and black[59] women, women with higher-grade tumors, estrogen receptor- and progesterone receptor-negative tumors, and tumors that overexpress HER-2.[71–77] Women with triple-negative breast cancer—e.g., hormone receptor and HER2-negative—are at the highest risk of developing brain metastasis, followed by women with HER2-positive tumors.[77]

In clinical studies, 40%–60% of the patients with metastatic breast cancer have a single-brain metastasis.[56,63] Posterior fossa metastasis without a supratentorial lesion is found in 13%.[63] Symptoms usually develop over 1–3 months and most commonly consist of headaches, altered mental status, behavioral changes, seizures, focal weakness, and/or focal sensory changes. The neurological symptoms and signs reflect the number, volume, and location of the lesions. Focal neurologic deficits, impairment of cognitive function, reduced level of consciousness, and papilledema may appear alone or in varying combinations. Gadolinium-enhanced MRI is the most sensitive method for detecting brain metastases (see Fig. 4).[63] A contrast-enhancing lesion in the brain is particularly likely to be a metastasis if the patient has active malignant disease outside the CNS, or if multiple lesions are present. If the diagnosis remains in doubt, gadolinium-enhanced MRI can be repeated following an interval of a few weeks.

The prognosis for symptomatic brain metastases is influenced in large part by the biological subtype of breast cancer. Patients with metastatic TNBC, for instance, survive a median of 27 months, while those with HER2 +/HR-, HER2-/HR +, and HER2 +/HR + have a median survival of 30, 47, and 58 months, respectively.[78] Without treatment, the prognosis for symptomatic brain metastases was historically poor and on the order of 2 months; however, in the current era with multimodality treatments, earlier detection, and modern systemic treatment options, some patients live 2 years or more.[64,79,80] Corticosteroids reduce symptoms related to vasogenic edema and are an important mainstay of symptomatic treatment.[63,66] Treatment approaches have become increasingly individualized in recent years, but typically involve radiotherapy, surgical resection, and/or medical therapy.[62]

The recommended approach depends on a number of factors including metastatic burden, lesion size, systemic disease control, and the patient's functional status. Surgery is indicated if there is a need for confirmation of tissue diagnosis or debulking for symptom relief and may often be considered for a single or dominant metastatic lesion larger than 3 cm not residing in a deep structure, or for lesions residing in the posterior fossa.[81] Postoperative adjuvant radiotherapy is typically standard, with either stereotactic radiosurgery (SRS) or fractionated stereotactic radiotherapy (FSRT).[82] For patients with oligometastatic disease—i.e., up to four lesions—numerous recent phase III trials have firmly established SRS alone as the favored approach, with its ability to deliver high doses of radiation to deep-seated lesions or those not amenable to surgical resection due to proximity to eloquent structures.[83–85] And though the maximum treatable metastatic burden for SRS to date remains somewhat controversial, there is some prospective evidence that for up to 10 metastatic lesions comprising less than 15 mL total tumor volume, SRS can be effective in providing local control.[86] SRS is therefore the preferred approach in patients with the otherwise well-controlled or stable systemic disease for which systemic therapy options exist.[87] SRS may also be used as salvage therapy

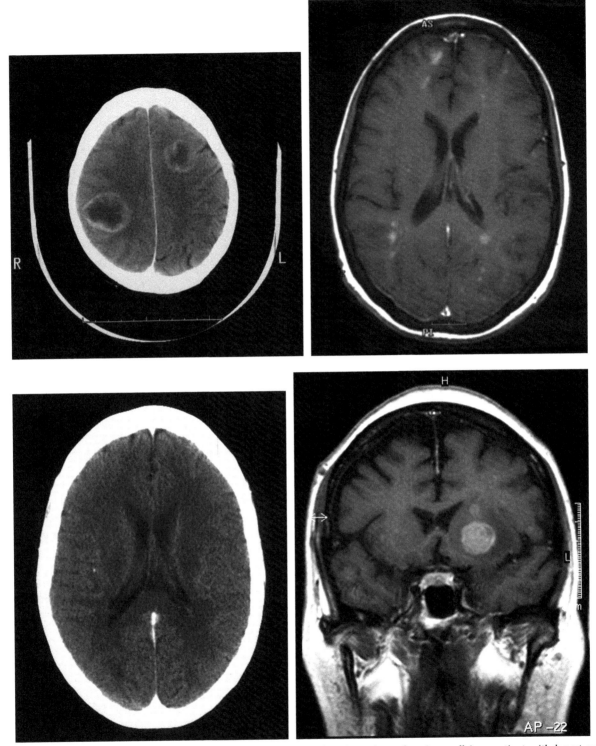

FIG. 4   (A) Brain metastases with a low attenuation central portion and an irregular, enhancing wall in a patient with breast cancer on contrast-enhanced CT. (B) Coronal T1-weighted MRI with gadolinium showing multiple brain metastases in a 58-year-old woman with breast cancer. (C and D) Multiple lesions at the junction of the gray and white matter of both cerebral hemispheres on gadolinium-enhanced T1-weighted MRI (C). These lesions are barely visible on contrast-enhanced CT (D).

following WBRT or neurosurgery. In an uncontrolled series of breast cancer patients, treatment of single- or multiple-brain metastases with SRS resulted in median survival rates of 7–13 months, 1-year survival of 30%, and 2-year survival of 13%.[88–91] Local tumor control was achieved in more than 90% of the lesions. SRS has not been compared with surgery in a randomized trial, but indirect comparison of surgical and radiosurgical case series suggests the outcomes are similar. It remains unclear if WBRT is required after surgical resection or radiosurgery of a single metastasis, as it does not prolong survival, but it does reduce recurrences in the brain.[92]

Patients with extensive metastatic burden or high metastatic volume (i.e., > 15–30 mL) may be considered for whole-brain radiotherapy (WBRT), which given its higher incidence of associated cognitive decline, should primarily be considered in patients for whom the risk of death due to distant failure is high, such as those with high intracranial disease burden and limited extracranial disease, and those with rapidly developing metastases in spite of SRS.[83,93] The high rate of long-term cognitive neurotoxicity (i.e., late onset of dementia, ataxia, and urinary incontinence)[94] with WBRT prompted investigation of hippocampal avoidance (HA-WBRT),[95] which was shown in the NRG CC001 phase III trial to significantly prevent cognitive decline and preserve the patient-reported quality of life.[96] HA-WBRT is therefore the preferred modality in most cases. After WBRT, 70% of the patients with breast cancer and brain metastases improve, 25% stabilize and 5% continue to deteriorate.[56] The median survival following WBRT is only 3–6 months and the survival rate at 1 year is about 20%.[56,63,64,66,97,98] Fifty percent experience progressive or recurrent disease at the original site or new metastases at another site in the brain.[98] Neurological disease is the cause of death or a major contributing factor in 45%–70% of the patients.[63,64] The prognosis after WBRT is affected by performance status, control of the primary tumor, control of systemic disease, menopausal status, and the size and number of the brain metastases.[66,98]

Recognition that the breakdown of the blood-brain barrier (BBB) within metastatic tumors may permit systemically administered drugs that are normally excluded from the brain to reach metastases has generated interest in systemic chemotherapy as a means of controlling extra- and intracranial disease burden. Some reports, for instance, have shown response of hormone receptor-positive brain metastases to tamoxifen and megestrol acetate.[99–101] Conventional cytotoxic chemotherapies such as cyclophosphamide, 5-fluorouracil, and methotrexate have shown activity in retrospective studies; however, prospective data are lacking.[102] Temozolomide and cisplatin were shown to have a 40% objective response rate in a phase II study including patients with brain metastasis from breast cancer primary,[103] but temozolomide as a single agent has failed to demonstrate significant activity despite its favorable BBB penetration.[104] Capecitabine, promisingly, readily crosses the BBB[105] and has demonstrated activity in patients with breast cancer with refractory intracranial metastatic as well as leptomeningeal disease.[106]

Brain metastases may respond to several different chemotherapeutic regimens.[102,103,107–110] Chemotherapy has not been compared with radiotherapy in a prospective randomized trial, but may have a role in patients with multiple cerebral metastases who already have received brain radiotherapy, or in patients with brain metastases and disseminated systemic disease. Targeted therapies hold great promise; a recent randomized, double-blinded, placebo-controlled trial of the oral HER2 inhibitor tucatinib in combination with trastuzumab and capecitabine for patients with refractory HER2 + metastatic breast cancer demonstrated a median overall survival of 21.9 versus 17.4 months with placebo.[111] A phase II study of epidermal growth factor receptor (EGFR) inhibitor lapatinib for patients with highly refractory HER2 + intracranial metastatic breast cancer did not meet its predefined efficacy criteria,[112] but investigators have shown antitumor activity in a phase I trial in combination with capecitabine.[113] Pertuzumab is a fully humanized monoclonal antibody targeting HER2 that is also being actively studied in a phase II trial in combination with high-dose trastuzumab.[114]

2.5 Spinal epidural

Spinal cord or cauda equina compression by metastatic tumor is a common complication of breast cancer, which causes progressive pain and paralysis unless a timely diagnosis is made and appropriate treatment is instituted. Breast, lung, and lymphoma constitute the three most common causes of malignant epidural spinal cord compression.[115–118] The cumulative probability of developing spinal cord compression during the 5 years preceding death from breast cancer is 5.5%.[119] Spinal cord compression typically occurs in patients with widespread metastatic disease; only 0.1% of the patients have malignant spinal cord compression at the time of diagnosis of breast cancer.[119] In contemporary series, the median overall survival from diagnosis of malignant epidural spinal cord compression due to metastatic breast cancer is 47 months.[120,121]

Spinal cord compression is usually caused by metastases to the vertebral body and less commonly to the posterior spinal elements. The spinal cord or nerve roots are compressed by expansion of the tumor into the epidural tissues or intervertebral foramina, or by collapse of the vertebral body and encroachment of the tumor and bone upon the epidural space. In some patients, the neurological deficit is caused by spinal cord ischemia. The thoracic spinal cord is affected in 75% and compression occurs at more than one level in about 30%.[115,118,120,122]

Back pain is usually the first symptom of malignant spinal cord compression, but later is accompanied by radicular pain and typically is exacerbated by Valsalva maneuver and the supine position.[115,118,123] The affected vertebrae are tender to percussion. Without treatment, weakness typically appears weeks later and usually progresses over several days, though a sudden deterioration occurs in about 15% of the patients.[118] Urinary and bowel symptoms typically ensue. A high index of suspicion is critical; in a modern series, 60.9% of the patients diagnosed with malignant epidural spinal cord compression had a minor or major neurologic deficit ranging from incomplete loss of sensory function to complete loss of motor function.[121,124] Encouragingly, 92.9% of those with preoperative neurologic deficits experienced complete recovery, highlighting the need for timely intervention.[121]

Back pain alone is an unreliable predictor of metastatic spinal cord compression, because vertebral metastases occur in two-thirds of patients with metastatic breast cancer.[13,117,123] Radiological investigations may identify patients who are developing spinal cord compression. Plain radiographs are abnormal at the symptomatic level in almost all patients with breast cancer and metastatic spinal cord compression.[118,125] Vertebral body collapse and erosion of the pedicles on plain films are associated with a greater likelihood of metastatic spinal cord compression.[123] In these patients, and those with abnormal neurological signs, MRI determines if there is compression of the spinal cord or nerve roots. Gadolinium-enhanced MRI of the whole spine is required to detect additional asymptomatic lesions. An algorithm for the evaluation of patients with breast cancer presenting with back pain has been developed.[126] Patients with a normal neurological examination, but progressive back pain, pain that is worse in the supine position, or pain made worse by the Valsalva maneuver, also should be evaluated with MRI if no medical contraindications exist. Patients with signs of a myelopathy or radiculopathy should be investigated immediately with MRI.

The main aims of treatment are to relieve pain and retain ambulation. Dexamethasone should be started immediately, because it reduces pain and may stabilize the neurological signs for a short time. Treatment with dexamethasone during radiotherapy increases the probability of ambulation following treatment.[127] The optimal dose of dexamethasone is uncertain, but best-quality evidence at this time supports an initial intravenous bolus of 10 mg followed by 16 mg daily dosing to be rapidly tapered after definitive treatment when there is a severe or rapidly progressive neurological deficit.[128]

In patients with breast cancer, treatment of metastatic spinal cord compression—particularly those who are not surgical candidates—usually consists of radiotherapy. No particular dose fractionation schedule has been conclusively shown to produce better pain relief or a better functional outcome than any other, but more protracted schedules have a lower rate of recurrence in the radiation field.[117,129,130(p1),131] This is important for patients with breast cancer, because they often survive longer and have a better functional outcome than patients with other primary tumors.[118,119] Treatment to 8 Gy over a single fraction or 20 Gy over five fractions is reasonable.[132]

More than 90% of the patients who are ambulatory before treatment remain so after treatment. If treatment is delayed until the patient cannot walk, only 45% regain ambulation.[120]

Recurrent compression at the same or a different site following radiotherapy occurs overall in 20% and is even more common in long-term survivors.[133] Recurrences may be managed with further radiotherapy,[134] but surgery and chemotherapy are other options. Other indications for surgical decompression include spinal instability due to fracture dislocation of the vertebral body and compression of the spinal cord by bony fragments.[117] Surgery is seldom required for tissue diagnosis. Decompressive laminectomy followed by radiotherapy does not have any advantage over radiotherapy alone. Laminectomy increases the risk of spinal instability if the vertebral body is already destroyed by tumor.[115,135]

Resolution of metastatic spinal cord compression may follow chemotherapy or hormonal therapy, but the response is often delayed and uncertain.[136] These treatments are usually not used alone in patients with spinal cord compression.

In patients with metastatic bone disease, bisphosphonates can reduce pain as well as the risk of nonvertebral fractures.[137,138]

2.6 Spinal cord

Intramedullary spinal cord metastases are uncommon compared with metastatic epidural spinal cord compression. Breast cancer accounts for 15% of the cases.[139–141] Most patients have progressive systemic cancer and concurrent brain or leptomeningeal metastases. Patients present with back pain and a rapidly progressive myelopathy.[140,141] The symptoms and signs do not reliably distinguish between an intramedullary spinal cord metastasis and metastatic spinal cord compression, but clues suggesting an intramedullary metastasis include asymmetric motor signs; atrophy and dissociated sensory loss at the level of the lesion; early involvement of bladder and bowel function; and the absence of metastases in adjacent vertebral bodies. Gadolinium-enhanced MRI is the best method for demonstrating an intramedullary metastasis (see Fig. 5).[141,142] The symptoms may improve or stabilize if dexamethasone and radiotherapy are commenced promptly,[140,141] but often, there is a poor outcome following treatment. Rarely, surgical resection of a well-circumscribed lesion is possible if the systemic disease is limited.[143,144]

2.7 Optic nerve

Breast cancer is one of the most common sources of optic nerve metastases.[145,146] An optic nerve metastasis presents with an acute or slowly progressive, painless,

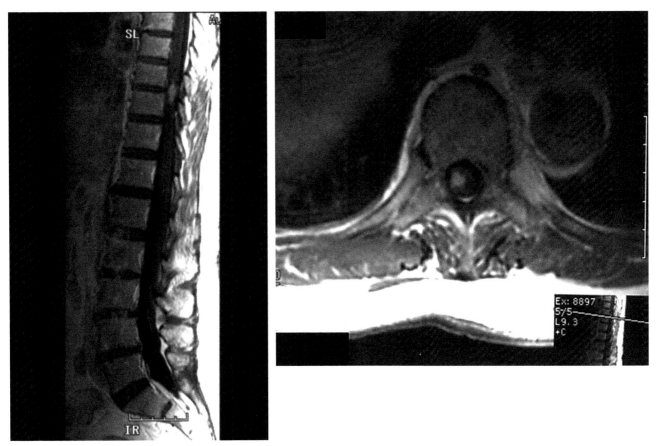

FIG. 5 An intramedullary metastasis in the lower thoracic spinal cord in a 74-year-old woman with breast cancer and widely disseminated metastases is seen on sagittal (A) and axial (B) gadolinium-enhanced MR images.

monocular loss of vision. Optic disk edema may be present. The lesion is usually visible with MR imaging. The clinical presentation can be confused with optic neuritis or ischemic optic neuropathy. Leptomeningeal metastases can present with an optic neuropathy.

2.8 Uvea

Breast cancer is the most common source of uveal metastases.[147] The choroid is most commonly affected, but metastases can lodge in the iris or ciliary body. Ocular screening examinations in patients with disseminated breast cancer identified choroidal metastases in 5%–27%.[148,149] The incidence in an autopsy study was 37%.[150] Bilateral choroidal metastases are present in 40% of the patients.[147,151] In 3%, the choroidal metastasis is the initial manifestation of breast cancer, and in 20%, it is the first site of metastasis.[147,151,152] In most patients, however, choroidal metastases are associated with systemic and brain metastases.[147]

Choroidal metastases have a preference for the macula. The most common symptom is blurred vision, but other symptoms include metamorphopsia, photopsia, floaters, and pain.[147,148,152,153] The diagnosis usually can be made by clinical examination. Treatment is required if the metastasis is affecting vision, associated with retinal detachment, or rapidly enlarging.[148] Choroidal metastases can be treated with radiotherapy, hormonal therapy, or chemotherapy.[147,152,153] Early radiotherapy usually prevents loss of vision and reduces pain.[148,151]

2.9 Pituitary

Metastases in the pituitary gland have been found at autopsy in 9% of the patients with breast cancer, and in 25% of those who have had a hypophysectomy as treatment for bony metastases.[154] The pituitary may be invaded by hematogenous spread, or direct extension of a bony metastasis in the sella.[155–157] The posterior lobe is more commonly involved than the anterior lobe.[158] Most pituitary metastases are asymptomatic, but when symptoms occur, headache, diabetes insipidus, and external ophthalmoplegia are the usual manifestations. Visual loss and hypopituitarism are uncommon.[156] A pituitary adenoma should be suspected if the patient presents with hypopituitarism and visual loss, and there are no other metastases. Typically, MRI shows bony erosion and soft tissue invasion. If the clinical features and imaging suggest there is a metastasis, corticosteroids and radiotherapy should be used. Symptoms may improve during systemic chemotherapy.[155] Resection of a pituitary metastasis is usually difficult, because infiltration of the cavernous sinus is common and the tumor can bleed profusely. However, an operation may be indicated if the clinical features suggest an adenoma may be present, there are no other metastases, and expected survival is greater than 6 months.

2.10 Brachial plexus

Cancers of the lung and breast are the most frequent causes of malignant brachial plexus infiltration.[159–161] Malignant brachial plexopathy occurs in 2.5% of the patients with breast cancer.[162] Unremitting severe pain in the shoulder and arm usually precedes other symptoms by several weeks or months, while weakness, wasting, and sensory symptoms appear later.[159,160] The motor and sensory signs reflect the location of the tumor in the brachial plexus. Breast cancer most commonly involves the lower trunk of the plexus, but the upper trunk can be affected by metastases in the supraclavicular lymph nodes. Horner's syndrome is common if the tumor extends into the paraspinal or epidural tissues. There may be a palpable mass in the axilla or supraclavicular fossa.

Malignant brachial plexopathy must be distinguished from radiation-induced plexopathy (see Table 2), complications of breast surgery, and brachial neuritis. Neurophysiological investigations show prolonged or absent F waves, low amplitude sensory and compound muscle action potentials, fibrillations, and neurogenic motor units. The paraspinal muscles are normal unless there is epidural extension of the tumor. Myokymia is uncommon unless the patient has received radiotherapy to the brachial plexus.[160] A mass within the brachial plexus on MRI or CT is highly predictive of malignant infiltration, but increased T2 signal adjacent to the brachial plexus and loss of fat planes is common in both malignant and radiation-induced plexopathy.[160,161,163,164] In some patients with malignant brachial plexopathy, the plexus appears to be normal on CT and MRI.[165,166] Positron emission tomography (PET) may show abnormal uptake of 18-fluorodeoxyglucose in the brachial plexus in the absence of a CT abnormality,[167] but the role of PET in distinguishing between malignant and radiation plexopathy is unclear. If the diagnosis remains uncertain, CT-guided biopsy[168] or surgical exploration of the plexus may be required.[159]

Radiotherapy is the treatment of choice for malignant brachial plexus infiltration. It may reduce pain initially, but chronic pain is common. In selected patients, anesthetic nerve blocks and neuroablative techniques are used.[169] The other symptoms of malignant brachial plexopathy usually worsen despite treatment. Chemotherapy and hormonal therapy can be tried if radiotherapy has been used in the past.

TABLE 2   Differentiation between malignant brachial plexopathy and radiation-induced brachial plexopathy.

	Malignant plexopathy	Radiation plexopathy
Prior radiotherapy	Sometimes	Always
Shoulder, arm pain	Early, severe	Late, milder
Palpable mass	Common	Nonexistent
Horner's syndrome	Sometimes	Uncommon
Myokymia	Rare[a]	Common
Paraspinal fibrillations	Rare	Common
MRI or CT	Mass	No mass
	Increased T2 signal	Increased T2 signal
	Loss of fat planes	Loss of fat planes

[a] Unless previous radiotherapy to the brachial plexus.

2.11 Lumbosacral plexus

Infiltration of the lumbosacral plexus by bony metastases in the sacrum and pelvis is a much less common complication of breast cancer. Breast cancer accounts for 10% of the cases of malignant lumbosacral plexopathy.[170] The typical presentation is insidious onset of severe pain in the low back, buttock, or thigh.[170] Unilateral or asymmetric weakness, sensory loss, reflex loss, and leg edema appear weeks or months later. Bowel and bladder function is usually normal unless there is epidural extension of the tumor. Compound muscle and sensory action potentials are reduced, and fibrillations and neurogenic motor units are present. Erosion of the sacrum is often visible on MRI or CT.[171] Radiotherapy and chemotherapy may stabilize the symptoms.

The differential diagnosis for malignant lumbosacral plexopathy should include leptomeningeal metastases, compression of the cauda equina by an epidural metastasis, and radiation-induced plexopathy. Leptomeningeal and epidural metastases usually affect bladder and bowel function. Early onset of bilateral leg weakness and sensory loss and myokymia are typical features of radiation-induced plexopathy and, while pain eventually develops in 50% of the patients, it rarely is a major problem.[171]

3 Nonmetastatic complications

3.1 Metabolic disorders

Metabolic encephalopathy—that is, altered consciousness and/or mentation due to a systemic cause—is one of the most common neurological manifestations of disseminated cancer. The condition is a clinical entity with many potential contributing factors, including electrolyte/osmolar or blood gas derangements, vital organ (e.g., liver or kidney) dysfunction, polypharmacy, and infection to name just a few.[172] The clinical features include confusion, inattention, behavioral changes, altered level of consciousness, and tremulousness.[173] The neurologic exam is typically nonfocal, though etiological hints may be gleaned from close examination of the pupils (i.e., miosis with opioid intoxication) or any abnormal movements (e.g., asterixis or myoclonus in patients with uremia, hyperammonemia, or hypercarbia). In patients with breast cancer, hypercalcemia is one of the most common causes of metabolic encephalopathy, though hyponatremia is also sometimes seen. The symptoms often begin acutely and often fluctuate. Comprehensive laboratory examination should include at a minimum a complete blood count with differential, metabolic panel with renal function testing, liver function testing, ammonia, blood gas and glucose measurements, and any measurable trough levels of maintenance medications. Electroencephalogram (EEG) may also be considered to rule out subclinical seizures, as even patients without structural intracranial lesions may develop these in the appropriate settings. If the history or aforementioned ancillary testing is supportive, obtaining CSF to evaluate for infectious, inflammatory/autoimmune or paraneoplastic disorders may be indicated.

3.2 Cerebrovascular disease

Knowledge of the various cerebrovascular complications that may result from breast cancer or its treatments is crucial. Patients with breast cancer accounted for 4% of the patients who were admitted to a large cancer hospital with a stroke.[174] While heart disease and atherosclerosis cause most of the strokes that occur in patients with cancer, other etiologies bear consideration in patients with breast cancer (see Table 3).[175–177] Certain chemotherapies increase the risk of cerebral infarction, hemorrhagic infarction, and other thromboembolic events in patients with breast cancer. Tamoxifen does not raise the overall risk of stroke or stroke mortality.[178–182] Raloxifene, a selective estrogen-receptor modulator, does not increase the overall incidence of stroke, but it is associated with a higher risk of fatal stroke.[183] Paradoxical embolism via a patent foramen ovale may be an underrecognized cause of ischemic stroke in patients with cancer, in addition to the hypercoagulability of malignancy and nonbacterial thrombotic (also known as "marantic") endocarditis.[184]

Any patient presenting with suspected stroke symptoms should undergo at minimum emergent noncontrast CT scan of the head, along with CT angiography of the head and neck[185] as part of their evaluation for candidacy for medical thrombolysis or mechanical thrombectomy. If time-sensitive decisions do not need to be made based on a patient's presentation, MRI—with consideration of gadolinium contrast if concern

TABLE 3 Causes of cerebrovascular disease in patients with breast cancer.

Ischemic strokes

Large artery disease

Atherosclerosis

Irradiation of extracranial carotid and vertebral arteries

Small vessel disease

Small vessel disease secondary to hypertension

Disseminated intravascular coagulopathy

Thrombotic microangiopathy

Leptomeningeal metastases

Intravascular mucin

Cardiac and other embolic causes

Common cardiac diseases (e.g., atrial fibrillation)

Nonbacterial thrombotic endocarditis

Paradoxical embolism through patent foramen ovale

Tumor embolism

Cerebral vein thrombosis

Hypercoagulable state

Metastatic compression of cerebral veins

Venous infiltration by tumor cells

Intracerebral Hemorrhage

Hemorrhagic metastases

Thrombocytopenia

Acute disseminated intravascular coagulopathy

Subdural hemorrhage secondary to dural metastasis

exists for intraparenchymal neoplastic involvement—of the brain with particular focus on diffusion-weighted imaging (DWI), apparent diffusion coefficient (ADC), and FLAIR sequences should be considered.[186] Once stroke or transient ischemic attack (TIA) is confirmed, a thorough workup of contributing vascular risk factors such as diabetes mellitus, hyperlipidemia, and cardiac arrhythmias should be undertaken. Transthoracic echocardiography should also be pursued to evaluate for intracardiac thrombi, patent foramen ovale, and valvular vegetations. The management and secondary prevention of ischemic stroke in patients with breast cancer depend, as in the general population, on the etiology. Antiplatelet agents should be considered for secondary prevention in patients with known cerebrovascular events such as TIAs or strokes secondary to atherosclerotic cerebrovascular disease, and anticoagulation is indicated in patients with stroke secondary to hypercoagulability of malignancy.

3.2.1 Disseminated intravascular coagulation

In an autopsy study, 1% of the patients with breast cancer had disseminated intravascular coagulation (DIC) with brain involvement.[187] Breast cancer accounts for 10% of the solid tumors complicated by DIC.[188] Overexpression of urokinase and the urokinase receptor in breast tumors may account for the occurrence of DIC in patients with breast cancer.[188] Patients typically present with an acute encephalopathy, but seizures and focal signs also may occur.[189] Death usually occurs within a few weeks. The neurological symptoms are caused by occlusion of small penetrating arteries and capillaries by fibrin thrombi, resulting in multiple cerebral infarcts and petechial hemorrhages. The neurological complications of DIC must be distinguished from other causes of diffuse or multifocal brain disease (see Table 4). Coagulation studies are often abnormal, but the abnormalities may be wrongly attributed to another cause.[187] CT and cerebral angiography are often normal. Heparin is usually not beneficial.

3.2.2 Thrombotic microangiopathy

Breast cancer may be complicated by a thrombotic microangiopathy leading to the formation of platelet-rich microvascular thrombi.[190] The cardinal features are a microangiopathic hemolytic anemia, thrombocytopenia, and renal failure. Neurological manifestations are common and may include headache, confusion, coma, and focal signs. Schistocytes are present in peripheral blood smears. The prothrombin and partial thromboplastin times are normal, unless there is a concomitant low-grade DIC. Chemotherapy has been implicated in the

TABLE 4 Differential diagnosis according to the localization of the lesion.

Headache

Metastases	Calvarium and skull base
	Dura
	Pituitary
	Leptomeninges
	Brain
Therapy	Bacterial meningitis 2° to Ommaya reservoir
	Acute meningoencephalopathy 2° to IT methotrexate
	Acute radiation encephalopathy
Vascular	Cerebral vein thrombosis
	Thrombotic microangiopathy
	Intracerebral hemorrhage
	Subdural hematoma

Encephalopathy

Metastases	Dura, leptomeninges, and brain
Metabolic	Hypercalcemia, liver failure, drugs, and others
Therapy	Radiation encephalopathy: acute and chronic
	Intrathecal methotrexate: acute and chronic
	High-dose tamoxifen
	Chemotherapy
Vascular	Disseminated intravascular coagulopathy
	Thrombotic microangiopathy
	Nonbacterial thrombotic endocarditis
	Intravascular mucin
	Cerebral vein thrombosis
	Intracerebral hemorrhage

TABLE 4 Differential diagnosis according to the localization of the lesion—cont'd

| Paraneoplastic | Limbic encephalitis |
| | Anti-Ri syndrome |

Seizures

Metastases	Dura, leptomeninges, and brain
Metabolic	Hyponatremia and others
Therapy	IT methotrexate leukoencephalopathy
	Inappropriate ADH secretion secondary to vincristine
	High-dose 5-fluorouracil
	Radiation encephalopathy
Vascular	Cerebral vein thrombosis
	Subdural hematoma
	Other causes
Paraneoplastic	Limbic encephalitis

Focal signs

Metastases	Dura, leptomeninges, and brain
Therapy	Methotrexate leukoencephalopathy
	Acute or chronic radiation encephalopathy
Vascular	Multiple causes

Ataxia

Metastases	Brain and leptomeninges
Therapy	Methotrexate leukoencephalopathy
	High-dose 5-fluorouracil
	High-dose tamoxifen
	Radiation
Vascular	
Paraneoplastic	Paraneoplastic cerebellar degeneration
	Opsoclonus-ataxia (anti-Ri)
	Anti-Ma2 encephalitis

Myelopathy

Metastatic	Epidural spinal cord compression
	Intramedullary spinal cord metastases
Therapy	Radiation myelopathy
	Intrathecal methotrexate
	Steroid-induced epidural lipomatosis

Arm pain and weakness

Metastases	Spinal epidural, leptomeninges, and brachial plexus
Therapy	Radiation plexopathy (early reversible or delayed)
	Radiation-induced brachial plexus tumors
	Ischemic brachial plexopathy
	Postmastectomy syndrome (intercostobrachial neuralgia)

Continued

III. Neurological complications of specific neoplasms

TABLE 4   Differential diagnosis according to the localization of the lesion—cont'd

Leg pain and weakness	
Metastatic	Spinal epidural, leptomeninges, and lumbosacral plexus
Therapy	Radiation plexopathy
Peripheral neuropathy	
Therapy	Taxanes and vincristine
Paraneoplastic	Sensory neuropathy
	Sensorimotor neuropathy
Unilateral visual loss	
Metastases	Orbit, optic nerve, choroid, and leptomeninges
Paraneoplastic	Optic neuritis
Bilateral visual loss	
Metastases	Leptomeninges, pituitary, and choroid
Therapy	Tamoxifen retinopathy
	Intrathecal methotrexate
	Radiation
Paraneoplastic	Paraneoplastic retinopathy
Ophthalmoplegia	
Metastases	Skull base, pituitary, and leptomeninges
Paraneoplastic	Anti-Ri
Cranial nerve palsies	
Metastases	Skull base, mandible, dura, leptomeninges, and brain stem
Therapy	Vincristine
Paraneoplastic	Anti-Ma2 encephalitis

ADH, antidiuretic hormone; *IT*, intrathecal.

pathogenesis of thrombotic microangiopathy in some patients.[191] Plasma exchange, immunoadsorption columns, and immunosuppressive drugs may be beneficial.

3.2.3 Nonbacterial thrombotic endocarditis

Five to 10% of the malignancies associated with non-bacterial thrombotic endocarditis (NBTE) are breast cancers.[192,193] These patients present with transient ischemic attacks and strokes in multiple vascular territories, diffuse encephalopathy, or both. The neurological manifestations result from embolization of valvular platelet-fibrin vegetations, or multifocal occlusion of cerebral arteries secondary to DIC, which commonly accompanies NBTE. Systemic emboli, venous thrombophlebitis, and a heart murmur sometimes are present. There are acute infarcts in multiple vascular territories on diffusion-weighted MRI and arteriography shows occlusion of multiple cerebral arteries.[194] Transesophageal echocardiography sometimes identifies the valvular vegetations.[195] Apart from the treatment of the underlying malignancy, indefinite anticoagulation is warranted;

in selected patients with severe valvular issues, surgery may be considered.[196–198]

3.2.4 Cerebral vein thrombosis

Cerebral vein thrombosis is a diagnostically challenging entity with few pathognomonic features that therefore requires a high index of suspicion to detect. It may be caused or contributed to by a baseline hypercoagulable state, venous compression by a dural or calvarial metastasis, or infiltration of veins by tumor cells or infections.[199–203] Headache, papilledema, seizures, focal signs, and/or a diffuse encephalopathy typically develop over a few days. Cerebral vein thrombosis may also be complicated by hemorrhagic venous infarction. MRI and MR venography confirm the diagnosis. Anticoagulation is the usual treatment, even in the presence of hemorrhagic infarction. Radiotherapy and chemotherapy may be warranted when cerebral vein thrombosis is caused by neoplastic venous compression. In select cases where patients clinically worsen despite appropriate anticoagulation, endovascular intervention may be indicated.[202,204]

3.2.5 Intravascular mucin

Production of intravascular mucin by breast carcinomas may cause strokes by a direct hypercoagulable effect of the mucinous proteins, hyperviscosity, concomitant development of NBTE, or occlusion of small cerebral arteries by mucinous material.[205–207] The typical clinical picture is a rapidly progressive encephalopathy and multiple acute strokes. At autopsy, there are multiple hemorrhagic cerebral infarcts. High serum levels of CA-125 suggest the diagnosis.[207]

3.2.6 Intracerebral hemorrhage

Intracerebral hemorrhage is relatively uncommon among patients with breast cancer. Less than 1% of the brain metastases are hemorrhagic and thrombocytopenia is a rare cause of intracerebral hemorrhage.[189] Dural metastases can cause—and, importantly, be obscured by—subdural hematoma. In contrast to patients with chronic low-grade DIC, intracranial hemorrhage is a common complication of an acute DIC.[208] Coagulation studies in acute DIC are severely deranged. These patients are managed with platelet infusions and replacement of clotting factors. The use of heparin is controversial.

3.3 Infections

Coagulase-negative staphylococci, *Propionibacterium acnes,* and *Staphylococcus aureus* are the most common causes of meningitis associated with the use of an Ommaya reservoir to administer intrathecal chemotherapy such as methotrexate, which occurs in 5%–8% of the patients who undergo Ommaya placement.[22,209–212] Although some cases can be successfully treated with intravenous and intrathecal antibiotics without the removal of the reservoir, it may be necessary to remove the reservoir if the patient has a persistent or recurrent infection. Other CNS infections are uncommon in patients with breast cancer, but chemotherapy can produce defects in the immune system, which increase susceptibility to infection.

3.4 Complications of chemotherapy

3.4.1 Intrathecal chemotherapy

A common iatrogenic cause of severe CNS complications in patients with breast cancer is intrathecal methotrexate. Acute meningitis consisting of headache, fever, neck stiffness, and vomiting can develop a few hours after an intrathecal injection of methotrexate.[21,22] There is a rise in the CSF leukocyte count and protein content above pretreatment values, but cultures are sterile. The symptoms resolve after 12–72h and they do not necessarily recur with subsequent administration of methotrexate.

A mild encephalopathy variably associated with neck stiffness or CSF abnormalities also can complicate intrathecal administration of methotrexate or other chemotherapy.[213] The symptoms generally resolve within a few days, but there may be a relationship between this syndrome and the late onset of a necrotizing leukoencephalopathy.

Leukoencephalopathy is the most serious late complication of repeated intrathecal injections of methotrexate, developing in more than 50% of those who survive 1 year or longer after starting treatment.[24] It usually occurs when intrathecal methotrexate is combined with whole-brain radiotherapy, but it may develop after intrathecal methotrexate alone. Behavioral abnormalities, confusion, dementia, somnolence, ataxia, seizures, hemiparesis, or quadriparesis begin insidiously or abruptly 3 months or more after treatment is commenced. There are periventricular white matter abnormalities on MRI and CT. The symptoms can stabilize when methotrexate is stopped, or they may progress to coma and death.

High concentrations of methotrexate adjacent to the Ommaya reservoir catheter can cause a focal leukoencephalopathy.[214] These lesions enhance on MRI after contrast injection and they may be associated with mass effect. Other rare complications attributed to intrathecal methotrexate include myelopathy, optic neuropathy, and sudden death.[215]

3.4.2 Ocular complications of tamoxifen

Tamoxifen can cause a retinopathy marked by the appearance of crystalline deposits in the nerve fiber and inner plexiform layers of the retina.[216,217] Extensive deposits may be associated with macular edema and impaired visual acuity. Most cases have occurred after high doses of tamoxifen, but retinopathy can occur with conventional doses administered over long periods. Macular edema and visual acuity improve after tamoxifen is withdrawn, but the retinal deposits persist. Corneal opacities also develop after prolonged treatment with tamoxifen, but they may not be clinically significant.[217] Optic neuropathy has been attributed to tamoxifen in a few patients, but the association may have been coincidental.[218] Other rare complications of tamoxifen include posterior reversible encephalopathy syndrome (PRES) and cerebellar ataxia.[219]

3.4.3 Other complications of chemotherapy

Retrospective and prospective studies have identified subtle cognitive impairments in up to 75% of the patients with breast cancer following systemic chemotherapy.[220,221] The most commonly affected cognitive domains are attention, learning, and processing speed. In 50% of these patients, cognitive impairment eventually improves and, in the others, it remains stable. Capecitabine can cause a reversible multifocal leukoencephalopathy in patients with breast cancer.[222] Many patients with metastatic breast cancer are treated with corticosteroids and, as in other patient populations, neurological complications, such as steroid myopathy, may occur. The vinca alkaloids and taxanes commonly cause peripheral neuropathy. Paclitaxel and docetaxel cause a predominantly sensory neuropathy by interfering with axonal microtubule function.

3.5 Complications of radiotherapy

Radiotherapy-related complications are generally divided into acute (days-weeks), early-delayed (1–6 months), and late-delayed (months to years) phases.[223] Acute complications can occur after standard conformal or stereotactic radiotherapy and are attributable to focal cerebral edema that can cause headache, nausea, seizures, and acute worsening of preexisting neurologic symptoms. These generally respond well to a short course of corticosteroids. Early-delayed—or subacute—complications are believed to be a consequence of radiation-induced demyelination and can take one of the various clinical syndromes including an encephalopathy with mental status depression, lethargy, and headache; this has been described as a "somnolence syndrome."[224,225] The best-known late-delayed complications of radiotherapy are radiation necrosis and progressive encephalopathy, the latter of which is historically associated with regimens including WBRT and has been discussed earlier.[87,226,227] Radiation necrosis can occur as early as 3 months to as late as several years after the receipt of radiotherapy, may have a "soap-bubble," "cut pepper," or "Swiss cheese" pattern of enhancement along with abundant T2 hyperintense signal usually within the prior radiation field, and, importantly, can be either symptomatic or asymptomatic. Increased relative cerebral blood volume on MR perfusion and MR spectroscopic findings of elevated lipid-lactate peak—as opposed to elevated choline-creatinine ratio—may also be helpful in distinguishing radiation necrosis from tumor recurrence. Patients with symptomatic radiation necrosis typically respond well to steroids, although in refractory cases medications such as bevacizumab may be indicated.[228]

3.5.1 Brachial plexus

Radiotherapy can affect the brachial plexus in several ways. A reversible brachial plexopathy develops in 1%–2% of the patients with breast cancer 1.5–14 months following irradiation of the axilla.[229,230] The initial symptoms are numbness and paresthesias in the hand and forearm, and mild shoulder and axillary pain. In some patients, there is wasting and weakness, but the scapular and rhomboid muscles are spared. The pathogenesis of this disorder is unknown, but it is not associated with a delayed, irreversible plexopathy.

The incidence of delayed, irreversible, radiation-induced brachial plexopathy in patients with breast cancer varies widely in different case series depending on the duration of follow-up and the use of different dose fractionation schedules.[231] Radiation plexopathy usually starts 6 months or more after radiotherapy.[231,232] Presentations more than 10 years after irradiation are not uncommon. The incidence is 2.9% per year.[233] The incidence increases as the total dose and the dose per fraction increase, and with overlapping treatment fields. Brachial plexopathy is more likely after total doses of 60 Gy or more, but it can occur after smaller doses, especially if they are delivered in fractions exceeding 2 Gy.[232,234,235] Initially, the upper trunk of the plexus is usually affected, but the lower trunk or the whole plexus may become involved. Paresthesias and numbness are the dominant presenting symptoms. Lymphedema, induration of the supraclavicular fossa, and slowly progressive arm weakness develop later.[159,160,231,236] The prevalence of pain increases over time, but it is seldom as severe as the pain associated with malignant plexopathy.[90,159] Myokymia and paraspinal fibrillations suggest radiation plexopathy, but nerve conduction studies and electromyography usually do not distinguish between malignant infiltration and radiation plexopathy. CT may show loss of normal tissue planes without a discrete mass.[160,163] Increased signal adjacent to the brachial plexus is typically present on T2-weighted MRI, but imaging may be normal.[161,164] Various treatments have been investigated throughout the years including steroids, hyperbaric oxygen, and antioxidants. Unfortunately, an effective therapy has yet to be identified.[237]

Peripheral nerve sheath tumors can develop many years after axillary radiotherapy, when patients are cured of their original tumors.[161,238] The cumulative incidence of sarcomas following irradiation of breast tumors is 0.2% at 10 years.[230,239] They present with a painful mass and a progressive brachial plexopathy. This complication may be more common in patients with neurofibromatosis, when radiation fields have overlapped.

Ischemic brachial plexopathy secondary to occlusion of the subclavian artery is a rare, late complication of radiotherapy. It presents with acute, nonprogressive, painless weakness of the arm, or a slowly progressive plexopathy.[240,241] There are reduced pulses and other signs of ischemia in the affected arm.

3.6 Complications of surgery

Chronic neuropathic pain develops in up to two-thirds of patients after breast surgery.[242] Three principal types of pain occur: intercostobrachial neuralgia, phantom breast pain, and pain associated with a neuroma.

3.6.1 Intercostobrachial neuralgia

Intercostobrachial neuralgia is the most common cause of arm pain after breast cancer surgery.[243] It is caused by damage to the intercostobrachial nerve and cutaneous branches of other intercostal nerves during axillary lymph node dissection.[244–246] This disorder was originally called postmastectomy pain syndrome, but the risk of damage to the intercostobrachial nerve is just as great after a lumpectomy with axillary dissection as after mastectomy.[242] Symptoms typically begin immediately after surgery.[246,247] Onset of pain beyond 6 months is unusual and, in this setting, recurrence of tumor in the chest wall should be excluded before a diagnosis of intercostobrachial neuralgia is made. Patients with intercostobrachial neuralgia have constricting, burning, and lancinating sensations, and sensory loss in the axilla, posteromedial upper arm, and anterior chest wall.[244,245,247] Shoulder movements increase the severity of the pain and a frozen shoulder can develop.[248] Attempts to prevent intercostobrachial neuralgia by modifying surgical technique are usually unsuccessful, because the surgeon cannot preserve the intercostobrachial nerve in one-third of axillary lymph node dissections.[249] Amitriptyline, venlafaxine, and topical capsaicin were helpful in managing intercostobrachial neuralgia in randomized, placebo-controlled trials.[242] Other therapeutic options include analgesics, carbamazepine, gabapentin, lidocaine patches, and nerve blocks. Despite treatment, 50% still have symptoms several years after surgery.[250]

3.6.2 Other complications of surgery

Phantom breast sensations develop in the first month in 15%–40% of the patients after a mastectomy.[251–254] Phantom breast pain is less common but still occurs in 10%–20% of the patients.[252–254] Intercostal neuromas can form in scar tissue after breast surgery and cause pain and trigger point sensitivity. Surgical resection of the neuroma may provide pain relief.[255] Carpal tunnel syndrome and brachial plexopathy have been attributed to lymphedema following radical mastectomy, but the evidence for a causal link is not convincing.[243,256]

3.7 Paraneoplastic syndromes

Neurological paraneoplastic syndromes are rare disorders that often develop before diagnosis of the tumor. Prompt recognition of a paraneoplastic syndrome may lead to early diagnosis of the primary tumor, though often extended surveillance is necessary to definitively diagnose a tumor. Fluorodeoxyglucose-PET can detect a breast tumor when mammography, echography, and MRI are normal, but normal PET does not exclude breast cancer.[257,258]

Small cell lung cancer, gynecological malignancies, and breast cancer are the tumors most commonly associated with neurological paraneoplastic syndromes.[259] Croft and Wilkinson found a carcinomatous neuromyopathy (abnormal neurological signs which could not be explained by metastases or another neurological disorder) in 4.4% of the women with breast cancer.[260] Several different neurological paraneoplastic syndromes can occur in patients with breast cancer, including paraneoplastic cerebellar degeneration, sensory peripheral neuropathy, opsoclonus-myoclonus, encephalomyelitis, stiff-person syndrome, limbic encephalitis, and retinopathy.[259,261]

3.7.1 Cerebellar degeneration

Paraneoplastic cerebellar degeneration (PCD) is the most common paraneoplastic syndrome encountered among patients with breast cancer.[257] The anti-Yo (anti-Purkinje cell) antibody is detected in the serum and the CSF in two-thirds of patients with breast cancer and PCD.[259] Most patients harboring an anti-Yo antibody have limited regional metastatic disease.[262] Anti-Yo is a marker for a pure cerebellar syndrome.[263–266] Truncal and bilateral limb ataxia, dysarthria, and nystagmus develop rapidly over a few days or weeks. The symptoms eventually stabilize, but only once the patient is severely incapacitated.[262] Spontaneous improvement does not occur. The CSF often shows a mild lymphocytic pleocytosis, increased protein and IgG levels, and oligoclonal bands. Initially, CT and MRI are normal, but later, they show cerebellar atrophy. The main pathological abnormality is a severe diffuse loss of Purkinje cells. Anti-Yo is a highly specific marker of PCD associated with breast, ovarian, and other gynecologic malignancies. It does not occur in breast cancer patients who do not have PCD.[259,267] Treatment of the breast cancer, corticosteroids, cyclophosphamide, and plasmapheresis does not improve PCD, although survival for several years is not unusual.[262]

Some patients with breast cancer and PCD do not have anti-Yo or any other onconeural antibody.[259] The clinical syndrome of seronegative PCD is heterogeneous. The clinical features may be indistinguishable from anti-Yo-associated PCD, but others have milder symptoms, which often improve after the removal of the primary tumor.[218,219]

3.7.2 Opsoclonus-myoclonus syndrome

Subacute onset of opsoclonus, which is often associated with truncal ataxia, dysarthria, myoclonus, vertigo, and encephalopathy, can occur with a variety of malignancies including breast cancer.[268,269] The CSF may show a mild pleocytosis and an elevated protein. MRI is usually normal. The symptoms usually stabilize and partial remissions can occur after the treatment of the primary tumor. Opsoclonus is caused by dysfunction of the brain stem omnipause neurons, but the neuropathology has not been well characterized. In some patients, there is no identifiable abnormality, while in others, the changes resemble PCD.[268]

The antineuronal antibody (anti-Ri) is most commonly found in patients with breast cancer and opsoclonus, but it can be present with other primary tumors.[270,271] The absence of anti-Ri does not rule out breast cancer in a patient presenting with opsoclonus.[259,269] Anti-Ri is not present in women with breast cancer without paraneoplastic disease.[271] A broad spectrum of neurological disorders other than opsoclonus has been observed in patients with an anti-Ri antibody including a subacute brain stem syndrome (vertigo, ophthalmoplegia, dysarthria, and ataxia), progressive polyradiculopathy, rigidity resembling stiff-person syndrome, and multifocal CNS disease.[259,272,273]

3.7.3 Limbic encephalitis

The main symptoms of limbic encephalitis are memory loss, confusion, seizures, psychiatric abnormalities, hypothalamic dysfunction, and hypersomnia.[274] The CSF often has a mild pleocytosis, increased protein, and oligoclonal bands. T2-weighted MRI shows abnormalities in the medial temporal lobes (see Fig. 6). Breast cancer accounts for 5% of the cases of paraneoplastic limbic encephalitis.[274] Most patients with breast cancer and limbic encephalitis do not have an onconeural antibody, but the anti-Ma2 antibody has been identified in several cases.[258,274–276] Anti-Ma2 is more commonly found in patients with testicular cancer and patients often have upper brain stem and diencephalic involvement, as well as limbic encephalitis.[277] Patients with breast cancer may have antibodies against Ma1 as well as Ma2.[277,278] Anti-Ma1 antibodies are associated with more severe

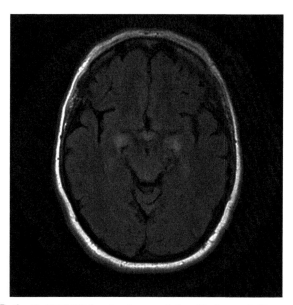

FIG. 6   FLAIR MRI showing increased signal in the medial temporal lobes in a 59-year-old woman with breast cancer and paraneoplastic limbic encephalitis. The patient presented with drowsiness, altered mood, amnesia, visual hallucinations, and somnolence. Antineuronal antibodies were not detected.

brain stem dysfunction and cerebellar ataxia. Atypical antibodies have been found in other patients with breast cancer and limbic encephalitis.[274,276] Paraneoplastic limbic encephalitis may improve when the tumor is treated or following treatment with corticosteroids.

3.7.4 Stiff-person syndrome

The stiff-person syndrome is characterized by progressive rigidity in the axial and proximal limb muscles and muscle spasms. The neurological examination and brain and spinal cord MRI are usually normal. Oligoclonal bands are often present in the CSF. The symptoms improve with high doses of benzodiazepines, corticosteroids, immunosuppressive drugs, or plasmapheresis. Antibodies to the synaptic protein amphiphysin are present in patients with breast cancer and paraneoplastic stiff-person syndrome.[279–282] Antiamphiphysin antibodies are not specific for breast cancer or stiff-person syndrome. They have been found in patients with other tumors and with other paraneoplastic diseases.[283–285] Many of the other clinical manifestations associated with antiamphiphysin antibodies may be explained by coexisting antibodies to other onconeural antigens expressed by the same tumor.[285] Rhabdomyolysis occurred in a patient with breast cancer, stiff-person syndrome, and amphiphysin antibodies.[286] IgG from a patient with breast cancer, antiamphiphysin antibody, atypical stiff-person syndrome, encephalopathy, and opsoclonus was used to passively transfer the disorder to mice.[287]

3.7.5 Sensory and sensorimotor neuropathy

A sensory neuropathy is among the most frequent paraneoplastic syndromes encountered in patients with breast cancer.[259] In a group of 12 patients with breast cancer and sensory neuropathy, three had an anti-Hu antibody and two had an antiamphiphysin antibody. The clinical picture in the seropositive patients resembled the subacute sensory neuropathy associated with small cell lung cancer and the anti-Hu antibody. The clinical features in the seronegative patients were heterogeneous. Some seronegative patients had a large fiber sensory neuropathy and gait ataxia. Two patients presented with paresthesias and itching in the legs. Similar patients with breast cancer and an axonal sensorimotor neuropathy presenting with paresthesias, itching, numbness, mild weakness, and cramps have been reported.[288,289] The initial sensory symptoms often occurred in an unusual distribution including the chest wall, perineum, and face. The symptoms developed over several weeks and then stabilized, progressed gradually, or fluctuated. Disability was minimal until late in the course. Peripheral nerve pathology has not been well characterized in these patients. Transient improvement followed treatment of the cancer in one-third. A sensorimotor neuropathy resembling Guillain-Barré syndrome or chronic inflammatory demyelinating polyradiculoneuropathy can occur in patients with breast cancer.[259]

3.7.6 Motor neuron disease

The occurrence of cancer and motor neuron disease in the same patient usually is coincidental. One possible exception is an association between a slowly progressive upper motor neuron disorder resembling primary lateral sclerosis and breast cancer.[290] Examination shows pyramidal weakness, hyperreflexia, extensor plantar responses, and pseudobulbar palsy, but a normal sensory examination. Antineuronal antibodies are not detected. Some patients eventually develop lower motor neuron dysfunction, but even after several years, there may be no lower motor neuron involvement. Women who present with primary lateral sclerosis probably should have mammography to rule out breast cancer.[290]

3.7.7 Other paraneoplastic syndromes

A few patients with breast cancer and Lambert-Eaton myasthenic syndrome have been reported, but the association may have been coincidental.[291] Dermatomyositis and polymyositis occur more frequently in patients with malignancy, including breast cancer.[267] Uveitis, macular edema, and optic disk swelling associated with a CSF pleocytosis have been described as a paraneoplastic syndrome in patients with breast cancer.[292,293] Breast cancer can cause a retinopathy, which presents with subacute loss of vision.[294] Antiretinal antibodies may be present.

4 Regional approach to the diagnosis of neurological symptoms

As such a wide variety of metastatic and nonmetastatic neurological complications of breast cancer exist, capable of affecting different parts of the central and peripheral nervous systems, a lesion at one site may have several potential causes. Once the location of the lesion has been determined by the history and the neurological examination, the etiology must be determined. Table 4 provides, for each type of localization or symptom, an initial set of differential diagnoses that may be considered in the appropriate context. In some patients, the history and examination suggest the likely cause but often further investigations are required.

5 Conclusion

From the preceding discussion, it may be concluded that although the neurological manifestations of breast cancer and cancer therapy are multifarious, a methodical approach based on careful characterization of symptoms and thorough physical examination informed by the per-

tinent historical framework will maximize the chances of accurate diagnosis and appropriate treatment.

References

1. Momenimovahed Z, Salehiniya H. Epidemiological characteristics of and risk factors for breast cancer in the world. *Breast Cancer (Dove Med Press).* 2019;11:151–164. https://doi.org/10.2147/BCTT.S176070.

2. Laigle-Donadey F, Taillibert S, Martin-Duverneuil N. Skull-base metastases. *J Neurooncol.* 2005;75(1):63–69.

3. Wright J, Gurney H, Glare P. Skull metastases masquerading as cerebral secondaries in patients with cancer. *Aust N Z J Med.* 1998;28(1):62.

4. Bullock JD, Yanes B. Ophthalmic manifestations of metastatic breast cancer. *Ophthalmology.* 1980;87(10):961–973.

5. Hall SM, Buzdar AU, Blumenschein GR. Cranial nerve palsies in metastatic breast cancer due to osseous metastasis without intracranial involvement. *Cancer.* 1983;52(1):180–184.

6. Greenberg HS, Deck MDF, Vikram B. Metastasis to the base of the skull: clinical findings in 43 patients. *Neurology.* 1981;31(5):530–537.

7. Jansen BPW, Pillay M, Bruin HG. 99mTc-SPECT in the diagnosis of skull base metastasis. *Neurology.* 1997;48(5):1326–1330.

8. Hudgins PA, Baugnon KL. Head and neck: skull base imaging. *Neurosurgery.* 2018;82(3):255–267. https://doi.org/10.1093/neuros/nyx492.

9. Lossos A, Siegal T. Numb chin syndrome in cancer patients: etiology, response to treatment, and prognostic significance. *Neurology.* 1992;42(6):1181–1184.

10. Horton J, Means ED, Cunningham TJ. The numb chin in breast cancer. *J Neurol Neurosurg Psychiatry.* 1973;36(2):211–216.

11. Li C, Yang W, Men Y, Wu F, Pan J, Li L. Magnetic resonance imaging for diagnosis of mandibular involvement from head and neck cancers: a systematic review and meta-analysis. *PLoS One.* 2014;9(11). https://doi.org/10.1371/journal.pone.0112267, e112267.

12. Posner JB, Chernik NL. In: Schoenberg BS, ed. *Intracranial Metastases From Systemic Cancer.* vol. 19. Raven Press; 1978:579–591.

13. Cifuentes N, Pickren JW. Metastases from carcinoma of mammary gland: an autopsy study. *J Surg Oncol.* 1979;11(3):193–205.

14. Tsukada Y, Fouad A, Pickren JW. Central nervous system metastasis from breast carcinoma. *Autopsy study. Cancer.* 1983;52(12):2349–2354.

15. Kleinschmidt-DeMasters BK. Dural metastases. A retrospective surgical and autopsy series. *Arch Pathol Lab Med.* 2001;125(7):880–887.

16. Minette SE, Kimmel DW. Subdural hematoma in patients with systemic cancer. *Mayo Clin Proc.* 1989;64(6):637–642.

17. Tseng SH, Liao CC, Lin SM. Dural metastasis in patients with malignant neoplasm and chronic subdural hematoma. *Acta Neurol Scand.* 2003;108(1):43–46.

18. Custer BS, Koepsell TD, Mueller BA. The association between breast carcinoma and meningioma in women. *Cancer.* 2002;94(6):1626–1635.

19. Johnson MD, Powell SZ, Boyer PJ. Dural lesions mimicking meningiomas. *Hum Pathol.* 2002;33(12):1211–1226.

20. Caroli E, Salvati M, Giangaspero F. Intrameningioma metastasis as first clinical manifestation of occult primary breast carcinoma. *Neurosurg Rev.* 2006;29(1):49–54.

21. Wasserstrom WR, Glass JP, Posner JB. Diagnosis and treatment of leptomeningeal metastases from solid tumors: experience with 90 patients. *Cancer.* 1982;49(4):759–772.

22. de Visser BWO, Somers R, Nooyen WH, van Heerde P, Hart AA, McVie JG. Intraventricular methotrexate therapy of leptomeningeal metastasis from breast carcinoma. *Neurology.* 1983;33(12):1565–1572.

23. Yap HY, Yap BS, Tashima CK. Meningeal carcinomatosis in breast cancer. *Cancer.* 1978;42(1):283–286.

24. Boogerd W, Hart AAM, Sande JJ. Meningeal carcinomatosis in breast cancer. Prognostic factors and influence of treatment. *Cancer.* 1991;67(6):1685–1695.

25. Lamovec J, Zidar A. Association of leptomeningeal carcinomatosis in carcinoma of the breast with infiltrating lobular carcinoma. An autopsy study. *Arch Pathol Lab Med.* 1991;115(5):507–510.

26. Kosmas C, Malamos NA, Tsavaris NB. Leptomeningeal carcinomatosis after major remission to taxane-based front-line therapy in patients with advanced breast cancer. *J Neurooncol.* 2002;56(3):265–273.

27. Jayson GC, Howell A, Harris M. Carcinomatous meningitis in patients with breast cancer. An aggressive disease variant. *Cancer.* 1994;74(12):3135–3141.

28. Smith DB, Howell A, Harris M. Carcinomatous meningitis associated with infiltrating lobular carcinoma of the breast. *Eur J Surg Oncol.* 1985;11(1):33–36.

29. Niwińska A, Rudnicka H, Murawska M. Breast cancer leptomeningeal metastasis: propensity of breast cancer subtypes for leptomeninges and the analysis of factors influencing survival. *Med Oncol.* 2013;30(1):408. https://doi.org/10.1007/s12032-012-0408-4.

30. Gleissner B, Chamberlain MC. Neoplastic meningitis. *Lancet Neurol.* 2006;5(5):443–452.

31. Freilich RJ, Krol G, DeAngelis LM. Neuroimaging and cerebrospinal fluid cytology in the diagnosis of leptomeningeal metastasis. *Ann Neurol.* 1995;38(1):51–57.

32. Collie DA, Brush JP, Lammie GA. Imaging features of leptomeningeal metastases. *Clin Radiol.* 1999;54(11):765–771.

33. van de Langerijt B, Gijtenbeek JM, de Reus HPM, et al. CSF levels of growth factors and plasminogen activators in leptomeningeal metastases. *Neurology.* 2006;67(1):114–119.

34. Twijnstra A, Zanten AP, Nooyen WJ. Sensitivity and specificity of single and combined tumour markers in the diagnosis of leptomeningeal metastasis from breast cancer. *J Neurol Neurosurg Psychiatry.* 1986;49(11):1246–1250.

35. Boire A, Brandsma D, Brastianos PK, et al. Liquid biopsy in central nervous system metastases: a RANO review and proposals for clinical applications. *Neuro Oncol.* 2019;21(5):571–584. https://doi.org/10.1093/neuonc/noz012.

36. Siravegna G, Marsoni S, Siena S, Bardelli A. Integrating liquid biopsies into the management of cancer. *Nat Rev Clin Oncol.* 2017;14(9):531–548. https://doi.org/10.1038/nrclinonc.2017.14.

37. DeAngelis LM. In: Harris JR, Lippman ME, Morrow M, Osborne CK, eds. *Leptomeningeal Metastasis.* 3rd ed. Lippincott Williams and Wilkins; 2004.

38. Yang TJ, Wijetunga NA, Yamada J, et al. Clinical trial of proton craniospinal irradiation for leptomeningeal metastases. *Neuro Oncol.* 2021. https://doi.org/10.1093/neuonc/noaa152.

39. Omuro AMP, Lallana EC, Bilsky MH. Ventriculoperitoneal shunt in patients with leptomeningeal metastasis. *Neurology.* 2005;64(9):1625–1627.

40. Yap HY, Yap BS, Rasmussen S. Treatment for meningeal carcinomatosis in breast cancer. *Cancer.* 1982;49(2):219–222.

41. Chamberlain MC, Kormanik PRN. Carcinomatous meningitis secondary to breast cancer: predictors of response to combined modality therapy. *J Neurooncol.* 1997;35(1):55–64.

42. Grant R, Naylor B, Greenberg HS. Clinical outcome in aggressively treated meningeal carcinomatosis. *Arch Neurol.* 1994;51(5):457–461.

43. Fizazi K, Asselain B, Vincent-Salomon A. Meningeal carcinomatosis in patients with breast carcinoma. Clinical features, prognostic factors, and results of a high-dose intrathecal methotrexate regimen. *Cancer.* 1996;77(7):1315–1323.

44. Glantz MJ, Cole BF, Recht L. High-dose intravenous methotrexate for patients with nonleukemic leptomeningeal cancer: is intrathecal chemotherapy necessary? *J Clin Oncol.* 1998;16(4):1561–1567.

45. Boogerd W, Bent MJ, Koehler PJ. The relevance of intraventricular chemotherapy for leptomeningeal metastasis in breast cancer: a randomised study. *Eur J Cancer.* 2004;40(18):2726–2733.

46. Groves MD, Glantz MJ, Chamberlain MC, et al. A multicenter phase II trial of intrathecal topotecan in patients with meningeal malignancies. *Neuro Oncol.* 2008;10(2):208–215. https://doi.org/10.1215/15228517-2007-059.

47. Park W-Y, Kim H-J, Kim K, et al. Intrathecal trastuzumab treatment in patients with breast cancer and leptomeningeal carcinomatosis. *Cancer Res Treat.* 2016;48(2):843–847. https://doi.org/10.4143/crt.2014.234.

48. Laakmann E, Witzel I, Müller V. Efficacy of liposomal cytarabine in the treatment of leptomeningeal metastasis of breast cancer. *Breast Care (Basel).* 2017;12(3):165–167. https://doi.org/10.1159/000464400.

49. Kumthekar P, Lassman AB, Lin N, et al. LPTO-02. Intrathecal (IT) trastuzumab (T) for the treatment of leptomeningeal disease (LM) in patients (PTS) with human epidermal receptor-2 positive (HER2+) cancer: a multicenter phase 1/2 study. *Neurooncol Adv.* 2019;1(Suppl 1):i6. https://doi.org/10.1093/noajnl/vdz014.025.

50. Brastianos PK, Lee EQ, Cohen JV, et al. Single-arm, open-label phase 2 trial of pembrolizumab in patients with leptomeningeal carcinomatosis. *Nat Med.* 2020;26(8):1280–1284. https://doi.org/10.1038/s41591-020-0918-0.

51. Grossman SA, Finkelstein DM, Ruckdeschel JC. Randomized prospective comparison of intraventricular methotrexate and thiotepa in patients with previously untreated neoplastic meningitis. *J Clin Oncol.* 1993;11(3):561–569.

52. Glantz MJ, Jaeckle KA, Chamberlain MC. A randomized controlled trial comparing intrathecal sustained-release cytarabine (Depo Cyt) to intrathecal methotrexate in patients with neoplastic meningitis from solid tumors. *Clin Cancer Res.* 1999;5(11):3394–3402.

53. Jaeckle KA, Phuphanich S, Bent MJ. Intrathecal treatment of neoplastic meningitis due to breast cancer with a slow-release formulation of cytarabine. *Br J Cancer.* 2001;84(2):157–163.

54. Orlando L, Curigliano G, Colleoni M. Intrathecal chemotherapy in carcinomatous meningitis from breast cancer. *Anticancer Res.* 2002;22(5):3057–3059.

55. ClinicalTrials.gov. *Tucatinib, Trastuzumab, and Capecitabine for the Treatment of HER2+ LMD—Full Text View.* ClinicalTrials.gov; 2020. https://clinicaltrials.gov/ct2/show/NCT03501979. Accessed 15 September 2020.

56. Cairncross JG, Kim J-H, Posner JB. Radiation therapy for brain metastases. *Ann Neurol.* 1980;7(6):529–541.

57. Lee YTN. Breast carcinoma: pattern of metastasis at autopsy. *J Surg Oncol.* 1983;23(3):175–180.

58. Schouten LJ, Rutten J, Huveneers HAM. Incidence of brain metastases in a cohort of patients with carcinoma of the breast, colon, kidney, and lung and melanoma. *Cancer.* 2002;94(10):2698–2705.

59. Barnholtz-Sloan JS, Sloan AE, Davis FG. Incidence proportions of brain metastases in patients diagnosed (1973 to 2001) in the metropolitan Detroit cancer surveillance system. *J Clin Oncol.* 2004;22(14):2865–2872.

60. Yawn BP, Wollan PC, Schroeder C. Temporal and gender-related trends in brain metastases from lung and breast cancer. *Minn Med.* 2003;86(12):32–37.

61. Carty NJ, Foggitt A, Hamilton CR. Patterns of clinical metastasis in breast cancer: an analysis of 100 patients. *Eur J Surg Oncol.* 1995;21(6):607–608.

62. Lin NU, Bellon JR, Winer EP. CNS metastases in breast cancer. *J Clin Oncol.* 2004;22(17):3608–3617.

63. Boogerd W, Vos VW, Hart AAM. Brain metastases in breast cancer; natural history, prognostic factors and outcome. *J Neurooncol.* 1993;15(2):165–174.

64. DiStefano A, Yap HY, Hortobagyi GN. The natural history of breast cancer patients with brain metastases. *Cancer.* 1979;44(5):1913–1918.

65. Paterson AHG, Agarwal M, Lees A. Brain metastases in breast cancer patients receiving adjuvant chemotherapy. *Cancer.* 1982;49(4):651–654.

66. Lentzsch S, Reichardt P, Weber F. Brain metastases in breast cancer: prognostic factors and management. *Eur J Cancer.* 1999;35(4):580–585.

67. Carey LA, Ewend MG, Metzger R. Central nervous system metastases in women after multimodality therapy for high risk breast cancer. *Breast Cancer Res Treat.* 2004;88(3):273–280.

68. Crivellari D, Pagani O, Veronesi A. High incidence of central nervous system involvement in patients with metastatic or locally advanced breast cancer treated with epirubicin and docetaxel. *Ann Oncol.* 2001;12(3):353–356.

69. Bendell JC, Domchek SM, Burstein HJ. Central nervous system metastases in women who receive trastuzumab-based therapy for metastatic breast carcinoma. *Cancer.* 2003;97(12):2972–2979.

70. Duchnowska R, Szczylik C. Central nervous system metastases in breast cancer patients administered trastuzumab. *Cancer Treat Rev.* 2005;31(4):312–318.

71. Evans AJ, James JJ, Cornford EJ. Brain metastases from breast cancer: identification of a high-risk group. *Clin Oncol.* 2004;16(5):345–349.

72. Maki DD, Grossman RI. Patterns of disease spread in metastatic breast carcinoma: influence of estrogen and progesterone receptor status. *Am J Neuroradiol.* 2000;21(6):1064–1066.

73. Miller KD, Weathers T, Haney LG. Occult central nervous system involvement in patients with metastatic breast cancer: prevalence, predictive factors and impact on overall survival. *Ann Oncol.* 2003;14(7):1072–1077.

74. Slimane K, Andre F, Delaloge S. Risk factors for brain relapse in patients with metastatic breast cancer. *Ann Oncol.* 2004;15(11):1640–1644.

75. Ryberg M, Nielsen D, Osterlind K. Predictors of central nervous system metastasis in patients with metastatic breast cancer. A competing risk analysis of 579 patients treated with epirubicin-based chemotherapy. *Breast Cancer Res Treat.* 2005;91(3):217–225.

76. Tabouret E, Chinot O, Metellus P, Tallet A, Viens P, Gonçalves A. Recent trends in epidemiology of brain metastases: an overview. *Anticancer Res.* 2012;32(11):4655–4662.

77. Witzel I, Oliveira-Ferrer L, Pantel K, Müller V, Wikman H. Breast cancer brain metastases: biology and new clinical perspectives. *Breast Cancer Res.* 2016;18(1):8. https://doi.org/10.1186/s13058-015-0665-1.

78. de Almeida Bastos DC, MVC M, Sawaya R, et al. Biological subtypes and survival outcomes in breast cancer patients with brain metastases in the targeted therapy era. *Neurooncol Pract.* 2018;5(3):161–169. https://doi.org/10.1093/nop/npx033.

79. Ogawa K, Yoshii Y, Nishimaki T, et al. Treatment and prognosis of brain metastases from breast cancer. *J Neurooncol.* 2008;86(2):231–238. https://doi.org/10.1007/s11060-007-9469-1.

80. Lee SS, Ahn J-H, Kim MK, et al. Brain metastases in breast cancer: prognostic factors and management. *Breast Cancer Res Treat.* 2008;111(3):523–530. https://doi.org/10.1007/s10549-007-9806-2.

81. Ewend MG, Morris DE, Carey LA, Ladha AM, Brem S. Guidelines for the initial management of metastatic brain tumors: role of surgery, radiosurgery, and radiation therapy. *J Natl Compr Canc Netw.* 2008;6(5):505–514. https://doi.org/10.6004/jnccn.2008.0038.

82. Palmer JD, Greenspoon J, Brown PD, Johnson DR, Roberge D. Neuro-Oncology Practice Clinical Debate: stereotactic radiosurgery or fractionated stereotactic radiotherapy following surgical resection for brain metastasis. *Neurooncol Pract.* 2020;7(3):263–267. https://doi.org/10.1093/nop/npz047.

83. Brown PD, Ballman KV, Cerhan JH, et al. Postoperative stereotactic radiosurgery compared with whole brain radiotherapy for resected metastatic brain disease (NCCTG N107C/

CEC·3): a multicentre, randomised, controlled, phase 3 trial. *Lancet Oncol.* 2017;18(8):1049–1060. https://doi.org/10.1016/S1470-2045(17)30441-2.

84. Aoyama H, Shirato H, Tago M, et al. Stereotactic radiosurgery plus whole-brain radiation therapy vs stereotactic radiosurgery alone for treatment of brain metastases: a randomized controlled trial. *JAMA.* 2006;295(21):2483–2491. https://doi.org/10.1001/jama.295.21.2483.

85. Kocher M, Soffietti R, Abacioglu U, et al. Adjuvant whole-brain radiotherapy versus observation after radiosurgery or surgical resection of one to three cerebral metastases: results of the EORTC 22952-26001 study. *J Clin Oncol Off J Am Soc Clin Oncol.* 2011;29(2):134–141. https://doi.org/10.1200/JCO.2010.30.1655.

86. Yamamoto M, Serizawa T, Shuto T, et al. Stereotactic radiosurgery for patients with multiple brain metastases (JLGK0901): a multi-institutional prospective observational study. *Lancet Oncol.* 2014;15(4):387–395. https://doi.org/10.1016/S1470-2045(14)70061-0.

87. Brown PD, Jaeckle K, Ballman KV, et al. Effect of radiosurgery alone vs radiosurgery with whole brain radiation therapy on cognitive function in patients with 1 to 3 brain metastases: a randomized clinical trial. *JAMA.* 2016;316(4):401–409. https://doi.org/10.1001/jama.2016.9839.

88. Amendola BE, Wolf AL, Coy S. Gamma knife radiosurgery in the treatment of patients with single and multiple brain metastases from carcinoma of the breast. *Cancer J.* 2000;6(2):88–92.

89. Firlik KS, Kondziolka D, Flickinger JC. Stereotactic radiosurgery for brain metastases from breast cancer. *Ann Surg Oncol.* 2000;7(5):333–338.

90. Lederman RJ, Wilbourn AJ. Brachial plexopathy: recurrent cancer or radiation? *Neurology.* 1984;34(10):1331–1335.

91. Muacevic A, Kreth FW, Tonn J-C. Stereotactic radiosurgery for multiple brain metastases from breast carcinoma. Feasibility and outcome of a local treatment concept. *Cancer.* 2004;100(8):1705–1711.

92. Patchell RA, Tibbs PA, Regine WF. Postoperative radiotherapy in the treatment of single metastases to the brain: a randomized trial. *JAMA.* 1998;280(17):1485–1489.

93. Palmer JD, Trifiletti DM, Gondi V, et al. Multidisciplinary patient-centered management of brain metastases and future directions. *Neurooncol Adv.* 2020;2(1). https://doi.org/10.1093/noajnl/vdaa034.

94. Wefel JS, Parsons MW, Gondi V, Brown PD. Neurocognitive aspects of brain metastasis. *Handb Clin Neurol.* 2018;149:155–165. https://doi.org/10.1016/B978-0-12-811161-1.00012-8.

95. Gondi V, Pugh SL, Tome WA, et al. Preservation of memory with conformal avoidance of the hippocampal neural stem-cell compartment during whole-brain radiotherapy for brain metastases (RTOG 0933): a phase ii multi-institutional trial. *J Clin Oncol.* 2014;32(34):3810–3816. https://doi.org/10.1200/JCO.2014.57.2909.

96. Brown PD, Gondi V, Pugh S, et al. Hippocampal avoidance during whole-brain radiotherapy plus memantine for patients with brain metastases: phase III trial NRG oncology CC001. *J Clin Oncol Off J Am Soc Clin Oncol.* 2020;38(10):1019–1029. https://doi.org/10.1200/JCO.19.02767.

97. Fokstuen T, Wilking N, Rutqvist LR. Radiation therapy in the management of brain metastases from breast cancer. *Breast Cancer Res Treat.* 2000;62(3):211–216.

98. Mahmoud-Ahmed AS, Suh JH, Lee S-Y. Results of whole brain radiotherapy in patients with brain metastases from breast cancer: a retrospective study. *Int J Radiat Oncol Biol Phys.* 2002;54(3):810–817.

99. Salvati M, Cervoni L, Innocenzi G, Bardella L. Prolonged stabilization of multiple and single brain metastases from breast cancer with tamoxifen. Report of three cases. *Tumori J.* 1993;79(5):359–362. https://doi.org/10.1177/030089169307900516.

III. Neurological complications of specific neoplasms

100. Stewart DJ, Dahrouge S. Response of brain metastases from breast cancer to megestrol acetate: a case report. *J Neurooncol.* 1995;24(3):299–301. https://doi.org/10.1007/BF01052847.

101. van der Gaast A, Alexieva-Figusch J, Vecht C, Verweij J, Stoter G. Complete remission of a brain metastasis to third-line hormonal treatment with megestrol acetate. *Am J Clin Oncol.* 1990;13(6):507–509.

102. Boogerd W, Dalesio O, Bais EM. Response of brain metastases from breast cancer to systemic chemotherapy. *Cancer.* 1992;69(4):972–980.

103. Christodoulou C, Bafaloukos D, Linardou H. Temozolomide (TMZ) combined with cisplatin (CDDP) in patients with brain metastases from solid tumors: a Hellenic Cooperative Oncology Group (HeCOG) phase II study. *J Neurooncol.* 2005;71(1):61–65.

104. Trudeau ME, Crump M, Charpentier D, et al. Temozolomide in metastatic breast cancer (MBC): a phase II trial of the National Cancer Institute of Canada—Clinical Trials Group (NCIC-CTG). *Ann Oncol.* 2006;17(6):952–956. https://doi.org/10.1093/annonc/mdl056.

105. Morikawa A, Peereboom DM, Thorsheim HR, et al. Capecitabine and lapatinib uptake in surgically resected brain metastases from metastatic breast cancer patients: a prospective study. *Neuro Oncol.* 2015;17(2):289–295. https://doi.org/10.1093/neuonc/nou141.

106. Ekenel M, Hormigo AM, Peak S, Deangelis LM, Abrey LE. Capecitabine therapy of central nervous system metastases from breast cancer. *J Neurooncol.* 2007;85(2):223–227. https://doi.org/10.1007/s11060-007-9409-0.

107. Colleoni M, Graiff C, Nelli P. Activity of combination chemotherapy in brain metastases from breast and lung adenocarcinoma. *Am J Clin Oncol.* 1997;20(3):303–307.

108. Franciosi V, Cocconi G, Michiara M. Front-line chemotherapy with cisplatin and etoposide for patients with brain metastases from breast carcinoma, nonsmall cell lung carcinoma, or malignant melanoma: a prospective study. *Cancer.* 1999;85(7):1599–1605.

109. Abrey LE, Olson JD, Raizer JJ. A phase II trial of temozolomide for patients with recurrent or progressive brain metastases. *J Neurooncol.* 2001;53(3):259–265.

110. Wang ML, Yung WK, Royce ME. Capecitabine for 5-fluorouracil-resistant brain metastases from breast cancer. *Am J Clin Oncol.* 2001;24(4):421–424.

111. Murthy RK, Loi S, Okines A, et al. Tucatinib, trastuzumab, and capecitabine for HER2-positive metastatic breast cancer. *N Engl J Med.* 2020;382(7):597–609. https://doi.org/10.1056/NEJMoa1914609.

112. Lin NU, Carey LA, Liu MC, et al. Phase II trial of lapatinib for brain metastases in patients with human epidermal growth factor receptor 2–positive breast cancer. *J Clin Oncol Off J Am Soc Clin Oncol.* 2008;26(12):1993–1999. https://doi.org/10.1200/JCO.2007.12.3588.

113. Morikawa A, de Stanchina E, Pentsova E, et al. Phase I study of intermittent high-dose lapatinib alternating with capecitabine for HER2-positive breast cancer patients with central nervous system metastases. *Clin Cancer Res.* 2019;25(13):3784–3792. https://doi.org/10.1158/1078-0432.CCR-18-3502.

114. Lin NU, Stein A, Nicholas A, et al. Planned interim analysis of PATRICIA: an open-label, single-arm, phase II study of pertuzumab (P) with high-dose trastuzumab (H) for the treatment of central nervous system (CNS) progression post radiotherapy (RT) in patients (pts) with HER2-positive metastatic breast cancer (MBC). *J Clin Oncol.* 2017;35(15_suppl):2074. https://doi.org/10.1200/JCO.2017.35.15_suppl.2074.

115. Gilbert RW, Kim J-H, Posner JB. Epidural spinal cord compression from metastatic tumor: diagnosis and treatment. *Ann Neurol.* 1978;3(1):40–51.

116. Chamberlain MC. Neoplastic meningitis and metastatic epidural spinal cord compression. *Hematol Oncol Clin North Am.* 2012;26(4):917–931. https://doi.org/10.1016/j.hoc.2012.04.004.

117. Loblaw DA, Perry J, Chambers A. Systematic review of the diagnosis and management of malignant extradural spinal cord compression: the Cancer Care Ontario Practice Guidelines Initiative's Neuro-oncology Disease Site Group. *J Clin Oncol.* 2005;23(9):2028–2037.

118. Stark RJ, Henson RA, Evans SJW. Spinal metastases. A retrospective survey from a general hospital. *Brain.* 1982;105(1):189–213.

119. Loblaw DA, Laperriere NJ, Mackillop WJ. A population-based study of malignant spinal cord compression in Ontario. *Clin Oncol.* 2003;15(4):211–217.

120. Hill ME, Richards MA, Gregory WM. Spinal cord compression in breast cancer: a review of 70 cases. *Br J Cancer.* 1993;68(5):969–973.

121. Pessina F, Navarria P, Riva M, et al. Long-term follow-up of patients with metastatic epidural spinal cord compression from breast cancer treated with surgery followed by radiotherapy. *World Neurosurg.* 2018;110:e281–e286. https://doi.org/10.1016/j.wneu.2017.10.156.

122. Boogerd W, Sande JJ, Kroger R. Early diagnosis and treatment of spinal epidural metastasis in breast cancer: a prospective study. *J Neurol Neurosurg Psychiatry.* 1992;55(12):1188–1193.

123. Lu C, Stomper PC, Drislane FW. Suspected spinal cord compression in breast cancer patients: a multidisciplinary risk assessment. *Breast Cancer Res Treat.* 1998;51(2):121–131.

124. Husband DJ. Malignant spinal cord compression: prospective study of delays in referral and treatment. *Br Med J.* 1998;317(7150):18–21.

125. Harrison KM, Muss HB, Ball MR. Spinal cord compression in breast cancer. *Cancer.* 1985;55(12):2839–2844.

126. Wen PY, McColl CD, Freilich RJ. In: Harris JR, Lippman ME, Morow M, Osborne CK, eds. *Epidural Metastases.* 3rd ed. Lippincott Williams and Wilkins; 2004.

127. Sørensen PS, Helweg-Larsen S, Mouridsen H. Effect of high-dose dexamethasone in carcinomatous metastatic spinal cord compression treated with radiotherapy: a randomised trial. *Eur J Cancer.* 1994;30A(1):22–27.

128. Kumar A, Weber MH, Gokaslan Z, et al. Metastatic spinal cord compression and steroid treatment: a systematic review. *Clin Spine Surg.* 2017;30(4):156–163. https://doi.org/10.1097/BSD.0000000000000528.

129. Rades D, Stalpers LJA, Veninga T. Evaluation of five radiation schedules and prognostic factors for metastatic spinal cord compression. *J Clin Oncol.* 2005;23(15):3366–3375.

130. Rades D, Lange M, Veninga T, et al. Preliminary results of spinal cord compression recurrence evaluation (score-1) study comparing short-course versus long-course radiotherapy for local control of malignant epidural spinal cord compression. *Int J Radiat Oncol Biol Phys.* 2009;73(1):228–234. https://doi.org/10.1016/j.ijrobp.2008.04.044.

131. Lee KA, Dunne M, Small C, et al. (ICORG 05-03): prospective randomized non-inferiority phase III trial comparing two radiation schedules in malignant spinal cord compression (not proceeding with surgical decompression); the quality of life analysis. *Acta Oncol.* 2018;57(7):965–972. https://doi.org/10.1080/0284186X.2018.1433320.

132. Hoskin PJ, Hopkins K, Misra V, et al. Effect of single-fraction vs multifraction radiotherapy on ambulatory status among patients with spinal canal compression from metastatic cancer: the SCORAD randomized clinical trial. *JAMA.* 2019;322(21):2084–2094. https://doi.org/10.1001/jama.2019.17913.

133. van der Sande JJ, Boogerd W, Kröger R, Kappelle AC. Recurrent spinal epidural metastases: a prospective study with a complete follow-up. *J Neurol Neurosurg Psychiatry.* 1999;66(5):623–627.

134. Maranzano E, Trippa F, Casale M, Anselmo P, Rossi R. Reirradiation of metastatic spinal cord compression: definitive results of two randomized trials. *Radiother Oncol.* 2011;98(2):234–237. https://doi.org/10.1016/j.radonc.2010.12.011.

135. Findlay GF. Adverse effects of the management of malignant spinal cord compression. *J Neurol Neurosurg Psychiatry.* 1984;47(8):761–768.

136. Boogerd W, Sande JJ, Kröger R. Effective systemic therapy for spinal epidural metastases from breast carcinoma. *Eur J Cancer Clin Oncol.* 1989;25(1):149–153.

137. Ross JR, Saunders Y, Edmonds PM, Patel S, Broadley KE, Johnston SRD. Systematic review of role of bisphosphonates on skeletal morbidity in metastatic cancer. *Br Med J.* 2003;327(7413):469.

138. Goldvaser H, Amir E. Role of bisphosphonates in breast cancer therapy. *Curr Treat Options Oncol.* 2019;20(4):26. https://doi.org/10.1007/s11864-019-0623-8.

139. Schiff D, O'Neill BP. Intramedullary spinal cord metastases: clinical features and treatment outcome. *Neurology.* 1996;47(4):906–912.

140. Villegas AE, Guthrie TH. Intramedullary spinal cord metastasis in breast cancer: clinical features, diagnosis, and therapeutic consideration. *Breast J.* 2004;10(6):532–535.

141. Kosmas C, Koumpou M, Nikolaou M. Intramedullary spinal cord metastases in breast cancer: report of four cases and review of the literature. *J Neurooncol.* 2005;71(1):67–72.

142. Crasto S, Duca S, Davini O. MRI diagnosis of intramedullary metastases from extra-CNS tumors. *Eur Radiol.* 1997;7(5):732–736.

143. Isla A, Paz JM, Sansivirini F. Intramedullary spinal cord metastasis. Case report. *J Neurosurg Sci.* 2000;44(2):99–101.

144. Gasser TG, Pospiech J, Stolke D. Spinal intramedullary metastases. Report of two cases and review of the literature. *Neurosurg Rev.* 2001;24(2-3):88–92.

145. Christmas NJ, Mead MD, Richardson EP. Secondary optic nerve tumors. *Surv Ophthalmol.* 1991;36(3):196–206.

146. Newman NJ, Grossniklaus HE, Wojno TH. Breast carcinoma metastatic to the optic nerve. *Arch Ophthalmol.* 1996;114(1):102–103.

147. Demirci H, Shields CL, Chao A-N. Uveal metastasis from breast cancer in 264 patients. *Am J Ophthalmol.* 2003;136(2):264–271.

148. Mewis L, Young SE. Breast carcinoma metastatic to the choroid. Analysis of 67 patients. *Ophthalmology.* 1982;89(2):147–151.

149. Wiegel T, Kreusel KM, Bornfeld N. Frequency of asymptomatic choroidal metastasis in patients with disseminated breast cancer: results of a prospective screening programme. *Br J Ophthalmol.* 1998;82(10):1159–1161.

150. Bloch RS, Gartner S. The incidence of ocular metastatic carcinoma. *Arch Ophthalmol.* 1971;85(6):673–675.

151. Thatcher N, Thomas PRM. Choroidal metastases from breast carcinoma: a survey of 42 patients and the use of radiation therapy. *Clin Radiol.* 1975;26(4):549–553.

152. Ratanatharathorn V, Powers WE, Grimm J. Eye metastasis from carcinoma of the breast: diagnosis, radiation treatment and results. *Cancer Treat Rev.* 1991;18(4):261–276.

153. Chan RVP, Young LH. Treatment options for metastatic tumors to the choroid. *Semin Ophthalmol.* 2005;20(4):207–216.

154. Gurling KJ, Scott GBD, Baron DN. Metastases in pituitary tissue removed at hypophysectomy in women with mammary carcinoma. *Br J Cancer.* 1957;11(4):519–523.

155. Yap HY, Tashima CK, Blumenschein GR. Diabetes insipidus and breast cancer. *Arch Intern Med.* 1979;139(9):1009–1011.

156. Max MB, Deck MDF, Rottenberg DA. Pituitary metastasis: incidence in cancer patients and clinical differentiation from pituitary adenoma. *Neurology.* 1981;31(8):998–1002.

157. Bobilev D, Shelef I, Lavrenkov K. Diabetes insipidus caused by isolated intracranial metastases in patient with breast cancer. *J Neurooncol.* 2005;73(1):39–42.

158. Kurkjian C, Armor JF, Kamble R. Symptomatic metastases to the pituitary infundibulum resulting from primary breast cancer. *Int J Clin Oncol.* 2005;10(3):191–194.

159. Kori SH, Foley KM, Posner JB. Brachial plexus lesions in patients with cancer: 100 cases. *Neurology.* 1981;31(1):45–50.

160. Harper CM, Thomas JE, Cascino TL. Distinction between neoplastic and radiation-induced brachial plexopathy, with emphasis on the role of EMG. *Neurology.* 1989;39(4):502–506.

161. Thyagarajan D, Cascino T, Harms G. Magnetic resonance imaging in brachial plexopathy of cancer. *Neurology.* 1995;45(3):421–427.

162. Son YH. Effectiveness of irradiation therapy in peripheral neuropathy caused by malignant disease. *Cancer.* 1967;20(9):1447–1451.

163. Cascino TL, Kori S, Krol G. CT of the brachial plexus in patients with cancer. *Neurology.* 1983;33(12):1553–1557.

164. Qayyum A, MacVicar AD, Padhani AR. Symptomatic brachial plexopathy following treatment for breast cancer: utility of MR imaging with surface-coil techniques. *Radiology.* 2000;214(3):837–842.

165. Meller I, Alkalay D, Mozes M. Isolated metastases to peripheral nerves. Report of five cases involving the brachial plexus. *Cancer.* 1995;76(10):1829–1832.

166. Lingawi SS, Bilbey JH, Munk PL. MR imaging of brachial plexopathy in breast cancer patients without palpable recurrence. *Skeletal Radiol.* 1999;28(6):318–323.

167. Ahmad A, Barrington S, Maisey M. Use of positron emission tomography in evaluation of brachial plexopathy in breast cancer patients. *Br J Cancer.* 1999;79(3/4):478–482.

168. Cole JW, Quint DJ, McGillicuddy JE. CT-guided brachial plexus biopsy. *Am J Neuroradiol.* 1997;18(8):1420–1422.

169. Cherny NI, Olsha O. In: Harris JR, Lippman ME, Morrow M, Osborne CK, eds. *Brachial Plexopathy in Patients With Breast Cancer.* 3rd ed. Lippincott Williams and Wilkins; 2004:1241–1255.

170. Jaeckle KA, Young DF, Foley KM. The natural history of lumbosacral plexopathy in cancer. *Neurology.* 1985;35(1):8–15.

171. Thomas JE, Cascino TL, Earle JD. Differential diagnosis between radiation and tumor plexopathy of the pelvis. *Neurology.* 1985;35(1):1–7.

172. Clouston PD, DeAngelis LM, Posner JB. The spectrum of neurological disease in patients with systemic cancer. *Ann Neurol.* 1992;31(3):268–273.

173. Plum F, Posner JB. *The Diagnosis of Stupor and Coma.* 3rd ed. F.A. Davis Company; 1980.

174. Cestari DM, Weine DM, Panageas KS. Stroke in patients with cancer. Incidence and etiology. *Neurology.* 2004;62(11):2025–2030.

175. Chaturvedi S, Ansell J, Recht L. Should cerebral ischemic events in cancer patients be considered a manifestation of hypercoagulability? *Stroke.* 1994;25(6):1215–1218.

176. Rogers LR. Cerebrovascular complications in cancer patients. *Neurol Clin.* 2003;21(1):167–192.

177. Katz JM, Segal AZ. Incidence and etiology of cerebrovascular disease in patients with malignancy. *Curr Atheroscler Rep.* 2005;7.

178. Wall JG, Weiss RB, Norton L. Arterial thrombosis associated with adjuvant chemotherapy for breast carcinoma: a Cancer and Leukemia Group B study. *Am J Med.* 1989;87(5):501–504.

179. Saphner T, Tormey DC, Gray R. Venous and arterial thrombosis in patients who received adjuvant therapy for breast cancer. *J Clin Oncol.* 1991;9(2):286–294.

180. Pritchard KI, Paterson AHG, Paul NA. Increased thromboembolic complications with concurrent tamoxifen and chemotherapy in a randomized trial of adjuvant therapy for women with breast cancer. *J Clin Oncol.* 1996;14(10):2731–2737.

181. Geiger AM, Fischberg GM, Chen W. Stroke risk and tamoxifen therapy for breast cancer. *J Natl Cancer Inst.* 2004;96(20):1528–1536.

182. Early Breast Cancer Trialists' Collaborative Group. Tamoxifen for early breast cancer: an overview of the randomised trials. *Lancet.* 1998;351(9114):1451–1467.

III. Neurological complications of specific neoplasms

183. Barrett-Connor E, Mosca L, Collins P. Effects of raloxifene on cardiovascular events and breast cancer in postmenopausal women. *N Engl J Med.* 2006;355(2):125–137.

184. Iguchi Y, Kimura K, Kobayashi K. Ischemic stroke with malignancy may be frequently caused by paradoxical embolism. *J Neurol Neurosurg Psychiatry.* 2006. https://doi.org/10.1136/jnnp.2006.092940.

185. Menon BK, Goyal M. Imaging paradigms in acute ischemic stroke: a pragmatic evidence-based approach. *Radiology.* 2015;277(1):7–12. https://doi.org/10.1148/radiol.2015151030.

186. Powers WJ, Rabinstein AA, Ackerson T, et al. Guidelines for the early management of patients with acute ischemic stroke: 2019 update to the 2018 guidelines for the early management of acute ischemic stroke: a guideline for healthcare professionals from the American heart association/American stroke association. *Stroke.* 2019;50(12). https://doi.org/10.1161/STR.0000000000000211.

187. Collins RC, Al-Mondhiry H, Chernik NL. Neurologic manifestations of intravascular coagulation in patients with cancer. A clinicopathologic analysis of 12 cases. *Neurology.* 1975;25(9):795–806.

188. Sallah S, Wan JY, Nguyen P, Hanrahan LR, Sigounas G. Disseminated intravascular coagulation in solid tumors: clinical and pathologic study. *Thromb Haemost.* 2001;86(3):828–833.

189. Graus F, Rogers LR, Posner JB. Cerebrovascular complications in patients with cancer. *Medicine (Baltimore).* 1985;64(1):16–35.

190. Kwaan HC, Gordon LI. Thrombotic microangiopathy in the cancer patient. *Acta Haematol.* 2001;106(1–2):52–56.

191. Fisher DC, Sherrill GB, Hussein A. Thrombotic microangiopathy as a complication of high-dose chemotherapy for breast cancer. *Bone Marrow Transplant.* 1996;18(1):193–198.

192. Biller J, Challa VR, Toole JF. Nonbacterial thrombotic endocarditis. A neurologic perspective of clinicopathologic correlations of 99 patients. *Arch Neurol.* 1982;39(2):95–98.

193. Rogers LR, Cho ES, Kempin S. Cerebral infarction from nonbacterial thrombotic endocarditis. Clinical and pathological study including the effects of anticoagulation. *Am J Med.* 1987;83(4):746–756.

194. Singhal AB, Topcuoglu MA, Buonanno FS. Acute ischemic stroke patterns in infective and nonbacterial thrombotic endocarditis. A diffusion-weighted magnetic resonance imaging study. *Stroke.* 2002;33(5):1267–1273.

195. Dutta T, Karas MG, Segal AZ. Yield of transesophageal echocardiography for nonbacterial thrombotic endocarditis and other cardiac sources of embolism in cancer patients with cerebral ischemia. *Am J Cardiol.* 2006;97(6):894–898.

196. Sutherland DE, Weitz IC, Liebman HA. Thromboembolic complications of cancer: epidemiology, pathogenesis, diagnosis, and treatment. *Am J Hematol.* 2003;72(1):43–52.

197. Salem DN, Stein PD, Al-Ahmad A. Antithrombotic therapy in valvular heart disease—native and prosthetic: the seventh ACCP conference on antithrombotic and thrombolytic therapy. *Chest.* 2004;126(3).

198. el-Shami K, Griffiths E, Streiff M. Nonbacterial thrombotic endocarditis in cancer patients: pathogenesis, diagnosis, and treatment. *Oncologist.* 2007;12(5):518–523. https://doi.org/10.1634/theoncologist.12-5-518.

199. Averback P. Primary cerebral venous thrombosis in young adults: the diverse manifestations of an underrecognized disease. *Ann Neurol.* 1978;3(1):81–86.

200. Sigsbee B, Deck MDF, Posner JB. Nonmetastatic superior sagittal sinus thrombosis complicating systemic cancer. *Neurology.* 1979;29(2):139–146.

201. Hickey WF, Garnick MB, Henderson IC. Primary cerebral venous thrombosis in patients with cancer—a rarely diagnosed paraneoplastic syndrome. Report of three cases and review of the literature. *Am J Med.* 1982;73(5):740–750.

202. Raizer JJ, DeAngelis LM. Cerebral sinus thrombosis diagnosed by MRI and MR venography in cancer patients. *Neurology.* 2000;54(6):1222–1226.

203. Finelli PF, Schauer PK. Cerebral sinus thrombosis with tamoxifen. *Neurology.* 2001;56(8):1113–1114.

204. Capecchi M, Abbattista M, Martinelli I. Cerebral venous sinus thrombosis. *J Thromb Haemost.* 2018;16(10):1918–1931. https://doi.org/10.1111/jth.14210.

205. Deck JHN, Lee MA. Mucin embolism to cerebral arteries: a fatal complication of carcinoma of the breast. *Can J Neurol Sci.* 1978;5(3):327–330.

206. Towfighi J, Simmonds MA, Davidson EA. Mucin and fat emboli in mucinous carcinomas. Cause of hemorrhagic cerebral infarcts. *Arch Pathol Lab Med.* 1983;107(12):646–649.

207. Jovin TG, Boosupalli V, Zivkovic SA. High titers of CA-125 may be associated with recurrent ischemic strokes in patients with cancer. *Neurology.* 2005;64(11):1944–1945.

208. Pasquini E, Gianni L, Aitini E. Acute disseminated intravascular coagulation syndrome in cancer patients. *Oncology.* 1995;52(6):505–508.

209. Obbens EAMT, Leavens ME, Beal JW. Ommaya reservoirs in 387 cancer patients: a 15-year experience. *Neurology.* 1985;35(9):1274–1278.

210. Lishner M, Perrin RG, Feld R. Complications associated with Ommaya reservoirs in patients with cancer. The Princess Margaret hospital experience and a review of the literature. *Arch Intern Med.* 1990;150:173–176.

211. Mead PA, Safdieh JE, Nizza P, Tuma S, Sepkowitz KA. Ommaya reservoir infections: a 16-year retrospective analysis. *J Infect.* 2014;68(3):225–230. https://doi.org/10.1016/j.jinf.2013.11.014.

212. Szvalb AD, Raad II, Weinberg JS, Suki D, Mayer R, Viola GM. Ommaya reservoir-related infections: clinical manifestations and treatment outcomes. *J Infect.* 2014;68(3):216–224. https://doi.org/10.1016/j.jinf.2013.12.002.

213. Boogerd W, Sande JJ, Moffie D. Acute fever and delayed leukoencephalopathy following low dose intraventricular methotrexate. *J Neurol Neurosurg Psychiatry.* 1988;51(10):1277–1283.

214. Lemann W, Wiley RG, Posner JB. Leukoencephalopathy complicating intraventricular catheters: clinical, radiographic and pathologic study of 10 cases. *J Neurooncol.* 1988;6(1):67–74.

215. Boogerd W, Moffie D, Smets LA. Early blindness and coma during intrathecal chemotherapy for meningeal carcinomatosis. *Cancer.* 1990;65(3):452–457.

216. Kaiser-Kupfer MI, Kupfer C, Rodrigues MM. Tamoxifen retinopathy: a clinicopathologic report. *Ophthalmology.* 1981;88(1):89–93.

217. Nayfield SG, Gorin MB. Tamoxifen-associated eye disease: a review. *J Clin Oncol.* 1996;14(3):1018–1026.

218. Pugesgaard T, Von Eyben FE. Bilateral optic neuritis evolved during tamoxifen treatment. *Cancer.* 1986;58(2):383–386.

219. Pluss JL, DiBella NJ. Reversible central nervous system dysfunction due to tamoxifen in a patient with breast cancer. *Ann Intern Med.* 1984;101(5):652.

220. Schagen SB, Muller MJ, Boogerd W. Late effects of adjuvant chemotherapy on cognitive function: a follow-up study in breast cancer patients. *Ann Oncol.* 2002;13(9):1387–1397.

221. Wefel JS, Lenzi R, Theriault RL. The cognitive sequelae of standard-dose adjuvant chemotherapy in women with breast carcinoma. Results of a prospective, randomized, longitudinal trial. *Cancer.* 2004;100(11):2292–2299.

222. Videnovic A, Semenov I, Chua-Adajar R. Capecitabine-induced multifocal leukoencephalopathy: a report of five cases. *Neurology.* 2005;65(11):1792–1794.

223. Sheline GE, Wara WM, Smith V. Therapeutic irradiation and brain injury. *Int J Radiat Oncol Biol Phys.* 1980;6(9):1215–1228. https://doi.org/10.1016/0360-3016(80)90175-3.

224. Helson L. Radiation-induced demyelination and remyelination in the central nervous system: a literature review. *Anticancer Res.* 2018;38(9):4999–5002. https://doi.org/10.21873/anticanres.12818.

225. Harjani RR, Gururajachar JM, Krishnaswamy U. Comprehensive assessment of Somnolence Syndrome in patients undergoing radiation to the brain. *Rep Pract Oncol Radiother.* 2016;21(6):560–566. https://doi.org/10.1016/j.rpor.2016.08.003.

226. Aoyama H, Tago M, Kato N, et al. Neurocognitive function of patients with brain metastasis who received either whole brain radiotherapy plus stereotactic radiosurgery or radiosurgery alone. *Int J Radiat Oncol Biol Phys.* 2007;68(5):1388–1395. https://doi.org/10.1016/j.ijrobp.2007.03.048.

227. Chang EL, Wefel JS, Hess KR, et al. Neurocognition in patients with brain metastases treated with radiosurgery or radiosurgery plus whole-brain irradiation: a randomised controlled trial. *Lancet Oncol.* 2009;10(11):1037–1044. https://doi.org/10.1016/S1470-2045(09)70263-3.

228. Vellayappan B, Tan CL, Yong C, et al. Diagnosis and management of radiation necrosis in patients with brain metastases. *Front Oncol.* 2018;8. https://doi.org/10.3389/fonc.2018.00395.

229. Salner AL, Botnick LE, Herzog AG. Reversible brachial plexopathy following primary radiation therapy for breast cancer. *Cancer Treat Rep.* 1981;65(9–10):797–802.

230. Pierce SM, Recht A, Lingos TI. Long-term radiation complications following conservative surgery (CS) and radiation therapy (RT) in patients with early stage breast cancer. *Int J Radiat Oncol Biol Phys.* 1992;23(5):915–923.

231. Fathers E, Thrush D, Huson SM. Radiation-induced brachial plexopathy in women treated for carcinoma of the breast. *Clin Rehabil.* 2002;16(2):160–165.

232. Powell S, Cooke J, Parson C. Radiation-induced brachial plexus injury: follow-up of two different fractionation schedules. *Radiother Oncol.* 1990;18(3):213–220.

233. Bajrovic A, Rades D, Fehlauer F. Is there a life-long risk of brachial plexopathy after radiotherapy of supraclavicular lymph nodes in breast cancer patients? *Radiother Oncol.* 2004;71(3):297–301.

234. Johansson S, Svensson H, Denekamp J. Dose response and latency for radiation-induced fibrosis, edema, and neuropathy in breast cancer patients. *Int J Radiat Oncol Biol Phys.* 2002;52(5):1207–1219.

235. Barr LC, Kissin MW. Radiation-induced brachial plexus neuropathy following breast conservation and radical radiotherapy. *Br J Surg.* 1987;74(9):855–856.

236. Rutherford R, Turley JJE. Carcinomatous versus radiation-induced brachial plexopathy in breast cancer. *Can J Neurol Sci.* 1983;10:154.

237. Pritchard J, Anand P, Broome J, et al. Double-blind randomized phase II study of hyperbaric oxygen in patients with radiation-induced brachial plexopathy. *Radiother Oncol.* 2001;58(3):279–286. https://doi.org/10.1016/s0167-8140(00)00319-4.

238. Foley KM, Woodruff JM, Ellis FT. Radiation-induced malignant and atypical peripheral nerve sheath tumors. *Ann Neurol.* 1980;7(4):311–318.

239. Taghian A, Vathaire F, Terrier P. Long-term risk of sarcoma following radiation treatment for breast cancer. *Int J Radiat Oncol Biol Phys.* 1991;21(2):361–367.

240. Gerard JM, Franck N, Moussa Z. Acute ischemic brachial plexus neuropathy following radiation therapy. *Neurology.* 1989;39(3):450–451.

241. Rubin DI, Schomberg PJ, Shepherd RFJ. Arteritis and brachial plexus neuropathy as delayed complications of radiation therapy. *Mayo Clin Proc.* 2001;76(8):849–852.

242. Jung BF, Ahrendt GM, Oaklander AL. Neuropathic pain following breast cancer surgery: proposed classification and research update. *Pain.* 2003;104(1/2):1–13.

243. Vecht CJ. Arm pain in the patient with breast cancer. *J Pain Symptom Manage.* 1990;5(2):109–117.

244. Granek I, Ashikari R, Foley K. The post-mastectomy pain syndrome: clinical and anatomical correlates. *Proc ASCO.* 1984;122.

245. Vecht CJ, Brand HJ, Wajer OJM. Post-axillary dissection pain in breast cancer due to a lesion of the intercostobrachial nerve. *Pain.* 1989;38(2):171–176.

246. Watson CPN, Evans RJ, Watt VR. The post-mastectomy pain syndrome and the effect of topical capsaicin. *Pain.* 1989;38(2):177–186.

247. Stevens PE, Dibble SL, Miaskowski C. Prevalence, characteristics, and impact of postmastectomy pain syndrome: an investigation of women's experiences. *Pain.* 1995;61(1):61–68.

248. Carpenter JS, Andrykowski MA, Sloan P. Postmastectomy/postlumpectomy pain in breast cancer survivors. *J Clin Epidemiol.* 1998;51(12):1285–1292.

249. Abdullah TI, Iddon J, Barr L. Prospective randomized controlled trial of preservation of the intercostobrachial nerve during axillary node clearance for breast cancer. *Br J Surg.* 1998;85(10):1443–1445.

250. Macdonald L, Bruce J, Scott NW. Long-term follow-up of breast cancer survivors with post-mastectomy pain syndrome. *Br J Cancer.* 2005;92(2):225–230.

251. Jamison K, Wellisch DK, Katz RL. Phantom breast syndrome. *Arch Surg.* 1979;114(1):93–95.

252. Downing R, Windsor CWO. Disturbance of sensation after mastectomy. *Br Med J.* 1984;288(6431):1650.

253. Krøner K, Krebs B, Skov J. Immediate and long-term phantom breast syndrome after mastectomy: incidence, clinical characteristics and relationship to pre-mastectomy breast pain. *Pain.* 1989;36(3):327–334.

254. Kroner K, Knudsen UB, Lundby L. Long-term phantom breast syndrome after mastectomy. *Clin J Pain.* 1992;8(4):346–350.

255. Wong L. Intercostal neuromas: a treatable cause of postoperative breast surgery pain. *Ann Plast Surg.* 2001;46(5):481–484.

256. Ganel A, Engel J, Sela M. Nerve entrapments associated with postmastectomy lymphedema. *Cancer.* 1979;44(6):2254–2259.

257. Younes-Mhenni S, Janier MF, Cinotti L. FDG-PET improves tumour detection in patients with paraneoplastic neurological syndromes. *Brain.* 2004;127(10):2331–2338.

258. Scheid R, Voltz R, Briest S. Clinical insights into paraneoplastic cerebellar degeneration. *J Neurol Neurosurg Psychiatry.* 2006;77(4):529–530.

259. Rojas-Marcos I, Rousseau A, Keime-Guibert F. Spectrum of paraneoplastic neurologic disorders in women with breast and gynecologic cancer. *Medicine (Baltimore).* 2003;82(3):216–223.

260. Croft PB, Wilkinson C. Carcinomatous neuromyopathy. Its incidence in patients with carcinoma of the ling and carcinoma of the breast. *Lancet.* 1963;i:184–188.

261. Fanous I, Dillon P. Paraneoplastic neurological complications of breast cancer. *Exp Hematol Oncol.* 2016;5. https://doi.org/10.1186/s40164-016-0058-x.

262. Rojas I, Graus F, Keime-Guibert F. Long-term clinical outcome of paraneoplastic cerebellar degeneration and anti-Yo antibodies. *Neurology.* 2000;55(5):713–715.

263. Anderson NE, Rosenblum MK, Posner JB. Paraneoplastic cerebellar degeneration: clinical-immunological correlations. *Ann Neurol.* 1988;24(4):559–567.

264. Hammack JE, Kimmel DW, O'Neill BP. Paraneoplastic cerebellar degeneration: a clinical comparison of patients with and without Purkinje cell cytoplasmic antibodies. *Mayo Clin Proc.* 1990;65(11):1423–1431.

265. Peterson K, Rosenblum MK, Kotanides H. Paraneoplastic cerebellar degeneration. I. A clinical analysis of 55 anti-Yo antibody-positive patients. *Neurology.* 1992;42(10):1931–1937.

266. Shams'íli S, Grefkens J, Leeuw B. Paraneoplastic cerebellar degeneration associated with antineuronal antibodies: analysis of 50 patients. *Brain.* 2003;126(6):1409–1418.

III. Neurological complications of specific neoplasms

267. Posner JB. *Neurologic Complications of Cancer.* FA Davis Company; 1995.
268. Anderson NE, Budde-Steffen C, Rosenblum MK. Opsoclonus, myoclonus, ataxia, and encephalopathy in adults with cancer: a distinct paraneoplastic syndrome. *Medicine (Baltimore).* 1988;67(2):100–109.
269. Bataller L, Graus F, Saiz A. Clinical outcome in adult onset idiopathic or paraneoplastic opsoclonus-myoclonus. *Brain.* 2001;124(2):437–443.
270. Budde-Steffen C, Anderson NE, Rosenblum MK. An antineuronal autoantibody in paraneoplastic opsoclonus. *Ann Neurol.* 1988;23(5):528–531.
271. Luque FA, Furneaux HM, Ferziger R. Anti-Ri: an antibody associated with paraneoplastic opsoclonus and breast cancer. *Ann Neurol.* 1991;29(3):241–251.
272. Escudero D, Barnadas A, Codina M. Anti-Ri-associated paraneoplastic disorder without opsoclonus in a patient with breast cancer. *Neurology.* 1993;43(8):1605–1606.
273. McCabe DJH, Turner NC, Chao D. Paraneoplastic "stiff person syndrome" with metastatic adenocarcinoma and anti-Ri antibodies. *Neurology.* 2004;62(8):1402–1404.
274. Gultekin SH, Rosenfeld MR, Voltz R. Paraneoplastic limbic encephalitis: neurological symptoms, immunological findings and tumour association in 50 patients. *Brain.* 2000;123(7):1481–1494.
275. Sutton I, Winer J, Rowlands D. Limbic encephalitis and antibodies to Ma2: a paraneoplastic presentation of breast cancer. *J Neurol Neurosurg Psychiatry.* 2000;69(2):266–268.
276. Scheid R, Honnorat J, Delmont E. A new anti-neuronal antibody in a case of paraneoplastic limbic encephalitis associated with breast cancer. *J Neurol Neurosurg Psychiatry.* 2004;75(2):338–340.
277. Dalmau J, Graus F, Villarejo A. Clinical analysis of anti-Ma2-associated encephalitis. *Brain.* 2004;127(8):1831–1844.
278. Rosenfeld MR, Eichen JG, Wade DF. Molecular and clinical diversity in paraneoplastic immunity to Ma proteins. *Ann Neurol.* 2001;50(3):339–348.
279. De Camilli P, Thomas A, Cofiell R, et al. The synaptic vesicle-associated protein amphiphysin is the 128-kD autoantigen of stiff-man syndrome with breast cancer. *J Exp Med.* 1993;178(6):2219–2223.
280. Folli F, Solimena M, Cofiell R. Autoantibodies to a 128-kd synaptic protein in three women with the stiff-man syndrome and breast cancer. *N Engl J Med.* 1993;328(8):546–551.
281. Rosin L, DeCamilli P, Butler M. Stiff-man syndrome in a woman with breast cancer: an uncommon central nervous system paraneoplastic syndrome. *Neurology.* 1998;50(1):94–98.
282. Schmierer K, Grosse P, De Camilli P. Paraneoplastic stiff-person syndrome: no tumor progression over 5 years. *Neurology.* 2002;58(1):148.
283. Antoine JC, Absi L, Honnorat J. Antiamphiphysin antibodies are associated with various paraneoplastic neurological syndromes and tumors. *Arch Neurol.* 1999;56(2):172–177.
284. Saiz A, Dalmau J, Husta BM. Anti-amphiphysin I antibodies in patients with paraneoplastic neurological disorders associated with small cell lung carcinoma. *J Neurol Neurosurg Psychiatry.* 1999;66(2):214–217.
285. Pittock SJ, Lucchinetti CF, Parisi JE. Amphiphysin autoimmunity: paraneoplastic accompaniments. *Ann Neurol.* 2005;58(1):96–107.
286. Petzold GC, Marcucci M, Butler MH. Rhabdomyolysis and paraneoplastic stiff-man syndrome with amphiphysin autoimmunity. *Ann Neurol.* 2004;55(2):286–290.
287. Sommer C, Weishaupt A, Brinkhoff J. Paraneoplastic stiff-person syndrome: passive transfer to rats by means of IgG antibodies to amphiphysin. *Lancet.* 2005;365(9468):1406–1411.
288. Peterson K, Forsyth PA, Posner JB. Paraneoplastic sensorimotor neuropathy associated with breast cancer. *J Neurooncol.* 1994;21(2):159–170.
289. Storstein A, Vedeler C. Neuropathy and malignancy: a retrospective survey. *J Neurol.* 2001;248(4):322–327.
290. Forsyth PA, Dalmau J, Graus F. Motor neuron syndromes in cancer patients. *Ann Neurol.* 1997;41(6):722–730.
291. O'Neill JH, Murray NMF, Newsom-Davis J. The Lambert-Eaton myasthenic syndrome. A review of 50 cases. *Brain.* 1988;111(3):577–596.
292. Rudge P. Optic neuritis as a complication of carcinoma of the breast. *Proc R Soc Med.* 1973;66(11):1106–1107.
293. Antoine JC, Honnorat J, Vocanson C. Posterior uveitis, paraneoplastic encephalomyelitis and auto-antibodies reacting with developmental protein of brain and retina. *J Neurol Sci.* 1993;117(1-2):215–223.
294. Klingele TG, Burde RM, Rappazzo JA. Paraneoplastic retinopathy. *J Clin Neuroophthalmol.* 1984;4(4):239–245.

Neurological complications of melanoma

David Gritsch and Maciej M. Mrugala

Department of Neurology, Mayo Clinic and Mayo Clinic Cancer Center, Phoenix/Scottsdale, AZ, United States

1 Introduction

Of the three major types of skin cancer—basal cell carcinoma, squamous cell carcinoma, and melanoma—only melanoma typically involves the central nervous system (CNS). Skin cancer and particularly melanoma can occasionally be detected early in its development, at a time when therapy can be curative. About 85% of patients with new diagnosis of cutaneous melanoma are present with clinically localized disease.[1] However, in many cases, a lack of appropriate screening leads to delayed diagnosis and results in poorer outcomes. In instances when initial therapy is not curative, malignant melanoma can commonly metastasize and produce neurological complications.

There has been a dramatic, 300% increase in the incidence of melanoma in the United States over the last 40 years, resulting in over 10,100 deaths annually.[2] Melanoma produces neurological complications by a variety of mechanisms, including direct invasion, compression from metastases to adjacent structures, vascular compromise, and systemic metabolic disturbances secondary to visceral metastases. Melanoma is the third most common cause of brain metastases, after lung and breast cancer, even though it accounts for only 1% of all cancer.[3–5] This chapter addresses the neurological complications of melanoma and discusses diagnostic and therapeutic considerations.

2 Neurological complications of malignant melanoma

2.1 Introduction

Cutaneous malignant melanoma is an increasingly common form of cancer, almost doubling in incidence every decade.[6] It has been estimated that over 76,000 new cases of melanoma were diagnosed in the United States in 2016 and that approximately 10,130 died from the disease.[1,2] There is a slight male predominance, with the lifetime risk of developing melanoma of 1 in 38.[7,8] The cause of the rising incidence is unknown, but higher levels of exposure to ultraviolet light, including the artificial sources, have been suggested as one possible factor. Cumulative exposure seems to play a significant role, hence the incidence of melanoma is higher in older individuals.[9] The median age at diagnosis of melanoma is in the early sixth decade. As a result of the rising incidence, the mortality rate from malignant melanoma is also increasing.

Melanomas arise from melanocytes, which are neural-crest-derived cells found in the skin, hair follicles, mesentery, iris and choroid, leptomeninges, and occasionally other sites in the body. Melanin, the pigment of melanocytes, is produced from tyrosine by the enzyme tyrosinase. Conversion of tyrosine to catecholamines occurs through a separate enzymatic pathway. Melanin is bound to proteins within intracellular organelles called melanosomes, where it absorbs ultraviolet light, partially protecting the skin from this form of irradiation. This may explain the fact that melanoma is more than 20 times less common in blacks than in whites.[8,10]

Cutaneous melanomas are clinically and pathologically classified into four groups: lentigo maligna, superficial spreading melanoma, nodular melanoma, and acral lentiginous melanoma. Initially, tumor cells extend in a radial (horizontal) plane. The development of vertical growth heralds an invasive phase in the progression of the tumor and the depth of invasion (Clark's or Breslow's classification; see Table 1) at the time of resection is highly correlated with the likelihood of metastatic disease and survival. The staging system by the American Joint Committee on Cancer (AJCC) is based on the tumor thickness, presence of primary ulcerations, and regional nodal tumor burden.[11] Regional nodal tumor burden is the most important prognostic factor for survival in patients without distant disease. Major changes to the 8th edition of the AJCC staging system include the use of a total of four M1 subgroups based on the location of the

TABLE 1   Clark level with approximately corresponding Breslow depth.

Clark level		Breslow depth
Level 1	Tumor is confined to the epidermis	≤1 mm
Level 2	Invasion of the papillary dermis	> 1.0–2.0
Level 3	Invasion of the dermis	> 2.0–4.0
Level 4	Invasion of the reticular dermis	
Level 5	Invasion of the subdermal fat.	> 4.0

Both staging systems grade the vertical progression of cutaneous malignant melanomas during the progressive phase of the disease and are highly correlated with the development of metastatic disease and survival. *Breslow A. Thickness, cross-sectional areas, and depth of invasion in the prognosis of cutaneous melanoma. Ann Surg. 1970;172:902; Clark WH Jr., From L, Bernardino EA, Mihm MC. The histogenesis and biologic behavior of primary human malignant melanomas of the skin. Cancer Res. 1969;29(3):705–727. PMID: 5773814.*

involved organ, with a new M1d designation being added for CNS metastases: distant skin, subcutaneous spread or distant lymph nodes, lung involvement, other visceral organ involvement excluding the CNS, and metastasis to the CNS. In addition, each of the four M categories is now subdivided by a notation indicating nonelevated or elevated serum lactate dehydrogenase, another important independent prognostic factor in metastatic melanoma.[12] Survival benefit was observed in patients with only subcutaneous or lymph node involvement.[13] Characteristics associated with a poor prognosis include nodular melanoma subtype, lesions located on the scalp, hands, or feet, and older age at diagnosis. When metastases occur, the organs most likely to be involved are the skin, lymph nodes, lung, and brain.

Neurological complications are common and grave events in the progression of malignant melanoma. In the majority of patients, these complications arise in the setting of disseminated, progressive systemic disease, thereby confirming an already poor prognosis. The incidence of central nervous system (CNS) metastases in patients with metastatic malignant melanoma ranges from 10% to 40% in clinical studies but is even higher in autopsy series where up to 90% of patients have CNS involvement.[14] About two-thirds of patients with metastatic melanoma develop clinically apparent neurological complications prior to death.[15,16]

While metastatic melanoma is notorious for its aggressive biological nature, it sometimes acts in an unpredictable manner, with metastatic disease becoming indolent, or on rare occasions, undergoing spontaneous regression. In fact, melanoma is one of the most common tumors to undergo spontaneous regression.[17] The rates of spontaneous regression are estimated to be between 10% and 20% based on histological studies that show that about 25% of melanomas exhibit partial regression.[17,18] More recently, the introduction and widespread use of molecularly targeted therapy and immunotherapies have ushered in a new era in the treatment of advanced melanomas and have led to a truly impressive improvement in outcomes.

2.2 Clinical presentation

Most of the neurological complications from malignant melanoma result from metastases directly to the nervous system (brain parenchyma, nerve roots) or adjacent structures with secondary neural compromise. Nonmetastatic complications, such as vascular disorders (e.g., nonbacterial thrombotic endocarditis and DIC) and paraneoplastic disorders, rarely occur in patients with melanoma.

Presenting symptoms are based on the location of the metastasis. Confusion (45%), headaches (27%), focal motor or sensory deficit (47%), and seizures (11%) are among the most common presentations.[16] Detection of CNS and other distant metastatic diseases has improved since the introduction of newer and more sophisticated imaging techniques, including gadolinium contrast-enhanced magnetic resonance imaging (MRI).[19]

2.2.1 Incidence and clinical features

The nervous system can be the initial site of metastasis in patients with melanoma in about 5% of cases.[16,20] The median interval from initial diagnosis of melanoma to the diagnosis of nervous system complications in one series was 37 months, but 21 patients (25%) were more than 5 years from diagnosis and 9 patients (11%) were more than 10 years out from initial diagnosis.[20] Frequently, the nervous system is not the sole site of metastatic disease at the time of diagnosis. Approximately 5%–20% of patients have no identifiable second site of disease at the time when brain metastases were diagnosed.[21,22]

Lung metastases are the most common type of extracranial disease in patients with brain metastases from melanoma, occurring twice as often as any other systemic site. This pattern has been noted with brain metastases from many types of cancer and suggests that the lungs are a common source for hematogenous spread to the brain. Some authors discovered that the incidence of CNS metastasis was higher if the primary lesions were located on the upper body (head, neck, or the trunk) as opposed to the lower extremities.[20,21] In the Vanderbilt series of 78 patients with neurological complication of melanoma, 82% had metastatic lesions in the brain while only 3% had lesions in the spinal cord. Seventeen percent of patients had lesions affecting cranial and peripheral nerves (see Table 2).

2.2.2 Brain metastases from malignant melanoma

The distribution of lesions between supratentorial and infratentorial compartments (see Table 2) is similar to that of brain metastases in general.[4] Postmortem studies have found a higher percentage (91%) of patients with multiple lesions than CT-based studies.[8]

TABLE 2   Site of nervous system involvement in 78 patients from a total sample of 208 patients with malignant melanoma.[a]

Location	N	(%)
Brain	61	(82%)[b]
Supratentorial	56	
Infratentorial	10	
Number of lesions		
1	26	(46%)[c]
2	10	(18%)
3 or more	20	(36%)
Spinal cord		
Epidural	2	(3%)
Intramedullary	0	
Meningeal		
Total	9	(12%)
Meningeal + brain	3	
Cranial and peripheral nerve	13	(17%)

[a] 12 patients had involvement of two nervous system sites.
[b] 5 patients had metastases above and below the tentorium.
[c] 46% of the 56 patients for whom lesion number is known.

Symptoms and signs

Headache, focal neurological deficits, cognitive changes, and seizures are the most common presenting features of brain metastases. In the Vanderbilt series, 15/56 (27%) patients with supratentorial brain metastases from melanoma developed seizures at some time during their course. Five of these patients had seizures as the presenting sign of their brain metastasis. Reports of the percentage of patients with seizures from brain metastases originating from other cancers and metastatic melanoma have described similar figures.[16,22,23] However, Hagen et al. found seizures at presentation in 37% of 35 patients with melanoma, and Byrne et al. reported a still higher figure of 48% of 80 patients.[24,25] Oberndorfer et al. looked at the frequency of seizures in patients with primary brain tumors and metastatic lesions.[26] In this series, metastatic melanoma was associated with a very high frequency of seizures (68%) during the course of the illness. A higher incidence of seizures in patients with CNS melanoma could be related to the propensity of lesions to become hemorrhagic.

The reported risk of developing seizures after diagnosis of brain metastases has varied between studies. In the Vanderbilt series, 10/51 (20%) of patients with brain metastasis developed seizures. Byrne et al. found late-onset seizures in 21/63 (37%) of melanoma patients, while Cohen et al. observed late-onset seizures in only

16/195 (8%) of patients with brain metastasis from a variety of solid tumors.[23,25] In the latter study, multiple cerebral metastases correlated with the development of late-onset seizures.

Patients with multiple small brain metastases may develop a diffuse (nonfocal) encephalopathy that suggests a metabolic etiology. While there usually are other clues to the presence of structural disease such as visual field loss, subtle focal weakness, or headache, these may be absent early in the course. Also, patients with brain metastases may have a lower threshold for the development of encephalopathy secondary to metabolic derangements or CNS depressant drugs than do those without such underlying brain abnormalities. Unusual presentations of brain metastases from melanoma may mimic subarachnoid hemorrhage, subdural hematoma, or cerebral sinus thrombosis.[27–29]

Diagnosis

The diagnosis of brain metastasis is usually confirmed by MRI or CT scanning. Because of the high frequency of brain metastases in patients with metastatic melanoma and because at autopsy patients have previously undetected brain metastases, a MRI or CT scan is appropriate for melanoma patients with headache, personality changes, or encephalopathy even when focal neurological signs are absent. MRI with gadolinium is more sensitive in the detection of metastatic deposits than CT.[30] Gadolinium-enhanced MRI is the study of choice, especially when resection of a presumed solitary metastasis is contemplated (Table 2).

Typical radiographic findings are single or multifocal parenchymal brain lesions that enhance following contrast administration, but subarachnoid or dural lesions are also seen.[31] Approximately 25% of pre-contrast CT scans demonstrate lesions with increased density, which reflects the presence of melanin or hemorrhage.

MRI may be able to differentiate among melanotic melanoma, amelanotic melanoma, and hemorrhagic melanoma metastases.[19,32] Melanotic melanomas may demonstrate increased intensity on T1-weighted sequences and mildly hypo- or isointense T2-weighted signals, which is the reverse of the findings with most metastases (see Fig. 1). This MRI finding is suggestive, although not diagnostic, of metastatic melanoma. Hemorrhagic lesions may be distinguished from melanotic metastases by the presence of susceptibility effect on gradient-echo or T2*-weighted images. Amelanotic melanoma metastases give MRI signal intensities that are similar to those seen with metastases from other primaries. In some instances, positron imaging tomography (PET) scanning can be helpful in the diagnostic process and for assessment of therapy; however, this technique has limitations, specifically for small CNS lesions.[33,34]

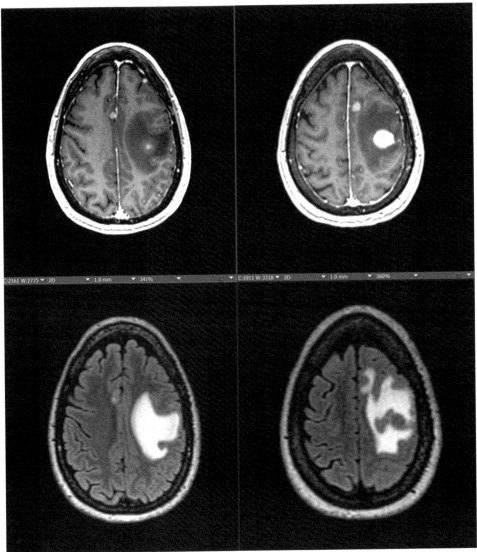

FIG. 1   MRI of brain metastases from malignant melanoma. The T1-weighted images after contrast administration (upper panel) show two cortically based lesions typical of melanoma. The T2/FLAIR images (lower panel) show classic "finger-like" projections of vasogenic edema (posterior frontal lesion) confined to the white matter and sparing the cortex. *Images courtesy of Dr. Leland Hu, Mayo Clinic Arizona.*

When radiological abnormalities consistent with brain metastases are seen in patients with known metastatic malignant melanoma, the diagnosis can be established with a high degree of confidence. However, the possibility of nonmetastatic diseases must always be kept in mind particularly with single lesions. In a prospective surgical series of cancer, patients with a single brain lesion thought to represent a metastasis, 6/54 (11%) were found to have other diseases at biopsy.[35] Second malignancies occur in about 5% of patients with metastatic melanoma. For these reasons, biopsy should be considered for patients with a solitary brain lesion and a distant history of melanoma or in the presence of stable or limited stage systemic disease.

In patients with neurological complaints but whose CT or MRI scan does not reveal a parenchymal abnormality, metastases to the leptomeninges and skull base must be considered. Hydrocephalus, leptomeningeal enhancement, and brain parenchymal lesions that abut the ventricles or the subarachnoid space suggest the possibility of leptomeningeal metastases.

Therapy of brain metastases

(i) Glucocorticoids. Symptoms and signs from melanoma brain metastases usually respond dramatically to dexamethasone. Dexamethasone should be tapered in patients whose neurological deficits are stable or improving, however, many patients will not be able to tolerate the withdrawal. Byrne et al. found that 86% of 80 patients with brain metastases from melanoma were steroid-dependent until death.[25]

(ii) Anticonvulsants. Given the high risk of recurrence and significant associated morbidity and mortality,

antiepileptic therapy is generally indicated in patients with seizures in the setting of primary and metastatic brain tumors. Some authors have recommended prophylactic anticonvulsants for patients with brain metastases from melanoma.[25] This recommendation is based on reports of a high risk of seizures in these patients (see above). Studies, however, have shown that prophylactic anticonvulsants may not be effective in reducing the incidence of first seizure in brain tumor patients.[36] In addition, initiation of antiepileptic therapy with valproate or levetiracetam at the start of chemoradiation did not result in improved survival in a meta-analysis of 1869 patients with glioblastoma.[37]

The goal of antiepileptic treatment is to achieve seizure freedom on the lowest effective dose of the anticonvulsant drug. In the absence of comparative randomized controlled trials, there is currently no clear evidence to suggest superiority of one anticonvulsant over the other. However, potential interactions with chemotherapeutic drugs and associated toxicities have to be considered when caring for patients with brain tumors. Therefore, anticonvulsants that do not interfere with the hepatic drug metabolism, such as levetiracetam, topiramate, or lacosamide, are generally preferred. Other good choices include oxcarbazepine, eslicarbazepine, perampanel, zonisamide, clobazam, or brivaracetam. Cost can be a factor that has to be considered with some of the newer 3rd generation antiepileptics. Because of its efficacy, ease of management, and low cost, levetiracetam is a common choice for antiepileptic therapy in patients with brain metastases. Levetiracetam is generally well tolerated; however, the risk of neuropsychiatric side effects may be increased in patients with brain tumors.[38] The risk of Stevens-Johnson syndrome may be increased in patients on phenytoin during radiation therapy so that special caution should be used when a rash occurs in this setting. Patients with structural brain disease may be more susceptible to the sedative effects of phenobarbital and lamotrigine is associated with the increased risk of severe skin rashes. Because most anticonvulsants are extensively bound by albumin, determination of free anticonvulsant levels may be useful in patients with hypoalbuminemia, in those taking numerous medications, and in those who develop signs of nervous system toxicity in the presence of a therapeutic total anticonvulsant level.[16] It has been suggested that around 50% of patients with brain tumors will suffer breakthrough seizures on first-line antiepileptic medication [B]. If patients continue to experience breakthrough seizures on monotherapy, despite dose escalation and adequate serum levels of antiepileptic drugs, it might be necessary to add a second anticonvulsant. Lacosamide has been demonstrated to be a safe and effective adjunctive treatment in patients with brain tumors and uncontrolled seizures.[39] Gabapentin and pregabalin are equally well tolerated

and can serve as useful adjunct to first-line antiepileptic therapy. Patients with brain metastases should be advised not to engage in potentially hazardous activities.

(iii) Targeted therapy and Immunotherapy. The introduction of BRAF-targeted therapy and immunotherapy has revolutionized the treatment of metastatic melanoma and resulted in an astonishing improvement in outcomes. BRAF and the downstream MEK protein kinase constitute part of the RAS/MAPK signaling pathway and are involved in the regulation of several cellular processes essential for oncogenesis such as proliferation, differentiation, migration, and survival. Cytotoxic T-lymphocyte antigen 4 (CTLA-4) and programmed cell death receptor 1 (PD-1) are two surface molecules expressed on T-cells that are part of a "checkpoint" system that limits T-cell activation and can be used by cancer cells to evade the immune response. Several phase II trials have evaluated the effectiveness and side-effect profile of single-agent BRAF inhibitors, BRAF inhibitors in combination with MEK inhibitors, single-agent PD-1 inhibitors, and PD-1 inhibitors combined with CTLA-4 inhibitors in the treatment of melanoma brain metastases.[40–46] While high response rates were achieved with both treatment modalities, immunotherapy appeared to produce more durable responses, especially when combining PD-1 and CTLA-4 inhibition.[42,45,46] Potential synergistic effects between radiotherapy and immunotherapy have been postulated based on both molecular and clinical data.[47–53] Comparative phase III studies and data regarding the combination of these modalities are unfortunately lacking.

Activating mutations in BRAF, resulting in an overactive mitogen-activated protein (MAP) kinase pathway, are important drivers of the malignant process and can be found in 40%–60% of advanced melanomas. The effectiveness of vemurafenib and dabrafenib in the treatment of advanced melanoma harboring one of the most common mutations in BRAF (V600E/K) has been demonstrated in randomized phase III trials. High initial response rates, unfortunately typically give way to disease progression. This likely represents reactivation of the MAP kinase pathway over alternative molecular mechanisms, including activation of MEK. Combination BRAF and MEK inhibitor therapy has therefore largely replaced single-agent BRAF inhibition.

The phase II BREAK-MB trial evaluated the effectiveness and safety of the BRAF inhibitor dabrafenib in 172 patients with melanoma brain metastases harboring the BRAF V600E mutation.[40] Intracranial response rate was 39% and 31% for patients with and without prior local treatment, respectively. The progression-free survival (PFS) was 16 weeks and overall survival (OS) 31 weeks. Overall toxicity was acceptable and no specific intracranial adverse events were observed.

Another phase II study investigated the BRAF inhibitor vemurafenib in patients with BRAF V600-mutated melanoma with brain metastases.[41] The observed intracranial response rate in patients without prior treatment was 18%. PFS was 4.0 months and 3.7 months and OS was 9.6 months and 8.9 months for patients with and without prior treatment, respectively. Notably adverse events were comparable to other studies of single-use vemurafenib and no significant CNS toxicity was observed.

The phase II COMBI-MB study evaluated the combination of dabrafenib with the MEK inhibitor trametinib in the treatment of BRAF V600-mutated melanoma with brain metastases.[42] The primary endpoint was intracranial response in 76 patients with BRAF V600E-mutated, asymptomatic melanoma brain metastases with no previous treatment and Eastern Cooperative Oncology Group (ECOG) performance status of 0 or 1. The study reported an acceptable side-effect profile for the combination of dabrafenib with trametinib and an intracranial response rate of 58%, PFS of 5.6 months, and OS of 10.8 months in the primary study population.

Cytotoxic T-lymphocyte antigen 4 (CTLA-4) and programmed cell death receptor 1 (PD-1) are molecules expressed on the surface of T-cells that are part of a physiologic break or "checkpoint" system that limits T-cell activation. Antibodies directed against PD-1 (nivolumab, pembrolizumab) or CTL-4 (ipilimumab), so-called checkpoint inhibitors, have shown to be effective and prolong survival in advanced melanoma.

A phase II study of 72 patients with advanced melanoma with brain metastases evaluated the safety and effectiveness of ipilimumab in this patient population.[43] The primary endpoint of complete intracranial response, partial intracranial response, or stable intracranial disease after 12 weeks was reached in 24% of asymptomatic patients without concurrent corticosteroid treatment and 10% of symptomatic patients on stable corticosteroid treatment. The side-effect profile was comparable to other studies evaluating the use of ipilimumab for solid malignancies. These results underline the importance of avoiding the use of corticosteroids in patients undergoing immunotherapy with checkpoint inhibitors.

Another phase II study evaluated the use of pembrolizumab in patients with melanoma or nonsmall-cell lung cancer and untreated brain metastases.[44] The study reported an intracranial response in four of 18 patients with melanoma brain metastases, and responses were durable throughout the study period (median follow-up of 7 months). Reported neurological adverse events included transient cognitive dysfunction and grade 1–2 seizures in the melanoma cohort.

Two open-label, multicenter, randomized phase 2 trials evaluated safety and efficacy of combination treatment with ipilimumab and nivolumab in patients with melanoma brain metastases. The ABC study evaluated the effect of nivolumab monotherapy and combination nivolumab plus ipilimumab in 79 immunotherapy-naive patients with active melanoma brain metastases.[45] Of 61 asymptomatic patients with no previous local brain therapy, intracranial response was reported in 46% and 20% for the combined treatment and monotherapy, respectively. Grade 3 or 4 treatment-related adverse events occurred in 54% of patients treated with the combination and 16% with the monotherapy. The CheckMate 204 study evaluated 94 patients with asymptomatic, nonirradiated melanoma brain metastases that were treated with the combination of nivolumab plus ipilimumab.[46] The overall rate of intracranial clinical benefit was 57%, with complete response in 26%, partial response in 30%, and stable disease for at least 6 months in 2% of the study population. The side-effect profile was similar to that reported in the treatment of extracranial melanoma. Grade 3 or 4 treatment-related adverse events occurred in 55% of patients, including central nervous system adverse events in 7%.

Taken together (see Table 3), the results of these studies demonstrate that both targeted therapy with BRAF inhibitors and immunotherapy with checkpoint inhibitors are safe and effective options in the treatment of advanced melanoma with brain metastases. Immunotherapy however appears to produce more durable results and the efficacy of immunotherapy appears to be higher when a combination of CTL-4 and PD1 inhibitors is used.

(iv) Radiation therapy. Melanoma is a radioresistant cancer. Whole brain radiation therapy (WBRT) has traditionally been used in melanoma brain metastases. However, a lack of survival benefit, high risk of relapse, and increasing awareness of neurotoxic side effects has largely limited its use to cases with numerous or extensive brain metastases and meningeal involvement.[54] Of particular concern are delayed neurocognitive side effects that can manifest over 12 months after completion of WBRT with a broad impact on cognition. Many studies of conventional, external beam radiation therapy for brain metastases from melanoma have been published, but the majorities are difficult to interpret since they are not prospective or controlled and because there is no direct measurement of the response of the brain metastases to radiation treatment. When neurological symptoms and signs are used to measure responses, glucocorticoid therapy may be responsible for any apparent response and rapidly progressive systemic disease can obscure any benefit of cranial irradiation on survival. Hagen et al. reported 35 patients with a single metastasis that was surgically removed.[24] Nineteen patients received postoperative RT and 16 did not. Relapse was measured by CT or MRI appearance of progressive disease. The irradiated group had a significantly longer median interval to progression than did the nonirradiated group (26 months versus 5.7 months, respectively).

TABLE 3   Overview of key studies of targeted therapy and immunotherapy in previously untreated and asymptomatic melanoma brain metastases.

Study	Treatment	RR[a]	PFS[b] (months)	OS[c] (months)
Targeted therapy: BRAF/MEK-inhibitors				
Long et al. 2012 (BREAK-MB)[37]	Dabrafenib	31%	16.1 Val600Glu; 8.1 Val600Lys	33.1 Val600Glu; 16.3 Val600Lys
McArthur et al. 2017[38]	Vemurafenib	18%	3.7	8.9
Davies et al. 2017[39] (COMBI-MB)	Dabrafenib and trametinib	58%	5.6	10.8
Immunotherapy: PD-L1 and CTLA-4 inhibitors				
Margolin et al. 2012[40]	Ipilimumab	24%	1.5	7.0
Goldberg et al. 2016[41]	Pembrolizumab	22%	2.3	NR
Long et al. 2018[42] (ABC)	Nivolumab vs nivolumab plus ipilimumab	20 vs 46%	2.5 vs NR	NR
Tawbi et al. 2018[43] (CheckMate 204)	Nivolumab plus ipilimumab	57%	59.5% 9-month PFS	82.8% 9-month OS

[a] *Intracranial response rate.*
[b] *Median intracranial progression-free survival.*
[c] *Median overall survival.*

Median survival times were the same for the two groups, however. Byrne et al. used discontinuation of glucocorticoids and CT scans to indicate responses to irradiation in 66 melanoma patients.[25] Using these criteria, approximately 10% to 15% of patients showed a response to irradiation. Accelerated fractionation schedules have been tested because of their beneficial effect on systemic sites of disease, but no evidence of benefit over conventional fractionation has been found.[22] Several authors have reported short-term toxicity in the form of headache and increase in focal neurological deficits in patients receiving high-dose fractions.[55,56] More recently, hippocampal-sparing protocols and pharmacological prophylaxis with memantine have been evaluated in an attempt to prevent or limit WBRT-induced neurocognitive decline.[57,58]

Experience with stereotactic external-beam RT (SRS, Gamma Knife radiosurgery) suggests that this approach is substantially more effective than conventional RT in the treatment of melanoma brain metastases.[59–63] In addition, current evidence suggests that SRS is equally effective as surgery when used in the appropriate clinical setting.[64,65] SRS is therefore currently preferred over WBRT in the treatment of patients with ≤4 melanoma brain metastases. Metastases are typically treated within a single fraction. Stable disease or partial responses are routinely obtained at the treatment site, and symptomatic radionecrosis has not been common. It appears that the radiobiology of a single high dose overcomes the radioresistance barrier of melanoma. SRS is considered of limited usefulness in patients with numerous or large (>3–4 cm) lesions.

However, the effectiveness of SRS in patients with 5–15 brain metastases is currently being evaluated in several clinical trials. In addition, fractionated SRS has been used successfully to limit the risk of radiotoxicity when managing larger lesions (>3 cm).[66] Several studies showed that higher Karnofsky performance status (KPS) (>90), female gender, lack of neurological symptoms prior to treatment, and supratentorial localization of the metastatic lesion are good prognostic factors.[61,63] The combination of surgery followed by adjuvant SRS has been shown to produce excellent results in cases where large metastases require urgent decompression.[67,68] The addition of WBRT to SRS has been demonstrated to improve local tumor control, but not survival and at the cost of a significantly higher risk of neurocognitive decline and an inferior quality of life.[69,70] The Alliance trial evaluated 213 patients with 1–3 brain metastases amenable to radiosurgery across 34 institutions in North America.[71] Patients were randomly assigned to receive SRS alone or in combination with WBRT (30 Gy in 12 fractions). At 3 months, there was significantly less cognitive deterioration (defined as decline >1 SD from baseline on at least 1 cognitive test) after SRS alone (63.5%) when compared to the combination therapy (91.7%; $P < .001$). In addition, overall quality of life was significantly higher at 3 months for SRS alone (mean change from baseline, −1.3 vs −10.9 points; $P = .002$). Time to intracranial failure was shorter following SRS alone compared to the combination treatment (hazard ratio, 3.6; $P < .001$) but no significant difference was observed in median overall survival and functional independence at 3 months.

(v) Combined treatment modalities. A synergistic effect of SRS with systemic therapies in the treatment of melanoma has been postulated. Putative mechanisms include activation of antitumoral immune response by SRS-induced apoptosis and a radiosensitizing effect of BRAF inhibition on melanoma cells.[47] Several studies have evaluated the clinical use of combination therapy with SRS and targeted therapy or immunotherapy in the treatment of melanoma brain metastases. While a number of retrospective studies have demonstrated excellent local control rates and safety profiles, prospective data are unfortunately lacking. Ahmed et al. evaluated the outcome of single-fraction SRS in combination with anti-PD-1 therapy, anti-CTLA-4 therapy, BRAF/MEK inhibitors, BRAF inhibitors, or conventional chemotherapy in a retrospective sample of 96 patients and a total of 314 melanoma brain metastases.[48] Combination of SRS with targeted therapy or immunotherapy significantly increased OS and distant control rates when compared with conventional chemotherapy. Knisley et al. performed a retrospective institutional review of 77 patients undergoing SRS for melanoma brain oligometastases, with 35% also receiving ipilimumab.[49] The use of SRS with ipilimumab was associated with significantly increased median survival (21.3 versus 4.9 months) and 2-year survival rate (47.2% versus 19.7%) when compared to SRS alone. In another retrospective study, Silk et al. reported 70 patients with melanoma brain metastases treated with WBRT or SRS, 33 of whom also received ipilimumab and 37 did not.[50] The authors reported significantly improved median survival (18.3 versus 5.3 months) and partial response rate (30% versus 9.1%) in patients who were treated with ipilimumab prior to radiotherapy compared to radiotherapy alone. In addition, the use of ipilimumab and SRS was each significant predictors of improved survival (hazard ratio = 0.43 and 0.45). Acharya et al. evaluated a cohort of 72 patients with a total of 233 melanoma brain metastases and reported decreased distant and local intracranial failure following SRS within 3 months of anti-CTLA-4 or anti-PD-1 therapy when compared to SRS alone.[51] Another retrospective multicenter study evaluated the combined use of SRS and BRAF inhibitors in 198 patients with known BRAF mutation status.[52] In their study, the authors reported improved survival when BRAF inhibitors were administered following SRS, compared to before or concurrent to SRS. Notably, an increased occurrence of ICH was observed in the group of patients receiving BRAF inhibition. Similarly, Hecht et al. explored the safety of concomitant radiation therapy with BRAF inhibitors in a retrospective cohort of 155 patients with known BRAF mutation status and reported improved survival and a reduction in radiation-induced toxicity when interrupting vemurafenib treatment during radiation.[53] Consensus guidelines from the Eastern Cooperative Oncology Group (ECOG) currently recommend holding BRAF and/or MEK inhibitors for at least 3 days before and after fractionated RT and at least 1 day before and after SRS.[52] An ongoing prospective phase II study is currently evaluating the concurrent continuous use of dabrafenib and trametinib with SRS in patients with melanoma brain metastases (NCT02974803).[72] Further prospective studies will be needed to evaluate the long-term toxicities and optimal timing of combined therapy with RT and systemic therapies.

(vi) Surgery. Resection of brain metastases is an increasingly common practice, and there is accumulating evidence to support this approach. In one prospective, randomized study comparing surgery plus conventional RT to RT alone in the treatment of solitary brain metastases from a variety of cancers, those patients undergoing surgery had a longer duration of survival, maintained functional independence for a longer period of time, and had a lower rate of local recurrence compared to the RT only group.[35] In addition, surgical resection can aid in establishing a tissue diagnosis and provide material for molecular testing. Following surgery, patients typically are treated with local radiation of the surgical bed or WBRT, depending on the size and number of metastatic lesions.

Several retrospective surgical series report longer survivals after surgery plus RT compared to RT alone in patients with metastatic melanoma. Wasif et al. reported significantly improved median and 5-year overall survival following metastasectomy in a population-based study of 4229 patients with stage IV melanoma.[73] However, selection bias may account for most of the reported differences, since patient age, performance status, extent of systemic disease, and rate of disease progression all affect the clinical decision to operate. Patients undergoing surgical resection of brain metastasis from melanoma often have survivals that are longer than would be expected from RT alone.[23] Furthermore, the only long-term survivals in patients with melanoma brain metastases are those undergoing surgery plus RT. In a series of 147 cases, Zacest et al. found that surgical treatment with adjuvant radiation therapy improves neurological symptoms and produces minimal morbidity.[74] Long-term survivals (>3 years) were observed in patients with single cerebral metastasis and no demonstrable extracranial disease. This suggests that in selected patients, complete surgical resection of brain metastases is beneficial. However, a series of surgically treated metastatic melanoma cases from Memorial Sloan-Kettering Cancer Center revealed that there was no difference in the length of survival in patients who underwent postoperative WBRT versus those who did not.[75] Authors of this report also found that patients who do not display preoperative neurological deficits have only one supratentorial lesion (infratentorial lesions are associated with poor prognosis) and no lung or visceral metastases may derive significant palliative benefit from surgical intervention.

Median survivals for patients undergoing surgical resection of a melanoma brain metastasis range from 5 to 10 months, with the longer durations appearing in series from recent years.[74–77] The extent of systemic disease is a key factor determining the duration of survival following surgery. Hagen et al. found a median survival of 19.2 months in those patients undergoing surgery with systemic disease that was undetectable or limited to the primary site.[24] In those with disseminated disease, the median postoperative survival was 3.7 months.

Survivals of 24 months or more have been reported in several series of patients undergoing resection of cerebral metastases from melanoma. Of 133 patients pooled from 6 surgical series, 16 (12%) survived 24 or more months. At least 3 (2%) of these patients survived for more than 5 years. In the Australian series described earlier, the 3-year survival was 9% while 5-year survival was 5%.[74] In MSKCC series, 3- and 5-year survivals were 13.2% and 6.6%, respectively.[75]

Several factors must be considered when surgical resection of brain metastasis from melanoma is contemplated. These factors include lesion number and surgical accessibility, status of systemic disease, and the physical condition of the patient (Karnofsky performance status). Gadolinium-enhanced MRI is the most sensitive way to assess the number of brain metastases.[30] Patients with one or two lesions that are surgically accessible should be considered for surgery. Surgical candidates should undergo a thorough staging evaluation. Only 5%–20% of patients with brain metastases from melanoma will not have identifiable systemic metastases.[22,25] Even in patients with limited systemic disease, brain metastases may herald a more aggressive phase of tumor growth which has yet to become clinically apparent. Some patients with brain metastases (10%, see later) will have clinically detectable, coexistent leptomeningeal metastases (LM).

(vii) Chemotherapy. Presently, chemotherapy for brain metastases from melanoma has rather limited role. It can be considered when the brain is a major symptomatic site of disease in patients whose lesions have not responded to prior therapy including RT. Cisplatin, administered intraarterially, has been one of the few agents capable of producing responses.[78,79] Systemic disease, however, may progress during an intracranial response to intraarterial cisplatin. Cisplatin in combination with temozolomide (TMZ), vinblastine, and subcutaneous interleukin-2 has produced responses in various metastatic locations including the brain.[80] Interferon alpha, DTIC, and the currently available nitrosoureas (BCNU, CCNU) have activity against systemic disease (25% response rate) but are not active against brain metastases.[81–84]

Fotemustine, an aminophosphonic acid-linked nitrosourea with high CNS penetration, has shown promising activity against brain and systemic metastases from melanoma in European trials.[85–87] In 39 patients treated with fotemustine prior to RT for brain metastases from melanoma, intracranial responses were measured in 11 (28%; 2 complete responses, 9 partial responses), but fotemustine was less effective when given after RT. Final results of the phase II trial evaluating 153 patients with metastatic melanoma treated with fotemustine revealed 3 complete responses and 34 partial responses overall with response rate (RR) of 24.2%. Response rate for intracerebral sites was reported to be 25%.[85,86] A randomized phase III trial compared fotemustine with dacarbazine in patients with disseminated malignant melanoma.[88] ORR was 15.2% in the fotemustine arm compared to 6.8% in the dacarbazine arm. In addition, a trend favoring fotemustine in terms of OS and time to brain metastasis was observed. Finally, another prospective randomized multicenter phase III trial evaluated the use of fotemustine plus WBRT vs fotemustine alone in melanoma brain metastases.[89] Addition of WBRT to fotemustine significantly delayed the time to cerebral progression in the study, without significantly affecting objective control or overall survival.

Temozolomide, used in combination therapies, has also been tried as a single agent. Given its known good penetration of the blood-brain barrier as well as activity against systemic melanoma this drug was an excellent candidate for therapy of metastatic melanoma. Complete response of multiple brain metastasis from melanoma after six cycles of therapy with TMZ was reported in a single case in 2001.[90] TMZ does have activity in both intra- and extra-CNS locations. Reported response rates range from 12% to 20%.[91,92] One study of TMZ-based chemotherapy for brain melanoma reported 24% overall RR, with 17% RR for monotherapy.[93] Another large multicenter study of TMZ in metastatic CNS melanoma without prior or concurrent radiotherapy revealed overall response rate to be only 6% with 26% of patients having stable disease 8 weeks into therapy. Median survival was 3.5 months.[94]

Therapies combining external-beam radiation with TMZ revealed encouraging response rates from 10% to 85%.[95,96] Larger studies are needed to confirm these results. An antiangiogenic agent thalidomide has been tested in several trials in combination with radiation therapy and TMZ. Response rates were rather poor (7%) with significant systemic toxicity.[97]

2.2.3 Spinal metastases from metastatic melanoma

Spinal metastases from malignant melanoma are rare. They can be divided into intramedullary and epidural. Intramedullary spinal cord metastases are usually discovered at autopsy and are present in less than 1%–2%

of cases.[98] Melanoma represents about 9% of all intramedullary spinal cord metastases.[99] Epidural spinal cord compression from metastasis (ESCC) is equally rare. In one series, it occurred in 2/78 (3%) of patients with neurological complications.[20] In both patients, ESCC occurred as a result of extension of a vertebral body metastasis into the spinal canal. This percentage is in the range reported by Amer et al., who found SCC in 4/56 (7%), and Bullard et al., who found SCC in 9.3% of 86 patients with neural metastases from melanoma.[16,21] These figures are also close to the 5% reported for SCC from all cancers.[100] Spinal cord compression from intradural masses secondary to leptomeningeal metastases may be more common in patients with melanoma than in patients with other types of cancer.

Hadden et al. performed a systematic literature review of two studies involving 39 patients to determine prognostic factors associated with metastatic spinal cord compression secondary to melanoma.[101] The most important variables associated with improved survival were ECOG performance status ≤ 2 and absence of visceral metastases and the most important predictors of good functional outcome were time between development of motor deficits and radiotherapy, performance status, and ability to ambulate prior to radiotherapy. Goodwin et al. performed a literature review of 65 studies to assess clinical outcomes in patients with spinal metastases from different types of skin cancers.[102] Median survival was reported as 4.0 months in patients with spinal metastases of malignant melanoma and squamous cell carcinoma, compared to 12.5 months for basal cell carcinoma, 3.0 months for pilomatrix carcinoma, and 1.5 months for Merkel cell carcinoma. The authors identified age > 65 years, sacral spinal involvement, presence of neurological deficits, and nonambulatory status as negative prognostic factors impacting survival. Notably the type of treatment received (surgical, medical, or combination) did not appear to significantly impact median survival in this study.

The optimal treatment approach for spinal metastases from melanoma remains unclear. Surgical resection, radiotherapy, and systemic therapies may all be valid treatment options when used in the appropriate clinical setting. Separation surgery followed by SRS has found increased use over en bloc resection in the treatment of extramedullary spinal metastases, given the higher risk of neurological injury and significant morbidity with the latter.[103] In addition, the benefit of SRS over conventional RT in the treatment of spinal metastases has been well established.[104,105] Laufer et al. performed a retrospective chart review of 186 patients with metastatic epidural spinal cord compression that were treated with surgical decompression followed by single-fraction SRS (24 Gy), high-dose hypofractionated SRS (24–30 Gy in 3 fractions), or low-dose hypofractionated SRS (18–36 Gy

in 5 or 6 fractions).[102] The authors reported a total rate of local progression of 16.4% 1 year following SRS (4.1% with high-dose hypofractionated SRS, 9.0% with single-fraction SRS, and 22.6% with low-dose hypofractionated SRS). While prospective data on the treatment of spinal metastases from malignant melanoma are unfortunately lacking, several small retrospective case series have reported encouraging results for SRS with local control rates ranging between 75% and 100%.[106–109]

Optimal treatment for intramedullary spinal cord metastases remains controversial. Microsurgical approaches with maximal removal of the lesion, with the goal of preserving existing function, can be considered when possible.[110] Radiation therapy can preserve existing spinal cord function or delay clinical progression. The total radiation doses vary from 30 to 50 Gy depending on the location and size of the lesion.[98] Immunotherapy and targeted therapies, alone or in combination with surgical resection and radiation, are promising treatment options based on the results with brain metastases and should be considered in the treatment of spinal disease.

2.2.4 Leptomeningeal metastases from metastatic melanoma

Leptomeningeal metastases (LM) were clinically diagnosed in 9/78 (12%) patients with neurological complications in the Vanderbilt series.[20] This is similar to the frequency reported by others in patients with metastatic melanoma and is higher than the frequency in most other cancers.[16] Autopsy series report LM in 24%–63% of patients.[111–113]

LM and brain metastases commonly coexist. Parenchymal brain lesions were noted in half of patients with LM in the Vanderbilt patients. Similar findings were noted by Amer et al.[16] Approximately 10% of patients with brain metastases from malignant melanoma will have clinically evident LM. Autopsy studies have reported that 50% of patients with brain metastases also have LM. This may reflect more disseminated disease by the time of death, or the detection of clinically inapparent cases at autopsy. Diagnosis can sometimes be challenging, especially in the absence of parenchymal disease. In one series, 41% of patients with leptomeningeal carcinomatosis proven at autopsy had normal CSF cytology prior to death. MRI is helpful and should be used if leptomeningeal disease is suspected.[113] Immunocytology with HMB-45 or MART1 antibodies may also be useful in the diagnosis of suspected LM from melanoma.[114]

The prognosis of LM from metastatic melanoma is unfortunately poor, with survival times typically ranging in the order of weeks.[115,116] There is little information available about the efficacy of treatments for LM from melanoma. Radiation therapy may be helpful when administered to clinically symptomatic areas. Two of 11 (18%) patients with LM from melanoma responded to RT plus intrathecal methotrexate in the study of Wasserstrom et al..[117] It is not known whether the addition of methotrexate to RT is beneficial. Experience with intrathecal injections of interferon alpha and dacarbazine is anecdotal and these therapies should be considered experimental.[118–120] There have been attempts to use intrathecal recombinant interleukin 2 (IL-2),[120] intrathecal cytotoxic T-cell immunotherapy,[121] as well as systemic combination of cisplatin and temozolomide[122]; however, despite good results, these represent single case reports and should be considered only if other modalities of treatment fail. A recently published retrospective series evaluated 43 patients treated with intrathecal IL-2 at the University of Texas MD Anderson Cancer Center between August 2006 and July 2014.[123] The diagnosis of LM was based on positive CSF cytology and radiographic evidence in 53%, CSF cytology alone in 19%, radiographic evidence alone in 21%, and histologic pathology in 7% of patients. The median OS following intrathecal IL-2 in the cohort was 7.8 months, with 1-year, 2-year, and 5-year OS rates of 36%, 26%, and 13%, respectively. Notably, all patients developed symptomatic intracranial hypertension that had to be managed with medication and/or CSF removal. No treatment-related deaths were reported.

Targeted therapy and immunotherapy are promising treatment options in metastatic melanoma, including patients with intracranial disease. Given the rarity of leptomeningeal disease, however, clinical data supporting the use of either of these therapies in this setting are currently limited to a small number of case reports.[124] Foppen et al. analyzed a retrospective case series of 39 patients with LM from melanoma treated at the Netherlands Cancer Institute between May 2010 and March 2015. Median OS was reported as 6.9 weeks (95% confidence interval 0.9–12.8) for the entire cohort of patients, 16.9 weeks for all 25 patients who had received therapy, and 21.7 weeks (range 2–235 weeks) for 21 patients that were treated with systemic targeted therapy and/or immunotherapy with or without RT.[125] Ferguson et al. reported outcomes and prognostic factors in a large retrospective cohort of 178 patients with LM from metastatic melanoma.[126] The majority of patients received at least one treatment for LM, including radiation (n=98), chemotherapy (n=89), targeted therapy (n=60), immunotherapy (n=12), or intrathecal therapy (n=64). Median OS was reported at 3.5 months, and 1-year, 2-year, and 5-year OS rates were 22%, 14%, and 9%, respectively. Factors significantly associated with OS on multivariate analysis were Eastern Cooperative Oncology Group (ECOG) performance status > 0 (HR 2.1), neurological symptoms (HR 1.6), absent systemic disease (HR 0.4), and LM treatment (HR 0.4) including targeted therapy (HR 0.6) and intrathecal therapy (HR 0.5). A phase I/Ib trial of intrathecal nivolumab in patients with LM, including from malignant melanoma, is currently ongoing (NCT03025256).

2.2.5 Peripheral nerve complications

Cranial nerve or peripheral nerve complications can occur in melanoma predominantly as a result of infiltration or compression of the surrounding structures, including cranial and peripheral nerves. Several syndromes resulting from melanoma invasion were reported. Jugular foramen syndrome with the involvement of cranial nerves IX-XI, sudden onset of complete hearing loss resulting from infiltration of the vestibulocochlear nerve, and blindness resulting from optic nerve involvement were all described in the setting of melanoma.[127–129] In Vanderbilt series, a total of 10/78 (13%) patients with neurological complications from metastatic melanoma had peripheral nerve involvement.[20] Each appeared to be secondary to metastases to nonneural sites (e.g., lymph nodes or bone) with compression of adjacent nerves. Cranial nerve complications included one patient with superior orbital fissure syndrome associated with a mass at the cavernous sinus, one with jugular foramen syndrome associated with a skull base metastasis, and one with an orbital mass with vision loss and pupillary paralysis. There was clinical evidence of brachial plexopathy in four patients. Cervical or axillary lymph nodes were enlarged in each. Leg pain, reflex loss, and weakness occurred in two patients. One had enlarged inguinal lymph nodes and the second had a pelvic mass. A seventh patient developed incontinence from a destructive sacral lesion.

A separate entity that deserves mentioning is spindle cell desmoplastic melanoma, and particularly its neurotropic variant called desmoplastic neurotropic melanoma (DNM).[130–132] These rare lesions, predominantly of the head and neck, frequently appear as a benign-looking nodule that later progresses to cranial nerve involvement. Pathologically a replacement of perineural and Schwann cells by tumor cells (neurotropism) is seen. Examples include invasion of inferior alveolar nerve,[133] and trigeminal nerve presenting as neuralgia.[134,135] DNM has also been described as involving the median nerve and nerves innervating the vulva.[136–138] Quinn et al. demonstrated that there was no statistically significant difference in survival for patients with desmoplastic melanoma and those with DNM, and overall survival for both was similar to that for patients with other cutaneous melanomas.[132]

2.2.6 Survival and cause of death

The prognosis of patients with melanoma brain metastases has traditionally been poor. Median survival ranged from 2.5 to 4 months.[16,25,139,140] In one series of 100 patients from Westminster Hospital in London median survival was 2.5 months. 8% of patients survived longer than 1 year and 4% survived longer than 2 years. Median survival from initial diagnosis of melanoma for the Vanderbilt series patients with all neurological complications was 45 months. Patients with systemic metastases, but not nervous system metastases, had the same duration of survival. Patients with CNS metastases had a median survival of 4 months from the diagnosis of their nervous system metastasis.[20] A large proportion of patients with melanoma metastatic to the CNS (20%–60%) would die as a result of their brain disease.[25,111,141] However, the recent introduction of BRAF-targeted therapy and immunotherapy has resulted in impressive improvements in response rates and survival for both extracranially and intracranially advanced melanoma. Combination therapy with dual anti-CTLA-4 and anti-PD-1/PD-L1 immune checkpoint inhibitors has shown especially encouraging results in phase II studies, with reported intracranial response rates of up to 57% and 6-month progression-free and overall survival of up to 64% and 92%, respectively.

2.2.7 l-dopa therapy and malignant melanoma

A temporal link between l-dopa therapy of Parkinson's disease (PD) and development or progression of malignant melanoma has been reported.[142,143] In addition, it has been suggested that there could be an association of melanoma with PD itself, regardless of treatment. Siple et al. conducted a survey of 34 published cases of melanoma in Parkinson's patients and concluded that the association between levodopa therapy and induction or exacerbation of malignant melanoma is unlikely.[144] A study of l-dopa pharmacokinetics in patients with PD and history of malignant melanoma did not provide evidence to support claims that l-dopa causes melanoma progression.[145] Letellier et al. in their prospective correlative study found that plasma l-dopa/l-tyrosine ratio reflects the tumor burden and correlates with the progression of malignant melanoma.[146] Overall, the association between levodopa therapy and malignant melanoma remains highly controversial in the absence of supportive data from randomized controlled trials. Olsen et al. performed a large retrospective study of 14,088 patients with PD in Denmark and found a nearly fourfold increased risk of malignant melanoma in patients with idiopathic PD when compared to patients with a diagnosis of questionable PD or other Parkinsonism.[147] However, no effect of levodopa therapy on the risk of malignant melanoma was observed in the study (OR of 1.0, 95% CI 0.8–1.3). Bertoni et al. evaluated 2106 patients with idiopathic PD for invasive malignant melanoma and found a 2.24-fold increased prevalence when compared to age- and sex-matched controls in the population.[148] Dalvin et al. performed a database review of 974 patients diagnosed with PD in Olmsted County MN and found a 3.8-fold increased probability of preexisting melanoma compared to controls.[149] The exact cause of the association between PD and melanoma is currently unknown, however,

genetic factors, pesticides, and abnormal cellular autophagy have all been discussed as potential mechanisms.[150]

2.2.8 Paraneoplastic syndromes

The National Institute of Neurological Disorders and Stroke defines paraneoplastic syndromes as a group of rare disorders in patients with malignancies that are caused by an aberrant immune response that targets normal cells. The immunologic response is thought to be mediated by cross-reactive antibodies and T-cells directed against shared antigens that are expressed on tumor cells and regular cells.[151] In a wider sense, paraneoplastic syndromes may also be caused by the production and secretion of humoral factors by the melanoma cells and other mechanisms that are not directly related to the location, bulk, and metabolic activity of the tumor or side effects of tumor-directed medications.[152] Melanoma-associated retinopathy (MAR) is the most well-recognized paraneoplastic syndrome in patients with malignant melanoma. MAR typically affects patients in the fourth to eighth decade of life and appears to be more common in men.[153] To date only around 90 cases of MAR have been reported since its first description in 1988 as a paraneoplastic syndrome.[154,155] Classic symptoms include acute onset of night blindness, visual field defects, and positive visual phenomena with relative preservation of visual acuity and color vision.[156,157] Pathogenic autoantibodies have been identified in patients with MAR and have been shown to be directed against melanoma antigens and cross-react with several retinal antigens, resulting in local tissue damage and dysfunction.[158] A diagnosis of MAR should be suspected based on the clinical presentation and supportive test results including abnormal visual field testing and scotopic response on electroretinogram.[159] The diagnosis can be confirmed by demonstration of serum autoantibodies reactive against retinal bipolar cells via western blot and immunohistochemistry.[160] Management of MAR is aimed at treating the underlying malignancy via a combination of cytoreductive surgery, radiotherapy, and chemotherapy approaches, as well as supportive immunosuppressive therapy targeted at reining in the autoimmune inflammatory process and limiting damage to retinal cells.[161] Limited data exist in the literature regarding the optimal choice and effectiveness of immunosuppressive therapy. The use of local or systemic corticosteroids, plasma exchange, and intravenous immunoglobulins has been described on a case-by-case basis with varying degrees of success.[157,161,162] Notably, it has been reported that the presence of autoantibodies in patients with advanced melanoma could be associated with improved prognosis.[157] The use of immunosuppressive therapy in mild or subclinical cases of MAR is therefore controversial.[156] Immune checkpoint inhibitors have been reported to exacerbate or induce autoimmunity and should therefore be avoided in patients with MAR.[163] Other paraneoplastic syndromes associated with malignant melanoma are rare but have been reported, including limbic encephalitis,[164,165] opsoclonus-myoclonus syndrome,[166,167] cerebellar degeneration,[168,169] demyelinating syndromes of the peripheral and central nervous system,[170–173] as well as cutaneous manifestations such as vitiligo,[174] dermatomyositis,[175–177] and bullous dermatoses.[178,179]

3 Neurological complications of noncutaneous melanomas

3.1 Primary nervous system melanoma

Two types of cutaneous melanotic lesions are associated with primary central nervous system melanoma: neurocutaneous melanosis and oculodermal melanosis. In the former, congenital, giant cutaneous hairy, and pigmented nevi are associated with abnormal meningeal melanocyte proliferation and malignant melanoma may arise in the skin or the meninges.[180] Cerebral hemispheric lesions and spinal cord lesions were also described in the setting of neurocutaneous melanosis.[181-184] In oculodermal melanosis (Nevus of Ota), pigmentation is usually unilateral and limited to the conjunctiva and periorbital skin but the involvement of sclera, nasopharynx, auricular mucosa, tympanic membrane, and dura can be seen. Glaucoma and malignant melanoma are the two known ocular complications in patients with Nevus of Ota. There may be an association with primary central nervous system melanoma with meningeal or parenchymal involvement.[185-187]

Melanoma can also arise in the brain without a predisposing neurocutaneous syndrome. Primary melanoma of the nervous system most commonly arises from the leptomeninges, sometimes with associated brain infiltration, but without a detectable systemic primary site.[188,189]

3.2 Metastatic ocular melanoma

Ocular melanomas are much less common than cutaneous melanomas and may metastasize throughout the body, including the nervous system. The organs most often involved are the liver, lung, and bone. An autopsy study of 92 patients who died from metastatic melanoma revealed a significantly lower frequency of brain metastases in patients with ocular melanoma (2/9, or 22%) compared to nonocular melanoma (48/73 or 66%).[190] In a retrospective study of 107 patients with metastatic ocular melanoma, 5 (4%) had clinically detectable neurological metastases.[191] An autopsy study that included four patients with ocular melanoma found brain metastases in three.[112] Skeletal metastases from metastatic ocular melanoma can lead to epidural spinal cord compression. Rarely, ocular melanoma extends intracranially along an optic nerve.[192] Metastatic ocular melanoma, like cutaneous melanoma, responds poorly to systemic chemotherapy. In addition, however, it should be noted that there are important biological and clinical differences between these two entities. Importantly, the mutation burden in uveal melanoma is significantly lower when compared to cutaneous melanoma,[193] and BRAF mutations are largely absent.[194] For this reason, no targeted therapy at the molecular level is currently available for the treatment of ocular melanoma. Several studies have evaluated the efficacy of immune checkpoint blockade in uveal melanoma. Unlike in cutaneous melanoma however, outcomes with single-agent immune checkpoint inhibition have largely been disappointing.[195-198] Most recently, preliminary data from two prospective studies evaluating combination therapy with ipilimumab and nivolumab in uveal melanoma have been reported, showing modestly more promising results.[199,200]

References

1. Siegel RL, Miller KD, Jemal A. Cancer statistics, 2020. *CA Cancer J Clin.* 2020;70(1):7–30. https://doi.org/10.3322/caac.21590.
2. Curti BD, Leachman S, Urba WJ. Cancer of the skin. In: Jameson J, Fauci AS, Kasper DL, Hauser SL, Longo DL, Loscalzo J, eds. *Harrison's Principles of Internal Medicine, 20e.* McGraw-Hill; 2021.
3. Zimm S, Wampler GL, Stablein D, et al. Intracerebral metastases in solid-tumor patients: natural history and results of treatment. *Cancer.* 1981;48(2):384–394.
4. Delattre JY, Krol G, Thaler HT, et al. Distribution of brain metastases. *Arch Neurol.* 1988;45(7):741–744.
5. Johnson JD, Young B. Demographics of brain metastasis. *Neurosurg Clin N Am.* 1996;7(3):337–344.
6. Koh HK. Cutaneous melanoma. *N Engl J Med.* 1991;325(3):171–182.
7. Zhang D, Wang Z, Shang D, Yu J, Yuan S. Incidence and prognosis of brain metastases in cutaneous melanoma patients: a population-based study. *Melanoma Res.* 2019;29(1):77–84.
8. National Cancer Institute. *SEER Cancer Stat Facts: Melanoma of the Skin;* 2020. Available at: https://seer.cancer.gov/statfacts/html/melan.html. [Accessed October 30].
9. Gilchrest BA, Eller MS, Geller AC, et al. The pathogenesis of melanoma induced by ultraviolet radiation. *N Engl J Med.* 1999;340(17):1341–1348.
10. Tsai T, Vu C, Henson DE. Cutaneous, ocular and visceral melanoma in African Americans and Caucasians. *Melanoma Res.* 2005;15(3):213–217.
11. Gershenwald JE, Scolyer RA, Hess KR, et al. Melanoma of the skin. In: Amin MB, ed. *AJCC Cancer Staging Manual.* 8th ed. Chicago: American Joint Committee on Cancer; 2017:563.
12. Kelderman S, Heemskerk B, van Tinteren H, et al. Lactate dehydrogenase as a selection criterion for ipilimumab treatment in metastatic melanoma. *Cancer Immunol Immunother.* 2014;63:449–458.
13. Homsi J, Kashani-Sabet M, Messina JL, et al. Cutaneous melanoma: prognostic factors. *Cancer Control.* 2005;12(4):223–229.
14. Bafaloukos D, Gogas H. The treatment of brain metastases in melanoma patients. *Cancer Treat Rev.* 2004;30(6):515–520.
15. Chason JL, Walker FB, Landers JW. Metastatic carcinoma in the central nervous system and dorsal root ganglia. A prospective autopsy study. *Cancer.* 1963;16:781–787.
16. Amer MH, Al-Sarraf M, Baker LH, et al. Malignant melanoma and central nervous system metastases: incidence, diagnosis, treatment and survival. *Cancer.* 1978;42(2):660–668.

17. Nathanson. Spontaneous regression of malignant melanoma: a review of the literature on incidence, clinical features, and possible mechanisms. *Natl Cancer Inst Monogr.* 1976;44:67–76.

18. Printz C. Spontaneous regression of melanoma may offer insight into cancer immunology. *J Natl Cancer Inst.* 2001;93(14):1047–1048.

19. Gaviani P, Mullins ME, Braga TA, et al. Improved detection of metastatic melanoma by t2*-weighted imaging. *AJNR Am J Neuroradiol.* 2006;27(3):605–608.

20. Henson JW. *Neurological Complications of Malignant Melanoma and Other Cutaneous Malignancies.* New York: Marcel Dekker; 1995.

21. Bullard DE, Cox EB, Seigler HF. Central nervous system metastases in malignant melanoma. *Neurosurgery.* 1981;8(1):26–30.

22. Choi KN, Withers HR, Rotman M. Intracranial metastases from melanoma. Clinical features and treatment by accelerated fractionation. *Cancer.* 1985;56(1):1–9.

23. Cohen N, Strauss G, Lew R, et al. Should prophylactic anticonvulsants be administered to patients with newly-diagnosed cerebral metastases? A retrospective analysis. *J Clin Oncol.* 1988;6(10):1621–1624.

24. Hagen NA, Cirrincione C, Thaler HT, et al. The role of radiation therapy following resection of single brain metastasis from melanoma. *Neurology.* 1990;40(1):158–160.

25. Byrne TN, Cascino TL, Posner JB. Brain metastasis from melanoma. *J Neurooncol.* 1983;1(4):313–317.

26. Oberndorfer S, Schmal T, Lahrmann H, et al. the frequency of seizures in patients with primary brain tumors or cerebral metastases. An evaluation from the Ludwig Boltzmann Institute of Neuro-Oncology and the Department of Neurology, Kaiser Franz Josef Hospital, Vienna. *Wien Klin Wochenschr.* 2002;114(21-22):911–916.

27. Clifford JR, Kirgis HD, Connolly ES. Metastatic melanoma of the brain presenting as subarachnoid hemorrhage. *South Med J.* 1975;68(2):206–208.

28. Palmer FJ, Poulgrain AP. Metastatic melanoma simulating subdural hematoma. Case report. *J Neurosurg.* 1978;49(2):301–302.

29. Akai T, Kuwayama N, Ogiichi T, et al. Leptomeningeal melanoma associated with straight sinus thrombosis—case report. *Neurol Med Chir (Tokyo).* 1997;37(10):757–761.

30. Davis PC, Hudgins PA, Peterman SB, et al. Diagnosis of cerebral metastases: double-dose delayed CT vs contrast-enhanced MR imaging. *AJNR Am J Neuroradiol.* 1991;12(2):293–300.

31. McGann GM, Platts A. Computed tomography of cranial metastatic malignant melanoma: features, early detection and unusual cases. *Br J Radiol.* 1991;64(760):310–313.

32. Atlas SW, Grossman RI, Gomori JM, et al. MR imaging of intracranial metastatic melanoma. *J Comput Assist Tomogr.* 1987;11(4):577–582.

33. Juweid ME, Cheson BD. Positron-emission tomography and assessment of cancer therapy. *N Engl J Med.* 2006;354(5):496–507.

34. Nguyen AT, Akhurst T, Larson SM, et al. Pet scanning with (18) f 2-fluoro-2-deoxy-d-glucose (FDG) in patients with melanoma. Benefits and limitations. *Clin Positron Imaging.* 1999;2(2):93–98.

35. Patchell RA, Tibbs PA, Walsh JW, et al. A randomized trial of surgery in the treatment of single metastases to the brain. *N Engl J Med.* 1990;322(8):494–500.

36. Glantz MJ, Cole BF, Forsyth PA, et al. Practice parameter: anticonvulsant prophylaxis in patients with newly diagnosed brain tumors. Report of the quality standards subcommittee of the American Academy of Neurology. *Neurology.* 2000;54(10):1886–1893.

37. Happold C, Gorlia T, Chinot O, et al. Does valproic acid or levetiracetam improve survival in glioblastoma? A pooled analysis of prospective clinical trials in newly diagnosed glioblastoma. *J Clin Oncol.* 2016;34(7):731–739. https://doi.org/10.1200/JCO.2015.63.6563. Epub 2016 Jan 19. PMID: 26786929; PMCID: PMC5070573.

38. Bedetti C, Romoli M, Maschio M, et al. Neuropsychiatric adverse events of antiepileptic drugs in brain tumour-related epilepsy: an Italian multicentre prospective observational study. *Eur J Neurol.* 2017;24(10):1283–1289. https://doi.org/10.1111/ene.13375. Epub 2017 Aug 10. PMID: 28796376.

39. Maschio M, Zarabla A, Maialetti A, et al. Quality of life, mood and seizure control in patients with brain tumor related epilepsy treated with lacosamide as add-on therapy: a prospective explorative study with a historical control group. *Epilepsy Behav.* 2017;73:83–89. https://doi.org/10.1016/j.yebeh.2017.05.031. Epub 2017 Jun 14. PMID: 28623754.

40. Long GV, Trefzer U, Davies MA, et al. Dabrafenib in patients with Val600Glu or Val600Lys BRAF-mutant melanoma metastatic to the brain (BREAK-MB): a multicentre, open-label, phase 2 trial. *Lancet Oncol.* 2012;13(11):1087–1095. https://doi.org/10.1016/S1470-2045(12)70431-X.

41. McArthur GA, Maio M, Arance A, et al. Vemurafenib in metastatic melanoma patients with brain metastases: an open-label, single-arm, phase 2, multicentre study. *Ann Oncol.* 2017;28(3):634–641. https://doi.org/10.1093/annonc/mdw641.

42. Davies MA, Saiag P, Robert C, et al. Dabrafenib plus trametinib in patients with BRAFV600-mutant melanoma brain metastases (COMBI-MB): a multicentre, multicohort, open-label, phase 2 trial. *Lancet Oncol.* 2017;18(7):863–873. https://doi.org/10.1016/S1470-2045(17)30429-1.

43. Margolin K, Ernstoff MS, Hamid O, et al. Ipilimumab in patients with melanoma and brain metastases: an open-label, phase 2 trial. *Lancet Oncol.* 2012;13(5):459–465. https://doi.org/10.1016/S1470-2045(12)70090-6.

44. Goldberg SB, Gettinger SN, Mahajan A, et al. Pembrolizumab for patients with melanoma or non-small-cell lung cancer and untreated brain metastases: early analysis of a non-randomised, open-label, phase 2 trial. *Lancet Oncol.* 2016;17(7):976–983. https://doi.org/10.1016/S1470-2045(16)30053-5.

45. Long GV, Atkinson V, Lo S, et al. Combination nivolumab and ipilimumab or nivolumab alone in melanoma brain metastases: a multicentre randomised phase 2 study. *Lancet Oncol.* 2018;19(5):672–681. https://doi.org/10.1016/S1470-2045(18)30139-6.

46. Tawbi HA, Forsyth PA, Algazi A, et al. Combined nivolumab and ipilimumab in melanoma metastatic to the brain. *N Engl J Med.* 2018;379(8):722–730. https://doi.org/10.1056/NEJMoa1805453.

47. Sambade MJ, Peters EC, Thomas NE, et al. Melanoma cells show a heterogeneous range of sensitivity to ionizing radiation and are radiosensitized by inhibition of B-RAF with PLX-4032. *Radiother Oncol.* 2011;98:394–399.

48. Ahmed KA, Abuodeh YA, Echevarria MI, et al. Clinical outcomes of melanoma brain metastases treated with stereotactic radiosurgery and anti-PD-1 therapy, anti-CTLA-4 therapy, BRAF/MEK inhibitors, BRAF inhibitor, or conventional chemotherapy. *Ann Oncol.* 2016;27(12):2288–2294. https://doi.org/10.1093/annonc/mdw417.

49. Knisely JP, Yu JB, Flanigan J, Sznol M, Kluger HM, Chiang VL. Radiosurgery for melanoma brain metastases in the ipilimumab era and the possibility of longer survival. *J Neurosurg.* 2012;117(2):227–233. https://doi.org/10.3171/2012.5.JNS111929.

50. Silk AW, Bassetti MF, West BT, Tsien CI, Lao CD. Ipilimumab and radiation therapy for melanoma brain metastases. *Cancer Med.* 2013;2(6):899–906. https://doi.org/10.1002/cam4.140.

51. Acharya S, Mahmood M, Mullen D, et al. Distant intracranial failure in melanoma brain metastases treated with stereotactic radiosurgery in the era of immunotherapy and targeted agents. *Adv Radiat Oncol.* 2017;2(4):572–580. https://doi.org/10.1016/j.adro.2017.07.003.

52. Mastorakos P, Xu Z, Yu J, et al. BRAF V600 mutation and BRAF kinase inhibitors in conjunction with stereotactic radiosurgery for intracranial melanoma metastases: a multicenter retrospective study. *Neurosurgery.* 2019;84(4):868–880. https://doi.org/10.1093/neuros/nyy203.

53. Hecht M, Meier F, Zimmer L, et al. Clinical outcome of concomitant vs interrupted BRAF inhibitor therapy during radiotherapy in melanoma patients. *Br J Cancer.* 2018;118(6):785–792. https://doi.org/10.1038/bjc.2017.489.

54. de la Fuente M, Beal K, Carvajal R, Kaley TJ. Whole-brain radiotherapy in patients with brain metastases from melanoma. *CNS Oncol.* 2014;3(6):401–406. https://doi.org/10.2217/cns.14.40.

55. Vlock DR, Kirkwood JM, Leutzinger C, et al. High-dose fraction radiation therapy for intracranial metastases of malignant melanoma: a comparison with low-dose fraction therapy. *Cancer.* 1982;49(11):2289–2294.

56. Ziegler JC, Cooper JS. Brain metastases from malignant melanoma: conventional vs. high-dose-per-fraction radiotherapy. *Int J Radiat Oncol Biol Phys.* 1986;12(10):1839–1842.

57. Brown PD, Pugh S, Laack NN, et al. Memantine for the prevention of cognitive dysfunction in patients receiving whole-brain radiotherapy: a randomized, double-blind, placebo-controlled trial. *Neuro Oncol.* 2013;15(10):1429–1437. https://doi.org/10.1093/neuonc/not114.

58. Brown PD, Gondi V, Pugh S, et al. Hippocampal avoidance during whole-brain radiotherapy plus memantine for patients with brain metastases: phase III trial NRG oncology CC001. *J Clin Oncol.* 2020;38(10):1019–1029. https://doi.org/10.1200/JCO.19.02767.

59. Loeffler JS, Kooy HM, Wen PY, et al. The treatment of recurrent brain metastases with stereotactic radiosurgery. *J Clin Oncol.* 1990;8(4):576–582.

60. Somaza S, Kondziolka D, Lunsford LD, et al. Stereotactic radiosurgery for cerebral metastatic melanoma. *J Neurosurg.* 1993;79(5):661–666.

61. Gonzalez-Martinez J, Hernandez L, Zamorano L, et al. Gamma knife radiosurgery for intracranial metastatic melanoma: a 6-year experience. *J Neurosurg.* 2002;97(5 Suppl):494–498.

62. Mingione V, Oliveira M, Prasad D, et al. Gamma surgery for melanoma metastases in the brain. *J Neurosurg.* 2002;96(3):544–551.

63. Koc M, McGregor J, Grecula J, et al. Gamma knife radiosurgery for intracranial metastatic melanoma: an analysis of survival and prognostic factors. *J Neurooncol.* 2005;71(3):307–313.

64. Muacevic A, Wowra B, Siefert A, Tonn JC, Steiger HJ, Kreth FW. Microsurgery plus whole brain irradiation versus Gamma Knife surgery alone for treatment of single metastases to the brain: a randomized controlled multicentre phase III trial. *J Neurooncol.* 2008;87(3):299–307. https://doi.org/10.1007/s11060-007-9510-4.

65. Rades D, Bohlen G, Pluemer A, et al. Stereotactic radiosurgery alone versus resection plus whole-brain radiotherapy for 1 or 2 brain metastases in recursive partitioning analysis class 1 and 2 patients. *Cancer.* 2007;109(12):2515–2521. https://doi.org/10.1002/cncr.22729.

66. Jeong WJ, Park JH, Lee EJ, Kim JH, Kim CJ, Cho YH. Efficacy and safety of fractionated stereotactic radiosurgery for large brain metastases. *J Korean Neurosurg Soc.* 2015;58(3):217–224. https://doi.org/10.3340/jkns.2015.58.3.217.

67. Soltys SG, Adler JR, Lipani JD, et al. Stereotactic radiosurgery of the postoperative resection cavity for brain metastases. *Int J Radiat Oncol Biol Phys.* 2008;70(1):187–193. https://doi.org/10.1016/j.ijrobp.2007.06.068.

68. Brown PD, Ballman KV, Cerhan JH, et al. Postoperative stereotactic radiosurgery compared with whole brain radiotherapy for resected metastatic brain disease (NCCTG N107C/CEC.3): a multicentre, randomised, controlled, phase 3 trial. *Lancet Oncol.* 2017;18(8):1049–1060. https://doi.org/10.1016/S1470-2045(17)30441-2.

69. Kocher M, Soffietti R, Abacioglu U, et al. Adjuvant whole-brain radiotherapy versus observation after radiosurgery or surgical resection of one to three cerebral metastases: results of the EORTC 22952-26001 study. *J Clin Oncol.* 2011;29(2):134–141. https://doi.org/10.1200/JCO.2010.30.1655.

70. Chang EL, Wefel JS, Hess KR, et al. Neurocognition in patients with brain metastases treated with radiosurgery or radiosurgery plus whole-brain irradiation: a randomised controlled trial. *Lancet Oncol.* 2009;10(11):1037–1044. https://doi.org/10.1016/S1470-2045(09)70263-3.

71. Brown PD, Jaeckle K, Ballman KV, et al. Effect of radiosurgery alone vs radiosurgery with whole brain radiation therapy on cognitive function in patients with 1 to 3 brain metastases: a randomized clinical trial. *JAMA.* 2016;316(4):401–409. https://doi.org/10.1001/jama.2016.9839.

72. Anker CJ, Grossmann KF, Atkins MB, Suneja G, Tarhini AA, Kirkwood JM. Avoiding severe toxicity from combined BRAF inhibitor and radiation treatment: consensus guidelines from the Eastern Cooperative Oncology Group (ECOG) [published correction appears in Int J Radiat Oncol Biol Phys. 2016 Oct 1;96(2):486]. *Int J Radiat Oncol Biol Phys.* 2016;95(2):632–646. https://doi.org/10.1016/j.ijrobp.2016.01.038.

73. Wasif N, Bagaria SP, Ray P, Morton DL. Does metastasectomy improve survival in patients with stage IV melanoma? A cancer registry analysis of outcomes. *J Surg Oncol.* 2011;104(2):111–115. https://doi.org/10.1002/jso.21903.

74. Zacest AC, Besser M, Stevens G, et al. Surgical management of cerebral metastases from melanoma: outcome in 147 patients treated at a single institution over two decades. *J Neurosurg.* 2002;96(3):552–558.

75. Wronski M, Arbit E. Surgical treatment of brain metastases from melanoma: a retrospective study of 91 patients. *J Neurosurg.* 2000;93(1):9–18.

76. Fell DA, Leavens ME, McBride CM. Surgical versus nonsurgical management of metastatic melanoma of the brain. *Neurosurgery.* 1980;7(3):238–242.

77. Brega K, Robinson WA, Winston K, et al. Surgical treatment of brain metastases in malignant melanoma. *Cancer.* 1990;66(10):2105–2110.

78. Weiden PL. Intracarotid cisplatin as therapy for melanoma metastatic to brain: ipsilateral response and contralateral progression. *Am J Med.* 1988;85(3):439–440.

79. Feun LG, Lee YY, Plager C, et al. Intracarotid cisplatin-based chemotherapy in patients with malignant melanoma and central nervous system (CNS) metastases. *Am J Clin Oncol.* 1990;13(5):448–451.

80. Gonzalez Cao M, Malvehy J, Marti R, et al. Biochemotherapy with temozolomide, cisplatin, vinblastine, subcutaneous interleukin-2 and interferon-alpha in patients with metastatic melanoma. *Melanoma Res.* 2006;16(1):59–64.

81. Moon JH, Gailani S, Cooper MR, et al. Comparison of the combination of 1,3-bis(2-chloroethyl)-1-nitrosourea (BCNU) and vincristine with two dose schedules of 5-(3,3-dimethyl-1-triazino)imidazole 4-carboxamide (DTIC) in the treatment of disseminated malignant melanoma. *Cancer.* 1975;35(2):368–371.

82. Beretta G, Bonadonna G, Cascinelli N, et al. Comparative evaluation of three combination regimens for advanced malignant melanoma: results of an international cooperative study. *Cancer Treat Rep.* 1976;60(1):33–40.

83. Costanzi JJ. DTIC (NSC-45388) studies in the southwest oncology group. *Cancer Treat Rep.* 1976;60(2):189–192.

84. Merimsky O, Inbar M, Reider-Groswasser I, et al. Brain metastases of malignant melanoma in interferon complete responders: clinical and radiological observations. *J Neurooncol.* 1992;12(2):137–140.

85. Jacquillat C, Khayat D, Banzet P, et al. Chemotherapy by fotemustine in cerebral metastases of disseminated malignant melanoma. *Cancer Chemother Pharmacol.* 1990;25(4):263–266.

86. Jacquillat C, Khayat D, Banzet P, et al. Final report of the French multicenter phase II study of the nitrosourea fotemustine in 153 evaluable patients with disseminated malignant melanoma including patients with cerebral metastases. *Cancer.* 1990;66(9):1873–1878.

III. Neurological complications of specific neoplasms

87. Khayat D, Avril MF, Gerard B, et al. Fotemustine: an overview of its clinical activity in disseminated malignant melanoma. *Melanoma Res.* 1992;2(3):147–151.

88. Avril MF, Aamdal S, Grob JJ, et al. Fotemustine compared with dacarbazine in patients with disseminated malignant melanoma: a phase III study. *J Clin Oncol.* 2004;22:1118–1125.

89. Mornex F, Thomas L, Mohr P, et al. A prospective randomized multicentre phase III trial of fotemustine plus whole brain irradiation versus fotemustine alone in cerebral metastases of malignant melanoma. *Melanoma Res.* 2003;13:97–103.

90. Biasco G, Pantaleo MA, Casadei S. Treatment of brain metastases of malignant melanoma with temozolomide. *N Engl J Med.* 2001;345(8):621–622.

91. Bleehen NM, Newlands ES, Lee SM, et al. Cancer research campaign phase II trial of temozolomide in metastatic melanoma. *J Clin Oncol.* 1995;13(4):910–913.

92. Middleton MR, Grob JJ, Aaronson N, et al. Randomized phase III study of temozolomide versus dacarbazine in the treatment of patients with advanced metastatic malignant melanoma. *J Clin Oncol.* 2000;18(1):158–166.

93. Bafaloukos D, Tsoutsos D, Fountzilas G, et al. The effect of temozolomide-based chemotherapy in patients with cerebral metastases from melanoma. *Melanoma Res.* 2004;14(4):289–294.

94. Agarwala SS, Kirkwood JM, Gore M, et al. Temozolomide for the treatment of brain metastases associated with metastatic melanoma: a phase II study. *J Clin Oncol.* 2004;22(11):2101–2107.

95. Dardoufas CMASC, Kouloulias V, et al. Concomitant temozolomide (TMZ) and radiotherapy (RT) followed by adjuvant treatment with temozolomide in patients with metastases from solid tumors. *Proc Am Soc Clin Oncol.* 2001;128:214–218.

96. Margolin KAB, Thompson A, et al. Temozolomide and whole brain irradiation in melanoma metastatic to the brain: a phase II trial of the cytokine working group. *J Cancer Res Clin Oncol.* 2002;128(4):214–218.

97. Atkins MBSJAS, Logan T, et al. A cytokine working group phase II study of temozolomide (TMZ), thalidomide (THAL) and whole brain radiotherapy (WBRT) for patients with brain metastases from melanoma. *Proc Am Soc Clin Oncol.* 2005;14(5):431–433.

98. Conill C, Sanchez M, Puig S, et al. Intramedullary spinal cord metastases of melanoma. *Melanoma Res.* 2004;14(5):431–433.

99. Connolly Jr ES, Winfree CJ, McCormick PC, et al. Intramedullary spinal cord metastasis: report of three cases and review of the literature. *Surg Neurol.* 1996;46(4):329–337. discussion 337–328.

100. Barron KD, Hirano A, Araki S, et al. Experiences with metastatic neoplasms involving the spinal cord. *Neurology.* 1959;9(2):91–106.

101. Hadden NJ, McIntosh JRD, Jay S, Whittaker PJ. Prognostic factors in patients with metastatic spinal cord compression secondary to melanoma: a systematic review. *Melanoma Res.* 2018;28(1):1–7. https://doi.org/10.1097/CMR.0000000000000411.

102. Goodwin CR, Sankey EW, Liu A, et al. A systematic review of clinical outcomes for patients diagnosed with skin cancer spinal metastases [published correction appears in J Neurosurg Spine. 2016 Nov;25(5):671]. *J Neurosurg Spine.* 2016;24(5):837–849. https://doi.org/10.3171/2015.4.SPINE15239.

103. Moussazadeh N, Laufer I, Yamada Y, Bilsky MH. Separation surgery for spinal metastases: effect of spinal radiosurgery on surgical treatment goals. *Cancer Control.* 2014;21(2):168–174. https://doi.org/10.1177/107327481402100210.

104. Chan NK, Abdullah KG, Lubelski D, et al. Stereotactic radiosurgery for metastatic spine tumors. *J Neurosurg Sci.* 2014;58(1):37–44.

105. Sahgal A, Larson DA, Chang EL. Stereotactic body radiosurgery for spinal metastases: a critical review [published correction appears in Int J Radiat Oncol Biol Phys. 2009 May 1;74(1):323]. *Int J Radiat Oncol Biol Phys.* 2008;71(3):652–665. https://doi.org/10.1016/j.ijrobp.2008.02.060.

106. Laufer I, Iorgulescu JB, Chapman T, et al. Local disease control for spinal metastases following "separation surgery" and adjuvant hypofractionated or high-dose single-fraction stereotactic radiosurgery: outcome analysis in 186 patients. *J Neurosurg Spine.* 2013;18(3):207–214. https://doi.org/10.3171/2012.11.SPINE12111.

107. Jahanshahi P, Nasr N, Unger K, Batouli A, Gagnon GJ. Malignant melanoma and radiotherapy: past myths, excellent local control in 146 studied lesions at Georgetown University, and improving future management. *Front Oncol.* 2012;2:167. https://doi.org/10.3389/fonc.2012.00167.

108. Gerszten PC, Burton SA, Ozhasoglu C, Welch WC. Radiosurgery for spinal metastases: clinical experience in 500 cases from a single institution. *Spine (Phila Pa 1976).* 2007;32(2):193–199. https://doi.org/10.1097/01.brs.0000251863.76595.a2.

109. Guckenberger M, Mantel F, Gerszten PC, et al. Safety and efficacy of stereotactic body radiotherapy as primary treatment for vertebral metastases: a multi-institutional analysis. *Radiat Oncol.* 2014;9:226. https://doi.org/10.1186/s13014-014-0226-2.

110. Hejazi N, Hassler W. Microsurgical treatment of intramedullary spinal cord tumors. *Neurol Med Chir (Tokyo).* 1998;38(5):266–271. discussion 271–263.

111. Patel JK, Didolkar MS, Pickren JW, et al. Metastatic pattern of malignant melanoma. A study of 216 autopsy cases. *Am J Surg.* 1978;135(6):807–810.

112. de la Monte SM, Moore GW, Hutchins GM. Patterned distribution of metastases from malignant melanoma in humans. *Cancer Res.* 1983;43(7):3427–3433.

113. Dupuis F, Sigal R, Margulis A, et al. Cerebral magnetic resonance imaging (MRI) in the diagnosis of leptomeningeal carcinomatosis in melanoma patients. *Ann Dermatol Venereol.* 2000;127(1):29–32.

114. Moseley RP, Davies AG, Bourne SP, et al. Neoplastic meningitis in malignant melanoma: diagnosis with monoclonal antibodies. *J Neurol Neurosurg Psychiatry.* 1989;52(7):881–886.

115. Le Rhun E, Taillibert S, Chamberlain MC. Carcinomatous meningitis: leptomeningeal metastases in solid tumors. *Surg Neurol Int.* 2013;4(Suppl. 4):S265–S288. https://doi.org/10.4103/2152-7806.111304.

116. Pape E, Desmedt E, Zairi F, et al. Leptomeningeal metastasis in melanoma: a prospective clinical study of nine patients. *In Vivo.* 2012;26:1079–1086.

117. Wasserstrom WR, Glass JP, Posner JB. Diagnosis and treatment of leptomeningeal metastases from solid tumors: experience with 90 patients. *Cancer.* 1982;49(4):759–772.

118. Champagne MA, Silver HK. Intrathecal dacarbazine treatment of leptomeningeal malignant melanoma. *J Natl Cancer Inst.* 1992;84(15):1203–1204.

119. Dorval T, Beuzeboc P, Garcia-Giralt E, et al. Malignant melanoma: treatment of metastatic meningitis with intrathecal interferon alpha-2b. *Eur J Cancer.* 1992;28(1):244–245.

120. Fathallah-Shaykh HM, Zimmerman C, Morgan H, et al. Response of primary leptomeningeal melanoma to intrathecal recombinant interleukin-2. A case report. *Cancer.* 1996;77(8):1544–1550.

121. Clemons-Miller AR, Chatta GS, Hutchins L, et al. Intrathecal cytotoxic T-cell immunotherapy for metastatic leptomeningeal melanoma. *Clin Cancer Res.* 2001;7(3 Suppl):917s–924s.

122. Salmaggi A, Silvani A, Eoli M, et al. Temozolomide and cisplatin in the treatment of leptomeningeal metastatic involvement from melanoma: a case report. *Neurol Sci.* 2002;23(5):257–258.

123. Glitza IC, Rohlfs M, Guha-Thakurta N, et al. Retrospective review of metastatic melanoma patients with leptomeningeal disease treated with intrathecal interleukin-2. *ESMO Open.* 2018;3(1):e000283. https://doi.org/10.1136/esmoopen-2017-000283.

124. Smalley KS, Fedorenko IV, Kenchappa RS, Sahebjam S, Forsyth PA. Managing leptomeningeal melanoma metastases in the era of immune and targeted therapy. *Int J Cancer.* 2016;139(6):1195–1201. https://doi.org/10.1002/ijc.30147.

125. Geukes Foppen MH, Brandsma D, Blank CU, van Thienen JV, Haanen JB, Boogerd W. Targeted treatment and immunotherapy in leptomeningeal metastases from melanoma. *Ann Oncol.* 2016;27(6):1138–1142. https://doi.org/10.1093/annonc/mdw134.

126. Ferguson SD, Bindal S, Bassett Jr RL, et al. Predictors of survival in metastatic melanoma patients with leptomeningeal disease (LMD). *J Neurooncol.* 2019;142(3):499–509. https://doi.org/10.1007/s11060-019-03121-2.

127. Schweinfurth JM, Johnson JT, Weissman J. Jugular foramen syndrome as a complication of metastatic melanoma. *Am J Otolaryngol.* 1993;14(3):168–174.

128. Currie L, Tomma A. Malignant melanoma presenting as sudden onset of complete hearing loss. *Ann Plast Surg.* 2001;47(3):336–337.

129. De Potter P, Shields CL, Eagle Jr RC, et al. Malignant melanoma of the optic nerve. *Arch Ophthalmol.* 1996;114(5):608–612.

130. Reed RJ, Leonard DD. Neurotropic melanoma. A variant of desmoplastic melanoma. *Am J Surg Pathol.* 1979;3(4):301–311.

131. Mack EE, Gomez EC. Neurotropic melanoma. A case report and review of the literature. *J Neurooncol.* 1992;13(2):165–171.

132. Quinn MJ, Crotty KA, Thompson JF, et al. Desmoplastic and desmoplastic neurotropic melanoma: experience with 280 patients. *Cancer.* 1998;83(6):1128–1135.

133. Lin D, Kashani-Sabet M, McCalmont T, et al. Neurotropic melanoma invading the inferior alveolar nerve. *J Am Acad Dermatol.* 2005;53(2 Suppl 1):S120–S122.

134. Hughes TA, McQueen IN, Anstey A, et al. Neurotropic malignant melanoma presenting as a trigeminal sensory neuropathy. *J Neurol Neurosurg Psychiatry.* 1995;58(3):381–382.

135. Newlin HE, Morris CG, Amdur RJ, et al. Neurotropic melanoma of the head and neck with clinical perineural invasion. *Am J Clin Oncol.* 2005;28(4):399–402.

136. Iyadomi M, Ohtsubo H, Gotoh Y, et al. Neurotropic melanoma invading the median nerve. *J Dermatol.* 1998;25(6):379–383.

137. Warner TF, Hafez GR, Buchler DA. Neurotropic melanoma of the vulva. *Cancer.* 1982;49(5):999–1004.

138. Byrne PR, Maiman M, Mikhail A, et al. Neurotropic desmoplastic melanoma: a rare vulvar malignancy. *Gynecol Oncol.* 1995;56(2):289–293.

139. Einhorn LH, Burgess MA, Vallejos C, et al. Prognostic correlations and response to treatment in advanced metastatic malignant melanoma. *Cancer Res.* 1974;34(8):1995–2004.

140. Retsas S, Gershuny AR. Central nervous system involvement in malignant melanoma. *Cancer.* 1988;61(9):1926–1934.

141. Budman DR, Camacho E, Wittes RE. The current causes of death in patients with malignant melanoma. *Eur J Cancer.* 1978;14(4):327–330.

142. Rampen FH. Levodopa and melanoma: three cases and review of literature. *J Neurol Neurosurg Psychiatry.* 1985;48(6):585–588.

143. Pfutzner W, Przybilla B. Malignant melanoma and levodopa: is there a relationship? Two new cases and a review of the literature. *J Am Acad Dermatol.* 1997;37(2 Pt 2):332–336.

144. Siple JF, Schneider DC, Wanlass WA, et al. Levodopa therapy and the risk of malignant melanoma. *Ann Pharmacother.* 2000;34(3):382–385.

145. Dizdar N, Granerus AK, Hannestad U, et al. L-dopa pharmacokinetics studied with microdialysis in patients with parkinson's disease and a history of malignant melanoma. *Acta Neurol Scand.* 1999;100(4):231–237.

146. Letellier S, Garnier JP, Spy J, et al. Development of metastases in malignant melanoma is associated with an increase in the plasma L-dopa/L-tyrosine ratio. *Melanoma Res.* 1999;9(4):389–394.

147. Olsen JH, Tangerud K, Wermuth L, Frederiksen K, Friis S. Treatment with levodopa and risk for malignant melanoma. *Mov Disord.* 2007;22(9):1252–1257. https://doi.org/10.1002/mds.21397. PMID: 17534943.

148. Bertoni JM, Arlette JP, Fernandez HH, et al. Increased melanoma risk in Parkinson disease: a prospective clinicopathological study. *Arch Neurol.* 2010;67(3):347–352. https://doi.org/10.1001/archneurol.2010.1.

149. Dalvin LA, Damento GM, Yawn BP, Abbott BA, Hodge DO, Pulido JS. Parkinson disease and melanoma: confirming and reexamining an association. *Mayo Clin Proc.* 2017;92(7):1070–1079. https://doi.org/10.1016/j.mayocp.2017.03.014.

150. Disse M, Reich H, Lee PK, Schram SS. A review of the association between Parkinson disease and malignant melanoma. *Dermatol Surg.* 2016;42(2):141–146. https://doi.org/10.1097/DSS.0000000000000591. PMID: 26771684.

151. Dalmau J, Gultekin HS, Posner JB. Paraneoplastic neurologic syndromes: pathogenesis and physiopathology. *Brain Pathol.* 1999;9(2):275–284. https://doi.org/10.1111/j.1750-3639.1999.tb00226.x. PMID: 10219745.

152. Wagner Jr RF, Nathanson L. Paraneoplastic syndromes, tumor markers, and other unusual features of malignant melanoma. *J Am Acad Dermatol.* 1986;14(2 Pt 1):249–256. https://doi.org/10.1016/s0190-9622(86)70029-7. PMID: 2869074.

153. Chan JW. Paraneoplastic retinopathies and optic neuropathies. *Surv Ophthalmol.* 2003;48:12–38.

154. Pfohler C, Preuss KD, Tilgen W, et al. Mitofilin and titin as target antigens in melanoma-associated retinopathy. *Int J Cancer.* 2007;120:788–795.

155. Berson EL, Lessell S. Paraneoplastic night blindness with malignant melanoma. *Am J Ophthalmol.* 1988;106:307–311.

156. Elsheikh S, Gurney SP, Burdon MA. Melanoma-associated retinopathy. *Clin Exp Dermatol.* 2020;45(2):147–152. https://doi.org/10.1111/ced.14095. Epub 2019 Nov 19. PMID: 31742740.

157. Chan C, O'Day J. Melanoma-associated retinopathy: does autoimmunity prolong survival? *Clin Experiment Ophthalmol.* 2001;29:235–238.

158. Milam AH, Saari JC, Jacobson SG, et al. Autoantibodies against retinal bipolar cells in cutaneous melanoma-associated retinopathy. *Invest Ophthalmol Vis Sci.* 1993;34:91–100.

159. Liu C-H, Wang N-K, Sun M-H. Melanoma-associated retinopathy. *Taiwan J Ophthalmol.* 2014;4:184–188.

160. Rahimy E, Sarraf D. Paraneoplastic and non-paraneoplastic retinopathy and optic neuropathy: evaluation and management. *Surv Ophthalmol.* 2013;58:430–458.

161. Powell SF, Dudek AZ. Treatment of melanoma-associated retinopathy. *Curr Treat Options Neurol.* 2010;12:54–63.

162. Boeck K, Hofmann S, Klopfer M, et al. Melanoma-associated paraneoplastic retinopathy: case report and review of the literature. *Br J Dermatol.* 1997;137:457–460.

163. Audemard A, de Raucourt S, Miocque S, et al. Melanoma-associated retinopathy treated with ipilimumab therapy. *Dermatology.* 2013;227:146–149.

164. Bartels F, Strönisch T, Farmer K, Rentzsch K, Kiecker F, Finke C. Neuronal autoantibodies associated with cognitive impairment in melanoma patients. *Ann Oncol.* 2019;30(5):823–829. https://doi.org/10.1093/annonc/mdz083. PMID: 30840061; PMCID: PMC6551450.

165. Becquart C, Ryckewaert G, Desmedt E, Defebvre L, Le Rhun E, Mortier L. Encéphalite limbique: une nouvelle manifestation auto-immune paranéoplasique associée au mélanome métastatique ? [Limbic encephalitis: a new paraneoplastic autoimmune manifestation associated with metastatic melanoma?]. *Ann Dermatol Venereol.* 2013;140(4):278–281. French https://doi.org/10.1016/j.annder.2013.01.424. Epub 2013 Feb 20. PMID: 23567229.

166. Dresco F, Aubin F, Deveza E, Revenco E, Tavernier L, Puzenat E. Paraneoplastic opsoclonus-myoclonus syndrome preceding a mucosal malignant melanoma. *Acta Derm Venereol.* 2019;99(3):337–338. https://doi.org/10.2340/00015555-3062. PMID: 30281137.

167. Berger JR, Mehari E. Paraneoplastic opsoclonus-myoclonus secondary to malignant melanoma. *J Neurooncol.* 1999;41(1):43–45. https://doi.org/10.1023/a:1006189210197. PMID: 10222421.

168. Valpione S, Zoccarato M, Parrozzani R, et al. Paraneoplastic cerebellar degeneration with anti-Yo antibodies associated with metastatic uveal melanoma. *J Neurol Sci.* 2013;335(1-2):210–212. https://doi.org/10.1016/j.jns.2013.08.026. Epub 2013 Aug 30. PMID: 24035275.

169. Jarius S, Steinmeyer F, Knobel A, et al. GABAB receptor antibodies in paraneoplastic cerebellar ataxia. *J Neuroimmunol.* 2013;256(1-2):94–96. https://doi.org/10.1016/j.jneuroim.2012.12.006. Epub 2013 Jan 14. PMID: 23332614.

170. Nicolae CD, Nicolae I. Antibodies against GM1 gangliosides associated with metastatic melanoma. *Acta Dermatovenerol Croat.* 2013;21(2):86–92. PMID: 24001415.

171. Palma JA, Martín-Algarra S. Chronic inflammatory demyelinating polyneuropathy associated with metastatic malignant melanoma of unknown primary origin. *J Neurooncol.* 2009;94(2):279–281. https://doi.org/10.1007/s11060-009-9848-x. Epub 2009 Mar 6. PMID: 19266164.

172. Kloos L, Sillevis Smitt P, Ang CW, Kruit W, Stoter G. Paraneoplastic ophthalmoplegia and subacute motor axonal neuropathy associated with anti-GQ1b antibodies in a patient with malignant melanoma. *J Neurol Neurosurg Psychiatry.* 2003;74(4):507–509. https://doi.org/10.1136/jnnp.74.4.507. PMID: 12640075; PMCID: PMC1738398.

173. Schoenberger SD, Kim SJ, Lavin P. Paraneoplastic optic neuropathy from cutaneous melanoma detected by positron emission tomographic and computed tomographic scanning. *Arch Ophthalmol.* 2012;130(9):1223–1225. https://doi.org/10.1001/archophthalmol.2012.449. PMID: 22965609.

174. Manganoni AM, Farfaglia R, Sereni E, Farisoglio C, Pavoni L, Calzavara-Pinton PG. Melanoma of unknown primary with nodal metastases, presenting with vitiligo-like depigmentation. *G Ital Dermatol Venereol.* 2012;147(2):210–211. PMID: 22481586.

175. Liakou AI, Trebing D, Zouboulis CC. Paraneoplastic dermatomyositis associated with metastatic melanoma. *J Dtsch Dermatol Ges.* 2012;10(1):63–64. https://doi.org/10.1111/j.1610-0387.2011.07774.x. Epub 2011 Aug 16. PMID: 21848981.

176. Tu J, Von Nida J. Metastatic malignant melanoma and dermatomyositis: a paraneoplastic phenomenon. *Australas J Dermatol.* 2011;52(2):e7–10. https://doi.org/10.1111/j.1440-0960.2010.00632.x. Epub 2010 Mar 31. PMID: 21605089.

177. Jouary T, Gracia C, Lalanne N, Vital A, Taieb A, Delaunay M. Rapidly lethal dermatomyositis associated with metastatic melanoma. *J Eur Acad Dermatol Venereol.* 2008;22(3):399–401. https://doi.org/10.1111/j.1468-3083.2007.02350.x. PMID: 18269627.

178. Kartono F, Shitabata PK, Magro CM, Rayhan D. Discohesive malignant melanoma simulating a bullous dermatoses. *J Cutan Pathol.* 2009;36(2):274–279. https://doi.org/10.1111/j.1600-0560.2007.01054.x. PMID: 19208079.

179. Meyer S, Kroiss M, Landthaler M, Vogt T. Thymoma, myasthenia gravis, eruptions of pemphigus vulgaris and a favourable course of relapsing melanoma: an immunological puzzle. *Br J Dermatol.* 2006;155(3):638–640. https://doi.org/10.1111/j.1365-2133.2006.07384.x. PMID: 16911302.

180. Reyes-Mugica M, Chou P, Byrd S, et al. Nevomelanocytic proliferations in the central nervous system of children. *Cancer.* 1993;72(7):2277–2285.

181. Sawamura Y, Abe H, Murai H, et al. An autopsy case of neurocutaneous melanosis associated with intracerebral malignant melanoma. *No To Shinkei.* 1987;39(8):789–795.

182. Jandro-Santel D, Popovic-Grle S, Cvitanovic L, et al. neurocutaneous melanosis and melanoma of the brain. *Acta Med Iugosl.* 1990;44(4):275–283.

183. Faillace WJ, Okawara SH, McDonald JV. Neurocutaneous melanosis with extensive intracerebral and spinal cord involvement. Report of two cases. *J Neurosurg.* 1984;61(4):782–785.

184. Poe LB, Roitberg D, Galyon DD. Neurocutaneous melanosis presenting as an intradural mass of the cervical canal: magnetic resonance features and the presence of melanin as a clue to diagnosis: case report. *Neurosurgery.* 1994;35(4):741–743.

185. Theunissen P, Spincemaille G, Pannebakker M, et al. Meningeal melanoma associated with nevus of OTA: case report and review. *Clin Neuropathol.* 1993;12(3):125–129.

186. Rahimi-Movaghar V, Karimi M. Meningeal melanocytoma of the brain and oculodermal melanocytosis (nevus of OTA): case report and literature review. *Surg Neurol.* 2003;59(3):200–210.

187. Sang DN, Albert DM, Sober AJ, et al. Nevus of OTA with contralateral cerebral melanoma. *Arch Ophthalmol.* 1977;95(10):1820–1824.

188. Rodriguez y Baena R, Gaetani P, Danova M, et al. Primary solitary intracranial melanoma: case report and review of the literature. *Surg Neurol.* 1992;38(1):26–37.

189. Kashiwagi N, Hirabuki N, Morino H, et al. Primary solitary intracranial melanoma in the sylvian fissure: MR demonstration. *Eur Radiol.* 2002;12(Suppl. 3):S7–10.

190. Zakka KA, Foos RY, Omphroy CA, et al. Malignant melanoma. Analysis of an autopsy population. *Ophthalmology.* 1980;87(6):549–556.

191. Lorigan JG, Wallace S, Mavligit GM. The prevalence and location of metastases from ocular melanoma: imaging study in 110 patients. *AJR Am J Roentgenol.* 1991;157(6):1279–1281.

192. Jones DR, Scobie IN, Sarkies NJ. Intracerebral metastases from ocular melanoma. *Br J Ophthalmol.* 1988;72(4):246–247.

193. Amaro A, Gangemi R, Piaggio F, et al. The biology of uveal melanoma. *Cancer Metastasis Rev.* 2017;36(1):109–140. https://doi.org/10.1007/s10555-017-9663-3.

194. Edmunds SC, Cree IA, Dí Nícolantonío F, Hungerford JL, Hurren JS, Kelsell DP. Absence of BRAF gene mutations in uveal melanomas in contrast to cutaneous melanomas. *Br J Cancer.* 2003;88(9):1403–1405. https://doi.org/10.1038/sj.bjc.6600919.

195. Royal RE, Levy C, Turner K, et al. Phase 2 trial of single agent Ipilimumab (anti-CTLA-4) for locally advanced or metastatic pancreatic adenocarcinoma. *J Immunother.* 2010;33(8):828–833. https://doi.org/10.1097/CJI.0b013e3181eec14c.

196. Zimmer L, Vaubel J, Mohr P, et al. Phase II DeCOG-study of ipilimumab in pretreated and treatment-naïve patients with metastatic uveal melanoma. *PLoS ONE.* 2015;10(3):e0118564. https://doi.org/10.1371/journal.pone.0118564.

197. Joshua AM, Monzon JG, Mihalcioiu C, Hogg D, Smylie M, Cheng T. A phase 2 study of tremelimumab in patients with advanced uveal melanoma. *Melanoma Res.* 2015;25(4):342–347. https://doi.org/10.1097/CMR.0000000000000175.

198. Rossi E, Pagliara MM, Orteschi D, et al. Pembrolizumab as first-line treatment for metastatic uveal melanoma. *Cancer Immunol Immunother.* 2019;68(7):1179–1185. https://doi.org/10.1007/s00262-019-02352-6.

199. Pelster M, et al. Phase II study of ipilimumab and nivolumab (IPI/NIVO) in metastatic uveal melanoma (UM). *J Clin Oncol.* 2019;37(15_suppl):9522.

200. Rodriguez JP, et al. Phase II multicenter, single arm, open label study of Nivolumab in combination with Ipilimumab in untreated patients with metastatic uveal melanoma. *Ann Oncol.* 2018;29(suppl_8):viii442–viii466.

18

Neurological complications of lymphoma

Amber Nicole Ruiz and Lynne P. Taylor

Departments of Neurology, Neurological Surgery and Medicine, University of Washington, Seattle Cancer Care Alliance, Seattle, WA, United States

1 Introduction

Lymphomas encompass a heterogenous group of malignant neoplasms of lymphoreticular origin. Based on 2017 data reported by the Centers for Disease Control and Prevention (CDC), Hodgkin lymphoma (HL), defined by the presence of Reed-Sternberg cells, accounts for less than 10% of new lymphoma cases with non-Hodgkin lymphoma (NHL) comprising the remainder.[1–3] Central nervous system (CNS) disease can be present at the time of diagnosis though is more often found at disease relapse. CNS infiltration is more common in aggressive lymphoma subtypes, including lymphoblastic and Burkitt lymphoma, with a CNS relapse rate around 30%–50% and rare in HL.[4,5] In comparison, diffuse large B-cell lymphoma (DLBCL), the most common form of NHL, has a more variable rate of CNS relapse depending on the presence or absence of high-risk features. Low-risk groups have an approximated CNS relapse rate of 2%–5%, in contrast to high-risk groups with reports of upward of 40% of CNS relapse.[4–12] Table 1 highlights the relative frequency of CNS involvement of both HL and NHL.

Lymphoma cells can reside anywhere within the central nervous system including the leptomeninges, brain, and spinal cord parenchyma, perivascular space, or the peripheral nerves; thus, a wide range of neurological signs and symptoms may develop as a consequence. These complications occur due to direct invasion from primary nodal or extranodal sites, compression, or secondarily due to paraneoplastic processes. This chapter delves into the potential neurological complications of lymphoma with an overview of the clinical features, imaging characteristics, and pathophysiology. We also provide a general approach to diagnosis, treatment, and symptom management based on current data.

1.1 Central nervous system prophylaxis

Risk models have been proposed as guidelines for CNS prophylactic strategies, though the role of CNS prophylaxis remains a debated topic with a wide variation in clinical practice. No molecular markers have been identified as a correlate of CNS relapse; however, several retrospective studies identified risk factors based on clinical features and patient performance status. The CNS International Prognostic Index (IPI) is the most robust, validated model to stratify DLBCL patients treated with R-CHOP into low-, medium-, and high-risk groups for CNS relapse.[7] They identified six features associated with increased risk of CNS disease including age greater than 60 years, elevated lactate dehydrogenase (LDH), involvement of more than one extranodal site, Eastern Cooperative Oncology Group performance scale (ECOG) greater than one, stage III/IV disease, and kidney and/or adrenal involvement. Site-specific involvement including breast, testes, and bone marrow has also been associated with increased risk of CNS relapse, though this is not consistent across studies.

Treatment of lymphoma subtypes with high risk of CNS involvement, such as lymphoblastic and Burkitt lymphoma, is modeled after childhood acute lymphoblastic leukemia that incorporates intrathecal chemotherapy (IT methotrexate and/or liposomal cytarabine) and systemic chemotherapy (some combination of cyclophosphamide, vincristine, doxorubicin, methotrexate, ifosfamide, mesna, etoposide, and cytarabine) with or without cranial radiotherapy.[13–15] Such aggressive treatment in adults, however, is prone to more pronounced acute treatment-related toxicities compared to pediatric populations. Several studies have demonstrated that radiation, systemic, and IT chemotherapies were more effective than IT chemotherapy alone in reducing CNS relapse in NHL. This suggests that IT administration is

TABLE 1   Hodgkin and non-Hodgkin lymphoma subtype relative propensity for central nervous system involvement.

Subtype	Relative frequency of CNS involvement
Hodgkin lymphoma	Rare
B-cell lymphomas	
Diffuse large B cell	Intermediate
Small lymphocytic	Rare
Follicular	Rare
Small cell	High
Marginal zone	Rare
Mantel cell	Rare
Burkitt	High
T-cell lymphomas	
Immunoblastic	Intermediate
Lymphoblastic	High
Anaplastic large cell	Rare

likely insufficient for uniform drug delivery within the CNS, particularly to the brain parenchyma.[4,6,15] Effective radiation-free CNS prophylaxis in the setting of combining high-dose methotrexate at doses of at least $5 g/m^2$ with IT cytarabine has been reported, suggesting that radiotherapy is not essential in CNS prophylaxis but rather an important consideration within a subset of the population.[14,16]

CNS prophylaxis in DLBCL is particularly contentious in the absence of randomized clinical trials, with neither a standardized method for patient identification nor a consensus on chemotherapeutic regimen—including specific agent(s), dose, duration of therapy, timing of therapy, or intrathecal versus systemic chemotherapy. Even with a validated model such as the CNS IPI, this tool is not uniformly applied to determine who should receive CNS prophylaxis. About 12%–23% of patients with DLBCL have a high-risk score based on the CSF IPI, though not all high-risk patients will experience CNS relapse; similarly, low-to-intermediate-risk patients may develop CNS relapse.[4,6] Aside from the lack of consensus on patient identification, there is no standardized CNS prophylactic regimen, with variations within and across clinical practices. In the rituximab era, a myriad of studies evaluated the impact of this anti-CD20 monoclonal antibody to standard DLBCL management (R-CHOP vs. CHOP alone) on CNS relapse. There is an overall trend in reduction in the frequency of CNS relapse with the use of rituximab, though this is not consistently appreciated.[8,9,17] The role of IT versus high-dose systemic methotrexate is similarly debated. High-dose methotrexate (MTX) can cross the blood-brain barrier; however, the

exact dose needed for consistent CNS penetration is unknown, with most favoring doses of at least $3 g/m^2$.[18] Intrathecal MTX has limitations as well, primarily in the ability to permeate the brain parenchyma. Liposomal formulations of intrathecal cytarabine had been used in conjunction with IT MTX as well as a single IT agent, showing promising results regarding prevention of CNS relapse, though this drug is no longer available due to technical issues with production.[17-20]

Though CNS prophylaxis remains a debated topic, it is clear that aggressive lymphomas and high-risk subgroups warrant evaluation for CNS involvement at the time of diagnosis and close monitoring for disease relapse. The CNS IPI is a useful tool in identifying high-risk patients, which will likely be further augmented by discovery of CNS biomarkers and tumor molecular characteristics. More research is needed for development of novel agents that penetrate the blood-brain barrier as well as agents that alter the immune landscape for effective prevention and treatment of CNS disease.

2 Intracranial metastases

2.1 Parenchymal disease

Intracranial involvement of lymphoma can manifest as an extradural mass, subdural mass, or less commonly as a parenchymal lesion. Intracranial lymphoma arises by direct extension of cervical lymph nodes, hematogenous spread, or via lymphatic spread. Direct CNS involvement is seen primarily in more aggressive histological subtypes of NHL. Parenchymal disease may be present with concurrent leptomeningeal metastases though this varies by lymphoma subtype. CNS relapse in DLBCL, for example, involves the brain parenchyma, either in isolation or in conjunction with leptomeningeal disease, in upward of 73% of cases.[4]

Clinical presentation is variable, ranging from asymptomatic or silent disease to development of focal neurological deficits. It is important to have a high index of suspicion, particularly in the presence of aggressive lymphomas, as initial signs and symptoms may be nonspecific. Diagnostic evaluation should involve neuroimaging with contrast-enhanced MRI and cerebrospinal fluid (CSF) evaluation. Parenchymal involvement can be variable as depicted in Fig. 1 that illustrates a small, solitary basal ganglia mass compared to multifocal disease as seen in Fig. 2. Cytological evaluation and/or flow cytometry immunophenotyping of CSF often reveals the presence of monoclonal, neoplastic cells. Definitive diagnosis rarely requires tissue confirmation with biopsy, though may be considered if there is a high clinical concern with otherwise unrevealing CSF or questionable imaging studies.

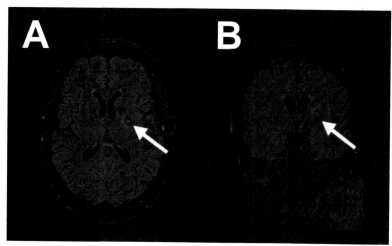

FIG. 1 (A) Axial T2 FLAIR image of a patient presenting with confusion and progressive gait instability several years after initial diagnosis of DLBCL revealing a solitary, heterogenous left basal ganglia mass with mild surrounding vasogenic edema. (B) Coronal T2 FLAIR of the left basal ganglia mass.

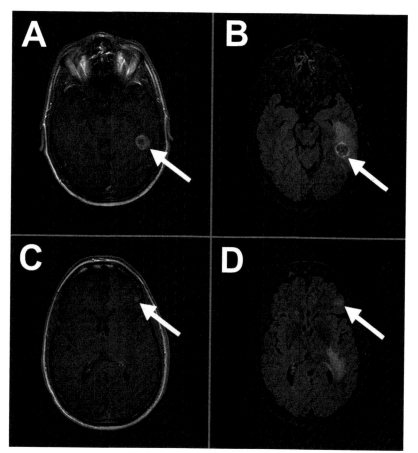

FIG. 2 (A) Axial T2 FLAIR image obtained during restaging of an asymptomatic patient with angioimmunoblastic T-cell lymphoma found to have a well-circumscribed, heterogenous lesion in the left temporal white matter with surrounding vasogenic edema. (B) Axial T1 postgadolinium image with ring-enhancement of the left temporal lesion. (C) Axial T2 FLAIR showing second, smaller hyperintense mass within the left frontal lobe with mild vasogenic edema. (D) Axial T1 postgadolinium with corresponding ring enhancement of the left frontal mass.

There is no standardized treatment for metastatic parenchymal disease. Treatment generally involves systemic chemotherapy with high-dose methotrexate or intrathecal chemotherapy with IT methotrexate or cytarbine.[2,21] As noted previously, liposomal cytarabine has shown promising results in treatment of CNS manifestations of lymphoma; however, this is no longer available due to complications with production. Whole brain radiation is a consideration in those with concurrent leptomeningeal disease though this is fraught with both short- and long-term complications. In young patients, treatment with high-dose methotrexate followed by high-dose systemic chemotherapy and stem cell transplantation has a 2-year overall survival of 54%–68%.[2] Even with treatment, prognosis is poor. Research has focused on the role and timing of CNS prophylaxis, which remains a controversial topic. As previously discussed, aggressive lymphoma subtypes and high-risk groups merit CNS evaluation at the time of diagnosis and close monitoring for disease relapse. Such individuals should be considered for CNS prophylaxis given the high mortality rate associated with CNS relapse in lymphoma.

2.2 Meningitic disease: Leptomeningeal metastases

Leptomeningeal metastasis, or lymphomatous meningitis, occurs through seeding of the leptomeninges by malignant cells. These malignant cells can penetrate the CNS either by hematogenous spread or direct exten-

sion. Leptomeningeal metastases occur in up to 25% of lymphoma patients.[22–25] Higher grade, more aggressive lymphomas carry an increased risk of developing leptomeningeal disease, with Burkitt and lymphoblastic lymphomas having higher rates of CNS relapse compared to HL or other NHL subtypes.[25,26] Leptomeningeal metastases, similar to parenchymal disease, are associated with a poor prognosis.

Malignant cells have a predilection for growth at the base of the brain, as demonstrated in Fig. 3A, and distal thecal sac; however, any portion of the neuroaxis can be involved, Fig. 4 demonstrates both cervical and thoracic leptomeningeal disease. [2,27] Patients may be asymptomatic, experience nonspecific symptoms such as headache, or develop more focal neurological symptoms including cranial neuropathies, radiculopathy, or signs of spinal cord dysfunction. Because symptomatology can be nonspecific, it is important to maintain a high index of suspicion and low threshold for evaluation.

Leptomeningeal metastases can be definitively diagnosed by identification of neoplastic cells by cytological evaluation of CSF. Flow cytometry immunophenotyping

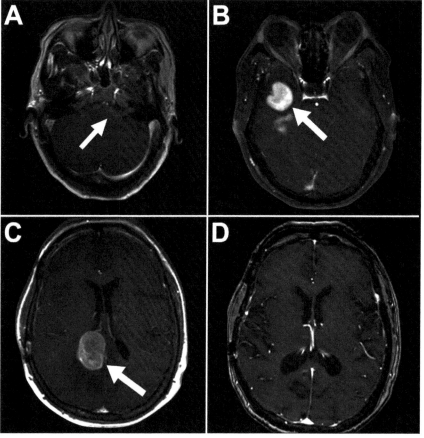

FIG. 3   (A) Axial postgadolinium image at initial presentation with a left facial droop showing dramatic enhancement around the brainstem and bilateral facial and vestibulocochlear nerves consistent with leptomeningeal lymphoma. (B) One year later with complete spontaneous resolution of symptoms, represented with headache found to have a large, homogeneously enhancing right temporal lobe mass. (C) One year later, presented with confusion, found to have an enhancing periventricular mass with biopsy-confirmed large B-cell lymphoma. (D) Two-year surveillance imaging following treatment with high-dose methotrexate and rituximab without clinical or radiographic evidence of lymphoma.

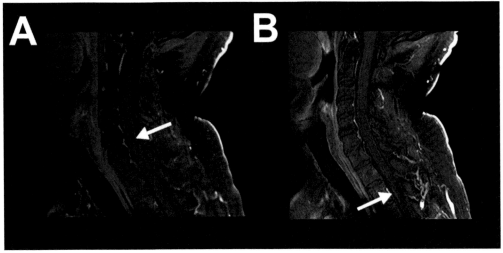

FIG. 4 (A) Sagittal and axial postgadolinium image of a patient with HIV and subsequent Burkitt lymphoma obtained for evaluation prior to bone marrow transplantation found to have both cervical (A) and thoracic (B) leptomeningeal enhancement.

is also a useful tool in establishing a diagnosis. CSF analysis should be included in disease staging in those with newly diagnosed non-Hodgkin's lymphoma or more aggressive histological subtypes. CNS invasion may be present even with negative cytology and flow studies. Neuroimaging with MRI is helpful when CSF studies are inconclusive or if one is unable to obtain CSF for evaluation. Gadolinium-enhanced MRI may demonstrate enhancing, nodular, leptomeningeal deposits. MRI of the complete neuroaxis should be obtained if there is concern for leptomeningeal disease as this may reveal asymptomatic drop metastases or spinal cord compression that requires urgent management.

As leptomeningeal metastasis is a diffuse process, treatment of the entire neuroaxis is required. Treatment generally includes a combination of corticosteroids, radiotherapy (such as focal or whole brain irradiation), and systemic and/or local chemotherapies. Local chemotherapy involves either high-dose methotrexate or cytarabine via intrathecal or intraventricular administration through an Ommaya reservoir.[25,28] Similar to management and prevention of parenchymal disease, CNS prophylaxis is an important consideration in aggressive and high-risk NHL subtypes.

2.3 Intravascular lymphomatosis

Intravascular lymphoma (IVL), or lymphomatoid granulomatosis, is a rare, rapidly fatal, complication of extranodal non-Hodgkin lymphoma. A meta-analysis of available literature on IVL suggests that most IVL's are of a B-cell origin, representing 88% of cases. T-cell and NK-cell origin represented much fewer cases at 6% and 2%, respectively.[29,30] Intravascular lymphomatosis is characterized by the presence of neoplastic cells within the lumen of small- and intermediate-sized blood vessels, often leading to vascular occlusion. Arterioles, capillaries, and venules of the skin and CNS are preferentially affected; however, the lungs, bone marrow, and spleen can be involved.

Clinical presentation is varied and often fluctuating though tends to present with stroke-like signs and symptoms. Diagnosis is difficult and is most frequently made postmortem due to delay in recognition/suspicion of IVL given the rarity of disease. Definitive diagnosis requires tissue evaluation with brain or meningeal biopsy, which further complicates timely diagnosis. Meta-analyses of IVL cohorts indicate that age, CNS involvement, and LDH level are important prognostic factors, with age > 70 years and LDH > 700 being associated with shortened survival.[29,31] IVL is sensitive to systemic chemotherapy, specifically rituximab with or without doxorubicin. Little is known regarding relapse, but data suggest higher rates of CNS relapse compared to other organ systems such as the skin, lungs, and bone marrow.[31]

3 Spinal cord processes

3.1 Intramedullary spinal cord metastases

Intramedullary spinal cord metastasis is a rare occurrence seen in both HL and NHL. Spinal cord involvement may be present at the time of initial lymphoma diagnosis or at disease relapse. There are reports that intramedullary spinal cord metastases account for 6% of myelopathies in solid tumors.[32,33] Intramedullary disease develops from hematogenous spread into the spinal cord parenchyma or via direct extension from spinal nerve roots or leptomeninges.[21,32] Lesions can be singular or multifocal. Fig. 5 demonstrates extensive, multifocal intramedullary metastases.

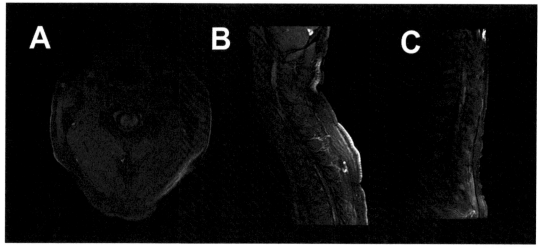

FIG. 5   (A) Axial postgadolinium image of the cervical spinal cord of a patient with a history of cutaneous T-cell lymphoma who presented with progressive, bilateral lower extremity weakness found to have homogenously enhancing intramedullary spinal cord metastases. (B–C) Sagittal postgadolinium imaging illustrating diffuse cervical, thoracic, and lumbar spine involvement.

Clinical presentation is usually a subacute, progressive myelopathy, with signs and symptoms dependent on the spinal level(s) involved. Pain may be a prominent component if nerve roots are also implicated. Patients are often wheelchair bound by the time of diagnosis.[32] Gadolinium-enhanced spinal MRI may demonstrate expansile, enhancing lesions within the spinal cord. It is important to note that patients may have concurrent brain lesions, so imaging the complete neuroaxis is essential. PET scans can aid in diagnosis with revelation of hypermetabolic spinal cord lesions.[32,33] Cytological evaluation of CSF may reveal the presence of neoplastic cells, with repeat sampling increasing diagnostic yield.

There is no standardized treatment of intramedullary spinal cord metastases. Initial treatment often involves corticosteroids that can provide symptomatic relief. Additional therapy may include intrathecal methotrexate and/or cytarabine, rituximab, or craniospinal radiation. Despite treatment, intramedullary spinal cord disease is associated with a poor prognosis, though median survival rates vary with reports of 3–11.5 months.[32,33]

3.2 Epidural lymphoma

Epidural lymphoma is a rare disease entity that occurs in both HL and NHL.[21,34–36] Epidural lymphoma accounts for 9% of epidural spinal tumors.[37] Spinal cord compression is reported in up to 10% of NHL cases with more variable reports in HL.[38] Epidural lymphoma is most frequently diagnosed when extranodal disease is already present.[21,39] Lymphomatous cells are thought to enter the epidural space through direct extension from the paraspinous region though this remains debated.

Patients may present with signs and symptoms of spinal cord compression based on the extent of disease and location of cord impingement. Lower extremity weakness,

bladder dysfunction, and localized back pain are the most common presenting symptoms.[34] As with other solid tumors, the thoracic spine is preferentially affected, followed by the cervical and lumbosacral spine, respectively.[38] Epidural lymphoma may be discovered during staging; though in some cases, the spinal column is the primary site of disease involvement. Delays in diagnosis can occur if steroids are administered prior to evaluation.

If spinal cord disease is suspected, imaging of the complete neuroaxis with gadolinium-enhanced MRI is warranted. MRI may reveal iso- or hypointense lesions with homogeneous enhancement within the epidural region. Lesions may extend over multiple vertebrae.[38] CSF evaluation is an important component of the initial workup, if not otherwise contraindicated. Treatment of epidural lymphoma often entails systemic chemotherapy, frequently including the R-CHOP regimen, employed in conjunction with local irradiation.[37,38] Though epidural lesions are radiation-sensitive, surgical intervention may be required if spinal cord compression is present, tumors are unresponsive to radiation, or less commonly to confirm diagnosis.[24,34,39] Analyses of primary spinal epidural lymphoma case series report the use of 3500–4000 cGy delivered in 20–25 fractions over 3–4 weeks.[34] Combined modality treatment has been associated with lower rates of recurrence and improved disease-free survival but not overall survival. Median survival rates vary with reports of upward of 12 months.[34,36,37]

4 Peripheral nervous system disease

4.1 Neurolymphomatosis

Neurolymphomatosis refers to infiltration of lymphoma into the peripheral nervous system. This is a rare, often painful and rapidly progressive neuropathy with a reported relative incidence of about 3% in newly diagnosed NHL.[40,41] Though neurolymphomatosis is primarily a disease of intermediate or high-grade B-cell lymphoma subtypes, there have been case reports of associated T-cell lymphoma; furthermore, acute leukemias also make up a considerable proportion, accounting for 10% of cases.[26,42,43] Neurolymphomatosis is thought to be due to hematogenous spread given the predilection for the perivascular space.[41,44,45] Lymphoma cells infiltrate the epineurium, perineurium, and endoneurium to varying degrees, leading to focal nerve fiber loss.[41,43] Both myelinated and unmyelinated fibers can be affected, leading to a variable degree of axonal degeneration.[46] Vessel walls, conversely, remain intact.

Clinical presentation is variable as any portion of the peripheral nervous system can be affected including cranial nerves, nerve roots, plexi, and autonomic structures. Peripheral nerves are the most common site involved, followed by spinal, cranial nerve, and plexus infiltration occurring at similar rates.[43] Pain is the most common symptom reported by patients. Clinical symptoms may precede diagnosis of neurolymphomatosis on the order of months to years; moreover, peripheral neurolymphomatosis precedes systemic disease in upward of 25% of patients.[46] Timely and accurate diagnosis is complicated by the phenotypic heterogeneity and frequently inconclusive diagnostic studies. This is an important consideration as early treatment is associated with a better prognosis.

Diagnostic workup includes imaging studies, CSF evaluation, electrodiagnostic studies, and/or nerve biopsy. MRI may reveal nerve or nerve root enlargement with variable reports of contrast enhancement.[43,47] MRI, however, is not specific and findings may represent other underlying neuropathic processes. PET-CT can assist in the diagnosis of neurolymphomatosis, though as with other imaging modalities, it lacks sensitivity. Imaging studies may have a role in identifying a target for biopsy with nerve biopsy boasting the highest diagnostic sensitivity of 80%–88%.[41,43] Histopathology reveals the presence of malignant cells within the epineurium, perineurium, and/or endoneurium with resultant nerve fiber damage.[41] Neurophysiologic studies may demonstrate a wide variety of abnormal, though nonspecific, findings with patterns consistent with a primary axonal neuropathy, demyelinating axonal neuropathy, or conduction block without demyelinating features.[46,48] As with imaging, neurophysiologic studies can aid in diagnosis by identifying a biopsy target but are likely not diagnostic in and of themselves.

Treatment of neurolymphomatosis can be difficult in terms of both symptomatic management and control of disease progression. Steroid therapy can provide short-term symptomatic relief but is limited by complications of long-term use. Treatment commonly involves systemic chemotherapy either with or without local radiation. Systemic chemotherapy with rituximab has a response rate of 82%, though this is not often durable and relapse is common.[2] High-dose methotrexate, alone or in combination with other systemic chemotherapeutic agents, is utilized by many centers. Radiotherapy is most effective when there is a limited field. It can provide significant neuropathic pain control when nerves, plexi, or roots are involved.[43] Unfortunately, prognosis is poor despite utilization of multimodal treatment. Most studies report a median survival of 10 months from diagnosis.[43,46,47]

4.2 Cranial neuropathies

Cranial neuropathies can develop as a consequence of direct invasion (neurolymphomatosis) as demonstrated in Fig. 6, compression, paraneoplastic process, or increased intracranial pressure. Leptomeningeal disease is the most common cause of multiple cranial neuropathies secondary to lymphoma. Cranial nerve palsies may also arise in the setting of a paraneoplastic process, though no specific antibody directed at cranial nerves has been identified. CNS complications of lymphoma leading to increased intracranial pressure can result in cranial nerve palsies, commonly abducens nerve palsies, resulting in pseudo-localization as opposed to direct cranial nerve infiltration by lymphoma.

Cranial nerve palsies may be the presenting symptom of an unknown lymphoma, neurological complication of a known lymphoma, or consequence of lymphoma treatment. One must have a high index of suspicion and consider urgent evaluation with advanced neuroimaging such as a contrast-enhanced MRI and possible CSF evaluation. Cranial nerve involvement may be subtle on imaging, as depicted in Fig. 6B, which reveals mild bilateral optic nerve enhancement. Treatment is dependent on underlying etiology—leptomeningeal metastases, neurolymphomatosis, paraneoplastic disease—described in detail within this chapter.

4.3 Nerve root disease, plexopathies, and other peripheral neuropathies

Similar to cranial neuropathies, the nerve roots and plexi can be implicated in lymphoma due to direct invasion, compression, or via paraneoplastic processes. Several antibodies including anti-Hu, anti-CV2/CRMP5, and antiamphiphysin antibodies have been associated with paraneoplastic peripheral sensory neuropathies secondary to HL. Radiographic evaluation, neurophysiologic studies, CSF analysis, or biopsy may be required to confirm diagnosis. Given the relative frequency of peripheral nervous system processes, care must be taken to rule out other possible etiologies. As with cranial

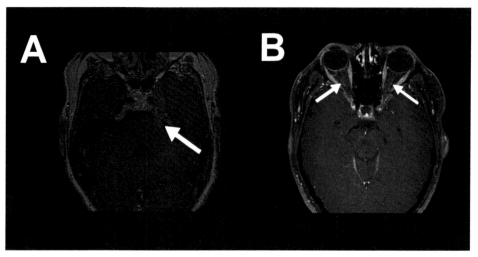

FIG. 6   (A) Axial postgadolinium image of a patient with a history of DLBCL who presented with left facial numbness and pain, intermittent diplopia, and nasal congestion found to have a homogenously enhancing lesion of the maxillary sinus extending into the pons involving the trigeminal nerve leading to nerve and expanding to the foramen rotundum and ovale. (B) Axial postgadolinium image of a different patient with peripheral T-cell lymphoma who presented with right eye vision loss was found to have subtle, bilateral optic nerve enhancement.

neuropathies, treatment is driven by the underlying pathologic process. Symptomatic management is also an important component of care as pain is a common feature shared with compressive processes and direct invasion.

5 Paraneoplastic neurological syndromes

Paraneoplastic neurological syndromes (PNS) are rare neurological disorders due to an underlying malignancy that are not attributable to direct neoplasm invasion, metastatic disease, coagulopathy, infection, metabolic derangements, nutritional deficiencies, or a consequence of chemotherapy. PNS can manifest in both the central and peripheral nervous systems. PNS is most commonly associated with solid tumors including thymoma, lung, testicular, and gynecologic malignancies with lymphomas accounting for a smaller percentage. The pathophysiology of PNS involves an immune-mediated response to onconeural antigens/proteins resulting in end-organ dysfunction. Current evidence suggests that tumors ectopically express antigens that are usually relegated to the central nervous system, these antigens are detected by the body's immune system leading to an inflammatory response mediated by both B and T cells. This immune response results in nervous system destruction.[49–53] Several well-defined autoantibodies to these onconeural proteins include Hu, Yo, Ri, CV2, amphiphysin, Ma, Ta, Tr, and NMDA.[53–57] Table 2 summarizes the identified paraneoplastic neurological syndromes and their associated autoantibodies and lymphoma subtypes.

PNS are rare, occurring in less than 1% of solid tumors and even less frequently in lymphoma; furthermore, the true incidence and prevalence of PNS are unknown.[49,50] A recent population-based epidemiologic

TABLE 2   Diagnostic criteria for paraneoplastic neurological syndromes (PNS) as defined by Graus et al.

Definite PNS

1. A classic syndrome and cancer that develops within 5 years of the diagnosis of the neurological disorder.

2. A nonclassic syndrome that resolves or significantly improves after cancer treatment without concomitant immunotherapy provided that the syndrome is not susceptible to spontaneous remission.

3. A nonclassic syndrome with onconeural antibodies (well-characterized or not) and cancer that develops within 5 years of the diagnosis of the neurological disorder.

4. A neurological syndrome (classic or not) with well-characterized onconeural antibodies (e.g., anti-Hu, Yo, CV2, Ri, Ma-2, or amphiphysin), and no cancer.

Possible PNS

1. A classic syndrome, no onconeural antibodies, no cancer but at high risk to have an underlying tumor.

2. A neurological syndrome (classic or not) with partially characterized onconeural antibodies and no cancer.

3. A nonclassic syndrome, no onconeural antibodies, and cancer present within 2 years of diagnosis.

From Graus F, et al. Recommended diagnostic criteria for paraneoplastic neurological syndromes. J Neurol Neurosurg Psychiatry. 2004;75:1135–1140.

study conducted in Italy calculated an incidence of 1/100,000 person-years and a prevalence of 4/100,000.[52] Several studies contend that the incidence of PNS is increasing over time due to increased clinical suspicion and advances in detection.[49,50,58] The recommended diagnostic criteria of PNS were described by Graus et al. in 2004, seen in Table 3, and are divided into definite or possible PNS based on clinical presentation (defined as

TABLE 3 Paraneoplastic neurological syndromes, identified autoantibodies, and associated lymphoma subtypes.

Paraneoplastic neurological syndrome	Antibodies	Lymphoma subtype
Limbic encephalitis	Hu	HL
	CRMP5/CV2	HL
	mGluR1	HL
	Ma2	NHL
Encephalomyelitis	Hu	HL
Cerebellar degeneration	Tr	HL
	GAD	NHL
Paraneoplastic chorea	CRMP5/CV2	NHL
Opsoclonus-myoclonus	None	NHL
Paraneoplastic myelopathy	None	HL and NHL
Sensory neuropathy	Hu	HL
	CRMP5/CV2	NHL
Sensorimotor neuropathy	None	HL and NHL
Autonomic ganglionopathy	AChR	HL and NHL
Stiff-person syndrome	GAD	NHL
	Amphiphysin	HL
Neuromyotonia	None	HL
Myasthenia gravis	AChR	HL and NHL
Lambert Eaton myasthenic syndrome	VGCC	NHL
Dermatomyositis	p155/p140	NHL

classical or nonclassical), presence or absence of underlying malignancy, and the presence or absence of autoantibodies to onconeural proteins.[51,59,60] Paraneoplastic neurological syndromes tend to develop rapidly and progress over the course of weeks to months, leading to significant neurological deficits. Spinal fluid studies often reveal a lymphocytic pleocytosis with an elevated protein level.[50,51] CSF IgG index may be high and oligoclonal bands may also be present. Notably, autoantibody specificity is high in serum assays. Imaging of the neuroaxis can be normal or have changes consistent with the clinical presentation (e.g., cerebellar atrophy seen in paraneoplastic cerebellar degeneration). PET scans may show focal or diffuse hypometabolism within the brain, though this is a nonspecific finding.[50,59]

Treatment is twofold and involves treatment of the underlying malignancy in addition to immune suppression.

Treatment of the underlying malignancy often prevents further PNS disease progression and can lead to substantial clinical improvement. As PNS is thought to be due to an aberrant immune response, a secondary treatment approach with immune suppression is usually employed. First-line immunosuppressive agents include corticosteroids, intravenous immunoglobulin (IVIG), and plasma exchange (PLEX). Cyclophosphamide and rituximab are also often used either in combination with these first-line agents or as second-line therapy.[57,59]

5.1 Cerebellar degeneration

Paraneoplastic cerebellar degeneration is associated with several malignancies, most notably HL. Development of symptoms often may precede the diagnosis of underlying malignancy; however, in the case of HL, symptoms tend to develop after the initial diagnosis of HL or during remission. Patients often develop vertiginous symptoms in conjunction with nausea and emesis followed by rapid progression of ataxia over the course of days to weeks with symptoms reaching a peak within months. Ataxia can be both axial and appendicular. Symptoms can be mild or more severe, leading to inability to ambulate without assistance, feed self, or perform other activities of daily living. Brain MRI in the early stages of disease is usually normal, but as the disease progresses, significant cerebellar atrophy is seen.

Several antibodies have been identified in association with paraneoplastic cerebellar degeneration including anti-Yo, anti-Tr, anti-Hu, and less commonly anti-Ma antibodies. Paraneoplastic cerebellar degenerations associated with HL most often express anti-Tr antibodies in contrast to anti-Yo antibodies, which are more often seen in breast and ovarian cancer. Anti-Tr antibody-associated cerebellar degeneration tends to be less severe than that of anti-Yo antibodies in terms of disease progression, symptom severity, and degree of cerebellar damage.[61,62] Anti-Hu antibodies are more often associated with an encephalomyelitis-type syndrome though there are reports of cerebellar degeneration. Antibody titers tend to be higher in the CSF compared to the serum, suggestive of intrathecal synthesis. These antibodies target proteins expressed on cerebellar Purkinje cells leading to cell death and cerebellar atrophy. On autopsy, there is cerebellar degeneration with severe loss of the Purkinje cell layer, which may be accompanied by molecular and granular layer thinning. Lymphocytic infiltrates are rarely found in the Purkinje cell layers.[50,62,63]

Treatment for paraneoplastic cerebellar degeneration is similar to that for all PNS and involves treatment of the underlying malignancy in conjunction with immunosuppression. Corticosteroid therapy generally involves 3–5 days of methylprednisolone 1000 mg daily, followed by an oral taper. If patients are steroid-responsive, there is consideration for more long-term treatment with IVIG at 2 g/kg. Other steroid-sparing agents that have been used include PLEX, cyclophosphamide (0.15–0.3 mg/kg/day), tacrolimus (0.15–0.3 mg/kg/day), rituximab (375 mg/m^2/week), and mycophenolate (1–1.5 g BID).[50] Anti-Tr antibody patients have a better prognosis compared to anti-Hu or anti-Yo antibody carriers, often experiencing improvement of disease or stabilization with immunosuppressive therapy. Recovery and response to treatment are likely related to the degree of Purkinje cell death.

5.2 Limbic encephalitis

Paraneoplastic limbic encephalitis is characterized by acute-to-subacute mood and personality changes, visual or auditory hallucinations, loss of short-term memory, and focal or generalized seizures. Symptoms tend to develop over the course of days to weeks. Encephalopathy can progress to coma in more severe cases. One should have a high index of suspicion in patients with acute personality or behavioral changes and hallucinations without a preexisting psychiatric history. Paraneoplastic limbic encephalitis is more often seen in HL as opposed to NHL, being the third most common cause after small-cell lung carcinoma and testicular germ cell tumors.[50] A lymphocytic pleocytosis is often seen in CSF studies. MRI brain can be normal or demonstrate T2 FLAIR hyperintensities of the medial temporal lobes sometimes with associated contrast enhancement. EEG may demonstrate increased seizure risk with epileptiform discharges, often over the temporal regions, or may show focal or generalized seizure activity.

The most common antibodies associated with paraneoplastic limbic encephalitis include anti-Hu, Ma2, CRMP5/CV2, NMDA, and GABA$_B$. There have been several case reports of anti-Tr antibodies associated with limbic encephalitis in HL. These autoantibodies tend to be elevated in both the serum and CSF, with CSF detection being more sensitive than serum.[50,64,65] Damage to the hippocampi occurs as a result of inflammation of the brain parenchyma, primarily consisting of T cells, and the perivascular space, involving both B and T cells.

Treatment, as with all PNS, involves immunosuppression and treatment of the underlying malignancy. The earlier treatment is initiated, the more likely the patient is to make a clinical recovery or stabilize the disease process. Acute treatment often involves 3–5 days of high-dose corticosteroids such as methylprednisolone 1000 mg daily with or without IVIG or PLEX. Long-term immunosuppression with steroid-sparing agents will depend on the patient's overall clinical course.[50,64,65] Symptomatic treatment of mood disturbances and seizures is also critical and will likely require ongoing close attention.

5.3 Neuromuscular junction syndromes

5.3.1 Lambert-Eaton and myasthenic syndrome

Paraneoplastic neurological syndromes of the neuromuscular junction are rare in lymphoma and include both Lambert-Eaton myasthenic syndrome (LEMS) and myasthenia gravis (MG). Several cases of paraneoplastic LEMS and MG have been reported in association with HL and T-cell NHL with involvement of the mediastinum.[49,51,66] In most cases, LEMS and MG develop around the time of diagnosis of lymphoma though symptoms may develop several years after initial diagnosis.

Though similar in clinical presentation, there are several clinical and pathophysiologic distinctions between LEMS and MG. Paraneoplastic LEMS is characterized by progressive proximal weakness (often involving the lower extremities greater than the upper extremities) that initially improves with repetitive use, diaphragmatic weakness, fatigability, and bulbar symptoms that are less severe than that of MG. Cholinergic dysautonomia develops in upward of 50% of cases later in the disease course.[49] Anti-VGCC (P/Q type voltage-gated calcium channel) antibodies are associated with LEMS. Characteristic electrophysiologic findings include small compound muscle action potentials, decrement of compound muscle action potentials at low stimulation rates compared to increment of action potentials at high stimulation rates. Antibodies react with presynaptic P/Q type VGCC leading to reduced acetylcholine release into the synaptic cleft though there are cases of LEMS associated with anti-Sox1 antibodies and lymphoma. Paraneoplastic MG similarly presents with fatigable muscle weakness though with an ocular-bulbar predilection as well as diaphragmatic weakness. Dysautonomia is not a common feature of MG. Electrophysiologic studies reveal decremental response to repetitive nerve stimulation. Anti-AchR (acetylcholine receptor) antibodies react to the postsynaptic muscle membrane leading to internalization and downregulation of AchR and thus reduced ion influx and impaired muscle contraction. There have been case reports of MG due to the presence of anti-MuSK (muscle-specific kinase) associated with NHL.[66]

Treatment of LEMS and MG is multifold, involving treatment of the underlying malignancy, immunosuppression, and symptomatic management. Targeted therapy of the underlying neuropathologic mechanism is key for clinical improvement. 3,4-Diaminopyridine (DAP), a potassium channel blocker, is used at doses of 80 mg/day for the management of LEMS. Pyridostigmine, an anticholinesterase, is used at 600 mg/day in divided doses for symptomatic management of MG. Other immunosuppressive agents that are utilized include corticosteroids (such as prednisone and prednisolone at various doses), azathioprine (up to 2.5 mg/kg/day), mycophenolate mofetil (1–3 g/day), rituximab (375 mg/m^2), cyclophosphamide (50 mg/kg/day for 4 days), PLEX, or IVIG (400–1000 mg/day for total 2–3 g).[49,66]

5.4 Dermatomyositis and polymyositis

Paraneoplastic inflammatory myopathies include dermatomyositis and polymyositis. Polymyositis is less likely than dermatomyositis to be paraneoplastic or related to an underlying malignancy. These paraneoplastic myopathic syndromes tend to precede the diagnosis of lymphoma, though often within the first year of diagnosis [67–69]. Dermatomyositis is characterized by skin changes and proximal muscle weakness. Skin changes include the classic, heliotrope rash of the upper eyelids; "shawl sign" defined by an erythematous rash of the face, neck, chest, and shoulders; and Gottron papules, scaly appearing lesions of the phalangeal joints. Polymyositis is an inflammatory myopathy similar to dermatomyositis though without dermatologic manifestations. Though inflammatory myopathies are more commonly associated with breast, ovarian, lung, and prostate cancers, there is a known association with NHL, primarily of B-cell lineage.[69]

Clinical presentation and laboratory findings of these paraneoplastic inflammatory myopathies are similar to idiopathic disease. Laboratories are notable for elevated serum creatine kinase (CK), liver enzymes (including AST and ALT), lactate dehydrogenase (LDH), and aldolase. Monitoring of serum CK is helpful in assessing response to therapy. Anti-p155/p140 (or anti-TIF1-γ) antibodies are associated with paraneoplastic dermatomyositis and polymyositis. Anti-p155/p140 (anti-TIF1-γ) is an antinuclear antibody with a proposed target of transcriptional intermediary factor 1-gamma leading to B- and T-cell-mediated inflammation of skeletal myocytes.[67,69] Electrophysiologic studies demonstrate fibrillations that suggest spontaneous muscle activity, complex repetitive discharges, and positive sharp waves. Muscle biopsy reveals a mixture of perivascular inflammatory infiltrates composed of both B and T cells as well as perifascicular muscle fiber atrophy.

Treatment often involves initiation of corticosteroids either in conjunction with or followed by steroid-sparing immunosuppressive agents such as azathioprine(up to 2.5 mg/kg/day), methotrexate (25 mg/week), cyclosporine A (100–150 mg BID), mycophenolate mofetil (2 g/day), cyclophosphamide (0.5–1 g/m²), or IVIG (400–1000 mg/day for total 2–3 g). [49,67,69] Despite treatment, up to 1/3 of patients have persistent motor impairment.

References

1. USCS Data Visualizations—CDC. Centers for Disease Control and Prevention; Published June 2020. https://gis.cdc.gov/Cancer/USCS/DataViz.html. Accessed 11 February 2021.

2. Mehta MP. *Principles and Practice of Neuro-Oncology: A Multidisciplinary Approach.* demosMedical; 2011. xxii, 951 p.

3. Louis DN, Ohgaki H, Wiestler OD, Cavenee WK, World Health Organization, International Agency for Research on Cancer. WHO classification of tumours of the central nervous system. In: *World Health Organization Classification of Tumours.* Revised 4th ed. International Agency For Research On Cancer; 2016. 408 p.

4. Kansara R. Central nervous system prophylaxis strategies in diffuse large B cell lymphoma. *Curr Treat Options Oncol.* 2018;19(11):52. https://doi.org/10.1007/s11864-018-0569-2.

5. Patrij K, Reiser M, Wätzel L, et al. Isolated central nervous system relapse of systemic lymphoma (SCNSL): clinical features and outcome of a retrospective analysis. *Ger Med Sci.* 2011;9. https://doi.org/10.3205/000134. Doc11.

6. Siegal T, Goldschmidt N. CNS prophylaxis in diffuse large B-cell lymphoma: if, when, how and for whom? *Blood Rev.* 2012;26(3):97–106. https://doi.org/10.1016/j.blre.2011.12.001.

7. Schmitz N, Zeynalova S, Nickelsen M, et al. CNS international prognostic index: a risk model for CNS relapse in patients with diffuse large B-cell lymphoma treated with R-CHOP. *J Clin Oncol.* 2016;34(26):3150–3156. https://doi.org/10.1200/JCO.2015.65.6520.

8. Savage KJ. Secondary CNS relapse in diffuse large B-cell lymphoma: defining high-risk patients and optimization of prophylaxis strategies. *Hematology Am Soc Hematol Educ Program.* 2017;2017(1):578–586. https://doi.org/10.1182/asheducation-2017.1.578.

9. Gleeson M, Counsell N, Cunningham D, et al. Central nervous system relapse of diffuse large B-cell lymphoma in the rituximab era: results of the UK NCRI R-CHOP-14 versus 21 trial. *Ann Oncol.* 2017;28(10):2511–2516. https://doi.org/10.1093/annonc/mdx353.

10. Cheah CY, George A, Giné E, et al. Central nervous system involvement in mantle cell lymphoma: clinical features, prognostic factors and outcomes from the European Mantle Cell Lymphoma Network. *Ann Oncol.* 2013;24(8):2119–2123. https://doi.org/10.1093/annonc/mdt139.

11. Ayanambakkam A, Ibrahimi S, Bilal K, Cherry MA. Extranodal marginal zone lymphoma of the central nervous system. *Clin Lymphoma Myeloma Leuk.* 2018;18(1):34–37.e8. https://doi.org/10.1016/j.clml.2017.09.012.

12. Drappatz J, Batchelor T. Neurologic complications of plasma cell disorders. *Clin Lymphoma.* 2004;5(3):163–171.

13. Gastwirt JP, Roschewski M. Management of adults with Burkitt lymphoma. *Clin Adv Hematol Oncol.* 2018;16(12):812–822.

14. Cortelazzo S, Ferreri A, Hoelzer D, Ponzoni M. Lymphoblastic lymphoma. *Crit Rev Oncol Hematol.* 2017;113:304–317. https://doi.org/10.1016/j.critrevonc.2017.03.020.

15. Bassan R, Maino E, Cortelazzo S. Lymphoblastic lymphoma: an updated review on biology, diagnosis, and treatment. *Eur J Haematol.* 2016;96(5):447–460. https://doi.org/10.1111/ejh.12722.

16. Bassan R, Masciulli A, Intermesoli T, et al. Randomized trial of radiation-free central nervous system prophylaxis comparing intrathecal triple therapy with liposomal cytarabine in acute lymphoblastic leukemia. *Haematologica.* 2015;100(6):786–793. https://doi.org/10.3324/haematol.2014.123273.

17. Ghose A, Elias HK, Guha G, Yellu M, Kundu R, Latif T. Influence of rituximab on central nervous system relapse in diffuse large B-cell lymphoma and role of prophylaxis—a systematic review of prospective studies. *Clin Lymphoma Myeloma Leuk.* 2015;15(8):451–457. https://doi.org/10.1016/j.clml.2015.02.026.

18. Goldschmidt N, Horowitz NA, Heffes V, et al. Addition of high-dose methotrexate to standard treatment for patients with high-risk diffuse large B-cell lymphoma contributes to improved freedom from progression and survival but does not prevent central nervous system relapse. *Leuk Lymphoma.* 2019;60(8):1890–1898. https://doi.org/10.1080/10428194.2018.1564823.

19. Jurczak W, Kroll-Balcerzak R, Giebel S, et al. Liposomal cytarabine in the prophylaxis and treatment of CNS lymphoma: toxicity analysis in a retrospective case series study conducted at Polish Lymphoma Research Group Centers. *Med Oncol.* 2015;32(4):90. https://doi.org/10.1007/s12032-015-0520-3.

20. González-Barca E, Canales M, Salar A, et al. Central nervous system prophylaxis with intrathecal liposomal cytarabine in a subset of high-risk patients with diffuse large B-cell lymphoma receiving first line systemic therapy in a prospective trial. *Ann Hematol.* 2016;95(6):893–899. https://doi.org/10.1007/s00277-016-2648-4.

21. Aminoff MJ, Josephson SA. *Aminoff's Neurology and General Medicine.* 5th ed. Elsevier/Academic Press; 2014. xxiii, 1368 p.

22. Palmieri D. Central nervous system metastasis, the biological basis and clinical considerations. In: *Cancer Metastasis—Biology and Treatment.* Springer; 2012. xi, 226 p.

23. Subirá D, Serrano C, Castañón S, et al. Role of flow cytometry immunophenotyping in the diagnosis of leptomeningeal carcinomatosis. *Neuro Oncol.* 2012;14(1):43–52. https://doi.org/10.1093/neuonc/nor172.

24. Bernstein M, Berger MS. *Neuro-Oncology : The Essentials.* 3rd ed. Thieme; 2015.

25. Murthy H, Anasetti C, Ayala E. Diagnosis and management of leukemic and lymphomatous meningitis. *Cancer Control.* 2017;24(1):33–41. https://doi.org/10.1177/107327481702400105.

26. Nayak L, Pentsova E, Batchelor TT. Primary CNS lymphoma and neurologic complications of hematologic malignancies. *Continuum (Minneap Minn).* 2015;21(2 Neuro-oncology):355–372. https://doi.org/10.1212/01.CON.0000464175.96311.0a.

27. Grimm S, Chamberlain M. Hodgkin's lymphoma: a review of neurologic complications. *Adv Hematol.* 2010;2011. https://doi.org/10.1155/2011/624578.

28. Brion A, Legrand F, Larosa F, et al. Intrathecal liposomal cytarabine (lipoCIT) administration in patients with leukemic or lymphomatous meningitis: efficacy and long-term safety in a single institution. *Invest New Drugs.* 2012;30(4):1697–1702. https://doi.org/10.1007/s10637-011-9632-6.

29. Fonkem E, Lok E, Robinson D, Gautam SG, Wong ET. The natural history of intravascular lymphomatosis. *Cancer Med.* 2014;3(4):1010–1024. https://doi.org/10.1002/cam4.269.

30. Lyden S, Dafer RM. Intravascular lymphomatosis presenting with spinal cord infarction and recurrent ischemic strokes. *J Stroke Cerebrovasc Dis.* 2019;28(9):e132–e134. https://doi.org/10.1016/j.jstrokecerebrovasdis.2019.06.009.

31. Fonkem E, Dayawansa S, Stroberg E, et al. Neurological presentations of intravascular lymphoma (IVL): meta-analysis of 654 patients. *BMC Neurol.* 2016;16:9. https://doi.org/10.1186/s12883-015-0509-8.

32. Flanagan EP, O'Neill BP, Habermann TM, Porter AB, Keegan BM. Secondary intramedullary spinal cord non-Hodgkin's lymphoma. *J Neurooncol.* 2012;107(3):575–580. https://doi.org/10.1007/s11060-011-0781-4.

33. Schiff D, O'Neill BP. Intramedullary spinal cord metastases: clinical features and treatment outcome. *Neurology.* 1996;47(4):906–912. https://doi.org/10.1212/wnl.47.4.906.

34. Xiong L, Liao LM, Ding JW, Zhang ZL, Liu AW, Huang L. Clinicopathologic characteristics and prognostic factors for primary spinal epidural lymphoma: report on 36 Chinese patients and review of the literature. *BMC Cancer.* 2017;17(1):131. https://doi.org/10.1186/s12885-017-3093-z.

35. Chapman S, Li J, Almiski M, Moffat H, Israels SJ. Epidural spinal mass as the presenting feature of B-acute lymphoblastic leukemia in a young child. *J Pediatr Hematol Oncol.* 2020;42(8):e845–e847. https://doi.org/10.1097/MPH.0000000000001609.

36. Nambiar RK, Nair SG, Prabhakaran PK, Mathew SP. Primary spinal epidural B-lymphoblastic lymphoma. *Proc (Baylor Univ Med Cent).* 2017;30(1):66–68. https://doi.org/10.1080/08998280.2017.11929533.

37. Mally R, Sharma M, Khan S, Velho V. Primary lumbo-sacral spinal epidural non-Hodgkin's lymphoma: a case report and review of literature. *Asian Spine J.* 2011;5(3):192–195. https://doi.org/10.4184/asj.2011.5.3.192.

38. Cho HJ, Lee JB, Hur JW, Jin SW, Cho TH, Park JY. A rare case of malignant lymphoma occurred at spinal epidural space: a case report. *Korean J Spine.* 2015;12(3):177–180. https://doi.org/10.14245/kjs.2015.12.3.177.

39. Berger MS, Prados M. *Textbook of Neuro-Oncology.* 1st ed. Elsevier Saunders; 2005. xx, 854 p.

40. Gan HK, Azad A, Cher L, Mitchell PL. Neurolymphomatosis: diagnosis, management, and outcomes in patients treated with rituximab. *Neuro Oncol.* 2010;12(2):212–215. https://doi.org/10.1093/neuonc/nop021.

41. Weis J. Neurolymphomatosis and rare focalor multifocal lesions. In: Jean-Michel Vallat JW, Gray F, Keohane K, eds. *Peripheral Nerve Disorders: Pathology and Genetics.* 1st ed. John Wiley & Sons, Ltd; 2014:291–293. [Chapter 38].

42. Sasannejad P, Azarpazhooh MR, Rahimi H, Ahmadi AM, Ardakani AM, Saber HR. Guillain-Barré-like syndrome, as a rare presentation of adult T-cell leukemia-lymphoma (ATLL): a case report. *Iran Red Crescent Med J.* 2012;14(8):497–498.

43. Grisariu S, Avni B, Batchelor TT, et al. Neurolymphomatosis: an International Primary CNS Lymphoma Collaborative Group report. *Blood.* 2010;115(24):5005–5011. https://doi.org/10.1182/blood-2009-12-258210.

44. Liang JJ, Singh PP, Witzig TE. Recurrent acute inflammatory demyelinating polyradiculoneuropathy following R-CHOP treatment for non-Hodgkin lymphoma. *Proc (Baylor Univ Med Cent).* 2013;26(2):156–158. https://doi.org/10.1080/08998280.2013.11928942.

45. Grisold W, Grisold A, Marosi C, Mengm S, Briani C. Neuropathies associated with lymphoma. *Neuro-Oncol Pract.* 2015;2(4):167–178. https://doi.org/10.1093/nop/npv025.

46. Keddie S, Nagendran A, Cox T, et al. Peripheral nerve neurolymphomatosis: clinical features, treatment, and outcomes. *Muscle Nerve.* 2020. https://doi.org/10.1002/mus.27045.

47. Kim JH, Jang JH, Koh SB. A case of neurolymphomatosis involving cranial nerves: MRI and fusion PET-CT findings. *J Neurooncol.* 2006;80(2):209–210. https://doi.org/10.1007/s11060-006-9164-7.

48. Stern BV, Baehring JM, Kleopa KA, Hochberg FH. Multifocal motor neuropathy with conduction block associated with metastatic lymphoma of the nervous system. *J Neurooncol.* 2006;78(1):81–84. https://doi.org/10.1007/s11060-005-9060-6.

49. Briani C, Vitaliani R, Grisold W, et al. Spectrum of paraneoplastic disease associated with lymphoma. *Neurology.* 2011;76(8):705–710. https://doi.org/10.1212/WNL.0b013e31820d62eb.

50. Grativvol RS, Cavalcante WCP, Castro LHM, Nitrini R, Simabukuro MM. Updates in the diagnosis and treatment of paraneoplastic neurologic syndromes. *Curr Oncol Rep.* 2018;20(11):92. https://doi.org/10.1007/s11912-018-0721-y.

51. Graus F, Delattre JY, Antoine JC, et al. Recommended diagnostic criteria for paraneoplastic neurological syndromes. *J Neurol Neurosurg Psychiatry.* 2004;75(8):1135–1140. https://doi.org/10.1136/jnnp.2003.034447.

52. Vogrig A, Gigli GL, Segatti S, et al. Epidemiology of paraneoplastic neurological syndromes: a population-based study. *J Neurol.* 2020;267(1):26–35. https://doi.org/10.1007/s00415-019-09544-1.

53. Jachiet V, Mekinian A, Carrat F, et al. Autoimmune manifestations associated with lymphoma: characteristics and outcome in a multicenter retrospective cohort study. *Leuk Lymphoma.* 2018;59(6):1399–1405. https://doi.org/10.1080/10428194.2017.1379075.

54. Rakocevic G, Hussain A. Stiff person syndrome improvement with chemotherapy in a patient with cutaneous T cell lymphoma. *Muscle Nerve.* 2013;47(6):938–939. https://doi.org/10.1002/mus.23706.

55. Kumar A, Lajara-Nanson WA, Neilson RW. Paraneoplastic opsoclonus-myoclonus syndrome: initial presentation of non-Hodgkins lymphoma. *J Neurooncol.* 2005;73(1):43–45. https://doi.org/10.1007/s11060-004-2465-9.

56. Gutmann B, Crivellaro C, Mitterer M, Zingerle H, Egarter-Vigl E, Wiedermann CJ. Paraneoplastic stiff-person syndrome, heterotopic soft tissue ossification and gonarthritis in a HLA B27-positive woman preceding the diagnosis of Hodgkin's lymphoma. *Haematologica.* 2006;91(12 Suppl):ECR59.

57. Valente M, Zhao H. Paraneoplastic myelopathy and ophthalmoplegia secondary to gray zone lymphoma with excellent response to immuno-chemotherapy: case report. *J Clin Neurosci.* 2017;43:128–130. https://doi.org/10.1016/j.jocn.2017.04.038.

58. Berger B, Dersch R, Ruthardt E, Rasiah C, Rauer S, Stich O. Prevalence of anti-SOX1 reactivity in various neurological disorders. *J Neurol Sci.* 2016;369:342–346. https://doi.org/10.1016/j.jns.2016.09.002.

59. Graus F, Dalmau J. Paraneoplastic neurological syndromes: diagnosis and treatment. *Curr Opin Neurol.* 2007;20(6):732–737. https://doi.org/10.1097/WCO.0b013e3282f189dc.

60. Greenlee JE. Recommended diagnostic criteria for paraneoplastic neurological syndromes. *J Neurol Neurosurg Psychiatry.* 2004;75(8):1090. https://doi.org/10.1136/jnnp.2004.038489.

61. Lakshmaiah KC, Viveka BK, Anil Kumar N, Saini ML, Sinha S, Saini KS. Gastric diffuse large B cell lymphoma presenting as paraneoplastic cerebellar degeneration: case report and review of literature. *J Egypt Natl Canc Inst.* 2013;25(4):231–235. https://doi.org/10.1016/j.jnci.2013.07.001.

62. Shimazu Y, Minakawa EN, Nishikori M, et al. A case of follicular lymphoma associated with paraneoplastic cerebellar degeneration. *Intern Med.* 2012;51(11):1387–1392. https://doi.org/10.2169/internalmedicine.51.7019.

63. Greene M. Antibodies to Delta/notch-like epidermal growth factor–related receptor in patients with anti-Tr, paraneoplastic cerebellar degeneration, and hodgkin lymphoma. *JAMA Neurol.* 2014;71(8):1003–1008. https://doi.org/10.1001/jamaneurol.2014.999.

64. Ju W, Qi B, Wang X, Yang Y. Anti-Ma2–associated limbic encephalitis with coexisting chronic inflammatory demyelinating polyneuropathy in a patient with non-Hodgkin lymphoma. *Medicine.* 2017;96(40):e8228. https://doi.org/10.1097/MD.0000000000008228.

65. Laffon M, Giordana C, Almairac F, Benchetrit M, Thomas P. Anti-Hu-associated paraneoplastic limbic encephalitis in Hodgkin lymphoma. *Leuk Lymphoma.* 2012;53(7):1433–1434. https://doi.org/10.3109/10428194.2011.645211.

66. Bhatt A, Farooq MU, Chang HT. Mantle cell lymphoma and anti-MuSK-positive myasthenia gravis. *Onkologie.* 2011;34(7):382–383. https://doi.org/10.1159/000329610.

67. Marie I, Guillevin L, Menard JF, et al. Hematological malignancy associated with polymyositis and dermatomyositis. *Autoimmun Rev.* 2012;11(9):615–620. https://doi.org/10.1016/j.autrev.2011.10.024.

68. Ohashi M, Shu E, Tokuzumi M, et al. Anti-p155/140 antibody-positive dermatomyositis with metastasis originating from an unknown site. *Acta Derm Venereol.* 2011;91(1):84–85. https://doi.org/10.2340/00015555-0955.

69. Stübgen JP. Inflammatory myopathies and lymphoma. *J Neurol Sci.* 2016;369:377–389. https://doi.org/10.1016/j.jns.2016.08.060.

Neurologic complications of the leukemias

Lynne P. Taylor

Departments of Neurology, Neurological Surgery and Medicine, University of Washington,
Seattle Cancer Care Alliance, Seattle, WA, United States

1 Introduction

The leukemias are a large group of hematologic malignancies first described in the 1800s. Virchow coined the term "leukåmie" in 1847 and was the first to understand that it represented a reversed balance between the red and white blood cells.[1] Previous to that time, it was thought that the "milky blood" of the transformed pale serum represented pus rather than a proliferation of leukocytes.

There were 61,090 new cases in 2021, which was 3% of all new cancer cases. The 5-year survival rate (2011–17) was 65%, a significant increase from the rate of 50% in the year 2000. [SEER Database. http://seer.cancer.gov. Accessed 18 September 2021]

With the 2016 publication of the WHO classification of leukemias and the expanding significance of new genetic information into the characterization of important subtypes,[2] it is more important than ever for neurologists to work together with our colleagues in hematology-oncology who are specialists in leukemia when caring for patients with neurologic complications. Leukemias are classified into the acute and chronic forms of lymphocytic and myelocytic subtypes. A progenitor stem cell matures into lymphoid and myeloid cells, with the lymphoid cell transforming into a lymphocyte and the myeloid precursor becoming red blood cells, platelets, and myeloblast white cells. Some of the neurologic complications of leukemia are distinct to this diagnosis, but many others share characteristics with the lymphomas and behave in such a way that they are indistinguishable in their effects on the central nervous system.

The factors that lead to leukemic infiltration into the nervous system are still poorly understood, but it is a characteristic of those leukemia subtypes with the worst prognosis, and involvement of the leptomeninges is associated with worsened overall survival and decreased quality of life,[3] particularly when it occurs as part of a recurrent leukemia.

Acute myeloid leukemia (AML) is more common in older, male adults. It is rare, with new cases representing 4.3/100,000 persons/year. Five-year relative survival is low at 29.7% (though increased from 17.2% in 2000), and it represents only 1.1% of new cancer cases.

Chronic myeloid leukemia (CML), on the contrary, is even less common, representing 0.5% of all new cancer cases. The 5-year survival, however, is 73.8% that is, again, a significant improvement over 2000 when it was 48.1%. The myeloid neoplasms, most common in adults, now have 11 distinct subtypes with most of them associated with unfavorable outcomes and distinct neurologic complications.

The acute lymphocytic leukemias (ALL), much more common in the pediatric population, do occur in adults where, unfortunately, they generally have a very poor prognosis. The median age at diagnosis is 17, with those under the age of 20 representing more than 50% of the new cases. The overall survival statistics hide a disparity, however, as 5-year survival rates near 90% in children but only 25% in those older than 50.[4] The rate of new cases is 1.8/100,000 persons/year. Five-year survival is higher than in AML at 70.3% (rising from 62.9% in 2000) and represents 0.3% of new cancer cases. One major change to the WHO classification system is that there are now two new recognized categories of acute leukemias, which are understood to act so much like lymphomas that they are given a hybrid name: B-cell and T-cell lymphoblastic leukemia/lymphomas; for which the reader is referred to the chapter on the neurologic complications of lymphomas for additional information, as their biological behavior is more like that larger group.

Chronic lymphocytic leukemia (CLL), most common in adult white male patients, represents 1.1% of all new cancer cases at 4.9/100,000 persons/year. The 5-year survival is the highest of all leukemia subtypes at 88.4%, a small increase over the rate of 79.8% in 2000.

Knowledge of a priori likelihood of direct neurologic invasion into the brain, dura, spinal cord, nerve roots, or cerebrospinal fluid (CSF) will also allow consideration of the possibility of indirect complications of the leukemias, such as infectious and vascular complications that are described in detail elsewhere in this volume.

2 Diagnostic considerations

With the exception of chloromas, CNS leukemia exists almost exclusively as leptomeningeal carcinomatosis. Given the wide distribution of the spinal fluid bathing the cerebral cortex, the cerebellum, spinal cord, cranial nerves, cervical, thoracic, and lumbar nerve roots; the clinical symptomatology can be diverse. When there are multiple symptoms spread across the neuraxis, especially if there is asymmetric involvement and signs of both upper and lower motor neuron dysfunction, the diagnosis of leptomeningeal leukemia can be easy to suspect but also very difficult to prove. Signs of cranial nerve or lumbar nerve root involvement on examination are quite helpful. Mild cognitive slowing and third, sixth, and seventh cranial nerve involvement are all quite common, but it is quite possible to have this diagnosis with no signs or symptoms of any kind. Standard examination of the CSF misses malignant cells in up to 45% of patients later found to be affected.[5] Flow cytometry is considered more sensitive for hematologic malignancies and should always be ordered along with evaluation of CSF glucose, protein, cell count, and cytology. In hematologic malignancies, immunohistochemistry has sensitivity of 89% (B-ALL and AML) while brain MRI has low sensitivity in the same patients of 39%–44%, compared to the 100% sensitivity seen in solid tumor patients.[6]

Given the known insensitivity of most brain MRI sequences, it is important to understand which to request in patients with hematologic malignancies in whom you suspect leptomeningeal disease. Enhanced T1-weighted images are more accurate for the diagnosis of brain metastases or foci of tumor that enters the brain parenchyma by the way of the Virchow-Robin perivascular spaces. However, delayed postcontrast FLAIR MRI improves the diagnostic sensitivity for tumor in the leptomeningeal space over other images, as the CSF suppression outlines the leptomeningeal space more clearly.[7] Both sequences should be ordered to ensure that both compartments are adequately visualized.

3 The acute leukemias

3.1 Acute lymphocytic leukemia

Acute lymphocytic leukemia has its peak incidence between the ages of 1–4 years. Intensified treatments and risk stratification have led to 5-year overall survival rates of 90% in children; however, the outcome in adults > 40 remains poor. B-cell acute lymphoblastic leukemia has many subtypes with complex alterations in chromosomes, which are beyond the scope of this study but does allow stratification into high-, medium-, and low-risk disease and allows the use of the more aggressive therapies to be focused on those expected to have the least favorable outcomes. There are genetic susceptibility syndromes such as Down's, Fanconi anemia, and Nijmegen breakage syndrome, as well as environmental factors such as prior exposure to ionizing radiation and pesticides. One important chromosomal abnormality is BCR-ABL (Philadelphia chromosome [t(9;22) (q34;q11)])-positive ALL that increases with age from 2% to 6% in children and young adults and > 25% in older adults. This gene fusion portends both poorer prognosis and more reliable development of leptomeningeal cancer. Other prognostic factors include white blood count at diagnosis, age, race, ethnicity, gender, cell lineage, and CNS involvement by CSF analysis. Despite recent advances, however, nearly one-quarter of patients lack a clear subtype and so evade risk stratification and are classified as "other."[8]

Frequency of CNS recurrence without prophylaxis is between 30% and 50% for ALL, making CNS prophylaxis routine. Treatment for newly diagnosed ALL is divided into four phases: induction, consolidation, dose intensification, and maintenance, which can take as long as 2–3 years total. Consideration of bone marrow transplant is kept for patients at high risk and has led to survival rates for children close to 90%. In adult patients beyond the age of 60, however, the 5-year overall survival is less than 20%.[4]

CNS prophylaxis is recommended for all ALL patients, as noted earlier. Ideally, CSF radionuclide flow studies should be performed prior to administration with normal CSF flow dynamics confirmed, though CSF blockage is less of an issue with hematologic malignancies than it is for solid tumor patients. Intrathecal methotrexate or cytarabine is commonly used, as the use of prophylactic cranial irradiation has fallen out of favor because of late neurocognitive effects and remote secondary cancers. Methotrexate is considered superior to ara-c due to a longer half-life in the leptomeninges and deeper penetration into the brain parenchyma. Pharmacokinetic studies suggest that doses higher than 6 mg are no better at maintaining cytotoxic doses within the CSF. Twelve mg in the lumbar route and 6 mg in the Ommaya route are considered bioequivalent. In many clinical trials, CNS prophylaxis consisted of 6–8 intrathecal (IT) injections, though in some protocols, there are as many as 16 IT treatments planned. CNS prophylaxis has been shown to result in significant increases in the interval before CNS relapses occur.[9] When using these drugs for treatment of active leptomeningeal leukemia, the methotrexate dose, chosen by route, is given twice a week until cytologic response as measured by flow cytometry, ideally with clearance of all leukemic cells. Note should be taken that methotrexate concentrations can be increased by coadministration with the following medications: sulfa drugs, penicillins, cyclosporine, and probenecid.[10] Cytosine arabinoside (ara-c) doses are generally 30 mg twice a week until clearance by flow or as part of standard CNS prophylaxis practice. Literature support of additional efficacy for the practice of triple-drug therapy with methotrexate, cytosine arabinoside, and hydrocortisone could not be documented (Fig. 1).

In adult patients with ALL, CNS involvement at recurrence portends a very grave prognosis. The incidence of CNS involvement with disease relapse is variably reported as between 7% and 15% and survival is very poor, with a median overall survival (OS) of 6 months and no survivors at 5 years. Mature B-cell subtype is associated with an increased risk of CNS disease as is an increase in lactate dehydrogenase (LDH) level, WBC count, and proliferative index. In one large study,[11] the differentiation between those ultimately found to have CNS disease and those who did not were higher WBC at diagnosis [25.6×10^9/L versus 13.5×10^9/L ($P < .001$)] and the presence of a mediastinal mass (22% versus 9%; $P < .001$). Among patients who have an isolated CNS recurrence, however, 88% subsequently developed recurrence of disease in the bone marrow.[12] This should make it clear that CNS involvement is rarely isolated and the argument that one can "clear the CNS compartment" by treatment is likely to fail due to early systemic relapse. Another study[13] found that there were no significant differences among patients with CNS involvement at diagnosis in complete remission (CR), OS, or disease-free survival (DFS). They also found that less than 10% of adult patients have CNS involvement at presentation (Figs. 2–4).

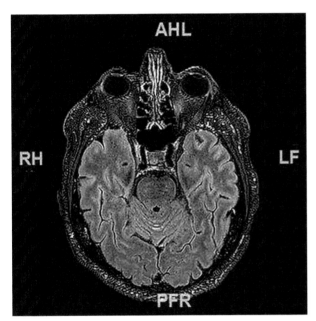

FIG. 1   Leptomeningeal leukemia (Axial 3D FLAIR + Gd). Images demonstrate scattered areas of leptomeningeal enhancement around the brainstem and within the cerebellar folia consistent with leptomeningeal leukemia. Clinical history: A 70-year-old man with a long history of leukemia. He had a systemic relapse diagnosed by PET-CT scan. He then presented within months with the subacute onset of saddle analgesia with severe constipation and urinary retention, left hand weakness, and diplopia. Neurologic examination revealed an alert man but with a hoarse and hypophonic voice. He had left third nerve palsy and a right sixth nerve palsy, palatal weakness, right arm C5 weakness, and sensory involvement and right leg sensory loss in an L5 distribution with a saddle sensory level. His examination was diagnostic of multiple cranial nerve, cervical and lumbar root involvement with associated leptomeningeal enhancement changes on brain MRI. CSF glucose was normal but protein was elevated at 171. CSF-nucleated cells were 116 and flow cytology was remarkable for atypical lymphoid cells. This was felt to be diagnostic of leptomeningeal leukemia meeting all clinical, radiographic, and laboratory criteria. Shortly after admission, he began having generalized tonic-clonic seizures. He was treated with dexamethasone, levetiracetam, and whole brain radiation therapy. Unfortunately, he died of leukemia several weeks after his presentation.

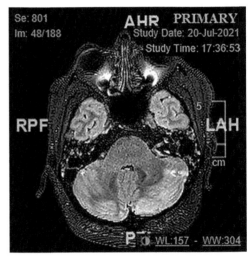

FIG. 2   AXIAL FLAIR. Cerebellar FLAIR hyperintensity involving the left cerebellar hemisphere. Clinical history: This was a 27-year-old man who was diagnosed with ALL when he presented with hemoptysis, fever, and dyspnea with a WBC of 84 and a hematocrit of 11. He had extensive mediastinal adenopathy. Systemic treatment led to a remission for several years but he had several isolated CNS recurrences in his spinal fluid treated with IT chemotherapy. He then went on to total body irradiation as part of a matched sibling peripheral stem cell transplant and was well without neurologic symptoms for some months. He then presented with diplopia and leg weakness. On examination, he had an unspecified diplopia, was somnolent and too weak to ambulate. CSF evaluation revealed a normal glucose and protein but blasts were found by flow cytometry. Given the clinical picture of widespread cerebral, cranial, and lumbar nerve root involvement, markedly abnormal brain MRI for leptomeningeal invasion and CSF diagnostic of leukemia, he was diagnosed with leptomeningeal ALL. He is now being treated with maximal supportive care with continued neurologic symptoms and some neuropathic pain.

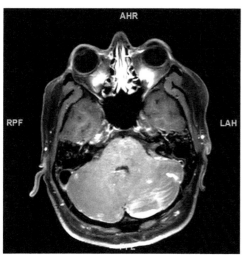

FIG. 3   AXIAL post GD. Extensive leptomeningeal enhancement of the left cerebellar hemisphere and vermis.

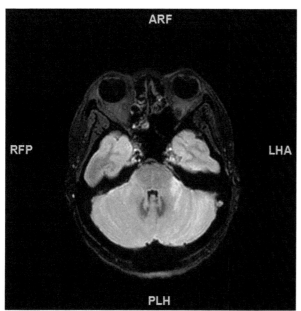

FIG. 4   Leptomeningeal leukemia, ALL. MRI description. Axial FLAIR Gd + Asymmetric foci of FLAIR signal abnormality involving the left brachium pontis, which was not felt to be diagnostic though she also had rounding of her third ventricle and enlarged lateral ventricles felt to be consistent with communicating hydrocephalus. Clinical history: A 50-year-old woman was diagnosed with Ph + precursor B-ALL. Several years later she presented with global HA, nausea and vomiting, and gait ataxia. CSF revealed 164 nucleated cells with normal glucose and protein. She had blasts in her CSF by flow cytometry and required a drain for communicating hydrocephalus. She had no obvious focal findings of cranial or lumbar root involvement though her symptoms were suggestive of increased intracranial pressure and likely cerebellar involvement. Despite normal CSF indices, she has blasts on flow cytometry so this is diagnostic of leptomeningeal leukemia. Note the subtlety of the MRI findings, the normal CSF indices and the nonlocalizing neurologic examination that could have led to a missed diagnosis without a high level of clinical suspicion and the flow cytometry.

Toxicities of treatment need to be carefully considered in the adult patient. Induction therapy is based on a combination of glucocorticoid, vincristine, L-asparaginase, and anthracycline as well as intrathecal chemotherapy. L-Asparaginase toxicities include pancreatitis, hepatotoxicity, and coagulation disorder. The coagulopathy, though uncommon, can cause both hemorrhagic and thrombotic complications. The reported incidence of thrombosis during treatment of childhood ALL can be quite low at 1% and range all the way to 37%, depending on the population studied.

3.2 Acute myeloid leukemia

In the pediatric population, CNS involvement is common in AML (6%–29%), though less is known about the incidence in adults. Frequency of CNS recurrence without prophylaxis is much lower than in ALL at < 5%, so CNS prophylaxis is not routine but considered for patients with hyperleukocytosis (HL).[14] Patients with CNS disease are presented with higher blast counts and younger age but also the same factors are seen in ALL that predispose to this complication: elevated LDH and overall WBC count. Cytogenetic abnormalities common to these patients are chromosome 11 abnormalities, inv(16), and trisomy 8 as well as a tendency toward complex cytogenetics.[15] Newly diagnosed patients who present with hyperleukocytosis (HL) with WBC in excess of 100,000 cells/ul can have diffuse intravascular coagulation (DIC) and leukostasis. Myeloid blasts are larger than immature lymphocytes and are less deformable, which is thought to explain the increased incidence of these complications in AML patients but not ALL. Leukostasis is a clinical diagnosis made when a patient with acute leukemia and hyperleukocytosis (HL) presents with respiratory and/or neurologic symptoms.

Tumor lysis syndrome is another dangerous complication, which can cause microscopic damage from sludging of cells within small arterioles to produce tissue damage in the lungs (39%), brain (27%), and kidneys (14%). Intracerebral bleeding and ischemic strokes are common in the brain. The treatment is both aggressive hydration and diuresis to reduce the elevated uric acid. Rapid attempts at cytoreduction through chemotherapy with hydroxyurea, cytarabine (ara-C), and danorubicin are effective.[3]

Leukapheresis to mechanically remove leukocytes can also be considered. This is somewhat controversial as the vast majority of the cells are in the bone marrow and a clear beneficial effect has never been adequately proven in clinical trials. In the absence of a significant reduction in circulating WBC, however, leukapheresis can be considered as long as the contraindications of cardiac disease or coagulation disorders are not present.[16]

In a large study of adults with AML, 32% were positive for CSF tumor at diagnosis and neurologic symptoms were seen in 4.8%. In their series of 103 adults, cytology was positive in 11% of cases and increased to 32% with the use of flow cytometry, a much more sensitive means of detection as we have already addressed. They note that NCCN guidelines suggest only performing an LP in adults with neurologic signs or symptoms. With their data, they suggested that diagnostic LP should be mandatory, especially in patients with the risk factors for higher levels of LDH or the myelomonocytic or monocytic subtypes, which are associated with an increased risk of CNS involvement.[17]

A myeloid sarcoma (chloroma) is an immature collection of myeloid cells, which appear in the extramedullary space. They can present anywhere in the body, but the most common sites are bones, soft tissues, lymph nodes, and skin. When they present intracranially, they are more often seen within the skull and the orbits. In 25% of the cases, the chloroma presents prior to the diagnosis of leukemia, often mimicking a meningioma. Radiographically they appear as hyperdense extra-axial masses on noncontrast head CT. Of the 21 reported cases of intracranial myeloid sarcoma in one series, they were found overlying all lobes of the brain and cerebellum, with one reported in the spine. There were intra-axial cases reported in 11/21 of the cases, or more than 50%.[18]

Other reports suggest that they are seen in 2%–8% of patients with AML as either single or multifocal tumors. Non-CNS sites are oral and nasal mucosa, breast, GI and GU tracts, chest wall, and pleura. There is no additional treatment recommended other than AML-directed systemic chemotherapies, as radiation therapy has not been shown to prolong OS (Fig. 5).[19]

4 The chronic leukemias

4.1 Chronic lymphocytic leukemia

Chronic lymphocytic leukemia (CLL) is the most common leukemia in the adult population, with diagnosis most commonly occurring in the sixth and seventh decades of life, with a median age of 69 at diagnosis, although it has been described in patients as young as 15 years of age. Prognosis varies widely, with some patients experiencing rapid progression, whereas others have clinically silent disease that is discovered incidentally and does not require treatment for many years. Significant progress has been made in treatment of this cancer, and 5-year overall survival rates have increased from 67.5% in 1975 to 83.2% in 2013.

Leptomeningeal carcinomatosis is rarely reported in CLL. Though as many as 8% of patients have leptomeningeal carcinomatosis in a large autopsy study, this

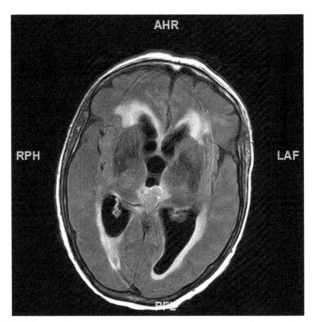

FIG. 5 Axial post Gd + FLAIR image. AML with pineal area chloroma. A 2 cm lobular enhancing pineal/posterior third ventricular mass causing obstructive hydrocephalus. CSF flow cytometry revealed an abnormal myeloid blast population with an immunophenotype similar to blood. Clinical history: A 67-year-old man with high-risk acute myeloid leukemia. He was treated with total body irradiation and an allogeneic stem cell transplant and did well for several years. He then presented with generalized weakness and upward gaze palsy and, in the setting of blasts in the CSF, the mass was felt to be a chloroma. He began treatment with whole brain radiation therapy that was discontinued due to clinical worsening.

previously had been thought to be so rare that it was diagnosed less than 1% of the time in living patients.[20]

Clinical case: Our patient was a 58-year-old man diagnosed with CLL after he was found to have lymphocytosis on routine lab studies. Subsequent evaluation revealed a monoclonal B-cell population. He was asymptomatic and remained well under observation only for 7 years. Thereafter he was treated with bendamustine, rituximab, and venetoclax, a small molecule targeted to the 17p deletion, as part of a clinical trial. He then again remained well until a full decade after his initial diagnosis when he began to develop muscular cramping and severe lancinating nerve pain in both legs. This eventually led to bilateral foot drop and his weakness progressed proximally over time until he was confined to a wheelchair. At this time, his symptoms broadened to include numbness in his left hand and isolated fingers of his right hand. LP revealed over 1000 nucleated white cells and an elevated protein of over 400 with a normal glucose. Flow cytometry was diagnostic with cell markers that matched his peripheral CLL, and he was diagnosed with leptomeningeal CLL.

He was ultimately treated with ibrutinib. Ibrutinib is a small molecule that acts as an irreversible inhibitor of Bruton's tyrosine kinase and has a CSF penetration of 21%–100%. Ibrutinib monotherapy results in 3-year response rates of 23% CR and 55% PR and is FDA approved for patients with CLL who have received at least one prior therapy. In line with the data for treatment response in systemic disease, we report that rituximab and ibrutinib were more frequently reported to achieve CR in leptomeningeal CLL than older agents. Given persistence of his CLL on flow cytometry, despite clinical improvement, IT rituximab targeted at his CD20 + cells was given weekly until CSF clearance. One year later, his neurologic exam had returned to normal and he was ambulating independently.

This case was so dramatic to us that we did a complete review of the literature to find all cases of leptomeningeal CLL, as it is widely thought that this type of leukemia rarely enters the leptomeningeal space. We were able to identify 136 cases published in 43 years since 1976, with many having OS greater than 9 months.[20] We concluded that, unlike for the acute leukemias, leptomeningeal CLL is potentially quite treatable with the possibility of neurologic recovery and prolonged survival with improved quality of life. Also, unlike other forms of leukemia, there do not seem to be any evident risk factors for the development of leptomeningeal disease.[21] Others have had similar findings including that, despite failure to clear CSF of tumor cells, prolonged survival from 23 + to 86 months was possible.[22]

4.2 Chronic myelogenous leukemia

Chronic myelogenous leukemia is a myeloproliferative disorder that presents with a chronic, accelerated, and blast phase. It was one of the first diseases in which a specific chromosome abnormality was identified, a translocation between chromosomes 22 and 9—the Philadelphia chromosome. The fusion gene that forms at the translocated site is known as the BCR-ABL mutation. Imatinib is a tyrosine kinase inhibitor that blocks the kinase activity of the oncoprotein BCR-ABL and produces remissions in patients treated for the chronic phase of CML.[23]

Studies in mice have shown that the central nervous system can become a sanctuary site for CML due to low levels of imatinib in the CSF. While quite rare, as in all other forms of leukemia, there have been reports of leptomeningeal enhancement and myeloblasts being discovered in patients, often in the blast phase.[24]

Though the outcome from leptomeningeal leukemia is usually considered grave, there were a small number of patients who had CSF recurrence from CML following hematopoietic stem cell transplant who were still alive three and 4 years later.[25] This is similar to the prognosis in the patients discussed with leptomeningeal CLL and suggests that the clinical spectrum is broader than previously considered.

5 Conclusion

While there are a very large number of leukemias described in the 2016 revision, the four broad categories previously mentioned adequately represent the predominant neurologic picture of leptomeningeal leukemia and chloromas. That leukemia can enter the nervous system is obvious, even in the rare instances of CLL and CML. The challenge will always be making the diagnosis early enough to allow disease control, hopefully eradication in some cases, and improved quality of life. Nothing will ever replace the initial suspicion of the diagnosis based on a careful history and neurologic examination, but there are considerations regarding flow cytometry and delayed FLAIR sequences with gadolinium that can improve diagnostic yield coupled with an understanding of which leukemias are more likely to present in the CNS.

With their cytopenias and exposure to drugs such as L-asparaginase, leukemia patients will also develop hematologic, infectious, and vascular complications, which are covered elsewhere in this volume. With increased survival and more frequent use of allogeneic stem cell transplants and chimeric antigen receptor T-cell therapy, late delayed and treatment-induced complications will also become more prevalent. Hopefully, as we learn more about these complications, we will also become more adept at therapies that will improve the lives of our patients.

References

1. Kampen KR. The discovery and early understanding of leukemia. *Leuk Res.* 2012;36(1):6–13. https://doi.org/10.1016/j.leukres.2011.09.028.
2. Leonard JP, Martin P, Roboz GJ. Practical implications of the 2016 revision of the World Health Organization classification of lymphoid and myeloid neoplasms and acute leukemia. *J Clin Oncol.* 2017;35(23). https://doi.org/10.1200/jco.2017.72.6745.
3. Berg S, Nand S. Neurological complications of the leukemias across the ages. *Curr Neurol Neurosci Rep.* 2017;17(2):13. https://doi.org/10.1007/s11910-017-0726-1.
4. Malard F, Mohty M. Acute lymphoblastic leukaemia. *Lancet.* 2020;395(10230):1146–1162.https://doi.org/10.1016/s0140-6736(19)33018-1.
5. Bromberg JEC, Breems DA, Kraan J, et al. CSF flow cytometry greatly improves diagnostic accuracy in CNS hematologic malignancies. *Neurology.* 2007;68(20):1674–1679. https://doi.org/10.1212/01.wnl.0000261909.28915.83.
6. Zeiser R, Burger JA, Bley TA, Windfuhr-Blum M, Schulte-Mönting J, Behringer DM. Clinical follow-up indicates differential accuracy of magnetic resonance imaging and immunocytology of the cerebral spinal fluid for the diagnosis of neoplastic meningitis—a single centre experience. *Br J Haematol.* 2004;124(6):762–768. https://doi.org/10.1111/j.1365-2141.2004.04853.x.
7. Kremer S, Eid MA, Bierry G, et al. Accuracy of delayed postcontrast flair MR imaging for the diagnosis of leptomeningeal infectious or tumoral diseases. *J Neuroradiol.* 2006;33(5):285–291. https://doi.org/10.1016/s0150-9861(06)77286-8.
8. Roberts KG. Genetics and prognosis of ALL in children vs adults. *Hematology.* 2018;2018(1):137–145. https://doi.org/10.1182/asheducation-2018.1.137.

9. Principe MID, Maurillo L, Buccisano F, et al. Central nervous system involvement in adult acute lymphoblastic leukemia: diagnostic tools, prophylaxis, and therapy. *Mediterr J Hematol Infect Dis.* 2014;6(1). https://doi.org/10.4084/mjhid.2014.075, e2014075.

10. Nagpal S, Recht L. Treatment and prophylaxis of hematologic malignancy in the central nervous system. *Curr Treat Options Neurol.* 2011;13(4):400–412. https://doi.org/10.1007/s11940-011-0128-7.

11. Lazarus HM. Central nervous system involvement in adult acute lymphoblastic leukemia at diagnosis: results from the international ALL trial MRC UKALL XII/ECOG E2993. *Blood.* 2006;108(2):465–472. https://doi.org/10.1182/blood-2005-11-4666.

12. Jabbour E, Thomas D, Cortes J, Kantarjian HM, O'Brien S. Central nervous system prophylaxis in adults with acute lymphoblastic leukemia. *Cancer.* 2010;116(10):2290–2300. https://doi.org/10.1002/cncr.25008.

13. Reman O, Pigneux A, Huguet F, et al. Central nervous system involvement in adult acute lymphoblastic leukemia at diagnosis and/or at first relapse: results from the GET-LALA group. *Leuk Res.* 2008;32(11):1741–1750. https://doi.org/10.1016/j.leukres.2008.04.011.

14. Nagpal S, Recht L. Treatment and prophylaxis of hematologic malignancy in the central nervous system. *Curr Treat Options Neurol.* 2011;13:400–412. https://doi.org/10.1007/s11940-011-0128-7.

15. Shihadeh F, Reed V, Faderl S, et al. Cytogenetic profile of patients with acute myeloid leukemia and central nervous system disease. *Cancer.* 2012;118(1):112–117. https://doi.org/10.1002/cncr.26253.

16. Röllig C, Ehninger G. How I treat hyperleukocytosis in acute myeloid leukemia. *Blood.* 2015;125(21):3246–3252. https://doi.org/10.1182/blood-2014-10-551507.

17. Principe MID, Buccisano F, Soddu S, et al. Involvement of central nervous system in adult patients with acute myeloid leukemia: incidence and impact on outcome. *Semin Hematol.* 2018;55(4):209–214. https://doi.org/10.1053/j.seminhematol.2018.02.006.

18. Cervantes GM, Cayci Z. Intracranial CNS manifestations of myeloid sarcoma in patients with acute myeloid leukemia: review of the literature and three case reports from the Author's institution. *J Clin Med.* 2015;4(5):1102–1113. https://doi.org/10.3390/jcm4051102.

19. Avni B, Koren-Michowitz M. Myeloid sarcoma: current approach and therapeutic options. *Ther Adv Hematol.* 2011;2(5):309–316. https://doi.org/10.1177/2040620711410774.

20. Naydenov AV, Taylor LP. Leptomeningeal carcinomatosis in chronic lymphocytic leukemia: a case report and review of the literature. *Oncologist.* 2019;24(9). https://doi.org/10.1634/theoncologist.2018-0619.

21. Moazzam AA, Drappatz J, Kim RY, Kesari S. Chronic lymphocytic leukemia with central nervous system involvement: report of two cases with a comprehensive literature review. *J Neurooncol.* 2012;106(1):185–200. https://doi.org/10.1007/s11060-011-0636-z.

22. Hanse MC, van't Veer MB, van Lom K, van den Bent MJ. Incidence of central nervous system involvement in chronic lymphocytic leukemia and outcome to treatment. *J Neurol.* 2008;255(6):828–830. https://doi.org/10.1007/s00415-008-0710-4.

23. Bornhauser M, Jenke A, Freiberg-Richter J, et al. CNS blast crisis of chronic myelogenous leukemia in a patient with a major cytogenetic response in bone marrow associated with low levels of imatinib mesylate and its N-desmethylated metabolite in cerebral spinal fluid. *Ann Hematol.* 2004;83(6):401–402. https://doi.org/10.1007/s00277-003-0829-4.

24. Altintas A, Cil T, Kilinc I, Kaplan MA, Ayyildiz O. Central nervous system blastic crisis in chronic myeloid leukemia on imatinib mesylate therapy: a case report. *J Neurooncol.* 2007;84(1):103–105. https://doi.org/10.1007/s11060-007-9352-0.

25. Oshima K, Kanda Y, Yamashita T, et al. Therapy K for. Central nervous system relapse of leukemia after allogeneic hematopoietic stem cell transplantation. *Biol Blood Marrow Transplant.* 2008;14(10):1100–1107. https://doi.org/10.1016/j.bbmt.2008.07.002.

20

Neurological complications of systemic cancer of the head and neck

Shreya Saxena[a], Patrick O'Shea[b], Karanvir Singh[a], Yasmeen Rauf[c], and Manmeet S. Ahluwalia[a]

[a]Miami Cancer Institute, Baptist Health South Florida, Miami, FL, United States, [b]Case Western Reserve Medical School, Cleveland, OH, United States, [c]University of North Carolina, Chapel Hill, NC, United States

1 Introduction

The number of head and neck cancer cases is rising in the United States. In 2021, the prevalence is estimated to be more than 60,000 and represents about 3% of the estimated total new cancer cases.[1] Increased substance abuse and smoking are some of the factors responsible for the increased cases. The oral cavity, pharynx, and larynx are the most frequent primary sites (94%). Alcohol usage and tobacco use, including cigarettes, cigars, and chewing tobacco, are common etiological factors. Human papillomavirus (HPV) infection is associated with about 70% of oropharyngeal cancers.

Head and neck cancer is more frequent in men after the age of 40 with estimated peak incidences in the 5th and 6th decades of life. There is a slight rise in the number of cases among the young and it may be related to nicotine use. The vast majority of cases is related to prolonged exposure to environmental factors, alcohol, and tobacco-related products. There is a possible linkage and causal association with pulmonary, hepatic, vascular, and nutritional disorders related to regular tobacco or alcohol exposure. These disorders are a potential source for neurological manifestations in patients with head and neck cancer. Anatomic, as well as histological, classification of head and neck cancers is done. The primary sites include the salivary glands, oral cavity, pharynx (naso-, oro-, or hypopharynx), larynx, paranasal sinuses, and the nasal cavity. The site of the origin of head and neck cancers mostly is the mucosal surfaces in the head and neck area. These are primarily squamous cell carcinomas or variants (lymphoepithelioma, spindle cell carcinoma, verrucous carcinoma, and undifferentiated carcinoma).[2] Sometimes, squamous cell carcinomas are found in the lymph nodes of the upper neck without evidence of a primary lesion (metastatic squamous neck cancer with unknown or occult primary). Less frequent histological tumor types accounting for < 10% include adenocarcinomas, olfactory neuroblastoma (esthesioneuroblastoma), and others.

Adenocarcinomas are typically seen in the paranasal sinuses, upper nasal cavity, or salivary glands. The olfactory neuroblastoma originates from the olfactory epithelium in the upper nasal cavity. Early detection and screening are associated with better outcomes in most cases. Conventionally, treatment has been based on surgery, radiotherapy, or a combination of both modalities. There are fewer complications with surgery and related modalities. Early-stage disease may be managed with the help of radiation,[3] while later stages of the disease may require a combination of therapeutic modalities. Chemotherapy still plays a significant role in the treatment of head and neck cancers. Recurrence is mainly noted in the cervical lymph nodes. Patients with clinically negative neck lymph nodes can be managed with radiation, functional neck dissection, or close observation and delayed treatment if cervical metastases develop. Patients with clinically positive cervical lymph nodes often require a combination of radical neck dissection and radiation.

The most well-recognized and frequently encountered complications in patients with head and neck cancer are neurological complications and have been reported often in the literature. Whether or not a neurological complication will occur depends on the histological variant and the staging of the tumor. The tumors with the most common neurological manifestations are mostly nasopharyngeal cancers, and a high number of neurological sequelae are reported.[4,5]

Metabolic or paraneoplastic disorders may accompany the neurological complications—the treatment modality is an important determining factor. Posttreatment sequelae may include neurological complications too.

Treatment of the disease is carried out in a multimodality manner with surgery, radiation, chemotherapy, or a combination of those, and itself may have acute or delayed neurological side effects. Advanced cases have a much higher likelihood of the development of neurological sequelae.

Table 1 summarizes neurological complications in head and neck cancer patients due to various etiologies and tumor types.

Various intracellular pathways are implicated in the understanding as well as studying the microenvironment of the tumor. It also helps identify the biomarkers to help stratify and divide patients into meaningful groups. If a patient presents with a new neurological complication, it would be suspicious for a second primary tumor or some treatment complications. The various manifestations also include secondary primary neoplastic diseases of the head and neck, the esophagus, and the lungs.

TABLE 1 Classification of neurological complications in head and neck cancer.

1. Direct	Growth limited to specific region (local)	
	Locoregional spread	– Ipsilateral and contralateral lymphatic spread
		– Perineural spread depending on tumor histologic type
	Distant spread	– Cerebral spread
		– Spinal epidural compression
		– Leptomeningeal spread
2. Indirect	Metabolic	– Wernicke encephalopathy
		– Alcohol withdrawal syndrome
	Paraneoplastic	– Neurologic symptoms
	Treatment complications	– Location of surgical resection may cause problems with speech, sight, language, etc.
		– Chemotherapy-induced peripheral neuropathy
		– Radiation-induced injury like necrosis

TABLE 2 Differential diagnosis of neurological symptoms due to local and locoregional complications in patients with head and neck cancer.

Symptoms	Differential diagnosis	Primary site/tumor
Diplopia (± proptosis)	Orbit invasion	– Frontal, maxillary, ethmoid sinuses; upper nasal cavity (SCC, ADC, ON)
	Cavernous sinus syndrome (III, IV, VI, V1)	– Sphenoid sinus; NSC
	LMD (VI, IV, III, or ICH)	– ADC (ethmoid, parotid); SCC (larynx, oral, oropharyngeal); NSC
Facial pain/ paresthesias/loss of sensations over the part of the face	Nerve V invasion and sinus obstruction	– Maxillary sinus – NSC – Lip (mental nerve neuropathy or "numb chin syndrome") – Nasal and paranasal sinuses
Facial weakness/ facial paralysis	Cranial nerve VII invasion	– Parotid
Otalgia	Cranial n. V, VII, IX, or X invasion	– Any primary site with PNS
	Pterygoid fossa invasion	– NSC, oropharyngeal, oral, parotid
	Otitis media obstruction	– NSC, nasal
Dysphonia[a]	Vocal cord invasion	– Laryngeal (SCC)
	Cranial n. X or recurrent laryngeal nerve invasion	– Any primary site with PNS or locoregional metastases
Odynophagia/ dysphagia/ dysarthria	Pharyngeal muscles and tongue invasion	– Oral, pharyngeal
	Cranial n. IX, X, XII invasion	– Pharyngeal
Headaches, anosmia, personality changes	Frontal lobe syndrome	– Frontal sinus, upper nasal cavity (SCC, ADC, ON); NSC

ADC, adenocarcinoma; ICH, intracranial hypertension; LMD, leptomeningeal disease; NSC, nasopharyngeal carcinoma; ON, olfactory neuroblastoma; PNS, perineural spread; SCC, squamous cell carcinoma; V1, ophthalmic division of V cranial nerve.
[a] *Most common tumors to cause dysphonia with vocal cord palsy are lung and thyroid carcinomas.[6]*

2 Local and locoregional complications

The anatomical location of the tumor determines the symptoms and signs of head and neck cancer. The pattern of spread is unique to each anatomical site.

Table 2 summarizes the differential diagnosis of the neurological symptoms most frequently seen in head and neck cancer patients due to local and locoregional complications.[6] Methods of the spread of the tumor are by local growth, perineural spread, or vascular spread.

2.1 Local growth

The local microenvironment and how the tumor spreads are dependent on the muscular or fascial planes affected and are a hallmark of head and neck cancer. The growth patterns of the tumor in specific anatomic sites may involve cranial nerves located in the vicinity of the primary tumor (e.g., facial weakness in the case of parotid tumors). A hallmark feature is the invasion of the parapharyngeal space which can lead to the spread of the tumor from the base of the skull to the root of the neck. This, in turn, compresses the nearby cranial nerves. Bone and cartilage are not usually invaded until later in the disease. Therefore, invasion of the skull base and brain is often an indication of advanced diseases. Sphenoid sinus tumors or nasopharyngeal cancers are associated with a cavernous sinus syndrome (ophthalmoplegia, loss of corneal reflex, sensory changes in the distribution of the ophthalmic division of the fifth cranial nerve, proptosis, and chemosis). In some cases, a frontal syndrome (anosmia, headache, and personality changes) may also essentially be developed due to the intracranial invasion by frontal sinus or upper nasal cavity tumors.

2.2 Lymphatic spread

The presence of lymph node invasion and lymphatic spread in the neck can be seen. These increase the risk of distant metastases. The frequency of lymph node involvement depends on the histological type plus the size of the tumor, the presence of vascular invasion, and the richness of the network of capillary lymphatics in the primary site.[2] Lymph node involvement may potentially cause neurologic symptoms by compression of neural or vascular structures in the neck.[7] Brachial plexopathy is a rather unusual neurologic complication of head and neck cancer and has been described in the literature.[2,6–8] The involvement of the brachial plexus may occur secondary to cervical lymph node metastases with capsular extension in the late stages of the disease. Differential diagnosis should include local growth or metastasis from primary lung carcinoma, a much more common cause of brachial plexopathy.[9]

2.3 Perineural spread

Perineural invasion (PNI) is defined as the presence of microscopic tumor cells in any of the three layers of the nerve sheath, meaning the endoneurium, epineurium, and perineurium.[10] Perineural tumor spread (PNTS) is, rather, a macroscopic growth of the primary tumor along the nerve and is generally radiographically or clinically apparent.[11]

PNI is reported to have a prevalence rate in head and neck cancer of 25%–80% but may vary according to the patient group or vary by source. It is a significant cause of morbidity and mortality and confers a worse prognosis and increased risk of local recurrence.[11] PNTS can be caused by various head and neck tumors and can cause weakness, pain, numbness, and paresthesias following the distribution of the affected nerve.[12] However, 40% of PNTS patients are asymptomatic at diagnosis, so clinical symptoms are not reliable.[13]

Perineural spread is an important and significant mechanism of tumor invasion in head and neck cancers based on certain histologies such as squamous cell carcinoma and salivary gland tumors of certain types. This is a common as well as a direct form of spread.[11,14]

Dissemination of tumor cells could occur along tissue planes of the neural sheath or within the lymphatics of the perineurium and/or epineurium. The imaging modality of choice is an MRI, and the failure to recognize early imaging findings in asymptomatic patients may lead to treatment failure or recurrence. Also, perineural spread[13] may indicate the need for more aggressive therapy, such as a wider surgical resection or an expansion of the radiation field.

The most typical imaging findings will include foraminal enlargement, foraminal destruction, nerve enlargement, obliteration of various fat planes and enhancement, neuropathic atrophy, and replacement of the trigeminal subarachnoid cistern with soft tissues.[15]

3 Distant metastases

The incidence of distant metastases in patients with head and neck cancer is less when compared to other malignancies.[16] The risk of distant metastases is related to the histologic type, location, and size of the primary tumor and neck stage and location of the involved lymph nodes.

The risk happens to be < 10% for N0–N1 disease and approximately 30% for N3 disease and N1–N2 nodes below the level of the thyroid notch. The risk varies based on the stage, etc., of the disease.

Most distant metastases occur during the first 3 years from diagnosis.[2] Treatment should be palliative for distant metastases. The most common site of distant metastases in head and neck cancer is the lung accounting for about > 50%, and almost all distant metastases are associated with lung involvement. It may be difficult to distinguish pulmonary metastases from a secondary lung malignancy, mainly if solitary. Bone metastases have been observed but are significantly less common (22%). Liver, skin, mediastinum, intracranial, and bone marrow metastases occur rarely but are mentioned in the literature.

3.1 Cerebral metastases

Head and neck cancers are uncommonly associated with intracranial metastases. The frequency has been reported as 3% of all intracranial metastases in a large autopsy study in patients with systemic cancers.[17] A direct cranial extension is more frequent, mainly in the case of nasopharyngeal carcinoma and some other types of head and neck cancers. The incidence of brain metastases seems to be higher in certain tumor types like neuroblastoma of the olfactory epithelium.[18,19]

The possibility of intracranial metastases from a second primary of the lung should be considered. Single brain metastasis is dealt with mainly by a surgical option or a combination of two or more modalities. Radiation therapy prolongs survival, although the overall prognosis is poor. Intracranial metastases are often associated with locoregional recurrence.

3.2 Spinal metastases

Metastases in the spinal epidural space are rarely seen in head and neck cancers as the incidence of hematogenous spread is low. The incidence has been reported as less than 2% or less.[20]

As with intracranial metastases, if an epidural metastasis is diagnosed in a patient without any other evidence of active disease, the possibility of a second primary neoplasm should be considered. Treatment outcome with high-dose steroids and radiation therapy seems similar to that observed with other carcinomas.[6] Surgical decompression may be considered as a viable option in selected patients with unstable spine, no improvement after radiotherapy, or in whom survival is expected to be more than 6 months.[21] Locoregional lymphatic spread of disease is also a potential etiology for the nerve root and/or spinal cord compression.[7] Palliative radiation may be difficult if the neck was previously irradiated and may have a limited effect due to the presence of radioresistant disease. Surgery is not feasible if there is diffuse recurrence in the neck, and it may be technically difficult in patients with previous surgery and/or radiotherapy to the same area.

4 Leptomeningeal carcinomatosis

Leptomeningeal carcinomatosis is a rare complication of head and neck cancer. The incidence has been estimated at about 1%–2% and varies in different age-groups and patient cohorts.

Perineural invasion and direct extension through the skull base are the predominant routes of the spread of malignant cells to the meninges; hematogenous spread is less frequent.[22,23]

The most common clinical finding is multiple cranial nerve dysfunction and is not seen very commonly. Spinal cord and spinal root involvement are uncommon. Diagnosis is based on clinical suspicion, CSF cytology, and imaging studies. The response to the treatment with intrathecal methotrexate, a chemotherapy agent with known activity against head and neck malignancies, needs further study.[24] Radiotherapy is indicated for the symptomatic or bulky disease. Systemic chemotherapy may also have a role in treating leptomeningeal disease.[18]

4.1 The nonmetastatic complications

The other types of complications include metabolic, paraneoplastic, etc. Nonmetastatic complications also include neurotoxicity due to the chemotherapy regimens and drugs. Sometimes, this becomes a dose-limiting factor in cancer treatments. Some alkaloid derivatives can result in toxicity signs like loss of ankle reflexes and others. Most of the toxicities are age and dose-dependent as well.[24]

4.2 Metabolic complications

4.2.1 Wernicke encephalopathy

Wernicke encephalopathy is a neurological disorder resulting from a deficiency of vitamin B1 (thiamine). Clinical manifestations include ophthalmoplegia, ataxia, and altered mental status.

The diagnosis is clinical, and differentials include metabolic abnormalities[25] and leptomeningeal disease. Most of the manifestations are generally reversible with the administration of vitamin B1 (100 mg i.v. daily until the patient resumes a regular diet). Nutritional support and counseling are essential.[26] Prophylactic supplementation with vitamin B1 must be started in patients at risk before the administration of I.V. glucose solutions. Other complications include alcohol withdrawal syndrome and may vary from a spectrum of mild-to-severe symptoms.[27]

4.3 Paraneoplastic syndromes

Squamous cell carcinoma and some histologic types. Neurological manifestations and chemotherapy-induced toxicity are noted. Confusion, hyperreflexia, and other such manifestations are commonly encountered (Table 3).

TABLE 3 Paraneoplastic syndromes in patients with head and neck cancer with neurological manifestations.

Neurological syndrome	Primary tumor type	Number of cases noted	Author (year)
Eaton-Lambert	Small cell neuroendocrine carcinoma of the larynx	1	Medina et al. (1984)
Dermatomyositis	Cancer of nasopharynx	10	Teo et al. (1989)
Ataxia	Cancer of larynx	1	Garcia et al. (1998)
Sensory neuropathy	Pharynx and tonsil SCC	1	Pericot et al. (2001)
Dermatomyositis	Cancer of nasopharynx	5	Mebazaa et al. (2003)
Dermatomyositis	Cancer of nasopharynx	1	Wang et al. (2003)
Dermatomyositis	Cancer of nasopharynx	1	Martini et al. (2005)
Dermatomyositis	Cancer of nasopharynx and cancer	2	Botsios et al. (2003)
Anti-Hu encephalomyelitis	Cancer of larynx	1	Baijens et al. (2006)

4.4 Complications associated with treatment

4.4.1 Surgery

The spinal accessory nerve (SAN) and its surrounding connections, often referred to as the spinal accessory nerve plexus, are an essential neurovascular pathway for shoulder and arm range of motion (ROM). Surgical neck dissection can negatively impact the SAN plexus, either directly or through causation of atherosclerosis of vessels supplying the plexus.[5] Eickmeyer et al. showed that when neck dissection was "nerve sacrificing," the ROM of the shoulder was lower than in "nerve-sparing" neck dissection. Both surgeries led to a lower ROM than patients undergoing no neck dissection. Decreased ROM in the affected shoulder can lead to a decrease in the quality of life for head and neck cancer survivors, including both physical and psychosocial detriment.

The most commonly injured nerve is the SAN though injuries to the vagus and hypoglossal nerves are also noted. Neuropathic pain can be seen after neck dissection from the interruption of the sensory branches.[28] Injury to the sympathetic fibers along the carotid artery will cause Horner's syndrome.[29] Lesions in the mediastinum or pulmonary apex may cause a preganglionic Horner's syndrome. Frey's syndrome, also known as auriculotemporal syndrome, is gustatory sweating and flushing. It is a well-known complication of parotid surgery and neck

dissection. It is due to iatrogenic injury to the auriculotemporal nerve.[30] As it pertains to head and neck cancer, there are specific side effects likely caused by radiation fibrosis in the common areas for head and neck tumors. The common sequelae seen in head and neck cancer survivors following radiation are dysphagia, dysphonia, dysgeusia, trismus, hearing loss, neuropathy, plexopathy, vasculopathy, and dystonia.[31,32] About 3%–5% of patients may suffer from a life-threatening emergency due to a carotid artery injury.[33,34] Severe neurological deficits[35] are seen in 16% to over 50% of patients who initially survive. The reported mortality varies from 3% to over 50%.[36] Bilateral radical neck dissection also results in intracranial hypertension with papilledema and visual symptoms due to the removal of both jugular veins.[37,38] This complication is also seen in unilateral neck dissection.[39]

4.4.2 Treatment/exercises

Swallowing exercises have been shown to be beneficial for long-term outcomes when dealing with dysphagia or mucositis when done before or during irradiation of the head and neck.[40,41] Working with a speech-language pathologist through the process can be very beneficial to head and neck cancer patients.[31]

4.4.3 Radiotherapy-linked complications

Radiation has long been a cornerstone treatment for head and neck cancer, often used with adjuvant surgery or chemotherapy. However, radiation-induced toxicity is a major cause of disability in head and neck cancer survivors.[31] Radiation is known to induce the creation of hydroxyl radicals, which is beneficial to patients because it damages the rapidly dividing tumor cells (Table 4).

However, radiation damage often extends to surrounding structures, leading to damaging side effects for head and neck cancer survivors.[42] Proinflammatory

TABLE 4 Classification of radiation therapy complications.

Acute complications	1. Lhermitte's sign (occurrence of electrical shock-like paresthesias which radiate from the neck to the lower spine and the extremities on flexion of the neck in patients) 2. Seizures 3. Loss of smell 4. Loss of taste 5. Transient cognitive impairment 6. Transient hearing loss 7. Brachial plexopathy 8. Paresthesias
Late complications	1. Focal brain/spine radionecrosis 2. Brainstem encephalopathy 3. Progressive myelopathy 4. Spinal hemorrhage 5. Lower motor neuron syndrome 6. Malignant nerve sheath tumors

states develop in radiated tissues, leading to a positive feedback loop of fibrosis and sclerosis in tissues and the providing microvasculature.[43] Over time, the fibrotic tissue plays a role in decreased function and adverse effects, as does insufficient neurovasculature, leading to atrophy.

Effects of radiation are divided into acute (during or immediately after treatment), early delayed (weeks after up to 3 months), and late delayed (more than 3 months following treatment).[3] There are not any well-established acute effects of radiation treatment in head and neck cancer, but delayed effects have been well documented.

Lhermitte's sign: Lhermitte's sign is an effect of cervical spinal cord radiation seen in the early delayed range.[44] Due to the location of head and neck cancer, cervical spinal cord radiation becomes a necessity at times, which can lead to this effect. Lhermitte's sign is characterized by a transient, short, shock-like feeling transmitting from the neck to the extremities. It is precipitated by flexion of the neck. If seen in the early delayed phase or up to 6 months, it often resolves on its own and requires no treatment. If seen in the late delayed phase or after a year, it could be a sign of more severe radiation myelopathy or spinal cord damage. The incidence of Lhermitte's sign in head and neck cancer increases with radiation to the cervical spinal cord.

Delayed late effects of radiation are a product of radiation fibrosis. Factors associated with a greater risk of radiation fibrosis include other treatment modalities (such as surgery/chemotherapy) being used with radiotherapy, large-volume or dose radiotherapy, and unusually high dose per fraction regimens.[45] Radiation fibrosis takes months to years to develop and is progressive but variable in its course, often progressing slowly.[43]

In addition to SAN damage from surgical neck dissection, radiation fibrosis can also cause damage to the SAN.[43] Radiation in head and neck cancer also can affect cervical nerve roots and the cervical plexus, leading to cervical dystonia, marked by pain, spasms, and contracture of the affected neck and shoulder muscles. Muscles affected include the sternocleidomastoid, scalenes, and trapezius.[33]

As it pertains to head and neck cancer, there are specific side effects likely caused by radiation fibrosis in the common areas for head and neck tumors. The common sequelae seen in head and neck cancer survivors following radiation are dysphagia, dysphonia, dysgeusia, trismus, hearing loss, neuropathy/myelopathy/plexopathy, vasculopathy, and dystonia.[31,32] These are all related to a similar damage and fibrosis created by radiation, but symptoms differ based upon the location of the tumor and the type of structures/nerves involved.

4.4.4 Dysphagia

Dysphagia is often seen in radiation-treated head and neck cancer, and it can be separated into early radiation injury and delayed radiation injury.[42] Early radiation injury is thought to be due to inflammation cascades and reactive oxygen species (ROS), which cause direct damage to structures involved in swallowing. Swallowing is often within normal limits, but when dysphagia is present, it is often due to various functional alterations, such as reduced retraction of the base of the tongue, poor epiglottic retroflexion, reduced laryngeal elevation, delay in pharyngeal transit, and/or poor coordination of swallowing muscles.[37,42,46] These issues can be accompanied by mucositis in the oral and pharyngeal cavities, causing pain and difficulty eating.[47]

Delayed radiation injury can be due to persistent effects of early damage, such as infection or an ulcer following inflammation of the mucosa.[42] Delayed injury is due to radiation fibrosis and is not necessarily preceded by early damage. Radiation-induced fibrosis is discussed earlier in this chapter.

Postradiation dysphagia can be due to early or delayed radiation injury. Early xerostomia and mucositis have been linked to later dysphagia 6–12 months posttreatment.[48] Dysphagia can also be present much later (2 + years) with no early signs of damage, presumably due to fibrosis and/or atrophy.[49,50] Radiation-caused ROS are able to damage the muscle tissue, particularly muscle with fewer mitochondria, as they exhibit worse ROS removal capacity. Further, inflammation and fibrosis from radiation can alter nerve transmission. This can lead to changes in swallowing reflexes and control and can ultimately lead to dysphagia.[42] Research has shown that limiting radiation dosage to certain structures in the head and neck can decrease both early and delayed radiation injury. Further research is attempting to elucidate how newer radiation techniques (IMRT, CyberKnife, and brachytherapy) can reduce dosage and minimize this damage.[42] However, these damaging outcomes of radiotherapy can be combated. Swallowing exercises have been shown to be beneficial for long-term outcomes when done before or during irradiation of the head and neck.[40,41] Working with a speech-language pathologist through the process can be very beneficial to head and neck cancer patients.[31]

5 Medical therapy

Chemotherapy and medical oncology remain a significant part of managing head and neck cancer patients either as single or combination drugs.[51] Despite various treatment modalities included nowadays, medical oncology and chemotherapy remain the mainstay of treatment and management. Cisplatin is one of the most important and efficacious agents. The most effective chemotherapy combinations are cisplatin with fluorouracil and paclitaxel or docetaxel with cisplatin or carboplatin.[2]

5.1 Cisplatin

The neurotoxicity associated with cisplatin use can cause ototoxicity, neuropathy and, in some cases, focal encephalopathy. Peripheral neuropathy is a dose-limiting toxicity and may occur after cumulative dosages of 300–600 mg/M^2. Sometimes, the dorsal root ganglia may be affected causing sensory neuropathy. Damage to the large sensory fibers results in tingling, numbness, loss of deep tendon reflexes, etc. Cisplatin-induced neuropathy may be reversible if the healing or recovery time is more than 1 year. Symptoms may even include sensory ataxia and an impaired sense of vibration. Cisplatin-induced neuropathy can be reversed if longer reversible or healing periods are allowed.[52]

5.2 Carboplatin

The reported cases of peripheral neuropathy are around 5% of the total patients on this therapy.

The incidence is significantly higher in people over 65 years of age. Toxicity is more in the high-dosage regimens.[53]

5.3 Paclitaxel and docetaxel

Taxanes are a group of drugs associated with sensory axonopathy and impaired paresthesias of the hand and feet,[54] which are most commonly encountered. Sensory neuropathy should be a dose-limiting toxicity.[55]

5.4 Fluorouracil (5-FU)

5-FU may cause sequelae like neurological toxicity and, in significantly high dosages, cerebellar syndrome. Sometimes, the deficiency of the DPD enzyme can result in severe 5-FU toxicity symptoms.

5.5 Immunotherapy

Patients with malignancies of the head and neck at an advanced stage treated with immune checkpoint inhibitors are at a much higher risk of immune-related neurological complications noted in the literature review.[59] Vigilance is essential at all stages of treatment and, in fact, even after treatment is completed. In some cases, there is a "double vulnerability" of the nervous system of a patient with head and neck cancer due to the paraneoplastic effect of the cancer cells and the neurotoxicity of the chemotherapeutic agent

as well.[59,60] Enhanced T-cell activation leads to autoimmune or other neurological complications affecting the neuromuscular junctions, neuronal routes, spinal cord, and even the brain.[61] The FDA had recently approved pembrolizumab for the first line of treatment of head and neck cancers.[61] It is extremely common for a patient with head and neck cancers to be at a greater risk of developing neuropathies, especially neuronal loss leading to sensorineural hearing loss and dysgeusia or dysosmia.[59,60] Radiation-induced cranial neuropathy is another common complication of head and neck cancers. Contouring of both the brachial and cranial nerve bundles can be useful for the prevention of radiation-induced brachial plexopathy and radiation-induced cranial neuropathy. Numerous ongoing clinical trials focus on the modulated management of radiation-induced peripheral neuropathy.[62]

6 Syncope in patients with head and neck cancer

When a patient presents with a transient loss of consciousness to the clinic, a detailed neurological consultation is a necessity. There is a broad differential diagnosis, including postural hypotension, metabolic derangements, or seizures. The patient requires a detailed physical and neurological examination, EKG, and electrolytes to understand the reason behind the loss of consciousness.[57] Once a cause is ascertained, more investigations like exercise treadmill tests and carotid ultrasounds are essential.[58] Recurrent syncopal attacks, if arising, would need hospitalization and proper workup and further treatment measures. Head and neck cancer-related syncope has its hallmarks like sweating, pallor, and bradycardia, and some of these events may be preceded by blacking out.[31,42] The patient may report upper or lower neck pain or head pain and a profound attack of hypotension. Three main types of syncopal attacks in patients with head and neck cancer with distinctive pathophysiology and manifestations are described: carotid sinus syndrome, glossopharyngeal neuralgia-asystole syndrome, and parapharyngeal space syndrome.[54]

In addition to these, the potential and significant cause of presyncope is the baroreflex failure secondary to surgery. Table 5 summarizes some of the reported characteristics of these syndromes, which may help with differential diagnosis and directly in choosing the right therapy. Treatment will include pharmacological therapy like vasopressors (dopamine and ephedrine), anticholinergics like propantheline or atropine, or the use of carbamazepine. Cardiac pacemakers and cardiac pacing may be needed in some patients. Surgical interventions include endarterectomy, sectioning of the nerve of Hering, and other modalities.

TABLE 5　Syncope in patients with head and neck cancer: proposed pathophysiology and differential diagnosis (D/D).

Syndrome	Pathophysiology	Distinctive symptoms
Carotid sinus	Invasion or pressure by the tumor on carotid artery bifurcation	– Carotid sinus massage/compression – Reduced heart rate and blood pressure
Glossopharyngeal neuralgia	Spontaneous afferent discharges in the cranial nerve IX	– Triggering factors (touch, swallowing, temperature, or taste in the posterior oropharynx) – Brief attacks of severe pain in the neck or throat, radiating to the ear, jaw, temple or occiput, preceding or accompanying syncope – Bradycardia and/or hypotension, asystole
Parapharyngeal region	Tumor compressing the parapharyngeal space with irritation of cranial n. IX	– No triggering factors – Not associated with pain – Episodes are more frequent, longer, and severe – Bradycardia and/or hypotension
Baroreflex failure	Surgery/radiotherapy	– Spontaneous or triggered by emotional and physical stress – Labile hyper/hypotension (sudden pressor and depressor episodes), brady- or tachycardia, headaches, diaphoresis, orthostatic hypotension

References

1　Ries L, Melbert D, Krapcho M, et al. *SEER Cancer Statistics Review, 1975–2005.* Bethesda, MD: National Cancer Institute; 2008:2999.

2　Mendenhall WM, Million RR. Elective neck irradiation for squamous cell carcinoma of the head and neck: analysis of time-dose factors and causes of failure. *Int J Radiat Oncol Biol Phys.* 1986;12(5):741–746.

3　Cross NE, Glantz MJ. Neurologic complications of radiation therapy. *Neurol Clin.* 2003;21(1):249–277.

4　Turgman J, Braham J, Modan B, Goldhammer Y. Neurological complications in patients with malignant tumors of the nasopharynx. *Eur Neurol.* 1978;17(3):149–154.

5　Leung S, Tsao S, Teo P, Foo W. Cranial nerve involvement by nasopharyngeal carcinoma: response to treatment and clinical significance. *Clin Oncol.* 1990;2(3):138–141.

6　Moots P, Wiley R. Neurological disorders in head and neck cancers. *Neurol Dis Ther.* 1995;37:353.

7　Mendes R, Nutting C, Harrington K. Residual or recurrent head and neck cancer presenting with nerve root compression affecting the upper limbs. *Br J Radiol.* 2004;77(920):688–690.

8　Kori SH, Foley KM, Posner JB. Brachial plexus lesions in patients with cancer = 100 cases. *Neurology.* 1981;31(1):45.

9　Wittenberg KH, Adkins MC. MR imaging of nontraumatic brachial plexopathies: frequency and spectrum of findings. *Radiographics.* 2000;20(4):1023–1032.

10　Liebig C, Ayala G, Wilks JA, Berger DH, Albo D. Perineural invasion in cancer: a review of the literature. *Cancer.* 2009;115(15):3379–3391.

11　Bakst RL, Glastonbury CM, Parvathaneni U, Katabi N, Hu KS, Yom SS. Perineural invasion and perineural tumor spread in head and neck cancer. *Int J Radiat Oncol Biol Phys.* 2019;103(5):1109–1124.

12　Agarwal M, Wangaryattawanich P, Rath TJ. Perineural tumor spread in head and neck malignancies. *Semin Roentgenol.* 2019;54(3):258–275.

13　Moonis G, Cunnane MB, Emerick K, Curtin H. Patterns of perineural tumor spread in head and neck cancer. *Magn Reson Imaging Clin N Am.* 2012;20(3):435–446.

14　Agarwal M, Wangaryattawanich P, Rath TJ. Perineural tumor spread in head and neck malignancies. In: *Paper Presented at: Seminars in Roentgenology;* 2019.

15　Ojiri H. Perineural spread in head and neck malignancies. *Radiat Med.* 2006;24(1):1–8.

16　Ferlito A, Shaha AR, Silver CE, Rinaldo A, Mondin V. Incidence and sites of distant metastases from head and neck cancer. *ORL J Otorhinolaryngol Relat Spec.* 2001;63(4):202–207.

17　Posner J. Intracranial metastases from systemic cancer. *Adv Neurol.* 1978;19:579–592.

18　Chamberlain MC. Treatment of intracranial metastatic esthesioneuroblastoma. *Cancer.* 2002;95(2):243–248.

19　Diaz Jr EM, Johnigan III RH, Pero C, et al. Olfactory neuroblastoma: the 22-year experience at one comprehensive cancer center. *Head Neck.* 2005;27(2):138–149.

20　Ampil FL, Nanda A, Aarstad RF, Hoasjoe DK, Chin HW, Hardjasudarma M. Spinal epidural compression in head and neck cancer: report of five cases. *J Craniomaxillofac Surg.* 1994;22(1):49–52.

21　Preciado DA, Sebring LA, Adams GL. Treatment of patients with spinal metastases from head and neck neoplasms. *Arch Otolaryngol Head Neck Surg.* 2002;128(5):539–543.

22　Lee O, Cromwell LD, Weider DJ. Carcinomatous meningitis arising from primary nasopharyngeal carcinoma. *Am J Otolaryngol.* 2005;26(3):193–197.

23　Thompson SR, Veness MJ, Morgan GJ, Shannon J, Kench JG. Leptomeningeal carcinomatosis from squamous cell carcinoma of the supraglottic larynx. *Australas Radiol.* 2003;47(3):325–330.

24　Redman BG, Tapazoglou E, Al-Sarraf M. Meningeal carcinomatosis in head and neck cancer: report of six cases and review of the literature. *Cancer.* 1986;58(12):2656–2661.

25　Brook I. Late side effects of radiation treatment for head and neck cancer. *Radiat Oncol J.* 2020;38(2):84.

26　Berry JA, Miulli DE, Lam B, et al. The neurosurgical wound and factors that can affect cosmetic, functional, and neurological outcomes. *Int Wound J.* 2019;16(1):71–78.

27　Sood S, Quraishi M, Bradley P. Frey's syndrome and parotid surgery. *Clin Otolaryngol Allied Sci.* 1998;23(4):291–301.

28　Kitagawa H, Iwabu J, Yokota K, Namikawa T, Hanazaki K. Intraoperative neurological monitoring during neck dissection for esophageal cancer with aberrant subclavian artery. *Anticancer Res.* 2019;39(6):3203–3205.

29　Kalani MYS, Kalb S, Martirosyan NL, et al. Cerebral revascularization and carotid artery resection at the skull base for treatment of advanced head and neck malignancies. *J Neurosurg.* 2013;118(3):637–642.

30　Brown H. Anatomy of the spinal accessory nerve plexus: relevance to head and neck cancer and atherosclerosis. *Exp Biol Med.* 2002;227(8):570–578.

31　Cohen EE, LaMonte SJ, Erb NL, et al. American Cancer Society head and neck cancer survivorship care guideline. *CA Cancer J Clin.* 2016;66(3):203–239.

32 Strojan P, Hutcheson KA, Eisbruch A, et al. Treatment of late se-quelae after radiotherapy for head and neck cancer. *Cancer Treat Rev.* 2017;59:79–92.

33 Nori P, Kline-Quiroz C, Stubblefield MD. Cancer rehabili-tation:: acute and chronic issues, nerve injury, radiation se-quelae, surgical and chemo-related, part 2. *Med Clin North Am.* 2020;104(2):251–262.

34 Khan MM, Ali H, Kazmi T, Iqbal H. Diagnostic, surgical, and post-operative challenges of neuroendocrine tumors of the neck: clini-cal experience and literature review. *Ann Vasc Surg.* 2017;45:92–97.

35 Botsios C, Ostuni P, Boscolo-Rizzo P, Da Mosto MC, Punzi L, Marchiori C. Dermatomyositis and malignancy of the pharynx in Caucasian patients: report of two observations. *Rheumatol Int.* 2003;23(6):309–311.

36 Barrett TF, Gill CM, Miles BA, et al. Brain metastasis from squa-mous cell carcinoma of the head and neck: a review of the litera-ture in the genomic era. *Neurosurg Focus.* 2018;44(6):E11.

37 Logemann JA, Rademaker AW, Pauloski BR, et al. Site of disease and treatment protocol as correlates of swallowing function in patients with head and neck cancer treated with chemoradiation. *Head Neck.* 2006;28(1):64–73.

38 Krekeler BN, Wendt E, Macdonald C, et al. Patient-reported dys-phagia after thyroidectomy: a qualitative study. *JAMA Otolaryngol Head Neck Surg.* 2018;144(4):342–348.

39 Bashjawish B, Patel S, Kılıç S, et al. Effect of elderly status on postoperative complications in patients with sinonasal cancer. In: *Paper Presented at: International Forum of Allergy & Rhinology*; 2019.

40 Hutcheson KA, Bhayani MK, Beadle BM, et al. Eat and exer-cise during radiotherapy or chemoradiotherapy for pharyn-geal cancers: use it or lose it. *JAMA Otolaryngol Head Neck Surg.* 2013;139(11):1127–1134.

41 Kulbersh BD, Rosenthal EL, McGrew BM, et al. Pretreatment, pre-operative swallowing exercises may improve dysphagia quality of life. *Laryngoscope.* 2006;116(6):883–886.

42 King SN, Dunlap NE, Tennant PA, Pitts T. Pathophysiology of radiation-induced dysphagia in head and neck cancer. *Dysphagia.* 2016;31(3):339–351.

43 Stubblefield MD. Radiation fibrosis syndrome: neuromuscular and musculoskeletal complications in cancer survivors. *PM R.* 2011;3(11):1041–1054.

44 Leung WM, Tsang NM, Chang FT, Lo CJ. Lhermitte's sign among nasopharyngeal cancer patients after radiotherapy. *Head Neck.* 2005;27(3):187–194.

45 O'Sullivan B, Levin W. Late radiation-related fibrosis: pathogene-sis, manifestations, and current management. *Semin Radiat Oncol.* 2003;13(3):274–289.

46 Logemann JA, Pauloski BR, Rademaker AW, et al. Swallowing dis-orders in the first year after radiation and chemoradiation. *Head Neck.* 2008;30(2):148–158.

47 Rademaker AW, Vonesh EF, Logemann JA, et al. Eating ability in head and neck cancer patients after treatment with chemoradia-tion: a 12-month follow-up study accounting for dropout. *Head Neck.* 2003;25(12):1034–1041.

48 van der Laan HP, Bijl HP, Steenbakkers RJ, et al. Acute symptoms during the course of head and neck radiotherapy or chemora-diation are strong predictors of late dysphagia. *Radiother Oncol.* 2015;115(1):56–62.

49 Hutcheson KA, Lewin JS, Barringer DA, et al. Late dysphagia af-ter radiotherapy-based treatment of head and neck cancer. *Cancer.* 2012;118(23):5793–5799.

50 Wall LR, Ward EC, Cartmill B, Hill AJ. Physiological changes to the swallowing mechanism following (chemo)radiother-apy for head and neck cancer: a systematic review. *Dysphagia.* 2013;28(4):481–493.

51 Eickmeyer SM, Walczak CK, Myers KB, Lindstrom DR, Layde P, Campbell BH. Quality of life, shoulder range of motion, and spinal accessory nerve status in 5-year survivors of head and neck cancer. *PM R.* 2014;6(12):1073–1080.

52 Colbert S, Ramakrishna S, Harvey J, Brennan P. Metastases in the cervical spine from primary head and neck cancers: current concepts of diagnosis and management. *Br J Oral Maxillofac Surg.* 2017;55(2):168–172.

53 Capatina C, Ntali G, Karavitaki N, Grossman AB. The manage-ment of head-and-neck paragangliomas. *Endocr Relat Cancer.* 2013;20(5):R291–R305.

54 Plitt A, El Ahmadieh TY, Bindal S, Myers L, White J, Gluf W. Hypoglossal schwannoma of neck: case report and review of liter-ature. *World Neurosurg.* 2018;110:240–243.

55 Baijens L, Manni J. Paraneoplastic syndromes in patients with pri-mary malignancies of the head and neck. Four cases and a review of the literature. *Eur Arch Otorhinolaryngol.* 2006;263(1):32–36.

56 Ghosh-Laskar S, Agarwal JP, Yathiraj PH, et al. Brain metasta-sis from nonnasopharyngeal head and neck squamous cell car-cinoma: a case series and review of literature. *J Cancer Res Ther.* 2016;12(3):1160.

57 Mobley SR, Miller BT, Astor FC, Fine B, Halliday NJ. Prone po-sitioning for head and neck reconstructive surgery. *Head Neck.* 2007;29(11):1041–1045.

58 Postow MA, Sidlow R, Hellmann MD. Immune-related adverse events associated with immune checkpoint blockade. *N Engl J Med.* 2018;378(2):158–168.

59 Fellner A, Makranz C, Lotem M, et al. Neurologic complications of immune checkpoint inhibitors. *J Neurooncol.* 2018;137(3):601–609.

60 Feng S, Coward J, McCaffrey E, Coucher J, Kalokerinos P, O'Byrne K. Pembrolizumab-induced encephalopathy: a review of neu-rological toxicities with immune checkpoint inhibitors. *J Thorac Oncol.* 2017;12(11):1626–1635.

61 Wick W, Hertenstein A, Platten M. Neurological sequelae of cancer immunotherapies and targeted therapies. *Lancet Oncol.* 2016;17(12):e529–e541.

Further reading

62 Van Wilgen CP, Dijkstra PU, van der Laan BF, Plukker JT, Roodenburg JL. Morbidity of the neck after head and neck cancer therapy. *Head Neck.* 2004;26(9):785–791.

21

Neurological complications of gynecological cancers

Susan C. Pannullo[a,b], Zhen Ni Zhou[c], Maricruz Rivera[a], Eseosa Odigie[d], Onyinye Balogun[e], Evan K. Noch[f], Jana Ivanidze[g], Jennifer Moliterno[h], and Eloise Chapman-Davis[c]

[a]Department of Neurological Surgery, Weill Cornell Medicine and New York Presbyterian Hospital, New York, NY, United States, [b]Department of Biomedical Engineering, Cornell University, Ithaca, NY, United States, [c]Department of Obstetrics and Gynecology, Division of Gynecologic Oncology, Weill Cornell Medicine and New York Presbyterian Hospital, New York, NY, United States, [d]Weill Cornell Medicine, New York, NY, United States, [e]Department of Radiation Oncology, Weill Cornell Medicine and New York Presbyterian Hospital, New York, NY, United States, [f]Department of Neurology, Division of Neuro-Oncology, Weill Cornell Medicine and New York Presbyterian Hospital, New York, NY, United States, [g]Department of Radiology, Divisions of Neuroradiology and Nuclear Medicine, Weill Cornell Medicine and New York Presbyterian Hospital, New York, NY, United States, [h]Department of Neurosurgery, Yale School of Medicine, New Haven, CT, United States

1 Introduction

Neurological complications occur in 20%–25% of patients with gynecological cancers.[1] For the purposes of this chapter, gynecological malignancies will include uterine (endometrial and cervical), ovarian, vulvar, vaginal, and fallopian tube cancers, as well as choriocarcinoma.[2] According to the American Cancer Society, approximately 114,000 women were diagnosed and 33,620 women died from gynecological cancer in the United States in 2020.[3] With approximately 58,000 cases diagnosed each year, endometrial cancer is the most common gynecological cancer and the fourth most common cancer in women in the United States.[4]

Central nervous system (CNS) metastases are rare in gynecological cancers. Fewer than 1% of patients with cervical, uterine, and endometrial gynecological tumors will develop CNS metastases; CNS metastases are more common in choriocarcinoma, with an incidence of 10%–20%.[5–7] However, neurological sequelae from local or regional invasion and local metastases or distant metastases involving neural structures can be problematic.[8] In addition, some gynecological cancers can produce neurological signs and symptoms without actual metastases to the central and/or peripheral nervous system structures through well-described paraneoplastic syndromes.

Finally, some treatments used for gynecological cancers, including surgery, radiation, and systemic agents (chemotherapy, targeted therapy, immunotherapy), can also result in neurological complications.

2 Local/regional spread of gynecological cancers to the lumbosacral plexus

The lumbosacral plexus is a neural network responsible for innervating the bladder and lower extremities. The lumbar plexus, formed by the ventral primary rami of L1–4, travels through the posterior psoas muscle and anterior to the L2–5 transverse processes.[9] The sacral plexus, located deeper within the pelvis between the pelvic fascia and piriformis muscle, connects to the lumbar plexus via the lumbosacral trunk. Given their proximity to the lumbosacral plexus, gynecological cancers may spread locally and directly invade or compress nearby neurological structures, producing neurological deficits.[10]

Most patients with metastatic disease to the lumbosacral plexus are present with exquisite pelvic and/or lumbosacral pain. Progressive weakness involving more than one nerve root, which ultimately results in focal muscle group paralysis, is the pathognomonic feature associated with lumbosacral plexus involvement.[11] Often, severe pain will precede neurological symptoms.[1,10] In patients with a history of gynecological cancer, prompt evaluation for locoregional metastatic disease must be considered in the setting of pain and/or neurological signs and symptoms, including motor and/or sensory deficits in the lower extremities and bowel and/or bladder dysfunction.

Metastatic disease may be visible on imaging studies such as computerized tomography (CT) and/or magnetic resonance imaging (MRI). However, these imaging modalities are unable to detect microscopic disease. Therefore, the diagnosis of locoregional spread to the lumbosacral plexus may be based upon a patient's clinical presentation. Irradiation is the cornerstone of treatment for metastatic disease involving the lumbosacral plexus. Although this treatment often leads to pain relief, improvement of neurological function generally does not occur.[12]

3 Distant CNS metastases from gynecological cancers

Gynecological cancers can metastasize to the spine, brain, calvarium, dura, and/or leptomeninges. Neurological symptoms often reflect tumor location, mass effect, and associated edema.

3.1 Spine metastases from gynecological cancers: Symptoms, testing, treatment

Spine metastases in patients with gynecological cancers are uncommon, with an incidence of less than 1%, and they account for less than 3% of all epidural metastases causing spinal cord or cauda equina compression.[13,14] However, back pain and/or motor/sensory changes in a patient with a history of gynecological cancer should be treated as spinal metastatic disease until proven otherwise. Most cancers metastasize to the spine through hematogenous venous spread, more commonly in the thoracic spine, followed by lumbosacral and cervical spine, respectively.[15,16] Pelvic tumors often spread to the lumbosacral spine primarily through Batson's venous plexus, causing neurological symptoms as the first sign of metastatic disease.[13–15] Spinal tumors are classified as extradural, intradural extramedullary, or intradural intramedullary, with the most common being extradural.[15] Intramedullary spinal cord metastases are extremely rare, with an incidence of less than 1% in all cancer types.[15]

The most common presenting symptoms of spinal metastases are back or radicular pain from mechanical instability or nerve compression, followed by motor/sensory changes, gait instability, myelopathy, and/or bladder/bowel dysfunction from compression of the spinal cord, conus medullaris, or cauda equina.[13,15] All patients with suspected spinal metastases should undergo MRI of the entire neuraxis to evaluate the extent of disease in the soft tissues and neural structures and assess for the presence of cord compression (Fig. 1). CT is

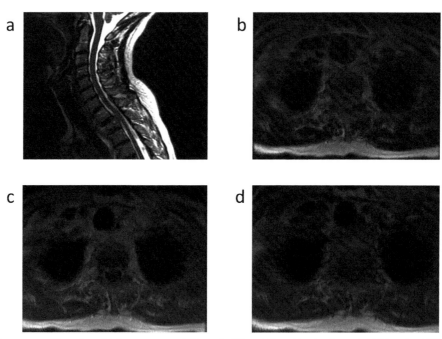

FIG. 1   A 62-year-old woman with endometrial cancer metastatic to the T2 vertebra. Cervicothoracic MRI sagittal T2, axial T2, and pre/post-axial T1 sequences are shown. There is epidural extension causing grade 2 epidural spinal cord compression and a pathologic fracture leading to mildly increased kyphosis. She had no posterior element involvement. The patient underwent a decompressive laminectomy and pedicle screw placement for stabilization.

helpful in the assessment of osseous lesions and to screen pathologic fractures and is further useful for operative planning should a decompression or instrumentation be needed. In the setting of MRI contraindications, CT myelography can be helpful in extent-of-disease evaluation.[15] Fluorodeoxyglucose (FDG) PET/CT can provide valuable additional information, especially in situations where there is a need to differentiate viable tumor from changes related to prior treatment (e.g., radiation necrosis). Once the presence of spinal cord metastases is confirmed, multimodal therapy consisting of radiation, chemo- or immunotherapy, and/or surgical decompression is tailored to each patient based on various factors.[17]

Historically, the management of spinal cord compression involved simple decompressive surgery with laminectomy (i.e., removing enough of the bony elements only to decompress the canal). However, depending on the extent of decompression, decompressive surgery can render patients more mechanically unstable than before surgery. In 2005, Patchell et al. conducted a randomized multiinstitutional study in which patients with spinal cord compression were treated with high-dose steroids and randomly assigned to receive either surgery followed by radiotherapy or radiation alone.[18] Surgical patients underwent spinal stabilization with fusion when deemed mechanically necessary. The authors found significantly more patients were able to maintain or regain ambulation posttreatment in the surgery group compared to those who received radiation alone, 84% and 57%, respectively.[18] The use of corticosteroids and analgesics was also less in the surgical cohort compared to the radiation group. The study was terminated early based on an interim analysis that demonstrated significant benefit and improved quality of life for the surgical decompression followed by radiotherapy group. The surgical group had improved quality of life but there was no significant effect on overall survival. Although these results argue in favor of surgical decompression with radiation, not every patient is a good surgical candidate given the often immune-compromised state and comorbidities associated with having metastatic cancer.[16,19] One study of 647 patients undergoing surgery for spine metastasis demonstrated a complication rate of 32% within 30 days of surgery, of which 12% were a major complication.[16]

Taking into account the complexities of oncology patients, Laufer et al. outlined a comprehensive multidisciplinary approach to the treatment of spinal metastases that includes neurologic, oncologic, mechanical, and systemic evaluation of every patient known as the NOMS framework.[20] Neurological evaluation includes both clinical and radiographic evidence of spinal cord compression. Patients are examined thoroughly for clinical signs of myelopathy, motor/sensory changes, saddle anesthesia, and/or bowel/bladder dysfunction.

Acute onset of these symptoms can indicate a neurosurgical emergency and urgent intervention may be necessary. Corticosteroids are often utilized to decrease spinal cord edema and inflammation at the site of compression, often resulting in a rapid improvement in symptoms. Vecht et al. evaluated the use of high (100 mg) or standard (10 mg) loading dose of dexamethasone followed by 16 mg daily with no difference in outcome.[21]

Epidural spinal cord compression (ESCC) is divided into six categories, with grade 2 and 3 considered high-grade and often resulting in myelopathy.[20] The gold standard for measuring the degree of spinal cord compression radiographically is through axial T2-weighted images of the most severe level (Fig. 1). The presence of high-grade cord compression and myelopathy is not the sole determining factors for surgical intervention. The oncologic portion of NOMS considers the type of metastatic tumor and predicted response to radiation therapy. Patients with high-grade cord compression from a radiosensitive tumor and without mechanical instability are treated with conventional external beam radiation therapy rather than surgery. Alternatively, high-grade cord compression from a radiosensitive tumor and without mechanical instability can be treated with surgical decompression and stereotactic radiosurgery. Surgical decompression is usually reserved for high-grade cord compression; however, surgical stabilization with instrumentation is offered for patients with mechanical instability regardless of ESCC grade.[20]

Mechanical instability is assessed through the Spinal Instability Neoplastic Score (SINS). Six factors are evaluated and assigned points. Location of the tumor at a junctional level receives a higher score compared to location in a rigid portion of the spine. Pain is evaluated as mechanical (pain with lying flat or changing position) or biological (pain at night). Additionally, the effect of the tumor on bone is considered: lytic vs. blastic, effect on spinal alignment, involvement of posterior bony elements, and degree of vertebral body collapse. A low SINS (0–6) is considered stable and a high SINS (13–18) is considered unstable and requires surgical intervention via percutaneous pedicle screws or open instrumentation. An intermediate SINS (7–12) needs to be further evaluated to determine the need for surgical intervention.[20]

Most importantly, NOMS evaluates each patient individually and considers his or her systemic extent of disease. Medical comorbidities, ability to withstand surgery, and oncologic prognosis are considered as essential in treatment planning. Patients with gynecological cancers rarely develop spinal metastases; however, once they do, the prognosis is poor.[13,14] Ultimately, the main objectives are to treat the spine tumor, control pain, and retain ambulation while keeping complications to a minimum. Although these measures may not affect overall survival, they allow the patient to maintain quality of life as long as possible.

3.2 Brain metastases: Symptoms, testing, treatment

Brain metastases from primary gynecological malignancies are extremely rare and associated with a poor prognosis. It has been postulated that carcinoma cells cross the blood-brain barrier (BBB) by disrupting the endothelial cells lining the central nervous system, thereby leading to increased vascular permeability.[22] This mechanism is not unique to gynecological malignancies.

Primary gynecological cancers associated with brain metastasis have a reported estimated incidence of less than 2% in ovarian cancer, 0.4%–1.2% in cervical cancer, and 0.3%–0.9% in endometrial cancer.[23] Recently, the incidence of brain metastases from gynecological malignancies has been rising.[2,24,25] The increased incidence may reflect an improvement in the ability to diagnose metastatic disease as more sensitive imaging modalities have become available, better access to imaging, and increased frequency of surveillance imaging. Additionally, with more effective treatment options available for gynecological malignancies, patients are surviving long enough to develop brain metastases.[24,26]

Of all the neurological complications of gynecological cancers, brain metastases are cited as being the most common.[12] In general, over two-thirds of patients with brain metastases will develop neurological symptoms including headache (40%–50%), focal neurological deficits (30%–40%), and seizures (15%–20%).[27,28] Brain metastases may also be an incidental finding made at the time of routine surveillance imaging. Choriocarcinomas, which account for approximately 35% of brain metastases stemming from a primary gynecological cancer, are associated with a high risk of hemorrhage; these patients may present with acute neurological deficits and symptoms.[8,12,28]

The optimal diagnostic imaging modality for detecting brain metastases in the setting of a primary gynecological cancer is contrast-enhanced MRI. FDG PET/CT and PET/MRI can aid in differentiating viable neoplasm and postradiation change in the postradiotherapy setting (Fig. 2). In most cancers, brain metastases are found at the gray-white matter junction in the supratentorial compartment. Gynecological brain metastases have a predilection for the infratentorial space.[2]

Regardless of location, brain metastases from gynecological malignancies appear radiographically similar to other brain metastases. The use of a susceptibility-weighted MRI may assist with

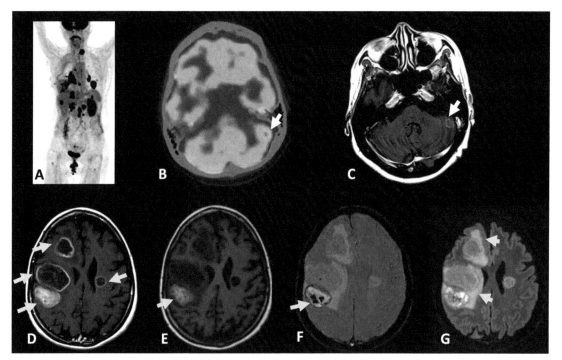

FIG. 2   A 50-year-old woman with metastatic cervical adenocarcinoma who presented for restaging and was found to have multiple brain metastases. Skull base-to-thigh FDG PET/CT MIP image (A) demonstrates extensive metastatic disease including nodal, hepatic, and soft tissue metastases. Axial-fused FDG PET/CT image from the same examination (B) reveals a focus of intense FDG avidity in the partially visualized left cerebellar hemisphere. Subsequent MRI confirms peripherally enhancing metastasis (C) and demonstrates numerous additional metastases for example in the bilateral corona radiata (D), with associated precontrast T1 shortening (E) and susceptibility hypointensity (F), suggestive of intralesional blood products, as well as regional mass effect and confluent surrounding T2-FLAIR hyperintensity (G) compatible with vasogenic edema.

differentiating areas of hemorrhage. Corticosteroids are often implemented to control peritumoral edema associated with brain metastases.[29] However, they should be avoided in patients with choriocarcinomas as steroids have been shown to stimulate their growth.[2,12] The use of anticonvulsants for seizure prophylaxis remains controversial in patients without a documented seizure. According to the American Academy of Neurology, the use of prophylactic anticonvulsants is not recommended as no substantial benefit, which was defined as a risk reduction of at least 26% for seizure-free survival, was noted in a study by Glantz et al.[30] In addition, anticonvulsants can be associated with serious adverse effects that can be life-threatening.[30] Some anticonvulsants, for example, phenytoin, carbamazepine, and phenobarbital, interfere with the cytochrome P450 enzyme complex and can thus affect the metabolism and efficacy of several commonly used chemotherapy agents used in the treatment of gynecological brain metastases.[28] Therefore, anticonvulsants are generally recommended in patients with brain metastases from gynecological cancers only in the setting of a documented seizure.

Overall disease burden must be taken into consideration when treating gynecological cancer brain metastases. Surgical resection is a viable option in those with symptomatic metastatic tumors that are surgically accessible. Tissue diagnosis prior to surgical resection is of utmost importance in patients with a known history of a gynecological malignancy and suspicion for brain metastases. To assess the effect of surgical resection on survival, Patchell et al. randomized patients with brain metastases to either surgery followed by radiation therapy or biopsy alone with radiation.[31] In patients with one metastatic brain lesion, surgical resection followed by radiation resulted in survival benefit with fewer subsequent recurrences and better quality of life compared to those who underwent biopsy followed by radiation.[31]

However, resection is not always feasible, especially if multiple metastases are present, or the location of the metastatic disease and/or the individual's comorbidities preclude safe surgery. In these circumstances, stereotactic radiosurgery with or without whole brain irradiation (WBRT) can effectively improve local disease control and survival when compared with WBRT alone.[28,32] In cases where resection and/or stereotactic radiosurgery are not possible, especially in situations where there are innumerable metastases, WBRT is considered standard of care, conferring a survival of about 18 weeks when used as a single agent treatment modality.[12,33]

Adjuvant systemic chemotherapy is often used upon the completion of surgery and radiation. While tumors are capable of disrupting the BBB, delivery of systemic chemotherapy to brain metastases has remained challenging. Certain chemotherapeutic agents, such as carboplatin, cisplatin, 5-fluorouracil, gemcitabine, temozolomide, and topotecan, have shown modest efficacy when given systemically.[34–37] However, lower response rates are often exhibited in patients who receive chemotherapy after failing radiation treatment for brain metastases.[37] Immune therapy may be a promising future additional option for treatment of brain metastases from gynecological cancers.

3.3 Calvarial, dural, and leptomeningeal metastases: Symptoms, testing, treatment

Symptomatic calvarial, dural, and leptomeningeal metastases from primary gynecological cancers are extraordinarily rare. However, the incidence of leptomeningeal metastases may be rising as a result of improvements in both the diagnosis and treatment of primary gynecological cancers.[2,12] Diagnosis is usually made using brain MRI (Fig. 3). Given the rarity of cerebrospinal fluid (CSF) dissemination of gynecological cancers, tissue sampling via biopsy or CSF analysis for cytology to confirm the suspected diagnosis of brain metastases is warranted. CA-125, a serum marker of ovarian cancer, can be measured in the CSF and has been used to diagnosis ovarian leptomeningeal metastases.[2] Calvarial metastases may be surgically excised, but the overall tumor burden must be considered before subjecting patients to an aggressive, invasive therapy with potential morbidity. Treatment of dural and leptomeningeal metastases consists of palliative radiation, which generally involves stereotactic radiosurgery or WBRT, intrathecal chemotherapy, and/or systemic chemotherapy. Development of leptomeningeal metastases portends a poor outcome; the majority of patients with leptomeningeal metastases typically die within months of diagnosis despite all efforts.[12]

4 Paraneoplastic syndromes due to gynecological cancers

Paraneoplastic neurological syndromes (PNS) are described as neurological complications related to an underlying neoplasm, but not directly caused by tumor invasion or metastatic involvement of the nervous system.[38,39] They can involve the central nervous system, peripheral nervous system, or both. It has been proposed that PNS are the result of an immune-mediated attack against nervous system antigens interpreted as foreign in non-nervous system cancers.[40] The best support for this theory comes from the identification of antineuronal antibodies in patients with these classical syndromes. Unfortunately, the neurological injury from the paraneoplastic syndrome may be more devastating and disabling than the cancer itself, even serving as the cause of death in some patients.[12]

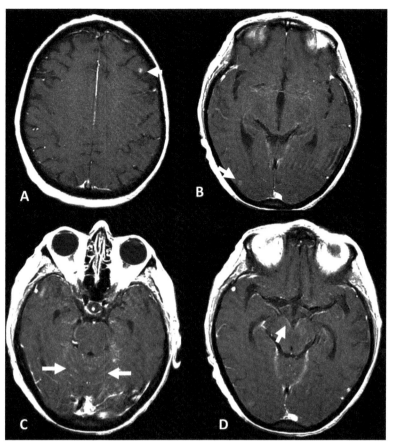

FIG. 3   A 68-year-old woman with endometrial cancer and headaches. MRI demonstrates leptomeningeal enhancing foci in the left frontal (A) and right occipital (B) sulci, as well as in the cerebellar foliae (C) and interpeduncular cistern (D), suspicious for leptomeningeal metastases.

Following small-cell lung cancer, gynecological and breast cancers are the most common solid tumors associated with PNS.[38,41] Although the spectrum of PNS associated with gynecological cancer includes limbic encephalitis, retinal degeneration, subacute sensory neuropathy, cancer-associated retinopathy, and opsoclonus associated with anti-Ri antibody, paraneoplastic cerebellar degeneration (PCD) is by far the most common and representative of PNS, and thus highlighted in this discussion.[12,42] Although PCD can occur with any malignancy, ovarian cancer is among the most common.[43]

Over two-thirds of patients with PCD experience symptoms months to years prior to a cancer diagnosis.[38,44] Although the exact pathogenesis remains unclear, PCD is associated with Purkinje cell loss, as well as inflammatory infiltration of the deep cerebellar nuclei[44,45] (Fig. 4). Clinically, affected patients present with pan-cerebellar dysfunction, including axial and appendicular ataxia, dysarthria, vertigo, downbeat nystagmus, and diplopia, in addition to dizziness, nausea, and vomiting.[12,44,46] Onset is often acute or subacute, with symptoms stabilizing after weeks to months. The gait ataxia can be so severe that patients can ultimately be confined to a wheelchair. Patients may also report an inability to

read or watch television secondary to diplopia and nystagmus, as well as difficulty with articulating words.[12] Any female with otherwise unexplained cerebellar signs and symptoms and evidence of cerebellar degeneration on brain imaging should undergo prompt workup for gynecological cancer, particularly ovarian cancer. Other studies, such as mammography, chest X-ray, CT, MRI, ultrasound, and PET should also be used to rule out other primary tumors.[38]

Antibodies against the cerebellar degeneration-related (CDR2) and CDR2-like (CDR2L) proteins expressed by cerebellar Purkinje cells, known as anti-Yo or anti-PCA1 antibodies, have been demonstrated in both the serum and CSF of patients with gynecological cancer-associated PCD.[47] Both CDR2 and CDR2L may contribute to Purkinje cell death in PCD and therefore may serve as therapeutic targets in this disease.[48,49] Thus, PCD appears to be immune-mediated, but it remains unclear whether it is a cell-mediated or a resultant humoral reaction.[12,50–53] Rojas-Marcos et al. reported that 88% of patients with PCD were positive for anti-Yo antibodies and all with gynecological cancer were positive for these antibodies.[41] Of the ovarian cancer patients who are positive for anti-Yo antibodies, 20% do not have

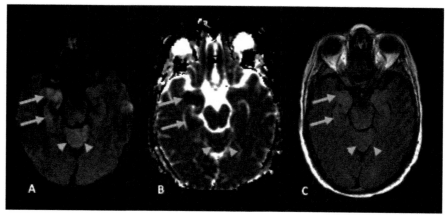

FIG. 4 A 57-year-old woman with malignant mixed Mullerian tumor of the cervix who presented with progressive confusion and altered mental status. MRI demonstrates abnormal diffusion restriction (hyperintensity on DWI sequence, (A), with corresponding hypointensity on ADC imaging, (B)) in the right medial temporal lobe (*arrow*) and the bilateral superior cerebellar hemispheres (*arrowheads*). There is corresponding T2-FLAIR hyperintensity (C). There was no associated contrast enhancement. Further workup confirmed paraneoplastic encephalitis.

neurological complications.[1] Animal models exploring the direct pathogenic effects of these antibodies have been largely inconclusive to date.[49,54–56] However, a recently developed animal model of PCD suggests that pro-inflammatory cerebellum-infiltrating T cells in PCD express pro-inflammatory cytokines, such as interferon gamma (IFN-gamma), that may serve as a therapeutic target in this disease.[57] Interestingly, patients with anti-Yo-associated PCD tend to live longer than those with anti-Hu-associated PCD.[58] In addition, it has been postulated that the immune response associated with PNS confers increased survival in cancer patients, suggesting that the hyperimmune response may play a role in tumor control.[59]

Our lack of understanding of the true pathology of PCD and other PNS has made treatment difficult. While some argue that treating the primary tumor is the best management for PCD, patients rarely improve neurologically.[38,46] This fact is echoed by the persistence of anti-Yo antibodies in the serum of patients years after disease remission.[12] Therapeutic efforts directed toward treating the neurological complications of paraneoplastic syndromes have been largely unsuccessful as well. Plasmapheresis, intravenous immunoglobulin (IVIG), and high-dose corticosteroids have generally shown limited success.[46] Supportive care and aggressive rehabilitation are often needed for a functional recovery.[12,38] Future therapies may be directed against CDR2- or CDR2L-specific cytotoxic T lymphocytes (CTLs) or IFN-gamma production in patients with PCD.[48,49,57]

5 Neurological complications associated with treatment of gynecological cancers

Treatment of gynecological cancer involves a multimodality approach that often includes surgery, radiation, and systemic therapies. Concurrent use of various therapies, such as surgery and radiotherapy, is common. Thus, it can become difficult to determine whether neurological complications are the result of one agent, or if they reflect more complex, synergistic toxic effects.

5.1 Complications of gynecological cancer surgery

One of the most common complications associated with surgical resection of gynecological malignancies is compression neuropathy, which is directly correlated with the length and complexity of the surgical procedure performed.[60] Although most neuropathies are transient and occur in the setting of pressure applied to a nerve from retractors or surgical positioning, postsurgical neuropathies may be permanent if a nerve is transected at the time of surgery. Neuromas, which can develop at the site of an injured peripheral nerve within the abdominal wall, are often diagnosed several weeks to months after the initial surgery. Patients may present with pain, numbness, and dysesthesia surrounding the incision site or in the pelvic or groin area.[61]

Among the compression neuropathies, femoral neuropathies are the most commonly reported neuropathy associated with gynecological surgeries.[60] They may occur in the setting of improper surgical positioning, compression from the use of surgical retractors, or from surgical dissection.[60] Additional neuropathies associated with gynecological surgery include those involving the genitofemoral, iliohypogastric, ilioinguinal, obturator, and peroneal nerves.[60] Injury to the lateral femoral cutaneous nerve due to external compression from a pelvic mass or retractor placement at the time of surgery can result in a rare complication called meralgia paraesthetica, which is characterized by pain and lateral thigh paresthesias.[60]

Extensive workup for postsurgical neuropathies is generally unwarranted when the diagnosis is suspected based upon the clinical history and findings. However, additional testing, including MRI or electrophysiologic studies, such as electromyography (EMG) or nerve conduction studies, may be indicated when symptoms have not improved and the diagnosis of neuropathy is unclear, to rule out recurrent disease.[62]

Treatment of neuropathies consists of symptom management and supportive care. Analgesics may be effective for pain control. Agents such as gabapentin, tramadol, and tricyclic antidepressants have also been utilized to treat dysesthesias and paraesthesias.[63] Rarely is additional surgery indicated to resect a postoperative neuroma or repair a transected nerve. A solid understanding of the anatomy of the pelvic organs in relation to surrounding neural structures and careful surgical positioning will help reduce the incidence of neurological complications from gynecological cancers.[64]

5.2 Neurological complications of gynecological radiation therapy

Primary and metastatic gynecological tumors are treated using various modalities including external beam radiation, brachytherapy, and stereotactic radiosurgery. Factors such as tumor location, size, and the proximity of normal organs influence the radiation modality that is selected. Radiation therapy, irrespective of modality, can cause adverse side effects. In some instances, radiation-induced symptoms may be similar to those of the actual tumor, making it difficult to distinguish radiation injury from damage due to local, regional invasion.

Irradiation of primary gynecological tumors in the pelvis can produce lumbosacral plexopathy due to inclusion of the lumbosacral plexus in the radiation field.[12] Radiation-induced plexopathy is generally a painless loss of neurological function, most often characterized by bilateral numbness and/or weakness, in the distribution of the nerve roots of the lumbosacral plexus. This phenomenon is extremely important to differentiate from metastatic disease involving the lumbosacral plexus, which is classically painful and unilateral, and lacking true neurological deficits.[65] Historically, radiation-induced plexopathy is estimated to occur in 1%–9% of women for all types of gynecological cancer and is usually a late effect, occurring, on average, 5 years after treatment.[65,66] Metastatic involvement of the plexus, however, typically has an earlier onset (< 1 year).[65] Nonetheless, imaging of the lumbosacral plexus is recommended in patients with lumbosacral plexopathy following radiation to exclude metastatic disease, although imaging may be limited in its diagnostic sensitivity.[12] EMG may also be helpful, as 50%–70% of radiation-induced lumbosacral plexopathies are associated with myokymia, a classic quivering movement that is nearly pathognomonic for radiation-induced plexopathy. Corticosteroids may be useful in ameliorating symptoms in cases that are not self-limiting.[65]

Due to technological advancements in radiation therapy, there is minimal risk of severe and/or chronic neurological toxicities due to irradiation of primary gynecological tumors in the pelvis. In the PORTEC-3 trial, which compared chemoradiotherapy to radiation therapy alone, there were no cases of motor or sensory neuropathy in the radiotherapy alone arm at 5 years postrandomization.[67] In contrast, 12 patients on the chemoradiotherapy arm (6%) reported Grade 2 sensory neuropathy, 1 patient reported Grade 2 motor neuropathy, and 1 patient experienced other Grade 2 neurological sequelae. In addition, 3 patients reported Grade 3 neurological side effects. Similarly, there were few instances of neurological side effects with radiotherapy alone in the GOG 249 study that compared pelvic radiotherapy with vaginal cuff brachytherapy followed by chemotherapy.[68] The rates of acute Grade 1, 2, and 3 toxicities were as follows: 3.9% vs. 41.7%, 0.7% vs. 6.9%, 0% vs. 1.5%, respectively. There were no Grade 4 or 5 acute neurological side effects.

Neurological complications related to cranial irradiation can, in part, reflect the type of radiation used. For instance, WBRT can result in nonspecific neurological complications, including headache, personality changes, and cognitive deficits. Radiation-related dementia is characterized by a decline in intellectual function and problems with short-term memory, lasting months to even decades after cranial irradiation.[65] Ensuring the total radiation dose is less than 5000 cGy helps decrease these side effects.[65] Although stereotactic radiosurgery minimizes global brain damage, it can result in focal acute, subacute, and chronic complications. Acute complications occur in less than 10% of patients within the first week of treatment and include the exacerbation of preexisting focal neurological deficits, headache, and seizure.[29,65] Lethargy is very common as well. The concurrent use of corticosteroids, however, may decrease the acute neurological complications associated with radiation.

Radiation necrosis is the best-defined late complication of cranial radiation, typically causing symptoms between 4 months and 4 years after radiation in approximately 8%–16% of patients.[65] The pathogenesis of radiation-induced necrosis involves two mechanisms: (1) via changes in the microvasculature, which results in infarction and coagulative necrosis and (2) by directly destroying astrocytes and oligodendrocytes thereby damaging brain tissue, particularly white matter.[69] Radiation necrosis can occur with all modalities of radiotherapy, though the frequency and time course is particularly heightened and accelerated with the more specific radiotherapies, namely stereotactic radiosurgery and brachytherapy.[65] In addition to headache and personality changes, seizures and focal deficits are common and reflect the tumor-like effect radiation necrosis can have. Patients can usually be managed with corticosteroids. Approximately 5%–10% of patients with radiation necrosis, however, require surgical intervention for symptom control.[29]

In addition to relieving symptoms, surgical intervention may also be necessary as it is the only way to reliably determine whether a mass is actually necrosis or recurrence of the previously treated tumor. It is not possible to differentiate reliably between radiation necrosis and tumor on routine CT or MRI, as both can appear as heterogeneously enhancing masses with associated edema. The use of more elaborate imaging, such as MR spectroscopy and positron emission tomography (PET), has gained support and popularity as these techniques provide insight by offering details regarding the metabolism and cellular activity of the areas in question.

Other less common side effects of brain irradiation include cerebral vasculopathies, including accelerated atherosclerosis and stroke (due to radiation-induced arterial stenosis), endocrinopathies (most commonly caused by radiation-induced primary hypothalamic dysfunction), and radiation-induced tumors, including meningiomas, sarcomas and, less frequently, gliomas.[69] The sensory cranial nerves, most notably the optic nerves and chiasm, are extremely sensitive to radiation and can develop neuropathy if not adequately protected from high doses of radiation.[69]

Radiation of the spinal cord, whether in the treatment field for targeting spinal metastases or as a secondary effect of treating the primary gynecological tumor, can lead to neurological complications by two different mechanisms. First, the spinal cord itself can be affected and patients can present with a Brown-Sequard syndrome, affecting half of the spinal cord, with eventual progression to paraparesis over weeks to months.[69] The pathogenesis is thought to be similar to that of radiation necrosis, such that the underlying process is coagulative necrosis of the spinal cord, particularly affecting the white matter. MRI is helpful in the evaluation of cord edema. Alternatively, and more commonly with radiation of pelvic tumors, patients can present with pure lower motor neuron signs, including flaccidity, atrophy, and areflexia after months or years of undergoing radiation. Sensation and bowel and bladder function are normally not affected. Such a constellation of signs and symptoms reflects the direct damage of the anterior horn cells in the gray matter of the spinal cord. These neurological complications usually stabilize after several months and patients are frequently able to retain their ability to walk.[69]

5.3 Neurological complications of gynecological cancer systemic therapy

Many chemotherapeutic agents used in the treatment of gynecological malignancies are associated with the development of distal sensorimotor peripheral neuropathy. Paclitaxel, especially when used in combination with cisplatin or carboplatin, can result in a dose-limiting sensory or sensorimotor neuropathy in up to 60% of patients.[2,69] While the symptoms appear to be dose-limited, some patients will continue to have sensorimotor peripheral neuropathy even with discontinuation of the offending chemotherapeutic agent.[70] Other complications associated with carboplatin, cisplatin, 5-fluorouracil, gemcitabine, temozolomide, and topotecan include headache, confusion, weakness, ataxia, acute encephalopathy, seizures, cranial neuropathy, hearing loss, myelopathy, and cerebellar syndrome.[71]

Recently, newer targeted therapies, such as targeted antibodies and poly (adenosine diphosphate-ribose) polymerase (PARP) inhibitors, are being increasingly used in the treatment of gynecological malignancies.[72,73] However, these therapies are associated with adverse neurological effects. For instance, the use of bevacizumab, a monoclonal antibody targeting human vascular endothelial growth factor A (VEGF-A), is associated with increased risk of hypertension and thrombotic events, which can increase the risk of posterior reversible leukoencephalopathy syndrome.[74,75] The overall incidence of neurotoxicity with PARP inhibitors, such as olaparib, niraparib, and rucaparib, is low; the most common documented neurotoxicities are fatigue and mild dysgeusia.[76,77]

Newer immunotherapies utilize various mechanisms to unleash antitumor immune responses. Neurological symptoms stemming from the use of immune checkpoint blockers, such as pembrolizumab, are typically mild and responsive to steroids; discontinuation of therapy is not often required. However, in rare circumstances, motor and sensory immune-mediated polyneuropathies have been noted.[78–80] Reports of Guillain-Barré syndrome, myasthenia gravis, and peripheral inflammatory neuropathy have been documented in patients treated with these agents.[79] Additionally, demyelinating polyradiculopathy has been noted with pembrolizumab use, an antiprogrammed death (PD)-1 therapy.[80] Neurological toxicities related to immunotherapy use may not appear different from the metastatic disease of the primary gynecological cancer. As a result, differentiation and recognition of these complications are imperative to the proper care and treatment of the patient.

6 Conclusion

Gynecological cancers can be associated with a wide range of neurological complications, either due to the underlying oncologic disease or as a consequence of aggressive therapy. Specifically, patients with gynecological cancers are at risk of tumoral invasion or compression of the lumbosacral plexus, as well as suffering the debilitating effects of classical paraneoplastic syndromes. Brain and spine metastases are becoming a more common manifestation as more effective treatment of the primary disease has enabled patients to survive long enough to develop such metastatic involvement. Multiparametric imaging including CT, MRI, and PET allows for earlier and clearer diagnosis of locoregional spread and distant involvement of neural structures. Treatments for gynecological cancer primary tumors and metastases, such as surgery, radiation, and systemic therapies, including newer targeted approaches and immune modulators, can themselves cause neurological sequelae. It is expected that as these therapeutic interventions continue to develop, their associated neurological complications will become even more apparent. Recognition of the neurological complications of gynecological cancers and their therapies may result in improvement in patient quality of life as long-term survival from these diseases increases.

References

1. Ramchandren S, Dalmau J. Metastases to the peripheral nervous system. *J Neurooncol*. 2005;75(1):101–110.
2. Wen PY, Schiff D. Neurologic complications of solid tumors. *Neurol Clin*. 2003;21(1):107–140. viii.
3. American Cancer Society. *Cancer Facts & Figures 2020*. Atlanta: American Cancer Society; 2020.
4. Braun MM, Overbeek-Wager EA, Grumbo RJ. Diagnosis and management of endometrial cancer. *Am Fam Physician*. 2016;93(6):468–474.
5. Hacker NF, Rao A. Surgical management of lung, liver and brain metastases from gynecological cancers: a literature review. *Gynecol Oncol Res Pract*. 2016;3:7.
6. Nasioudis D, Persaud A, Taunk NK, Latif NA. Brain metastases from gynecologic malignancies: prevalence and management. *Am J Clin Oncol*. 2020;43(6):418–421.
7. Dadlani R, Furtado SV, Ghosal N, Prasanna KV, Hegde AS. Unusual clinical and radiological presentation of metastatic choriocarcinoma to the brain and long-term remission following emergency craniotomy and adjuvant EMA-CO chemotherapy. *J Cancer Res Ther*. 2010;6(4):552–556.
8. Mahmoud-Ahmed AS, Kupelian PA, Reddy CA, Suh JH. Brain metastases from gynecological cancers: factors that affect overall survival. *Technol Cancer Res Treat*. 2002;1(4):305–310.
9. Planner AC, Donaghy M, Moore NR. Causes of lumbosacral plexopathy. *Clin Radiol*. 2006;61(12):987–995.
10. Jaeckle KA, Young DF, Foley KM. The natural history of lumbosacral plexopathy in cancer. *Neurology*. 1985;35(1):8–15.
11. Saphner T, Gallion HH, Van Nagell JR, Kryscio R, Patchell RA. Neurologic complications of cervical cancer. A review of 2261 cases. *Cancer*. 1989;64(5):1147–1151.
12. Abrey LE. Female reproductive tract cancers. In: Schiff D, Wen PY, eds. *Cancer Neurology in Clinical Practice*. Totowa, NJ: Humana Press; 2003:397–403.
13. Abrey LE. Neurologic complications of female reproductive tract cancer. In: Schiff D, Kesari S, Wen PY, eds. *Cancer Neurology in*

Clinical Practice: Neurologic Complications of Cancer and its Treatment. Totowa, NJ: Humana Press; 2008:449–458.
14. Liu A, Sankey EW, Goodwin CR, et al. Postoperative survival and functional outcomes for patients with metastatic gynecological cancer to the spine: case series and review of the literature. *J Neurosurg Spine*. 2016;24(1):131–144.
15. Hussain I, Pennicooke BH, Baaj AA. Introduction to spinal metastases. In: Ramakrishna R, Magge RS, Baaj AA, Knisely JPS, eds. *Central Nervous System Metastases: Diagnosis and Treatment*. Cham: Springer International Publishing; 2020:487–494.
16. Paulino Pereira NR, Ogink PT, Groot OQ, et al. Complications and reoperations after surgery for 647 patients with spine metastatic disease. *Spine J*. 2019;19(1):144–156.
17. Yamada S, Tsuyoshi H, Yamamoto M, et al. Prognostic value of 16α-[18F]-fluoro-17β-estradiol positron emission tomography as a predictor of disease outcome in endometrial cancer: a prospective study. *J Nucl Med*. 2020;62(5):636–642.
18. Patchell RA, Tibbs PA, Regine WF, et al. Direct decompressive surgical resection in the treatment of spinal cord compression caused by metastatic cancer: a randomised trial. *Lancet*. 2005;366(9486):643–648.
19. Yahanda AT, Buchowski JM, Wegner AM. Treatment, complications, and outcomes of metastatic disease of the spine: from Patchell to PROMIS. *Ann Transl Med*. 2019;7(10):216.
20. Laufer I, Rubin DG, Lis E, et al. The NOMS framework: approach to the treatment of spinal metastatic tumors. *Oncologist*. 2013;18(6):744–751.
21. Vecht CJ, Haaxma-Reiche H, van Putten WL, de Visser M, Vries EP, Twijnstra A. Initial bolus of conventional versus high-dose dexamethasone in metastatic spinal cord compression. *Neurology*. 1989;39(9):1255–1257.
22. Stewart PA, Hayakawa K, Farrell CL, Maestro RFD. Quantitative study of microvessel ultrastructure in human peritumoral brain tissue. *J Neurosurg*. 1987;67(5):697.
23. Kim SB, Hwang K, Joo JD, Han JH, Kim YB, Kim CY. Outcomes in 20 gynecologic cancer patient with brain metastasis: a single institution retrospective study. *Brain Tumor Res Treat*. 2017;5(2):87–93.
24. Ogawa K, Yoshii Y, Aoki Y, et al. Treatment and prognosis of brain metastases from gynecological cancers. *Neurol Med Chir (Tokyo)*. 2008;48(2):57–62. discussion 62–53.
25. Robinson JB, Morris M. Cervical carcinoma metastatic to the brain. *Gynecol Oncol*. 1997;66(2):324–326.
26. Wong ET, Berkenblit A. The role of topotecan in the treatment of brain metastases. *Oncologist*. 2004;9(1):68–79.
27. Parker RG, Janjan NA, Selch MT. Intracranial metastases. In: *Radiation Oncology for Cure and Palliation*. Berlin, Heidelberg: Springer Berlin Heidelberg; 2003:29–35.
28. Soffietti R, Cornu P, Delattre JY, et al. EFNS guidelines on diagnosis and treatment of brain metastases: report of an EFNS Task Force. *Eur J Neurol*. 2006;13(7):674–681.
29. Wen PY, Loeffler JS. Brain metastases. *Curr Treat Options Oncol*. 2000;1(5):447–457.
30. Glantz MJ, Cole BF, Forsyth PA, et al. Practice parameter: anticonvulsant prophylaxis in patients with newly diagnosed brain tumors. Report of the Quality Standards Subcommittee of the American Academy of Neurology. *Neurology*. 2000;54(10):1886–1893.
31. Patchell RA, Tibbs PA, Walsh JW, et al. A randomized trial of surgery in the treatment of single metastases to the brain. *N Engl J Med*. 1990;322(8):494–500.
32. Breneman JC, Warnick RE, Albright Jr RE, et al. Stereotactic radiosurgery for the treatment of brain metastases. *Cancer*. 1997;79(3):551–557.
33. Kurtz JM, Gelber R, Brady LW, Carella RJ, Cooper JS. The palliation of brain metastases in a favorable patient population: a randomized clinical trial by the Radiation Therapy Oncology Group. *Int J Radiat Oncol Biol Phys*. 1981;7(7):891–895.

34. Chura JC, Shukla K, Argenta PA. Brain metastasis from cervical carcinoma. *Int J Gynecol Cancer.* 2007;17(1):141–146.

35. Melichar B, Urminská H, Kohlová T, Nová M, Česák T. Brain metastases of epithelial ovarian carcinoma responding to cisplatin and gemcitabine combination chemotherapy: a case report and review of the literature. *Gynecol Oncol.* 2004;94(2):267–276.

36. Cormio G, Gabriele A, Maneo A, Zanetta G, Bonazzi C, Landoni F. Complete remission of brain metastases from ovarian carcinoma with carboplatin. *Eur J Obstet Gynecol Reprod Biol.* 1998;78(1):91–93.

37. van den Bent MJ. The role of chemotherapy in brain metastases. *Eur J Cancer.* 2003;39(15):2114–2120.

38. Santillan A, Bristow RE. Paraneoplastic cerebellar degeneration in a woman with ovarian cancer. *Nat Clin Pract Oncol.* 2006;3(2):108–112.

39. Dalmau J, Rosenfeld MR. Paraneoplastic syndromes of the CNS. *Lancet Neurol.* 2008;7(4):327–340.

40. Posner JB, Dalmau J. Paraneoplastic syndromes. *Curr Opin Immunol.* 1997;9(5):723–729.

41. Rojas-Marcos I, Rousseau A, Keime-Guibert F, et al. Spectrum of paraneoplastic neurologic disorders in women with breast and gynecologic cancer. *Medicine (Baltimore).* 2003;82(3):216–223.

42. Viau M, Renaud MC, Grégoire J, Sebastianelli A, Plante M. Paraneoplastic syndromes associated with gynecological cancers: a systematic review. *Gynecol Oncol.* 2017;146(3):661–671.

43. Vogrig A, Gigli GL, Segatti S, et al. Epidemiology of paraneoplastic neurological syndromes: a population-based study. *J Neurol.* 2020;267(1):26–35.

44. Zaborowski MP, Spaczynski M, Nowak-Markwitz E, Michalak S. Paraneoplastic neurological syndromes associated with ovarian tumors. *J Cancer Res Clin Oncol.* 2015;141(1):99–108.

45. Albert ML, Darnell JC, Bender A, Francisco LM, Bhardwaj N, Darnell RB. Tumor-specific killer cells in paraneoplastic cerebellar degeneration. *Nat Med.* 1998;4(11):1321–1324.

46. Leypoldt F, Wandinger KP. Paraneoplastic neurological syndromes. *Clin Exp Immunol.* 2014;175(3):336–348.

47. Cao Y, Abbas J, Wu X, Dooley J, van Amburg AL. Anti-Yo positive paraneoplastic cerebellar degeneration associated with ovarian carcinoma: case report and review of the literature. *Gynecol Oncol.* 1999;75(1):178–183.

48. Kråkenes T, Herdlevaer I, Raspotnig M, Haugen M, Schubert M, Vedeler CA. CDR2L is the major Yo antibody target in paraneoplastic cerebellar degeneration. *Ann Neurol.* 2019;86(2):316–321.

49. Greenlee JE, Clawson SA, Hill KE, et al. Anti-Yo antibody uptake and interaction with its intracellular target antigen causes Purkinje cell death in rat cerebellar slice cultures: a possible mechanism for paraneoplastic cerebellar degeneration in humans with gynecological or breast cancers. *PLoS One.* 2015;10(4):e0123446.

50. Vialatte de Pémille C, Berzero G, Small M, et al. Transcriptomic immune profiling of ovarian cancers in paraneoplastic cerebellar degeneration associated with anti-Yo antibodies. *Br J Cancer.* 2018;119(1):105–113.

51. Storstein A, Krossnes BK, Vedeler CA. Morphological and immunohistochemical characterization of paraneoplastic cerebellar degeneration associated with Yo antibodies. *Acta Neurol Scand.* 2009;120(1):64–67.

52. Small M, Treilleux I, Couillault C, et al. Genetic alterations and tumor immune attack in Yo paraneoplastic cerebellar degeneration. *Acta Neuropathol.* 2018;135(4):569–579.

53. Yshii L, Bost C, Liblau R. Immunological bases of paraneoplastic cerebellar degeneration and therapeutic implications. *Front Immunol.* 2020;11:991.

54. Graus F, Illa I, Agusti M, Ribalta T, Cruz-Sanchez F, Juarez C. Effect of intraventricular injection of an anti-Purkinje cell antibody (anti-Yo) in a guinea pig model. *J Neurol Sci.* 1991;106(1):82–87.

55. Schubert M, Panja D, Haugen M, Bramham CR, Vedeler CA. Paraneoplastic CDR2 and CDR2L antibodies affect Purkinje cell calcium homeostasis. *Acta Neuropathol.* 2014;128(6):835–852.

56. Tanaka K, Tanaka M, Igarashi S, Onodera O, Miyatake T, Tsuji S. Trial to establish an animal model of paraneoplastic cerebellar degeneration with anti-Yo antibody. 2. Passive transfer of murine mononuclear cells activated with recombinant Yo protein to paraneoplastic cerebellar degeneration lymphocytes in severe combined immunodeficiency mice. *Clin Neurol Neurosurg.* 1995;97(1):101–105.

57. Yshii L, Pignolet B, Mauré E, et al. IFN-γ is a therapeutic target in paraneoplastic cerebellar degeneration. *JCI Insight.* 2019;4(7).

58. Shams'ili S, Grefkens J, de Leeuw B, et al. Paraneoplastic cerebellar degeneration associated with antineuronal antibodies: analysis of 50 patients. *Brain.* 2003;126(Pt 6):1409–1418.

59. Dalmau J, Posner J. Neurological paraneoplastic syndromes. *Neuroscientist.* 1998;4:443–453.

60. Abdalmageed OS, Bedaiwy MA, Falcone T. Nerve injuries in gynecologic laparoscopy. *J Minim Invasive Gynecol.* 2017;24(1):16–27.

61. Ducic I, Moxley M, Al-Attar A. Algorithm for treatment of postoperative incisional groin pain after cesarean delivery or hysterectomy. *Obstet Gynecol.* 2006;108(1):27–31.

62. Herrera-Ornelas L, Tolls RM, Petrelli NJ, Piver S, Mittelman A. Common peroneal nerve palsy associated with pelvic surgery for cancer. *Dis Colon Rectum.* 1986;29(6):392–397.

63. Stubblefield MD, Vahdat LT, Balmaceda CM, Troxel AB, Hesdorffer CS, Gooch CL. Glutamine as a neuroprotective agent in high-dose paclitaxel-induced peripheral neuropathy: a clinical and electrophysiologic study. *Clin Oncol (R Coll Radiol).* 2005;17(4):271–276.

64. Irvin W, Andersen W, Taylor P, Rice L. Minimizing the risk of neurologic injury in gynecologic surgery. *Obstet Gynecol.* 2004;103(2):374–382.

65. Tasdemiroglu E, Kaya A, Bek S, et al. Neurologic complications of cancer: part 2: vascular, infectious, paraneoplastic, neuromuscular, and treatment-related complications. *Neurosurg Q.* 2004;14:133–153.

66. Cross NE, Glantz MJ. Neurologic complications of radiation therapy. *Neurol Clin.* 2003;21(1):249–277.

67. de Boer SM, Powell ME, Mileshkin L, et al. Adjuvant chemoradiotherapy versus radiotherapy alone for women with high-risk endometrial cancer (PORTEC-3): final results of an international, open-label, multicentre, randomised, phase 3 trial. *Lancet Oncol.* 2018;19(3):295–309.

68. Randall ME, Filiaci V, McMeekin DS, et al. Phase III trial: adjuvant pelvic radiation therapy versus vaginal brachytherapy plus paclitaxel/carboplatin in high-intermediate and high-risk early stage endometrial cancer. *J Clin Oncol.* 2019;37(21):1810–1818.

69. Keime-Guibert F, Napolitano M, Delattre JY. Neurological complications of radiotherapy and chemotherapy. *J Neurol.* 1998;245(11):695–708.

70. Loprinzi CL, Lacchetti C, Bleeker J, et al. Prevention and management of chemotherapy-induced peripheral neuropathy in survivors of adult cancers: ASCO guideline update. *J Clin Oncol.* 2020;38(28):3325–3348.

71. Plotkin SR, Wen PY. Neurologic complications of cancer therapy. *Neurol Clin.* 2003;21(1):279–318. x.

72. Tewari KS, Burger RA, Enserro D, et al. Final overall survival of a randomized trial of bevacizumab for primary treatment of ovarian cancer. *J Clin Oncol.* 2019;37(26):2317–2328.

73. Pothuri B, O'Cearbhaill R, Eskander R, Armstrong D. Frontline PARP inhibitor maintenance therapy in ovarian cancer: a Society of Gynecologic Oncology practice statement. *Gynecol Oncol.* 2020;159(1):8–12.

74. Tlemsani C, Mir O, Boudou-Rouquette P, et al. Posterior reversible encephalopathy syndrome induced by anti-VEGF agents. *Target Oncol.* 2011;6(4):253–258.

75. Seet RC, Rabinstein AA. Clinical features and outcomes of posterior reversible encephalopathy syndrome following bevacizumab treatment. *QJM.* 2012;105(1):69–75.

III. Neurological complications of specific neoplasms

76. Shaw HM, Hall M. Emerging treatment options for recurrent ovarian cancer: the potential role of olaparib. *Onco Targets Ther.* 2013;6:1197–1206.

77. Swisher EM, Lin KK, Oza AM, et al. Rucaparib in relapsed, platinum-sensitive high-grade ovarian carcinoma (ARIEL2 Part 1): an international, multicentre, open-label, phase 2 trial. *Lancet Oncol.* 2017;18(1):75–87.

78. Liao B, Shroff S, Kamiya-Matsuoka C, Tummala S. Atypical neurological complications of ipilimumab therapy in patients with metastatic melanoma. *Neuro Oncol.* 2014;16(4):589–593.

79. Thaipisuttikul I, Chapman P, Avila EK. Peripheral neuropathy associated with ipilimumab: a report of 2 cases. *J Immunother.* 2015;38(2):77–79.

80. de Maleissye MF, Nicolas G, Saiag P. Pembrolizumab-induced demyelinating polyradiculoneuropathy. *N Engl J Med.* 2016;375(3):296–297.

CHAPTER

22

Neurological complications of GI cancers

Denise Leung[a], Moh'd Khushman[b], and Larry Junck[a]

[a]Department of Neurology, University of Michigan, Ann Arbor, MI, United States, [b]Department of Hematology-Oncology, The University of Alabama at Birmingham, Birmingham, AL, United States

1 Overview of gastrointestinal cancers

Gastrointestinal (GI) cancers are common, accounting for 28% of cancer deaths.[1] They are led in frequency by colorectal cancer, the third leading cause of cancer death in the United States, and pancreatic cancer, the fourth leading cause (see Table 1).[1] These are followed by gastric and esophageal cancer. Most are adenocarcinomas.

TABLE 1   Estimated new cases and deaths from GI cancers in the United States in 2020.

Type of cancer	# of cases/ year	# of deaths/ year	% of cancer deaths
Total from GI tract	333,680	167,790	27.7
Specific GI sites:			
Esophageal	18,440	16,170	2.7
Stomach	27,600	11,010	1.8
Small intestine	11,110	1700	0.3
Colorectal	147,950	53,200	8.8
Anus, anal canal, anorectum	8590	1350	0.2
Liver and intrahepatic bile duct	42,810	30,160	5.0
Gallbladder and other biliary	11,980	4090	0.7
Pancreas	57,600	47,050	7.8
Other digestive organs	7600	3060	0.5
Total from all sites	1,806,590	606,520	100.0

From Siegel RL, Miller KD, Jemal A. Cancer statistics, 2020. CA Cancer J Clin. 2020;70(1):7–30.

1.1 Esophageal cancer

Esophageal carcinoma usually presents with difficulty in swallowing, often accompanied by weight loss. It occurs with two different histologies, squamous cell carcinoma and adenocarcinoma; both are more common in men.[2] Esophageal cancer rates have been declining over the last decade.

Squamous cell carcinoma occurs throughout the esophagus and is strongly related to alcohol and tobacco use.[2] Squamous cell carcinoma was previously more common but now makes up only 30% of esophageal cancer cases. The initial site of recurrence is typically local or regional.

In contrast, adenocarcinoma predominates in the distal esophagus and is increasing in frequency.[2] It is also linked to alcohol and tobacco use, and also with obesity. Adenocarcinoma can arise from metaplasia associated with gastroesophageal reflux disease (Barrett's esophagus). Recurrence can be distant.

For both types of esophageal carcinoma, surgery, radiation, chemotherapy, targeted therapy, and immunotherapy can all play important roles in treatment.[2] There has been improved treatment over the years and survival rates have improved. The 5-year survival rate has improved from only 5% in the 1970s to 20% in recent years.[2]

1.2 Gastric cancer

Gastric cancer is a larger cause of deaths worldwide than in the United States.[3] Gastric cancer is nearly always adenocarcinoma.[2] Carcinoid tumors and gastrointestinal stromal tumors (GIST) can also start in the stomach but are much rarer. Gastric cancer is twice as common in smokers as in nonsmokers. *Helicobacter pylori* infection is a significant cause of gastric cancer, especially in the distal stomach. Atrophic gastritis, which may be associated with pernicious anemia, is also a risk factor. The incidence of gastric cancer varies with geographic location, thought related to dietary factors and prevalence of *H. pylori* infection.

Hereditary diffuse gastric cancer syndrome, sometimes caused by mutations in the CDH1 gene,[4] is rare, but the lifetime risk of gastric cancer in this population is up to 80%.[2] Women with this syndrome are also at risk of breast cancer. People with BRCA1 or BRCA2 mutations or Li-Fraumeni syndrome may also be at higher risk of gastric cancer.[5–7] Those with Li-Fraumeni syndrome may have early-onset gastric cancer, especially if there is a family history of this cancer.[6]

In North America, the predominant locus of disease has shifted from the distal to the proximal stomach.[2] Gastric cancer often presents with weight loss (due to anorexia, nausea, dysphagia, or early satiety) and abdominal pain. Diagnosis is usually by endoscopic biopsy. CT is used to assess extent of disease. If tumor is localized to the stomach, surgery offers a chance for cure, but most surgically treated patients relapse. Most patients are present in an advanced stage. Metastasis is most common to regional lymph nodes, liver, and peritoneum.

1.3 Hepatocellular cancer

There are several subtypes of hepatocellular carcinoma but identifying the subtype does not change management or prognosis.[2] However, the fibrolamellar subtype is of note because, although rare, it occurs most commonly in women younger than age 35 and has a better prognosis.[2] Most other hepatocellular carcinomas arise in livers affected by chronic liver disease and cirrhosis.[2] Chronic viral hepatitis B and C and alcohol are the most important risk factors because they can cause cirrhosis. Tobacco use also raises this risk. The incidence is much higher in certain sub-Saharan Africa and Southeast Asian countries than in the United States[2].

It usually presents with right upper quadrant pain, abdominal enlargement, or weight loss.[8] Some patients have indirect manifestations such as hypoglycemia, erythrocytosis, hypercalcemia, and hypercholesterolemia. Diagnosis is established by percutaneous biopsy or by characteristic imaging supported by elevated serum alpha-fetoprotein (AFP). Surgical resection or liver transplant is possible only in a small minority. Various ablation or embolization techniques, or radiation can be used in appropriate cases.[9] Chemotherapy is ineffective in advanced disease. The last few years have witnessed several advances in the systemic treatment of patients with hepatocellular carcinoma where immunotherapy and targeted therapy have improved survival outcomes.[10–16] Metastasis is most common to regional lymph nodes, lungs, and bones.

Hepatoblastoma is a rare cancer that develops in children younger than age 4.[2] The tumor cells are much like fetal liver cells. Most cases are successfully treated with surgery and chemotherapy.[2]

1.4 Gallbladder and biliary tract cancer

Carcinomas of the gallbladder and bile ducts (cholangiocarcinoma) are usually adenocarcinomas.[17] Gallbladder cancer is often an unexpected finding at cholecystectomy for cholelithiasis; if the disease is localized, cholecystectomy is often curative. Advanced tumors may present with jaundice or weight loss. It is much more common in certain Asian and South American countries because of the high liver fluke infection rate in those areas. Risk factors include cholelithiasis, chronic inflammation, and chronic infection, including *H. pylori*.[17] Cancers arising from the intra- or extrahepatic bile ducts are uncommon. Surgical therapy can be curative, although radiation and/or chemotherapy can also be used afterward to decrease recurrence risk.[18] Many patients are present with advanced, unresectable disease and this has poor prognosis.[3] Targeted therapies and immunotherapy can be used in select patients. Metastasis occurs to the nodes, liver, and lungs.[2]

1.5 Pancreatic cancer

Pancreatic cancer is responsible for 7% of cancer deaths in the United States.[1] The most important risk factor is tobacco use.[2] Other risk factors include obesity, diabetes, and chronic pancreatitis. Patients with familial pancreatitis caused by mutations in the PRSS1 gene are also at higher risk.[19]

The most common presentations of pancreatic cancer include pain in the upper abdomen radiating to the back, weight loss, and jaundice.[2] The tumor is usually visualized on CT, MR, ultrasound, or cholangiopancreatography. A minority of tumors are resectable with pancreaticoduodenectomy (Whipple procedure). In most cases, tissue diagnosis is achieved by endoscopic ultrasound-guided fine needle aspirate or CT-guided biopsy of primary or metastatic lesions. Surgery, ablation, radiation, chemotherapy, targeted therapy, or immunotherapy can be reasonable considerations for select patients. Pancreatic cancer commonly spreads by invasion of adjacent structures and by metastasis to the liver. The 5-year relative survival rate (all stages combined) is 9%.[2]

1.6 Colorectal cancer

Risk factors for colorectal cancer include colonic polyps, diabetes mellitus, obesity, low physical activity, and diet high in red meats and processed meats.[20] The most common presenting symptoms are abdominal pain, change in bowel habits, hematochezia, or melena, followed by fatigue, anemia, and weight loss. Diagnosis is usually by colonoscopic biopsy. Patients with potentially surgically curable tumors may go straight to resection. Treatment is surgical for localized disease but includes chemotherapy for disease that has spread to lymph nodes and for distant metastatic disease. The most common metastatic sites are regional lymph nodes, liver, lungs, and peritoneum.

1.6.1 Genetic syndromes of colorectal and CNS neoplasms

Historically, Turcot's syndrome applies to families with colorectal cancer and CNS neoplasms.[21,22] It is associated with two different types of germ-line defects that increase the risk of colorectal cancer. Both are autosomal dominant.

The first type is familial adenomatous polyposis (FAP), which includes a variety of germ-line mutations in the *adenomatous polyposis coli (APC)* gene, all of which lead to a truncated protein.[23] The mutation location within the gene is associated with the age of cancer onset, degree of cancer risk, survival, and extracolonic manifestations. Affected individuals develop multiple adenomatous polyps in their childhood and teens and have nearly a 100% incidence of colorectal cancer. The average age of diagnosis is 39, which is much younger than patients without FAP. The young age combined with the sheer number of adenomas confers the increased colorectal cancer risk; the adenomas themselves are not intrinsically more malignant than those that are sporadic. Prophylactic colectomy is recommended in identified gene carriers with classic FAP. FAP is responsible for approximately 1% of colorectal cancer. Dependent upon the mutation site in the APC gene, some families share additional features, including desmoid tumors, thyroid tumors, hepatoblastoma, skin cysts, pigmented ocular fundus lesions, osteomas, and osteosclerotic jaw lesions (Gardner's syndrome). Patients with FAP have a low but increased risk of cerebellar medulloblastoma (median age of onset, 14 years; range, 5–26); those so affected are considered to have Turcot's syndrome.[23]

Another type of germ line mutation leading to Turcot's syndrome is hereditary nonpolyposis colorectal cancer (HNPCC) or Lynch syndrome.[24] The genetic defect in this syndrome occurs in a DNA mismatch repair gene, several of which have been identified (*MLH1, MSH2, MSH6, PMS2, EPCAM*), leading to DNA replication errors, microsatellite instability, and a predisposition to colorectal neoplasia. These families typically have few polyps but develop colorectal cancer at a young age, 15–20 years earlier than sporadically occurring disease. HNPCC is estimated to account for 3%–5% of colorectal cancer. Extracolonic cancer may also occur in some families, often uterine or upper GI malignancy. CNS neoplasms in HNPCC are typically glioblastomas (median age, 33; range, 4–35).[24]

In a person with colonic polyposis or colorectal carcinoma, a brain tumor occurring at a young age, or with an appearance or location atypical for metastasis should lead to consideration of Turcot's syndrome. Family history should be explored, and medical genetics consultation should be offered.

Note that there are also other genetic syndromes associated with polyposis without CNS neoplasms. One such example is *MUTYH*-associated polyposis.[25] The phenotype of this disease is variable and depends on the genotype, epigenetic, and environmental factors. Another example is Peutz-Jeghers syndrome. These patients develop hamartomas of the digestive tract and are at higher risk of colorectal cancer and other cancers, like breast, ovary, pancreas.[2]

1.7 Neuroendocrine tumors (carcinoid tumors)

Neuroendocrine neoplasms (NENs) are mixed groups of malignancies with cells that have dense core granules similar to those in serotonergic neurons (the "neuro" property) and the ability to make and secrete monoamines (the "endocrine" property).[26] Historically, gastroenteropancreatic (GEP) NENs that were well differentiated were called carcinoid tumors. In current practice, they are now categorized as low-grade or high-grade carcinomas, which are more useful for prognostication and treatments decisions.[27] They originate most commonly in the GI tract (62%–67%), followed by the lungs (22%–27%).[26] GI NENs occur more commonly in African Americans and lung NENs in Caucasians. NENs are more common in women (2.5:1) and are increasing in incidence, felt to be from improved awareness. The prevalence is low and meets the criteria for orphan disease status (< 200,000 in the United States). Of those NENs originating in the GI tract, the origin is most often in the small intestine and rectum.[2] Other common locations include the colon, appendix, and stomach. Only a small minority (< 1%) are accompanied by the carcinoid syndrome of episodic diarrhea and flushing, caused by serotonin release from the tumors, usually in the setting of advanced liver metastasis.[26] More common symptoms of GI carcinoids are intestinal obstruction, hematochezia, and melena. Many NENs are asymptomatic, including some that have metastasized. Their behavior is much less aggressive than that of the more common adenocarcinomas of the GI tract.

Most NENs occur sporadically but others are associated with some hereditary syndromes.[26] These include von Hippel-Lindau syndrome, multiple endocrine neoplasia type 1 (MEN-1), MEN-2, tuberous sclerosis, and neurofibromatosis. Mutations in the mTOR pathway genes occur in 15% of patients and make mTOR inhibitors a rational choice for treatment. NENs are very vascular, making vascular endothelial growth factor inhibition also a rational treatment.

The female preponderance may be from selection bias because of more frequent diagnostic laparoscopy performed to investigate possible gynecological conditions in premenopausal women with lower abdominal pain.[26] Several imaging methods can be used for diagnosis, including endoscopy, CT, MRI, ultrasound, and PET. Surgery is the best treatment for nonmetastatic disease, so localization is essential. Common tumor markers include urinary 5-HIAA, the final secreted product of serotonin, and serum CgA, which is more sensitive. Octreotide and somatostatin can be labeled with radionucleotides for diagnosis by scintigraphy as GEP NENs generally express somatostatin receptors.

Surgery can be curative in nonmetastatic tumors.[26] The role of adjuvant therapy in these patients is not clear. Ablations, embolization, or selective internal radiation may be possible for liver NENs. There are various chemotherapy or targeted therapy options for systemic treatment, including somatostatin analogs.

Patchell and Posner[28] reported 36 neurological complications among 219 patients with carcinoid tumors. They found 14 patients with epidural spinal cord compression (ESCC), 13 with brain metastasis, 1 with base-of-skull metastasis, 1 with leptomeningeal metastasis, and 5 with plexus or nerve involvement. One patient had carcinoid myopathy, a recognized complication thought due to serotonin effects on muscle; the patient responded to cyproheptadine, a serotonin antagonist. ESCC has been reported but is very rare.[29–32] The carcinoid syndrome may be considered a paraneoplastic syndrome. Neurological paraneoplastic syndromes have been reported with bronchial carcinoid tumors but are virtually unknown with GI carcinoid tumors.[28]

2 Neurological complications of GI cancers

Although adenocarcinomas of the lung and breast commonly spread to the nervous system as brain metastases and leptomeningeal metastases, those of the gastrointestinal tract do so less commonly. The low incidence of brain and leptomeningeal metastases with GI adenocarcinomas suggests that the high rate seen with adenocarcinomas of the lung and breast is a property specific to tumors originating in these organs, not a property of adenocarcinomas in general. Bony metastases and, hence, spinal cord compression are also fairly uncommon. Colorectal carcinoma is the most common cause of lumbosacral plexus involvement. Paraneoplastic syndromes are fairly uncommon with GI malignancies. Neurological toxicities of chemotherapy regimens used to treat GI malignancies are a fairly common and important problem.

GI cancers have interesting differences in their patterns of metastasis. Gastric cancer appears to have a greater risk of leptomeningeal metastasis, while colorectal cancer has more tendencies to metastasize to the brain. Cancers of the liver and pancreas rarely spread to either site. The risk of epidural spinal cord compression appears highest with colorectal cancer and gastric cancer. Liver cancer probably has the greatest likelihood of spreading to bone, yet it rarely causes epidural spinal cord compression.

2.1 Brain metastasis

Nearly 20% of cancer patients will develop brain metastasis, but it is felt that the true incidence is higher as these estimates often include only those who are considered for treatment.[33] Autopsy studies suggest that the incidence may be up to 40%. This is likely to increase further as overall survival outcomes continue to improve after initial cancer diagnosis and with increased screening. There is increased risk of brain metastasis with more advanced primary disease. Brain metastasis is less common from GI primaries than from certain other cancers such as lung cancer, breast cancer, and melanoma.[33] It is not clear why some cancers have a predilection for brain metastasis, but there are some interesting new insights from various molecular or genetic analyses. For example, there is a link between hypomethylation and metastatic invasiveness in colorectal cancer. Other studies have shown that there are genetic differences in brain metastases when compared to their matched primary tumors.

Brain metastasis occurs in less than 4% of those with GI cancers,[3] but this rate is increasing.[33] Of the GI cancers, colorectal cancer has the highest incidence of brain metastasis (Table 2[34]; see also Esmaeilzadeh et al.[35]).

TABLE 2 Proportion of patients with primary GI cancers among 2382 patients with brain metastasis from all primary tumors.

Type of cancer	# of cases	% of all patients with brain metastasis
Total from GI tract	119	5.0
Specific GI sites:		
Esophageal	13	0.5
Gastric	8	0.3
Liver	2	0.1
Gallbladder and biliary	0	0.0
Pancreas	4	0.2
Colorectal	92	3.9

Reproduced with permission from Junck L, Zalupski MM. Neurological complications of GI cancers. In: Newton HB, Malkin MG, eds. Neurological Complications of Systemic Cancer and Antineoplastic Therapy. 1st ed. Boca Raton, FL: CRC Press; 2010:312–335.

Nussbaum et al.[36] reported that brain metastases from GI cancers are more often single (67%) than those from other tumors (53% of all patients with metastases). Metastases from GI cancers may involve the cerebellum more often (31% of cases with single metastases) than other cancers (18% of all cases with single metastases).

The optimal treatment for patients with primary GI cancers and brain metastasis is identical to that of patients with other primary cancers. The National Comprehensive Cancer Network (NCCN) guidelines for brain metastasis make little differentiation among the primary cancers.[37] This includes surgical resection if the metastasis is single and resectable, and the patient has favorable surgical risk and controlled systemic disease. Resection may also be appropriate for the management of mass effect or symptoms. Resection is usually followed by radiation. The decision for whole brain radiation (WBRT) or stereotactic radiosurgery (SRS) may depend on several factors, including total brain tumor volume, number of metastases, and reasonable systemic treatment options. In general, SRS is preferred for low tumor number and volume. Hippocampal avoidance is preferred when possible if WBRT is pursued. Chemotherapy is of modest benefit for patients with a GI primary cancer and brain metastasis. Some targeted agents or immunotherapies have good brain penetration and may be good options in select patients. Dexamethasone offers substantial relief of edema-related symptoms.

Gaspar et al.[38] delineated three prognostic classes among patients with brain metastases from all primary tumors based on a scoring system using recursive partitioning analysis (RPA). They used age, Karnofsky performance score (KPS), primary tumor control status, and presence/absence of extracranial metastases for scoring. The three different RPA classes had median survivals of 2.3, 4.2, and 7.1 months. More recent attempts to define diagnosis-specific indices have resulted in the graded prognostic assessment (GPA); it was noted that for GI cancer, age, KPS, presence/absence of extracranial metastases, and the number of brain metastases were prognostic.[39] There were four groups defined; median overall survival by groups was 3, 7, 11, and 17 months, with more than 30% of patients in the worst prognostic group.

2.1.1 Brain metastasis from esophageal cancer

It is likely that brain metastasis from esophageal cancer occurs via Batson's vertebral venous plexus as this provides a communication between the central nervous system and the esophagus.[3] It is rare, occurring in less than 2% of patients. Go et al.[3] note that there are only about 100 clinical or autopsy cases reported in the literature around the world. The median interval between diagnosis of esophageal cancer and brain metastasis ranged from 5.6 to 12.3 months. Patients with more advanced stage and size of the primary tumor were more

likely to have brain metastasis. They suggest that tumor histology may not be an independent risk factor for brain metastasis as Western series showed that most patients had adenocarcinoma, but in Japan, most patients had squamous cell carcinoma; these results mirror the higher incidence of adenocarcinoma in the United States and squamous cell carcinoma in Asia.

Gabrielson et al.[40] reported a series of 334 patients who underwent esophagectomy for esophageal carcinoma and documented 12 with brain metastasis (3.6%), including 10 of 230 (4.3%) with adenocarcinoma and 2 of 104 (2%) with squamous carcinoma. Another series found 2 instances of brain metastasis among 293 patients with esophageal carcinoma, but no instances of metastasis were detected on 240 patients undergoing head computed tomography (CT) prior to esophagectomy.

Weinberg et al.[41] reported 27 cases of brain metastasis among 1588 patients with esophageal carcinoma (1.7%). The pathology was adenocarcinoma in 82% and squamous cell carcinoma in 7% and 48% were single metastases. Median time from diagnosis of the primary cancer was 5.6 months and 70% had systemic metastases. Median survival after brain metastasis was only 3.8 months despite aggressive treatment. Liver metastasis was an adverse predictor.

Khuntia et al.[42] reported 27 patients with esophageal carcinoma and brain metastasis, and 3.1% of 837 patients were seen with esophageal carcinoma. Average latency from the diagnosis of esophageal carcinoma was 10 months, and only one had both diagnoses made at the same time. Moreover, 92% of the patients had adenocarcinoma and 8% squamous cell carcinoma. Median survival was 3.6 months overall. Performance status and "aggressive treatment" (primarily surgical) were favorable predictors in multivariate analysis.

In a series from Japan, Ogawa et al.[43] reported 36 patients with brain metastasis of 2554 patients seen at their hospitals with esophageal carcinoma (1.4%). Only 31% had lung metastasis at the time of brain metastasis, indicating that seeding of the brain does not require lung metastasis as an intermediate step. Thirty-three had squamous cell carcinoma and only 1 had adenocarcinoma, probably reflecting a difference from the United States in epidemiology of esophageal cancer. Seventeen percent had single metastases. Median survival was 3.9 months overall, 9.6 months with surgery and RT, and 1.8 months for RT alone. In multivariate analysis, treatment and performance status were predictive.

Also from Japan, Yoshida[44] reported 17 patients with brain metastasis out of 1141 patients treated for esophageal carcinoma (1.5%) and out of 803 patients treated for brain metastasis (2.1%). As in the series of Ogawa et al.,[43] squamous cell carcinoma was the most common pathology. Treatment included resection (10 patients), WBRT (5), and stereotactic radiosurgery (2). Treatment results were impressive, including median survival of 38 months among 10 patients with a single metastasis and 16 months among 7 with multiple metastases.

Kothari et al.[45] identified 49 patients with brain metastasis from esophageal cancer, stage I–IV between 1998 and 2015 at a single tertiary referral center. Eighty-two percent of patients had adenocarcinoma. Median survival after identification of brain metastasis was 5 months. Multivariate analysis showed that patients in RPA class I or II fared better than those in class III, and those with one to two brain metastases fared better than those with three or more. Patients with these favorable characteristics had a median survival of 11.1 months.

Go et al.[3] note that some small studies have shown a survival benefit with surgical resection, with or without WBRT. The patients with the longest survival were those with excellent KPS (90%–100%), no extracranial metastasis, and solitary brain metastasis.

2.1.2 Brain metastasis from gastric cancer

The incidence of brain metastasis from gastric cancer is less than 1%.[3] There have been two large single-institution retrospective reviews.

York et al.[46] reported on 3320 patients with gastric cancer, with only 9 of these patients known to have brain metastasis diagnosed during life. Another 5 patients were found to have brain metastasis at autopsy. Median survival was 2.4 months after diagnosis of brain metastasis for the whole group, 1.7 months for 4 given no specific treatment, 2.1 months for 11 treated with radiation, and 12.5 months for 3 treated with resection.

Kasakura et al.[47] found only 11 patients with brain metastasis out of 2322 patients. These 11 patients had advanced-stage disease. The patients who had resection had better survival outcomes; several of these patients also had chemotherapy and/or radiation, suggesting that aggressive multidisciplinary treatment (or perhaps those patients who have good enough KPS to qualify for this treatment) likely improves outcomes.

2.1.3 Brain metastasis from hepatocellular cancer

Retrospective case series and case reports indicate an overall incidence of about 1%[48–51]; however, autopsy reports indicate the incidence may be as high as 7%.[52] Patients with brain metastasis from hepatocellular cancer usually have RPA II-III, poor KPS, diffuse systemic disease (especially lung metastases), and neurological symptoms from frontal or occipital lesions. Some patients may have intracranial hemorrhage. Murakami et al.[53] reported 16 patients from Japan; 14 were hemorrhagic. All occurred in patients previously diagnosed with hepatocellular carcinoma (median latency, 20 months). Eighty-one percent had lung metastases; median survival was 6 weeks. Among 7 patients in Korea with brain metastasis, Kim et al.[48] reported that lung metastases were already present in 5. Median latency after diagnosis of the primary tumor was 13 months; in 1 patient, brain metastasis was the initial manifestation.

2.1.4 Brain metastasis from gallbladder and biliary tract cancer

Brain metastasis is very rare from these primary tumors but has been noted in case reports with adenocarcinoma of the gallbladder and bile ducts.[54–58]

2.1.5 Brain metastasis from pancreatic cancer

Most reports indicate that brain metastasis is rare with pancreatic cancer possibly because this cancer is locally aggressive and so patients do not survive long enough to develop them, although this may be changing.[3] In 1964, Aronson et al.[59] reported 7 patients with pancreatic cancer among 250 patients with brain metastases (2.8%), a higher incidence than more recent series, suggesting a change in pattern of metastasis or diagnosis. Park et al.[60] reported 4 cases of brain metastasis occurring among 1229 patients in Korea (0.3%); in 2, it was the presenting manifestation of the cancer. Median survival was less than 3 months.

2.1.6 Brain metastasis from colorectal cancer

There are four molecular subtypes of colorectal cancer based on extensive molecular testing[33]: CMS1 has strong immune activation and microsatellite instability; CMS2 is characterized by WNT and MYC upregulation; CMS3 has metabolic dysregulation; and CMS4 shows TGFβ activation, cellular invasion, and angiogenesis. RAS family mutations and PIK3CA mutations have also been described in brain metastases from colorectal cancer.

Cascino et al.[61] reported 40 patients with brain metastases from colon cancer, representing 4% of total patients with colon cancer. Brain metastasis was discovered prior to cancer diagnosis in only one patient; in others, the diagnosis of colon cancer preceded the neurological presentation by a median of 24 months. Extensive systemic metastasis was present in 92% including 85% with lung metastases. Twenty patients had a single metastasis, while 20 had multiple metastases as assessed by CT. Of the single tumors, 13 were supratentorial (65%) and 7 infratentorial (35%). Median survival was only 9 weeks in 32 patients treated with RT; follow-up scans showed improvement in only 2 of 11.

Alden et al.[62] reported 19 patients with brain metastasis from colorectal cancer. The median time from diagnosis of colorectal cancer to brain metastasis was 32 months (range, 0–100 months). Seventy-nine percent had metastases at other sites. A single tumor was present in 63%, located in the cerebellum in 32%. Median survival was 2.8 months in the whole group, 2.6 months in 5 patients treated with RT, and 4.9 months in 5 who underwent resection.

Hammoud et al.[63] reported 100 patients with brain metastases out of nearly 9000 patients with colorectal cancer (1.2%). The median interval between primary diagnosis and diagnosis of metastasis was 26 months. A rectal primary cancer was associated with brain metastasis; 33% among those with brain metastasis and 24% of the overall group had primary rectal tumors. Sixty-four percent had a single metastasis. Median survival with brain metastasis was 3 months without specific treatment, 3 months after radiation, and 9 months after surgery. In multivariate analysis, predictors of survival included surgical treatment of the brain metastasis and location of the primary tumor in proximal colon.

Nieder et al.[64] reported 20 patients with colorectal cancer and brain metastasis, treated with RT. Among 10 patients with a solitary metastasis, 5 were infratentorial, and among 9 with multiple metastases, 5 had at least one infratentorial tumor. Forty-five percent of patients had lung metastases when brain metastasis was detected, but in 25%, the brain was the first site of metastasis and in 20%, brain metastasis preceded the diagnosis of the primary cancer. Median survival after resection (6 cases) and RT (all 20) was only 7 weeks, and only 1 patient was alive at 1 year.

Wronski and Arbit[65] reported that among 709 patients undergoing resection of brain metastases, 73 (10%) had colorectal carcinoma as the primary diagnosis. The primary tumor had been resected in all patients, and median time to brain metastasis was 28 months. Fifty-four patients (74%) had pulmonary metastases. Median survival was 8 months, while survival was 32% at 1 year, 7% at 2 years, and 4% at 5 years. Death was due mainly or entirely to brain metastasis in 30 (43%) and to systemic metastasis without brain recurrence in 33 (47%). The mode of surgery for the brain metastasis (en bloc in 40, piecemeal in 11, and by ultrasonic aspiration in 22) was unrelated to survival. Metastases were supratentorial in 47 (64%) and cerebellar in 26 (36%). Cerebellar location of metastasis was an adverse prognostic feature.

Ko et al.,[66] in a series from Taiwan, reported 53 patients with colorectal cancer and brain metastasis. Of these, 37 were from a series of 3773 patients with colorectal cancer (1.0%). Adding 16 patients to their analysis, they reported that the primary tumor was in the rectum in 28 of 53 (53%). Twenty-nine had concurrent metastases in the lung and 10 in the liver, while 12 had no metastases. Median survival was an astounding 87 months among 6 patients undergoing resection of the metastasis but only 2.9 months among the other patients.

Schouten et al.[67] reported brain metastasis in 10 patients (1.4%) from a cohort of 720 patients with colorectal carcinoma.

Sundermeyer et al.[68] reported 3% of 1020 patients with colorectal cancer to have brain metastases; the incidence was 6.2% in patients with lung metastasis and 1.2% in those without.

Amichetti et al.[69] reported an Italian series of 23 patients with brain metastases from colorectal cancer treated with RT. The interval from primary diagnosis to diagnosis of brain metastasis was 28 months. Ninety-one percent had metastases at other sites, including the lung in 74% and the liver in 52%. Fourteen patients (61%) had multiple brain metastases. The location was supratentorial in 14, infratentorial in 7, and both in 2. Median survival was 3 months, but the authors indicated that brain metastasis was the primary cause of death in only 3.

These series of patients with colorectal carcinoma and brain metastasis, as well as other data, suggest that brain metastasis is (1) more often solitary, (2) more often infratentorial, (3) relatively more common with rectal cancer than with colon cancer, and (4) often helped by surgical resection. Factors for prolonged survival include age less than 65 years, absence of systemic disease, solitary brain metastasis, bone metastasis, and memory loss at presentation.[3]

2.2 Leptomeningeal metastasis

The numbers of patients with various GI primaries among patients with leptomeningeal carcinoma is depicted in Table 3.[34] Comparing these figures with the incidence and death figures in Table 1, it appears that gastric carcinoma has a higher risk of leptomeningeal metastasis among GI cancers, while hepatocellular and pancreatic cancer have distinctly lower risk. Giglio et al.[71] reported 21 patients with leptomeningeal metastasis from GI primaries, including 7 from esophageal carcinoma (0.25% of

TABLE 3   Proportion of patients with primary GI cancers among 2765 patients in 49 series with leptomeningeal metastasis from all primary tumors.[70]

Type of cancer	% of cases
Total from GI tract	2.6
Specific GI sites:	
Esophageal	0.3
Gastric	1.3
Liver	0
Gallbladder and biliary	0.1
Pancreas	0.3
Colorectal	0.6
Other/unknown	0.1

their patients with metastatic esophageal cancer), 8 from gastric (0.19% of those with metastases), 1 from pancreatic (0.023%), and 5 from colorectal carcinoma (0.027%). Their data suggest that esophageal cancer shares gastric cancer's risk of metastasis to leptomeninges, but this is not reflected in the data in Table 3.[34]

Eleven of the 21 patients of Giglio et al.[71] had signet ring carcinomas. Other series have reported 5 of 8 patients[72] and 8 of 17 patients[73] with gastric carcinoma and leptomeningeal metastasis to have signet ring morphology. The "signet ring" is a cell with a characteristic appearance caused by central mucin pushing the nucleus aside. Signet ring cancers are known to behave aggressively and metastasize early.[74] Signet ring morphology is present in only 1%–2% of colorectal cancers[74] but has increased in gastric carcinomas in recent years to approximately 20%.[75] The signet ring morphology is a subtype of the diffuse type of gastric carcinomas, which generally have mutations in the *CDH1* gene that encodes E-cadherin, a cell adhesion molecule.[76] Perhaps the loss of E-cadherin plays a role in leptomeningeal metastasis.

The management of patients with leptomeningeal metastasis from GI primaries is similar to that of patients with other primaries.[37,70,77] Manifestations are often related to multiple levels of the neuraxis, including mental status abnormalities, cranial neuropathies, myelopathy, and radiculopathy. Pain is often a feature but typically not severe. Seizures can be present. The diagnosis is usually by MRI showing leptomeningeal deposits (especially with a nodular appearance) or by CSF cytology, in the setting of a person with a known GI malignancy, usually in the advanced state. Limiting the management to supportive measures is appropriate for most patients, especially those with poor performance status and uncontrolled systemic disease. Hydrocephalus is common and responds to shunting. Radiotherapy (RT) may be offered to symptomatic sites such as the brain or sections of the spinal canal. Intra-CSF chemotherapy (e.g., with liposomal cytarabine or methotrexate) is of limited benefit and probably should be offered only to those with the most favorable risk factors.[70] Dexamethasone may offer symptom relief, especially for tumors affecting the brain or spinal cord function. Giglio et al.[78] reported a response to oral capecitabine in a patient with gastroesophageal carcinoma, while another with esophageal carcinoma did not respond. Targeted agents can also be considered in appropriate patients, either given systemically or even intra-CSF, depending on the agent and the target.[79]

In their series of 21 patients with leptomeningeal metastasis from a primary GI cancer, Giglio et al.[71] reported the median latency after diagnosis of the primary cancer to be 39 weeks (range, 0–870). Median survival after various treatments was only 7 weeks.

2.2.1 *Leptomeningeal metastasis from esophageal cancer*

The series reported by Giglio,[71] including 7 patients with esophageal carcinoma, suggests that esophageal carcinoma may share with gastric carcinoma a risk of leptomeningeal metastasis that is higher than that of other GI cancers. Median survival was only 8 weeks. A series by Lukas et al.[80] reported 7 cases of leptomeningeal disease from esophageal cancer. Onset of leptomeningeal involvement was variable, in relation to primary disease diagnosis, ranging from 5 months to 3 years. However, progression was rapid in all of those patients, ranging from 2.5 to 16 weeks. Patients in this series received different therapies but there was no clearly beneficial treatment. Fig. 1 shows imaging from a patient with leptomeningeal metastasis from esophageal cancer.

2.2.2 *Leptomeningeal metastasis from gastric cancer*

The risk of gastric carcinoma for leptomeningeal metastasis is discussed earlier. Delaunoit et al.[81] have proposed that the aggressive linitis plastica variant of gastric adenocarcinoma is especially susceptible to leptomeningeal metastasis, observing this in 4 patients, 5% of their patients with this variant. These patients often have signet ring cells in their CSF supporting this diagnosis. Another series reported 5 of 8 patients with gastric carcinoma and leptomeningeal metastasis to have the linitis plastica variant[72]; 4 of these 5 also had signet ring morphology. Lisenko et al.[72] reported a paucity and low volume of liver metastases, frequent atypical sites of metastasis, and the diffuse nature of metastases among their patients.

In a series from Korea, where gastric carcinoma is common, Lee et al.[82] reported median survival of 0.9 months for 18 patients with leptomeningeal metastasis, 2.7 months for 10 receiving intra-CSF chemotherapy, and 0.3 months for 8 receiving only supportive care. Among the 8 patients of Giglio et al.,[71] median survival was 6 weeks. The patients of Delaunoit et al.[81] had surprisingly good survival with intra-CSF methotrexate, with 2 of 4 living 9 + months.

2.2.3 *Leptomeningeal metastasis from other GI cancers*

Leptomeningeal metastasis is quite rare from other GI cancers. From primary tumors of the biliary tract, at least 6 cases have been reported with adenocarcinoma of the gallbladder[83] and one with cholangiocarcinoma.[84] Giglio et al.[71] reported 5 cases of leptomeningeal metastasis from colorectal cancer; median survival was 7 weeks.

2.3 Spinal cord compression

The term *epidural spinal cord compression (ESCC)* will be used in this chapter to include compression of either the spinal cord or the cauda equina.

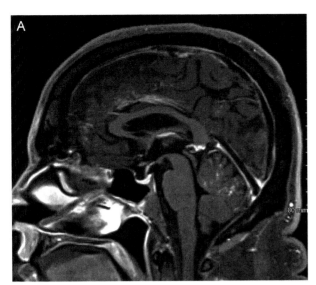
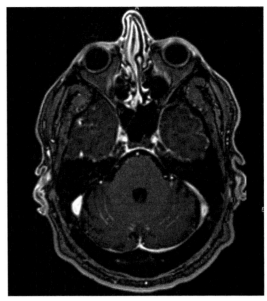

FIG. 1   Leptomeningeal metastases in a patient with known poorly differentiated esophageal adenocarcinoma. MRI T1 with contrast shows enhancement in the cerebellar folia in the (A) sagittal and (B) axial views. The patient had received treatment with FOLFOX and was on capecitabine maintenance therapy when these metastases were discovered. The MRI was pursued because the patient presented with progressive headaches, back pain, difficulty walking, nausea, and dizziness. His symptoms improved with dexamethasone but did not completely resolve. At the time of this publication, the patient was receiving WBRT with a plan to initiate pembrolizumab. Alternatively, the patient was also considering hospice instead, given the poor prognosis and his worsening symptoms.

Investigation and management of ESCC from GI cancers should be similar to that of ESCC from other cancers.[37,85] Nearly all patients present with severe pain[86]; many have radiculopathy, and some have myelopathy. Motor deficits are more common than sensory deficits, and bowel and/or bladder dysfunctions are often a late finding. It is important to image the entire spine, as epidural deposits will be found at multiple levels in about one-third, even when only one level is symptomatic.[37,87] MRI is nearly 100% sensitive in detecting ESCC, except in some ambulatory patients with only mild motor symptoms.[88] In patients with myelopathy, or in whom the pain is severe, treatment should begin with high-dose dexamethasone.[86,89] Direct surgical resection has been shown to preserve neurological function and should be considered in appropriately selected patients.[90] There are now several prognostic scoring systems to help surgeons make these decisions.[86] RT should be offered urgently when decompressive surgery will not be performed. No RT regimen has been proven superior to another, but most centers offer a treatment course of 2 weeks or less. Spine radiosurgery may be preferred for oligometastatic disease, especially if the tumor is suspected to be radioresistant, which is the case in some colorectal cancers.[37] Pain control (including palliative radiation to symptomatic metastases), bladder management, and rehabilitative measures are important.[86] Treatment with a biphosphonate such as zoledronic acid should be considered in patients with bone metastases to prevent fractures and other skeletal complications.[91]

The data in Table 4[34] suggest that most cases of ESCC from GI primaries are associated with colon cancer, followed by gastric cancer. Larger numbers, preferably from a population-based study or large cohort study, would be needed to be confident of differences in risk of ESCC among primary GI cancers. When a patient presents with ESCC by tumor without an established diagnosis of cancer, a GI cancer will be found in only about 3% of cases.[92]

ESCC is distinctly rare with esophageal carcinoma (see Table 4).[34] One case has been reported of thoracic myelopathy caused by direct extension of large-cell carcinoma of the esophagus.[93] As per Table 4, ESCC is known to occur occasionally in patients with gastric cancer. It is uncommon in pancreatic cancer.

Bone metastasis is moderately common in hepatocellular cancer, occurring in 5%–17% of patients, generally as osteolytic lesions.[49,50,94,95] Vertebral metastases have been reported in 22 of 403 patients (5.5%)[49] and 18 of 482 patients (3.7%).[51] ESCC, however, has been uncommon (see Table 4).[34] Natsuizaka et al.[51] reported 8 patients with ESCC among 18 with vertebral metastases and 482 with hepatocellular cancer. Doval et al.[96] reported 4 patients with ESCC; in 2, this was the presenting manifestation of their cancer. Treatment included RT in all four and laminectomy in two, but median survival was 4 months or less.

In colorectal cancer, up to 5% have bone metastasis at the time of primary tumor diagnosis and up to 27% have bone metastasis during their disease course.[97] The most

TABLE 4 Proportion of patients with primary GI cancers among 1104 patients in 12 series with epidural spinal cord compression from all primary tumors.[85]

Type of cancer	% of cases
Total from GI tract	**6.1**
Specific GI sites:	
Esophageal	0.1
Gastric	0.7
Liver	0.5
Gallbladder and biliary	0.0
Pancreas	0.4
Colorectal	4.3

Reproduced with permission from Junck L, Zalupski MM. Neurological complications of GI cancers. In: Newton HB, Malkin MG, eds. Neurological Complications of Systemic Cancer and Antineoplastic Therapy. 1st ed. Boca Raton, FL: CRC Press; 2010:312–335.

common location for bone metastasis in these patients is the vertebral column.

Brown et al.[98] reported 45 episodes of ESCC in 39 patients with colorectal cancer. Of 34 patients treated with RT, 17 had colon cancer and 17 rectal cancer. All patients had metastases at other sites. Metastases were in the lumbar spine in 55%, thoracic spine in 32%, and cervical spine in 12%, a difference from the distribution of metastases in most cancers, which are uncommon in lumbosacral spine (24%) and predominate in thoracic spine (65%). Of 63% who were ambulatory at presentation, 20 maintained ambulation (95%). Local failure occurred in 3 (8%). Median survival was 4.1 months, fairly typical for ESCC with the more unfavorable cancers[90,98] but worse than that with more favorable primary cancers (e.g., breast, prostate, lymphoma, and myeloma). Prognostic factors in multivariate analysis were primary tumor in the rectum rather than colon and RT dose > 30 Gy, both favorable.

Rades et al.[99] reported 81 patients with colorectal cancer and ESCC, treated with RT alone. Eleven (14%) improved neurologically, including 5 nonambulatory patients who regained ambulation. Median survival was 4 months. Three patients had recurrent ESCC within the RT field.

Bostel et al.[97] identified 162 bone lesions in the thoracic (60%) and lumber (40%) spine of 94 patients with colorectal cancer treated with RT. Only 39% of patients were still alive 6 months after RT. Median survival after diagnosis of bone metastasis was 4.2 months.

2.4 Plexopathy

Among the GI cancers, colorectal cancer is by far the most likely to cause lumbosacral plexopathy because of its proximity to the lumbosacral and coccygeal plexuses.[100,101] Plexopathy, however, is probably less common than brain metastasis and epidural metastasis. Plexopathy is caused by invasion from the primary tumor or regional adenopathy, often at recurrence.

Lumbosacral plexopathy usually presents with pain referred to the distribution of the involved trunks.[101] The pain is usually unilateral (but may be bilateral) and affects the low back, hip, and thigh.[102–104] The pain may worsen in the recumbent position at night. Isolated tumor involvement at the L1 level can affect the ilioinguinal, iliohypogastric, or genitofemoral nerves while sparing motor trunks of the plexus; these patients report pain or paresthesias in the groin or lower abdomen with no motor or sensory symptoms in the lower extremities.[102] The pain may progress relentlessly and is often refractory to narcotics. After several weeks or months, weakness and sensory symptoms ensue in a pattern indicating involvement of more than one nerve root or peripheral nerve. Bladder dysfunction is uncommon unless the metastasis extends medially, compressing the cauda equina, or inferiorly, reaching the S2, S3, and S4 nerve roots or the inferior hypogastric plexus. Involvement of the coccygeal plexus usually occurs in patients with a known invasive rectal cancer, either at initial presentation or, more commonly, at recurrence. These patients present with perineal pain, sensory loss, and early bowel symptoms.

Examination reveals weakness, sensory loss, and decreased reflexes in a pattern indicating involvement of more than one nerve root.[102] Approximately half of the patients have sacral or sciatic notch tenderness or a positive straight leg raising test. With involvement of the sacral or coccygeal plexus, tumor may be palpable on rectal exam.

The differential diagnosis of lumbosacral plexopathy includes the closely related condition of radiculopathy due to direct tumor extension or vertebral metastasis.[104] With epidural compression of the spinal cord or the cauda equina, back pain is the predominant early symptom. With leptomeningeal metastasis involving the conus or cauda equina, manifestations are usually bilateral, pain is less prominent, and bladder and bowel dysfunction occur early. Plexopathy can occur as a complication of intra-arterial chemotherapy using drugs such as cisplatin and 5-FU.[103,105] Conditions unrelated to cancer such as lumbar disk disease and lumbar stenosis may be considerations.

Radiation plexopathy should be considered in patients who previously received RT to the area of the plexus.[101] It rarely begins earlier than 3 months after RT, and the median latency after RT is 5 years. These patients typically have a more indolent presentation with weakness and atrophy involving L2 through S1 muscles with an L5–S1 predominance.[103,104] One-third of patients have prominent early numbness and paresthesias. Approximately half of the patients experience pain, but it usually occurs later, and rarely it is as severe or refractory to treatment as the pain of tumor invasion.

Diagnostic testing begins with MRI or CT. MRI is more sensitive and is the modality of choice.[101] It is also useful to look for involvement of conus, cauda, or nerve roots.[102–104,106] PET can be considered to confirm metastases in those with indeterminate MRI findings.[101,107] The most reliable finding is the demonstration of a tumor mass. Scans often show bilateral plexus involvement, even in patients with unilateral symptoms and signs. Epidural tumor extension is common, reported in 45% of patients with lumbosacral plexopathy from cancer.[102] In the absence of an overt mass, nerve trunks can be thickened by tumor involvement. For radiation plexopathy, imaging may be normal or may show poor definition of normal tissue planes without an identifiable mass.[104] Gadolinium enhancement can be seen with radiation injury and need not imply tumor. When the initial imaging is normal but symptoms persist or worsen, repeat imaging in 3–6 months may disclose tumor not seen earlier.

When the diagnosis is not clear, EMG may be helpful, especially when radiation plexopathy is a consideration.[102] EMG shows evidence of severe acute and chronic axonal loss, often bilateral involvement, even when symptoms and signs are unilateral. Sensory nerve action potentials may be absent. Myokymia is the EMG finding that most reliably differentiates radiation plexopathy from tumor involvement. Myokymic discharges are present in approximately 60% of patients with radiation lumbosacral plexopathy, but rarely with tumor involving lumbosacral plexus.

For lumbosacral plexopathy due to tumor invasion, RT produces significant pain relief in 30%–80%.[102,103,108] Improvement of neurological deficits is less likely, occurring in no more than one-third of patients. Unlike with bone metastases, higher doses of radiation are required for tumor regression that is causing neuropathic pain from nerve root compression; thus, a protracted course is more effective.[101] The prognosis for survival depends on the overall status of the malignancy. Patients with lumbosacral plexus tumor involvement often survive no more than several months.[104]

For radiation plexopathy, no treatment is known to be effective. The symptoms are often progressive, including weakness, lymphedema, and pain, which can be substantial.[109] A trial of hyperbaric oxygen therapy for radiation brachial plexopathy showed no benefit.[110] Treatment usually emphasizes supportive care. Pain treatment options include local nerve blocks, intrathecal narcotic infusion, transcutaneous electrical nerve stimulation, percutaneous cervical cordotomy, and chemical sympathectomy. Some patients obtain temporary pain relief from dexamethasone.[103]

Patients with motor disability may benefit from physical therapy.

2.5 Paraneoplastic neurological syndromes

Paraneoplastic neurological syndromes (PNS) are a heterogeneous group of clinical manifestations and are designated as rare by the Office of Rare Diseases Research of the National Institutes of Health.[111] They affect only 0.01%–1% of cancer patients, although the true incidence may be underestimated. There are some case reports among the GI cancers.[112]

No particular syndrome predominates nor does a particular antineuronal antibody. Dermatomyositis and polymyositis are mentioned because there have been population-based studies on these topics. A number of cases of other syndromes are mentioned later under each of the specific tumor types, but in the absence of clear association between a particular syndrome or antineuronal antibody and a specific cancer, one cannot be confident in isolated patients that the cancer has caused the paraneoplastic syndrome. Cases of nonspecific polyneuropathy and motor neuron disease are not listed because the association of these diagnoses with cancer is questionable.

It is well accepted that dermatomyositis is associated with malignancy, and it is considered a classical PNS.[111] The risk of cancer in patients with dermatomyositis can be as high as 30%–42%.[111] Population-based studies have shown patients with dermatomyositis to have relative incidence rates for cancer of 2.4 (males) to 3.4 (females),[113] 3.8,[114] 6.2,[115] and 7.7,[116] compared to patients without dermatomyositis. Patients older than 40 years with rapid onset of symptoms, higher levels of inflammatory markers, and the presence of autoantibodies against nuclear antitranscription intermediary factor (TIF)-γ are more likely to have associated malignancy.[111] Hill et al.[117] found the relative risk of patients with dermatomyositis to have digestive tract malignancies to be 2.9 for esophageal cancer, 3.5 for stomach cancer, 3.8 for pancreatic cancer, and 2.5 for colorectal cancer, similar to that of malignancies in general (3.0) but lower than that for ovarian and lung cancer. The number of cases of dermatomyositis associated with GI cancers in the literature is at least 8 for esophageal (mostly adenocarcinomas), 21 for gastric, 1 for hepatic, 3 for gallbladder, 1 for biliary ducts, 6 for pancreatic, and 34 for colorectal cancer (most references not provided).[117]

The population-based studies of polymyositis show a much lower relative risk of 1.7–1.8,[113] 1.7,[114] 2.0,[115] and 2.1.[116] The relative risk of cancer is also lower with inclusion body myositis 2.4.[115] It is possible that medical screening in patients with these conditions accounts for the incidental discovery of cancers that are actually unrelated.

2.5.1 PNS from esophageal cancer

These are rare. Reported cases include 4 patients with limbic encephalitis,[118] including 3 identified in a review of 50 patients with this paraneoplastic syndrome,[119] brainstem encephalitis associated with GABA-B-receptor antibodies,[120] cerebellar degeneration associated with anti-Yo antibody in 2 male patients with adenocarcinoma,[121,122] opsoclonus-myoclonus from squamous cell carcinoma,[123] encephalomyelitis associated with anti-Hu antibody,[124] necrotizing myelopathy,[125] motor neuronopathy associated with adenocarcinoma,[126] sensory motor polyneuropathy in a patient with small-cell carcinoma,[127] and acute inflammatory demyelinating polyradiculopathy (AIDP),[128] including one associated with antiganglioside antibodies.[129]

2.5.2 PNS from gastric cancer

These are also rare. Reported cases include encephalopathy and sensorimotor polyneuropathy with anti-α-enolase antibody,[130] encephalopathy with opsoclonus-myoclonus without identified antineuronal antibody,[131] refractory focal epilepsy with antibodies against the glutamate receptor,[132] cerebellar degeneration associated with anti-Yo antibody in patients with adenocarcinoma,[133–136] cerebellar degeneration associated with the anti-Ri antibody in a patient with neuroendocrine carcinoma of the stomach,[137] brachial plexopathy (Parsonage-Turner syndrome),[138] subacute sensory neuropathy associated with the anti-Hu antibody,[139] and necrotizing arteritis of the peripheral nervous system.[140]

2.5.3 PNS from hepatocellular carcinoma

The most common paraneoplastic syndromes with hepatocellular carcinoma are probably endocrine and cutaneous syndromes.[141] It has been suggested that clear-cell hepatocellular carcinoma is associated with more neurological paraneoplastic syndromes.[142] Reported neurological paraneoplastic syndromes include encephalomyelitis with clear-cell hepatocellular carcinoma,[142] multifocal necrotizing leukoencephalopathy in a 19-year-old male, unassociated with treatment,[143] noninflammatory vasculopathy with cerebral infarcts in a 23-year-old female,[143] multifocal demyelinating encephalopathy with radiculopathy,[144] cancer-associated retinopathy,[145] opsoclonus-myoclonus in an infant with Beckwith-Wiedemann syndrome and hepatoblastoma,[146] subacute motor neuronopathy,[147] AIDP,[148] 3 cases of chronic inflammatory demyelinating polyneuropathy (CIDP),[149–151] and demyelinating neuropathy with the PR3-antineutrophil cytoplasmic antibody.[152] There are also several case reports of polymyositis.[153–155]

2.5.4 PNS from cancers of the gallbladder and biliary tract

These are quite rare. There are documented cases of opsoclonus and cerebellar ataxia with adenocarcinoma of the gallbladder without antineuronal antibody,[156] AIDP with adenocarcinoma of the gallbladder,[157] and CIDP with cholangiocarcinoma.[158]

2.5.5 PNS from pancreatic cancer

Endocrine paraneoplastic syndromes are very common with neuroendocrine tumors of the pancreas.[159] Of these, the most common syndrome is the Zollinger-Ellison syndrome of gastric hypersecretion stimulated by gastrin. Others include insulinomas causing hypoglycemia, which is commonly neurological in presentation, and glucagonoma causing weight loss, dermatitis, stomatitis, and diabetes mellitus.

Neurological paraneoplastic syndromes reported with pancreatic cancer are rare. Reported cases include encephalomyelitis with anti-GAD antibodies,[160] Miller-Fisher syndrome,[161] cerebellar degeneration and the anti-Hu antibody with pancreatic small-cell carcinoma,[162] CIDP,[158] and acquired gastroparesis.[163] There are 2 cases of opsoclonus, 1 with encephalopathy and ataxia associated with an atypical anti-Purkinje cell antibody,[164] the other with rhombencephalitis and meningoencephalitis without anti-Ri or other autoantibodies.[165]

2.5.6 PNS from colorectal cancer

Perhaps because of the frequency of colorectal cancer, a number of patients with PNS have been reported:

- Retinopathy with antibipolar antibodies[166]
- Optic neuropathy[167,168]
- Tonic pupils with focal weakness and dysmetria[169]
- Limbic encephalitis,[170,171] including 2 patients identified in a review of 50 patients with paraneoplastic limbic encephalitis[119]
- Cerebellar degeneration with atypical anti-Purkinje cell antibody[172]
- Cerebellar degeneration and limbic encephalitis associated with an atypical antibody to a 4 kDa protein[173]
- Cerebellar or brainstem syndrome with anti-Ma1, -Ma2, and -Ma3 antibodies[174]
- Lambert-Eaton myasthenic syndrome in a patient with rectal cancer[175]
- Subacute sensory neuronopathy[176]
- CIDP[158]
- Myositis with an antibody to a 34-kDA sarcoplasmic protein[177]
- Two cases of myotonia, one followed by myasthenia and myositis[178,179]
- Polymyositis[180]
- Stiff person syndrome[181]

2.6 Metabolic and cerebrovascular complications

Vitamin B12 deficiency is a recognized complication of gastrectomy due to the role of the stomach in producing intrinsic factors.[182] Symptoms include unsteady gait, weakness, numbness, paresthesia, and mental status alterations, while exam findings include loss of proprioception and vibration as well as Romberg sign. Prevention is key because response to B12 replacement is variable. These patients can also be deficient in iron.

Hepatic encephalopathy is uncommon with GI malignancies that have metastasized to the liver, with the exception of persons with underlying cirrhosis. Diagnosis is clinical and includes a history of worsening liver disease, usually with precipitating causes such as hypovolemia, GI bleeding, hypokalemia, sedating medications (especially benzodiazepines), infection, and thrombosis of the hepatic or portal vein.[183] Examination shows encephalopathy with asterixis and hyperventilation. The diagnosis is supported by prolongation of prothrombin time and, when necessary, electroencephalography showing triphasic waves. When hepatic encephalopathy occurs in the late stage of malignancies, supportive care without specific treatment is often appropriate. When treatment is indicated, the first step in treatment is to identify and treat the precipitating causes. There are several strategies that can be used to treat chronic and/or acute disease.[184] This is usually accompanied by ammonia-lowering therapy, which may include purgatives, probiotics, medications to modulate interorganic ammonia, and nonabsorbable antibiotics. Dietary changes are also important.

Cerebrovascular complications are moderately common in cancer, especially GI cancers.[185] Intravascular mucinosis associated with mucin-producing adenocarcinomas may cause vascular thrombosis and may contribute to nonbacterial thrombotic endocarditis (NBTE), which is the most common cause of ischemic strokes from embolic infarcts. The first cases of this were described in patients with pancreatic cancer. The treatment of choice is subcutaneous low molecular weight heparin or intravenous unfractionated heparin.

GI malignancies notoriously are associated with hypercoagulability and disseminated intravascular coagulation (DIC).[185] Venous occlusion may be related to the hypercoagulable state, especially in colon cancer, gallbladder cancer, and pancreatic cancer. Cerebral hemorrhage may be caused by thrombocytopenia from treatment, DIC, therapeutic anticoagulation, or hemorrhage into brain metastases. Gastric cancer is often associated with subdural hematomas. Cerebral sinus thromboses have been noted in gallbladder carcinoma. Certain chemotherapy treatments (notably bevacizumab with GI cancers) may contribute to ischemic and hemorrhagic strokes. The focus of treatment is to determine the underlying cause and correct it if possible.[185]

2.7 Pain management

Pain specifically from radiation plexopathy was discussed earlier in the chapter. Pain can otherwise be a significant problem with GI cancers, most notably pancreatic cancer.[186] Cancer pain management begins with analgesics, including opiates as indicated, and often this suffices.[187,188]

Another approach is celiac plexus block (neurolysis).[186,188–190] This blocks pain pathways from the upper abdominal viscera, including the stomach to the proximal colon, but it is most commonly used for the pain of pancreatic cancer.[186,190,191] Invasion of the celiac plexus probably contributes to the pain of some pancreatic cancers, but it is difficult to diagnose clinically, and involvement of the plexus need not be a criterion for celiac plexus block. Injection of alcohol may be guided by CT, fluoroscopy, or X-ray, or performed endoscopically with ultrasound guidance. Pain relief can be substantial within the first 1–2 weeks but often wanes, but many patients still have partial relief at 3 months.[190] Many patients, however, are not able to reduce their dose of opioid analgesics.

3 Common systemic treatment regimens for GI cancers and their neurological complications

3.1 Chemotherapy

3.1.1 Fluoropyrimidines

These are a class of chemotherapy drugs that include intravenous 5-fluorouracil (5-FU) and orally active fluorouracil. 5-FU is a pyrimidine analog antimetabolite that interferes with DNA and RNA syntheses. Orally active FUs are prodrugs of 5-FU and include capecitabine (Xeloda) and ftorafur (tegafur).[192,193] FUs are widely used in various GI malignancies. 5-FU by infusion is not myelosuppressive and therefore is easily combined with other chemotherapeutic agents, such as combining 5-FU with *leucovorin* and *oxaliplatin* (FOLFOX) or leucovorin and *irinotecan* (FOLFIRI).[194,195] FU crosses the blood-brain barrier well and may cause a spectrum of neurotoxicity. This is more commonly reported in patients with dihydropyrimidine dehydrogenase deficiency,[196] the enzyme responsible for more than 85% of pyrimidine catabolism. Partial or complete DPD deficiency is estimated to occur in 3% of the adult cancer population.[197] 5-FU produced acute cerebellar toxicity in as many as 2%–5% of patients when given by rapid infusion.[198–200] It typically appears after weeks or months of treatment but begins acutely with ataxia, dysarthria, and nystagmus. MRI and CSF studies are unrevealing. Patients usually make a good recovery. Rechallenge may result in recurrence. This syndrome is thought to be due to direct toxicity to granular cells, Purkinje cells, and deep nuclei.[198] 5-FU can also cause encephalopathy, sometimes associated with hyperammonemia[200,201]; one case has been reported to improve with thymidine infusion.[196] Other reported neurotoxicities include optic neuropathy, dystonia, parkinsonism, and seizures.[202,203] All of these complications are thought to be less common with the prolonged infusions generally in use today.

Five cases of multifocal leukoencephalopathy have been reported in patients on capecitabine,[204] presenting with various combinations of nausea, mental changes, dysarthria, and ataxia in the first 3–7 days. Brain MRI shows multiple white matter lesions with restricted diffusion that are hyperintense on T2 and fluid-attenuated inversion recovery (FLAIR) images and without mass effect. Hand-foot syndrome (erythrodysesthesia) is associated with prolonged capecitabine use and may mimic peripheral neuropathy.

3.1.2 Platinum

Platinum chemotherapy drugs (carboplatin, cisplatin, and oxaliplatin) are alkylating agents that inhibit DNA synthesis. They are used in the treatment of patients with various GI malignancies. Platinum chemotherapy drugs are often combined with other chemotherapeutic agents such as combining 5-FU with leucovorin and oxaliplatin (FOLFOX) or docetaxel, cisplatin, and 5-FU (DCF).[194,205] Peripheral neuropathy caused by platinum chemotherapy is usually cumulative and has common features to cisplatin, carboplatin, and oxaliplatin with some distinctive patterns. It is often described as paresthesia in stocking-glove distribution, areflexia, and loss of proprioception and vibratory sensation. Unfortunately, neurotoxicity may progress or appear after treatment disconsolation for up to several months (the "coasting" phenomenon).[206] Patients may experience gradual improvement; however, severe cases may have incomplete recovery and may be related to cumulative dose.[206] Loss of motor strength has been reported with cisplatin. The neurotoxicity resulting from carboplatin administration is less frequent (4%–6%) than that observed with cisplatin or oxaliplatin (15%–60%). The risk of carboplatin peripheral neurotoxicity increases in patients older than 65 and particularly with high doses.[207] Cisplatin can also lead to hearing loss and vestibular toxicity.

There are two patterns of neuropathy that are caused by oxaliplatin: an acute cold-aggravated but transient condition and a chronic form that usually starts after multiple exposures.[208] A less common aspect is acute sensory disturbance in the pharyngolaryngeal structures, characterized by subjective dyspnea, difficulty swallowing, and cold sensitivity of the oropharynx. The pharyngolaryngeal syndrome occurs in 0.3%–2.5% of patients. The cumulative neurosensory subacute syndrome is one of distal paresthesias, hypesthesia, and dysesthesia, with

decreased proprioception and incoordination, which can also include a component of impaired coordination. Motor symptoms may be present and may include fasciculations, tetanic spasms, and prolonged muscle contractions.[209] Reflexes are typically decreased.[210] Unusual features include Lhermitte's sign and urinary retention. It has been reported to worsen acutely after surgery.[211] Some patients experience "coasting."[209]

Electromyography (EMG) in the acute phase (< 3 weeks) shows repetitive compound muscle action potentials (CMAPs) and myotonic discharges that resolve within 3 weeks.[210] In the chronic phase, EMG shows decreased sensory amplitude of sensory nerve action potentials (SNAPs) with normal conduction velocity.[210] Generally, the diagnosis can be established clinically, and EMG is rarely needed.

Laboratory evidence indicates that oxaliplatin alters the kinetics of sodium movement at dorsal root ganglia neurons, probably due to an effect on voltage-gated Na^+ channels, providing a potential explanation for these toxicities.[212] Several polymorphisms of different genes, some related to these Na^+ channels, have been associated with increased incidence of this neuropathy.[209]

Since the neuropathy associated with oxaliplatin can be severe and irreversible, there have been efforts to prevent onset.[209] One study suggests that calcium and magnesium infusion before and after oxaliplatin protects against neuropathy, but there is concern that they may also interfere with its efficacy.[213] Other treatments that have been suggested to diminish the severity include exercise, amifostine, and alpha-lipoic acid,[214] but these are not established as standard treatments. Duloxetine has been recommended for treatment of oxaliplatin neuropathy.[209] Some evidence suggests that the effects for combined nortriptyline and gabapentin are additive for symptom relief.[214] Topical agents can also be considered; their efficacy is not known but they are generally felt to be low risk.[214]

3.1.3 Taxanes

Taxane chemotherapy drugs (docetaxel, paclitaxel, and albumin-bound paclitaxel) suppress the dynamic assembly of microtubules. They are used in patients with various GI malignancies. Taxane chemotherapy drugs are often combined with other chemotherapeutic agents such as combining paclitaxel and ramucirumab or gemcitabine and albumin-bound paclitaxel.[215,216] Taxanes can cause peripheral neuropathy and autonomic neuropathy.[217]

3.1.4 Gemcitabine

Gemcitabine is a pyrimidine antimetabolite that inhibits DNA synthesis. It is used to treat patients with pancreatic and bile duct cancer. Gemcitabine is often combined with other chemotherapeutic agents such as cisplatin and albumin-bound paclitaxel. Paresthesia has been reported in up to 10%–20%, probably due to mild sensory neuropathy. Autonomic neuropathy has been reported that improved 4 weeks after gemcitabine treatment was stopped.[218] Gemcitabine can cause posterior reversible encephalopathy syndrome (PRES), discussed later.[216,219,220]

3.1.5 Irinotecan

Irinotecan and its liposomal formulation are topoisomerase I inhibitors. They are used to treat patients with various GI malignancies. They are often combined with other chemotherapeutic agents such as 5-FU. They are not commonly associated with neurotoxicity.[221]

3.1.6 Trifluridine/tipiracil

Trifluridine, the active cytotoxic component of trifluridine/tipiracil, is a thymidine-based nucleic acid analog and it interferes with DNA synthesis. Tipiracil is a potent thymidine phosphorylase inhibitor that prevents the rapid degradation of trifluridine. It is used to treat patients with colorectal cancer and gastric cancer. It is not commonly associated with neurotoxicity.[222,223]

3.2 Targeted therapy

3.2.1 Vascular endothelial growth factor (VEGF) inhibitor

VEGF inhibitors (bevacizumab and ramucirumab) are recombinant monoclonal antibodies that bind to VEGF leading to antiangiogenesis effect. They are used to treat patients with various GI malignancies. They are often combined with other chemotherapeutic agents such as FOLFIRI and FOLFOX. The main toxicities are nonneurologic, including hypertension, proteinuria, thromboembolism, intratumoral hemorrhage, poor healing of surgical wounds, and GI organ perforation.[224] In clinical trials, patients on bevacizumab had a higher risk of cerebrovascular events (relative risk of 3.28), including ischemic strokes and cerebral hemorrhage.[225] Transient ischemic attacks have also been reported.

Like gemcitabine, they can cause PRES.[226,227] Typically, patients are present with any combination of headache, visual disturbance, abnormal mental function, and seizures, usually accompanied by hypertension.[228,229] The standard approach should be to aggressively lower the blood pressure and treat seizures if appropriate. The risk of cerebral hemorrhage and PRES argue for aggressively treating hypertension in patients on bevacizumab.

3.2.2 Human epidermal growth receptor 2 (HER2/neu) inhibitor

The monoclonal antibodies, trastuzumab and pertuzumab, and the tyrosine kinase inhibitor lapatinib target tumor cells that overexpress HER2/neu. In patients with tumors that overexpress HER2/neu, trastuzumab

is often combined with pertuzumab or lapatinib in patients with colorectal cancer and with chemotherapeutic agents in patients with gastric and esophageal cancers. They can cause peripheral neuropathy.[230,231]

3.2.3 Multitarget tyrosine kinase inhibitor (TKI)

Multitarget TKIs (sorafenib, lenvatinib, regorafenib, and cabozantinib) inhibit tumor growth by inhibiting multiple targets including intracellular RAF kinases (CRAF, BRAF, and mutant BRAF), cell surface kinase receptors (VEGFR-1, VEGFR-2, VEGFR-3, PDGFR-beta, cKIT, FLT-3, RET, and RET/PTC), and others. Sorafenib, lenvatinib, regorafenib, and cabozantinib are approved in patients with hepatocellular carcinoma. Regorafenib is also approved in patients with colorectal cancer. TKIs can cause peripheral sensory neuropathy and PRES.[10-13]

3.2.4 Epidermal growth factor receptor (EGFR) kinase inhibitor

EGFR inhibitor erlotinib inhibits EFGR tyrosine kinase activity. Erlotinib is approved in patients with pancreatic cancer in combination with gemcitabine.[232] Cetuximab and panitumumab are monoclonal antibodies that bind to EGFR (EGFR, HER1, c-ErbB-1) and inhibit the binding of EGF and other ligands. They are approved in patients with KRAS wild-type colorectal cancer. They can cause peripheral neuropathy, dizziness, paresthesia, and confusion.[233,234]

3.2.5 BRAF inhibitor

Encorafenib targets BRAF V600E and suppresses the MAPK pathway.[235] In BRAF-mutant colorectal cancer, EGFR-mediated pathway activation is a resistance mechanism to BRAF inhibition; the combination of a BRAF inhibitor and anti-EGFR agent has been shown to overcome this resistance mechanism. Encorafenib is approved in patients with BRAF V600E-mutant colorectal cancer in combination with cetuximab or panitumumab. Encorafenib can cause peripheral neuropathy.[236]

3.2.6 Fibroblast growth factor receptor (FGFR) kinase inhibitor

FGFR kinase inhibitor (pemigatinib) binds and inhibits FGFR1, FGFR2, and FGFR3. Pemigatinib is approved in patients with cholangiocarcinoma with FGFR2 fusion. Pemigatinib is not commonly associated with neurotoxicity.[237]

3.2.7 Isocitrate dehydrogenase 1 (IDH 1) inhibitor

IDH 1 inhibitor (ivosidenib) is an oral small-molecule inhibitor of the mutant isocitrate dehydrogenase 1 (IDH 1). Ivosidenib is listed as a treatment option in patients with cholangiocarcinoma with IDH 1 mutation. It can cause peripheral neuropathy and Guillain-Barre syndrome.[238]

3.2.8 Neurotrophic tropomyosin-related kinase (NTRK) inhibitor

NTRK inhibitors (larotrectinib and entrectinib) are potent inhibitors of the three tropomyosin receptor kinase (TRK) proteins, TRKA, TRKB, and TRKC that are encoded by NTRK1, NTRK2, and NTRK3 genes, respectively. Entrectinib also inhibits proto-oncogenic tyrosine-protein kinase ROS1 and anaplastic lymphoma kinase (ALK). They are approved in patients with cancers with NTRK gene fusion. Larotrectinib can cause delirium, dysarthria, dizziness, gait disturbance, paresthesia, memory impairment, tremor, and encephalopathy.[239,240]

3.3 Immunotherapy

3.3.1 Immune checkpoint (ICP) inhibitors

Checkpoint inhibitor drugs that target PD-1 or PD-L1: programmed death-1 (PD-1) inhibitors (nivolumab and pembrolizumab) are human monoclonal antibodies that inhibit PD-1 activity by binding to the PD-1 receptor to block the ligands PD-L1 and PD-L2 from binding. This leads to disrupting the negative PD-1 receptor signaling that regulates T-cell activation, reverses T-cell suppression, and induces antitumor responses.[241,242] Programmed death ligand-1 (PD-L1) inhibitor (atezolizumab) is a human monoclonal antibody that binds PD-L1 and prevents its interaction with B7.1 (CD80) receptor and that restores antitumor T-cell function.[243] Nivolumab and pembrolizumab are approved in various GI malignancies.[14,15] Atezolizumab is approved in patients with HCC.[16] PD-1 and PD-L1 inhibitors can cause headache, dizziness, peripheral neuropathy, neuritis, peripheral nerve palsy, facial and abducens nerve palsy, demyelinating disease, encephalitis (limbic/lymphocytic/viral; maybe immune-mediated), aseptic meningitis, myasthenia gravis, and Guillain-Barre syndrome. Some of the neurological adverse events are immune-mediated.[244,245]

Checkpoint inhibitor drugs that target CTLA-4: cytotoxic T-lymphocyte-associated antigen (CTLA-4) inhibitor (ipilimumab) is a recombinant human monoclonal antibody that binds to the CTLA-4 and that allows enhanced T-cell activation and response. Combining nivolumab (anti-PD-1) and ipilimumab (anti-CTLA-4) enhances T-cell function resulting in improved antitumor response.[246] Ipilimumab, in combination with nivolumab, is effective in patients with microsatellite instability-high colorectal cancer and HCC.[246] Ipilimumab can cause headache, peripheral neuropathy, meningitis, encephalitis, myasthenia gravis, and Guillain-Barre syndrome. Some of the neurological adverse events are immune-mediated.

Permissions summary

One figure, not previously published, original material. No permissions needed.

References

1. Siegel RL, Miller KD, Jemal A. Cancer statistics, 2020. *CA Cancer J Clin.* 2020;70(1):7–30.
2. American Cancer Society. *Cancer Facts & Figures 2020.* Atlanta, GA: American Cancer Society; 2020.
3. Go PH, Klaassen Z, Meadows MC, Chamberlain RS. Gastrointestinal cancer and brain metastasis: a rare and ominous sign. *Cancer.* 2011;117(16):3630–3640.
4. Colvin H, Yamamoto K, Wada N, Mori M. Hereditary gastric cancer syndromes. *Surg Oncol Clin N Am.* 2015;24(4):765–777.
5. Chen W, Wang J, Li X, et al. Prognostic significance of BRCA1 expression in gastric cancer. *Med Oncol.* 2013;30(1):423.
6. Masciari S, Dewanwala A, Stoffel EM, et al. Gastric cancer in individuals with Li-Fraumeni syndrome. *Genet Med.* 2011;13(7):651–657.
7. Jakubowska A, Nej K, Huzarski T, Scott RJ, Lubinski J. BRCA2 gene mutations in families with aggregations of breast and stomach cancers. *Br J Cancer.* 2002;87(8):888–891.
8. National Cancer Institute. *Physician Query Data (PDQ). Adult Primary Liver Cancer Symptoms, Tests, Prognosis, and Stages—Patient Version;* 2020. https://www.cancer.gov/types/liver/patient/about-adult-liver-cancer-pdq. Accessed 18 October 2020.
9. National Comprehensive Cancer Network. *NCCN Clinical Practice Guidelines in Oncology: Hepatobiliary Cancers;* 2020. https://www.nccn.org/professionals/physician_gls/pdf/hepatobiliary.pdf. Accessed 20 June 2020.
10. Abou-Alfa GK, Meyer T, Cheng AL, et al. Cabozantinib in patients with advanced and progressing hepatocellular carcinoma. *N Engl J Med.* 2018;379(1):54–63.
11. Kudo M, Finn RS, Qin S, et al. Lenvatinib versus sorafenib in first-line treatment of patients with unresectable hepatocellular carcinoma: a randomised phase 3 non-inferiority trial. *Lancet.* 2018;391(10126):1163–1173.
12. Bruix J, Qin S, Merle P, et al. Regorafenib for patients with hepatocellular carcinoma who progressed on sorafenib treatment (RESORCE): a randomised, double-blind, placebo-controlled, phase 3 trial. *Lancet.* 2017;389(10064):56–66.
13. Llovet JM, Ricci S, Mazzaferro V, et al. Sorafenib in advanced hepatocellular carcinoma. *N Engl J Med.* 2008;359(4):378–390.
14. Zhu AX, Finn RS, Edeline J, et al. Pembrolizumab in patients with advanced hepatocellular carcinoma previously treated with sorafenib (KEYNOTE-224): a non-randomised, open-label phase 2 trial. *Lancet Oncol.* 2018;19(7):940–952.
15. Overman MJ, McDermott R, Leach JL, et al. Nivolumab in patients with metastatic DNA mismatch repair-deficient or microsatellite instability-high colorectal cancer (CheckMate 142): an open-label, multicentre, phase 2 study. *Lancet Oncol.* 2017;18(9):1182–1191.
16. Finn RS, Qin S, Ikeda M, et al. Atezolizumab plus bevacizumab in unresectable hepatocellular carcinoma. *N Engl J Med.* 2020;382(20):1894–1905.
17. Petrick JL, Yang B, Altekruse SF, et al. Risk factors for intrahepatic and extrahepatic cholangiocarcinoma in the United States: a population-based study in SEER-Medicare. *PLoS One.* 2017;12(10):e0186643.
18. National Comprehensive Cancer Network. *NCCN Clinical Practice Guidelines in Oncology: Hepatobiliary Cancers;* 2020. https://www.nccn.org/professionals/physician_gls/pdf/hepatobiliary.pdf. Accessed 18 October 2020.
19. Whitcomb DC. Hereditary pancreatitis: new insights into acute and chronic pancreatitis. *Gut.* 1999;45(3):317–322.
20. Siegel RL, Miller KD, Goding Sauer A, et al. Colorectal cancer statistics, 2020. *CA Cancer J Clin.* 2020;70(3):145–164.
21. Hamilton SR, Liu B, Parsons RE, et al. The molecular basis of Turcot's syndrome. *N Engl J Med.* 1995;332(13):839–847.

22. Cavenee W, Hawkins C, Burger P, Leung S, Van Meir E, Tabori U. Turcot syndrome. In: Louis DLOH, Wiestler OD, Cavenee WK, eds. *World Health Classification of Tumours*. 4th ed. Lyon: International Agency for Research on Cancer (IARC); 2016:317–318.

23 Chung D, Rodgers L. Clinical manifestations and diagnosis of familial adenomatous polyposis; 2020. http://www.uptodate.com. Accessed 1 October 2020.

24 Win AK. Lynch syndrome (hereditary nonpolyposis colorectal cancer): clinical manifestations and diagnosis; 2020. http://www.uptodate.com. Accessed 1 October 2020.

25 Grover S, Stoffel E. MUTYH-associated polyposis; 2020. http://www.uptodate.com. Accessed 1 October 2020.

26. Oronsky B, Ma PC, Morgensztern D, Carter CA. Nothing but NET: a review of neuroendocrine tumors and carcinomas. *Neoplasia*. 2017;19(12):991–1002.

27. Kim JY, Hong SM, Ro JY. Recent updates on grading and classification of neuroendocrine tumors. *Ann Diagn Pathol*. 2017;29:11–16.

28. Patchell RA, Posner JB. Neurologic complications of carcinoid. *Neurology*. 1986;36(6):745–749.

29. Scott S, Antwi-Yeboah Y, Bucur S. Metastatic carcinoid tumour with spinal cord compression. *J Surg Case Rep*. 2012;2012(7):5.

30. Gray JA, Nishikawa H, Jamous MA, Grahame-Smith DG. Spinal cord compression due to carcinoid metastasis. *Postgrad Med J*. 1988;64(755):703–705.

31. Kirkpatrick DB, Dawson E, Haskell CM, Batzdorf U. Metastatic carcinoid presenting as a spinal tumor. *Surg Neurol*. 1975;4(3):283–287.

32. Tanabe M, Akatsuka K, Umeda S, et al. Metastasis of carcinoid to the arch of the axis in a multiple endocrine neoplasia patient: a case report. *Spine J*. 2008;8(5):841–844.

33. Achrol AS, Rennert RC, Anders C, et al. Brain metastases. *Nat Rev Dis Primers*. 2019;5(1):5.

34. Junck L, Zalupski MM. Neurological complications of GI cancers. In: Newton HB, Malkin MG, eds. *Neurological Complications of Systemic Cancer and Antineoplastic Therapy*. 1st ed. Boca Raton, FL: CRC Press; 2010:312–335.

35. Esmaeilzadeh M, Majlesara A, Faridar A, et al. Brain metastasis from gastrointestinal cancers: a systematic review. *Int J Clin Pract*. 2014;68(7):890–899.

36. Nussbaum ES, Djalilian HR, Cho KH, Hall WA. Brain metastases. Histology, multiplicity, surgery, and survival. *Cancer*. 1996;78(8):1781–1788.

37. National Comprehensive Center Network. *National Comprehensive Center Network Clinical Practice Guidelines in Oncology: Central Nervous System Cancers*; 2020. https://www.nccn.org/professionals/physician_gls/pdf/cns.pdf. Accessed 17 October 2020.

38. Gaspar L, Scott C, Rotman M, et al. Recursive partitioning analysis (RPA) of prognostic factors in three Radiation Therapy Oncology Group (RTOG) brain metastases trials. *Int J Radiat Oncol Biol Phys*. 1997;37(4):745–751.

39. Sperduto PW, Fang P, Li J, et al. Estimating survival in patients with gastrointestinal cancers and brain metastases: an update of the graded prognostic assessment for gastrointestinal cancers (GI-GPA). *Clin Transl Radiat Oncol*. 2019;18:39–45.

40. Gabrielsen TO, Eldevik OP, Orringer MB, Marshall BL. Esophageal carcinoma metastatic to the brain: clinical value and cost-effectiveness of routine enhanced head CT before esophagectomy. *AJNR Am J Neuroradiol*. 1995;16(9):1915–1921.

41. Weinberg JS, Suki D, Hanbali F, Cohen ZR, Lenzi R, Sawaya R. Metastasis of esophageal carcinoma to the brain. *Cancer*. 2003;98(9):1925–1933.

42. Khuntia D, Sajja R, Chidel MA, et al. Factors associated with improved survival in patients with brain metastases from esophageal cancer: a retrospective review. *Technol Cancer Res Treat*. 2003;2(3):267–272.

43. Ogawa K, Toita T, Sueyama H, et al. Brain metastases from esophageal carcinoma: natural history, prognostic factors, and outcome. *Cancer*. 2002;94(3):759–764.

44. Yoshida S. Brain metastasis in patients with esophageal carcinoma. *Surg Neurol*. 2007;67(3):288–290.

45. Kothari N, Mellon E, Hoffe SE, et al. Outcomes in patients with brain metastasis from esophageal carcinoma. *J Gastrointest Oncol*. 2016;7(4):562–569.

46. York JE, Stringer J, Ajani JA, Wildrick DM, Gokaslan ZL. Gastric cancer and metastasis to the brain. *Ann Surg Oncol*. 1999;6(8):771–776.

47. Kasakura Y, Fujii M, Mochizuki F, Suzuki T, Takahashi T. Clinicopathological study of brain metastasis in gastric cancer patients. *Surg Today*. 2000;30(6):485–490.

48. Kim M, Na DL, Park SH, Jeon BS, Roh JK. Nervous system involvement by metastatic hepatocellular carcinoma. *J Neurooncol*. 1998;36(1):85–90.

49. Katyal S, Oliver 3rd JH, Peterson MS, Ferris JV, Carr BS, Baron RL. Extrahepatic metastases of hepatocellular carcinoma. *Radiology*. 2000;216(3):698–703.

50. Ishii H, Furuse J, Kinoshita T, et al. Extrahepatic spread from hepatocellular carcinoma: who are candidates for aggressive anti-cancer treatment? *Jpn J Clin Oncol*. 2004;34(12):733–739.

51. Natsuizaka M, Omura T, Akaike T, et al. Clinical features of hepatocellular carcinoma with extrahepatic metastases. *J Gastroenterol Hepatol*. 2005;20(11):1781–1787.

52. Menis J, Fontanella C, Follador A, Fasola G, Aprile G. Brain metastases from gastrointestinal tumours: tailoring the approach to maximize the outcome. *Crit Rev Oncol Hematol*. 2013;85(1):32–44.

53. Murakami K, Nawano S, Moriyama N, et al. Intracranial metastases of hepatocellular carcinoma: CT and MRI. *Neuroradiology*. 1996;38(Suppl. 1):S31–S35.

54. Agrawal A, Agrawal CS, Kumar A, Tiwari A, Lakshmi R, Yadav R. Gall bladder carcinoma: stroke as first manifestation. *Indian J Gastroenterol*. 2006;25(6):316.

55. Gudesblatt MS, Sencer W, Sacher M, Lanzieri CF, Song SK. Cholangiocarcinoma presenting as a cerebellar metastasis: case report and review of the literature. *J Comput Tomogr*. 1984;8(3):191–195.

56. Takano S, Yoshii Y, Owada T, Shirai S, Nose T. Central nervous system metastasis from gallbladder carcinoma—case report. *Neurol Med Chir (Tokyo)*. 1991;31(12):782–786.

57. Smith WD, Sinar J, Carey M. Sagittal sinus thrombosis and occult malignancy. *J Neurol Neurosurg Psychiatry*. 1983;46(2):187–188.

58. Kawamata T, Kawamura H, Kubo O, Sasahara A, Yamazato M, Hori T. Central nervous system metastasis from gallbladder carcinoma mimicking a meningioma. Case illustration. *J Neurosurg*. 1999;91(6):1059.

59. Aronson SM, Garcia JH, Aronson BE. Metastatic neoplasms of the brain: their frequency in relation to age. *Cancer*. 1964;17:558–563.

60. Park KS, Kim M, Park SH, Lee KW. Nervous system involvement by pancreatic cancer. *J Neurooncol*. 2003;63(3):313–316.

61. Cascino TL, Leavengood JM, Kemeny N, Posner JB. Brain metastases from colon cancer. *J Neurooncol*. 1983;1(3):203–209.

62. Alden TD, Gianino JW, Saclarides TJ. Brain metastases from colorectal cancer. *Dis Colon Rectum*. 1996;39(5):541–545.

63. Hammoud MA, McCutcheon IE, Elsouki R, Schoppa D, Patt YZ. Colorectal carcinoma and brain metastasis: distribution, treatment, and survival. *Ann Surg Oncol*. 1996;3(5):453–463.

64. Nieder C, Niewald M, Schnabel K. Hirnmetastasen von kolon- und rektumkarzinomen. *Wien Klin Woc*. 1997;109:239–243.

65. Wronski M, Arbit E. Resection of brain metastases from colorectal carcinoma in 73 patients. *Cancer*. 1999;85(8):1677–1685.

66. Ko FC, Liu JM, Chen WS, Chiang JK, Lin TC, Lin JK. Risk and patterns of brain metastases in colorectal cancer: 27-year experience. *Dis Colon Rectum*. 1999;42(11):1467–1471.

67. Schouten LJ, Rutten J, Huveneers HA, Twijnstra A. Incidence of brain metastases in a cohort of patients with carcinoma of the breast, colon, kidney, and lung and melanoma. *Cancer.* 2002;94(10):2698–2705.

68. Sundermeyer ML, Meropol NJ, Rogatko A, Wang H, Cohen SJ. Changing patterns of bone and brain metastases in patients with colorectal cancer. *Clin Colorectal Cancer.* 2005;5(2):108–113.

69. Amichetti M, Lay G, Dessi M, et al. Results of whole brain radiation therapy in patients with brain metastases from colorectal carcinoma. *Tumori.* 2005;91(2):163–167.

70. Junck L. Leptomeningeal metastasis. In: Gilman S, Goldstein GW, eds. *Medlink Neurology.* San Diego, CA: MedLink Corporation; 2008.

71. Giglio P, Weinberg JS, Forman AD, Wolff R, Groves MD. Neoplastic meningitis in patients with adenocarcinoma of the gastrointestinal tract. *Cancer.* 2005;103(11):2355–2362.

72. Lisenko Y, Kumar AJ, Yao J, Ajani J, Ho L. Leptomeningeal carcinomatosis originating from gastric cancer: report of eight cases and review of the literature. *Am J Clin Oncol.* 2003;26(2):165–170.

73. Lee JL, Kang YK, Kim TW, et al. Leptomeningeal carcinomatosis in gastric cancer. *J Neurooncol.* 2004;66(1–2):167–174.

74. Psathakis D, Schiedeck TH, Krug F, Oevermann E, Kujath P, Bruch HP. Ordinary colorectal adenocarcinoma vs. primary colorectal signet-ring cell carcinoma: study matched for age, gender, grade, and stage. *Dis Colon Rectum.* 1999;42(12):1618–1625.

75. Henson DE, Dittus C, Younes M, Nguyen H, Albores-Saavedra J. Differential trends in the intestinal and diffuse types of gastric carcinoma in the United States, 1973–2000: increase in the signet ring cell type. *Arch Pathol Lab Med.* 2004;128(7):765–770.

76. Lauwers G, Kumarasinghe P. Gastric cancer: pathology and molecular pathogenesis. In: Goldberg RM, ed. *UpToDate.* Waltham, MA: UptoDate; 2020.

77. Batool A, Kasi A. Leptomeningeal carcinomatosis. Treasure Island, FL: StatPearls Publishing; 2020.

78. Giglio P, Tremont-Lukats IW, Groves MD. Response of neoplastic meningitis from solid tumors to oral capecitabine. *J Neurooncol.* 2003;65(2):167–172.

79. Thomas KH, Ramirez RA. Leptomeningeal disease and the evolving role of molecular targeted therapy and immunotherapy. *Ochsner J.* 2017;17(4):362–378.

80. Lukas RV, Mata-Machado NA, Nicholas MK, Salgia R, Antic T, Villaflor VM. Leptomeningeal carcinomatosis in esophageal cancer: a case series and systematic review of the literature. *Dis Esophagus.* 2015;28(8):772–781.

81. Delaunoit T, Boige V, Belloc J, et al. Gastric linitis adenocarcinoma and carcinomatous meningitis: an infrequent but aggressive association—report of four cases. *Ann Oncol.* 2001;12(6):869–871.

82. Lee JH, Shin JH, Kim DS, et al. A case of Lambert-Eaton myasthenic syndrome associated with atypical bronchopulmonary carcinoid tumor. *J Korean Med Sci.* 2004;19(5):753–755.

83. Miyagui T, Luchemback L, Teixeira GH, de Azevedo KM. Meningeal carcinomatosis as the initial manifestation of a gallbladder adenocarcinoma associated with a Krukenberg tumor. *Rev Hosp Clin Fac Med Sao Paulo.* 2003;58(3):169–172.

84. Huffman JL, Yeatman TJ, Smith JB. Leptomeningeal carcinomatosis: a sequela of cholangiocarcinoma. *Am Surg.* 1997;63(4):310–313.

85. Junck L. Metastases to the meninges, spine, and plexus. In: Noseworthy JH, ed. *Neurological Therapeutics: Principles and Practice.* London: Informa Healthcare; 2006:912–933.

86. Al-Qurainy R, Collis E. Metastatic spinal cord compression: diagnosis and management. *BMJ.* 2016;353:i2539.

87. Ropper AE, Ropper AH. Acute spinal cord compression. *N Engl J Med.* 2017;376(14):1358–1369.

88. Switlyk MD, Hole KH, Skjeldal S, et al. MRI and neurological findings in patients with spinal metastases. *Acta Radiol.* 2012;53(10):1164–1172.

89. Sorensen S, Helweg-Larsen S, Mouridsen H, Hansen HH. Effect of high-dose dexamethasone in carcinomatous metastatic spinal cord compression treated with radiotherapy: a randomised trial. *Eur J Cancer.* 1994;30A(1):22–27.

90. Patchell RA, Tibbs PA, Regine WF, et al. Direct decompressive surgical resection in the treatment of spinal cord compression caused by metastatic cancer: a randomised trial. *Lancet.* 2005;366(9486):643–648.

91. Ross JR, Saunders Y, Edmonds PM, Patel S, Broadley KE, Johnston SR. Systematic review of role of bisphosphonates on skeletal morbidity in metastatic cancer. *BMJ.* 2003;327(7413):469.

92. Schiff D, O'Neill BP, Suman VJ. Spinal epidural metastasis as the initial manifestation of malignancy: clinical features and diagnostic approach. *Neurology.* 1997;49(2):452–456.

93. Tibble JA, Ireland AC. Carcinoma of the oesophagus causing paraparesis by direct extension to the spinal cord. *Eur J Gastroenterol Hepatol.* 1995;7(10):1003–1004.

94. Kudo M, Izumi N, Kubo S, et al. Report of the 20th Nationwide follow-up survey of primary liver cancer in Japan. *Hepatol Res.* 2020;50(1):15–46.

95. Yoon KT, Kim JK, Kim DY, et al. Role of 18F-fluorodeoxyglucose positron emission tomography in detecting extrahepatic metastasis in pretreatment staging of hepatocellular carcinoma. *Oncology.* 2007;72(Suppl. 1):104–110.

96. Doval DC, Bhatia K, Vaid AK, et al. Spinal cord compression secondary to bone metastases from hepatocellular carcinoma. *World J Gastroenterol.* 2006;12(32):5247–5252.

97. Bostel T, Forster R, Schlampp I, et al. Spinal bone metastases in colorectal cancer: a retrospective analysis of stability, prognostic factors and survival after palliative radiotherapy. *Radiat Oncol.* 2017;12(1):115.

98. Brown PD, Stafford SL, Schild SE, Martenson JA, Schiff D. Metastatic spinal cord compression in patients with colorectal cancer. *J Neurooncol.* 1999;44(2):175–180.

99. Rades D, Dahm-Daphi J, Rudat V, et al. Is short-course radiotherapy with high doses per fraction the appropriate regimen for metastatic spinal cord compression in colorectal cancer patients? *Strahlenther Onkol.* 2006;182(12):708–712.

100. Mendez JS, DeAngelis LM. Metastatic complications of cancer involving the central and peripheral nervous systems. *Neurol Clin.* 2018;36(3):579–598.

101. Schiff D, Sherman J, Brown PD. Metastatic tumours: spinal cord, plexus, and peripheral nerve. In: Batchelor TT, Nishikawa R, Tarbell NJ, Weller M, eds. *Oxford Textbook of Neuro-Oncology.* Oxford: Oxford University Press; 2017:223–232.

102. Jaeckle KA, Young DF, Foley KM. The natural history of lumbosacral plexopathy in cancer. *Neurology.* 1985;35(1):8–15.

103. Pettigrew LC, Glass JP, Maor M, Zornoza J. Diagnosis and treatment of lumbosacral plexopathies in patients with cancer. *Arch Neurol.* 1984;41(12):1282–1285.

104. Thomas JE, Cascino TL, Earle JD. Differential diagnosis between radiation and tumor plexopathy of the pelvis. *Neurology.* 1985;35(1):1–7.

105. Castellanos AM, Glass JP, Yung WK. Regional nerve injury after intra-arterial chemotherapy. *Neurology.* 1987;37(5):834–837.

106. Taylor BV, Kimmel DW, Krecke KN, Cascino TL. Magnetic resonance imaging in cancer-related lumbosacral plexopathy. *Mayo Clin Proc.* 1997;72(9):823–829.

107. Hathaway PB, Mankoff DA, Maravilla KR, et al. Value of combined FDG PET and MR imaging in the evaluation of suspected recurrent local-regional breast cancer: preliminary experience. *Radiology.* 1999;210(3):807–814.

108. Ampil FL. Palliative irradiation of carcinomatous lumbosacral plexus neuropathy. *Int J Radiat Oncol Biol Phys.* 1986;12(9):1681–1686.

109. Warade AC, Jha AK, Pattankar S, Desai K. Radiation-induced brachial plexus neuropathy: a review. *Neurol India.* 2019;67(Supplement):S47–S52.

110. Pritchard J, Anand P, Broome J, et al. Double-blind randomized phase II study of hyperbaric oxygen in patients with radiation-induced brachial plexopathy. *Radiother Oncol.* 2001;58(3):279–286.

111. Martel S, De Angelis F, Lapointe E, Larue S, Speranza G. Paraneoplastic neurologic syndromes: clinical presentation and management. *Curr Probl Cancer.* 2014;38(4):115–134.

112. Rodríguez Páez LR, Yurgaky SJ, Otero Regino W, Faizal M. Síndromes paraneoplásicos en tumores gastrointestinales. Revisión de tema. *Rev Colomb Gastroenterol.* 2017;32(3):230–244.

113. Sigurgeirsson B, Lindelof B, Edhag O, Allander E. Risk of cancer in patients with dermatomyositis or polymyositis. A population-based study. *N Engl J Med.* 1992;326(6):363–367.

114. Chow WH, Gridley G, Mellemkjaer L, McLaughlin JK, Olsen JH, Fraumeni Jr JF. Cancer risk following polymyositis and dermatomyositis: a nationwide cohort study in Denmark. *Cancer Causes Control.* 1995;6(1):9–13.

115. Buchbinder R, Hill CL. Malignancy in patients with inflammatory myopathy. *Curr Rheumatol Rep.* 2002;4(5):415–426.

116. Stockton D, Doherty VR, Brewster DH. Risk of cancer in patients with dermatomyositis or polymyositis, and follow-up implications: a Scottish population-based cohort study. *Br J Cancer.* 2001;85(1):41–45.

117. Hill CL, Zhang Y, Sigurgeirsson B, et al. Frequency of specific cancer types in dermatomyositis and polymyositis: a population-based study. *Lancet.* 2001;357(9250):96–100.

118. Menezes RB, de Lucena AF, Maia FM, Marinho AR. Limbic encephalitis as the presenting symptom of oesophageal adenocarcinoma: another cancer to search? *BMJ Case Rep.* 2013;2013.

119. Gultekin SH, Rosenfeld MR, Voltz R, Eichen J, Posner JB, Dalmau J. Paraneoplastic limbic encephalitis: neurological symptoms, immunological findings and tumour association in 50 patients. *Brain.* 2000;123(Pt 7):1481–1494.

120. Mundiyanapurath S, Jarius S, Probst C, Stocker W, Wildemann B, Bosel J. GABA-B-receptor antibodies in paraneoplastic brainstem encephalitis. *J Neuroimmunol.* 2013;259(1–2):88–91.

121. Xia K, Saltzman JR, Carr-Locke DL. Anti-Yo antibody-mediated paraneoplastic cerebellar degeneration in a man with esophageal adenocarcinoma. *MedGenMed.* 2003;5(3):18.

122. Sutton IJ, Fursdon Davis CJ, Esiri MM, Hughes S, Amyes ER, Vincent A. Anti-Yo antibodies and cerebellar degeneration in a man with adenocarcinoma of the esophagus. *Ann Neurol.* 2001;49(2):253–257.

123. Rossor AM, Perry F, Botha A, Norwood F. Opsoclonus myoclonus syndrome due to squamous cell carcinoma of the oesophagus. *BMJ Case Rep.* 2014;2014.

124. Shirafuji T, Kanda F, Sekiguchi K, et al. Anti-Hu-associated paraneoplastic encephalomyelitis with esophageal small cell carcinoma. *Intern Med.* 2012;51(17):2423–2427.

125. Urai Y, Matsumoto K, Shimamura M, et al. Paraneoplastic necrotizing myelopathy in a patient with advanced esophageal cancer: an autopsied case report. *J Neurol Sci.* 2009;280(1–2):113–117.

126. Khealani BA, Qureshi R, Wasay M. Motor neuronopathy associated with adenocarcinoma of esophagus. *J Pak Med Assoc.* 2004;54(3):165–166.

127. Shimoda T, Koizumi W, Tanabe S, et al. Small-cell carcinoma of the esophagus associated with a paraneoplastic neurological syndrome: a case report documenting a complete response. *Jpn J Clin Oncol.* 2006;36(2):109–112.

128. Tola-Arribas MA, Canibano-Gonzalez MA. Guillain-Barre syndrome associated with gastric adenocarcinoma. Paraneoplastic origin or coincidence? *Rev Neurol.* 2001;33(8):797–798.

129. Mostoufizadeh S, Souri M, de Seze J. A case of paraneoplastic demyelinating motor polyneuropathy. *Case Rep Neurol.* 2012;4(1):71–76.

130. Tojo K, Tokuda T, Yazaki M, et al. Paraneoplastic sensorimotor neuropathy and encephalopathy associated with anti-alpha-enolase antibody in a case of gastric adenocarcinoma. *Eur Neurol.* 2004;51(4):231–233.

131. Bataller L, Graus F, Saiz A, Vilchez JJ, Spanish Opsoclonus-Myoclonus Study Group. Clinical outcome in adult onset idiopathic or paraneoplastic opsoclonus-myoclonus. *Brain.* 2001;124(Pt 2):437–443.

132. Tanaka Y, Nishida H, Yamada O, Takahashi Y, Moriwaki H. Intrathecal glutamate receptor antibodies in a patient with elderly-onset refractory epilepsy. *Rinsho Shinkeigaku.* 2003;43(6):345–349.

133. Kurokawa T, Taniwaki T, Arakawa K, et al. A case of paraneoplastic cerebellar degeneration with resting tremor. *Rinsho Shinkeigaku.* 2001;41(1):24–30.

134. Meglic B, Graus F, Grad A. Anti-Yo-associated paraneoplastic cerebellar degeneration in a man with gastric adenocarcinoma. *J Neurol Sci.* 2001;185(2):135–138.

135. Debes JD, Lagarde SM, Hulsenboom E, et al. Anti-Yo-associated paraneoplastic cerebellar degeneration in a man with adenocarcinoma of the gastroesophageal junction. *Dig Surg.* 2007;24(5):395–397.

136. Goto A, Kusumi M, Wakutani Y, Nakaso K, Kowa H, Nakashima K. Anti-Yo antibody associated paraneoplastic cerebellar degeneration with gastric adenocarcinoma in a male patient: a case report. *Rinsho Shinkeigaku.* 2006;46(2):144–147.

137. Kikuchi H, Yamada T, Okayama A, et al. Anti-Ri-associated paraneoplastic cerebellar degeneration without opsoclonus in a patient with a neuroendocrine carcinoma of the stomach. *Fukuoka Igaku Zasshi.* 2000;91(4):104–109.

138. Cayla J, Bouchacourt P, Rondier J. Parsonage-turner syndrome associated with superficial cancer of the stomach. A coincidence or paraneoplastic syndrome? *Rev Rhum Mal Osteoartic.* 1984;51(5):281–282.

139. Ichimura M, Yamamoto M, Kobayashi Y, et al. Tissue distribution of pathological lesions and Hu antigen expression in paraneoplastic sensory neuronopathy. *Acta Neuropathol.* 1998;95(6):641–648.

140. Naka T, Yorifuji S, Fujimura H, Takahashi M, Tarui S. A case of paraneoplastic neuropathy with necrotizing arteritis localized in the peripheral nervous system. *Rinsho Shinkeigaku.* 1991;31(4):427–432.

141. Kassianides C, Kew MC. The clinical manifestations and natural history of hepatocellular carcinoma. *Gastroenterol Clin North Am.* 1987;16(4):553–562.

142. Coeytaux A, Kressig R, Zulian GB. Hepatocarcinoma with concomitant paraneoplastic encephalomyelitis. *J Palliat Care.* 2001;17(1):59–60.

143. Norris S, Rajendiran S, Sheahan K, et al. Noncirrhotic hepatoma presenting with paraneoplastic neurologic manifestations: two cases. *Am J Gastroenterol.* 1997;92(10):1923–1926.

144. Phanthumchinda K, Rungruxsirivorn S. Encephaloradiculopathy: a non-metastatic complication of hepatocellular carcinoma. *J Med Assoc Thai.* 1991;74(5):288–291.

145. Chang PY, Yang CH, Yang CM. Cancer-associated retinopathy in a patient with hepatocellular carcinoma: case report and literature review. *Retina.* 2005;25(8):1093–1096.

146. Wilfong AA, Parke JT, McCrary 3rd JA. Opsoclonus-myoclonus with Beckwith-Wiedemann syndrome and hepatoblastoma. *Pediatr Neurol.* 1992;8(1):77–79.

147. Turgut N, Karagol H, Celik Y, Uygun K, Reyhani A. Subacute motor neuronopathy associated with hepatocellular carcinoma. *J Neurooncol.* 2007;83(1):95–96.

148. Calvey HD, Melia WM, Williams R. Polyneuropathy: an unreported non-metastatic complication of primary hepatocellular carcinoma. *Clin Oncol.* 1983;9(3):199–202.

149. Abe K, Sugai F. Chronic inflammatory demyelinating polyneuropathy accompanied by carcinoma. *J Neurol Neurosurg Psychiatry.* 1998;65(3):403–404.

150. Sugai F, Abe K, Fujimoto T, et al. Chronic inflammatory demyelinating polyneuropathy accompanied by hepatocellular carcinoma. *Intern Med.* 1997;36(1):53–55.

151. Arguedas MR, McGuire BM. Hepatocellular carcinoma presenting with chronic inflammatory demyelinating polyradiculoneuropathy. *Dig Dis Sci.* 2000;45(12):2369–2373.

152. Walcher J, Witter T, Rupprecht HD. Hepatocellular carcinoma presenting with paraneoplastic demyelinating polyneuropathy and PR3-antineutrophil cytoplasmic antibody. *J Clin Gastroenterol.* 2002;35(4):364–365.

153. Hasegawa K, Uesugi H, Kubota K, et al. Polymyositis as a paraneoplastic manifestation of hepatocellular carcinoma. *Hepatogastroenterology.* 2000;47(35):1425–1427.

154. Kishore D, Khurana V, Raj A, Gambhir IS, Diwaker A. Hepatocellular carcinoma presenting as polymyositis: a paraneoplastic syndrome. *Ann Saudi Med.* 2011;31(5):533–535.

155. Tekaya R, Abdelghni K, Abdelmoula L, Ben Hadj Yahia C, Chaabouni L, Zouari R. Hepatocellular carcinoma with polymyositis as an initial symptom: a case report. *Acta Clin Belg.* 2011;66(1):53–54.

156. Corcia P, De Toffol B, Hommet C, Saudeau D, Autret A. Paraneoplastic opsoclonus associated with cancer of the gall bladder. *J Neurol Neurosurg Psychiatry.* 1997;62(3):293.

157. Phan TG, Hersch M, Zagami AS. Guillain-Barre syndrome and adenocarcinoma of the gall bladder: a paraneoplastic phenomenon? *Muscle Nerve.* 1999;22(1):141–142.

158. Antoine JC, Mosnier JF, Lapras J, et al. Chronic inflammatory demyelinating polyneuropathy associated with carcinoma. *J Neurol Neurosurg Psychiatry.* 1996;60(2):188–190.

159. Kaltsas G, Androulakis II, de Herder WW, Grossman AB. Paraneoplastic syndromes secondary to neuroendocrine tumours. *Endocr Relat Cancer.* 2010;17(3):R173–R193.

160. Hernandez-Echebarria L, Saiz A, Ares A, et al. Paraneoplastic encephalomyelitis associated with pancreatic tumor and anti-GAD antibodies. *Neurology.* 2006;66(3):450–451.

161. Tahrani AA, Sharma S, Rangan S, Macleod AF. A patient with worsening mobility: a diagnostic challenge. *Eur J Intern Med.* 2008;19(4):292–294.

162. Salmeron-Ato P, Medrano V, Morales-Ortiz A, et al. Paraneoplastic cerebellar degeneration as initial presentation of a pancreatic small-cell carcinoma. *Rev Neurol.* 2002;35(12):1112–1115.

163. Caras S, Laurie S, Cronk W, Tompkins W, Brashear R, McCallum RW. Case report: pancreatic cancer presenting with paraneoplastic gastroparesis. *Am J Med Sci.* 1996;312(1):34–36.

164. Honnorat J, Trillet M, Antoine JC, Aguera M, Dalmau J, Graus F. Paraneoplastic opsomyoclonus, cerebellar ataxia and encephalopathy associated with anti-Purkinje cell antibodies. *J Neurol.* 1997;244(5):333–335.

165. Aggarwal A, Williams D. Opsoclonus as a paraneoplastic manifestation of pancreatic carcinoma. *J Neurol Neurosurg Psychiatry.* 1997;63(5):687–688.

166. Jacobson DM, Adamus G. Retinal anti-bipolar cell antibodies in a patient with paraneoplastic retinopathy and colon carcinoma. *Am J Ophthalmol.* 2001;131(6):806–808.

167. Chao D, Chen WC, Thirkill CE, Lee AG. Paraneoplastic optic neuropathy and retinopathy associated with colon adenocarcinoma. *Can J Ophthalmol.* 2013;48(5):e116–e120.

168. Rahimy E, Sarraf D. Paraneoplastic and non-paraneoplastic retinopathy and optic neuropathy: evaluation and management. *Surv Ophthalmol.* 2013;58(5):430–458.

169. Maitland CG, Scherokman BJ, Schiffman J, Harlan JW, Galdi AP. Paraneoplastic tonic pupils. *J Clin Neuroophthalmol.* 1985;5(2):99–104.

170. Aggarwal I, Beller J, Tzimas D. A case of confusion. *Gastroenterology.* 2014;147(6):e5–e6.

171. Sio TT, Paredes M, Uzair C. Neurological manifestation of colonic adenocarcinoma. *Rare Tumors.* 2012;4(2):e32.

172. Anderson NE, Rosenblum MK, Posner JB. Paraneoplastic cerebellar degeneration: clinical-immunological correlations. *Ann Neurol.* 1988;24(4):559–567.

173. Tsukamoto T, Mochizuki R, Mochizuki H, et al. Paraneoplastic cerebellar degeneration and limbic encephalitis in a patient with adenocarcinoma of the colon. *J Neurol Neurosurg Psychiatry.* 1993;56(6):713–716.

174. Rosenfeld MR, Eichen JG, Wade DF, Posner JB, Dalmau J. Molecular and clinical diversity in paraneoplastic immunity to Ma proteins. *Ann Neurol.* 2001;50(3):339–348.

175. Macdonell RA, Rich JM, Cros D, Shahani BT, Ali HH. The Lambert-Eaton myasthenic syndrome: a cause of delayed recovery from general anesthesia. *Arch Phys Med Rehabil.* 1992;73(1):98–100.

176. Kiylioglu N, Meydan N, Barutca S, Akyol A. Sub-acute sensory neuronopathy as a preceding sign of recurrence in colon carcinoma. *Int J Gastrointest Cancer.* 2003;34(2–3):135–137.

177. Ueyama H, Kumamoto T, Araki S. Circulating autoantibody to muscle protein in a patient with paraneoplastic myositis and colon cancer. *Eur Neurol.* 1992;32(5):281–284.

178. Soichot P, Audry-Chaboud D, Martin F, Bady B. Paraneoplastic neuromuscular manifestations—apropos of a case presenting successively a picture of myotonia, then myasthenia with myositis 4 years before the diagnosis of a colonic neoplasm. *Rev Electroencephalogr Neurophysiol Clin.* 1982;12(2):147–152.

179. Pascual J, Sanchez-Pernaute R, Berciano J, Calleja J. Paraneoplastic myotonia. *Muscle Nerve.* 1994;17(6):694–695.

180. Pautas E, Cherin P, Wechsler B. Polymyositis as a paraneoplastic manifestation of rectal adenocarcinoma. *Am J Med.* 1999;106(1):122–123.

181. Liu YL, Lo WC, Tseng CH, Tsai CH, Yang YW. Reversible stiff person syndrome presenting as an initial symptom in a patient with colon adenocarcinoma. *Acta Oncol.* 2010;49(2):271–272.

182. Williams JA, Hall GS, Thompson AG, Cooke WT. Neurological disease after partial gastrectomy. *Br Med J.* 1969;3(5664):210–212.

183. Shawcross DL, Dunk AA, Jalan R, et al. How to diagnose and manage hepatic encephalopathy: a consensus statement on roles and responsibilities beyond the liver specialist. *Eur J Gastroenterol Hepatol.* 2016;28(2):146–152.

184. Wright G, Chattree A, Jalan R. Management of hepatic encephalopathy. *Int J Hepatol.* 2011;2011:841407.

185. Mehta D, El-Hunjul M, Leary MC. Cerebrovascular complications of cancer. In: Caplan LR, Leary MC, Thomas AJ, et al., eds. *Primer on Cerebrovascular Diseases.* 2nd ed. Academic Press; 2017:573–579.

186. Wyse JM, Carone M, Paquin SC, Usatii M, Sahai AV. Randomized, double-blind, controlled trial of early endoscopic ultrasound-guided celiac plexus neurolysis to prevent pain progression in patients with newly diagnosed, painful, inoperable pancreatic cancer. *J Clin Oncol.* 2011;29(26):3541–3546.

187. Portenoy RK, Mehta Z, Ahmed E. Cancer pain management: general principles and risk management for patients receiving opioids. In: Abraham J, ed. *UpToDate.* Waltham, MA: UpToDate; 2020.

188. Network. NCC. *NCCN Clinical Practice Guidelines in Oncology: Adult Cancer Pain.* https://www.nccn.org/professionals/physician_gls/pdf/pain.pdf.

189. Portenoy RK, Copenhaver DJ. Cancer pain management: interventional therapies. In: Abraham J, Fishman S, eds. *UpToDate.* Waltham, MA: UptoDate; 2020.

190. Wong GY, Schroeder DR, Carns PE, et al. Effect of neurolytic celiac plexus block on pain relief, quality of life, and survival in patients with unresectable pancreatic cancer: a randomized controlled trial. *JAMA*. 2004;291(9):1092–1099.

191. Eisenberg E, Carr DB, Chalmers TC. Neurolytic celiac plexus block for treatment of cancer pain: a meta-analysis. *Anesth Analg*. 1995;80(2):290–295.

192. Hoff PM, Cassidy J, Schmoll HJ. The evolution of fluoropyrimidine therapy: from intravenous to oral. *Oncologist*. 2001;6(Suppl. 4):3–11.

193. Huang WY, Ho CL, Lee CC, et al. Oral tegafur-uracil as metronomic therapy following intravenous FOLFOX for stage III colon cancer. *PLoS One*. 2017;12(3):e0174280.

194. Andre T, Boni C, Mounedji-Boudiaf L, et al. Oxaliplatin, fluorouracil, and leucovorin as adjuvant treatment for colon cancer. *N Engl J Med*. 2004;350(23):2343–2351.

195. Saltz LB, Cox JV, Blanke C, et al. Irinotecan plus fluorouracil and leucovorin for metastatic colorectal cancer. Irinotecan Study Group. *N Engl J Med*. 2000;343(13):905–914.

196. Takimoto CH, Lu ZH, Zhang R, et al. Severe neurotoxicity following 5-fluorouracil-based chemotherapy in a patient with dihydropyrimidine dehydrogenase deficiency. *Clin Cancer Res*. 1996;2(3):477–481.

197. Franco DA, Greenberg HS. 5-FU multifocal inflammatory leukoencephalopathy and dihydropyrimidine dehydrogenase deficiency. *Neurology*. 2001;56(1):110–112.

198. Moertel CG, Reitemeier RJ, Bolton CF, Shorter RG. Cerebellar ataxia associated with fluorinated pyrimidine therapy. *Cancer Chemother Rep*. 1964;41:15–18.

199. Riehl JL, Brown WJ. Acute cerebellar syndrome secondary to 5-fluorouracil therapy. *Neurology*. 1964;14:961–967.

200. Pirzada NA, Ali II, Dafer RM. Fluorouracil-induced neurotoxicity. *Ann Pharmacother*. 2000;34(1):35–38.

201. Shehata N, Pater A, Tang SC. Prolonged severe 5-fluorouracil-associated neurotoxicity in a patient with dihydropyrimidine dehydrogenase deficiency. *Cancer Invest*. 1999;17(3):201–205.

202. Adams JW, Bofenkamp TM, Kobrin J, Wirtschafter JD, Zeese JA. Recurrent acute toxic optic neuropathy secondary to 5-FU. *Cancer Treat Rep*. 1984;68(3):565–566.

203. Brashear A, Siemers E. Focal dystonia after chemotherapy: a case series. *J Neurooncol*. 1997;34(2):163–167.

204. Videnovic A, Semenov I, Chua-Adajar R, et al. Capecitabine-induced multifocal leukoencephalopathy: a report of five cases. *Neurology*. 2005;65(11):1792–1794. discussion 1685.

205. Cunningham D, Allum WH, Stenning SP, et al. Perioperative chemotherapy versus surgery alone for resectable gastroesophageal cancer. *N Engl J Med*. 2006;355(1):11–20.

206. Windebank AJ, Grisold W. Chemotherapy-induced neuropathy. *J Peripher Nerv Syst*. 2008;13(1):27–46.

207. McWhinney SR, Goldberg RM, McLeod HL. Platinum neurotoxicity pharmacogenetics. *Mol Cancer Ther*. 2009;8(1):10–16.

208. Kanat O, Ertas H, Caner B. Platinum-induced neurotoxicity: a review of possible mechanisms. *World J Clin Oncol*. 2017;8(4):329–335.

209. Salat K. Chemotherapy-induced peripheral neuropathy-part 2: focus on the prevention of oxaliplatin-induced neurotoxicity. *Pharmacol Rep*. 2020;72(3):508–527.

210. Lehky TJ, Leonard GD, Wilson RH, Grem JL, Floeter MK. Oxaliplatin-induced neurotoxicity: acute hyperexcitability and chronic neuropathy. *Muscle Nerve*. 2004;29(3):387–392.

211. Cersosimo RJ. Oxaliplatin-associated neuropathy: a review. *Ann Pharmacother*. 2005;39(1):128–135.

212. Adelsberger H, Quasthoff S, Grosskreutz J, Lepier A, Eckel F, Lersch C. The chemotherapeutic oxaliplatin alters voltage-gated Na(+) channel kinetics on rat sensory neurons. *Eur J Pharmacol*. 2000;406(1):25–32.

213. Grothey A, Nikcevich DA, Sloan JA, et al. Intravenous calcium and magnesium for oxaliplatin-induced sensory neurotoxicity in adjuvant colon cancer: NCCTG N04C7. *J Clin Oncol*. 2011;29(4):421–427.

214. Guo Y, Jones D, Palmer JL, et al. Oral alpha-lipoic acid to prevent chemotherapy-induced peripheral neuropathy: a randomized, double-blind, placebo-controlled trial. *Support Care Cancer*. 2014;22(5):1223–1231.

215. Wilke H, Muro K, Van Cutsem E, et al. Ramucirumab plus paclitaxel versus placebo plus paclitaxel in patients with previously treated advanced gastric or gastro-oesophageal junction adenocarcinoma (RAINBOW): a double-blind, randomised phase 3 trial. *Lancet Oncol*. 2014;15(11):1224–1235.

216. Von Hoff DD, Ervin T, Arena FP, et al. Increased survival in pancreatic cancer with nab-paclitaxel plus gemcitabine. *N Engl J Med*. 2013;369(18):1691–1703.

217. Zajaczkowska R, Kocot-Kepska M, Leppert W, Wrzosek A, Mika J, Wordliczek J. Mechanisms of chemotherapy-induced peripheral neuropathy. *Int J Mol Sci*. 2019;20(6).

218. Dormann AJ, Grunewald T, Wigginghaus B, Huchzermeyer H. Gemcitabine-associated autonomic neuropathy. *Lancet*. 1998;351(9103):644.

219. Kabre RS, Kamble KM. Gemcitabine and cisplatin induced posterior reversible encephalopathy syndrome: a case report with review of literature. *J Res Pharm Pract*. 2016;5(4):297–300.

220. Valle J, Wasan H, Palmer DH, et al. Cisplatin plus gemcitabine versus gemcitabine for biliary tract cancer. *N Engl J Med*. 2010;362(14):1273–1281.

221. Wang-Gillam A, Li CP, Bodoky G, et al. Nanoliposomal irinotecan with fluorouracil and folinic acid in metastatic pancreatic cancer after previous gemcitabine-based therapy (NAPOLI-1): a global, randomised, open-label, phase 3 trial. *Lancet*. 2016;387(10018):545–557.

222. Mayer RJ, Van Cutsem E, Falcone A, et al. Randomized trial of TAS-102 for refractory metastatic colorectal cancer. *N Engl J Med*. 2015;372(20):1909–1919.

223. Girard PM, de Broucker T, Fryer DG, Netter JM, Saimot AG, Coulaud JP. Cerebellar syndrome in mild *Plasmodium falciparum* malaria. *Trans R Soc Trop Med Hyg*. 1988;82(2):204.

224. Kabbinavar FF, Schulz J, McCleod M, et al. Addition of bevacizumab to bolus fluorouracil and leucovorin in first-line metastatic colorectal cancer: results of a randomized phase II trial. *J Clin Oncol*. 2005;23(16):3697–3705.

225. Zuo PY, Chen XL, Liu YW, Xiao CL, Liu CY. Increased risk of cerebrovascular events in patients with cancer treated with bevacizumab: a meta-analysis. *PLoS One*. 2014;9(7):e102484.

226. Seet RC, Rabinstein AA. Clinical features and outcomes of posterior reversible encephalopathy syndrome following bevacizumab treatment. *QJM*. 2012;105(1):69–75.

227. Hurwitz H, Fehrenbacher L, Novotny W, et al. Bevacizumab plus irinotecan, fluorouracil, and leucovorin for metastatic colorectal cancer. *N Engl J Med*. 2004;350(23):2335–2342.

228. Glusker P, Recht L, Lane B. Reversible posterior leukoencephalopathy syndrome and bevacizumab. *N Engl J Med*. 2006;354(9):980–982. discussion 980–982.

229. Ozcan C, Wong SJ, Hari P. Reversible posterior leukoencephalopathy syndrome and bevacizumab. *N Engl J Med*. 2006;354(9):980–982. discussion 980–982.

230. Meric-Bernstam F, Hurwitz H, Raghav KPS, et al. Pertuzumab plus trastuzumab for HER2-amplified metastatic colorectal cancer (MyPathway): an updated report from a multicentre, open-label, phase 2a, multiple basket study. *Lancet Oncol*. 2019;20(4):518–530.

231. Sartore-Bianchi A, Trusolino L, Martino C, et al. Dual-targeted therapy with trastuzumab and lapatinib in treatment-refractory, KRAS codon 12/13 wild-type, HER2-positive metastatic colorectal cancer (HERACLES): a proof-of-concept, multicentre, open-label, phase 2 trial. *Lancet Oncol*. 2016;17(6):738–746.

III. Neurological complications of specific neoplasms

232. Moore MJ, Goldstein D, Hamm J, et al. Erlotinib plus gemcitabine compared with gemcitabine alone in patients with advanced pancreatic cancer: a phase III trial of the National Cancer Institute of Canada Clinical Trials Group. *J Clin Oncol.* 2007;25(15):1960–1966.

233. Douillard JY, Siena S, Cassidy J, et al. Randomized, phase III trial of panitumumab with infusional fluorouracil, leucovorin, and oxaliplatin (FOLFOX4) versus FOLFOX4 alone as first-line treatment in patients with previously untreated metastatic colorectal cancer: the PRIME study. *J Clin Oncol.* 2010;28(31):4697–4705.

234. Pfeiffer P, Nielsen D, Bjerregaard J, Qvortrup C, Yilmaz M, Jensen B. Biweekly cetuximab and irinotecan as third-line therapy in patients with advanced colorectal cancer after failure to irinotecan, oxaliplatin and 5-fluorouracil. *Ann Oncol.* 2008;19(6):1141–1145.

235. Dummer R, Ascierto PA, Gogas HJ, et al. Encorafenib plus binimetinib versus vemurafenib or encorafenib in patients with BRAF-mutant melanoma (COLUMBUS): a multicentre, open-label, randomised phase 3 trial. *Lancet Oncol.* 2018;19(5):603–615.

236. Kopetz S, Grothey A, Yaeger R, et al. Encorafenib, binimetinib, and cetuximab in BRAF V600E-mutated colorectal cancer. *N Engl J Med.* 2019;381(17):1632–1643.

237. Abou-Alfa GK, Sahai V, Hollebecque A, et al. Pemigatinib for previously treated, locally advanced or metastatic cholangiocarcinoma: a multicentre, open-label, phase 2 study. *Lancet Oncol.* 2020;21(5):671–684.

238. Abou-Alfa GK, Macarulla T, Javle MM, et al. Ivosidenib in IDH1-mutant, chemotherapy-refractory cholangiocarcinoma (ClarIDHy): a multicentre, randomised, double-blind, placebo-controlled, phase 3 study. *Lancet Oncol.* 2020;21(6):796–807.

239. Doebele RC, Drilon A, Paz-Ares L, et al. Entrectinib in patients with advanced or metastatic NTRK fusion-positive solid tumours: integrated analysis of three phase 1-2 trials. *Lancet Oncol.* 2020;21(2):271–282.

240. Drilon A, Laetsch TW, Kummar S, et al. Efficacy of larotrectinib in TRK fusion-positive cancers in adults and children. *N Engl J Med.* 2018;378(8):731–739.

241. Robert C, Long GV, Brady B, et al. Nivolumab in previously untreated melanoma without BRAF mutation. *N Engl J Med.* 2015;372(4):320–330.

242. Hamid O, Robert C, Daud A, et al. Safety and tumor responses with lambrolizumab (anti-PD-1) in melanoma. *N Engl J Med.* 2013;369(2):134–144.

243. Rosenberg JE, Hoffman-Censits J, Powles T, et al. Atezolizumab in patients with locally advanced and metastatic urothelial carcinoma who have progressed following treatment with platinum-based chemotherapy: a single-arm, multicentre, phase 2 trial. *Lancet.* 2016;387(10031):1909–1920.

244. Spain L, Walls G, Julve M, et al. Neurotoxicity from immune-checkpoint inhibition in the treatment of melanoma: a single centre experience and review of the literature. *Ann Oncol.* 2017;28(2):377–385.

245. de Maleissye MF, Nicolas G, Saiag P. Pembrolizumab-induced demyelinating polyradiculoneuropathy. *N Engl J Med.* 2016;375(3):296–297.

246. Overman MJ, Lonardi S, Wong KYM, et al. Durable clinical benefit with nivolumab plus ipilimumab in DNA mismatch repair-deficient/microsatellite instability-high metastatic colorectal cancer. *J Clin Oncol.* 2018;36(8):773–779.

23

Neurologic complications associated with genitourinary cancer

Lalanthica V. Yogendran[a], Marc S. Ernstoff[b], and Camilo E. Fadul[a]

[a]Department of Neurology, Division of Neuro-Oncology, University of Virginia, Charlottesville, VA, United States,
[b]Immuno-Oncology Branch, Developmental Therapeutics Program, Division of Cancer Therapy & Diagnosis, National Cancer Institute, Bethesda, MD, United States

1 Introduction

Genitourinary (GU) malignancies comprise a diverse group of cancers that have distinct biological behavior. Although direct or indirect involvement of the nervous system by these tumors usually appears in the setting of advanced neoplastic disease, it varies significantly and in some cases may be the first manifestation of the disease. Understanding the metastatic patterns will assist the clinician in the evaluation and treatment of patients with neurologic complications from GU malignancies. Because of their low frequency and subtle initial symptoms (Table 1), the clinician should be aware of the possibility of neurologic compromise, since prompt evaluation and treatment will improve patient outcome.

This chapter will describe general clinical concepts and treatment modalities that apply to the most frequent neurologic complications observed in patients suffering from these types of cancers. Evaluation of neurologic signs and symptoms must first determine if the complication is the result of direct or remote nervous system involvement. Although direct tumor infiltration is more common, remote complications have become relevant as new treatment modalities with the potential for neurologic complications become available, and previously unrecognized paraneoplastic neurologic syndromes (PNSs) are being described (Tables 2 and 3). For each specific type of cancer, we review the most frequently associated nervous system complications, including a succinct description of their purported causal relationship, cardinal clinical manifestations, treatment, and prognosis.

2 Prostate cancer

Prostate cancer is the most prevalent nondermatologic cancer in men in the United States.[1,2] Because prostate cancer has a propensity to spread to the axial skeleton, most neurologic complications are secondary to compression of neural structures from bone metastases. In the case of epidural spinal cord compression, the vertebra eroded by metastatic spread may collapse or have an extension resulting in cord compression or nerve root entrapment. Likewise, metastatic involvement of the pelvis may result in invasion of the lumbosacral plexus, while metastases to the base of the skull may compress the cranial nerves. In rare cases, prostate cancer metastasizes to the brain, dura, or cerebrospinal fluid as leptomeningeal metastases (LMs) (Fig. 1).

2.1 Epidural spinal cord compression

At least one episode of malignant epidural spinal cord compression will occur in 7% of men dying from prostate cancer.[3] Tumor grade, metastatic burden, and longer duration of hormone therapy are associated with the increased risk of spinal cord compression.[4] Involvement of the spinal cord is usually secondary to local extension of the bone metastasis into the spinal canal, although metastatic lesions occasionally compress venous flow resulting in clinical manifestations secondary to edema.[5]

The most frequent initial symptom is back pain that may be focal, radicular, or referred. Patients with only back pain at the time of diagnosis of vertebral bone metastases and early treatment will rarely develop

TABLE 1   Incidence (%) of neurologic complications among GU cancers.

	Prostate	Cervix	Endometrium	Ovarian	Kidney	Testicle	Bladder
Brain metastases	< 1.0	< 1.2	< 1	0.49–6.1	3.5–17	< 2	Rare

TABLE 2   Complications of therapy for GU cancer.

Agent	Indication—cancer site	Type of complication
Cisplatin/ carboplatin	Uterine cervix, ovarian	Peripheral neuropathy, ototoxicity, paresthesia, visual impairment, and Lhermitte's phenomenon
Taxanes	Ovarian	Peripheral neuropathy
PARP inhibitors	Ovarian	Headache, insomnia, dizziness
Bevacizumab	Renal	PRES, uncontrolled hypertension
TKI	Renal	PRES, hallucinations

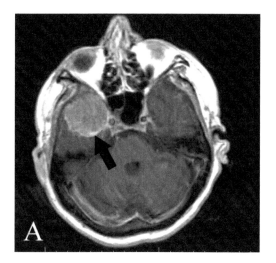

TABLE 3   Paraneoplastic neurologic syndromes associated with GU cancer.

Primary cancer site	Paraneoplastic syndromes	Antibodies
Ovary	PCD, encephalitis, dermatomyositis	Anti-Yo, anti-cdr2L, anti-Ri, antiamphiphysin, anti-NMDAR
Testicle	Limbic, hypothalamic, brainstem/cerebellar encephalitis, dermatomyositis	Anti-Ma, anti-Ta, anti-KLH11
Prostate	PCD, limbic encephalitis, peripheral neuropathy, Lambert-Eaton Syndrome[a]	Anti-Hu
Bladder	Cerebellar degeneration, dermatomyositis, opsoclonus-ataxia[b]	Anti-Ri, anti-Yo

a   Seen more with small cell carcinoma of prostate.

b   More common in transitional cell carcinoma.

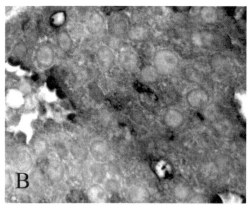

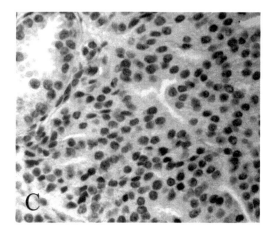

cord compression. Unfortunately, at the time of diagnosis, most patients have some degree of myelopathy that limits their ambulation. Therefore, new onset of back pain in a man with the diagnosis of prostate cancer should raise the suspicion of metastatic disease to the spine. Patients with metastatic prostate carcinoma should be alerted about the risk of cord compression and instructed to seek medical attention promptly after symptoms appear, to improve the likelihood of preserving or restoring function. It is essential for the clinician to be aware that a normal neurologic examination does not exclude the possibility of an impending spinal cord compression. The delay in diagnosis may result in significant and permanent neurologic impairment, in

FIG. 1   A man with advanced hormone refractory prostate cancer presents with a 1-week history of left side weakness. Brain MRI revealed a right temporal lobe mass observed on (A) axial T1 weighted image after contrast. Patient had surgical resection and immunohistochemistry of pathology revealed (B) metastatic adenocarcinoma that is PSA positive on and (C) was the control.

which the ability to walk after treatment is the most important predictor of survival.[6]

Serum PSA is a sensitive marker for prostate cancer staging and for monitoring response to therapy. When back pain is the first manifestation of prostate cancer, evaluation should include a serum PSA measurement, physical examination, and use of imaging studies as warranted. For patients with a normal neurologic examination and back pain, plain films of the spine and bone scan may be adequate to detect bone metastases to the region. If the neurologic examination is abnormal or the clinician has a high degree of suspicion, magnetic resonance imaging (MRI) is the study of choice to make the diagnosis of metastatic epidural spinal cord compression. The study should include the total spine, because there are other areas of metastatic disease frequently identified requiring treatment. The above diagnostic process is pertinent, because as previously emphasized, delay in diagnosis and treatment will lead to poor outcomes primarily because of loss of ambulation at the time of treatment.[7]

The standard first-line treatment of metastatic prostate cancer is androgen deprivation therapy (ADT), which includes luteinizing hormone-releasing hormone (LHRH) analogs, orchiectomy, and more recently gonadotropin releasing hormone antagonist therapy. LHRH treatment can cause the adverse event of flare phenomenon, which is a transient increase in luteinizing hormone and testosterone resulting in the increase of symptoms including spinal cord compression. Therefore, first-line single-agent LHRH is not routinely recommended for spinal cord involvement in prostate cancer. Studies are now looking at the concomitant use of LHRH with antiandrogens to reduce the risk of this flare phenomenon, but there is currently no consensus on these recommendations.[8] Neuroendocrine prostate cancer remains particularly challenging as they are hormone refractory due to low PSA levels. Therefore, surgery is a key for pathology diagnosis and treatment.[9]

Zoledronic acid (ZA) is still the only Food and Drug Administration (FDA)-approved bisphosphonate to delay skeletal-related events (SREs). However, there are several new bone-targeting drugs to reduce the risk of SREs secondary to prostate cancer, including denosumab, which has shown decreased risk of SREs in castration-resistant prostate cancer (CRPC) and Radium-223 that has shown to improve overall survival in metastatic CRPC.[2,10]

Regardless of preventive SRE measures, spinal cord compression still occurs. Pain management consists of utilizing opioids, and corticosteroids, which also reduces edema and retains neurologic function while diagnostic studies are performed to assess the best course of treatment. There is still controversy about the total steroid dose to be used.[3,11] The benefits of corticosteroids were not remarkably different among various doses and even placebo, but those with high dose had more side effects.[11,12] The neurologic, oncologic, mechanical, and systemic (NOMS) decision framework has been proposed to guide the decision-making process for optimal treatment.[13] Radiation remains the backbone of treatment with surgery and corticosteroids having additive benefits. The current focus of surgery is to decompress the spinal cord to deliver the optimal radiation dose.[13] In the most recent Cochrane review, surgery and radiation versus radiation alone did not result in more adverse events.[12] Prognosis of metastatic epidural spinal cord compression is dependent on many factors. Overall, patients who are ambulatory, have no additional metastases, have a longer interval between primary diagnosis and metastatic cancer, lack of bone or visceral metastases, and a slower course of motor dysfunction have improved survival.[14,15] For those who receive decompressive surgery, the best predictor of survival was preoperative Karnofsky Performance Scale (KPS) > 80.[6]

2.2 Brain metastases

Intracranial metastasis is a rare event estimated to occur between 0.16% and 0.63% of patients with metastatic prostate cancer.[16] The histologic examination of brain metastases denotes that most are moderately or poorly differentiated adenocarcinomas.[17,18] Three mechanisms for prostate cancer metastasizing to the brain have been proposed, including secondary spread from lung metastases, tumor bypassing the lung and embolizing through a patent foramen ovale, and cancer cells accessing the capillary system into the left heart.[19]

The clinical picture is nonspecific and will depend on the location of the metastatic lesions. Metastases from prostate carcinoma, in contrast with other solid tumors, appear to have a preference for dural location.[19] MRI establishes the presence of a mass, but because it is so rare, pathologic confirmation is frequently necessary to establish the diagnosis (Fig. 1). Immunostaining for prostatic acid phosphatase and PSA in the metastatic tissue confirms the origin of the lesion. This technique is especially useful for the exceptional patient, whose first manifestation of prostate cancer is brain metastases.[17]

With improved treatment options, the prognosis has improved in recent years.[16,20,21] Treatment for brain metastases includes corticosteroids, surgery, stereotactic radiosurgery (SRS), and radiotherapy. As with brain metastases from other types of cancer, treatment has to be individualized according to the functional status, number of metastases, and extent of systemic disease. Craniotomy and resection of solitary metastasis in good surgical candidates, combined with SRS, has been increasing in clinical practice and can result in longer survival, although prognosis continues to be poor.

2.3 Base of the skull metastases

While still uncommon, prostate cancer is the most frequent type of malignancy affecting the skull base, and it usually occurs in patients who have hormone refractory disease with other bone metastases.[22] Usually, pain precedes involvement of the cranial nerves and follows any of the five skull base syndromes that have been described: orbital (diplopia, visual impairment), parasellar (diplopia), middle-fossa (facial numbness, facial palsy), jugular foramen (hoarseness, dysphagia), and occipital condyle (tongue weakness).[23,24] The diagnosis may be difficult if imaging studies are not guided by the clinical presentation. MRI is the most useful examination to establish the diagnosis, but frequently computed tomography (CT) scans with bone windows and bone scan are required to demonstrate bone involvement. Palliative therapy with radiation and concomitant corticosteroids results in improvement of symptoms in most patients, and the prognosis will depend on the systemic disease and the timing of the treatment.

2.4 Paraneoplastic syndromes

Following renal cancer, prostate cancer is the most common urologic malignancy associated with PNS[25,26] and usually occurs in the setting of advanced cancer. Specific neurologic syndromes that may be associated with prostate cancer are paraneoplastic cerebellar degeneration (PCD), limbic encephalitis, Eaton-Lambert myasthenic syndrome, and peripheral neuropathy.[26–28] One review found a strong association with anti-Hu antibody in a 37-case-series evaluation, in which a majority of patients had adenocarcinoma.[29] Recognition of these syndromes is important as it may identify an underlying malignancy and influence treatment options. Early treatment of the cancer and immunosuppressive therapies may stabilize or improve the neurologic impairment, but in most cases, the likelihood of recovery is poor.

2.5 Treatment-related complications

In rare cases, robotic-assisted laparoscopic prostatectomy may result in debilitating peripheral neuropathies. One retrospective study showed a 0.16% associated incidence with these surgeries in prostate cancer.[30] It usually occurs after time-consuming surgeries in the Trendelenburg position, in which compression, stretching, and ischemia of peripheral nerves can lead to loss of function.[30] Erectile dysfunction is common after radiation therapy for prostate cancer and has been shown to be radiation dose related.[31] Brachytherapy preserves function at a higher rate than external beam radiation. Prostate surgery has also been shown to affect erectile dysfunction, in which bilateral nerve sparing is a predictor of preserving baseline erectile function.[32]

3 Uterus

3.1 Uterine cervix

Cervical cancer has the second highest morbidity among cancers in women, and it rarely metastasizes to the brain, with an incidence ranging between 0.4% and 1.2%.[33,34] Metastasis is most commonly found supratentorially[33] and arises from hematogenous spread from the lungs, the most common site of cervical metastasis.[33,34] A review of the histologic types shows that, although most are pure squamous cell carcinoma (SCC), approximately one third of cases are adenocarcinomas alone or mixed with SCC.[35] The signs and symptoms are nonspecific for the type of primary cancer, and MRI establishes the presence of brain metastases. Treatment is dependent on multiple factors, including number of metastases, and is a combination of surgery with or without radiotherapy. Surgery, when feasible and followed with radiation, seems to improve survival.[33–35] If chemotherapy is utilized, cisplatin is the most frequently used and has been noted to show response in brain metastases and systemic tumor burden.[35,36]

The frequency of lumbosacral plexopathy associated with cervical cancer ranges from 0.3% to 1.3%,[37] and may be secondary to tumor infiltration or as a complication from radiotherapy. Tumor perineural spread will depend on the level of stromal, parametrium, and lymphovascular space invasion, tumor size, and lymph node metastasis.[38,39] Although most GU neoplasms cause a panplexopathy, cervical cancer usually involves the lower plexus (L4-S1). Fifty percent of patients will present with moderate pelvic pain, which can be asymmetric at onset but becomes bilateral. As the pain becomes progressive, motor and sensory impairment appears.

CT scan or MRI of the lumbosacral plexus, including visualization from the upper segment of the plexus and extending down to the sciatic notch, will help to establish the diagnosis. However, the complexity of the lumbosacral plexus makes it challenging to accurately diagnose the specific areas of involvement. High-resolution magnetic resonance neurography allows for improved visualization and can differentiate tumor invasion versus radiation-induced plexopathy.[40] When it is determined to be radiation-induced, a wide range of radiation doses have been implicated.[37,41,42] No effective treatment exists for plexopathy secondary to tumor invasion, although radiation may help with symptom management. For radiation plexus injury, pentoxifylline and tocopherol may decrease symptoms and can be synergistically effective with the addition of clodronate, a bisphosphonate.[37] Additional symptomatic treatment includes medications to reduce neuropathic pain.

Metastatic epidural spinal cord compression is unusual in patients with carcinoma of the uterine cervix.

The clinical manifestations are similar to those seen with epidural metastases arising from other solid tumors in other locations. MRI is indicated for the accurate diagnosis of the spinal involvement. Although the prognosis is poor, treatment with radiotherapy may offer palliation.

Cisplatin, which is often used in cervical cancer, can result in peripheral neuropathy and ototoxicity. Sensory peripheral neuropathy is a common complication, and treatment is symptomatic.[43] Ototoxicity is identified as sensorineural hearing loss, which is bilateral and also irreversible.[43]

Encephalopathy is rarely seen, but is one of the most frequent remote neurologic complications due to cervical cancer. It can be a result of metabolic deterioration secondary to cervical cancer or from a platinum-based treatment regimen. Seizures and metabolic encephalopathy may develop in patients with renal failure secondary to bilateral ureteral obstruction in advanced-stage cervical cancer.

3.2 Uterine body

Adenocarcinoma of the endometrium may be responsible for several types of neurologic complications, including lumbosacral plexopathy, spinal cord compression, and brain metastases. Uterine small cell carcinoma, although very rare, has also been seen to metastasize to the brain in late stages of the disease.[44,45] In general, neurologic complications are even more uncommon than those caused by cancer of the cervix.

Brain metastases from endometrial carcinoma are very rare with a reported incidence of 0.3%–0.9%[46,47] and occur at a median interval of 17 months after primary diagnosis.[48] Brain metastasis may also be an initial manifestation when endometrial carcinoma is undifferentiated, with deep myometrial and vascular invasion.[49]

Purported risk factors for the development of brain metastasis include high tumor grade, advanced stage, and lymphovascular space invasion. Because this remains a very rare event, treatment options are similar to treatment of other solid tumors with brain metastases. Although the median survival after diagnosis of brain metastases is 5 months,[48] aggressive treatment with surgical resection combined with radiation therapy in selected cases may improve survival.[50]

4 Ovarian cancer

Ovarian cancer is the fifth most common cancer and the leading cause of death from gynecologic malignancies among women in the United States. Neurologic complications associated with ovarian cancer are rare with a reported incidence of 0.29%–11.6%,[51] although their incidence may be increasing due to improved survival associated with more effective therapies.

4.1 Brain metastases

The incidence of brain metastases from ovarian cancer is estimated between 0.49% and 6.1%.[52] Metastasis has been thought to occur via direct hematogenous spread through Virchow-Robin perivascular spaces or by direct invasion into the nervous system after bone involvement. Retrograde lymphatic spread has also been suggested when there is leptomeningeal involvement.[53] More recently, BRCA1 and BRCA2 gene mutations and androgen receptor expression have been associated with increased risk of developing brain metastases.[52] Positive prognostic factors include well-controlled primary disease at the time of central nervous system (CNS) involvement, low-grade tumor, platinum-responsive treatment, and a high KPS score.[54]

Serous cystadenocarcinoma is the most common histologic subtype of ovarian cancer and accounts for 90% of malignant ovarian tumors,[54] thus most frequently associated with CNS involvement. The most common presenting symptoms are attributed to increased intracranial pressure, although other focal neurologic symptoms may occur. The median survival after diagnosis is 6 months.[55]

Established treatment for brain metastases from ovarian cancer does not exist and is not different to other solid tumors. Combination treatment, with surgery, radiation, and systemic chemotherapy, is preferred over a single modality treatment and shows upward of 18–33 months of survival.[51] If WBRT is chosen, hippocampal sparing is preferred as it better preserves cognitive function and patient-reported symptoms.[56]

4.2 Leptomeningeal metastases

Ovarian cancer with LM is still rare, but reported more frequently because of more effective treatments and longer survival. When present, it is often seen with prior or concurrent brain metastases, and the clinical manifestations are the same as observed with LM from other cancers.[57] MRI of the brain and spine and CSF cytology establish the diagnosis. Palliative radiation to symptomatic areas remains the staple treatment, with an overall survival of a few months, which is similar to other solid tumors.[51]

4.3 Paraneoplastic syndromes

Patients may have PNS prior to the diagnosis as often as after the diagnosis of ovarian cancer.[58] About 50% of patients with nonfamilial subacute cerebellar degeneration have PNS associated with either ovarian or lung cancer[59] (Fig. 2). A group of disorders secondary to neuronal surface antibodies that frequently are present as an encephalitis have been more recently described, but

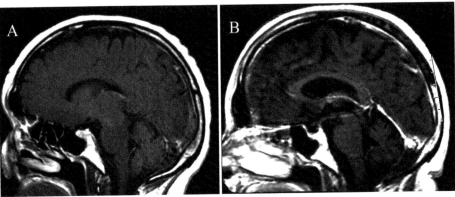

FIG. 2   Patient with paraneoplastic cerebellar degeneration associated with ovarian cancer. Sagittal T1 weighted MRI images reveal (A) normal cerebellum 2 weeks after onset of symptoms and (B) cerebellar atrophy 14 months after initial study.

they rarely associate with malignancies. The exception to this rule being, *N*-methyl-ᴅ-aspartate receptor antibody (NMDAR-Abs) encephalitis which appears in young women with benign ovarian teratomas. The etiology of PNS is considered to be an autoimmune process.

Diagnosis is based on neurologic syndrome, tumor diagnosis, and associated antibodies, which are detected in serum or CSF through indirect immunofluorescence and confirmatory Western blot. Additionally, CSF may show pleocytosis, elevated protein, and oligoclonal bands.[59] A recent study has even proposed that paraneoplastic antibody surveillance can be used for early ovarian cancer detection, even in the setting of normal CA-125 concentrations, using the combination of four antigens, Ro52, CDR2, HARS, and 5H6.[60] Regardless of the detection, treatment of the primary tumor improves overall survival, and early detection and treatment can improve the overall neurologic outcome.

Patients with PCD present with ataxia, dysarthria, and often downbeat nystagmus. The most common antibody associated with PCD and ovarian cancer is against the surface protein cdr2, also known as anti-Yo or Purkinje cell cytoplasmic antibody type 1 (PCA-1). The co-existence of anti-cdr2-like (cdr2L) and anti-Yo antibodies in patients with PCD carries a high probability of underlying ovarian or breast cancer.[59] Specifically, the presence in CSF of anti-Yo, anti-Ri, or antiamphiphysin antibodies should trigger suspicion for ovarian cancer. MRI can show cerebellar atrophy, and FDG-PET will commonly reveal cerebellar hypometabolism. In woman with PCD without evidence of cancer, the NCCN Guidelines recommend transvaginal ultrasound combined with CA-125 measurements every 6 months.[61] In women with anti-Yo cerebellar degeneration and worsening neurologic status, especially if they are postmenopausal, surgical exploration and oophorectomy are suggested.[61] The presence of PCD does not seem to affect the ovarian cancer prognosis, but the neurologic syndrome is frequently severely incapacitating and irreversible.

Patients with NMDAR-Abs present with an encephalitis syndrome, characterized by psychosis and memory changes with subsequent seizures, dyskinesia, and autonomic instability. Multiple studies have now shown a strong correlation of NMDAR-Abs with the presence of ovarian teratoma. CSF findings include inflammatory markers and NMDAR-Abs. The strong link between the NMDAR-Abs encephalitis and the ovarian teratomas is explained by the cross-presentation of the same antigen.[59] CNS imaging is not always helpful, as normal MRI findings are common. If a lesion is present, they are most commonly found in the hippocampus.[62] Seizures, with corresponding EEG abnormalities, resistant to antiepileptic treatment are observed in many patients with NMDAR-Abs. MRI or CT imaging of the pelvis is a key to identifying a teratoma, but even if a teratoma is not found on imaging an exploratory surgery may prove to be informative (Fig. 3). In those patients with an identified teratoma, surgical removal in combination with plasma exchange, steroids, and intravenous immunoglobulin (IVIG) will provide best recovery.

Between 20% and 25% of patients with dermatomyositis have an associated malignancy,[61] and when associated with

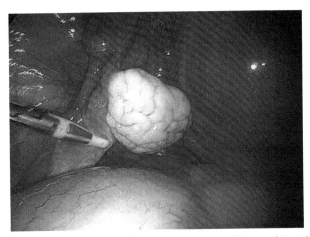

FIG. 3   Surgical finding of an ovarian teratoma in a patient with anti-NMDAR encephalitis.

ovarian cancer, it is commonly in women over the age of 40. In fact, women of 40–60 years with dermatomyositis have an ovarian cancer relative risk of 16.7.[63] It may precede the development of ovarian cancer by months and even years. In women with dermatomyositis without cancer, the recommendation is for pelvic ultrasound screening annually for the first 3 years after diagnosis.[61] Rarely, will it appear after an established diagnosis of ovarian cancer. It has been suggested that the concurrent presence of Raynaud's syndrome with dermatomyositis may exclude an underlying ovarian cancer.[61] The diagnosis is usually established by muscle biopsy that reveals typical pathologic changes of myositis. In most cases, symptoms of dermatomyositis regress after treatment of the underlying ovarian cancer. Corticosteroids, IVIG, and other immunosuppressive treatment approaches may be effective in some instances.[64]

4.4 Treatment-related complications

Cisplatin and paclitaxel, alone or in combination, produce the majority of the neurologic complications associated with ovarian cancer treatment (Table 2). The most common complication is peripheral neuropathy.[65] Paclitaxel-induced neuropathy predominantly affects sensory fibers, with sparing of motor nerves. The symptoms appear rapidly after paclitaxel administration and usually involve the hands and feet simultaneously. Although neurotoxicity is reversible once treatment is suspended, symptoms usually persist and may increase, affecting the quality of life.

Cisplatin, on the other hand, affects primarily large myelinated fibers of the peripheral sensory nerves. Although the nerve injury affects most patients, it is symptomatic in approximately 50%. Ototoxicity (hearing loss and tinnitus) is also seen with cisplatin administration. It is usually bilateral, symmetric, and predominantly affects high-frequency audition. The likelihood of developing this complication is higher in older patients and those with preexisting hearing loss. Carboplatin has replaced cisplatin in most chemotherapy regimens since it provides similar benefit while it carries less nephrotoxic and neurotoxic effects. More recently, poly ADP-ribose polymerase (PARP) inhibitors have become the front-line treatment for Stages III–IV ovarian cancer, especially with BRCA1 or BRCA2 mutations. Rare neurologic toxicities from these inhibitors include headaches, insomnia, and dizziness.[66]

To date, there are no preventive measures or effective treatments for chemotherapy-induced painful neuropathy. Duloxetine is the only recommended treatment for painful neuropathy, which has been shown to improve pain.[67] One study showed promise of venlafaxine as a successful preventive treatment, but long-term follow-up did not hold up its use.[68] Gabapentinoids and tricyclic antidepressants are reasonable to try due to their reported efficacy in other neuropathies. However, two trials failed

to show benefit from the former and no prospective trials exist for the later; therefore, the use of either has waned.[68] Topical agents (including baclofen, amitriptyline, and ketamine) also have shown little efficacy in a few trials and no FDA approved product exists.[68]

5 Renal carcinoma

Kidney cancer is the sixth and eighth most common cancers in men and women in the United States. Most kidney cancers derive from the renal cortex with a primary histologic type being clear cell.[69] Brain metastases and epidural spinal cord compression are the most frequent neurologic complications related to renal cell carcinoma (RCC). Metastases from kidney cancer are highly vascularized and have been considered less sensitive to radiation therapy than metastases from other primary cancers.

5.1 Brain metastases

The incidence of brain metastases from RCC is estimated between 3.5% and 17% of all patients with metastatic disease. The median interval between diagnosis of the primary cancer and the neurologic involvement is 10.7 months,[69,70] with a 5-year survival rate after diagnosis of 12%.[69,70] Metastases seed the brain through hematogenous spread and appear in advanced stages of the disease.[71] The clinical manifestations are similar to those observed with brain metastases from other malignancies. Because metastases of RCC are highly vascularized, there is an increased risk for hemorrhage. In some cases, the clinical presentation can be abrupt, with an intracerebral hemorrhage.[70,72]

The Renal Graded Prognostic Assessment (GPA) tool is a diagnosis-specific prognostic index that helps in treatment selection. It found KPS, number of brain metastases, extracranial metastases, and hemoglobin to be the most specific factors associated with survival at 1 year.[73] Standard of care for many years now has been WBRT for patients with multiple lesions: however, mean survival is approximately 4.4 months, and overall not considered effective.[69] If a lone metastasis larger than 3 cm within a noneloquent area exists, surgery is favored.[74] If less than 10 lesions and smaller than 3 cm with a good performance status, then SRS is preferred.[69] It is also used as an adjuvant after surgical resection. SRS is favored to WBRT because of the potential result of cognitive decline. A retrospective study noted 11% of patients alive 12 months after WBRT developed dementia.[75]

Although anti-VEGFR tyrosine kinase inhibitors (TKIs) (specifically sorafenib and sunitinib)[76] and immune checkpoint inhibitors are the current mainstay treatment for metastatic RCC, their benefit for brain metastasis is unknown as no Phase III trials have included patients with brain metastasis.[77] Many trials excluded RCC patients with brain metastasis because of safety concerns that included risk of pseudo- and hyperprogression, the limited ability to cross the blood brain barrier (BBB), steroid use that can alter the immune system activity with immune checkpoint inhibitors, and increased risk of hemorrhage with anti-VEGF treatments.[74] A recent multicenter retrospective study showed promising results of brain metastases from RCC patients effectively treated with checkpoint inhibitors (ipilimumab and nivolumab).[78]

Although anti-VEGFR, TKIs, and checkpoint inhibitors show responses in extra-cranial disease, the intracranial response is moderate at best.[74] Furthermore, the median time to response using checkpoint inhibitors or anti-VEGF therapy is about 3 months, which is another limitation of using systemic treatments.[79] Checkpoint inhibitors attack brain lesions by activating peripheral T-cells: however, the systemic treatment is not enough to control intracranial disease. Prior studies using anti-VEGF treatments and checkpoint inhibitors in patients with intracranial RCC did show an improvement in overall survival despite the lack of intracranial control, which is attributed to control of extracranial disease.[79] Ongoing trials are assessing for the use of immunotherapy with radiation, which is postulated to be effective because radiation results in necrosis of the tumor tissue, resulting in release and exposure of tumor antigens to systemic therapy.[74] Some retrospective studies using this combination suggest an increase in overall survival.[77,79]

5.2 Epidural spinal cord compression

Kidney cancer is the second most common GU malignancy, after prostate cancer, to cause epidural spinal cord compression. It is estimated that up to 30% of patients with advanced RCC will develop spinal metastases, presenting with clinical features that are indistinguishable from other malignancies.[80] One study found 28% of SREs associated with bone metastases from RCC had a high incidence of spinal cord and nerve root compression.[81] Further neurologic impairment worsens overall quality of life and survival. Other associated symptoms include change in bladder and bowel function, weakness, and loss of sensation that can imply spinal nerve root or cord compression.[82] The diagnosis is established by MRI, which demonstrates low signal intensity of the involved vertebral bodies and the extent of the cord compression.

The treatment options for spinal cord compressions from renal carcinoma include decompression with steroids, radiotherapy, surgery, interventional radiology, and systemic therapy. Surgical interventions are used to decrease the risk of fracture, treat a pathologic fracture, and decrease spinal instability in the setting of spinal cord compression.[83] Radiotherapy plays a primary role in the palliative treatment of spinal metastases.[81,83] It successfully helps reduce bone pain and improves spinal cord compression. However, spinal cord compression from renal carcinoma seems to be less responsive to radiation than other primary malignancies, and ultimately the treatment outcome and survival time after radiotherapy may be not as good as with other primaries. There is still a lack of literature that reveals the outcomes from radiation treatment, in which a few studies only report a short course of pain improvement and reduction in fractures.[83] SRS can deliver high doses with a small number of fractions, allowing for accurate targeting and avoiding other critical structures.[82] SRS is also an option for patients who progress in spite of external beam fractionated radiation therapy, often achieving local tumor control, with minimal toxicity.[82] Surgery followed by radiation would be recommended in patients who have a low surgical risk with limited systemic disease and a single area of cord compression. Because surgery carries a high complication rate due to the high vascularity of RCC, arterial embolization has been used to reduce the blood loss during the procedure.[82] Most recently, a combination of radiation and drug therapy is being assessed. Small-powered studies have looked at combining ZA with radiotherapy and show improved and longer SRE survival compared to radiation alone.[81,84] As with cord compression from other cancers, the life expectancy depends on multiple factors such as extent of disease, ambulatory status before radiotherapy, performance status, and the number of involved vertebrae.[15,80]

5.3 Paraneoplastic syndromes and leptomeningeal metastases

Though rare, paraneoplastic neurologic conditions occur in RCC having a variety of presentations including sensory or motor neuropathies, nonspecific myalgias, phrenic nerve paralysis, and one case of amyotrophic lateral sclerosis.[85] As these cases were associated with localized neoplastic disease, nephrectomy is considered curative.[85] Even less frequent is leptomeningeal metastasis from RCC. The clinical presentation is similar to other cancers with LM. Treatment options include TKI, checkpoint inhibitors, and radiation,[86–88] but the prognosis remains poor.

5.4 Treatment-related complications

Bevacizumab, a monoclonal antibody that targets and inhibits the vascular endothelial growth factor, significantly prolongs survival in patients with metastatic RCC. Several cases of a posterior reversible leukoencephalopathy syndrome (PRES), similar to that observed with chemotherapy and immunosuppressive therapies, have been reported after bevacizumab.[89,90] Uncontrolled hypertension is thought to be a predisposing factor, and therefore, in addition to discontinuing the offending agent, strict blood pressure control is recommended when this complication appears.[90] There are also rare but severe neurologic adverse events associated with immune checkpoint inhibitor (ICI) therapy, with myasthenia gravis cases most frequently reported.[91] Given the increased use of ICI therapies, we expect the frequency of neurologic adverse events to increase as well. The cornerstones of managing ICI-induced neurologic adverse events are early discontinuation of ICI therapy, and initiation of immunosuppressive therapy including steroids, IVIG, or plasmapheresis.[91] Specific TKIs also carry neurologic risks, including axitinib and pazopanib that are associated with PRES.[92]

6 Germ cell testicular cancer

Seminomas and nonseminomatous germ cell tumors account for more than 95% of all testicular cancers, with the latter having more aggressive growth behavior. Testicular cancer continues to be the most common primary malignant tumor in young males in the United States. Although frequently there is evidence of systemic metastases at the time of diagnosis, seeding of the CNS is unusual.

6.1 Brain metastases

Brain metastases occur in 1%–2% of all testicular germ cell tumors and 10%–15% of those with advanced disease with pulmonary metastases.[93] Specifically, 1.3% of men with disseminated nonseminomatous and 1.2% with disseminated seminomas have brain metastasis at the time of diagnosis.[94] Brain metastases should be suspected in patients with neurologic symptoms, excessively elevated serum human chorionic gonadotrophin (HCG) levels, and/or malignant teratoma histology. In addition, brain imaging should be considered in all advanced diseases to diagnose clinically asymptomatic metastases.

Currently, the standard of care for testicular germ cell brain metastases is the combination of cisplatin followed by radiation.[95] Those receiving chemotherapy, especially the first cycle, should be monitored for risk of intracranial hemorrhage.[96] Additionally, for patients with one or a few brain metastases, surgery can be considered followed by chemotherapy.[95–97] For patients with advanced disease and identified to have a poor prognosis, cisplatin and etoposide with autologous stem cell transplant are viable. In recent years, new chemotherapies such as docetaxel, paclitaxel, gemcitabine, and oxaliplatin are used as second-line or salvage treatment in those with poor prognosis.[93] One retrospective review looking at 27 patients with brain metastases showed the 5-year survival and 10-year survival were both 35.9%,[93] which is consistent with other studies.[97]

Patients who develop brain metastases after initial chemotherapy for testicular germ cell tumors have a poor prognosis. This is hypothesized to occur because of incomplete penetrance of chemotherapy through the BBB.[95] A clinical trial showed brain metastases appear to be the singular site of relapse, occurring early and frequently in some patients.[98] Furthermore, if brain metastases relapse, it is often associated with rising serum tumor markers (α-fetoprotein concentration $> 1000\,ng/mL$ or HCG concentration $> 5000\,IU/L$); in this setting, the likelihood of survival is significantly decreased.[94,99] Studies show that platinum refractory-relapsed disease with brain metastases can be cured with high-dose chemotherapy in 40% of the cases.[99] Therefore, cure is possible and aggressive therapy is usually indicated.

6.2 Epidural spinal cord compression

Testicular cancer may involve the retroperitoneal lymph nodes, and from there directly grow into the spinal canal or cause radiculopathies. When there is spinal canal invasion without bone involvement, bone scan or plain X-rays may be misleadingly negative, and MRI should be performed. Rarely, the spinal cord compression is the result of vertebral bone metastasis. Patients present with classical signs and symptoms of spinal cord compression and, like with other tumors, early diagnosis and aggressive management are essential for improved outcomes.

Radiation and surgery remain the mainstays of treatment for epidural spinal cord compression. In general, primary treatment has been urgent radiotherapy for most patients. In a small four-case series, patients treated with radiotherapy had a 75% 2-year survival rate.[100] For patients with late relapse of nonseminomatous germ cell cancer, resection is a mainstay of treatment and can be followed by aggressive local radiotherapy and adjuvant chemotherapy.[101] Surgical decompression and stabilization of the spine is also utilized when there is acute neurologic deterioration, instability, or intractable pain. In certain instances, cisplatin-based combination chemotherapy is added to radiation and surgery once diagnosis is confirmed.

6.3 Paraneoplastic syndromes

The PNS of limbic, hypothalamic, and brain stem encephalitis associated with the anti-Ma antibodies has been well described in patients with testicular cancer (Table 3). Characteristically, the neurologic syndrome is more often associated with the anti-Ta (formerly known as anti-Ma2) antibody subtype in patients who responded well to tumor treatment.[102] Patients will have symptoms referable to the area of involvement that could be limbic-hypothalamic system, brainstem, or both. Symptoms include personality and mood changes that develop over days or weeks, associated with severe impairment of recent memory, confusion, and occasionally seizures. Early treatment of testicular cancer should improve the outcome, and therefore, it is necessary to make an early diagnosis in patients in which PNS is the first and sole manifestation of cancer.

Recently, an autoantibody against Kelch-like protein 11 (KLHL11-abs) was described in 13 patients with paraneoplastic brainstem/cerebellar encephalitis and testicular seminoma without anti-Ma/Ta antibodies.[103] Based on this and other reports, three clinical settings have been associated with KLHL11-abs. First, brainstem and cerebellar syndromes with KLHL11-abs occurring with benign teratoma or testicular tumor (+/− anti-Ma/Ta antibodies) respond moderately to treatment.[103] Second, some patients with NMDAR-Abs encephalitis have been found to also simultaneously have KLHL11-abs. In these cases, the presence of KLH11-abs has no neurologic implications; therefore, these cases still present as a classical NMDAR-Abs encephalitis.[103] Third, about 5% presented with KLH11-abs and a neurologic syndrome that was not paraneoplastic.[103] Patients with KLHL11-abs present with rhomboencephalitis, ataxia, diplopia, dysarthria, vertigo, hearing loss, and tinnitus, in which the latter two symptoms are unique manifestations.[104] Brain MRI shows T2/FLAIR abnormalities in the brainstem or limbic system. In one study, only 25% of 32 patients showed neurologic improvement following cancer treatment and/or immunotherapy, but ataxia, hearing loss, and diplopia persisted.[104] Independently, there has also been identification of NMDAR-Abs encephalitis associated with a testicular teratoma[102] as well as a few sporadic cases of dermatomyositis in which subsequently testicular germ cell tumors were diagnosed.[105]

6.4 Treatment-related complications

Cisplatin-based chemotherapy may cure many patients with testicular cancer, but neurotoxicity is the most important dose-limiting side effect. Cisplatin-induced neuropathy is characterized by numbness and tingling of the extremities, which can occasionally be painful, as well as ototoxicity. A less frequent manifestation of cisplatin neurotoxicity is the development of Lhermitte's phenomenon, a manifestation of pathology involving the posterior columns of the spinal cord. Clinically, relevant neuropathy symptoms persist in approximately 20%–25% of long-term testicular cancer survivors,[106] affecting their quality of life.

7 Bladder carcinoma

Bladder cancer is the ninth most common cancer diagnosed worldwide.[107] CNS involvement is rare, but can occur as brain metastases, LM, spinal cord compression, and, rarely, as a PNS. In contrast to many malignancies, nonmetastatic neurologic complications occur more frequently than metastatic nervous system involvement.[108]

7.1 Brain metastases

Brain metastases remain a rare event, with a three- to ninefold higher incidence in small cell bladder cancer than transitional-cell carcinoma (TCC) of the bladder.[109] When brain metastases occur, they are frequently diagnosed concurrently with the initial presentation of bladder carcinoma. Some studies suggest that the incidence of CNS metastases is associated with disease extension and is higher among patients receiving aggressive therapy.[109] Additionally, the prolonged systemic remission due to chemotherapy and inability of the usual chemotherapy regimen, including methotrexate, vinblastine, doxorubicin, and cisplatin (M-VAC), to penetrate the blood-brain barrier may contribute to the increased incidence of brain metastases.[110]

Optimal management is still uncertain, and the efficacy of chemotherapy and immune checkpoint inhibitors in brain metastases remains unclear.[111] Historically, WBRT and steroids have been the standard treatment for multiple brain metastases, with an average survival of 2–7 months[112]; as such, SRS is now utilized more often. It is reasonable to resect single lesions in noneloquent areas and follow with adjuvant SRS.[111] Good prognostic features of brain metastases include high KPS, solitary lesion, and absent extracranial metastatic disease.

7.2 Leptomeningeal metastases

There are only a few reported cases of LM secondary to bladder cancer.[57,107,113] The majority of patients developing LM have been previously treated with M-VAC chemotherapy.[107] Prolonged survival after bladder cancer diagnosis, selection of tumor cells resistant to treatment, and poor penetration of chemotherapeutic agents into the CNS have been some of the explanations proposed for the higher frequency of this neurologic complication in recent years.[107,113] Prognosis is poor with a median survival of 2.2 months.[107]

7.3 Spinal cord complications

Spinal cord or cauda equina complications secondary to metastatic bladder cancer are rare and may occur through direct invasion of tumor mass.[114,115] Furthermore, secondary direct extension of the tumor into the epidural space from bone structures or, more commonly, soft tissues may cause similar complications. Diagnostic approaches and treatment are similar to metastatic epidural spinal cord compression from other cancers.

7.4 Lumbosacral plexopathy

The incidence of lumbosacral plexopathy in bladder cancer patients is estimated to be 2%.[116] Lumbosacral plexus dysfunction occurs in patients with bladder carcinoma by direct extension of tumor into the posterior pelvis, thought to be the result of perineural spread.[38,116] In one study, perineural invasion was noted in 47.7% of pathological analyses of radical cystectomy specimens for bladder cancer.[38] Pain is usually the initial symptom, although

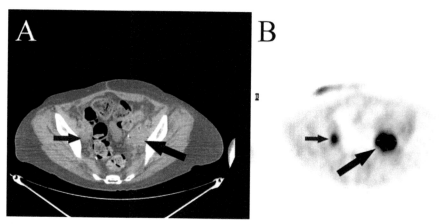

FIG. 4 (A) CT scan and (B) PET scan of a patient who had a history of bladder carcinoma and developed a left lumbosacral plexopathy, showing a left pelvic mass (*large arrow*). *Small arrow* reveals a smaller mass on the right pelvis.

paresthesias, numbness, and weakness are frequently reported. Treatments with analgesics, chemotherapy, and radiotherapy have all been used to alleviate symptoms, since it is difficult to surgically remove the tumor given the proximity of the bladder to the plexus (Fig. 4).[108]

7.5 Paraneoplastic syndromes

PNSs have been diagnosed in a few cases of bladder cancer, with opsoclonus-ataxia and cerebellar degeneration, associated with anti-Ri and anti-Yo antibodies, being the most common.[108] Dermatomyositis has also been reported in bladder cancer, specifically TCC.[117,118] In a few cases, small cell carcinoma of the bladder has also been presented as an aggressive neuroendocrine cancer and dermatomyositis.[118]

8 Conclusion

Although neurologic complications associated with GU cancers are infrequent, their incidence is on the rise as more effective therapies improve the survival of patients with these neoplasms. The first step in establishing the neurologic diagnosis is to determine if the complication is secondary to direct or metastatic involvement of neural structures by the cancer, neurotoxicity associated with the oncologic treatment, or a remote effect of cancer (PNS). The prognosis of the neurologic complications usually depends on early diagnosis and treatment. Some are medical emergencies, like spinal cord compression, while others like a plexopathy may have a more indolent presentation. Although most have a poor prognosis, a thoughtful stepwise treatment approach is necessary to improve the neurologic and oncologic outcomes of patients with neurologic disease associated with GU malignancies.

References

1. Coleman RE. Clinical features of metastatic bone disease and risk of skeletal morbidity. *Clin Cancer Res.* 2006;12(20 Pt. 2):6243s–6249s.
2. Albany C, Hahn NM. Novel bone-targeting agents in prostate cancer. *Prostate Cancer Prostatic Dis.* 2014;17(2):112–118.
3. Loblaw A, Mitera G. Malignant extradural spinal cord compression in men with prostate cancer. *Curr Opin Support Palliat Care.* 2011;5(3):206–210.
4. Sutcliffe P, Connock M, Shyangdan D, Court R, Kandala NB, Clarke A. A systematic review of evidence on malignant spinal metastases: natural history and technologies for identifying patients at high risk of vertebral fracture and spinal cord compression. *Health Technol Assess.* 2013;17(42):1–274.
5. Benjamin R. Neurologic complications of prostate cancer. *Am Fam Physician.* 2002;65(9):1834–1840.
6. Bakar D, Tanenbaum JE, Phan K, et al. Decompression surgery for spinal metastases: a systematic review. *Neurosurg Focus.* 2016;41(2):E2.
7. Crnalic S, Hildingsson C, Bergh A, Widmark A, Svensson O, Lofvenberg R. Early diagnosis and treatment is crucial for neurological recovery after surgery for metastatic spinal cord compression in prostate cancer. *Acta Oncol.* 2013;52(4):809–815.
8. Clinton TN, Woldu SL, Raj GV. Degarelix versus luteinizing hormone-releasing hormone agonists for the treatment of prostate cancer. *Expert Opin Pharmacother.* 2017;18(8):825–832.
9. Patel GK, Chugh N, Tripathi M. Neuroendocrine differentiation of prostate cancer-an intriguing example of tumor evolution at play. *Cancers.* 2019;11(10).
10. Smith MR, Coleman RE, Klotz L, et al. Denosumab for the prevention of skeletal complications in metastatic castration-resistant prostate cancer: comparison of skeletal-related events and symptomatic skeletal events. *Ann Oncol.* 2015;26(2):368–374.
11. Kumar A, Weber MH, Gokaslan Z, et al. Metastatic spinal cord compression and steroid treatment: a systematic review. *Clin Spine Surg.* 2017;30(4):156–163.
12. George R, Jeba J, Ramkumar G, Chacko AG, Tharyan P. Interventions for the treatment of metastatic extradural spinal cord compression in adults. *Cochrane Database Syst Rev.* 2015;(9):CD006716.
13. Laufer I, Rubin DG, Lis E, et al. The NOMS framework: approach to the treatment of spinal metastatic tumors. *Oncologist.* 2013;18(6):744–751.
14. Gao ZY, Zhang T, Zhang H, Pang CG, Jiang WX. Prognostic factors for overall survival in patients with spinal metastasis secondary to prostate cancer: a systematic review and meta-analysis. *BMC Musculoskelet Disord.* 2020;21(1):388.

15. Rades D, Walz J, Stalpers LJ, et al. Short-course radiotherapy (RT) for metastatic spinal cord compression (MSCC) due to renal cell carcinoma: results of a retrospective multi-center study. *Eur Urol.* 2006;49(5):846–852. discussion 52.

16. Bhambhvani HP, Greenberg DR, Srinivas S, Hayden Gephart M. Prostate cancer brain metastases: a single-institution experience. *World Neurosurg.* 2020;138:e445–e449.

17. de Oliveira Barros EG, Meireles Da Costa N, Palmero CY, Ribeiro Pinto LF, Nasciutti LE, Palumbo Jr A. Malignant invasion of the central nervous system: the hidden face of a poorly understood outcome of prostate cancer. *World J Urol.* 2018;36(12):2009–2019.

18. Hatzoglou V, Patel GV, Morris MJ, et al. Brain metastases from prostate cancer: an 11-year analysis in the MRI era with emphasis on imaging characteristics, incidence, and prognosis. *J Neuroimaging.* 2014;24(2):161–166.

19. Ganau M, Gallinaro P, Cebula H, et al. Intracranial metastases from prostate carcinoma: classification, management, and prognostication. *World Neurosurg.* 2020;134:e559–e565.

20. Berg KD, Thomsen FB, Mikkelsen MK, et al. Improved survival for patients with de novo metastatic prostate cancer in the last 20 years. *Eur J Cancer.* 2017;72:20–27.

21. Nevedomskaya E, Baumgart SJ, Haendler B. Recent advances in prostate cancer treatment and drug discovery. *Int J Mol Sci.* 2018;19(5).

22. McDermott RS, Anderson PR, Greenberg RE, Milestone BN, Hudes GR. Cranial nerve deficits in patients with metastatic prostate carcinoma: clinical features and treatment outcomes. *Cancer.* 2004;101(7):1639–1643.

23. Chamoun RB, DeMonte F. Management of skull base metastases. *Neurosurg Clin N Am.* 2011;22(1):61–66. vi-ii.

24. Laigle-Donadey F, Taillibert S, Martin-Duverneuil N, Hildebrand J, Delattre JY. Skull-base metastases. *J Neurooncol.* 2005;75(1):63–69.

25. Hong MK, Kong J, Namdarian B, et al. Paraneoplastic syndromes in prostate cancer. *Nat Rev Urol.* 2010;7(12):681–692.

26. Sacco E, Pinto F, Sasso F, et al. Paraneoplastic syndromes in patients with urological malignancies. *Urol Int.* 2009;83(1):1–11.

27. Aly R, Emmady PD. *Paraneoplastic Cerebellar Degeneration.* Treasure Island, FL: StatPearls; 2020.

28. Bost C, Chanson E, Picard G, et al. Malignant tumors in autoimmune encephalitis with anti-NMDA receptor antibodies. *J Neurol.* 2018;265(10):2190–2200.

29. Storstein A, Raspotnig M, Vitaliani R, et al. Prostate cancer, Hu antibodies and paraneoplastic neurological syndromes. *J Neurol.* 2016;263(5):1001–1007.

30. Maerz DA, Beck LN, Sim AJ, Gainsburg DM. Complications of robotic-assisted laparoscopic surgery distant from the surgical site. *Br J Anaesth.* 2017;118(4):492–503.

31. Faiena I, Patel N, Seftel AD. Prevention of erectile dysfunction after radiotherapy for prostate cancer. *Asian J Androl.* 2014;16(6):805–806.

32. Salonia A, Castagna G, Capogrosso P, Castiglione F, Briganti A, Montorsi F. Prevention and management of post prostatectomy erectile dysfunction. *Transl Androl Urol.* 2015;4(4):421–437.

33. Branch BC, Henry J, Vecil GG. Brain metastases from cervical cancer—a short review. *Tumori.* 2014;100(5):e171–e179.

34. Chura JC, Shukla K, Argenta PA. Brain metastasis from cervical carcinoma. *Int J Gynecol Cancer.* 2007;17(1):141–146.

35. Cordeiro JG, Prevedello DM, da Silva Ditzel LF, Pereira CU, Araujo JC. Cerebral metastasis of cervical uterine cancer: report of three cases. *Arq Neuropsiquiatr.* 2006;64(2A):300–302.

36. Hwang JH, Yoo HJ, Lim MC, et al. Brain metastasis in patients with uterine cervical cancer. *J Obstet Gynaecol Res.* 2013;39(1):287–291.

37. Bourhafour I, Benoulaid M, El Kacemi H, El Majjaoui S, Kebdani T, Benjaafar N. Lumbosacral plexopathy: a rare long term complication of concomitant chemo-radiation for cervical cancer. *Gynecol Oncol Res Pract.* 2015;2:12.

38. Capek S, Howe BM, Amrami KK, Spinner RJ. Perineural spread of pelvic malignancies to the lumbosacral plexus and beyond: clinical and imaging patterns. *Neurosurg Focus.* 2015;39(3):E14.

39. Gupta L, Yadav M, Thulkar S. 'Trident sign' in pelvis: sinister sign with poor prognosis. *BMJ Case Rep.* 2017;2017.

40. Muniz Neto FJ, Kihara Filho EN, Miranda FC, Rosemberg LA, Santos DCB, Taneja AK. Demystifying MR neurography of the lumbosacral plexus: from protocols to pathologies. *Biomed Res Int.* 2018;2018:9608947.

41. Klimek M, Kosobucki R, Luczynska E, Bieda T, Urbanski K. Radiotherapy-induced lumbosacral plexopathy in a patient with cervical cancer: a case report and literature review. *Contemp Oncol.* 2012;16(2):194–196.

42. Rash D, Durbin-Johnson B, Lim J, et al. Dose delivered to the lumbosacral plexus from high-dose-rate brachytherapy for cervical cancer. *Int J Gynecol Cancer.* 2015;25(5):897–902.

43. Santos N, Ferreira RS, Santos ACD. Overview of cisplatin-induced neurotoxicity and ototoxicity, and the protective agents. *Food Chem Toxicol.* 2020;136, 111079.

44. Koo YJ, Kim DY, Kim KR, et al. Small cell neuroendocrine carcinoma of the endometrium: a clinicopathologic study of six cases. *Taiwan J Obstet Gynecol.* 2014;53(3):355–359.

45. Sawada M, Matsuzaki S, Yoshino K, et al. Long-term survival in small-cell carcinoma of the endometrium with liver and brain metastases. *Anticancer Drugs.* 2016;27(2):138–143.

46. Kasper E, Ippen F, Wong E, Uhlmann E, Floyd S, Mahadevan A. Stereotactic radiosurgery for brain metastasis from gynecological malignancies. *Oncol Lett.* 2017;13(3):1525–1528.

47. Gien LT, Kwon JS, D'Souza DP, et al. Brain metastases from endometrial carcinoma: a retrospective study. *Gynecol Oncol.* 2004;93(2):524–528.

48. Hacker NF, Rao A. Surgical management of lung, liver and brain metastases from gynecological cancers: a literature review. *Gynecol Oncol Res Pract.* 2016;3:7.

49. Kottke-Marchant K, Estes ML, Nunez C. Early brain metastases in endometrial carcinoma. *Gynecol Oncol.* 1991;41(1):67–73.

50. Gressel GM, Lundsberg LS, Altwerger G, et al. Factors predictive of improved survival in patients with brain metastases from gynecologic cancer: a single institution retrospective study of 47 cases and review of the literature. *Int J Gynecol Cancer.* 2015;25(9):1711–1716.

51. Teckie S, Makker V, Tabar V, et al. Radiation therapy for epithelial ovarian cancer brain metastases: clinical outcomes and predictors of survival. *Radiat Oncol.* 2013;8:36.

52. Borella F, Bertero L, Morrone A, et al. Brain metastases from ovarian cancer: current evidence in diagnosis, treatment, and prognosis. *Cancers.* 2020;12(8).

53. Kumar A, Kaushik S, Tripathi RP, Kaur P, Khushu S. Role of in vivo proton MR spectroscopy in the evaluation of adult brain lesions: our preliminary experience. *Neurol India.* 2003;51(4):474–478.

54. Wohl A, Kimchi G, Korach J, et al. Brain metastases from ovarian carcinoma: an evaluation of prognostic factors and treatment. *Neurol India.* 2019;67(6):1431–1436.

55. Cohen ZR, Suki D, Weinberg JS, et al. Brain metastases in patients with ovarian carcinoma: prognostic factors and outcome. *J Neurooncol.* 2004;66(3):313–325.

56. Brown PD, Gondi V, Pugh S, et al. Hippocampal avoidance during whole-brain radiotherapy plus memantine for patients with brain metastases: Phase III Trial NRG Oncology CC001. *J Clin Oncol.* 2020;38(10):1019–1029.

57. Yust-Katz S, Mathis S, Groves MD. Leptomeningeal metastases from genitourinary cancer: the University of Texas MD Anderson Cancer Center experience. *Med Oncol.* 2013;30(1):429.

58. Chatterjee M, Hurley LC, Tainsky MA. Paraneoplastic antigens as biomarkers for early diagnosis of ovarian cancer. *Gynecol Oncol Rep.* 2017;21:37–44.

59. Zaborowski MP, Spaczynski M, Nowak-Markwitz E, Michalak S. Paraneoplastic neurological syndromes associated with ovarian tumors. *J Cancer Res Clin Oncol.* 2015;141(1):99–108.

60. Chatterjee M, Hurley LC, Levin NK, Stack M, Tainsky MA. Utility of paraneoplastic antigens as biomarkers for surveillance and prediction of recurrence in ovarian cancer. *Cancer Biomark.* 2017;20(4):369–387.

61. Titulaer MJ, Soffietti R, Dalmau J, et al. Screening for tumours in paraneoplastic syndromes: report of an EFNS task force. *Eur J Neurol.* 2011;18(1):19–e3.

62. Zhang T, Duan Y, Ye J, et al. Brain MRI characteristics of patients with anti-N-methyl-D-aspartate receptor encephalitis and their associations with 2-year clinical outcome. *AJNR Am J Neuroradiol.* 2018;39(5):824–829.

63. Scheinfeld N. A review of the cutaneous paraneoplastic associations and metastatic presentations of ovarian carcinoma. *Clin Exp Dermatol.* 2008;33(1):10–15.

64. Ben-Zvi N, Shani A, Ben-Baruch G, et al. Dermatomyositis following the diagnosis of ovarian cancer. *Int J Gynecol Cancer.* 2005;15(6):1124–1126.

65. Abrey LE, Dalmau JO. Neurologic complications of ovarian carcinoma. *Cancer.* 1999;85(1):127–133.

66. LaFargue CJ, Dal Molin GZ, Sood AK, Coleman RL. Exploring and comparing adverse events between PARP inhibitors. *Lancet Oncol.* 2019;20(1):e15–e28.

67. Smith EM, Pang H, Cirrincione C, et al. Effect of duloxetine on pain, function, and quality of life among patients with chemotherapy-induced painful peripheral neuropathy: a randomized clinical trial. *JAMA.* 2013;309(13):1359–1367.

68. Loprinzi CL, Lacchetti C, Bleeker J, et al. Prevention and management of chemotherapy-induced peripheral neuropathy in survivors of adult cancers: ASCO guideline update. *J Clin Oncol.* 2020;38(28):3325–3348.

69. Ramalingam S, George DJ, Harrison MR. How we treat brain metastases in metastatic renal cell carcinoma. *Clin Adv Hematol Oncol.* 2018;16(2):110–114.

70. Maria B, Antonella V, Michela R, et al. Multimodality treatment of brain metastases from renal cell carcinoma in the era of targeted therapy. *Ther Adv Med Oncol.* 2016;8(6):450–459.

71. Muacevic A, Siebels M, Tonn JC, Wowra B. Treatment of brain metastases in renal cell carcinoma: radiotherapy, radiosurgery, or surgery? *World J Urol.* 2005;23(3):180–184.

72. Shirotake S. Management of brain metastases from renal cell carcinoma. *Ann Transl Med.* 2019;7(suppl 8):S369.

73. Sperduto PW, Deegan BJ, Li J, et al. Estimating survival for renal cell carcinoma patients with brain metastases: an update of the Renal Graded Prognostic Assessment tool. *Neuro Oncol.* 2018;20(12):1652–1660.

74. Matsui Y. Current multimodality treatments against brain metastases from renal cell carcinoma. *Cancers.* 2020;12(10).

75. DeAngelis LM, Delattre JY, Posner JB. Radiation-induced dementia in patients cured of brain metastases. *Neurology.* 1989;39(6):789–796.

76. Khan M, Zhao Z, Arooj S, Liao G. Impact of tyrosine kinase inhibitors (TKIs) combined with radiation therapy for the management of brain metastases from renal cell carcinoma. *Front Oncol.* 2020;10:1246.

77. Yekeduz E, Arzu Yasar H, Utkan G, Urun Y. A systematic review: role of systemic therapy on treatment and prevention of brain metastasis in renal cell carcinoma. *J Oncol Pharm Pract.* 2020;26(4):972–981.

78. Brown LC, Desai K, Kao C, et al. A multicenter retrospective study to evaluate real-world clinical outcomes in patients with metastatic renal cell carcinoma (mRCC) and brain metastasis treated with ipilimumab and nivolumab. *J Clin Oncol.* 2020;38(6_suppl):637.

79. Fallah J, Ahluwalia MS. The role of immunotherapy in the management of patients with renal cell carcinoma and brain metastases. *Ann Transl Med.* 2019;7(suppl 8):S313.

80. Gerszten PC, Burton SA, Ozhasoglu C, et al. Stereotactic radiosurgery for spinal metastases from renal cell carcinoma. *J Neurosurg Spine.* 2005;3(4):288–295.

81. Woodward E, Jagdev S, McParland L, et al. Skeletal complications and survival in renal cancer patients with bone metastases. *Bone.* 2011;48(1):160–166.

82. Teyssonneau D, Gross-Goupil M, Domblides C, et al. Treatment of spinal metastases in renal cell carcinoma: a critical review. *Crit Rev Oncol Hematol.* 2018;125:19–29.

83. Wood SL, Brown JE. Skeletal metastasis in renal cell carcinoma: current and future management options. *Cancer Treat Rev.* 2012;38(4):284–291.

84. Chen SC, Kuo PL. Bone metastasis from renal cell carcinoma. *Int J Mol Sci.* 2016;17(6).

85. Palapattu GS, Kristo B, Rajfer J. Paraneoplastic syndromes in urologic malignancy: the many faces of renal cell carcinoma. *Rev Urol.* 2002;4(4):163–170.

86. Chilkulwar A, Pottimutyapu R, Wu F, Padooru KR, Pingali SR, Kassem M. Leptomeningeal carcinomatosis associated with papillary renal cell carcinoma. *Ecancermedicalscience.* 2014;8:468.

87. Dalhaug A, Haukland E, Nieder C. Leptomeningeal carcinomatosis from renal cell cancer: treatment attempt with radiation and sunitinib (case report). *World J Surg Oncol.* 2010;8:36.

88. Bonomi L, Bettini AC, Arnoldi E, et al. Nivolumab efficacy in leptomeningeal metastasis of renal cell carcinoma: a case report. *Tumori.* 2020;, 300891620904411.

89. Glusker P, Recht L, Lane B. Reversible posterior leukoencephalopathy syndrome and bevacizumab. *N Engl J Med.* 2006;354(9):980–982. discussion 980–982.

90. Ozcan C, Wong SJ, Hari P. Reversible posterior leukoencephalopathy syndrome and bevacizumab. *N Engl J Med.* 2006;354(9):980–982. discussion 980–982.

91. Ernstoff MS, Puzanov I, Robert C, Diab AM, Hersey PM. *SITC's Guide to Managing Immunotherapy Toxicity.* Springer Publishing Company; 2019.

92. Zukas AM, Schiff D. Neurological complications of new chemotherapy agents. *Neuro Oncol.* 2018;20(1):24–36.

93. Nonomura N, Nagahara A, Oka D, et al. Brain metastases from testicular germ cell tumors: a retrospective analysis. *Int J Urol.* 2009;16(11):887–893.

94. Gilligan T. Decision making in a data-poor environment: management of brain metastases from testicular and extragonadal germ cell tumors. *J Clin Oncol.* 2016;34(4):303–306.

95. Oechsle K, Kollmannsberger C, Honecker F, Boehlke I, Bokemeyer C. Cerebral metastases in non-seminomatous germ cell tumour patients undergoing primary high-dose chemotherapy. *Eur J Cancer.* 2008;44(12):1663–1669.

96. Lavoie JM, Kollmannsberger CK. Current management of disseminated germ cell tumors. *Urol Clin North Am.* 2019;46(3):377–388.

97. Kopp HG, Kuczyk M, Classen J, et al. Advances in the treatment of testicular cancer. *Drugs.* 2006;66(5):641–659.

98. Loriot Y, Pagliaro L, Flechon A, et al. Patterns of relapse in poor-prognosis germ-cell tumours in the GETUG 13 trial: implications for assessment of brain metastases. *Eur J Cancer.* 2017;87:140–146.

99. Adra N, Einhorn LH. Testicular cancer update. *Clin Adv Hematol Oncol.* 2017;15(5):386–396.

100. Bolm L, Janssen S, Bartscht T, Rades D. Radiotherapy alone for malignant spinal cord compression in young men with seminoma. *Anticancer Res.* 2016;36(4):2033–2034.

101. Biswas A, Puri T, Goyal S, et al. Spinal intradural primary germ cell tumour—review of literature and case report. *Acta Neurochir.* 2009;151(3):277–284.

102. Hoffmann LA, Jarius S, Pellkofer HL, et al. Anti-Ma and anti-Ta associated paraneoplastic neurological syndromes: 22 newly diagnosed patients and review of previous cases. *J Neurol Neurosurg Psychiatry.* 2008;79(7):767–773.

103. Maudes E, Landa J, Munoz-Lopetegi A, et al. Clinical significance of Kelch-like protein 11 antibodies. *Neurol Neuroimmunol Neuroinflamm.* 2020;7(3).

104. Dubey D, Wilson MR, Clarkson B, et al. Expanded clinical phenotype, oncological associations, and immunopathologic insights of paraneoplastic Kelch-like Protein-11 encephalitis. *JAMA Neurol.* 2020;77:1420–1429.

105. Raghavan D. Cutaneous manifestations of genitourinary malignancy. *Semin Oncol.* 2016;43(3):347–352.

106. Fossa SD, de Wit R, Roberts JT, et al. Quality of life in good prognosis patients with metastatic germ cell cancer: a prospective study of the European organization for research and treatment of Cancer Genitourinary Group/Medical Research Council Testicular Cancer Study Group (30941/TE20). *J Clin Oncol.* 2003;21(6):1107–1118.

107. Swallow TW, Mabbutt S, Bell CR. Muscle invasive bladder cancer culminating with leptomeningeal carcinomatosis. *Can Urol Assoc J.* 2015;9(11–12):E903–E904.

108. Anderson TS, Regine WF, Kryscio R, Patchell RA. Neurologic complications of bladder carcinoma: a review of 359 cases. *Cancer.* 2003;97(9):2267–2272.

109. Bex A, Sonke GS, Pos FJ, Brandsma D, Kerst JM, Horenblas S. Symptomatic brain metastases from small-cell carcinoma of the urinary bladder: the Netherlands Cancer Institute experience and literature review. *Ann Oncol.* 2010;21(11):2240–2245.

110. Mahmoud-Ahmed AS, Kupelian PA, Reddy CA, Suh JH. Brain metastases from gynecological cancers: factors that affect overall survival. *Technol Cancer Res Treat.* 2002;1(4):305–310.

111. Brenneman RJ, Gay HA, Christodouleas JP, et al. Brain metastases in bladder cancer. *Bladder Cancer.* 2020;1–12 [Preprint].

112. Fokas E, Henzel M, Engenhart-Cabillic R. A comparison of radiotherapy with radiotherapy plus surgery for brain metastases from urinary bladder cancer: analysis of 62 patients. *Strahlenther Onkol.* 2010;186(10):565–571.

113. Teyssonneau D, Daste A, Dousset V, Hoepffner JL, Ravaud A, Gross-Goupil M. Metastatic non-muscle invasive bladder cancer with meningeal carcinomatosis: case report of an unexpected response. *BMC Cancer.* 2017;17(1):323.

114. Bothig R, Kurze I, Fiebag K, et al. Clinical characteristics of bladder cancer in patients with spinal cord injury: the experience from a single centre. *Int Urol Nephrol.* 2017;49(6):983–994.

115. Gui-Zhong L, Li-Bo M. Bladder cancer in individuals with spinal cord injuries: a meta-analysis. *Spinal Cord.* 2017;55(4):341–345.

116. Aghion DM, Capek S, Howe BM, et al. Perineural tumor spread of bladder cancer causing lumbosacral plexopathy: an anatomic explanation. *Acta Neurochir.* 2014;156(12):2331–2336.

117. Requena C, Alfaro A, Traves V, et al. Paraneoplastic dermatomyositis: a study of 12 cases. *Actas Dermosifiliogr.* 2014;105(7):675–682.

118. Sagi L, Amichai B, Barzilai A, et al. Dermatomyositis and small cell carcinoma of the bladder. *Can Fam Physician.* 2009;55(10):997–999.

Sarcoma and the nervous system

Karan S. Dixit[a,b], Jean-Paul Wolinsky[b,c], Priya Kumthekar[a,b],
Craig Horbinski[b,c,d], and Rimas V. Lukas[a,b]

[a]Department of Neurology, Northwestern University, Chicago, IL, United States, [b]Lou & Jean Malnati Brain Tumor Institute, Northwestern University, Chicago, IL, United States, [c]Department of Neurological Surgery, Northwestern University, Chicago, IL, United States, [d]Department of Pathology, Northwestern University, Chicago, IL, United States

1 Introduction

Sarcomas are neoplasms arising from mesenchymal progenitor cells. Mesenchymal cells are of mesodermal origin, as opposed to ectodermal or endodermal origin. This differentiates them phylogenetically from other central nervous system (CNS) tumors, such as gliomas, which arise from cells that are ectodermal in origin. The dura mater is thought to arise from a combination of cells of ectodermal and mesodermal origin. As cells of mesenchymal origin are not a substantial component of the CNS, it is unsurprising that primary tumors arising from those cell types are infrequently seen in the CNS. Many CNS sarcomas, when they do occur, most commonly arise in the pediatric population. Our understanding of their epidemiology is lacking. In the Central Brain Tumor Registry of the United States (CBTRUS), arguably the most comprehensive national brain tumor registry, these tumors can be classified along with other neoplasms under the following categories: mesenchymal tumors (under tumors of meninges), neoplasm unspecified (under unclassified tumors), and all others (also under unclassified tumors).[1] As these categories can also include other tumor types, it is difficult to ascertain the incidence of CNS sarcomas. While gliosarcoma remains a term in the WHO classification system, we will refrain from covering it in this chapter, as it is viewed more akin to the infiltrating gliomas arising from cells of ectodermal origin, as opposed to the CNS sarcomas arising from cells of mesodermal origin.

This chapter will detail sarcomas which arise within the CNS, as well as those which occur adjacent to the CNS, in bony tissue for example, and can unfavorably impact the CNS. These rare tumors are often managed by multidisciplinary teams which include neuro-oncologists, pediatric neuro-oncologists, oncologists, neurosurgeons, orthopedic surgeons, radiation oncologists, general surgeons, neuroradiologists, and pathologists. The composition of the teams depends upon the specific disease, its location, and a number of institutional factors. The optimal management of these tumors is less clear than that of other CNS tumors. The rarity of these diseases makes randomized studies untenable. Most of what we know arises from retrospective single-institution case series. These are hampered by all of the drawbacks intrinsic to this study type.

The management of CNS sarcoma involves obtaining a histologic diagnosis via a surgical procedure. There are no pathognomonic radiographic findings which allow for a reliable diagnosis without pathology. As other CNS tumor types are far more common, CNS sarcoma typically does not lay at the top of any differential diagnosis. The therapeutic management across all CNS sarcoma subtypes begins with maximal surgical resection, ideally with histologically negative margins. Postoperative management is less clearly defined, but is specific to sarcomatous subtype, and may involve focal radiation therapy and/or systemic therapy. Specific radiation dosing regimens have limited evidence of superiority over others. A similar scenario is noted with respect to systemic therapies. Across all of the sarcomatous CNS tumors discussed within, there is a range of predominantly cytotoxic agents that have been used. The level of evidence to support one over another is severely limited.

We will begin our discussion highlighting a number of rare primary CNS sarcomas. This will be followed by discussion of primary sarcomas of the peripheral nervous system. This will be followed by sarcomas adjacent to the CNS and negatively impacting it. Next, sarcoma metastases to the CNS will be covered. Discussion of the common neurologic toxicities associated with extra-CNS sarcoma treatments will conclude this chapter.

2 Primary CNS sarcomas

Primary CNS sarcomas are very rare tumors. As noted earlier, the management of primary CNS sarcoma typically involves an attempt at gross total resection. This is often followed by radiation therapy (RT) to residual disease or the resected tumor bed. Systemic treatments with chemotherapies can also be employed.

Radiographically, the majority of these tumors enhance. Cysts, calcification, and intratumoral hemorrhage may be present. Approximately half are adherent to or arise from the dura, which is unsurprising due to the partial mesodermal origin of the dura. One-third of these tumors demonstrate a distinct plane differentiating them from normal brain. As with sarcomas elsewhere in the body, immunohistochemistry (IHC) demonstrates staining for the Type III intermediate filament vimentin.[2] Other specific tumor markers will be discussed with each tumor type. Survival remains poor with these tumors. Median overall survival (OS) has been reported as 4.6 years with a range up to 16 years. A number of factors, including elevated proliferation index, very young age (< 1 year old) at diagnosis, and cerebrospinal fluid (CSF) dissemination, have all been associated with poorer outcomes.[2,3]

2.1 DICER1-associated central nervous system sarcoma

At least some primary CNS sarcomas are associated with the tumor predisposition DICER1 syndrome. This has been recently termed as DICER1-associated central nervous system sarcoma (DCS). DICER1 encodes a protein, RNase III endoribonuclease, involved in the cleaving of precursor RNA for the creation of micro-RNA (miRNA). miRNA in turn regulates the expression of messenger RNA (mRNA). Dysregulation of this system leads to a predisposition toward development of rare tumors. A majority of mutations are inherited in an autosomal dominant manner, with a smaller number appearing to be de novo. An even smaller percentage of patients exhibit a somatic mosaicism.[4] In addition to primary CNS sarcoma, patients with DICER1 syndrome are at risk for developing other rare CNS tumors and extra-CNS tumors. These include pineoblastoma, pituitary blastoma, ciliary body medulloepithelioma, and infantile embryonal cerebellar tumors among the CNS tumors. Non-CNS tumors include pleuropulmonary blastoma, Sertoli-Leydig cell tumor of the ovary, cystic nephroma, thyroid carcinoma, renal sarcoma, embryonal rhabdomyosarcoma, and nasal chondromesenchymal hamartoma.[4]

Primary CNS sarcoma in DICER1 syndrome tends to affect children and adolescents ranging from ages of 3 to 15.[4,5] When associated with DICER1 syndrome, these tumors typically demonstrate a rhabdomyoblastic differentiation, which is in line with the other "blastic" tumors found in DICER1 syndrome.[6] IHC has demonstrated patchy staining for the Type III intermediate filament desmin and the muscle-specific transcription factor myogenin, as well as patchy loss of H3K27me3, a finding which can be seen in other sarcomas such as malignant peripheral nerve sheath tumors (MPNSTs).[5] Molecular studies reveal TP53 inactivation and *RAS* pathway activation.[5] These tumors also appear to harbor a specific methylation profile which may be of benefit in diagnostic evaluation.[6]

As with non-DICER1-associated CNS sarcomas, surgical resection followed by focal RT is the mainstay of treatment. Systemic chemotherapies are often also utilized with agents including ifosfamide, doxorubicin, vincristine, and dactinomycin.[7]

2.2 Solitary fibrous tumor of the nervous system/hemangiopericytoma

The term solitary fibrous tumor (SFT) of the nervous system/hemangiopericytoma was codified in the 2016 revision of the World Health Organization (WHO) classification system.[8] These tumors are analogous to SFTs in other non-CNS locations. Previously, this entity, when presenting in the CNS, had been called hemangiopericytoma, although the original report describing them did utilize the term SFT.[9] These tumors arise from perivascular cells. They most often involve the dura intracranially although they can be seen in the dural tissue surrounding the spine. Radiographically, their appearance is similar to that of meningiomas, another dural-based tumor. They may harbor cystic components, a more heterogeneous enhancement pattern, or evidence of extra-cranial extension through the skull.[10] The mean age of diagnosis is 44 years old. In younger patients, there is a higher incidence in males and in older patients a higher incidence in females.[11] While diffusion-weighted imaging and MR perfusion have not been demonstrated to be of value in differentiating SFT from meningiomas, MR spectroscopy may be of use. Elevated myo-inositol, particularly in patients with a high pretest probability, is suggestive of SFT.[12] An elevated choline-to-creatine ratio has been described on MR spectroscopy, although it is unknown how common this finding is.[10] Histologically, these tumors have been described as having a "staghorn" branching pattern of vasculature.[13] The defining molecular hallmark of SFTs is a fusion between *NGFI-A binding protein 2* (*NAB2*) and *signal transducers and activators 6* (*STAT6*), caused by a paracentric inversion of chromosome 12q13.[9,14]

The presence of the NAB2-STAT6 fusion leads to abnormally high nuclear STAT6 expression, which plays a role as a transcriptional transactivator in the inflammatory response to cytokines such as IL-4 and IL-13. Thus, strong nuclear STAT6 immunostaining in nearly 100% of tumor cells is a reliable marker of *NAB2-STAT6* fusion.[15] Like many other kinds of tumors, necrosis and elevated mitotic index have been associated with more aggressive SFT behavior.[16] Other markers, such as the decreased expression of CD34 and Bcl-2, or TERT promoter mutations, may be suggestive of a shorter progression-free survival, but require further validation.[17]

These tumors are usually treated with surgical resection, with a gross total resection as the goal. Extent of resection appears to be associated with survival.[10,11] Due to the risk of postoperative recurrence, surgery may be followed by postoperative radiotherapy. The radiotherapeutic approaches have included both focal fractionated radiation and stereotactic radiosurgery with radiosurgery being used in both the newly diagnosed and progressive/recurrent disease settings. No standard dosing has been established in this rare disease. Survival data are limited in these small case series. Five-year overall survival ranges from 61% to 100%. A number of nontreatment-related factors likely influence this.[18] Even with an aggressive upfront approach, these tumors have the potential to recur locally, as well as at intracranial locations distant from the original site. Additionally, they may metastasize outside of the nervous system with the lungs and bones being the most frequently affected tissues. These tumors can be differentiated from most malignancies with a high metastatic potential by the latency between initial diagnosis and treatment and subsequent development of recurrence usually measured in years and at times greater than a decade.[10,11,19,20] This aspect underlines the importance of continued long-term surveillance. Chemotherapy has limited activity in SFTs, and there is no established chemotherapy regimen for refractory or metastatic disease. Systemic chemotherapy regimens using cyclophosphamide, doxorubicin, vincristine, interferon, ifosfamide, carboplatin, and etoposide have been studied with only a fraction of patients demonstrating radiographic response, but a majority had stable disease.[21] Temozolomide and bevacizumab, two agents used frequently in brain tumors, were also studied in patients with recurrent and metastatic disease with almost 80% of patients achieving a partial response with a median progression-free survival of 9.7 months.[22] Targeted agents such as sunitinib, sorafenib, and pazopanib have also been studied with evidence of response.[23] Pazopanib, a multitargeted receptor tyrosine kinase inhibitor, has been the most recently studied agent with promising activity in both frontline and recurrent disease.[24,25]

2.3 Primary CNS histiocytic sarcoma

Primary CNS histiocytic sarcoma is a very rare manifestation of a hematopoietic malignancy more often found in the gastrointestinal tract, skin, lymph nodes, or other soft tissue locations. Only a very limited number of cases of this tumor arising in the CNS have been reported.[26] There do not appear to be clear predisposing factors for development of this disease, although at least one case was deemed to be radiation induced.[26] Pathology reveals large atypical round/ovoid cells with presence of spindling in some tumors. Abundant cytoplasm with round/ovoid nuclei is seen. As would be expected of a tumor with histiocytic differentiation, CD68-positive lysosomes are frequently noted in the cytoplasm. The cells are often positive for markers of histiocytic lineage such as cluster of differentiation (CD) 4, CD11c, CD14, CD68 (KP-1), CD163, lysozyme, and fascin.[26] They lack the markers typically seen in Langerhans cell histiocytosis such as CD1a, CD21, CD23, and CD35. Molecular alterations in the RAS/RAF/MAPK pathway are highly characteristic of histiocytic sarcomas.[27]

As only a limited number of cases have been reported, it is uncertain what is the optimal management. These tumors have been treated with surgical resection and/or radiation therapy. Prognosis remains poor, with survival usually measured only in months.[28]

2.4 Kaposi's sarcoma

Kaposi's sarcoma is a neoplasm of mesenchymal cells lining the blood vessels and lymphatics. Tumorigenesis is driven by the Kaposi's sarcoma-associated herpes virus (KSHV). KSHV has been detected via polymerase chain reaction (PCR) in the neurons of the CNS in HIV infected patients.[29]

Although usually affecting mucocutaneous tissues, Kaposi's sarcoma can rarely affect the central and peripheral nervous system. Peripheral nervous system involvement has primarily been reported only on autopsy studies in patients who experienced radicular symptoms with evidence of tumor infiltrating the nerve roots, plexus, and nerves.[30]

Central nervous system involvement has been reported more often, but it is still very rare with poor characterization of its presenting symptoms and radiographic appearance. Histologically, it looks like a highly vascular tumor with brisk inflammation. Immunohistochemical markers for factor VIII-related antigen, CD31, and CD34 are usually positive, but the single best differentiator from other histologic mimickers is probably LNA-1 immunopositivity, which indicates the presence of HHV8.[31] In an autopsy case of a patient who developed hemiparesis, several hemorrhagic and necrotic brain lesions were found.[32] In patients with HIV and Kaposi's sarcoma, it is important to rule out other potential infectious intracranial diseases rather than presume they are metastases given the overall rarity.[33]

2.5 Ewing's sarcoma

Ewing's sarcoma is a small round blue cell tumor deemed to represent a primary bone tumor arising from neuroectodermal tissue found most frequently in the pediatric/young adult population. Primary CNS Ewing's sarcoma is exceedingly rare. Histopathologically, it resembles other primary CNS small round blue cell tumors, such as medulloblastoma and the primitive neuroectodermal tumors (PNETs), a classification retired in 2016 by the WHO. When Ewing's sarcoma does appear to arise from an intracranial primary, it typically manifests with dural involvement.[34,35] Primary dural Ewing's sarcoma represents only 1%–4% of extraosseous Ewing's sarcoma.[34] It is not always clear if this represents a primary bone tumor with invasion of the dura or the inverse scenario. Regardless of location, pathology is similar. Homer-Wright rosettes are characteristic. Vascular invasion and necrosis have been reported.[34] CD99/MIC-2, Friend Leukemia Virus Integration 1 (Fli-1), NXK2.2, EMA, INI-1/SMARCB1, and neurofilament protein expression have all been noted to varying degrees. The proliferative index in these tumors is often quite high.[35] The *EWSR1-Fli-1* fusion is present in about 85% of these tumors. Less common fusion partners with *EWSR1* include *ERG* and *ETV1*. A break-apart fluorescence in situ hybridization probe targeting *EWSR1* is a reliable test for all these fusions, although it cannot tell which fusion partner is involved.

Management of typical Ewing's sarcoma consists of intensive neoadjuvant chemotherapy followed by either *en bloc* surgical resection or definitive radiotherapy for inoperable tumors followed by maintenance chemotherapy.[36] Management of intracranial disease consists of surgical resection with an attempt at a disease-free margin. If bone is also involved, this is also resected. Postoperative radiation and chemotherapy have also been utilized. Their effects on outcome are difficult to parcel out due to the lack of randomized clinical data as well as reliable historical controls.[35]

2.6 Leiomyosarcoma

Leiomyosarcoma is a neoplasm arising from smooth-muscle-derived tissue. CNS leiomyosarcomas are thought to arise from the intracranial vasculature or a subpopulation of dural cells.[37] These tumors typically manifest outside the CNS although cases of primary CNS involvement have been reported. It is important to screen for disease outside the CNS as a primary site elsewhere may be detected. When involving the CNS, it is

usually within the context of an immunocompromised state. In the setting of HIV+, it usually occurs when CD4 counts are markedly diminished.[38,39] Oftentimes, in this setting, the tumor is positive for Epstein-Barr virus (EBV). Prior cranial radiation is also a potential risk factor for developing these tumors.[40]

Radiographically, the lesion is likely to be enhancing and can arise from the meninges, being suggestive of a meningioma.[41] They may also arise within the brain parenchyma. These lesions can appear relatively well circumscribed.[42–44] There may be a predisposition for the parieto-occipital region.[45] Perineural extension, along cranial nerves, has been described.[46] Once a tissue diagnosis is obtained, immunohistochemistry can reveal staining for α-smooth muscle actin, the Type III intermediate filaments vimentin and desmin, the calcium-binding protein calponin, the cell-cell adhesion factor CD34, the antiapoptotic protein BCL2, and progesterone receptor.[37,42,45]

As with other CNS sarcomas, the management revolves around surgical resection. However, it is thought that the risk of local recurrence remains high. As with other CNS sarcomas, postoperative RT and/or chemotherapy have been used although their value is not clearly established. Chemotherapeutic agents which have been reported include procarbazine, vincristine, ifosfamide, doxorubicin, etoposide, cyclophosphamide, as well as the antiangiogenic bevacizumab, the EGFR-targeting antibody nimotuzumab, and the immunotherapeutic interferon-α.[39,42,44]

2.7 Angiosarcoma

Angiosarcomas are rare sarcomas thought to arise from endothelial cells of the vasculature or lymphatics. Their presence as primary CNS neoplasms is particularly uncommon. Radiographically, what is seen are enhancing parenchymal tumors. Cystic components may be present.[47] There are no radiographic features pathognomonic for CNS angiosarcoma. Histologically, angiosarcomas show anastomosing vascular channels lined by mitotically active, malignant-appearing cells. Immunopositivity for the usual vascular markers, including CD31, CD34, ERG, and FLI1, is typical. Approximately quarter of angiosarcomas express insulin-like growth factor 2 messenger RNA-binding protein 3 (IGF2BP3), which is not found in nonneoplastic lesions.[48] Angiosarcomas broadly speaking arise in the elderly population. However, the age distribution for CNS angiosarcomas is quite broad ranging from congenital tumors to older patients. A number of the reported cases have been congenital in nature. There appears to be dichotomy with respect to outcomes. A substantial number of patients progress rapidly to death while others demonstrate no evidence of recurrence after gross total resection.[47]

2.8 Radiation-induced sarcomas affecting the CNS

Radiation exposure to the CNS increases the incidence of CNS tumors manyfold.[49] Radiation-induced sarcomas of the CNS can be due to incidental radiation exposure or can be iatrogenically induced. Radiation-induced sarcomas are far rarer than radiation-induced meningiomas and gliomas.[49] These tumors arise within the radiation field and typically follow a latency period of years to decades.[50] Of note, these tumors can be induced over much shorter intervals in primate models.[51] Other factors including radiation dose and patient age at the time of radiation likely influence the risk of developing these secondary malignancies. It is premature to make any claims regarding mitigation of risk of developing secondary malignancies including CNS sarcomas by utilizing techniques to limit entry/exit doses with the use of protons as opposed to photons.[52] Imaging findings are similar to spontaneous CNS sarcomas.[50] At this time, a histologic or molecular signature is not known for differentiating radiation-induced CNS sarcomas from spontaneous ones. Outcomes in patients with radiation-induced CNS sarcoma are poor with a median OS of 11 months and few 5-year survivors. There is a suggestion that systemic chemotherapy may be associated with improved survival.[50]

3 Primary sarcomas of the peripheral nervous system

As in the central nervous system, primary sarcomas of the peripheral nervous system are uncommon tumors. When they do occur, they are predominantly MPNSTs. These can occur spontaneously or within the background of a neurocutaneous syndrome. Sarcomas involving the peripheral nervous system can manifest with motor and sensory symptomatology. At times, they can grow in a fashion which impinges on the adjacent CNS causing additional symptoms. MPNSTs will be discussed in detail below.

3.1 Malignant peripheral nerve sheath tumors

MPNSTs are aggressive soft tissue sarcomas arising from peripheral nerves. MPNSTs are rare, with an incidence of 1 in 100,000 with 50% occurring in patients with neurofibromatosis 1 (NF1), 10% being radiation induced (often occurring 10–20 years later), and the rest occurring sporadically.[53] Patients with NF1 have an 9%–13% lifetime risk of developing MPNST, most often from pre-existing plexiform neurofibromas.[54] NF1 patients with internal plexiform neurofibromas are 20 times more likely to develop MPNSTs compared to those without them.[55] Patients with NF1 tend to develop MPNSTs at a younger age in the 3rd–4th decade of life compared to the 7th decade of life in patients without NF1.[54]

The most common location for MPNSTs is within the extremities and trunk, often along the course of major nerves and nerve trunks such as the sciatic and brachial plexus, and less often along the head and neck (Fig. 1).[56] Clinical presentation often is a combination of new or worsening pain, rapid growth of a known mass, and sensorimotor deficits in the distribution of the affected nerves.[57] Any of these symptoms in a patient with NF1 should prompt urgent workup for MPNST which includes MRI, PET/CT, and tissue sampling. MRI features suggestive of MPNST include large tumor size, heterogeneous enhancement, ill-defined margins, invasion of fat planes, and surrounding edema.[58] There is mounting evidence that diffusion-weighted imaging (DWI) and apparent diffusion coefficient (ADC) mapping can also reliably distinguish MPNST from benign nerve sheath tumors.[59] Fluorodeoxyglucose positron emission tomography (FDG-PET) is also a sensitive and specific modality to differentiate between MPNST and benign tumors with areas of FDG uptake allowing for targeted biopsies to maximize diagnostic yield.[60] An SUV_{max} threshold of ≥6.1 can differentiate MPNST from benign nerve sheath tumors with a sensitivity of 94% and specificity of 91%.[61] Full body PET/CT is an ideal imaging modality because it also provides systemic staging given

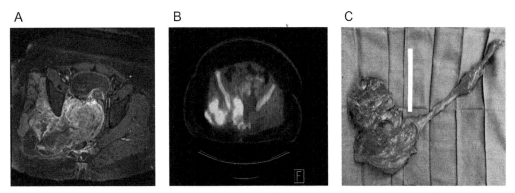

FIG. 1   Malignant peripheral nerve sheath tumor of right sciatic nerve. (A) MRI with gadolinium with avidly enhancing lesion, (B) PET/CT demonstrates avid hypermetabolic activity, and (C) *en bloc* specimen of right sciatic nerve and tumor.

potential metastases to the lung, liver, bone, regional lymph nodes, and retroperitoneum.[62]

MPNSTs are usually high-grade tumors with histopathologic appearance notable for mitotically active spindle cells, which are weak or negative for S100 and SOX10, indicating dedifferentiation from Schwann cells.[63,64] A subset of MPNSTs feature epithelioid morphology, or even other mesenchymal patterns like smooth muscle, bone, and cartilage. In addition to NF1, inactivating mutations in the PRC2 complex, such as *SUZ12* and *EED*, are common.[65] As a result of PRC2 disruption, complete loss of H3K27me3 immunopositivity has been suggested as a marker of MPNST though its reliability is controversial.[66–68] Staging is performed using the American Joint Committee on Cancer staging system including tumor grade, size ($<5\,cm$, $\geq 5\,cm$), deep versus superficial location, and presence of metastases. High tumor grade and tumor size $\geq 5\,cm$ is associated with poorer outcomes. Although the prevailing notion was that NF1 patients with MPNST experienced worse survival, a large metaanalysis of over 1800 patients with MPNST demonstrated no significant difference in outcome between the groups.[69]

Management of MPNSTs is like that of other soft tissue sarcomas and can be heterogeneous. Complete *en bloc* surgical resection with negative margins may possibly be curative and should be the primary goal if possible.[70] Patients with metastatic disease are usually not considered for *en bloc* resection unless the risk of surgical morbidity is low. Neoadjuvant chemotherapy (with ifosfamide and epirubicin) or radiotherapy to downsize tumors to allow for a better surgical target can be considered and is often employed.[71,72] Local recurrence can occur in nearly 50% of cases; thus, adjuvant therapy with radiation is considered.[73] Adjuvant radiation is often used when MPNSTs are not amenable to surgical resection and can improve local control to improve disease specific survival.[74] The role of chemotherapy in localized disease is not well defined with treatment extrapolated from other soft tissue sarcomas including the use of doxorubicin, dacarbazine, and ifosfamide with variable response ranging from 20% to 46%.[75] With the advent of molecularly targeted therapies, selumetinib, a MEK inhibitor, was granted US Breakthrough Therapy Designation in April 2020 for progressive or inoperable MPNSTs in NF1 patients.

4 Extra-CNS sarcoma affecting the nervous system

The CNS is enveloped by protective skeletal structures, specifically the skull and the vertebral column. Sarcomas which develop in these structures have the potential to compress the underlying CNS leading to potential neurologic deficits, injury, and death. The specific type of symptomatology will relate predominantly to neuroanatomic localization but can also lead to generalized symptoms (e.g., increased intracranial pressure symptoms) depending on tumor size and rate of growth. We will discuss three specific types of sarcomas arising from the bone: chordoma, chondrosarcoma, and osteosarcoma. Similarly, sarcomas directly involving the CNS management involve surgical resection, often radiation therapy, and to a limited degree systemic medical therapy.

4.1 Chordoma

Chordomas are rare tumors of the bone, which arise from remnants of the notochord. They are slowly growing but locally aggressive and often occur in the axial, or midline, skeleton. Almost half occur within the sacrococcygeal region, 35% occur in the skull base, and the remaining 15% occur along the vertebral column.[76] Overall annual incidence is 1 per 1,000,000, and they typically occur in the 4th–6th decades of life affecting males at a nearly 2:1 ratio.[77] Neurologic symptoms depend upon the location of the tumor. Skull base chordomas, often affecting the clivus, can impinge surrounding neural structures and present with multiple cranial neuropathies with symptoms including diplopia, dysphagia, hearing loss, vertigo, headache, and neck pain. Sacral chordomas expectedly can cause progressive lower extremity sensory or motor deficits, bowel and bladder dysfunction, and neuropathic pain (Fig. 2).

Radiological studies demonstrate a destructive osseous lesion on CT with T1 hypointensity, T2 hyperintensity, and variable contrast enhancement on MRI.[78] Histologic appearance is notable for relatively bland-appearing vacuolated cells (physaliphorous cells) surrounded by a myxoid matrix. Immunohistochemistry is often positive for S100 and brachyury, though the single best marker differentiating chordoma from chondrosarcoma is a cytokeratin antibody.[79,80]

Chordomas demonstrate locally aggressive growth, with over 50% of patients developing loco-regional recurrence necessitating aggressive upfront management.[81] Risk factors for poorer outcome include female sex, older age at diagnoses, larger tumors, subtotal resection, and presence of metastases.[82] Histologically, elevated Ki67 proliferation index has also been associated with more aggressive behavior, as has homozygous deletion of *CDKN2A/B* and 1p36.[83,84] Surgery is the primary treatment modality for chordoma with the goal for complete surgical resection if possible, with patients undergoing subtotal resection having a nearly four times risk of recurrence within 5 years.[85] Wide *en bloc* resections often are not possible in the skull base but may be possible in the sacrum and spine and are associated with superior

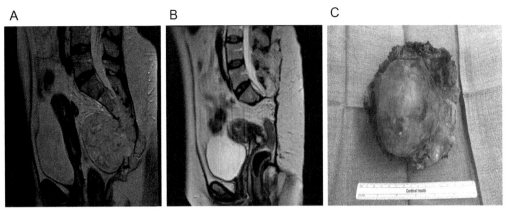

FIG. 2 Sacral chordoma. (A) Preoperative T2-weighted MRI with large sacral mass, (B) 2-year follow-up, T2-weighted MRI with no evidence of disease, and (C) *en bloc* specimen with negative margin.

outcomes.[86] Surgical technique has evolved over time, especially for clival chordomas, with the endoscopic endonasal approach becoming a preferred method.[87] Cranial neuropathies are often the presenting symptom for cranial chordomas and can often worsen immediately after surgery though surgically related deficits tend to improve through recovery.[88] Given the high risk of locoregional recurrence, adjuvant postoperative radiation is employed if an *en bloc* resection with negative margins is not achievable. Particle radiation, most often with proton beam therapy, is favored given its ability to deliver larger doses of radiation to a highly conformal field as there are often critical neurovascular structures surrounding chordomas.[89] Image-guided intensity-modulated radiotherapy (IG-IMRT) is an another modality that is being studied given the relatively few proton radiation centers available.[90] There is currently no role for chemotherapy in the management of chordoma. Based on the increasing knowledge of the molecular pathogenesis of chordomas, multiple targeted agents for specific pathways, including EGFR, PDGFR, mTOR, and VEGF, have been tried with modest response.[91] There are cases of refractory chordoma responding to immunotherapy though much translational and clinical research is needed.[92]

4.2 Chondrosarcoma

Chondrosarcomas are rare malignant tumors arising from cartilaginous matrix which often affect the pelvis or long bone; however, they can rarely also occur in the skull base and spine which ultimately can cause neurologic symptoms.[93–95]

Intracranial chondrosarcomas are rare and account for 6% of skull base tumors and 0.2% of intracranial tumors.[96] In a single case series, median age was 37 years and nearly 50% of patients were classified as Grade I tumors while 30% were classified as mesenchymal type, which are the most malignant subtype. The most

common bones affected were the petrosal bone, followed by occipital, clivus, and then the sphenoid. The most common presenting symptoms were diplopia, headache, hearing loss, dizziness, and tinnitus.[97] Spinal chondrosarcomas usually will involve the spine and the rib head.[98] Pelvic osteosarcomas usually involve the sacroiliac joint and can extend into the sacrum or into the pelvis.[99]

Chondrosarcomas can be challenging to distinguish from chordomas based on radiographic appearance. Immunohistochemistry can be used to differentiate the two as chondrosarcomas do not express cytokeratin while chordomas do.[100]

The primary management of chondrosarcomas is centered around surgical resection. However, similar to chordoma, the extent of resection can be limited by surrounding neurologic structures. With surgery alone, 5-year recurrence is estimated at 44% while patients who received adjuvant radiation had a 5-year recurrence rate of only 9%.[94] Chondrosarcomas of the skull base treated with surgical resection followed by radiotherapy have excellent local control rates of greater than 90% at 5 years.[101] Spinal and pelvic chondrosarcomas, however, often behave in a more aggressive manner with higher rates of local recurrence, up to 35%, with local recurrence being associated with chondrosarcoma-related death.[102] There is limited role for chemotherapy in chondrosarcomas. Overall, outcomes with chondrosarcomas are often superior to chordomas with a 5- and 10-year survival of close to 99% compared to 51% and 35%, respectively.[100]

An inherited genetic disorder of particular note in relation to chondrosarcomas is Maffucci syndrome, due to germ line isocitrate dehydrogenase (IDH) mutation. Somatic mutation of IDH is noted across a range of cancers including leukemias and infiltrating gliomas. Its germ line mutation is far rarer. Maffucci syndrome is associated with the development of infiltrating gliomas, multiple endochondromas, cutaneous hemangiomas, and in some patients chondrosarcoma. The related

Ollier disease, also due to germ line IDH mutation, is associated with endochondromas, chondrosarcomas, and other malignancies, but not the cutaneous hemangiomas. The role of IDH mutation in tumorigenesis in chondrosarcomas associated with Maffucci/Ollier syndromes raises the possibility of therapeutically targeting this abnormality in chondrosarcomas with both germ line and somatic IDH mutations.[103]

4.3 Osteosarcoma

Osteosarcomas are the most common primary malignant tumors of the bone and have a bimodal distribution affecting either adolescents or adults over 65 years of age (Fig. 3).[104] They often occur in the long bones and pelvis with rare involvement of the spine, in approximately 3%–5% of cases, which present as spinal cord compression and pain.[105,106] Metastases to the CNS are also very rare occurring in 2%–6% of cases which are often preceded by the development of pulmonary metastases.[107,108] In one small series of five patients, two patients presented with significant intracranial hemorrhage and three patients developed status epilepticus.[109] The increased rate of brain metastases in patients with pulmonary lesions is theorized to occur from lung tumor emboli, and CNS surveillance has been proposed for this

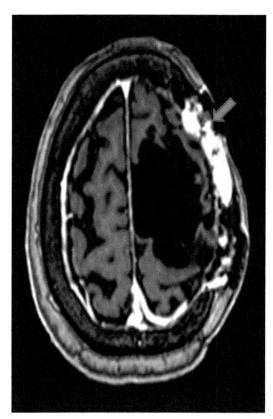

FIG. 3   A low-grade osteosarcoma of the skull *(arrow)*. This tumor arose many years after the treatment of an adjacent glioma which was treated with focal radiation.

group though it is not validated.[110] There is no consensus on optimal management though surgical resection, especially for solitary lesions, followed by radiation is frequently employed with modest response.[111] Stereotactic radiosurgery has also been used with favorable local control and without notable toxicity.[112]

5 CNS metastases from extra-CNS sarcoma

As first noted in the Section 1, the majority of sarcomas affecting the nervous system do so either via direct compression caused by extra-CNS disease or via metastasis of sarcoma to the CNS. The incidence of CNS metastases due to sarcomas is uncertain but is deemed to be an uncommon pattern of spread for sarcomas and is thought to comprise a small component of CNS metastases overall. When CNS metastases occur, they are often single, rarely oligometastatic, and unlikely to be multiple. The metastases frequently involve the dura with associated invasion of surrounding structures including overlying bone and underlying brain. The radiographic appearance lacks features which are uniquely definitive for CNS sarcoma metastases. However, as CNS metastases are not typically early manifestations of disease, a preexisting history of sarcoma is often noted.

The management of CNS metastases from sarcoma follows a similar paradigm as that of metastases from other solid tumors as well as the paradigm for primary CNS sarcomas. As the lesions are often single, maximum, and ideally gross total, resection is frequently the initial therapeutic step. Due to tumor location, gross total resection is not always feasible. For residual tumor, postoperatively focal radiation is often employed. This is usually delivered either in a fractionated manner or with stereotactic radiosurgery with doses of 15–20 Gy to the target which can lead to durable local control.[112] Development of CNS metastases is often a poor prognostic sign with extent of systemic disease burden and overall functional status being important factors to guide prognosis and therapy. Overall survival can range from 7 to 16 months with the majority of patients living less than 12 months. Surgical intervention to improve neurologic function in patients with well-controlled systemic disease and favorable performance status is a feasible option. Palliative whole brain radiotherapy is an option for patients with multiple brain metastases and concurrent active systemic disease.[111]

6 Neurologic toxicities from sarcoma treatments

As the adverse outcomes related to the use of surgery and radiation for the treatment of sarcomas are not substantially different from those to treat other malignancies, they will not be highlighted here. While radiation is used for the treatment of sarcomas and can be associated with neurologic toxicity, we will focus specifically on the neurologic toxicities associated with the systemic therapies commonly used to treat sarcomas.

Some of the most common chemotherapies used to treat soft tissue sarcoma and osteosarcomas include cytotoxic chemotherapies such as anthracyclines, taxanes, platinum agents, alkylating agents as well as topoisomerase inhibitors. These agents can cause a variety of known neurologic complications which are reviewed below.

6.1 Peripheral neuropathy

Chemotherapy-induced peripheral neuropathy is a common dose-limiting toxicity of chemotherapy and can affect quality of life for patients. Several classes of chemotherapy used routinely in sarcoma management can cause peripheral neuropathy.

Platinum agents, namely cisplatin and carboplatin to a lesser degree, primarily cause a large fiber sensory neuropathy that affects the dorsal root ganglion in a dose-dependent manner. Clinical symptoms typically include sensory loss, proprioception deficits, and loss of deep tendon reflexes.[113] Vinca alkaloids, such as vincristine, and taxanes, such as paclitaxel, affect microtubules which cause a mixed small and large fiber sensorimotor neuropathy secondary to its effects on the dorsal root ganglion, microtubules, and nerve terminals. Symptoms often are painful paresthesias and myalgias.[114] Vincristine can also cause weakness, cranial neuropathies, and autonomic dysfunction manifesting as orthostatic hypotension, dizziness, and constipation.[115] Chemotherapy-induced neuropathy is often reversible with cessation of the offending agent. There are no well-defined agents for prevention of chemotherapy-induced neuropathy though calcium and magnesium may be of benefit. Symptom management with tricyclic antidepressants and anticonvulsants has been studied with limited benefit, though duloxetine, a serotonin/norepinephrine reuptake inhibitor, has demonstrated activity compared to placebo for managing neuropathic pain.[116]

6.2 Hearing loss

Ototoxicity is another dose-limiting toxicity and is typically associated with cisplatin. Cisplatin-induced hearing loss first affects higher frequencies, then progresses to lower ones, is bilateral, and can be permanent.[117] Symptoms include hearing loss and tinnitus, and can also cause vestibular dysfunction.[118] The underlying mechanisms are not fully well understood though apoptosis of hair cells in the organ of Corti is thought to play a significant role.[119] Patients on cisplatin should have routine audiometric testing for early detection of hearing loss to guide therapy.[120] Dose reduction, drug cessation, and changing to a less ototoxic chemotherapy such as carboplatin should be considered.

6.3 Central nervous system complications

In addition to peripheral nervous system toxicities, chemotherapies can also cause CNS complications including acute encephalopathy, stroke, and seizures.

Acute encephalopathy has been described after administration of cisplatin. Other acute-to-subacute toxicities related to cisplatin include seizures, stroke, cortical blindness, language disturbance, and motor deficits which usually resolve without any specific therapy and may not occur with repeat exposure to the drug.[121] Ifosfamide has a higher incidence of CNS toxicity with acute encephalopathy occurring in up to 10%–40% of patients. Hypoalbuminemia has been associated with increased risk of developing ifosfamide-related encephalopathy. Symptoms can include somnolence, agitation, visual hallucinations, aphasia, seizures, and even coma.[122] EEG can show diffuse slowing without epileptiform discharges.[121] Intravenous administration of methylene blue is often used with ifosfamide encephalopathy though it is unclear if patients resolve without any specific intervention.[123]

7 Conclusions

Sarcomas have the potential to affect both the CNS and the PNS. They may do so via the development and progression of primary sarcomas of these tissues. More commonly, what is seen are extra-CNS sarcomas affecting the nervous system. This occurs most frequently with the impingement and invasion of the nervous system by sarcomas arising from the adjacent bony structures. Extra-CNS sarcomas also in rare circumstances may metastasize to the nervous system. This often includes dural involvement. Finally, the systemic therapies, particularly cytotoxic chemotherapies, used to treat sarcomas can manifest neurologic toxicities. A multidisciplinary approach can be utilized to address all of these scenarios.

References

1. Ostrom QT, Cioffi G, Gittleman H, et al. CBTRUS statistical report: primary brain and other central nervous system tumors diagnosed in the United States in 2012-2016. *Neuro-Oncology.* 2019;21(Suppl. 5):v1–v100. https://doi.org/10.1093/neuonc/noz150.
2. Al-Gahtany M, Shroff M, Bouffet E, et al. Primary central nervous system sarcomas in children: clinical, radiological, and pathological features. *Childs Nerv Syst.* 2003;19(12):808–817. https://doi.org/10.1007/s00381-003-0839-5.

3. Wang K, Ma XJ, Guo TX, et al. Intracranial mesenchymal chondrosarcoma: report of 16 cases. *World Neurosurg.* 2018;116:e691–e698. https://doi.org/10.1016/j.wneu.2018.05.069.

4. de Kock L, Priest JR, Foulkes WD, Alexandrescu S. An update on the central nervous system manifestations of DICER1 syndrome. *Acta Neuropathol.* 2020;139(4):689–701. https://doi.org/10.1007/s00401-019-01997-y.

5. Kamihara J, Paulson V, Breen MA, et al. DICER1-associated central nervous system sarcoma in children: comprehensive clinicopathologic and genetic analysis of a newly described rare tumor. *Mod Pathol.* 2020;33(10):1910–1921. https://doi.org/10.1038/s41379-020-0516-1.

6. Koelsche C, Mynarek M, Schrimpf D, et al. Primary intracranial spindle cell sarcoma with rhabdomyosarcoma-like features share a highly distinct methylation profile and DICER1 mutations. *Acta Neuropathol.* 2018;136(2):327–337. https://doi.org/10.1007/s00401-018-1871-6.

7. In GK, Hu JS, Tseng WW. Treatment of advanced, metastatic soft tissue sarcoma: latest evidence and clinical considerations. *Ther Adv Med Oncol.* 2017;9(8):533–550. https://doi.org/10.1177/1758834017712963.

8. Louis DN, Perry A, Reifenberger G, et al. The 2016 World Health Organization classification of tumors of the central nervous system: a summary. *Acta Neuropathol.* 2016;131(6):803–820. https://doi.org/10.1007/s00401-016-1545-1.

9. Carneiro SS, Scheithauer BW, Nascimento AG, Hirose T, Davis DH. Solitary fibrous tumor of the meninges: a lesion distinct from fibrous meningioma. A clinicopathologic and immunohistochemical study. *Am J Clin Pathol.* 1996;106(2):217–224. https://doi.org/10.1093/ajcp/106.2.217.

10. Ma L, Wang L, Fang X, Zhao CH, Sun L. Diagnosis and treatment of solitary fibrous tumor/hemangiopericytoma of central nervous system. Retrospective report of 17 patients and literature review. *Neuro Endocrinol Lett.* 2018;39(2):88–94.

11. Ghose A, Guha G, Kundu R, Tew J, Chaudhary R. CNS hemangiopericytoma: a systematic review of 523 patients. *Am J Clin Oncol.* 2017;40(3):223–227. https://doi.org/10.1097/COC.0000000000000146.

12. Ohba S, Murayama K, Nishiyama Y, et al. Clinical and radiographic features for differentiating solitary fibrous tumor/hemangiopericytoma from meningioma. *World Neurosurg.* 2019;130:e383–e392. https://doi.org/10.1016/j.wneu.2019.06.094.

13. Stout AP, Murray MR. Hemangiopericytoma: a vascular tumor featuring Zimmermann's pericytes. *Ann Surg.* 1942;116(1):26–33. https://doi.org/10.1097/00000658-194207000-00004.

14. Doyle LA, Vivero M, Fletcher CD, Mertens F, Hornick JL. Nuclear expression of STAT6 distinguishes solitary fibrous tumor from histologic mimics. *Mod Pathol.* 2014;27(3):390–395. https://doi.org/10.1038/modpathol.2013.164.

15. Berghoff AS, Kresl P, Bienkowski M, et al. Validation of nuclear STAT6 immunostaining as a diagnostic marker of meningeal solitary fibrous tumor (SFT)/hemangiopericytoma. *Clin Neuropathol.* 2017;36(2):56–59. https://doi.org/10.5414/NP300993.

16. Fritchie K, Jensch K, Moskalev EA, et al. The impact of histopathology and NAB2-STAT6 fusion subtype in classification and grading of meningeal solitary fibrous tumor/hemangiopericytoma. *Acta Neuropathol.* 2019;137(2):307–319. https://doi.org/10.1007/s00401-018-1952-6.

17. Bertero L, Anfossi V, Osella-Abate S, et al. Pathological prognostic markers in central nervous system solitary fibrous tumour/hemangiopericytoma: evidence from a small series. *PLoS One.* 2018;13(9):e0203570. https://doi.org/10.1371/journal.pone.0203570.

18. Spina A, Boari N, Gagliardi F, Donofrio CA, Franzin A, Mortini P. The current role of Gamma Knife radiosurgery in the management of intracranial haemangiopericytoma. *Acta*

19. *Neurochir.* 2016;158(4):635–642. https://doi.org/10.1007/s00701-016-2742-3.

19. Melone AG, D'Elia A, Santoro F, et al. Intracranial hemangiopericytoma—our experience in 30 years: a series of 43 cases and review of the literature. *World Neurosurg.* 2014;81(3–4):556–562. https://doi.org/10.1016/j.wneu.2013.11.009.

20. Gonzalez-Vargas PM, Thenier-Villa JL, Sanroman Alvarez P, et al. Hemangiopericytoma/solitary fibrous tumor in the central nervous system. experience with surgery and radiotherapy as a complementary treatment: a 10-year analysis of a heterogeneous series in a single tertiary center. *Neurocirugia (Astur).* 2020;31(1):14–23. https://doi.org/10.1016/j.neucir.2019.06.001.

21. Chamberlain MC, Glantz MJ. Sequential salvage chemotherapy for recurrent intracranial hemangiopericytoma. *Neurosurgery.* 2008;63(4):720–726. Author reply 726–727 https://doi.org/10.1227/01.NEU.0000325494.69836.51.

22. Park MS, Patel SR, Ludwig JA, et al. Activity of temozolomide and bevacizumab in the treatment of locally advanced, recurrent, and metastatic hemangiopericytoma and malignant solitary fibrous tumor. *Cancer.* 2011;117(21):4939–4947. https://doi.org/10.1002/cncr.26098.

23. Park MS, Ravi V, Conley A, et al. The role of chemotherapy in advanced solitary fibrous tumors: a retrospective analysis. *Clin Sarcoma Res.* 2013;3(1):7. https://doi.org/10.1186/2045-3329-3-7.

24. Maruzzo M, Martin-Liberal J, Messiou C, et al. Pazopanib as first line treatment for solitary fibrous tumours: the Royal Marsden Hospital experience. *Clin Sarcoma Res.* 2015;5:5. https://doi.org/10.1186/s13569-015-0022-2.

25. Apra C, Alentorn A, Mokhtari K, Kalamarides M, Sanson M. Pazopanib efficacy in recurrent central nervous system hemangiopericytomas. *J Neuro-Oncol.* 2018;139(2):369–372. https://doi.org/10.1007/s11060-018-2870-0.

26. Wu W, Tanrivermis Sayit A, Vinters HV, Pope W, Mirsadraei L, Said J. Primary central nervous system histiocytic sarcoma presenting as a postradiation sarcoma: case report and literature review. *Hum Pathol.* 2013;44(6):1177–1183. https://doi.org/10.1016/j.humpath.2012.11.002.

27. Egan C, Nicolae A, Lack J, et al. Genomic profiling of primary histiocytic sarcoma reveals two molecular subgroups. *Haematologica.* 2020;105(4):951–960. https://doi.org/10.3324/haematol.2019.230375.

28. So H, Kim SA, Yoon DH, et al. Primary histiocytic sarcoma of the central nervous system. *Cancer Res Treat.* 2015;47(2):322–328. https://doi.org/10.4143/crt.2013.163.

29. Tso FY, Sawyer A, Kwon EH, et al. Kaposi's sarcoma-associated herpesvirus infection of neurons in HIV-positive patients. *J Infect Dis.* 2017;215(12):1898–1907. https://doi.org/10.1093/infdis/jiw545.

30. Gonzalez-Crussi F, Mossanen A, Robertson DM. Neurological involvement in Kaposi's sarcoma. *Can Med Assoc J.* 1969;100(10):481–484.

31. Dourmishev LA, Dourmishev AL, Palmeri D, Schwartz RA, Lukac DM. Molecular genetics of Kaposi's sarcoma-associated herpesvirus (human herpesvirus-8) epidemiology and pathogenesis. *Microbiol Mol Biol Rev.* 2003;67(2):175–212. table of contents https://doi.org/10.1128/mmbr.67.2.175-212.2003.

32. Rwomushana RJ, Bailey IC, Kyalwazi SK. Kaposi's sarcoma of the brain. A case report with necropsy findings. *Cancer.* 1975;36(3):1127–1131. https://doi.org/10.1002/1097-0142(197509)36:3<1127::aid-cncr2820360344>3.0.co;2-i.

33. Pantanowitz L, Dezube BJ. Kaposi sarcoma in unusual locations. *BMC Cancer.* 2008;8:190. https://doi.org/10.1186/1471-2407-8-190.

34. Panagopoulos D, Themistocleous M, Apostolopoulou K, Sfakianos G. Primary, dural-based, ewing sarcoma manifesting with seizure activity: presentation of a rare tumor entity with literature review.

World Neurosurg. 2019;129:216–220. https://doi.org/10.1016/j.wneu.2019.06.036.

35. Chen J, Jiang Q, Zhang Y, et al. Clinical features and long-term outcome of primary intracranial Ewing sarcoma/peripheral primitive neuroectodermal tumors: 14 cases from a single institution. *World Neurosurg.* 2019;122:e1606–e1614. https://doi.org/10.1016/j.wneu.2018.11.151.

36. Gaspar N, Hawkins DS, Dirksen U, et al. Ewing sarcoma: current management and future approaches through collaboration. *J Clin Oncol Off J Am Soc Clin Oncol.* 2015;33(27):3036–3046. https://doi.org/10.1200/JCO.2014.59.5256.

37. Aeddula NR, Pathireddy S, Samaha T, Ukena T, Hosseinnezhad A. Primary intracranial leiomyosarcoma in an immunocompetent adult. *J Clin Oncol Off J Am Soc Clin Oncol.* 2011;29(14):e407–e410. https://doi.org/10.1200/JCO.2010.33.4805.

38. Litofsky NS, Pihan G, Corvi F, Smith TW. Intracranial leiomyosarcoma: a neuro-oncological consequence of acquired immunodeficiency syndrome. *J Neuro-Oncol.* 1998;40(2):179–183. https://doi.org/10.1023/a:1006167629968.

39. Francisco CN, Alejandria M, Salvana EM, Andal VMV. Primary intracranial leiomyosarcoma among patients with AIDS in the era of new chemotherapeutic and biological agents. *BMJ Case Rep.* 2018;2018. https://doi.org/10.1136/bcr-2018-225714.

40. Zhang H, Dong L, Huang Y, et al. Primary intracranial leiomyosarcoma: review of the literature and presentation of a case. *Onkologie.* 2012;35(10):609–616. https://doi.org/10.1159/000342676.

41. Li XL, Ren J, Niu RN, et al. Primary intracranial leiomyosarcoma in an immunocompetent patient: case report with emphasis on imaging features. *Medicine (Baltimore).* 2019;98(17):e15269. https://doi.org/10.1097/MD.0000000000015269.

42. Torihashi K, Chin M, Yoshida K, Narumi O, Yamagata S. Primary intracranial leiomyosarcoma with intratumoral hemorrhage: case report and review of literature. *World Neurosurg.* 2018;116:169–173. https://doi.org/10.1016/j.wneu.2018.05.004.

43. Gautam S, Meena RK. Primary intracranial leiomyosarcoma presenting with massive peritumoral edema and mass effect: case report and literature review. *Surg Neurol Int.* 2017;8:278. https://doi.org/10.4103/sni.sni_219_17.

44. Polewski PJ, Smith AL, Conway PD, Marinier DE. Primary CNS leiomyosarcoma in an immunocompetent patient. *J Oncol Pract.* 2016;12(9):827–829. https://doi.org/10.1200/JOP.2016.012310.

45. Gallagher SJ, Rosenberg SA, Francis D, Salamat S, Howard SP, Kimple RJ. Primary intracranial leiomyosarcoma in an immunocompetent patient: case report and review of the literature. *Clin Neurol Neurosurg.* 2018;165:76–80. https://doi.org/10.1016/j.clineuro.2017.12.014.

46. Barbiero FJ, Huttner AJ, Judson BL, Baehring JM. Leiomyosarcoma of the infratemporal fossa with perineurial spread along the right mandibular nerve: a case report. *CNS Oncol.* 2017;6(4):281–285. https://doi.org/10.2217/cns-2017-0004.

47. Jerjir N, Lambert J, Vanwalleghem L, Casselman J. Primary angiosarcoma of the central nervous system: case report and review of the imaging features. *J Belg Soc Radiol.* 2016;100(1):82. https://doi.org/10.5334/jbr-btr.1087.

48. Okabayshi M, Kataoka TR, Oji M, et al. IGF2BP3 (IMP3) expression in angiosarcoma, epithelioid hemangioendothelioma, and benign vascular lesions. *Diagn Pathol.* 2020;15(1):26. https://doi.org/10.1186/s13000-020-00951-x.

49. Lee JW, Wernicke AG. Risk and survival outcomes of radiation-induced CNS tumors. *J Neuro-Oncol.* 2016;129(1):15–22. https://doi.org/10.1007/s11060-016-2148-3.

50. Yamanaka R, Hayano A. Radiation-induced sarcomas of the central nervous system: a systematic review. *World Neurosurg.* 2017;98:818–828 e7. https://doi.org/10.1016/j.wneu.2016.11.008.

51. Kent SP, Pickering JE. Neoplasms in monkeys (*Macaca mulatta*): spontaneous and irradiation induced. *Cancer.* 1958;11(1):138–147. https://doi.org/10.1002/1097-0142(195801/02)11:1<138::aid-cncr2820110125>3.0.co;2-p.

52. Weber DC, Habrand JL, Hoppe BS, et al. Proton therapy for pediatric malignancies: fact, figures and costs. A joint consensus statement from the pediatric subcommittee of PTCOG, PROS and EPTN. *Radiother Oncol.* 2018;128(1):44–55. https://doi.org/10.1016/j.radonc.2018.05.020.

53. Le Guellec S, Decouvelaere AV, Filleron T, et al. Malignant peripheral nerve sheath tumor is a challenging diagnosis: a systematic pathology review, immunohistochemistry, and molecular analysis in 160 patients from the French sarcoma group database. *Am J Surg Pathol.* 2016;40(7):896–908. https://doi.org/10.1097/PAS.0000000000000655.

54. Evans DG, Baser ME, McGaughran J, Sharif S, Howard E, Moran A. Malignant peripheral nerve sheath tumours in neurofibromatosis 1. *J Med Genet.* 2002;39(5):311–314. https://doi.org/10.1136/jmg.39.5.311.

55. Tucker T, Wolkenstein P, Revuz J, Zeller J, Friedman JM. Association between benign and malignant peripheral nerve sheath tumors in NF1. *Neurology.* 2005;65(2):205–211. https://doi.org/10.1212/01.wnl.0000168830.79997.13.

56. Stucky CC, Johnson KN, Gray RJ, et al. Malignant peripheral nerve sheath tumors (MPNST): the Mayo Clinic experience. *Ann Surg Oncol.* 2012;19(3):878–885. https://doi.org/10.1245/s10434-011-1978-7.

57. Baehring JM, Betensky RA, Batchelor TT. Malignant peripheral nerve sheath tumor: the clinical spectrum and outcome of treatment. *Neurology.* 2003;61(5):696–698. https://doi.org/10.1212/01.wnl.0000078813.05925.2c.

58. Pilavaki M, Chourmouzi D, Kiziridou A, Skordalaki A, Zarampoukas T, Drevelengas A. Imaging of peripheral nerve sheath tumors with pathologic correlation: pictorial review. *Eur J Radiol.* 2004;52(3):229–239. https://doi.org/10.1016/j.ejrad.2003.12.001.

59. Ahlawat S, Blakeley JO, Rodriguez FJ, Fayad LM. Imaging biomarkers for malignant peripheral nerve sheath tumors in neurofibromatosis type 1. *Neurology.* 2019;93(11):e1076–e1084. https://doi.org/10.1212/WNL.0000000000008092.

60. Brahmi M, Thiesse P, Ranchere D, et al. Diagnostic accuracy of PET/CT-guided percutaneous biopsies for malignant peripheral nerve sheath tumors in neurofibromatosis type 1 patients. *PLoS One.* 2015;10(10):e0138386. https://doi.org/10.1371/journal.pone.0138386.

61. Benz MR, Czernin J, Dry SM, et al. Quantitative F18-fluorodeoxyglucose positron emission tomography accurately characterizes peripheral nerve sheath tumors as malignant or benign. *Cancer.* 2010;116(2):451–458. https://doi.org/10.1002/cncr.24755.

62. Ducatman BS, Scheithauer BW, Piepgras DG, Reiman HM, Ilstrup DM. Malignant peripheral nerve sheath tumors. A clinicopathologic study of 120 cases. *Cancer.* 1986;57(10):2006–2021. https://doi.org/10.1002/1097-0142(19860515)57:10<2006::aid-cncr2820571022>3.0.co;2-6.

63. Rodriguez FJ, Folpe AL, Giannini C, Perry A. Pathology of peripheral nerve sheath tumors: diagnostic overview and update on selected diagnostic problems. *Acta Neuropathol.* 2012;123(3):295–319. https://doi.org/10.1007/s00401-012-0954-z.

64. Miettinen MM, Antonescu CR, Fletcher CDM, et al. Histopathologic evaluation of atypical neurofibromatous tumors and their transformation into malignant peripheral nerve sheath tumor in patients with neurofibromatosis 1-a consensus overview. *Hum Pathol.* 2017;67:1–10. https://doi.org/10.1016/j.humpath.2017.05.010.

65. Sahm F, Reuss DE, Giannini C. WHO 2016 classification: changes and advancements in the diagnosis of miscellaneous primary CNS tumours. *Neuropathol Appl Neurobiol.* 2018;44(2):163–171. https://doi.org/10.1111/nan.12397.

III. Neurological complications of specific neoplasms

66. Cleven AH, Sannaa GA, Briaire-de Bruijn I, et al. Loss of H3K27 tri-methylation is a diagnostic marker for malignant peripheral nerve sheath tumors and an indicator for an inferior survival. *Mod Pathol.* 2016;29(6):582–590. https://doi.org/10.1038/modpathol.2016.45.

67. Lu VM, Marek T, Gilder HE, et al. H3K27 trimethylation loss in malignant peripheral nerve sheath tumor: a systematic review and meta-analysis with diagnostic implications. *J Neuro-Oncol.* 2019;144(3):433–443. https://doi.org/10.1007/s11060-019-03247-3.

68. Lyskjaer I, Lindsay D, Tirabosco R, et al. H3K27me3 expression and methylation status in histological variants of malignant peripheral nerve sheath tumours. *J Pathol.* 2020;252(2):151–164. https://doi.org/10.1002/path.5507.

69. Kolberg M, Holand M, Agesen TH, et al. Survival meta-analyses for >1800 malignant peripheral nerve sheath tumor patients with and without neurofibromatosis type 1. *Neuro-Oncology.* 2013;15(2):135–147. https://doi.org/10.1093/neuonc/nos287.

70. Dunn GP, Spiliopoulos K, Plotkin SR, et al. Role of resection of malignant peripheral nerve sheath tumors in patients with neurofibromatosis type 1. *J Neurosurg.* 2013;118(1):142–148. https://doi.org/10.3171/2012.9.JNS101610.

71. Hirbe AC, Cosper PF, Dahiya S, Van Tine BA. Neoadjuvant ifosfamide and epirubicin in the treatment of malignant peripheral nerve sheath tumors. *Sarcoma.* 2017;2017:3761292. https://doi.org/10.1155/2017/3761292.

72. Wang D, Zhang Q, Eisenberg BL, et al. Significant reduction of late toxicities in patients with extremity sarcoma treated with image-guided radiation therapy to a reduced target volume: results of radiation therapy oncology group RTOG-0630 trial. *J Clin Oncol Off J Am Soc Clin Oncol.* 2015;33(20):2231–2238. https://doi.org/10.1200/JCO.2014.58.5828s.

73. Watson KL, Al Sannaa GA, Kivlin CM, et al. Patterns of recurrence and survival in sporadic, neurofibromatosis Type 1-associated, and radiation-associated malignant peripheral nerve sheath tumors. *J Neurosurg.* 2017;126(1):319–329. https://doi.org/10.3171/2015.12.JNS152443.

74. Ferner RE, Gutmann DH. International consensus statement on malignant peripheral nerve sheath tumors in neurofibromatosis. *Cancer Res.* 2002;62(5):1573–1577.

75. Verma S, Bramwell V. Dose-intensive chemotherapy in advanced adult soft tissue sarcoma. *Expert Rev Anticancer Ther.* 2002;2(2):201–215. https://doi.org/10.1586/14737140.2.2.201.

76. Stacchiotti S, Sommer J. Chordoma global consensus G. building a global consensus approach to chordoma: a position paper from the medical and patient community. *Lancet Oncol.* 2015;16(2):e71–e83. https://doi.org/10.1016/S1470-2045(14)71190-8.

77. George B, Bresson D, Herman P, Froelich S. Chordomas: a review. *Neurosurg Clin N Am.* 2015;26(3):437–452. https://doi.org/10.1016/j.nec.2015.03.012.

78. Meyers SP, Hirsch Jr WL, Curtin HD, Barnes L, Sekhar LN, Sen C. Chordomas of the skull base: MR features. *AJNR Am J Neuroradiol.* 1992;13(6):1627–1636.

79. Vujovic S, Henderson S, Presneau N, et al. Brachyury, a crucial regulator of notochordal development, is a novel biomarker for chordomas. *J Pathol.* 2006;209(2):157–165. https://doi.org/10.1002/path.1969.

80. Oakley GJ, Fuhrer K, Seethala RR. Brachyury, SOX-9, and podoplanin, new markers in the skull base chordoma vs chondrosarcoma differential: a tissue microarray-based comparative analysis. *Mod Pathol.* 2008;21(12):1461–1469. https://doi.org/10.1038/modpathol.2008.144.

81. Stacchiotti S, Gronchi A, Fossati P, et al. Best practices for the management of local-regional recurrent chordoma: a position paper by the Chordoma Global Consensus Group. *Ann Oncol.* 2017;28(6):1230–1242. https://doi.org/10.1093/annonc/mdx054.

82. Bakker SH, Jacobs WCH, Pondaag W, et al. Chordoma: a systematic review of the epidemiology and clinical prognostic factors predicting progression-free and overall survival. *Eur Spine J.* 2018;27(12):3043–3058. https://doi.org/10.1007/s00586-018-5764-0.

83. Zenonos GA, Fernandez-Miranda JC, Mukherjee D, et al. Prospective validation of a molecular prognostication panel for clival chordoma. *J Neurosurg.* 2018;1:1–10. https://doi.org/10.3171/2018.3.JNS172321.

84. Horbinski C, Oakley GJ, Cieply K, et al. The prognostic value of Ki-67, p53, epidermal growth factor receptor, 1p36, 9p21, 10q23, and 17p13 in skull base chordomas. *Arch Pathol Lab Med.* 2010;134(8):1170–1176. https://doi.org/10.1043/2009-0380-OA.1.

85. Di Maio S, Temkin N, Ramanathan D, Sekhar LN. Current comprehensive management of cranial base chordomas: 10-year meta-analysis of observational studies. *J Neurosurg.* 2011;115(6):1094–1105. https://doi.org/10.3171/2011.7.JNS11355.

86. Walcott BP, Nahed BV, Mohyeldin A, Coumans JV, Kahle KT, Ferreira MJ. Chordoma: current concepts, management, and future directions. *Lancet Oncol.* 2012;13(2):e69–e76. https://doi.org/10.1016/S1470-2045(11)70337-0.

87. Zoli M, Milanese L, Bonfatti R, et al. Clival chordomas: considerations after 16 years of endoscopic endonasal surgery. *J Neurosurg.* 2018;128(2):329–338. https://doi.org/10.3171/2016.11.JNS162082.

88. Sen C, Triana AI, Berglind N, Godbold J, Shrivastava RK. Clival chordomas: clinical management, results, and complications in 71 patients. *J Neurosurg.* 2010;113(5):1059–1071. https://doi.org/10.3171/2009.9.JNS08596.

89. Matloob SA, Nasir HA, Choi D. Proton beam therapy in the management of skull base chordomas: systematic review of indications, outcomes, and implications for neurosurgeons. *Br J Neurosurg.* 2016;30(4):382–387. https://doi.org/10.1080/02688697.2016.1181154.

90. Sahgal A, Chan MW, Atenafu EG, et al. Image-guided, intensity-modulated radiation therapy (IG-IMRT) for skull base chordoma and chondrosarcoma: preliminary outcomes. *Neuro-Oncology.* 2015;17(6):889–894. https://doi.org/10.1093/neuonc/nou347.

91. Di Maio S, Yip S, Al Zhrani GA, et al. Novel targeted therapies in chordoma: an update. *Ther Clin Risk Manag.* 2015;11:873–883. https://doi.org/10.2147/TCRM.S50526.

92. Migliorini D, Mach N, Aguiar D, et al. First report of clinical responses to immunotherapy in 3 relapsing cases of chordoma after failure of standard therapies. *Oncoimmunology.* 2017;6(8):e1338235. https://doi.org/10.1080/2162402X.2017.1338235.

93. Sangma MM, Dasiah S. Chondrosarcoma of a rib. *Int J Surg Case Rep.* 2015;10:126–128. https://doi.org/10.1016/j.ijscr.2015.03.052.

94. Bloch OG, Jian BJ, Yang I, et al. Cranial chondrosarcoma and recurrence. *Skull Base.* 2010;20(3):149–156. https://doi.org/10.1055/s-0029-1246218.

95. Strike SA, McCarthy EF. Chondrosarcoma of the spine: a series of 16 cases and a review of the literature. *Iowa Orthop J.* 2011;31:154–159.

96. Bingaman KD, Alleyne Jr CH, Olson JJ. Intracranial extraskeletal mesenchymal chondrosarcoma: case report. *Neurosurgery.* 2000;46(1):207–211. discussion 211–212.

97. Korten AG, ter Berg HJ, Spincemaille GH, van der Laan RT, Van de Wel AM. Intracranial chondrosarcoma: review of the literature and report of 15 cases. *J Neurol Neurosurg Psychiatry.* 1998;65(1):88–92. https://doi.org/10.1136/jnnp.65.1.88.

98. Arshi A, Sharim J, Park DY, et al. Chondrosarcoma of the osseous spine: an analysis of epidemiology, patient outcomes, and prognostic factors using the SEER registry from 1973 to 2012. *Spine (Phila Pa 1976).* 2017;42(9):644–652. https://doi.org/10.1097/BRS.0000000000001870.

99. Lex JR, Evans S, Stevenson JD, Parry M, Jeys LM, Grimer RJ. Dedifferentiated chondrosarcoma of the pelvis: clinical outcomes and current treatment. *Clin Sarcoma Res.* 2018;8:23. https://doi.org/10.1186/s13569-018-0110-1.

100. Rosenberg AE, Nielsen GP, Keel SB, et al. Chondrosarcoma of the base of the skull: a clinicopathologic study of 200 cases with emphasis on its distinction from chordoma. *Am J Surg Pathol.* 1999;23(11):1370–1378. https://doi.org/10.1097/00000478-199911000-00007.

101. Hug EB, Loredo LN, Slater JD, et al. Proton radiation therapy for chordomas and chondrosarcomas of the skull base. *J Neurosurg.* 1999;91(3):432–439. https://doi.org/10.3171/jns.1999.91.3.0432.

102. Fisher CG, Versteeg AL, Dea N, et al. Surgical management of spinal chondrosarcomas. *Spine (Phila Pa 1976).* 2016;41(8):678–685. https://doi.org/10.1097/BRS.0000000000001485.

103. Golub D, Iyengar N, Dogra S, et al. Mutant isocitrate dehydrogenase inhibitors as targeted cancer therapeutics. *Front Oncol.* 2019;9:417. https://doi.org/10.3389/fonc.2019.00417.

104. Mirabello L, Troisi RJ, Savage SA. Osteosarcoma incidence and survival rates from 1973 to 2004: data from the surveillance, epidemiology, and end results program. *Cancer.* 2009;115(7):1531–1543. https://doi.org/10.1002/cncr.24121.

105. Korovessis P, Repanti M, Stamatakis M. Primary osteosarcoma of the L2 lamina presenting as "silent" paraplegia: case report and review of the literature. *Eur Spine J.* 1995;4(6):375–378. https://doi.org/10.1007/BF00300304.

106. Katonis P, Datsis G, Karantanas A, et al. Spinal osteosarcoma. *Clin Med Insights Oncol.* 2013;7:199–208. https://doi.org/10.4137/CMO.S10099.

107. Kebudi R, Ayan I, Gorgun O, Agaoglu FY, Vural S, Darendeliler E. Brain metastasis in pediatric extracranial solid tumors: survey and literature review. *J Neuro-Oncol.* 2005;71(1):43–48. https://doi.org/10.1007/s11060-004-4840-y.

108. Chaigneau L, Patrikidou A, Ray-Coquard I, et al. Brain metastases from adult sarcoma: prognostic factors and impact of treatment. a retrospective analysis from the French Sarcoma Group (GSF/GETO). *Oncologist.* 2018;23(8):948–955. https://doi.org/10.1634/theoncologist.2017-0136.

109. Baram TZ, van Tassel P, Jaffe NA. Brain metastases in osteosarcoma: incidence, clinical and neuroradiological findings and management options. *J Neuro-Oncol.* 1988;6(1):47–52. https://doi.org/10.1007/BF00163540.

110. Yonemoto T, Tatezaki S, Ishii T, Osato K, Takenouchi T. Longterm survival after surgical removal of solitary brain metastasis from osteosarcoma. *Int J Clin Oncol.* 2003;8(5):340–342. https://doi.org/10.1007/s10147-003-0341-9.

111. Shweikeh F, Bukavina L, Saeed K, et al. Brain metastasis in bone and soft tissue cancers: a review of incidence, interventions, and outcomes. *Sarcoma.* 2014;2014:475175. https://doi.org/10.1155/2014/475175.

112. Flannery T, Kano H, Niranjan A, et al. Gamma knife radiosurgery as a therapeutic strategy for intracranial sarcomatous metastases. *Int J Radiat Oncol Biol Phys.* 2010;76(2):513–519. https://doi.org/10.1016/j.ijrobp.2009.02.007.

113. Park SB, Goldstein D, Krishnan AV, et al. Chemotherapy-induced peripheral neurotoxicity: a critical analysis. *CA Cancer J Clin.* 2013;63(6):419–437. https://doi.org/10.3322/caac.21204.

114. Carlson K, Ocean AJ. Peripheral neuropathy with microtubule-targeting agents: occurrence and management approach. *Clin Breast Cancer.* 2011;11(2):73–81. https://doi.org/10.1016/j.clbc.2011.03.006.

115. van de Velde ME, Kaspers GL, Abbink FCH, Wilhelm AJ, Ket JCF, van den Berg MH. Vincristine-induced peripheral neuropathy in children with cancer: a systematic review. *Crit Rev Oncol Hematol.* 2017;114:114–130. https://doi.org/10.1016/j.critrevonc.2017.04.004.

116. Piccolo J, Kolesar JM. Prevention and treatment of chemotherapy-induced peripheral neuropathy. *Am J Health Syst Pharm.* 2014;71(1):19–25. https://doi.org/10.2146/ajhp130126.

117. Rybak LP. Mechanisms of cisplatin ototoxicity and progress in otoprotection. *Curr Opin Otolaryngol Head Neck Surg.* 2007;15(5):364–369. https://doi.org/10.1097/MOO.0b013e3282eee452.

118. Watts KL. Ototoxicity: visualized in concept maps. *Semin Hear.* 2019;40(2):177–187. https://doi.org/10.1055/s-0039-1684046.

119. Clerici WJ, DiMartino DL, Prasad MR. Direct effects of reactive oxygen species on cochlear outer hair cell shape in vitro. *Hear Res.* 1995;84(1–2):30–40. https://doi.org/10.1016/0378-5955(95)00010-2.

120. Simpson TH, Schwan SA, Rintelmann WF. Audiometric test criteria in the detection of cisplatin ototoxicity. *J Am Acad Audiol.* 1992;3(3):176–185.

121. Newton HB. Neurological complications of chemotherapy to the central nervous system. *Handb Clin Neurol.* 2012;105:903–916. https://doi.org/10.1016/B978-0-444-53502-3.00031-8.

122. Sweiss KI, Beri R, Shord SS. Encephalopathy after high-dose Ifosfamide: a retrospective cohort study and review of the literature. *Drug Saf.* 2008;31(11):989–996. https://doi.org/10.2165/00002018-200831110-00003.

123. Patel PN. Methylene blue for management of Ifosfamide-induced encephalopathy. *Ann Pharmacother.* 2006;40(2):299–303. https://doi.org/10.1345/aph.1G114.

III. Neurological complications of specific neoplasms

25

Neurological complications of multiple myeloma

Ankush Bhatia[a] and Nina A. Paleologos[b]

[a]Department of Neurology, The University of Houston Health Science Center at Houston, McGovern Medical School, Houston, TX, United States, [b]Department of Neurology, Advocate Medical Group, Advocate Healthcare, Rush University Medical School, Chicago, IL, United States

1 Introduction

Multiple myeloma is a neoplastic disease involving proliferation of terminally differentiated plasma cells which, in most cases, produce a monoclonal immunoglobulin protein (also known as M protein). Clonal plasma cells primarily reside in the bone marrow, but they can also be seen in the peripheral blood and other extramedullary sites, such as the central and peripheral nervous system, especially late in the disease course.[1] Multiple myeloma is the second most common hematological malignancy after non-Hodgkin lymphoma, comprising 1.8% of all neoplasms and approximately 10% of all hematologic malignancies, with an annual incidence rate of approximately 4.6 per 100,000 population.[2–5] The average age of onset is 66 years, and the incidence is higher in black males.[6,7] The most common presenting symptoms are fatigue, weight loss, bone pain, and recurrent infections, driven by monoclonal protein (most commonly IgG kappa). CRAB features, integral to the recently revised diagnostic criteria, describe signs of end-organ damage and include hypercalcemia, renal insufficiency, anemia, and/or bone disease.[8,9]

Although the first reports of this disease appeared in the literature in the 1840s, it was not until the 1920s that neurological complications were described.[2,3,10,11] Multiple possible etiologies exist to account for the neurological symptoms and signs common to multiple myeloma patients; no symptoms or signs are exclusive to one etiology. Both the central and peripheral nervous systems can be affected. Mechanisms of neurological dysfunction include the following: (1) direct infiltration of tumor cells, for example, spinal cord compression and leptomeningeal myelomatosis (LMM); (2) indirect effects via toxic metabolic derangements due to end-organ damage, autoimmune processes such as paraneoplastic disease, or amyloid deposition; and (3) iatrogenic effects related to chemotherapeutic drugs. Since multiple myeloma is a condition within the larger group of disorders of monoclonal gammopathies, the neurological symptomology must be carefully considered within the context of the laboratory, imaging, and bone marrow biopsy findings. The various neuropathies associated with plasma cell dyscrasias are summarized in Table 1 and will be reviewed in detail in the following sections.

2 Central nervous system complications

2.1 Spinal cord compression

Given that multiple myeloma is a malignancy of the bone marrow, it follows that one of the most frequent neurological complications is a direct consequence of this bony involvement. Spinal cord or cauda equina compression may result in severe pain and permanent, profound neurological deficits in patients with multiple myeloma. Extradural spinal cord or cauda equina compression (ESCC) occurs in approximately 10%–24% of patients with multiple myeloma and may be the presenting manifestation, with an incidence varying across series from 5% to 85%.[10,12–16] Multiple myeloma is the attributable malignancy in 5%–10% of all cases of ESCC.[15,17,18] Patients with multiple myeloma have a higher probability of developing ESCC (7.9%) in the 5 years preceding death than patients with other cancers, including lung, breast, and prostate.[19]

The rate of onset of the neurological signs and symptoms of ESCC secondary to multiple myeloma varies

TABLE 1   Neuropathies associated with plasma cell dyscrasias.

Diagnosis	Monoclonal protein	Clinical presentation	Neuropathy phenotype	Autonomic involvement	Treatment
Multiple myeloma					
No evidence of amyloid	IgG > IgA	Varies	Sensory > motor	Rare	Underlying disorder
Associated with amyloid		Painful	Sensory = motor	Significant	Consider SCT
Treatment-emergent (bortezomib, thalidomide)	N/A	Painful	Sensory predominant	Minor	Dose reduction or discontinuation of offending drug
Monoclonal gammopathy of undetermined significance	IgM kappa	Varies	Sensory predominant, demyelinating, anti-MAG antibodies	None	Varied, depends on severity
Waldenström's macroglobulinemia	IgM kappa	Numbness, sensory ataxia	Sensory predominant, axonal or demyelinating	None	Rituximab, cyclophosphamide, and dexamethasone, plasmapheresis, SCT
Immunoglobulin light-chain amyloidosis	Lambda > kappa	Painful	Sensory = motor, axonal	Significant	SCT vs melphalan-dexamethasone or cyclophosphamide/bortezomib/dexamethasone
POEMS syndrome	IgG or IgA, lambda	Varies, distal, symmetric	Polyradiculopathy, sensory = motor, demyelinating	Possible	Underlying disorder

SCT, stem cell transplant.
Modified from Drappatz J., Batchelor T. Neurologic complications of plasma cell disorders. Clin Lymphoma. 2004;5(3):163–171.

greatly, occurring suddenly in some cases, or in rare cases developing over a decade.[12,13,15] An average duration of symptoms of 1–4 weeks prior to diagnosis has been reported.[12,15] Many patients (58%–80%) with multiple myeloma have bone pain at the time of diagnosis,[2,3,20] and it is a prominent symptom in almost all patients at some point in the course of their illness.[10,13,21] Relatively, few of these patients will concurrently have or later develop ESCC. While the pain experienced by patients with ESCC is often impossible to differentiate from that experienced by patients without spinal cord compression, some characteristic features of the pain associated with ESCC do exist.[13] Pain that is persistently localized or of sudden onset may be due to a pathologic fracture.[2] Severe and persistent pain, particularly of a radicular nature, often precedes the neurological symptoms of spinal cord compression.[13,22] In several series of patients with myeloma and ESCC, 70%–100% had prolonged back pain that preceded the diagnosis by weeks to, in rare cases, years.[12–15,22]

Weakness is also a frequent presenting symptom of ESCC.[22] Weakness and motor deficits occur more commonly (90%–100%) in patients with ESCC due to myeloma than in patients with ESCC due to metastases from other systemic malignancies (60%–85%).[12,14,15] Some report that 20%–65% of patients with ESCC will present with complete paraplegia.[12,14] Sensory deficits (usually a sensory level occurring at or below the involved segment of spinal cord) have been found in 60%–100% of patients, and bowel and bladder dysfunction in 40%–60% of patients with ESCC due to myeloma.[10,14]

The most frequent site for ESCC in patients with multiple myeloma is the mid- to lower-thoracic spine, with the lumbar spine as the second most common location.[12–15,17,22] Multiple explanations have been proposed to explain the predilection of ESCC for the thoracic spine. Since it is the longest spinal segment with the highest volume of bone, myelomatous infiltration is most likely to occur in this area.[14,17] Thoracic myelomatous infiltration is more likely to cause a symptomatic cord compression because the anteroposterior distance of the thoracic spinal canal is narrower than in the cervical or lumbar segments, resulting in the smallest proportion of spinal canal to spinal cord diameter.[17,23] Additionally, the mid-thoracic spinal cord is more prone to ischemia secondary to its comparatively small blood supply.[13,14]

Myelomatous infiltration with subsequent vertebral collapse and extension of plasma cells directly into the extradural space is the most common mechanism of myeloma-associated spinal cord or cauda equina compression.[10,12–15,24] Extradural myeloma causing ESCC in the absence of bony involvement is considered to be a relatively uncommon cause by most, but reached 20% in one reported series.[10,13,24] The rarity of compression without bony involvement may be due to inherent resistance of the dura to infiltration by myeloma cells.[10] Intradural metastasis and amyloidosis are unusual causes of myelopathy.[10,13,24]

Magnetic resonance imaging (MRI) is the most appropriate initial imaging modality for the evaluation of patients in whom the diagnosis of ESCC is being considered.[25–28] It provides precise localization of the compression and assessment of the extent of the underlying disease process, allowing for informed therapeutic decisions (Fig. 1A).[17] Myelography may also be used to confirm the presence of ESCC and was the imaging modality of choice prior to the era of MRI.[10,14] Computed tomography (CT) scan with soft tissue windowing is also recommended to delineate whether bone or soft tissue is causing the spinal cord, cauda equina, or nerve root compression.

The clinical examination, specifically at the sensory level, is known to be unreliable as a localization tool, and diffuse involvement of the entire spine may be missed if only the suspected level is imaged.[17] Therefore, the entire spine should be imaged with and without gadolinium contrast and with T2-weighted images.[17] Subsequent histopathological assessment can help not only determine the extent of involvement of the spinal bone and epidural space, but also to allow for definitive diagnosis after appropriate immunohistochemical staining (Fig. 1B–D).

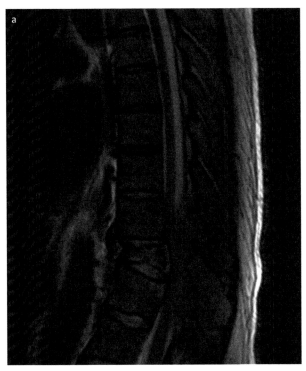

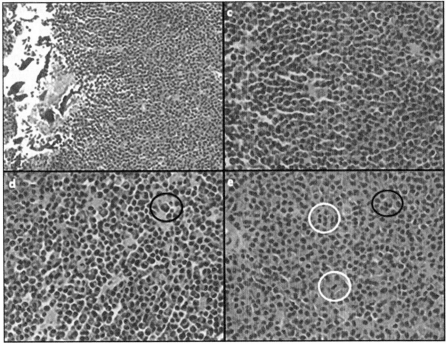

FIG. 1 (A) T_2-weighted sagittal image of severe spinal cord compression in a patient with multiple myeloma due to neoplastic involvement of the T8-9 to T11-12 vertebra with marked pathologic compression fracture of the T10 vertebral body. There is a large soft tissue mass that arises from the vertebrae, effacing the lower thoracic canal and extending into the posterior paraspinal soft tissues. T2 hyperintense signal/edema is present within the cord at the above levels. (B–E) Hematoxylin and eosin (H&E) stain of plasma cell neoplasm infiltrating spinal bone and epidural space with mitoses *(white circles)* and binucleate cells *(black circles)*. The tumor shows diffuse sheets of well- to moderately differentiated plasma cells that are monoclonal, with kappa restriction, and are diffusely positive for CD138 immunostaining. (B) 200 × mag; (C–E) 400 × mag. *Figure courtesy of Meenakshi B. Bhattacharjee, M.D., University of Texas Health Science Center at Houston.*

III. Neurological complications of specific neoplasms

The mainstays of acute management are corticosteroids, chemotherapy, radiotherapy (RT), and surgery.[29] Decisions on optimal management should be individualized using a multidisciplinary approach. Myelomatous deposits of soft tissue in the spinal canal can be treated effectively with chemotherapy, RT, and steroids, without the need for surgical decompression. However, any significant spinal cord compromise or cauda equina compression may still require surgical decompression. Corticosteroids improve pain and result in better clinical outcomes, and therefore should be started promptly as first-line therapy.[17] No prospective trials have been conducted to determine the most advantageous schedule and dose.[17] Regimens using initial doses of 10–100 mg, with daily dosing of 16–96 mg, have been reported.[17] Patients should be followed closely for adverse effects such as myopathy and myalgias, increased appetite and weight gain, elevated blood glucose, lower extremity edema, and Cushingoid features, with the steroids being tapered and discontinued as quickly as neurological symptoms allow.[17]

Low-dose RT (up to 30 Gy) is recommended for impending spinal cord compression.[29] Upfront external beam RT should also be considered for patients with spinal cord compression.[29] However, RT should be given sparingly and in fractionated doses to limit the deleterious effects on normal spinal cord and to spare marrow function.[17,29] The majority of studies examining the role of RT to date have been retrospective with small patient cohorts, but there is a clear benefit of RT in providing pain relief, decreasing analgesic use, improving neurological symptoms and motor function, and increasing quality of life in patients with multiple myeloma.[29–32] A retrospective study of 172 patients reported that long-course RT (10 fractions of 3 Gy, 15 fractions of 2.5 Gy, or 20 fractions of 2 Gy) resulted in more frequent improvement of motor function compared to short-course RT (1 fraction of 8 Gy, or 5 fractions of 4 Gy).[30] Techniques such as intensity-modulated RT and tomotherapy can decrease the amount of radiation that the surrounding healthy spinal cord receives.[17] A large retrospective series demonstrated RT alone improved motor function in 75% of patients with multiple myeloma and spinal cord compression with 1-year local control of 100% and 1-year survival of 94%.[33] In addition to its efficacy in the acute treatment of ESCC, RT has been shown to have other significant benefits when used in the management of patients with myelomatous vertebral involvement. Significantly, fewer new fractures and focal lesions developed in irradiated vertebrae compared to nonirradiated vertebrae in patients followed prospectively with MRI in one series.[34] Additionally, especially when given concurrently with chemotherapy, RT has been shown to relieve pain secondary to spinal involvement.[21,35]

The role of surgery in the treatment of ESCC due to multiple myeloma, a radiosensitive neoplasm, has not yet been clearly defined.[12,17,18,29,36,37] Orthopedic and/or neurosurgical consultation is recommended, and consideration for surgery should be made in consultation with the treating oncologists.[29] While surgery can still be directed toward preventing or repairing axial fractures, unstable spinal fractures, or spinal cord compression, decompressive laminectomy is rarely required in patients with multiple myeloma, unless needed in rare cases of radioresistant multiple myeloma or retropulsed bony fragments.[29,38] A multicenter, randomized, prospective trial of surgery followed by RT versus RT alone for the treatment of ESCC from metastatic cancer showed that combined therapy was superior to RT alone; however, patients with multiple myeloma were excluded.[36] In several retrospective series, a greater number of patients with myeloma-associated ESCC treated with both surgery (usually decompressive laminectomy) and RT had significant clinical improvement compared to those treated with either modality alone.[10,12] While these historical studies support the use of surgery in some patients, it is important to note that surgical management is not appropriate in the majority of patients due to the often extensive burden of disease.[37,39,40] The diffuse nature of bony disease impairs the strength of vertebrae, which can increase the risk of complications when inserting rods and screws safely.[37,39,40] Furthermore, the majority of these patients have significant comorbidities and are poor candidates for surgical decompression, increasing the risk of morbidity, mortality, and longer recovery times.[37,39,40]

Following successful treatment of spinal cord or cauda equina compression with steroids, chemotherapy, and/or radiation, cement augmentation is often required to relieve pain associated with the fracture or reestablish stability of the spine.[41] Balloon kyphoplasty and percutaneous vertebroplasty are minimally invasive procedures that have demonstrated consistency in relieving pain and restoring physical functioning.[42] Several studies have demonstrated the benefit of these procedures in the treatment of symptomatic vertebral compression fractures, with reports of significant improvements in physical functioning, back pain, and quality of life.[29,42–45]

Bisphosphonates, which inhibit osteoclastic activity, have been shown to decrease bone turnover, decrease bone pain, and reduce skeletal complications, including ESCC, due to vertebral compression fractures.[20,46–49] The American Society of Clinical Oncology Practice Guidelines recommend that bisphosphonates should be given concurrently with antimyeloma therapy to patients with multiple myeloma with or without detectable bone lesions on imaging.[29] Intravenous (IV) zoledronic acid or pamidronate is recommended for preventing skeletal-related events in patients with multiple myeloma. Zoledronic acid has the advantage of a much shorter infusion time (15 min compared to 2 h for pamidronate), but is more expensive. In a large phase III, randomized, double-blinded study comparing pamidronate and zoledronic acid, there was no difference in time to the first skeletal-related event between the two groups.[50] Denosumab is a novel monoclonal antibody that was recently shown to be noninferior to zoledronic acid for time to skeletal-related events.[51] Several other novel targets are under study in the treatment of patients with multiple myeloma bone disease.[52]

Neurological function at the time of diagnosis is the most significant prognostic factor in patients with ESCC, irrespective of the causative underlying malignancy.[17,18] Patients who are ambulatory at the time of initiation of treatment have a much higher likelihood of retaining this function than patients who are not ambulatory at the time of treatment.[17] Regardless of the modality of therapy, patients with a shorter duration of symptoms prior to treatment, especially those symptomatic for less than 24 h, have a greatly improved prognosis compared to patients whose symptoms were present longer.[12,14,15,22] Response to treatment also has significant prognostic implications. Patients with partial or no improvement in their neurological deficits have a significantly shorter survival from time of diagnosis of ESCC than those who have complete resolution of their neurological deficits.[12,14]

2.2 Central nervous system myelomatosis

The incidence of extramedullary disease involving the central nervous system (CNS) in multiple myeloma patients continues to increase. This is likely due to advances in imaging techniques and improved survival of myeloma patients resulting from novel treatments, including stem cell transplantation, proteasome inhibitors, and immunomodulatory drugs.[53–56] Although myelomatous involvement of the CNS by direct plasma cell infiltration of the parenchyma, leptomeninges, or cerebrospinal fluid (CSF) has been described in the literature since the early 20th century, only 172 reported cases were found in the largest multiinstitutional retrospective study published in 2016.[57] CNS involvement has been reported to occur in approximately 1% of patients with multiple myeloma, and after the diagnosis is made, the prognosis is extremely poor, with a reported median overall survival of 7 months or less.[57–66] However, intracerebral plasmacytomas can be successfully treated with radiation, in stark contrast to aggressive myelomatous meningitis.[56] The rarity of this complication is even more striking when compared to the frequency with which other hematologic (incidence of 20%–75%) and solid tumor (5%–8%) malignancies spread to the CNS.[57–66] CNS involvement of multiple myeloma can occur at any stage of the disease process, with up to a quarter of cases discovered at initial diagnosis.[57]

The skull is one of the most frequent sites of bony lesions in patients with multiple myeloma,[2] with the most common skull locations being the sella, the body of the sphenoid ridge, and the apex of the petrous bones.[10,67–70] Direct extension from erosion of contiguous skull-based myeloma lesions into the dura is a well-documented mechanism of initial entry into the meninges and cerebral parenchyma.[60,71–79] Extension may also occur from myeloma lesions located in the oropharynx and paranasal sinuses.[80] LMM without evidence of direct invasion was not reported until the late 20th century.[81] A proposed mechanism for LMM in the absence of direct invasion is hematogenous spread by circulating malignant plasma cells or by lymphocytes that may be the progenitors of myeloma cells.[72–74,76,82–84] This mechanism of entry is supported by the increased prevalence of circulating plasma cells and of plasma cell leukemia in myeloma patients with CNS involvement.[72–74,76] Hematogenous spread, whether by circulating plasma cells or by lymphocytes, would also account for intraparenchymal myeloma lesions that occur in the absence of direct extension from a skull-based lesion or meningeal involvement.

The location of the skull-based lesion dictates which neurological signs and symptoms occur.[24,28] The clinical picture may be that of an isolated endocrinological abnormality, such as diabetes insipidus or hypopituitarism, cranial nerve palsies, focal brainstem findings, seizures, headache, or papilledema and other signs of increased intracranial pressure.[10,24,28,82–87] Compression and distortion of cranial nerves from lytic lesions of the skull, especially those of the sphenoid or petrous bones, and by foraminal involvement by skull-based lesions, are the usual etiology of palsies encountered in patients with multiple myeloma, even without diffuse leptomeningeal involvement.[10,28,88–91] The most commonly affected cranial nerves are II, III, V, VI, VII, and VIII; the lower cranial nerves may also be affected.[10,28,77–79,87] Alterations in the level of consciousness, with coma as the most severe manifestation, may also be seen with intraparenchymal involvement of multiple myeloma.[24,28] Myelomatous meningitis presents most commonly with mental status changes, leg weakness, and gait disorders.[74,76,77,92,93] LMM may also present with many other neurological signs or symptoms, including cranial nerve palsies, headache, speech disorder, seizures, meningismus, hemiparesis, hemisensory loss, vertigo, dysmetria, muscle atrophy, pain of the limbs or face, fecal incontinence, and Lhermitte's sign.[72–74,76,77,92,93] Obstructive hydrocephalus with papilledema secondary to LMM has been reported.[94] The interval between the time of initial diagnosis of multiple myeloma and the onset of symptoms secondary to LMM was reported in early studies as being approximately 9 months[73,76,95]; more recent reviews report a median interval of 2–3 years.[57,61–63] This longer interval may be due to improved treatment and the use of high-dose chemotherapy followed by peripheral blood stem cell or bone marrow transplant.

The age at presentation varies between reported studies, with a median age of onset of CNS multiple myeloma around 50–60 years old, younger than the median age for multiple myeloma diagnosis (approximately 70 years old).[57,96] Taking into account the limitations of retrospective case series, there is a bias toward a lower M:F ratio with a mean of 57% males in all studies combined.[97] LMM is most frequently seen in patients with an aggressive underlying malignancy, characterized by a high disease burden, stage III disease, extramedullary manifestations, circulating plasma cells in the peripheral blood, and plasma cell leukemia.[60,72–74,76] However, LMM has been reported to occur in patients in apparent remission, with low tumor burden after high-dose chemotherapy and stem cell transplant.[72,73,77]

There is no consensus for associations between light-chain restriction and Ig class in patients with CNS involvement of multiple myeloma, as studies are small and limited by their retrospective nature. In the largest retrospective series to date, IgA myeloma represented 27% of CNS multiple myeloma cases compared to 21% in a large case series of newly diagnosed multiple myeloma patients.[3,57] Also, several reports show a higher proportion of cases of lambda rather than kappa light-chain expression in CNS multiple myeloma compared with non-CNS multiple myeloma.[54,57]

Cytogenetic and immunohistologic studies of plasma cells found in the CSF and bone marrow of patients with CNS multiple myeloma have shown complex abnormalities including translocations and deletions of chromosome 13, a high incidence of p53 deletions, and low expression of CD56, a neural adhesion molecule, compared to multiple myeloma patients without CNS involvement.[57,60,98,99] These genetic and immunohistologic characteristics have been hypothesized to be related to the metastatic potential of the myeloma cells in patients with LMM.[57,99] Conversely, the absence of these and other, as yet undefined, biologic and genetic characteristics in multiple myeloma patients without LMM has been hypothesized to explain the low incidence of this complication.[77,98,99]

Intraparenchymal plasmacytomas that do not directly extend from the bone or dura are a very rare manifestation of myelomatous CNS involvement, occurring less frequently than meningeal involvement (Fig. 2).[10,57,72,100] Cerebral plasmacytomas may occur in the absence of systemic disease.[21,28,72] As with skull- and dural-based lesions, the clinical picture is based on the location and size of the lesion; focal deficits, seizures, and signs of increased intracranial pressure may all occur.[24,28]

Neuroimaging studies are not diagnostic in patients with LMM, as they may be negative for any abnormalities.[72,73,101,102] Skull-based myeloma lesions may appear

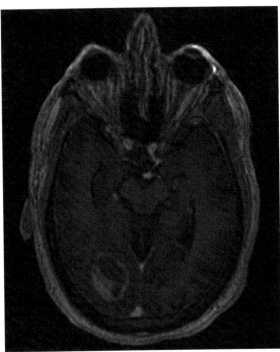

FIG. 2 Axial T1-weighted postcontrast imaging of an oval, well-circumscribed internally complex enhancing mass in the right occipital lobe, demonstrating thick slightly nodular peripheral enhancement and mild septation and measuring 2.8×3.1×2.7 cm. There is moderate surrounding vasogenic edema with sulcal narrowing and mild anterior displacement of the posterior horn of the right lateral ventricle. Pathology was consistent with an intracerebral plasmacytoma, in keeping with the patient's history of multiple myeloma status post stem cell transplant. The patient was subsequently treated with postoperative radiation.

as destructive masses on CT and MRI.[100] LMM may appear on MRI as diffuse thickness of the cranial meninges with focal or diffuse enhancement; multiple intraparenchymal lesions also may be present.[72,77,103,104] LMM can appear as extra-axial leptomeningeal-based mass lesions.[72,77] On CT, LMM may appear as slightly hyperdense subdural masses.[72]

The diagnosis of LMM is made by the presence of plasma cells in the CSF.[71–74,77,105–107] Since plasma cells may be seen in the CSF in a variety of infectious and inflammatory conditions,[75,108,109] a definitive diagnosis should be based not only on the traditional cell differential, but with immunocytochemical studies to determine the monoclonality of the plasma cell population.[72–74,106,107] The monoclonality of the plasma cells and the paraprotein found in the CSF should match that found in the serum, urine, or bone marrow; however, rare exceptions have been reported.[60,74,76] The presence of paraprotein in CSF is not sufficient to diagnose LMM, as a direct relationship exists between the amount of paraprotein found in the serum and the CSF, and patients with a heavy myeloma burden will have abnormal amounts of paraprotein in the CSF even in the absence of CNS involvement.[60,73,74,76,105] Other common CSF findings in patients with LMM are increased protein, hypoglycorrhachia (though not as common as in other forms of leptomeningeal malignancies), and an elevated opening pressure.[60,73,74,76]

Radiation therapy, systemic intermediate- and high-dose chemotherapy, allogeneic and autologous stem cell transplant, and intrathecal chemotherapy have all been used to treat myelomatous CNS involvement in various combinations and with varying degrees of success.[55,65–67] Systemic chemotherapy, including agents such as melphalan, cyclophosphamide, high-dose methotrexate or cytarabine, is thought to be limited in efficacy due to poor blood-brain barrier and/or CSF penetration or inactivity against tumor-resistant myeloma cells.[56,110,111] One exception may be bendamustine, which is capable of penetrating the blood-brain barrier, and has shown some efficacy in two cases of LMM in combination with thalidomide, dexamethasone, and craniospinal irradiation.[111] Jurczyszyn et al. in 2016[57] reported a significantly longer median overall survival (OS) in patients with CNS multiple myeloma who received systemic therapy than those who did not (OS 12 vs 3 months), although these data may be biased due to the likelihood that patients with a better performance status are more likely to receive systemic therapy. Thalidomide and lenalidomide, while reported to safely cross the blood-brain barrier in LMM, did not demonstrate any survival benefit in multiple reports,[57,110,112,113] unless they were combined with intrathecal therapy and cranial-spinal irriadiation.[96] A durable CSF response has been reported in one case using the combination of pomalidomide and dexamethasone.[114] Novel protease inhibitors including bortezomib and marizomib have demonstrated potential efficacy in a small number of CNS multiple myeloma patients.[115–117] The most common regimen of intrathecal chemotherapy used in CNS multiple myeloma is the combination of hydrocortisone, methotrexate, and/or cytarabine. There have been reports of successful clearing of CSF plasma cells, but the resultant clinical response was transient.[59,96,106,118] Not one of these therapies seems clearly more beneficial than others, and there are no prospective studies of treatment in patients with LMM. Treatment of LMM in patients with multiple myeloma may rapidly produce an excellent, though often transient, response, resulting in symptomatic relief and improved quality of life, supporting an aggressive treatment approach in some patients.[76,77] The limited number of cases in the literature of LMM in patients with multiple myeloma, and the lack of a clearly successful treatment regimen, precludes the formation of generic treatment recommendations. Since responses are rarely durable, decisions regarding treatment should be made on an individual basis.

Cranial or cranial-spinal irradiation is an effective treatment for solitary plasmacytomas of bone, parenchymal CNS lesions, and other extramedullary plasmacytomas, given the known radiosensitivity of malignant plasma cells.[54,119–121] Targeted radiation may also improve muscle weakness from intramedullary spinal cord lesions.[120] There are reports of stem cell transplantation as a method to overcome a poor prognosis due to CNS involvement in one study with a 25-month long-term survivor.[53,60,122,123]

The prognosis of myeloma patients with CNS involvement is remarkably poor, with reported median survival from the time of diagnosis of CNS involvement of 7 months. Patients who respond to treatment tend to have a survival advantage of several months over those who do not respond.[57,76] Patients with LMM who respond well to therapy often succumb to their typically aggressive systemic disease.[57,76] In nine patients with CNS multiple myeloma, the three longest survivors were those who underwent a stem cell transplant.[106] Another interesting caveat was that all nine patients were not receiving maintenance therapy prior to detection of CNS involvement.[97,106]

2.3 Hyperviscosity syndrome

The classic hyperviscosity syndrome (HVS) triad of bleeding of the mucous membranes, visual disturbances, and neurological abnormalities was first described in a landmark article by Fahey et al. in 1965.[124] All of the patients reported had Waldenström's macroglobulinemia, and the authors emphasized the rarity of this syndrome in patients with multiple myeloma.[124,125] Subsequent studies found an incidence of approximately 2%–10% in multiple myeloma patients.[14,126–128] In a few small case series of multiple myeloma patients with HVS, 50%–100% of patients had neurological manifestations.[127–129] In one series, greater than 25% of the patients had a severe neurological deficit as the presenting feature of their HVS.[127]

The neurological manifestations of HVS are extensive in their breadth and degree of severity. HVS is considered an emergent condition with potentially catastrophic neurological manifestations. It may result in coma, seizures, and stroke with irreversible deficits.[88,124,130–133] If untreated, initially mild deficits may rapidly progress. Symptoms may appear at any time during the course of the myeloma. One retrospective series found a higher frequency of HVS symptoms at the time of initial clinical presentation of the myeloma and at the time of recurrence of disease.[127] The authors attributed this to the elevated paraprotein concentrations at these times.[127]

HVS often manifests neurologically as alterations in level of consciousness and mentation.[124,127,128,131,132] Drowsiness, lethargy, and confusion are common and may rapidly progress to somnolence and stupor.[127,131,133]

Alternatively, symptoms may be subacute at onset, consistent with a dementia syndrome.[134] Headaches (often persistent) and dizziness are also frequent, and hallucinations may occur.[127,128,132] Other neurological manifestations include hearing loss, ataxia, vertigo, nystagmus, gait impairment, diplopia, and dysarthria.[124,127,128] Paresthesias may occur from the effects of hyperviscosity on the CNS or on peripheral vasculature.[130] Deficits may be transient or constant. Neurological symptoms are thought to occur due to the effect of increased serum viscosity on the cerebral circulation.[88,132] Increased serum viscosity results in slowing of blood flow.[88,132,135] This impairment in flow leads to occlusion of blood vessels, with those of the microcirculation being most involved.[132] Neurological complications occur as a consequence of ischemia.[88,132]

Decreased visual acuity is characteristic of HVS and may range in severity from mild blurring to blindness, and may be acute or gradual in onset.[124,127,128] Slow blood flow through the retinal veins will initially result in venous distention, an early finding on funduscopic examination.[124] Marked congestion, tortuosity, and sausage-shaped dilatation and constriction of the retinal veins follow, with flame-shaped hemorrhages occurring later.[14,124,127,128,132] Capillary microaneurysms and retinal vein thromboses may also be seen.[124] In severe cases, the funduscopic examination will show exudates, striking congestion, and numerous hemorrhages.[124,127]

Bleeding in various forms (e.g., epistaxis, bleeding of the mucous membranes, and/or gastrointestinal bleeding) is the most frequent presenting symptom and most common clinical feature of HVS[124,127,128]; intracerebral hemorrhage rarely occurs.[88] Serum viscosity is a function of multiple factors, including the quantity, shape, molecular size, and interactions of the paraproteins present.[124,127–129,132] HSV occurs most commonly in patients with multiple myeloma subtypes IgA and IgG-3, due to the propensity of these subtypes to form polymers and aggregates, respectively.[2,124,126,127,129,132] HVS can occur in patients with pure light-chain myeloma due to the ability of kappa light chains to form aggregates.[126,132] No absolute value of serum viscosity exists at which myeloma patients become symptomatic; however, the clinical manifestations of serum hyperviscosity are not usually seen below a relative viscosity of 4.0 (normal 1.4–1.8). Patients may be asymptomatic with a viscosity of 8–10.[124,128,132] The value at which symptoms of hyperviscosity occur is a particular characteristic of each patient, not of the disease process as a whole, and is dependent on several factors including patient age and the presence of concomitant diseases and other hematologic abnormalities.[2,124,128,134]

Chemotherapeutic agents reduce the formation of plasma paraproteins, thereby reducing their concentration, resulting in a decrease in serum hyperviscos-

ity and improvement of clinical symptoms.[88] However, these outcomes are delayed, and therefore, chemotherapy is not useful as first-line treatment for this emergent condition. Plasmapheresis quickly clears intravascular paraproteins and has been shown to rapidly and markedly improve signs and symptoms of HVS.[124,127,128,131,134] While controversy exists regarding the appropriate frequency of plasmapheresis, overall, it is regarded as a relatively safe, well-tolerated, and effective initial treatment of HVS.[124,131,132]

Treatment should be initiated promptly. Given the high potential for poor neurological outcomes and the effectiveness of early treatment, HVS should be considered in all multiple myeloma patients with neurological or visual symptoms.[131]

2.4 Toxic metabolic encephalopathy

Uremia and hypercalcemia are commonly encountered in patients with multiple myeloma. These may result in neurological abnormalities that are as varied and potentially severe as those caused by direct involvement of the neoplasm. In contrast to many of the neurological complications caused by direct involvement of multiple myeloma, those secondary to metabolic derangements will usually have a rapid and marked response to intervention if therapy is initiated promptly and prior to progression of symptoms. Patients with myeloma should therefore be monitored closely for the clinical manifestations and laboratory abnormalities of these metabolic processes, and a metabolic etiology should be considered in patients with headaches and other seemingly minor neurological symptoms, including mild changes in mental status.

2.4.1 Uremia

Renal dysfunction is a major source of morbidity and mortality in patients with multiple myeloma.[136] Approximately 25%–50% of myeloma patients will develop renal insufficiency during the course of their disease. In a large series of 1027 patients, 48% had a creatinine level of 1.3 mg/dL or higher at the time of diagnosis.[2,136–139] The high frequency of renal insufficiency in myeloma patients is explained by the multiple mechanisms of renal injury that are intrinsic to the neoplastic process. The clinical syndrome of uremia is usually seen once the creatinine clearance is less than 10–20 mL/min; every organ system can be affected.[140] In a retrospective series of 277 patients with multiple myeloma, approximately 4% had neurological manifestations of uremia.[10]

Uremic encephalopathy is characterized by fluctuating changes in mood, level of orientation, and degree of consciousness, at times progressing to coma.[99,100] Other neurological abnormalities associated with uremic encephalopathy include hallucinations, headache, fatigue, asterixis, tremor, dysarthria, myoclonus, tetany, ataxia, hemiparesis, generalized weakness, hypertonia, seizures, meningeal signs, and restless leg syndrome.[19,99,100] Dialysis rapidly and effectively improves uremic encephalopathy, but rare adverse neurological events, such as disequilibrium syndrome, can result.[99,100] Treatment of the underlying malignancy with chemotherapy is effective at improving renal function in 50% of patients with mild renal insufficiency.[136,141] Avoidance of uremia, by close monitoring and prevention of its multiple causes, is the best management strategy.[136,141] This includes adequate hydration, rapid detection, and treatment of hyperviscosity, hypercalcemia, hyperuremia, and urological infections.[136,141]

2.4.2 Hypercalcemia

Thirteen percent of newly diagnosed multiple myeloma patients had hypercalcemia in a large retrospective series.[3] It eventually occurs in approximately 20%–40% of patients.[10,20,88] Hypercalcemia results not only from direct metastasis to the bone by plasma cells, but also from the generalized amplification of osteoclast activity by myeloma cell-released cytokines, such as tumor necrosis factor B, interleukin (IL)-1β, and IL-6.[2,10,20,46,88] It is exacerbated by renal insufficiency, specifically decreased glomerular function, and possibly the increased reabsorption of calcium by the renal tubules.[20,88] Hypercalcemia is more commonly associated with a large tumor burden and an advanced stage of disease, but may occur at any stage of disease and with any degree of tumor burden.[20]

In two studies of the neurological complications of multiple myeloma, approximately 1%–5% of patients had complications thought to be attributable to hypercalcemia.[10,14] Hypercalcemia, especially at levels above 14 mg/dL, commonly results in mental status changes, including confusion, changes in mood, and alteration in the level of consciousness.[2,10,14,24,88,141] Other neurological manifestations include fatigue, headache, asterixis, muscle weakness, and myalgias secondary to decreased neuromuscular excitability, and rarely, seizures.[10,14,24,88,141] If untreated, patients may progress to coma. Lumbar puncture may show an elevated CSF protein level.[88] Nonneurological symptoms mainly involve the gastrointestinal system, including nausea, vomiting, anorexia, constipation, and abdominal pain; their presence may provide clues to the etiology of the patient's neurological symptoms.[2,88,141]

The first line of therapy in myeloma patients with symptomatic hypercalcemia is urgent, aggressive (but metered) hydration, usually in association with glucocorticoids.[14,20,88,136,141] This combination improves renal function, thereby decreasing calcium reabsorption and improving calcium excretion.[88] Treatment with a loop diuretic, such as furosemide, should also be initiated.[88,141]

Initial therapeutic intervention often results in a swift and significant improvement in the neurological manifestations. Patients treated early have better outcomes.[2,20,141,142] Calcitonin inhibits bone resorption and may be transiently beneficial; however, it is typically not necessary due to the effectiveness of first-line therapy.[88,141] Bisphosphonates inhibit osteoclastic activity and will improve hypercalcemia in myeloma patients.[46] This response does not occur rapidly, and these drugs should not be considered part of the initial or urgent management strategy in these patients.[88,141]

3 Peripheral nervous system complications

3.1 Compression, infiltration, and numb chin syndrome

Multiple myeloma can affect the cranial nerves and nerve roots via leptomeningeal infiltration, compression from adjacent bony lesions, or plasmacytomas. The most classic example is that of numb chin syndrome (NCS), most commonly due to lytic lesions of the mandible.[143] NCS, or mental neuropathy, is numbness in the distribution of the mental or inferior alveolar nerve, involving the chin, lower lip, and rarely the buccal gingiva and lower teeth.[144–147] Although NCS may result from a benign etiology, such as trauma to the face or dental disease, a significant association exists with the presence of an underlying malignancy, most commonly breast cancer or lymphoma.[144–149] The exact incidence of NCS in patients with multiple myeloma is not known.[80] NCS may be the presenting sign of a previously occult malignancy or signify tumor progression or relapse.[144,146–149] In NCS from any etiology, the sensory deficits occur more unilaterally than bilaterally, and as the nerves affected are solely sensory nerves, taste and motor function are spared.[144–147]

Metastases to the mandible may cause compression of the mental or inferior alveolar nerve resulting in NCS in patients with multiple myeloma. The reported range of mandibular involvement in multiple myeloma patients is broad, occurring in 5%–70% of cases.[80,150,151] In addition to mandibular metastases, other multiple myeloma-related etiologies of NCS include leptomeningeal spread and intracranial involvement of the mandibular nerve through the base of the skull lytic lesions.[144,146–148]

Patients with multiple myeloma who present with numbness in the mental nerve distribution should undergo a thorough evaluation for all possible causes, including careful neurological examination for the presence of other cranial nerve palsies. These include CT of the head, base of the skull, and mandible, MRI of the brain with and without gadolinium, and lumbar puncture.[144–148] The etiology dictates the therapeutic course in these patients.[147] RT of base-of-skull lesions, intrathecal chemotherapy for leptomeningeal metastasis, and chemotherapy for control of systemic disease in patients with mandibular lesions have all been shown to have some efficacy in patients with neoplastic-related NCS.[145–147]

NCS due to any neoplastic etiology is associated with a very poor prognosis, with one study reporting a median survival of 8 months after diagnosis.[144,145,147–149] The prognostic implications of mandibular involvement in the absence of NCS in myeloma patients are not as clear.[149,151]

3.2 Multiple myeloma-associated peripheral neuropathy

Peripheral neuropathy is a well-recognized neurological complication of multiple myeloma, and in rare cases may be the presenting manifestation.[152–155] The incidence of multiple myeloma-associated peripheral neuropathy has been reported as 3%–5% in several large retrospective series.[10,14,28,152,155,156] It reached 13% in one prospective series of 23 patients when based on neurological examination, and 39% using electrophysiologic studies.[156] The clinical presentation is variable. It is usually a symmetric, mixed-sensorimotor neuropathy, with greater involvement of the lower extremities, manifested symptomatically by distal numbness and weakness.[10,14,155,157] Symptoms are usually mild and the course stable or progressive, though it may be acute, monophasic, or relapsing and remitting in some cases.[14,28,155,158] Pure sensory or motor neuropathies (some with associated cranial nerve or respiratory involvement) are well documented, but less common.[14,155,158] Patients with a pure sensory neuropathy may have a marked sensory ataxia secondary to loss of joint position sense.[155] In the absence of amyloidosis, autonomic symptoms and neuropathic pain are not significant features.[14,155] In patients with the most common clinical variant, examination may reveal mild distal, symmetric weakness, and hyporeflexia (most frequently at the ankles), with mild decrease of all sensory modalities.[155]

Results of electrophysiologic examination correspond to the heterogeneous clinical picture and may be consistent with an axonal neuropathy, demyelinating neuropathy, or an axonal neuropathy with demyelinating features.[14,153,155,158–161] Nerve conduction velocity (NCV) tests usually show mild slowing of motor conduction velocity, slightly low-amplitude compound muscle action potentials, and low-to-absent sensory nerve compound action potentials.[14,155–157] Fibrillations and denervation potentials, often of the distal muscles, are commonly seen on electromyography (EMG).[14,155,157]

Histopathological studies are also varied and may show evidence of active axonal degeneration, demy-elination, and loss of both myelinated and unmyelinated fibers, either as individual findings or in the same biopsy.[155–157,161,162] Infiltration of peripheral nerves by myeloma cells has been reported, but is not a finding common to all patients with myeloma-associated neuropathy.[155,163] Neurogenic atrophy may be seen on muscle biopsy.[157]

Etiology is most commonly the toxic effect of treatment for the underlying malignancy.[153] However, many patients develop a neuropathy without having ever received therapy. An immunological pathogenesis has been proposed, possibly involving antineuronal antibodies, but specific antibodies and their antigen targets have not been definitively identified.[154,158,161,164] Antibodies against myelin-associated glycoprotein (MAG) and peripheral nerve myelin (PNM) are often found in patients with a demyelinating neuropathy associated with IgM monoclonal gammopathy of undetermined significance (MGUS).[154,159,164] However, IgM multiple myeloma is rare, and patients with multiple myeloma seldom have anti-PNM or anti-MAG antibodies.[3,165] Direct invasion of peripheral nerves by plasma cells may also result in neuropathy.[10,153,156,161,163] Relative nutritional deficiencies and the metabolic derangements commonly seen in patients with multiple myeloma, such as chronic renal failure, are also likely contributing factors.[88,156] The exact mechanism of nontreatment-related multiple myeloma-associated peripheral neuropathy has yet to be clearly delineated; all of the earlier-mentioned etiologies likely contribute in some way to the pathogenesis of this complication.

There is no specific treatment for the neuropathy associated with multiple myeloma other than treatment of the underlying multiple myeloma. Treatment options may include autologous stem cell transplant, as well as a variety of chemotherapeutic regimens. Victor et al. reported in 1958 that the course of multiple myeloma-associated peripheral neuropathy appeared distinct from that of the malignancy and from the malignancy's response to treatment.[162] This has been subsequently supported by several retrospective series and reviews.[14,154,155,159,164] There are reports describing patients treated with RT and/or chemotherapy for the underlying myeloma whose neuropathy improved; however, patients with osteosclerotic myeloma and osteolytic myeloma were not reported separately.[157,158] Treatments targeting an immunological mechanism, such as plasmapheresis and intravenous immunoglobulin, have not been adequately studied. Medications such as pregabalin and gabapentin used to treat painful neuropathies of other etiologies may be useful if patients are experiencing neuropathic pain. There are no studies evaluating these drugs in this patient population. Dialysis may improve or stabilize neuropathy due to uremia from chronic renal failure. Treatment of other underlying metabolic derangements or nutritional deficiencies may also help. Therapy is mainly supportive and includes physical therapy, orthotics, and ambulatory aids.[166]

3.3 Treatment-emergent peripheral neuropathy

The leading cause of peripheral neuropathy in multiple myeloma patients is now secondary to novel therapeutics such as bortezomib and thalidomide. These agents have significantly improved outcomes, but can often lead to neurotoxicity requiring dose reductions or discontinuation of the offending agent. Since patients with multiple myeloma often have preexisting peripheral neuropathy, it is extremely important to screen patients for symptoms and signs of peripheral neuropathy, and even consider baseline nerve conduction studies prior to starting treatment of the disease.

3.3.1 Proteasome inhibitors

As part of a new generation of anticancer drugs, bortezomib, a dipeptidyl boronic acid, is a reversible proteasome inhibitor that primarily targets the chymotrypsin-like and caspase-like active sites of the proteasome. This results in suppression of tumor survival pathways and arrests tumor growth, spread, and angiogenesis.[167,168] While bortezomib has significantly improved median overall survival in multiple myeloma,[169,170] many patients are unable to tolerate treatment due to side effects, including the peripheral neuropathy that occurs in up to two-thirds of patients.[171,172] Switching the dosing schedule of bortezomib to subcutaneous or weekly IV dosing may reduce the incidence of peripheral neuropathy to approximately 40%.[173,174] Bortezomib-associated peripheral neuropathy is described as a painful, distal, and sensory-predominant polyneuropathy. Nerve conduction studies may show low sensory nerve action potentials. Other reports have demonstrated severe, motor-predominant, polyradiculoneuropathy with demyelinating features on nerve conduction studies with elevated CSF protein.[175,176] Dose reduction or discontinuation of bortezomib often reverses the neuropathic symptoms in most patients. A number of new second-generation proteasome inhibitors (carfilzomib, marizomib, and ixazomib) report a lower-incidence of treatment-emergent peripheral neuropathy with superior efficacy, especially in those patients with myeloma resistant to bortezomib.[167,177,178]

3.3.2 Thalidomide

Thalidomide was the first immune-modulatory drug in its class and demonstrated efficacy in the treatment of multiple myeloma in 1999.[179] Through multiple mechanisms of action, thalidomide causes dorsal root ganglia degeneration.[180] Treatment was quickly associated with treatment-emergent peripheral neuropathy,

which was found to be cumulative, dose-dependent, and linked to treatment duration. One study reported thalidomide-associated neuropathy in 38% and 73% of patients at 6 and 12 months, respectively.[181] The clinical characteristics and nerve conduction study findings of thalidomide-associated neuropathy are similar to those of bortezomib. However, symptoms can begin after treatment discontinuation, may persist for several months, and can be irreversible. Newer agents such as lenalidomide and pomalidomide have a lower association with treatment-emergent peripheral neuropathy.[177]

3.4 Other plasma cell dyscrasias

3.4.1 Amyloidosis

Amyloidosis occurs concurrently with multiple myeloma in 6%–15% of patients.[182] Patients with myeloma who have neuropathy have been found to have amyloidosis in 20%–30% of cases.[10,14,155,164] In those patients, the peripheral neuropathy is homogeneous, and typically indistinguishable from that seen in patients with primary systemic amyloidosis without multiple myeloma.[159,164,182] Patients with primary amyloidosis and peripheral neuropathy are subdivided into primary systemic amyloidosis and familial amyloidosis. Both are defined by the subunit of protein that alters the amyloid fibril. Familial amyloidosis is due to mutations in the transthyretin (TTR) gene and less commonly apolipoprotein A-1 or gelsolin genes.[183]

Amyloid light-chain (AL) amyloidosis causes neuropathy due to deposits of monoclonal light chains, kappa or lambda, in the peripheral nerves, but is rarely associated with heavy-chain fragments.[184] The incidence of peripheral neuropathy in these patients is estimated to be 17%.[185] In both AL and familial forms of amyloidosis, the neuropathy is chronic and progressive, usually associated with autonomic failure and a painful, length-dependent, sensorimotor peripheral neuropathy.[155,159,182,186] Sensory findings, especially those of a small fiber neuropathy (pain and temperature loss), tend to precede motor abnormalities, and the lower extremities are usually affected before the upper extremities. Pain is commonly a prominent and early complaint, and may be burning or lancinating.[155,159,187,188] Autonomic symptoms include decreased sweating, orthostatic hypotension (which may be profound), gastrointestinal dysfunction with malabsorption and diarrhea, impotence, and loss of bladder control.[155,159,187] Superimposed bilateral carpal tunnel syndrome is seen in approximately 25% of patients, secondary to amyloid infiltration of the flexor retinaculum.[10,154,155,159,164,182] EMG/NCV studies typically are consistent with an axonal neuropathy.[28,155,160] Amyloidoma, a localized deposit of amyloid on a peripheral nerve, can cause a focal neuropathy via a compressive mechanism.[189,190]

If small-fiber involvement is suspected, quantitative sudomotor axon reflex testing, thermoregulatory sweat testing, and/or skin biopsy to determine epidermal nerve fiber density are useful for diagnosis. Serum and urine immunofixation and immunoglobulin free light-chain assays should be ordered; but tissue confirmation is often necessary if a diagnosis of AL amyloidosis is suspected.[191] Rectal or abdominal fat biopsy in combination with a sural nerve biopsy can confirm the diagnosis. If the diagnosis is made expeditiously, stem cell transplant is the treatment of choice, with survival rates of 53% in patients with a complete response and 43% for all treated patients. If the patient is not eligible for transplant, chemotherapy is recommended, which can include melphalan-dexamethasone or cyclophosphamide-bortezomib-dexamethasone.[192,193]

3.4.2 POEMS syndrome

POEMS syndrome is a rare paraneoplastic syndrome associated with a plasma cell disorder. Historically, it has also been named osteosclerotic myeloma, Takatsuki syndrome, or Crow-Fukase syndrome. POEMS is an acronym that describes the most common features of the syndrome, including peripheral neuropathy, organomegaly, endocrinopathy, monoclonal plasma cell disorder, and skin changes.[194] In 2012, the diagnostic criteria were further refined.[8] In addition to the requirement that the patients have a plasma cell disorder and neuropathy, one of three major criteria must be met, which includes Castleman disease, sclerotic bone lesions, and elevated vascular endothelial growth factor.[8] Additionally, one of six minor criteria must be met: organomegaly, endocrinopathy, edema, skin changes, papilledema, or thrombocytosis/erythrocytosis.[8]

Neuropathy is required for the diagnosis of POEMS syndrome.[154,155,187,188] It is a predominantly motor, demyelinating neuropathy that sometimes begins in the upper extremities and is frequently the presenting symptom of the underlying neoplastic process.[24,154,155,159,188] Progression is gradual and occurs distally to proximally.[155,159,164] Sensory loss occurs to a lesser degree than weakness, with the large fiber modalities of vibration and proprioception usually more impaired than pain and temperature.[155,159] Muscle stretch reflexes are absent or significantly reduced.[155,157,159] EMG/NCV findings are usually consistent with a predominantly demyelinating process, with moderate-to-marked slowing of conduction velocities; CSF protein is usually elevated.[155,159,164,187] The characteristics of this neuropathy are similar to chronic inflammatory demyelinating polyradiculopathy, and patients are often misdiagnosed.[155,159,164] POEMS-associated neuropathy occurs in younger patients than in those with multiple myeloma-associated neuropathy.[154,155,157–159,188] The evaluation of POEMS syndrome requires evaluation for serum and urine monoclonal proteins, where IgG or IgA lambda monoclonal gammopathy makes the diagnosis more likely. Skeletal survey for osteosclerotic lesions followed by bone marrow biopsy is necessary, as two-thirds of patients demonstrate plasma cell disorders with 91% lambda, megakaryocyte hyperplasia, and megakaryocyte clustering.[195] CT of the chest, abdomen, and pelvis might reveal organomegaly and lymphadenopathy. Vascular endothelial growth factor (VEGF) is increased in POEMS and correlates with disease activity.[196]

The approach to treatment depends on the plasma cell involvement of the bone marrow and the number of bone lesions. If the patient has no bone marrow involvement, the bone lesions can be radiated directly with curative intent, often resulting in a better overall survival.[197] If there is involvement of the bone marrow, treatment involves systemic chemotherapy or autologous stem cell transplantation. Lenalidomide, thalidomide, and bortezomib are often used in conjunction with chemotherapy.[198–200]

3.4.3 Waldenström's macroglobulinemia

Waldenström's macroglobulinemia (WM) is a lymphoplasmacytic lymphoma associated with IgM monoclonal gammopathy. Bone marrow biopsy findings demonstrate a lymphoplasmacytic infiltration with an intertrabecular pattern.[201] The clinical manifestations of WM include peripheral neuropathy, hepatosplenomegaly, lymphadenopathy, and HSV (as discussed previously). Another CNS manifestation is Bing-Neel syndrome, which is characterized by direct infiltration of lymphoplasmacytic cells. This is often diagnosed radiographically with an MRI of the brain and supporting CSF studies. While the pathogenesis of Bing-Neel syndrome is unclear, it is thought to be caused by IgM deposits in the parenchyma.[202] Peripheral neuropathy from WM is indistinguishable from neuropathy associated with IgM-MGUS.[202] With an estimated 5-year overall survival rate of about 78%, treatment regimens should weigh benefit and toxicity.[203] For asymptomatic patients, observation is reasonable. Patients with symptoms including anemia, thrombocytopenia, or signs of end-organ damage may be treated with single-agent rituximab. Patients with severe symptoms and/or HSV may require a more aggressive regimen including dexamethasone, rituximab, and cyclophosphamide with autologous stem cell transplantation for relapsing or refractory cases.[204]

4 Conclusion

Multiple myeloma may involve the nervous system through a variety of mechanisms, and many patients with multiple myeloma will have neurological manifestations of their disease. No symptom or sign is unique to one neurological complication. The risk of a catastrophic outcome with delay in treatment, and the excellent response achieved if therapy is initiated promptly that is shared by the many possible neurological complications, strongly encourages careful consideration of all diagnostic possibilities.

References

1. Kumar SK, Rajkumar V, Kyle RA, et al. Multiple myeloma. *Nat Rev Dis Primers*. 2017;3:17046.
2. Kyle RA. Multiple myeloma: review of 869 cases. *Mayo Clin Proc*. 1975;50(1):29–40.
3. Kyle RA, Gertz MA, Witzig TE, et al. Review of 1027 patients with newly diagnosed multiple myeloma. *Mayo Clin Proc*. 2003;78(1):21–33.
4. Kyle RA, Therneau TM, Rajkumar SV, Larson DR, Plevak MF, Melton 3rd LJ. Incidence of multiple myeloma in Olmsted County, Minnesota: trend over 6 decades. *Cancer*. 2004;101(11):2667–2674.
5. Siegel RL, Miller KD, Jemal A. Cancer statistics, 2019. *CA Cancer J Clin*. 2019;69(1):7–34.
6. Huang SY, Yao M, Tang JL, et al. Epidemiology of multiple myeloma in Taiwan: increasing incidence for the past 25 years and higher prevalence of extramedullary myeloma in patients younger than 55 years. *Cancer*. 2007;110(4):896–905.
7. Waxman AJ, Mink PJ, Devesa SS, et al. Racial disparities in incidence and outcome in multiple myeloma: a population-based study. *Blood*. 2010;116(25):5501–5506.
8. Rajkumar SV. Multiple myeloma: 2020 update on diagnosis, risk-stratification and management. *Am J Hematol*. 2020;95(5):548–567.
9. Rajkumar SV, Dimopoulos MA, Palumbo A, et al. International Myeloma Working Group updated criteria for the diagnosis of multiple myeloma. *Lancet Oncol*. 2014;15(12):e538–e548.
10. Silverstein A, Doniger DE. Neurologic complications of myelomatosis. *Arch Neurol*. 1963;9:534–544.
11. Clamp JR. Some aspects of the first recorded case of multiple myeloma. *Lancet*. 1967;2(7530):1354–1356.
12. Brenner B, Carter A, Tatarsky I, Gruszkiewicz J, Peyser E. Incidence, prognostic significance and therapeutic modalities of central nervous system involvement in multiple myeloma. *Acta Haematol*. 1982;68(2):77–83.
13. Clarke E. Spinal cord involvement in multiple myelomatosis. *Brain*. 1956;79(2):332–348.
14. Camacho J, Arnalich F, Anciones B, et al. The spectrum of neurological manifestations in myeloma. *J Med*. 1985;16(5–6):597–611.
15. Woo E, Yu YL, Ng M, Huang CY, Todd D. Spinal cord compression in multiple myeloma: who gets it? *Aust N Z J Med*. 1986;16(5):671–675.
16. Wallington M, Mendis S, Premawardhana U, Sanders P, Shahsavar-Haghighi K. Local control and survival in spinal cord compression from lymphoma and myeloma. *Radiother Oncol*. 1997;42(1):43–47.
17. Schiff D. Spinal cord compression. *Neurol Clin*. 2003;21(1):67–86. viii.
18. Gilbert RW, Kim JH, Posner JB. Epidural spinal cord compression from metastatic tumor: diagnosis and treatment. *Ann Neurol*. 1978;3(1):40–51.
19. Loblaw DA, Laperriere NJ, Mackillop WJ. A population-based study of malignant spinal cord compression in Ontario. *Clin Oncol (R Coll Radiol)*. 2003;15(4):211–217.
20. Mundy GR. Myeloma bone disease. *Eur J Cancer*. 1998;34(2):246–251.
21. Bosch A, Frias Z. Radiotherapy in the treatment of multiple myeloma. *Int J Radiat Oncol Biol Phys*. 1988;15(6):1363–1369.
22. Dahlstrom U, Jarpe S, Lindstrom FD. Paraplegia in myelomatosis—a study of 20 cases. *Acta Med Scand*. 1979;205(3):173–178.
23. Chan L, Snyder HS, Verdile VP. Cervical fracture as the initial presentation of multiple myeloma. *Ann Emerg Med*. 1994;24(6):1192–1194.
24. Davies-Jones GAB, Michael J, Josephson SA. Neurological manifestations of hematological disorders. In: *Aminoff's Neurology and General Medicine*. 5th ed. Amsterdam: Elsevier/AP; 2014.
25. Rahmouni A, Divine M, Mathieu D, et al. Detection of multiple myeloma involving the spine: efficacy of fat-suppression and contrast-enhanced MR imaging. *AJR Am J Roentgenol*. 1993;160(5):1049–1052.
26. Lecouvet FE, Malghem J, Michaux L, et al. Vertebral compression fractures in multiple myeloma. Part II. Assessment of fracture risk with MR imaging of spinal bone marrow. *Radiology*. 1997;204(1):201–205.
27. Leeds NE, Kumar A. Diagnostic imaging. In: Levin VA, ed. *Cancer in the Nervous System*. New York: Churchill Livingstone; 1996:13–49.
28. Pollard JD, Young GA. Neurology and the bone marrow. *J Neurol Neurosurg Psychiatry*. 1997;63(6):706–718.
29. Terpos E, Morgan G, Dimopoulos MA, et al. International Myeloma Working Group recommendations for the treatment of multiple myeloma-related bone disease. *J Clin Oncol*. 2013;31(18):2347–2357.
30. Rades D, Hoskin PJ, Stalpers LJ, et al. Short-course radiotherapy is not optimal for spinal cord compression due to myeloma. *Int J Radiat Oncol Biol Phys*. 2006;64(5):1452–1457.
31. Balducci M, Chiesa S, Manfrida S, et al. Impact of radiotherapy on pain relief and recalcification in plasma cell neoplasms: long-term experience. *Strahlenther Onkol*. 2011;187(2):114–119.
32. Hirsch AE, Jha RM, Yoo AJ, et al. The use of vertebral augmentation and external beam radiation therapy in the multimodal management of malignant vertebral compression fractures. *Pain Physician*. 2011;14(5):447–458.
33. Rades D, Veninga T, Stalpers LJ, et al. Outcome after radiotherapy alone for metastatic spinal cord compression in patients with oligometastases. *J Clin Oncol*. 2007;25(1):50–56.
34. Lecouvet F, Richard F, Vande Berg B, et al. Long-term effects of localized spinal radiation therapy on vertebral fractures and focal lesions appearance in patients with multiple myeloma. *Br J Haematol*. 1997;96(4):743–745.
35. Adamietz IA, Schober C, Schulte RW, Peest D, Renner K. Palliative radiotherapy in plasma cell myeloma. *Radiother Oncol*. 1991;20(2):111–116.
36. Patchell RA, Tibbs PA, Regine WF, et al. Direct decompressive surgical resection in the treatment of spinal cord compression caused by metastatic cancer: a randomised trial. *Lancet*. 2005;366(9486):643–648.
37. Molloy S, Lai M, Pratt G, et al. Optimizing the management of patients with spinal myeloma disease. *Br J Haematol*. 2015;171(3):332–343.
38. Wedin R. Surgical treatment for pathologic fracture. *Acta Orthop Scand Suppl*. 2001;72(302):1–29. 2p.
39. Gokaraju K, Butler JS, Benton A, Suarez-Huerta ML, Selvadurai S, Molloy S. Multiple myeloma presenting with acute bony spinal cord compression and mechanical instability successfully managed nonoperatively. *Spine J*. 2016;16(8):e567–e570.
40. Cawley DT, Butler JS, Benton A, et al. Managing the cervical spine in multiple myeloma patients. *Hematol Oncol*. 2019;37(2):129–135.
41. Kyriakou C, Molloy S, Vrionis F, et al. The role of cement augmentation with percutaneous vertebroplasty and balloon kyphoplasty for the treatment of vertebral compression fractures in multiple myeloma: a consensus statement from the International Myeloma Working Group (IMWG). *Blood Cancer J*. 2019;9(3):27.

42. Patel MS, Ghasem A, Greif DN, Huntley SR, Conway SA, Al Maaieh M. Evaluating treatment strategies for spinal lesions in multiple myeloma: a review of the literature. *Int J Spine Surg.* 2018;12(5):571–581.

43. Yeh HS, Berenson JR. Treatment for myeloma bone disease. *Clin Cancer Res.* 2006;12(20 Pt 2):6279s–6284s.

44. Dudeney S, Lieberman IH, Reinhardt MK, Hussein M. Kyphoplasty in the treatment of osteolytic vertebral compression fractures as a result of multiple myeloma. *J Clin Oncol.* 2002;20(9):2382–2387.

45. Berenson J, Pflugmacher R, Jarzem P, et al. Balloon kyphoplasty versus non-surgical fracture management for treatment of painful vertebral body compression fractures in patients with cancer: a multicentre, randomised controlled trial. *Lancet Oncol.* 2011;12(3):225–235.

46. Berenson JR, Hillner BE, Kyle RA, et al. American Society of Clinical Oncology clinical practice guidelines: the role of bisphosphonates in multiple myeloma. *J Clin Oncol.* 2002;20(17):3719–3736.

47. Rosen LS, Gordon D, Kaminski M, et al. Long-term efficacy and safety of zoledronic acid compared with pamidronate disodium in the treatment of skeletal complications in patients with advanced multiple myeloma or breast carcinoma: a randomized, double-blind, multicenter, comparative trial. *Cancer.* 2003;98(8):1735–1744.

48. Terpos E, Sezer O, Croucher PI, et al. The use of bisphosphonates in multiple myeloma: recommendations of an expert panel on behalf of the European Myeloma Network. *Ann Oncol.* 2009;20(8):1303–1317.

49. Berenson JR, Lichtenstein A, Porter L, et al. Efficacy of pamidronate in reducing skeletal events in patients with advanced multiple myeloma. Myeloma Aredia Study Group. *N Engl J Med.* 1996;334(8):488–493.

50. Rosen LS, Gordon D, Kaminski M, et al. Zoledronic acid versus pamidronate in the treatment of skeletal metastases in patients with breast cancer or osteolytic lesions of multiple myeloma: a phase III, double-blind, comparative trial. *Cancer J.* 2001;7(5):377–387.

51. Raje N, Terpos E, Willenbacher W, et al. Denosumab versus zoledronic acid in bone disease treatment of newly diagnosed multiple myeloma: an international, double-blind, double-dummy, randomised, controlled, phase 3 study. *Lancet Oncol.* 2018;19(3):370–381.

52. Terpos E, Christoulas D, Gavriatopoulou M. Biology and treatment of myeloma related bone disease. *Metabolism.* 2018;80:80–90.

53. Wirk B, Wingard JR, Moreb JS. Extramedullary disease in plasma cell myeloma: the iceberg phenomenon. *Bone Marrow Transplant.* 2013;48(1):10–18.

54. Nieuwenhuizen L, Biesma DH. Central nervous system myelomatosis: review of the literature. *Eur J Haematol.* 2008;80(1):1–9.

55. Tirumani SH, Shinagare AB, Jagannathan JP, Krajewski KM, Munshi NC, Ramaiya NH. MRI features of extramedullary myeloma. *AJR Am J Roentgenol.* 2014;202(4):803–810.

56. Gertz MA. Pomalidomide and myeloma meningitis. *Leuk Lymphoma.* 2013;54(4):681–682.

57. Jurczyszyn A, Grzasko N, Gozzetti A, et al. Central nervous system involvement by multiple myeloma: a multi-institutional retrospective study of 172 patients in daily clinical practice. *Am J Haematol.* 2016;91(6):575–580.

58. Abdallah AO, Atrash S, Shahid Z, et al. Patterns of central nervous system involvement in relapsed and refractory multiple myeloma. *Clin Lymphoma Myeloma Leuk.* 2014;14(3):211–214.

59. Lee D, Kalff A, Low M, et al. Central nervous system multiple myeloma—potential roles for intrathecal therapy and measurement of cerebrospinal fluid light chains. *Br J Haematol.* 2013;162(3):371–375.

60. Fassas AB, Ward S, Muwalla F, et al. Myeloma of the central nervous system: strong association with unfavorable chromosomal abnormalities and other high-risk disease features. *Leuk Lymphoma.* 2004;45(2):291–300.

61. Gangatharan SA, Carney DA, Prince HM, et al. Emergence of central nervous system myeloma in the era of novel agents. *Hematol Oncol.* 2012;30(4):170–174.

62. Paludo J, Painuly U, Kumar S, et al. Myelomatous involvement of the central nervous system. *Clin Lymphoma Myeloma Leuk.* 2016;16(11):644–654.

63. Rasche L, Bernard C, Topp MS, et al. Features of extramedullary myeloma relapse: high proliferation, minimal marrow involvement, adverse cytogenetics: a retrospective single-center study of 24 cases. *Ann Hematol.* 2012;91(7):1031–1037.

64. Varga G, Mikala G, Gopcsa L, et al. Multiple myeloma of the central nervous system: 13 cases and review of the literature. *J Oncol.* 2018;2018:3970169.

65. Varettoni M, Corso A, Pica G, Mangiacavalli S, Pascutto C, Lazzarino M. Incidence, presenting features and outcome of extramedullary disease in multiple myeloma: a longitudinal study on 1003 consecutive patients. *Ann Oncol.* 2010;21(2):325–330.

66. Weberpals J, Pulte D, Jansen L, et al. Survival of patients with lymphoplasmacytic lymphoma and solitary plasmacytoma in Germany and the United States of America in the early 21(st) century. *Haematologica.* 2017;102(6):e229–e232.

67. Yang W, Zheng J, Li R, et al. Multiple myeloma with pathologically proven skull plasmacytoma after a mild head injury: case report. *Medicine (Baltimore).* 2018;97(39):e12327.

68. D'Arena G, Pietrantuono G, Mansueto G, Villani O, Zandolino A, Cammarota A. Parietal skull extramedullary relapse in multiple myeloma. *Postgrad Med J.* 2020;96(1136):360.

69. Chertok Shacham E, Brikman S, Chap Marshak D, Denisov V, Dori G. A rare case of IgM multiple myeloma with a skull neoplasm. *Isr Med Assoc J.* 2019;21(9):632–633.

70. Bitelman VM, Lopes JA, Nogueira AB, Frassetto FP, Duarte-Neto AN. "Punched out" multiple myeloma lytic lesions in the skull. *Autops Case Rep.* 2016;6(1):7–9.

71. Sham RL, Phatak PD, Kouides PA, Janas JA, Marder VJ. Hematologic neoplasia and the central nervous system. *Am J Hematol.* 1999;62(4):234–238.

72. Patriarca F, Zaja F, Silvestri F, et al. Meningeal and cerebral involvement in multiple myeloma patients. *Ann Hematol.* 2001;80(12):758–762.

73. Petersen SL, Wagner A, Gimsing P. Cerebral and meningeal multiple myeloma after autologous stem cell transplantation. A case report and review of the literature. *Am J Hematol.* 1999;62(4):228–233.

74. Cavanna L, Invernizzi R, Berte R, Vallisa D, Buscarini L. Meningeal involvement in multiple myeloma: report of a case with cytologic and immunocytochemical diagnosis. *Acta Cytol.* 1996;40(3):571–575.

75. Truong LD, Kim HS, Estrada R. Meningeal myeloma. *Am J Clin Pathol.* 1982;78(4):532–535.

76. Leifer D, Grabowski T, Simonian N, Demirjian ZN. Leptomeningeal myelomatosis presenting with mental status changes and other neurologic findings. *Cancer.* 1992;70(7):1899–1904.

77. Schluterman KO, Fassas AB, Van Hemert RL, Harik SI. Multiple myeloma invasion of the central nervous system. *Arch Neurol.* 2004;61(9):1423–1429.

78. Qu XY, Fu WJ, Xi H, Zhou F, Wei W, Hou J. Clinical features of multiple myeloma invasion of the central nervous system in Chinese patients. *Chin Med J (Engl).* 2010;123(11):1402–1406.

79. Marjanovic S, Mijuskovic Z, Stamatovic D, et al. Multiple myeloma invasion of the central nervous system. *Vojnosanit Pregl.* 2012;69(2):209–213.

80. Hogan MC, Lee A, Solberg LA, Thome SD. Unusual presentation of multiple myeloma with unilateral visual loss and numb chin syndrome in a young adult. *Am J Hematol.* 2002;70(1):55–59.

81. Spiers AS, Halpern R, Ross SC, Neiman RS, Harawi S, Zipoli TE. Meningeal myelomatosis. *Arch Intern Med.* 1980;140(2):256–259.

82. Sekhri A, Khattar P, Islam H, Liu D. Multiple myeloma with leptomeningeal involvement and positive CSF. *Stem Cell Investig.* 2014;1:21.

83. Mourad AR, Kharfan-Dabaja MA, Benson K, Moscinski LC, Baz RC. Leptomeningeal myeloma as the sole manifestation of relapse: an unusual presentation. *Am J Med Sci.* 2010;339(1):81–82.

84. Bommer M, Kull M, Teleanu V, et al. Leptomeningeal myelomatosis: a rare but devastating manifestation of multiple myeloma diagnosed using cytology, flow cytometry, and fluorescent in situ hybridization. *Acta Haematol.* 2018;139(4):247–254.

85. Yellu MR, Engel JM, Ghose A, Onitilo AA. Overview of recent trends in diagnosis and management of leptomeningeal multiple myeloma. *Hematol Oncol.* 2016;34(1):2–8.

86. Ren H, Zou Y, Zhao Y, et al. Cerebrospinal fluid cytological diagnosis in multiple myeloma with leptomeningeal involvement: a report of two cases. *Diagn Cytopathol.* 2017;45(1):66–68.

87. Pak N, Shakki Katouli F, Radmard AR, Shaki Katuli MH, Rezwanifar MM, Sepehri Boroujeni N. Multiple cranial nerve palsy concomitant with leptomeningeal involvement in multiple myeloma: a case report and review of literature. *Int J Hematol Oncol Stem Cell Res.* 2018;12(1):8–13.

88. Drappatz J, Batchelor T. Neurologic complications of plasma cell disorders. *Clin Lymphoma.* 2004;5(3):163–171.

89. Knapp AJ, Gartner S, Henkind P. Multiple myeloma and its ocular manifestations. *Surv Ophthalmol.* 1987;31(5):343–351.

90. Chin KJ, Kempin S, Milman T, Finger PT. Ocular manifestations of multiple myeloma: three cases and a review of the literature. *Optometry.* 2011;82(4):224–230.

91. Bellan LD, Cox TA, Gascoyne RD. Parasellar syndrome caused by plasma cell leukemia. *Can J Ophthalmol.* 1989;24(7):331–334.

92. Kusano Y, Terui Y, Nishimura N, Yokoyama M, Ueda K, Hatake K. Myelomatous meningitis: a case report. *Int J Hematol.* 2016;104(2):149–150.

93. Brum M, Antonio AS, Guerreiro R. Myelomatous meningitis: a rare neurological involvement in complete remission of multiple myeloma. *J Neurol Sci.* 2014;340(1–2):241–242.

94. Dennis M, Chu P. A case of meningeal myeloma presenting as obstructive hydrocephalus—a therapeutic challenge. *Leuk Lymphoma.* 2000;40(1–2):219–220.

95. Damaj G, Mohty M, Vey N, et al. Features of extramedullary and extraosseous multiple myeloma: a report of 19 patients from a single center. *Eur J Haematol.* 2004;73(6):402–406.

96. Chen CI, Masih-Khan E, Jiang H, et al. Central nervous system involvement with multiple myeloma: long term survival can be achieved with radiation, intrathecal chemotherapy, and immunomodulatory agents. *Br J Haematol.* 2013;162(4):483–488.

97. Egan PA, Elder PT, Deighan WI, O'Connor SJM, Alexander HD. Multiple myeloma with central nervous system relapse. *Haematologica.* 2020;105(7):1780–1790.

98. Chang H, Bartlett ES, Patterson B, Chen CI, Yi QL. The absence of CD56 on malignant plasma cells in the cerebrospinal fluid is the hallmark of multiple myeloma involving central nervous system. *Br J Haematol.* 2005;129(4):539–541.

99. Chang H, Sloan S, Li D, Keith Stewart A. Multiple myeloma involving central nervous system: high frequency of chromosome 17p13.1 (p53) deletions. *Br J Haematol.* 2004;127(3):280–284.

100. Montalban C, Martin-Aresti J, Patier JL, Millan JM, Cosio MG. Unusual cases in multiple myeloma and a dramatic response in metastatic lung cancer: case 3. Intracranial plasmacytoma with cranial nerve neuropathy in multiple myeloma. *J Clin Oncol.* 2005;23(1):233–235.

101. Moran CC, Anderson CC, Caldemeyer KS, Smith RR. Meningeal myelomatosis: CT and MR appearances. *AJNR Am J Neuroradiol.* 1995;16(7):1501–1503.

102. van Ginkel S, Snijders TJ, van de Donk NW, Klijn CJ, Broekman ML. Progressive neurological deficits in multiple myeloma: meningeal myelomatosis without MRI abnormalities. *J Neurol.* 2012;259(6):1231–1233.

103. Silva N, Delamain M, Duarte G, Reis F. Meningeal myelomatosis illustrated on FLAIR post-contrasted images. *Can J Neurol Sci.* 2019;46(4):477–479.

104. Gascon N, Perez-Montero H, Guardado S, D'Ambrosi R, Cabeza MA, Perez-Regadera JF. Dural plasmacytoma with meningeal myelomatosis in a patient with multiple myeloma. *Case Rep Hematol.* 2018;2018:6730567.

105. de la Fuente J, Prieto I, Albo C, Sopena B, Somolinos N, Martinez C. Plasma cell myeloma presented as myelomatous meningitis. *Eur J Haematol.* 1994;53(4):244–245.

106. Majd N, Wei X, Demopoulos A, Hormigo A, Chari A. Characterization of central nervous system multiple myeloma in the era of novel therapies. *Leuk Lymphoma.* 2016;57(7):1709–1713.

107. Marini A, Carulli G, Lari T, et al. Myelomatous meningitis evaluated by multiparameter flow cytometry: report of a case and review of the literature. *J Clin Exp Hematop.* 2014;54(2):129–136.

108. Peter A. The plasma cells of the cerebrospinal fluid. *J Neurol Sci.* 1967;4(2):227–239.

109. Sarid N, Katz BZ. Dividing plasma cells in the cerebrospinal fluid of a patient with refractory multiple myeloma. *Blood.* 2015;126(18):2162.

110. Anderson KC. Lenalidomide and thalidomide: mechanisms of action—similarities and differences. *Semin Hematol.* 2005;42(4 Suppl. 4):S3–S8.

111. Nahi H, Svedmyr E, Lerner R. Bendamustine in combination with high-dose radiotherapy and thalidomide is effective in treatment of multiple myeloma with central nervous system involvement. *Eur J Haematol.* 2014;92(5):454–455.

112. Katodritou E, Terpos E, Kastritis E, et al. Lack of survival improvement with novel anti-myeloma agents for patients with multiple myeloma and central nervous system involvement: the Greek Myeloma Study Group experience. *Ann Hematol.* 2015;94(12):2033–2042.

113. Vicari P, Ribas C, Sampaio M, et al. Can thalidomide be effective to treat plasma cell leptomeningeal infiltration? *Eur J Haematol.* 2003;70(3):198–199.

114. Mussetti A, Dalto S, Montefusco V. Effective treatment of pomalidomide in central nervous system myelomatosis. *Leuk Lymphoma.* 2013;54(4):864–866.

115. Badros A, Singh Z, Dhakal B, et al. Marizomib for central nervous system-multiple myeloma. *Br J Haematol.* 2017;177(2):221–225.

116. Gozzetti A, Cerase A, Lotti F, et al. Extramedullary intracranial localization of multiple myeloma and treatment with novel agents: a retrospective survey of 50 patients. *Cancer.* 2012;118(6):1574–1584.

117. Harrison SJ, Spencer A, Quach H. Myeloma of the central nervous system—an ongoing conundrum! *Leuk Lymphoma.* 2016;57(7):1505–1506.

118. Chang WJ, Kim SJ, Kim K. Central nervous system multiple myeloma: a different cytogenetic profile? *Br J Haematol.* 2014;164(5):745–748.

119. Kauffmann G, Buerki RA, Lukas RV, Gondi V, Chmura SJ. Case report of bone marrow-sparing proton therapy craniospinal irradiation for central nervous system myelomatosis. *Cureus.* 2017;9(11):e1885.

120. Riley JM, Russo JK, Shipp A, Alsharif M, Jenrette JM. Central nervous system myelomatosis with optic neuropathy and intramedullary spinal cord compression responding to radiation therapy. *Jpn J Radiol.* 2011;29(7):513–516.

III. Neurological complications of specific neoplasms

121. Tsang RW, Campbell BA, Goda JS, et al. Radiation therapy for solitary plasmacytoma and multiple myeloma: guidelines from the International Lymphoma Radiation Oncology Group. *Int J Radiat Oncol Biol Phys.* 2018;101(4):794–808.

122. Wu P, Davies FE, Boyd K, et al. The impact of extramedullary disease at presentation on the outcome of myeloma. *Leuk Lymphoma.* 2009;50(2):230–235.

123. Lee SE, Kim JH, Jeon YW, et al. Impact of extramedullary plasmacytomas on outcomes according to treatment approach in newly diagnosed symptomatic multiple myeloma. *Ann Hematol.* 2015;94(3):445–452.

124. Fahey JL, Barth WF, Solomon A. Serum hyperviscosity syndrome. *JAMA.* 1965;192:464–467.

125. Mehta J, Singhal S. Hyperviscosity syndrome in plasma cell dyscrasias. *Semin Thromb Hemost.* 2003;29(5):467–471.

126. Khan P, Roth MS, Keren DF, Foon KA. Light chain disease associated with the hyperviscosity syndrome. *Cancer.* 1987;60(9):2267–2268.

127. Preston FE, Cooke KB, Foster ME, Winfield DA, Lee D. Myelomatosis and the hyperviscosity syndrome. *Br J Haematol.* 1978;38(4):517–530.

128. Pruzanski W, Watt JG. Serum viscosity and hyperviscosity syndrome in IgG multiple myeloma. Report on 10 patients and a review of the literature. *Ann Intern Med.* 1972;77(6):853–860.

129. Alkner U, Hansson UB, Lindstrom FD. Factors affecting IgA related hyperviscosity. *Clin Exp Immunol.* 1983;51(3):617–623.

130. Crawford J, Cox EB, Cohen HJ. Evaluation of hyperviscosity in monoclonal gammopathies. *Am J Med.* 1985;79(1):13–22.

131. Gertz MA. Acute hyperviscosity: syndromes and management. *Blood.* 2018;132(13):1379–1385.

132. Kwaan HC, Bongu A. The hyperviscosity syndromes. *Semin Thromb Hemost.* 1999;25(2):199–208.

133. Park MS, Kim BC, Kim IK, et al. Cerebral infarction in IgG multiple myeloma with hyperviscosity. *J Korean Med Sci.* 2005;20(4):699–701.

134. Mueller J, Hotson JR, Langston JW. Hyperviscosity-induced dementia. *Neurology.* 1983;33(1):101–103.

135. Ovadia S, Lysyy L, Floru S. Emergency plasmapheresis for unstable angina in a patient with hyperviscosity syndrome. *Am J Emerg Med.* 2005;23(6):811–812.

136. Martinez-Maldonado M, Yium J, Suki WN, Eknoyan G. Renal complications in multiple myeloma: pathophysiology and some aspects of clinical management. *J Chronic Dis.* 1971;24(4):221–227.

137. Fotiou D, Dimopoulos MA, Kastritis E. Managing renal complications in multiple myeloma. *Expert Rev Hematol.* 2016;9(9):839–850.

138. Faiman BM, Mangan P, Spong J, Tariman JD, The International Myeloma Foundation Nurse Leadership Board. Renal complications in multiple myeloma and related disorders: survivorship care plan of the International Myeloma Foundation Nurse Leadership Board. *Clin J Oncol Nurs.* 2011;15(Suppl):66–76.

139. Niesvizky R, Badros AZ. Complications of multiple myeloma therapy, part 2: risk reduction and management of venous thromboembolism, osteonecrosis of the jaw, renal complications, and anemia. *J Natl Compr Cancer Netw.* 2010;8(Suppl. 1):S13–S20.

140. Barbano RL. Structure and function of the kidneys. In: Goldman L, Schafer AI, eds. *Cecil-Goldman Medicine.* 26th ed. Philadelphia: Elsevier; 2019.

141. Avigan D, Rosenblatt J. Current treatment for multiple myeloma. *N Engl J Med.* 2014;371(10):961–962.

142. Mikhael JR, Dingli D, Roy V, et al. Management of newly diagnosed symptomatic multiple myeloma: updated Mayo Stratification of Myeloma and Risk-Adapted Therapy (mSMART) consensus guidelines 2013. *Mayo Clin Proc.* 2013;88(4):360–376.

143. Tejani N, Cooper A, Rezo A, Pranavan G, Yip D. Numb chin syndrome: a case series of a clinical syndrome associated with malignancy. *J Med Imaging Radiat Oncol.* 2014;58(6):700–705.

144. Laurencet FM, Anchisi S, Tullen E, Dietrich PY. Mental neuropathy: report of five cases and review of the literature. *Crit Rev Oncol Hematol.* 2000;34(1):71–79.

145. Bruyn RP, Boogerd W. The numb chin. *Clin Neurol Neurosurg.* 1991;93(3):187–193.

146. Burt RK, Sharfman WH, Karp BI, Wilson WH. Mental neuropathy (numb chin syndrome). A harbinger of tumor progression or relapse. *Cancer.* 1992;70(4):877–881.

147. Lossos A, Siegal T. Numb chin syndrome in cancer patients: etiology, response to treatment, and prognostic significance. *Neurology.* 1992;42(6):1181–1184.

148. Sweet JM. The numb chin syndrome: a critical sign for primary care physicians. *Arch Intern Med.* 2004;164(12):1347–1348.

149. Miera C, Benito-Leon J, dela Fuente M, de la Serna J. Numb chin syndrome heralding myeloma relapse. *Muscle Nerve.* 1997;20(12):1603–1606.

150. Witt C, Borges AC, Klein K, Neumann HJ. Radiographic manifestations of multiple myeloma in the mandible: a retrospective study of 77 patients. *J Oral Maxillofac Surg.* 1997;55(5):450–453. discussion 454–455.

151. Furutani M, Ohnishi M, Tanaka Y. Mandibular involvement in patients with multiple myeloma. *J Oral Maxillofac Surg.* 1994;52(1):23–25.

152. Faiman B, Doss D, Colson K, Mangan P, King T, Tariman JD. Renal, GI, and peripheral nerves: evidence-based recommendations for the management of symptoms and care for patients with multiple myeloma. *Clin J Oncol Nurs.* 2017;21(5 Suppl):19–36.

153. Denier C, Lozeron P, Adams D, et al. Multifocal neuropathy due to plasma cell infiltration of peripheral nerves in multiple myeloma. *Neurology.* 2006;66(6):917–918.

154. Meier C. Polyneuropathy in paraproteinaemia. *J Neurol.* 1985;232(4):204–214.

155. Kelly Jr JJ, Kyle RA, Miles JM, O'Brien PC, Dyck PJ. The spectrum of peripheral neuropathy in myeloma. *Neurology.* 1981;31(1):24–31.

156. Walsh JC. The neuropathy of multiple myeloma. An electrophysiological and histological study. *Arch Neurol.* 1971;25(5):404–414.

157. Driedger H, Pruzanski W. Plasma cell neoplasia with peripheral polyneuropathy. A study of five cases and a review of the literature. *Medicine (Baltimore).* 1980;59(4):301–310.

158. Davis LE, Drachman DB. Myeloma neuropathy. Successful treatment of two patients and review of cases. *Arch Neurol.* 1972;27(6):507–511.

159. Bosch EP, Smith BE. Peripheral neuropathies associated with monoclonal proteins. *Med Clin North Am.* 1993;77(1):125–139.

160. Kelly Jr JJ. The electrodiagnostic findings in peripheral neuropathy associated with monoclonal gammopathy. *Muscle Nerve.* 1983;6(7):504–509.

161. Ohi T, Kyle RA, Dyck PJ. Axonal attenuation and secondary segmental demyelination in myeloma neuropathies. *Ann Neurol.* 1985;17(3):255–261.

162. Victor M, Banker BQ, Adams RD. The neuropathy of multiple myeloma. *J Neurol Neurosurg Psychiatry.* 1958;21(2):73–88.

163. Barron KD, Rowland LP, Zimmerman HM. Neuropathy with malignant tumor metastases. *J Nerv Ment Dis.* 1960;131:10–31.

164. Kelly Jr JJ. Peripheral neuropathies associated with monoclonal proteins: a clinical review. *Muscle Nerve.* 1985;8(2):138–150.

165. Vrethem M, Cruz M, Wen-Xin H, Malm C, Holmgren H, Ernerudh J. Clinical, neurophysiological and immunological evidence of polyneuropathy in patients with monoclonal gammopathies. *J Neurol Sci.* 1993;114(2):193–199.

166. Wicklund MP, Kissel JT. Paraproteinemic neuropathy. *Curr Treat Options Neurol.* 2001;3(2):147–156.

167. Moreau P, Richardson PG, Cavo M, et al. Proteasome inhibitors in multiple myeloma: 10 years later. *Blood.* 2012;120(5):947–959.

168. Staff NP, Podratz JL, Grassner L, et al. Bortezomib alters microtubule polymerization and axonal transport in rat dorsal root ganglion neurons. *Neurotoxicology.* 2013;39:124–131.

169. Richardson PG, Sonneveld P, Schuster MW, et al. Bortezomib or high-dose dexamethasone for relapsed multiple myeloma. *N Engl J Med.* 2005;352(24):2487–2498.

170. Richardson PG, Barlogie B, Berenson J, et al. A phase 2 study of bortezomib in relapsed, refractory myeloma. *N Engl J Med.* 2003;348(26):2609–2617.

171. Richardson PG, Xie W, Mitsiades C, et al. Single-agent bortezomib in previously untreated multiple myeloma: efficacy, characterization of peripheral neuropathy, and molecular correlations with response and neuropathy. *J Clin Oncol.* 2009;27(21):3518–3525.

172. San Miguel JF, Schlag R, Khuageva NK, et al. Bortezomib plus melphalan and prednisone for initial treatment of multiple myeloma. *N Engl J Med.* 2008;359(9):906–917.

173. Bringhen S, Larocca A, Rossi D, et al. Efficacy and safety of once-weekly bortezomib in multiple myeloma patients. *Blood.* 2010;116(23):4745–4753.

174. Moreau P, Pylypenko H, Grosicki S, et al. Subcutaneous versus intravenous administration of bortezomib in patients with relapsed multiple myeloma: a randomised, phase 3, non-inferiority study. *Lancet Oncol.* 2011;12(5):431–440.

175. Mauermann ML, Blumenreich MS, Dispenzieri A, Staff NP. A case of peripheral nerve microvasculitis associated with multiple myeloma and bortezomib treatment. *Muscle Nerve.* 2012;46(6):970–977.

176. Ravaglia S, Corso A, Piccolo G, et al. Immune-mediated neuropathies in myeloma patients treated with bortezomib. *Clin Neurophysiol.* 2008;119(11):2507–2512.

177. Richardson PG, Delforge M, Beksac M, et al. Management of treatment-emergent peripheral neuropathy in multiple myeloma. *Leukemia.* 2012;26(4):595–608.

178. Chen D, Frezza M, Schmitt S, Kanwar J, Dou QP. Bortezomib as the first proteasome inhibitor anticancer drug: current status and future perspectives. *Curr Cancer Drug Targets.* 2011;11(3):239–253.

179. Singhal S, Mehta J, Desikan R, et al. Antitumor activity of thalidomide in refractory multiple myeloma. *N Engl J Med.* 1999;341(21):1565–1571.

180. Giannini F, Volpi N, Rossi S, Passero S, Fimiani M, Cerase A. Thalidomide-induced neuropathy: a ganglionopathy? *Neurology.* 2003;60(5):877–878.

181. Mileshkin L, Stark R, Day B, Seymour JF, Zeldis JB, Prince HM. Development of neuropathy in patients with myeloma treated with thalidomide: patterns of occurrence and the role of electrophysiologic monitoring. *J Clin Oncol.* 2006;24(27):4507–4514.

182. Verghese JP, Bradley WG, Nemni R, McAdam KP. Amyloid neuropathy in multiple myeloma and other plasma cell dyscrasias. A hypothesis of the pathogenesis of amyloid neuropathies. *J Neurol Sci.* 1983;59(2):237–246.

183. Plante-Bordeneuve V, Said G. Familial amyloid polyneuropathy. *Lancet Neurol.* 2011;10(12):1086–1097.

184. Solomon A, Weiss DT, Murphy C. Primary amyloidosis associated with a novel heavy-chain fragment (AH amyloidosis). *Am J Hematol.* 1994;45(2):171–176.

185. Kyle RA, Gertz MA. Primary systemic amyloidosis: clinical and laboratory features in 474 cases. *Semin Hematol.* 1995;32(1):45–59.

186. Wang AK, Fealey RD, Gehrking TL, Low PA. Patterns of neuropathy and autonomic failure in patients with amyloidosis. *Mayo Clin Proc.* 2008;83(11):1226–1230.

187. Kyle RA. Monoclonal proteins in neuropathy. *Neurol Clin.* 1992;10(3):713–734.

188. Rajkumar SV, Dispenzieri A, Kyle RA. Monoclonal gammopathy of undetermined significance, Waldenstrom macroglobulinemia, AL amyloidosis, and related plasma cell disorders: diagnosis and treatment. *Mayo Clin Proc.* 2006;81(5):693–703.

189. Ladha SS, Dyck PJ, Spinner RJ, et al. Isolated amyloidosis presenting with lumbosacral radiculoplexopathy: description of two cases and pathogenic review. *J Peripher Nerv Syst.* 2006;11(4):346–352.

190. Sadek I, Mauermann ML, Hayman SR, Spinner RJ, Gertz MA. Primary systemic amyloidosis presenting with asymmetric multiple mononeuropathies. *J Clin Oncol.* 2010;28(25):e429–e432.

191. Gertz MA. Immunoglobulin light chain amyloidosis: 2020 update on diagnosis, prognosis, and treatment. *Am J Hematol.* 2020;95(7):848–860.

192. Sanchorawala V, Sun F, Quillen K, Sloan JM, Berk JL, Seldin DC. Long-term outcome of patients with AL amyloidosis treated with high-dose melphalan and stem cell transplantation: 20-year experience. *Blood.* 2015;126(20):2345–2347.

193. Cordes S, Dispenzieri A, Lacy MQ, et al. Ten-year survival after autologous stem cell transplantation for immunoglobulin light chain amyloidosis. *Cancer.* 2012;118(24):6105–6109.

194. Bardwick PA, Zvaifler NJ, Gill GN, Newman D, Greenway GD, Resnick DL. Plasma cell dyscrasia with polyneuropathy, organomegaly, endocrinopathy, M protein, and skin changes: the POEMS syndrome. Report on two cases and a review of the literature. *Medicine (Baltimore).* 1980;59(4):311–322.

195. Dao LN, Hanson CA, Dispenzieri A, Morice WG, Kurtin PJ, Hoyer JD. Bone marrow histopathology in POEMS syndrome: a distinctive combination of plasma cell, lymphoid, and myeloid findings in 87 patients. *Blood.* 2011;117(24):6438–6444.

196. D'Souza A, Hayman SR, Buadi F, et al. The utility of plasma vascular endothelial growth factor levels in the diagnosis and follow-up of patients with POEMS syndrome. *Blood.* 2011;118(17):4663–4665.

197. Dispenzieri A, Kyle RA, Lacy MQ, et al. POEMS syndrome: definitions and long-term outcome. *Blood.* 2003;101(7):2496–2506.

198. D'Souza A, Lacy M, Gertz M, et al. Long-term outcomes after autologous stem cell transplantation for patients with POEMS syndrome (osteosclerotic myeloma): a single-center experience. *Blood.* 2012;120(1):56–62.

199. Humeniuk MS, Gertz MA, Lacy MQ, et al. Outcomes of patients with POEMS syndrome treated initially with radiation. *Blood.* 2013;122(1):68–73.

200. Karam C, Klein CJ, Dispenzieri A, et al. Polyneuropathy improvement following autologous stem cell transplantation for POEMS syndrome. *Neurology.* 2015;84(19):1981–1987.

201. Owen RG, Treon SP, Al-Katib A, et al. Clinicopathological definition of Waldenstrom's macroglobulinemia: consensus panel recommendations from the Second International Workshop on Waldenstrom's Macroglobulinemia. *Semin Oncol.* 2003;30(2):110–115.

202. Fintelmann F, Forghani R, Schaefer PW, Hochberg EP, Hochberg FH. Bing-Neel Syndrome revisited. *Clin Lymphoma Myeloma.* 2009;9(1):104–106.

203. Gertz MA. Waldenstrom macroglobulinemia: 2015 update on diagnosis, risk stratification, and management. *Am J Hematol.* 2015;90(4):346–354.

204. Ansell SM, Kyle RA, Reeder CB, et al. Diagnosis and management of Waldenstrom macroglobulinemia: Mayo stratification of macroglobulinemia and risk-adapted therapy (mSMART) guidelines. *Mayo Clin Proc.* 2010;85(9):824–833.

III. Neurological complications of specific neoplasms

26

Neurologic complications in the treatment of childhood malignancies

Angela Liou[a], Cassie Kline[a], and Sabine Mueller[b,c]

[a]Children's Hospital of Philadelphia, Philadelphia, PA, United States, [b]University of California, San Francisco, San Francisco, CA, United States, [c]University Children's Hospital Zurich, Zurich, Switzerland

1 Introduction

Childhood malignancy imposes a significant burden on the lives of children and their families. Pediatric cancer is the second leading cause of death in children between 1 and 14 years, with an estimated 11,050 children under the age of 15 to be diagnosed with cancer in the United States in 2020.[1] According to Surveillance, Epidemiology and End Results (SEER) data from 2017, the most common malignancy is acute leukemia, accounting for 29% of all pediatric cancers, followed by brain tumors (26%), lymphoma (predominantly non-Hodgkin lymphoma; 12%), soft tissue sarcoma (predominantly rhabdomyosarcoma; 6%), neuroblastoma (6%), and Wilms tumors (5%).[2,3] Remarkably, this past decade has heralded major advances in medicine and scientific research that have catalyzed the development of curative therapies. Historical survival rates, once dismal with a 5-year overall survival (OS) rate of only 58%–62% in the 1970s, are now in sharp contrast to the current survival rates of > 80%–85% for many pediatric cancers. However, as survivorship increases, the acute and late effects of cancer and its treatment are becoming ever apparent in children.

Neurologic complications of childhood cancers are unfortunately common. The central nervous system (CNS) is uniquely susceptible to the direct and indirect impacts of malignancy and its treatments. Direct effects of cancer include tumor mass effect with potentially serious cases of herniation, seizures, intracranial hemorrhage, hydrocephalus, and focal neurologic injury. These effects can also be seen in both primary CNS tumors and brain metastases from solid malignancies, such as neuroblastoma, bony or soft tissue sarcomas, as well as leukemic infiltration into the CNS. Indirect neurologic complications, on the other hand,

are often a consequence of paraneoplastic syndromes, tumor-induced coagulopathies, and cancer-directed therapies.[4] Specifically, long-term neurotoxicity is a well-recognized complication in cancer treatment. This is especially true in the pediatric population due to the unique vulnerabilities of the developing brain. Owing to an immature nervous system, differences in tumor types, and responses to therapies, the neurologic complications in children are often different from those in the adult population. Furthermore, with the advent of genomic technologies expanding our understanding of molecular drivers of disease, the landscape of childhood cancer treatment is rapidly evolving and has given rise to distinct treatment-specific neurotoxicities.

In this chapter, we will explore the specific neurologic complications seen in conventional cancer treatment, as well as those in novel cancer therapies (Fig. 1). This chapter is organized into two major sections: acute and late effects common to each treatment strategy (Table 1). As more children with cancer are surviving into adulthood, it becomes increasingly important to highlight both the acute and chronic consequences of therapy in order to increase recognition and improve the care of childhood cancer survivors.

2 Part I: Acute complications of cancer treatment

It is well known that many treatment modalities in pediatric cancer are associated with significant acute complications that can exact a substantial toll, particularly on the neurologic development of children. These acute neurologic morbidities are often debilitating and require early recognition and acute management. In the literature, there exists a vast body of knowledge on this

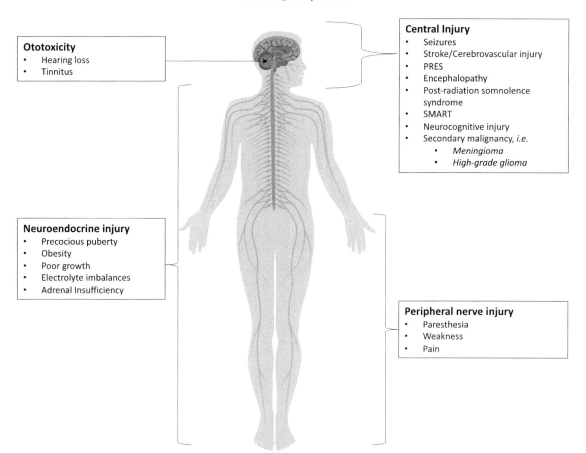

FIG. 1   Potential therapy-related toxicities delineated according to the affected region of the central nervous system.

subject. Recently, there have been several outstanding review papers published on the acute neurologic toxicities of cancer treatment.[4–8] Herein, we aim to provide a comprehensive compilation of what has been published and the perspectives well established in the pediatric oncology community.

2.1 Systemic chemotherapy

2.1.1 Seizures

Seizure is a widely known complication in children receiving systemic chemotherapy. It is a manifestation of exaggerated, synchronous discharges in the cerebral cortex that can cause an altered state of consciousness. Seizure-like activity often varies widely from transient subclinical events to generalized tonic-clonic episodes that, when prolonged, can evolve into status epilepticus. A number of chemotherapy agents are known culprits. Among the most well known are methotrexate (MTX), platinum-based compounds, and vincristine (VCR). Regimens used in bone marrow transplant, such as busulfan and cyclosporine, can also provoke seizures.

MTX neurotoxicity is commonly observed and is one of the most well-implicated agents associated with seizures, both when given systemically and intrathecally

(IT). MTX is a folate antagonist that prevents de novo purine synthesis and DNA repair. How MTX induces seizure formation is not clear, but it has been linked to an aberrant production of glutamate, damaging glutamatergic synapses and inducing neuronal cell death.[9] For childhood cancers, MTX is commonly incorporated into the backbone for acute lymphoblastic leukemia (ALL), lymphoma, and solid tumor regimens. MTX neurotoxicities have been observed across a range of doses whether at low-dose systemic therapy in ALL therapies or at high-dose IV administration in solid malignancies,[10] and the risks may be even higher following the repeat IT delivery.[11] Seizures are most commonly associated with intense IV and IT regimens. In a 2014 study by Bhojwani et al. investigating MTX-induced neurotoxicity, the authors found a strong correlation between MTX exposure and risk of seizures in a cohort of 369 children with ALL diagnosed between 2000 and 2007 at St. Jude Children's Research Hospital.[10] Patients received upfront treatment consisting of high-dose MTX 1 g/m^2 during induction, followed by 2.5 g/m^2 or 5 g/m^2 MTX based on either low risk vs standard/high risk assignment during consolidation. Leucovorin rescue was administered during all treatment periods (leucovorin is an active folate analog that can selectively replete folic acid stores in normal

TABLE 1   Specific cancer-directed therapies with impacts on neurologic domains and toxicities associated with each affected domain.

Type of therapy	Neurologic domain affected	Toxicity
Chemotherapy		
Intrathecal e.g., Methotrexate Cytarabine	Neurocognition	Executive function Attention Processing speed Memory
	Nerves	Peripheral neuropathy
Systemic e.g., Methotrexate Cyclophosphamide Ifosfamide Busulfan Thiotepa Asparaginase	Consciousness	Seizures Encephalopathy Fatigue
	Vascular	Stroke Moyamoya syndrome Thrombosis/hemorrhage PRES
	Hearing	Tinnitus Hearing loss
	Infection	Viral/bacterial infection
Immune-based therapy		
T-cell-based therapy e.g., CAR-T TCR	Pain	Headaches Edema
	Nerves	Autoimmune neuropathies
Checkpoint inhibition e.g., PD-1 inhibitors IDO-1 inhibitors CTLA-4 inhibitors	Consciousness	Confusion Encephalopathy Somnolence Fatigue Disorientation
Vaccines	Speech	Aphasia
	Vision	Vision deficits
Oncolytic viruses	Motor	Tremor Apraxia
Radiation		
	Neurocognition	Executive function Attention Processing speed Memory
	Consciousness	Seizures Encephalopathy Somnolence Fatigue
	Vascular	Stroke Moyamoya syndrome Arteriopathy
	Vision	Vision deficits
	Hearing	Tinnitus Hearing deficits
	Neuroendocrine	Specific hormone deficiencies—growth hormone, thyroid, gonadotropins, adrenocorticotropic decreased linear growth Obesity
	Cancer	Secondary malignancies

III. Neurological complications of specific neoplasms

cells, thereby abrogating MTX toxicities). Additionally, patients received IT MTX in the form of triple therapy consisting of MTX, hydrocortisone, and cytarabine. During maintenance, patients received low-dose intravenous MTX (40 mg/m^2). MTX-associated neurologic toxicities developed in 14/369 patients (3.8%), of which 7 patients presented with seizures (1.9%). Clinical manifestations ranged from complex partial to generalized tonic-clonic events. Most events were brief, achieving complete resolution in less than 24 h. A majority of the patients developed their first seizure during consolidation and within 1 week of IV/IT MTX. After resolution, most patients were rechallenged with subsequent doses of MTX without further complications.

In another well-cited historic study by Mahoney et al., the authors examined the incidence of acute neurologic complications in >1000 children with B-precursor ALL who received MTX during a 6-month intensification regimen.[12] A total of 864 pediatric patients received an intermediate dose IV MTX regimen of 1000 mg/m^2 vs 345 patients who received a low-dose oral MTX regimen of 30 mg/m^2. IT MTX was also provided for CNS prophylaxis. The authors found that 9.5% of the patients developed neurotoxicities in the intermediate-dose arm vs 2.8% in the low-dose arm. Seizure events were the most prevalent complications and presented within a median of 10–11 days after IV or IT MTX. In the intermediate MTX arm, 7.9% of patients developed seizures compared to only 3.7% of patients in the low-dose regimen. In comparison with current protocols, the overall MTX:leucovorin ratio in this study was much higher. This suboptimal leucovorin rescue may have contributed to a higher incidence of MTX-induced seizure activity.

Other agents such as cisplatin and VCR can also potentiate seizures by causing electrolyte derangements.[6,13] However, the incidence is rare. Cisplatin is a platinum-based alkylating agent that cross-links DNA to form adducts and subsequent cell apoptosis. It is widely used in the treatment of pediatric sarcomas and CNS tumors. Cisplatin causes renal electrolyte wasting, which can lead to hypomagnesemia, hypokalemia, and hyponatremia and predispose the onset of generalized seizures. Furthermore, cisplatin can traverse the blood-brain barrier and cause direct neuronal injury. Cortical white matter gliosis correlating with epileptiform discharges has been observed after cisplatin therapy.[14] Albeit rare, VCR has also been linked to seizures. VCR belongs to the class of vinca alkaloids that acts to prevent microtubule polymerization and subsequent cell division arrest. It is broadly used in the treatment of many pediatric cancers including liquid and solid malignancies. VCR is believed to lower the seizure threshold by causing hyponatremia through the syndrome of inappropriate antidiuretic hormone or SIADH.[14–16] Concurrent administration of VCR with antifungal azoles, such as fluconazole, is known to augment VCR neurotoxicity and increase seizure frequency,[17–20] and in such cases, cessation of VCR and azole therapy leads to full resolution of seizures.

In general, patients with seizures regardless of the offending agent are treated with antiepileptic medications (AED). Common abortive therapies include lorazepam, levetiracetam, and valproic acid. For patients requiring long-term therapy or prophylaxis, newer agents are favored that do not interfere with chemotherapy metabolic clearance via cytochrome P450.

2.1.2 Stroke

Cerebrovascular disease as a cancer treatment complication is documented in children. Certain intravenous chemotherapies can significantly alter the coagulation cascade and precipitate cerebral infarction or hemorrhage. Asparaginase (ASP) is one widely accepted offending agent in this regard. ASP constitutes the backbone of ALL therapy and has proven indispensable in transforming pediatric leukemia into a curable disease. However, ASP is also associated with a number of adverse effects, including the risk of thromboembolism. ASP catalyzes the hydrolysis of asparagine amino acid and decreases the proliferation and growth of cancer cells. This cytotoxic effect is specific to leukemia because leukemia cells lack asparagine synthetase to reconstitute the asparagine pool. Notably, ASP also perturbs the delicate homeostasis between hemostatic and anticoagulation pathways. Although ASP is known to reduce both thrombotic and antithrombotic proteins, its complication is predominately seen in the form of thromboembolism due to reductions in proteins C, S, and antithrombin III.[21,22]

In a large meta-analysis evaluating the association between pediatric ALL and its thrombotic complications, the authors analyzed the incidence of thrombosis in 1752 children across 17 prospective studies.[23] The study identified a 5.2% rate of thrombosis. The highest incidence of thromboembolism occurred during the induction phase, and interestingly, risk was associated with lower doses of < 6000 U/m^2 of ASP administered daily over longer periods of > 8 days. There was an incidence rate of 9.6% in patients who received 9 or more days and 6000 U/m^2 or less of ASP. Thrombotic events presented as central-line-associated deep vein thrombosis and stroke. A majority of CNS thrombotic events were classified as cerebral venous thrombosis (CVST). In a recent 2019 retrospective study of 778 children with ALL treated per DCOG ALL-10 protocol across 6 Dutch pediatric centers, the authors identified that 59 of 778 patients developed venous thromboembolism (VTE) at an incidence rate of 7.6%.[24] Of note, 44.1% of these patients presented with CVST diagnosed on contrast-enhanced CT or MR venography. CVST often occurred after a median of 3 days post-ASP therapy and mostly in conjunction with concomitant corticosteroid use (87.5%).

Emergent management of CVST entails prompt, but thoughtful, initiation of anticoagulation. The decision to start anticoagulation is often balanced by the risk of clot extension vs hemorrhagic conversion. Therapy typically begins with continuous infusion of unfractionated heparin, which is favored due to its rapid reversibility with protamine. Later, patients are transitioned to low molecular weight heparin. Direct oral anticoagulants may be considered in young adults. A full treatment course is at least 3 months, and the decision to discontinue hinges on radiographic evidence of clot stability, the presence of ongoing ASP therapy, or any additional hypercoagulation risk factors. Studies have demonstrated that continuing ASP in patients with history of CVST is feasible and safe when on anticoagulation prophylaxis.[5,25] Prognosis after CVST treatment is widely variable depending on the extent of thrombosis and the presence of hemorrhagic conversion. In the 2019 Dutch study, 34.6% of the patients with CVST acquired irreversible neurologic sequelae that included symptomatic epilepsy, focal motor deficits, and cognitive disabilities.[24]

2.1.3 Encephalopathy

Encephalopathy can be seen in children receiving systemic chemotherapy and includes symptoms of increased sleepiness, confusion, lack of cooperativity, and agitation. Several intravenous chemotherapies such as ifosfamide, cytarabine, MTX, ASP, VCR, cisplatin, and etoposide have been implicated.[6] Ifosfamide is an alkylating agent that is administered as a prodrug and requires metabolic activation by hepatic cytochrome P450 to impact DNA cross-linking, leading to impaired DNA synthesis and repair. This drug has a role in the treatment of solid tumors, including CNS tumors, neuroblastoma, Ewing's sarcoma, and salvage regimens of recurrent/resistant tumors. In the literature, the incidence of ifosfamide-associated neurotoxicities is reported in 10%–80% of all patients.[26] The precise pathophysiology remains poorly understood. It is hypothesized that the generation of toxic metabolites, in particular chloroacetaldehyde, leads to neuronal damage and altered neuronal signaling. In children, the spectrum of neurologic changes can involve disorientation, psychosis, focal motor deficits, paresis, and incontinence. Seizure-like activity is a rare phenomenon, mostly seen in adults.[27,28] Aberrant EEG changes have been captured postdrug exposure, but are often short lived. Risk factors for ifosfamide encephalopathy include hypoalbuminemia, concomitant antiemetic use (notably aprepitant, a selective antiemetic that antagonizes substance P/neurokinin 1 receptors while also competing for cytochrome P450

system clearance), elevated creatinine, and prior cisplatin exposure.[29–32] Hypoalbuminemia is often cited as one of the greatest risk factors, yet prophylaxis with albumin infusion has rendered little clinical benefit.[33] From a management perspective, prompt discontinuation of ifosfamide often leads to complete spontaneous reversal and return to neurologic baseline within 72 h. The use of methylene blue as a reversal agent and as neurologic prophylaxis can also be considered, and recovery with or without methylene blue is promising.[34] Re-initiation of ifosfamide after toxicity is administered with caution, but not contraindicated.

Intravenous high-dose cytarabine (HDARC) given at $> 3 \text{ g/m}^2$ is another drug closely associated with encephalopathy.[35] HDARC neurotoxicity can include confusion, altered consciousness, lethargy, seizures, and acute cerebellar syndrome. ARC is a nucleoside analog that blocks DNA synthesis by inhibiting DNA polymerase. In pediatrics, HDARC forms a mainstay of acute myelogenous leukemia therapy. After ARC administration, acute cerebellar syndrome can occur within days and is characterized by classic cerebellar signs such as dysdiadochokinesia, dysarthria, dysmetria, ataxia, and nystagmus. In a historic study, Herzig et al. analyzed 418 patients aged between 2 and 74 years who received cumulative HDARC doses between 36 and 48 g/m^2 for the treatment of leukemia or lymphoma.[36] The study identified that the risk of cerebellar impairment was age dependent, and 19% of the patients > 50 years developed neurotoxicity compared to only 3% in younger and pediatric patients. The mechanism behind this neurotoxicity is attributed to direct injury to the Purkinje cells.[35] In postmortem studies, significant Purkinje cell loss with infiltration and recruitment of glial cells evoking an inflammatory response has been observed. Management is usually the cessation of HDARC. There is no standard treatment, although CSF removal with saline replacement has been attempted. Pellier et al. report a case of treating a 17-year-old patient who developed neurologic toxicities after HDARC with a rapid improvement after removing and replacing CSF with intrathecal infusions of normal saline.[37] This approach is not common practice, and little is known about its potential harm. In contrast, the use of corticosteroids has been successful in the treatment of HDARC neurotoxicity, implicating an inflammatory component as contributory.[38,39]

2.1.4 Posterior reversible encephalopathy syndrome (PRES)

Posterior reversible encephalopathy syndrome (PRES) is a complication closely linked to acute hypertension.[40,41] PRES is diagnosed by the constellation of clinical presentation and radiographic findings on MRI FLAIR sequences.[42,43] Neurologic changes involve altered mental status, seizures, headaches, cortical blindness, or visual disturbances. These symptoms are almost invariably accompanied by elevated blood pressure and cardinal radiographic evidence of symmetric T2 hyperintense signals within the bilateral parieto-occipital lobes. It is accepted that pathogenesis owes to a rapid rise in blood pressure causing dysregulation of the autoregulatory system within cerebral vasculature. Consequently, this results in vasogenic edema, predominately in the posterior lobes due to limited sympathetic innervation and lack of appropriate vasoconstriction. In children undergoing chemotherapy, PRES is suspected in patients who present with a constellation of seizure-like activity, headaches, and sudden elevations in blood pressure. In a 2011 retrospective study, de Laat et al. performed a comprehensive literature review of 56 children aged between 2 and 17 years who developed PRES during the course of their cancer treatment.[44] There was a broad array of primary diagnoses, of which 55% were ALL and 23% were solid tumors. Given different treatment regimens of varied tumor types, PRES was not significantly linked to a particular agent. Multiple agents appeared contributory, especially corticosteroids. The authors found that 87% of the patients were acutely hypertensive at the time of PRES and highlighted this as the most critical risk factor. The etiology of hypertension was multifactorial, including the primary disease and its treatment, renal insufficiency, distress or pain, and systemic inflammation. Of interest, the study concluded that it was not the underlying cause of hypertension that precipitated PRES, but the actual presence of hypertension. PRES and its radiographic changes were reversible after blood pressures normalize. MRI lesions typically resolved within 1–6 months. However, long-term neurologic sequelae can occur, and de Laat et al. reported that 7% of patients ultimately developed epilepsy. Re-challenging with the offending drug can be safe once patients are managed on antihypertensive agents and seizure prophylaxis.

2.1.5 Peripheral neuropathy

Peripheral neuropathy is a common dose-limiting adverse effect of chemotherapy, more formally termed chemotherapy-induced peripheral neuropathy or CIPN. CIPN results from medication injury to the peripheral nervous system either due to direct cytotoxicity or due to inflammation.[45] Both the somatic and autonomic nervous systems are affected. As such, CIPN manifests in a spectrum of neurologic presentations, including numbness/tingling, motor weakness in extremities, jaw pain, altered thermoregulation, and intestinal dysmotility seen as functional constipation.[46] Common culprits include vinca alkaloids, platinum agents, taxanes, and bortezomib.[5,46,47] VCR is a vinca alkaloid that restricts the tubulin elongation. It not only causes microtubule arrest during cell division, but also causes arrest of axonal transport within neurons, thus resulting in a sensory/motor defect. In a 2016 meta-analysis on pediatric CIPN, the authors identified 1580 articles with > 200 children receiving cancer-directed therapy and 61 studies met inclusion criteria.[48] The study found that VCR was the single most important agent associated with peripheral neurotoxicity. Half of the studies attributed CIPN to VCR exposure compared to < 5–10 for all other agents. VCR induces both sensorimotor and autonomic changes with motor impairment. Neuropathy in toe extensors and ankle dorsiflexion, followed by foot drop, is most apparent. Additionally, cranial nerve involvement presenting as voice hoarseness from vocal cord paresis, jaw paresis/pain, ptosis, and ocular nerve palsies is well described.[48] Typically, VCR-related neurologic toxicities occur in the first 1–2 months of therapy and worsen with continuation.

Other agents strongly associated with CIPN are platinum-based compounds, notably cisplatin and oxaliplatin. As a platinum agent used in pediatric solid and CNS tumors, cisplatin causes sensory neuropathy due to preferential drug binding to the dorsal root ganglion (DRG).[49,50] Patients often develop sensory neuropathy, including paresthesia, pain, and muscle spasms/myalgia. Compared to other agents within the same class, oxaliplatin induces more potent cytotoxicity to the DRG, leading to long-term neuropathies.[51]

Although none are FDA-approved to date, several medications have provided excellent benefits for children with CIPN, including anticonvulsants (i.e., gabapentin, carbamazepine) and antidepressants (amitriptyline, duloxetine).[5] Depending on severity, symptoms often resolve with discontinuation of the offending agent; however, persistence of motor polyneuropathy can be seen.[52]

2.1.6 Ototoxicity

Ototoxicity is another well-established neurotoxicity associated with platinum agents, in particular cisplatin. These agents constitute the backbone of many pediatric solid tumors and brain tumors, including osteosarcoma, hepatoblastoma, neuroblastoma, and medulloblastoma. Although platinum agents have an excellent antineoplastic profile, they can cause debilitating acute and long-term hearing impairment, tinnitus, or vertigo. In children, cisplatin-associated hearing loss is more common than carboplatin. High-frequency hearing is often impacted first, before low-frequency hearing, due to the exquisite sensitivity of the outer cochlear cells.[53–55] Continued drug exposure then further breaks down the blood-labyrinth barrier, whereby hearing insensitivity results from injury to the inner cells. Clinically, this is appreciated by the inability to discern subtle differences in sounds, and in children, it can interfere with speech and noise recognition. Incidence of cisplatin-induced ototoxicity is reported in > 40%–60% of pediatric patients.[56,57] In a recent COG study measuring the hearing outcomes of children and young adults treated on cisplatin-based protocols, the authors found an ototoxicity prevalence rate of 40%–50% for any degree of hearing impairment and a rate of 7%–22% for severe ototoxicity.[58] Median cumulative dose of cisplatin was 395 mg/m^2. Clinically apparent hearing impairment can be seen after only one cycle of platinum exposure, but the median time to toxicity from diagnosis is 135 days.[56] Comparatively, carboplatin-associated toxicity is less frequently seen in children, although the combined rate of both cisplatin and carboplatin can be as high as 75%.[59]

Susceptibility to platinum ototoxicity may have a genetic basis. Several genetic variants involved in drug metabolism increase vulnerability. Recently, genetic alterations in the thiopurine S-methyltransferase gene (TPMT) were found to be strongly associated with cisplatin ototoxicity with an odds ratio of 17.[60] However, it is noteworthy that, in general, most children who develop platinum ototoxicity do not have the TPMT variant. Other variants implicated include COMT, GST, and ACYPT2.[60–62] In the future, further pharmacogenomics studies can inform clinical decision making on adapting dose-lowering therapies for individuals with these genetic variants. Moreover, literature cites cumulative dose, history of cranial radiation, and concurrent use of other ototoxic mediations (loop diuretics, aminoglycosides) as additional risk factors.[55,59]

Diagnosis of ototoxicity is based on sequential audiologic testing. Two modalities are frequently used in detecting hearing loss: (1) extended high-frequency audiometry (EHF) and (2) otoacoustic emissions (OAEs). Children receiving platinum therapy are often tested at baseline prior to the initiation of therapy, then serially through their treatment course. Most therapy protocols include specific guidelines for dose modification of the offending agents without compromising curative intent. Reassuringly, a Children's Oncology Group (COG) study showed that dose modifications of cisplatin-based regimens for medulloblastoma did not lead to inferior survival outcomes.[63] From a treatment perspective, children with hearing loss are fitted with hearing aids, which often become a life-long intervention. Currently, there are no FDA-approved agents for otoprotection or ototoxicity reversal. Several preclinical drugs are showing promising results, including amifostine, sodium thiosulfate, N-acetyl cysteine, D-methionine, and ebselen.[53] These drugs have been integrated into clinical guidelines and practice, although large-scale clinical trials in pediatrics are still needed to gain FDA approval for standard of therapy.[64]

2.2 Radiation therapy

Historically, conventional photon-based radiation therapy (RT) constituted the cornerstone of numerous cancer treatments and has been fundamental in improving cure rates. In the 1970s, cranial RT (CRT) was used in primary CNS malignancies, secondary brain metastasis, as well as acute lymphoblastic leukemia/acute myeloid leukemia with CNS infiltration. Over the past decade, there has been a sharp decline in using cranial RT for non-CNS tumors and a paradigm shift toward restricting its use in very young children.[65] This not only owes to the advent of more effective chemotherapeutics, but also to greater appreciation of RT complications. As cancer therapy improves and survivorship increases, we are learning more about the debilitating consequences of CRT, in its acute and long-term complications. Neurologic complications following radiotherapy are categorized by their timing of presentation, specifically, into early (< 1 month of therapy), early-delayed (1–6 months), and late effects of therapy (> 6 months). In the immediate period, headaches, nausea and vomiting, somnolence, and mental fatigue are seen. These effects are mostly transient and reversible, as opposed to the late complications of CRT. Long-term side effects include increased stroke risk due to cerebral vasculopathies, endocrine disorders, neurocognitive deficits, and secondary malignancies. These complications derive from off-target effects causing cell death in healthy, adjacent brain tissue. Dose reductions, sparing of brain regions important for memory and cognition such as the hippocampus, and the use of proton-based RT (PRT) are frequently used to combat long-term injury. Compared to photon-based RT, PRT delivers relatively lower radiation doses outside the target range and is found to result in improved neurological sequelae without curtailing antineoplastic effect.[66,67]

2.2.1 Early effects (< 1 month)—Encephalopathy

The acute neurotoxicity of RT is mostly attributed to intracerebral vascular damage. In the acute phase, ionizing radiation evokes a significant inflammatory response. Clinically, children can present with signs of mild encephalopathy, including generalized headaches, nausea and vomiting, drowsiness, and fatigue. Symptoms often present within the first 2 weeks of RT. They are typically transient and responsive to symptomatic management. In addition to analgesia, antiemetics and corticosteroids are used. Steroids render an anti-inflammatory effect that can ameliorate the clinical manifestations of intracranial edema.[68] Similarly, bevacizumab is now also used routinely to counteract the effects of radiation-induced inflammation and necrosis.[69–71]

2.2.2 Subacute effects (1–6 months)—Postradiation somnolence syndrome

Early-delayed neurologic side effects are present between 1 and 6 months. During this period, a number of complications have been described, including transient encephalopathy, myelopathy from spinal radiation, and postradiation somnolence syndrome. First reported several decades ago, postradiation somnolence syndrome is a well-documented phenomenon, preferentially found in children after whole brain radiation for hematologic malignancies.[72,73] Postradiation somnolence syndrome is mostly a consequence of whole brain radiation and whether it can be caused by focal RT is not precisely known. Children present with marked sleepiness, lethargy, slow mental processing, and irritability. The pathophysiology is not entirely clear, but it has been attributed to impaired myelin synthesis by oligodendrocytes and diffuse cortical demyelination.[72] In the literature, somnolence syndrome can be seen in both adults and children undergoing cranial RT with an incidence ranging from 13% to 79%.[74] More specifically, in children with ALL receiving whole brain RT, the incidence has been reported to be between 13% and 58% following 18 Gy and 60% and 79% following 24 Gy.[75] Symptoms typically present by 4–8 weeks, but are often short-lived and self-resolve within 2–3 weeks. In a 2009 retrospective report by Vern et al., out of 27 pediatric patients with hematologic malignancies who received CRT between 1981 and 2007, 71% presented with somnolence syndrome over a median time to presentation of 28 days from RT completion and with symptoms lasting between 7 and 21 days.[75]

2.3 Hematopoietic stem cell transplantation

Hematopoietic stem cell transplantation (HSCT) is a treatment modality used across a spectrum of diseases and malignancies. In childhood cancers, allogenic transplantation involving the infusion of donor hematopoietic stem cells is used in pediatric leukemia/lymphoma, whereas autologous transplantation involving harvesting and infusion of the patient's own stem cells is used in pediatric solid tumors. Complications from transplant are common, often due to chemo-cytotoxicity of conditioning regimens, severe immunosuppression, and graft-vs-host disease (GVHD). Moreover, there is a high incidence of neurologic complications, reported in the range of 10%–60%.[76–78] Neurologic complications can be serious, leading to high mortality and morbidity after transplantation.[76–80] The majority of neurotoxicities are present in the immediate period after HST, but late effects can be seen, mostly in the setting of GVHD complications. Here, we discuss HST-related neurologic sequelae based on timing of onset: acute (< 6 months) vs late effects (> 6 months) as well as the common underlying causes.

2.3.1 Acute effects (< 6 months)

Neurologic complications in the immediate posttransplant period (< 1 month) frequently manifest as encephalopathy, seizures, PRES, cranial nerve deficits, and also intracranial hemorrhage.[76,78] The etiology is often multifactorial, but can be linked to the well-known culprits,

including busulfan in conditioning regimens, calcineurin inhibitors for GVHD prophylaxis, CNS infections from severe neutropenia, and hemorrhage or VTE in the setting of thrombocytopenia or coagulation dysfunction.

In the pretransplant phase, many neurologic changes are caused by the high-dose myeloablative chemotherapies, and of these, busulfan is most widely implicated. Busulfan is an alkylating agent that interferes with DNA replication and transcription. It was first introduced > 30 years ago as an alternative to total body irradiation (TBI)[81] and now forms the backbone of most preconditioning regimens for both allogeneic and autologous transplants. As a lipophilic agent with minimal plasma protein binding, busulfan freely traverses the blood-brain barrier (BBB; a selective barrier surrounding the CNS, which functions to limit the permeability of circulating solutes in the blood from reaching the protected space of the CNS) and can induce direct neuronal injury.[81,82] Seizures, typically generalized tonic-clonic in semiology, are the most common manifestation of busulfan-related neurologic complications. Historically in the 1980–1990s, the incidence of seizures in children receiving high-dose busulfan was seen in approximately 7% of patients.[82] Literature now supports the use of seizure prophylaxis in patients receiving busulfan.[83–85] A number of antiepileptic drugs are used, including benzodiazepines (i.e., lorazepam, diazepam), second-generation agents (i.e., levetiracetam, lamotrigine) in addition to older drugs such as valproic acid.[83] However, there is controversy regarding the true utility of seizure prophylaxis as newer case studies have challenged the incidence of busulfan-associated seizures.[86] Regardless, seizure prophylaxis remains standard practice and is associated with a low incidence of drug-related seizures. In a 2014 multiinstitutional analysis of pediatric patients that underwent 954 busulfan-based conditioning transplants using prophylactic AEDs, only 1.3% of patients developed seizures and this dropped to 0.75% when eliminating children with predisposing CNS conditions.[87] In this cohort, 40% of patients developed seizures during the course of busulfan administration, while the remaining developed events between day + 1 and day + 86 from transplant.

Calcineurin inhibitor (CI)-associated neurologic complication is another highly recognized event following transplant in children. It has been reported in almost 48% of pediatric patients posttransplant and is commonly seen within 4–6 weeks from transplantation.[76] Cyclosporine A (CSA) is widely used after HSCT to mitigate the risk of transplant rejection and GVHD and is frequently the offending agent. CSA inhibits calcineurin phosphatase activity, which in turn impairs the interleukin production in T cells and immune activation.[88] Although CSA is highly efficacious as an immunosuppressive agent, neurologic complication is a serious and widely reported toxicity.[88,89] CSA is often started posttransplant, and if without evidence of GVHD, CSA is tapered starting at 6 months before discontinuation at 1 year. The pathophysiology of CSA neurologic injury is believed to result from mitochondrial dysregulation. In animal models, CSA can impair metabolism and energy production in brain cells, leading to the exaggerated activation of nitric oxide synthase and the production of reactive oxygen species.[88] CSA-related changes can manifest as headaches, confusion, tremor, cortical blindness, aphasia, ataxia, and seizures.[88,90] Clinical symptoms correlate with MRI findings of cerebral edema/gyral swelling and signal aberrations in the white matter.[88] Incidence of neurologic changes after CSA exposure has been reported in approximately 4%–11% of patients.[90] In 2010, Noe et al. examined the incidence of CSA-associated neurotoxicities in 67 pediatric patients who underwent HSCT and found that almost 9% of children developed mental status changes.[90] Generalized seizures were prevalent (66.6%), and the onset of symptoms ranged from day + 1 to day + 40 after HSCT. Successful management entailed prompt discontinuation of CSA with resolution within 2–3 days. Patients were subsequently switched to an alternate CI (tacrolimus with or without mycophenolate). Patients tolerated re-challenging with a CI without further seizures. Similarly in a larger study by Straathof et al., re-introduction of CI proved to be a viable option.[91] Of interest however, even following the re-introduction of immunosuppression, the majority of these patients experienced poor outcomes related to GVHD-associated morbidity. This may be due to general inability to tolerate CSA-based immunosuppression. The complexities affecting the outcomes of these patients, though, remain to be fully resolved.

Myeloablative conditioning regimens and ongoing GVHD prophylaxis in HSCT place patients in a precarious state of severe immunosuppression. Children become highly susceptible to life-threatening infections from bacteria, viruses, fungi, and parasites. Vulnerability to specific species of pathogens changes over time. During the initial month, profound neutropenia increases the risk of common bacterial or nosocomial infections, such as methicillin-resistant staphylococcus, multiresistant enterococci, and gram-negative organisms (*Escherichia coli*, *Pseudomonas*, *Klebsiella*).[78,79] During this period, CNS candidiasis and septic embolism from aspergillus species also account for a substantial percentage of CNS infections.[78,79] In the class of viral infections, cytomegalovirus can cause encephalitis and ventriculitis, as well as peripheral nervous system disorders, such as chorioretinitis and myelitis.[78] Additional viral culprits include human herpesvirus, BK virus, Epstein-Barr virus, varicella zoster, and JC polyomavirus.[92] In the 1- to 6-month period following HSCT, there is a high incidence of opportunistic invasive fungal diseases and viral infections. In particular, CNS aspergillosis is the most common fungal disease and HHV-6 encephalitis is the most common viral disease.[78,92] After 6 months, when immunosuppression for GVHD prophylaxis is tapered, the risk of opportunistic CNS infection declines. However, for patients on chronic immunosuppression for GVHD treatment, the risk of cryptococcal meningitis and progressive multifocal leukoencephalopathy from JC viral infections remains substantial.[92] Initial treatment involves broad-spectrum antimicrobials that can be narrowed pending clinical improvement, determination of the pathogenic isolate, and its antimicrobial sensitivities. Historically, patients with fungal CNS infections, such as cerebral aspergillosis and toxoplasmosis, have fared poorly despite aggressive antifungal therapy.[80]

Broadly speaking, cerebrovascular disease as a posttransplant complication is associated with inferior outcomes. Intracranial hemorrhage is more commonly reported in adults than in children,[93–95] occurring in only 2% of children who had undergone HSCT.[77] Causes of ICH owe to a myriad of risk factors, including thrombocytopenia, coagulation dysfunction, hypertension, and a history of TBI or cranial radiation. Separately, two other prominent vascular disorders, veno-occlusive disease (VOD) or sinusoidal obstruction syndrome (SOS) and transplant-associated thrombotic microangiopathy (TMA), are also well known to cause altered mental status among a constellation of diagnostic features. In both disorders, confusion, altered consciousness, hallucinations, and PRES have all been reported.[5,96,97] For treatment, defibrotide is used in VOD and eculizumab in TMA.

2.4 Novel agents: Immune-based therapies

Over the recent decade, our understanding of how to harness the body's own immune surveillance to antagonize neoplastic cells has grown drastically. Adoptive cellular T-cell therapy, checkpoint inhibitors, vaccines, and oncolytic viruses have transformed the current landscape of cancer treatment, promising cure especially in children with refractory/relapsed disease. Given their novelty, our appreciation for their unique adverse effects is still evolving, but a number of acute neurologic complications are well recognized. Conversely, late effects remain to be seen as long-term follow-up data mature.

2.4.1 T-cell-based immunotherapy

In little more than half a decade, there has been an unprecedented rise in T-cell-based therapies. To our repertoire of antineoplastic agents, we have now added several T-cell-based immunotherapies. Their therapeutic efficacy is best recognized in the treatment of pediatric leukemia. To date, three agents have received full FDA approval for children with relapsed/refractory B-cell ALL: tisagenlecleucel, blinatumomab, and inotuzumab.[98,99]

New COG clinical trials are also underway to integrate these therapies into frontline regimens for newly diagnosed B-cell ALL patients.[99] Tisagenlecleucel is a chimeric antigen receptor-modified T cell (CAR-T 19) that targets CD-19 residing on B cells. This novel therapy genetically modifies one's own cytotoxic T cells to express an extracellular domain that binds CD19 in order to engage and subsequently kill CD-19-expressing B-cell ALL cells. The first use of CAR-T in pediatrics was in 2012 through a phase I/IIA trial in which two children with relapsed/refractory B-ALL were treated with tisagenlecleucel.[100] Intriguingly, both children achieved negative minimal residual disease status (MRD < 0.01%) in 1 month. These results paved the way for larger international trials such as the pivotal phase II ELIANA study in 2015.[101] From this cohort of children, we have learned a tremendous amount regarding the acute toxicities associated with CAR-T 19, in particular the neurotoxicities from cytokine release syndrome (CRS). CRS results from the exaggerated immune activation caused by a cytokine storm. High levels of cytokines such as IL-6, IL-10, and soluble IL-2 receptors are associated with life-threatening cardiovascular and organ damage. The spectrum of CRS manifestations ranges from low-grade fever to multisystem organ failure, including neurotoxicities. In the ELIANA trial, 79% of the patients developed CRS with a notable portion of these children presenting with acute neurologic symptoms, including headaches, confusion, and somnolence.[101] In a 2018 study by Gofshteyn et al., the authors analyzed the acute neurotoxicities developing within 30 days post-CAR-T 19 infusion in 51 pediatric patients.[102] The study found that 45% of the patients demonstrated acute neurologic changes, most in association with CRS. Symptoms presented at a median of 6 days postinfusion or 3 days after CRS onset. Encephalopathy was the most common presentation, while seizures occurred in 8% of the patients. Most neurologic complications were transient and resolved after the treatment of CRS. Similar incidences have been reported in other studies.[103,104] Mild CRS self-resolves with supportive measures, while severe CRS is treated with targeted IL-6 blockade, tocilizumab, while corticosteroids are reserved for a subset of patients with severe CRS refractory to tocilizumab.[105]

There exists another CAR-T-associated neurologic syndrome termed "CAR-T-cell-related encephalopathy syndrome" (CRES) that is an entity distinct from CRS and can occur in isolation.[106,107] CRES encompasses a constellation of neurologic changes, including headache, disorientation, focal neurologic deficits, hallucinations, aphasia, somnolence, and, less commonly, seizures. CRES can present within the first few days after CAR-T infusion or after several months of receiving therapy. In both CRS and CRES, the pathophysiology of these neurotoxicities is poorly understood, but has been attributed to glial injury secondary to systemic inflammation. In a 2019 prospective study by Gust et al., the authors obtained inflammatory CSF and blood biomarkers on 43 pediatric and young adult patients treated with CAR-T 19.[108] The study identified that 19 of 43 patients developed neurotoxicities. Delirium and confusion were the leading complaints. Severity correlated with CRS, elevations in IL-6, IFN-γ, IL-10, granzyme B in CSF and serum, and high peak levels of serum CAR-T-T. Intriguingly, similar concentrations of CAR-T cells in the CSF were identified in both patients with or without neurologic changes, suggesting that direct neuronal injury from CNS penetration was not the primary mechanism. Systemic inflammation, however, was associated with glial injury on neuropathology. Analysis of postmortem brain samples of patients who received CAR-T infusion often demonstrates moderate cortical gliosis and microglial activation with perivascular changes and hemosiderin deposition, suggesting a history of microbleeds.[108]

Blinatumomab is a bispecific T-cell engager (BiTE) targeting CD-19, which links CD-19-expressing B-cell ALL cells to cytotoxic T cells. Following success in a pediatric phase II trial in 2016, it gained accelerated FDA approval for use in children with relapsed/refractory B-cell ALL.[98,109] Of the 70 patients who received the recommended blinatumomab dosage, 24% developed neurologic events consisting of dizziness, tremor, and somnolence, and 2 patients developed seizures that prompted temporary suspension of therapy.[109] No neurologic event caused permanent discontinuation. Symptoms typically resolved shortly after stopping therapy, and patients tolerated re-challenge with blinatumomab. In light of its efficacy and favorable toxicity profile, blinatumomab is now incorporated into the backbone of newly diagnosed standard risk B-cell ALL patients in recent COG trials.[99]

2.4.2 Immune checkpoint inhibitors

Immune checkpoint inhibition is another mechanism that exploits immune surveillance for tumor eradication. Cytotoxic T cells defend our body against cancer cells; however, in the setting of tumor progression, T-cell exhaustion is frequently observed due to upregulation of checkpoint proteins that suppress cytotoxic T-cell activity. A hallmark of cancer is the activation of inhibitory pathways, such as programmed death receptor-1 (PD-1)/programmed death receptor-1 ligand (PD-L1), cytotoxic T lymphocyte antigen-4 (CTLA-4), B7H3, indoleamine 2,3 dioxygenase (IDO), among others. Physiologically, checkpoint proteins permit necessary and normal autoimmune tolerance. However, when expressed on tumor cells, they provide a protective mechanism that facilitates immune evasion. In adult trials, several agents have proven efficacious in eliminating adult solid tumors and are now FDA-approved for patient use. These results have translated to the introduction of checkpoint inhibitors into clinical trials for children with extracranial solid tumors and brain cancer.

PD-L1 is expressed on the membrane of malignant cells, and upon interaction with PD-1 receptors on T cells, it can downregulate the cytotoxic immune response, leading to tumor cell survival. Several agents targeting the PD-L1/PD-1 pathway exist, including PD-1 inhibitors, nivolumab and pembrolizumab. Several studies are currently ongoing, including phase I trials using cemiplimab for recurrent solid tumors and in combination with radiation for diffuse intrinsic pontine glioma and high-grade glioma (NCT03690869), and pembrolizumab in recurrent/refractory pediatric high-grade glioma and diffuse intrinsic pontine glioma (NCT02359565). There is also a COG-based phase I/II KEYNOTE-051 trial assessing the use of pembrolizumab in pediatric sarcomas and CNS tumors (NCT02332668). Additionally, several European phase I/II trials are underway determining the effect of nivolumab in children.[110,111] Similar to PD-1, IDO is an intracellular enzyme and a critical regulator of the tryptophan catabolism pathway; it has been shown that T cells and natural killer cells are inhibited with high levels of IDO. Currently, there is phase I clinical trial using indoximod, a targeted INO inhibitor, in children aged 3–21 years with malignant brain tumors (NCT02502708). Finally, CTLA-4 is a membrane-bound protein found on cytotoxic T cells and regulatory T cells. In children, the first trial investigating the use of ipilimumab for patients < 21 years old found that 55% of the patients developed therapy-associated toxicities of any grade, but none with clinically significant neurotoxicity.[112]

Given the early stages of immune checkpoint inhibition in pediatrics, the neurological complications are still being appreciated. A number of retrospective meta-analyses have highlighted a unique set of neurologic consequences related to PD-L1/PD1 and CTLA-4 inhibitors,[113,114] including headaches, aseptic meningitis, or autoimmune encephalitis. The peripheral nervous system is also impacted with sensory/motor neuropathies reminiscent of Guillain-Barre-like syndromes, chronic inflammatory demyelinating polyneuropathy, and cranial nerve deficits as well as neuromuscular junction involvement presenting as myasthenia gravis. Most neurologic toxicities develop within 6 weeks after the onset of therapy and often resolve after the cessation of treatment. Near-complete or full reversal of side effects can be augmented by steroid therapy.[113,115] One unique consequence of checkpoint inhibition to carefully consider is pseudoprogression in primary brain tumors. In this instance, there is immune cell infiltration into the tumor bed, causing focal mass expansion, edema, and potential compression of vital brain structures. Focal neurological deficits, seizures, hydrocephalus, and herniation can result. Judicious use of steroids can quell inflammation, but also negatively impact the efficacy of checkpoint inhibitor therapy. Newer protocols are now incorporating VEGF inhibitors as steroid-sparing alternatives.[115]

2.4.3 Vaccines

Therapeutic cancer vaccines are another modality first trialed in adult solid tumors that have expanded their use to include pediatric cancers, in particular pediatric brain tumors and neuroblastoma. This strategy harnesses the body's adaptive immune system to promote tumor cell killing by introducing immune stimulants. Currently, there are multiple forms of vaccines, including peptide-based vaccines, that display a specific tumor antigen or tumor-associated antigens and DNA and RNA-based vaccines that activate dendritic cells.[116] The Children's Hospital of Pittsburgh recently conducted a pilot trial using peptide vaccines against common glioma antigens, EphA2, survivin, and IL-13Ra2.[117] Another example of vaccine therapy is the H3.3K27M peptide vaccine trial undertaken by the Pediatric Neuro-Oncology Consortium (PNOC; NCT02960230). This vaccine targets the H3.3K27M mutation, which is a known oncogenic driver and present in over 70% of diffuse midline gliomas.[118] Introducing a synthetic molecule that harbors a component of the H3.3K27M protein can entrain the immune system to specifically target these cancer cells. Similar tumor-based antigen vaccines are also being explored in the treatment of pediatric neuroblastoma and Ewing sarcoma (NCT01192555, NCT02511132, NCT03495921). Acute neurologic complications are due to systemic inflammation. Most commonly seen are flu-like symptoms presenting within 1 week postadministration.[115] Accompanying complaints also include headaches and fatigue. As is common in immunomodulating therapies, pseudoprogression can be seen following vaccine therapy with cranial neuropathies and cerebellar signs seen for posterior fossa tumors.[119,120]

2.4.4 Viral therapies

Finally, oncolytic viral therapy is a form of therapy that takes advantage of viruses capable of destroying tumor cells and tumor cell lysis release of cancer-specific antigens that can trigger an adaptive immune response. Several viruses, including the adenovirus, herpes simplex virus, polio, and measles, have been engineered to or intrinsically infect cancer cells. These therapies have been best studied in adult glioblastoma with promising results and demonstrable life-prolonging benefits.[115,121] Gleaning from adult studies, neurologic complications can be significant, including headaches, confusion, seizures, and focal neurologic deficits. The frequency of these side effects is often related to the type of oncolytic virus, CNS tumor type, and the patient's baseline neurologic function. Bevacizumab, as a VEGF inhibitor, and steroids are frequently used to reduce local tumor inflammation across all immune-based therapies, and the same is true for viral-based therapies.[122]

3 Part II: Late effects of cancer treatment

Long-term survivorship has improved significantly over the past several decades. Prior to the 1970s, many pediatric cancer patients succumbed to their primary disease. However, in the modern era of multimodal therapies and prompt supportive care, cure rates of pediatric cancers have markedly increased. It is estimated that the survival rate of children is currently about 80% with more than 500,000 childhood cancer survivors living in the United States alone.[123–125] This has resulted in an unprecedented and burgeoning population of long-term childhood cancer survivors, each uniquely impacted by cancer and its treatment. Late complications can impact every organ system, and in growing children, its impact on neurocognitive development can be devastating. Never before has recognition of these late neurological complications become so relevant. Currently, the Childhood Cancer Survivor Study (CCSS) group has published a number of seminal papers on the largest and most extensively analyzed cohort of 5-year pediatric cancer survivors in the United States.[123,124] Additionally, larger consortia such as COG (Table 2) and PNOC have put forth comprehensive sets of guidelines to facilitate the systematic follow-up and screening of childhood cancer survivors.[126–128] Similar guidelines also exist from European groups, such as the Scottish Intercollegiate Guidelines Network (SIGN), the Late Effects Group of the United Kingdom Children's Cancer and Leukaemia Group (UKCCLG), and the International Late Effects of Childhood Cancer Guideline Harmonization Group (IGHG).[129,130] In this section, we present the key late neurologic complications commonly associated with each treatment modality and potential interventions (Table 3).

3.1 Systemic chemotherapy

3.1.1 Neurocognitive sequelae

During childhood, the developing brain is particularly vulnerable to toxicities and the relationship between chemotherapeutics and adverse neurodevelopmental effects is well documented. Chemotherapy can indelibly impact the pediatric brain and impart long-lasting cognitive defects. Survivors tend to perform poorly on measures of executive function, attention, processing speed, and memory. Several reviews have identified agents commonly implicated in neurocognitive injury,

TABLE 2   A listing of recommended survivorship guidelines for neurologic complications commonly seen in pediatric cancer survivors.

Type of injury	Survivorship guidelines/considerations[a]
Neurocognition	• Regular neuropsychology evaluation • Routine screening for school performance • Involvement of school liaison
Nerves	• Routine physical/neurologic examinations • Consideration of physical therapy/occupational therapy referral/consultation
Vascular	• Blood pressure screening • Diabetes screening • Cholesterol screening • Magnetic resonance angiography screening • Color Doppler ultrasound (*for patients for significant radiation to the neck*) • Consideration of neurosurgery referral/consultation or other services to assist with the above potential contributing factors (i.e., cardiologist for blood pressure management)
Function	• Routine vision/hearing screens • Routine screen for developmental milestones • Consideration of formal audiology/ophthalmology assessments • Consideration of speech therapy referral/consultation • Involvement of school liaison
Neuroendocrine	• Routine physical examination/tanner staging • Growth monitoring • Hormonal function screening • Screening for menses onset/sexual function • Consideration of endocrine referral/consultation

[a] *Guidelines adapted from and additional details available—Children's Oncology Group. Long-Term Follow-Up Guidelines for Survivors of Childhood, Adolescent and Young Adult Cancers, Version 5.0. Monrovia, CA: Children's Oncology Group; October 2018; Available online: www.survivorshipguidelines.org. Modified from Children's Oncology Group guidelines.*

TABLE 3 Neurocognitive domains that are commonly affected by cancer-directed therapies in pediatric patients, with potential therapeutic interventions for prevention or treatment.

Type of injury	Domain of intervention	Intervention/therapy
Neurocognition	Physical	• Physical exercise
	Pharmacologic	• NMDA receptor antagonists • Stimulants
	Therapeutic support	• School interventions • Educational support
Nerves	Physical	• Physical therapy • Occupational therapy
	Pharmacologic	• Analgesia • Anticonvulsants • Antidepressants
	Therapeutic support	• Bracing • Mobility support
Vascular	Surgical	• Neurosurgical evaluation for affected vasculature
	Pharmacologic	• Blood pressure management • Diabetic management • Cholesterol management • Anticoagulants
Function	Physical	• Hearing aids • Visual aids
	Pharmacologic	• Protectants (i.e., amifostine, sodium thiosulfate, *N*-acetylcysteine, D-methionine, ebselen)
	Therapeutic support	• Speech/language therapy • School interventions • Educational support
Neuroendocrine	Physical	• Sperm/egg preservation measures
	Pharmacologic	• Hormone replacement • Gonadotropin-releasing hormone agonists
	Therapeutic support	• Endocrine subspecialist support

including MTX, DNA cross-linking agents (cisplatin, cyclophosphamide, ifosfamide, thiotepa), taxanes, and vincristine.[131,132] Of these, MTX appears to be the most consequential. In animal studies, these agents are known to produce widespread injury to the cortex, thalamus, and hippocampus; pronounced synaptic remodeling, neuronal apoptosis with gliosis, and increased blood-brain barrier permeability are all seen.[133] Furthermore, the impact of MTX can be long-lasting, related to general disruption of brain tissue homeostasis and the inability of oligodendrocytes to recover after injury.[134]

For at least three decades, IT and high-dose IV MTX have been known to cause acute neurotoxicities and leukoencephalopathy on MRI, as defined by abnormal T2-weighted hyperintensities within the periventricular and deep white matter. However, as follow-up data on cancer survivors mature, we now have a greater appreciation for the persistence of MTX-associated brain lesions. Evident in leukemia patients treated on historic trials incorporating CRT for CNS prophylaxis, it is

unsurprising that MTX in combination with irradiation leads to worse neurocognitive outcomes. In a 1990 COG study, the authors performed IQ testing on 70 leukemia survivors with a median of 5 years after receiving 24-Gy CRT with or without IT MTX.[135] The study identified a pronounced decline in IQ in the cohort who received MTX with CRT. Intriguingly, timing of IT MTX also mattered. In a retrospective study of 72 pediatric patients, Balsom et al. found that MTX given prior to CRT seemed to confer less neurotoxicity than when given either concomitantly or post-CRT.[136]

Even with the transition away from CNS prophylaxis, ALL survivors treated on only systemic chemotherapy continue to demonstrate worse cognitive function compared to healthy controls. In data taken from the St. Jude Lifetime Cohort Study, > 10% of ALL survivors have impairment across multiple cognitive domains, most significantly in processing speed (16.8%) and executive function (15.9%).[137] The cause is chiefly attributed to MTX. Duffner et al. conducted a 2014 study comparing two historic ALL protocols that differed largely in the intensity of MTX administration.[138] None of the patients analyzed underwent CRT. The authors found that the subset of patients treated with intensive CNS-directed MTX therapy performed worse than their counterparts on a battery of cognitive assessments. Testing was conducted more than 7 years postinitial therapy. Of interest, the study assessed for radiographic evidence of leukoencephalopathy and found a significant proportion of patients (68%) who received intensive MTX therapy demonstrated persistent leukoencephalopathy. These MRI findings correlated with poor neurocognitive performance. Similarly, in another study based on the St. Jude cohort, the authors report an intimate association between leukoencephalopathy and neurobehavioral (i.e., behavioral regulation-impulse control, attention shifting/transition, regulation of emotions, and metacognition) deficits.[139] Cheung et al. performed a longitudinal analysis of 190 evaluable patients who were initially treated during 2000–10 then participated in follow-up studies on neurocognitive assessments and MRI obtained at least 5 years from diagnosis.[139] Survivors showed persistent leukoencephalopathy on MRI that correlated with low neurocognitive scores.

The potential mechanism of MTX-related injury has recently been investigated in postmortem frontal lobe biospecimens from children and young adults and in mouse models of MTX-related neurologic injury.[134] This study identified that MTX injury is associated with decreased (i) white matter oligodendrocyte precursors and recovery post-MTX-exposure, (ii) oligodendrocyte differentiation, and (iii) overall myelination. Other studies have expanded on MTX depletion of folate and associated perturbed methylation of neurotransmitters and cytosine bases. These effects cause aberrant neuronal biosynthesis and circuitry remodeling following cell apoptosis.[140] Additionally, the absence of folate places a higher demand on choline to serve as the methyl group donor and depletes choline pools, which can limit the myelin synthesis. Myelin's role in quick transmission of electrical impulses may explain why patients exhibit such impaired processing speeds.

In patients with ALL, the effect of MTX is likely compounded by other agents such as VCR, cyclophosphamide, and ASP. Genschaft et al. analyzed the MRI of 27 ALL survivors paired with 27 age-matched healthy individuals. Survivors were in remission for a median of 12.5 years at the time of evaluation.[141] There were significantly smaller volumes in the bilateral hippocampi, nucleus accumbens, amygdala, and thalamus on the MRI of survivors compared to paired controls. These radiographic findings translated to lower scores in hippocampus-dependent function such as memory. Follow-up work has confirmed this relationship between chemotherapy and hippocampal dysfunction.[142]

In response, efforts among collaborative groups are now burgeoning to determine effective preventative and treatment measures. Much emphasis is placed on identifying patients at risk of neurocognitive defects and implementing long-term surveillance. Currently, COG is leading a large prospective trial called the "Neuropsychological, Social, Emotional and Behavioral Outcomes in Children with Cancer," in which the investigators are integrating routine cognitive testing into survivorship follow-up (ALTE07C1). PNOC, too, incorporates a battery of quality of life and neurocognitive outcomes within their interventional clinical trials. This battery is notable for utilizing Cogstate, a validated, computer-based program that can be done in person or remotely.[143,144] From an intervention perspective, there are physical fitness-based approaches, cognitive rehabilitation programs,[145] and a pharmacologic-based approach with a vast body of work pointing to routine exercise as neuroprotective and promoting reconstitution of damaged neural networks. Several national and international clinical trials are addressing the use of physical exercise as remediation against treatment-induced neurocognitive injury (NCT02153957; NCT02749877). As evidenced on patient MRIs, exercise can increase cortical thickness and hippocampal volume.[146] From a pharmacologic approach, methylphenidate has been used with excellent success as a stimulant agent for survivors who have cognitive impairments.[147] Conklin et al. performed a randomized, cross-over trial of children previously treated for ALL or brain tumors who developed learning disabilities and found that methylphenidate treatment leads to higher attention and processing speeds.[147] These promising results paved the way for a follow-up study that extends the use of methylphenidate to all cancer survivors regardless of a preexisting learning disability (NCT01100658).

3.1.2 Chemotherapy-induced peripheral neuropathy (CIPN)

CIPN, once considered an acute complication that resolves with discontinuation, is now known to have late effects. Prominent culprits are vinca alkaloids (VCR) and platinum agents (cisplatin, carboplatin). In a recent 2018 large observational study, Kandula et al. evaluated 121 childhood cancer survivors, 107 of which were exposed to at least one neurotoxic chemotherapy agent.[148] The authors conducted electrical neurophysiologic testing and functional physical assessments, and obtained patient-reported functional outcomes. The median time to testing from initial diagnosis was 8.5 years. Over half of the patients with history of vinca alkaloids and/or platinum exposure scored abnormally on the pediatric-modified Total Neuropathy Scoring Scale. They also demonstrated reduced sensory electrical amplitudes and self-reported decreased function. Intriguingly, there was also a distinct difference between VCR vs cisplatin-induced CIPN. Cisplatin conferred a greater neurotoxicity profile and was seen to have longer-term neurologic deficits and sensory changes compared to VCR. Other studies also report that up to 72% of survivors treated with cisplatin have chronic sensory changes, some for more than 10 years. This contrasts to the only 30%–40% of patients who received VCR and have ongoing neuropathy for up to 3 years posttreatment.[148] Unlike cisplatin, VCR-associated CIPN typically improves over time with incremental normalization on electrophysiology testing. It is posited that VCR damages retrograde axons that are more amenable to repair, while cisplatin injures neuronal bodies in the dorsal root ganglion.[149]

From a management perspective, there currently exist no evidence-based recommendations for CIPN. In 2014, a large meta-analysis was conducted with the intent to construct American Society of Clinical Oncology Clinical practice guidelines.[150] However, on careful evaluation of 48 total RCTs, the authors found paucity of high-quality evidence to support practice recommendations. Only a moderate recommendation could be made for duloxetine as treatment. For nonpharmacologic measures, physical training is encouraged. Although to date there are no clinical trials investigating the benefits of physiotherapy and occupational therapy, the merits of these approaches to counter deconditioning and improve functional outcomes are incontrovertible.

3.1.3 Ototoxicity

Platinum-related hearing loss is a chronic complication well reported in pediatric cancer survivors. The prevalence of hearing loss among cancer survivors is high, reaching more than 50% in some reports.[56,151,152] As previously discussed, the two agents most commonly associated with ototoxicity are cisplatin and carboplatin. After the initial onset, the condition is usually permanent and stable after completion of therapy. However, recent reports indicate that hearing loss can be progressive even after stopping therapy. In a well-cited study by Bertolini et al., the authors followed a cohort of 120 survivors who received platinum-based therapy for the treatment of pediatric solid tumors and performed serial hearing assessments.[153] The median follow-up period was 7 years. Median cumulative dose was $400\,mg/m^2$ for cisplatin and $1600\,mg/m^2$ for carboplatin. Compellingly, the study reported that only 5% of audiograms were positive for > grade 1 toxicity at the end of treatment compared to 11% within 2 years of diagnosis and 44% after > 2 years.

The burden of hearing impairment is substantial and impacts all facets of life from childhood and beyond. Indisputably, hearing loss can affect speech and language development, communication and social interaction, psycho-behavioral development, and overall quality of life. The inability to detect high-frequency sounds hinders proper language acquisition, especially in young children. The constructs of many languages abound with high-frequency phonemes and their recognition are crucial to proper language development. After attaining language competency, speech impediments in young adults tend to not develop until losses in the 3000- to 4000-Hz range. Young adults use syntactic cues as a compensatory mechanism.[56] Furthermore, it is also reported that high-frequency losses at > 2000 Hz increase children's risk of poor academic achievement and indirectly influence their emotional growth.[153–155] There are now concerted efforts to formalize hearing surveillance internationally. Most recently in 2019, Clemens et al. published a report summarizing ototoxicity monitoring recommendations put forth by IGHG and the European Union-funded PanCare Consortium.[152] This panel of 32 experts from 10 countries performed a systemic search of MEDLINE for literature on platinum-based ototoxicity published between 1980 and 2017. From this effort, the group compiled evidence-based surveillance guidelines on ototoxicity.[152] For management, the guidelines are less well defined. Currently, there are no FDA-approved agents for otoprotection. The mainstay of management relies on a multidisciplinary approach, including referral to audiology, otolaryngology, speech and language pathology, school-based educational groups, and social work support.[156]

3.2 Radiation therapy

3.2.1 Neurocognitive defects

Neurocognitive compromise as a late effect after cranial radiation (CRT) is well established with reports dating back to the 1960–1970s. Neuropathologically, CNS radiation induces a number of irreversible CNS microvascular and direct cellular changes that cause tissue infarction, necrosis, and breakdown of the blood-brain barrier.[157] Calcifications in the cortical neurons and multifocal white matter lesions can be seen on histology and on MRIs post-CRT.[158–160] In historic studies when children with ALL were treated with whole brain CRT as CNS prophylaxis, an overwhelming majority of patients developed life-long neurocognitive decline.[161] In an early study, Rowland et al. assessed children treated for leukemia at > 1 year post-CRT using an array of neurocognitive achievement tests.[162] The authors found compelling evidence that CRT was inextricably linked to compromised executive function. On Full Scale Intelligent Quotient (IQ) testing, < 15% of the children in the CRT group scored above the cutoff for average IQ compared to > 55% of children in MTX arms. Children in the CRT cohort also performed poorly across a wide range of neuro-psychometric testing from verbal and mathematical processing to motor function.

These results have now been upheld in many other studies.[163,164] Similarly, for brain tumor survivors, CRT also exacts a significant neurocognitive cost. It leads to profound cognitive decline independent of confounding variables, such as surgical injury, primary tumor location, and associated morbidities such as hydrocephalus.[138,165] Alarmingly, approximately 30%–70% of brain tumor survivors score below-average intelligence on IQ testing and only < 10% of children score above-average IQ.[166-168] In a large meta-analysis of 22 studies, Mulhern showed that across all age groups, there was a notable decline of 12–14 IQ points in children treated with CRT compared to matched controls.[169] In a cohort of 120 children with medulloblastoma, Hoppe-Hirsch et al. showed that this decline can be progressive. At the start of the study, 58% had IQ > 80 at 5 years post-CRT, which dropped precipitously to 15% at 10 years with 36% of the patients incapable of living independently.[168]

Critical predictors of CRT and its neurocognitive impact include age at the time of CRT, radiation dose, CNS tumor location, and adjunctive MTX. Young age at the time of CRT is undoubtedly a key risk factor. Although the expansion of neural gray matter peaks around 4–5 years of age, white matter growth and axonal myelination continue well into the second and early third decades of life.[157] Early injury from CRT causes demyelination and white matter volume reduction. Waber et al. assessed 51 children with ALL who received 24-Gy CRT and saw a clear association between young age at the time of CRT and microcephaly.[170] Similarly, Jannoun et al. assessed a cohort of brain tumor patients who received CRT.[171] The total radiation dose ranged between 50 and 55 Gy for children > 3 years old and 45 Gy for children < 3 years old. On serial cognitive testing 3–20 years posttreatment, the study found that IQ scores were most impacted in the very young age group. The mean IQ was 72 in children who received CRT at 1–5 years of age, 93 in the 6–10 age group, and 107 in the 11–15 age group. CRT dosage is another important risk factor. High-dose CRT positively correlates with greater cognitive morbidity.[172] Several studies have attempted to study this relationship through neuraxial dose reduction in the treatment of medulloblastoma. In a landmark study based on the Pediatric Oncology Group (POG) and the Children's Study Group (CCC) treatment protocols for low-risk medulloblastoma, the study randomized 126 patients to receive either 36 Gy or 23.4 Gy.[173] This study was terminated at 16 months due to unacceptable rates of early relapse associated with neuraxial RT dose reduction. However, in follow-up analyses of this cohort, Mulhern et al. found that younger children treated with 23.4 Gy had a 10- to 15-point increase in IQ points over their matched cohorts who received 36 Gy.[174] Additionally, irradiation to particular brain regions (notably, the cortico-hemispheric region and the

hypothalamus) and adjunctive MTX (predisposing to leukoencephalopathy) are other known risk factors for cognitive impairment.[175-177]

Several strategies have been adopted to mitigate the impact of CRT on neurocognition, including the elimination of CNS prophylaxis in select ALL patients, dose reduction, and use of different radiation strategies in CNS tumors. Standard ALL protocols have moved away from routine administration of CNS prophylaxis, and radiotherapy is now reserved for high-risk patients and those with CNS relapse or refractory to chemotherapy. Although withholding radiotherapy is not an option for the majority of brain tumors, transitioning to intensity-modulated photon radiation therapy (IMRT) and proton-based radiation are increasingly favored options. Hyperfractionation is another modality whereby the total radiation dose is partitioned into multiple small fractions with no more than 5 fractions per week to allow healthy neurons time to self-repair.[178,179] While proton therapy generates less radiation scatter than conventional photon-based CRT and theoretically causes less damage to the surrounding healthy tissue, few studies have investigated the neurocognitive consequences postproton CRT. In a recent study published by Ventura et al., the authors followed a cohort of 65 children treated with proton therapy for a primary brain tumor.[180] Neurocognitive assessments were made at least 1 year from the time of CRT. Compared to photon outcomes reported in the literature, the authors observed improved preservation of intellectual capacity following the proton therapy. The overall academic achievements and quality of life of the children were within the normal limits, although processing speed and memory capacity were both negatively impacted compared to population norms. This was posited to be attributed to concomitant risk factors such as history of CSI exposure, surgical resection, and chemotherapy.

Several pharmacologic therapies for neuroprotection or cognitive stimulation are under active investigation, including memantine, modafinil, and donepezil. Memantine is a NMDA receptor antagonist used in Alzheimer's disease that has proven neurocognitive-stimulating effects. Currently, there is a trial assessing the use of memantine in pediatrics to prevent late cognitive effects after CRT (NCT03194906). Modafinil is another agent with neurologic-stimulating properties used in pediatric trials. Although distinct from sympathomimetic agents like amphetamine, modafinil is shown to have neuro-awakening benefits and is commonly used in patients with narcolepsy.[181] A pediatric trial is ongoing, evaluating its use to treat children with memory and attention deficit problems following brain tumor therapy (NCT01381718). Finally, the general management of impaired neurocognition necessarily takes a multidisciplinary approach focused on cognitive rehabilitation, creating school-based individualized education programs and supporting the child's psycho-behavioral needs.

3.2.2 Cardiovascular injury: Stroke, moyamoya

Cerebrovascular disease is a known late effect of cranial radiation in conventional photon RT. There is a robust body of literature reporting late vasculopathies in cancer survivors postcranial radiation.[182–184] These cerebrovascular changes include strokes, veno-occlusive disease/moyamoya, cavernous malformations and aneurysms, and stroke-like migraines. The pathophysiology is attributed to ionizing radiation causing damage to medium and large cerebral vessels.[185] Encased by the thick muscular tunica media, large arteries initially protected from acute radiation injury over time become weakened by remodeling and chronic inflammation. This leads to excessive intimal fibrosis and narrowing of large vessels, reminiscent of severe atherosclerosis.[186]

In cancer survivors with a history of CRT, several key studies have documented an unequivocal association between increased stroke risk and CRT.[182] In a large, multi-institutional study conducted by the CCSS group, Bowers et al. analyzed the incidence of self-reported strokes in > 10,000 survivors of childhood cancers treated with CRT during 1970–86.[183] The authors compared stroke risk between 4828 leukemia survivors and 1871 brain tumor survivors vs 3846 sibling controls. In the leukemia cohort, late-occurring strokes presented at a median of 9.8 years post-CRT exposure. The rate of occurrence was 58/100,000 person-years in the leukemia group, with a cumulative incidence of 0.73% at 25 years. Compared to sibling controls, stroke among leukemia survivors had a 6.4 relative risk (RR) ratio. Comparatively higher in the brain tumor vs leukemia cohort, the rate of occurrence was 267.6 per 100,000 person-years with a cumulative incidence of 5.58% at 25 years. The RR of stroke among brain tumor survivors was 29.0. Strikingly, there was a clear difference in stroke risk between the brain tumor and leukemia group that was largely attributed to RT exposure and dosage. To investigate the correlation between stroke risk and CRT dose, the authors also performed a combined analysis of leukemia and brain tumor survivors and then determined the rate of stroke stratified by RT dose. The study found statistically significant differences between survivors who have received < 29 Gy vs 30–49 Gy and > 50 Gy. Furthermore, a follow-up CCSS study using the same cohort of stroke patients in the Bower's study found that CRT is a strong predictor of recurrent stroke.[184] The 10-year cumulative incidence of recurrence was 33% for patients treated with high-dose CRT > 50 Gy vs 11% for patients with no prior CRT exposure.[184] The CCSS cohort has also been utilized to highlight the detrimental impact of stroke and stroke recurrence on mortality and quality of life in childhood cancer survivors, showing over double the rate of mortality after a single vs recurrent stroke and a substantial increase in odds ratio (OR 5.3) of living with a caregiver after a recurrent stroke.[187] A key limitation of the CCSS studies is that outcomes are based on patient reports and details on type of strokes could not be evaluated. However, additional studies using imaging review supported the observed associations between CRT and stroke and stroke recurrence.[188] One retrospective chart review of > 430 pediatric brain tumor patients treated with CRT between 1993 and 2002[189] assessed the incidence of medically diagnosed stroke or transient ischemic attack (TIA) as a late complication of CRT over a median follow-up period of 6.3 years. The group observed a 10-fold increase in stroke/TIA incidence of 584 per 100,000 survivors compared to the general population of 2–8 per 100,000. The median time between CRT and the first event was 4.9 years. Of interest, this study also hypothesized that radiation to large vessels involving the circle of Willis would confer the greatest stroke risk. The circle of Willis resides near the prepontine cistern, providing cerebral blood flow to the entire brain. Indeed, the group uncovered a stroke hazard ratio of 9.0 following irradiation of the circle of Willis compared to 3.4 for radiation outside this region.[189]

Taken together, these studies make evident that CRT poses a significant risk to the late development of stroke, a risk modulated by RT dosage and involvement of large vessels. The literature also cites alkylating therapy, young age, and comorbid atherosclerotic risk factors as additional risk-modifying factors.[190] Furthermore, in a multitude of studies, neurofibromatosis 1 (NF1) has emerged as another important risk factor in the development of cerebral steno-occlusive disease, especially moyamoya. Similarly, radiation injury can cause large cerebral artery fibrosis and create a compensatory network of wispy anastomoses characteristic of moyamoya disease. In a well-cited study by Ullrich et al., the authors analyzed the incidence of moyamoya in 345 pediatric patients who received CRT for primary CNS tumors.[191] The median follow-up period was 4.5 years. Using serial MRI/MRA, the study identified 12 patients (3.5%) with radiographic evidence of moyamoya. Nine (75%) out of 12 patients subsequently developed symptomatic stroke and/or seizure. Of importance, this is also one of the first studies to highlight NF1 as a critical risk factor. NF1 is a genetic disorder that leads to the loss of neurofibromin, a GTP-ase protein that negatively regulates the RAS pathway, but of relevance is also expressed in the endothelium that when absent may lead to abnormal vascular proliferation after injury.[192] In NF1 patients, Ullrich et al., saw a markedly shorted latency to moyamoya, 38 months compared to 55 months in non-NF1 patients.

For proton-based RT, this form of irradiation can also confer a significant risk of stroke, despite having less off-target effects. In a 2017 retrospective study by Kralik et al., the authors analyzed a cohort of 75 pediatric patients who have undergone proton therapy for the treatment of primary brain tumors.[193] The median follow-up period was 4.3 years. The authors identified that 6.7% (5/75 patients) developed MRI findings of large cerebral vessel stenosis with a median time to development of 1.5 years. Four of the five patients subsequently presented with a clinically significant stroke. In a large 2018 case series, Hall et al., similarly reported proton therapy to be a significant predictor of cerebral vasculopathy.[194] The authors evaluated 644 pediatric patients with primary brain tumor treated with proton therapy between 2006 and 2015. The median period of follow-up was 3 years. The study identified a 3-year cumulative rate of vasculopathy to be 6.4% and of clinically significant stroke/TIA to be 2.6%. Age was a significant risk factor. Patients < 5 years old were significantly more vulnerable to proton-induced vasculopathy. This aligns with earlier reports in the literature.[195,196] Moyamoya, on the other hand, is less well described in proton therapy with only a small number of case reports to date.[197–199] One child presented with acute intermittent left focal deficits and a history of posterior fossa ependymoma treated with gross total resection and proton therapy.[198] MRI consequently revealed hallmark findings of moyamoya, including dilated lenticulostriate arteries.

3.2.3 SMART syndrome

Stroke-like migraine attacks have been reported in adult and pediatric patients several years after CRT. Patients frequently present between 2 and 10 years from CRT exposure with neurological complaints reminiscent of a migraine attack.[5,182] Symptoms include severe, recurrent headaches, focal neurologic deficits, and seizures—all signs initially worrisome for CNS tumor recurrence. However, these neurologic changes are transient and reversible. MRI often shows cortical swelling and enhancement within the parieto-occipital region and hypermetabolism on PET. EEG demonstrates a diffuse cortical slowing. The exact pathophysiology is not clear, but CRT-induced vasculopathy, injury to the trigeminovascular system, and ion channel deficiencies have been proposed.[5,182] Regardless of its underlying mechanism, this disorder often presents late and only lasts a couple of weeks before full spontaneous resolution.[200] As a mimicker of recurrent CNS disease, this entity merits attention to provide provider reassurance and limit unnecessary interventions.

3.2.4 Vascular malformations: Cavernous malformation, aneurysms

Vascular malformations following the photon-based radiation therapy is a complication more commonly seen in pediatrics than in adults.[201,202] Radiation injury compromises the integrity of vessels that become prone to rupture. These malformations often take the form of intracerebral cavernous malformation (ICM), aneurysms, and telangiectasia. ICMs are a known late complication of RT and are defined as dilated vascular sinusoids encased by a single layer of endothelial cells with no intervening brain parenchyma. The natural prevalence of ICM is reported to be < 0.5%–0.6% in the general population, but becomes markedly higher after CRT.[203] In a large 2015 retrospective study by Gastelum et al., the authors assessed the incidence of radiographic ICM in children who received CRT between 1980 and 2009.[203] Within this single institutional study of 362 patients, 10 were found to have imaging-proven ICMs. The median time to onset was 12 years after CRT with a 10-year cumulative incidence of 3%. Similarly in another large retrospective series from Germany, the authors analyzed all pediatric patients who received CRT between 1980 and 2003.[204] The study included 171 patients who underwent CRT for a spectrum of malignancies, including primary brain tumors and leukemia, as well as TBI for transplant. Nearly 5% developed ICM with a median latency time of 14.6 years. Consistent with other studies, the cumulative incidence 10 years post-CRT was 3.8%. Of interest, none of the patients developed ICM-associated complications or intracranial hemorrhage. Several patients underwent prophylactic surgical resection. On risk factor analysis, it was striking that most ICM survivors were treated with CRT during the first 10 years of life. In this cohort, the 10-year cumulative incidence almost doubled at 7%, highlighting young age as a critical predictor of ICM development. In another notable study assessing ICM in all patients indiscriminate of age or initial diagnosis, Heckl et al. evaluated 189 cases of ICM and found that 40 patients had a history of CRT.[205] Remarkably, a vast majority of these patients received CRT during childhood; in fact, 63% of these patients received CRT when < 10 years old. Although composed of tenuous blood vessels, ICM was rarely reported to cause intracranial hemorrhage, likely preempted by early surgical resection. Aneurysms are another vascular complication. Although far less frequent than ICM, they are linked to greater mortality and prone to life-threatening rupture. Patients often present > 10 years from the initial CRT and after acute-onset aneurysmal subarachnoid hemorrhage with a high mortality rate of > 40%.[206] Finally, small vessel diseases such as cerebral microbleeds are also linked to prior CRT exposure. Patients with cerebral microbleeds often demonstrate worse neurocognitive outcomes, including executive function, particularly for patients with microbleeds localized within the frontal or temporal lobes.[207] Little is known about proton-based CRT and the associated incidence of vascular malformation, but vasculopathies have been seen in large cerebral arteries with stroke-like manifestations as previously discussed.

3.2.5 Endocrinopathies

Endocrine abnormalities are common after radio-therapy. The central hypothalamic-pituitary axis is remarkably radiosensitive, in particular the hypothalamus where injury can be seen with doses as low as 18–24 Gy.[208] The pituitary gland is slightly less vulnerable, and deficiencies such as growth hormone (GH), thyroid-stimulating hormone, ACTH, and gonadotropin occur at doses > 50 Gy.[179] In survivors of childhood cancers, endocrine derangements not only directly impact growth, energy, and pubertal maturation, but also have far-reaching consequences on normal psychosocial/behavioral development.

Growth hormone deficiency

Growth hormone deficiency (GHD) is the most common CRT-related endocrinopathy.[179] Clinically, patients present with cardinal features of GHD, including blunted growth velocity, delayed skeletal maturation, and insufficient GH provocation by pharmacologic secretagogues. In early ALL studies involving 24-Gy whole brain CRT, investigators found pronounced height differences between survivors compared to their age-matched peers.[209] Katz et al. evaluated the heights of 51 long-term survivors who received CNS prophylaxis and found significantly compromised linear growth.[209] For the 51 patients, the mean adult height was greater than one standard deviation below the average height of the general population. Similarly, in an older study, 46 patients treated with craniospinal radiation for ALL were followed over a course of 6 years and underwent serial GH assessments.[210] The authors found that the height of over 71% of patients was greater than one standard deviation below average and most survivors demonstrated partial or complete GH deficiency. Of interest, the average height in the remainder of patients without GH aberrations on testing was also markedly lower than in the general population. The authors attributed this observation to the limitations of the GH provocative assay since a portion of these patients subsequently responded to GH replacement, despite normal testing. CRT-induced GHD is uniquely characterized by diminished spontaneous GH secretion with the preservation of peak GH responses on pharmacological assays.[211] Following the removal of CRT as CNS prophylaxis from COG-based standard-risk ALL regimens, growth hormone deficiencies in ALL survivors are no longer common. For CNS tumors however, CRT remains the mainstay of treatment. Literature cites that as many as 95% of survivors of pediatric brain tumors experience GHD.[212,213] In a 2011 prospective study, Merchant et al. performed endocrine testing on a cohort of 192 pediatric patients treated with conformal radiation therapy.[214] Interval provoked GH stimulation testing was provided between 6 and 60 months from the time of CRT. The authors found that CRT was indeed a critical predictor of GH deficiency, as well as injury to the hypothalamus. The authors determined that the risk of GHD exceeds 50% at 5 years post-CRT after > 16 Gy dose to hypothalamus. Unsurprisingly, several studies also support that increased radiation dose and inclusion of the hypothalamus within the radiation field confer high GHD risk.[215,216] In a large Childhood Cancer Survivor study conducted by Gurney et al., the authors obtained treatment records and survey responses from 921 adult survivors of pediatric brain tumors.[217] Approximately 40% of participants had an adult height markedly below population norms. Young age is another important risk. When stratified by age, 53% of adults treated for CNS malignancy before age 5 were below the 10th percentile for height. As seen in the ALL studies, hypothalamus exposure and CRT dose significantly correlated with short adult stature. In fact, hypothalamic-pituitary axis (HPA) exposure of 20 Gy or greater was linked to a threefold increase in risk of short stature and > 60 Gy was linked to a fivefold risk increase.[218,219] There is a myriad of alternative causes that further contribute to the development of poor growth in the pediatric cancer population, including poor nutrition, chemotherapy, other endocrinopathies, early puberty, and stunted spinal growth from spinal radiation. Spinal irradiation in particular can cause significant vertebral body damage especially when exposure occurs during puberty at a time of accelerated spinal growth.[220]

GH replacement forms the cornerstone of GHD treatment. Although GH replacement can improve the growth velocity, it does not completely reverse the growth impediment conferred by radiotherapy. In general, the final height gain in treated children is still less than the predicted height for age. This is also true when these children are compared to individuals with idiopathic GHD (iGHD). In a study undertaken by the Pfizer International Growth Study Group, the investigators analyzed the response to GH in children treated for medulloblastoma (MB) vs children with iGHD.[221] Final height attainment in MB patients was significantly lower compared to the iGHD cohort despite GH therapy.[221] To ameliorate the height disadvantage, gonadotropin-releasing hormone (GnRH) analogs in conjunction with GH replacement have also been used. One of the first studies analyzing the efficacy of combined GH and GnRH treatment is a 2000 report by Adan et al.[222] The study included 56 survivors of childhood cancer indiscriminate of initial diagnosis who received CRT and GH replacement plus GnRH analogs between 1986 and 1997.[222] In treated patients, there was a significant improvement in adult heights compared to untreated individuals. Improvements in height were attributed to earlier testing for GHD, prompt initiation of replacement therapy, and the combined use of GnRH analogs. However, the final heights of this cohort were still shorter than population norms. Overall, it is reasonable to conclude that final height in children with GHD following CRT will be less than normal predicted height despite GH replacement.

Historically, there has been much controversy regarding the use of GH due to safety concerns of inducing disease recurrence or secondary malignancy. Earlier studies have reported a heightened incidence of leukemia in pediatric patients treated on GH.[223–225] However, numerous follow-up studies have disproven this conclusion.[226–228] Large cohort follow-up studies also did not show an increased leukemia incidence in solid tumor or brain tumor relapse.[229–231] The risk of secondary malignancy, on the other hand, is less clear, and there is data to support a minimal increase, especially of meningiomas, but the data is subjected to statistical bias given the small number of events.[232,233] Overall, GH administration in cancer survivors is deemed safe. There is a common practice, however, to defer GH therapy until at least 1 year from remission because the risk of recurrence is generally highest during the first year after the end of treatment.

Gonadotrophin deficiency

The second-most common clinically significant endocrinopathy is gonadotropin deficiency. Clinical presentation of this hormonal aberration appears highly dependent on RT dosage. Doses > 50 Gy result in GnRH deficiency, whereas low- to moderate-dose RT in the range of 20–50 Gy paradoxically results in early puberty.[179,234] When ALL patients were historically treated with 24 Gy for CNS prophylaxis, early puberty was common, particularly in girls.[234,235] This gender dichotomy was no longer evident when patients were exposed to higher RT doses in the range of 30–50 Gy.[236] Literature cites the cessation of negative feedback by the cortex on the hypothalamus as the cause for early puberty.[234] On the other hand, when RT doses reach an excess of 50 Gy, absolute suppression of the hypothalamic-pituitary-gonadal axis is seen. The clinical presentation of GnRH deficiency is wide and can vary from subtle decreases in sex hormones to severe drops in circulating hormones, resulting in severe sexual maturity and secondary sexual development.

Adrenocorticotropic hormone deficiencies

Adrenocorticotropic hormone (ACTH) deficiency in children after CRT is a rare occurrence, but one that merits close attention due to its potential life-threatening complications. When severe, ACTH deficiency can manifest as labile blood pressure, organ hypoperfusion, hypoglycemia, or the impaired ability to mount an appropriate physiologic response to systemic illness. Radiation dosage is closely associated with increasing incidence of this disorder. With low-dose RT of 18–24 Gy previously used in ALL patients, few to no cases of RT-associated ACTH

deficiency were observed. However, when higher doses of RT were administered to brain tumor patients, ACTH deficiencies were detected at greater frequencies.[213,234,237] This disorder often presents as a late effect and has been detected in patients > 10 years post-CRT. Vigilance in surveillance and awareness of this complication is advised to continue well into adulthood. Treatment involves mineralocorticoid replacement such as hydrocortisone to maintain physiologic levels with higher doses used during periods of illness.

Hypothyroidism

Hypothyroidism is another known complication of RT. Clinical manifestations include an array of metabolic derangements from growth impairment to weight gain, as well as to neurobehavioral disturbances coupled with poor school performance and academic outcomes. Both central and primary hypothyroidism are seen but each is caused by different RT modalities (cranial vs spinal). The hypothalamic-pituitary-thyroid axis is relatively radioresistant, and central hypothyroidism only becomes clinically appreciated at RT doses in excess of 50 Gy. Thyroid dysfunction can be seen in ALL survivors after 24 Gy cranial RT, but clinical symptoms are subtle and mostly detected on laboratory testing of thyroid-stimulating hormones free thyroxine (free T4), thyroxine (T4), and triiodothyronine (T3).[220] In contrast, central hypothyroidism is clinically relevant in children with brain tumors. In a well-cited study by Rose et al., the authors followed 208 pediatric cancer survivors with a history of various tumor diagnosis, including brain tumors, head/neck cancers, and ALL.[238] The study found that central hypothyroidism was discerned in approximately 50% of patients with suprasellar or nasopharyngeal tumors and approximately 35% of patients who had supratentorial or posterior fossa tumors. Incidence was significantly correlated with total cranial radiation dose. Cumulative incidence of hypothyroidism at 10 years postdiagnosis after a history of > 30 Gy RT exposure was 39% compared to only 8% in survivors with history of 15- to 29-Gy RT exposure.[238] Primary hypothyroidism is comparatively more common due to injury to the thyroid gland that lies within the radiation field of CSI. CSI forms the backbone of treatment in medulloblastoma to mitigate the risk of leptomeningeal dissemination, and primary hypothyroidism is best studied in this group of children. A prevalence rate of 20%–60% has been reported in survivors of medulloblastoma following CSI.[239,240] Thyroid replacement therapy with thyroxine forms the mainstay of therapy and is generally well tolerated. Appropriate thyroxine therapy often ameliorates symptoms; however, complete reversal of linear growth and neurologic-behavioral complications from hypothyroidism is debated. Separately, the risk of thyroid carcinoma increases proportionally with increasing dosage of RT within the low-moderate range; however, the risk plateaus at higher doses. In fact, cancer risk declines at high doses of > 30 Gy and is a reflection of RT-induced irreparable damage or cell death.[212] Most thyroid cancers consist of mature papillary carcinomas, and timing of presentation is usually at > 5 years from therapy.[241,242]

Obesity

Obesity is a recognized late effect in cancer survivors following CRT exposure. The mechanism is multifactorial and largely attributed to hypothalamic obesity, untreated GHD, and steroid administration. Direct cellular damage to the hypothalamus, in particular to the ventromedial hypothalamus (VMH), causes dysregulated energy intake and expenditure. In a 1999 study by Lustig et al., the authors found significant increases in the BMI of brain tumor survivors 10 years postdiagnosis compared to population norms.[243] Key risk factors include hypothalamic injury, CRT > 50 Gy, and young age at the time of treatment.[244] Effects of hypothalamic obesity can be debilitating with pervasive physical and social impacts. Given the relationship between hyperinsulinism from VMH damage and hyperphagia, octreotide as a somatostatin analog has been trialed in pediatric brain tumor survivors with promising results. In a 2003 randomized, placebo-controlled trial, investigators found that the administration of octreotide therapy improved BMI with a mean change in BMI by − 0.2 in the treatment group vs + 2.2 in the placebo group.[245]

To mitigate these adverse endocrinopathies from photon-based irradiation, proton CRT is now increasingly favored given its tissue-sparing properties. Indeed, proton CRT is seen to be associated with fewer late endocrine effects. In a report by Eaton et al., the study showed a statistically significant association between improved final heights and reduced incidences of hypothyroidism and sex hormone deficiency in patients treated with proton-based CRT compared to photon-based CRT.[246] However, proton therapy did not eliminate all radiation-associated late endocrine defects, in particular GHD.

3.2.6 Secondary malignancy

As survivorship increases with advanced antineoplastic therapies, there is a concurrent rise in treatment-associated secondary malignancy. These secondary cancers are distinct from primary recurrence and now form the leading cause of morbidity and mortality in cancer survivors.[124,247] Secondary CNS malignancy after cranial RT is well documented. Specifically, CRT increases the risk of secondary meningiomas, malignant gliomas, and embryonal tumors.[247,248] In the early 2010s, several CCSS studies highlight radiotherapy as a critical factor for the subsequent neoplasms. Friedman et al. performed a large study consisting of 14,359 survivors within the CCSS cohort and report an overall 20.5% 30-year

cumulative incidence for all secondary malignancies.[249] In the group, 1402 survivors developed a total of 2703 neoplasms, 9.6% of which were CNS tumors. Among the CNS tumors, meningioma and glioma were identified with the highest frequency. When stratified by secondary malignancy subtype, the cumulative 30-year incidence of meningioma was 3.1% and was highest among survivors of medulloblastoma at 16.4%. Across all secondary malignancies, history of radiotherapy exposure was found to be the key risk factor (RR 2.7). Consistent with an earlier nested case-control study on the CCSS cohort, Neglia et al. investigated the occurrence of secondary neoplasms of the CNS in 14,361 long-term survivors and performed a detailed assessment of RT dose-risk relationships.[250] The most common neoplasms were meningioma ($n = 66$), glioma ($n = 40$), and primitive neuro-ectodermal tumor ($n = 6$). Median time to development was 9 years for gliomas, but a much longer latency for meningiomas with a median time of 17 years. On risk factor assessment, the authors found that radiation therapy was significantly correlated with the development of meningioma (OR 9.9) and glioma (OR 6.8) with a combined OR for all CNS tumors of 7.1. There is also a linear correlation between dose of radiation and rates of CNS tumors. For both meningioma and gliomas, OR for development peaked at exposures of 30–44 Gy (OR 96.3 for meningioma and OR 21 for gliomas). This is consistent with later studies.[251] In a 2013 retrospective meta-analysis by Bowers et al., the authors found that meningioma and high-grade glioma ranked as the most frequent secondary neoplasms in which nearly all patients (95%–100%) had a positive history of cranial irradiation.[252]

In light of increasing awareness between risk of secondary neoplasm and RT dose, there has been a push to dose reduce CRT or eliminate CRT entirely in case of low/standard risk patients. In a large 2017 CCSS study assessing changing trends of antineoplastic therapy and temporal correlation with incidence of secondary cancers, Turcotte et al. found that risk of subsequent neoplasms became markedly lower due to therapy modifications.[253] The authors performed a retrospective multicenter cohort analysis on 23,603 patients diagnosed between 1970 and 1999 with a mean follow-up of 20.5 years. The analysis revealed distinct differences in the 15-year cumulative incidences of subsequent neoplasms between patients treated in the 1970s (2.1%) vs those treated in the 1990s (1.3%). For all secondary cancers, including meningioma, RR declined for every 5-year increment of treatment era.

For childhood cancer survivors who subsequently develop secondary neoplasms, genetic susceptibility plays a key role. There are a number of predisposition syndromes that exacerbate the risk of subsequent brain tumors following cranial radiation. Sensitivity to radiotherapy is well described in patients with NF1 or 2, von Hippel-Lindau disease, Li-Fraumeni syndrome, Gorlin syndrome, and ataxia telangiectasia, among others.[254,255] It is no surprise that many of these syndromes are linked to DNA repair defects with cells poised to undergo malignant transformation. Radiation induces innumerable double-strand breaks that can overwhelm an already-defective DNA repair system. Of late, efforts to identify children with predisposition syndrome and perform early surveillance is gaining traction. In 2017, the Pediatric Cancer Working Group of the American Association for Cancer Research (AACR) assembled a large multinational group of medical professionals to publish a set of consensus guidelines aimed to standardize approaches to early cancer detection for children with predisposition syndromes.[256] However, for the children who have concurrent or prior history of primary malignancies, guidelines on therapy modifications or implications on secondary cancer risk were not addressed. The authors emphasized that in these unique cases, management and subsequent surveillance are highly personalized and will depend on the underlying disorder and the primary cancer therapy received.

3.3 Hematopoietic stem cell transplant

Late neurologic complications in HSCT are rare and mostly pertain to the development of chronic graft-vs-host disease (GVHD). Chronic GVHD is mediated by donor T-cell attack of host tissue antigens and multiorgan damage, including the skin, gastrointestinal tract, liver, and lungs.[79,257] Chronic GVHD is known to affect the nervous system; however, its neurologic sequelae are best recognized in the peripheral nervous system (PNS).[79,258,259] This is due to the difficulty isolating chronic GVHD as the primary cause in the presence of confounding factors, such as history of cytotoxic drug exposure, radiation, and infections. Neurologic manifestations occur from several months after transplantation in parallel with the development of chronic GVHD[79,258] and most literature is based on adult survivors.[258–260]

PNS manifestations from GVHD is an established phenomenon.[258] Chronic GVHD can affect all anatomic structures within the PNS from the nerve bundle, nerve sheath, and ganglion to the neuromuscular junction and associated muscle fascia. It has been reported that chronic GVHD can result in disorders, including immune-mediated neuropathies such as Guillain-Barré syndrome (GBS), neuromuscular junction disorders such as myasthenia gravis and myopathies, polymyositis, and myopathies.[258] In the 2014 National Institutes of Health (NIH) consensus criteria for chronic GVHD however, only peripheral neuropathy, polymyositis, and myositis were recognized as distinct neurologic sequelae caused by chronic GVHD.[261]

In contrast, CNS involvement in GVHD is controversial. Although CNS manifestations were not included within the 2014 NIH scoring criteria, several groups do ascribe CNS complications to chronic GVHD. Both Openshaw et al. and Grauer et al. have proposed a set of diagnostic criteria to facilitate the diagnosis of potential GHVD-associated CNS changes.[258,259] These guidelines take into consideration the presence of chronic GVHD concomitant with neurologic findings not otherwise explained, abnormal brain MRI, abnormal CSF studies such as pleocytosis of elevated immunoglobulin, and response to immunosuppressive therapy.[258] Although definitive diagnosis necessitates histologic confirmation of GVHD-induced changes, this is seldom obtained. Neurologic consequences from chronic GVHD can be quite variable in presentation, encompassing a spectrum of disorders that includes encephalopathy, seizures, cerebrovascular disease, demyelinating diseases, focal neurologic deficits, and immune-mediated encephalitis.[79,258,260] Brain MRIs often show multifocal white matter changes, areas of ischemic infarction, and small hemorrhages.[79]

Treatment of neurologic manifestations is directed to treating the underlying GVHD. This often entails a regimen of corticosteroids, calcineurin inhibitors as well as newer immunomodulatory therapies.[79,258] Neurologic complications may improve with treatment; however, chronic GVHD is a significant predictor of increased mortality and morbidity. A 2015 study by Kang et al., assessing a total of 383 pediatric patients who received allogenic transplantation, found that 70 patients subsequently developed neurologic complications.[76] One-third of the patients developed symptoms > 100 days posttransplantation. In a multivariate analysis of critical predictors for mortality, the study identified that extensive chronic GVHD was a statistically significant risk factor with a hazard risk of 5.98.

3.4 Novel agents: Immune-based therapies

The advent of immunotherapy and targeted therapies marks a new era of cancer treatment and personalized care. These therapies hold remarkable potential for cure with hypothetically, less off-target side effects. However, these regimens are still in their infancy. In contrast to our considerable experience with conventional agents, XRT and BMT, the long-term tolerability of novel therapeutics is largely unknown.[262,263] In our short experience administering targeted therapies, a small number of potential late effects in children have been seen, including endocrine abnormalities (predominately in thyroid and bone mineral metabolism), cardiomyopathy, and chronic immune suppression.[262] However, a complete understanding of the late effects and their impact on children's neurocognitive development and psycho-behavioral maturity will require longitudinal follow-up. Several phase I/II SIOP and COG trials are now underway to study the tolerability profile and disease response of novel therapeutics and long-term follow-up data from these cohorts will be indispensable. It is also evident that to formulate an accurate appreciation of the late effects, rigorous, adequately powered, and prospective clinical trials evaluating the chronic complications and neurologic changes over time are necessary in children. Fortunately, there is concerted movement toward this effort. The FDA has recently instituted a mandate encouraging sponsors and investigators to include late effect correlative study aims, particularly for cellular therapy and gene therapy products.[264] In the future, understanding the late effects of novel agents can provide a more complete picture of the risk and benefits and help providers and families better navigate the intricacies between old vs new treatment modalities.

4 Conclusion

Remarkable advances in cancer research have accelerated the development of effective cures. Excitingly, these breakthroughs in science and medicine have translated to dramatic increases in the OS of children and young adults with malignancy. In less than 3 decades, the average 5-year survival rate for childhood cancers has gone from 50% in the 1970s to now over 80%–85%.[1,2,265] However, concurrent with the rise in survival, the number of patients experiencing acute and late complications from treatment has also risen. In children, the nervous system is particularly susceptible to toxicities that can result in devastating and sometimes irreversible consequences. The breadth of potential neurologic injury can impact all facets of children's lives from physical growth to psychosocial impairment, all of which can persist well into adulthood. Given the improved treatments and a growing population of childhood cancer survivors, it has become ever important to understand the acute and late effects of therapy along with the development of effective survivorship management. Future studies will need to focus on neurological changes during the acute and posttreatment period and how these changes affect the long-term neurocognitive growth and quality of life. Such investigations will be of special importance in the setting of new therapy modalities such as targeted and immune-based therapies, which continue to actively evolve.

References

1. Siegel RL, Miller KD, Jemal A. Cancer statistics, 2020. *CA Cancer J Clin.* 2020;70(1):7–30.
2. Howlader N, Noone A, Krapcho M, et al. *SEER Cancer Statistics Review, 1975–2018.* Bethesda, MD: National Cancer Institute; 2018:1–12.
3. Siegel RL, Miller KD, Jemal A. Cancer statistics, 2018. *CA Cancer J Clin.* 2018;68(1):7–30.
4. Giglio P, Gilbert MR. Neurologic complications of cancer and its treatment. *Curr Oncol Rep.* 2010;12(1):50–59.
5. Sun LR, Cooper S. Neurological complications of the treatment of pediatric neoplastic disorders. *Pediatr Neurol.* 2018;85:33–42.
6. Reddy AT, Witek K. Neurologic complications of chemotherapy for children with cancer. *Curr Neurol Neurosci Rep.* 2003;3(2):137–142.
7. Neil EC, Hanmantgad S, Khakoo Y. Neurological complications of pediatric cancer. *J Child Neurol.* 2016;31(12):1412–1420.
8. Armstrong C, Sun LR. Neurological complications of pediatric cancer. *Cancer Metastasis Rev.* 2020;1–21.
9. Leke R, Oliveira D, Schmidt A, et al. Methotrexate induces seizure and decreases glutamate uptake in brain slices: prevention by ionotropic glutamate receptors antagonists and adenosine. *Life Sci.* 2006;80(1):1–8.
10. Bhojwani D, Sabin ND, Pei D, et al. Methotrexate-induced neurotoxicity and leukoencephalopathy in childhood acute lymphoblastic leukemia. *J Clin Oncol.* 2014;32(9):949.
11. Cheung YT, Khan RB, Liu W, et al. Association of cerebrospinal fluid biomarkers of central nervous system injury with neurocognitive and brain imaging outcomes in children receiving chemotherapy for acute lymphoblastic leukemia. *JAMA Oncol.* 2018;4(7), e180089.
12. Mahoney Jr DH, Shuster JJ, Nitschke R, et al. Acute neurotoxicity in children with B-precursor acute lymphoid leukemia: an association with intermediate-dose intravenous methotrexate and intrathecal triple therapy—a Pediatric Oncology Group study. *J Clin Oncol.* 1998;16(5):1712–1722.
13. Singh G, Rees JH, Sander JW. Seizures and epilepsy in oncological practice: causes, course, mechanisms and treatment. *J Neurol Neurosurg Psychiatry.* 2007;78(4):342–349.
14. Steeghs N, De Jongh F, Smitt PS, Van den Bent M. Cisplatin-induced encephalopathy and seizures. *Anticancer Drugs.* 2003;14(6):443–446.
15. Hurwitz RL, Mahoney Jr DH, Armstrong DL, Browder TM. Reversible encephalopathy and seizures as a result of conventional vincristine administration. *Med Pediatr Oncol.* 1988;16(3):216–219.
16. Lennon AS, Norales G, Armstrong MB. Cardiac arrest and possible seizure activity after vincristine injection. *Am J Health Syst Pharm.* 2012;69(16):1394–1397.
17. Mahapatra M, Kumar R, Choudhry VP. Seizures as an adverse drug reaction after therapeutic dose of vincristine. *Ann Hematol.* 2007;86(2):153–154.
18. Eiden C, Palenzuela G, Hillaire-buys D, et al. Posaconazole-increased vincristine neurotoxicity in a child: a case report. *J Pediatr Hematol Oncol.* 2009;31(4):292–295.
19. van Schie RM, Brüggemann RJ, Hoogerbrugge PM, Te Loo D. Effect of azole antifungal therapy on vincristine toxicity in childhood acute lymphoblastic leukaemia. *J Antimicrob Chemother.* 2011;66(8):1853–1856.
20. Hamdy DA, El-Geed H, El-Salem S, Zaidan M. Posaconazole-vincristine coadministration triggers seizure in a young female adult: a case report. *Case Rep Hematol.* 2012;2012.
21. Goyal G, Bhatt VR. L-asparaginase and venous thromboembolism in acute lymphocytic leukemia. *Future Oncol.* 2015;11(17):2459–2470.
22. Truelove E, Fielding A, Hunt B. The coagulopathy and thrombotic risk associated with L-asparaginase treatment in adults with acute lymphoblastic leukaemia. *Leukemia.* 2013;27(3):553–559.
23. Caruso V, Iacoviello L, Di Castelnuovo A, et al. Thrombotic complications in childhood acute lymphoblastic leukemia: a meta-analysis of 17 prospective studies comprising 1752 pediatric patients. *Blood.* 2006;108(7):2216–2222.
24. Klaassen IL, Lauw MN, Fiocco M, et al. Venous thromboembolism in a large cohort of children with acute lymphoblastic leukemia: risk factors and effect on prognosis. *Res Pract Thromb Haemost.* 2019;3(2):234–241.

25. Qureshi A, Mitchell C, Richards S, Vora A, Goulden N. Asparaginase-related venous thrombosis in UKALL 2003-re-exposure to asparaginase is feasible and safe. *Br J Haematol.* 2010;149(3):410–413.

26. Ajithkumar T, Parkinson C, Shamshad F, Murray P. Ifosfamide encephalopathy. *Clin Oncol.* 2007;19(2):108–114.

27. Feyissa AM, Tummala S. Ifosfamide related encephalopathy: the need for a timely EEG evaluation. *J Neurol Sci.* 2014;336(1–2):109–112.

28. Taupin D, Racela R, Friedman D. Ifosfamide chemotherapy and nonconvulsive status epilepticus: case report and review of the literature. *Clin EEG Neurosci.* 2014;45(3):222–225.

29. Kataria PS, Kendre PP, Patel AA. Ifosfamide-induced encephalopathy precipitated by aprepitant: a rarely manifested side effect of drug interaction. *J Pharmacol Pharmacother.* 2017;8(1):38.

30. Howell JE, Szabatura AH, Hatfield Seung A, Nesbit SA. Characterization of the occurrence of ifosfamide-induced neurotoxicity with concomitant aprepitant. *J Oncol Pharm Pract.* 2008;14(3):157–162.

31. Ide Y, Yanagisawa R, Kubota N, et al. Analysis of the clinical characteristics of pediatric patients who experience ifosfamide-induced encephalopathy. *Pediatr Blood Cancer.* 2019;66(12), e27996.

32. Szabatura AH, Cirrone F, Harris C, et al. An assessment of risk factors associated with ifosfamide-induced encephalopathy in a large academic cancer center. *J Oncol Pharm Pract.* 2015;21(3):188–193.

33. Kettle JK, Grauer D, Folker TL, O'Neal N, Henry DW, Williams CB. Effectiveness of exogenous albumin administration for the prevention of ifosfamide-induced encephalopathy. *Pharmacotherapy.* 2010;30(8):812–817.

34. Pelgrims J, De Vos F, Van den Brande J, Schrijvers D, Prové A, Vermorken J. Methylene blue in the treatment and prevention of ifosfamide-induced encephalopathy: report of 12 cases and a review of the literature. *Br J Cancer.* 2000;82(2):291–294.

35. Baker WJ, Royer Jr GL, Weiss RB. Cytarabine and neurologic toxicity. *J Clin Oncol.* 1991;9(4):679–693.

36. Herzig RH, Hines JD, Herzig GP, et al. Cerebellar toxicity with high-dose cytosine arabinoside. *J Clin Oncol.* 1987;5(6):927–932.

37. Pellier I, Leboucher B, Rachieru P, Ifrah N, Rialland X. Flushing out of cerebrospinal fluid as a therapy for acute cerebellar dysfunction caused by high dose of cytosine arabinoside: a case report. *J Pediatr Hematol Oncol.* 2006;28(12):837–839.

38. Dotson JL, Jamil MO. Successful treatment of cytarabine-related neurotoxicity with corticosteroids, a case series. *Int J Hematol.* 2018;108(5):554–557.

39. Malhotra P, Mahi S, Lal V, Kumari S, Jain S, Varma S. Cytarabine-induced neurotoxicity responding to methyl prednisolone and research, Chandigarh, India. *Am J Hematol.* 2004;77(4):416.

40. Anastasopoulou S, Eriksson MA, Heyman M, et al. Posterior reversible encephalopathy syndrome in children with acute lymphoblastic leukemia: clinical characteristics, risk factors, course, and outcome of disease. *Pediatr Blood Cancer.* 2019;66(5), e27594.

41. Morris EB, Laningham FH, Sandlund JT, Khan RB. Posterior reversible encephalopathy syndrome in children with cancer. *Pediatr Blood Cancer.* 2007;48(2):152–159.

42. Khan SJ, Arshad AA, Fayyaz MB, Ud Din Mirza I. Posterior reversible encephalopathy syndrome in pediatric cancer: clinical and radiologic findings. *J Glob Oncol.* 2017;4:1–8.

43. Raman R, Devaramane R, Jagadish GM, Chowdaiah S. Various imaging manifestations of posterior reversible encephalopathy syndrome (PRES) on magnetic resonance imaging (MRI). *Pol J Radiol.* 2017;82:64.

44. de Laat P, te Winkel ML, Devos A, Catsman-Berrevoets C, Pieters R, van den Heuvel-Eibrink M. Posterior reversible encephalopathy syndrome in childhood cancer. *Ann Oncol.* 2011;22(2):472–478.

45. Park SB, Goldstein D, Krishnan AV, et al. Chemotherapy-induced peripheral neurotoxicity: a critical analysis. *CA Cancer J Clin.* 2013;63(6):419–437.

46. Gilchrist L. Chemotherapy-induced peripheral neuropathy in pediatric cancer patients. In: *Paper presented at: Seminars in Pediatric Neurology*; 2012.

47. Mora E, Smith EML, Donohoe C, Hertz DL. Vincristine-induced peripheral neuropathy in pediatric cancer patients. *Am J Cancer Res.* 2016;6(11):2416.

48. Kandula T, Park SB, Cohn RJ, Krishnan AV, Farrar MA. Pediatric chemotherapy induced peripheral neuropathy: a systematic review of current knowledge. *Cancer Treat Rev.* 2016;50:118–128.

49. McDonald ES, Randon KR, Knight A, Windebank AJ. Cisplatin preferentially binds to DNA in dorsal root ganglion neurons in vitro and in vivo: a potential mechanism for neurotoxicity. *Neurobiol Dis.* 2005;18(2):305–313.

50. Staff NP, Grisold A, Grisold W, Windebank AJ. Chemotherapy-induced peripheral neuropathy: a current review. *Ann Neurol.* 2017;81(6):772–781.

51. Argyriou AA, Polychronopoulos P, Iconomou G, Chroni E, Kalofonos HP. A review on oxaliplatin-induced peripheral nerve damage. *Cancer Treat Rev.* 2008;34(4):368–377.

52. Jain P, Gulati S, Seth R, Bakhshi S, Toteja G, Pandey R. Vincristine-induced neuropathy in childhood ALL (acute lymphoblastic leukemia) survivors: prevalence and electrophysiological characteristics. *J Child Neurol.* 2014;29(7):932–937.

53. Brock PR, Knight KR, Freyer DR, et al. Platinum-induced ototoxicity in children: a consensus review on mechanisms, predisposition, and protection, including a new International Society of Pediatric Oncology Boston ototoxicity scale. *J Clin Oncol.* 2012;30(19):2408.

54. Sheth S, Mukherjea D, Rybak LP, Ramkumar V. Mechanisms of cisplatin-induced ototoxicity and otoprotection. *Front Cell Neurosci.* 2017;11:338.

55. Brooks B, Knight K. Ototoxicity monitoring in children treated with platinum chemotherapy. *Int J Audiol.* 2018;57(Suppl. 4):S62–S68.

56. Knight KRG, Kraemer DF, Neuwelt EA. Ototoxicity in children receiving platinum chemotherapy: underestimating a commonly occurring toxicity that may influence academic and social development. *J Clin Oncol.* 2005;23(34):8588–8596.

57. Peleva E, Emami N, Alzahrani M, et al. Incidence of platinum-induced ototoxicity in pediatric patients in Quebec. *Pediatr Blood Cancer.* 2014;61(11):2012–2017.

58. Knight KR, Chen L, Freyer D, et al. Group-wide, prospective study of ototoxicity assessment in children receiving cisplatin chemotherapy (ACCL05C1): a report from the Children's Oncology Group. *J Clin Oncol.* 2017;35(4):440.

59. Clemens E, de Vries AC, Pluijm SF, et al. Determinants of ototoxicity in 451 platinum-treated Dutch survivors of childhood cancer: a DCOG late-effects study. *Eur J Cancer.* 2016;69:77–85.

60. Ross CJ, Katzov-Eckert H, Dubé M-P, et al. Genetic variants in TPMT and COMT are associated with hearing loss in children receiving cisplatin chemotherapy. *Nat Genet.* 2009;41(12):1345–1349.

61. Tserga E, Nandwani T, Edvall NK, et al. The genetic vulnerability to cisplatin ototoxicity: a systematic review. *Sci Rep.* 2019;9(1):3455.

62. Wheeler HE, Gamazon ER, Frisina RD, et al. Variants in WFS1 and other Mendelian deafness genes are associated with cisplatin-associated ototoxicity. *Clin Cancer Res.* 2017;23(13):3325–3333.

63. Nageswara Rao AA, Wallace DJ, Billups C, Boyett JM, Gajjar A, Packer RJ. Cumulative cisplatin dose is not associated with event-free or overall survival in children with newly diagnosed average-risk medulloblastoma treated with cisplatin based adjuvant chemotherapy: report from the Children's Oncology Group. *Pediatr Blood Cancer.* 2014;61(1):102–106.

64. Freyer DR, Brock PR, Chang KW, et al. Prevention of cisplatin-induced ototoxicity in children and adolescents with cancer: a clinical practice guideline. *Lancet Child Adolesc Health.* 2020;4(2):141–150.

65. Jairam V, Roberts KB, James BY. Historical trends in the use of radiation therapy for pediatric cancers: 1973–2008. *Int J Radiat Oncol Biol Phys.* 2013;85(3):e151–e155.

66. Baumann BC, Hallahan DE, Michalski JM, Perez CA, Metz JM. Concurrent chemo-radiotherapy with proton therapy: reduced toxicity with comparable oncological outcomes vs photon chemo-radiotherapy. *Br J Cancer*. 2020;123:1–2.

67. Baumann BC, Mitra N, Harton JG, et al. Comparative effectiveness of proton vs photon therapy as part of concurrent chemoradiotherapy for locally advanced cancer. *JAMA Oncol*. 2020;6(2):237–246.

68. Keime-Guibert F, Napolitano M, Delattre J-Y. Neurological complications of radiotherapy and chemotherapy. *J Neurol*. 1998;245(11):695–708.

69. Foster KA, Ares WJ, Pollack IF, Jakacki RI. Bevacizumab for symptomatic radiation-induced tumor enlargement in pediatric low grade gliomas. *Pediatr Blood Cancer*. 2015;62(2):240–245.

70. Levin VA, Bidaut L, Hou P, et al. Randomized double-blind placebo-controlled trial of bevacizumab therapy for radiation necrosis of the central nervous system. *Int J Radiat Oncol Biol Phys*. 2011;79(5):1487–1495.

71. Liu AK, Macy ME, Foreman NK. Bevacizumab as therapy for radiation necrosis in four children with pontine gliomas. *Int J Radiat Oncol Biol Phys*. 2009;75(4):1148–1154.

72. Freeman J, Johnston P, Voke J. Somnolence after prophylactic cranial irradiation in children with acute lymphoblastic leukaemia. *Br Med J*. 1973;4(5891):523–525.

73. Littman P, Rosenstock J, Gale G, et al. The somnolence syndrome in leukemic children following reduced daily dose fractions of cranial radiation. *Int J Radiat Oncol Biol Phys*. 1984;10(10):1851–1853.

74. Powell C, Guerrero D, Sardell S, et al. Somnolence syndrome in patients receiving radical radiotherapy for primary brain tumours: a prospective study. *Radiother Oncol*. 2011;100(1):131–136.

75. Vern TZ, Salvi S. Somnolence syndrome and fever in pediatric patients with cranial irradiation. *J Pediatr Hematol Oncol*. 2009;31(2):118–120.

76. Kang J-M, Kim Y-J, Kim JY, et al. Neurologic complications after allogeneic hematopoietic stem cell transplantation in children: analysis of prognostic factors. *Biol Blood Marrow Transplant*. 2015;21(6):1091–1098.

77. Uckan D, Cetin M, Yigitkanli I, et al. Life-threatening neurological complications after bone marrow transplantation in children. *Bone Marrow Transplant*. 2005;35(1):71–76.

78. Dulamea AO, Lupescu IG. Neurological complications of hematopoietic cell transplantation in children and adults. *Neural Regen Res*. 2018;13(6):945.

79. Saiz A, Graus F. Neurological complications of hematopoietic cell transplantation. In: *Paper presented at: Seminars in Neurology*; 2004.

80. Weber C, Schaper J, Tibussek D, et al. Diagnostic and therapeutic implications of neurological complications following paediatric haematopoietic stem cell transplantation. *Bone Marrow Transplant*. 2008;41(3):253–259.

81. Ciurea SO, Andersson BS. Busulfan in hematopoietic stem cell transplantation. *Biol Blood Marrow Transplant*. 2009;15(5):523–536.

82. Vassal G, Gouyette A, Hartmann O, Pico J, Lemerle J. Pharmacokinetics of high-dose busulfan in children. *Cancer Chemother Pharmacol*. 1989;24(6):386–390.

83. Eberly AL, Anderson GD, Bubalo JS, McCune JS. Optimal prevention of seizures induced by high-dose busulfan. *Pharmacotherapy*. 2008;28(12):1502–1510.

84. De RLC, Tomas J, Figuera A, Berberana M, Fernandez-Ranada J. High dose busulfan and seizures. *Bone Marrow Transplant*. 1991;7(5):363–364.

85. Chan K, Mullen C, Worth LL, et al. Lorazepam for seizure prophylaxis during high-dose busulfan administration. *Bone Marrow Transplant*. 2002;29(12):963–965.

86. Ruiz-Argüelles GJ, Gomez-Almaguer D, Steensma DP. Outdated dogma? Busulfan, seizure prophylaxis, and stem cell allografting. *Am J Hematol*. 2012;87(9):941.

87. Caselli D, Rosati A, Faraci M, et al. Risk of seizures in children receiving busulphan-containing regimens for stem cell transplantation. *Biol Blood Marrow Transplant*. 2014;20(2):282–285.

88. Serkova NJ, Christians U, Benet LZ. Biochemical mechanisms of cyclosporine neurotoxicity. *Mol Interv*. 2004;4(2):97.

89. Gijtenbeek J, Van den Bent M, Vecht CJ. Cyclosporine neurotoxicity: a review. *J Neurol*. 1999;246(5):339–346.

90. Noè A, Cappelli B, Biffi A, et al. High incidence of severe cyclosporine neurotoxicity in children affected by haemoglobinopathies undergoing myeloablative haematopoietic stem cell transplantation: early diagnosis and prompt intervention ameliorates neurological outcome. *Ital J Pediatr*. 2010;36(1):14.

91. Straathof K, Anoop P, Allwood Z, et al. Long-term outcome following cyclosporine-related neurotoxicity in paediatric allogeneic haematopoietic stem cell transplantation. *Bone Marrow Transplant*. 2017;52(1):159–162.

92. Zunt JR. Central nervous system infection during immunosuppression. *Neurol Clin*. 2002;20(1):1–22.

93. Najima Y, Ohashi K, Miyazawa M, et al. Intracranial hemorrhage following allogeneic hematopoietic stem cell transplantation. *Am J Hematol*. 2009;84(5):298–301.

94. Graus F, Saiz A, Sierra J, et al. Neurologic complications of autologous and allogeneic bone marrow transplantation in patients with leukemia: a comparative study. *Neurology*. 1996;46(4):1004–1009.

95. Zhang X, Han W, Chen Y, et al. *Intracranial Hemorrhage and Mortality in 1461 Patients After Allogeneic Hematopoietic Stem Cell Transplantation for 6-Year Follow-Up: Study of 44 Cases*. Washington, DC: American Society of Hematology; 2013.

96. Ho VT, Cutler C, Carter S, et al. Blood and marrow transplant clinical trials network toxicity committee consensus summary: thrombotic microangiopathy after hematopoietic stem cell transplantation. *Biol Blood Marrow Transplant*. 2005;11(8):571–575.

97. Mohty M, Malard F, Abecassis M, et al. Sinusoidal obstruction syndrome/veno-occlusive disease: current situation and perspectives—a position statement from the European Society for Blood and Marrow Transplantation (EBMT). *Bone Marrow Transplant*. 2015;50(6):781–789.

98. Teachey DT, Hunger SP. Immunotherapy for ALL takes the world by storm. *Nat Rev Clin Oncol*. 2018;15(2):69–70.

99. Gupta S, Maude SL, O'Brien MM, Rau RE, McNeer JL. How the COG is approaching the high-risk patient with ALL: incorporation of immunotherapy into frontline treatment. *Clin Lymphoma Myeloma Leuk*. 2020;20:S8–S11.

100. Grupp SA, Kalos M, Barrett D, et al. Chimeric antigen receptor–modified T cells for acute lymphoid leukemia. *N Engl J Med*. 2013;368(16):1509–1518.

101. Buechner J, Grupp SA, Maude SL, et al. Global registration trial of efficacy and safety of CTL019 in pediatric and young adult patients with relapsed/refractory (R/R) acute lymphoblastic leukemia (ALL): update to the interim analysis. *Clin Lymphoma Myeloma Leuk*. 2017;17:S263–S264.

102. Gofshteyn JS, Shaw PA, Teachey DT, et al. Neurotoxicity after CTL019 in a pediatric and young adult cohort. *Ann Neurol*. 2018;84(4):537–546.

103. Lee DW, Kochenderfer JN, Stetler-Stevenson M, et al. T cells expressing CD19 chimeric antigen receptors for acute lymphoblastic leukaemia in children and young adults: a phase 1 dose-escalation trial. *Lancet*. 2015;385(9967):517–528.

104. Gardner RA, Finney O, Annesley C, et al. Intent-to-treat leukemia remission by CD19 CAR T cells of defined formulation and dose in children and young adults. *Blood*. 2017;129(25):3322–3331.

105. DiNofia AM, Maude SL. Chimeric antigen receptor T-cell therapy clinical results in pediatric and young adult B-ALL. *HemaSphere*. 2019;3(4).

106. Shimabukuro-Vornhagen A, Gödel P, Subklewe M, et al. Cytokine release syndrome. *J Immunother Cancer*. 2018;6(1):56.

III. Neurological complications of specific neoplasms

107. Neelapu SS, Tummala S, Kebriaei P, et al. Chimeric antigen receptor T-cell therapy—assessment and management of toxicities. *Nat Rev Clin Oncol.* 2018;15(1):47–62.

108. Gust J, Finney OC, Li D, et al. Glial injury in neurotoxicity after pediatric CD19-directed chimeric antigen receptor T cell therapy. *Ann Neurol.* 2019;86(1):42–54.

109. von Stackelberg A, Locatelli F, Zugmaier G, et al. Phase I/ phase II study of blinatumomab in pediatric patients with relapsed/refractory acute lymphoblastic leukemia. *J Clin Oncol.* 2016;34(36):4381–4389.

110. Ring EK, Markert JM, Gillespie GY, Friedman GK. Checkpoint proteins in pediatric brain and extracranial solid tumors: opportunities for immunotherapy. *Clin Cancer Res.* 2017;23(2):342–350.

111. Foster JB, Madsen PJ, Hegde M, et al. Immunotherapy for pediatric brain tumors: past and present. *Neuro Oncol.* 2019;21(10):1226–1238.

112. Merchant MS, Wright M, Baird K, et al. Phase I clinical trial of ipilimumab in pediatric patients with advanced solid tumors. *Clin Cancer Res.* 2016;22(6):1364–1370.

113. Cuzzubbo S, Javeri F, Tissier M, et al. Neurological adverse events associated with immune checkpoint inhibitors: review of the literature. *Eur J Cancer.* 2017;73:1–8.

114. Johnson DB, Manouchehri A, Haugh AM, et al. Neurologic toxicity associated with immune checkpoint inhibitors: a pharmacovigilance study. *J Immunother Cancer.* 2019;7(1):134.

115. Finch EA, Duke E, Hwang EI, Packer RJ. Immunotherapy approaches for pediatric CNS tumors and associated neurotoxicity. *Pediatr Neurol.* 2020;107:7–15.

116. Hutzen B, Ghonime M, Lee J, et al. Immunotherapeutic challenges for pediatric cancers. *Mol Ther Oncolytics.* 2019;15:38–48.

117. Pollack IF, Jakacki RI, Butterfield LH, et al. Antigen-specific immunoreactivity and clinical outcome following vaccination with glioma-associated antigen peptides in children with recurrent high-grade gliomas: results of a pilot study. *J Neurooncol.* 2016;130(3):517–527.

118. Mueller S, Taitt JM, Villanueva-Meyer JE, et al. Mass cytometry detects H3.3K27M-specific vaccine responses in diffuse midline glioma. *J Clin Invest.* 2020;130:6325–6337.

119. Pollack IF, Jakacki RI, Butterfield LH, et al. Antigen-specific immune responses and clinical outcome after vaccination with glioma-associated antigen peptides and polyinosinic-polycytidylic acid stabilized by lysine and carboxymethylcellulose in children with newly diagnosed malignant brainstem and nonbrainstem gliomas. *J Clin Oncol.* 2014;32(19):2050.

120. Ceschin R, Kurland BF, Abberbock SR, et al. Parametric response mapping of apparent diffusion coefficient as an imaging biomarker to distinguish pseudoprogression from true tumor progression in peptide-based vaccine therapy for pediatric diffuse intrinsic pontine glioma. *Am J Neuroradiol.* 2015;36(11):2170–2176.

121. Stepanenko AA, Chekhonin VP. Recent advances in oncolytic virotherapy and immunotherapy for glioblastoma: a glimmer of hope in the search for an effective therapy? *Cancer.* 2018;10(12):492.

122. Desjardins A, Gromeier M, Herndon JE, et al. Recurrent glioblastoma treated with recombinant poliovirus. *N Engl J Med.* 2018;379(2):150–161.

123. American Academy of Pediatrics Section on Hematology/ Oncology Children's Oncology Group. Long-term follow-up care for pediatric cancer survivors. *Pediatrics.* 2009;123(3):906.

124. Armstrong GT, Liu Q, Yasui Y, et al. Late mortality among 5-year survivors of childhood cancer: a summary from the Childhood Cancer Survivor Study. *J Clin Oncol.* 2009;27(14):2328.

125. Jones RM, Pattwell SS. Future considerations for pediatric cancer survivorship: translational perspectives from developmental neuroscience. *Dev Cogn Neurosci.* 2019;38:100657.

126. Children's Oncology Group. *Long-Term Follow-Up Guidelines for Survivors of Childhood, Adolescent, and Young Adult Cancers Version 5.0*; 2018, October.

127. Landier W, Bhatia S, Eshelman DA, et al. Development of risk-based guidelines for pediatric cancer survivors: the Children's Oncology Group Long-term Follow-up Guidelines from the children's Oncology Group Late Effects Committee and Nursing Discipline. *J Clin Oncol.* 2004;22(24):4979–4990.

128. Eshelman-Kent D, Kinahan KE, Hobbie W, et al. Cancer survivorship practices, services, and delivery: a report from the Children's Oncology Group (COG) nursing discipline, adolescent/young adult, and late effects committees. *J Cancer Surviv.* 2011;5(4):345–357.

129. Landier W, Skinner R, Wallace WH, et al. Surveillance for late effects in childhood cancer survivors. *J Clin Oncol.* 2018;36(21):2216.

130. Kremer LC, Mulder RL, Oeffinger KC, et al. A worldwide collaboration to harmonize guidelines for the long-term follow-up of childhood and young adult cancer survivors: a report from the International Late Effects of Childhood Cancer Guideline Harmonization Group. *Pediatr Blood Cancer.* 2013;60(4):543–549.

131. Ikonomidou C. Chemotherapy and the pediatric brain. *Mol Cell Pediatr.* 2018;5(1):8.

132. Kline CN, Mueller S. Neurocognitive outcomes in children with brain tumors. In: *Paper presented at: Seminars in Neurology*; 2020.

133. Rzeski W, Pruskil S, Macke A, et al. Anticancer agents are potent neurotoxins in vitro and in vivo. *Ann Neurol.* 2004;56(3):351–360.

134. Gibson EM, Nagaraja S, Ocampo A, et al. Methotrexate chemotherapy induces persistent tri-glial dysregulation that underlies chemotherapy-related cognitive impairment. *Cell.* 2019;176(1–2):43–55.e13.

135. Bleyer W, Fallavollita J, Robison L, et al. Influence of age, sex, and concurrent intrathecal methotrexate therapy on intellectual function after cranial irradiation during childhood: a report from the Children's Cancer Study Group. *Pediatr Hematol Oncol.* 1990;7(4):329–338.

136. Balsom WR, Bleyer WA, Robison LL, et al. Intellectual function in long-term survivors of childhood acute lymphoblastic leukemia: protective effect of pre-irradiation methotrexate a Children's Cancer Study Group study. *Med Pediatr Oncol.* 1991;19(6):486–492.

137. Krull KR, Hardy KK, Kahalley LS, Schuitema I, Kesler SR. Neurocognitive outcomes and interventions in long-term survivors of childhood cancer. *J Clin Oncol.* 2018;36(21):2181.

138. Duffner PK, Armstrong FD, Chen L, et al. Neurocognitive and neuroradiologic central nervous system late effects in children treated on Pediatric Oncology Group (POG) P9605 (standard risk) and P9201 (lesser risk) acute lymphoblastic leukemia protocols (ACCL0131): a methotrexate consequence? A report from the Children's Oncology Group. *J Pediatr Hematol Oncol.* 2014;36(1):8.

139. Cheung YT, Sabin ND, Reddick WE, et al. Leukoencephalopathy and long-term neurobehavioural, neurocognitive, and brain imaging outcomes in survivors of childhood acute lymphoblastic leukaemia treated with chemotherapy: a longitudinal analysis. *Lancet Haematol.* 2016;3(10):e456–e466.

140. van der Plas E, Nieman BJ, Butcher DT, et al. Neurocognitive late effects of chemotherapy in survivors of acute lymphoblastic leukemia: focus on methotrexate. *J Can Acad Child Adolesc Psychiatry.* 2015;24(1):25.

141. Genschaft M, Huebner T, Plessow F, et al. Impact of chemotherapy for childhood leukemia on brain morphology and function. *PLoS One.* 2013;8(11), e78599.

142. Dietrich J, Prust M, Kaiser J. Chemotherapy, cognitive impairment and hippocampal toxicity. *Neuroscience.* 2015;309:224–232.

143. Cromer JA, Harel BT, Yu K, et al. Comparison of cognitive performance on the cogstate brief battery when taken in-clinic, in-group, and unsupervised. *Clin Neuropsychol.* 2015;29(4):542–558.

144. Heitzer AM, Ashford JM, Harel BT, et al. Computerized assessment of cognitive impairment among children undergoing radiation therapy for medulloblastoma. *J Neurooncol.* 2019;141(2):403–411.

145. Richard NM, Bernstein LJ, Mason WP, et al. Cognitive rehabilitation for executive dysfunction in brain tumor patients: a pilot randomized controlled trial. *J Neurooncol*. 2019;142(3):565–575.

146. Szulc-Lerch KU, Timmons BW, Bouffet E, et al. Repairing the brain with physical exercise: cortical thickness and brain volume increases in long-term pediatric brain tumor survivors in response to a structured exercise intervention. *NeuroImage*. 2018;18:972–985.

147. Conklin HM, Khan RB, Reddick WE, et al. Acute neurocognitive response to methylphenidate among survivors of childhood cancer: a randomized, double-blind, cross-over trial. *J Pediatr Psychol*. 2007;32(9):1127–1139.

148. Kandula T, Farrar MA, Cohn RJ, et al. Chemotherapy-induced peripheral neuropathy in long-term survivors of childhood cancer: clinical, neurophysiological, functional, and patient-reported outcomes. *JAMA Neurol*. 2018;75(8):980–988.

149. Bjornard KL, Gilchrist LS, Inaba H, et al. Peripheral neuropathy in children and adolescents treated for cancer. *Lancet Child Adolesc Health*. 2018;2(10):744–754.

150. Hershman DL, Lacchetti C, Dworkin RH, et al. Prevention and management of chemotherapy-induced peripheral neuropathy in survivors of adult cancers: American Society of Clinical Oncology clinical practice guideline. *J Clin Oncol*. 2014;32(18):1941–1967.

151. Bass JK, Hua C-H, Huang J, et al. Hearing loss in patients who received cranial radiation therapy for childhood cancer. *J Clin Oncol*. 2016;34(11):1248.

152. Clemens E, van den Heuvel-Eibrink MM, Mulder RL, et al. Recommendations for ototoxicity surveillance for childhood, adolescent, and young adult cancer survivors: a report from the International Late Effects of Childhood cancer Guideline Harmonization Group in collaboration with the PanCare Consortium. *Lancet Oncol*. 2019;20(1):e29–e41.

153. Bertolini P, Lassalle M, Mercier G, et al. Platinum compound-related ototoxicity in children: long-term follow-up reveals continuous worsening of hearing loss. *J Pediatr Hematol Oncol*. 2004;26(10):649–655.

154. Bess FH, Dodd-Murphy J, Parker RA. Children with minimal sensorineural hearing loss: prevalence, educational performance, and functional status. *Ear Hear*. 1998;19(5):339–354.

155. Gurney JG, Tersak JM, Ness KK, Landier W, Matthay KK, Schmidt ML. Hearing loss, quality of life, and academic problems in long-term neuroblastoma survivors: a report from the Children's Oncology Group. *Pediatrics*. 2007;120(5):e1229–e1236.

156. Grewal S, Merchant T, Reymond R, McInerney M, Hodge C, Shearer P. Auditory late effects of childhood cancer therapy: a report from the Children's Oncology Group. *Pediatrics*. 2010;125(4):e938–e950.

157. Mulhern RK, Palmer SL. Neurocognitive late effects in pediatric cancer. *Curr Probl Cancer*. 2003;27(4):177–197.

158. Mulhern RK, Palmer SL, Reddick WE, et al. Risks of young age for selected neurocognitive deficits in medulloblastoma are associated with white matter loss. *J Clin Oncol*. 2001;19(2):472–479.

159. Reddick WE, Shan ZY, Glass JO, et al. Smaller white-matter volumes are associated with larger deficits in attention and learning among long-term survivors of acute lymphoblastic leukemia. *Cancer*. 2006;106(4):941–949.

160. Reddickaij WE, Russell JM, Glass JO, et al. Subtle white matter volume differences in children treated for medulloblastoma with conventional or reduced dose craniospinal irradiation. *Magn Reson Imaging*. 2000;18(7):787–793.

161. Rubenstein CL, Varni JW, Katz ER. Cognitive functioning in long-term survivors of childhood leukemia: a prospective analysis. *J Dev Behav Pediatr*. 1990;11(6):301–305.

162. Rowland JH, Glidewell O, Sibley R, et al. Effects of different forms of central nervous system prophylaxis on neuropsychologic function in childhood leukemia. *J Clin Oncol*. 1984;2(12):1327–1335.

163. Moss HA, Nannis ED, Poplack DG. The effects of prophylactic treatment of the central nervous system on the intellectual functioning of children with acute lymphocytic leukemia. *Am J Med*. 1981;71(1):47–52.

164. Tamaroff M, Miller D, Murphy M, Salwen R, Ghavimi F, Nir Y. Immediate and long-term posttherapy neuropsychologic performance in children with acute lymphoblastic leukemia treated without central nervous system radiation. *J Pediatr*. 1982;101(4):524–529.

165. Mulhern RK, Merchant TE, Gajjar A, Reddick WE, Kun LE. Late neurocognitive sequelae in survivors of brain tumours in childhood. *Lancet Oncol*. 2004;5(7):399–408.

166. Duffner PK, Cohen ME, Thomas P. Late effects of treatment on the intelligence of children with posterior fossa tumors. *Cancer*. 1983;51(2):233–237.

167. Hirsch J, Renier D, Czernichow P, Benveniste L, Pierre-Kahn A. Medulloblastoma in childhood. Survival and functional results. *Acta Neurochir*. 1979;48(1–2):1–15.

168. Hoppe-Hirsch E, Renier D, Lellouch-Tubiana A, Sainte-Rose C, Pierre-Kahn A, Hirsch J. Medulloblastoma in childhood: progressive intellectual deterioration. *Childs Nerv Syst*. 1990;6(2):60–65.

169. Mulhern RK, Hancock J, Fairclough D, Kun L. Neuropsychological status of children treated for brain tumors: a critical review and integrative analysis. *Med Pediatr Oncol*. 1992;20(3):181–191.

170. Waber DP, Urion DK, Tarbell NJ, Niemeyer C, Gelber R, Sallan SE. Late effects of central nervous system treatment of acute lymphoblastic leukemia in childhood are sex-dependent. *Dev Med Child Neurol*. 1990;32(3):238–248.

171. Jannoun L, Bloom H. Long-term psychological effects in children treated for intracranial tumors. *Int J Radiat Oncol Biol Phys*. 1990;18(4):747–753.

172. Silber JH, Radcliffe J, Peckham V, et al. Whole-brain irradiation and decline in intelligence: the influence of dose and age on IQ score. *J Clin Oncol*. 1992;10(9):1390–1396.

173. Deutsch M, Thomas PR, Krischer J, et al. Results of a prospective randomized trial comparing standard dose neuraxis irradiation (3,600 cGy/20) with reduced neuraxis irradiation (2,340 cGy/13) in patients with low-stage medulloblastoma. *Pediatr Neurosurg*. 1996;24(4):167–177.

174. Mulhern RK, Kepner JL, Thomas PR, Armstrong FD, Friedman HS, Kun LE. Neuropsychologic functioning of survivors of childhood medulloblastoma randomized to receive conventional or reduced-dose craniospinal irradiation: a Pediatric Oncology Group study. *J Clin Oncol*. 1998;16(5):1723–1728.

175. Ellenberg L, McComb GJ, Siegel SE, Stowe S. Factors affecting intellectual outcome in pediatric brain tumor patients. *Neurosurgery*. 1987;21(5):638–644.

176. Decker AL, Szulc KU, Bouffet E, et al. Smaller hippocampal subfield volumes predict verbal associative memory in pediatric brain tumor survivors. *Hippocampus*. 2017;27(11):1140–1154.

177. Acharya S, Wu S, Ashford JM, et al. Association between hippocampal dose and memory in survivors of childhood or adolescent low-grade glioma: a 10-year neurocognitive longitudinal study. *Neuro Oncol*. 2019;21(9):1175–1183.

178. Hopewell JW. Radiation injury to the central nervous system. *Med Pediatr Oncol*. 1998;30(S1):1–9.

179. Mostoufi-Moab S, Grimberg A. Pediatric brain tumor treatment: growth consequences and their management. *Pediatr Endocrinol Rev*. 2010;8(1):6.

180. Ventura LM, Grieco JA, Evans CL, et al. Executive functioning, academic skills, and quality of life in pediatric patients with brain tumors post-proton radiation therapy. *J Neurooncol*. 2018;137(1):119–126.

181. Minzenberg MJ, Carter CS. Modafinil: a review of neurochemical actions and effects on cognition. *Neuropsychopharmacology*. 2008;33(7):1477–1502.

182. Morris B, Partap S, Yeom K, Gibbs I, Fisher P, King A. Cerebrovascular disease in childhood cancer survivors: a Children's Oncology Group Report. *Neurology*. 2009;73(22):1906–1913.

III. Neurological complications of specific neoplasms

183. Bowers DC, Liu Y, Leisenring W, et al. Late-occurring stroke among long-term survivors of childhood leukemia and brain tumors: a report from the Childhood Cancer Survivor Study. *J Clin Oncol.* 2006;24(33):5277–5282.

184. Fullerton HJ, Stratton K, Mueller S, et al. Recurrent stroke in childhood cancer survivors. *Neurology.* 2015;85(12):1056–1064.

185. Louis EL, McLoughlin MJ, Wortzman G. Chronic damage to medium and large arteries following irradiation. *J Can Assoc Radiol.* 1974;25(2):94–104.

186. Fajardo LF. The pathology of ionizing radiation as defined by morphologic patterns. *Acta Oncol.* 2005;44(1):13–22.

187. Mueller S, Kline CN, Buerki RA, et al. Stroke impact on mortality and psychologic morbidity within the Childhood Cancer Survivor Study. *Cancer.* 2020;126(5):1051–1059.

188. Mueller S, Sear K, Hills NK, et al. Risk of first and recurrent stroke in childhood cancer survivors treated with cranial and cervical radiation therapy. *Int J Radiat Oncol Biol Phys.* 2013;86(4):643–648.

189. Campen CJ, Kranick SM, Kasner SE, et al. Cranial irradiation increases risk of stroke in pediatric brain tumor survivors. *Stroke.* 2012;43(11):3035–3040.

190. Mueller S, Fullerton HJ, Stratton K, et al. Radiation, atherosclerotic risk factors, and stroke risk in survivors of pediatric cancer: a report from the Childhood Cancer Survivor Study. *Int J Radiat Oncol Biol Phys.* 2013;86(4):649–655.

191. Ullrich N, Robertson R, Kinnamon D, et al. Moyamoya following cranial irradiation for primary brain tumors in children. *Neurology.* 2007;68(12):932–938.

192. Kaas B, Huisman TA, Tekes A, Bergner A, Blakeley JO, Jordan LC. Spectrum and prevalence of vasculopathy in pediatric neurofibromatosis type 1. *J Child Neurol.* 2013;28(5):561–569.

193. Kralik SF, Watson GA, Shih C-S, Ho CY, Finke W, Buchsbaum J. Radiation-induced large vessel cerebral vasculopathy in pediatric patients with brain tumors treated with proton radiation therapy. *Int J Radiat Oncol Biol Phys.* 2017;99(4):817–824.

194. Hall MD, Bradley JA, Rotondo RL, et al. Risk of radiation vasculopathy and stroke in pediatric patients treated with proton therapy for brain and skull base tumors. *Int J Radiat Oncol Biol Phys.* 2018;101(4):854–859.

195. Wang C, Roberts KB, Bindra RS, Chiang VL, James BY. Delayed cerebral vasculopathy following cranial radiation therapy for pediatric tumors. *Pediatr Neurol.* 2014;50(6):549–556.

196. Tsang DS, Murphy ES, Merchant TE. Radiation therapy for optic pathway and hypothalamic low-grade gliomas in children. *Int J Radiat Oncol Biol Phys.* 2017;99(3):642–651.

197. Reynolds MR, Haydon DH, Caird J, Leonard JR. Radiation-induced moyamoya syndrome after proton beam therapy in the pediatric patient: a case series. *Pediatr Neurosurg.* 2016;51(6):297–301.

198. Zwagerman NT, Foster K, Jakacki R, Khan FH, Yock TI, Greene S. The development of Moyamoya syndrome after proton beam therapy. *Pediatr Blood Cancer.* 2014;61(8):1490–1492.

199. Scala M, Vennarini S, Garrè ML, et al. Radiation-induced moyamoya syndrome after proton therapy in child with clival chordoma: natural history and surgical treatment. *World Neurosurg.* 2019;123:306–309.

200. Kerklaan JP, á Nijeholt GJL, Wiggenraad RG, Berghuis B, Postma TJ, Taphoorn MJ. SMART syndrome: a late reversible complication after radiation therapy for brain tumours. *J Neurol.* 2011;258(6):1098–1104.

201. Nimjee SM, Powers CJ, Bulsara KR. Review of the literature on de novo formation of cavernous malformations of the central nervous system after radiation therapy. *Neurosurg Focus.* 2006;21(1):1–6.

202. Burn S, Gunny R, Phipps K, Gaze M, Hayward R. Incidence of cavernoma development in children after radiotherapy for brain tumors. *J Neurosurg Pediatr.* 2007;106(5):379–383.

203. Gastelum E, Sear K, Hills N, et al. Rates and characteristics of radiographically detected intracerebral cavernous malformations after cranial radiation therapy in pediatric cancer patients. *J Child Neurol.* 2015;30(7):842–849.

204. Strenger V, Sovinz P, Lackner H, et al. Intracerebral cavernous hemangioma after cranial irradiation in childhood. *Strahlenther Onkol.* 2008;184(5):276–280.

205. Heckl S, Aschoff A, Kunze S. Radiation-induced cavernous hemangiomas of the brain: a late effect predominantly in children. *Cancer.* 2002;94(12):3285–3291.

206. Sciubba DM, Gallia GL, Recinos P, Garonzik IM, Clatterbuck RE. Intracranial aneurysm following radiation therapy during childhood for a brain tumor. Case report and review of the literature. *J Neurosurg.* 2006;105(2 suppl):134–139.

207. Roddy E, Sear K, Felton E, et al. Presence of cerebral microbleeds is associated with worse executive function in pediatric brain tumor survivors. *Neuro Oncol.* 2016;18(11):1548–1558.

208. Follin C, Erfurth EM. Long-term effect of cranial radiotherapy on pituitary-hypothalamus area in childhood acute lymphoblastic leukemia survivors. *Curr Treat Options Oncol.* 2016;17(9):50.

209. Katz J, Pollock BH, Jacaruso D, Morad A. Final attained height in patients successfully treated for childhood acute lymphoblastic leukemia. *J Pediatr.* 1993;123(4):546–552.

210. Kirk J, Stevens M, Menser M, et al. Growth failure and growth-hormone deficiency after treatment for acute lymphoblastic leukaemia. *Lancet.* 1987;329(8526):190–193.

211. Bercu BB, Diamond Jr FB. Growth hormone neurosecretory dysfunction. *Clin Endocrinol Metab.* 1986;15(3):537–590.

212. Roddy E, Mueller S. Late effects of treatment of pediatric central nervous system tumors. *J Child Neurol.* 2016;31(2):237–254.

213. Livesey E, Hindmarsh P, Brook C, et al. Endocrine disorders following treatment of childhood brain tumours. *Br J Cancer.* 1990;61(4):622–625.

214. Merchant TE, Rose SR, Bosley C, Wu S, Xiong X, Lustig RH. Growth hormone secretion after conformal radiation therapy in pediatric patients with localized brain tumors. *J Clin Oncol.* 2011;29(36):4776–4780.

215. Schmiegelow M, Lassen S, Poulsen H, et al. Cranial radiotherapy of childhood brain tumours: growth hormone deficiency and its relation to the biological effective dose of irradiation in a large population based study. *Clin Endocrinol (Oxf).* 2000;53(2):191–197.

216. Darzy KH, Shalet SM. Hypopituitarism following radiotherapy. *Pituitary.* 2009;12(1):40–50.

217. Gurney JG, Ness KK, Stovall M, et al. Final height and body mass index among adult survivors of childhood brain cancer: Childhood Cancer Survivor Study. *J Clin Endocrinol Metabol.* 2003;88(10):4731–4739.

218. Diller L, Chow EJ, Gurney JG, et al. Chronic disease in the Childhood Cancer Survivor Study cohort: a review of published findings. *J Clin Oncol.* 2009;27(14):2339–2355.

219. Gurney JG, Kadan-Lottick NS, Packer RJ, et al. Endocrine and cardiovascular late effects among adult survivors of childhood brain tumors: Childhood Cancer Survivor Study. *Cancer.* 2003;97(3):663–673.

220. Duffner PK. Long-term effects of radiation therapy on cognitive and endocrine function in children with leukemia and brain tumors. *Neurologist.* 2004;10(6):293–310.

221. Ranke MB, Price DA, Lindberg A, Wilton P, Darendeliler F, Reiter EO. Final height in children with medulloblastoma treated with growth hormone. *Horm Res Paediatr.* 2005;64(1):28–34.

222. Adan L, Sainte-Rose C, Souberbielle J, Zucker J, Kalifa C, Brauner R. Adult height after growth hormone (GH) treatment for GH deficiency due to cranial irradiation. *Med Pediatr Oncol.* 2000;34(1):14–19.

223. Watanabe S, Tsunematsu Y, Fujimoto J, et al. Leukaemia in patients treated with growth hormone. *Lancet.* 1988;331(8595):1159–1160.

224. Stahnke N, Zeisel H. Growth hormone therapy and leukaemia. *Eur J Pediatr.* 1989;148(7):591–596.

225. Fradkin JE, Mills JL, Schonberger LB, et al. Risk of leukemia after treatment with pituitary growth hormone. *JAMA.* 1993;270(23):2829–2832.

226. Stahnke N. Leukemia in growth-hormone treated patients: an update, 1992. *Horm Res Paediatr.* 1992;38(suppl 1):56–62.

227. Nishi Y, Tanaka T, Takano K, et al. Recent status in the occurrence of leukemia in growth hormone-treated patients in Japan. *J Clin Endocrinol Metabol.* 1999;84(6):1961–1965.

228. Bell J, Parker K, Swinford R, Hoffman A, Maneatis T, Lippe B. Long-term safety of recombinant human growth hormone in children. *J Clin Endocrinol Metabol.* 2010;95(1):167–177.

229. Moshang Jr T, Rundle AC, Graves DA, Nickas J, Johanson A, Meadows A. Brain tumor recurrence in children treated with growth hormone: the national cooperative growth study experience. *J Pediatr.* 1996;128(5):S4–S7.

230. Swerdlow A, Reddingius R, Higgins C, et al. Growth hormone treatment of children with brain tumors and risk of tumor recurrence. *J Clin Endocrinol Metabol.* 2000;85(12):4444–4449.

231. Ogilvy-Stuart AL, Ryder W, Gattamaneni H, Clayton PE, Shalet SM. Growth hormone and tumour recurrence. *Br Med J.* 1992;304(6842):1601–1605.

232. Sklar CA, Mertens AC, Mitby P, et al. Risk of disease recurrence and second neoplasms in survivors of childhood cancer treated with growth hormone: a report from the Childhood Cancer Survivor Study. *J Clin Endocrinol Metabol.* 2002;87(7):3136–3141.

233. Ergun-Longmire B, Mertens AC, Mitby P, et al. Growth hormone treatment and risk of second neoplasms in the childhood cancer survivor. *J Clin Endocrinol Metabol.* 2006;91(9):3494–3498.

234. Gleeson HK, Shalet SM. The impact of cancer therapy on the endocrine system in survivors of childhood brain tumours. *Endocr Relat Cancer.* 2004;11(4):589–602.

235. Leiper A, Stanhope R, Kitching P, Chessells J. Precocious and premature puberty associated with treatment of acute lymphoblastic leukaemia. *Arch Dis Child.* 1987;62(11):1107–1112.

236. Ogilvy-Stuart AL, Clayton PE, Shalet SM. Cranial irradiation and early puberty. *J Clin Endocrinol Metabol.* 1994;78(6):1282–1286.

237. Constine LS, Woolf PD, Cann D, et al. Hypothalamic-pituitary dysfunction after radiation for brain tumors. *N Engl J Med.* 1993;328(2):87–94.

238. Rose SR, Lustig RH, Pitukcheewanont P, et al. Diagnosis of hidden central hypothyroidism in survivors of childhood cancer. *J Clin Endocrinol Metabol.* 1999;84(12):4472–4479.

239. Paulino AC. Hypothyroidism in children with medulloblastoma: a comparison of 3600 and 2340 cGy craniospinal radiotherapy. *Int J Radiat Oncol Biol Phys.* 2002;53(3):543–547.

240. Ricardi U, Corrias A, Einaudi S, et al. Thyroid dysfunction as a late effect in childhood medulloblastoma: a comparison of hyperfractionated versus conventionally fractionated craniospinal radiotherapy. *Int J Radiat Oncol Biol Phys.* 2001;50(5):1287–1294.

241. Schneider AB, Ron E, Lubin J, Stovall M, Gierlowski TC. Dose-response relationships for radiation-induced thyroid cancer and thyroid nodules: evidence for the prolonged effects of radiation on the thyroid. *J Clin Endocrinol Metabol.* 1993;77(2):362–369.

242. Sigurdson AJ, Ronckers CM, Mertens AC, et al. Primary thyroid cancer after a first tumour in childhood (the Childhood Cancer Survivor Study): a nested case-control study. *Lancet.* 2005;365(9476):2014–2023.

243. Lustig RH, Rose SR, Burghen GA, et al. Hypothalamic obesity caused by cranial insult in children: altered glucose and insulin dynamics and reversal by a somatostatin agonist. *J Pediatr.* 1999;135(2):162–168.

244. Lustig RH, Post SR, Srivannaboon K, et al. Risk factors for the development of obesity in children surviving brain tumors. *J Clin Endocrinol Metabol.* 2003;88(2):611–616.

245. Lustig RH, Hinds PS, Ringwald-Smith K, et al. Octreotide therapy of pediatric hypothalamic obesity: a double-blind, placebo-controlled trial. *J Clin Endocrinol Metabol.* 2003;88(6):2586–2592.

246. Eaton BR, Esiashvili N, Kim S, et al. Endocrine outcomes with proton and photon radiotherapy for standard risk medulloblastoma. *Neuro Oncol.* 2016;18(6):881–887.

247. Morton LM, Onel K, Curtis RE, Hungate EA, Armstrong GT. The rising incidence of second cancers: patterns of occurrence and identification of risk factors for children and adults. *Am Soc Clin Oncol Educ Book.* 2014;34(1):e57–e67.

248. Braganza MZ, Kitahara CM, Berrington de González A, Inskip PD, Johnson KJ, Rajaraman P. Ionizing radiation and the risk of brain and central nervous system tumors: a systematic review. *Neuro Oncol.* 2012;14(11):1316–1324.

249. Friedman DL, Whitton J, Leisenring W, et al. Subsequent neoplasms in 5-year survivors of childhood cancer: the Childhood Cancer Survivor Study. *J Natl Cancer Inst.* 2010;102(14):1083–1095.

250. Neglia JP, Robison LL, Stovall M, et al. New primary neoplasms of the central nervous system in survivors of childhood cancer: a report from the Childhood Cancer Survivor Study. *J Natl Cancer Inst.* 2006;98(21):1528–1537.

251. Taylor AJ, Little MP, Winter DL, et al. Population-based risks of CNS tumors in survivors of childhood cancer: the British Childhood Cancer Survivor Study. *J Clin Oncol.* 2010;28(36):5287.

252. Bowers DC, Nathan PC, Constine L, et al. Subsequent neoplasms of the CNS among survivors of childhood cancer: a systematic review. *Lancet Oncol.* 2013;14(8):e321–e328.

253. Turcotte LM, Liu Q, Yasui Y, et al. Temporal trends in treatment and subsequent neoplasm risk among 5-year survivors of childhood cancer, 1970-2015. *JAMA.* 2017;317(8):814–824.

254. Evans DGR, Birch JM, Ramsden R, Sharif S, Baser ME. Malignant transformation and new primary tumours after therapeutic radiation for benign disease: substantial risks in certain tumour prone syndromes. *J Med Genet.* 2006;43(4):289–294.

255. Travis LB, Rabkin CS, Brown LM, et al. Cancer survivorship—genetic susceptibility and second primary cancers: research strategies and recommendations. *J Natl Cancer Inst.* 2006;98(1):15–25.

256. Brodeur GM, Nichols KE, Plon SE, Schiffman JD, Malkin D. Pediatric cancer predisposition and surveillance: an overview, and a tribute to Alfred G Knudson Jr. *Clin Cancer Res.* 2017;23(11):e1–e5.

257. Filipovich AH, Weisdorf D, Pavletic S, et al. National Institutes of Health consensus development project on criteria for clinical trials in chronic graft-versus-host disease: I. Diagnosis and Staging Working Group report. *Biol Blood Marrow Transplant.* 2005;11(12):945–956.

258. Grauer O, Wolff D, Bertz H, et al. Neurological manifestations of chronic graft-versus-host disease after allogeneic haematopoietic stem cell transplantation: report from the Consensus Conference on Clinical Practice in chronic graft-versus-host disease. *Brain.* 2010;133(10):2852–2865.

259. Openshaw H. Neurological manifestations of chronic graft versus host disease. In: Vogelsang GB, Pavletic SZ, eds. *Chronic Graft Versus Host Disease.* New York: Cambridge University Press; 2009:243–251.

260. Sostak P, Padovan C, Yousry T, Ledderose G, Kolb H-J, Straube A. Prospective evaluation of neurological complications after allogeneic bone marrow transplantation. *Neurology.* 2003;60(5):842–848.

261. Jagasia MH, Greinix HT, Arora M, et al. National Institutes of Health consensus development project on criteria for clinical trials in chronic graft-versus-host disease: I. The 2014 Diagnosis and Staging Working Group report. *Biol Blood Marrow Transplant.* 2015;21(3):389–401.e381.

III. Neurological complications of specific neoplasms

262. Chow EJ, Antal Z, Constine LS, et al. New agents, emerging late effects, and the development of precision survivorship. *J Clin Oncol.* 2018;36(21):2231.

263. Cooney T, Yeo KK, Kline C, et al. Neuro-oncology practice clinical debate: targeted therapy vs conventional chemotherapy in pediatric low-grade glioma. *Neuro Oncol Pract.* 2020;7(1):4–10.

264. U.S. Food and Drug Administration. *Cellular & Gene Therapy Guidances*; 2020, February 14.

265. Curtin SC, Minino AM, Anderson RN. Declines in cancer death rates among children and adolescents in the United States, 1999–2014. *NCHS Data Brief.* 2016;257:1–8.

Neurological complications of antineoplastic therapy

27

Neurological complications of radiation therapy

Kailin Yang[a], Erin S. Murphy[a], Simon S. Lo[b], Samuel T. Chao[a], and John H. Suh[a]

[a]Department of Radiation Oncology, Taussig Cancer Center, Cleveland Clinic, Cleveland, OH, United States,
[b]Department of Radiation Oncology, University of Washington School of Medicine, Seattle, WA, United States

1 Introduction

Radiation therapy is one of the main treatment approaches for primary and secondary malignancies of the central nervous system (CNS). Prophylactic radiation to the brain or craniospinal axis is also performed in cancers with propensity to spread into the neuraxis, in order to prevent development of future brain or spinal metastasis. Radiation to the neurological system causes a series of temporal injuries to the normal tissue and associated blood vessels, leading to transient and/or permanent changes.[1] Recent progress from basic and clinical studies demonstrates that the neurological complications of radiation treatment are mediated through cellular injuries sustained by a number of cell types.[2] Traditionally, neurological complications are classified based on the temporal relationship with the delivery of radiation: acute (during radiation), early-delayed (up to 6 months after radiation), and late-delayed (>6 months to years postradiation).[2,3] With the improved efficacy of anticancer therapies, the clinical manifestations of neurological complications and efforts to minimize radiation-related toxicity have received more attention during recent decades. It is generally accepted that total radiation dose, dose per fraction, treatment volume, use of concurrent chemotherapy, age, and predisposing mutations are risk factors of neurological complications.[4,5] Therefore, a thorough understanding of the molecular mechanisms and clinical manifestations of radiation-induced neurological toxicity is essential to the development of novel approaches to reduce the incidence and severity of treatment-related complications.

2 Pathophysiology of neurological complications from radiation

In the modern era, several forms of radiation, including photon, electron, proton, and other particles, are commonly used in the clinic. In spite of the various physical properties of the radiation modality, the induction of cellular death results from an overwhelming number of double-stranded DNA breaks, which is thought to be the main mechanism for neurological complications.[4] In addition, direct damage to the cellular machinery, such as lipid membranes and protein complexes, also contributes to the severity of toxicity. Traditionally, endothelial cells and oligodendrocytes have been proposed as the main target cells of radiation-induced CNS toxicity.[1] However, recent evidence suggests a more complex picture focusing on the functional integrity of the neurological system, as summarized in Table 1.[2]

In the acute phase after radiation treatment, an increase in the permeability of the blood–brain barrier (BBB) is observed, which is often dependent on the radiation dose and recovers in a few weeks.[6] Apoptosis of vascular endothelial cells is considered to be the main driver of such transient but acute leakage of the BBB. The death of endothelial cells is mediated through the activation of acid sphingomyelinase and can be inhibited by fibroblast growth factor.[7] Overexpression of intercellular adhesion molecules (ICAM-1) in irradiated endothelial cells also correlates with the disruption of the BBB, both spatially and temporarily.[8] Secretion of tumor necrosis factor-α (TNF-α) from microglial cells may contribute to the changes of BBB permeability.[9] Apoptosis of oligo-progenitor cells also occurs in the acute phase, though mature oligodendrocytes are still preserved.[10] Neuronal stem cells undergo apoptosis after acute radiation, leading to a decrease in the proliferating repertoire in a mouse model.[11]

After the transient disruption to endothelial function during the acute phase, restoration of BBB integrity is seen during the early phase postradiation. Proliferation of endothelial cells, at least partially, reverses the increased barrier permeability.[6] Paradoxically, such temporary repair processes are usually followed by a second

TABLE 1   Pathophysiology of radiation-induced neurological complications.

Phase	Timeline (after radiation)	Key events
Acute	During or days	Transient disruption of blood–brain barrier
		Apoptosis of endothelial cells, oligoprogenitor cells, and neuronal stem cells
Early-delayed	Weeks to 6 months	Proliferation of endothelial cells
		Temporary repair of blood–brain barrier
		Demyelination and loss of mature oligodendrocytes
Late-delayed	>6 months to years	Vasogenic edema and necrosis Progressive demyelination Neuronal degeneration and impaired neurogenesis

wave of barrier disruption, more commonly seen in the early-delayed phase. Spatial correlation between the second phase of barrier breakdown and production of TNF-α by microglial cells and astrocytes reveals mechanistic insight into the dynamics of BBB function.[9] The exact relationship between the first and second phases of deterioration of barrier function remains elusive, though normal aging process and incomplete repair during the first phase may contribute to the development of the second wave of barrier dysfunction. Along with the malfunctioning BBB, hypoxia and accumulation of reactive oxygen species start to occur. Another hallmark of radiation toxicity in the early and early-delayed phases is demyelination. In the rat model, demyelination starts to occur 2 weeks after radiation and persists for months. At the cellular level, loss of myelin correlates with the apoptosis of oligodendroglial precursor cells during early phase, along with subsequent loss of mature oligodendrocytes during the early-delayed phase. A shift of neuronal differentiation is also concomitant with the loss of myelin, with neural precursor cells preferentially differentiating into glial cells.[12]

During the late-delayed phase after radiation treatment, necrosis follows the second wave of disruption of the BBB. This process is thought to be regulated by astrocytes, which exhibit high expression levels of vascular endothelial growth factor (VEGF) and hypoxia-inducible factor-1α (HIF-1α). Vasogenic edema occurs in the area of barrier dysfunction, leading to hypoxia and stabilization of HIF-1α, which in turn stimulates the expression and secretion of VEGF from surrounding astrocytes.[13] In animal models, VEGF is shown as the mediator of radiation-induced neurological injury, possibly through modulating ICAM-1 expression on endothelial cells.[14] Loss of mature oligodendrocytes

and associated demyelination also continues during the late-delayed phase. The impact on neuronal function emerges during this phase, though mature neurons are usually terminally differentiated with low likelihood to undergo apoptosis from radiation damage. In contrast, axonal dysfunction and synaptic changes have been proposed as the primary mechanisms of neuronal toxicity from radiation.[15] The reservoir of neural stem cells in the subventricular zone is low at 15 months after radiation treatment.[16] The shift of neuronal differentiation from early and delayed-early phases may also compromise neurogenesis, and cause a decreased number of newly formed neurons in the hippocampus.[12]

Demyelination and vascular changes are the main histological events across different phases after radiation.[1,17] These processes are coordinated by diverse cell populations, with complex spatial and temporal relationships. Understanding the cellular and molecular mechanisms of such neurological complications forms the foundation to develop new therapeutic approaches for radiation-induced toxicity.[2]

3 Radiation-induced brain toxicity

The types of radiation-induced toxicity on the brain highly correlate with the time interval from radiation treatment, which is detailed in Table 2. Acute reactions typically occur within the first day of radiation, early and early-delayed complications happen between one day and six months, and late-delayed toxicities occur within the time frame of months and even years postradiation.[1,2] Both radiation factors and patient factors contribute to the incidence and severity of such neurological complications.[4]

TABLE 2   Radiation-induced brain toxicity.

Phase	Type	Pathogenesis	Prognosis
Acute	Acute encephalopathy	Disruption of blood-brain barrier	Reversible
Early-delayed	Fatigue/somnolence syndrome	Demyelination	Reversible
	Focal encephalopathy	Demyelination	Likely reversible
	Pseudoprogression	Vasogenic edema	Reversible vs. disease progression
Late-delayed	Radiation necrosis	Necrosis	Response to intervention
	Leukoencephalopathy	Demyelination	Permanent
	Cognitive decline	Impaired neurogenesis	Permanent

3.1 Acute complications

Acute encephalopathy includes symptoms such as somnolence, headache, nausea, vomiting, and exacerbation of existing neurological deficits, which can occur as soon as a few hours after the first radiation treatment. With modern radiation techniques, most patients present with a mild form of symptoms, though death associated with acute encephalopathy has been reported with large radiation doses.[18] Increased intracranial pressure is considered a predisposing factor for acute encephalopathy with worse symptoms.[19] Mechanistically, acute perturbation of the BBB leads to increased permeability, which usually responds to corticosteroids.

3.2 Early-delayed complications

Early-delayed complications include a diverse range of clinical symptoms that can occur from a few weeks up to 6 months postradiation therapy. These symptoms are typically reversible, and depending on the severity, medical intervention might be indicated. Appropriate patient education regarding the incidence and extent of early-delayed complications is essential to the management during the postradiation period.

Fatigue is a common side effect after radiation treatment, which can manifest in up to 90% of patients undergoing radiation treatment for primary brain malignancy.[20] Being mild among most patients, severe fatigue may limit the daily activities of patients and may not be relieved with rest. Fatigue typically occurs during the first few weeks of radiation treatment and peaks approximately 2 weeks postcompletion. Among children younger than 3 years old receiving whole-brain radiation, a more severe form—somnolence syndrome—can occur with up to 50% incidence. Somnolence syndrome can manifest as severe lethargy, associated with nausea, irritability, hypersomnia, fever, and papilledema, which typically appear ~ 3–8 weeks after completion of radiation and has a favorable prognosis.[21] In most patients, fatigue can be managed with supportive care, focusing on regular sleep hygiene with proper naps. Corticosteroids can be given to facilitate the recovery from somnolence syndrome.[22] For refractory or severe fatigue, psychostimulants may be considered.[23]

Focal encephalopathy is more common among patients receiving radiation to the eye, ear, pituitary gland, and brainstem.[24] It consists of ataxia, diplopia, hearing loss, and dysarthria. The onset of symptoms is typically a few months postradiation, correlating with the loss of myelin resulting from apoptosis of oligodendroglial precursor cells and loss of mature oligodendrocytes.

Pseudoprogression can occur in up to 25% of patients receiving definitive radiation for malignant gliomas, and its incidence is associated with the use of concurrent chemotherapy such as temozolomide.[25] On magnetic resonance imaging (MRI), pseudo-progression manifests as increased T1 contrast enhancement with increased surrounding fluid-attenuated inversion recovery (FLAIR) changes. Some patients may present with worsening of preexisting neurological symptoms, making it difficult to discern between early tumor progression and pseudo-progression. Response Assessment in Neuro-Oncology (RANO) criteria have been proposed to define disease progression as increased enhancement beyond 80% isodose line during the first 12 weeks after completion of chemoradiation.[26,27] Both clinical symptoms and radiographic changes typically improve over the course of a few months. Repeat MRI and/or positron emission tomography (PET) can be performed to assess resolution of pseudo-progression.

More severe neurological complications such as peritumoral edema and seizure may occur, particularly with stereotactic radiosurgery for both primary and metastatic brain tumors.[28] The incidence is dependent on the total dose and per fraction dose. For patients with meningioma, the parasagittal location is associated with a higher incidence of postradiation edema.[29] Symptoms typically resolve in the next few months after onset, though steroids and antiepileptic drugs can be used for management.

3.3 Late-delayed complications

Late-delayed complications are typically irreversible changes, which start to appear clinically 6 months to years after radiation treatment.[1,2] Common toxicities include focal radiation necrosis, leukoencephalopathy, and cognitive decline. Factors related to the severity of late-delayed complications include the total radiation dose, per fraction dose, radiation volume, concurrent chemotherapy, age, genetic predisposition, and comorbidities.[30]

Focal radiation necrosis occurs after radiation treatment involving partial or whole-brain RT for primary or secondary neurological malignancies (Fig. 1). Its incidence is 10%–15% of patients after standard dosing for high-grade gliomas (typical dose of 59.4–60 Gy in 1.8–2.0 Gy per fraction), though higher among patients receiving concurrent chemotherapy.[30,31] The peak time of onset is around 12 months postradiation, though the time interval can be as long as 5 years or more.[30] Modern dose constraints to normal brain tissue are mainly designed to minimize risk for radiation necrosis.[32] After stereotactic radiosurgery, symptomatic radiation necrosis can occur in up to 5% of patients.[33] The risk is associated with radiation dose and volume of radiation field, based on the results from RTOG 9005 study.[34]

Temporal lobe necrosis has been reported in patients with primary head and neck cancer such as nasopharyngeal cancer treated with definitive radiation.[35] Similarly,

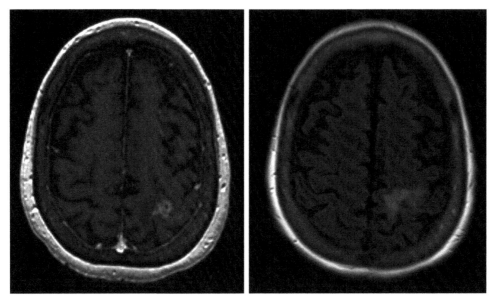

FIG. 1   Radiographic appearance of radiation necrosis. T1-weighted MRI image (left panel) shows a lesion with heterogeneous contrast enhancement, and T2/FLAIR-weighted MRI image (right panel) delineates surrounding vasogenic edema.

necrosis of the frontal lobe has been reported in patients receiving radiation treatment to the orbit and maxillary sinuses.[36] Such complications result from radiation exposure to the inferior aspects of the temporal lobe, frontal lobe, and brainstem in the treatment field. Microvascular injury from radiation treatment to the hippocampal area may prompt the development of cognitive dysfunction.[37] Modern radiation delivery methods such as three-dimensional conformal radiation therapy (3D-CRT) and intensity-modulated radiation therapy (IMRT) can significantly reduce the radiation dose to the temporal lobes and therefore decrease the incidence of temporal lobe necrosis.

The pathology of focal radiation necrosis is characterized by coagulative necrosis involving primarily the white matter.[36] Cerebral cortex and deep gray matter are usually spared from radiation necrosis. Widespread vascular injury is common in the areas of radiation necrosis, with hyalinization of the vessel walls, fibrinoid necrosis, vascular thrombosis, and telangiectatic changes. Loss of oligodendrocytes, dystrophic calcifications, fibrillary gliosis, and axonal dysfunction are also seen in surrounding regions.

The radiographic appearance of radiation necrosis is usually a mass within the previous radiation field with vasogenic edema and contrast enhancement on T1 series of MRI (Fig. 1, left).[38] The increased T2 signal change on MRI is a manifestation of edema caused by disruption of the BBB (Fig. 1, right). Using conventional computed tomography (CT) and MRI, it is difficult to differentiate radiation necrosis from tumor recurrence. Alternate radiological modalities have been proposed to accurately diagnose focal radiation necrosis. Using PET imaging with ^{18}F-fluorodeoxyglucose (FDG), radiation necrosis is associated with low FDG avidity, consistent with its hypometabolic nature.[39] PET scans using ^{11}C-methionine have also been shown to differentiate radiation necrosis with low avidity from tumor recurrence, though this tracer is not routinely available for clinical use.[40] The lesions of radiation necrosis typically have increased capacity of fluid diffusion, and therefore, low apparent diffusion coefficient (ADC) is diagnostic on diffusion-weighted MRI series.[41] On perfusion-weighted MRI, radiation necrosis has a lower absolute cerebral blood volume (CBV) compared to tumor recurrence.[42] Using MR spectroscopy, elevated ratios of choline/creatinine and choline/N-aspartyl acetate (NAA) are associated with tumor recurrence, while a lactate peak is indicative of radiation necrosis.[43]

Corticosteroids have been the first-line therapy for clinical management of symptomatic radiation necrosis, while some patients may become steroid-dependent requiring advanced management options.[44] Alternate medical therapies include hyperbaric oxygen, pentoxifylline, and vitamin E, which either increase oxygen delivery to the necrotic tissue and/or normalize vascular changes.[45,46] Recent but small-scale studies also demonstrate the efficacy of bevacizumab, a monoclonal antibody targeting VEGF, in the management of radiation necrosis.[47] Surgical resection remains the definitive option for radiation necrosis, but it may be particularly helpful among patients who need to establish pathological diagnosis or become refractory to medical intervention.[44] An alternative surgical method, laser interstitial thermal therapy (LITT), shows efficacy for patients who are steroid dependent.[48] LITT uses a laser to generate high temperatures within the necrotic tissue, inducing thermal coagulation.

Leukoencephalopathy and associated cerebral atrophy are more common among patients receiving whole-brain radiation therapy (Fig. 2).[49] These complications may also occur in patients with low-grade gliomas who are young and have a more favorable prognosis. On imaging, typical abnormalities include widespread white matter changes (more obvious in periventricular areas), cortical atrophy, and ventriculomegaly. In severe cases, patients may present with significant co-morbidities including progressive dementia, ataxia, and urinary incontinence, consistent with normal pressure hydrocephalus.[50] Cognitive decline is another long-term toxicity after radiation treatment that may occur in as many as 50% of patients receiving whole-brain radiation therapy, with a substantial impact on daily function.[51] The Hopkins Verbal Learning Test-Revised (HVLT-R) is commonly used to assess cognitive decline in clinical practice and trials. Several therapeutic agents, such as donepezil, methylphenidate, and modafinil, in addition to cognitive rehabilitation, have been attempted for symptomatic relief and preservation of normal daily function after receiving radiation.[52] Interventions including hippocampal avoidance and memantine have been shown to reduce the incidence of cognitive decline in prospective trials.[53]

Decline of the intelligence quotient (IQ) is a common long-term toxicity in survivors who have received cranial radiation therapy during childhood for primary brain tumors.[54] Decline of the IQ score typically occurs

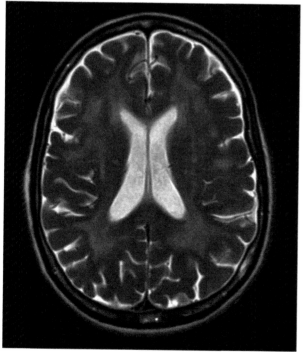

FIG. 2   Leukoencephalopathy after radiation treatment. T2-weighted MRI image depicts widespread white matter changes in a periventricular pattern after whole-brain radiation therapy.

at a rate of 2–4 points per year, and associated factors include younger age at onset of radiation (particularly younger than 3 years old), total radiation dose, and radiation volume.[55] Mechanistically, these survivors encounter difficulty with acquiring new skills and information, leading to a delay compared to healthy peers. Similar to adult patients, diagnostic imaging would reveal diffuse white matter change and cerebral atrophy. A unique feature of mineralizing microangiopathy has been reported in one-third of pediatric patients, which is commonly characterized by calcified deposits at small blood vessels within gray/white matter junction.[56]

4 Radiation-induced complications of spinal cord

4.1 Acute and early-delayed myelopathy

Acute myelopathy after radiation treatment has not been supported by clinical or experimental evidence. Historic use of 10 Gy as a single-fraction to spinal metastasis did not cause any acute myelopathy.[57] A transient form of radiation-induced spinal myelopathy typically occurs 2–6 months after treatment (Table 3).[19] It usually presents with mild symptoms including paresthesias and sensation of electric shock that spreads along the spine inferiorly and through extremities, consistent with Lhermitte's sign that is commonly seen in patients with multiple sclerosis. The incidence is up to 10% of patients, and risk factors include high per fraction dose (>2 Gy per day), high total dose (>50 Gy), and concurrent chemotherapy.[58] These signs are usually symmetrical and induced by physical exertion including neck flexion, and resolve within a few months after initial occurrence. In contrast to multiple sclerosis, MRI scans are not typically revealing for patients with early-delayed spinal myelopathy.[59] Mechanistically, transient demyelination of the posterior columns has been proposed as the etiology, and this is not associated with progressive disease or more severe forms of permanent myelopathy. Given the mild and reversible nature of the clinical presentation, no therapeutic intervention is indicated for most patients with transient myelopathy, though carbamazepine or gabapentin may be considered for patients with severe symptoms.

4.2 Late-delayed myelopathy

Late-delayed myelopathy typically presents with progressive and irreversible symptoms that can range from mild deficits of sensory and motor function to complete paraplegia (Table 3). Its clinical course usually follows an insidious pattern initially, with mild loss of proprioception or temperature sensation. There is steady progression over the ensuing months to years, with debilitating symptoms including weakness, incontinence, loss of

TABLE 3   Complications of spinal cord after radiation treatment.

Phase	Type	Pathogenesis	Prognosis
Early-delayed	Early myelopathy (Lhermitte's sign)	Transient demyelination	Reversible
Late-delayed	Late myelopathy (progressive motor and sensory deficit)	Demyelination, vascular necrosis, axonal injury	Progressive/permanent
	Lower motor neuron syndrome	Vasculopathy	Permanent
	Hemorrhage	Telangiectasia	Favorable/subacute

bowel function, hyperreflexia, and even quadriplegia or paraplegia. Pain is rare in late-delayed myelopathy. MRI imaging of late-delayed myelopathy reveals an intramedullary T2 hyperintensity consistent with edema, with T1 hypointensity at the affected spinal level, which may progress into spinal atrophy later in the course.[60]

Pathological changes of late-delayed myelopathy include demyelination with axonal degeneration and focal necrosis, more commonly seen in the posterior column. Vascular change is another characterizing feature, ranging from telangiectasia, wall thickening, hyalinization, to fibrinoid necrosis.[61] In animal models, expression of VEGF induced by HIF-1α has been shown to mediate the pathogenesis, likely through increasing vascular permeability leading to edema.[13]

With conventional fraction size of 1.8–2 Gy, the risk of myelopathy is <1% and <10% at 54 Gy and 61 Gy, respectively.[62] Spinal cord regeneration after radiation has been suggested from re-irradiation series, though in general at least a 6-month interval is adopted in most studies prior to the administration of repeat spinal radiation.[63] With the emergence of stereotactic body radiation therapy (SBRT), several forms of dose constraints have been proposed.[62,64] Current protocols for clinical trials involving spinal SBRT typically limit maximal doses to the spinal cord at 13–14 Gy for single fraction.[62,65]

Management options are limited for patients who develop late-delayed myelopathy after spinal radiation treatment. Use of high-dose corticosteroids and hyperbaric oxygen has been reported.[66] Anticoagulation therapy such as heparin and enoxaparin, as well as the use of bevacizumab, can be considered for patients with refractory disease after corticosteroids.[67,68]

Other rare late-delayed toxicities include lower motor neuron syndrome and spinal cord hemorrhage. Lower motor neuron syndrome involves symmetric flaccid weakness of lower extremities, with atrophy, fasciculation, and areflexia.[69] Deficits in sensory function may also be present. Chronic denervation, particularly at the level of prior radiation, is evident with needle electromyography. Contrast enhancement in the cauda equina can be detected on MRI scan, likely resulting from vasculopathy of proximal spinal roots induced by radiation. Spinal cord hemorrhage from telangiectasia is extremely rare, which may occur decades after radiation therapy.[70]

It is characterized by subacute hemorrhage on MRI, which tends to recover in most patients.

5 Radiation-induced nerve injury

5.1 Toxicity of cranial nerves after radiation

Injury to cranial nerves is a common complication after radiation treatment. Direct damage to axons and radiation-induced fibrosis are the possible pathological mechanisms of nerve injury.[19] Total dose to the nerve, dose per fraction, and the extent of the cranial nerve in the radiation field are important risk factors.

The most commonly affected cranial nerve is the optic nerve (CN II), along with the optic chiasm, which has been reported among patients receiving radiation to the nasopharynx, nasal/paranasal sinuses, orbit, and sella. The average time interval to develop optic neuropathy is 1.5 years after radiation.[71] Clinically, patients commonly present with painless loss of visual acuity in one or both eyes. Visual field deficits may also be present. Patchy enhancement can be seen along the affected nerve on MRI, associated with nerve enlargement during the early phase, followed by atrophy in the later phase.[72] In general, optic nerve injury occurs in patients with a total dose higher than 50 Gy, though the dose per fraction has been reported to be more critical than total dose.[73] Radiation modality is another risk factor given the anatomic location, with a high incidence (50%) being reported for proton radiation to clival chordomas.[74] Radiation-induced optic neuropathy has an irreversible and progressive course, which may lead to eventual blindness on the affected side. Most patients do not respond to medical intervention, though rare case reports with bevacizumab or hyperbaric oxygen have shown some promise.[75,76]

Hypoglossal nerve (CN XII) followed by the vagus nerve (CN X) are the next most commonly affected by radiation therapy.[77] Unilateral palsy of the hypoglossal nerve typically causes paralysis and atrophy of the ipsilateral side of the tongue, leading to partial dysphagia, though bilateral palsy is associated with more devastating symptoms, including severe dysphagia and dysarthria. The pathogenesis of injury to the hypoglossal and vagus nerves is thought to be fibrosis of soft tissue and muscle induced by radiation, typically after high doses (>50Gy) to the head and neck area such as for nasopharyngeal cancer.[78] Therefore, minimizing hot spots in the era of IMRT would be essential. Alternatively, disease progression/recurrence may be another contributing factor to nerve palsy, and therefore, MRI scans of the head and neck are valuable to exclude recurrence and confirm the extent of fibrosis. Needle electromyography (EMG) is a commonly used diagnostic test, as denervation of the affected nerve can be readily detected.[79] A complex pattern of spontaneous bursts of electric discharges, termed myokymia, has been seen on EMG in patients with radiation-induced nerve injury.[80] In general, this is an irreversible process, with some therapeutic success using corticosteroids.[81]

Sensation of an unpleasant odor during radiation treatment may be from direct stimulation of the olfactory nerve (CN I), and long-term injury to the olfactory nerve, particularly deterioration in olfactory threshold scores, has also been reported with high doses of radiation.[82] Palsy of the oculomotor nerve (CN III) or trigeminal nerve (CN V) after radiation is very rare and is usually a reversible process.[83] Facial nerve (CN VII) dysfunction has been studied in patients undergoing stereotactic radiosurgery for vestibular schwannoma, and its incidence can be as high as 6%, linked to single fraction treatment.[84] A fractionated regimen has been suggested to minimize the risk of nerve injury.[85] Sensorineural hearing loss is a delayed complication after radiation, though direct toxicity to the cochlea rather than the acoustic nerve is the main driving event.

5.2 Radiation-induced brachial plexopathy

Injury to the brachial plexus may occur in some patients receiving radiation treatment for breast cancer, lung cancer, and Hodgkin's lymphoma. Radiation doses to the supraclavicular, infraclavicular, and axillary nodal areas are associated with a higher incidence of brachial plexopathy.[19] Transient but reversible paresthesia has been reported during acute or early phases of radiation treatment, but has been rare.[86]

Delayed brachial plexopathy peaks approximately a few years after radiation treatment, though the risk continues decades after.[87] Patients present with initial paresthesia and numbness of the ipsilateral hand and fingers, followed by progressive weakness of the shoulder and proximal arm. Pain is rare in radiation-induced brachial plexopathy. The extensive fibrosis from radiation likely contributes to the injury of the brachial plexus, though direct injury to axonal ends and demyelination are also plausible causes.[19] The use of large daily fractions and the presence of hot spots are associated risk factors. With conventional per fraction dose of 1.8–2Gy, the commonly accepted constraint for maximal dose is 60–66Gy.[88]

Certain extent of repair of brachial plexus has been suggested from the patients receiving re-irradiation, and in general, the cumulative maximal dose should be lower than 95Gy to minimize the risk for neural injury from repeat radiation treatment.[89] Brachial plexopathy induced by SBRT has also been reported, which may lead to symptoms including paresthesia, weakness, neuropathic pain, and paralysis.[90] An early study from Indiana University showed that grade 2+ brachial plexopathy occurred at a median time of 7months after SBRT for apical lung cancers with an incidence of 19%, and suggested to keep to the maximal dose to the brachial plexus <26Gy in 3 or 4 fractions.[91] A separate study from Sweden recommended the maximal dose to the brachial plexus to be kept <30Gy for 3-fraction SBRT.[92]

Similar to cranial nerve injury, needle EMG has been used to diagnose brachial plexopathy after radiation. Myokymia is a characteristic feature of radiation-induced plexopathy, present in as many as 70% of the patients.[93] This is particularly helpful as myokymia is rarely present from alternate etiologies such as recurrence or metastasis. MRI of the brachial plexus can be considered to confirm the presence of radiation-induced fibrosis (low-intensity signal on T2-weighted image) and the absence of a mass.[94] Diffuse thickening and contrast enhancement of the brachial plexus have also been seen on MRI. PET/CT can be useful to rule out hypermetabolic metastasis.[95]

Management options are very limited for radiation-induced injury to the brachial plexus. No neurological improvement was seen in a randomized phase II study using hyperbaric oxygen.[96,97] Surgical neurolysis to remove epineural scar tissue has been explored, with some benefit for pain relief but minimal effect on neurological function.[98] Neuropathic pain can be managed symptomatically with neurotropins. For patients with advanced symptoms such as intractable pain and flaccid paralysis of the ipsilateral limb, periscapular amputation can be considered as an aggressive form of surgical management.[99]

5.3 Radiation-induced lumbosacral plexopathy

Injury to the lumbosacral plexus is commonly associated with radiation to the lower abdominal and pelvic area.[19] Mild forms of lumbosacral plexopathy during early phase have been described among patients receiving low-dose radiation for early-stage testicular seminoma.[100] Most patients have a favorable prognosis with symptomatic resolution in 1–3months.

Late-delayed toxicity of lumbosacral plexus usually starts with an insidious course, at a median interval of a few years after radiation. Patients experience asymmetric weakness in both lower extremities with progressive sensory loss. Pain is relatively rare, though can be an accompanying sign during the late phase. Fibrosis compressing on small vessels and nerve ends and direct vascular injury from radiation are the possible etiologies. Needle EMG can be used, as the presence of myokymia is diagnostic for radiation-induced plexopathy.[101] CT, MRI, and PET/CT can be used to rule out a metastatic etiology, though increased signal intensity can be seen in radiated nerves on T2-weighted MRI series.[102]

Symptoms for the majority of patients will continue to progress in a step-wise manner, with episodes of stabilization.[103] Treatment options are limited for late-delayed lumbosacral plexopathy. Combined pentoxifylline-tocopherol (PE) is known to reduce radiation fibrosis. Adding clodronate to PE has shown some benefits to improve sensorimotor symptoms in case reports.[104]

6 Endocrinopathy after radiation therapy

Endocrine dysfunction is a common complication after radiation therapy to the CNS. With doses starting at 20 Gy, impairment of the endocrine functions of the hypothalamus and pituitary gland has been extensively studied.[105–107] As early as 1 year after radiation treatment, patients may develop a subclinical abnormality in hormonal levels, long before the emergence of any clinical symptoms.[108] The incidence continues to rise over time, as high as 80%. Patients receiving cranial radiation due to glioma and nasopharyngeal cancer are at increased risk for endocrinopathy in the hypothalamic-pituitary axis.[108,109] Deficiency in growth hormone is the most commonly observed. Deficiencies in gonadotropin and adrenocorticotrophic hormone, hyperprolactinemia, and central hypothyroidism have also been reported. Patients receiving radiation to the hypothalamus or pituitary gland should have baseline endocrine evaluation within 1 year after completion of treatment. They should be followed regularly with blood tests to screen for abnormalities in hormonal levels along the hypothalamic–pituitary axis.[4]

Deficiency in thyroid hormone is another common endocrine dysfunction after radiation therapy. The incidence can be as high as 20%, mainly due to central hypothyroidism from cranial radiation or primary hypothyroidism from direct radiation to the thyroid gland.[106] Regular screening with serum levels of thyroid stimulating hormone and free T4 would be indicated for long-term survivors. Thyroid hormone replacement therapy such as levothyroxine can be started when hormonal levels become low and/or patients develop symptoms.

7 Cerebrovascular complications after radiation treatment

Vascular complications, including vasculopathy, stroke, and cavernous malformation, are commonly seen after cranial radiation treatment.[110] The internal carotid artery and circle of Willis are the most susceptible sites for radiation-induced vascular injuries.[111] Large blood vessels tend to develop atherosclerosis, thromboembolism, and aneurysm, while small blood vessels are at risk for capillary loss and ischemic necrosis after radiation.[112,113] The time interval for vascular complications is typically years after radiation, with radiation dose, fractionation, and field of exposure affecting the incidence and severity.[114] The Childhood Cancer Survivor Study reported a risk of 3.4% to develop stroke at a median of 14 years after diagnosis among survivors with brain tumors.[115] The risk of stroke is the highest (6.5%) for patients who received both cranial radiation and chemotherapy. The circle of Willis is particularly susceptible to develop long-term stroke after radiation, with risk approaching 11.3% at age 45 among a large cohort of childhood cancer survivors who have received 10 Gy or more.[116] The observed 20-year mortality rate for cerebrovascular events after radiation is ~1%, with increased risk for prior radiation to the central aspect of the brain.[117] Conservative approaches to reduce risk factors for cerebrovascular events, such as tight management of hypertension, diabetes, and hypercholesterolemia, are recommended for cancer survivors who have undergone cranial radiation.[114] Angioplasty, stenting, and arterial repair have also been attempted with optimal outcome for patients who develop transient ischemic attack (TIA) or stroke.[118,119]

Moyamoya arteriopathy is commonly associated with survivors of childhood brain tumor therapy, after receiving cranial radiation.[120] Clinically, it manifests as an abnormal network of collateral vessels, with close proximity to spontaneously occluded vessels around the internal carotid artery or the circle of Willis. Risk factors include younger age at the time of radiation, neurofibromatosis type 1, and exposure of the circle of Willis to radiation.[121] The median time of onset is within the first 5 years after radiotherapy. Patients of Moyamoya typically present with recurrent episodes of TIA or stroke, headache, seizure, and progressive cognitive deterioration. Surgical revascularization and carotid bypass have been used to manage radiation-induced Moyamoya arteriopathy.[122,123]

8 Radiation-induced secondary malignancy

Secondary malignancy is a well-characterized complication after radiation exposure.[124] Cahan's criteria were previously proposed to define radiation-induced sarcoma.[125] In general, several conditions need to be met to attribute the incidence of a second malignancy to the effect of prior radiation: First, the second malignancy must arise in the field of prior radiation; second, a sufficient latent period between radiation and subsequent diagnosis of second malignancy; third, different histology between previously radiated malignancy and the second malignancy; and fourth, prior to radiation, no metabolic or genetic abnormality has been detected in the tissue in which the second malignancy arises. Common histologies of secondary malignancies associated with radiation to neural tissues include meningioma, glioma, nerve sheath tumor, and sarcoma.[126]

The risk of secondary malignancy is associated with radiation dose and treatment field.[127] A large historical cohort study on patients receiving childhood radiation for tinea capitis demonstrates that radiation doses as low as 1–2 Gy can significantly increase the risk for secondary malignancy in the nervous system.[128] The relative risks to develop meningioma, glioma, and nerve sheath tumor are reported as 9.5, 2.6, and 18.8, respectively. Meningioma is the most common neurological secondary malignancy, accounting for approximately 70% of reported cases.[129] The risk is associated with radiation dose, and radiation-induced meningiomas are characterized by aggressive pathological features with complex cytogenetic aberrations.[130]

Gliomas are the second most common type (~ 20%) of neurological malignancy induced by radiation treatment.[129] There can be a wide range of latency from radiation to the development of a secondary glioma, with a median time of ~ 11 years.[131] The majority of secondary gliomas are high-grade, such as glioblastoma, WHO grade 4.[132] Profiling of the genetic landscape of radiation-induced gliomas reveals a high frequency of TP53 mutation, CDK4 amplification, CDKN2A homozygous deletion, and amplifications or rearrangements involving receptor tyrosine kinase genes, supporting the pathogenesis mediated by radiation-induced double-strand breaks.[133] Patients with high-grade secondary gliomas are associated with a very poor clinical prognosis. Nerve sheath tumors and sarcomas of the nervous system are rare complications of radiation therapy. Osteosarcoma and undifferentiated pleomorphic sarcoma (historically named as malignant fibrous histiocytoma) are the most common histologies of secondary sarcoma.[134] Overall prognosis remains poor for these rare types of secondary malignancies.[135]

Genetic predisposition is another risk factor for radiation-induced secondary malignancy.[124] Germ line mutations in tumor suppressor genes likely contribute to the development of secondary malignancies after radiation. Cancer-prone conditions, including neurofibromatosis 1, neurofibromatosis 2, von Hippel-Lindau disease, Li-Fraumeni syndrome, and retinoblastoma, have been associated with increased risk for radiation-induced neoplasm, though the quality of data and follow-up vary depending on the study.[136] A careful balance between the safety and the efficacy of radiation treatment needs to be conducted prior to initiation of radiation treatment for these patients.

9 Conclusion

With the improvement of modern cancer care and longer patient survival, the short-term and long-term side effects of anticancer treatment have received increasing attention during recent decades.[2] As one of the major modalities, radiation therapy is known to cause a series of potential side effects, which are related to treatment factors including total radiation dose, dose per fraction, treatment volume, use of concurrent chemotherapy, age, and predisposing mutations.[5,19] Recognition of these neurological complications from radiation and appropriate counseling of patients are essential to clinical management. The evolution of precise radiation techniques such as IMRT has led to increased accuracy through limiting radiation dose to organs at risk and reducing toxicity.[4] Advances in basic research have discovered novel molecular mechanisms of the underlying pathological processes mediating radiation-induced complications, revealing new avenues for interventional strategies to minimize side effects from radiation.[2] Future research should focus on the improved understanding between patient-specific genetic factors and refined radiation delivery to achieve precise radiation planning with optimal efficacy and minimal toxicity.

References

1. Sheline GE, Wara WM, Smith V. Therapeutic irradiation and brain injury. *Int J Radiat Oncol Biol Phys.* 1980;6:1215–1228.
2. Soussain C, Ricard D, Fike JR, Mazeron JJ, Psimaras D, Delattre JY. CNS complications of radiotherapy and chemotherapy. *Lancet.* 2009;374:1639–1651.
3. Sheline GE. Radiation therapy of brain tumors. *Cancer.* 1977;39:873–881.
4. Tanguturi SK, Alexander BM. Neurologic complications of radiation therapy. *Neurol Clin.* 2018;36:599–625.
5. Rogers LR. Neurologic complications of radiation. *Continuum (Minneap Minn).* 2012;18:343–354.
6. Li YQ, Chen P, Jain V, Reilly RM, Wong CS. Early radiation-induced endothelial cell loss and blood-spinal cord barrier breakdown in the rat spinal cord. *Radiat Res.* 2004;161:143–152.
7. Pena LA, Fuks Z, Kolesnick RN. Radiation-induced apoptosis of endothelial cells in the murine central nervous system: protection by fibroblast growth factor and sphingomyelinase deficiency. *Cancer Res.* 2000;60:321–327.
8. Nordal RA, Wong CS. Intercellular adhesion molecule-1 and blood-spinal cord barrier disruption in central nervous system radiation injury. *J Neuropathol Exp Neurol.* 2004;63:474–483.
9. Daigle JL, Hong JH, Chiang CS, McBride WH. The role of tumor necrosis factor signaling pathways in the response of murine brain to irradiation. *Cancer Res.* 2001;61:8859–8865.
10. Atkinson S, Li YQ, Wong CS. Changes in oligodendrocytes and myelin gene expression after radiation in the rodent spinal cord. *Int J Radiat Oncol Biol Phys.* 2003;57:1093–1100.
11. Mizumatsu S, Monje ML, Morhardt DR, Rola R, Palmer TD, Fike JR. Extreme sensitivity of adult neurogenesis to low doses of X-irradiation. *Cancer Res.* 2003;63:4021–4027.
12. Monje ML, Mizumatsu S, Fike JR, Palmer TD. Irradiation induces neural precursor-cell dysfunction. *Nat Med.* 2002;8:955–962.

13. Nordal RA, Nagy A, Pintilie M, Wong CS. Hypoxia and hypoxia-inducible factor-1 target genes in central nervous system radiation injury: a role for vascular endothelial growth factor. *Clin Cancer Res.* 2004;10:3342–3353.

14. Proescholdt MA, Heiss JD, Walbridge S, et al. Vascular endothelial growth factor (VEGF) modulates vascular permeability and inflammation in rat brain. *J Neuropathol Exp Neurol.* 1999;58:613–627.

15. Pellmar TC, Schauer DA, Zeman GH. Time- and dose-dependent changes in neuronal activity produced by X radiation in brain slices. *Radiat Res.* 1990;122:209–214.

16. Panagiotakos G, Alshamy G, Chan B, et al. Long-term impact of radiation on the stem cell and oligodendrocyte precursors in the brain. *PLoS ONE.* 2007;2, e588.

17. van der Kogel AJ. Radiation-induced damage in the central nervous system: an interpretation of target cell responses. *Br J Cancer Suppl.* 1986;7:207–217.

18. Young DF, Posner JB, Chu F, Nisce L. Rapid-course radiation therapy of cerebral metastases: results and complications. *Cancer.* 1974;34:1069–1076.

19. Dropcho EJ. Neurotoxicity of radiation therapy. *Neurol Clin.* 2010;28:217–234.

20. Faithfull S, Brada M. Somnolence syndrome in adults following cranial irradiation for primary brain tumours. *Clin Oncol (R Coll Radiol).* 1998;10:250–254.

21. Ch'ien LT, Aur RJ, Stagner S, et al. Long-term neurological implications of somnolence syndrome in children with acute lymphocytic leukemia. *Ann Neurol.* 1980;8:273–277.

22. Mandell LR, Walker RW, Steinherz P, Fuks Z. Reduced incidence of the somnolence syndrome in leukemic children with steroid coverage during prophylactic cranial radiation therapy. Results of a pilot study. *Cancer.* 1989;63:1975–1978.

23. Breitbart W, Alici Y. Psychostimulants for cancer-related fatigue. *J Natl Compr Canc Netw.* 2010;8:933–942.

24. Greene-Schloesser D, Robbins ME, Peiffer AM, Shaw EG, Wheeler KT, Chan MD. Radiation-induced brain injury: a review. *Front Oncol.* 2012;2:73.

25. Brandsma D, Stalpers L, Taal W, Sminia P, van den Bent MJ. Clinical features, mechanisms, and management of pseudoprogression in malignant gliomas. *Lancet Oncol.* 2008;9:453–461.

26. Linhares P, Carvalho B, Figueiredo R, Reis RM, Vaz R. Early Pseudoprogression following chemoradiotherapy in glioblastoma patients: the value of RANO evaluation. *J Oncol.* 2013;2013, 690585.

27. Wen PY, Chang SM, Van den Bent MJ, Vogelbaum MA, Macdonald DR, Lee EQ. Response assessment in neuro-oncology clinical trials. *J Clin Oncol.* 2017;35:2439–2449.

28. Chang JH, Chang JW, Choi JY, Park YG, Chung SS. Complications after gamma knife radiosurgery for benign meningiomas. *J Neurol Neurosurg Psychiatry.* 2003;74:226–230.

29. Cai R, Barnett GH, Novak E, Chao ST, Suh JH. Principal risk of peritumoral edema after stereotactic radiosurgery for intracranial meningioma is tumor-brain contact interface area. *Neurosurgery.* 2010;66:513–522.

30. Ruben JD, Dally M, Bailey M, Smith R, McLean CA, Fedele P. Cerebral radiation necrosis: incidence, outcomes, and risk factors with emphasis on radiation parameters and chemotherapy. *Int J Radiat Oncol Biol Phys.* 2006;65:499–508.

31. Kumar AJ, Leeds NE, Fuller GN, et al. Malignant gliomas: MR imaging spectrum of radiation therapy- and chemotherapy-induced necrosis of the brain after treatment. *Radiology.* 2000;217:377–384.

32. Emami B, Lyman J, Brown A, et al. Tolerance of normal tissue to therapeutic irradiation. *Int J Radiat Oncol Biol Phys.* 1991;21:109–122.

33. Swinson BM, Friedman WA. Linear accelerator stereotactic radiosurgery for metastatic brain tumors: 17 years of experience at the University of Florida. *Neurosurgery.* 2008;62:1018–1031 [discussion 1031–1012].

34. Shaw E, Scott C, Souhami L, et al. Single dose radiosurgical treatment of recurrent previously irradiated primary brain tumors and brain metastases: final report of RTOG protocol 90-05. *Int J Radiat Oncol Biol Phys.* 2000;47:291–298.

35. Chen J, Dassarath M, Yin Z, Liu H, Yang K, Wu G. Radiation induced temporal lobe necrosis in patients with nasopharyngeal carcinoma: a review of new avenues in its management. *Radiat Oncol.* 2011;6:128.

36. Rottenberg DA, Chernik NL, Deck MD, Ellis F, Posner JB. Cerebral necrosis following radiotherapy of extracranial neoplasms. *Ann Neurol.* 1977;1:339–357.

37. Cheung M, Chan AS, Law SC, Chan JH, Tse VK. Cognitive function of patients with nasopharyngeal carcinoma with and without temporal lobe radionecrosis. *Arch Neurol.* 2000;57:1347–1352.

38. Walker AJ, Ruzevick J, Malayeri AA, et al. Postradiation imaging changes in the CNS: how can we differentiate between treatment effect and disease progression? *Future Oncol.* 2014;10:1277–1297.

39. Chao ST, Suh JH, Raja S, Lee SY, Barnett G. The sensitivity and specificity of FDG PET in distinguishing recurrent brain tumor from radionecrosis in patients treated with stereotactic radiosurgery. *Int J Cancer.* 2001;96:191–197.

40. Tsuyuguchi N, Sunada I, Iwai Y, et al. Methionine positron emission tomography of recurrent metastatic brain tumor and radiation necrosis after stereotactic radiosurgery: is a differential diagnosis possible? *J Neurosurg.* 2003;98:1056–1064.

41. Shah R, Vattoth S, Jacob R, et al. Radiation necrosis in the brain: imaging features and differentiation from tumor recurrence. *Radiographics.* 2012;32:1343–1359.

42. Wang B, Zhao B, Zhang Y, et al. Absolute CBV for the differentiation of recurrence and radionecrosis of brain metastases after gamma knife radiotherapy: a comparison with relative CBV. *Clin Radiol.* 2018;73:758 e751–758 e757.

43. Sundgren PC. MR spectroscopy in radiation injury. *AJNR Am J Neuroradiol.* 2009;30:1469–1476.

44. Vellayappan B, Tan CL, Yong C, et al. Diagnosis and Management of Radiation Necrosis in patients with brain metastases. *Front Oncol.* 2018;8:395.

45. Ohguri T, Imada H, Kohshi K, et al. Effect of prophylactic hyperbaric oxygen treatment for radiation-induced brain injury after stereotactic radiosurgery of brain metastases. *Int J Radiat Oncol Biol Phys.* 2007;67:248–255.

46. Williamson R, Kondziolka D, Kanaan H, Lunsford LD, Flickinger JC. Adverse radiation effects after radiosurgery may benefit from oral vitamin E and pentoxifylline therapy: a pilot study. *Stereotact Funct Neurosurg.* 2008;86:359–366.

47. Levin VA, Bidaut L, Hou P, et al. Randomized double-blind placebo-controlled trial of bevacizumab therapy for radiation necrosis of the central nervous system. *Int J Radiat Oncol Biol Phys.* 2011;79:1487–1495.

48. Ahluwalia M, Barnett GH, Deng D, et al. Laser ablation after stereotactic radiosurgery: a multicenter prospective study in patients with metastatic brain tumors and radiation necrosis. *J Neurosurg.* 2018;130:804–811.

49. Swennen MH, Bromberg JE, Witkamp TD, Terhaard CH, Postma TJ, Taphoorn MJ. Delayed radiation toxicity after focal or whole brain radiotherapy for low-grade glioma. *J Neurooncol.* 2004;66:333–339.

50. DeAngelis LM, Delattre JY, Posner JB. Radiation-induced dementia in patients cured of brain metastases. *Neurology.* 1989;39:789–796.

51. Meyers CA, Smith JA, Bezjak A, et al. Neurocognitive function and progression in patients with brain metastases treated with whole-brain radiation and motexafin gadolinium: results of a randomized phase III trial. *J Clin Oncol.* 2004;22:157–165.

52. Cramer CK, Cummings TL, Andrews RN, et al. Treatment of radiation-induced cognitive decline in adult brain tumor patients. *Curr Treat Options Oncol.* 2019;20:42.

53. Brown PD, Gondi V, Pugh S, et al. Hippocampal avoidance during whole-brain radiotherapy plus memantine for patients with brain metastases: phase III trial NRG oncology CC001. *J Clin Oncol.* 2020;38:1019–1029.

54. Palmer SL, Goloubeva O, Reddick WE, et al. Patterns of intellectual development among survivors of pediatric medulloblastoma: a longitudinal analysis. *J Clin Oncol.* 2001;19:2302–2308.

55. Fouladi M, Gilger E, Kocak M, et al. Intellectual and functional outcome of children 3 years old or younger who have CNS malignancies. *J Clin Oncol.* 2005;23:7152–7160.

56. Davis PC, Hoffman Jr JC, Pearl GS, Braun IF. CT evaluation of effects of cranial radiation therapy in children. *AJR Am J Roentgenol.* 1986;147:587–592.

57. Tefft M, Mitus A, Schulz MD. Initial high dose irradiation for metastases causing spinal cord compression in children. *Am J Roentgenol Radium Ther Nucl Med.* 1969;106:385–393.

58. Fein DA, Marcus Jr RB, Parsons JT, Mendenhall WM, Million RR. Lhermitte's sign: incidence and treatment variables influencing risk after irradiation of the cervical spinal cord. *Int J Radiat Oncol Biol Phys.* 1993;27:1029–1033.

59. Pak D, Vineberg K, Feng F, Ten Haken RK, Eisbruch A. Lhermitte sign after chemo-IMRT of head-and-neck cancer: incidence, doses, and potential mechanisms. *Int J Radiat Oncol Biol Phys.* 2012;83:1528–1533.

60. Komachi H, Tsuchiya K, Ikeda M, Koike R, Matsunaga T, Ikeda K. Radiation myelopathy: a clinicopathological study with special reference to correlation between MRI findings and neuropathology. *J Neurol Sci.* 1995;132:228–232.

61. Lengyel Z, Reko G, Majtenyi K, et al. Autopsy verifies demyelination and lack of vascular damage in partially reversible radiation myelopathy. *Spinal Cord.* 2003;41:577–585.

62. Kirkpatrick JP, van der Kogel AJ, Schultheiss TE. Radiation dose-volume effects in the spinal cord. *Int J Radiat Oncol Biol Phys.* 2010;76:S42–S49.

63. Abbatucci JS, Delozier T, Quint R, Roussel A, Brune D. Radiation myelopathy of the cervical spinal cord: time, dose and volume factors. *Int J Radiat Oncol Biol Phys.* 1978;4:239–248.

64. Sahgal A, Ma L, Gibbs I, et al. Spinal cord tolerance for stereotactic body radiotherapy. *Int J Radiat Oncol Biol Phys.* 2010;77:548–553.

65. Ryu S, Pugh SL, Gerszten PC, et al. RTOG 0631 phase II/III study of image-guided stereotactic radiosurgery for localized (1-3) spine metastases: phase II results. *Int J Radiat Oncol Biol Phys.* 2011;81:S131–S132.

66. Graber JJ, Nolan CP. Myelopathies in patients with cancer. *Arch Neurol.* 2010;67:298–304.

67. Liu CY, Yim BT, Wozniak AJ. Anticoagulation therapy for radiation-induced myelopathy. *Ann Pharmacother.* 2001;35:188–191.

68. Chamberlain MC, Eaton KD, Fink J. Radiation-induced myelopathy: treatment with bevacizumab. *Arch Neurol.* 2011;68:1608–1609.

69. van der Sluis RW, Wolfe GI, Nations SP, et al. Post-radiation lower motor neuron syndrome. *J Clin Neuromuscul Dis.* 2000;2:10–17.

70. Allen JC, Miller DC, Budzilovich GN, Epstein FJ. Brain and spinal cord hemorrhage in long-term survivors of malignant pediatric brain tumors: a possible late effect of therapy. *Neurology.* 1991;41:148–150.

71. Danesh-Meyer HV. Radiation-induced optic neuropathy. *J Clin Neurosci.* 2008;15:95–100.

72. Archer EL, Liao EA, Trobe JD. Radiation-induced optic neuropathy: clinical and imaging profile of twelve patients. *J Neuroophthalmol.* 2019;39:170–180.

73. Parsons JT, Bova FJ, Fitzgerald CR, Mendenhall WM, Million RR. Radiation optic neuropathy after megavoltage external-beam irradiation: analysis of time-dose factors. *Int J Radiat Oncol Biol Phys.* 1994;30:755–763.

74. Bowyer J, Natha S, Marsh I, Foy P. Visual complications of proton beam therapy for clival chordoma. *Eye (Lond).* 2003;17:318–323.

75. Farooq O, Lincoff NS, Saikali N, Prasad D, Miletich RS, Mechtler LL. Novel treatment for radiation optic neuropathy with intravenous bevacizumab. *J Neuroophthalmol.* 2012;32:321–324.

76. Borruat FX, Schatz NJ, Glaser JS, Feun LG, Matos L. Visual recovery from radiation-induced optic neuropathy. The role of hyperbaric oxygen therapy. *J Clin Neuroophthalmol.* 1993;13:98–101.

77. Lin YS, Jen YM, Lin JC. Radiation-related cranial nerve palsy in patients with nasopharyngeal carcinoma. *Cancer.* 2002;95:404–409.

78. Janssen S, Glanzmann C, Yousefi B, et al. Radiation-induced lower cranial nerve palsy in patients with head and neck carcinoma. *Mol Clin Oncol.* 2015;3:811–816.

79. Alqahtani SA, Agha C, Rothstein T. Isolated unilateral tongue atrophy: a possible late complication of juxta cephalic radiation therapy. *Am J Case Rep.* 2016;17:535–537.

80. Tiftikcioglu BI, Bulbul I, Ozcelik MM, Piskin-Demir G, Zorlu Y. Tongue myokymia presenting twelve years after radiation therapy. *Clin Neurophysiol Pract.* 2016;1:41–42.

81. Rigamonti A, Lauria G, Mantero V, Stanzani L, Salmaggi A. Bilateral radiation-induced hypoglossal nerve palsy responsive to steroid treatment. *J Clin Neurol.* 2018;14:244–245.

82. Jalali MM, Gerami H, Rahimi A, Jafari M. Assessment of olfactory threshold in patients undergoing radiotherapy for head and neck malignancies. *Iran J Otorhinolaryngol.* 2014;26:211–217.

83. Grabau O, Leonhardi J, Reimers CD. Recurrent isolated oculomotor nerve palsy after radiation of a mesencephalic metastasis. Case report and mini review. *Front Neurol.* 2014;5:123.

84. Badakhshi H, Graf R, Bohmer D, Synowitz M, Wiener E, Budach V. Results for local control and functional outcome after linac-based image-guided stereotactic radiosurgery in 190 patients with vestibular schwannoma. *J Radiat Res.* 2014;55:288–292.

85. Kalapurakal JA, Silverman CL, Akhtar N, Andrews DW, Downes B, Thomas PR. Improved trigeminal and facial nerve tolerance following fractionated stereotactic radiotherapy for large acoustic neuromas. *Br J Radiol.* 1999;72:1202–1207.

86. Salner AL, Botnick LE, Herzog AG, et al. Reversible brachial plexopathy following primary radiation therapy for breast cancer. *Cancer Treat Rep.* 1981;65:797–802.

87. Johansson S, Svensson H, Larsson LG, Denekamp J. Brachial plexopathy after postoperative radiotherapy of breast cancer patients—a long-term follow-up. *Acta Oncol.* 2000;39:373–382.

88. Thomas TO, Refaat T, Choi M, et al. Brachial plexus dose tolerance in head and neck cancer patients treated with sequential intensity modulated radiation therapy. *Radiat Oncol.* 2015;10:94.

89. Chen AM, Yoshizaki T, Velez MA, Mikaeilian AG, Hsu S, Cao M. Tolerance of the brachial plexus to high-dose reirradiation. *Int J Radiat Oncol Biol Phys.* 2017;98:83–90.

90. Schaub SK, Tseng YD, Chang EL, et al. Strategies to mitigate toxicities from stereotactic body radiation therapy for spine metastases. *Neurosurgery.* 2019;85:729–740.

91. Forquer JA, Fakiris AJ, Timmerman RD, et al. Brachial plexopathy from stereotactic body radiotherapy in early-stage NSCLC: dose-limiting toxicity in apical tumor sites. *Radiother Oncol.* 2009;93:408–413.

92. Lindberg K, Grozman V, Lindberg S, et al. Radiation-induced brachial plexus toxicity after SBRT of apically located lung lesions. *Acta Oncol.* 2019;58:1178–1186.

93. Olsen NK, Pfeiffer P, Johannsen L, Schroder H, Rose C. Radiation-induced brachial plexopathy: neurological follow-up in 161 recurrence-free breast cancer patients. *Int J Radiat Oncol Biol Phys.* 1993;26:43–49.

94. Wittenberg KH, Adkins MC. MR imaging of nontraumatic brachial plexopathies: frequency and spectrum of findings. *Radiographics.* 2000;20:1023–1032.

95. Ho L, Henderson R, Luong T, Malkhassian S, Wassef H. 18F-FDG PET/CT appearance of metastatic brachial plexopathy involving epidural space from breast carcinoma. *Clin Nucl Med.* 2012;37:e263–e264.

96. Yarnold J. Double-blind randomised phase II study of hyperbaric oxygen in patients with radiation-induced brachial plexopathy. *Radiother Oncol.* 2005;77:327.

97. Pritchard J, Anand P, Broome J, et al. Double-blind randomized phase II study of hyperbaric oxygen in patients with radiation-induced brachial plexopathy. *Radiother Oncol.* 2001;58:279–286.

98. Lu L, Gong X, Liu Z, Wang D, Zhang Z. Diagnosis and operative treatment of radiation-induced brachial plexopathy. *Chin J Traumatol.* 2002;5:329–332.

99. Behnke NK, Crosby SN, Stutz CM, Holt GE. Periscapular amputation as treatment for brachial plexopathy secondary to recurrent breast carcinoma: a case series and review of the literature. *Eur J Surg Oncol.* 2013;39:1325–1331.

100. Brydoy M, Storstein A, Dahl O. Transient neurological adverse effects following low dose radiation therapy for early stage testicular seminoma. *Radiother Oncol.* 2007;82:137–144.

101. Ko K, Sung DH, Kang MJ, et al. Clinical, electrophysiological findings in adult patients with non-traumatic plexopathies. *Ann Rehabil Med.* 2011;35:807–815.

102. Bourhafour I, Benoulaid M, El Kacemi H, El Majjaoui S, Kebdani T, Benjaafar N. Lumbosacral plexopathy: a rare long term complication of concomitant chemo-radiation for cervical cancer. *Gynecol Oncol Res Pract.* 2015;2:12.

103. Delanian S, Lefaix JL, Pradat PF. Radiation-induced neuropathy in cancer survivors. *Radiother Oncol.* 2012;105:273–282.

104. Delanian S, Lefaix JL, Maisonobe T, Salachas F, Pradat PF. Significant clinical improvement in radiation-induced lumbosacral polyradiculopathy by a treatment combining pentoxifylline, tocopherol, and clodronate (Pentoclo). *J Neurol Sci.* 2008;275:164–166.

105. Constine LS, Woolf PD, Cann D, et al. Hypothalamic-pituitary dysfunction after radiation for brain tumors. *N Engl J Med.* 1993;328:87–94.

106. Appelman-Dijkstra NM, Malgo F, Neelis KJ, Coremans I, Biermasz NR, Pereira AM. Pituitary dysfunction in adult patients after cranial irradiation for head and nasopharyngeal tumours. *Radiother Oncol.* 2014;113:102–107.

107. Pai HH, Thornton A, Katznelson L, et al. Hypothalamic/pituitary function following high-dose conformal radiotherapy to the base of skull: demonstration of a dose-effect relationship using dose-volume histogram analysis. *Int J Radiat Oncol Biol Phys.* 2001;49:1079–1092.

108. Taphoorn MJ, Heimans JJ, van der Veen EA, Karim AB. Endocrine functions in long-term survivors of low-grade supratentorial glioma treated with radiation therapy. *J Neurooncol.* 1995;25:97–102.

109. Woo E, Lam K, Yu YL, Ma J, Wang C, Yeung RT. Temporal lobe and hypothalamic-pituitary dysfunctions after radiotherapy for nasopharyngeal carcinoma: a distinct clinical syndrome. *J Neurol Neurosurg Psychiatry.* 1988;51:1302–1307.

110. Murphy ES, Xie H, Merchant TE, Yu JS, Chao ST, Suh JH. Review of cranial radiotherapy-induced vasculopathy. *J Neurooncol.* 2015;122:421–429.

111. Nordstrom M, Felton E, Sear K, et al. Large vessel arteriopathy after cranial radiation therapy in pediatric brain tumor survivors. *J Child Neurol.* 2018;33:359–366.

112. Stewart FA, Heeneman S, Te Poele J, et al. Ionizing radiation accelerates the development of atherosclerotic lesions in ApoE−/− mice and predisposes to an inflammatory plaque phenotype prone to hemorrhage. *Am J Pathol.* 2006;168:649–658.

113. Price RA, Birdwell DA. The central nervous system in childhood leukemia. III. Mineralizing microangiopathy and dystrophic calcification. *Cancer.* 1978;42:717–728.

114. Xu J, Cao Y. Radiation-induced carotid artery stenosis: a comprehensive review of the literature. *Interv Neurol.* 2014;2:183–192.

115. Bowers DC, Liu Y, Leisenring W, et al. Late-occurring stroke among long-term survivors of childhood leukemia and brain tumors: a report from the childhood cancer survivor study. *J Clin Oncol.* 2006;24:5277–5282.

116. El-Fayech C, Haddy N, Allodji RS, et al. Cerebrovascular diseases in childhood cancer survivors: role of the radiation dose to Willis circle arteries. *Int J Radiat Oncol Biol Phys.* 2017;97:278–286.

117. Aizer AA, Du R, Wen PY, Arvold ND. Radiotherapy and death from cerebrovascular disease in patients with primary brain tumors. *J Neurooncol.* 2015;124:291–297.

118. Kashyap VS, Moore WS, Quinones-Baldrich WJ. Carotid artery repair for radiation-associated atherosclerosis is a safe and durable procedure. *J Vasc Surg.* 1999;29:90–96 [discussion 97–99].

119. DeZorzi C. Radiation-induced coronary artery disease and its treatment: a quick review of current evidence. *Cardiol Res Pract.* 2018;2018:8367268.

120. Desai SS, Paulino AC, Mai WY, Teh BS. Radiation-induced moyamoya syndrome. *Int J Radiat Oncol Biol Phys.* 2006;65:1222–1227.

121. Ullrich NJ, Robertson R, Kinnamon DD, et al. Moyamoya following cranial irradiation for primary brain tumors in children. *Neurology.* 2007;68:932–938.

122. Kornblihtt LI, Cocorullo S, Miranda C, Lylyk P, Heller PG, Molinas FC. Moyamoya syndrome in an adolescent with essential thrombocythemia: successful intracranial carotid stent placement. *Stroke.* 2005;36:E71–E73.

123. Taniguchi M, Taki T, Tsuzuki T, Tani N, Ohnishi Y. EC-IC bypass using the distal stump of the superficial temporal artery as an additional collateral source of blood flow in patients with Moyamoya disease. *Acta Neurochir.* 2007;149:393–398.

124. Dracham CB, Shankar A, Madan R. Radiation induced secondary malignancies: a review article. *Radiat Oncol J.* 2018;36:85–94.

125. Cahan WG, Woodard HQ, Higinbotham NL, Stewart FW, Coley BL. Sarcoma arising in irradiated bone: report of eleven cases. 1948. *Cancer.* 1998;82:8–34.

126. Kleinschmidt-Demasters BK, Kang JS, Lillehei KO. The burden of radiation-induced central nervous system tumors: a single institution s experience. *J Neuropathol Exp Neurol.* 2006;65:204–216.

127. Galloway TJ, Indelicato DJ, Amdur RJ, Morris CG, Swanson EL, Marcus RB. Analysis of dose at the site of second tumor formation after radiotherapy to the central nervous system. *Int J Radiat Oncol Biol Phys.* 2012;82:90–94.

128. Ron E, Modan B, Boice Jr JD, et al. Tumors of the brain and nervous system after radiotherapy in childhood. *N Engl J Med.* 1988;319:1033–1039.

129. Amirjamshidi A, Abbassioun K. Radiation-induced tumors of the central nervous system occurring in childhood and adolescence. Four unusual lesions in three patients and a review of the literature. *Childs Nerv Syst.* 2000;16:390–397.

130. Al-Mefty O, Topsakal C, Pravdenkova S, Sawyer JR, Harrison MJ. Radiation-induced meningiomas: clinical, pathological, cytokinetic, and cytogenetic characteristics. *J Neurosurg.* 2004;100:1002–1013.

131. Pettorini BL, Park YS, Caldarelli M, Massimi L, Tamburrini G, Di Rocco C. Radiation-induced brain tumours after central nervous system irradiation in childhood: a review. *Childs Nerv Syst.* 2008;24:793–805.

132. Paulino AC, Mai WY, Chintagumpala M, Taher A, Teh BS. Radiation-induced malignant gliomas: is there a role for reirradiation? *Int J Radiat Oncol Biol Phys.* 2008;71:1381–1387.

133. Lopez GY, Van Ziffle J, Onodera C, et al. The genetic landscape of gliomas arising after therapeutic radiation. *Acta Neuropathol.* 2019;137:139–150.

134. Patel SR. Radiation-induced sarcoma. *Curr Treat Options Oncol.* 2000;1:258–261.

135. Yamanaka R, Hayano A. Radiation-induced malignant peripheral nerve sheath tumors: a systematic review. *World Neurosurg.* 2017;105(961–970), e968.

136. Evans DG, Birch JM, Ramsden RT, Sharif S, Baser ME. Malignant transformation and new primary tumours after therapeutic radiation for benign disease: substantial risks in certain tumour prone syndromes. *J Med Genet.* 2006;43:289–294.

IV. Neurological complications of antineoplastic therapy

28

Neurological complications of systemic cancer and antineoplastic therapy

Gilbert Youssef[a,b,c], Patrick Y. Wen[a,b,c], and Eudocia Q. Lee[a,b,c]

[a]Center for Neuro-Oncology, Department of Medical Oncology, Dana-Farber Cancer Institute, Boston, MA, United States, [b]Division of Cancer Neurology, Brigham and Women's Hospital, Boston, MA, United States, [c]Harvard Medical School, Boston, MA, United States

1 Introduction

The spectrum of cancer-targeted therapies has been expanding in the last 2 decades and includes the traditional cytotoxic chemotherapeutic and hormonal agents, and the newer modalities such as monoclonal antibodies and small-molecule signal transduction inhibitors. In addition, a new approach aimed at stimulating the antitumoral immune response has arisen in the form of immunologic modulators and immunotherapy. As patients are diagnosed earlier in their disease course and are treated aggressively more often with multiple agents, alone or in combination, the range of possible neurological toxicities developing from anticancer therapy, both in the short and long terms, has expanded as well. This chapter will review the spectrum, mechanism, and treatment of neurotoxicity due to conventional chemotherapies (Table 1), monoclonal antibodies, and targeted therapies (Table 2) used in the treatment of patients with cancer.

2 Mechanisms of neurotoxicity

Conventional cytotoxic agents generally target rapidly dividing cells. The central nervous system (CNS) has no to little active cell turnover, with the exception of glia, macrophages, and endothelial cells that have a faster turnover, though still slow compared to other cell lineages. Also, the blood-brain barrier (BBB) prevents many large molecules and hydrophilic agents from entering the cerebrospinal fluid (CSF) and the CNS parenchyma. Although this would theoretically protect the CNS from the toxicity associated with systemic chemotherapy, toxicity even from agents with poor CNS penetration has been reported. The absence of the BBB in some areas in the CNS, namely the area postrema, allows these agents to penetrate the CNS and explains the nausea and vomiting experienced with most cytotoxic agents.[1] However, other factors have been identified as mediators for increased susceptibility of the CNS to chemotherapy-related toxicity, including direct administration of drug into the CSF by lumbar puncture or intraventricular reservoir injection, disruption of the BBB by concurrent radiation to the CNS, or by direct or indirect vascular injury induced by chemotherapy, disruption of the BBB efflux pumps in some individuals, and augmentation of intravascular dose level by intra-arterial administration.[2–4] The CNS can also be subject to indirect toxicity through hormonal changes, systemic hypertension, and coagulopathies induced by chemotherapy drugs. Advanced age, organ dysfunction leading to alteration in drug metabolism, concomitant administration of other cytotoxic agents, coexisting medical morbidities, and metabolic abnormalities can also potentiate neurotoxicity.[1,5–7] CNS toxicity can be acute or chronic and can affect the brain and spinal cord. In the brain, it can range from acute confusion, visual and/or auditory hallucinations, drowsiness, seizures, and cerebellar symptoms, to a chronic leukoencephalopathy and "chemobrain," a condition characterized by cognitive and behavioral changes. Spinal cord toxicity is often acute and mainly manifests as transverse myelitis or aseptic meningitis occurring after intrathecal chemotherapy administration.

Toxic effects on the peripheral nervous system (PNS) are more common and often dose-limiting. Chemotherapy-induced peripheral neuropathy (CIPN) is the most common complication. It is usually dose-dependent, cumulative, and sometimes persists long after treatment cessation. Different mechanisms have been

TABLE 1   Common neurologic toxicities of conventional chemotherapy.

Drug class	Specific drug	Neurologic toxicity	
		CNS	PNS
Alkylating agents	Nitrosoureas	Leukoencephalopathy, seizures, optic neuropathy	
	Busulfan	Seizures	
	Chlorambucil	Seizures	
	Cyclophosphamide	Confusion, blurred vision	
	Ifosfamide	Encephalopathy	Painful axonal sensorimotor PN
	Procarbazine	Headaches, depression, psychosis	
	Temozolomide	Headaches	
	Thiotepa	Encephalopathy, myelopathy[a]	
	Dacarbazine	Headaches, seizures	
	Estramustine	Cerebral infarctions	
Antimetabolites	Cladribine	Headaches	Sensorimotor PN, Guillain-Barré-like syndrome
	Capecitabine	Multifocal leukoencephalopathy, cerebellar ataxia, hypertonia	
	Cytarabine	Acute cerebellar syndrome, aseptic meningitis[a]	
	Fludarabine	Headaches, confusion, acute leukoencephalopathy	
	5-Fluorouracil	Acute cerebellar syndrome	
	Gemcitabine	Encephalopathy	Sensory and autonomic PN, acute inflammatory myopathy
	Hydroxyurea	Headaches, confusion, sedation, seizures	
	Methotrexate	Aseptic meningitis[a], transverse myelopathy[a], acute encephalopathy, delayed leukoencephalopathy	
Platinum compounds	Cisplatin	Headaches, encephalopathy, cortical blindness, transient demyelination of posterior spinal cord columns	Ototoxicity, sensory axonal PN
	Carboplatin	Retinopathy	
	Oxaliplatin		Transient paresthesias, muscle spasms, cold hypersensitivity, sensorimotor axonal PN
Antineoplastic antibiotics	Doxorubicin	Cerebral infarcts, subacute ascending myelopathy[a], encephalopathy[a]	
	Daunorubicin	Subacute ascending myelopathy[a], encephalopathy[a]	
Vinca alkaloids	Vincristine	Fatal myeloencephalopathy[a]	Sensorimotor small-fiber axonal PN, mononeuropathies, autonomic neuropathy
	Vinorelbine		Sensory predominant axonal PN
Taxanes	Paclitaxel	Infusion-related photopsia	Sensorimotor axonal PN, transient acute myalgias, and arthralgias
	Docetaxel		Large-fiber sensory PN
Topoisomerase inhibitors	Irinotecan	Dizziness, dysarthria	
	Topotecan	Headaches	Paresthesias
	Etoposide	Headaches, encephalopathy	Sensorimotor PN

Continued

TABLE 1 Common neurologic toxicities of conventional chemotherapy—cont'd

Drug class	Specific drug	Neurologic toxicity	
		CNS	PNS
Biologic agents	Interferons Interleukins Tumor necrosis factors	Tremor, confusion, personality changes, encephalopathy, hypertonia, seizures	Sensorimotor axonal PN
	Enzalutamide	Depression, seizures	
	L-asparaginase	Dural sinus thrombosis, cerebral infarctions	
Immune modulators	Thalidomide	Transient somnolence	Sensorimotor axonal PN

[a] When administered intrathecally.
PN, peripheral neuropathy.

TABLE 2 Common neurologic toxicities with targeted therapies.

Drug class	Specific drug	Neurologic toxicities	
		CNS	PNS
Monoclonal antibodies	Alemtuzumab	PML, encephalopathy, seizures, speech disorders	
	Bevacizumab	Intracranial hemorrhage, strokes, PRES, optic neuropathy, cognitive impairment	
	Brentuximab	PML	Sensorimotor axonal PN, Guillain-Barré syndrome, CIDP
	Cetuximab	Somnolence	Muscle cramps, paresthesias
	Polatuzumab		Sensorimotor PN
	Rituximab	Headaches, dizziness, PML, PRES[a]	Myalgias, lumbosacral paresthesias[a]
	Trastuzumab	Headaches, dizziness, insomnia	Painful sensory PN
	Ado-trastuzumab	Headaches	Sensory PN
Small-molecule tyrosine kinase inhibitors	BCR-ABL1 inhibitors (imatinib, nilotinib, dasatinib)	Cerebrovascular events, transverse myelitis, headaches	Reversible proximal myopathy, sensorimotor axonal PN
	EGFR inhibitors (afatinib, erlotinib, gefitinib)	Headaches, seizures	Sensory PN
	VEGFR inhibitors (sunitinib, sorafenib)	PRES, ischemic strokes	
	ALK inhibitors (crizotinib, alectinib, ceritinib, cabozantinib)	Headaches, optic neuropathy	Myalgias, sensory PN
	Ibrutinib	Headaches, dizziness	
	BRAF inhibitors (vemurafenib, dabrafenib)	Cerebral edema, intraparenchymal hemorrhage	Bilateral facial palsy (vemurafenib)
	MEK inhibitors (trametinib, cobimetinib)		Rhabdomyolysis, serous retinopathy
	Palbociclib		Myalgias, dysgeusia
	NTRK inhibitors (larotrectinib, entrectinib)	Encephalopathy, dizziness, gait disturbances, sleep disturbances	
Proteasome inhibitors	Bortezomib Carfilzomib Ixazomib		Painful sensory PN

Continued

IV. Neurological complications of antineoplastic therapy

TABLE 2	Common neurologic toxicities with targeted therapies—cont'd

Drug class	Specific drug	Neurologic toxicities	
		CNS	PNS
mTOR inhibitors	Rapamycin Everolimus Temsirolimus	PRES	Reversible muscle weakness
SMO inhibitors	Sonidegib Vismodegib		Dysgeusia, myalgias, muscle spasms

[a] When administered intrathecally.

CIDP, chronic inflammatory demyelinating polyradiculopathy; PML, progressive multifocal leukoencephalopathy; PN, peripheral neuropathy; PRES, posterior reversible encephalopathy syndrome.

implicated in the development of CIPN. Drugs that alter microtubule polymerization, such as the vinca alkaloids and the taxanes, inhibit fast axonal transport and lead to a length-dependent axonal injury.[8,9] Other agents, like thalidomide, may affect the blood supply to the nerves, leading to Wallerian degeneration, while a reversible transient demyelination of posterior columns has been observed with cisplatin.[1,10] Acute reversible neuropathy, cumulative CIPN, and less commonly visual changes, voice changes, and perioral paresthesias, have been observed with oxaliplatin, and have been attributed to the disruption of the voltage-gated sodium channels due to the chelation of calcium and magnesium by the oxalate metabolites.[11] The dorsal root ganglia of primary sensory neurons have fenestrated blood vessels and lack a BBB, rendering them more susceptible than the motor neurons to the effect of chemotherapy agents. Hence, sensory symptoms are predominant, although motor and autonomic symptoms can occur. Neuromuscular complications are much less common than CIPN and manifest mainly by cramps and myalgias with tyrosine-kinase inhibitors, and exacerbation or rarely development of a myasthenic syndrome with doxorubicin, cisplatin, etoposide, or immunotherapies.[12,13]

3 Neurotixicity of specific agents

3.1 Alkylating agents

3.1.1 Nitrosoureas

The nitrosourea drug class includes carmustine (BCNU), lomustine (CCNU), semustine (methyl-CCNU), nimustine (ACNU), fotemustine, and streptozotocin. They are used in the treatment of high-grade gliomas, melanoma, and lymphoma. They are highly lipid soluble and easily go into the CSF where they reach concentrations of 15%–30% that of the plasma. They are administered orally, intravenously, or intra-arterially, and BCNU wafers are implanted locally into the resection cavity. Neurotoxicity is rare at conventional doses, but high doses have been associated with ocular symptoms (acute

orbital pain during infusion, optic neuropathy, and retinal injury) and leukoencephalopathy manifesting with focal or generalized seizures, focal weakness, and rarely stroke and coma.[14] Intra-arterial delivery of BCNU has been associated with a subacute necrotizing leukoencephalopathy characterized by focal necrosis and mineralizing axonopathy and potentiated in patients who received prior or concurrent radiation.[15,16] A mild incidence of neurotoxicity is observed with BCNU wafers, and it usually manifests with increased edema, seizure activity, and contralateral focal deficits.

3.1.2 Busulfan

Busulfan is a bifunctional alkylating agent that acts by cross-linking nucleic acid strands and interfering with transcription and translation. It is an oral drug that is often used in the treatment of chronic myelogenous leukemia, and is one of the most frequently used drugs in the conditioning regimens of hematopoietic stem cell transplant.[17] It freely crosses the BBB and achieves high concentrations in the CSF that are similar to those in plasma.[18,19] Approximately 10% of patients receiving high-dose therapy experience generalized tonic-clonic seizures that typically occur between the second day of therapy initiation and within 24h after the final dose administration.[20] Prophylactic antiepileptic therapy decreases the incidence of seizures and has become standard in the pre-transplant conditioning regimen. The most studied antiepileptic medications are phenytoin and lorazepam, but levetiracetam has emerged as a safe and effective option.[21–23]

3.1.3 Chlorambucil

Chlorambucil is a well-tolerated oral nitrogen mustard derivative that is approved as a single agent or in a combinatorial regimen for the treatment of leukemias, lymphomas, and some solid tumors. Generalized seizures have been reported with high-dose therapy, especially in children with nephrotic syndrome. Low-dose therapy was also associated with seizures, but only in patients with a prior history of seizures.[24] Seizures typically occur within 24h up to 3weeks after initiation of therapy, but spells occurring as early as few hours and as late as 90 days have been reported. Diplopia, retinal hemorrhages, and progressive multifocal leukoencephalopathy (PML) have also been reported.[25,26]

3.1.4 Cyclophosphamide

Cyclophosphamide requires metabolic activation in the liver and has broad spectrum activity against leukemias, lymphomas, and solid tumors. Neurotoxicity is rare at conventional doses, but reversible blurred vision, dizziness, and confusion have been reported with high doses.[27,28]

3.1.5 Ifosfamide

Ifosfamide is an analog of cyclophosphamide that is hydroxylated by the liver to active metabolites 4-hydroxy-ifosfamide and isofosforamide mustard. It is approved for the treatment of sarcomas, and testicular and lung cancers. 15% to 40% of patients receiving high-dose therapy develop an idiosyncratic encephalopathy characterized by confusion, hallucinations, cerebellar dysfunction, extrapyramidal symptoms, cranial neuropathies, seizures, and occasionally coma.[7,29] Symptoms usually develop within 24 h of the infusion and are more likely to occur in patients with hepatic or renal impairment, hypoalbuminemia, hyponatremia, concurrent use of phenobarbital, and prior treatment with cisplatin.[30,31] They usually resolve spontaneously within a few days, although irreversible deficits and deaths have been reported, and rechallenge is uncommon given the high risk of recurrence. Methylene blue has been used to treat severe ifosfamide-induced encephalopathy and helps improve the symptoms in a matter of days as well; however, spontaneous recovery can also occur, and the lack of its evaluation in randomized clinical trials and its potential of causing headaches and dizziness make its usefulness unclear. Benzodiazepines and thiamine have also been used to reverse the neurotoxicity symptoms.[9] Painful axonal peripheral neuropathy has also been reported with ifosfamide, mainly in patients with an underlying neuropathy.[32,33]

3.1.6 Procarbazine

Procarbazine is an oral alkylating agent used in the treatment of lymphomas and gliomas. Its outdated intravenous formulation used to cause severe somnolence. Due to its weak monoamine oxidase inhibitor activity, its main neurological adverse events are headaches and hypertensive encephalopathy and are mainly observed with concurrent use of sympathomimetic agents or after consumption of tyramine-containing food. Therefore, dietary restrictions are recommended during the treatment course. Other neurotoxicities include depression and psychosis.[34]

3.1.7 Temozolomide

Temozolomide is an oral alkylating agent with relatively high BBB penetration that is widely used in the treatment of malignant glioma. Headaches have been reported in up to 40% of patients, although their etiology cannot be distinguished from the underlying malignancy. Severe neurological complications are rare.[35]

3.1.8 Thiotepa

Thiotepa is related to nitrogen mustard and easily crosses the BBB.[36] It is used intravenously in the treatment of breast, ovarian and bladder cancer, lymphoma, as part of a multi-agent chemotherapy regimen prior to hematopoietic stem cell transplantation, and intrathecally for the treatment of leptomeningeal disease. Neurotoxicity is rare at conventional intravenous doses, but encephalopathy and an acute confusional state can develop at high doses. Toxicity is mainly observed with intrathecal use, including mild reversible chemical meningitis, and in rare cases a myelopathy that can seldom be permanent.[37]

3.1.9 Other alkylating agents

Dacarbazine is an intravenous alkylating agent used in the treatment of lymphoma, melanoma, and other solid tumors. Mild self-limited headaches and malaise have been reported, but severe neurotoxicity is rare, manifesting as encephalopathy and seizures.[38]

3.2 Antimetabolites

3.2.1 2-Chlorodeoxyadenosine (cladribine)

Cladribine is a purine nucleoside analog used to treat lymphoid proliferative disorders, notably hairy cell leukemia, chronic lymphocytic leukemia, and low-grade lymphoma. Neurotoxicity is uncommon at low doses, although headaches and distal motor weakness have been reported. The incidence of neurotoxicity increases with incremental doses, and the main adverse event is a severe peripheral sensorimotor neuropathy characterized by axonal degeneration that might be irreversible.[39] Myelopathy and a Guillain-Barré-like syndrome have also been reported in a case series.[40]

3.2.2 Capecitabine

Capecitabine is an oral pro-drug that is metabolized to its cytotoxic form, 5-fluorouracil, by thymidine phosphorylase. It is used to treat breast, lung, head and neck, pancreatic and colorectal cancer. Neurotoxicity is uncommon, but cases of multifocal leukoencephalopathy have been reported.[41] Symptoms usually develop within a week after treatment, and consist of headaches, vertigo, confusion, ataxia, dysarthria, and generalized increased muscle tone. Brain MRI often shows a hyperintense T2/FLAIR restricting signal in the splenium of the corpus callosum, and less often in the brachium pontis and deep periventricular white matter. Symptoms usually resolve over a few days after treatment discontinuation. Cerebellar ataxia, acute reversible trismus, ocular abnormalities and mild peripheral neuropathy have also been reported[42–44] (Fig. 2).

3.2.3 Cytosine arabinoside (Cytarabine, Ara-C)

Cytarabine is a pyrimidine analog whose active metabolite inhibits DNA-polymerase in the S phase of the cell cycle and causes premature chain termination. It is used in the treatment of leukemias, lymphomas and leptomeningeal metastases. The incidence and severity of neurotoxicity depends on the dose and route of administration. Intravenous administration is mostly associated with cerebellar toxicity, especially at doses higher than $1 g/m^2$. An acute cerebellar syndrome is observed in up to 25% of patients treated with doses higher than $3 g/m^2$.[45] The syndrome usually occurs within 24 h and starts with somnolence and confusion that are then followed by cerebellar signs, including truncal ataxia and gait instability, along with dysarthria and nystagmus.[46] The infusion should be stopped, and symptoms usually resolve spontaneously within 2 weeks. Risk factors for the development of cerebellar toxicity include age greater than 40, abnormal kidney function, underlying neurological disease, and cumulative doses greater than 30 g.[47] CSF studies are usually normal. Brain imaging usually shows reversible white matter changes, but cerebellar atrophy subsequent to Purkinje fiber loss can be observed on long-term follow-up.[9] Blurred vision, burning eye pain, reversible reduced visual acuity, pseudobulbar palsy, Horner's syndrome and anosmia have been occasionally reported.[48–50]

Intrathecal administration of Cytarabine produces high levels of the drug in the CSF for at least 24 h, leading to potential adverse events occurring mainly within the first 24 h after its administration, but which onset can be delayed up to days of the dose. Aseptic meningitis is the most common complication and has been reported in up to 10% of patients. Patients typically report lower back pain, and often complain of headaches.[51] Myelopathy is a rare complication and can occur within days to weeks after intrathecal dose administration; it manifests as lower back pain followed by rapidly ascending flaccid paraparesis, sensory level and sphincter function abnormalities.[52] Spine MRI can be normal initially, but then focal cord swelling and intramedullary enhancement can be apparent. Other rare complications include locked in syndrome reported in one patient, acute encephalopathy and seizures.[32,53] The sustained-release formulation of cytarabine for intrathecal administration, liposomal cytarabine, is no longer commercially available but used to produce cytotoxic levels in the CSF for up to 10 to

14 days, leading to the development of aseptic meningitis in up to 25% of treatment cycles.[54] Coadministration of corticosteroids greatly reduced the likelihood of this complication.

PNS toxicity is rare, but painful sensory neuropathy has been reported when intravenous cytarabine was used in combination to other cytotoxic agents.[55,56] Brachial plexopathy and lateral rectus muscle palsy have also been reported.[57]

3.2.4 Fludarabine

Fludarabine is a purine analog that acts via inhibition of DNA polymerase and ribonucleotide reductase and is used for the treatment of B-cell chronic lymphoblastic leukemia and in some conditioning regimens for allogeneic transplants. Neurotoxicity is uncommon and usually manifests with headaches, confusion, blurred vision, and proprioceptive deficits.[58] At doses greater than $50 mg/m^2/day$, an acute toxic leukoencephalopathy may develop and can be severe and rarely fatal.[59] Polymorphisms in genes encoding for concentrative nucleoside transporters and deoxycytidine kinase have been correlated with increased incidence of neurotoxicity.[60]

3.2.5 5-Fluorouracil (5-FU)

5-FU is a fluorinated pyrimidine that interferes with DNA synthesis through inhibition of thymidylate synthetase. Like capecitabine, it is widely used in the treatment of solid cancers, including colorectal, head and neck and breast cancers. It readily crosses the BBB, with highest concentrations found in the cerebellum. The main neurotoxicity is an acute cerebellar syndrome that occurs within weeks to months after drug administration in approximately 5% of patients, and is characterized by the sudden onset of ataxia, dysarthria and nystagmus.[61–63] A rare syndrome of acute encephalopathy has been reported with cumulative doses of 5-FU, mainly with concurrent use of levamisole.[64] Patients experience a variety of symptoms, 6 to 19 weeks after the initiation of therapy, including confusion, altered mental status, depressed mood, aphasia, dysarthria, ataxia, diplopia, vertigo, hiccups, gait imbalance, akinesia, paresthesias and seizures. The pathophysiology of this syndrome is an inflammatory reaction reflected by perivascular lymphocyte infiltration with myelin loss and relative axonal sparing on pathological examination and leading to multifocal demyelination in the periventricular white matter. Most patients improve clinically and radiographically after discontinuation of therapy and a course of corticosteroids. Posterior reversible encephalopathy syndrome (PRES) has also been reported as a delayed toxicity from 5-FU.

Toxicity from 5-FU and its prodrug capecitabine can be severe, and sometimes fatal, in patients with dihydropyrimidine dehydrogenase (DPD) deficiency.[65,66]

DPD is the initial and rate-limiting enzyme of the 5-FU catabolic pathway and is responsible for the catabolism of 85% of 5-FU. Reduced DPD activity leads to the accumulation of 5-FU and has been reported in up to 59% of patients who experienced unanticipated 5-FU toxicity. Polymorphisms in the genes encoding thymidylate synthetase, methylene tetrahydrofolate reductase, and orotate phosphoribosyltransferase have been also associated with increased incidence of 5-FU neurotoxicity.[67]

3.2.6 Gemcitabine

Gemcitabine is a cytidine analog that is activated intracellularly to two active cytotoxic metabolites and is used in the treatment of pancreatic, bladder, and non-small-cell lung cancers. Neurotoxicity is uncommon, although sensory and autonomic neuropathies have been reported in up to 10% of patients.[68] An encephalopathy syndrome, characterized by headaches, confusion, visual impairment, aphasia, ataxia and seizures can occur 1–2 weeks after a chemotherapy regimen that combines gemcitabine to platinum-based agents and/or taxanes.[69] This combination has also been associated with the development of an acute inflammatory myopathy characterized by painful and focal weakness, typically involving proximal muscles, muscles in previously irradiated areas, and the muscles of the abdominal wall.[70] Symptoms usually resolve on steroids after discontinuation of the treatment.

3.2.7 Hydroxyurea

Hydroxyurea is a ribonucleotide reductase inhibitor that is used to treat myeloproliferative disorders and some solid tumors, including head and neck cancers. Neurological complications are rarely observed with its use. Sedation can be seen with high doses. Headache, confusion, hallucinations and seizures have been reported, especially when hydroxyurea is combined with highly active antiretroviral therapy in patients with acquired immunodeficiency syndrome.[71,72]

3.2.8 Methotrexate

Methotrexate is an irreversible inhibitor of dihydrofolate reductase and inhibits the biosynthesis of both purines and thymidine. It is the most widely used antimetabolite agent in the treatment of both hematologic and solid tumors. Its poor BBB penetration is enhanced by high-dose intravenous ($\geq 1\,g/m^2$) and intrathecal administration.[1] Its toxic adverse events affect mainly the CNS, and can be acute, subacute or chronic.

Aseptic meningitis is the most common acute toxicity and can occur in up to 60% of patients receiving intrathecal therapy. Symptoms start within 4h and usually resolve spontaneously within 72h; they consist of headaches, fever, nuchal rigidity and lethargy.[73] Development of such a toxicity is idiosyncratic and is not predictive of similar reactions with subsequent doses.[1] Transverse myelopathy is another idiosyncratic acute complication of intrathecal administration of methotrexate. It manifests with acute back and radicular leg pain, with possible sensory loss, paraplegia and sphincter dysfunction in severe cases, and is pathologically characterized by vacuolar necrotizing demyelination of the spinal cord and often brainstem.[1,74] Its incidence increases with prior or concurrent radiation to the spine, leptomeningeal disease, and advanced age. Clinical improvement is variable over time. When administered through an intraventricular reservoir, methotrexate can penetrate the white matter directly, through CSF flow obstruction or through a misplaced catheter, and cause an acute encephalopathy characterized by sudden delirium, somnolence, and rarely seizures.[75] This complication has been also reported with intravenous high doses in the setting of prior or concurrent brain radiation.

A stroke-like syndrome, with fluctuating focal findings (alternating hemiparesis, aphasia or seizures) can develop days to weeks after intravenous administration of high-dose methotrexate, typically occurring after the third or fourth dose. The mental status can be variable and change from extreme somnolence to inappropriate laughter. Symptoms typically resolve within days, and do not preclude further treatment. Posterior reversible encephalopathy syndrome (PRES) has been reported in patients receiving intravenous or intrathecal methotrexate. It can develop acutely, or within days of treatment, and is characterized by headaches, confusion and visual disturbances.[76,77] Its incidence increases with concurrent hypertension and hypomagnesemia.

Delayed leukoencephalopathy is the most devastating complication of methotrexate, and can occur in up to 26% of long-term adult CNS lymphoma survivors who have received multiple high-dose intravenous treatments, especially if the treatment was preceded by or accompanied with cranial irradiation[78]; progressive dementia is the hallmark symptom and can develop up to 6 years after treatment. Gait disturbances, hemiparesis and aphasia have also been reported.[79] Brain MRI typically shows cerebral atrophy, with diffuse white matter abnormalities, ventricular enlargement, and sometimes cortical calcifications.[80] Loss of oligodendrocytes, gliosis, and necrotizing leukoencephalopathy are typically observed on autopsy.

3.2.9 Pentostatin

Pentostatin is a purine analog used in the treatment of hairy cell leukemia. Neurotoxicity is rare with conventional doses. Seizures and encephalopathy were observed with higher doses.[81]

3.3 Platinum compounds

3.3.1 Cisplatin

Cisplatin is a platinum-based agent that binds to DNA, producing inter- and intrastrand cross-links, impairing DNA synthesis and transcription. It is used to treat a wide variety of solid tumors including ovarian, testicular, lung, gastrointestinal, bladder, head and neck cancers, and medulloblastoma. It has minor BBB penetration, but high concentrations were found in dorsal root ganglia cells on autopsies.[82] Although the CSF concentration is relatively low, CNS toxicity can occur, especially after intra-arterial administration into the carotid arteries in patients with brain tumors.[83] Symptoms can result from direct toxicity of cisplatin or from cisplatin-induced electrolyte abnormalities, like hyponatremia as a result of the syndrome of inappropriate antidiuretic hormone (SIADH), or renal toxicity, hypomagnesemia or hypocalcemia.[9] Headaches, encephalopathy, cortical blindness, focal deficits, and seizures have been reported and are usually reversible after cessation of treatment or correction of the metabolic disturbances. Delayed vascular toxicity leading to strokes has also been described. A transient demyelination of posterior spinal cord columns can be seen in up to 40% of patients and manifests with Lhermitte phenomenon during or after the infusion.[10] High concentrations of cisplatin-induced platinated DNA accumulates in the nuclei of outer hair cells, leading to accumulation of reactive oxygen species, and activation of mitochondrial pathways that lead to apoptosis of these cells.[84] This results in ototoxicity and increase in the hearing threshold on audiometric studies in the majority of patients. Tinnitus is reported in up to 36% of patients, and permanent bilateral hearing loss can affect 20% of patients at low cumulative doses, and up to 50% of patients at high cumulative doses.[85] The incidence of ototoxicity increases with incremental doses, noise exposure, exposure to other cytotoxic drugs, malnutrition state and cranial irradiation.[86–88] Patients treated at a younger age are more likely to develop ototoxicity as well.

Peripheral neuropathy is the major dose-limiting toxicity of cisplatin, typically developing in more than 30% of patients receiving cumulative doses >400 mg/m^2.[89] Axonal damage predominantly affects the large myelinated sensory fibers, and patients usually develop numbness and paresthesia in their distal extremities, but more severe deficits in proprioception leading to ataxic gait can develop. Impaired pain and temperature sensation and dysautonomia are less common, and motor fibers are usually spared. Symptom progression may occur for months after cessation of therapy (so-called coasting phenomenon), but most patients eventually improve, although recovery is often incomplete.[82]

3.3.2 Carboplatin

Carboplatin is a second-generation platinum complex with similar activity to cisplatin but with fewer neurologic side effects. It crosses the BBB poorly but achieves higher CSF concentrations than cisplatin.[90] CNS toxicity was mainly reported with intra-arterial administration, and retinopathy is the most common adverse event; cortical blindness and cortical infarcts are much less frequent.[91] Rare cases of peripheral neuropathy have been reported in patients previously treated with cisplatin or receiving high-dose carboplatin therapy.

3.3.3 Oxaliplatin

Oxaliplatin is a third-generation platinum compound that is approved for the treatment of colorectal cancer. Peripheral neuropathy is the most common nonhematologic adverse event.[92] Almost all patients experience acute transient paresthesias, muscular spasms involving the limbs and perioral region, and voice changes that are triggered or exacerbated by cold. Patients often report throat discomfort and difficulty swallowing cold liquids.[93] Symptoms can occur during the drug infusion, but they typically get worse 2 to 3 days after each dose, and their severity often increases with subsequent doses. These symptoms are believed to be related to chelation of calcium and magnesium by the oxalate metabolites, leading to altered neuronal voltage-gated sodium channels.[94,95] Later in the course of the treatment, a chronic dose-dependent sensorimotor axonal neuropathy may develop in 10% to 20% of patients receiving cumulative doses greater than 750 mg/m^2.[96]

3.4 Antineoplastic antibiotics

Several antibiotics have antineoplastic effects by causing single-strand breaks in DNA, and inhibition of DNA-directed RNA polymerase and topoisomerase II. Anthracyclines (doxorubicin and daunorubicin) are classic examples, along with bleomycin, mitomycin C, and mitoxantrone. Most of these agents have little to no CNS penetration; hence, neurotoxicity is infrequent with their use. Cerebral infarcts and hemorrhagic necrosis have been reported with intra-arterial administration of doxorubicin, and inadvertent intrathecal injection of doxorubicin or daunorubicin may lead to a fatal acute or subacute ascending myelopathy and encephalopathy.[97,98] Rare visual blurriness, seizures, encephalopathy, and cerebral infarcts have been observed with bleomycin, while a disseminated intravascular coagulation and thrombotic microangiopathy can occur with mitomycin C, leading to headaches and cerebral infarcts.[38,99]

3.5 Vinca alkaloids

Vinca alkaloids are plant-derived compounds that inhibit the assembly of microtubules in the mitotic spindle, hence causing cell cycle arrest. Vincristine and vinblastine are natural alkaloids, whereas vindesine and vinorelbine are semi-synthetic compounds, among others.

Vincristine is the most widely used for the treatment of leukemias, lymphomas, myelomas, various sarcomas, and brain tumors, and is the most neurotoxic in this drug class (particularly affecting the PNS). CNS toxicity is rare, although an accidental massive intrathecal overdose may lead to a fatal myeloencephalopathy.[100] Encephalopathy, cortical blindness, athetosis, ataxia, parkinsonism, and seizures have been infrequently reported and are usually reversible without any sequelae.[101] Cranial neuropathies have also been described, including ptosis, ophthalmoplegia, diplopia, vocal cord paresis, sensorineural hearing loss, and facial palsy.[102–104]

PNS toxicity is common with vincristine and is thought to be caused by inhibition of fast axonal transport by microtubules in the peripheral nerves. A length-dependent, sensory more than motor, small fiber polyneuropathy is the most common complication and occurs in the majority of patients, but femoral and peroneal mononeuropathies have also been described. Paresthesia in the fingers and toes is the initial symptom and is often followed by loss of ankle reflexes. Sural nerve biopsy shows axonal degeneration with segmental demyelination. Weakness might develop, and it usually involves the extensor muscles of the wrist and dorsiflexors of the toes.[105] The severity of peripheral neuropathy is dose-related and tends to be more severe in older or malnourished patients, those with prior irradiation to the peripheral nerves, a history of hepatic insufficiency or a demyelinating neuropathy like Charcot-Marie-Tooth, or receiving concurrent hematopoietic colony-stimulating factors or inhibitors of the CYP3A4 enzyme.[103,106–108] Up to one-third of patients develop autonomic neuropathy predominantly affecting the gastrointestinal tract, leading to constipation, and in rare cases to paralytic ileus or megacolon, although bladder atony, impotence, and orthostatic hypotension can occur.[109] Peripheral neuropathy is usually reversible, and its development does not preclude continuing the treatment. However, therapy should be discontinued if symptoms are very severe and affecting the patient's quality of life. A dose equal to or less than $1.4 \, mg/m^2$ is recommended to minimize the risk of neuropathy.

Vinblastine does not cause CNS toxicity, but can cause a milder form of neuropathy than vincristine. Vinorelbine and vindesine might also cause a mild distal sensory predominant neuropathy with loss of ankle reflexes.[110,111]

3.6 Taxanes

3.6.1 Paclitaxel

Paclitaxel is a plant derivative that blocks the cell cycle in the late G2 mitotic phase through polymerization and stabilization of microtubules. It is often combined with platinum compounds and used in the treatment of a variety of solid tumors including breast, ovarian, and lung cancers. CNS toxicity is rare and occurs with very high doses ($> 600 \, mg/m^2$), mainly in the form of an acute severe encephalopathy.[112] Some patients also report photopsia (a sensation of flashing light across their visual fields) during the infusion.[113] PNS toxicity is more common and is actually the most predominant adverse event observed with paclitaxel. The abundance of rigid microtubules in the peripheral nerves inhibits axonal transport and leads to a predominantly large-fiber polyneuropathy that may develop within 48h after treatment, and usually occurs after cumulative doses of $100–200 \, mg/m^{2}$.[114] Patients start experiencing paresthesias, numbness, and sometimes burning pain in the hands and feet. Symptoms may progress to impaired dexterity, and unsteadiness, particularly in the dark, due to proprioceptive defects.[115] Some patients develop a mild proximal muscle weakness that usually resolves spontaneously.[116] Transient acute arthralgias and myalgias have been reported, and they are believed to be a manifestation of neuropathy.[117,118] The coadministration of granulocyte colony stimulating factor (G-CSF) at doses greater than $250 \, mg/m^2$ may worsen the neuropathy, making it a dose-limiting toxicity.[119] High-dose cycles administered in a short infusion, high cumulative doses, diabetes mellitus, and preexisting neuropathy increase the incidence of CIPN. Sural nerve biopsy shows fiber loss, lack of axonal sprouting, and axonal atrophy, with secondary demyelination and remyelination.[120] The neuropathy is seldom permanent.

3.6.2 Docetaxel

Docetaxel is a semisynthetic analog of paclitaxel used to treat a variety of solid tumors. It is a more potent inhibitor of cell replication than paclitaxel.[9] Like paclitaxel, it can cause a predominant large-fiber sensory peripheral neuropathy that is usually mild to moderate and resolves within weeks of treatment discontinuation. Neuropathy is more severe with cumulative doses greater than $400 \, mg/m^2$, and is histologically characterized by loss of large, myelinated fibers, and occasional axonal degeneration.[121] Reversible Lhermitte's phenomenon during infusion and a mild form of proximal motor weakness have been occasionally reported.[116,122]

3.7 Topoisomerase inhibitors

Topoisomerases are DNA enzymes that control the configuration of the DNA double helix during its transcription and replication. Topoisomerase I initiates the cleavage of a single strand of the DNA, while topoisomerase II cleaves both DNA strands. Topoisomerase inhibitors induce irreversible DNA replication defects leading to subsequent cell cycle arrest and cell death.

Irinotecan is a water-soluble semisynthetic analog of camptothecin (CPT) that is metabolized in the liver to its active metabolite. It is a topoisomerase I inhibitor used in the treatment of lymphomas, leukemias, and progressive or metastatic colorectal cancer. Severe neurotoxicity has not been observed. Nonspecific dizziness and occasional transient dysarthria have been reported.[38]

3.7.1 Topotecan

Topotecan is also a topoisomerase I inhibitor with activity against ovarian and small cell lung cancer. Neurotoxicity is uncommon; headache and paresthesias are infrequently reported.

3.7.2 Etoposide and teniposide

Both of these agents are semisynthetic derivatives of podophyllotoxin and inhibit topoisomerase II. They have poor CNS penetration, reaching a CSF concentration only 0.3%–5% that of plasma. Etoposide is more frequently used, mainly in the treatment of lung cancer, germ cell tumors, and lymphomas. CNS toxicity is rare with conventional doses. Headaches and encephalopathy are occasionally reported during or shortly after high dose intravenous infusions. A mild-to-moderate sensorimotor neuropathy has also been described, particularly in patients with advanced age, poor nutritional status, and prior neurotoxic chemotherapy treatment.[38,123]

4 Targeted therapies

4.1 Monoclonal antibodies

4.1.1 Bevacizumab

Bevacizumab is a humanized antibody against vascular endothelial growth factor (VEGF) that is approved for treatment of metastatic colorectal cancer and glioblastoma. Neurotoxicity is related to its anti-VEGF activity, and it mainly consists of cerebrovascular events.[124,125] VEGF has a protective effect on the endothelial cells, through promoting their proliferation, suppressing the pathways involved in their activation and coagulation, and induction of nitric oxide (NO) and prostacyclin.[126,127] Blockade of VEGF impairs the endothelial cell regenerative capacity and leads to a proinflammatory state characterized by expression of proinflammatory cytokines and inhibition of NO and prostacyclin, resulting in an increased incidence of hypertension, damage of the blood vessels in the brain, and eventually intracranial thrombosis and hemorrhage.[126,127] PRES occurs

in less than 0.1% of patients, and symptom onset can range from 1 day to 1 year after initiation of therapy. Discontinuation of the therapy and control of any underlying hypertension leads to resolution of the symptoms.[128] Severe optic neuropathy has been reported in some patients receiving bevacizumab in the setting of prior radiation to the optic apparatus.[129] Long-term cognitive impairment has also been described in patients who received prior bevacizumab, and it is thought to be caused by reduced synaptic plasticity in the hippocampus through inhibition of the VEGF expressed by CA1 neurons in the hippocampus.[130]

4.1.2 Cetuximab

Cetuximab is a chimeric mouse-human antibody directed against epithelial growth factor receptor (EGFR). It is used in the treatment of colorectal cancer and head and neck cancer. It can cause symptomatic hypomagnesemia, manifesting with fatigue, somnolence, muscle cramps, and paresthesias. It was also reported to aggravate symptoms of oxaliplatin-induced neuropathy through the same mechanism.[131] Two cases of chronic inflammatory demyelinating polyradiculoneuropathy have also been reported with cetuximab use.[132]

4.1.3 Iodine-131 tositumomab

Iodine-131 tositumomab is a radiolabeled anti-CD20 murine monoclonal antibody. It delivers targeted radiotherapy to malignant B cells and is approved for the treatment of indolent non-Hodgkin lymphoma. Significant neurotoxicity has not been reported, but few patients may experience headache, fatigue, or dizziness during infusion.[133,134]

4.1.4 Rituximab

Rituximab is a chimeric monoclonal antibody directed against the cell surface antigen CD20, a nonglycosylated phosphoprotein expressed on the surface of mature B cells. It is used to treat systemic B-cell lymphoma and primary CNS lymphoma. Neurotoxicity is rare with rituximab. The most common neurologic side effects are headaches, nonspecific dizziness, and myalgias that have been reported during infusion, but rare cases of ischemic or hemorrhagic strokes, seizures, serotonin syndrome, and PRES have been described.[135–137] The most serious complication, though rare, is the reactivation of JC virus, leading to PML.[138] Intrathecal rituximab is usually well tolerated, but rare cases of acute and reversible encephalopathy and lumbosacral paresthesias have been reported.[139,140]

4.1.5 Trastuzumab (Herceptin)

Trastuzumab is a humanized monoclonal antibody directed against human epidermal growth factor receptor-2 (HER2). It is used alone or in combination with standard chemotherapeutic drugs for the treatment of HER2-positive breast cancer. The most common infusion-related neurologic side effects reported include headache, dizziness, fatigue, and insomnia. A sensory neuropathy, that can be sometimes painful, has been reported with high cumulative doses and usually starts 1 to 2 years after initiation of therapy.[141]

T-DM1, also called ado-trastuzumab, is an antibody-drug conjugate (ADC) that combines trastuzumab to emtansine, a cytotoxic antimicrotubule agent, and has been also associated with a sensory polyneuropathy, and headaches, especially in patients with brain metastases.[142–144] It also potentiates the effect of stereotactic radiation and leads to a higher incidence of radiation necrosis and brain edema.[145,146]

Trastuzumab deruxtecan is another ADC, where trastuzumab is conjugated to a topoisomerase I inhibitor, that also caused a painful sensory neuropathy in a minority of patients.[147]

4.1.6 Brentuximab vedotin

Brentuximab vedotin is an ADC that is used in the treatment of CD30-positive lymphomas. The most common neurological adverse event is peripheral neuropathy that usually develops 24 to 30 weeks after initiation of therapy.[148] It has been reported in up to 80% of patients and is more frequent in the elderly with predisposition to

neuropathy. Symptoms are predominantly sensory, but motor symptoms can develop in up to 15% of cases.[149,150] Therapy should be held for neuropathy of a grade greater than 2, until improvement to grade 1, then can be restarted at lower doses. However, it should be discontinued if a grade 4 neuropathy develops. Neuropathy is attributed to the internalization of CD30 receptor, leading to intracellular release of monomethyl auristatin E which disrupts the microtubule network. A few cases of PML, Guillain-Barré syndrome, and CIDP have been reported.[151,152] Myalgias and mild reversible proximal muscle weakness can also be observed (Figs. 1 and 2).

4.1.7 Polatuzumab vedotin

This ADC combines the monoclonal anti-CD79b antibody polatuzumab to the microtubule disrupting agent monomethyl auristatin E. It is used in the treatment of large B-cell lymphoma. The main neurotoxicity is a sensorimotor polyneuropathy that can develop in up to 40% of the patients; it is usually mild and resolves after dose reduction.[153,154]

4.1.8 Other monoclonal antibodies

Alemtuzumab is a humanized monoclonal anti-CD52 antibody and is approved for the treatment of B-cell chronic lymphocytic leukemia. It has been associated with PML in a few patients.[155]

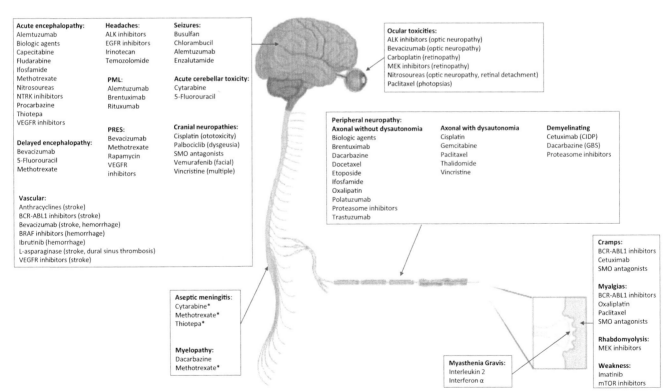

FIG. 1 Common neurologic toxicities of chemotherapy and targeted therapy agents (created in BioRender.com). CIDP: chronic inflammatory demyelinating polyradiculopathy; GBS: Guillain-Barré syndrome; PML: progressive multifocal leukoencephalopathy; PRES: posterior reversible encephalopathy syndrome; *when administered intrathecally.

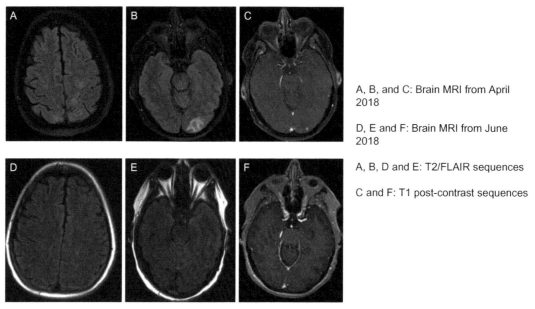

A, B, and C: Brain MRI from April 2018

D, E and F: Brain MRI from June 2018

A, B, D and E: T2/FLAIR sequences

C and F: T1 post-contrast sequences

FIG. 2   Posterior reversible encephalopathy syndrome after treatment with gemcitabine. This is the case of a 52-year-old lady with a poorly differentiated peritoneal carcinoma, with liver metastases, who was treated with gemcitabine after her disease had progressed on doxorubicin and bevacizumab. One week after the first dose of gemcitabine, she started developing occipital headaches, nausea, and elevated blood pressure. A few days later, she was brought to the emergency department with a generalized tonic-clonic seizure. Brain MRI showed T2/FLAIR hyperintense changes in the cortical and subcortical areas of the bilateral parieto-occipital lobes, along with punctate enhancement in the left occipital lobe. Lumbar puncture was negative for any infections or leptomeningeal involvement by her disease. After discontinuation of gemcitabine, and antihypertensive and antiepileptic treatment, the symptoms gradually improved and then resolved. A follow-up brain MRI 2 months later showed resolution of the abnormal T2/FLAIR signal and enhancement.

Blinatumomab is an anti-CD19/CD3 bispecific T-cell engager that is approved for the treatment of acute lymphocytic leukemia. Neurotoxicity has been observed in up to 50% of patients, and the reported adverse events were encephalopathy, seizures, speech disorders, coordination, and balance disorders, and were attributed to the activation of the immune system.[156–158] Similar adverse events, and many more involving the CNS and PNS, have also been observed with immune checkpoint blockade agents. The immune-related neurological adverse events of these agents will be discussed in another chapter.

4.2 Small-molecule tyrosine kinase inhibitors

4.2.1 BCR-Abl inhibitors

BCR-Abl tyrosine kinase inhibitors are oral agents used in the treatment of chronic myeloid leukemia (CML) and Philadelphia-positive acute lymphoblastic leukemia (pH + ALL). Imatinib is the first engineered agent in this class and was approved in 2001 for CML. Second and third generation agents were developed and include bosutinib, nilotinib, ponatinib, and dasatinib. An increased risk of arterial occlusive disease and cerebrovascular events has been reported with these agents, particularly with ponatinib, due to its VEGFR inhibitory activity, where up to 48% develop such complications.[159] Lipid profile and glucose levels should be monitored

during therapy, not to further increase the risk of strokes. ABL kinases have also been implicated in the development of neuromuscular junctions, and myalgias and cramps are not infrequent with these agents; they are usually mild and improve with calcium and magnesium administration.[12] However, a more severe reversible proximal myopathy has been described with imatinib.[151] A case series of patients treated with such agents identified one patient who developed transverse myelitis on imatinib, that recurred with dasatinib, as well as a length-dependent sensorimotor axonal neuropathy in one patient, and an optic neuropathy with visual field defects in another patient.[160] Headaches and nonspecific dizziness have also been reported with these agents.

4.2.2 Epidermal growth factor receptor inhibitors

The human epidermal growth factor receptor (HER) belongs to the receptor-tyrosine kinase superfamily and comprises epidermal growth factor receptor (EGFR) and HER2 among others. The oral EGFR inhibitors are afatinib, erlotinib, gefitinib, and the third-generation agent osimertinib, and they are mainly used in the treatment of non-small-cell lung cancers expressing EGFR pathway alterations. Lapatinib is an oral HER2 inhibitor and is used in breast cancer. Neurotoxicity is infrequent with these agents, except for mild headaches that can occur in up to 40% of patients.[161] They can all cause depletion in a variety of electrolytes, leading to potential seizures, especially in patients with brain metastases.[151] Peripheral neuropathy was occasionally observed with lapatinib.

4.2.3 Vascular endothelial growth factor receptor (VEGFR) inhibitors

Sorafenib and sunitinib are multikinase inhibitors that mainly inhibit VEGFR and platelet-derived growth factor receptor (PDGFR) and are used in a variety of solid tumors. Due to their main effect on VEGFR, the most common neurologic adverse events are PRES and increased incidence of ischemic strokes.[162,163] A reversible encephalopathy was also reported with both agents.[12] Sunitinib can also cause hypothyroidism and may induce myxedema coma.[164,165]

4.2.4 Anaplastic lymphoma kinase (ALK) inhibitors

Crizotinib, alectinib, ceritinib, and lorlatinib are oral ALK inhibitors used in the treatment of ALK-rearranged non-small-cell lung cancer, whereas cabozantinib is used in the treatment of renal cell carcinoma and medullary thyroid cancer. Neurotoxicity is not common with these agents. The reported adverse events are headaches and optic neuropathy that can be permanent with crizotinib,[166] myalgias with alectinib,[167] and peripheral neuropathy with ceritinib.[168] Also, more than 50% of patients receiving lorlatinib develop neurologic adverse events, the most common of which are changes in cognitive function, mood and speech, that usually occur within the first 2 months of treatment initiation and are typically reversible.[169]

4.2.5 Proteasome inhibitors

Bortezomib, carfilzomib, and ixazomib are proteasome inhibitors that are used in the treatment of multiple myeloma. Bortezomib and carfilzomib are intravenous agents, whereas ixazomib is a novel oral agent. Neurotoxicity is most common with bortezomib and is mainly in the form of a painful sensory polyneuropathy that can occur in up to 80% of patients.[170] The exact pathophysiology of the neuropathy is not known, but is thought to be related to the direct inhibitory effect of proteasomes on the dorsal root ganglion cells that lead to endoplasmic reticulum stress and mitochondrial disruption of intracellular calcium homeostasis and, subsequently, to a demyelinating or mixed demyelinating and axonal neuropathy.[151] Symptoms usually occur after cumulative doses of $> 26\,mg/m^2$ and are more likely to develop in heavily pretreated patients and in those with preexisting baseline neuropathy.[171–173] Motor neuropathy consisting of distal lower extremity weakness has been reported in 10% of patients. Symptoms usually resolve within 3 months after treatment discontinuation, but may persist up to 2 years.[174] Peripheral neuropathy is less common and

less severe with carfilzomib and ixazomib than with bortezomib.[175,176]

4.2.6 Neurotrophic tyrosine receptor kinase (NTRK) inhibitors

Larotrectinib and entrectinib are oral agents that inhibit the neurotrophic tyrosine receptor kinase (NTRK) and are approved for the treatment of sarcomas, lung cancer, and malignant salivary gland tumors. Neurologic adverse events are common and often mild. The most common complications reported are encephalopathy, dizziness, gait disturbances, and sleep disturbances.

4.2.7 Other tyrosine kinase (TKI) inhibitors

Ibrutinib is a Bruton tyrosine kinase (BTK) inhibitor used in the treatment of chronic lymphocytic leukemia and some lymphomas. The expression of BTK on the platelet surface has led to increased incidence of intracranial hemorrhage associated with its use.[177] It has also been associated with an increased incidence of opportunistic infections with CNS involvement, particularly aspergillosis.[178] It may also cause headaches and nonspecific dizziness.[179]

Vemurafenib and dabrafenib are BRAF kinase inhibitors that are used mainly in melanomas harboring BRAF V600 mutations. Their good CNS penetration has made them particularly useful in the treatment of melanoma metastatic to the CNS. Neurotoxicity has been rarely seen with them. However, given that they are radiosensitizers, their concurrent use with stereotactic radiation in brain metastases has led to worsening intraparenchymal edema and hemorrhage.[180] Facial palsy has also been described with vemurafenib and was bilateral in one patient; it is usually reversible on steroids.[181,182]

BRAF inhibitors are often used in combination with mitogen-activated protein kinase (MEK) inhibitors like trametinib and cobimetinib. These agents have been associated with rhabdomyolysis and retinal detachment with serous retinopathy in less than 1% of patients.[183]

Cyclin-dependent kinase (CDK) inhibitors are other small oral TKIs used in hormone receptor-positive breast cancer, namely palbociclib. Palbociclib-associated neurotoxicity is mild and limited to myalgias and dysgeusia.[184]

Mechanistic target of rapamycin (mTOR) inhibitors is approved for the treatment of various solid tumors, like breast, lung, kidney, and pancreatic cancer, and subependymal giant cell astrocytoma in tuberous sclerosis and in posttransplant settings. The main mTOR inhibitors are rapamycin, everolimus, and temsirolimus. Although neurotoxicity is rare with them, they may cause hypophosphatemia, inducing a symptomatic reversible muscle weakness, and PRES has been described with rapamycin in organ transplant patients.

Sonidegib and vismodegib are smoothened (SMO) antagonists that target the Hedgehog pathway and are approved for the treatment of some solid and hematologic malignancies. They can cause dysgeusia, myalgias, and muscle spasms in 20% to 40% of patients.[185] Zinc supplementation can help improve the dysgeusia.[186]

5 Biological agents

Until recently, biologic response modifiers, like interferons, interleukins, and tumor necrosis factors, have been used in conjunction with chemotherapeutic agents for the treatment of some hematologic and solid tumors. Interferon alpha is used in the treatment of hairy cell leukemia, Kaposi sarcoma, chronic myeloid leukemia, and melanoma. Neurotoxicity is dose dependent and is thought to be related to overwhelming inflammatory reactions. The most frequent adverse event at low doses is tremor, whereas confusion, lethargy, personality changes, cognitive impairment, hallucinations, hypertonia, and seizures have been reported with high doses.[187–189] Cognitive changes consist mainly of impairment in verbal memory and executive functions. The reported personality changes include depression, increased somatic concern, and stress reactions, and are more likely to occur in patients with preexisting psychiatric disorders.[190] Cases of oculomotor palsy, sensorimotor axonal polyneuropathy, myasthenia gravis, and polyradiculopathy have been described.[191,192]

Neuropsychiatric adverse events, including amnesia, delusion, hallucinations, and depression, have also been reported with interleukin 2 in up to 50% of patients with melanoma and renal cell carcinoma, and they resolved after discontinuation of therapy.[193]

Treatment with tumor necrosis factor alpha has been associated with transient aphasia, encephalopathy, and mild sensory neuropathy.[194]

Depression has also been observed with androgen deprivation therapy that reduces the plasma level of testosterone.[195] Enzalutamide, an androgen receptor signaling inhibitor, was also associated with an increased incidence of seizures in patients with castration-resistant prostate cancer.[196]

L-asparaginase is an enzyme used in the treatment of acute lymphoblastic leukemia. It catalyzes the amino acid L-asparagine, depriving the leukemic cells of it, thus inhibiting protein synthesis in these cells. Neurotoxicity is rare. However, the induced inhibition of synthesis of coagulation factors can lead to a coagulopathy, which typically occurs weeks after initiation of treatment, causing hemorrhagic and thrombotic complications, including dural sinus thrombosis and cerebral infarction.[197]

5.1 Immune modulators

Thalidomide is a potent antiangiogenic agent with immunomodulatory properties that is used for the treatment of plasma cell dyscrasias. It has been associated with a distal peripheral neuropathy in which incidence increases with higher cumulative doses and may develop in about 50% of patients.[198] It is a predominantly sensory neuropathy, but motor symptoms have been reported in 30% of patients, and autonomic dysfunction, manifesting with constipation, bradycardia, and impotence, was rarely observed.[199–201] The pathophysiology of the neuropathy is unknown, but antiangiogenesis and dysregulation of neurotrophin activity through nuclear factor kappa B are plausible mechanisms. A transient somnolence is common with therapy initiation in elderly patients, but it progressively improves and eventually wears off in most patients after few weeks.[202]

Lenalidomide and pomalidomide are newer antiangiogenic and immunomodulator agents that are used for the treatment of multiple myeloma. Peripheral neuropathy is much less frequent with these agents and has been reported in less than 2% of patients.

6 Chemotherapy-induced cognitive impairment

With advances in the treatment of cancer, the number of cancer survivors continues to increase, and the median overall survival is also increasing. This has been associated with an emerging long-term complication from chemotherapy: chemotherapy-related cognitive impairment (CICI), sometimes referred to as "chemobrain" or "chemofog." Studies associating cognitive changes with chemotherapy date back to the 1970s,[203] but as cancer patients are living longer, this complication has been more frequently reported and can affect up to 70% of long-term cancer survivors.[204,205] Cognitive changes are transient in most patients and often resolve within months following treatment cessation but might persist for years and even get worse in some patients.[206] They are typically subtle and are often only noticed by the patient, but a score below average can be detected in at least one cognitive domain. The largest studies have been conducted in breast and lung cancer survivors. Verbal memory and executive function are particularly affected, but impairment in attention, episodic and working memory, visual memory, processing speed, and visuospatial and constructional ability have also been reported. Retrieval and remote memory are relatively preserved.[207–210] Patients are not affected equally, and the severity of symptoms does not relate to the types of chemotherapy agents received remotely or to the duration from completion of therapy. However, the risk of CICI increases with heavier treatment regimens. Higher regimen doses, advanced age at the time of treatment, low pre-treatment intelligence quotient, and education level are other risk factors.[1] CICI is also more often reported in female patients, which may be related to hormonal status. Symptoms were initially attributed to depression and mood changes that can accompany a cancer diagnosis. However, comprehensive neuropsychological testing data suggest a pattern of cognitive impairment that is different from the one seen in depression. Detection of chemotherapy agents in the CSF and brain parenchyma, even though at low levels, has further supported the CICI theory.[211–213] Also, a positron emission tomography (PET) study evaluating the F-18 fluorodeoxyglucose (FDG) uptake and cerebral blood flow in subjects treated for breast cancer with remote chemotherapy, during performance of control and memory-related tasks, showed reduced cerebral blood volume and FDG uptake in the basal ganglia, inferior frontal cortex, and cerebellum in the pre-treated subjects compared to controls.[214] Another study assessing the diffuse tensor imaging (DTI) sequences in the same cancer population resulted in reduced fractional anisotropy in the corona radiata, corpus callosum, and frontal, parietal and occipital white matter tracts, suggesting a global functional impairment of the brain.[215]

The exact pathophysiology of CICI is unknown, but multiple potential mechanisms have been suggested. Polymorphisms in the multidrug resistance 1 gene (MDR1), encoding the protein P-glycoprotein responsible for the extracellular efflux of drugs, make some subjects more susceptible to having higher chemotherapy agents' concentration in the CSF, and hence to their effects on the brain parenchyma, especially on the glial progenitor cells and nondividing oligodendrocytes.[216,217] Once in the CSF and brain parenchyma, the chemotherapeutic drugs lead to oxidative damage of the cells, particularly neurons, leading to impaired DNA transcription and loss of essential gene products, and hence to apoptosis and cognitive impairment—a process shared by multiple neurodegenerative disorders.[218,219] Also, chemotherapeutic agents have shown to generate reactive oxygen species and induce the release of pro-inflammatory cytokines, which leads to mitochondrial damage, increased rate of telomere shortening, accelerating the glial aging process, and to a disruption of the BBB, allowing for a higher preponderance of the toxic agents in the CSF.[220–223] Decrease in hormone levels after prolonged hormonal therapy has also been implicated in CICI, since sex hormones, particularly estrogen, have neurotrophic and neuroprotective effects and are involved in speech and memory function.[224,225] Finally, alteration of the microbiota-gut-brain axis is the subject of extensive current research as a culprit of CICI. The gastrointestinal microbiota are heavily connected to the nervous system through the vagus nerve and produce short-chain fatty acids and neurotransmitters that are involved in the modulation of maturation of glial cells, CNS innate immunity, and BBB permeability.[226–228] Mucositis is a frequent adverse event of chemotherapy and leads to significant disruption of the gut fauna, indirectly altering normal brain functioning.[229,230] Polymorphisms in transmembrane protein Toll-like receptor 4 (TLR4) can increase this risk.[231] Apolipoprotein E, brain-derived neurotrophic factor, and catechol-O-methyltransferase have been implicated in memory, attention, and motor speed. Single-nucleotide polymorphisms in these genes can also lead to a neuronal genetic predisposition to CICI.[232–234]

There are no specific guidelines for the treatment of CICI, and there are no approved therapies. However, some nonpharmaceutical and pharmaceutical approaches have been showing promising results. The best initial approach is to encourage physical and mental interventions, as exercise has shown to improve cerebral blood flow and neuroplasticity and decrease neuroinflammation.[235–237] A medium-intensity exercise performed on a regular basis 2 to 5 times per week can help preserve surviving cells. A study evaluating tai-chi for 10 weeks in 23 female patients with breast cancer showed significant improvement in attention, memory, executive function, and processing speed.[238] Cognitive training has also been shown to extend the survival of neurons in the dentate gyrus for months in animal studies, and helps eliminate negative emotions and unsuitable behaviors, while improving sleep quality.[239,240] A balanced diet, rich in fruits and vegetables, was associated with improved verbal fluency and executive function in patients with breast cancer and is recommended by some researchers.[241] However, no randomized clinical trials have been conducted so far evaluating the benefit of such interventions.

Some medications have also been suggested for the treatment of CICI. Memantine is an N-methyl-D-aspartate (NMDA) antagonist, used in Alzheimer's disease, that reduced the incidence of cognitive impairment in patients with brain metastases after whole brain radiation.[242] Donepezil, a reversible acetylcholinesterase (AChE) inhibitor also used in Alzheimer's disease, has shown improvement in some of the cognitive domains in patients who had received remote chemotherapy.[243,244] A better bioavailability of AChE leading to reduced atrophy of the cholinergic system of the basal ganglia, and improved cerebral perfusion are possible mechanisms of the clinical benefit observed. CNS stimulants, like methylphenidate and modafinil, have also been studied in CICI. Methylphenidate showed controversial results among studies, while modafinil was associated with improved attention and memory in patients with breast cancer complaining of fatigue, but it is not clear whether the effects are only related to improvement of the underlying fatigue.[245–247]

Other treatment modalities have failed to show any benefit in CICI, including erythropoietin, *Ginkgo biloba*, transdermal nicotine patches, and repeated transcranial magnetic stimulation, whereas body acupuncture, combined with electroacupuncture trigeminal nerve stimulation, showed improved quality of life and working memory in a double-blind randomized control trial.

Finally, probiotics and granulocyte colony-stimulating factor (G-CSF) can be potentially beneficial but have not been studied in humans yet.

7 Conclusion

Neurologic disorders affect up to 15% of cancer patients and can affect the patient's quality of life and sometimes survival. Some of these disorders can be direct complications of the primary cancer, while others are induced by the treatment and have become more frequent as the number of survivors is increasing. Radiation, conventional chemotherapy, and newer tumor-targeted agents have all been associated with neurotoxicity. Their toxicity profiles are different and depend on the mechanism of action, dosage, and treatment duration of tumor-targeted agents. Patients' individual predisposition and comorbid conditions, like hypertension and diabetes mellitus, also come into play. Recognition of these factors may help identify patients at highest risk for neurologic complications of cancer therapy, and lead to an early diagnosis of these complications to preserve neurologic function and prevent avoidable sequelae. However, some permanent neurologic injuries can be delayed. Vigorous measures to identify genetic predispositions to, and preventive and therapeutic strategies of, neurotoxicity are necessary to improve cancer survivors' quality of life.

References

1. Nolan CP, LM DA. Neurologic complications of chemotherapy and radiation therapy. *Continuum (Minneap Minn)*. 2015;23.
2. Qin D, Ma J, Xiao J, Tang Z. Effect of brain irradiation on blood-CSF barrier permeability of chemotherapeutic agents. *Am J Clin Oncol*. 1997;20(3):263–265.
3. Ahles TA, Saykin A. Cognitive effects of standard-dose chemotherapy in patients with cancer. *Cancer Invest*. 2001;19(8):812–820.
4. Mk T, Sw H. Neurotoxicity secondary to antineoplastic drugs. *Cancer Treat Rev*. 1994;20(2):191–214.
5. Ahles TA, Saykin AJ, McDonald BC, et al. Longitudinal assessment of cognitive changes associated with adjuvant treatment for breast cancer: impact of age and cognitive reserve. *J Clin Oncol Off J Am Soc Clin Oncol*. 2010;28(29):4434–4440.
6. Joly F, Rigal O, Noal S, Giffard B. Cognitive dysfunction and cancer: which consequences in terms of disease management? *Psychooncology*. 2011;20(12):1251–1258.
7. Szabatura AH, Cirrone F, Harris C, et al. An assessment of risk factors associated with ifosfamide-induced encephalopathy in a large academic cancer center. *J Oncol Pharm Pract Off Publ Int Soc Oncol Pharm Pract*. 2015;21(3):188–193.
8. Antoine J-C, Camdessanché J-P. Peripheral nervous system involvement in patients with cancer. *Lancet Neurol*. 2007;6(1):75–86.

9. Verstappen CCP, Heimans JJ, Hoekman K, Postma TJ. Neurotoxic complications of chemotherapy in patients with cancer: clinical signs and optimal management. *Drugs*. 2003;63(15):1549–1563.
10. DeAngelis LM, Posner JB. *Neurologic Complications of Cancer*. Oxford University Press; 2013. [cited 2020 Oct 4]. Available from: https://oxfordmedicine.com/view/10.1093/med/9780195366747.001.0001/med-9780195366747.
11. Grolleau F, Gamelin L, Boisdron-Celle M, Lapied B, Pelhate M, Gamelin E. A possible explanation for a neurotoxic effect of the anticancer agent oxaliplatin on neuronal voltage-gated sodium channels. *J Neurophysiol*. 2001;85(5):2293–2297.
12. Schiff D, Wen PY, van den Bent MJ. Neurological adverse effects caused by cytotoxic and targeted therapies. *Nat Rev Clin Oncol*. 2009;6(10):596–603.
13. Wolf S, Barton D, Kottschade L, Grothey A, Loprinzi C. Chemotherapy-induced peripheral neuropathy: prevention and treatment strategies. *Eur J Cancer*. 2008;44(11):1507–1515.
14. Shingleton BJ, Bienfang DC, Albert DM, Ensminger WD, Chandler WF, Greenberg HS. Ocular toxicity associated with high-dose carmustine. *Arch Ophthalmol*. 1982;100(11):1766–1772.
15. Cascino TL, Byrne TN, Deck MDF, Posner JB. Intra-arterial BCNU in the treatment of metastatic brain tumors. *J Neurooncol*. 1983;1(3):211–218.
16. Tsuboi K, Yoshii Y, Hyodo A, Takada K, Nose T. Leukoencephalopathy associated with intra-arterial ACNU in patients with gliomas. *J Neurooncol*. 1995;23(3):223–231.
17. Deeg HJ, Maris MB, Scott BL, Warren EH. Optimization of allogeneic transplant conditioning: not the time for dogma. *Leukemia*. 2006;20(10):1701–1705.
18. Vassal G, Deroussent A, Hartmann O, et al. Dose-dependent neurotoxicity of high-dose busulfan in children: a clinical and pharmacological study. *Cancer Res*. 1990;50(19):6203–6207.
19. Hassan M, Ehrsson H, Smedmyr B, et al. Cerebrospinal fluid and plasma concentrations of busulfan during high-dose therapy. *Bone Marrow Transplant*. 1989;4(1):113–114.
20. Murphy CP, Harden EA, Thompson JM. Generalized seizures secondary to high-dose Busulfan therapy. *Ann Pharmacother*. 2016. [cited 2020 Oct 4]; Available from: https://journals.sagepub.com/doi/10.1177/106002809202600107.
21. Chan K, Mullen C, Worth L, et al. Lorazepam for seizure prophylaxis during high-dose busulfan administration. *Bone Marrow Transplant*. 2002;29(12):963–965.
22. Eberly AL, Anderson GD, Bubalo JS, McCune JS. Optimal prevention of seizures induced by high-dose Busulfan. *Pharmacotherapy*. 2008;28(12):1502–1510.
23. Akiyama K, Kume T, Fukaya M, et al. Comparison of levetiracetam with phenytoin for the prevention of intravenous busulfan-induced seizures in hematopoietic cell transplantation recipients. *Cancer Chemother Pharmacol*. 2018;82(4):717–721.
24. Salloum E, Khan KK, Cooper DL. Chlorambucil-induced seizures. *Cancer*. 1997;79(5):1009–1013.
25. Crews SJ. Drug-induced ocular side effects and drug interactions. *Br J Ophthalmol*. 1984;68(5):371.
26. Danovska M, Ovcharova E, Marinova-Trifonova D, Mladenovski I. Chlorambucil-induced progressive multifocal leucoencephalopathy - a case report. *J Neurol Sci*. 2019;405:166.
27. Kende G, Sirkin SR, Thomas PR, Freeman AI. Blurring of vision: a previously undescribed complication of cyclophosphamide therapy. *Cancer*. 1979;44(1):69–71.
28. Tashima CK. Immediate cerebral symptoms during rapid intravenous administration of cyclophosphamide (NSC-26271). *Cancer Chemother Rep*. 1975;59(2 Pt 1):441–442.
29. Dimaggio JR, Baile WF, Brown R, Schapira D. Hallucinations and ifosfamide-induced neurotoxicity. *Cancer*. 1994;73(5):1509–1514.
30. Pratt CB, Green AA, Horowitz ME, et al. Central nervous system toxicity following the treatment of pediatric patients

with ifosfamide/mesna. *J Clin Oncol Off J Am Soc Clin Oncol.* 1986;4(8):1253–1261.

31. Curtin JP, Koonings PP, Gutierrez M, Schlaerth JB, Morrow CP. Ifosfamide-induced neurotoxicity. *Gynecol Oncol.* 1991;42(3):193–196 [discussion 191–192].

32. Patel SR, Vadhan-Raj S, Papadopolous N, et al. High-dose ifosfamide in bone and soft tissue sarcomas: results of phase II and pilot studies- -dose-response and schedule dependence. *J Clin Oncol Off J Am Soc Clin Oncol.* 1997;15(6):2378–2384.

33. Frisk P, Stålberg E, Strömberg B, Jakobson A. Painful peripheral neuropathy after treatment with high-dose ifosfamide. *Med Pediatr Oncol.* 2001;37(4):379–382.

34. Postma TJ, van Groeningen CJ, Witjes RJ, Weerts JG, Kralendonk JH, Heimans JJ. Neurotoxicity of combination chemotherapy with procarbazine, CCNU and vincristine (PCV) for recurrent glioma. *J Neurooncol.* 1998;38(1):69–75.

35. Middleton MR, Grob JJ, Aaronson N, et al. Randomized phase III study of temozolomide versus dacarbazine in the treatment of patients with advanced metastatic malignant melanoma. *J Clin Oncol Off J Am Soc Clin Oncol.* 2000;18(1):158–166.

36. Maanen MJ, Smeets CJ, Beijnen JH. Chemistry, pharmacology and pharmacokinetics of N,N',N'' -triethylenethiophosphoramide (ThioTEPA). *Cancer Treat Rev.* 2000;26(4):257–268.

37. Gutin PH, Levi JA, Wiernik PH, Walker MD. Treatment of malignant meningeal disease with intrathecal thiotepa: a phase II study. *Cancer Treat Rep.* 1977;61(5):885–887.

38. Newton HB. Neurological complications of chemotherapy to the central nervous system. In: *Handbook of Clinical Neurology.* Elsevier; 2012:903–916. [cited 2020 Oct 8]. Available from: https://linkinghub.elsevier.com/retrieve/pii/B9780444535023000318.

39. Vahdat L, Wong ET, Wile MJ, Rosenblum M, Foley KM, Warrell RJ. Therapeutic and neurotoxic effects of 2-chlorodeoxyadenosine in adults with acute myeloid leukemia. *Blood.* 1994;84(10):3429–3434.

40. Saven A, Kawasaki H, Carrera CJ, et al. 2-Chlorodeoxyadenosine dose escalation in nonhematologic malignancies. *J Clin Oncol Off J Am Soc Clin Oncol.* 1993;11(4):671–678.

41. Videnovic A, Semenov I, Chua-Adajar R, et al. Capecitabine-induced multifocal leukoencephalopathy: a report of five cases. *Neurology.* 2005;65(11):1792–1794.

42. Renouf D, Gill S. Capecitabine-induced cerebellar toxicity. *Clin Colorectal Cancer.* 2006;6(1):70–71.

43. Couch LSB, Groteluschen DL, Stewart JA, Mulkerin DL. Capecitabine-related neurotoxicity presenting as trismus. *Clin Colorectal Cancer.* 2003;3(2):121–123.

44. Saif MW, Wood TE, McGee PJ, Diasio RB. Peripheral neuropathy associated with capecitabine. *Anticancer Drugs.* 2004;15(8):767–771.

45. Gottlieb D, Bradstock K, Koutts J, Robertson T, Lee C, Castaldi P. The neurotoxicity of high-dose cytosine arabinoside is age-related. *Cancer.* 1987;60(7):1439–1441.

46. Herzig RH, Hines JD, Herzig GP, et al. Cerebellar toxicity with high-dose cytosine arabinoside. *J Clin Oncol Off J Am Soc Clin Oncol.* 1987;5(6):927–932.

47. Kannarkat G, Lasher EE, Schiff D. Neurologic complications of chemotherapy agents. *Curr Opin Intern Med.* 2008;7(1):88–94.

48. Ritch PS, Hansen RM, Heuer DK. Ocular toxicity from high-dose cytosine arabinoside. *Cancer.* 1983;51(3):430–432.

49. Shaw PJ, Procopis PG, Menser MA, Bergin M, Antony J, Stevens MM. Bulbar and pseudobulbar palsy complicating therapy with high-dose cytosine arabinoside in children with leukemia. *Med Pediatr Oncol.* 1991;19(2):122–125.

50. Hoffman DL, Howard JR, Sarma R, Riggs JE. Encephalopathy, myelopathy, optic neuropathy, and anosmia associated with intravenous cytosine arabinoside. *Clin Neuropharmacol.* 1993;16(3):258–262.

51. Kwong Y-L, Yeung DYM, Chan JCW. Intrathecal chemotherapy for hematologic malignancies: drugs and toxicities. *Ann Hematol.* 2009;88(3):193–201.

52. Dunton SF, Nitschke R, Spruce WE, Bodensteiner J, Krous HF. Progressive ascending paralysis following administration of intrathecal and intravenous cytosine arabinoside. A pediatric oncology group study. *Cancer.* 1986;57(6):1083–1088.

53. Resar LM, Phillips PC, Kastan MB, Leventhal BG, Bowman PW, Civin CI. Acute neurotoxicity after intrathecal cytosine arabinoside in two adolescents with acute lymphoblastic leukemia of B-cell type. *Cancer.* 1993;71(1):117–123.

54. Glantz MJ, Jaeckle KA, Chamberlain MC, et al. A randomized controlled trial comparing intrathecal sustained-release cytarabine (DepoCyt) to intrathecal methotrexate in patients with neoplastic meningitis from solid tumors. *Clin Cancer Res Off J Am Assoc Cancer Res.* 1999;5(11):3394–3402.

55. Russell JA, Powles RL. Letter: neuropathy due to cytosine arabinoside. *Br Med J.* 1974;4(5945):652–653.

56. Powell BL, Capizzi RL, Lyerly ES, Cooper MR. Peripheral neuropathy after high-dose cytosine arabinoside, daunorubicin, and asparaginase consolidation for acute nonlymphocytic leukemia. *J Clin Oncol Off J Am Soc Clin Oncol.* 1986;4(1):95–97.

57. Scherokman B, Filling-Katz MR, Tell D. Brachial plexus neuropathy following high-dose cytarabine in acute monoblastic leukemia. *Cancer Treat Rep.* 1985;69(9):1005–1006.

58. Annaloro C, Costa A, Fracchiolla NS, et al. Severe fludarabine neurotoxicity after reduced intensity conditioning regimen to allogeneic hematopoietic stem cell transplantation: a case report. *Clin Case Rep.* 2015;3(7):650–655.

59. Beitinjaneh A, McKinney AM, Cao Q, Weisdorf DJ. Toxic leukoencephalopathy following Fludarabine-associated hematopoietic cell transplantation. *Biol Blood Marrow Transplant.* 2011;17(3):300–308.

60. Plunkett W, Gandhi V, Huang P, et al. Fludarabine: pharmacokinetics, mechanisms of action, and rationales for combination therapies. *Semin Oncol.* 1993;20(5 Suppl 7):2–12.

61. Koenig H, Patel A. Biochemical basis of the acute cerebellar syndrome in 5-fluorouracil chemotherapy. *Trans Am Neurol Assoc.* 1969;94:290–292.

62. Barbieux C, Patri B, Cerf I, de Parades V. Acute cerebellar syndrome after treatment with 5-fluorouracil. *Bull Cancer (Paris).* 1996;83(1):77–80.

63. Noguerón E, Berrocal A, Albert A, Camps C, Vicent JM. Acute cerebellar syndrome due to 5-fluorouracil. *Rev Neurol.* 1997;25(148):2053–2054.

64. Niemann B, Rochlitz C, Herrmann R, Pless M. Toxic encephalopathy induced by Capecitabine. *Oncology.* 2004;66(4):331–335.

65. Takimoto CH, Lu ZH, Zhang R, et al. Severe neurotoxicity following 5-fluorouracil-based chemotherapy in a patient with dihydropyrimidine dehydrogenase deficiency. *Clin Cancer Res Off J Am Assoc Cancer Res.* 1996;2(3):477–481.

66. Saif MW. Dihydropyrimidine dehydrogenase gene (DPYD) polymorphism among Caucasian and non-Caucasian patients with 5-FU- and capecitabine-related toxicity using full sequencing of DPYD. *Cancer Genomics Proteomics.* 2013;10(2):89–92.

67. Saif MW. Capecitabine-induced cerebellar toxicity and TYMS pharmacogenetics. *Anticancer Drugs.* 2019;30(4):431–434.

68. Dormann AJ, Grünewald T, Wigginghaus B, Huchzermeyer H. Gemcitabine-associated autonomic neuropathy. *Lancet.* 1998;351(9103):644.

69. Larsen FO, Hansen SW. Severe neurotoxicity caused by gemcitabine treatment. *Acta Oncol.* 2004;43(6):590–591.

70. Pentsova E, Liu A, Rosenblum M, O'Reilly E, Chen X, Hormigo A. Gemcitabine induced myositis in patients with pancreatic cancer: case reports and topic review. *J Neurooncol.* 2012;106(1):15–21.

71. Donehower RC. An overview of the clinical experience with hydroxyurea. *Semin Oncol.* 1992;19(3 Suppl 9):11–19.

72. Barry M, Clarke S, Mulcahy F, Back D. Hydroxyurea-induced neurotoxicity in HIV disease. *AIDS.* 1999;13(12):1592.

73. Mott MG, Stevenson P, Wood CB. Methotrexate meningitis. *Lancet.* 1972;2(7778):656.

74. Bates S, McKeever P, Masur H, et al. Myelopathy following intrathecal chemotherapy in a patient with extensive Burkitt's lymphoma and altered immune status. *Am J Med.* 1985;78(4):697–702.

75. Bhojwani D, Sabin ND, Pei D, et al. Methotrexate-induced neurotoxicity and leukoencephalopathy in childhood acute lymphoblastic leukemia. *J Clin Oncol Off J Am Soc Clin Oncol.* 2014;32(9):949–959.

76. Pavlidou E, Pavlou E, Anastasiou A, et al. Posterior reversible encephalopathy syndrome after intrathecal methotrexate infusion: a case report and literature update. *Quant Imaging Med Surg.* 2016;6(5):605–611.

77. Aradillas E, Arora R, Gasperino J. Methotrexate-induced posterior reversible encephalopathy syndrome. *J Clin Pharm Ther.* 2011;36(4):529–536.

78. Correa DD, Shi W, Abrey LE, et al. Cognitive functions in primary CNS lymphoma after single or combined modality regimens. *Neuro-Oncol.* 2012;14(1):101–108.

79. Blay JY, Conroy T, Chevreau C, et al. High-dose methotrexate for the treatment of primary cerebral lymphomas: analysis of survival and late neurologic toxicity in a retrospective series. *J Clin Oncol Off J Am Soc Clin Oncol.* 1998;16(3):864–871.

80. Lien HH, Blomlie V, Saeter G, Solheim O, Fosså SD. Osteogenic sarcoma: MR signal abnormalities of the brain in asymptomatic patients treated with high-dose methotrexate. *Radiology.* 1991;179(2):547–550.

81. Cheson BD, Vena DA, Foss FM, Sorensen JM. Neurotoxicity of purine analogs: a review. *J Clin Oncol Off J Am Soc Clin Oncol.* 1994;12(10):2216–2228.

82. Gregg RW, Molepo JM, Monpetit VJ, et al. Cisplatin neurotoxicity: the relationship between dosage, time, and platinum concentration in neurologic tissues, and morphologic evidence of toxicity. *J Clin Oncol Off J Am Soc Clin Oncol.* 1992;10(5):795–803.

83. Tfayli A, Hentschel P, Madajewicz S, et al. Toxicities related to intraarterial infusion of cisplatin and etoposide in patients with brain tumors. *J Neurooncol.* 1999;42(1):73–77.

84. Clerici WJ, Yang L. Direct effects of intraperilymphatic reactive oxygen species generation on cochlear function. *Hear Res.* 1996;101(1–2):14–22.

85. Reddel RR, Kefford RF, Grant JM, Coates AS, Fox RM, Tattersall MH. Ototoxicity in patients receiving cisplatin: importance of dose and method of drug administration. *Cancer Treat Rep.* 1982;66(1):19–23.

86. Bokemeyer C, Berger CC, Hartmann JT, et al. Analysis of risk factors for cisplatin-induced ototoxicity in patients with testicular cancer. *Br J Cancer.* 1998;77(8):1355–1362.

87. Huang E, Teh BS, Strother DR, et al. Intensity-modulated radiation therapy for pediatric medulloblastoma: early report on the reduction of ototoxicity. *Int J Radiat Oncol Biol Phys.* 2002;52(3):599–605.

88. Kopelman J, Budnick AS, Sessions RB, Kramer MB, Wong GY. Ototoxicity of high-dose cisplatin by bolus administration in patients with advanced cancers and normal hearing. *Laryngoscope.* 1988;98(8 Pt 1):858–864.

89. Boogerd W, ten Bokkel Huinink WW, Dalesio O, Hoppenbrouwers WJ, van der Sande JJ. Cisplatin induced neuropathy: central, peripheral and autonomic nerve involvement. *J Neurooncol.* 1990;9(3):255–263.

90. Riccardi R, Riccardi A, Di Rocco C, et al. Cerebrospinal fluid pharmacokinetics of carboplatin in children with brain tumors. *Cancer Chemother Pharmacol.* 1992;30(1):21–24.

91. O'Brien ME, Tonge K, Blake P, Moskovic E, Wiltshaw E. Blindness associated with high-dose carboplatin. *Lancet.* 1992;339(8792):558.

92. Cvitkovic E, Bekradda M. Oxaliplatin: a new therapeutic option in colorectal cancer. *Semin Oncol.* 1999;26(6):647–662.

93. Loprinzi CL, Lacchetti C, Bleeker J, et al. Prevention and management of chemotherapy-induced peripheral neuropathy in survivors of adult cancers: ASCO guideline update. *J Clin Oncol.* 2020;38(28):3325–3348.

94. Gamelin E, Gamelin L, Bossi L, Quasthoff S. Clinical aspects and molecular basis of oxaliplatin neurotoxicity: current management and development of preventive measures. *Semin Oncol.* 2002;29(5 Suppl 15):21–33.

95. Grothey A. Oxaliplatin-safety profile: neurotoxicity. *Semin Oncol.* 2003;30(4 Suppl 15):5–13.

96. André T, Boni C, Mounedji-Boudiaf L, et al. Oxaliplatin, fluorouracil, and leucovorin as adjuvant treatment for colon cancer. *N Engl J Med.* 2004;350(23):2343–2351.

97. Neuwelt EA, Glasberg M, Frenkel E, Barnett P. Neurotoxicity of chemotherapeutic agents after blood-brain barrier modification: neuropathological studies. *Ann Neurol.* 1983;14(3):316–324.

98. Mortensen ME, Cecalupo AJ, Lo WD, Egorin MJ, Batley R. Inadvertent intrathecal injection of daunorubicin with fatal outcome. *Med Pediatr Oncol.* 1992;20(3):249–253.

99. Pisoni R, Ruggenenti P, Remuzzi G. Drug-induced thrombotic microangiopathy: incidence, prevention and management. *Drug Saf.* 2001;24(7):491–501.

100. Bain PG, Lantos PL, Djurovic V, West I. Intrathecal vincristine: a fatal chemotherapeutic error with devastating central nervous system effects. *J Neurol.* 1991;238(4):230–234.

101. Hurwitz RL, Mahoney DH, Armstrong DL, Browder TM. Reversible encephalopathy and seizures as a result of conventional vincristine administration. *Med Pediatr Oncol.* 1988;16(3):216–219.

102. Delaney P. Vincristine-induced laryngeal nerve paralysis. *Neurology.* 1982;32(11):1285–1288.

103. Sandler SG, Tobin W, Henderson ES. Vincristine-induced neuropathy. A clinical study of fifty leukemic patients. *Neurology.* 1969;19(4):367–374.

104. Lugassy G, Shapira A. Sensorineural hearing loss associated with vincristine treatment. *Blut.* 1990;61(5):320–321.

105. DeAngelis LM, Gnecco C, Taylor L, Warrell RP. Evolution of neuropathy and myopathy during intensive vincristine/corticosteroid chemotherapy for non-Hodgkin's lymphoma. *Cancer.* 1991;67(9):2241–2246.

106. Naumann R, Mohm J, Reuner U, Kroschinsky F, Rautenstrauss B, Ehninger G. Early recognition of hereditary motor and sensory neuropathy type 1 can avoid life-threatening vincristine neurotoxicity. *Br J Haematol.* 2001;115(2):323–325.

107. Weintraub M, Adde MA, Venzon DJ, et al. Severe atypical neuropathy associated with administration of hematopoietic colony-stimulating factors and vincristine. *J Clin Oncol Off J Am Soc Clin Oncol.* 1996;14(3):935–940.

108. Teusink AC, Ragucci D, Shatat IF, Kalpatthi R. Potentiation of vincristine toxicity with concomitant fluconazole prophylaxis in children with acute lymphoblastic leukemia. *Pediatr Hematol Oncol.* 2012;29(1):62–67.

109. Haim N, Epelbaum R, Ben-Shahar M, Yarnitsky D, Simri W, Robinson E. Full dose vincristine (without 2-mg dose limit) in the treatment of lymphomas. *Cancer.* 1994;73(10):2515–2519.

110. Focan C, Olivier R, Le Hung S, Bays R, Claessens JJ, Debruyne H. Neurological toxicity of vindesine used in combination chemotherapy of 51 human solid tumors. *Cancer Chemother Pharmacol.* 1981;6(2):175–181.

111. Romero A, Rabinovich MG, Vallejo CT, et al. Vinorelbine as first-line chemotherapy for metastatic breast carcinoma. *J Clin Oncol Off J Am Soc Clin Oncol.* 1994;12(2):336–341.

112. Nieto Y, Cagnoni PJ, Bearman SI, et al. Acute encephalopathy: a new toxicity associated with high-dose paclitaxel. *Clin Cancer Res Off J Am Assoc Cancer Res*. 1999;5(3):501–506.

113. Seidman AD, Barrett S, Canezo S. Photopsia during 3-hour paclitaxel administration at doses > or = 250 mg/m2. *J Clin Oncol Off J Am Soc Clin Oncol*. 1994;12(8):1741–1742.

114. van Gerven JM, Moll JW, van den Bent MJ, et al. Paclitaxel (Taxol) induces cumulative mild neurotoxicity. *Eur J Cancer*. 1994;30A(8):1074–1077.

115. Postma TJ, Vermorken JB, Liefting AJ, Pinedo HM, Heimans JJ. Paclitaxel-induced neuropathy. *Ann Oncol Off J Eur Soc Med Oncol*. 1995;6(5):489–494.

116. Freilich RJ, Balmaceda C, Seidman AD, Rubin M, DeAngelis LM. Motor neuropathy due to docetaxel and paclitaxel. *Neurology*. 1996;47(1):115–118.

117. Lorenz E, Hagen B, Himmelmann A, et al. A phase II study of biweekly administration of paclitaxel in patients with recurrent epithelial ovarian cancer. *Int J Gynecol Cancer Off J Int Gynecol Cancer Soc*. 1999;9(5):373–376.

118. Loprinzi CL, Maddocks-Christianson K, Wolf SL, et al. The paclitaxel acute pain syndrome: sensitization of nociceptors as the putative mechanism. *Cancer J*. 2007;13(6):399–403.

119. Schiller JH, Storer B, Tutsch K, et al. Phase I trial of 3-hour infusion of paclitaxel with or without granulocyte colony-stimulating factor in patients with advanced cancer. *J Clin Oncol Off J Am Soc Clin Oncol*. 1994;12(2):241–248.

120. Sahenk Z, Barohn R, New P, Mendell JR. Taxol neuropathy. Electrodiagnostic and sural nerve biopsy findings. *Arch Neurol*. 1994;51(7):726–729.

121. New PZ, Jackson CE, Rinaldi D, Burris H, Barohn RJ. Peripheral neuropathy secondary to docetaxel (Taxotere). *Neurology*. 1996;46(1):108–111.

122. Hilkens PH, Verweij J, Stoter G, Vecht CJ, van Putten WL, van den Bent MJ. Peripheral neurotoxicity induced by docetaxel. *Neurology*. 1996;46(1):104–108.

123. Falkson G, van Dyk JJ, van Eden EB, van der Merwe AM, van den Bergh JA, Falkson HC. A clinical trial of the oral form of 4'-demethyl-epipodophyllotoxin-beta-D ethylidene glucoside (NSC 141540) VP 16-213. *Cancer*. 1975;35(4):1141–1144.

124. Seet RCS, Rabinstein AA, Lindell PE, Uhm JH, Wijdicks EF. Cerebrovascular events after bevacizumab treatment: an early and severe complication. *Neurocrit Care*. 2011;15(3):421–427.

125. Fraum TJ, Kreisl TN, Sul J, Fine HA, Iwamoto FM. Ischemic stroke and intracranial hemorrhage in glioma patients on antiangiogenic therapy. *J Neurooncol*. 2011;105(2):281–289.

126. Kamba T, McDonald DM. Mechanisms of adverse effects of anti-VEGF therapy for cancer. *Br J Cancer*. 2007;96(12):1788–1795.

127. Zachary I. Signaling mechanisms mediating vascular protective actions of vascular endothelial growth factor. *Am J Physiol Cell Physiol*. 2001;280(6):C1375–C1386.

128. Fugate JE, Rabinstein AA. Posterior reversible encephalopathy syndrome: clinical and radiological manifestations, pathophysiology, and outstanding questions. *Lancet Neurol*. 2015;14(9):914–925.

129. Sherman JH, Aregawi DG, Lai A, et al. Optic neuropathy in patients with glioblastoma receiving bevacizumab. *Neurology*. 2009;73(22):1924–1926.

130. Fathpour P, Obad N, Espedal H, et al. Bevacizumab treatment for human glioblastoma. Can it induce cognitive impairment? *Neuro-Oncol*. 2014;16(5):754–756.

131. Kono T, Satomi M, Asama T, et al. Cetuximab-induced hypomagnesaemia aggravates peripheral sensory neurotoxicity caused by oxaliplatin. *J Gastrointest Oncol*. 2010;1(2):97–101.

132. Beydoun SR, Shatzmiller RA. Chronic immune-mediated demyelinating polyneuropathy in the setting of cetuximab treatment. *Clin Neurol Neurosurg*. 2010;112(10):900–902.

133. Vose JM, Wahl RL, Saleh M, et al. Multicenter phase II study of iodine-131 tositumomab for chemotherapy-relapsed/refractory low-grade and transformed low-grade B-cell non-Hodgkin's lymphomas. *J Clin Oncol Off J Am Soc Clin Oncol*. 2000;18(6):1316–1323.

134. Kaminski MS, Estes J, Zasadny KR, et al. Radioimmunotherapy with iodine 131I tositumomab for relapsed or refractory B-cell non-Hodgkin lymphoma: updated results and long-term follow-up of the University of Michigan experience. *Blood*. 2000;96(4):1259–1266.

135. Stone JB, DeAngelis LM. Cancer-treatment-induced neurotoxicity—focus on newer treatments. *Nat Rev Clin Oncol*. 2016;13(2):92–105.

136. Glusker P, Recht L, Lane B. Reversible posterior leukoencephalopathy syndrome and bevacizumab. *N Engl J Med*. 2006;354(9):980–982. discussion 980–982.

137. Emery P, Fleischmann R, Filipowicz-Sosnowska A, et al. The efficacy and safety of rituximab in patients with active rheumatoid arthritis despite methotrexate treatment: results of a phase IIB randomized, double-blind, placebo-controlled, dose-ranging trial. *Arthritis Rheum*. 2006;54(5):1390–1400.

138. Carson KR, Evens AM, Richey EA, et al. Progressive multifocal leukoencephalopathy after rituximab therapy in HIV-negative patients: a report of 57 cases from the research on adverse drug events and reports project. *Blood*. 2009;113(20):4834–4840.

139. Rubenstein JL, Fridlyand J, Abrey L, et al. Phase I study of intraventricular administration of rituximab in patients with recurrent CNS and intraocular lymphoma. *J Clin Oncol Off J Am Soc Clin Oncol*. 2007;25(11):1350–1356.

140. Bromberg JEC, Doorduijn JK, Baars JW, van Imhoff GW, Enting R, van den Bent MJ. Acute painful lumbosacral paresthesia after intrathecal rituximab. *J Neurol*. 2012;259(3):559–561.

141. Soffietti R, Trevisan E, Rudà R. Neurologic complications of chemotherapy and other newer and experimental approaches. In: *Handbook of Clinical Neurology*. Elsevier; 2014:1199–1218. [cited 2020 Oct 8]. Available from: https://linkinghub.elsevier.com/retrieve/pii/B9780702040887000808.

142. Verma S, Miles D, Gianni L, et al. Trastuzumab emtansine for HER2-positive advanced breast cancer. *N Engl J Med*. 2012;367(19):1783–1791.

143. Perez EA, Barrios C, Eiermann W, et al. Trastuzumab emtansine with or without Pertuzumab versus Trastuzumab plus Taxane for human epidermal growth factor receptor 2-positive, advanced breast cancer: primary results from the phase III MARIANNE study. *J Clin Oncol Off J Am Soc Clin Oncol*. 2017;35(2):141–148.

144. Jin J, Wang B, Gao Y, et al. Exposure–safety relationship of trastuzumab emtansine (T-DM1) in patients with HER2-positive locally advanced or metastatic breast cancer (MBC). *J Clin Oncol*. 2013;31(15_suppl):646.

145. Carlson JA, Nooruddin Z, Rusthoven C, et al. Trastuzumab emtansine and stereotactic radiosurgery: an unexpected increase in clinically significant brain edema. *Neuro-Oncol*. 2014;16(7):1006–1009.

146. Stumpf PK, Cittelly DM, Robin TP, et al. Combination of Trastuzumab emtansine and stereotactic radiosurgery results in high rates of clinically significant Radionecrosis and dysregulation of Aquaporin-4. *Clin Cancer Res*. 2019;25(13):3946–3953.

147. Iwata H, Tamura K, Doi T, et al. Trastuzumab deruxtecan (DS-8201a) in subjects with HER2-expressing solid tumors: long-term results of a large phase 1 study with multiple expansion cohorts. *J Clin Oncol*. 2018;36(15_suppl):2501.

148. Foyil KV, Bartlett NL. Brentuximab vedotin for the treatment of CD30+ lymphomas. *Immunotherapy*. 2011;3(4):475–485.

149. Younes A, Gopal AK, Smith SE, et al. Results of a pivotal phase II study of brentuximab vedotin for patients with relapsed or refractory Hodgkin's lymphoma. *J Clin Oncol Off J Am Soc Clin Oncol*. 2012;30(18):2183–2189.

150. Forero-Torres A, Holkova B, Goldschmidt J, et al. Phase 2 study of frontline brentuximab vedotin monotherapy in Hodgkin lymphoma patients aged 60 years and older. *Blood*. 2015;126(26):2798–2804.

151. Wick W, Hertenstein A, Platten M. Neurological sequelae of cancer immunotherapies and targeted therapies. *Lancet Oncol*. 2016;17(12):e529–e541.

152. Fargeot G, Dupel-Pottier C, Stephant M, et al. Brentuximab vedotin treatment associated with acute and chronic inflammatory demyelinating polyradiculoneuropathies. *J Neurol Neurosurg Psychiatry*. 2020;91(7):786–788.

153. Palanca-Wessels MCA, Czuczman M, Salles G, et al. Safety and activity of the anti-CD79B antibody-drug conjugate polatuzumab vedotin in relapsed or refractory B-cell non-Hodgkin lymphoma and chronic lymphocytic leukaemia: a phase 1 study. *Lancet Oncol*. 2015;16(6):704–715.

154. Sehn LH, Herrera AF, Flowers CR, et al. Polatuzumab Vedotin in relapsed or refractory diffuse large B-cell lymphoma. *J Clin Oncol*. 2019;38(2):155–165.

155. Piccinni C, Sacripanti C, Poluzzi E, et al. Stronger association of drug-induced progressive multifocal leukoencephalopathy (PML) with biological immunomodulating agents. *Eur J Clin Pharmacol*. 2010;66(2):199–206.

156. Topp MS, Gökbuget N, Stein AS, et al. Safety and activity of blinatumomab for adult patients with relapsed or refractory B-precursor acute lymphoblastic leukaemia: a multicentre, single-arm, phase 2 study. *Lancet Oncol*. 2015;16(1):57–66.

157. Topp MS, Gökbuget N, Zugmaier G, et al. Phase II Trial of the Anti-CD19 bispecific T cell–engager blinatumomab shows hematologic and molecular remissions in patients with relapsed or refractory B-precursor acute lymphoblastic leukemia. *J Clin Oncol*. 2014;32(36):4134–4140.

158. Maude SL, Frey N, Shaw PA, et al. Chimeric antigen receptor T cells for sustained remissions in leukemia. *N Engl J Med*. 2014;371(16):1507–1517.

159. Gainor JF, Chabner BA. Ponatinib: accelerated disapproval. *Oncologist*. 2015;20(8):847–848.

160. Rafei H, Jabbour EJ, Kantarjian H, et al. Neurotoxic events associated with BCR-ABL1 tyrosine kinase inhibitors: a case series. *Leuk Lymphoma*. 2019;60(13):3292–3295.

161. Welslau M, Diéras V, Sohn J-H, et al. Patient-reported outcomes from EMILIA, a randomized phase 3 study of trastuzumab emtansine (T-DM1) versus capecitabine and lapatinib in human epidermal growth factor receptor 2-positive locally advanced or metastatic breast cancer. *Cancer*. 2014;120(5):642–651.

162. Martín G, Bellido L, Cruz JJ. Reversible posterior leukoencephalopathy syndrome induced by sunitinib. *J Clin Oncol Off J Am Soc Clin Oncol*. 2007;25(23):3559.

163. Choueiri TK, Schutz FAB, Je Y, Rosenberg JE, Bellmunt J. Risk of arterial thromboembolic events with sunitinib and sorafenib: a systematic review and meta-analysis of clinical trials. *J Clin Oncol Off J Am Soc Clin Oncol*. 2010;28(13):2280–2285.

164. Lele AV, Clutter S, Price E, De Ruyter ML. Severe hypothyroidism presenting as myxedema coma in the postoperative period in a patient taking sunitinib: case report and review of literature. *J Clin Anesth*. 2013;25(1):47–51.

165. Mannavola D, Coco P, Vannucchi G, et al. A novel tyrosine-kinase selective inhibitor, sunitinib, induces transient hypothyroidism by blocking iodine uptake. *J Clin Endocrinol Metab*. 2007;92(9):3531–3534.

166. Chun SG, Iyengar P, Gerber DE, Hogan RN, Timmerman RD. Optic neuropathy and blindness associated with crizotinib for non-small-cell lung cancer with EML4-ALK translocation. *J Clin Oncol Off J Am Soc Clin Oncol*. 2015;33(5):e25–e26.

167. Ou S-HI, Ahn JS, De Petris L, et al. Alectinib in crizotinib-refractory ALK-rearranged non-small-cell lung cancer: a phase II global study. *J Clin Oncol Off J Am Soc Clin Oncol*. 2016;34(7):661–668.

168. Cooper MR, Chim H, Chan H, Durand C. Ceritinib: a new tyrosine kinase inhibitor for non-small-cell lung cancer. *Ann Pharmacother*. 2015;49(1):107–112.

169. Bauer TM, Felip E, Solomon BJ, et al. Clinical management of adverse events associated with lorlatinib. *Oncologist*. 2019;24(8):1103–1110.

170. Richardson PG, Briemberg H, Jagannath S, et al. Frequency, characteristics, and reversibility of peripheral neuropathy during treatment of advanced multiple myeloma with bortezomib. *J Clin Oncol Off J Am Soc Clin Oncol*. 2006;24(19):3113–3120.

171. Richardson PG, Sonneveld P, Schuster MW, et al. Reversibility of symptomatic peripheral neuropathy with bortezomib in the phase III APEX trial in relapsed multiple myeloma: impact of a dose-modification guideline. *Br J Haematol*. 2009;144(6):895–903.

172. Argyriou AA, Iconomou G, Kalofonos HP. Bortezomib-induced peripheral neuropathy in multiple myeloma: a comprehensive review of the literature. *Blood*. 2008;112(5):1593–1599.

173. Lanzani F, Mattavelli L, Frigeni B, et al. Role of a pre-existing neuropathy on the course of bortezomib-induced peripheral neurotoxicity. *J Peripher Nerv Syst*. 2008;13(4):267–274.

174. Cavaletti G, Jakubowiak AJ. Peripheral neuropathy during bortezomib treatment of multiple myeloma: a review of recent studies. *Leuk Lymphoma*. 2010;51(7):1178–1187.

175. Dimopoulos MA, Moreau P, Palumbo A, et al. Carfilzomib and dexamethasone versus bortezomib and dexamethasone for patients with relapsed or refractory multiple myeloma (ENDEAVOR): a randomised, phase 3, open-label, multicentre study. *Lancet Oncol*. 2016;17(1):27–38.

176. Moreau P, Masszi T, Grzasko N, et al. Oral ixazomib, lenalidomide, and dexamethasone for multiple myeloma. *N Engl J Med*. 2016;374(17):1621–1634.

177. Seiter K, Stiefel MF, Barrientos J, et al. Successful treatment of ibrutinib-associated central nervous system hemorrhage with platelet transfusion support. *Stem Cell Invest*. 2016;3:27.

178. Rogers KA, Mousa L, Zhao Q, et al. Incidence of opportunistic infections during ibrutinib treatment for B-cell malignancies. *Leukemia*. 2019;33(10):2527–2530.

179. Dreyling M, Jurczak W, Jerkeman M, et al. Ibrutinib versus temsirolimus in patients with relapsed or refractory mantle-cell lymphoma: an international, randomised, open-label, phase 3 study. *Lancet*. 2016;387(10020):770–778.

180. Kroeze SGC, Fritz C, Hoyer M, et al. Toxicity of concurrent stereotactic radiotherapy and targeted therapy or immunotherapy: a systematic review. *Cancer Treat Rev*. 2017;53:25–37.

181. Shailesh FNU, Singh M, Tiwari U, Hutchins LF. Vemurafenib-induced bilateral facial palsy. *J Postgrad Med*. 2014;60(2):187.

182. Klein O, Ribas A, Chmielowski B, et al. Facial palsy as a side effect of vemurafenib treatment in patients with metastatic melanoma. *J Clin Oncol Off J Am Soc Clin Oncol*. 2013;31(12):e215–e217.

183. Robert C, Karaszewska B, Schachter J, et al. Improved overall survival in melanoma with combined dabrafenib and trametinib. *N Engl J Med*. 2015;372(1):30–39.

184. DeMichele A, Clark AS, Tan KS, et al. CDK 4/6 inhibitor palbociclib (PD0332991) in Rb+ advanced breast cancer: phase II activity, safety, and predictive biomarker assessment. *Clin Cancer Res Off J Am Assoc Cancer Res*. 2015;21(5):995–1001.

185. Villani A, Fabbrocini G, Costa C, Scalvenzi M. Sonidegib: safety and efficacy in treatment of advanced basal cell carcinoma. *Dermatol Ther*. 2020;10(3):401–412.

186. Ramelyte E, Amann VC, Dummer R. Sonidegib for the treatment of advanced basal cell carcinoma. *Expert Opin Pharmacother*. 2016;17(14):1963–1968.

187. Caraceni A, Gangeri L, Martini C, et al. Neurotoxicity of interferon-α in melanoma therapy. *Cancer*. 1998;83(3):482–489.

188. Pavol MA, Meyers CA, Rexer JL, Valentine AD, Mattis PJ, Talpaz M. Pattern of neurobehavioral deficits associated with interferon alpha therapy for leukemia. *Neurology*. 1995;45(5):947–950.

189. Valentine AD, Meyers CA, Kling MA, Richelson E, Hauser P. Mood and cognitive side effects of interferon-alpha therapy. *Semin Oncol.* 1998;25(1 Suppl 1):39–47.

190. Hensley ML, Peterson B, Silver RT, Larson RA, Schiffer CA, Szatrowski TP. Risk factors for severe neuropsychiatric toxicity in patients receiving interferon alfa-2b and low-dose cytarabine for chronic myelogenous leukemia: analysis of cancer and leukemia group B 9013. *J Clin Oncol Off J Am Soc Clin Oncol.* 2000;18(6):1301–1308.

191. Bora I, Karli N, Bakar M, Zarifoğlu M, Turan F, Oğul E. Myasthenia gravis following IFN-alpha-2a treatment. *Eur Neurol.* 1997;38(1):68.

192. Rutkove SB. An unusual axonal polyneuropathy induced by low-dose interferon alfa-2a. *Arch Neurol.* 1997;54(7):907–908.

193. Denicoff KD, Rubinow DR, Papa MZ, et al. The neuropsychiatric effects of treatment with interleukin-2 and lymphokine-activated killer cells. *Ann Intern Med.* 1987;107(3):293–300.

194. Drory VE, Lev D, Groozman GB, Gutmann M, Klausner JM. Neurotoxicity of isolated limb perfusion with tumor necrosis factor. *J Neurol Sci.* 1998;158(1):1–4.

195. Pirl WF, Siegel GI, Goode MJ, Smith MR. Depression in men receiving androgen deprivation therapy for prostate cancer: a pilot study. *Psychooncology.* 2002;11(6):518–523.

196. Scher HI, Fizazi K, Saad F, et al. Increased survival with enzalutamide in prostate cancer after chemotherapy. *N Engl J Med.* 2012;367(13):1187–1197.

197. Feinberg WM, Swenson MR. Cerebrovascular complications of L-asparaginase therapy. *Neurology.* 1988;38(1):127–133.

198. Mileshkin L, Stark R, Day B, Seymour JF, Zeldis JB, Prince HM. Development of neuropathy in patients with myeloma treated with thalidomide: patterns of occurrence and the role of electrophysiologic monitoring. *J Clin Oncol Off J Am Soc Clin Oncol.* 2006;24(27):4507–4514.

199. Dimopoulos MA, Eleutherakis-Papaiakovou V. Adverse effects of thalidomide administration in patients with neoplastic diseases. *Am J Med.* 2004;117(7):508–515.

200. Rajkumar SV, Rosiñol L, Hussein M, et al. Multicenter, randomized, double-blind, placebo-controlled study of thalidomide plus dexamethasone compared with dexamethasone as initial therapy for newly diagnosed multiple myeloma. *J Clin Oncol Off J Am Soc Clin Oncol.* 2008;26(13):2171–2177.

201. Murphy PT, O'Donnell JR. Thalidomide induced impotence in male hematology patients: a common but ignored complication? *Haematologica.* 2007;92(10):1440.

202. Waage A, Gimsing P, Fayers P, et al. Melphalan and prednisone plus thalidomide or placebo in elderly patients with multiple myeloma. *Blood.* 2010;116(9):1405–1412.

203. Silberfarb PM. Chemotherapy and cognitive defects in cancer patients. *Annu Rev Med.* 1983;34:35–46.

204. Runowicz CD, Leach CR, Henry NL, et al. American cancer society/American society of clinical oncology breast cancer survivorship care guideline. *J Clin Oncol Off J Am Soc Clin Oncol.* 2015;34(6):611–635.

205. Castellino SM, Ullrich NJ, Whelen MJ, Lange BJ. Developing interventions for cancer-related cognitive dysfunction in childhood cancer survivors. *J Natl Cancer Inst.* 2014;106(8).

206. Koppelmans V, Breteler MMB, Boogerd W, Seynaeve C, Gundy C, Schagen SB. Neuropsychological performance in survivors of breast cancer more than 20 years after adjuvant chemotherapy. *J Clin Oncol Off J Am Soc Clin Oncol.* 2012;30(10):1080–1086.

207. Komaki R, Meyers CA, Shin DM, et al. Evaluation of cognitive function in patients with limited small cell lung cancer prior to and shortly following prophylactic cranial irradiation. *Int J Radiat Oncol Biol Phys.* 1995;33(1):179–182.

208. Tannock IF, Ahles TA, Ganz PA, Van Dam FS. Cognitive impairment associated with chemotherapy for cancer: report of a workshop. *J Clin Oncol Off J Am Soc Clin Oncol.* 2004;22(11):2233–2239.

209. Ferguson RJ, Ahles TA. Low neuropsychologic performance among adult cancer survivors treated with chemotherapy. *Curr Neurol Neurosci Rep.* 2003;3(3):215–222.

210. Anderson-Hanley C, Sherman ML, Riggs R, Agocha VB, Compas BE. Neuropsychological effects of treatments for adults with cancer: a meta-analysis and review of the literature. *J Int Neuropsychol Soc.* 2003;9(7):967–982.

211. Ginos JZ, Cooper AJ, Dhawan V, et al. [13N]cisplatin PET to assess pharmacokinetics of intra-arterial versus intravenous chemotherapy for malignant brain tumors. *J Nucl Med Off Publ Soc Nucl Med.* 1987;28(12):1844–1852.

212. Mitsuki S, Diksic M, Conway T, Yamamoto YL, Villemure J-G, Feindel W. Pharmacokinetics of 11C-labelled BCNU and SarCNU in gliomas studied by PET. *J Neurooncol.* 1991;10(1):47–55.

213. Gangloff A, Hsueh W-A, Kesner AL, et al. Estimation of paclitaxel biodistribution and uptake in human-derived xenografts in vivo with (18)F-fluoropaclitaxel. *J Nucl Med Off Publ Soc Nucl Med.* 2005;46(11):1866–1871.

214. Silverman DHS, Dy CJ, Castellon SA, et al. Altered frontocortical, cerebellar, and basal ganglia activity in adjuvant-treated breast cancer survivors 5–10 years after chemotherapy. *Breast Cancer Res Treat.* 2007;103(3):303–311.

215. Deprez S, Amant F, Smeets A, et al. Longitudinal assessment of chemotherapy-induced structural changes in cerebral White matter and its correlation with impaired cognitive functioning. *J Clin Oncol.* 2012;30(3):274–281.

216. Jamroziak K, Robak T. Pharmacogenomics of MDR1/ABCB1 gene: the influence on risk and clinical outcome of haematological malignancies. *Hematology.* 2004;9(2):91–105.

217. Seigers R, Schagen SB, Van Tellingen O, Dietrich J. Chemotherapy-related cognitive dysfunction: current animal studies and future directions. *Brain Imaging Behav.* 2013;7(4):453–459.

218. Fishel ML, Vasko MR, Kelley MR. DNA repair in neurons: so if they don't divide what's to repair? *Mutat Res.* 2007;614(1–2):24–36.

219. Caldecott KW. DNA single-strand breaks and neurodegeneration. *DNA Repair.* 2004;3(8–9):875–882.

220. Ahles TA, Saykin AJ. Candidate mechanisms for chemotherapy-induced cognitive changes. *Nat Rev Cancer.* 2007;7(3):192–201.

221. Gaman AM, Uzoni A, Popa-Wagner A, Andrei A, Petcu E-B. The role of oxidative stress in etiopathogenesis of chemotherapy induced cognitive impairment (CICI)-"Chemobrain". *Aging Dis.* 2016;7(3):307–317.

222. Schröder CP, Wisman GBA, de Jong S, et al. Telomere length in breast cancer patients before and after chemotherapy with or without stem cell transplantation. *Br J Cancer.* 2001;84(10):1348–1353.

223. Bolzán AD, Bianchi MS. DNA and chromosome damage induced by bleomycin in mammalian cells: an update. *Mutat Res.* 2018;775:51–62.

224. Shilling V, Jenkins V, Fallowfield L, Howell T. The effects of hormone therapy on cognition in breast cancer. *J Steroid Biochem Mol Biol.* 2003;86(3–5):405–412.

225. Liu J, Lin H, Huang Y, Liu Y, Wang B, Su F. Cognitive effects of long-term dydrogesterone treatment used alone or with estrogen on rat menopausal models of different ages. *Neuroscience.* 2015;290:103–114.

226. Rhee SH, Pothoulakis C, Mayer EA. Principles and clinical implications of the brain-gut-enteric microbiota axis. *Nat Rev Gastroenterol Hepatol.* 2009;6(5):306–314.

227. Cryan JF, O'Riordan KJ, Cowan CSM, et al. The microbiota-gut-brain axis. *Physiol Rev.* 2019;99(4):1877–2013.

228. Erny D, de Angelis ALH, Jaitin D, et al. Host microbiota constantly control maturation and function of microglia in the CNS. *Nat Neurosci.* 2015;18(7):965–977.

229. Pedroso SHSP, Vieira AT, Bastos RW, et al. Evaluation of mucositis induced by irinotecan after microbial colonization in germ-free mice. *Microbiol Read Engl.* 2015;161(10):1950–1960.

230. Secombe KR, Coller JK, Gibson RJ, Wardill HR, Bowen JM. The bidirectional interaction of the gut microbiome and the innate immune system: implications for chemotherapy-induced gastrointestinal toxicity. *Int J Cancer.* 2019;144(10):2365–2376.

231. Wardill HR, Gibson RJ, Van Sebille YZA, et al. Irinotecan-induced gastrointestinal dysfunction and pain are mediated by common TLR4-dependent mechanisms. *Mol Cancer Ther.* 2016;15(6):1376–1386.

232. Ahles TA, Saykin AJ, Noll WW, et al. The relationship of APOE genotype to neuropsychological performance in long-term cancer survivors treated with standard dose chemotherapy. *Psychooncology.* 2003;12(6):612–619.

233. Cheng H, Li W, Gan C, Zhang B, Jia Q, Wang K. The COMT (rs165599) gene polymorphism contributes to chemotherapy-induced cognitive impairment in breast cancer patients. *Am J Transl Res.* 2016;8(11):5087–5097.

234. Dooley LN, Ganz PA, Cole SW, Crespi CM, Bower JE. Val66Met BDNF polymorphism as a vulnerability factor for inflammation-associated depressive symptoms in women with breast cancer. *J Affect Disord.* 2016;197:43–50.

235. Oh B, Butow PN, Mullan BA, et al. Effect of medical Qigong on cognitive function, quality of life, and a biomarker of inflammation in cancer patients: a randomized controlled trial. *Support Care Cancer.* 2012. [cited 2020 Oct 14]. Available from: https://pubmed.ncbi.nlm.nih.gov/21688163/.

236. Hillman CH, Erickson KI, Kramer AF. Be smart, exercise your heart: exercise effects on brain and cognition. *Nat Rev Neurosci.* 2008;9. [cited 2020 Oct 14]. Available from https://pubmed.ncbi.nlm.nih.gov/18094706/.

237. Derry HM, Jaremka LM, Bennett JM, et al. Yoga and self-reported cognitive problems in breast cancer survivors: a randomized controlled trial. *Psychooncology.* 2015;24(8):958.

238. Reid-Arndt SA, Matsuda S, Cox CR. Tai Chi effects on neuropsychological, emotional, and physical functioning following cancer treatment: a pilot study. Vol. 18, Complementary therapies in clinical practice. *Complement Ther Clin Pract.* 2012. [cited 2020 Oct 14]. Available from: https://pubmed.ncbi.nlm.nih.gov/22196570/.

239. Ferguson RJ, McDonald BC, Rocque MA, et al. Development of CBT for chemotherapy-related cognitive change: results of a waitlist control trial. *Psychooncology.* 2012;21(2):176–186.

240. Dos Santos M, Rigal O, Léger I, et al. Cognitive rehabilitation program to improve cognition of cancer patients treated with chemotherapy: a randomized controlled multicenter trial. *J Clin Oncol.* 2019;37(15_suppl):11521.

241. Zuniga KE, Moran NE. Low serum carotenoids are associated with self-reported cognitive dysfunction and inflammatory markers in breast cancer survivors. *Nutrients.* 2018. [cited 2020 Oct 14];10(8). Available from: https://www.ncbi.nlm.nih.gov/pmc/articles/PMC6116006/.

242. Brown PD, Pugh S, Laack NN, et al. Memantine for the prevention of cognitive dysfunction in patients receiving whole-brain radiotherapy: a randomized, double-blind, placebo-controlled trial. *Neuro-Oncol.* 2013;15(10):1429.

243. Rapp SR, Case LD, Peiffer A, et al. Donepezil for irradiated brain tumor survivors: a phase III randomized placebo-controlled clinical trial. *J Clin Oncol.* 2015;33(15):1653.

244. Lawrence JA, Griffin L, Balcueva EP, et al. A study of donepezil in female breast cancer survivors with self-reported cognitive dysfunction 1 to 5 years following adjuvant chemotherapy. *J Cancer Surviv Res Pract.* 2016;10(1):176.

245. Escalante CP, Meyers C, Reuben JM, et al. A randomized, double-blind, 2-period, placebo-controlled crossover trial of a sustained-release methylphenidate in the treatment of fatigue in cancer patients. *Cancer J.* 2014;20(1):8–14.

246. Mar Fan HG, Clemons M, Xu W, et al. A randomised, placebo-controlled, double-blind trial of the effects of d-methylphenidate on fatigue and cognitive dysfunction in women undergoing adjuvant chemotherapy for breast cancer. *Support Care Cancer.* 2008;16. [cited 2020 Oct 14]. Available from: https://pubmed.ncbi.nlm.nih.gov/17972110/.

247. Kohli S, Fisher SG, Tra Y, et al. The effect of modafinil on cognitive function in breast cancer survivors. *Cancer.* 2009;115. [cited 2020 Oct 14]. Available from: https://pubmed.ncbi.nlm.nih.gov/19309747/.

Neurological complications of hematopoietic stem cell transplantation

Eudocia Q. Lee[a,b,c,d]

[a]Center for Neuro-Oncology, Department of Medical Oncology, Dana-Farber Cancer Institute, Boston, MA, United States,
[b]Division of Cancer Neurology, Brigham and Women's Hospital, Boston, MA, United States, [c]Department of Neurology,
Brigham and Women's Hospital, Boston, MA, United States, [d]Harvard Medical School, Boston, MA, United States

1 Introduction

Hematopoietic stem cell transplantation (HSCT) is a treatment used for many hematologic and nonhematologic diseases in which hematopoietic stem cells are infused following a chemotherapy and/or radiation conditioning regimen.[1] More than 50,000 transplants are performed annually worldwide.[2] HSCTs are potentially curative for many hematologic malignancies and nonmalignant bone marrow disorders, but are associated with significant early and late treatment related mortality.[3] The main causes of death in HSCT recipients are disease relapse, infections, and GVHD (in allogeneic only).[3]

The purpose and risks of the transplant depend on the type of transplant. When the stem cells are derived from the patient's own bone marrow (BM) or peripheral blood (PBSC), this is called an autologous transplant. In this type of transplant, patients are given a conditioning regimen to rid the body of the chemosensitive malignancy, which also ablates or significantly depletes the bone marrow. The stem cell infusion is intended to rescue the bone marrow, as opposed to treating the disease itself. Because the donor is self, neither graft versus host disease (GVHD) nor graft versus tumor (GvT) effect is anticipated. The most common indications for autologous HSCT are multiple myeloma and lymphoma.[1] When the stem cells are derived from a healthy donor, this is called an allogeneic transplant. A conditioning regimen is administered to kill malignant cells, ablate the bone marrow, and suppress the patient's immune system. The donor stem cells replace the immune system and help rid the body of malignancy. Indications for allogeneic HSCT include acute myeloid leukemia, acute lymphoblastic leukemia, myelodysplastic syndromes, myeloproliferative neoplasms, and select solid tumors.[1]

This type of transplant has the lowest risk of disease relapse due to the GvT effect. However, downsides associated with allogeneic transplants include GVHD, graft failure, and immunodeficiency. Another possible source for stem cells is partially-matched umbilical cord blood. However, the amount of available cord blood is very limited, and therefore mainly used for children.

The timeline of the HSCT process can help provide a framework when considering potential complications (Fig. 1). When a patient is a candidate for transplant, he or she first undergoes a conditioning or preparative regimen. Conditioning is followed by stem cell infusion. The timing of immune reconstitution varies but occurs gradually following the transplant. Innate or natural immunity recovers weeks to months after transplant and includes physical barriers (skin and mucosal surfaces), neutrophils, macrophages, and natural killer cells. Neutropenia can last for 7–14 days in autologous transplant and for 20–30 days in allogeneic transplant, although neutrophil function may not fully recover for a few months. In allogenic transplant patients, adaptive immunity through B and T lymphocytes can take years to return and can be delayed by GVHD. Day 100 generally marks a milestone for allogeneic HSCT patients as engraftment (the first 3 consecutive days with a neutrophil count greater than 0.5×10^9/L) generally has occurred by this date and the risk of critical complications is decreased. Day 100 also divides early postengraftment (within the first 30–100 days) from late postengraftment (> 100 days). The neurologic complications seen during these different phases of treatment may be slightly different. Following allogeneic transplants, early complications are usually related to the conditioning regimen, pancytopenia, infection, and drug toxicities.[4] Later,

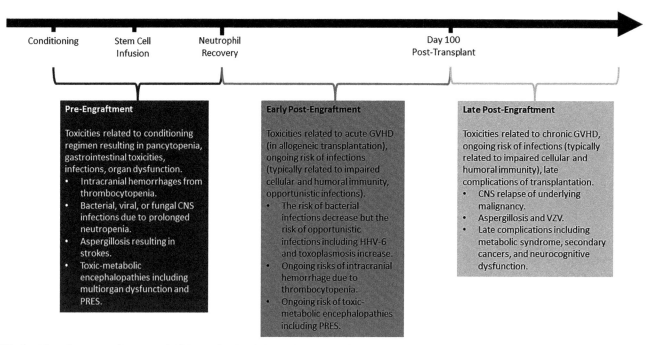

FIG. 1   Neurologic complications of HSCT can be divided into three phases based on timing: preengraftment period (from the start of the conditioning regimen to neutrophil recovery), the early postengraftment period (from neutrophil recovery to posttransplantation day 100), and the late postengraftment period (day 100 and beyond).

complications are usually related to GVHD, immunosuppressive therapy, or disease relapse.

The frequency of neurologic complications varies according to the study, types of neurotoxicity included, patient populations, and types of transplant, ranging from 2.8% to 56%.[5] In a prospective study evaluating central nervous system (CNS) disorders in HSCT recipients, the incidence of CNS infections was 0.1% after autologous HSCT and 2.6% after allogeneic HSCT, whereas the incidence of noninfectious CNS disorders was 0.4% after autologous HSCT and 4.5% after allogeneic HSCT.[6] Fungi including *Aspergillus* species (spp.) were the most common cause of CNS infections. Metabolic and drug-induced disorders, most commonly cyclosporine, were the most common cause of noninfectious CNS toxicity. In another study, the cumulative incidence of CNS complications in allogeneic HSCT patients was 9% at day 30, 18% at day 100, 20% at day 180, and 23% at day 365.[7] Studies also suggest that patients who develop CNS neurotoxicity have a higher mortality rate.[4,8] In the aforementioned prospective study, 50% of the HSCT recipients who developed a CNS disorder died within 30 days, with the highest rates of mortality reported for CNS fungal infections and cerebrovascular events.[6] Based on an evaluation of 33,047 adult patients who died following an allogenic HSCT performed in the 2010s, 345 (1.04%) died from CNS toxicity at a median of 96 days posttransplant.[9]

2 Neurologic complications associated with conditioning regimens

Conditioning regimens (typically chemotherapy or radiation or both) are classified as myeloablative, reduced intensity, or nonmyeloablative. As the name suggests, myeloablative regimens cause more profound long-lasting pancytopenia within 1–3 weeks from administration, possibly even irreversible pancytopenia until hematopoiesis is restored by stem cell infusion. Reduced intensity regimens cause prolonged cytopenias, but not as intense as myeloablative regimens. Nonmyeloablative regimens cause minimal cytopenia (with significant lymphopenia). Below is a list of agents commonly included in conditioning regimens, listed in alphabetical order.

2.1 Alemtuzumab

Alemtuzumab is a humanized anti-CD52 monoclonal antibody that depletes B and T cells from the circulating blood without a significant effect on hematopoietic progenitor cells.[10] When used in conditioning regimens for allogeneic HSCT, alemtuzumab can help reduce the risk of acute and chronic GVHD and prevent graft rejection. Immune reconstitution, particularly CD4 + lymphocytes, is slowed and thus there is an increased risk of opportunistic infections, including progressive multifocal leukoencephalopathy (PML).

2.2 Antithymocyte globulin (ATG)

ATG is used in some conditioning regimens for HSCT from HLA mismatched donors to reduce the risk of rejection and GVHD. With equine ATG, seizures, paresthesias, confusion, dizziness, syncope, and headaches have been reported in less than 5% of patients.[11] With rabbit ATG, headaches have been reported in 18%–40% of patients.[12]

2.3 Busulfan

Busulfan is an alkylating agent used in high-dose chemotherapy based conditioning regimens.[13] Seizures can occur in patients receiving high-dose busulfan and caution is recommended in patients with a history of a seizure disorder or head trauma or in patients receiving potentially epileptogenic drugs.[14] Premedication with anticonvulsants is advised, starting 12 hours prior to bulsufan and up to 24 hours after the last dose of bulsufan. Without premedication, seizures frequency ranges from 1% to 40%.[15] A retrospective review performed in patients who received intravenous busulfan noted a seizure prevention rate of 98.6% with phenytoin 125 mg/day to 300 mg/day and 100% with levetiracetam 1000 mg/day, although there were no statistically significant differences in seizure prevention rate or occurrence of adverse events between the two regimens.[15]

2.4 Cyclophosphamide

Cyclophosphamide is used in combination with total body irradiation (TBI) or other chemotherapies as conditioning and is used to manage chronic GVHD posttransplant.[13] In HSCT patients, Guillain Barre syndrome has rarely been reported.[16] Other neurotoxicities reported with cyclophosphamide in postmarketing surveillance include encephalopathy, peripheral neuropathy, seizures, posterior reversible encephalopathy syndrome (PRES), and myelopathy.[17]

2.5 Cytosine arabinoside

Cytosine arabinoside (also known as cytarabine or Ara-C) has been combined with TBI or other chemotherapies for conditioning.[13] CNS toxicity is dose-related and is the dose limiting toxicity of cytarabine.[18] In patients receiving high-dose cytarabine, the incidence of neurotoxicity is approximately 10%.[19] Cerebellar dysfunction is the most common CNS manifestation, although seizures or encephalopathy (somnolence, confusion, disorientation, memory loss, cognitive dysfunction, psychosis) can also occur in association with cerebellar dysfunction or as the sole manifestation.[19] Cerebellar symptoms such as dysarthria, dysdiadochokinesia, dysmetria, and ataxia typically occur 3–8 days after initiating treatment. These symptoms usually resolve, typically within 5 days of stopping cytarabine, although approximately 30% will not regain normal cerebellar function. Pathologic studies reveal Purkinje cell loss. Older age, higher cumulative doses, and renal insufficiency are risk factors for neurotoxicity.

2.6 Fludarabine

Fludarabine is a nucleoside analog with immunosuppressive properties and synergizing effect with alkylators. Dose-dependent severe CNS toxicity was reported in early dose-finding studies of fludarabine. Chun et al. reported that 13 of 36 (36%) patients treated with high-dose fludarabine $\geq$ 96 mg/m^2/day for 5–7 days (doses approximately four times greater than the current recommended dose) developed a syndrome characterized by blindness, progressive encephalopathy, and even death.[20,21] A lower rate of neurotoxicity has been reported with lower doses of fludarabine.[21] Nonetheless, neurotoxicity has been reported even with standard dose fludarabine in reduced intensity conditioning regimens (Fig. 2).[22] In a single institution retrospective review, Beitinjaneh and colleagues reported

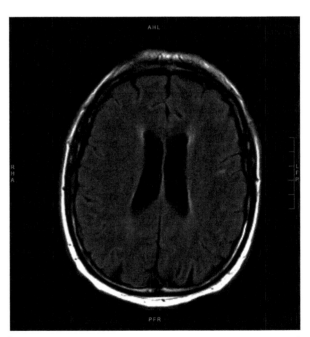

FIG. 2   A 64-year-old man with T-cell large granular lymphocyte leukemia complicated by pure red cell aplasia was admitted for allogenic stem cell transplant. As part of his reduced intensity conditioning regimen, he received fludarabine, cyclophosphamide, antithymocyte globulin, and total body irradiation. One month later, he developed low back pain radiating down both legs, sensory loss, and weakness followed by cortical blindness. Initial brain MRI was unremarkable but repeat brain MRI 7 days later showed subtle patchy T2 hyperintensities in periventricular white matter, mesial temporal lobes, and brainstem. Repeat CSF studies demonstrated no evidence of inflammation or infection. His syndrome was ultimately felt to be most consistent with fludarabine neurotoxicity.

severe fludarabine-associated leukoencephalopathy in 39 of 1596 (2.4%) of adult and pediatric patients undergoing HSCT.[22] They described three clinical syndromes: (1) PRES characterized by seizures, headache, vision changes, and variable mental status alterations, as well as cortical and subcortical white matter involvement on brain MRI, (2) acute toxic leukoencephalopathy (ATL) characterized by cognitive dysfunction, depressed consciousness, and vision changes as well as deep white matter changes on MRI, and (3) other leukoencephalopathy which is similar to ATL but less prominent deep white matter changes on MRI. CNS toxicity presented approximately 2 months after starting fludarabine and plateaued in severity over the next month. Fourteen patients (36%) died from neurotoxicity, all during the 6 months after onset of neurotoxicity. Of the 10 patients with fludarabine-related neurotoxicity who survived more than 1 year after HSCT and were not lost to follow-up, 6 had recovered neurologically, 3 had substantial but incomplete recovery, and 1 had persistent neurologic deficits.

2.7 Melphalan

Melphalan is used in some conditioning regimens for autologous and allogeneic transplants due to its ability to ablate the bone marrow with minimal extramedullary toxicity and potent immunosuppressive effects.[13,23] Dizziness was reported in 38% patients who received melphalan for autologous transplantation.[24] Severe encephalopathy with or without tonic-clonic seizures has rarely been reported in patients with renal failure who received high dose intravenous melphalan, although hypercytokinemia was also present in some of these cases.[25–27]

2.8 Thiotepa

Thiotepa is used in combination with other chemotherapies for conditioning.[13] Dose-dependent neurologic complications occurring during or shortly after administration include headache, confusion, amnesia, hallucinations, drowsiness, seizures, coma, and forgetfulness.[28] Fatal encephalopathy has rarely been reported in patients treated with high-dose thiotepa.[28,29] In a clinical trial of 20 patients with oligodendroglioma who underwent thiotepa consolidation (300 mg/m^2/day for 3 days) with stem cell transplantation, 4 patients (20%) died of delayed toxicities related to thiotepa.[29] Three of these 4 had developed a severe progressive encephalopathy characterized by apathy, disorientation, and psychomotor retardation. Another patient died from a fatal intratumoral hemorrhage occurring 48 hours after stem cell reinfusion.

In clinical practice, thiotepa neurotoxicity is relevant but generally not fatal. A more recent retrospective review of 307 courses of high-dose thiotepa in children undergoing HSCT reported 57 (18.6%) neurologic adverse events at least possibly related to thiotepa.[30] Most adverse events were mild (grade 1) or moderate (grade 2) and included headache, dizziness, or confusion, although six events were considered severe including seizures, tremor, and cerebellar syndrome. All the neurologic complications resolved without sequelae at a median interval of 3 days, and some patients were able to tolerate a re-challenge with thiotepa.

2.9 Total body irradiation (TBI)

TBI is widely used in conditioning regimens to induce immunosuppression to prevent rejection of the donor marrow (in allogeneic transplants) and to eradicate malignant cells, particularly in sanctuary sites such as the CNS and testes.[13,31] Acute neurotoxicities associated with TBI may include headache and fatigue.[32] Late neurocognitive dysfunction as a consequence of TBI is speculated but studies (most of them small) have demonstrated mixed results in adults and pediatric populations. A meta-analysis of 404 HSCT patients did not demonstrate an association between TBI and changes in cognitive function.[33] However, a longitudinal study of pediatric stem cell transplant recipients revealed that patients < 3 years old who received TBI had a significantly lower intelligence quotient (IQ) than those who did not receive TBI.[34] A TBI dose over 12 Gy has also been associated with increased risk of neurologic complications within the first 100 days after transplant.[35]

3 Toxic-metabolic encephalopathy

Metabolic encephalopathy affects approximately 5% of allogeneic HSCT recipients[36] and more commonly occurs in the preengraftment or early postengraftment periods.[37] The underlying etiology is often multifactorial including multiorgan dysfunction (i.e., liver, kidneys), electrolyte abnormalities, endocrine disorders, infection, and drug toxicity (including the previously mentioned conditioning agents). Patients may develop Wernicke encephalopathy from prolonged total parenteral nutrition and thiamine deficiency. Central pontine myelinolysis results from osmotic changes and rapid correction of hyponatremia. Hepatic veno-occlusive disease (VOD), also known as sinusoidal obstructive syndrome (SOS), can develop in a subset of patients following myeloablative or (less commonly) reduced intensity HSCT and is characterized by confusion, encephalopathy, renal failure, pleural effusions, and hypoxia due to multiorgan failure.[38] Posterior reversible leukoencephalopathy (PRES) can be caused by a variety of drugs used in HSCT patients, including calcineurin inhibitors discussed further below. Antibiotics cause a reversible encephalopathy in less than 1% of patients.[39] Cephalosporins and penicillins tend to cause an encephalopathy with myoclonus or seizures. Procaine penicillin, sulfonamides, fluoroquinolones, and macrolides are associated with psychosis. Metronidazole induced encephalopathy is characterized by cerebellar dysfunction and rare seizures.

4 Cerebrovascular complications

4.1 Intracranial hemorrhage (ICH)

HSCT recipients can develop intraparenchymal (IPH), subdural (SDH), or subarachnoid (SAH) hemorrhages or a combination of different types of ICH. In studies, the incidence of ICH among HSCT recipients varies from 1.5% to 32.2%.[40] Based on autopsy studies, hemorrhages can range in size from petechiae to focal bleeds to large hemorrhages.[41] Thrombocytopenia is an important underlying cause, although platelet refractoriness, infections, coagulopathy, hypertension, and PRES among other factors may also contribute. In a retrospective review of 622 patients who underwent allogeneic HSCT between 2004 and 2014, 21 patients (3.4%) developed an intracranial hemorrhage (15 intraparenchymal, 2 subarachnoid, 4 subdural).[42] Median time from transplant to ICH was 63 days (range 6–3488 days). The mortality rate of IPH is high. In the aforementioned review, the majority of patients with IPH died, whereas all those with SAH or SDH survived. Modern practice is to provide prophylactic platelet transfusions in patients with a platelet count of 10×10^9 cells/L or less to reduce the risk for spontaneous bleeding including ICH.[43]

4.2 Ischemic strokes

Ischemic strokes are much less common than intracranial hemorrhages in HSCT recipients. When strokes occur in the peri-engraftment period, possible causes include dimethyl sulfoxide (DMSO) and fungal infections, particularly Aspergillosis (discussed further below). DMSO is a cryoprotectant to prevent freezing injury of stem cells in transplants that exhibits a dose-dependent vasoconstrictor effect in animal models.[44] Neurologic toxicities including PRES, seizures, and stroke have been reported with DMSO. The incidence of toxicity may be reduced when cells are washed before transplantation or when lower concentrations of DMSO are used.

An immune-mediated cerebral vasculitis presenting as large or small vessel disease has also been described in HSCT patients, possibly related to chronic GVHD.[45] Medium to large vessel vasculitis may present with focal neurologic deficits due to ischemic stroke or hemorrhagic stroke. Small vessel vasculitis is characterized by a relapsing progressive course of headache, cognitive impairment, seizures, and (in children) fever, malaise, or flu-like symptoms. MRI may demonstrate multifocal or confluent white matter changes and ischemic strokes of varying ages. Cerebral blood vessel imaging is recommended, although a brain biopsy may be needed to confirm vasculitis. Management of cerebral vasculitis includes immunosuppressive agents such as corticosteroids and cyclophosphamide.

Stroke can also occur as a long-term consequence of HSCT. In general, long-term survivors of HSCT are at high risk for developing metabolic syndrome with an estimated prevalence of 31%–49%. Compared to the general population, HSCT survivors are more likely to report strokes (4.8% vs. 3.3%), cardiomyopathy (4.0% vs. 2.6%), dyslipidemia (33.9% vs. 22.3%), and diabetes (14.3% vs. 11.7%).[46] The cumulative incidence of stroke after adult HSCT is 1%–5% at a median of 4–10 years following the transplant.[47] When comparing autologous versus allogeneic donors, the magnitude of stroke risk for autologous survivors was lower than that for allogeneic survivors, but still elevated compared to the general population. Screening and management of cardiovascular risk factors is key to prevention of ischemic strokes in HSCT survivors.

4.3 Transplant-associated thrombotic microangiopathy (TA-TMA)

TA-TMA is a multisystem disorder caused by endothelial injury and microthrombus formation, most commonly affecting the kidneys, but can also involve the lungs, gastrointestinal tract, brain, and heart.[48,49] This complication is more common after allogeneic HSCT, but has been reported following autologous HSCT. The incidence varies widely in studies, in part due to the lack of standardized diagnostic criteria. In a registry based retrospective analysis of 23,665 allo-HSCT recipients, the 3 year cumulative incidence of TA-TMA was 3%.[50] Although the exact pathogenesis is not known, potential triggers for the development of TA-TMA include high-dose chemotherapy, radiation, calcineurin inhibitors, GVHD, and infections. Neurologic manifestations have been reported in up to half of patients with TA-TMA and include confusion, headache, hallucinations, or seizures.[49] Although cerebral vasculature may be affected, most of the neurologic injury arises from acute uncontrolled TMA-associated hypertension, including PRES and possibly hemorrhage. Management includes withdrawal of medication triggers, treatment of co-morbidities that may promote TA-TMA such as infections, and aggressive hypertension management. TA-TMA is associated with high mortality.

5 CNS infections

Immunocompromised patients, particularly those who receive allogeneic transplants, are at elevated risk for infections. In one study, fungal infections were the most common cause of CNS infection in HSCT recipients, followed by viral, parasitic, and bacterial infections (Table 1).[6] Different infections are likely to occur in different phases of the transplant timeline (Fig. 1).[37] During preengraftment, neutropenia and damaged mucosal barriers lead to high risk for gram-positive and gram-negative bacterial infections, some fungal infections (*Aspergillus* and *Candida*), and viral infections (herpes simplex virus). With severe neutropenia, classic signs of infection may be absent due to impaired inflammatory responses, and thus brain lesions demonstrate less rim enhancement, mass effect, and cerebral edema. During early postengraftment, impaired cellular and humoral immunity, determined in part by extent of GVHD and immunosuppressant therapy, results in opportunistic infections (human herpes virus 6, cytomegalovirus, toxoplasmosis). During postengraftment, as immunity slowly recovers, patients are at risk for *Aspergillus* and varicella zoster virus.

Management requires a high level of awareness as neurologic symptoms can be nonspecific and require prompt intervention, as delayed treatment can lead to poor outcomes.[51] If a CNS infection is suspected, neuroimaging (preferably MRI), cerebrospinal fluid analysis,

TABLE 1   CNS infections in HSCT patients.

Organism	Clinical manifestation	MRI findings	Diagnostic testing	Antimicrobial treatment
Parasitic infections				
Toxoplasma spp.	Cerebral toxoplasmosis	Hyperintense or isointense lesions in basal ganglia, posterior fossa, or gray-white matter junction. Compared to toxoplasmosis in HIV, lesions in HSCT patients may demonstrate hemorrhage and may not enhance or exhibit surrounding edema.	CSF: demonstration of tachyzoites or cysts, Toxoplasma PCR	Pyrimethamine and sulfadiazine Alternative options: pyrimethamine and clindamycin, trimethoprim-sulfamethoxazole
Fungal infections				
Aspergillus spp.	Brain abscess Vasculopathy with infarctions or hemorrhages Cavernous sinus thrombosis Cerebral artery pseudoaneurysms	Typically multiple, located in basal ganglia, thalamus, subcortical white matter. Evidence of hemorrhage or infarction. Ring-like enhancement may be absent.	CSF: galactomannan, Aspergillus PCR	Voriconazole In select cases, surgical removal of the abscess may be beneficial
Mucor spp. and *Rhizopus* spp.	Rhinocerebral mucormycosis Disseminated infection with CNS involvement	In rhinocerebral mucormycosis, infiltration from adjacent rhino-sinu-orbital regions, bone destruction of paranasal sinuses and hard palate. Possible cavernous sinus thrombosis, parenchymal invasion, cerebral infarction.	Pathology of infected tissue demonstrating irregular fungal hyphae with wide-angle branching	L-AmB Surgical resection of necrotic tissue
Candida spp.	Meningitis (typically in the presence of disseminated disease, ventricular draining device, or as isolated chronic meningitis)	Hydrocephalus (particularly with infected CNS shunts), microabscesses (multiple, small ring enhancing lesions with possible hemorrhagic component)	CSF: yeast stain and fungal culture	L-AmB and 5-FC Fluconazole with or without 5-FC as oral consolidation therapy Removal of indwelling catheter if source of infection (ventricular drain or central venous line)
Cryptococcus spp.	Meningitis	Dilated Virchow-Robin spaces, cyst like structures, granuloma of choroid plexus	CSF: India Ink staining and fungal cultures, Cryptococcus PCR	L-AmB and 5-FC In select patients, increased ICP may be managed by repeated LPs

TABLE 1   CNS infections in HSCT patients.—cont'd

Organism	Clinical manifestation	MRI findings	Diagnostic testing	Antimicrobial treatment
Bacterial infections				
Gram-positive bacteria (*Staphylococcus aureus, Staphylococcus epidermidis, Streptococcus pneumoniae, Listeria monocytogenes*)	Meningitis Ventriculitis	In meningitis, leptomeningeal enhancement In ventriculitis, hydrocephalus, abnormal ependymal enhancement, intraventricular debris with diffusion restriction	CSF: gram stain and bacterial culture	Depends on the causative organism
Viral infections				
VZV	Vasculitis causing thrombosis, infarctions, and rarely hemorrhage	Multiple infarctions throughout the brain, cerebral vessel imaging may demonstrate stenosis or occlusion	CSF: VZV PCR, anti-VZV IgG antibody levels	Acyclovir
HHV-6	Encephalitis	Symmetric, bilateral T2-hyperintense lesions in medial temporal lobe, hippocampus, amygdala	CSF: HHV-6 PCR (be aware of chromosomally integrated HHV6)	Foscarnet or ganciclovir
CMV	Ventriculoencephalitis	May be normal, may demonstrate patchy white matter T2 hyperintensities with periventricular enhancement	CSF: CMV PCR, CMV antigen	Ganciclovir and foscarnet
HSV	Encephalitis	T2 hyperintensities in medial and inferior temporal lobe, insula, cingulate	CSF: HSV PCR	Acyclovir
JC virus	PML	Asymmetric subcortical white matter abnormalities without mass effect or contrast enhancement	CSF: JC virus PCR Brain biopsy	No established treatment, restoration of the immune system
EBV	Meningoencephalitis	MRI may be normal or may demonstrate multifocal areas of T2 hyperintensity in basal ganglia and cortex	CSF: EBV PCR	Unclear which antiviral should be used, reduction of immunosuppression
	PTLD	Variable appearance, multiple enhancing masses	CSF: EBV PCR in many Brain biopsy	Reduction of immunosuppression, anticancer therapy

Abbreviations: *5-FC*, flucytosine; *CMV*, cytomegalovirus; *CSF*, cerebrospinal fluid; *EBV*, Epstein-Barr virus; *HHV-6*, human herpes virus 6; *HSV*, herpes simplex virus; *JC*, John Cunningham; *L-AmB*, liposomal amphotericin B; *MRI*, magnetic resonance imaging; *PCR*, polymerase chain reaction; *PML*, progressive multifocal leukoencephalopathy; *PTLD*, posttransplant lymphoproliferative disorder; *VZV*, varicella zoster virus.

and (in select cases) biopsy of the lesion are warranted. Empiric antimicrobial treatment should be started after the collection of blood and CSF studies and subsequently modified after isolating the causative organism.

5.1 Fungal

With the widespread use of fluconazole prophylaxis, the mortality rates due to invasive *Candida albicans* infections have decreased in HSCT patients.[52] Molds instead emerged as a major source of infection related mortality. *Aspergillus* causes more than 90% of the cerebral abscesses in HSCT recipients.[53] Invasive aspergillosis spreads to the brain either via hematogenous route from the lungs (resulting in an infectious vasculopathy causing strokes or hemorrhages, or extension into tissue causing cerebritis, meningitis, or abscesses) or via spread from the paranasal sinuses (causing brain abscesses, cavernous sinus thrombosis, or carotid pseudoaneurysm).[53] Galactomannan, a component of the cell wall in *Aspergillus* species, can be detected in the serum, CSF, or bronchoalveolar lavage. Definitive diagnosis is made by biopsy demonstrating septate hyphae. Voriconazole is recommended for the management of CNS aspergillosis.[51] Surgical resection may be beneficial in select CNS cases. Mortality rates are unfortunately high despite treatment.

5.2 Parasitic

Toxoplasmosis is a rare but potentially fatal opportunistic infection seen in HSCT recipients.[54] Reactivation of a latent *Toxoplasma gondii* infection is the most common cause of toxoplasmosis in HSCT. The prevalence of seropositivity and severity vary by geographic area. Immunocompromised patients can develop severe toxoplasmosis manifesting as encephalitis, pneumonitis, or myocarditis. The risk of reactivation is highest for a seropositive allo-HSCT patient receiving a seronegative graft.[55] Allogeneic HSCT recipients generally receive trimethoprim-sulfamethoxazole (TMP-SMX) for *Pneumocystis jirovecii* prophylaxis, which is also effective prophylaxis against toxoplasmosis.[56] Cerebral toxoplasmosis may present with mental status changes, fatigue, seizures, and fever.[51] Brain MRI reveals lesions, often nodular or ring enhancing, mostly involving the basal ganglia and the frontal lobe. Diagnosis can be made by demonstrating tachyzoites or cysts in the CSF or by CSF PCR. Treatment includes pyrimethamine and sulfadiazine.

5.3 Viral

Many viruses including herpes simplex virus (HSV), varicella zoster virus (VZV), Epstein-Barr virus (EBV), cytomegalovirus (CMV), human herpes virus 6 (HHV-6), Adenovirus (ADV), John Cunningham (JC) virus, and West Nile virus can cause CNS infection in HSCT recipients (Table 1).[57] CNS viral infections occur within 6 months after HSCT in 0.6% of patients. Appropriate antiviral prophylaxis is recommended for select transplant patients who are seropositive for CMV, HSV, or VZV.[56]

HHV-6B is the most common cause of viral encephalitis after allogeneic HSCT.[57] HHV-6B remains latent after primary infection (typically during childhood), but HHV-6B reactivates when the immune system is suppressed. Poor T cell function (as can be seen in cord blood transplants, T-cell depleted allografts, mismatched or unrelated donors, acute GVHD, and treatment with glucocorticosteroids) is a risk factor for HHV-6B encephalitis.[58] The cumulative incidence of HHV-6 encephalitis at day 70 is 7.9% in cord blood transplant recipients versus 1.2% in recipients of transplants from other donor sources.[59] HHV-6B encephalitis presents as a post-transplant acute limbic encephalitis with confusion, encephalopathy, short-term memory loss, syndrome of inappropriate antidiuretic hormone secretion (SIADH), seizures, and insomnia.[58] Brain MRI is initially normal but 60% will develop nonenhancing, T2 hyperintense lesions in the medial temporal lobes. EEG is often diffusely abnormal and temporal lobe seizures are relatively common. Diagnosis can be made by demonstrating HHV-6 DNA by PCR in the CSF in combination with a concerning clinical picture, as well as the exclusion of other causes. Patients with HHV-6B encephalitis typically have plasma HHV-6B viral loads ≥ 100,000 copies. Caution should be exercised when interpreting HHV-6 DNA test results to distinguish HHV-6 reactivation from chromosomally integrated HHV-6 (CIHHV-6), which does not generally cause encephalitis. Approximately 1% of the population will have CIHHV-6 and thus viral DNA in latent form originating from human chromosomal DNA will be persistently detected at high levels in whole blood, serum, and CSF.[58] Treatment options for HHV-6B encephalitis include intravenous foscarnet or ganciclovir.[58] Even so, HHV-6 encephalitis is associated with a poor prognosis. One study of allogeneic transplant recipients reported an overall survival rate of 58.3% at day 100 in patients who developed HHV-6 encephalitis versus 80.5% in those who did not.[60] Among survivors, 57% retained neuropsychological dysfunction after completing antiviral therapy, most often memory disturbance.

5.4 Bacterial

Bacteremia is a common infectious complication in HSCT, with an increasing concern for multidrug resistant strains.[61,62] Bacteria used to be the main cause of infection-related mortality during neutropenia in the 1960s, but strategies including fluoroquinolone prophylaxis for neutropenia and prompt empiric therapy for neutropenic fever have led to improvements in the mortality rate. Most centers report a higher frequency of gram-positive bacteremia compared to gram-negative bacteremia.

Bacterial meningitis is relatively uncommon in HSCT recipients compared to other types of CNS infections. Nonetheless, the incidence of bacterial meningitis in HSCT patients is 30-fold higher than the general population.[63] The risk is higher in allogeneic than autologous HSCT and is higher in patients with intraventricular devices (such as external ventricular drain) or following a neurosurgical procedure. *Streptococcus pneumoniae* and *Neisseria meningitidis* were the most common causative organism in a series from the Netherlands.[63] The median interval from transplant to onset of bacterial meningitis was 4 years in this series although the range was wide. The presentation of bacterial meningitis in HSCT is similar to the general population, although the CSF white blood cell count can be deceptively low due to leukopenia. Empiric antimicrobial treatment is recommended in suspected meningitis and then narrowed when the causative organism is isolated. Patients can experience additional neurologic complications in the setting of bacterial meningitis including cerebral infarction, venous sinus thrombosis, and hemorrhage. In the Netherland series, 21% of HSCT recipients who developed bacterial meningitis died and 36% of the survivors were left with moderate disabilities.[63]

6 Neurologic complications associated with chronic graft versus host disease (GVHD)

Chronic GVHD is a late complication of allogeneic HSCT and is an immune-mediated disorder that can affect a variety of organs including the skin, eyes, oral mucosa, lungs, intestinal tract, and liver.[45] Neurologic involvement is rare, particularly in children, but when it does occur, GVHD typically affects the peripheral nervous system. Central nervous system involvement is less common. Based on a meta-analysis, the pooled incidence of immune-mediated neurologic disorders including GVHD was 0.6%.[8] Based on the 2014 National Institutes of Health consensus criteria for chronic GVHD, myositis is a distinctive but nondiagnostic manifestation of chronic GVHD.[64] Other neurologic syndromes such as immune neuropathies (i.e., acute or chronic inflammatory demyelinating polyneuropathy), myasthenia gravis, cerebral vasculitis, demyelination, and encephalitis are less established and therefore attribution to GVHD is a diagnosis of exclusion.

Myositis occurs in 2%–3% patients after allogeneic HSCT.[45] Symptom onset ranges from 3 to 69 months after transplant and can present with or without other signs of chronic GVHD. Patients report moderate to severe symmetrical proximal muscle weakness, particularly involving the neck flexors and limb girdle. Dysphagia and respiratory dysfunction can occur. Creatine kinase is typically elevated 5–50 times above normal. Electromyography may demonstrate a myopathic pattern including fibrillation potentials. Pathology from muscle biopsy can help confirm diagnosis. Management should follow general guidelines for chronic GVHD. Corticosteroids are commonly used, although patients may require combination therapy with other immunosuppressive drugs. In general, neurologic function substantially returns in the first 2–7 months of treatment.

Guillain Barre Syndrome (GBS) is reported in 1%–2% of patients following allogeneic HSCT and usually develops within the first 3 months after transplant.[45] GBS has been reported to occur in association with acute and chronic GVHD, and in the absence of GVHD. GBS is classically described as rapidly progressive symmetric, ascending motor weakness with numbness and hyporeflexia that peaks within 2–4 weeks. Other possible manifestations include cranial neuropathies, respiratory insufficiency, and autonomic involvement. Most GBS cases after allogeneic HSCT follow infections, including those with coxsackie viruses, CMV, *Chlamydia pneumoniae*, *Mycoplasma pneumoniae*, *Campylobacter jejuni*, EBV, hepatitis C virus, and VZV. All subtypes of GBS have been reported in association with chronic GVHD including acute inflammatory demyelinating polyneuropathy, acute motor sensory axonal neuropathy, acute motor axonal neuropathy, and clinical variants such as Miller-Fisher syndrome. Nerve conduction studies are important for diagnosis. CSF studies typically demonstrate albumino-cytologic dissociation. Plasma exchange or intravenous gamma globulin can be used to treat GBS after allogeneic HSCT.

7 Neurologic complications associated with immunosuppressive therapy

7.1 Corticosteroids

Corticosteroids are often used to manage acute and chronic GVHD. Side effects include neuropsychiatric symptoms ranging from mood lability to memory impairment to psychosis. Steroid myopathy can occur with long-term steroid use and the risk is elevated at doses equivalent to 30 mg/day or higher of prednisone.[45] Symptoms including proximal muscle weakness and atrophy begin insidiously. Serum creatine kinase and electromyography are generally normal. Pathology is not necessary for diagnosis; when biopsy is obtained, type II fiber atrophy is the primary histologic finding. Management includes tapering of corticosteroids and switching to another immunosuppressant for the management of GVHD.

7.2 Calcineurin inhibitors

The calcineurin inhibitors cyclosporine and tacrolimus are frequently used to treat chronic GVHD. Neurotoxicities seen with calcineurin inhibitors include a fine tremor involving the upper extremities that is most apparent with arm extension, migraine occipital headache, peripheral neuropathy and paresthesia, and neuropsychiatric symptoms ranging from agitation to psychosis.[65] PRES has been reported in 5.5%–7.2% of allogenic HSCT patients receiving calcineurin inhibitors for GVHD prophylaxis.[7] PRES is a (mostly) reversible syndrome characterized by neurologic symptoms (such as seizures, encephalopathy, headache, and visual disturbances) and subcortical vasogenic edema on imaging.[66] Onset of neurologic symptoms is acute or subacute, developing over hours to days. Encephalopathy can range from mild confusion to stupor. Generalized tonic-clonic seizures occur in 60%–75% of patients with PRES. Unexplained mental status changes should prompt an EEG to rule out status epilepticus. Headaches are usually dull and diffuse, although thunderclap headaches can occur with cerebral vasoconstriction and should prompt cerebrovascular imaging. Visual disturbances seen with PRES include decreased visual acuity, visual field deficits, cortical blindness, and visual hallucinations. Brain MRI typically demonstrates T2 signal abnormality consistent with subcortical vasogenic edema. The three main patterns of involvement include predominantly parieto-occipital regions of the cerebral hemispheres, holo-hemispheric watershed areas between the territories of the anterior and middle cerebral arteries, or superior frontal sulci. Edema can even affect basal ganglia, brainstem, or cerebellum. Involvement may be asymmetric but should be bilateral. Intracranial hemorrhage can occur in PRES and may occur in higher frequency when PRES occurs after allogeneic transplantation as compared to other disease settings. PRES is usually (but not always) reversible with removal or treatment of the underlying cause. There are no consensus guidelines on how to manage PRES in HSCT. Hypertension management is important. Removal of the precipitating immunosuppressive drug may reverse PRES, but increases the risk of graft rejection or development of GVHD. Strategies for management have not been studied in prospective trials, but could include medication adjustment or changing to an alternate immunosuppressive agent.[67] Elevated serum levels of the calcineurin inhibitor may be associated with neurotoxicity, but PRES has been reported even with normal drug levels.

8 Neurocognitive dysfunction

The extent of neurocognitive dysfunction in HSCT is not fully understood, given the heterogeneity between studies on definition and assessment of neurocognitive function, small sample sizes, and varied patient populations.[35] Risk factors for neurocognitive dysfunction include pretransplant chemotherapy, total body irradiation in conditioning, immunosuppressive therapies, length of hospital stay, and GVHD.[35]

Across studies, many adult HSCT patients have some neurocognitive dysfunction prior to transplant, worsen in the first few months after HSCT, and partially recover over time; but up to 60% of adults will have some degree of neurocognitive deficits at 22–82 months post-HSCT.[35] Syrjala and colleagues performed a prospective evaluation of adult allogeneic HSCT recipients using standardized neuropsychological tests at baseline (prior to transplant) as well as at 80 days, 1 year, and 5 years after transplant.[68,69] At baseline, 15%–32% of patients had some impairment (pretransplant impairments in motor dexterity, verbal memory, and verbal fluency were about twice the expected rate).[69] Performance on all tests declined from baseline to 80 days, but then recovered at 1 year for most survivors (with the exception of motor speed and dexterity as measured by the Pegboard). Further recovery in information processing speed and executive function occurred between 1 and 5 years posttransplant.[68] However, the Global Deficit Score indicated some neurocognitive dysfunction (mostly mild) at 5 years in 41.5% of long-term survivors compared to

19.7% of matched controls. In addition, transplant recipients showed persistent deficits in motor speech and dexterity at 5 years compared to their baseline and compared to matched healthy controls.

In children, pre- and posttransplant testing can reveal neurocognitive dysfunction and associated decrements in intelligence quotient (IQ).[35] Younger age at diagnosis and treatment seems to be associated with more significant declines.[34] One longitudinal study of pediatric HSCT patients demonstrated that the youngest patients (age < 3 at baseline) experienced declines in cognitive function during the first year.[34] Those that did not receive TBI largely recovered functioning in subsequent years, whereas those who did receive TBI failed to recover those losses.

There are no prospective studies on the prevention or management of neurocognitive dysfunction in HSCT recipients. Strategies to mitigate neurocognitive dysfunction include minimizing neurotoxic regimens such as TBI, management of acute CNS toxicities after transplantation, nonpharmacologic interventions such as cognitive rehabilitation, and pharmacologic interventions (although data are limited).[70]

9 Secondary malignancies

The risk of solid cancer is increased after HSCT compared to the general population. The standardized incidence ratio (the ratio of observed cancer cases in a HSCT cohort to expected cancer cases in the general population of similar age and gender) of CNS tumors is 3.8–9.5.[71,72] CNS tumors reported to have occurred following HSCT include astrocytoma (including glioblastoma), meningioma (Fig. 2), ependymoma, medulloblastoma, primitive neuro-ectodermal tumor (PNET), and CNS lymphoma. Younger age at transplant, cranial irradiation as part of prior treatment, and TBI for pretransplant conditioning may be risk factors.[72]

HSCT recipients are also at risk for developing posttransplant lymphoproliferative disorder (PTLD), which represents a heterogenous group of malignant diseases that can occur in the setting of suppression of T cell function (typically from immunosuppressants) after transplantation.[73,74] The most common type is related to EBV (EBV-PTLD), which occurs in 3.2% of allogeneic HSCT recipients.[74] The clinical presentation ranges from incidental, asymptomatic findings to fulminant disease, and can affect the CNS in 5%–20% of patients.[73] In CNS-PTLD, brain MRI typically demonstrates multifocal ring-enhancing lesions, mostly involving the lobes, basal ganglia, and thalamus.[75] Positive CSF PCR for EBV is highly suggestive of CNS-PTLD in the correct clinical setting, although biopsy is generally required for definitive diagnosis. CNS-PTLD pathology usually resembles aggressive diffuse large cell lymphomas. Treatment includes reduction of immunosuppression and anticancer treatment. Involvement of extra-lymphoid tissue including the CNS is associated with worse outcomes in PTLD.[74]

10 Conclusion

Neurologic complications of HSCT are associated with morbidity and mortality. Patients undergoing allogeneic transplantation are at higher risk for complications than patients undergoing autologous transplantation. Neurotoxicities can arise from treatments related to HSCT itself including the conditioning regimen, as a complication of HSCT such as infections and GVHD, or even from treatments used to manage those complications including antibiotics and calcineurin inhibitors.

References

1. Bazinet A, Popradi G. A general practitioner's guide to hematopoietic stem-cell transplantation. *Curr Oncol.* 2019;26(3):187–191.
2. World Health Organization. *Haematopoietic Stem Cell Transplantation HSCtx.* https://www.who.int/transplantation/hsctx/en/. Accessed 7 October 2020.
3. Styczyński J, Tridello G, Koster L, et al. Death after hematopoietic stem cell transplantation: changes over calendar year time, infections and associated factors. *Bone Marrow Transplant.* 2020;55(1):126–136.
4. Balaguer-Rosello A, Bataller L, Piñana JL, et al. Noninfectious neurologic complications after allogeneic hematopoietic stem cell transplantation. *Biol Blood Marrow Transplant.* 2019;25(9):1818–1824.
5. Sheikh MA, Toledano M, Ahmed S, Gul Z, Hashmi SK. Noninfectious neurologic complications of hematopoietic cell transplantation: a systematic review. *Hematol Oncol Stem Cell Ther.* 2021;14(2):87–94.
6. Schmidt-Hieber M, Engelhard D, Ullmann A, et al. Central nervous system disorders after hematopoietic stem cell transplantation: a prospective study of the Infectious Diseases Working Party of EBMT. *J Neurol.* 2020;267(2):430–439.
7. Siegal D, Keller A, Xu W, et al. Central nervous system complications after allogeneic hematopoietic stem cell transplantation: incidence, manifestations, and clinical significance. *Biol Blood Marrow Transplant.* 2007;13(11):1369–1379.
8. Gavriilaki M, Mainou M, Gavriilaki E, et al. Neurologic complications after allogeneic transplantation: a meta-analysis. *Ann Clin Transl Neurol.* 2019;6(10):2037–2047.
9. Schultze-Florey CR, Peczynski C, de Marino I, et al. Frequency of lethal central nervous system neurotoxicity in patients undergoing allogeneic stem cell transplantation: a retrospective registry analysis. *Bone Marrow Transplant.* 2020;55(8):1642–1646.
10. Poiré X, van Besien K. Alemtuzumab in allogeneic hematopoietic stem cell transplantation. *Expert Opin Biol Ther.* 2011;11(8):1099–1111.
11. *Product Information: ATGAM(R) IV Injection, Lymphocyte Immune Globulin, Anti-Thymocyte Globulin [Equine] Sterile IV Injection.* Kalamazoo, MI: Pharmacia & Upjohn Company; 2005. https://www.fda.gov/media/78206/download.
12. *Product Information: THYMOGLOBULIN(R) Intravenous Injection, Anti-Thymocyte Globulin (Rabbit) Intravenous Injection.* Cambridge, MA: Genzyme Corporation (per manufacturer); 2017. https://www.fda.gov/media/74641/download.
13. Gyurkocza B, Sandmaier BM. Conditioning regimens for hematopoietic cell transplantation: one size does not fit all. *Blood.* 2014;124(3):344–353.
14. *Product Information: BUSULFEX(R) Intravenous Injection, Busulfan Intravenous Injection.* Rockville, MD: Otsuka America Pharmaceutical, Inc (per FDA); 2015. https://www.accessdata.fda.gov/drugsatfda_docs/label/2015/020954s014lbl.pdf.
15. Akiyama K, Kume T, Fukaya M, et al. Comparison of levetiracetam with phenytoin for the prevention of intravenous busulfan-induced seizures in hematopoietic cell transplantation recipients. *Cancer Chemother Pharmacol.* 2018;82(4):717–721.
16. Rodriguez V, Kuehnle I, Heslop HE, Khan S, Krance RA. Guillain-Barré syndrome after allogeneic hematopoietic stem cell transplantation. *Bone Marrow Transplant.* 2002;29(6):515–517.
17. *Product Information: CYCLOPHOSPHAMIDE Intravenous Injection, Oral Tablets, Cyclophosphamide Intravenous Injection, Oral Tablets.* Deerfield, IL: Baxter Healthcare Corporation (per FDA); 2013. https://www.accessdata.fda.gov/drugsatfda_docs/label/2013/012141s090,012142s112lbl.pdf.
18. Herzig RH, Hines JD, Herzig GP, et al. Cerebellar toxicity with high-dose cytosine arabinoside. *J Clin Oncol.* 1987;5(6):927–932.
19. Baker WJ, Royer Jr GL, Weiss RB. Cytarabine and neurologic toxicity. *J Clin Oncol.* 1991;9(4):679–693.
20. Chun HG, Leyland-Jones BR, Caryk SM, Hoth DF. Central nervous system toxicity of fludarabine phosphate. *Cancer Treat Rep.* 1986;70(10):1225–1228.
21. Ding X, Herzlich AA, Bishop R, Tuo J, Chan C-C. Ocular toxicity of fludarabine: a purine analog. *Expert Rev Ophthalmol.* 2008;3(1):97–109.
22. Beitinjaneh A, McKinney AM, Cao Q, Weisdorf DJ. Toxic leukoencephalopathy following fludarabine-associated hematopoietic cell transplantation. *Biol Blood Marrow Transplant.* 2011;17(3):300–308.
23. Bayraktar UD, Bashir Q, Qazilbash M, Champlin RE, Ciurea SO. Fifty years of Melphalan use in hematopoietic stem cell transplantation. *Biol Blood Marrow Transplant.* 2013;19(3):344–356.
24. *Product Information: EVOMELA Intravenous Injection, Melphalan Intravenous Injection.* Irvine, CA: Spectrum Pharmaceuticals, Inc. (per FDA); 2016. https://www.accessdata.fda.gov/drugsatfda_docs/label/2016/207155s000lbl.pdf.
25. Schuh A, Dandridge J, Haydon P, Littlewood TJ. Encephalopathy complicating high-dose melphalan. *Bone Marrow Transplant.* 1999;24(10):1141–1143.
26. Freeman M, Dubey D, Neeley O, Carter G. Hypercytokinemia induced encephalopathy following Melphalan administration (P5.102). *Neurology.* 2015;84(14 suppl). P5.102.
27. Alayón-Laguer D, Alsina M, Ochoa-Bayona JL, Ayala E. Melphalan culprit or confounder in acute encephalopathy during autologous hematopoietic stem cell transplantation? *Case Rep Transplant.* 2012;2012:942795.
28. *Product Information: TEPADINA(R) Intravenous, Intracavitary, Intravesical Injection, Thiotepa Intravenous, Intracavitary, Intravesical Injection.* Cedar Park, TX: ADIENNE SA (per FDA); 2017. https://www.accessdata.fda.gov/drugsatfda_docs/label/2017/208264s000lbl.pdf.
29. Cairncross G, Swinnen L, Bayer R, et al. Myeloablative chemotherapy for recurrent aggressive oligodendroglioma. *Neuro Oncol.* 2000;2(2):114–119.
30. Maritaz C, Lemare F, Laplanche A, Demirdjian S, Valteau-Couanet D, Dufour C. High-dose thiotepa-related neurotoxicity and the role of tramadol in children. *BMC Cancer.* 2018;18(1):177.
31. Paix A, Antoni D, Waissi W, et al. Total body irradiation in allogeneic bone marrow transplantation conditioning regimens: a review. *Crit Rev Oncol Hematol.* 2018;123:138–148.
32. American College of Radiology and American Society for Radiation Oncology. *ACR–Astro Practice Parameter for the*

Performance of Total Body Irradiation. https://www.astro.org/uploadedFiles/_MAIN_SITE/Patient_Care/Clinical_Practice_Statements/Content_Pieces/ACRASTROPracticeParameterTBI.pdf. Accessed 8 October 2020.

33. Phillips KM, McGinty HL, Cessna J, et al. A systematic review and meta-analysis of changes in cognitive functioning in adults undergoing hematopoietic cell transplantation. *Bone Marrow Transplant.* 2013;48(10):1350–1357.

34. Willard VW, Leung W, Huang Q, Zhang H, Phipps S. Cognitive outcome after pediatric stem-cell transplantation: impact of age and total-body irradiation. *J Clin Oncol.* 2014;32(35):3982–3988.

35. Buchbinder D, Kelly DL, Duarte RF, et al. Neurocognitive dysfunction in hematopoietic cell transplant recipients: expert review from the late effects and Quality of Life Working Committee of the CIBMTR and complications and Quality of Life Working Party of the EBMT. *Bone Marrow Transplant.* 2018;53(5):535–555.

36. Dowling MR, Li S, Dey BR, et al. Neurologic complications after allogeneic hematopoietic stem cell transplantation: risk factors and impact. *Bone Marrow Transplant.* 2018;53(2):199–206.

37. Bonardi M, Turpini E, Sanfilippo G, Mina T, Tolva A, Zappoli Thyrion F. Brain imaging findings and neurologic complications after allogenic hematopoietic stem cell transplantation in children. *Radiographics.* 2018;38(4):1223–1238.

38. Dalle JH, Giralt SA. Hepatic veno-occlusive disease after hematopoietic stem cell transplantation: risk factors and stratification, prophylaxis, and treatment. *Biol Blood Marrow Transplant.* 2016;22(3):400–409.

39. Bhattacharyya S, Darby RR, Raibagkar P, Gonzalez Castro LN, Berkowitz AL. Antibiotic-associated encephalopathy. *Neurology.* 2016;86(10):963–971.

40. Zhang XH, Wang QM, Chen H, et al. Clinical characteristics and risk factors of intracranial hemorrhage in patients following allogeneic hematopoietic stem cell transplantation. *Ann Hematol.* 2016;95(10):1637–1643.

41. Bleggi-Torres LF, Werner B, Gasparetto EL, de Medeiros BC, Pasquini R, de Medeiros CR. Intracranial hemorrhage following bone marrow transplantation: an autopsy study of 58 patients. *Bone Marrow Transplant.* 2002;29(1):29–32.

42. Najima Y, Ohashi K, Miyazawa M, et al. Intracranial hemorrhage following allogeneic hematopoietic stem cell transplantation. *Am J Hematol.* 2009;84(5):298–301.

43. Kaufman RM, Djulbegovic B, Gernsheimer T, et al. Platelet transfusion: a clinical practice guideline from the AABB. *Ann Intern Med.* 2015;162(3):205–213.

44. Windrum P, Morris TC, Drake MB, Niederwieser D, Ruutu T. Variation in dimethyl sulfoxide use in stem cell transplantation: a survey of EBMT centres. *Bone Marrow Transplant.* 2005;36(7):601–603.

45. Grauer O, Wolff D, Bertz H, et al. Neurological manifestations of chronic graft-versus-host disease after allogeneic haematopoietic stem cell transplantation: report from the Consensus Conference on Clinical Practice in chronic graft-versus-host disease. *Brain.* 2010;133(10):2852–2865.

46. Chow EJ, Baker KS, Lee SJ, et al. Influence of conventional cardiovascular risk factors and lifestyle characteristics on cardiovascular disease after hematopoietic cell transplantation. *J Clin Oncol.* 2014;32(3):191–198.

47. DeFilipp Z, Duarte RF, Snowden JA, et al. Metabolic syndrome and cardiovascular disease after hematopoietic cell transplantation: screening and preventive practice recommendations from the CIBMTR and EBMT. *Biol Blood Marrow Transplant.* 2016;22(8):1493–1503.

48. Khosla J, Yeh AC, Spitzer TR, Dey BR. Hematopoietic stem cell transplant-associated thrombotic microangiopathy: current paradigm and novel therapies. *Bone Marrow Transplant.* 2018;53(2):129–137.

49. Jodele S, Laskin BL, Dandoy CE, et al. A new paradigm: diagnosis and management of HSCT-associated thrombotic microangiopathy as multi-system endothelial injury. *Blood Rev.* 2015;29(3):191–204.

50. Epperla N, Li A, Logan B, et al. Incidence, risk factors for and outcomes of transplant-associated thrombotic microangiopathy. *Br J Haematol.* 2020;189(6):1171–1181.

51. Schmidt-Hieber M, Silling G, Schalk E, et al. CNS infections in patients with hematological disorders (including allogeneic stem-cell transplantation)-Guidelines of the Infectious Diseases Working Party (AGIHO) of the German Society of Hematology and Medical Oncology (DGHO). *Ann Oncol.* 2016;27(7):1207–1225.

52. Marr KA, Seidel K, Slavin MA, et al. Prolonged fluconazole prophylaxis is associated with persistent protection against candidiasis-related death in allogeneic marrow transplant recipients: long-term follow-up of a randomized, placebo-controlled trial. *Blood.* 2000;96(6):2055–2061.

53. Concepcion NDP, Romberg EK, Phillips GS, Lee EY, Laya BF. Imaging assessment of complications from transplantation from pediatric to adult patients: part 2: hematopoietic stem cell transplantation. *Radiol Clin North Am.* 2020;58(3):569–582.

54. Martino R, Maertens J, Bretagne S, et al. Toxoplasmosis after hematopoietic stem cell transplantation. *Clin Infect Dis.* 2000;31(5):1188–1194.

55. Robert-Gangneux F, Meroni V, Dupont D, et al. Toxoplasmosis in transplant recipients, Europe, 2010–2014. *Emerg Infect Dis J.* 2018;24(8):1497.

56. Tomblyn M, Chiller T, Einsele H, et al. Guidelines for preventing infectious complications among hematopoietic cell transplant recipients: a global perspective. *Bone Marrow Transplant.* 2009;44(8):453–455.

57. Abidi MZ, Hari P, Chen M, et al. Virus detection in the cerebrospinal fluid of hematopoietic stem cell transplant recipients is associated with poor patient outcomes: a CIBMTR contemporary longitudinal study. *Bone Marrow Transplant.* 2019;54(8):1354–1360.

58. Ward KN, Hill JA, Hubacek P, et al. Guidelines from the 2017 European Conference on Infections in Leukaemia for management of HHV-6 infection in patients with hematologic malignancies and after hematopoietic stem cell transplantation. *Haematologica.* 2019;104(11):2155–2163.

59. Ogata M, Satou T, Kadota J, et al. Human herpesvirus 6 (HHV-6) reactivation and HHV-6 encephalitis after allogeneic hematopoietic cell transplantation: a multicenter, prospective study. *Clin Infect Dis.* 2013;57(5):671–681.

60. Ogata M, Oshima K, Ikebe T, et al. Clinical characteristics and outcome of human herpesvirus-6 encephalitis after allogeneic hematopoietic stem cell transplantation. *Bone Marrow Transplant.* 2017;52(11):1563–1570.

61. Mikulska M, Del Bono V, Viscoli C. Bacterial infections in hematopoietic stem cell transplantation recipients. *Curr Opin Hematol.* 2014;21(6):451–458.

62. Misch EA, Andes DR. Bacterial infections in the stem cell transplant recipient and hematologic malignancy patient. *Infect Dis Clin North Am.* 2019;33(2):399–445.

63. van Veen KEB, Brouwer MC, van der Ende A, van de Beek D. Bacterial meningitis in hematopoietic stem cell transplant recipients: a population-based prospective study. *Bone Marrow Transplant.* 2016;51(11):1490–1495.

64. Jagasia MH, Greinix HT, Arora M, et al. National institutes of health consensus development project on criteria for clinical trials in chronic graft-versus-host disease: I. The 2014 diagnosis and staging working group report. *Biol Blood Marrow Transplant.* 2015;21(3):389–401.e381.

65. Coe CL, Horst SN, Izzy MJ. Neurologic toxicities associated with tumor necrosis factor inhibitors and calcineurin inhibitors. *Neurol Clin.* 2020;38(4):937–951.

66. Fugate JE, Rabinstein AA. Posterior reversible encephalopathy syndrome: clinical and radiological manifestations, pathophysiology, and outstanding questions. *Lancet Neurol.* 2015;14(9):914–925.

67. Cerejo MC, Barajas RF, Cha S, Logan AC. Management strategies for posterior reversible encephalopathy syndrome (PRES) in patients receiving calcineurin-inhibitor or sirolimus therapy for hematologic disorders and allogeneic transplantation. *Blood.* 2014;124(21):1144.

68. Syrjala KL, Artherholt SB, Kurland BF, et al. Prospective neurocognitive function over 5 years after allogeneic hematopoietic cell transplantation for cancer survivors compared with matched controls at 5 years. *J Clin Oncol.* 2011;29(17):2397–2404.

69. Syrjala KL, Dikmen S, Langer SL, Roth-Roemer S, Abrams JR. Neuropsychologic changes from before transplantation to 1 year in patients receiving myeloablative allogeneic hematopoietic cell transplant. *Blood.* 2004;104(10):3386–3392.

70. Kelly DL, Buchbinder D, Duarte RF, et al. Neurocognitive dysfunction in hematopoietic cell transplant recipients: expert review from the Late Effects and Quality of Life Working committee of the Center for International Blood and Marrow Transplant Research and Complications and Quality of Life Working Party of the European Society for Blood and Marrow Transplantation. *Biol Blood Marrow Transplant.* 2018;24(2):228–241.

71. Inamoto Y, Shah NN, Savani BN, et al. Secondary solid cancer screening following hematopoietic cell transplantation. *Bone Marrow Transplant.* 2015;50(8):1013–1023.

72. Curtis RE, Rowlings PA, Deeg HJ, et al. Solid cancers after bone marrow transplantation. *N Engl J Med.* 1997;336(13):897–904.

73. Dierickx D, Habermann TM. Post-transplantation lymphoproliferative disorders in adults. *N Engl J Med.* 2018;378(6):549–562.

74. Styczynski J, Gil L, Tridello G, et al. Response to rituximab-based therapy and risk factor analysis in Epstein Barr virus-related lymphoproliferative disorder after hematopoietic stem cell transplant in children and adults: a study from the Infectious Diseases Working Party of the European Group for Blood and Marrow Transplantation. *Clin Infect Dis.* 2013;57(6):794–802.

75. White ML, Moore DW, Zhang Y, Mark KD, Greiner TC, Bierman PJ. Primary central nervous system post-transplant lymphoproliferative disorders: the spectrum of imaging appearances and differential. *Insights Imaging.* 2019;10(1):46.

Chemotherapy-induced peripheral neuropathy

Zhi-Jian Chen[a,b] *and Mark G. Malkin*[a,b,c]

[a]Neuro-Oncology Program at Massey Cancer Center, Virginia Commonwealth University School of Medicine, Richmond, VA, United States, [b]Neuro-Oncology Division, Department of Neurology, Virginia Commonwealth University School of Medicine, Richmond, VA, United States, [c]Neurology and Neurosurgery, William G. Reynolds, Jr. Chair in Neuro-Oncology, Virginia Commonwealth University School of Medicine, Richmond, VA, United States

1 Introduction

Peripheral neuropathy is the most common clinical neurologic syndrome caused by chemotherapy.[1] The overall incidence of chemotherapy-induced peripheral neuropathy (CIPN) is estimated to be approximately 38% in patients treated with multiple agents,[2] although this percentage varies depending on chemotherapy regimens, duration of exposure, and assessment methods.[3,4] Chemotherapy combinations with higher incidence include those that involve platinum drugs, vinca alkaloids, bortezomib, and taxanes.[5] In the US, more than 400,000 patients develop CIPN annually at a cost of approximately 2.5 billion dollars per year.[6] Symptoms range from early post treatment pain to paresthesia, sensory ataxia, and mechanical and cold allodynia that can have a significant impact on quality of life, often persisting long after the chemotherapy has been discontinued. Based on a self-report data study using the Quality-of-Life Questionnaire—CIPN20,[7] the most common symptoms were tingling in the hands and feet (30%), numbness in the toes and feet (19%), tingling in the hands or fingers (15%), and burning or shooting pain in the toes or feet (13%).[8] The actual clinical pattern of neuropathy differs among drugs, depending on the predominant site and mechanism(s) of nerve injury. Of note, survivors reported symptoms for up to 11 years after the completion of chemotherapy.[8] Importantly, CIPN can limit the dosage and selection of chemotherapeutic agents, delay further treatment cycles, and is the most common reason for discontinuation of chemotherapy. In early stage breast cancer patients undergoing taxane-based chemotherapy for adjuvant and neoadjuvant settings, 17% CIPN-associated dose reduction was reported.[9] Studies have shown that even modest reductions in planned total dose of chemotherapy can have a negative impact on cancer survival.[10,11]

2 Pathophysiology of CIPN

The pathophysiology of CIPN can be more complicated than other types of peripheral neuropathy because of the variety of mechanisms of different chemotherapeutic agents, e.g., microtubule-targeting agents, platinum-containing agents, proteasome inhibitors, angiogenesis inhibitors, the newer class of chemotherapeutics known as immune checkpoint inhibitors, and different locations of the injury involving the entire peripheral nervous system, and even some of the central nervous system or spinal cord. The mechanisms of action of chemotherapeutic agents that lead to potent effects on tumor cell proliferation and cell death are well-studied and relatively well understood. However, it is not entirely clear whether these (mostly) desirable effects on rapidly proliferating cells are also responsible for causing undesirable effects on nonproliferating sensory neurons, or whether additional pharmacological effects contribute to the development of CIPN. Axonopathy, primarily of the sensory neurons, is the hallmark of CIPN.[12,13] It is associated with DNA damage, oxidative stress, mitochondrial toxicity, and ion channel remodeling in the neurons of the peripheral nervous system. The events that trigger the onset of peripheral neurotoxicity and neuropathic pain can differ depending on the chemotherapy. Dorsal root ganglia, sensory neurons, satellite cells, Schwann cells, as well as neuronal and glial cells in the spinal cord are the preferential sites at which chemotherapy neurotoxicity occurs.

The vinca alkaloids (vincristine, vinblastine, vindesine) are most frequently used in the treatment of hematological and lymphatic malignancies as well as some solid tumors of the breast and lungs. They inhibit the assembly of microtubules, promote their disassembly, and lead to alterations in the neuronal cytoskeleton and axonal transport. The affinity for tubulin differs among the vinca alkaloids, with vincristine having the highest affinity and, consequently, producing the severest neuropathy, followed by vinblastine and vinorelbine.[14]

Paclitaxel and docetaxel belong to the family of taxanes: chemotherapeutic agents used in the treatment of breast, prostate, lungs, pancreatic, gynecological, and other solid tumors that act by inhibiting disassembly of tubulin from the microtubule polymer. Neither compound crosses the blood-brain-barrier (BBB), although paclitaxel is accumulated in dorsal root ganglion neurons—which lie outside the BBB—via largely unknown mechanisms.[15,16] Both paclitaxel and docetaxel are extensively metabolized, via cytochrome P 2C8 and 3A4, respectively, albeit pharmacological activity of these metabolites is modest at best and their contribution to the development of CIPN is unclear.

Platinum-based chemotherapeutic agents (oxaliplatin, cisplatin, and carboplatin) are widely used in the treatment of several types of solid tumors. Oxaliplatin is indicated for the treatment of digestive tract tumors (advanced colorectal, esophageal, stomach, liver, and pancreatic cancers), while cisplatin and carboplatin are indicated for the treatment of other types of tumors (small-cell lung cancer, testicular, ovarian, brain, uterine, and bladder cancers). Chronic platinum induced a toxic effect on DNA binding in dorsal root ganglion sensory neurons. Oxaliplatin is unique in that it causes acute cold-induced hyperalgesia, which likely arises because of increased neuronal excitability via oxaliplatin's effect on voltage-gated sodium channels.[17]

Bortezomib and carfilzomib are reversible proteasome inhibitors used for the treatment of multiple myeloma and certain types of lymphoma. Compelling evidence is emerging that ceramide and sphingosine-1 phosphate (S1P) share potent inflammatory and nociceptive actions.[18] Within astrocytes, bortezomib causes an increase in sphingolipid metabolism, leading to an increase in the production of ceramide as well as sphingosine-1 phosphate (S1P) and dihydrosphingosine-1-phosphate (DH-S1P). Released S1P binds to the S1P receptor (S1PR1) on astrocytes, which ultimately leads to an increase in the release of presynaptic glutamate at the level of the dorsal horn of the spinal cord and to the development of neuropathic pain.[19,20]

Immune checkpoint inhibitors (ICIs) are increasingly used and are becoming the standard of care in the treatment of various tumor types. Ipilimumab (an antibody targeting cytotoxic T lymphocyte associated antigen 4,

CTLA-4), nivolumab, pembrolizumab, and cemiplimab (antibodies directed against the programmed cell death 1 ligand, PD-1), atezolizumab, durvalumab, and avelumab (anti-PD-1 ligand, PD-L1) have led to improved survival in patients with various cancer types including lung, kidney, liver, bladder, and breast cancer, melanoma, and lymphoma.[21,22] With the fast development of immune checkpoint inhibitors in cancer treatment, the immune-mediated neuropathies are becoming more recognized.[23] They are mostly demyelinating peripheral nerve lesions, such as demyelinating polyradiculoneuropathy. Potential mechanisms may relate to antigens shared between affected tissue and tumor leading to a cross reactivity between tumor neo-antigens and normal tissue antigens favoring the development of immune-related adverse events by immunotherapy.[24,25] Additionally, direct binding of ICIs to targets expressed in normal tissue (antibody deposition in the pituitary gland as a consequence of CTLA-4 expression at this level) may induce an antibody dependent toxicity and complement-mediated inflammation.[26] The reactions thought to be involved are through type IV (T-cell dependent) and type II (IgG-dependent) immune mechanisms..[27]

Similar to the studies of many diseases, especially cancers, entering the era of molecular biology/genetics in recent years, a fair number of studies have identified genetic risk factors associated with the development of CIPN in cancer patients. Many of these are pharmacogenomic in nature, affecting either the absorption, distribution, metabolism, or excretion of these chemotherapeutic agents. They are: (1) Charcot-Marie-Tooth related genes, such as FGD4, PRX, SBF2; (2) DNA repair-related genes, e.g., ERCC1, XPC, KIAA0146-PRKDb; (3) Drug metabolism-related genes, e.g., CYP2C8, GSTP1, CYP1B1, CYP3A4, CYP3A5, GSTM1, GSTM3; (4) Inflammatory response-related genes, e.g., ITGA1, IL-1b; (5) Nerve-related genes, e.g., ARHGEF10, EPHA5, EPHA6; (6) Transport protein genes, e.g., ABCB1. ABCC2, ABCC4, SLCO1B1c; (7) other biomarkers, such as GPX7, KLC3, TAC1, etc..[28]

Genetic variants associated with the development and severity of taxane-induced neuropathy include low-frequency variants in the ephrin receptor genes EPHA6, EPHA5, and EPHA8, the Charcot-Marie-Tooth disease gene ARHGEF10, the glycogen synthase kinase-3b (GSK3b) gene, the DNA repair pathway genes XPC, the congenital peripheral neuropathy gene FGD4, the b-tubulin IIa gene (TUBB2A), and VAC14, a gene coding for a component of a trimolecular complex that tightly regulates the level of phosphatidylinositol 3,5-bisphosphate.[29] The polymorphism Ile105Val of the GSTP1 gene, encoding glutathione transferase P1, has been associated with a decreased risk of developing severe oxaliplatin-related cumulative neuropathy.[30] A SNP (single nucleotide polymorphism) in the centrosome protein encoded by CEP72 gene has been associated with the development of vincristine-induced neuropathy in the later phases of therapy.[31]

The discovery of biomarkers capable of reliably determining the likelihood of CIPN development in a patient would allow for chemotherapeutic regimens to be tailored to the unique needs for individual patients and perhaps avoided completely for patients at high CIPN risk. But so far, there is no sensitive and ideal biomarker available for either early diagnosis or longitudinal monitoring of the disease course of CIPN. There are some clinical trials underway that are focused on natural history and biomarker validation, such as ECOG-ACRIN's prospective validation trial of taxane therapy and risk of chemotherapy-induced peripheral neuropathy in African American women, and an industry sponsored CIPN natural history study (EPIPHANY). As the costs of DNA sequencing continue to decrease, it becomes more likely that physicians will begin to use genome sequencing as a tool to gain insights into individual patients.

3 Clinical presentation and diagnostic evaluation of CIPN

Approximately 70% of patients using paclitaxel[10] and 31%–64% of patients treated with vinca alkaloids and proteasome inhibitors for hematological malignancy treatment reported CIPN symptoms.[11] Between 21% and 63% of patients using platin-based regimens for colorectal cancers and 63% of patients using thalidomide reported CIPN cases.[10] CIPN is dose dependent, occurring as early as 24–72h following drug administration but often as much as 3 months after completion of chemotherapy, a phenomenon known as coasting. In many cases, it persists after discontinuation of treatment. Symptoms persist in 58%–64% of breast cancer patients treated with taxanes who finished chemotherapy within 5 years[12] and in 30% of lymphoma patients treated with vincristine more than 5 years postchemotherapy. The spectrum of CIPN severity varies from mild to severe, and its risk factors include cumulative chemotherapy dose, history of neuropathy, and genetic polymorphisms.[12] Classical symptoms of CIPN are peripheral neuropathy with a stocking and glove distribution characterized by sensory loss, paresthesia, dysesthesia, and numbness, especially with neuropathic pain in the most serious cases. Neuropathic symptoms typically appear within 6 months of chemotherapy initiation.[7,13] CIPN can be painful and/or disabling, causing significant loss of functional abilities and decreasing quality of life.[3,6] In addition to sensory neuropathic symptoms, CIPN can also cause muscle cramps, spasm, weakness, as well as autonomic symptoms such as constipation, ileus, urinary retention, and orthostatic hypotension. The long-term effects are associated with comorbidities including depression, insomnia, and falls in cancer survivors.[11]

Since the exact cause of CIPN remains elusive, it is difficult to reach consensus as to which assessment can be considered the "gold-standard" for CIPN.[1] Consequently, without reliable and reproducible assessment tools, the prevalence of CIPN can be misestimated in certain clinics, and the evaluation of treatment for CIPN may also not be sensitive or reliable enough to identify efficacious therapy. In clinical practice, diagnosis is generally made by the oncologist first. Rarely, nerve conduction velocity, skin biopsy, or nerve biopsies are used. Validated patient and clinician-reported outcome measures have been developed. Clinician reported outcomes (ClinROs) used for CIPN grading include the physician assessed common toxicity criteria of the National Cancer Institute-Common Terminology Criteria for Adverse Events (NCI-CTCAE) and the Total Neuropathy Score clinical version (TNSc). Patient-reported outcomes (PROs) include the European Organization for Research and Treatment of Cancer-Quality of Life Questionnaire-CIPN-20 (EORTC QLQ-CIPN-20), and the Treatment-Induced Neuropathy Assessment Scale (TNAS).[2–4] It is of note that selecting an effective and sensitive assessment tool for CIPN is one of the most important components of a successful clinical study. When analyzing the seven published clinical trials of patients with chronic CIPN, only four of them evaluated therapeutic pain as the first line treatment effect. Among these four, the one that successfully demonstrated the efficacy of the treatment on neuropathic pain (i.e., duloxetine) was the only one that not only used a primary outcome measure that assessed only pain, but also used a minimum pain intensity for the principal inclusion criterion.[5] Besides the PROs and ClinROs, functional measures are the tests that can potentially cover both subjective and objective components affected by CIPN. Being potentially useful and easy to administer, different function tests to assess peripheral neuropathy have been developed and used in neuropathy assessment for more than a decade, but they have become more attractive recently due to the technologic advance in either measurement device development or better recognition of CIPN related specific clinical symptoms to be evaluated and tested by the digital functional test device.

A recent retrospective cohort study of ~ 70,000 cancer patients demonstrated twice the fall rate when patients were treated with neurotoxic versus nonneurotoxic agents.[32,33] Since gait disturbance and falls have been demonstrated to be one of the most functionally serious complications of CIPN, these are perfect clinical symptoms for applying wearable device technology for characterizing disease progression.

4 Clinical management of CIPN

ASCO (American Society of Clinical Oncology) has published two versions of guidelines on prevention and management of chemotherapy-induced peripheral neuropathy in survivors of adult cancers which comprehensively summarize the conducted clinical trials and provide evidence-based guidance.[34,35] The identified data reconfirmed that no agents are recommended for the prevention of CIPN. The use of acetyl-L-carnitine for the prevention of CIPN in patients with cancer should be discouraged. Furthermore, clinicians should assess the appropriateness of dose delaying, dose reduction, substitutions, or stopping chemotherapy in patients who develop intolerable neuropathy and/or functional impairment. Duloxetine is the only agent that has appropriate evidence to support its use for patients with established painful CIPN.[36]

In practice, for patients with neuropathic pain related to cancer and its treatment, the consensus is to use antidepressants and anticonvulsants as first-line adjuvant analgesics in patients whose pain is only partially responsive to opioids. So, for patients with cancer who are experiencing CIPN, it is suggested to use duloxetine which is supported by Class 2b evidence. Gabapentin/pregabalin or a tricyclic antidepressant (amitriptyline/nortriptyline) is a reasonable alternative given the limited therapeutic option and the demonstrated efficacy of these drugs for other neuropathic pain conditions. It is also sensible to try a compounded topical agent containing baclofen, amitriptyline, ketamine, or lidocaine. In addition to the above choices, exercise, acupuncture, and scrambler therapy have attracted more attention for managing neuropathic symptoms from CIPN.

There is a great need to investigate promising treatments either for preventive or therapeutic efficacy in CIPN. Better understanding of the pathophysiological mechanism of CIPN, natural disease course, and pharmacogenomic profiles will be crucial to predict specific effective management of CIPN.

References

1. Brewer JR, Morrison G, Dolan ME, Fleming GF. Chemotherapy-induced peripheral neuropathy: current status and progress. *Gynecol Oncol.* 2016;140(1):176–183.
2. Cavaletti G, Zanna C. Current status and future prospects for the treatment of chemotherapy-induced peripheral neurotoxicity. *Eur J Cancer.* 2002;38(14):1832–1837.
3. Cavaletti G, Frigeni B, Lanzani F, et al. Chemotherapy-induced peripheral neurotoxicity assessment: a critical revision of the currently available tools. *Eur J Cancer.* 2010;46(3):479–494.
4. Cavaletti G, Cornblath DR, Merkies ISJ, et al. The chemotherapy-induced peripheral neuropathy outcome measures standardization study: from consensus to the first validity and reliability findings. *Ann Oncol.* 2013;24(2):454–462.
5. Staff NP, Podratz JL, Grassner L, et al. Bortezomib alters microtubule polymerization and axonal transport in rat dorsal root ganglion neurons. *Neurotoxicology.* 2013;39:124–131.
6. Pachman DR, Barton DL, Watson JC, Loprinzi CL. Chemotherapy-induced peripheral neuropathy: prevention and treatment. *Clin Pharmacol Ther.* 2011;90(3):377–387.
7. Postma TJ, Aaronson NK, Heimans JJ, Muller MJ, Hildebrand JG, Delattre JY. The development of an EORTC quality of life questionnaire to assess chemotherapy-induced peripheral neuropathy: the QLQ-CIPN20. *Eur J Cancer.* 2005;41(8):1135–1139.
8. Beijers AJ, Mols F, Tjan-Heijnen VC, Faber CG, van de Poll-Franse LV, Vreugdenhil G. Peripheral neuropathy in colorectal cancer survivors: the influence of oxaliplatin administration. Results from the population-based PROFILES registry. *Acta Oncol.* 2015;54(4):463–469.
9. Bhatnagar B, Gilmore S, Goloubeva O, et al. Chemotherapy dose reduction due to chemotherapy induced peripheral neuropathy in breast cancer patients receiving chemotherapy in the neoadjuvant or adjuvant settings: a single-center experience. *Springerplus.* 2014;3:366–371.
10. Denduluri N, Lyman GH, Wang Y, et al. Chemotherapy dose intensity and overall survival among patients with advanced breast or ovarian cancer. *Clin Breast Cancer.* 2018;18(5):380–386.
11. Pettengell R, Schwenkglenks M, Bosly A. Association of reduced relative dose intensity and survival in lymphoma patients receiving CHOP-21 chemotherapy. *Ann Hematol.* 2008;87(5):429–430.
12. Miltenburg NC, Boogerd W. Chemotherapy-induced neuropathy: a comprehensive survey. *Cancer Treat Rev.* 2014;40(7):872–882.
13. Lema MJ, Foley KM, Hausheer FH. Types and epidemiology of cancer-related neuropathic pain: the intersection of cancer pain and neuropathic pain. *Oncologist.* 2010;15(Suppl. 2):3–8.
14. Sisignano M, Baron R, Scholich K, Geisslinger G. Mechanism-based treatment for chemotherapy-induced peripheral neuropathic pain. *Nat Rev Neurol.* 2014;10(12):694–707.
15. Cavaletti G, Cavalletti E, Montaguti P, Oggioni N, De Negri O, Tredici G. Effect on the peripheral nervous system of the short-term intravenous administration of paclitaxel in the rat. *Neurotoxicology.* 1997;18(1):137–145.
16. Wozniak KM, Vornov JJ, Wu Y, et al. Sustained accumulation of microtubule-binding chemotherapy drugs in the peripheral nervous system: correlations with time course and neurotoxic severity. *Cancer Res.* 2016;76(11):3332–3339.
17. Zajączkowska R, Kocot-Kępska M, Leppert W, Wrzosek A, Mika J, Wordliczek J. Mechanisms of chemotherapy-induced peripheral neuropathy. *Int J Mol Sci.* 2019;20(6):1451–1480.
18. Salvemini D, Doyle T, Kress M, Nicol G. Therapeutic targeting of the ceramide-to-sphingosine 1-phosphate pathway in pain. *Trends Pharmacol Sci.* 2013;34(2):110–118.
19. Stockstill K, Doyle TM, Yan X, et al. Dysregulation of sphingolipid metabolism contributes to bortezomib-induced neuropathic pain. *J Exp Med.* 2018;215(5):1301–1313.
20. Emery EC, Wood JN. Gaining on pain. *N Engl J Med.* 2018;379(5):485–487.
21. Emens LA, Ascierto PA, Darcy PK, et al. Cancer immunotherapy: opportunities and challenges in the rapidly evolving clinical landscape. *Eur J Cancer.* 2017;81:116–129.
22. Hargadon KM, Johnson CE, Williams CJ. Immune checkpoint blockade therapy for cancer: an overview of FDA-approved immune checkpoint inhibitors. *Int Immunopharmacol.* 2018;62:29–39.
23. Psimaras D, Velasco R, Birzu C, et al. Immune checkpoint inhibitors-induced neuromuscular toxicity: from pathogenesis to treatment. *J Peripher Nerv Syst.* 2019;24(Suppl. 2):S74–S85.
24. Johnson DB, Balko JM. Biomarkers for immunotherapy toxicity: are cytokines the answer? *Clin Cancer Res.* 2019;25(5):1452–1454.
25. Michot JM, Bigenwald C, Champiat S, et al. Immune-related adverse events with immune checkpoint blockade: a comprehensive review. *Eur J Cancer.* 2016;54:139–148.

26. Thompson JA, Schneider BJ, Brahmer J, et al. Management of immunotherapy-related toxicities, version 1.2019. *J Natl Compr Canc Netw.* 2019;17(3):255–289.

27. Caturegli P, Di Dalmazi G, Lombardi M, et al. Hypophysitis secondary to cytotoxic T-lymphocyte-associated protein 4 blockade: insights into pathogenesis from an autopsy series. *Am J Pathol.* 2016;186(12):3225–3235.

28. Diaz PL, Furfari A, Wan BA, et al. Predictive biomarkers of chemotherapy-induced peripheral neuropathy: a review. *Biomark Med.* 2018;12(8):907–916.

29. Starobova H, Vetter I. Pathophysiology of chemotherapy-induced peripheral neuropathy. *Front Mol Neurosci.* 2017;10:174–195.

30. Lecomte T, Landi B, Beaune P, Laurent-Puig P, Loriot MA. Glutathione S-transferase P1 polymorphism (Ile105Val) predicts cumulative neuropathy in patients receiving oxaliplatin-based chemotherapy. *Clin Cancer Res.* 2006;12(10):3050–3056.

31. Diouf B, Crews KR, Lew G, et al. Association of an inherited genetic variant with vincristine-related peripheral neuropathy in children with acute lymphoblastic leukemia. *JAMA.* 2015;313(8):815–823.

32. Tofthagen C, Overcash J, Kip K. Falls in persons with chemotherapy-induced peripheral neuropathy. *Support Care Cancer.* 2012;20(3):583–589.

33. Ward PR, Wong MD, Moore R, Naeim A. Fall-related injuries in elderly cancer patients treated with neurotoxic chemotherapy: a retrospective cohort study. *J Geriatr Oncol.* 2014;5(1):57–64.

34. Hershman DL, Lacchetti C, Dworkin RH, et al. American Society of Clinical Oncology. Prevention and management of chemotherapy-induced peripheral neuropathy in survivors of adult cancers: American Society of Clinical Oncology clinical practice guideline. *J Clin Oncol.* 2014;32(18):1941–1967.

35. Loprinzi CL, Lacchetti C, Bleeker J, et al. Prevention and management of chemotherapy-induced peripheral neuropathy in survivors of adult cancers: ASCO guideline update. *J Clin Oncol.* 2020;38(28):3325–3348.

36. Smith EM, Pang H, Cirrincione C, et al, Alliance for Clinical Trials in Oncology. Effect of duloxetine on pain, function, and quality of life among patients with chemotherapy-induced painful peripheral neuropathy: a randomized clinical trial. *JAMA.* 2013;309(13):1359–1367.

Neurological complications of immunotherapy and monoclonal antibody therapy

Alberto Picca[a,b] *and Dimitri Psimaras*[a,b]

[a]Service de Neurologie 2—Mazarin, Neurology Department, Pitié-Salpêtrière Hospital, APHP, Paris, France,
[b]OncoNeuroTox Group, Center for Patients with Neurological Complications of Oncologic Treatments, Pitié-Salpetrière Hospital, Paris, France

1 Introduction

It is long recognized that the immune system plays an important role in preventing the development and growth of tumors, and in the last 10 years, the use of immunotherapies has evolved enormously in cancer patients. The introduction of immune checkpoint inhibitors represented a breakthrough in the management of several solid tumors,[1] similar to chimeric antigen receptor T (CAR T) cells in refractory or relapsing B-cell malignancies.[2] At the same time, the use of other immunomodulatory agents such as interleukins and interferons reduced progressively.

Monoclonal antibodies (mAbs) are engineered antibodies that may exert an antineoplastic effect in several different ways.[3] Some of them may bind and subsequently block transmembrane receptors implied in pro-growth signaling pathways in tumor cells, such as the antiepidermal growth factor receptor (EGFR) mAb cetuximab[4] and the antihuman epidermal growth factor receptor 2 (HER2) mAb trastuzumab,[5] while others counteract pro-angiogenic signals in the tumor microenvironment, including the antivascular endothelial growth factor (VEGF) mAb bevacizumab.[6] They may otherwise cause the activation of a host immune response against tumor cells via the complement-dependent cytotoxicity (CDC), antibody-dependent cellular phagocytosis (ADCP), and antibody-dependent cellular cytotoxicity (ADCC), as it has been proposed for the anti-CD20 mAb rituximab.[3] Those mechanisms are not mutually exclusive, and it is likely that more than one mechanism takes place to a different extent in explaining the antineoplastic effects of each compound. More recently, mAbs have been used as vehicles to deliver effector molecules, as cytotoxic compounds or radionuclides that are specifically released in target cells after antigen-specific cell internalization.[7] The

first antibody-drug conjugate (ADC) approved by FDA was brentuximab vedotin, composed of an anti-CD30 mAb coupled to the spindle poison monomethyl auristatin E (MMSE). Other recent developments include engineered bispecific mAbs that concurrently target a tumor antigen and an activating receptor on effector cells (e.g., CD3 on T cells), resulting in a strong immune reaction against tumor cells.[8] They are usually indicated with the term of Bispecific T cell Engagers (BiTEs); the first receiving FDA approval was the anti-CD19, anti-CD3 blinatumomab, for acute lymphoblastic leukemia.[9] Monoclonal antibodies may also exert an antineoplastic effect without directly targeting tumor cells or the tumor microenvironment. The anti-cytotoxic T-lymphocyte-associated protein 4 (CTLA4) and anti-programmed cell death protein 1 (PD1)/programmed death-ligand 1 (PDL1) checkpoint inhibitors bind to host immune cells, enhancing the immunologic response against the tumor.[1]

Monoclonal antibodies may cause neurological toxicities according to different mechanisms, including on-target vascular effects (e.g., cerebrovascular events and reversible posterior leukoencephalopathy syndrome in bevacizumab), off-target release of neurotoxic molecules (e.g., MMSE in brentuximab vedotin), induction of an immunodepressed state (e.g., anti-CD20 mAbs), or other poorly characterized mechanisms, such as cetuximab-associated aseptic meningitis or blinatumomab associated-encephalopathy.

In the following paragraphs, we will discuss the most relevant neurological side effects of monoclonal antibodies that target tumor cells or the tumor microenvironment (summarized in Table 1). Immune checkpoint inhibitors may also induce immune-related neurological adverse events[10]; they are further discussed in Chapter 32. The main neurological toxicities of cytokines used in cancer treatment (interleukin-2 and interferon alpha) are also briefly reviewed.

TABLE 1   Most relevant neurological toxicities associated with monoclonal antibodies targeting tumor cells.

Compound	Target	Relevant neurological toxicities
Bevacizumab	VEGF	Cerebrovascular events RPLS Optic neuropathy (?)
Cetuximab	EGFR	Aseptic meningitis Symptomatic hypomagnesemia RPLS
Panitumumab	EGFR	Symptomatic hypomagnesemia RPLS
Trastuzumab	HER2	RPLS
Rituximab	CD20	PML RPLS
Obinutuzumab	CD20	PML
Ofatumumab	CD20	PML
Dinutuximab	GD2	Pain
Antibody-drug conjugates		
Brentuximab vedotin	CD30	Peripheral neuropathy
Polatuzumab vedotin	CD79b	Peripheral neuropathy
Enfortumab vedotin	Nectin-4	Peripheral neuropathy
Trastuzumab emtansine	HER2	Peripheral neuropathy
Bi-specific T-cell engagers		
Blinatumomab	CD19, CD3	Encephalopathy Seizures Focal symptoms Tremor Headache

PML, progressive multifocal leukoencephalopathy; *RPLS*, reversible posterior leukoencephalopathy syndrome.

2 Bevacizumab

Bevacizumab is an antiangiogenic drug consisting of a recombinant humanized monoclonal antibody that targets the vascular endothelial growth factor (VEGF).[11] It is currently FDA approved for the treatment of metastatic colorectal cancer (mCRC), advanced nonsmall cell lung cancer, cervical cancer, metastatic renal cell carcinoma, and progressive glioblastoma.[12]

2.1 Bevacizumab and cerebrovascular events

Bevacizumab is known to increase the risk of vascular ischemic and hemorrhagic events,[13,14] including cerebrovascular events. In a metaanalysis of 12,917 patients from 17 randomized clinical trials, Zuo and colleagues reported a 3.28 risk ratio (RR) for any cerebrovascular events in bevacizumab-treated patients compared to control arms.[15] The RR for central nervous system (CNS) ischemic events was 3.22, and 3.09 for CNS bleeding. The risk of cerebrovascular events in bevacizumab-treated patients appeared to be dose-dependent. Nevertheless, the overall incidence of CNS ischemic events in bevacizumab-treated patients was 0.5%, and CNS hemorrhagic stroke incidence was 0.3%.[15] In a single-center retrospective analysis from the Memorial Sloan-Kettering Cancer Center, Khasraw and colleagues reported a similar incidence of CNS bleeding in patients with primary or secondary CNS tumors treated with or without bevacizumab, when stratified for similar histologies.[16] Bevacizumab may also increase the risk of venous thromboembolism,[17] including cerebral venous thrombosis.

Multiple underlying mechanisms are involved in these vascular processes. VEGF acts as a protective vascular factor; bevacizumab may induce endothelial cell damage, increase pro-inflammatory cytokine levels, and reduce nitric oxide production, leading to increased blood clot formation and vessel wall fragility. Additionally, bevacizumab induces arterial hypertension in a relevant proportion of patients.[18] Pathology-induced or therapy-induced factors, such as thrombocytopenia, curative anticoagulation, cancer-induced hypercoagulability states, and radiation-induced vessel damage, may further play a role.

Presenting symptoms do not differ from those of cerebrovascular events from other causes; ischemic stroke typically manifests with abrupt focal deficits (e.g., hemiparesis, sensory deficits, aphasia, or dysarthria), while intracranial hemorrhage usually presents with headaches, focal deficits, decreased consciousness, and rarely seizures. Median time from treatment initiation to cerebrovascular events in two retrospective series was 3–4 months, but ranged widely from one to 50 weeks.[19,20] They appear to be serious events; only around a half of patients recover their baseline neurological status; 26%–60% have been reported to expire within 3 months.[19,20] Management consists primarily in supportive treatments. Few data are available about the role of antiplatelet therapy in bevacizumab-induced arterial ischemic strokes. In patients with cerebral veinous thrombosis, anticoagulation with low molecular weight heparins is the preferred option.[21]

2.2 Bevacizumab and RPLS

Reversible posterior leukoencephalopathy syndrome (RPLS), also indicated as posterior reversible encephalopathy syndrome (PRES), is a rare, severe, but potentially reversible adverse class-effect of anti-VEGF compounds, including bevacizumab.[22–24] This nosological entity clinically manifests with headaches, nausea and vomiting, seizures, visual disturbances (including typically cortical blindness), confusion, and decreased consciousness.[25] Radiological correlates include white matter hyperintensities prevalently affecting the posterior temporal, parietal, and occipital lobes (Fig. 1), but may also be present in the anterior hemispheres and posterior fossa structures. Contrast enhancement may be present in a subset of cases. The condition is caused by a disruption of the blood-brain barrier and failure of the autoregulatory mechanisms of the cerebral vasculature, resulting in vasogenic edema.[26]

RPLS onset has been reported after a median of 10 weeks of treatment with bevacizumab[24]; symptoms do not differ from those of RPLS from other causes. Proposed possible risk factors are female sex, history of hypertension, and the development of proteinuria and/ or uncontrolled hypertension while under bevacizumab treatment.[24] Almost all reported patients had elevated blood pressure values at RPLS diagnosis.[24,27] A causative role of bevacizumab may be difficult to discern when the compound is administered concomitantly with other chemotherapies associated with the development of RPLS, such as platinum compounds.

Bevacizumab should be immediately discontinued in patients suspected of developing RPLS; treatment includes essentially systemic blood pressure control with antihypertensive drugs. Antiepileptic drugs should be administered in patients who experience seizures. As for RPLS from other causes, patient outcome is usually favorable, with symptom resolution after a median of 9 days[24]; nevertheless, permanent sequelae may be present in patients developing ischemic or hemorrhagic events. Successful reintroduction of bevacizumab without any recurrence of RPLS has been reported,[28] and it may be considered in patients with limited therapeutic options, under the condition of complete resolution of the first event and adequate control of systemic blood pressure.

2.3 Bevacizumab and optic neuropathy

Sherman et al. raised concern in 2009 about the possible induction of optic neuropathy in six patients treated with bevacizumab for glioblastoma.[29] The patients developed unilateral ($n = 3$) or bilateral ($n = 3$) optic

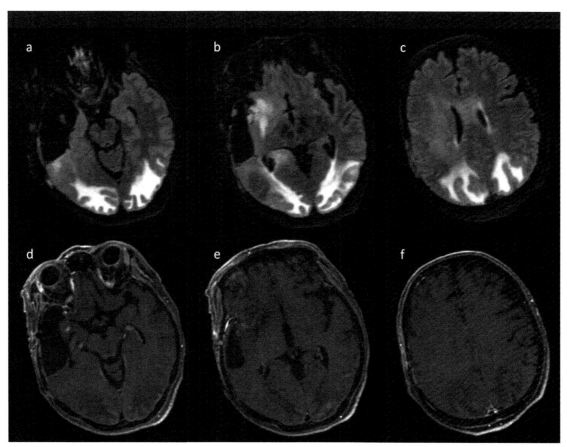

FIG. 1 Reversible posterior leukoencephalopathy syndrome in a 54-year old woman with a right temporal resected glioblastoma under treatment with bevacizumab. (A–C) T2-weighted Fluid Attenuated Inversion Recovery (FLAIR) sequence shows bilateral white matter hyperintensity prevailing in occipital and posterior parietal lobes, corresponding to vasogenic edema. (D–F) T1-weighted sequence after gadolinium injection shows faint cortical enhancement of the posterior regions. Right temporo-insular FLAIR hyperintensity (A–C) and postgadolinium enhancement (D–F) correspond to the area infiltrated by the underlying glioma.

neuropathy after a median of 7.5 doses of bevacizumab. All patients received previous irradiation involving optic nerve structures. Histopathologic findings in one case showed postradiation changes. The authors hypothesized a priming effect of radiation for optic nerve injury. Nevertheless, data have never been replicated since this first report, and both intravitreal and systemic bevacizumab have shown promising effects in treating radiation-induced optic neuropathies.[30–32]

3 Cetuximab and other anti-EGFR mAbs

Cetuximab is a chimeric monoclonal antibody directed against the epidermal growth factor receptor (EGFR).[4] It is currently approved by the FDA for the treatment of recurrent or advanced squamous cell carcinoma of the head and neck, as well as for Ras wild-type, EGFR-expressing, mCRC.[33]

3.1 Cetuximab and aseptic meningitis

Aseptic meningitis associated with cetuximab was first described in a patient undergoing a phase I trial in 2000 by Baselga and colleagues.[34] Postmarketing reports confirmed this entity as a rare but insidious side effect of the treatment with cetuximab.[35–41] Aseptic meningitis typically occurs after the first administration of a loading dose of cetuximab at $400 mg/m^2$. Exceptionally, however, it may occur also after several administrations of the drug.[40] Symptoms present in the first 24 hours following the infusion (most commonly 2–12 hours later), and consist of fever, severe headache, photophobia, neck stiffness, nausea, vomiting, and, rarely, altered mental status, mimicking bacterial meningitis. Imaging (brain CT or MRI) is normal. CSF analysis discloses a marked elevation of white blood cells (usually $\geq 1000/\mu L$, up to $5000/\mu L$) with a neutrophilic predominance ($> 80\%$) and an associated increase in protein level. The diagnosis of drug induced aseptic meningitis remains a diagnosis of exclusion, and infectious causes, along with carcinomatous meningitis, should be always ruled out. An empiric antibiotic therapy is mandatory until the negative results of bacterial cultures. Symptoms typically resolve in a few days with supportive care only; rarely, symptom resolution requires up to two weeks.[35] Re-challenge with cetuximab is usually safe; reducing the infusion rate and premedicating the patient with glucocorticoids may reduce the risk of relapses.[35,42]

3.2 Cetuximab and hypomagnesemia

Hypomagnesemia is a class-effect of anti-EGFR therapies[43] that may manifest with prevailing neurological symptoms. Hypomagnesemia in patients treated with anti-EGFR compounds is the consequence of impaired renal magnesium reabsorption.[44] A prospective study demonstrated that almost all patients under treatment with EGFR-targeting antibodies have a decrease in serum magnesium concentration.[44] Retrospective monocentric series estimated a 24%–27% incidence of grade ≥ 3 hypomagnesemia in cetuximab-treated patients[45,46]; the incidence rate grows up to 47% in patients receiving cetuximab for more than 6 months.[45]

Marked hypomagnesemia may manifest with the neurological symptoms of fatigue, paresthesias, muscle cramps, confusion, and lethargy. Hyperreflexia, along with Chvostek and Trousseau signs, may be present. In more severe cases, profound muscle weakness, decreased consciousness, psychosis, tetany, and convulsions may occur.[43,47] Concurrent hypocalcemia may be present as a consequence of impaired parathyroid hormone release and activity; correction of hypomagnesemia usually reverses the secondary hypocalcemia.[46]

Due to renal wasting, high doses of parenteral magnesium sulfate are required to reverse cetuximab-induced hypomagnesemia.[44–46] Oral supplementation is ineffective and poorly tolerated because of secondary diarrhea. In grade 3–4 hypomagnesemia, intravenous supplementation (up to 8–10 g per day) is beneficial but short lasting (48–72 hours)[44–46]; several administrations per week are necessary.[43] The side effect is usually reversible after cetuximab discontinuation,[44,45] with a return to normal magnesium values after 30–90 days. Nevertheless, in a minority of cases, hypomagnesemia may last several months after the end of anti-EGFR treatment.[45]

3.3 Cetuximab and RPLS

Although anecdotal, cetuximab has also been associated with the development of typical reversible posterior leukoencephalopathy syndrome.[48,49] Outcomes have been reported as favorable.[48] Pathogenesis of cetuximab-associated RPLS remains uncharacterized, but drug-induced hypomagnesemia, an electrolyte disorder that has been associated with RPLS,[50] may play a role as a predisposing or precipitating factor.

3.4 Other anti-EGFR mAbs

Similar to cetuximab, hypomagnesemia is a typical side effect of other FDA-approved monoclonal antibodies targeting the EGF receptor, including panitumumab[51] and necitumumab.[52]

4 Trastuzumab

Trastuzumab is a monoclonal antibody targeting the human epidermal growth factor receptor 2 (HER2), approved by FDA for the treatment of HER2 overexpressing breast cancer, metastatic gastric, or gastroesophageal junction adenocarcinoma.[53]

Trastuzumab has been anecdotally associated with the development of RPLS.[54–56] Patients presented with typical signs of RPLS, including seizures and cortical blindness, and corresponding T2/FLAIR hyperintensities affecting the occipital lobes on MRI. All cases of trastuzumab-associated RPLS had elevated blood pressure; the antiangiogenic effect of trastuzumab[57] has been postulated as a possible mechanism of action.[54] In one case, trastuzumab was re-administered after the resolution of symptoms and MRI abnormalities, without further relapses.[55]

5 Rituximab and other anti-CD20 mAbs

Rituximab is a chimeric anti-CD20 monoclonal antibody currently approved by the FDA for the treatment of patients with non-Hodgkin lymphoma (NHL) or chronic lymphocytic leukemia (CLL), along with nononcological indications in rheumatoid arthritis[58] and a wide off-label use in several other autoimmune disorders. In 1997, it became the first monoclonal antibody approved for the treatment of cancer.

5.1 Rituximab and PML

Progressive multifocal leukoencephalopathy (PML) is a rare but severe brain disease caused by local reactivation of the double-stranded DNA JC polyomavirus (JCV).[59–61] JC virus is a ubiquitous virus that after a silent infection in childhood remains latent in target organs including the kidney, lymphoid structures, and bone marrow. More than 50% of the adult population is seropositive for JCV. Under conditions of immunosuppression, JC virus may reactivate and migrate to the brain where it induces death of oligodendrocytes and consequent demyelination.[59–61] AIDS has been the most frequent condition associated with the development of PML, but other situations, such as hematological malignancies, chronic autoimmune disorders, and the use of immunosuppressive treatments, are known risk factors.[59,60]

Rituximab has been associated with the development of fatal PML since 2006, warranting an FDA boxed warning.[58] The exact incidence of PML in patients with hematological malignancies treated with rituximab remains nonetheless poorly defined, because of the lack of prospective data and the presence of several confounders, including concomitant chemotherapies and the underlying cancer that are per se predisposing factors. A single-center retrospective study estimated a PML incidence of 7.8 cases per 10,000 non-HIV NHL patients exposed to rituximab compared to an incidence of 1.5

cases per 10,000 nonexposed patients, resulting in a relative risk of 5.4.[62] Even more marked, in a subsequent study from the same hospital, the relative risk was of 19.9 in non-HIV CLL patients treated with rituximab versus those nontreated (estimated incidence of 16.5 versus 0.8 per 10,000 patients, respectively).[63] A relatively higher risk in CLL compared to NHL patients treated with rituximab was also noted in data obtained from the Roche company safety database and published in 2019 by Focosi et al.[64] Patients treated with regimens including bendamustine may be at higher risk.[64] In a monocentric retrospective series of 47 patients treated with a bendamustine plus rituximab regimen (BR) for high tumor burden follicular lymphoma, D'Alò and colleagues reported three cases of fatal PML.[65] Two of them had peripheral blood immunophenotyping that showed a severe CD4$^+$ cells decrease ($\leq 200/\mu L$).[65]

CD4$^+$ lymphopenia is a well-known risk factor for PML,[66,67] and has been postulated as a mechanism of rituximab-associated PML.[64] Rituximab may indeed cause a decrease in CD4$^+$ lymphocyte counts.[68] The rate of severe CD4$^+$ lymphopenia is further increased in regimens including bendamustine,[64] possibly explaining the increased risk. Another hypothesis provides that rituximab-induced B-cell depletion causes the mobilization of hematopoietic progenitor cells infected with latent JCV, resulting in hematogenous spread to the brain.[69]

Symptoms of PML depend on the affected brain regions, and usually include cognitive deterioration and focal signs such as hemiparesis, sensory loss, visual field deficits, aphasia, or ataxia.[59–61] Presentation is typically subacute in onset, over several days or weeks. The diagnosis of PML should always be evoked in rituximab-treated patients developing new neurological signs or symptoms. The MRI features are those of T2-hyperintense white matter lesions, usually located in the proximity of the white-gray matter junction, with little or no mass effect, and absent or faint contrast enhancement.[59,61] The brainstem and cerebellum may be less frequently involved, while optic nerves and the spinal cord are typically spared. Definitive diagnosis requires the demonstration of JC virus in the CSF or histopathological specimens from patients with compatible clinical and imaging features.[61]

In the largest case series published to date, median time to PML diagnosis was 16 months from the first rituximab dose[69]; outcomes are usually poor. The majority of patients in Carson and colleagues' series died of PML after a median time of 2 months from PML diagnosis.[69] No effective direct treatment against JCV is currently available. Rituximab should be immediately discontinued once the suspicion of PML is confirmed. Similar to management of natalizumab-related PML, plasma exchange may be used to reduce the rituximab half-life and accelerate the restoration of immunocompetence in patients that received the last dose less than 2–3 months before.[70]

When occurring, subsequent immune restoration often results in local inflammation causing symptom exacerbation, a condition named IRIS (immune reconstitution inflammatory syndrome).[59] MRI may disclose an increase in lesion size and the appearance of contrast enhancement. Although a marker of an ongoing immune response against the JCV, IRIS may cause a severe degradation of the patient's neurological condition, necessitating a short course of high-dose glucocorticoids (i.e., 1 g of methylprednisolone per day for 3–5 days).

5.2 Rituximab and RPLS

Rituximab exposure has been sporadically associated with the development of RPLS in both onco-hematological and rheumatological patients.[71,72] A direct causative role of rituximab is difficult to ascertain, due to several other potential factors (concomitant chemotherapies, hypertension, end-stage renal failure). Nevertheless, at least in a subset of cases, the temporal relationship strongly points toward rituximab as the precipitating element.[72,73] The presence of classical neurological symptoms (cortical blindness, headache, seizure) and a typical radiological pattern warranted the diagnosis. Management did not differ from that of RPLS from other causes, including an aggressive correction of any accompanying hypertension, secondary antiepileptic prophylaxis, and supportive care. The outcome is usually favorable, with a complete clinical and radiological remission.[71]

5.3 Other anti-CD20 mAbs

Obinutuzumab is a humanized anti-CD20 monoclonal antibody with FDA approbation in CLL and follicular lymphoma.[74] Ofatumumab is a human CD20-directed cytolytic monoclonal antibody, currently FDA-approved for the treatment of CLL.[75] Both compounds have been associated with the development of progressive multifocal leukoencephalopathy,[76–78] and their FDA labels include a boxed warning about the PML risk.[74,75] However, the experience with these agents is still limited and further knowledge is needed to fully estimate the associated risk of JCV reactivation.

6 Dinutuximab

Dinutuximab is a chimeric human-mouse monoclonal antibody directed against the GD2 ganglioside.[79,80] It is currently FDA-approved for the treatment of high-risk neuroblastoma, in combination with GM-CSF and IL2.[81] The drug exerts its antineoplastic effect by inducing antibody-dependent cytotoxicity in neuroblastoma cells

that overexpress the target ganglioside.[79,80] Nevertheless, GD2 is also normally expressed by healthy cells of neuroectodermal origin, including neurons and peripheral nerve fibers, explaining the frequent neurological side effects seen with dinutuximab treatment.

The most relevant adverse event of dinutuximab administration is neuropathic pain without described neuropathy.[82–84] Pain occurred in 85% of patients treated with dinutuximab in a pivotal Children's Oncology Group (COG ANBL0032) trial, being severe (grade 3–4) in 52%.[82] Pain has been linked to antibody-mediated local activation of the complement cascade on peripheral nerve fibers and subsequent inflammation.[85] It manifests as spontaneous abdominal pain and/or allodynia.[85] Pain reactions are usually in close temporal relation with drug infusions, and their rate decrease with repeated cycles, being maximal at the first infusion.[82,84] In order to reduce infusion reactions, including pain, infusion time has been extended from 5 to 10 hours in recent studies.[83,84] The management recommended by FDA also requires the administration of intravenous opioids prior to, during, and for 2 hours after the completion of drug administration.[81]

7 Brentuximab vedotin

Brentuximab vedotin (BV) is an antibody-drug conjugate (ADC) in which a chimeric anti-CD30 monoclonal antibody is associated with the spindle poison monomethyl auristatin E (MMAE). Its mechanism of action relies on its internalization in CD30-expressing cells and the subsequent release of the antimicrotubule agent MMAE that inhibits microtubule polymerization, thus blocking the progression of the cell cycle in target cells.[86] It is currently FDA-approved for the treatment of classic Hodgkin lymphoma, systemic anaplastic large cell lymphoma, primary cutaneous anaplastic large cell lymphoma, or other CD30-expressing peripheral T-cell lymphomas.[87]

7.1 Brentuximab vedotin and peripheral neuropathy

Peripheral neuropathy (PN) is the main side effect of BV. In pivotal phase III trials, the incidence of treatment-emergent PN was 67%, with grade 3 events in 9%–13% of patients, significantly higher compared to the control arms.[88,89] Peripheral neuropathy was the most frequent adverse event causing treatment discontinuation, in 14%–23% of cases.[88,89] Nevertheless, PN improved or resolved in most patients (82%–85%) after treatment discontinuation or completion.

Median time from the start of treatment to PN onset is around 15 weeks,[88,90] and the adverse event is dose-dependent.[90] Peripheral neuropathy is most commonly of sensory type, but motor neuropathies have been reported in up to 23% of patients.[88] Most common symptoms include distal numbness, paresthesia, neuropathic pain, and proprioceptive ataxia.[90,91] When present, motor symptoms usually consist in mild distal limb weakness. Neurological examination discloses loss of deep tendon reflexes and impaired vibratory sensation. Neurophysiological studies typically demonstrate a selective or predominant length-dependent axonal injury,[90,91] consistent with the proposed mechanism of disruption of the axonal cytoskeleton.[92] The recommended management for grade 2–3 PN is to withhold BV treatment until improvement to grade ≤1, and then restart at a lower dose; treatment discontinuation is warranted for grade 4 PN.[87]

Moreover, it has been recently reported that BV may also, although less frequently, induce inflammatory polyradiculoneuropathies fulfilling the criteria for Guillain-Barré syndrome or chronic inflammatory demyelinating polyradiculoneuropathy (CIDP).[93] The presenting symptom is frequently represented by distal sensory impairment, but patients may often develop proprioceptive ataxia and mild-to-severe predominantly distal weakness. In the reported series, at the worst of symptoms intensity, three-quarters of the patients necessitated a support for walking.[93] Nerve conduction studies are useful in demonstrating demyelinating features, in a sural-sparing pattern. Due to symptom severity, most patients developing inflammatory polyradiculoneuropathies require BV discontinuation. Intravenous immunoglobulins at 2 g/kg over 3–5 days appeared beneficial, although patients are typically left with mild-to-moderate residual deficits.[93]

7.2 Brentuximab vedotin and PML

In 2012, FDA issued a warning after the first reports of JC virus reactivation in patients under treatment with BV.[94,95] A subsequent series described the main features of five cases of PML associated with BV therapy.[96] Three patients received BV because of relapsing Hodgkin's lymphoma, while two had an underlying cutaneous lymphoma. All of the patients received previous lines of chemotherapy, and the three cases of HL also underwent autologous stem cell transplantation (ASCT). Differing from cases of PML associated with other monoclonal antibodies (e.g., rituximab or natalizumab), time to onset was less than 3 months from first BV administration in most cases. Presenting symptoms were confusion, gait impairment, and focal neurological deficits. Most patients rapidly worsened to death. Only one case of immune reconstitution inflammatory syndrome (IRIS) had a positive evolution.[96] Although in most reported cases, the temporal relationship was in favor of a causative

role of BV in JCV reactivation, previously administered chemotherapy regimens and ASCT are known potential causes of PML as well. Underlying lymphomas are per se risk factors for the development of PML.

All patients with newly presenting CNS symptoms such as focal symptoms or altered mental status under treatment with BV should be evaluated for PML. MRI may reveal white matter lesions.[59,61] The diagnosis is confirmed with the detection of JC virus in the CSF or histopathological specimens.[61]

8 Polatuzumab vedotin

Polatuzumab vedotin (PoV) is an antibody-drug conjugate consisting of a humanized anti-CD79b monoclonal antibody bound to the antimicrotubule compound MMAE.[97] It recently received accelerated approval by the FDA in association with rituximab and bendamustine for the treatment of relapsed or refractory (r/r) diffuse large B-cell lymphoma (DLBCL) after at least two prior therapies[98]; it is currently under investigation for the treatment of other r/r non-Hodgkin lymphomas.[97]

Similarly, to the anti-CD30 ADC brentuximab vedotin, the main dose-limiting toxicity of PoV is peripheral neuropathy. Peripheral neuropathy is usually sensory and mild to moderate (grade 1–2) in severity, but motor or combined PN has also been seen.[99–101] The hypothesized mechanism is the off-target toxicity of MMAE on nerve axons.

In the phase II ROMULUS trial evaluating PoV (at the dose of 2.4 mg/kg every 3 weeks) plus rituximab in r/r DLBCL or follicular lymphoma (FL), the rate of all grade PN was 36% in the DLBCL group and 65% in the FL group.[100] Nevertheless, incidence data should be interpreted with caution, as patients previously received neurotoxic compounds (e.g., vinca alkaloids) and about one third of cases had grade 1 PN at study entry. Three cases (8%) of grade 3 sensory PN and one case (3%) of grade 3 motor PN were reported in the DLBCL group, and none in the FL group. Peripheral neuropathy was the reason for treatment discontinuation in 18% of PoV-rituximab treated patients in the DLBCL group and in 55% in the FL group.[100] A subsequent analysis of data from phase I and II trials led to the reduction of PoV dose to 1.8 mg/kg in order to reduce the incidence of grade ≥2 PN.[102] In the GO29365 phase Ib/II trial, the reported incidence of all-grades PN in the PoV (at 1.8 mg/kg every 3 weeks) plus bendamustine-rituximab arm was 40%.[101] Median time to onset was 2 months. Most cases were grade 1–2 toxicities, and only four (2.3%) patients suffered from grade 3 PN.[101]

The recommended management of PoV-induced grade 2–3 peripheral neuropathy is to withhold treatment until improvement to grade 1 or less and to restart at a reduced dose (1.4 mg/kg). In cases of grade 4 PN, if there is no improvement to grade 1 or less within 14 days, or down to grade 2–3 PN during treatment at a reduced dose, treatment discontinuation is recommended.[98]

9 Enfortumab vedotin

Enfortumab vedotin (EV) is a fully humanized antibody-drug conjugate directed against the adhesion molecule Nectin-4 that delivers to targeted cells the antimicrotubule compound MMAE.[103] It received accelerated approval in 2019 for the treatment of locally advanced or metastatic urothelial cancer in patients previously treated with an anti-PD1/PDL1 and a platinum-containing chemotherapy.[104]

As for other MMAE-conjugated mAbs, peripheral neuropathy is a frequent side effect of EV, mostly mild to moderate in intensity and predominantly sensory. In the phase I EV-101 trial,[105] PN incidence increased with increased EV doses. In patients treated with the target 1.25 mg/kg dose, peripheral sensory neuropathy occurred in 38% of cases, but was mainly mild in severity; only one patient (1%) experienced grade 3 sensory PN. Nevertheless, PN was the main treatment-related adverse event leading to EV discontinuation, in 3% of exposed patients.[105]

In the phase 2 EV-201 trial,[106] all-grades treatment-related PN affected 50% of patients; the sensory form was more frequent (44%) than the motor form (14%). Two patients (2%) suffered from grade ≥3 sensory PN. Again, sensory PN was identified as the main toxicity leading to dose reduction (in 9% of patients) or drug discontinuation (6%). Median time to onset for any grade PN was 2.43 months (3.8 months for grade ≥2).[104,106] Of note, the trial included patients previously treated with platinum-based compounds; cases with ongoing grade ≥2 PN were excluded from the study. Forty-two patients were reported with PN at enrolment; 52% of this subgroup presented a worsening of their PN. In the phase 3 EV-301 trial, treatment-related PN was reported in 46.3% of patients in the EV arm (versus 30.6% in the chemotherapy group). Grade 3 neuropathy was seen in 3.7% of patients treated with EV, versus 2.4% of patients receiving chemotherapy. Most cases were sensory, and motor neuropathy was seen in 22 of 296 patients in the EV arm (7.4%), being grade 3 in five (1.7%). Again, sensory PN was the main adverse event resulting in dose reduction, interruption, of discontinuation of the treatment (in 7.1%, 15.5%, and 2.4% of patients, respectively).[107]

Current FDA recommendations include the temporary interruption of EV administration for grade 2 PN, and resumption after improvement to at least grade 1 (at the same dose if first occurrence, at a reduced dose if recurring PN). Treatment discontinuation is recommended for grade ≥3 PN.[104] In the EV-201 trial, 76% of patients with PN improved to grade 1 or less at last follow-up (reported median time to improvement 1.18 months).[106]

10 Trastuzumab emtansine

Trastuzumab emtansine (also known as T-DM1) is an ADC consisting of a HER2-targeted monoclonal antibody (trastuzumab) bound to the antimicrotubule agent mertansine (also known as DM1). It is currently FDA approved for the treatment of patients with HER2-positive breast cancer.[108]

The most frequently encountered neurological side effects are nonspecific and mild, including headache and myalgias. Headache has been reported in 23%–32% of patients treated with T-DM1,[109–111] but usually of grade 1–2. The incidence of grade 3–4 headache has been less than 2%–3% in landmark phase III trials of T-DM1.[109,111–113] Relatively higher rates of all-grades (28% versus 21%) and grade 3 (2.5% versus 0.7%) headache have been reported in patients with known brain metastases at baseline.[110] Similarly, myalgias occur in 10%–18% of patients treated with T-DM1 but are usually not clinically relevant.

Peripheral neuropathy due to axonal damage is an expected toxicity of the antimicrotubule compound DM1.[114] Indeed, peripheral neuropathy has been reported in 5%–32% of patients treated with trastuzumab entamsine,[109,111,112,115] with grade 3–4 events in 1.6%–2.2% of cases.[112,115] FDA label recommends to temporarily discontinue the drug in patients with grade ≥3 PN, until improvement to at least grade 2.[108]

11 Blinatumomab

Blinatumomab is a bi-specific T cell engager (BiTE) antibody that concurrently binds the CD3 antigen expressed by cytotoxic T cells and the CD19 antigen expressed by B lineage cells, inducing the T cell mediated killing of CD19 expressing cells.[116] It is currently approved by FDA for the treatment of r/r B-cell precursor acute lymphoblastic leukemia (ALL) and underwent accelerated approval for B-cell precursor ALL in first or second complete remission with minimal residual disease ≥0.1%[117]; it is under investigation for the treatment of other CD19-expressing malignancies, such as r/r B-cell NHL.

In patients treated for ALL, neurological toxicity is the main dose limiting adverse event,[116] and warranted a boxed warning in the blinatumomab FDA package insert.[117] Flu-like symptoms including headache (in 29%–47%) and fatigue (in up to 50% of cases) are common, but are likely due to the expression of a mild cytokine release syndrome rather than direct CNS involvement. They are usually of grade 1–2.[9,118,119] Tremor is another frequently reported grade 1–2 adverse event, in 10%–36% of patients.[9,118,119]

The reported incidence of serious (grade ≥3) neurological adverse events was 13%–22% in phase II trials[9,118,120] and 9.4% (25/267) in the phase III TOWER trial.[119] They are typically represented by encephalopathic symptoms (confusion, decreased consciousness, ataxia, speech disorders, tremor) and/or seizures.[120] One case of fatal encephalopathy has been reported in a patient with a history of CNS involvement from his ALL and intense preblinatumomab intrathecal treatment.[120]

Of note, patients with active CNS leukemia or clinically relevant CNS pathologies have been excluded from trials, due to the fear of an increased risk of neurological toxicities; the safety of blinatumomab treatment in this population is yet to be defined.

Most neurological adverse events (both all-grade and grade ≥ 3) occur after the first cycle; their incidence decreases in subsequent cycles.[120] Median time to onset from blinatumomab initiation is 9 days for all-grade events, and 16.5 days for grade ≥ 3 events.[120] The possible risk factors for developing neurological adverse events during blinatumomab treatment remain poorly defined; patients at higher risk may be those ≥ 65 years of age, with preexisting neurological co-morbidities, who have received more than two prior salvage therapies, and of ethnicity other than white.[120]

Recommended management of grade 3 neurological adverse events from blinatumomab is to discontinue drug administration until the event resolves to ≤ grade 1, and in any case for not less than 3 days, then restart with blinatumomab infusion at the reduced dose of 9 μg/day; if the event does not resolve to at least grade 1 within 7 days, or occurs at the 9 μg/day dose, or in case of grade 4 events, treatment discontinuation is mandated.[117] Dexamethasone administration (at least 8 mg three times daily with subsequent scaling over the next 4 days) should be also administered in patients with serious neurological adverse events.[120] Patients experiencing seizures should start secondary antiepileptic prophylaxis. Recurring neurological events seem to respond well to treatment interruption and corticosteroid administration.[120] Nevertheless, blinatumomab discontinuation is recommended for patients experiencing more than one seizure.[117]

Neurological adverse events were the dose limiting toxicities in blinatumomab phase I and II trials for r/r NHL.[121,122] All-grade neurological adverse events were seen in about 70% of treated patients, and consisted mostly of grade 1–2 tremor, headache, and dizziness.[121,122] Grade 3 events occurred in 22% of patients, and included encephalopathy, confusion, somnolence, speech disorders (aphasia, dysarthria), ataxia, and seizures.[121,122] No grade 4 or 5 neurological events were seen. Most cases occurred within 2 days from treatment onset or dosage increase.[122] Treatment interruption was recommended per protocol in all patients with grade 3 toxicity; permanent discontinuation was indicated if no resolution occurred within 14 days or with grade 4 toxicities. Neurological toxicities were the most common reason for treatment discontinuation in 17%–22% of patients. Nevertheless, almost all neurological events

resolved after blinatumomab interruption or discontinuation.[121,122] To reduce the incidence of neurological side effects, a stepwise dose escalation and early dexamethasone administration before the first dose and every dose step have been recommended.

Long-term monitoring of patients included in the phase I MT103-104 trial showed no neurological abnormalities attributable to blinatumomab in surviving patients; no major cognitive impairment was identified in nine cases evaluated after a median of 5 years (range 4.2–7.2 years) since the completion of blinatumomab therapy, including four patients who previously experienced neurological adverse events.[123]

Preclinical experiments showed that blinatumomab increases T cells adhesiveness to blood vessel endothelium, with consequent T lymphocytes redistribution to endothelial activation.[124] Klinger and colleagues proposed a model in which T cell adhesion to brain endothelium is followed by T lymphocytes transmigration in perivascular spaces, even in the presence of an intact blood-brain barrier. Consistently, CD3+/CD8+ T cells have been found in perivascular and leptomeningeal spaces at autopsy in a patient with r/r ALL treated with blinatumomab and recurring neurological adverse events.[124] In the proposed model, extravasated T lymphocytes are then activated by the encounter with rare B cells in the CNS and in the presence of blinatumomab; this initiates a local inflammatory cascade that is further maintained by the recruitment of other immune cells (e.g., monocytes), local production of pro-inflammatory cytokines, and the secondary disruption of blood-brain barrier. A secondary hypothesis suggests an impaired cerebral microcirculation and consequent hypoxia due to the massive leukocyte adhesion to brain endothelium as a cause of transient blinatumomab-induced neurological symptoms.[124] Antiadhesive compounds have been tested in vitro as potential preventive treatments to reduce blinatumomab neurological toxicity.[124] The P-selectin inhibitor pentosan polysulfate (PPS) appears to reduce T cell redistribution after blinatumomab infusion.[124] In a phase I trial of blinatumomab for r/r NHL, three patients considered at high risk for neurological adverse events received concomitant PPS at treatment initiation and dosage increases, resulting in no dose discontinuations.[122]

12 Interleukin-2 (IL2)

Aldesleukin is a recombinant IL2-derived product with antineoplastic and immunomodulatory properties, currently approved for the treatment of metastatic melanoma and metastatic renal cell carcinoma.[125] The theoretical mechanism is an enhanced antitumoral activity of cytotoxic T lymphocytes and natural killer cells. IL2 therapy has significant neurotoxicity, possibly due to the disruption of the blood-brain barrier that may be severely disabling and treatment-limiting.

12.1 IL2 and acute/subacute encephalopathy

This complication occurs in 30%–50% of patients who receive IL2 in an IV bolus, usually after 5–6 days of treatment.[126] The incidence is reduced by a continuous infusion of IL2. The toxicity appears to be dose dependent. The clinical picture is characterized by behavioral changes (agitation, irritability, aggressiveness, depression, hallucinations, rare paranoid delusions), cognitive changes (disorientation, memory loss), and somnolence. Sometimes, seizures occur and lethargy may progress to coma.[127–130] Symptoms usually resolve over a few days after discontinuation of treatment. Neuroleptics are useful in case of aggressive behavior or delusions.[130] One fatal case has been reported, with autopsy findings suggestive of acute disseminated encephalomyelitis.[131]

Interleukin-2 treatment may also induce RPLS with typical clinical (i.e., seizures, cortical blindness, decreased consciousness) and radiological (i.e., predominant involvement of occipital lobes and posterior fossa structures) patterns.[132] Outcome is usually favorable after drug discontinuation.

12.2 IL2 and peripheral neuropathies

A few cases of brachial plexopathy, sometimes bilateral, have been reported. These cases completely resolved over several weeks.[133] Nerve entrapment syndromes, and particularly carpal tunnel syndrome, are a classic complication of IL2; it is thought to be a direct consequence of interstitial edema compressing the affected nerve.[134–136]

13 Interferon alpha

Interferons are cytokines that have antiviral, antineoplastic, and immunomodulatory effects.[137] Interferon alpha (IFN-alpha) is used in oncology to treat hematologic malignancies (hairy cell leukemia, follicular lymphoma), malignant melanoma, and AIDS-related Kaposi's sarcoma.

The neurotoxic effects of systemically administered interferon alpha can be acute, subacute, or chronic. They occur in about a third of treated patients with severe complications occurring in less than 10%.[138] The most frequent neurological side effect is an influenza-like syndrome (headache, myalgia, fever, fatigue, arthralgia, malaise, and rarely seizures) that is transient (4–8 hours) and rarely treatment limiting.[139] The syndrome is usually worse after the first administrations of the drug, and

tends to improve during the course of the treatment. To improve tolerance, a premedication with paracetamol or nonsteroidal inflammatory drugs is helpful. Tremor is another frequent side effect of IFN-alpha treatment.[140] Beta-blockers may be useful in these cases. Seizures occur in 1%–4% of patients[137] and have been reported following both high doses[141] and standard doses of interferon alpha.[142] Seizures resolve once interferon administration is stopped and, even without chronic anticonvulsant therapy, they do not recur. If the treatment with interferon needs to be continued after seizures, the use of an anticonvulsant drug may be required. Noncontrolled epilepsy and a past history of epilepsy are contraindications for the administration of interferon alpha-2b and interferon alpha-2a.

13.1 Interferon alpha and subacute encephalopathy

Symptoms typically appear between the second and fourth week of treatment and are characterized by fatigue, behavioral (lethargy, affective disorders, psychosis) and cognitive changes (disorientation, impaired memory, speech difficulties, attention deficits, and deficits in verbal abstraction and visuographic skills).[143–145] Focal neurological symptoms are rare. Progression of severe encephalopathy to coma may occur. EEG shows diffuse slowing, but these abnormalities do not always correlate with the clinical status of the patient. No evocative radiological pattern has been described, and the CSF is usually normal. The appearance of subacute CNS toxicity is largely dose-dependent.[143,146] Almost all patients develop severe encephalopathy with high doses of IFN-alpha, while the incidence is reduced (4%–20%) with the common dose of 3 MIU three times per week.[143,147] The symptoms improve within 2–3 weeks after dose reduction or discontinuation of IFN treatment. However, some neurotoxic side effects may persist long after treatment is discontinued, and a few patients may get worse despite discontinuation of IFN treatment, resulting in chronic encephalopathy.

13.2 Interferon alpha and chronic neuropsychiatric manifestations

The incidence of chronic complications is unknown, but was estimated by Meyers et al. at 14/1300 at 3-year posttreatment.[148] Psychiatric symptoms, mostly depression, are seen in 10%–40% of patients treated with IFN-alpha and may be a cause for treatment discontinuation.[149] Severe psychiatric manifestations, permanent dementia, or even a vegetative state may result.[148,150] EEG shows diffuse nonspecific slow activity. Brain CT or MRI reveals cerebral atrophy in about half of cases, while CSF is reported to be normal.[148,150] Predisposing factors for severe neuropsychiatric side effects include older age, higher IFN doses, and intense treatment schedules.

13.3 Interferon alpha and peripheral neurotoxicities

Peripheral nervous system side effects are rare and poorly characterized adverse events of IFN-alpha treatment.[151] Reported cases include axonal polyneuropathy,[152] chronic inflammatory demyelinating polyneuropathy,[153,154] cranial neuropathies,[155] myasthenia gravis,[156] and polymyositis.[157–159] Nevertheless, a direct causative role is difficult to ascertain, and IFN-alpha has also been successfully used to treat immune-mediated neuropathies.[160] It is unclear whether IFN alpha can induce myasthenia or simply reveals or aggravates a preexisting subclinical disease. However, IFN therapy should be avoided in patients with myasthenia gravis, as symptoms may deteriorate despite the cessation of IFN therapy. In chronic hepatitis patients with cryoglobulinemia, IFN-alpha may exacerbate an underlying neuropathy, and should be thus avoided.[151,161]

References

1. Ribas A, Wolchok JD. Cancer immunotherapy using checkpoint blockade. *Science*. 2018;359(6382):1350–1355. https://doi.org/10.1126/science.aar4060.
2. June CH, Sadelain M. Chimeric antigen receptor therapy. *N Engl J Med*. 2018;379(1):64–73. https://doi.org/10.1056/NEJMra1706169.
3. Zahavi D, Weiner L. Monoclonal antibodies in cancer therapy. *Antibodies*. 2020;9:3. https://doi.org/10.3390/antib9030034.
4. Kirkpatrick P, Graham J, Muhsin M. Cetuximab. *Nat Rev Drug Discov*. 2004;3(7):549–550. https://doi.org/10.1038/nrd1445.
5. Spector NL, Blackwell KL. Understanding the mechanisms behind trastuzumab therapy for human epidermal growth factor receptor 2–positive breast cancer. *J Clin Oncol*. 2009;27(34):5838–5847. https://doi.org/10.1200/JCO.2009.22.1507.
6. Garcia J, Hurwitz HI, Sandler AB, et al. Bevacizumab (Avastin®) in cancer treatment: a review of 15 years of clinical experience and future outlook. *Cancer Treat Rev*. 2020;86. https://doi.org/10.1016/j.ctrv.2020.102017.
7. Lambert JM, Berkenblit A. Antibody–drug conjugates for cancer treatment. *Annu Rev Med*. 2018;69(1):191–207. https://doi.org/10.1146/annurev-med-061516-121357.
8. Krishnamurthy A, Jimeno A. Bispecific antibodies for cancer therapy: a review. *Pharmacol Ther*. 2018;185:122–134. https://doi.org/10.1016/j.pharmthera.2017.12.002.
9. Topp MS, Gökbuget N, Stein AS, et al. Safety and activity of blinatumomab for adult patients with relapsed or refractory B-precursor acute lymphoblastic leukaemia: a multicentre, single-arm, phase 2 study. *Lancet Oncol*. 2015;16(1):57–66. https://doi.org/10.1016/S1470-2045(14)71170-2.
10. Berzero B, Picca A, Psimaras D. Neurological complications of chimeric antigen receptor T cells and immune-checkpoint inhibitors: ongoing challenges in daily practice. *Curr Opin Oncol*. 2020;32(6):603–612. https://doi.org/10.1097/CCO.0000000000000681.
11. Ferrara N, Hillan KJ, Gerber H-P, Novotny W. Discovery and development of bevacizumab, an anti-VEGF antibody for treating cancer. *Nat Rev Drug Discov*. 2004;3(5):391–400. https://doi.org/10.1038/nrd1381.
12. Genentech, Inc. *AVASTIN® (bevacizumab). [package insert]*. U.S. Food and Drug Administration website; 2020. https://www.accessdata.fda.gov/drugsatfda_docs/label/2020/125085s332lbl.pdf. Accessed 30 November 2020.
13. Schutz FAB, Je Y, Azzi GR, Nguyen PL, Choueiri TK. Bevacizumab increases the risk of arterial ischemia: a large study in cancer patients with a focus on different subgroup outcomes. *Ann Oncol*. 2011;22(6):1404–1412. https://doi.org/10.1093/annonc/mdq587.
14. Hapani S, Sher A, Chu D, Wu S. Increased risk of serious hemorrhage with bevacizumab in cancer patients: a meta-analysis. *Oncology*. 2010;79(1–2):27–38. https://doi.org/10.1159/000314980.
15. Zuo P-Y, Chen X-L, Liu Y-W, Xiao C-L, Liu C-Y. Increased risk of cerebrovascular events in patients with cancer treated with bevacizumab: a meta-analysis. *PLoS One*. 2014;9(7):e102484. https://doi.org/10.1371/journal.pone.0102484.
16. Khasraw M, Holodny A, Goldlust SA, DeAngelis LM. Intracranial hemorrhage in patients with cancer treated with bevacizumab: the Memorial Sloan-Kettering experience. *Ann Oncol*. 2012;23(2):458–463. https://doi.org/10.1093/annonc/mdr148.
17. Nalluri SR, Chu D, Keresztes R, Zhu X, Wu S. Risk of venous thromboembolism with the angiogenesis inhibitor bevacizumab in cancer patients: a meta-analysis. *JAMA*. 2008;300(19):2277–2285. https://doi.org/10.1001/jama.2008.656.
18. Ranpura V, Pulipati B, Chu D, Zhu X, Wu S. Increased risk of high-grade hypertension with bevacizumab in cancer patients: a meta-analysis. *Am J Hypertens*. 2010;23(5):460–468. https://doi.org/10.1038/ajh.2010.25.
19. Seet RCS, Rabinstein AA, Lindell PE, Uhm JH, Wijdicks EF. Cerebrovascular events after bevacizumab treatment: an early and severe complication. *Neurocrit Care*. 2011;15(3):421–427. https://doi.org/10.1007/s12028-011-9552-5.
20. Tlemsani C, Mir O, Psimaras D, et al. Acute neurovascular events in cancer patients receiving anti-vascular endothelial growth factor agents: clinical experience in Paris University Hospitals. *Eur J Cancer*. 2016;66:75–82. https://doi.org/10.1016/j.ejca.2016.07.008.
21. Brandes AA, Bartolotti M, Tosoni A, Poggi R, Franceschi E. Practical management of bevacizumab-related toxicities in glioblastoma. *Oncologist*. 2015;20(2):166–175. https://doi.org/10.1634/theoncologist.2014-0330.
22. Glusker P, Recht L, Lane B. Reversible posterior leukoencephalopathy syndrome and bevacizumab. *N Engl J Med*. 2006;354(9):980–982. https://doi.org/10.1056/NEJMc052954.
23. Ozcan C, Wong SJ, Hari P. Reversible posterior leukoencephalopathy syndrome and bevacizumab. *N Engl J Med*. 2006;354(9):980–982.
24. Tlemsani C, Mir O, Boudou-Rouquette P, et al. Posterior reversible encephalopathy syndrome induced by anti-VEGF agents. *Target Oncol*. 2011;6(4):253–258. https://doi.org/10.1007/s11523-011-0201-x.
25. Hinchey J, Chaves C, Appignani B, et al. A reversible posterior leukoencephalopathy syndrome. *N Engl J Med*. 1996;334(8):494–500. https://doi.org/10.1056/NEJM199602223340803.
26. Fugate JE, Rabinstein AA. Posterior reversible encephalopathy syndrome: clinical and radiological manifestations, pathophysiology, and outstanding questions. *Lancet Neurol*. 2015;14(9):914–925. https://doi.org/10.1016/S1474-4422(15)00111-8.
27. Seet RCS, Rabinstein AA. Clinical features and outcomes of posterior reversible encephalopathy syndrome following bevacizumab treatment. *QJM Mon J Assoc Physicians*. 2012;105(1):69–75. https://doi.org/10.1093/qjmed/hcr139.

28. Lou E, Turner S, Sumrall A, et al. Bevacizumab-induced reversible posterior leukoencephalopathy syndrome and successful retreatment in a patient with glioblastoma. *J Clin Oncol.* 2011;6. https://doi.org/10.1200/JCO.2011.36.1865. Published online September.

29. Sherman JH, Aregawi DG, Lai A, et al. Optic neuropathy in patients with glioblastoma receiving bevacizumab. *Neurology.* 2009;73(22):1924–1926. https://doi.org/10.1212/WNL.0b013e3181c3fd00.

30. Finger PT, Chin KJ. Antivascular endothelial growth factor bevacizumab for radiation optic neuropathy: secondary to plaque radiotherapy. *Int J Radiat Oncol Biol Phys.* 2012;82(2):789–798. https://doi.org/10.1016/j.ijrobp.2010.11.075.

31. Farooq O, Lincoff NS, Saikali N, Prasad D, Miletich RS, Mechtler LL. Novel treatment for radiation optic neuropathy with intravenous bevacizumab. *J Neuroophthalmol.* 2012;32(4):321–324. https://doi.org/10.1097/WNO.0b013e3182607381.

32. Finger PT. Anti-VEGF bevacizumab (Avastin) for radiation optic neuropathy. *Am J Ophthalmol.* 2007;143(2):335–338. https://doi.org/10.1016/j.ajo.2006.09.014.

33. ImClone LLC. *ERBITUX® (cetuximab). [package insert].* U.S. Food and Drug Administration website; 2020. https://www.accessdata.fda.gov/drugsatfda_docs/label/2019/125084s273lbl.pdf. Accessed 30 November 2020.

34. Baselga J, Pfister D, Cooper MR, et al. Phase I studies of anti-epidermal growth factor receptor chimeric antibody C225 alone and in combination with cisplatin. *J Clin Oncol.* 2000;18(4):904–914. https://doi.org/10.1200/JCO.2000.18.4.904.

35. Feinstein TM, Gibson MK, Argiris A. Cetuximab-induced aseptic meningitis. *Ann Oncol.* 2009;20(9):1609–1610. https://doi.org/10.1093/annonc/mdp382.

36. Emani MK, Zaiden RA. Aseptic meningitis: a rare side effect of cetuximab therapy. *J Oncol Pharm Pract.* 2013;19(2):178–180. https://doi.org/10.1177/1078155212447973.

37. Nagovskiy N, Agarwal M, Allerton J. Cetuximab-induced aseptic meningitis. *J Thorac Oncol.* 2010;5(5):751. https://doi.org/10.1097/JTO.0b013e3181d408bc.

38. Kaur S, Khatun H, Kumar A, Ellis M, Mehta D, Guron G. Cetuximab induced aseptic meningitis: a rare side effect. *J Cancer Sci Ther.* 2018;10:211–213. https://doi.org/10.4172/1948-5956.1000546.

39. Prasanna D, Elrafei T, Shum E, Strakhan M. More than a headache: a case of cetuximab-induced aseptic meningitis. *BMJ Case Rep.* 2015;2015. https://doi.org/10.1136/bcr-2015-209622.

40. Rohrer CL, Grullon Z, George SK, Castillo R, Karasiewicz K. A case of aseptic meningitis in a cetuximab-experienced patient with metastatic colon cancer. *J Oncol Pharm Pract.* 2018;24(8):632–633. https://doi.org/10.1177/1078155217739685.

41. Vulsteke CA, Joosens E. Aseptic meningitis as a rare but serious side effect of cetuximab therapy. *Belg J Med Oncol.* 2010;4(6):257–259.

42. Maritaz C, Metz C, Baba-Hamed N, Jardin-Szucs M, Deplanque G. Cetuximab-induced aseptic meningitis: case report and review of a rare adverse event. *BMC Cancer.* 2016;16:384. https://doi.org/10.1186/s12885-016-2434-7.

43. Fakih M, Vincent M. Adverse events associated with anti-EGFR therapies for the treatment of metastatic colorectal cancer. *Curr Oncol.* 2010;17(Suppl. 1):S18–S30. https://doi.org/10.3747/co.v17is1.615.

44. Tejpar S, Piessevaux H, Claes K, et al. Magnesium wasting associated with epidermal-growth-factor receptor-targeting antibodies in colorectal cancer: a prospective study. *Lancet Oncol.* 2007;8(5):387–394. https://doi.org/10.1016/S1470-2045(07)70108-0.

45. Fakih MG, Wilding G, Lombardo J. Cetuximab-induced hypomagnesemia in patients with colorectal cancer. *Clin Colorectal Cancer.* 2006;6(2):152–156. https://doi.org/10.3816/CCC.2006.n.033.

46. Schrag D, Chung KY, Flombaum C, Saltz L. Cetuximab therapy and symptomatic hypomagnesemia. *J Natl Cancer Inst.* 2005;97(16):1221–1224. https://doi.org/10.1093/jnci/dji242.

47. Riggs JE. Neurologic manifestations of electrolyte disturbances. *Neurol Clin.* 2002;20(1):227–239. vii https://doi.org/10.1016/s0733-8619(03)00060-4.

48. Palma J-A, Gomez-Ibañez A, Martin B, Urrestarazu E, Gil-Bazo I, Pastor MA. Nonconvulsive status epilepticus related to posterior reversible leukoencephalopathy syndrome induced by cetuximab. *Neurologist.* 2011;17(5):273–275. https://doi.org/10.1097/NRL.0b013e3182173655.

49. Kamiya-Matsuoka C, Paker AM, Chi L, Youssef A, Tummala S, Loghin ME. Posterior reversible encephalopathy syndrome in cancer patients: a single institution retrospective study. *J Neurooncol.* 2016;128(1):75–84. https://doi.org/10.1007/s11060-016-2078-0.

50. Chardain A, Mesnage V, Alamowitch S, et al. Posterior reversible encephalopathy syndrome (PRES) and hypomagnesemia: a frequent association? *Rev Neurol.* 2016;172(6):384–388. https://doi.org/10.1016/j.neurol.2016.06.004.

51. Douillard JY, Siena S, Cassidy J, et al. Final results from PRIME: randomized phase III study of panitumumab with FOLFOX4 for first-line treatment of metastatic colorectal cancer. *Ann Oncol.* 2014;25(7):1346–1355. https://doi.org/10.1093/annonc/mdu141.

52. Thatcher N, Hirsch FR, Luft AV, et al. Necitumumab plus gemcitabine and cisplatin versus gemcitabine and cisplatin alone as first-line therapy in patients with stage IV squamous non-small-cell lung cancer (SQUIRE): an open-label, randomised, controlled phase 3 trial. *Lancet Oncol.* 2015;16(7):763–774. https://doi.org/10.1016/S1470-2045(15)00021-2.

53. Genentech, Inc. *HERCEPTIN® (trastuzumab) [package insert].* U.S. Food and Drug Administration website; 2020. https://www.accessdata.fda.gov/drugsatfda_docs/label/2010/103792s5250lbl.pdf. Accessed 30 November 2020.

54. Kaneda H, Okamoto I, Satoh T, Nakagawa K. Reversible posterior leukoencephalopathy syndrome and trastuzumab. *Invest New Drugs.* 2012;30(4):1766–1767. https://doi.org/10.1007/s10637-011-9696-3.

55. Ladwa R, Peters G, Bigby K, Chern B. Posterior reversible encephalopathy syndrome in early-stage breast cancer. *Breast J.* 2015;21(6):674–677. https://doi.org/10.1111/tbj.12502.

56. Abughanimeh O, Abu Ghanimeh M, Qasrawi A, Al Momani LA, Madhusudhana S. Trastuzumab-associated posterior reversible encephalopathy syndrome. *Cureus.* 2018;10(5). https://doi.org/10.7759/cureus.2686.

57. Izumi Y, Xu L, di Tomaso E, Fukumura D, Jain RK. Tumour biology: herceptin acts as an anti-angiogenic cocktail. *Nature.* 2002;416(6878):279–280. https://doi.org/10.1038/416279b.

58. Genentech, Inc. *RITUXAN® (rituximab) [package insert].* U.S. Food and Drug Administration website; 2020. https://www.accessdata.fda.gov/drugsatfda_docs/label/2010/103705s5311lbl.pdf. Accessed November 30, 2020.

59. Aksamit AJ. Progressive multifocal leukoencephalopathy. *Continuum.* 2012;18:1374–1391. https://doi.org/10.1212/01.CON.0000423852.70641.de [6 Infectious Disease].

60. Tan CS, Koralnik IJ. Progressive multifocal leukoencephalopathy and other disorders caused by JC virus: clinical features and pathogenesis. *Lancet Neurol.* 2010;9(4):425–437. https://doi.org/10.1016/S1474-4422(10)70040-5.

61. Berger JR, Aksamit AJ, Clifford DB, et al. PML diagnostic criteria: consensus statement from the AAN Neuroinfectious Disease Section. *Neurology.* 2013;80(15):1430–1438. https://doi.org/10.1212/WNL.0b013e31828c2fa1.

62. Norris LB, Georgantopoulos P, Rao GA, Haddock KS, Bennett CL. Association between rituximab use and progressive multifocal leukoencephalopathy among non-HIV, non-Hodgkin lymphoma Veteran's Administration patients. *J Clin Oncol.* 2014;32(15_Suppl):e19540. https://doi.org/10.1200/jco.2014.32.15_suppl.e19540.

63. Norris LB, Georgantopoulos P, Rao GA, Sartor AO, Bennett CL. Rituximab is associated with increased risk of progressive multifocal leukoencephalopathy developing among non-HIV-infected veterans with chronic lymphocytic leukemia. *J Clin Oncol.* 2015;33(15_Suppl):e18033. https://doi.org/10.1200/jco.2015.33.15_suppl.e18033.

64. Focosi D, Tuccori M, Maggi F. Progressive multifocal leukoencephalopathy and anti-CD20 monoclonal antibodies: what do we know after 20 years of rituximab. *Rev Med Virol.* 2019;29(6):e 2077. https://doi.org/10.1002/rmv.2077.

65. D'Alò F, Malafronte R, Piludu F, et al. Progressive multifocal leukoencephalopathy in patients with follicular lymphoma treated with bendamustine plus rituximab followed by rituximab maintenance. *Br J Haematol.* 2020;189(4):e140–e144. https://doi.org/10.1111/bjh.16563.

66. Gasnault J, Kahraman M, De Goër de Herve MG, Durali D, Delfraissy J-F, Taoufik Y. Critical role of JC virus-specific CD4 T-cell responses in preventing progressive multifocal leukoencephalopathy. *AIDS.* 2003;17(10):1443–1449.

67. Delgado-Alvarado M, Sedano MJ, González-Quintanilla V, de Lucas EM, Polo JM, Berciano J. Progressive multifocal leukoencephalopathy and idiopathic CD4 lymphocytopenia. *J Neurol Sci.* 2013;327(1):75–79. https://doi.org/10.1016/j.jns.2013.02.002.

68. Mélet J, Mulleman D, Goupille P, Ribourtout B, Watier H, Thibault G. Rituximab-induced T cell depletion in patients with rheumatoid arthritis: association with clinical response. *Arthritis Rheum.* 2013;65(11):2783–2790. https://doi.org/10.1002/art.38107.

69. Carson KR, Evens AM, Richey EA, et al. Progressive multifocal leukoencephalopathy after rituximab therapy in HIV-negative patients: a report of 57 cases from the Research on Adverse Drug Events and Reports project. *Blood.* 2009;113(20):4834–4840. https://doi.org/10.1182/blood-2008-10-186999.

70. Clifford DB, Ances B, Costello C, et al. Rituximab-associated progressive multifocal leukoencephalopathy in rheumatoid arthritis. *Arch Neurol.* 2011;68(9):1156–1164. https://doi.org/10.1001/archneurol.2011.103.

71. Mustafa KN, Qasem U, Al-Ryalat NT, Bsisu IK. Rituximab-associated posterior reversible encephalopathy syndrome. *Int J Rheum Dis.* 2019;22(1):160–165. https://doi.org/10.1111/1756-185X.13427.

72. Mizutani M, Nakamori Y, Sakaguchi H, et al. Development of syndrome of inappropriate secretion of ADH and reversible posterior leukoencephalopathy during initial rituximab-CHOP therapy in a patient with diffuse large B-cell lymphoma. *Rinsho Ketsueki.* 2013;54(3):269–272.

73. Mavragani CP, Vlachoyiannopoulos PG, Kosmas N, Boletis I, Tzioufas AG, Voulgarelis M. A case of reversible posterior leucoencephalopathy syndrome after rituximab infusion. *Rheumatology (Oxford).* 2004;43(11):1450–1451. https://doi.org/10.1093/rheumatology/keh305.

74. Genentech, Inc. *GAZYVA® (obinutuzumab) [package insert].* U.S. Food and Drug Administration website; 2020. https://www.accessdata.fda.gov/drugsatfda_docs/label/2017/125486s017s018lbl.pdf. Accessed 30 November 2020.

75. Novartis Pharmaceuticals Corporation. *ARZERRA® (ofatumumab) [package insert].* U.S. Food and Drug Administration website; 2020. https://www.accessdata.fda.gov/drugsatfda_docs/label/2016/125326s062lbl.pdf. Accessed 30 November 2020.

76. Raisch DW, Rafi JA, Chen C, Bennett CL. Detection of cases of progressive multifocal leukoencephalopathy associated with new biologicals and targeted cancer therapies from the FDA's adverse event reporting system. *Expert Opin Drug Saf.* 2016;15(8):1003–1011. https://doi.org/10.1080/14740338.2016.1198775.

77. Pejsa V, Lucijanic M, Jonjic Z, Prka Z, Vukorepa G. Progressive multifocal leukoencephalopathy developing after obinutuzumab treatment for chronic lymphocytic leukemia. *Ann Hematol.* 2019;98(6):1509–1510. https://doi.org/10.1007/s00277-018-3552-x.

78. Forryan J, Yong J. Rapid cognitive decline in a patient with chronic lymphocytic leukaemia: a case report. *J Med Case Reports.* 2020;14(1):1–6. https://doi.org/10.1186/s13256-020-2360-9.

79. Ploessl C, Pan A, Maples KT, Lowe DK. Dinutuximab: an anti-GD2 monoclonal antibody for high-risk neuroblastoma. *Ann Pharmacother.* 2016. https://doi.org/10.1177/1060028016632013. Published online February 25.

80. Keyel ME, Reynolds CP. Spotlight on dinutuximab in the treatment of high-risk neuroblastoma: development and place in therapy. *Biol Targets Ther.* 2018;13:1–12. https://doi.org/10.2147/BTT.S114530.

81. United Therapeutics Corp. *UNITUXIN® (dinutuximab) [package insert].* U.S. Food and Drug Administration website; 2020. https://www.accessdata.fda.gov/drugsatfda_docs/label/2015/125516s000lbl.pdf. Accessed 30 November 2020.

82. Yu AL, Gilman AL, Ozkaynak MF, et al. Anti-GD2 antibody with GM-CSF, interleukin-2, and isotretinoin for neuroblastoma. *N Engl J Med.* 2010;363(14):1324–1334. https://doi.org/10.1056/NEJMoa0911123.

83. Mody R, Naranjo A, Van Ryn C, et al. Irinotecan-temozolomide with temsirolimus or dinutuximab in children with refractory or relapsed neuroblastoma (COG ANBL 1221): an open-label, randomised, phase 2 trial. *Lancet Oncol.* 2017;18(7):946–957. https://doi.org/10.1016/S1470-2045(17)30355-8.

84. Ozkaynak MF, Gilman AL, London WB, et al. A comprehensive safety trial of chimeric antibody 14.18 with GM-CSF, IL-2, and isotretinoin in high-risk neuroblastoma patients following myeloablative therapy: children's oncology group study ANBL 0931. *Front Immunol.* 2018;9. https://doi.org/10.3389/fimmu.2018.01355.

85. Sorkin LS, Otto M, Baldwin WM, et al. Anti-GD2 with an FC point mutation reduces complement fixation and decreases antibody-induced allodynia. *Pain.* 2010;149(1):135–142. https://doi.org/10.1016/j.pain.2010.01.024.

86. Oak E, Bartlett NL. A safety evaluation of brentuximab vedotin for the treatment of Hodgkin lymphoma. *Expert Opin Drug Saf.* 2016;15(6):875–882. https://doi.org/10.1080/14740338.2016.1179277.

87. Seattle Genetics, Inc. *ADCETRIS® (brentuximab vedotin) [package insert].* U.S. Food and Drug Administration website; 2020. https://www.accessdata.fda.gov/drugsatfda_docs/label/2014/125388_S056S078lbl.pdf. Accessed 30 November 2020.

88. Moskowitz CH, Nademanee A, Masszi T, et al. Brentuximab vedotin as consolidation therapy after autologous stem-cell transplantation in patients with Hodgkin's lymphoma at risk of relapse or progression (AETHERA): a randomised, double-blind, placebo-controlled, phase 3 trial. *Lancet.* 2015;385(9980):1853–1862. https://doi.org/10.1016/S0140-6736(15)60165-9.

89. Prince HM, Kim YH, Horwitz SM, et al. Brentuximab vedotin or physician's choice in CD30-positive cutaneous T-cell lymphoma (ALCANZA): an international, open-label, randomised, phase 3, multicentre trial. *Lancet.* 2017;390(10094):555–566. https://doi.org/10.1016/S0140-6736(17)31266-7.

90. Corbin ZA, Nguyen-Lin A, Li S, et al. Characterization of the peripheral neuropathy associated with brentuximab vedotin treatment of Mycosis Fungoides and Sézary Syndrome. *J Neurooncol*. 2017;132(3):439–446. https://doi.org/10.1007/s11060-017-2389-9.

91. Mariotto S, Tecchio C, Sorio M, et al. Clinical and neurophysiological serial assessments of brentuximab vedotin-associated peripheral neuropathy. *Leuk Lymphoma*. 2019;60(11):2806–2809. https://doi.org/10.1080/10428194.2019.1605068.

92. Mariotto S, Ferrari S, Sorio M, et al. Brentuximab vedotin: axonal microtubule's Apollyon. *Blood Cancer J*. 2015;5:e343. https://doi.org/10.1038/bcj.2015.72.

93. Fargeot G, Dupel-Pottier C, Stephant M, et al. Brentuximab vedotin treatment associated with acute and chronic inflammatory demyelinating polyradiculoneuropathies. *J Neurol Neurosurg Psychiatry*. 2020;91(7):786–788. https://doi.org/10.1136/jnnp-2020-323124.

94. Wagner-Johnston ND, Bartlett NL, Cashen A, Berger JR. Progressive multifocal leukoencephalopathy in a patient with Hodgkin lymphoma treated with brentuximab vedotin. *Leuk Lymphoma*. 2012;53(11):2283–2286. https://doi.org/10.3109/10428194.2012.676170.

95. von Geldern G, Pardo CA, Calabresi PA, Newsome SD. PML-IRIS in a patient treated with brentuximab. *Neurology*. 2012;79(20):2075–2077. https://doi.org/10.1212/WNL.0b013e3182749f17.

96. Carson KR, Newsome SD, Kim EJ, et al. Progressive multifocal leukoencephalopathy associated with brentuximab vedotin therapy: a report of 5 cases from the Southern Network on Adverse Reactions (SONAR) project. *Cancer*. 2014;120(16):2464–2471. https://doi.org/10.1002/cncr.28712.

97. Deeks ED. Polatuzumab vedotin: first global approval. *Drugs*. 2019;79(13):1467–1475. https://doi.org/10.1007/s40265-019-01175-0.

98. Genentech, Inc. *POLIVY® (polatuzumab vedotin-piiq) [package insert]*. US Food and Drug Administration website; 2020. https://www.accessdata.fda.gov/drugsatfda_docs/label/2019/761121s000lbl.pdf. Accessed 30 November 2020.

99. Palanca-Wessels MCA, Czuczman M, Salles G, et al. Safety and activity of the anti-CD79B antibody–drug conjugate polatuzumab vedotin in relapsed or refractory B-cell non-Hodgkin lymphoma and chronic lymphocytic leukaemia: a phase 1 study. *Lancet Oncol*. 2015;16(6):704–715. https://doi.org/10.1016/S1470-2045(15)70128-2.

100. Morschhauser F, Flinn IW, Advani R, et al. Polatuzumab vedotin or pinatuzumab vedotin plus rituximab in patients with relapsed or refractory non-Hodgkin lymphoma: final results from a phase 2 randomised study (ROMULUS). *Lancet Haematol*. 2019;6(5):e254–e265. https://doi.org/10.1016/S2352-3026(19)30026-2.

101. Sehn LH, Herrera AF, Flowers CR, et al. Polatuzumab vedotin in relapsed or refractory diffuse large B-cell lymphoma. *J Clin Oncol*. 2020;38(2):155–165. https://doi.org/10.1200/JCO.19.00172.

102. Lu D, Gillespie WR, Girish S, et al. Time-to-event analysis of polatuzumab vedotin-induced peripheral neuropathy to assist in the comparison of clinical dosing regimens. *CPT Pharmacomet Syst Pharmacol*. 2017;6(6):401–408. https://doi.org/10.1002/psp4.12192.

103. Challita-Eid PM, Satpayev D, Yang P, et al. Enfortumab vedotin antibody-drug conjugate targeting nectin-4 is a highly potent therapeutic agent in multiple preclinical cancer models. *Cancer Res*. 2016;76(10):3003–3013. https://doi.org/10.1158/0008-5472.CAN-15-1313.

104. Astellas Pharma US, Inc. *PADCEV® (enfortumab vedotin) [package insert]*. U.S. Food and Drug Administration website; 2020. https://www.accessdata.fda.gov/drugsatfda_docs/label/2019/761137s000lbl.pdf. Accessed 30 November 2020.

105. Rosenberg J, Sridhar SS, Zhang J, et al. EV-101: a phase I study of single-agent enfortumab vedotin in patients with nectin-4-positive solid tumors, including metastatic urothelial carcinoma. *J Clin Oncol*. 2020;38(10):1041–1049. https://doi.org/10.1200/JCO.19.02044.

106. Rosenberg JE, O'Donnell PH, Balar AV, et al. Pivotal trial of enfortumab vedotin in urothelial carcinoma after platinum and anti-programmed death 1/programmed death ligand 1 therapy. *J Clin Oncol*. 2019;37(29):2592–2600. https://doi.org/10.1200/JCO.19.01140.

107. Powles T, Rosenberg JE, Sonpavde GP, et al. Enfortumab vedotin in previously treated advanced urothelial carcinoma. *N Engl J Med*. 2021;384(12):1125–1135. https://doi.org/10.1056/NEJMoa2035807.

108. Genentech, Inc. *KADCYLA® (ado-trastuzumab emtansine) [package insert]*. U.S. Food and Drug Administration website; 2020. https://www.accessdata.fda.gov/drugsatfda_docs/label/2019/125427s105lbl.pdf. Accessed 30 November 2020.

109. Montemurro F, Ellis P, Anton A, et al. Safety of trastuzumab emtansine (T-DM1) in patients with HER2-positive advanced breast cancer: primary results from the KAMILLA study cohort 1. *Eur J Cancer*. 2019;109:92–102. https://doi.org/10.1016/j.ejca.2018.12.022.

110. Montemurro F, Delaloge S, Barrios CH, et al. Trastuzumab emtansine (T-DM1) in patients with HER2-positive metastatic breast cancer and brain metastases: exploratory final analysis of cohort 1 from KAMILLA, a single-arm phase IIIb clinical trial. *Ann Oncol*. 2020;31(10):1350–1358. https://doi.org/10.1016/j.annonc.2020.06.020.

111. Perez EA, Barrios C, Eiermann W, et al. Trastuzumab emtansine with or without pertuzumab versus trastuzumab plus taxane for human epidermal growth factor receptor 2-positive, advanced breast cancer: primary results from the phase III MARIANNE study. *J Clin Oncol*. 2017;35(2):141–148. https://doi.org/10.1200/JCO.2016.67.4887.

112. Verma S, Miles D, Gianni L, et al. Trastuzumab emtansine for HER2-positive advanced breast cancer. *N Engl J Med*. 2012;367(19):1783–1791. https://doi.org/10.1056/NEJMoa1209124.

113. Krop IE, Kim S-B, González-Martín A, et al. Trastuzumab emtansine versus treatment of physician's choice for pretreated HER2-positive advanced breast cancer (TH3RESA): a randomised, open-label, phase 3 trial. *Lancet Oncol*. 2014;15(7):689–699. https://doi.org/10.1016/S1470-2045(14)70178-0.

114. Poon KA, Flagella K, Beyer J, et al. Preclinical safety profile of trastuzumab emtansine (T-DM1): mechanism of action of its cytotoxic component retained with improved tolerability. *Toxicol Appl Pharmacol*. 2013;273(2):298–313. https://doi.org/10.1016/j.taap.2013.09.003.

115. von Minckwitz G, Huang C-S, Mano MS, et al. Trastuzumab emtansine for residual invasive HER2-positive breast cancer. *N Engl J Med*. 2019;380(7):617–628. https://doi.org/10.1056/NEJMoa1814017.

116. Goebeler M-E, Bargou RC. T cell-engaging therapies—BiTEs and beyond. *Nat Rev Clin Oncol*. 2020;17(7):418–434. https://doi.org/10.1038/s41571-020-0347-5.

117. Amgen Inc. *BLINCYTO® (blinatumomab) [package insert]*. U.S. Food and Drug Administration website; 2020. https://www.accessdata.fda.gov/drugsatfda_docs/label/2018/125557s013lbl.pdf. Accessed 30 November 2020.

118. Topp MS, Gökbuget N, Zugmaier G, et al. Phase II trial of the anti-CD19 bispecific T cell–engager blinatumomab shows hematologic and molecular remissions in patients with relapsed or refractory B-precursor acute lymphoblastic leukemia. *J Clin Oncol*. 2014;32(36):4134–4140. https://doi.org/10.1200/JCO.2014.56.3247.

119. Kantarjian H, Stein A, Gökbuget N, et al. Blinatumomab versus chemotherapy for advanced acute lymphoblastic leukemia. *N Engl J Med.* 2017;376(9):836–847. https://doi.org/10.1056/NEJMoa1609783.

120. Stein AS, Schiller G, Benjamin R, et al. Neurologic adverse events in patients with relapsed/refractory acute lymphoblastic leukemia treated with blinatumomab: management and mitigating factors. *Ann Hematol.* 2019;98(1):159–167. https://doi.org/10.1007/s00277-018-3497-0.

121. Viardot A, Goebeler M-E, Hess G, et al. Phase 2 study of the bispecific T-cell engager (BiTE) antibody blinatumomab in relapsed/refractory diffuse large B-cell lymphoma. *Blood.* 2016;127(11):1410–1416. https://doi.org/10.1182/blood-2015-06-651380.

122. Goebeler M-E, Knop S, Viardot A, et al. Bispecific T-cell engager (BiTE) antibody construct blinatumomab for the treatment of patients with relapsed/refractory non-Hodgkin lymphoma: final results from a phase I study. *J Clin Oncol.* 2016;34(10):1104–1111. https://doi.org/10.1200/JCO.2014.59.1586.

123. Dufner V, Sayehli CM, Chatterjee M, et al. Long-term outcome of patients with relapsed/refractory B-cell non-Hodgkin lymphoma treated with blinatumomab. *Blood Adv.* 2019;3(16):2491–2498. https://doi.org/10.1182/bloodadvances.2019000025.

124. Klinger M, Zugmaier G, Nägele V, et al. Adhesion of T cells to endothelial cells facilitates blinatumomab-associated neurologic adverse events. *Cancer Res.* 2020;80(1):91–101. https://doi.org/10.1158/0008-5472.CAN-19-1131.

125. Prometheus Laboratories Inc. *PROLEUKIN® (aldesleukin) [package insert].* U.S. Food and Drug Administration website; 2020. https://www.accessdata.fda.gov/drugsatfda_docs/label/2012/103293s5130lbl.pdf. Accessed 30 November 2020.

126. Denicoff KD, Rubinow DR, Papa MZ, et al. The neuropsychiatric effects of treatment with interleukin-2 and lymphokine-activated killer cells. *Ann Intern Med.* 1987;107(3):293–300. https://doi.org/10.7326/0003-4819-107-2-293.

127. Dillman RO, Oldham RK, Barth NM, et al. Continuous interleukin-2 and tumor-infiltrating lymphocytes as treatment of advanced melanoma. A national biotherapy study group trial. *Cancer.* 1991;68(1):1–8. https://doi.org/10.1002/1097-0142(19910701)68:1<1::aid-cncr2820680102>3.0.co;2-k.

128. Siegel JP, Puri RK. Interleukin-2 toxicity. *J Clin Oncol.* 1991;9(4):694–704. https://doi.org/10.1200/JCO.1991.9.4.694.

129. Rosenberg SA, Lotze MT, Muul LM, et al. A progress report on the treatment of 157 patients with advanced cancer using lymphokine-activated killer cells and interleukin-2 or high-dose interleukin-2 alone. *N Engl J Med.* 1987;316(15):889–897. https://doi.org/10.1056/NEJM198704093161501.

130. Margolin KA, Rayner AA, Hawkins MJ, et al. Interleukin-2 and lymphokine-activated killer cell therapy of solid tumors: analysis of toxicity and management guidelines. *J Clin Oncol.* 1989;7(4):486–498. https://doi.org/10.1200/JCO.1989.7.4.486.

131. Vecht CJ, Keohane C, Menon RS, Punt CJ, Stoter G. Acute fatal leukoencephalopathy after interleukin-2 therapy. *N Engl J Med.* 1990;323(16):1146–1147. https://doi.org/10.1056/nejm199010183231616.

132. Karp BI, Yang JC, Khorsand M, Wood R, Merigan TC. Multiple cerebral lesions complicating therapy with interleukin-2. *Neurology.* 1996;47(2):417–424. https://doi.org/10.1212/wnl.47.2.417.

133. Loh FL, Herskovitz S, Berger AR, Swerdlow ML. Brachial plexopathy associated with interleukin-2 therapy. *Neurology.* 1992;42(2):462–463. https://doi.org/10.1212/wnl.42.2.462.

134. Heys SD, Mills KL, Eremin O. Bilateral carpal tunnel syndrome associated with interleukin 2 therapy. *Postgrad Med J.* 1992;68(801):587–588. https://doi.org/10.1136/pgmj.68.801.587.

135. Puduvalli VK, Sella A, Austin SG, Forman AD. Carpal tunnel syndrome associated with interleukin-2 therapy. *Cancer.* 1996;77(6):1189–1192. https://doi.org/10.1002/(sici)1097-0142(19960315)77:6<1189::aid-cncr27>3.0.co;2-x.

136. Sikora SS, Samsonov ME, Dookeran KA, Edington H, Lotze MT. Peripheral nerve entrapment: an unusual adverse event with high-dose interleukin-2 therapy. *Ann Oncol.* 1996;7(5):535–536. https://doi.org/10.1093/oxfordjournals.annonc.a010647.

137. Williams CD, Linch DC. Interferon alfa-2a. *Br J Hosp Med.* 1997;57(9):436–439.

138. Vial T, Choquet-Kastylevsky G, Liautard C, Descotes J. Endocrine and neurological adverse effects of the therapeutic interferons. *Toxicology.* 2000;142(3):161–172. https://doi.org/10.1016/s0300-483x(99)00141-9.

139. Ravaud A, Bedane C, Geoffrois L, Lesimple T, Delaunay M. Toxicity and feasibility of adjuvant high-dose interferon alpha-2b in patients with melanoma in clinical oncologic practice. *Br J Cancer.* 1999;80(11):1767–1769. https://doi.org/10.1038/sj.bjc.6690595.

140. Caraceni A, Gangeri L, Martini C, et al. Neurotoxicity of interferon-alpha in melanoma therapy: results from a randomized controlled trial. *Cancer.* 1998;83(3):482–489. https://doi.org/10.1002/(sici)1097-0142(19980801)83:3<482::aid-cncr17>3.0.co;2-s.

141. Kirkwood JM, Ernstoff MS, Davis CA, Reiss M, Ferraresi R, Rudnick SA. Comparison of intramuscular and intravenous recombinant alpha-2 interferon in melanoma and other cancers. *Ann Intern Med.* 1985;103(1):32–36. https://doi.org/10.7326/0003-4819-103-1-32.

142. Janssen HLA, Berk L, Vermeulen M, Schalm SW. Seizures associated with low-dose α-interferon. *Lancet.* 1990;336(8730):1580. https://doi.org/10.1016/0140-6736(90)93356-T.

143. Rohatiner AZ, Prior PF, Burton AC, Smith AT, Balkwill FR, Lister TA. Central nervous system toxicity of interferon. *Br J Cancer.* 1983;47(3):419–422. https://doi.org/10.1038/bjc.1983.63.

144. Adams F, Quesada JR, Gutterman JU. Neuropsychiatric manifestations of human leukocyte interferon therapy in patients with cancer. *JAMA.* 1984;252(7):938–941. https://doi.org/10.1001/jama.1984.03350070056026.

145. Quesada JR, Talpaz M, Rios A, Kurzrock R, Gutterman JU. Clinical toxicity of interferons in cancer patients: a review. *J Clin Oncol.* 1986;4(2):234–243. https://doi.org/10.1200/JCO.1986.4.2.234.

146. Merimsky O, Chaitchik S. Neurotoxicity of interferon-alpha. *Anticancer Drugs.* 1992;3(6):567–570. https://doi.org/10.1097/00001813-199212000-00002.

147. Färkkilä M, Iivanainen M, Roine R, et al. Neurotoxic and other side effects of high-dose interferon in amyotrophic lateral sclerosis. *Acta Neurol Scand.* 1984;70(1):42–46. https://doi.org/10.1111/j.1600-0404.1984.tb00801.x.

148. Meyers CA, Scheibel RS, Forman AD. Persistent neurotoxicity of systemically administered interferon-alpha. *Neurology.* 1991;41(5):672–676. https://doi.org/10.1212/wnl.41.5.672.

149. Gool ARV, Kruit WHJ, Stoter G, Engels FK, Eggermont AMM. Neuropsychiatric side effects of interferon-alfa therapy. *Pharm World Sci.* 2003;25(1):11–20. https://doi.org/10.1023/A:1022449613907.

150. Merimsky O, Reider-Groswasser I, Inbar M, Chaitchik S. Interferon-related mental deterioration and behavioral changes in patients with renal cell carcinoma. *Eur J Cancer.* 1990;26(5):596–600. https://doi.org/10.1016/0277-5379(90)90086-9.

151. Stübgen J-P. Interferon alpha and neuromuscular disorders. *J Neuroimmunol.* 2009;207(1):3–17. https://doi.org/10.1016/j.jneuroim.2008.12.008.

152. Rutkove SB. An unusual axonal polyneuropathy induced by low-dose interferon alfa-2a. *Arch Neurol.* 1997;54(7):907–908. https://doi.org/10.1001/archneur.1997.00550190093020.

153. Meriggioli MN, Rowin J. Chronic inflammatory demyelinating polyneuropathy after treatment with interferon-alpha. *Muscle Nerve.* 2000;23(3):433–435. https://doi.org/10.1002/(sici)1097-4598(200003)23:3<433::aid-mus17>3.0.co;2-o.

154. Anthoney DA, Bone I, Evans TR. Inflammatory demyelinating polyneuropathy: a complication of immunotherapy in malignant melanoma. *Ann Oncol.* 2000;11(9):1197–1200. https://doi.org/10.1023/a:1008362714023.

155. Bauherz G, Soeur M, Lustman F. Oculomotor nerve paralysis induced by alpha II-interferon. *Acta Neurol Belg.* 1990;90(2):111–114.

156. Batocchi AP, Evoli A, Servidei S, Palmisani MT, Apollo F, Tonali P. Myasthenia gravis during interferon alfa therapy. *Neurology.* 1995;45(2):382–383. https://doi.org/10.1212/wnl.45.2.382.

157. Arai H, Tanaka M, Ohta K, Kojo T, Niijima K, Imawari M. Symptomatic myopathy associated with interferon therapy for chronic hepatitis C. *Lancet.* 1995;345(8949):582. https://doi.org/10.1016/s0140-6736(95)90490-5.

158. Cirigliano G, Della Rossa A, Tavoni A, Viacava P, Bombardieri S. Polymyositis occurring during alpha-interferon treatment for malignant melanoma: a case report and review of the literature. *Rheumatol Int.* 1999;19(1–2):65–67. https://doi.org/10.1007/s002960050103.

159. Falcone A, Bodenizza C, Musto P, Carotenuto M. Symptomatic myopathy during interferon alfa therapy for chronic myelogenous leukemia. *Leukemia.* 1998;12:1329. https://doi.org/10.1038/sj.leu.2401104.

160. Gorson KC, Allam G, Simovic D, Ropper AH. Improvement following interferon-alpha 2A in chronic inflammatory demyelinating polyneuropathy. *Neurology.* 1997;48(3):777–780. https://doi.org/10.1212/wnl.48.3.777.

161. Ammendola A, Sampaolo S, Ambrosone L, et al. Peripheral neuropathy in hepatitis-related mixed cryoglobulinemia: electrophysiologic follow-up study. *Muscle Nerve.* 2005;31(3):382–385. https://doi.org/10.1002/mus.20184.

32

Neurologic complications of immune modulatory therapy

Brian M. Andersen and David A. Reardon

Center for Neuro-Oncology, Department of Medical Oncology, Dana-Farber Cancer Institute, Boston, MA, United States

1 Introduction

Immune modulation is a novel therapeutic strategy targeting mechanisms of tolerance established by malignancies. In contrast to most chemotherapeutics, immunotherapies manipulate nonneoplastic cells, specifically within the adaptive immune system. Immune checkpoint blockade (ICB), initially FDA approved in 2011, has become the most common immune modulatory therapy. Chimeric antigen receptor (CAR) T cells and bispecific T cell engagers (BiTEs) are other types of immune modulatory therapies approved for several hematologic malignancies and under investigation in many solid tumors. Immune modulatory therapies cause adverse effects that are unique from those of conventional chemotherapy and sometimes affect the nervous system.

ICB agents are monoclonal antibodies that block immune checkpoint molecules expressed on the surface of lymphocytes. ICB therapies currently approved target the following molecules on T lymphocytes: cytotoxic T-lymphocyte antigen-4 (CTLA-4; ipilimumab), programmed death-1 (PD-1: pembrolizumab, nivolumab, and cemiplimab), and programmed death ligand-1 (PD-L1: atezolizumab, durvalumab, and avelumab). Numerous other T, B, NK cell and other innate immune checkpoints are under investigation. Upon binding of these drugs to their targets, inhibitory signals are hindered, allowing for the activation of T cells and killing of antigen-expressing tumor cells. When ICB is given to cancer patients, checkpoint blockade occurs in all T cells, which include T cells that recognize unique cancer neoantigen-derived epitopes, cancer-associated antigen-derived epitopes present on malignant and normal cells, and self-antigens present only on nonmalignant cells. ICB can also indirectly trigger B cell activation and autoimmune antibody responses through the inhibition of the same immune checkpoints on helper T cells. Toxicities due to ICB-inducing autoimmunity are therefore termed immune-related adverse events (irAEs).

Paraneoplastic neurologic syndromes (PNSs) are a rare group of disorders that deserve mention due to their relationship to neurological irAEs (N-irAEs). PNSs are spontaneous autoimmune responses to neuronal antigens expressed by neoplastic cells, resulting in damage to nonneoplastic neuronal tissues. PNSs are characterized by a combination of distinct clinical signs and/or symptoms, with or without the presence of specific autoantibodies in the sera of affected cancer patients, or the presence of nonclassic neurologic symptoms that improve with cancer therapy or immune suppression. Across all cancers, the prevalence of PNSs in the absence of immunotherapy is less than 1%, but there is considerable variability across cancer types. N-irAEs encompass a wide range or diagnoses, some of which are PNSs; therefore, discussion of PNSs in the context of N-irAEs is relevant. This chapter will describe neurologic irAEs (N-irAEs) seen in patients receiving ICB, which include disorders of central and peripheral nervous systems that occasionally resemble PNSs.[1]

This chapter also discusses neurologic toxicity induced by CAR T cell therapy, also known as immune effector cell-associated neurotoxicity syndrome (ICANS), whose mechanisms are poorly understood but are distinct from those of ICB. CAR T cells are T lymphocytes expressing an engineered receptor whose external structure resembles an antibody and internal structure resembles the internal domains of T cell and co-stimulatory receptors. CAR T cells (FDA-approved in 2017) are produced by the isolation of autologous T cells from the patient, transduction to express the gene of the chimeric receptor, and ex vivo expansion. CAR T cells are then infused into patients, where they proliferate further and

Neurological Complications of Systemic Cancer and Antineoplastic Therapy
https://doi.org/10.1016/B978-0-12-821976-8.00025-6

lyse tumor cells expressing the receptor target. The success of CAR T cell therapies in non-Hodgkin lymphoma, acute lymphoblastic leukemia, and multiple myeloma is attributed to their expression of a surface antigen that is expressed by all tumor cells such as CD19, CD20, or B cell maturation antigen (BCMA).[2] CAR T cells can induce a profound proinflammatory state termed cytokine release syndrome (CRS), in which high-level cytokines such as interleukin (IL-6) and interleukin 1 beta (IL-1b) are produced. CRS can evolve into hemophagocytic lymphohistiocytosis (HLH), also known as macrophage activation syndrome, in which there is multiorgan dysfunction.[3] Some patients with CRS also exhibit ICANS, ranging from mild confusion or encephalopathy to severe aphasia, seizures, cerebral edema, coma, or death. The underlying mechanisms of CAR T cell-induced neurotoxicity are poorly understood. Current evidence argues against direct actions by T cells but rather indirect effects such as brain endothelial cell dysfunction, proinflammatory cytokine penetration, and macrophage activity in the brain parenchyma.[2,4,5]

A third class of novel immunotherapy product, bispecific T cell engagers (BiTEs) such as blinatumomab, which is FDA-approved for acute lymphoblastic leukemia, consists of two single-chain variable fragments of antibody structures binding a tumor-specific antigen and a T cell-specific activation molecule. BiTEs can induce a neurotoxicity syndrome similar to ICANS.[6]

Oncologists, neurologists, and neuro-oncologists should be familiar with the diagnosis and treatment of N-irAEs and ICANS as rapid diagnosis and treatment are crucial to preventing deterioration, permanent morbidity, and mortality.

2 Epidemiology

2.1 Neurologic immune-related adverse events from immune checkpoint blockade therapy

Since 2011, hundreds of thousands of cancer patients in the United States have received ICB therapy. While there are now dozens of indications for the use of ICB, anti-CTLA4 and anti-PD1 antibodies have been prescribed most commonly, with the majority of data on N-irAEs existing in patients with metastatic melanoma, renal cell carcinoma, small-cell and nonsmall-cell lung cancer, and microsatellite instability-high cancers. The incidence of irAEs and N-irAEs depends on the type of cancer, the agent used (with dual-agent therapy causing more irAEs and N-irAEs), and the dose given. Severe irAEs of any type occur in up to 43% of patients receiving ipilimumab[7] and up to 20% of patients receiving anti-PD1/PD-L1 antibodies.[8,9] The presence of irAEs from anti-PD1 and/or anti-PD-L1 antibodies, if properly treated, is associated with improved outcomes,[9–11]

suggesting that irAEs may indicate that the immune system is more likely to exert an antitumor activity. Fatal irAEs occur in 0.3%–1.3% of patients, a rate comparable to that of many conventional chemotherapies.[12] Regardless of ICB agent, fatal toxicities tend to occur early in the course of treatment and evolve rapidly.[13]

Data indicate that the incidence of N-irAEs indicate it may not be as rare as initially estimated (Table 1). Initially, N-irAEs were described in case reports and in large-scale clinical trials. A systematic review of N-irAEs cases up to February 2016 reported an incidence ranging 3.8%–12%; the majority of cases included minor nonspecific symptoms such as headache (55% of cases).[19] Initial clinical trials reported that 0.4%–2.8% of cases developed grade III or IV N-irAEs.[20,21] A recent retrospective cohort study assessed the rate of N-irAEs in 1834 patients who received ICB over a six-year period at a single academic institution. 1.5% of patients (28) experienced N-irAEs limiting activities of daily living. The kinetics of antitumor responses and irAE onset from ICB tend to

TABLE 1   Neurologic immune-related adverse events by incidence (percent of patients receiving immune checkpoint blockade).[14–18]

Diagnosis	Incidence	Clinical features
Peripheral nervous system		
Peripheral neuropathy	0.7%–1.16%	Distal weakness and/or sensory loss
Cranial neuropathies	0.46%	Most frequently facial neuropathy
Guillain-Barré Syndrome	0.25%	Rapidly progressive weakness, back pain, respiratory compromise
Myasthenia gravis	0.12%–0.47%	Fatigable weakness, ptosis, respiratory compromise
Myositis	0.06%–0.55%	Proximal muscle weakness, ptosis, sometimes myalgias
Chronic (CIDP-like) polyneuropathies	0.051%	Indolent progression of distal, usually symmetric weakness and or numbness/paresthesias
Central nervous system		
Hypophysitis	8%–13%	Headache, diffuse weakness, generalized fatigue
Encephalitis	0.5%	Confusion, aphasia, inattention
Meningitis	0.15%	Meningismus, headache
Demyelinating disease	0.08%	Focal neurologic deficits corresponding to new enhancing CNS lesion
Cerebral vasculopathy	0.07%	Focal neurologic deficits corresponding to new diffusion-restricting CNS lesion

vary more widely than responses and adverse effects of conventional chemotherapies. The majority N-irAEs developed within the first six cycles (12–18 weeks) of therapy. Patients receiving dual ICB with anti-CTLA4 and anti-PD1 antibodies had the highest incidence rate of N-irAEs (2.8%), followed by anti-CTLA4 monotherapy (2.2%), and the lowest rate (1.0%) occurred in patients receiving anti-PD1 or anti-PD-L1 antibodies. Sixty-eight percent of patients (19) had co-occurring nonneurologic irAEs such as colitis, pneumonitis, thyroiditis, vitiligo, or myocarditis. The frequency of central versus peripheral nervous system N-irAEs also varied by therapy, with anti-CTLA4 with a greater tendency to cause irAEs in the CNS and anti-PD1/anti-PD-L1 therapy more likely to cause peripheral nervous system irAEs. Eighteen percent of patients had N-irAE of the CNS and PNS.[14] In summary, earlier studies estimated a slightly lower rate of serious N-irAEs, with a recent meta-analysis and retrospective cohort study reporting incidence between one and 2%. With increasing numbers of approved indications for ICB, the prevalence of N-irAEs is expected to increase further.

2.2 Neurotoxicity from chimeric antigen-receptor T cell therapy (ICANS)

In comparison with ICB, CAR T cell therapies have been given to fewer patients, making the calculation of incidence more challenging. Cytokine release syndrome (CRS) is the most common adverse effect of CAR T cell therapy, comprising a diffuse inflammatory reaction resulting in fever, hypotension, and multiorgan dysfunction. In some trials, the incidence of CRS has been as high as 100%, with high-grade CRS occurring in as many as 46% of patients.[22,23] ICANS is a related but distinct adverse effect occurring less frequently. Incidence of any degree or type of neurotoxicity ranges between 20% and 64%, with severe neurotoxicity occurring in 11%–52% of patients.[2,22,24–29]

3 Headache

Headache occurred in more than half of the patients receiving ICB in a recent review.[19] The vast majority of cases were mild and treated symptomatically with acetaminophen or nonsteroidal therapies. As with any new-onset headache in a cancer patient, important history and examination are vital to determining whether an underlying secondary cause is present. Certain serious adverse events mentioned below such as hypophysitis, aseptic meningitis, and cerebral edema (from ICB or CAR T cell neurotoxicity) may present initially as a mild headache. Red flags that should prompt brain imaging and/or neurologic evaluation include new onset,

crescendo nature, unilateral location, high severity, positional nature (worse either supine or standing), and accompanying symptoms. Accompanying nausea, vomiting, visual disturbance, aphasia, confusion, or other new neurologic symptoms should prompt an emergent evaluation. New-onset headache after ICB or CAR T therapy in the absence of other concerning features should be monitored closely and palliated with medications that minimize neurologic side effects or immune suppression.

4 Malignant cerebral edema

Brain metastasis is a common complication of cancer, which is traditionally treated with radiation and sometimes surgical resection. The potential for ICB to induce antitumor immune responses in multiple tumor types within the central nervous system is now irrefutable, with response rates to brain metastases in melanoma and nonsmall-cell lung cancer equivalent to that of systemic disease.[30–34] There are also small numbers of patients receiving CAR T cell therapy with tumor regression in the brain and leptomeninges.[35,36] Neurologic symptoms from brain tumors, whether focal or generalized, can worsen in the setting of an ongoing local antitumor immune response. Cerebral edema from metastases can transiently worsen due to increased tumor lysis, leukocyte infiltration, and local cytokine production.

Moreover, patients who have received CAR T cell therapies can develop diffuse cerebral edema in the absence of central nervous system metastases.[15] The range of symptoms from cerebral edema, with or without metastases, as discussed in greater detail elsewhere in this text, which can occur de novo following immune modulation, recur, or worsen, include: (1) headache, especially following positional changes and accompanied by nausea and/or vomiting; (2) intermittent drowsiness or confusion; (3) seizures; (4) focal neurologic deficits such as hemiparesis, cranial neuropathies, hemineglect or hemianopia, sensory loss, or ataxia; or (5) herniation syndromes, which warrant emergent noncontrast head CT and consideration for monitoring in an intensive care unit. Signs of herniation syndromes due to malignant cerebral edema include:

(1) Transfalcine herniation, which results from displacement of the cingulate gyrus under the falx, compressing both anterior frontal lobes. The ipsilateral anterior cerebral artery is particularly at risk for ischemia due to compression. Signs of transfalcine herniation include bilateral Babinski responses, urgency incontinence, and unilateral leg and proximal arm weakness;

(2) Uncal herniation, resulting from displacement of the uncus of the temporal lobe into the ipsilateral midbrain causing contralateral hemiparesis, third cranial nerve palsy causing ipsilateral pupillary dilation, and potentially compression of the posterior cerebral artery causing occipital lobe infarction resulting in contralateral visual field deficits;

(3) Transtentorial herniation due to a supratentorial cerebral hemispheric mass can also lead to lateral and caudal displacement of the thalamus, hypothalamus, midbrain, and progressive loss of consciousness. Supratentorial masses also cause the obstruction of CSF drainage at the third ventricle or the foramen of Monro, causing ventricular entrapment and hydrocephalus, resulting in gait difficulty, headache, nausea or vomiting, and progressive loss of consciousness;

(4) Upward transtentorial herniation occurs from posterior fossa masses causing displacement of brainstem and/ or cerebellar tissue to shift upward, exerting pressure on the midbrain, causing pupillary or eye movement abnormalities, and can also lead to hydrocephalus; and

(5) Cerebellar tonsillar herniation through the foramen magnum can occur from large supratentorial masses or posterior fossa masses, causing syncope or sudden apnea.[37]

Malignant cerebral edema is managed by a combination of hyperosmolar therapy such as hypertonic saline or mannitol, maintaining high-normal to elevated serum sodium concentration, and corticosteroids. The location of brain tumors, the acuity of the neurologic change, timing from onset of immunotherapy, extent of ongoing treatment response, and overall goals of care of the patient should be accounted for when determining the aggressiveness of treatment, in particular corticosteroid dosage. While there are no trials to guide therapy of malignant cerebral edema, most often this scenario is grave enough to warrant medium- to high-dose dexamethasone (usually a bolus of 10 mg intravenously, followed by at least 8 mg twice a day) for several days. Given that immune responses can be persistent or durable, and do not necessarily cease immediately after interrupting ICB, symptoms of cerebral edema may persist for weeks. In the case of combined immune-related and tumor growth-related cerebral edema, deficits may be permanent.

5 Peripheral nervous system

Adverse effects that impact the peripheral nervous system are seen in ICB but not CAR T cell therapy. Peripheral N-irAEs are roughly twice as common as central N-irAEs, particularly in patients receiving anti-PD1/anti-PD-1 antibodies, and often resemble peripheral nervous system disorders in noncancer patients.[14,16] In roughly one-fifth of cases, N-irAEs are present in the CNS and PNS simultaneously.[14] The majority of PNS N-irAEs are antibody and B cell-driven; however, there is evidence for the role of T cells in certain disorders. While the spectrum of severity and prognosis with ICB-induced neuropathy is in general favorable, ICB-induced myasthenia gravis and myositis can portend a worse prognosis.[17] A recent systematic review of neuromuscular complications of anti-PD1 therapy determined that evidence did not support the stated diagnosis in 13% of cases,[38] highlighting the need for thorough, multidisciplinary workup of these disorders.

5.1 Neuropathy

Neuropathy is a broad category of syndromes causing peripheral nerve dysfunction with diverse causes: trauma, toxins, nutritional deficiency, infection, hereditary syndromes, microvascular dysfunction, and autoimmunity. Estimated incidence of immune-mediated neuropathy in patients receiving ICB is 0.7%–1.16%.[16,18] In general, neuropathic findings of progressive muscle weakness and/or sensory loss can be noticed by most clinicians, but etiology is difficult to ascertain. Immune-mediated neuropathy in the absence of cancer is routinely diagnosed and treated by neurologists who specialize in neuromuscular disorders. Diagnosis is made by recognition of clinical syndromes in combination with laboratory or nerve function testing, which most often lack a pathognomonic autoantibody or antigen. Diagnosis often requires a neurologist, as weakness is sometimes difficult to distinguish from fatigue, preexisting cytotoxic chemotherapy-induced neuropathy is often present, and localization of the peripheral nerve abnormalities is key. Workup includes a thorough neurologic examination, serologic and/or cerebrospinal fluid testing, electromyography-nerve conduction study (EMG-NCS), and occasionally MRI of the brain or spine.

The structures and basic anatomy of the peripheral neuron and its myelin are important to understanding the diverse locations, mechanisms of pathogenesis, and signs and symptoms of autoimmune neuropathies (Fig. 1). Sensory and motor neurons enter and exit the spinal cord at or near the level of the body where they serve to function. The cell bodies of motor neurons reside within the anterior horns of the spinal cord, while the cell bodies of sensory neurons reside in the dorsal root ganglia outside the spine. Moreover, autonomic neurons are sometimes selectively affected by autoimmune attack in their locations in the periphery. Sensory, motor, and autonomic neurons share some but not all antigens, and their axons, surrounded by Schwann cells,

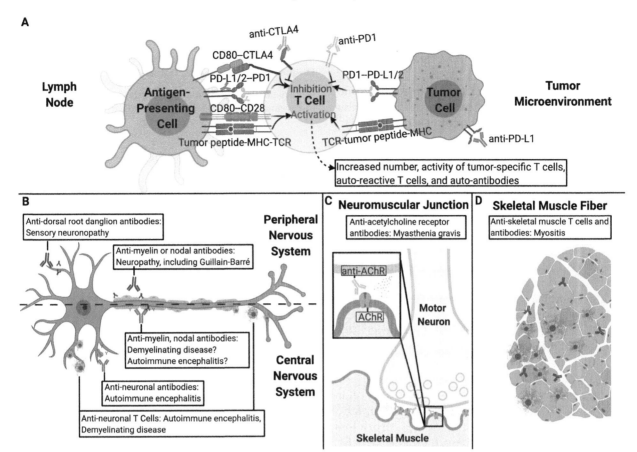

FIG. 1 Mechanisms of ICB-induced encephalopathy, neuropathy, neuromuscular disease, and myopathy. (A) T cell immune checkpoints induce inhibitory signals that counteract the activation provided by T cell receptor signaling upon antigen detection. Immune checkpoint blockade (ICB) antibodies such as anti-CTLA4, anti-PD1, and anti-PD-L1 bind extracellularly and prevent ligand binding from tumor cells and antigen-presenting cells. ICB targets all T cells, including tumor and self-reactive T cells, and can lead to autoimmune T cell and antibody-mediated diseases. (B) Numerous regions of neurons and myelin can be targeted by autoreactive T cells, which become activated by ICB. (C) ICB induces myasthenia gravis through the induction of antibodies to postsynaptic neuromuscular junction proteins such as the nicotinic acetylcholine receptor (AChR), although a significant proportion of patients are negative for anti-AChR antibodies. (D) ICB-induced myositis, which frequently co-occurs with myasthenia gravis, is associated with autoreactive T cells and antibodies to skeletal muscle antigens. Figure created with BioRender.com.

consist of the myelin sheath interrupted by nodes of Ranvier. Antibodies bind to targets on nodes of Ranvier (on the axon), myelin, or to cell bodies. Therefore, clinical syndromes mirror the neuron type under attack, and sometimes the diagnosis is made by EMG-NCS through the detection of the type of neurophysiologic dysfunction induced by the structure of the neuron or myelin under attack.[39]

The most practical division of neuropathy type is by acuity, indicating which conditions require urgent inpatient versus outpatient neurologic evaluation (e.g., acute inflammatory demyelinating polyradiculoneuropathy, AIDP, also known as Guillain-Barré syndrome, GBS). In cases with subacute or chronic onset, it is also reasonable to consider other common causes of neuropathy that may be the only underlying etiology or co-occurring with ICB-induced neuropathy. Patients should be asked about alcohol use, and basic serologic testing could be performed to assess for diabetes or numerous

vitamin deficiencies depending on their nutritional status. Nutritional causes of neuropathy include deficiencies in B12, B1 (thiamine), B6, vitamin E, copper, or zinc. Thyroid-stimulating hormone (TSH), HIV, serum immunofixation, and light chains should also be tested.[40,41] Given occasional overlap with myasthenia gravis and myositis, it would also be reasonable to test for creatine phosphokinase, perform an EKG to rule out myocardial involvement, and perform EMG-NCS.

Immune-mediated neuropathies can be divided into more than a dozen types[39]; many but not all types have been described following the administration of ICB. Below, the principal types are divided by the timing of onset (acute versus nonacute) and the type or part of neuron affected. While the below diagnoses are typically applied, patient variability, disease heterogeneity, and potentially differing mechanisms of immune pathogenesis most likely lead to the distinct features of ICB-induced neuropathies described below.

5.1.1 Subacute to chronic onset: Cranial neuropathy; sensory and motor neuropathy resembling chronic inflammatory demyelinating polyneuropathy

Impairment of cranial nerves occurs in roughly one-third of patients with ICB-induced neuropathy, making it the most common type of neuropathy seen (Table 1).[16] Most patients present with upper and lower facial nerve weakness resembling Bell's palsy, with a subacute onset. Other cranial mononeuropathies occur, such as trigeminal neuralgia. MRI of the brain demonstrates gadolinium contrast enhancement of the affected cranial nerves in the vast majority of cases; often, numerous cranial nerves can exhibit enhancement despite the lack of clinical evidence of multiple cranial neuropathies. CSF frequently shows a lymphocytic pleocytosis with elevated protein.[18,42-44] Following the interruption of ICB and initiation of prednisone, resolution of cranial neuropathies following completion of the prednisone taper is expected.[18,45] Facial weakness can also be seen in GBS; however, it most often occurs with much greater severity than isolated facial weakness and is preceded by ascending limb weakness.

Sensory neuropathy can develop in isolation or in combination with motor or autonomic neuropathy. Patients present with indolent, progressive sensory loss or pain, often accompanied by paresthesias in the distribution of affected dermatomes. Sensory dysfunction can be length-dependent or nonlength-dependent (the latter distinguishes it from chemotherapy-induced neuropathy). Small sensory nerve fiber dysfunction affects nociception and temperature sensation, manifesting as painful paresthesias most commonly in a length-dependent fashion. Large sensory fiber dysfunction manifests with vibration or proprioception deficits, leading to gait ataxia. In rare cases, the brachial or lumbosacral plexus is selectively involved, which can cause severe pain or weakness that involves multiple sensory neurons. Sensory nerve conduction studies, which can only detect large fiber conduction, most frequently show diminished velocities, with or without conduction block, indicating a demyelinating process.[18,46,47] Diminished amplitudes are also seen, indicating axonal loss, more commonly than among patients without cancer (Table 2).

Motor neuropathies most frequently present with limb weakness, which is most commonly distal. If present for weeks to months, muscle atrophy and fasciculations may be seen. Nerve conduction studies can show decreased velocities in the case of demyelination and decreased amplitudes in cases of axonal damage. EMG can show evidence of denervation such as increased insertional activity, fibrillation potentials, and reduced recruitment of motor unit potentials. As mentioned above, weakness in multiple nerve distributions suggests selective dysfunction at the plexus.[14,18,46,50,51]

TABLE 2   Electrophysiological properties of neuropathic, neuromuscular, and myopathic N-irAEs.[17,18,48,49]

Diagnosis	Nerve conduction study features	Electromyogram features
Sensory neuropathy (demyelinating)	Prolonged sensory nerve distal latency; slowed sensory conduction velocity	Absence of fibrillations or fasciculations
Sensory neuropathy (axonal)	Decreased SNAP amplitude	Absence of fibrillations or fasciculations
Motor neuropathy (demyelinating; classic idiopathic Guillain-Barré)	Normal CMAP amplitude	Prolonged distal motor latency, slowed conduction velocity, prolonged F wave
Motor neuropathy (axonal; often ICB-induced Guillain-Barré)	Reduced CMAP amplitude	Occasional fibrillations or fasciculations (weeks to months)
Myasthenia gravis	> 10% decrease in CMAP amplitude with repetitive nerve stimulation	Increased variability/"jitter" in latency of muscle fiber activation in single-fiber EMG[a]
Myositis	Normal SNAP and sensory latency; normal to reduced CMAP amplitude	Fibrillations, positive sharp waves, spontaneous activity; early recruitment; increased jitter in single-fiber EMG[a]

[a] Jitter is not specific for MG or myositis; it also occurs in amyotrophic lateral sclerosis and Lambert-Eaton myasthenic syndrome.
Abbreviations: CMAP, compound motor action potential; SNAP, sensory nerve action potential; Note: Electrophysiological abnormalities for neuropathy (including GBS) usually do not appear until at least 10 days after disease onset.

Interventions depend on the severity of symptoms based on the National Cancer Institute Common Terminology Criteria for Adverse Events (CTCAE) version 5.0 grade.[52] CTCAE was developed to standardize the severity of adverse events in clinical trials for cancer-directed therapies but is widely used to grade severity outside of trials. Grade 1 refers to AEs in which the patient is asymptomatic and symptoms are only noticed on clinical examination or diagnostic testing, and continuation of ICB can be considered with careful weekly re-examination. Grade 2 events occur when persons notice symptoms with mild impact on activities of daily living. Grade 3 and higher events are considered severe and often lead to hospitalization for workup and monitoring. Grade 3 events are defined by the need for assistance with some activities of daily living. Grade 4 events are marked by severe impairments such as inability to walk, with possible breathing and/or swallowing impairment.[40,41,53,54]

Treatment guidelines of higher-grade neuropathy are based on consensus from case reports and literature series rather than randomized trials. Grade 2 neuropathies can be treated with oral prednisone, started at roughly 50–60 mg daily and tapered over 6–8 weeks; the vast majority of patients experience improvement or resolution of symptoms.[14,18] Once symptoms resolve, resumption of ICB could be considered. Clinicians have described cases that begin in a chronic form, then accelerate and resemble GBS once a second ICB agent is added, or when ICB is switched.[18] Grade 3 or higher cases should be treated with intravenous methylprednisolone and discontinuation of ICB therapy. If no improvement, plasmapheresis is performed, or intravenous immune globulin or rituximab is given, as is done for sporadic chronic inflammatory demyelinating polyneuropathy (CIDP).[55,56] Severely refractory cases after guideline-based therapies have led some clinicians to use nonsteroid cytotoxic immune-suppressive mediations such as mycophenolate mofetil.[57] Duloxetine, pregabalin, or gabapentin could be prescribed for neuropathic pain.[40]

5.1.2 Autonomic neuropathy

Autonomic neuropathy has been observed in a small number of persons following ICB. Any combination of autonomic symptoms can occur from this neuropathy, such as anhidrosis, gastroparesis or bowel dysmotility, pupillary mydriasis, bowel, bladder, or sexual dysfunction, or orthostatic hypotension.[18,58] Cases of autonomic neuropathy may feature co-occurring sensory neuropathy, may be responsive to corticosteroids, or may be severe and refractory to all immune suppressive efforts. Autonomic neuropathy can sometimes be a feature of GBS (as discussed later in this chapter), or due to a paraneoplastic neurologic syndrome (PNS), and paraneoplastic antibody serological testing can be considered. A full neurologic examination should be performed, and EMG-NCS should be considered for concomitant peripheral sensorimotor neuropathy (Table 2). Patients should be assessed for orthostatic hypotension and, if present, provided counseling and supportive care such as compression stockings, physical therapy, and fludrocortisone, midodrine, droxidopa, atomoxetine, or pyridostigmine.[59–62]

5.1.3 Mononeuritis multiplex

Cases in which the pattern of sensorimotor neuropathy appears in multiple locations in an irregular and asymmetric distribution, known as mononeuritis multiplex, should be evaluated for evidence of vasculitis. Vasculitic neuropathy can occur in isolation or in the presence of systemic vasculitis. Serologic testing is particularly helpful in assessing for vasculitis as well as other causes of mononeuritis multiplex: tests to consider include ESR, CRP, C3, C4, ANA, anti-dsDNA antibodies, SSA, SSB, RF, c and p-ANCA, vitamin B12, and methylmalonic acid.

5.1.4 Guillain-Barré syndrome—Acute inflammatory demyelinating polyradiculoneuropathy (AIDP)

Distal limb sensory loss, paresthesias, weakness, with or without dyspnea progressing over hours in patients receiving ICB, should prompt an evaluation for GBS. Other clinical features of GBS may be present such as back pain, tachycardia, gastroparesis, cranial neuropathies, ataxia, or vertigo. Cranial nerve and vestibular dysfunction signify the less common GBS variant Miller-Fisher syndrome. GBS is roughly five times as common in patients receiving ICB than CIDP-type variants, and given the dangers of respiratory failure, treating clinicians need to be aware that admission for expedited workup, monitoring, and treatment is often warranted when symptoms are acute.[16,18]

Patients under evaluation for GBS should have a regular assessment of negative inspiratory force and vital capacity every 3–4 h by a respiratory therapist. Evaluation includes spine MRI, and potentially also a brain MRI if cranial neuropathies are also present. Enhancement of lumbar nerve roots including the cauda equina is common, and many cases demonstrate diffuse leptomeningeal (including cranial nerve) enhancement. CSF should be sent for cell count, protein, glucose and Gram stain, culture, PCR for HSV1 and 2, cytology, oligoclonal bands, and autoimmune encephalopathy panels. CSF analysis most frequently shows a moderate lymphocytic pleocytosis, which distinguishes ICB-induced GBS from most classic cases of GBS where an elevation of WBCs is typically not seen. Elevated CSF protein is nearly universal.[18,46,47,50,55,57,63–65] EMG-NCS can be normal if performed within the first 10 days of disease; therefore, empiric treatment must precede EMG-NCS (Table 2). As mentioned previously for chronic neuropathy, demyelination manifests as diminished nerve conduction velocity while axonal damage leads to diminished amplitudes.[66] Whereas classic GBS is most commonly demyelinating, ICB-induced GBS more frequently shows diminished amplitude that signifies axonal loss. Numerous ganglioside antigens have been associated with non-ICB-related cases of GBS, and serum antibodies to these gangliosides should be tested. Most cases of ICB-induced GBS are seronegative; seronegativity in combination with numerous pathology specimens showing T cell infiltration argues for direct cytolytic T cell activity as a likely contributing mechanism of many cases of disease.[18]

GBS is by definition grade 2 or higher and warrants holding ICB once diagnosed. Treatment of ICB-induced GBS, in contrast to other cases of GBS, should include high-dose methylprednisolone at 1 mg/kg given for 5 days, in tandem with plasmapheresis or IVIG.[40,41,53,54] Patients should continue to be monitored as inpatients where ICU transfer can occur if needed. Patients with severe disease may require intubation and symptomatic management of autonomic dysfunction. As with more indolent peripheral neuropathy, neuropathic pain can be significant but can be mitigated with duloxetine, pregabalin, or gabapentin.[40]

5.2 Myasthenia gravis and myositis

Idiopathic myasthenia gravis (MG) is an autoimmune disorder of the neuromuscular junction occurring sporadically and associated with thymoma in a substantial proportion of cases. Idiopathic MG is characterized by fluctuating, fatigable weakness of the limbs or bulbar muscles that worsen over the course of the day or with repetitive limb use. In practice, fatigability is not always reported or found on examination. Serum antibodies that bind and/or block or modulate the postsynaptic neuromuscular junction proteins are found in 65%–95% of cases, most commonly the nicotinic acetylcholine receptor (nAChR).[67] Antibodies against the muscle-specific tyrosine kinase (MuSK) or lipoprotein receptor–related protein 4 (LRP4) protein may be present in cases seronegative for nAChR.[68,69] MG often but not always leads to abnormalities on EMG-NCS (Table 2), most notably a decremental response on repetitive nerve stimulation. The Tensilon test is also occasionally performed, whereby edrophonium or neostigmine (acetylcholinesterase inhibitors) is intravenously administered and patients are serially examined for transient improvement in weakness over minutes. All patients are imaged to determine the presence of thymoma and, if present, resection is performed.[70] Rarely, myasthenic crisis can occur, with acute respiratory or diffuse weakness. Myasthenic crisis can be treated effectively with plasma exchange or IVIG given the disease is antibody-mediated. Chronically, MG is managed with pyridostigmine (which is not immune modulatory but improves neuromuscular transmission) and immune suppression in the form of corticosteroids, IVIG, and/or immune suppressive medications such as azathioprine.[71] While a spectrum of outcomes exists following treatment, prognosis in most patients with idiopathic MG is positive.

Lambert-Eaton myasthenic syndrome (LEMS) is a paraneoplastic neurologic disorder 10 times as rare as idiopathic MG and is strongly associated with small-cell lung adenocarcinoma. While small-cell lung carcinoma is often treated with anti-PD-L1 antibodies, incidence of LEMS remains extremely low, with two cases being reported in the literature.[72,73] In contrast to MG, LEMS is characterized by anti-P/Q-type calcium channel antibodies acting in the presynaptic membrane, increasing strength with use, and incremental response with repetitive nerve stimulation/nerve conduction study.[72]

Myositis, sometimes referred to as idiopathic inflammatory myopathy outside of the context of ICB, classically includes dermatomyositis (DM), polymyositis (PM), and inclusion body myositis (IBM). More recently, there has been movement to eliminate PM as a diagnosis and re-classify inclusion body myositis as a noninflammatory myositis. A revised categorization based on newly found autoantibodies divides myositis into four diagnoses: dermatomyositis (DM), immune-mediated necrotizing myopathy, antisynthetase syndrome, and overlap myositis.[74] All types classically present with insidious onset symmetric proximal limb weakness, elevated muscle enzymes, and inflammatory muscle biopsies. Occasionally, patients present with bulbar weakness including ptosis. Muscle atrophy and/or tenderness to palpation may be present on examination.[75] There are numerous autoantibodies identified, including antihistidyl-ARN-t-synthetase (Jo1), antithreonine-ARN-t-synthetase (PL7), antialanine-ARN-t-synthetase (PL12), anticomplex nucleosome remodeling histone deacetylase (Mi2), antiKu, antipolymyositis/systemic scleroderma (PMScl), antitopoisomerase 1 (Scl70), and antisignal recognition particle (SRP) antibodies. Diagnosis is most often achieved through a combination of clinical examination, serologic testing, MRI, and muscle biopsy data.[76] EMG-NCS may show myopathic changes (Table 2), which include normal sensory nerve conduction studies, normal or low amplitude motor nerve conduction studies, and abnormal spontaneous activity with or without short-duration, low-amplitude motor unit potentials with an early recruitment pattern on EMG. Muscle biopsy shows a lymphoplasmacytic infiltrate of the perimysium and perivascular regions in dermatomyositis and anti-synthetase, but a lack of lymphocytic infiltrate and the presence of necrosis in muscle fibers in necrotizing immune-mediated myopathy. Treatment, which has rarely been substantiated by randomized studies, consists of corticosteroids such as prednisone, started at 0.5 mg/kg/day and increased to 80 mg/day. In severe cases, methylprednisolone is given intravenously for 5 days. Afterward, prednisone is prescribed and tapered after 4–6 weeks, and nonsteroidal immune suppressants such as IVIG, rituximab, methotrexate, mycophenolate mofetil, or azathioprine are used.[77–79] Due to the association between immune-mediated myopathy and cancer, thorough screening for underlying malignancy is also performed.[80]

Myasthenia gravis is estimated to occur in 0.12%–0.47% of patients following ICB, roughly 10-fold higher than the general population.[16,17] Although rare, it is associated with polyneuropathy, thyroiditis, myositis and myocarditis, the latter being the most fatal of immune-related adverse events.[14,17,48,81] ICB-induced MG in isolation can also present with a fulminant myasthenic crisis leading to respiratory failure and rapid decline. Several important clinical observations should be kept in mind while assessing for potential ICB-induced MG, regardless of the acuity of onset. In comparison with idiopathic MG, ICB-induced MG more frequently causes bulbar or respiratory weakness earlier in the clinical course. ICB-induced MG also co-presents with myositis and myocarditis. Symptoms usually begin within one to four cycles of treatment.[14,17,82] Workup should include rapid respiratory and neurologic assessment if onset is acute. Testing should include parallel assessment for MG, myositis, thyroiditis, myocarditis, and potentially neuropathy. Serum should be tested for the anti-AChR antibody-blocking or modulating panel (for MG), ESR and CRP, creatine kinase (myositis), anti-striated muscle antibodies (myositis), TSH and free T4 (thyroiditis), troponin (myocarditis), and EKG (myocarditis). If creatine kinase is elevated, serum should be tested for autoantibodies associated with inflammatory myopathies (anti-Jo1, PL7, PL12, Mi2, PMScl, Scl70, and SRP). Electrophysiology is crucial in order to establish the diagnosis, as MG antibody testing is both less sensitive and less specific in ICB-induced MG, and clinically distinguishing MG from neuropathy or myositis can be difficult. If standard EMG-NCS is negative, single-fiber EMG and potentially muscle MRI or biopsy should be performed (Table 2).[49,83]

Treatment for ICB-induced MG, in contrast to idiopathic MG, should begin with corticosteroids. In addition to holding further ICB, grade 2 ICB-MG should be treated with pyridostigmine 30 mg three times a day, and low-dose prednisone 20–100 mg daily. Low dose (0.5–1 mg/kg/day prednisone) should also be started in cases of grade 2 myositis, and tapering should be performed over weeks based on improvement in creatinine kinase. Cases of grade 3 ICB-induced MG or myositis should be treated inpatient, with intravenous methylprednisolone at 1 mg/kg, for 5 days. Plasma exchange or IVIG can be considered for MG or myositis if there is no improvement after 2–3 days of methylprednisolone. Care should be taken to avoid medications that may worsen MG such as beta blockers, ciprofloxacin, and intravenous magnesium.[40,41,53,54] Many patients respond to corticosteroids, and some improve rapidly; however, the fatality rate in patients with ICB-induced MG has been reported to be as high as 20%.[17,38] The guidelines of the American Society of Clinical Oncology and the National Cancer Institute state that ICB should be permanently discontinued if patients develop grade 3 or higher MG or myositis[40,54]; however, some oncology and supporting teams have administered rechallenge of ICB with the same or different agents under close surveillance if MG or other AEs resolve rapidly.[17,83] The presence of myocarditis nearly uniformly leads to permanent discontinuation of ICB.

6 Central nervous system

Immune-checkpoint blockade-induced disorders of the central nervous system are roughly half as common as those of the peripheral nervous system, are more frequently induced by anti-CTLA4 blockade with ipilimumab and resemble previously classified CNS disorders induced by autoimmunity. Hypophysitis is most frequently considered an endocrine disorder but will be briefly described below. Aseptic meningitis, encephalitis, and myelitis may occur in isolation or in combination with one another or with ICB-induced PNS disorders. Rarely vascular events such as posterior reversible encephalopathy syndrome or vasculitic changes in the CNS occur, which are thought to be due to autoimmune activity within cerebral arteries and capillaries.

6.1 Hypophysitis

Autoimmune hypophysitis occurs in roughly one in nine million persons sporadically[84]; in contrast, 8%–13% of oncology patients receiving anti-CTLA4 antibodies alone or in combination with anti-PD1 or PD-L1 antibodies develop hypophysitis.[85,86] Onset typically occurs six to 12 weeks after starting therapy,[86,87] and in 89% of patients, initial symptoms are headache, diffuse weakness, and generalized fatigue.[87] Ten to twenty percent of patients report symptoms more telling of pituitary or endocrine dysfunction such as nausea, vomiting, anorexia, weight loss, visual changes, heat or cold intolerance, and occasionally confusion.[86] Half of patients with hypophysitis have hyponatremia, an indicator of the major consequence of hypophysitis, secondary adrenal insufficiency. Other common laboratory abnormalities are adrenocorticotrophic hormone (ACTH) and thyroid-stimulating hormone (TSH) deficiencies, followed by other anterior pituitary hormone deficiencies. Diabetes insipidus from posterior pituitary dysfunction is extremely rare.[88,89] Clinicians must be highly vigilant for symptoms of adrenal crisis such as hypotension, severe electrolyte derangements, and dehydration. On neurologic examination, visual field deficits are rare; there should be a low threshold to test sera for electrolyte disturbances and hormone levels in collaboration with oncologists and endocrinologists. MR of the brain often but not always demonstrates avid gadolinium enhancement and enlargement of the pituitary, and such abnormalities typically develop on average 1 week prior to detectible blood endocrine disturbances.[86] Treatment with corticosteroids is highly effective, often inducing clinical responses within days.[87] Most often, adrenal insufficiency is permanent, necessitating chronic low-dose corticosteroids.[86,90,91] Thyroid dysfunction, which is treatable with levothyroxine, is more likely to reverse but still only does so in a minority of patients.[86,91]

Hypogonadotropic hypogonadism leading to low estradiol levels in women and low testosterone levels in men as well as disturbance in prolactin secretion can occur with ICB therapy.[92]

6.2 Aseptic meningitis

Autoimmune inflammation of the meninges, aseptic meningitis, occurs in approximately 0.15% of ICB-treated patients,[16] although mild forms of meningitis may underlie mild idiopathic headache.[19] Aside from headache, other presenting symptoms include nausea/vomiting, fever, meningismus, papilledema, or signs of cerebral cortical irritation such as confusion (raising high suspicion for concomitant encephalitis), seizures, or focal neurologic deficits.[40,53,54] The mainstay of testing should be MR of the brain (with pituitary protocol to assess for hypophysitis) and total spine with and without gadolinium contrast, serological testing for morning cortisol and ACTH, followed by lumbar puncture. CSF analysis should include the measurement of opening pressure, cell count, protein, glucose, and Gram stain, culture, PCR for HSV1 and 2 and other viral encephalitis, cytology, and circulating tumor cell analysis if available, oligoclonal bands, and autoimmune encephalopathy.[40,54] CSF most often contains a lymphocytic pleocytosis and elevated protein with no further abnormalities. Occasionally, CSF oligoclonal bands are seen, indicating intrathecal production of antibodies. Often patients are treated empirically with acyclovir to cover for possible false-negative HSV PCR and are serially examined. Imaging and/or lumbar puncture may also be repeated to assess for improvement or resolution. Treatment for ICB-induced meningitis can be less emergent than for other N-irAEs as disease is often less severe. Most often patients may be monitored, but once infection has been ruled out, oral (0.5–1 mg/kg prednisone) or intravenous (1 mg/kg methylprednisolone daily for 5 days) corticosteroids should be considered if symptoms are moderate or severe.[40,53,54] Immune checkpoint blockade is held until symptoms of meningitis or radiographic appearance improves. Meningeal enhancement on MR brain or spine may persist after the resolution of symptoms.

Two cases of aseptic meningitis following ICB provided compelling evidence that the underlying etiology of the meningitis was de novo central nervous system sarcoidosis. Both patients had de novo, ICB-induced, active biopsy-proven pulmonary sarcoidosis and severe meningitis. While pathologic diagnosis was not definitively confirmed by meningeal biopsy, the exclusion of other etiologies was rigorous, radiographic appearance of meningitis was bulkier than typical aseptic meningitides, and symptoms and radiographic meningitis improved following ICB discontinuation and immune suppressive treatment.[93,94] While extremely rare, new pulmonary symptoms in patients following ICB concomitant with headache or other signs of meningitis may require additional workup (serum and CSF angiotensin-converting enzyme) for systemic and CNS sarcoidosis.

6.3 Encephalitis, demyelination syndromes, and cerebellitis

Lymphocyte or antibody-mediated damage to the brain parenchyma can manifest as a focal or diffuse syndrome depending on the location and extent of the immune response. N-irAEs of the brain parenchyma occur in roughly 0.5%–0.9% of patients receiving ICB,[16,95] have a myriad of presentations, including diffuse cerebral, brainstem or limbic encephalitis, cerebellitis, focal demyelination syndromes, or paraneoplastic encephalitides.[96] Twenty percent of brain parenchymal N-irAEs co-occur with other N-irAEs such as meningitis and peripheral neuropathy.[14,95] N-irAEs of the brain parenchyma are slightly more common in patients treated with anti-PD1 therapy but have been reported in anti-CTLA4 or anti-PD-L1 monotherapy.[96] While a range of severities can be seen with ICB-induced encephalitis, and disease is usually treatable, rare fatalities have been reported, with no clear difference in odds of severity among ICB agents.[95]

ICB-induced encephalitis symptoms begin slightly later on average in comparison with other N-irAEs, a median of 80 days after starting ICB. Presenting symptoms are diverse and most commonly include confusion and headache in 50% and 36% of patients, respectively. Seizures occur in roughly 20% of patients, and fever is less common (18%). Patients should undergo systemic infectious workup, MR of the brain and often spine, electroencephalography (EEG) and CSF analysis for infection, malignancy, and specific autoimmune diseases including paraneoplastic neurologic disorders. In particular, EEG should be performed early in workup to rapidly treat nonconvulsive seizures. Specific tests from the CSF should include cell count, protein, glucose and Gram stain, culture, PCR for HSV1 and 2 and other viral encephalitis, bacterial culture and rapid testing, cytology, oligoclonal bands, and autoimmune encephalopathy panels should be sent. In one series, the most common MR brain abnormality reported was temporal or occipital lobe enhancement, occurring in 22% of patients. Sixty-eight percent of patients recovered completely, 22% of patients had persistent neurologic abnormalities, and 9% died.[95] Cases have been described in terms of the locations of brain that appear most affected based on neurologic findings and MR of the brain. Cases range from limbic encephalitis (including paraneoplastic antibody-positive and negative cases),[95,96] diffuse encephalitis satisfying diagnostic criteria for Hashimoto's encephalopathy,[14] diffuse fulminant encephalitis extending to the brainstem,[97] or cerebellitis.[46,98]

While some cases of encephalitis definitively appear to be antibody-mediated,[96] autopsy from two lethal cases showed T lymphocyte-predominant infiltration, with no B or plasma cells.[95,97] Encephalitis can be due to paraneoplastic neurologic syndromes such as NMDA receptor or anti-Ma2 encephalitis,[96,99] in which antigen expression is suspected to be shared between the involved region of brain parenchyma and the tumor. As is the case for myasthenia gravis antibody testing, paraneoplastic antibody positivity can be more common after the administration of ICB without the accompanying clinical syndrome. While determining if a given encephalopathy is paraneoplastic may seem academic, rapidly ruling out other potential causes of encephalopathy is important and often challenging.

Aside from encephalitis, ICB-induced brain parenchymal damage has occurred from focal immune infiltration resembling demyelination,[14,100,101] in addition to damage from immune vascular attack causing posterior reversible encephalopathy syndrome (PRES).[14,50,102] Moreover, demyelination syndromes from ICB can also cause myelitis with the involvement of the spinal cord.[90,101,103] MR of the total spine should be performed on all persons with new focal neurologic deficits, but spine imaging should be prioritized over brain imaging when patients develop a clear cervical, thoracic, or lumbar level deficit, severe back pain, saddle anesthesia, urinary or bowel incontinence, a central or hemicord syndrome, or paraplegia. Workup for demyelination should otherwise include the same CSF studies as mentioned for encephalitis and serum testing for antibodies against aquaportin-4 (AQP4) and myelin oligodendrocyte glycoprotein (MOG). Workup and treatment for PRES should include reviewing medications, blood pressure, and renal function to eliminate all contributors.

Treatment for ICB-induced encephalitis includes holding ICB, considering concurrent iv acyclovir until PCR results return negative, and administering 0.5 mg/kg-1.0 mg/kg prednisone for mild disease, 1–2 mg/kg methylprednisolone for moderate disease, or 1 g intravenous methylprednisolone daily for 3–5 days with or without IVIG for 5 days in severe disease. If no improvement or if autoantibodies have been detected, then treating clinicians should consider administering rituximab or plasmapheresis.[40,54] Cases of transverse myelitis most often warrant high-dose methylprednisolone and permanent discontinuation of ICB as mentioned below.[40,54,90]

7 CAR T cell-mediated neurotoxicity (immune effector cell-associated neurotoxicity syndrome)

CAR T cell therapies have been administered to over a thousand patients with hematologic and lymphoid malignancies, for which they are FDA-approved, in addition to many solid tumor patients in clinical trials. CAR T cell-mediated neurotoxicity, hereafter referred to as immune effector cell-associated neurotoxicity syndrome (ICANS), manifests with focal or diffuse neurologic symptoms which are not from direct activity of CAR T cells but rather macrophage and monocyte activation, endothelial cell disruption, cytokine penetration into the brain, and potentially activation of astrocytes.[4,5] Timing of onset of ICANS can vary; a small number of patients have concomitant cytokine release syndrome (CRS) with ICANS in the first 7 days, while the majority of patients with ICANS present seven to 21 days postinfusion.[25,26,28,104] Numerous factors have been associated with increased risk of ICANS, including younger age, preexisting neurologic comorbidities, extramedullary disease, fludarabine-containing conditioning, higher total CAR T cell dose, higher peak CAR T cell counts, early or severe CRS, cytopenia (in particular thrombocytopenia), and diffuse intravascular coagulation.[104,105]

Due to a frequency as high as 60% in patients receiving CAR T cell therapy,[104] baseline neurologic examination and brain MRI are often performed prior to infusion and examination is systematically repeated over the high-risk period after infusion in all patients. Common ICANS symptoms include confusion (~ 90%), which manifests with signs such as fluctuating drowsiness, inattention, agitation, calculation deficits, or apraxia; aphasia (50%), headache (40%), tremor (33%), difficulty with handwriting, frontal release signs, memory loss, meningismus, dysarthria, focal weakness (10%), ataxia, and seizures.[24,104,106] Of note, headache in isolation occurs frequently in CRS and is not normally predictive of ICANS.[3] Radiographic findings are similarly diverse: diffuse cerebral edema is the most common finding, followed by focal areas of restricted diffusion and T2/FLAIR abnormalities observed in various locations including bilateral thalami, the splenium of the corpus callosum, or may be multifocal.[104,106,107] Microhemorrhages and leptomeningeal enhancement can also develop.[105]

Severity of ICANS is determined by the consensus statement from the American Society for Blood and Bone Marrow Transplantation, who proposed the Immune Effector Cell-Associated Encephalopathy (ICE) score (Table 3).[3] In addition to the ICE score, each patient should undergo neurologic examination to assess for other neurologic domains that count toward the CTCAE grade. These other domains include level of consciousness, motor symptoms, seizures, and signs of elevated ICP/cerebral edema, which may occur with or without encephalopathy (Table 3). Before proceeding to treatment, proper workup for systemic and/or CNS infection should be performed expeditiously as recommended for ICB-induced encephalitis, often requiring serologic and CSF testing.

TABLE 3 Immune effector cell-associated encephalopathy ICE score.[3]

Exam component	Mental status features examined	Total points possible
Orientation	Orientation to year, month, city, hospital	4
Naming	Ability to name three objects (e.g., clock, pen button)	3
Following commands	Ability to follow simple commands (e.g., "Show me your thumb")	1
Writing	Ability to write a standard sentence (e.g., "Today is a sunny day")	1
Attention	Ability to count backward from 100 by 10	1
Total score		10

Scoring: 10, no impairment; 7–9 grade 1 ICANS; 3–6 grade 2 ICANS; 0–2 grade 3 ICANS; 0 due to the inability of patient to complete assessment, grade 4 ICANS.

Prior to the administration of CAR T cell therapy, patients should undergo brain MRI to establish a baseline; many clinicians also start a prophylactic dose (500 mg twice daily) of the antiseizure medication levetiracetam, to be continued for several weeks. Patients with early signs of encephalopathy (ICE score 7–9; grade 1 if no further findings on neurologic assessment) suspicious for ICANS should be considered strongly for noncontrast head CT to assess for cerebral edema, potentially followed by continuous EEG monitoring. If further neurologic deterioration occurs, there should be a low threshold to perform continuous EEG monitoring and brain MRI with and without gadolinium contrast, and start dexamethasone, usually one or two empiric doses followed by repeat neurologic examination. Initial doses vary widely from 4 to 10 mg twice daily. Worsening neurologic function may warrant starting dexamethasone per the clinician's discretion, up to 10 mg every 6 h. Except in cases of malignant cerebral edema, in which intensive care unit monitoring and other measures for malignant edema are simultaneously used (see Section 4), it is usually reasonable to dose dexamethasone every 12 rather than 6 or 8 h due to its long half-life.[108] There is a significant variability regarding the management of ICANS beyond basic workup and the use of corticosteroids. Current recommendations are based on expert opinion and institutional experience, with one working group primarily composed of authors from one center[109] being received with controversy.[110] Anticytokine therapies that are given for CRS such as tocilizumab (an anti-IL6 receptor antibody) and anakinra (an anti-IL1 receptor antibody) are used less often for ICANS as efficacy has not been established, but use is institution dependent. While prognosis is highly variable,

the majority of patients make a full recovery. Roughly 4%–10% of patients have lasting neurologic deficits.[2,25,26]

8 Conclusion: Diverse presentation, kinetics, importance of multidisciplinary and early treatment, consideration of re-challenge

Immune modulatory therapies have become a mainstay for many malignancies, in particular in advanced disease. Immune modulation and its complications will continue to grow, making knowledge of these adverse effects crucial for safe navigation through immunotherapy. N-irAEs often resemble idiopathic autoimmune diseases, which can impact all components of the nervous system. In contrast, ICANS is less well understood, presents clinically less often with localizable signs, and treatment is less uniform across institutions. While each entity on its own may be rare, the development of neurologic symptoms as a result of aberrant immune modulation is rather common. Unlike other cancer-directed therapies, the kinetics of immune responses range widely; however, most patients present within the first four to six doses of ICB and roughly within 7–21 days of CAR T cell infusion. Rapid workup and treatment are paramount for acute-onset symptoms given the potential for rapid neurologic deterioration with inaction and complete recovery with early treatment. In certain situations, patients with grade 2 or 3 N-irAEs may be considered for re-challenge with the same or another ICB agent. While guidelines recommend permanent discontinuation for grade 4 events, or any grade of certain N-irAEs such as transverse myelitis and myocarditis,[40,54] the majority of cases will be decided through conversation with the patient regarding risk of ICB re-challenge versus treatment alternatives.

References

1. Graus F, Dalmau J. Paraneoplastic neurological syndromes in the era of immune-checkpoint inhibitors. *Nat Rev Clin Oncol.* 2019;16(9):535–548.
2. Schuster SJ, Svoboda J, Chong EA, et al. Chimeric antigen receptor T cells in refractory B-cell lymphomas. *N Engl J Med.* 2017;377(26):2545–2554.
3. Lee DW, Santomasso BD, Locke FL, et al. ASTCT consensus grading for cytokine release syndrome and neurologic toxicity associated with immune effector cells. *Biol Blood Marrow Transplant.* 2019;25(4):625–638.
4. Hunter BD, Jacobson CA. CAR T-cell associated neurotoxicity: mechanisms, clinicopathologic correlates, and future directions. *J Natl Cancer Inst.* 2019;111(7):646–654.
5. Gust J, Finney OC, Li D, et al. Glial injury in neurotoxicity after pediatric CD19-directed chimeric antigen receptor T cell therapy. *Ann Neurol.* 2019;86(1):42–54.
6. Topp MS, Gokbuget N, Stein AS, et al. Safety and activity of blinatumomab for adult patients with relapsed or refractory B-precursor acute lymphoblastic leukaemia: a multicentre, single-arm, phase 2 study. *Lancet Oncol.* 2015;16(1):57–66.

7. Bertrand A, Kostine M, Barnetche T, Truchetet ME, Schaeverbeke T. Immune related adverse events associated with anti-CTLA-4 antibodies: systematic review and meta-analysis. *BMC Med.* 2015;13:211.

8. Maughan BL, Bailey E, Gill DM, Agarwal N. Incidence of immune-related adverse events with program death receptor-1- and program death receptor-1 ligand-directed therapies in genitourinary cancers. *Front Oncol.* 2017;7:56.

9. Ricciuti B, Genova C, De Giglio A, et al. Impact of immune-related adverse events on survival in patients with advanced non-small cell lung cancer treated with nivolumab: long-term outcomes from a multi-institutional analysis. *J Cancer Res Clin Oncol.* 2019;145(2):479–485.

10. Rogado J, Sanchez-Torres JM, Romero-Laorden N, et al. Immune-related adverse events predict the therapeutic efficacy of anti-PD-1 antibodies in cancer patients. *Eur J Cancer.* 2019;109:21–27.

11. Toi Y, Sugawara S, Kawashima Y, et al. Association of immune-related adverse events with clinical benefit in patients with advanced non-small-cell lung cancer treated with nivolumab. *Oncologist.* 2018;23(11):1358–1365.

12. Wang DY, Salem JE, Cohen JV, et al. Fatal toxic effects associated with immune checkpoint inhibitors: a systematic review and meta-analysis. *JAMA Oncol.* 2018;4(12):1721–1728.

13. Postow MA, Sidlow R, Hellmann MD. Immune-related adverse events associated with immune checkpoint blockade. *N Engl J Med.* 2018;378(2):158–168.

14. Dubey D, David WS, Reynolds KL, et al. Severe neurological toxicity of immune checkpoint inhibitors: growing spectrum. *Ann Neurol.* 2020;87(5):659–669.

15. JCAR015 in ALL: A root-cause investigation. *Cancer Discov.* 2018;8(1):4–5.

16. Johnson DB, Manouchehri A, Haugh AM, et al. Neurologic toxicity associated with immune checkpoint inhibitors: a pharmacovigilance study. *J Immunother Cancer.* 2019;7(1):134.

17. Suzuki S, Ishikawa N, Konoeda F, et al. Nivolumab-related myasthenia gravis with myositis and myocarditis in Japan. *Neurology.* 2017;89(11):1127–1134.

18. Dubey D, David WS, Amato AA, et al. Varied phenotypes and management of immune checkpoint inhibitor-associated neuropathies. *Neurology.* 2019;93(11):e1093–e1103.

19. Cuzzubbo S, Javeri F, Tissier M, et al. Neurological adverse events associated with immune checkpoint inhibitors: review of the literature. *Eur J Cancer.* 2017;73:1–8.

20. Spain L, Walls G, Julve M, et al. Neurotoxicity from immune-checkpoint inhibition in the treatment of melanoma: a single centre experience and review of the literature. *Ann Oncol.* 2017;28(2):377–385.

21. Robert C, Long GV, Brady B, et al. Nivolumab in previously untreated melanoma without BRAF mutation. *N Engl J Med.* 2015;372(4):320–330.

22. Grupp SA, Kalos M, Barrett D, et al. Chimeric antigen receptor-modified T cells for acute lymphoid leukemia. *N Engl J Med.* 2013;368(16):1509–1518.

23. Maude SL, Frey N, Shaw PA, et al. Chimeric antigen receptor T cells for sustained remissions in leukemia. *N Engl J Med.* 2014;371(16):1507–1517.

24. Kochenderfer JN, Dudley ME, Kassim SH, et al. Chemotherapy-refractory diffuse large B-cell lymphoma and indolent B-cell malignancies can be effectively treated with autologous T cells expressing an anti-CD19 chimeric antigen receptor. *J Clin Oncol.* 2015;33(6):540–549.

25. Maude SL, Laetsch TW, Buechner J, et al. Tisagenlecleucel in children and young adults with B-cell lymphoblastic leukemia. *N Engl J Med.* 2018;378(5):439–448.

26. Neelapu SS, Locke FL, Bartlett NL, et al. Axicabtagene Ciloleucel CAR T-cell therapy in refractory large B-cell lymphoma. *N Engl J Med.* 2017;377(26):2531–2544.

27. Park JH, Riviere I, Gonen M, et al. Long-term follow-up of CD19 CAR therapy in acute lymphoblastic leukemia. *N Engl J Med.* 2018;378(5):449–459.

28. Schuster SJ, Bishop MR, Tam CS, et al. Tisagenlecleucel in adult relapsed or refractory diffuse large B-cell lymphoma. *N Engl J Med.* 2019;380(1):45–56.

29. Gardner RA, Finney O, Annesley C, et al. Intent-to-treat leukemia remission by CD19 CAR T cells of defined formulation and dose in children and young adults. *Blood.* 2017;129(25):3322–3331.

30. Borghaei H, Paz-Ares L, Horn L, et al. Nivolumab versus docetaxel in advanced nonsquamous non-small-cell lung cancer. *N Engl J Med.* 2015;373(17):1627–1639.

31. Gauvain C, Vauleon E, Chouaid C, et al. Intracerebral efficacy and tolerance of nivolumab in non-small-cell lung cancer patients with brain metastases. *Lung Cancer.* 2018;116:62–66.

32. Goldberg SB, Gettinger SN, Mahajan A, et al. Pembrolizumab for patients with melanoma or non-small-cell lung cancer and untreated brain metastases: early analysis of a non-randomised, open-label, phase 2 trial. *Lancet Oncol.* 2016;17(7):976–983.

33. Long GV, Atkinson V, Lo S, et al. Combination nivolumab and ipilimumab or nivolumab alone in melanoma brain metastases: a multicentre randomised phase 2 study. *Lancet Oncol.* 2018;19(5):672–681.

34. Tawbi HA, Forsyth PA, Algazi A, et al. Combined nivolumab and ipilimumab in melanoma metastatic to the brain. *N Engl J Med.* 2018;379(8):722–730.

35. Brown CE, Alizadeh D, Starr R, et al. Regression of glioblastoma after chimeric antigen receptor T-cell therapy. *N Engl J Med.* 2016;375(26):2561–2569.

36. Abramson JS, McGree B, Noyes S, et al. Anti-CD19 CAR T cells in CNS diffuse large-B-cell lymphoma. *N Engl J Med.* 2017;377(8):783–784.

37. Posner JB. Neurologic complications of systemic cancer. *Dis Mon.* 1978;25(2):1–60.

38. Johansen A, Christensen SJ, Scheie D, Hojgaard JLS, Kondziella D. Neuromuscular adverse events associated with anti-PD-1 monoclonal antibodies: systematic review. *Neurology.* 2019;92(14):663–674.

39. Kieseier BC, Mathey EK, Sommer C, Hartung HP. Immune-mediated neuropathies. *Nat Rev Dis Primers.* 2018;4(1):31.

40. Brahmer JR, Lacchetti C, Schneider BJ, et al. Management of immune-related adverse events in patients treated with immune checkpoint inhibitor therapy: American Society of Clinical Oncology clinical practice guideline. *J Clin Oncol.* 2018;36(17):1714–1768.

41. Puzanov I, Diab A, Abdallah K, et al. Managing toxicities associated with immune checkpoint inhibitors: consensus recommendations from the Society for Immunotherapy of Cancer (SITC) Toxicity Management Working Group. *J Immunother Cancer.* 2017;5(1):95.

42. Ali S, Lee SK. Ipilimumab therapy for melanoma: a mimic of leptomeningeal metastases. *AJNR Am J Neuroradiol.* 2015;36(12):E69–E70.

43. Altman AL, Golub JS, Pensak ML, Samy RN. Bilateral facial palsy following Ipilimumab infusion for melanoma. *Otolaryngol Head Neck Surg.* 2015;153(5):894–895.

44. Yost MD, Chou CZ, Botha H, Block MS, Liewluck T. Facial diplegia after pembrolizumab treatment. *Muscle Nerve.* 2017;56(3):E20–E21.

45. Zimmer L, Goldinger SM, Hofmann L, et al. Neurological, respiratory, musculoskeletal, cardiac and ocular side-effects of anti-PD-1 therapy. *Eur J Cancer.* 2016;60:210–225.

46. Kao JC, Liao B, Markovic SN, et al. Neurological complications associated with anti-programmed death 1 (PD-1) antibodies. *JAMA Neurol.* 2017;74(10):1216–1222.

47. Thaipisuttikul I, Chapman P, Avila EK. Peripheral neuropathy associated with ipilimumab: a report of 2 cases. *J Immunother.* 2015;38(2):77–79.

48. Takamatsu K, Nakane S, Suzuki S, et al. Immune checkpoint inhibitors in the onset of myasthenia gravis with hyperCKemia. *Ann Clin Transl Neurol.* 2018;5(11):1421–1427.

49. Shah M, Tayar JH, Abdel-Wahab N, Suarez-Almazor ME. Myositis as an adverse event of immune checkpoint blockade for cancer therapy. *Semin Arthritis Rheum.* 2019;48(4):736–740.

50. Liao B, Shroff S, Kamiya-Matsuoka C, Tummala S. Atypical neurological complications of ipilimumab therapy in patients with metastatic melanoma. *Neuro Oncol.* 2014;16(4):589–593.

51. Tanaka R, Maruyama H, Tomidokoro Y, et al. Nivolumab-induced chronic inflammatory demyelinating polyradiculoneuropathy mimicking rapid-onset Guillain-Barre syndrome: a case report. *Jpn J Clin Oncol.* 2016;46(9):875–878.

52. 2017. https://ctep.cancer.gov/protocoldevelopment/electronic_applications/docs/CTCAE_v5_Quick_Reference_5x7.pdf. [Accessed].

53. Haanen J, Carbonnel F, Robert C, et al. Management of toxicities from immunotherapy: ESMO clinical practice guidelines for diagnosis, treatment and follow-up. *Ann Oncol.* 2017;28(suppl 4):iv119–iv142.

54. Thompson JA, Schneider BJ, Brahmer J, et al. Management of immunotherapy-related toxicities, version 1.2019. *J Natl Compr Canc Netw.* 2019;17(3):255–289.

55. de Maleissye MF, Nicolas G, Saiag P. Pembrolizumab-induced demyelinating polyradiculoneuropathy. *N Engl J Med.* 2016;375(3):296–297.

56. Sepulveda M, Martinez-Hernandez E, Gaba L, et al. Motor polyradiculopathy during pembrolizumab treatment of metastatic melanoma. *Muscle Nerve.* 2017;56(6):E162–E167.

57. Gu Y, Menzies AM, Long GV, Fernando SL, Herkes G. Immune mediated neuropathy following checkpoint immunotherapy. *J Clin Neurosci.* 2017;45:14–17.

58. Appelbaum J, Wells D, Hiatt JB, et al. Fatal enteric plexus neuropathy after one dose of ipilimumab plus nivolumab: a case report. *J Immunother Cancer.* 2018;6(1):82.

59. Byun JI, Moon J, Kim DY, et al. Efficacy of single or combined midodrine and pyridostigmine in orthostatic hypotension. *Neurology.* 2017;89(10):1078–1086.

60. Okamoto LE, Shibao CA, Gamboa A, et al. Synergistic pressor effect of atomoxetine and pyridostigmine in patients with neurogenic orthostatic hypotension. *Hypertension.* 2019;73(1):235–241.

61. Strassheim V, Newton JL, Tan MP, Frith J. Droxidopa for orthostatic hypotension: a systematic review and meta-analysis. *J Hypertens.* 2016;34(10):1933–1941.

62. Eschlbock S, Wenning G, Fanciulli A. Evidence-based treatment of neurogenic orthostatic hypotension and related symptoms. *J Neural Transm (Vienna).* 2017;124(12):1567–1605.

63. Manousakis G, Koch J, Sommerville RB, et al. Multifocal radiculoneuropathy during ipilimumab treatment of melanoma. *Muscle Nerve.* 2013;48(3):440–444.

64. Gaudy-Marqueste C, Monestier S, Franques J, Cantais E, Richard MA, Grob JJ. A severe case of ipilimumab-induced Guillain-Barre syndrome revealed by an occlusive enteric neuropathy: a differential diagnosis for ipilimumab-induced colitis. *J Immunother.* 2013;36(1):77–78.

65. Wilgenhof S, Neyns B. Anti-CTLA-4 antibody-induced Guillain-Barre syndrome in a melanoma patient. *Ann Oncol.* 2011;22(4):991–993.

66. Chen X, Haggiagi A, Tzatha E, DeAngelis LM, Santomasso B. Electrophysiological findings in immune checkpoint inhibitor-related peripheral neuropathy. *Clin Neurophysiol.* 2019;130(8):1440–1445.

67. Chan KH, Lachance DH, Harper CM, Lennon VA. Frequency of seronegativity in adult-acquired generalized myasthenia gravis. *Muscle Nerve.* 2007;36(5):651–658.

68. McConville J, Farrugia ME, Beeson D, et al. Detection and characterization of MuSK antibodies in seronegative myasthenia gravis. *Ann Neurol.* 2004;55(4):580–584.

69. Higuchi O, Hamuro J, Motomura M, Yamanashi Y. Autoantibodies to low-density lipoprotein receptor-related protein 4 in myasthenia gravis. *Ann Neurol.* 2011;69(2):418–422.

70. Wolfe GI, Kaminski HJ, Aban IB, et al. Randomized trial of thymectomy in myasthenia gravis. *N Engl J Med.* 2016;375(6):511–522.

71. Verschuuren J, Strijbos E, Vincent A. Neuromuscular junction disorders. *Handb Clin Neurol.* 2016;133:447–466.

72. Agrawal K, Agrawal N. Lambert-Eaton myasthenic syndrome secondary to nivolumab and ipilimumab in a patient with small-cell lung cancer. *Case Rep Neurol Med.* 2019;2019:5353202.

73. Manson G, Maria ATJ, Poizeau F, et al. Worsening and newly diagnosed paraneoplastic syndromes following anti-PD-1 or anti-PD-L1 immunotherapies, a descriptive study. *J Immunother Cancer.* 2019;7(1):337.

74. Selva-O'Callaghan A, Pinal-Fernandez I, Trallero-Araguas E, Milisenda JC, Grau-Junyent JM, Mammen AL. Classification and management of adult inflammatory myopathies. *Lancet Neurol.* 2018;17(9):816–828.

75. Pinal-Fernandez I, Parks C, Werner JL, et al. Longitudinal course of disease in a large cohort of myositis patients with autoantibodies recognizing the signal recognition particle. *Arthritis Care Res.* 2017;69(2):263–270.

76. Allenbach Y, Benveniste O, Goebel HH, Stenzel W. Integrated classification of inflammatory myopathies. *Neuropathol Appl Neurobiol.* 2017;43(1):62–81.

77. Dalakas MC, Illa I, Dambrosia JM, et al. A controlled trial of high-dose intravenous immune globulin infusions as treatment for dermatomyositis. *N Engl J Med.* 1993;329(27):1993–2000.

78. Mahler EA, Blom M, Voermans NC, van Engelen BG, van Riel PL, Vonk MC. Rituximab treatment in patients with refractory inflammatory myopathies. *Rheumatology (Oxford).* 2011;50(12):2206–2213.

79. Vencovsky J, Jarosova K, Machacek S, et al. Cyclosporine A versus methotrexate in the treatment of polymyositis and dermatomyositis. *Scand J Rheumatol.* 2000;29(2):95–102.

80. Selva-O'Callaghan A, Grau JM, Gamez-Cenzano C, et al. Conventional cancer screening versus PET/CT in dermatomyositis/polymyositis. *Am J Med.* 2010;123(6):558–562.

81. Chen JH, Lee KY, Hu CJ, Chung CC. Coexisting myasthenia gravis, myositis, and polyneuropathy induced by ipilimumab and nivolumab in a patient with non-small-cell lung cancer: a case report and literature review. *Medicine (Baltimore).* 2017;96(50), e9262.

82. Guidon AC. Lambert-eaton myasthenic syndrome, botulism, and immune checkpoint inhibitor-related myasthenia gravis. *Continuum (Minneap Minn).* 2019;25(6):1785–1806.

83. Reynolds KL, Guidon AC. Diagnosis and management of immune checkpoint inhibitor-associated neurologic toxicity: illustrative case and review of the literature. *Oncologist.* 2019;24(4):435–443.

84. Caturegli P, Newschaffer C, Olivi A, Pomper MG, Burger PC, Rose NR. Autoimmune hypophysitis. *Endocr Rev.* 2005;26(5):599–614.

85. Albarel F, Gaudy C, Castinetti F, et al. Long-term follow-up of ipilimumab-induced hypophysitis, a common adverse event of the anti-CTLA-4 antibody in melanoma. *Eur J Endocrinol.* 2015;172(2):195–204.

86. Faje AT, Sullivan R, Lawrence D, et al. Ipilimumab-induced hypophysitis: a detailed longitudinal analysis in a large cohort of patients with metastatic melanoma. *J Clin Endocrinol Metab.* 2014;99(11):4078–4085.

87. Ryder M, Callahan M, Postow MA, Wolchok J, Fagin JA. Endocrine-related adverse events following ipilimumab in patients with advanced melanoma: a comprehensive retrospective review from a single institution. *Endocr Relat Cancer.* 2014;21(2):371–381.

88. Dillard T, Yedinak CG, Alumkal J, Fleseriu M. Anti-CTLA-4 antibody therapy associated autoimmune hypophysitis: serious immune related adverse events across a spectrum of cancer subtypes. *Pituitary.* 2010;13(1):29–38.

89. Juszczak A, Gupta A, Karavitaki N, Middleton MR, Grossman AB. Ipilimumab: a novel immunomodulating therapy causing autoimmune hypophysitis: a case report and review. *Eur J Endocrinol.* 2012;167(1):1–5.

90. Abou Alaiwi S, Xie W, Nassar AH, et al. Safety and efficacy of restarting immune checkpoint inhibitors after clinically significant immune-related adverse events in metastatic renal cell carcinoma. *J Immunother Cancer.* 2020;8(1):e000144.

91. Min L, Hodi FS, Giobbie-Hurder A, et al. Systemic high-dose corticosteroid treatment does not improve the outcome of ipilimumab-related hypophysitis: a retrospective cohort study. *Clin Cancer Res.* 2015;21(4):749–755.

92. Chang LS, Barroso-Sousa R, Tolaney SM, Hodi FS, Kaiser UB, Min L. Endocrine toxicity of cancer immunotherapy targeting immune checkpoints. *Endocr Rev.* 2019;40(1):17–65.

93. Dunn-Pirio AM, Shah S, Eckstein C. Neurosarcoidosis following immune checkpoint inhibition. *Case Rep Oncol.* 2018;11(2):521–526.

94. Tan I, Malinzak M, Salama AKS. Delayed onset of neurosarcoidosis after concurrent ipilimumab/nivolumab therapy. *J Immunother Cancer.* 2018;6(1):77.

95. Johnson DB, McDonnell WJ, Gonzalez-Ericsson PI, et al. A case report of clonal EBV-like memory CD4(+) T cell activation in fatal checkpoint inhibitor-induced encephalitis. *Nat Med.* 2019;25(8):1243–1250.

96. Williams TJ, Benavides DR, Patrice KA, et al. Association of autoimmune encephalitis with combined immune checkpoint inhibitor treatment for metastatic cancer. *JAMA Neurol.* 2016;73(8):928–933.

97. Bossart S, Thurneysen S, Rushing E, et al. Case report: encephalitis, with brainstem involvement, following checkpoint inhibitor therapy in metastatic melanoma. *Oncologist.* 2017;22(6):749–753.

98. Zurko J, Mehta A. Association of immune-mediated cerebellitis with immune checkpoint inhibitor therapy. *Mayo Clin Proc Innov Qual Outcomes.* 2018;2(1):74–77.

99. Vogrig A, Fouret M, Joubert B, et al. Increased frequency of anti-Ma2 encephalitis associated with immune checkpoint inhibitors. *Neurol Neuroimmunol Neuroinflamm.* 2019;6(6):e604.

100. Duraes J, Coutinho I, Mariano A, Geraldo A, Macario MC. Demyelinating disease of the central nervous system associated with pembrolizumab treatment for metastatic melanoma. *Mult Scler.* 2019;25(7):1005–1008.

101. Maurice C, Schneider R, Kiehl TR, et al. Subacute CNS demyelination after treatment with nivolumab for melanoma. *Cancer Immunol Res.* 2015;3(12):1299–1302.

102. Maur M, Tomasello C, Frassoldati A, Dieci MV, Barbieri E, Conte P. Posterior reversible encephalopathy syndrome during ipilimumab therapy for malignant melanoma. *J Clin Oncol.* 2012;30(6):e76–e78.

103. Narumi Y, Yoshida R, Minami Y, et al. Neuromyelitis optica spectrum disorder secondary to treatment with anti-PD-1 antibody nivolumab: the first report. *BMC Cancer.* 2018;18(1):95.

104. Santomasso BD, Park JH, Salloum D, et al. Clinical and biological correlates of neurotoxicity associated with CAR T-cell therapy in patients with B-cell acute lymphoblastic leukemia. *Cancer Discov.* 2018;8(8):958–971.

105. Gust J, Hay KA, Hanafi LA, et al. Endothelial activation and blood-brain barrier disruption in neurotoxicity after adoptive immunotherapy with CD19 CAR-T cells. *Cancer Discov.* 2017;7(12):1404–1419.

106. Rubin DB, Danish HH, Ali AB, et al. Neurological toxicities associated with chimeric antigen receptor T-cell therapy. *Brain.* 2019;142(5):1334–1348.

107. Karschnia P, Jordan JT, Forst DA, et al. Clinical presentation, management, and biomarkers of neurotoxicity after adoptive immunotherapy with CAR T cells. *Blood.* 2019;133(20):2212–2221.

108. Lim-Fat MJ, Bi WL, Lo J, et al. Letter: When less is more: dexamethasone dosing for brain tumors. *Neurosurgery.* 2019;85(3):E607–E608.

109. Neelapu SS, Tummala S, Kebriaei P, et al. Chimeric antigen receptor T-cell therapy—assessment and management of toxicities. *Nat Rev Clin Oncol.* 2018;15(1):47–62.

110. Teachey DT, Bishop MR, Maloney DG, Grupp SA. Toxicity management after chimeric antigen receptor T cell therapy: one size does not fit 'ALL'. *Nat Rev Clin Oncol.* 2018;15(4):218.

33

Neurological complications of steroids and of supportive care

Shannon Fortin Ensign[a] and Alyx B. Porter[b]

[a]Department of Hematology and Oncology, Mayo Clinic, Phoenix, AZ, United States, [b]Department of Neurology, Mayo Clinic, Phoenix, AZ, United States

1 Introduction

Corticosteroids are commonly employed within oncology primarily as either part of a direct treatment regimen or more broadly a supportive care modality. However, despite the widespread and well-established benefits of corticosteroids in these roles, a myriad of side effects can develop in correlation with the chronicity and/or dosage required in use. In particular, neurological side effects can be pronounced, leading to a limitation in their cumulative use and a concern for the potential significant impact on patient quality of life. Neurologic complications also arise during the care of patients with cancer as a direct consequence of additional supportive care medications, including antiepileptics, analgesics, antiemetics, and antidepressants in particular, or during the management of tumor-associated CNS vascular damage, including ischemic infarction or hemorrhagic injury. Here, we review the spectrum of corticosteroid-induced neurologic complications as well as the neurologic complications that arise secondary to supportive care measures of patients within oncology.

2 Corticosteroids

Corticosteroids have a multifaceted role within oncology care. Corticosteroids are commonly utilized directly as part of treatment regimens within many hematologic malignancies, including non-Hodgkin lymphomas. In the case of brain metastases, primary brain and spinal cord tumors, and/or of spinal cord compression, corticosteroids are the mainstay of treatment to reduce edema and improve neurologic function. Other applications of corticosteroids in oncology include the alleviation of pain, as an antiemetic, or to prevent infusion reactions to various chemotherapies. While the systemic use of glucocorticoids can enact toxicities across multiple organ systems, including dermatologic, ophthalmologic, cardiovascular, gastrointestinal, musculoskeletal, endocrine/metabolic, hematologic, and immune system adverse events, including increased susceptibility to infection, here we will focus on the side effects resulting in dysfunction of the central or peripheral nervous system.

2.1 Steroid-induced myopathy

Steroid-induced myopathy may result in significant morbidity with the typical clinical presentation characterized by painless proximal motor weakness in both the upper and lower extremities. Over time, the proximal muscle groups atrophy. Some patients only report subjective feelings of weakness, without clinical signs. However, up to 35% of patients require major assistance on walking.[1] Steroid-induced myopathy can also affect the muscles involved with respiration.[2-4] Of note, asymmetrical distribution of the weakness is rarely observed.[5]

The incidence of this complication varies in the literature between 6% and 60% for patients treated with steroids for an extended period of time.[2,6] Most likely, a clinically significant myopathy develops in 10%–20% of patients treated with steroids.[1,7] This may be impacted by age and gender, with studies suggesting higher risk in the elderly and females, as well as by the concurrent use of other medications, such as phenytoin, that alter the metabolism of corticosteroids.[1,7,8]

In general, steroid-induced myopathy is found more commonly when corticosteroids are used in higher doses and/or for more prolonged duration, without a clear threshold above which myopathy is consistently documented.[1,9] Most commonly, steroid-induced myopathy develops gradually over weeks to months and rarely develops within the first 4 weeks of treatment.[7] A retrospective study of patients with primary brain tumors who received at least 2 weeks of continuous daily dexamethasone noted that two-thirds of patients developed weakness by weeks 9–12 of continuous use.[1] However, a prospective study found that a myopathy can develop much faster, i.e., within 3 weeks of treatment, with doses of 4–16 mg of dexamethasone per day.[2,10] A study in asthma patients to assess dose relationships of systemic prednisone found that the development of myopathy was rare at a dose under 30 mg per day, and moreover, the risk for myopathy is possibly smaller with nonfluorinated steroids, such as prednisone or prednisolone.[1,6]

EMG and laboratory investigations are often normal.[1,7,9] However, the EMG may demonstrate brief motor unit potentials of low amplitude and polyphasic activity in the affected muscles.[1,7] Typically, no spontaneous muscle fiber activity is recorded using EMG investigations.[7] Creatine kinase, LDH, and SGOT may be elevated and urinary creatinine excretion may be increased.[3,4,9] Muscle biopsies in steroid-induced myopathy are often normal and have little clinical utility in establishing the diagnosis, although they may show selective-type IIb fiber atrophy and increased muscle glycogen may be present in the type IIb fibers.[7,8] Given that systemic steroid use results in multisystem side effects, myopathy in a patient utilizing corticosteroids is more likely to be presumed a causal relationship when observed in combination with other well-established steroid-induced toxicities.

There are no specific treatment options to improve strength in patients with steroid-induced myopathy. If possible, the steroid dose should be decreased, and usually, this leads to a gradual improvement in the severity of weakness. Recovery, after discontinuation of steroids, may take weeks to months.[1,6,7] Some patients may experience permanent residual weakness.[6] Physiotherapy can help to foster the mobility of the patient and may possibly prevent muscle weakness to some degree.[2,6,7]

2.2 Epidural lipomatosis

Chronic steroid therapy may result in an unusual deposition of fat in the epidural space, known as epidural lipomatosis. The majority of the reported cases involve male patients.[11] This side effect is rare and usually occurs only after years of treatment. Some patients, however, may develop this complication after only 5 months of treatment.[12] As a result of compression of the spinal cord or the cauda equine, patients may experience back pain, radiculopathy, paraparesis, or neurogenic claudication.[12,13] The epidural deposits are most often situated at the thoracic levels of the spinal column and less frequently lumbosacral.[11] Diagnosis is made by MRI. Treatment of epidural lipmatosis includes reduction and taper of corticosteroid dose. In the case of obesity, weight

loss may also improve symptoms.[12] If the neurological deficit is severe, laminectomy can be considered, which generally leads to neurological recovery.[11] Patients using high-dose steroids may benefit following surgery.[11]

2.3 Cognitive impairment

Cognitive impairment characterized by declarative or verbal memory deficits can be associated with both the long- and short-term use of corticosteroids.[14,15] Steroid use results in hippocampal dysfunction with the reversible atrophy of hippocampal neurons, with declarative memory deficits seen as early as 4–5 days following the treatment of dexamethasone or prednisone therapy.[15–17] Significantly smaller hippocampal volumes as compared to controls have been observed in patients using chronic steroid treatment as well,[18] and clinically, these deficits may remain persistent with cognitive impairment after starting corticosteroids notably stable in six patients with a mean follow-up of 4 years in one study.[18] This more severe cognitive impairment may be more consistent with corticosteroid-induced dementia.[15] Acute memory disturbance is associated with dose of steroid and appears to be completely reversible after discontinuation of corticosteroids; however, recovery of cerebral atrophy following chronic corticosteroid use is less clear.[17]

2.4 Major psychiatric disturbance

Psychiatric symptoms as side effects of corticosteroids have been reported in various studies ranging in incidence from 13% to 62%, with a weighted-average incidence across a meta-analysis of 11 studies of 27.6%. Severe symptoms occurred with a mean incidence of 5.7%.[19] Psychiatric syndromes included depression (41%), mania (28%), depression and mania (8%), psychosis (14%), and delirium (10%). These syndromes can occur at any time during treatment, mostly early after starting therapy.[15] The risk of developing steroid-induced psychiatric symptoms correlates with the dosage of corticosteroid.[15] Symptoms were rarely seen with daily doses up to 40 mg prednisone, while they occur in over 18% of patients using 80 mg of prednisone or more.[19] Manic episodes are more likely provoked with high-dose steroid use, whereas depression more often occurs during chronic treatment in a relatively low dose.[18] Depending on the severity of symptoms, further management by a neurologist or psychiatrist may be warranted. If possible, steroid dose should be decreased.

2.5 Minor neuropsychiatric symptoms

Irritability, insomnia, anxiety, and labile mood are commonly encountered side effects in patients requiring corticosteroid use. In addition, patients may observe steroid-induced tremor and hyperkinesia.[15,20]

The frequency of these symptoms is not well reported. In one study in which patients received up to 6000 to 9000 mg of prednisolone for oral pemphigus, seven out of 12 patients reported insomnia.[21] In a retrospective study of 88 patients with brain metastases treated with dexamethasone, 24% had complaints of insomnia.[22]

2.6 Steroid dosing

As corticosteroids may result in many side effects that reduce quality of life, duration of treatment should be as short as possible utilizing the lowest effective dose. Short steroid bursts and/or tapers can often be employed as part of a strategy to control acute cancer-associated pain, or prevent nausea and/or infusion reactions before systemic chemotherapy treatments where the cumulative steroid exposure and resulting steroid-associated toxicity is thus limited. Corticosteroids utilized in standard treatment regimens for hematologic malignancies will often require dose reductions over time during the therapy course to minimize toxicity. In patients where corticosteroid use is supportive, for example, to control edema in patients with brain tumors, the dose can be tailored to the severity of clinical findings. The standard initial dose in patients with brain edema associated with brain tumors is historically 16 mg per day (4 mg every 6 h).[23] This initial dose can be considered for patients with signs of increased intracranial pressure, or for patients with a tumor deposit in the posterior fossa of the brain. Without clinical signs of increased intracranial pressure, a prospective randomized study has shown that a daily dose of 4 mg of dexamethasone is usually enough to control neurological signs and symptoms, being equally effective to a dose of 16 mg of dexamethasone per day.[10] Twice daily dosing of corticosteroids should be employed when able, given the insomnia often caused by dosing every 6 and 8 h.[24] In practice, one may start during the first 48 h with a loading dose of 8 mg per day, followed by a maintenance dose of 4 mg per day, and dosages may be further tapered down following more definitive tumor therapies, including radiation and/or surgery. It is also important to recognize that the baseline use of corticosteroids greater than 10 mg of prednisone equivalent per day is associated with poorer outcomes in patients receiving immune-checkpoint inhibitors for their systemic cancer-directed therapy.[25] Thus, alternative strategies to manage complications of brain metastases in these patients are prudent if definitive control with radiation and/or surgery is not possible. Recently, bevacizumab has gained in popularity for use as a steroid-sparing strategy to control cerebral edema, and it should be considered as a steroid alternative in these cases.[26]

Care should be taken, as sudden tapering of corticosteroids may lead to a steroid withdrawal syndrome, most commonly characterized by depression, anxiety, and fatigue.[15] Alternatively, corticosteroid withdrawal may be characterized by headache, fatigue, low-grade fever, malaise, nausea, bilateral arthralgias in hips and knees, myalgias, and other symptoms secondary to suppression of the hypothalamic–pituitary–adrenal axis (HPA) leading to secondary adrenal insufficiency.[15,27–29] Dose reduction of steroids should, in general, be done gradually. Some reports suggest that a full discontinuation within 72 h is safe in patients who have used corticosteroids for no longer than 2 weeks.[8] Patients with a longer duration of treatment may reduce the dose in a period of four to 12 weeks. A commonly employed strategy is to perform a 25% reduction in dose of the steroid each 3–5 days. This period is based on the long biological half-life of the glucocorticoids, often up to 72 h or more, despite a plasma half-life of about 4 h.[22]

3 Seizures

About two-third of the patients with primary brain tumors and one-third of patients with brain metastases suffer from seizures.[30,31] The prevalence of epilepsy is notably higher among lower-grade gliomas reaching up to 90%, which may be related to the tumor biology accompanying isocitrate dehydrogenase (IDH) mutations.[32] In the case of cerebral metastases, seizures occur as the presenting symptom in 18% of the patients and another 15% develop seizures later in the disease course.[30] Seizures may also occur in cancer patients with or without a cerebral localization of their malignancy as a result of medication, metabolic disturbances, radiation toxicity, or the reversible posterior leukoencephalopathy syndrome.[33,34] Therefore, it is important to evaluate all medications and cancer-directed treatments a patient has received to ascertain a potential contribution of each modality to the occurrence of the seizures. As there is no evidence to support prophylactic treatment with antiepileptic drugs (AED) in patients with brain tumors without seizures, treatment should only be started in patients with clinically evident epilepsy, and possibly in those patients with no seizures, but with additional risk factors.[35] Antitumor therapy by surgery, radiotherapy, or chemotherapy often leads to a substantial decrease in the frequency of seizures.[30,31] The choice of AED is primarily based on provider preference in combination with an assessment of anticipated drug-specific toxicities reviewed within the context of patient comorbidities or laboratory abnormalities. Earlier studies of valproate suggested the additional benefit of antineoplastic activity; however, subsequent analyses have not confirmed this and the use of valproate remains with seizure control alone in patients with malignancy-induced epilepsy.[32] Levetiracetam remains one of the most commonly utilized AEDs in the setting of tumor-induced epilepsy.[32]

3.1 Side effects of antiepileptic drugs

Antiepileptic drugs (AEDs) play an essential part within the multimodal treatment of patients with brain tumors and seizures, however, with well-recognized toxicity. Skin reactions, neurocognitive disturbances, bone marrow suppression, and liver abnormalities are all commonly seen with AEDs.[36–38] Severe skin reactions such as Stevens-Johnson syndrome are rare but can occur mainly during the first 4–8 weeks after starting carbamazepine, phenobarbital, phenytoin, or lamotrigine.[39,40] Additionally, these reactions have also been observed during cranial radiotherapy while patients were on phenytoin.[41] This complication can be fatal. Neurocognitive deficits represent a common toxicity of antiepileptic drug use.[36,42] Adverse effects include somnolence, dizziness, ataxia, blurred vision, diplopia, irritability, nausea, headache, tremor, depression, insomnia, memory disturbance, apathy, or aggression and hostility.[32] The impairment of cognition is stronger at higher levels of the AEDs.[43] In patients with low-grade gliomas, both the tumor and the use of antiepileptic drugs were associated with changes in cognition.[44] In addition, it can be difficult to remove patients off AED use with low-grade gliomas due to the high underlying seizure incidence, and there are no large prospective studies to adequately address the optimal withdrawal timeline during the treatment of these tumors.[32]

3.2 Drug interactions with AEDs

Antiepileptic drugs have many potential interactions with other medications that may be used in oncology, including chemotherapeutic regimens. These interactions have potential consequences for both tumor and seizure control and may lead to toxic effects of one or both groups of agents. Changes in the activity of the cytochrome-P450 pathway of protein breakdown are the main cause of these interactions. The classical AEDs (i.e., phenobarbital, primidone, carbamazepine, and phenytoin) do all induce cytochrome-P450 coenzymes such as 3A4, 2C9, or 2C19, leading to a faster metabolism and thus to lower plasma concentrations of agents that share the same metabolic isoenzyme.[45]

Enzyme-inducing AEDs will impair the efficacy of several chemotherapeutic agents such as nitrosoureas, paclitaxel, cyclophosphamide, etoposide, topotecan, irinotecan, thiotepa, doxorubicin, and methotrexate.[45] Valproic acid is a broad-spectrum enzyme-inhibiting AED and may decrease the metabolism of a second drug, thus leading to an increase of the plasma concentrations of this drug. Valproic acid may thus cause both

enhanced activity and toxic effects of a concomitantly given drug. Toxic effects of nitrosoureas, cisplatin, and etoposide have also been observed in association with the use of valproic acid.[46] Vice versa, many chemotherapeutic agents can also induce the isoenzymes of the cytochrome-P450 pathway and may thus influence the plasma concentration of concomitantly prescribed antiepileptic drugs, which share the same coenzyme for their metabolism. Methotrexate, cisplatin, and doxorubicin can reduce the plasma concentration of valproic acid. Doxorubicin and cisplatin can decrease the plasma concentration of carbamazepine or valproic acid.[46] Combination of phenytoin with fluoropyrimidines (i.e., fluorouracil, tegafur, and capecitabine) increases phenytoin's toxic effects.[47]

Phenytoin and phenobarbital shorten the half-life of dexamethasone and prednisone and thus the activity of these agents.[48] Phenytoin concentrations may either increase, possibly by a lesser availability of protein-binding sites, or alternatively decrease due to interference with hepatic metabolism by concomitantly administered dexamethasone.[49] Therefore, phenytoin concentrations in patients treated with dexamethasone should be monitored closely—particularly during withdrawal of dexamethasone, which subsequently may lead to toxic levels of phenytoin.

4 Vascular complications

Patients with cancer are at an increased risk of cerebrovascular complications.[50] In one retrospective study, autopsy reports of 3426 patients with systemic cancer were reviewed.[51] Evidence of cerebrovascular events was found in 500 patients, of which 255 patients had experienced clinically significant signs and symptoms during life. Ischemic stroke was present in 223, intracerebral hemorrhage in 157, subdural hematoma in 63, sinus thrombosis in 33, and subarachnoid hemorrhage in 24 patients. In a group of 69 oncological patients with a stroke, the majority (63.8%) developed stroke within a year after the cancer diagnosis.[50]

Most of these events are related to the tumor, such as tumor-associated coagulation disorders or tumor embolism.[50,52] Besides, several chemotherapeutic agents may induce coagulation disorders and atherosclerotic abnormalities. Patients treated with cisplatin, and to a lesser extent other platinum compounds, have the highest risk for developing ischemic stroke.[53] Thrombocytopenia, which is frequently encountered in patients on antineoplastic therapy, may further increase the risk of intracerebral hemorrhage as well as the use of anticoagulation in order to treat malignancy-associated venous thromboembolism (VTE).[50] Glioma patients, in particular, have been shown to have one of the highest risks of VTE with rates up to 25%–39%, and concurrent use of bevacizumab in these patients can further complicate the risk of bleed.[32]

5 Analgesics

5.1 Opioids

Opioids are widely used in oncology, as they generally provide adequate control of pain.[54] However, because of their side effects, opioids are not always the first choice as an analgesic agent. Commonly used opioids include codeine, tramadol, oxycodone, fentanyl, and morphine. Opioids have several side effects involving the central nervous system, including sedation, fatigue, dizziness, euphoria, hallucinations, sleep disorders, depression, agitation, seizures, and respiratory depression.[55–58] Rarely, the paradoxical response of opioid-induced hyperalgesia may be observed, typically in patients who have used high doses of opioids for a long time.[59,60] Opioids may also induce nausea by a direct effect on the brainstem.[61] A systematic review of placebo-controlled randomized controlled trials of opioid use in patients with chronic noncancer pain established a prevalence of nausea of 32%.[62] Among the most common neurologic side effects, the prevalence of somnolence/sedation, dizziness, and headache was 29%, 20%, and 8%, respectively.

Of the most frequently encountered neurologic side effects of opioids, sedation can have a severe impact on quality of life. It is important to realize that chronic fatigue in patients with cancer may also sometimes reflect opioid-induced sedation. Sedation occurs early during the start of opioid treatment for which tolerance typically develops, as well as later during chronic treatment as a result of the need for higher doses for adequate pain control. Many patients do not develop tolerance to this latter sedation, limiting the treatment of pain.[55] The prevalence of sedation in cancer patients receiving chronic treatment of long-acting opioids in combination with fentanyl varies from 7% to 13%.[63] Some studies have investigated the treatment options for the chronic type of sedation. Beneficial effects of daily doses of 10 and 15 mg methylphenidate have been described.[54,58] One non-blinded study with six cancer patients indicated that the sedation improves with donepezil.[63] The starting dose of donepezil was 2.5 to 5.0 mg, which was in some patients subsequently raised to a maximum of 10 mg, and in one patient to 15 mg. One retrospective study evaluated the effect of modafinil on sedation in 11 patients.[55] The mean starting dose of modafinil was 264 mg (SD 112 mg) and the mean dose at maintenance was 427 mg (SD 156 mg). This study found a beneficial effect of modafinil on opioid-induced sedation. Respiratory depression can be treated with naloxone or nalbuphine.[61] Sedation can also be a confounding complication of opioid-associated sleep-disordered breathing, which is important to recognize in patients with underlying comorbidities of chronic obstructive pulmonary disease or obesity.[64] Moreover, elderly patients may have an increased susceptibility to opioids, and particular care in monitoring for adverse neurologic toxicity should be taken in this population.[65]

Many patients develop tolerance to opioids, necessitating an increase of the daily dose to maintain adequate analgesia. It is important to realize that tolerance to one opioid does not implicate tolerance to all opioids.[58] Therefore, changing of one opioid for another may achieve better pain control. However, care should be taken as changing to an equivalent dose of another opioid may result in an overdose, as tolerance may only have developed for the first opioid. Patients can also become physically dependent on opioids, resulting in signs and symptoms of autonomic and somatic hyperactivity after withdrawal.[58] Switching from one opioid to another may also cause signs and symptoms of withdrawal.[66]

5.2 Nonsteroidal antiinflammatory drugs

Nonsteroidal antiinflammatory drugs (NSAIDs) are often prescribed for arthritic pains and other pain originating from inflammation. Inherent with the mechanism of action of NSAIDs is a reduced aggregation of thrombocytes.[67] This may lead to increased bleeding times, especially with the use of acetylsalicylic acid. Nevertheless, the risk for intracerebral hemorrhage is low, except in case of thrombocytopenia or in combination with other factors influencing hemostasis.[68] The risk of hemorrhagic stroke in patients using NSAIDs other than aspirin seems not to be increased.[69] In elderly patients, NSAIDs may induce renal insufficiency, sometimes acutely, and patients who already have impairment of renal perfusion are at higher risk[70]; thus, these populations should be routinely assessed for bleeding risk on NSAIDs.

5.3 Antiepileptic drugs and antidepressants

Neuropathic pain is commonly treated with antiepileptic drugs, antidepressants, and variants of these drugs. Carbamazepine, gabapentin, pregabalin, duloxetine, and amitriptyline are among the most frequently employed.[71] Carbamazepine may give rise to multiple neurological side effects as previously reviewed in the section regarding seizures.

Gabapentin is effective in reducing neuropathic pain in gradually increasing doses up to 1200 mg, three times daily.[72] In higher doses, including one study utilizing 2400 mg in patients with postherpetic neuralgia, the most prominent neurological side effects were dizziness (33%) and somnolence (20%).[73] Similar to gabapentin, the most frequent neurological side effects of pregabalin are dizziness and somnolence.[74] These side effects appear to be dose related.

Amitriptyline has anticholinergic properties, which may induce mydriasis, fatigue, tremor, convulsions, and delirium.[75,76] The occurrence and severity of delirium correlates with the anticholinergic activity in plasma.[77] Hence, higher doses of amitriptyline are more likely to induce delirium. Severe anticholinergic side effects can be countered by 1–2 mg intravenous physostigmine.[76] A positive effect of a daily oral dose of 5 mg donepezil for signs of amitriptyline toxicity has been reported.[76]

Duloxetine has been shown to decrease the pain associated with chemotherapy-induced peripheral neuropathy and lessen the impact of pain on daily functional ability in patients with cancer. Duloxetine was reported to be well tolerated overall, with fatigue reported as the most prominent adverse event (7%), followed by insomnia (5%) and nausea (5%).[78]

6 Antiemetics

Nausea occurs frequently in patients with cancer, especially as a result of chemotherapy and radiation. One of the most common antiemetic classes utilized are the selective 5-HT$_3$-receptor antagonists.[79,80] Currently, these antagonists, which include ondansetron or longer-acting palonosetron, are the first choice in the prevention and treatment of nausea in cancer patients. Although severe neurological side effects are rare, a common neurological side effect is headache.[80] Some patients may experience dizziness or fatigue.[81] Rarely transient blindness has been reported after bolus injections of ondansetron.[82] Ondansetron may also decrease the analgesic properties of tramadol.[83]

Dopamine antagonists are also commonly utilized in the management of chemotherapy-induced nausea and vomiting. Prochlorperazine, an antipsychotic that blocks postsynaptic mesolimbic dopaminergic D1 and D2 receptors in the brain, is routinely used to treat nausea. Headache, insomnia, sedation, and more rarely extrapyramidal reactions can occur. In addition, the dopamine antagonists, such as metoclopramide, are available for use as subsequent line therapy if needed; however, in practice, this remains infrequent. Metoclopramide may also result in extrapyramidal disorders, mainly dystonia, parkinsonism, tardive dyskinesia, and akathisia.[80,84–86] The incidence of extrapyramidal side effects of metoclopramide is estimated at 0.2% for the adult population; however, in children and elderly, the risk may increase to 25%.[85] Symptoms usually develop between 4 and 36 h after administration, although occasionally after 6 days of use.[87] After the withdrawal of metoclopramide, the extrapyramidal symptoms usually resolve within 12 h.[87]

Neurokinin-1-receptor antagonists, such as aprepitant, constitute a newer instrument in the treatment of nausea. Aprepitant is most commonly utilized in combination with other antiemetics in the preventative setting for pretreatment nausea control. The most common neurological side effects of aprepitant are fatigue and asthenia.[80,88]

Oral cannabinoids (i.e., synthetic pharmaceutical-grade THC products including dronabinol and nabilone) have become increasingly utilized for the treatment of chemotherapy-induced nausea and vomiting. The most common adverse events of these medications included disorientation, dizziness, euphoria, confusion, and drowsiness.[89]

Steroids are commonly utilized for both nausea prevention and treatment, and the toxicities of these medications are detailed previously above.

7 Depression

The frequency of major depression in cancer patients is approximately 15%, and less severe symptoms of depression are even more frequent.[90] Depressive symptoms usually have a profound impact on the quality of life.[91] A depression may be the result of the emotional impact of the primary disease or of the associated treatment. Impairment of cerebral function by the brain cancer itself, resulting in surgery and radiation, and/or by metabolic disturbances can confound the exact precipitant. Once identified, it is important to exclude the presence of cerebral metastases, infection, and toxic/metabolic encephalopathy. The majority of patients are treated with antidepressants.

Severe adverse drug reactions as a result of antidepressants were found in 734 patients, out of a group of 53,042 patients treated in psychiatric hospitals.[92] In 360 of these 734 patients, the antidepressant drugs were thought to be the sole cause of these adverse events. In patients using tricyclic antidepressants, toxic delirium and neurological adverse drug reactions were found in 0.15%–0.21%. The neurological adverse events involved mostly seizures. SSRIs can induce other psychiatric symptoms, most notably agitation, and neurological adverse events occur in 0.15% of the patients. Furthermore, 0.05% of patients had electrolyte disturbances, most notably including hyponatremia. In this study, mirtazapine, mianserin, venlafaxine, nefazodone, and reboxetine were grouped as "other antidepressant drugs," and their most frequent neurological adverse events were toxic delirium and seizures, both occurring in about 0.05% of patients. No significant relation was found for the dosage of the antidepressant and the occurrence of adverse events. As this study did not involve oncological patients and only investigated severe adverse events, the type and frequency of adverse events may be different in an oncological population. One study investigated the effect of a treatment algorithm for major depression in 59 patients with advanced cancer.[91] In the first week, 19 patients dropped out, eight of whom as a result of delirium. In five patients, delirium was probably caused by the use of an antidepressant: alprazolam in three patients and amitriptyline and amoxapine both in one patient.

A meta-analysis indicated that benzodiazepines in combination with the use of antidepressant may lead to a faster improvement of depression.[93] Also, the dropout rate for discontinuing medication was lower in patients using such combination therapy. The addition of benzodiazepines may improve insomnia and anxiety and therefore improve compliance, thus leading to a quicker resolution of symptoms. Current management strategies are centered around SSRI and SNRI medications in combination with cognitive behavioral therapy.

8 Summary

Corticosteroids are routinely employed in oncology practice, including their direct use in treatment regimens or more broadly as a supportive care modality. Despite their clear benefit, corticosteroids are associated with many toxicities, including dysfunction of the central or peripheral nervous system. Moreover, additional supportive care treatments employed within the multimodal care of patients with cancer, including antiepileptics, analgesics, antiemetics, and antidepressants, also enact neurologic toxicity. A detailed understanding, anticipation, and recognition of the spectrum of potential neurologic adverse events is paramount in order to plan a mitigation strategy to lessen the adverse impact on quality of life that patients with cancer may experience during therapy. Here, we highlight the neurologic complications that may be encountered and require intervention during steroid use and supportive care of patients in oncology.

Acknowledgment

This chapter contains an updated version of the previous work by Willem-Johan van de Beek and Charles J. Vecht, Neuro-Oncology Unit, Dept. of Neurology, Medical Center, The Hague.

References

1. Dropcho EJ, Soong SJ. Steroid-induced weakness in patients with primary brain tumors. *Neurology*. 1991;41:1235–1239.
2. Batchelor TT, Taylor LP, Thaler HT, Posner JB, DeAngelis LM. Steroid myopathy in cancer patients. *Neurology*. 1997;48:1234–1238.
3. Janssens S, Decramer M. Corticosteroid-induced myopathy and the respiratory muscles. Report of two cases. *Chest*. 1989;95:1160–1162.
4. van Balkom RH, van der Heijden HF, van Herwaarden CL, Dekhuijzen PN. Corticosteroid-induced myopathy of the respiratory muscles. *Neth J Med*. 1994;45:114–122.
5. Sun DY, Edgar M, Rubin M. Hemiparetic acute myopathy of intensive care progressing to triplegia. *Arch Neurol*. 1997;54:1420–1422.
6. Bowyer SL, LaMothe MP, Hollister JR. Steroid myopathy: incidence and detection in a population with asthma. *J Allergy Clin Immunol*. 1985;76:234–242.
7. Ubogu EE, Kaminski HJ. Endocrine myopathies. In: Engel AG, Franzini-Armstrong C, eds. *Myology—Basic and Clinical*. vol. 2. 3rd ed. New York: McGraw-Hill Companies; 2004:1713–1738.
8. Wen PY, et al. Medical management of patients with brain tumors. *J Neurooncol*. 2006;80:313–332.
9. Askari A, Vignos Jr PJ, Moskowitz RW. Steroid myopathy in connective tissue disease. *Am J Med*. 1976;61:485–492.
10. Vecht CJ, Hovestadt A, Verbiest HB, van Vliet JJ, van Putten WL. Dose-effect relationship of dexamethasone on Karnofsky performance in metastatic brain tumors: a randomized study of doses of 4, 8, and 16 mg per day. *Neurology*. 1994;44:675–680.
11. Roy-Camille R, Mazel C, Husson JL, Saillant G. Symptomatic spinal epidural lipomatosis induced by a long-term steroid treatment. Review of the literature and report of two additional cases. *Spine (Phila Pa 1976)*. 1991;16:1365–1371.
12. Gupta R, Kumar AN, Gupta V, Madhavan SM, Sharma SK. An unusual cause of paraparesis in a patient on chronic steroid therapy. *J Spinal Cord Med*. 2007;30:67–69.
13. Burkhardt N, Hamann GF. Extradural lipomatosis after long-term treatment with steroids. *Nervenarzt*. 2006;**77**:1477–1479.
14. Wolkowitz OM, Lupien SJ, Bigler E, Levin RB, Canick J. The "steroid dementia syndrome": an unrecognized complication of glucocorticoid treatment. *Ann N Y Acad Sci*. 2004;1032:191–194.
15. Warrington TP, Bostwick JM. Psychiatric adverse effects of corticosteroids. *Mayo Clin Proc*. 2006;81:1361–1367.
16. Coluccia D, et al. Glucocorticoid therapy-induced memory deficits: acute versus chronic effects. *J Neurosci*. 2008;28:3474–3478.
17. Brown ES, Rush AJ, McEwen BS. Hippocampal remodeling and damage by corticosteroids: implications for mood disorders. *Neuropsychopharmacology*. 1999;21:474–484.
18. Brown ES, Vera E, Frol AB, Woolston DJ, Johnson B. Effects of chronic prednisone therapy on mood and memory. *J Affect Disord*. 2007;99:279–283.
19. Lewis DA, Smith RE. Steroid-induced psychiatric syndromes. A report of 14 cases and a review of the literature. *J Affect Disord*. 1983;5:319–332.
20. Stiefel FC, Breitbart WS, Holland JC. Corticosteroids in cancer: neuropsychiatric complications. *Cancer Invest*. 1989;7:479–491.
21. Mignogna MD, et al. High-dose intravenous 'pulse' methylprednisone in the treatment of severe oropharyngeal pemphigus: a pilot study. *J Oral Pathol Med*. 2002;31:339–344.
22. Sturdza A, et al. The use and toxicity of steroids in the management of patients with brain metastases. *Support Care Cancer*. 2008;16:1041–1048.
23. Galicich JH, French LA, Melby JC. Use of dexamethasone in treatment of cerebral edema associated with brain tumors. *J Lancet*. 1961;81:46–53.
24. Lim-Fat MJ, et al. Letter: when less is more: dexamethasone dosing for brain tumors. *Neurosurgery*. 2019;85:E607–E608.
25. Arbour KC, et al. Impact of baseline steroids on efficacy of programmed cell death-1 and programmed death-ligand 1 blockade in patients with non-small-cell lung cancer. *J Clin Oncol*. 2018;36:2872–2878.
26. Meng X, et al. Efficacy and safety of bevacizumab treatment for refractory brain edema: case report. *Medicine (Baltimore)*. 2017;96:e8280.
27. Wolfson AH, et al. The role of steroids in the management of metastatic carcinoma to the brain. A pilot prospective trial. *Am J Clin Oncol*. 1994;17:234–238.
28. Cooper MS, Stewart PM. Corticosteroid insufficiency in acutely ill patients. *N Engl J Med*. 2003;348:727–734.

29. Coursin DB, Wood KE. Corticosteroid supplementation for adrenal insufficiency. *JAMA*. 2002;287:236–240.

30. Junck L. Supportive management in neuro-oncology: opportunities for patient care, teaching, and research. *Curr Opin Neurol*. 2004;17:649–653.

31. Hildebrand J, Lecaille C, Perennes J, Delattre JY. Epileptic seizures during follow-up of patients treated for primary brain tumors. *Neurology*. 2005;65:212–215.

32. Schiff D, Alyahya M. Neurological and medical complications in brain tumor patients. *Curr Neurol Neurosci Rep*. 2020;20:33.

33. Lee VH, Wijdicks EF, Manno EM, Rabinstein AA. Clinical spectrum of reversible posterior leukoencephalopathy syndrome. *Arch Neurol*. 2008;65:205–210.

34. Minisini AM, Pauletto G, Andreetta C, Bergonzi P, Fasola G. Anticancer drugs and central nervous system: clinical issues for patients and physicians. *Cancer Lett*. 2008;267:1–9.

35. Sirven JI, Wingerchuk DM, Drazkowski JF, Lyons MK, Zimmerman RS. Seizure prophylaxis in patients with brain tumors: a meta-analysis. *Mayo Clin Proc*. 2004;79:1489–1494.

36. Aldenkamp AP, et al. Cognitive side-effects of phenytoin compared with carbamazepine in patients with localization-related epilepsy. *Epilepsy Res*. 1994;19:37–43.

37. Koenig SA, et al. Valproic acid-induced hepatopathy: nine new fatalities in Germany from 1994 to 2003. *Epilepsia*. 2006;47:2027–2031.

38. Gerstner T, Bauer MO, Longin E, Bell N, Koenig SA. Reversible hepatotoxicity, pancreatitis, coagulation disorder and simultaneous bone marrow suppression with valproate in a 2-year-old girl. *Seizure*. 2007;16:554–556.

39. Rzany B, et al. Risk of Stevens-Johnson syndrome and toxic epidermal necrolysis during first weeks of antiepileptic therapy: a case-control study. Study group of the international case control study on severe cutaneous adverse reactions. *Lancet*. 1999;353:2190–2194.

40. Mockenhaupt M, Messenheimer J, Tennis P, Schlingmann J. Risk of Stevens-Johnson syndrome and toxic epidermal necrolysis in new users of antiepileptics. *Neurology*. 2005;64:1134–1138.

41. Khafaga YM, et al. Stevens-Johnson syndrome in patients on phenytoin and cranial radiotherapy. *Acta Oncol*. 1999;38:111–116.

42. Riva D, Devoti M. Discontinuation of phenobarbital in children: effects on neurocognitive behavior. *Pediatr Neurol*. 1996;14:36–40.

43. Thompson PJ, Trimble MR. Anticonvulsant serum levels: relationship to impairments of cognitive functioning. *J Neurol Neurosurg Psychiatry*. 1983;46:227–233.

44. Klein M, et al. Effect of radiotherapy and other treatment-related factors on mid-term to long-term cognitive sequelae in low-grade gliomas: a comparative study. *Lancet*. 2002;360:1361–1368.

45. Vecht CJ, Wagner GL, Wilms EB. Interactions between antiepileptic and chemotherapeutic drugs. *Lancet Neurol*. 2003;2:404–409.

46. Bourg V, Lebrun C, Chichmanian RM, Thomas P, Frenay M. Nitroso-urea-cisplatin-based chemotherapy associated with valproate: increase of haematologic toxicity. *Ann Oncol*. 2001;12:217–219.

47. Gilbar PJ, Brodribb TR. Phenytoin and fluorouracil interaction. *Ann Pharmacother*. 2001;35:1367–1370.

48. Ruegg S. Dexamethasone/phenytoin interactions: neurooncological concerns. *Swiss Med Wkly*. 2002;132:425–426.

49. Lackner TE. Interaction of dexamethasone with phenytoin. *Pharmacotherapy*. 1991;11:344–347.

50. Zhang YY, Chan DK, Cordato D, Shen Q, Sheng AZ. Stroke risk factor, pattern and outcome in patients with cancer. *Acta Neurol Scand*. 2006;114:378–383.

51. Graus F, Rogers LR, Posner JB. Cerebrovascular complications in patients with cancer. *Medicine (Baltimore)*. 1985;64:16–35.

52. Cestari DM, Weine DM, Panageas KS, Segal AZ, DeAngelis LM. Stroke in patients with cancer: incidence and etiology. *Neurology*. 2004;62:2025–2030.

53. Li SH, et al. Incidence of ischemic stroke post-chemotherapy: a retrospective review of 10,963 patients. *Clin Neurol Neurosurg*. 2006;108:150–156.

54. Bruera E, Fainsinger R, MacEachern T, Hanson J. The use of methylphenidate in patients with incident cancer pain receiving regular opiates. A preliminary report. *Pain*. 1992;50:75–77.

55. Webster L, Andrews M, Stoddard G. Modafinil treatment of opioid-induced sedation. *Pain Med*. 2003;4:135–140.

56. Trescot AM, et al. Opioid guidelines in the management of chronic non-cancer pain. *Pain Physician*. 2006;9:1–39.

57. Daniell HW. DHEAS deficiency during consumption of sustained-action prescribed opioids: evidence for opioid-induced inhibition of adrenal androgen production. *J Pain*. 2006;7:901–907.

58. Benyamin R, et al. Opioid complications and side effects. *Pain Physician*. 2008;11:S105–S120.

59. Angst MS, Clark JD. Opioid-induced hyperalgesia: a qualitative systematic review. *Anesthesiology*. 2006;104:570–587.

60. DuPen A, Shen D, Ersek M. Mechanisms of opioid-induced tolerance and hyperalgesia. *Pain Manag Nurs*. 2007;8:113–121.

61. Ruan X. Drug-related side effects of long-term intrathecal morphine therapy. *Pain Physician*. 2007;10:357–366.

62. Kalso E, Edwards JE, Moore RA, McQuay HJ. Opioids in chronic non-cancer pain: systematic review of efficacy and safety. *Pain*. 2004;112:372–380.

63. Slatkin NE, Rhiner M, Bolton TM. Donepezil in the treatment of opioid-induced sedation: report of six cases. *J Pain Symptom Manage*. 2001;21:425–438.

64. Rosen IM, et al. Chronic opioid therapy and sleep: an American Academy of Sleep Medicine position statement. *J Clin Sleep Med*. 2019;15:1671–1673.

65. Rao A, Cohen HJ. Symptom management in the elderly cancer patient: fatigue, pain, and depression. *J Natl Cancer Inst Monogr*. 2004;150–157. https://doi.org/10.1093/jncimonographs/lgh031.

66. McMunnigall F, Welsh J. Opioid withdrawal syndrome on switching from hydromorphone to alfentanil. *Palliat Med*. 2008;22:191–192.

67. Scharbert G, et al. Point-of-care platelet function tests: detection of platelet inhibition induced by nonopioid analgesic drugs. *Blood Coagul Fibrinolysis*. 2007;18:775–780.

68. Thrift AG, McNeil JJ, Forbes A, Donnan GA. Risk of primary intracerebral haemorrhage associated with aspirin and non-steroidal anti-inflammatory drugs: case-control study. *BMJ*. 1999;318:759–764.

69. Choi NK, Park BJ, Jeong SW, Yu KH, Yoon BW. Nonaspirin nonsteroidal anti-inflammatory drugs and hemorrhagic stroke risk: the Acute Brain Bleeding Analysis study. *Stroke*. 2008;39:845–849.

70. Pannu N, Nadim MK. An overview of drug-induced acute kidney injury. *Crit Care Med*. 2008;36:S216–S223.

71. Finnerup NB, et al. Pharmacotherapy for neuropathic pain in adults: a systematic review and meta-analysis. *Lancet Neurol*. 2015;14:162–173.

72. Bennett MI, Simpson KH. Gabapentin in the treatment of neuropathic pain. *Palliat Med*. 2004;18:5–11.

73. Rice AS, Maton S, Postherpetic G. Neuralgia Study, Gabapentin in postherpetic neuralgia: a randomised, double blind, placebo controlled study. *Pain*. 2001;94:215–224.

74. Frampton JE, Foster RH. Pregabalin: in the treatment of postherpetic neuralgia. *Drugs*. 2005;65:111–118. discussion 119–120.

75. Raethjen J, et al. Amitriptyline enhances the central component of physiological tremor. *J Neurol Neurosurg Psychiatry*. 2001;70:78–82.

76. Noyan MA, Elbi H, Aksu H. Donepezil for anticholinergic drug intoxication: a case report. *Prog Neuropsychopharmacol Biol Psychiatry*. 2003;27:885–887.

77. Flacker JM, et al. The association of serum anticholinergic activity with delirium in elderly medical patients. *Am J Geriatr Psychiatry*. 1998;6:31–41.

78. Smith EM, et al. Effect of duloxetine on pain, function, and quality of life among patients with chemotherapy-induced painful peripheral neuropathy: a randomized clinical trial. *JAMA.* 2013;309:1359–1367.

79. Noble A, Bremer K, Goedhals L, Cupissol D, Dilly SG. A double-blind, randomised, crossover comparison of granisetron and ondansetron in 5-day fractionated chemotherapy: assessment of efficacy, safety and patient preference. The Granisetron Study Group. *Eur J Cancer.* 1994;30A:1083–1088.

80. Hesketh PJ. Chemotherapy-induced nausea and vomiting. *N Engl J Med.* 2008;358:2482–2494.

81. Shi Y, et al. Ramosetron versus ondansetron in the prevention of chemotherapy-induced gastrointestinal side effects: a prospective randomized controlled study. *Chemotherapy.* 2007;53:44–50.

82. Cherian A, Maguire M. Transient blindness following intravenous ondansetron. *Anaesthesia.* 2005;60:938–939.

83. Arcioni R, et al. Ondansetron inhibits the analgesic effects of tramadol: a possible 5-HT(3) spinal receptor involvement in acute pain in humans. *Anesth Analg.* 2002;94:1553–1557. table of contents.

84. Putnam PE, Orenstein SR, Wessel HB, Stowe RM. Tardive dyskinesia associated with use of metoclopramide in a child. *J Pediatr.* 1992;121:983–985.

85. Yis U, Ozdemir D, Duman M, Unal N. Metoclopramide induced dystonia in children: two case reports. *Eur J Emerg Med.* 2005;12:117–119.

86. Basch E, et al. Antiemetic use in oncology: updated guideline recommendations from ASCO. *Am Soc Clin Oncol Educ Book.* 2012;532–540. https://doi.org/10.14694/EdBook_AM.2012.32.230.

87. Low LC, Goel KM. Metoclopramide poisoning in children. *Arch Dis Child.* 1980;55:310–312.

88. Osorio-Sanchez JA, Karapetis C, Koczwara B. Efficacy of aprepitant in management of chemotherapy-induced nausea and vomiting. *Intern Med J.* 2007;37:247–250.

89. Badowski ME. A review of oral cannabinoids and medical marijuana for the treatment of chemotherapy-induced nausea and vomiting: a focus on pharmacokinetic variability and pharmacodynamics. *Cancer Chemother Pharmacol.* 2017;80:441–449.

90. Hotopf M, Chidgey J, Addington-Hall J, Ly KL. Depression in advanced disease: a systematic review Part 1. Prevalence and case finding. *Palliat Med.* 2002;16:81–97.

91. Okamura M, et al. Clinical experience of the use of a pharmacological treatment algorithm for major depressive disorder in patients with advanced cancer. *Psychooncology.* 2008;17:154–160.

92. Degner D, et al. Severe adverse drug reactions of antidepressants: results of the German multicenter drug surveillance program AMSP. *Pharmacopsychiatry.* 2004;37(Suppl 1):S39–S45.

93. Furukawa TA, Streiner DL, Young LT. Is antidepressant-benzodiazepine combination therapy clinically more useful? A meta-analytic study. *J Affect Disord.* 2001;65:173–177.

IV. Neurological complications of antineoplastic therapy

Psychiatric, pain, psychosocial, and supportive care issues

Psychiatric aspects of care in the cancer patient

William S. Breitbart[a,b,c], *Yesne Alici*[b], *and Mark Kurzrok*[b]

[a]Department of Psychiatry, Weill Medical College of Cornell University, New York, NY, United States, [b]Department of Psychiatry and Behavioral Sciences, Memorial Sloan-Kettering Cancer Center, New York, NY, United States, [c]Department of Medicine, Pain and Palliative Care Service, Memorial Sloan-Kettering Cancer Center, New York, NY, United States

1 Introduction

Psycho-oncology, the study and treatment of the psychological and psychiatric aspects of cancer along the continuum from prevention to cure, is a dynamic field that addresses the psychological response to cancer of patients, their families, and their care providers, as well as the psychological, behavioral, and social factors that influence cancer risk, detection, and survival.

Psycho-oncology clinicians and researchers exist throughout the world, and they are represented by several national and international organizations. A more recent accreditation standard set forth in 2012 by the American College of Surgeons Commission on Cancer requires that cancer centers have an on-site program that identifies distressed patients and provides resources for them to seek psychological care. Most cancer centers have an identified member or members responsible for providing psychosocial support. Clinicians and researchers from psychiatry, psychology, social work, nursing, and clergy offer unique perspectives and tools to alleviate distress as members of diverse interdisciplinary psycho-oncology teams.

The relatively young discipline of psycho-oncology continues to grow. With the development of novel psychotherapy modalities for cancer patients and the proliferation of communication skills training, providers can be trained in a host of evidence-based interventions to manage the complex psychiatric and psychological needs of patients with cancer.

This chapter reviews the history of psycho-oncology, the unique concerns that arise from adaptation to cancer diagnosis, the assessment and management of major psychiatric disorders in cancer, and psychiatric aspects in the management of fatigue. Issues related to the psychosocial care of families, bereavement, and cancer survivorship are also discussed.

2 Historical background

For centuries, physicians were reluctant to discuss the diagnosis of cancer with patients and their families. Cancer represented inevitable death due to the lack of effective treatment; consequently, revealing a diagnosis of cancer was regarded as cruel and destructive.[1]

By the early 1970s, with improvements in survival due to advances in surgical technique, chemotherapy, and radiation coinciding with a cultural shift toward patient-centered care, the discussion of cancer diagnoses became a more customary part of oncological treatment. Clinicians' enhanced comfort with communicating a diagnosis of cancer, increased concern for providing the best pain management and palliative care, and growing interest in patient autonomy and quality of life spurred clinicians and researchers to devote specific focus to supportive and psychological aspects of care. In the 1980s, psycho-oncology units began to develop in larger cancer centers and the first prevalence studies of psychiatric and psychological sequelae in cancer were published.[1–4] In the 1990s, behavioral research in changing habits (such as smoking), diet, and lifestyle improved education of the public on cancer prevention.[1] Health-related quality of life assessments and more recently patient-reported outcomes have become a routine part of outcome measures in clinical trials.[5] Despite the developments in treatment outlined above, pervasive stigmata regarding cancer and mental illness remain an important barrier to accessing care, especially among underserved populations and in various parts of the world.[1]

3 Common psychological responses to cancer

The diagnosis of cancer creates a crisis that requires patients to quickly adapt to catastrophic news while often experiencing significant distress. In addition to fear of death, major concerns after cancer diagnosis include dependency, disfigurement, disability, and abandonment, as well as disruptions in relationships, role functioning, and financial status.[6] The diagnosis of cancer induces distress and individuals show an initial characteristic response of shock and denial, which usually resolves within 1 week.[7] Patients must manage significant emotional distress while making crucial treatment decisions. The presence of a relative or friend can help with the processing of important information. Research has shown that the way news is conveyed by healthcare professionals can influence a patient's beliefs, emotions, and attitudes toward the future. There are evidence-based guidelines and recommendations available that describe effective ways to share the diagnosis, treatment plan, and prognosis to patients with cancer.[3,6–9]

The second phase of response to cancer diagnosis is characterized by a period of emotional turmoil with mixed symptoms of anxiety, depression, irritability, insomnia, and poor concentration. At times, symptoms become so severe that patients become unable to function. These symptoms usually begin to resolve with support from family, friends, and their care provider, who outlines a treatment plan that offers hope. This phase usually lasts for 1–2 weeks.[3]

During the third phase and the final phase, patients adapt to the diagnosis of cancer and its treatment. Patients return to previously utilized coping strategies that are helpful in reducing distress. These coping strategies are influenced by maturation and adaptation.[3]

The legal requirement of informed consent has improved communication between physicians and patients about illness, treatment options, and prognosis. However, increased involvement with care may represent an additional burden to some patients due to the awareness of the severity of their illness.[6] Psychological effects of cancer persist long after remission; after patients complete their treatment, they are frequently monitored for recurrence, which often precipitates anxiety and fear.

Patient responses to the diagnosis of cancer are modulated by societal factors, individual factors, and factors related to the specific diagnosis of cancer received. Societal factors relate to the society's perception and knowledge of cancer and its treatment. Individual factors include preexisting character traits, coping ability, ego strength, developmental stage of life, and the impact and meaning of the cancer at the given developmental stage of life.[3,10,11] Lower socioeconomic status has been shown to be a potential barrier to access health care, and

a risk factor for mental illness after cancer diagnosis.[12] Finally, adaptation to cancer is related to characteristics of the disease itself, such as stage of the disease, symptoms, site, prognosis, type of treatment, and impact on functionality.[3,6] Consideration of the factors that predict poor adjustment to cancer allows for earlier intervention for the most vulnerable among this population.

4 Epidemiology of psychiatric disorders in cancer patients

The prevalence of psychiatric disorders in cancer patients is approximately 50%.[4,5,13–18] Most psychiatric disturbances in patients with cancer relate to the cancer itself or adverse effects of treatment. The prevalence is highest among patients with advanced disease and poor prognosis.[4,13–19] Poor social support, low income, and prior history of mental illness are significant risk factors for psychiatric morbidity.[20] More than two-thirds of psychiatric diagnoses represent adjustment disorders, 4%–15% depression, and approximately 10% delirium.[4,14,20,21] Inpatient studies show a higher incidence of both depression (20%–45%) and delirium (15%–75%) with advanced disease.[14,22] Treatable syndromes, such as major depression and delirium, continue to be underdiagnosed and undertreated despite their known prevalence in cancer patients.[13,14,23,24]

5 General principles of psychiatric assessment and treatment in patients with cancer

Psychiatric assessment of cancer patients consists of a thorough medical evaluation of cancer site, stage, treatment, and any associated medical conditions or treatments, and a comprehensive assessment of the patient's past psychiatric history, current mental status, and the patient's understanding of their illness and prognosis. Another important area of assessment is the role of the environment, with particular focus on the patient's family, their interface with the healthcare system, and financial concerns.[2,6,25]

The Panel for Management of Psychosocial Distress, appointed by the National Comprehensive Cancer Network (NCCN), developed and recently released revised, evidence-based standard-of-care guidelines to assist oncology teams in attending to patients' distress.[25,26] The Panel initially proposed the use of the word "distress" due to stigma among patients, families, and healthcare providers associated with terms such as "psychiatric" or "psychological." The Panel defines distress as a "multifactorial, unpleasant experience of a psychologic (i.e., cognitive, behavioral, emotional), social, spiritual, and/or physical nature that may interfere with the ability to cope effectively with cancer, its physical symptoms, and its treatment." Per patient surveys, 20%–52% of patients with cancer have a significant amount of distress. Distress is associated with nonadherence to cancer treatment, longer hospital stays, and poorer quality of life.[26–29]

The guidelines include a simple Distress Thermometer to rapidly screen patients for distress. To use the Distress Thermometer, patients are asked to rate their level of distress on a scale of 0–10. The thermometer is accompanied by a problem list that patients use to indicate major sources of distress, namely, physical, psychological, social, spiritual, or practical problems such as finances. Widespread use of this scale for distress screening in outpatient settings has led to improved integration of psychosocial support and psychiatric care into the treatment of cancer patients.[3,24,25]

The hallmark of treatment in psycho-oncology is the simultaneous use of several modalities in an aggressive attempt for rapid relief of psychiatric symptoms and psychological suffering.[6,18] The physician must also address the family, as well as the family-staff interface. Physicians must remain active through all stages of illness.[6]

A growing number of psychotherapy modalities have been shown to reduce distress and improve quality of life in patients with cancer. The family should be involved with the therapeutic process, especially as disease progresses and patients lose function and ability to advocate for themselves.

Cognitive-behavioral therapy (CBT) is a time-limited, structured psychotherapeutic modality that encourages patients to identify and correct negative or inaccurate thought patterns that lead to changes in mood, anxiety, and behavior. CBT has been shown to be effective in reducing distressing psychological and physical symptoms in cancer patients.[30–32]

Evidence-based stress reduction and relaxation exercises designed to help patients become mentally and physically relaxed include passive relaxation, progressive muscle relaxation, medication, mindfulness meditation, biofeedback, and guided imagery. Patients may find some rather than others helpful; any method that is effective for reducing stress can be helpful and should be reinforced.[33]

Physical exercise is a safe and effective evidence-based treatment for patients with cancer, both during and after treatment. Physical activity promotes numerous physical and psychological benefits, including improved energy, sleep, fatigue, pain, depression, anxiety, immune function, blood and lymph flow, as well as overall quality of life. Exercise modality and intensity level should be individualized to patients' current physical state and titrated such that patients feel more energized after physical activity, not more fatigued. Patients should avoid any exercises that precipitate pain, shortness of breath, or other problems related to preexisting health conditions.[34,35]

Support groups are backed by three decades of research that shows positive effects on coping with cancer and quality of life. Support groups of various types have been found to improve mood, quality of life, coping skills, pain management, self-esteem, and interpersonal relationships in cancer patients.[8] Positive effects are often more pronounced among patients who are initially more distressed.[36–38] Emerging research in digital support groups has found that such therapeutic platforms decrease depression and negative reaction to pain while increasing enthusiasm for life and spirituality.[39]

Meaning Centered Psychotherapy (MCP) is a manualized adaptation of Victor Frankel's work that addresses existential distress.[40] Utilizing a multimodal approach that includes didactic, discussion, and experiential elements, sessions help patients identify and enhance sources of meaning. MCP in both individual and group formats has been shown to improve quality of life and spiritual well-being, and to reduce hopelessness, wish for hastened death, and somatic distress among patients with cancer.[41]

Managing Cancer and Living Meaningfully (CALM) is a manualized psychotherapeutic intervention intended to treat and prevent depression and end-of-life distress in patients with advanced cancer. Areas of therapeutic focus include symptom management and communication with healthcare providers, changes in self and personal relationships, spiritual well-being and the sense of meaning and purpose, and mortality and future-oriented concerns.[41]

The COPE Model (Creativity, Optimism, Planning and Expert Information) is an evidence-based conceptual model that was designed to help family caregivers of patients with cancer by applying the principles of problem-solving training to stressful problems experienced by caregivers. Though this modality is targeted toward caregivers, research has also shown direct benefit to patients via a reduction of symptom distress.[41,42]

In the following section, we will briefly review the psychiatric disorders commonly encountered in cancer patients.

6 Delirium

Delirium is a common neuropsychiatric syndrome characterized by an abrupt onset of disturbances in consciousness, attention, cognition, and perception that fluctuate over time. Symptoms are often highly distressing to patients, their families, and care providers.[43] Delirium is indicative of an underlying physiological disturbance, commonly including infection, metabolic disturbance, drug withdrawal, or adverse medication effect.[43]

Delirium is the most common neuropsychiatric complication of medical illness. It is a medical emergency that must be prevented, identified, and thoughtfully treated.

Unfortunately, delirium is frequently underdiagnosed and untreated in the medical setting, leading to increased morbidity and mortality, interference in the management of symptoms such as pain, increased length of hospitalization stay, increased healthcare costs, and increased distress for patients and their caregivers.[22,23,43–51] In a study of hospitalized patients with cancer, 54% were found to have recalled their delirium experience after recovery from delirium.[52] Factors predicting delirium recall included the degree of short-term memory impairment, delirium severity, and the presence of perceptual disturbances. The most significant factor predicting distress for patients was the presence of delusions. Patients with hypoactive delirium were found to be as distressed as patients with hyperactive delirium.[53,54]

Cancer is a risk factor for delirium. Among cancer patients, delirium is present in 25%–85% depending on the stage of illness.[22,23,49,50,54] Estimates of delirium in cancer patients range broadly due to varied clinical settings and contexts.[55]

Cancer-related factors such as uncontrolled pain, central nervous system tumors, brain metastases, paraneoplastic syndromes, and several immunotherapies and chemotherapies are common causes of delirium in patients with cancer. While prevalence rates of delirium range from 15% to 30% in hospitalized cancer patients, delirium becomes more prevalent in the last weeks of life, affecting 40%–85%.[22,23,50] Delirium is often a harbinger of death, as elderly patients admitted to a hospital who develop delirium have an estimated mortality rate of 22%–76% during their hospitalization.[50,56]

Predisposing risk factors such as age, functional impairment, and nature and severity of illness increase the risk of delirium during hospitalization.[23,49,50,56,57] Unplanned hospital admission with metabolic abnormalities strongly correlates with an increased risk of delirium in cancer patients.[58] Dementia is an important independent risk factor for developing delirium.[49] Sensory impairment, low body mass index (BMI), and malnutrition have also been associated with an increased risk for delirium.[23,49,50]

In patients with advanced cancer, direct effects of cancer on the central nervous system (CNS) as well as indirect CNS effects of the disease or treatments can precipitate symptoms of delirium.[49] Given frequent polypharmacy, older age, and diminished physiological state among cancer patients, even routinely ordered hypnotics are enough to induce delirium.[23,58] Narcotic analgesics are common causes of confusional states, particularly in the elderly and terminally ill. Benzodiazepines, an independent risk factor of delirium, pose an even greater risk in the elderly due to an increased sensitivity to sedation and

extended duration of action given their lipid solubility and accumulation in adipose tissue.[56] Chemotherapeutic agents known to cause delirium include ifosfamide, methotrexate, fluorouracil, vincristine, vinblastine, bleomycin, BCNU, cis-platinum, asparaginase, procarbazine, and corticosteroids.[23,59] Immunotherapies such as IL-2, interferon-alpha, and the newer chimeric antigen receptor (CAR) T-cell therapy are associated with delirium as well.[57,58] Drug withdrawal (benzodiazepines or opioids) may be a common cause of delirium, especially in postoperative patients whose drugs have been interrupted for surgery and have not been promptly restarted.[23,50,53]

The diagnosis of delirium is primarily clinical. Clinical features of delirium include rapidly fluctuating course, attention disturbance, altered level of alertness and arousal, increased or decreased psychomotor activity, disturbance of sleep-wake cycle, affective symptoms, perceptual disturbances, disorganized thinking, paranoia, incoherent speech, disorientation, and memory impairment.[41,50] Neurological abnormalities, including cortical (aphasia, apraxia, dysnomia, dysgraphia) and motor symptoms (tremor, asterixis, myoclonus, changes in muscle tone and deep tendon reflexes), can also be present.[23,50]

Delirium is classified into three clinical subtypes, based on arousal disturbance and psychomotor behavior: hyperactive, hypoactive, and mixed.[60,61] Approximately two-thirds are either of the hypoactive or of the mixed subtype.[60,62] The hyperactive form is most often characterized by hallucinations, delusions, agitation, and disorientation, while the hypoactive form is characterized by disorientation, sedation, but is rarely accompanied by hallucinations or delusions.[23,61]

Many of the clinical features and symptoms of delirium can also be associated with other psychiatric disorders such as depression, mania, psychosis, and dementia. Delirium, particularly the hypoactive subtype, is often initially misdiagnosed as depression. In distinguishing delirium from depression, particularly in the context of advanced cancer, an evaluation of the onset and temporal sequencing of depressive and cognitive symptoms is particularly helpful.[23,50]

While the "gold-standard" of diagnostic evaluation for delirium remains clinical interview with the *Diagnostic and Statistical Manual of Mental Disorders* (*DSM-5*), several scales and instruments have been developed for the screening of delirium by various providers, and some have been evaluated for use with cancer patients. These include the Delirium Rating Scale—Revised 98, the Confusion Assessment Method for ICU, the Delirium Observation Scale, and the Memorial Delirium Assessment Scale.[63–70]

The Memorial Delirium Assessment Scale (MDAS) is a 10-item delirium screening and assessment tool, validated among hospitalized patients with advanced cancer and AIDS.[66] Scale items assess disturbances in arousal and level of consciousness, as well as in several areas of cognitive functioning, including memory, attention, orientation, disturbances in thinking, and psychomotor activity. A cutoff score of 13 is diagnostic of delirium. MDAS was further examined for clinical utility and validation in a population of advanced cancer patients admitted to a palliative care unit. Investigators found the MDAS to be useful in this population, with a cutoff score of 7 out of 30, yielding the highest sensitivity (98%) and specificity (76%) for a delirium diagnosis in this palliative care population.[67,68]

The Confusion Assessment Method for ICU (CAM-ICU) was developed from the longer Confusion Assessment Method for the rapid screening of delirium in critically ill patients.[68,69] The instrument does not require verbal participation from the patient and includes four-staged criteria: acute change or fluctuating mental status, inattention, altered level of consciousness, and disorganized thinking. It is important to limit the use of the CAM-ICU to critical care settings, as a study of this instrument's use in hospitalized, noncritically ill cancer patients demonstrated poor sensitivity below 20%.[69]

The Delirium Rating Scale Revised 98 (DRS-R-98) has been evaluated for use with cancer patients and may be used as both a diagnostic and an assessment tool.[70] This instrument is a 16-item, clinician-related scale, with 13 severity items, namely, attention, delusions, sleep-wake cycle disturbance, perceptual disturbances and hallucinations, lability of affect, language, orientation, psychomotor agitation or retardation, short/long term memory, thought process abnormalities, and visuospatial ability, and three diagnostic items: fluctuation of symptom severity, physical disorder, and temporal onset of symptoms.[70-72]

The standard approach to managing delirium in the medically ill, including those with advanced disease, includes a search for underlying causes, correction of those factors, and management of symptoms with pharmacologic and nonpharmacologic interventions.[23,50] The desired and often achievable outcome is a patient who is awake, alert, calm, cognitively intact, and communicating coherently. In the terminally ill patient who develops delirium in the last days of life (referred to as terminal delirium), the management of delirium is unique. Management of terminal delirium presents several dilemmas, and the desired clinical outcome may be significantly altered by the dying process.[22,23,50,73-76] When confronted with delirium in the terminally ill or dying patient, a differential diagnosis should always be formulated as to the likely etiology or etiologies. Diagnostic workup in pursuit of an etiology for delirium may be limited by either practical constraints such as the setting (home, hospice) or the focus on patient comfort.[23,50] Most often, the etiology of terminal delirium is multifactorial or may not be determined. Etiology is discovered in less than 50% of terminally ill patients with delirium.

Unfortunately, when a distinct cause is found for delirium in the terminally ill, it is often irreversible or difficult to treat. One study found that 68% of delirious cancer patients could be improved, despite a 30-day mortality of 31%. In a recent prospective study of delirium in patients on a palliative care unit, investigators reported that the etiology of delirium was multifactorial in the great majority of cases. Although delirium occurred in 88% of dying patients in the last week of life, delirium was reversible in approximately 50% of episodes. Causes of delirium that were most associated with reversibility included dehydration and medications. Hypoxic and metabolic encephalopathies were less likely to be reversed in terminal delirium.[23,46,50,75]

Since the publication of the American Psychiatric Association guidelines for the management of delirium in 1999, various systematic review articles and guidelines have been published, offering clinicians evidence-based best clinical practices. The most recently published set of guidelines, the first to be published by the United Kingdom in a decade, is the 2019 Scottish Intercollegiate Guidelines Network (SIGN) publication on delirium. The guidelines emphasize the importance of multimodal nonpharmacological strategies, care-provider education, communication and follow-up, avoidance of high-risk medications, and medical evaluation of underlying cases.[77,78] The guidelines also advise the limited use of pharmacotherapy save for specific clinical situations, including patient distress and risk of harm to patient or staff. Specifically, the guidelines recommend avoidance of benzodiazepines except in the setting of withdrawal or Lewy-body dementia, consideration of cumulative anticholinergic burden, and careful titration of pain medications to avoid undertreated pain or opioid toxicity. The guidelines state that there "may" be a role for a short course of low-dose antipsychotics for patients with intractable distress.[79]

Diagnostic evaluation includes an assessment of potentially reversible causes of delirium. A full physical examination should assess for evidence of sepsis, dehydration, or major organ failure. Medications that could contribute to delirium should be reviewed and then tapered or discontinued as possible. Screening common laboratory values allows for assessment of the possible role of metabolic abnormalities or infection. Imaging studies of the brain and assessment of the cerebrospinal fluid may be appropriate in some cases to identify sources of delirium should the primary workup fail to reveal a source.[23,50]

Prevention is the most effective strategy for reducing morbidity and mortality associated with delirium. A delirium prevention protocol that targeted orientation, mobilization, medication reconciliation, sleep-wake cycle preservation, sensory impairment, and dehydration reduced the incidence of delirium by 6% on a general medical unit. This protocol has been shown to be effective when adapted to other settings, including nursing homes and surgical units. Avoidance or minimization of deliriogenic medications such as benzodiazepines, opioids, and anticholinergics, especially among the elderly and physically frail, is another effective intervention to prevent delirium.

While current evidence does not support the use of antipsychotic medications for the prevention of delirium in higher-risk patients, they may be useful for preventing postoperative delirium.[79,80] Some studies support the use of the selective alpha-2 adrenergic receptor agonist dexmedetomidine for preventing delirium in preoperative and intensive care settings.[81] Two smaller studies suggest melatonin and the melatonin analog, ramelteon, are effective in reducing the incidence of delirium in older hospitalized patients with medical illness.[82,83]

Nonpharmacologic and supportive treatment approaches are a critical part of effective delirium treatment. Evidence suggests that nonpharmacological interventions can hasten the improvement of symptoms, but not necessarily mortality or health-related quality of life when compared to usual care. Nonpharmacologic and supportive treatment approaches prioritize sleep-wake cycle regulation, nutrition and hydration, vision and hearing optimization, early mobilization, and orientation. Effective interventions include creating a calm, comfortable, well-lit environment, using orienting objects such as calendars and clocks, having family members or photos of family at bedside, limiting room and staff changes, and allowing patients to have uninterrupted periods of rest at night to promote healthy sleep-wake cycle. One-to-one nursing observation may also be useful and sometimes necessary to monitor for behavioral disturbances that place the patient and staff at risk of harm.[22,23,49,50,84–87]

Clinician-patient communication plays an especially crucial role in delirium, a disorienting, often disturbing state that can cause significant distress in both patients and their caregiver. Evidence suggests that patients' knowledge about delirium and plans for care can help them feel safer and more reassured.[88]

Supportive techniques alone are often not effective in controlling the symptoms of delirium, and pharmacologic treatment is necessary for dangerous or distressing symptoms.[23] Antipsychotics constitute the primary pharmacological intervention.[50] Haloperidol has been effectively used in the treatment of delirium for many years.[50] Atypical antipsychotic medications, including olanzapine, risperidone, quetiapine, ziprasidone, and aripiprazole, are increasingly being used in the treatment of delirium due to their favorable side-effect profile.[89–95]

None of the antipsychotic medications have been approved by the Food and Drug administration (FDA) for the treatment of delirium.[89] Most studies are limited to open-label trials, case reports, and retrospective reviews. Comparison trials have not identified any antipsychotic medication as superior to another in terms of efficacy.[89–95]

The past 10 years has yielded a surge in treatment trials for delirium. Several recent studies found antipsychotics to have no benefit in terms of delirium duration or severity of delirium. A 2016 systematic review examined antipsychotic drugs, including oral risperidone, oral olanzapine, oral quetiapine, intramuscular ziprasidone, and oral, intravenous, and intramuscular haloperidol, and concluded that the current evidence does not support the use of antipsychotics for the treatment or prevention of delirium in hospitalized older adults.[96] There was no significant decrease in delirium incidence among 19 studies and no change in delirium duration or severity, hospital or intensive care length of stay, or mortality. Potential harm was demonstrated in two studies in which more patients required institutionalization after treatment with antipsychotic medications. Another recent systematic review of antipsychotics in the treatment of delirium, among 16 randomized controlled trials and 10 observational studies, found no difference in delirium duration, length of hospital stay, sedation status, or mortality between haloperidol and atypical antipsychotic medications versus placebo.[97] This study also found no difference in delirium severity and cognitive functioning for haloperidol versus the second-generation antipsychotic medications. The authors noted that heterogeneity was present among the studies in terms of dose, administration route of antipsychotics, outcomes, and measurement instruments, as well as insufficient or no evidence regarding multiple clinically important outcomes. In a randomized controlled trial of atypical antipsychotic medications in palliative care settings, participants receiving oral risperidone or haloperidol had higher delirium symptom scores and were more likely to require breakthrough treatment compared with participants receiving placebo.[98] Limitations of this analysis that must be considered into account include that survival was not one of the primary study outcomes and that patients taking haloperidol were more medically ill, required more opioids, and had more severe delirium.[98]

The choice of medication in the treatment of delirium depends on multiple factors, including the degree of agitation, subtype of delirium, the available route of administration, concurrent medical conditions, and consideration of adverse-effect profile.[23,50,89] A recent Cochrane review on drug therapy for delirium in the terminally ill concluded that based on a double-blind randomized controlled study, haloperidol is the most suitable medication for the treatment of patients with delirium near the end of life, with chlorpromazine as an acceptable alternative as long as a small risk of cognitive impairment is not a concern. However, it is also emphasized that due to the small number of patients in that study ($n = 30$), the evidence is insufficient to draw any conclusion about the role of pharmacotherapy in terminal delirium.[95,99] Another Cochrane Review, comparing the efficacy and the incidence of adverse effects between haloperidol and atypical antipsychotics, concluded that haloperidol and selected newer atypical antipsychotics (risperidone, olanzapine) are effective in managing the symptoms of delirium.[94] The American Psychiatric Association guidelines for the treatment of delirium recommend low-dose haloperidol (i.e., 1–2 mg PO every 4 h as needed or 0.25–0.5 mg PO every 4 h for the elderly) as the treatment of choice in cases where medications are necessary.[50] Lorazepam (0.5–1.0 mg q 1–2 h PO or IV) along with haloperidol may be effective in rapidly sedating the agitated delirious patient and may minimize extrapyramidal symptoms associated with haloperidol. However, benzodiazepine monotherapy should be avoided unless the delirium is due to alcohol or benzodiazepine withdrawal, as benzodiazepines can cause disinhibition and oversedation. An alternative strategy in agitated patients is to switch from haloperidol to a more sedating antipsychotic such as chlorpromazine.[50,100] It is important to note the anticholinergic and hypotensive side effects of chlorpromazine, particularly in elderly patients.[100] Table 1 lists the antipsychotic medications used in the treatment of delirium with recommended dose ranges.

In light of recent literature on the use of antipsychotics for the treatment of delirium, it is important to emphasize that for patients with hypoactive delirium, or delirium of any subtype with mild to moderate severity, antipsychotics are best avoided or only used if the benefits clearly outweigh the risks, such as when symptoms interfere with medical care or pose danger to the patient or staff. As noted in a debate article by Meagher et al., evidence-based concerns must be applied to all interventions, both pharmacological and nonpharmacological, with equal vigilance.[101] Nonpharmacological interventions can be effective in the prevention of delirium in certain settings and may be a helpful tool in the treatment of delirium, though interventions such as early mobilization

TABLE 1   Antipsychotics in the treatment of delirium.

Medication	Dose range*a*	Available route of administration*b*	Side effects	Comments
Typical antipsychotics				
Haloperidol	0.5–2 mg every 2–12 h	PO, IV, IM, SC	> 4.5 mg/day has been associated with extrapyramidal side effects. Monitor QTc interval.	Remains to be the gold standard medication in the treatment of terminal delirium. May add lorazepam (0.5–1 mg every 2–4 h) for agitated patients.
Chlorpromazine	12.5–50 mg every 4–6 h	PO, IV, IM, SC, PR	More sedating and anticholinergic compared to haloperidol. Monitor blood pressure for hypotension.	Preferred in agitated patients due to its sedative effect.
Atypical antipsychotics				
Olanzapine	2.5–5 mg every 12–24 h	PO^c	Sedation is the main dose-limiting side effect in short-term use.	Older age, preexisting dementia, and hypoactive subtype seem to be associated with poor response.
Risperidone	0.25–1 mg every 12–24 h	PO^c	> 6 mg/day is associated with an increased risk of extrapyramidal symptoms. Orthostatic hypotension.	Clinical experience suggests better results in patients with hypoactive delirium.
Quetiapine	12.5–100 mg every 12–24 h	PO	Sedation, orthostatic hypotension.	Preferred in patients with Parkinson's disease or Lewy body dementia due to lower risk of parkinsonian side effects.
Ziprasidone	10–40 mg every 12–24 h	PO	Monitor QTc interval on ECG.	Evidence limited to case reports. Least preferred in the medically ill due to risk of QT prolongation compared to other atypical antipsychotics.
Aripiprazole	5–30 mg every 24 h	PO^c	Monitor for akathisia.	Clinical experience suggests better results in hypoactive delirium.

a *Lower doses with slow titration are recommended in older patients and in patients with multiple medical comorbidities.*

b *Olanzapine, aripiprazole, and ziprasidone are available in IM formulations; however, there are no case reports or studies on their use in the management of delirious patients.*

c *Risperidone, olanzapine, and aripiprazole are available in orally disintegrating tablet forms.*

or cognitive remediation may be agitating to patients with severe delirium rather than being beneficial.

The APA guidelines for the treatment of delirium recommend the use of low-dose haloperidol (i.e., 1–2 mg PO every 4 h as needed or 0.25–0.5 mg PO every 4 h for the elderly) as the treatment of choice in cases where medications are necessary.[98] Existing literature supports the use of risperidone for the treatment of delirium, starting at doses ranging from 0.125 to 1 mg and titrated as necessary to a maximum of 4 mg daily with particular attention to an increased risk of extrapyramidal symptoms, orthostatic hypotension, and sedation at higher doses.[98] Olanzapine can be started between 2.5 and 5 mg nightly and titrated with sedation often a major limiting factor, though this medication effect may be favorable in the treatment of hyperactive delirium. Recent literature also supports the use of olanzapine for the management of chemotherapy-induced nausea and vomiting, and this medication is frequently used in both outpatient and inpatient cancer settings.[99] Literature supports the use of quetiapine with a starting dose of 12.5–50 mg and titration up to 100–200 mg daily (often in divided doses). Sedation and orthostatic hypotension are often dose-limiting factors.[89] Quetiapine is the preferred agent for patients at an increased risk of extrapyramidal symptoms including those with Parkinson's disease or Lewy Body dementia given it is the antipsychotic medication with the lowest risk of extrapyramidal effects.[102] Recent research and clinical experience suggest a starting dose of 2–5 mg daily for aripiprazole, with a maximum dose of 30 mg daily. Clinical evidence supports its use in hypoactively delirious patients.[89]

Important considerations in starting treatment with any antipsychotic for delirium may include extrapyramidal symptom risk, sedation, anticholinergic effects, cardiac arrhythmias, and drug-drug interactions. The FDA released a public health advisory on increased risk of death related to the use of antipsychotics in the treatment of behavioral disturbances of patients with dementia. This advisory was followed by studies showing increased mortality rates in elderly patients using antipsychotics. A recent meta-analysis found that the relative risk for all-cause mortality is close to 2 (HR = 1.9–2.19), with mortality remaining high in patients both with and without dementia. This study also found that the period of greatest mortality risk is the initial 180 days of treatment and that all-cause mortality risk is dose dependent. Little difference in mortality risk was found between typical and atypical antipsychotics.[103]

A retrospective cohort of elderly people on antipsychotics found that the typical antipsychotics were associated with higher rates of mortality as compared to atypical antipsychotics.[104,105] Increase in mortality was seen most often shortly after medication initiation. It is unknown whether those warnings apply to the short-term use (i.e., 1–2 weeks) of antipsychotics in a medically ill elderly population. The FDA has recently issued another warning about the risk of QT prolongation and torsades de pointes with intravenous haloperidol; thus, QTc intervals should be monitored regularly.[106,107] Given the inherent risks of antipsychotic use, clinicians should only use them to treat symptoms that are distressing to the patient or that put the patient or staff at risk of harm. Antipsychotic medications should be used at the lowest possible dosages for the shortest duration possible. These medications should be discontinued at discharge unless there is a clear plan for taper or outpatient psychiatric care for longer-term follow up.

Limited evidence supports the use of some other agents in addition to antipsychotics, which may be helpful for patients with antipsychotic-refractory agitation or elevated QT interval on EKG.

A few retrospective studies support the use of valproic acid as a second-line agent for agitation in the setting of delirium.[108–110] Theoretically, valproic acid may modulate delirium via effects on neurotransmitter systems, inflammation, oxidative stress, and transcriptional changes implicated in the pathophysiology of delirium. Of note, no current literature studies the use of valproic acid in cancer patients, and some characteristics of this medication may proscribe its use in this population, including risk of thrombocytopenia, and medication interactions through a number of mechanisms, including increased clearance by protein displacement (methotrexate, cisplatin, and irinotecan), and via the inhibition of CYP2C9, CYP3A4 (temsirolimus and imatinib), UGT1A4 (lamotrigine), UGT2B7, and epoxide hydrolase.[111–113]

Dexmedetomidine is an alpha-2 agonist that has been studied in postoperative and critical settings, often with mechanically ventilated patients. A recent systematic review and meta-analysis demonstrated that dexmedetomidine reduces the incidence of delirium and agitation in intensive care patients.[114] Adverse effects include bradycardia and hypotension, and limitations for its use include the need for high-level care and current expense.

Melatonin and their analogues are an emerging target of therapeutic interest for patients with delirium. One recent meta-analysis showed that melatonin supplementation decreased the incidence of delirium among elderly patients on general medical wards.[115] Another recent meta-analysis supported the efficacy of prophylactic melatonin supplementation to prevent postoperative delirium in older patients.[116]

Delirium in pediatric oncology settings is managed similarly, with careful assessment for and treatment of underlying medical etiologies and minimizing iatrogenic factors in care. If symptoms of delirium persist and the child demonstrates agitated behavior that is distressing or interfering with medical care, pharmacologic therapy may be considered. Most experts recommend the use of the atypical antipsychotic medications, such as quetiapine.[117]

7 Adjustment disorders

Adjustment disorders with depressed mood, anxious mood, mixed emotional features, and other variants represent the most common group of psychiatric diagnoses found in cancer patients.[43,115] Diagnosis of cancer may invoke a normal reaction to stress that can affect mood. Patients with cancer are on a continuum, with acute stress responses at one end and mood and anxiety disorders at the other.[3,6] Adjustment disorders represent the development of emotional or behavioral symptoms due to an identifiable stressor within 3 months of stressor onset, with distress out of proportion to the severity or intensity of the stressor, or impairment in functioning in social, occupational, or one or more psychosocial areas. Symptoms must not meet criteria for another psychiatric disorder and cannot be compatible with normal bereavement. The DSM-5 states that environmental context and cultural factors that may impact symptom severity and presentation must be considered as well. This disorder may resolve when the stressor resolves but the condition may become chronic, requiring medications along with therapy.[3,6,43,118,119]

Prevalence estimates vary due to the lack of a standardized screening instrument for adjustment disorder. A meta-analysis by Mitchell et al. found the prevalence of adjustment disorder to be 19% in cancer patients.[16]

8 Anxiety disorders

Anxiety is the most common response in the setting of cancer, a normal adaptive response to a significant threat, but one that can become maladaptive. Anxiety is manifested by a broad array of physical signs and symptoms, changes in thinking (i.e., intrusive thoughts), and behavior.

In this section, we will review the anxiety disorders standardized in the Diagnostic and Statistical Manual of Mental Disorders, 5th edition (DSM-5), commonly encountered in patients with cancer.[43]

Generalized anxiety disorder is characterized by excessive worry across different settings occurring more days than not for at least 6 months, with at least three of the following six symptoms: restlessness or feeling keyed up or on edge, being easily fatigued, difficulty concentrating or mind going blank, irritability, muscle tension, and sleep disturbance.[43]

Panic disorder is characterized by recurrent, expected, or unexpected panic attacks followed by worry and concern about panic attacks and behavior changes to avoid

precipitating future attacks.[43] Panic symptoms have specific features, including rapid onset of symptoms in various situations such that anticipatory avoidance is not possible.[16]

Specific phobias are characterized by persistent and excessive fear elicited by the presence or anticipation of a specific object or situation. Phobias of blood, needles, hospitals, magnetic resonance imaging machines, and radiation simulators may complicate treatment adherence. Patients undergoing chemotherapy have nausea and vomiting and about 25%–30% develop anticipatory nausea and vomiting despite adequate antiemetic regimens. Patients experience nausea when they are exposed to stimuli that remind them of drug infusions.[6,23,120]

Anxiety disorder due to a General Medical Condition is meant to describe cases in which the symptoms are a direct physiological consequence of the illness. Anxiety in cancer can be caused by different medical conditions, including pulmonary embolism, hypoxia, hypoglycemia, and cardiac disorders. Hormone-secreting neoplasms and paraneoplastic syndromes can also cause anxiety.

Treatment aims include reducing the patient's overall level of emotional distress as well as reducing specific target symptoms that may impair social or occupational functioning. Treatment of anxiety in cancer patients depends on the etiology and the timing of onset of symptoms. Delirium can present with anxiety; therefore, it must be considered when evaluating cancer patients presenting with an acute change in symptoms or fluctuating course.[6]

Nonpharmacologic treatment involves several approaches, including interpersonal psychotherapy, supportive psychotherapy, and cognitive behavioral therapy. Several behavioral methods have been used such as progressive muscle relaxation, breathing exercises, meditation, biofeedback, and guided imagery with good results in cancer patients with anxiety disorders.[3,6,16]

Medications used to treat anxiety in cancer patients are also used in primary anxiety disorders. These medications are benzodiazepines (such as lorazepam, clonazepam, diazepam) for short-term relief of acute anxiety and antidepressants (typically selective serotonin reuptake inhibitors such as escitalopram and sertraline or serotonin norepinephrine reuptake inhibitors such as venlafaxine or duloxetine due to less toxicity and medication interactions than TCAs) for long-term use. Low-dose atypical antipsychotic medications can be used for short periods of time—given the risk of tardive dyskinesia, metabolic syndrome, and increased all-cause mortality—particularly in frequently monitored inpatient settings with anxiety as the presenting symptom for delirium or in comfort care settings.[3,6,16] Beta blockers can be considered for physical symptoms of anxiety such as tremor and palpitations.[121]

9 Trauma and stressor-related disorders

Trauma- and stressor-related disorders, reclassified from acute stress disorder in the DSM-5, involve direct or indirect exposure to an actual or threatened death, serious injury, or sexual violation. The diagnostic criteria of trauma- and stressor-related disorders include the presence of nine or more symptoms from five categories, namely, intrusion, negative mood, dissociation, avoidance, and arousal, that begin or worsen after the traumatic event.[43] Trauma- and stressor-related disorders should be considered for symptoms that develop within 3 days to 1 month of a traumatic event; posttraumatic stress disorder (PTSD) should be considered for symptoms that have persisted for more than 1 month after exposure to a traumatic event.[119] The prevalence of trauma- and stressor-related disorders has not been established in cancer patients. Having cancer is perceived as a life-threatening event. For those patients who have significant psychological trauma, the fear can result in dissociative experiences, avoidance to everything related to cancer, nightmares, irritability, hypervigilance, and poor concentration. PTSD has been found in up to 32% of cancer patients. Patients with cancer who are younger, who are diagnosed with more advanced disease, or who recently completed cancer treatment may be at greater risk of developing PTSD. The literature also reports that up to 80% of patients are likely to experience PTSD symptoms following cancer.[3,6,120,122-125] It is difficult to assess classical PTSD symptoms in cancer patients. Evaluation of intrusive symptoms is dubious since cancer patients often report fears related to the future and rarely flashbacks or intrusive memories. Avoidance behavior is difficult to ascertain, as patients are frequently confronted with the disease and its treatment with potential trauma-related stressors.[6,119-126]

10 Depressive disorders

Depression is a common psychiatric complication of cancer, though it is often not properly recognized in clinical care.[127] Depression is a common psychiatric complication of cancer and an important risk factor for suicide. Cancer patients are vulnerable to these symptoms at all stages of illness. It is important to identify when normal sadness or distress associated with cancer becomes overwhelming or impairing, potentially indicating the presence of a depressive disorder rather than a normal emotional response. Improved recognition and treatment of depressive disorders leads to increased adherence to cancer treatment, improved quality of life, and reduced serious consequences such as the desire for hastened death and suicide.[127]

The prevalence estimates of major depressive disorder in cancer patients have varied widely in different studies, from a low of 1% to a high of 50%.[14] The use of different diagnostic measures and cutoff criteria impacts the prevalence estimates. Physical symptoms and the site and stage of cancer are additional factors that contribute to differences in prevalence. Although it is well established that depression is more prevalent among women than among men in the general population, this gender difference is not evident among cancer patients.[13,14,127-133]

There are several risk factors for depression in cancer patients.[131,134-138] Chemotherapy regimens including vinblastine, vincristine, interferon, procarbazine, asparaginase, tamoxifen, cyproterone, and corticosteroids are associated with greater risk. A higher prevalence of depressive disorders has been found among patients with pancreatic, head and neck, breast, and lung cancer with relatively lower rates observed among those patients with lymphoma, colon, and gynecological cancers.[136] Depression may also develop secondary to organ failure, or nutritional, endocrine, and neurological complications of cancer. Other risk factors for depressive disorders are advanced disease stage, physical disability, presence of other chronic medical illnesses, a previous history of depression, family history of depression, uncontrolled pain, low social support, social isolation, and recent experience of a significant loss.[3,16,131]

Diagnosis of depression is challenging in cancer patients due to the neurovegetative symptoms that mimic many symptoms caused by cancer and/or its treatment, such as loss of appetite, fatigue, sleep disturbances, psychomotor retardation, apathy, and poor concentration. The assessment of depressive symptoms in cancer patients should focus on the presence of dysphoria, anhedonia, hopelessness, worthlessness, excessive or inappropriate guilt, and suicidal ideation. Changes in appetite, sleep disturbances, fatigue, and loss of energy can still be used to diagnose the cancer patient who is physically well, although one must bear in mind that chemotherapy and radiation therapy-induced fatigue and other vegetative disturbances can continue for a long time. Poor memory and impaired concentration are more likely to be the initial chief complaints in elderly depressed patients. In medically ill depressed patients, the presence of delusions and hallucinations may be reflective of a diagnosis of delirium, which should be ruled out first.[3,6,131,139]

Several subcategories of DSM-5 depression diagnoses are found in the context of cancer. A diagnosis of "depressive disorder due to cancer with depressive features" pertains to depressive symptoms due to cancer that do not meet criteria for a depressive episode. "Depressive disorder due to cancer with major-depressive like episode" should be used when symptoms meet full criteria for a major depressive episode. When symptoms of mania or hypomania due to cancer are present concurrently with depressive symptoms, "depressive disorder due to cancer with mixed features" is most appropriate. When a medication (such as interferon) is the underlying cause of a depressive disorder, the diagnosis of "substance/medication-induced depressive disorder" is used. Corticosteroids are used in a wide range of clinical situations that can cause depressive, manic, and psychotic symptoms.[19,132,139] Establishing the etiology of symptoms of depression in cancer patients is difficult, and usually, multiple factors contribute to depressive symptoms in most of these patients. Self-report symptoms inventories are useful for screening purposes but are not discriminating between a major depressive disorder and a mood disorder due to general medical condition with depressive features.[139–141] Major depressive disorder is defined as an episode of clinically significant persistent and pervasive depressed mood and/or anhedonia, accompanied by cognitive and behavioral symptoms. Patients who have had a major depressive episode in the past are at a risk of a recurrence in the context of cancer.[139,142]

Demoralization is an important depressive syndrome common among patients with cancer that is not recognized in the DSM-5. This syndrome has been described and explored in the literature for over 20 years. De Figueiredo and Fava et al. each delineate demoralization from major depressive disorder, finding that those with demoralization feel subjectively incompetent, believe they have failed to meet expectations of themselves or others, or are unable to cope with their current situation, resulting in feelings of despair, helplessness, hopelessness, or giving up.[143,144] Demoralized patients perceive their source of distress as outside of themselves, do not suffer from anhedonia or guilt, and are able to rationally evaluate their problems. Studies have demonstrated that demoralization can negatively affect patients with cancer, as the condition has been associated with poorer quality of life, increased preoccupation regarding cancer, loss of dignity, and suicidal ideation.[144] Fava et al. proposed the Diagnostic Criteria for Psychosomatic Research (DCPR) for demoralization, revised for improved utility in research and clinical care.[142] A recent study using DCPR to identify demoralized patients found a prevalence of over 30%, while prevalence of major depression was less than 17%.[144,145] Growing literature that documents patient experiences of demoralization, and recent systematic reviews that summarize growing knowledge about the symptoms and consequences, convincingly assert this syndrome's importance in the setting of cancer.[146–150]

10.1 Table for diagnostic criteria

Management of depression in cancer patients requires a comprehensive approach that addresses evaluation,

treatment, and follow-up.[140,141] Initial management begins with the establishment of a therapeutic alliance with the clinician and recruiting support from family or friends. The APA practice guidelines for the treatment of depressive disorders in physically healthy individuals have been applied to the treatment of depression in cancer patients by the National Comprehensive Cancer Network (NCCN).[25,140] There are several pharmacologic and psychotherapeutic strategies available. Prior to selecting an appropriate treatment, the site of cancer, current cancer treatment, comorbid medical conditions, and medications should be taken into consideration, any of which may contribute to depressive symptoms. If the depressive disorder is believed to be caused by a medical condition or by a drug, the clinician should treat the underlying condition or change the drug; however, antidepressants are usually started concurrently to relieve patient's suffering as quickly as possible.[6,132,142]

The use of antidepressant medications in cancer patients creates unique challenges. A rapid onset of action is preferable in cancer patients, especially in the terminally ill; however, antidepressants may take several weeks to have a therapeutic effect due to their delayed onset of action.[141] An appropriate antidepressant should be selected based on the potential side effects of each antidepressant, drug-drug interactions, patients' prognoses, primary symptoms of depression, and comorbid condition. Antidepressants should be started at low doses and titrated slowly in medically frail and elderly cancer patients.[132]

Selective serotonin reuptake inhibitors (SSRIs) have become the first line of treatment for depressive disorders in the medically ill.[132,140] They are efficacious, generally well tolerated, and are not as toxic in overdose as tricyclic antidepressants. Some SSRIs, such as fluoxetine and fluvoxamine, are inhibitors of cytochrome P450 isoenzymes. It is therefore important to monitor for the possibility of drug-drug interactions.[143,144] Sertraline, citalopram, and escitalopram are less protein-bound and may have a lower risk of drug interactions with the P450 system. Many of the SSRIs are available in liquid forms, making it easier for patients who cannot swallow pills.[132]

Tricyclic antidepressants (TCAs) have been around for many years and are less expensive than many of the SSRIs. Because of their anticholinergic, antiadrenergic, and antihistaminergic side effects, they are less frequently used in cancer patients. Their role as adjunct pain medications, especially for neuropathic pain, has become the most common indication for their use in cancer patients.[132,140]

Monoamine oxidase inhibitors (MAOIs) are rarely used in the treatment of cancer patients with depression due to the risk of fatal hypertensive crisis when concurrently used with tyramine-rich food or sympathomimetic drugs.[132,140]

Of the newer antidepressants, bupropion acts primarily on the dopamine system and may have a mild stimulant effect, which can be beneficial for cancer patients with fatigue or psychomotor retardation. It is generally tolerated well in the medically ill. Bupropion is associated with an increased risk of seizures at higher doses and should be used with extreme caution in individuals with central nervous system tumors or seizure disorders. Venlafaxine and duloxetine work as reuptake inhibitors of serotonin and norepinephrine (SNRI). They are generally well tolerated, with a benign side-effect profile like that of SSRIs. Blood pressure monitoring is recommended for patients on an SNRI due to the potentiation of noradrenergic neurotransmission. It is important to note that venlafaxine primarily inhibits serotonin reuptake at low doses; its effect on norepinephrine reuptake inhibition is seen at doses higher than 150 mg per day. Venlafaxine and duloxetine are preferably used for patients with comorbid depression and neuropathic pain due to their effects as adjunct pain medications. Mirtazapine acts by blocking the 5-HT_2, 5-HT_3, and $\alpha2$ adrenergic receptor sites. Its side effects of sedation and weight gain are beneficial for many cancer patients with insomnia and weight loss. Mirtazapine also has antiemetic properties via 5-HT_3 blockade. It is also available in a dissolvable tablet form, which is particularly useful for patients who cannot swallow or who have difficulty with nausea and vomiting.[132,140,151,152]

Psychostimulants and wakefulness-promoting agents can be helpful to treat depressed cancer patients' symptoms of fatigue, psychomotor retardation, and poor concentration. Psychostimulants used in cancer patients include dextroamphetamine and methylphenidate. They have a major advantage over antidepressants due to their rapid onset of action, and their role in relieving fatigue, promoting wakefulness, and countering opioid-related sedation. Adverse effects include anorexia, anxiety, insomnia, euphoria, irritability, and mood lability. Side effects are not common at low doses and can be avoided by slow titration. Hypertension and cardiac complications can occur; thus, it is advisable to monitor cardiac function. Modafinil is a nonstimulant wakefulness-promoting agent. It produces increased alertness, wakefulness, and energy. It is presumably better tolerated than the psychostimulants, but may also cause anxiety, restlessness, and insomnia. This medication should be used with caution in patients with poorly controlled hypertension.[132,140,153] Table 2 includes a list of medications used in the treatment of cancer patients with depression.

Several different psychotherapeutic techniques have been successfully employed with depressed cancer patients. Psychotherapy is often combined with a pharmacologic intervention. The most utilized forms of psychotherapy are supportive psychotherapy and cognitive-behavioral therapy.[132,140]

TABLE 2   Medications used to treat depression in cancer patients.

Medication	Starting dose/ therapeutic range	Common side effects/ comments
Selective serotonin reuptake inhibitors (SSRIs)		
Fluoxetine[a]	10–20 mg/20–60 mg	Varying degrees of gastrointestinal distress, nausea, headache, insomnia, increased anxiety, sexual dysfunction. Sertraline, citalopram, and escitalopram are associated with the fewest p450 system interactions.
Sertraline[a]	25–50 mg/50–200 mg	
Paroxetine	10–20 mg/20–50 mg	
Fluvoxamine	50 mg/100–300 mg	
Citalopram[a]	10–20 mg/20–60 mg	
Escitalopram[a]	10 mg/10–20 mg	
Serotonin and norepinephrine reuptake inhibitors (SNRIs)		
Duloxetine	20–30 mg/30–60 mg bid	Activating, anxiety, nausea. Can be helpful with chronic pain.
Venlafaxine ER	37.5 mg bid/ 75–225 mg	Activating, nausea, anxiety, sedation, sweating, hypertension.
Desvenlafaxine	25 mg/50 mg	Activating, nausea, anxiety, sedation, sweating, hypertension.
Tricyclic antidepressants (TCAs)		
Amitriptyline	10–25 mg/50–150 mg	Sedating
Imipramine	10–25 mg/50–300 mg	Sedating
Desipramine	25 mg/75–200 mg	Minimal sedation or orthostasis; moderate anticholinergic effects.
Nortriptyline[a]	10–25 mg/50–150 mg	Sedation; minimal anticholinergic effects or orthostasis.
Doxepin[a]	25 mg/75–300 mg	Sedating; anticholinergic effects; orthostasis.
Atypical antidepressants		
Bupropion	75 mg/150–450 mg	Activating. Seizures if predisposed; Less sexual dysfunction. Can be helpful with smoking cessation.
Trazodone	50 mg/150–200 mg	Sedation; useful as a sleep aid; priapism.
Nefazodone	100/150–300 mg	Risk of liver failure; sedation, dizziness, constipation. Sexual dysfunction unlikely.
Mirtazapine	7.5–15 mg/15–45 mg	Sedation, weight gain; dissolvable tablet form available.
Vilazodone		Gastrointestinal distress, nausea, insomnia. Rare weight gain, less sexual dysfunction.

TABLE 2 Medications used to treat depression in cancer patients—cont'd

Medication	Starting dose/ therapeutic range	Common side effects/ comments
Vortioxetine		Gastrointestinal distress, nausea, insomnia. Rare weight gain, less sexual dysfunction.
Stimulants and wakefulness-promoting agents		
Dextroamphet-amine	2.5 mg/5–30 mg	Possible cardiac complications; Agitation, anxiety, agitation, nausea.
Methylphenidate	2.5 mg/5–15 mg bid	
Modafinil	50 mg/100–400 mg	Activating, nausea, cardiac side effects, usually well tolerated.

[a] *Available in liquid form. Adapted from Coups, Winell, Holland,[132] Holland and Friedlander.[2]*

Group therapy can be helpful to improve social networks, connecting the patient with others who have the same diagnosis and/or treatment, decreasing the patient's sense of isolation. Supportive-expressive and cognitive-existential group psychotherapies have been studied and used successfully for cancer patients.[132,154–156]

Electroconvulsive therapy (ECT) is an effective treatment modality for depressed patients. ECT should be considered in patients who are refractory to psychopharmacologic treatment, have severe weight loss secondary to depression, exhibit acute psychosis, or have a high suicide risk. Although there are no absolute contraindications to ECT, it is used with caution among individuals with central nervous system tumors or cardiac problems.[157,158]

11 Suicide assessment and management in cancer patients

The incidence of suicide is higher in cancer patients compared with the general population. Studies suggest that although a small number of cancer patients commit suicide, the relative risk of suicide is almost four and a half times than that of the general population.[159–162] Suicide is more likely to occur in patients with advanced cancer who suffer escalating depression, hopelessness, and poorly controlled symptoms, particularly pain. Patients with bladder, lung, head and neck, and testicular cancers are at an increased risk for completing suicide. Suicidal thoughts in patients with advanced disease, poor prognosis, or poorly controlled symptoms should not be viewed as rational.[161–165] It is important to note that those patients may have a treatable major depressive episode precipitating their suicidal ideation. Clinicians should evaluate for hopelessness and a diagnosis of depression in terminally ill patients with persistent desire for death or suicidal intention.[162–164]

Prior history of psychiatric illness, previous history of depression or suicide attempts, recent bereavement, history of alcohol or other substance abuse or dependence, male gender, family history of depression or suicide, lack of family or social support, and recent losses are common risk factors for suicide.[161,166] Untreated delirium may lead to unpredictable suicide attempts due to impaired judgment and impulse control. Older patients and individuals with head and neck, lung, breast, urogenital, gastrointestinal cancers, and myeloma seem to have an increased risk of suicide.[161,165,166] An international population-based study from Denmark, Finland, Norway, Sweden, and the United States has shown a small, but statistically significant, increased long-term risk of suicide, 25 or more years after a breast cancer diagnosis.[167] A Finnish survey of suicides in 1 year revealed that 4.3% had cancer. Half of the patients were in remission at the time of suicide. Patients were noted to have a history of psychiatric illness prior to cancer diagnosis, particularly those related to substance abuse.[167]

Evaluation of suicidal thoughts should consider the disease stage and prognosis. It is helpful to consider the issue of suicidality from four perspectives: suicidal thoughts that occur transiently in all patients with cancer; suicidal thoughts in patients who are in remission with a good prognosis; suicidal thoughts in patients with poor prognosis/poor symptom control; and suicidal thoughts in the terminally ill. Suicidal ideation or plans to commit suicide in cancer patients with good prognosis or those in remission also require careful assessment.[161]

It is important to recognize and aggressively treat high-risk patients for depression and address suicidal risk with psychiatric hospitalization, if necessary. Maintaining a supportive relationship, symptom control (e.g., pain, nausea, depression), involving the family or friends are the initial steps in the management of a suicidal patient. A recent study examining the suicidal ideation and past suicidal attempt in adult survivors of childhood cancer found a strong correlation between physical health and suicidality, which underscores the importance of symptom control in cancer patients.[168] Early psychiatric involvement with high-risk individuals can often avert suicide in cancer patients. A careful evaluation includes an exploration of the reasons for suicidal thoughts and the seriousness of the risk. The clinician should listen empathically, without appearing critical or judgmental. Allowing the patient to discuss suicidal thoughts often decreases the risk of suicide despite common belief to the contrary. Patients often reconsider and reject the idea of suicide when the physician acknowledges the legitimacy of their option and the need to retain a sense of control over aspects of their death.[161]

12 Psychiatric considerations in cancer-related fatigue

Cancer-related fatigue is a persistent, subjective sense of tiredness related to cancer or cancer treatment that interferes with usual functioning. Physical, emotional, and motivational components may contribute to the interrelated symptoms caused by cancer-related fatigue. It is associated with reduced quality of life and considerable psychological and functional morbidity.[169] Fatigue can be more distressing than pain or nausea and vomiting, which are manageable by medications in most patients.[157,169–171] Cancer-related fatigue is often more distressing than fatigue in individuals without cancer, and it is usually refractory to sleep and rest. Cancer-related fatigue is quite common; the reported prevalence of cancer-related fatigue ranges from 4% to 91% depending on the specific cancer population studied and the method of assessment utilized.[157,169,170] Fatigue is present at the time of cancer diagnosis in approximately 50% of cancer patients. It occurs in up to 75% of patients with bone metastases. Fatigue is also prevalent in long-term cancer survivors and can continue to have a serious impact on quality of life long after remission.[157,169–174]

A group of expert clinicians proposed a set of diagnostic criteria which have been included in the 10th edition of International Classification of Disease (ICD-10).[170] A standardized interview guide has been designed and studied for its reliability and validity to assess the proposed clinical syndrome, with results suggesting its utility in identifying patients experiencing clinically significant cancer-related fatigue.

Fatigue is difficult to quantify. There are a variety of standardized self-report scales to measure fatigue, most of which have been developed in the context of cancer.[169,170] Given the multifactorial nature of fatigue, accessory scales (e.g., depression scales) and measurements of certain biological parameters should be used in addition to fatigue assessment tools in order to obtain a more complete evaluation of a patient's fatigue.

Fatigue and depression often coexist in cancer patients, and there is considerable overlap of symptoms in these two conditions. It is necessary to clarify the relationship between depression and fatigue in order to effectively treat cancer-related fatigue. Depressive symptoms due to fatigue are typically less severe, and patients tend to attribute such symptoms to the consequences of fatigue. Depression, on the other hand, is more likely in the presence of hopelessness, feelings of worthlessness and/or guilt, suicidal ideation, and a family history of depression. The nature of any causal relationship between cancer-related fatigue and depression is unclear.

The pathophysiology of cancer-related fatigue is likely multifactorial, caused by several physical and psychosocial mechanisms. These include tumor by-products, opioids or other drugs (such as antidepressants, beta-blockers, benzodiazepines, antihistamines), hypogonadism, hypothyroidism, cachexia, anemia, serotonin dysregulation, circadian rhythm dysregulation, chemotherapy, radiation therapy, bone marrow transplantation, and treatment with biological response modifiers.[175] Cancer-related fatigue has been related to pain, depression, emotional distress, sleep deprivation, and reduced physical activity. Cytokine dysregulation (primarily IL-1, IL-6, TNF-α) has a role in the development of cancer-related fatigue.[169,175]

All patients should be screened for fatigue at their initial visit and then at regular intervals during cancer treatment and following remission. The NCCN practice guidelines on cancer-related fatigue recommend the use of numerical self-report scales or verbal scales to assess the severity of fatigue.[169] If the severity of fatigue is moderate to severe (i.e., a score greater than 4 on a 0–10 scale with higher numbers indicating increased severity), a focused history and clinical examination, evaluation of the pattern of fatigue, associated symptoms, and any interference with functioning is recommended. Potentially reversible causes of fatigue (such as pain, emotional distress, sleep disturbance, anemia, hypothyroidism) should be identified and treated, and nonessential centrally acting drugs should be eliminated.

Nonpharmacological approaches have been recommended by NCCN guidelines for the treatment of cancer-related fatigue.[169,176,177] Activity enhancement and psychosocial interventions (i.e., education, support groups, individual counseling, stress management training) are well supported by research. Dietary management, attention-restoring therapy, and sleep therapy have also been recommended in the treatment of cancer-related fatigue.[169,176,177]

Psychostimulants (i.e., methylphenidate, dextroamphetamine), wakefulness-promoting agents (i.e., modafinil and armodafinil), and antidepressants have been studied for use in the treatment of cancer-related fatigue.[178–184] Modafinil has been shown to improve cancer-related fatigue in patients with severe rather than mild-to-moderate symptoms, though current evidence is currently limited.[185]

Antidepressants have only been found helpful for treating cancer-related fatigue in the presence of comorbid depression.[182,184]

If anemia is determined to be a major contributing cause of cancer-related fatigue, the physician should determine the necessity of a transfusion in severely symptomatic patients.[169] Erythropoietin and other erythropoiesis-stimulating agents may be considered for anemia-related fatigue with caution given ample clinical and laboratory evidence of its antiapoptotic, growth-promoting effect in malignant and nonmalignant cells. Given the risk of worsened malignancy, these agents may be most suitable in palliative rather than curative care settings.[186]

Corticosteroids are frequently used to treat cancer-related fatigue in many parts of the world. Three randomized controlled trials have shown corticosteroids to improve cancer-related fatigue versus placebo for approximately 2 weeks, but the trials had high attrition rates and serious adverse effects, including insulin resistance, myopathy, and more frequent infections. Given their limited duration of effect and significant side-effect profile, corticosteroids for the treatment of cancer-related fatigue should be reserved for end-of-life care for those with comorbid anorexia, or for the treatment of pain related to metastases to the brain or bone.[187]

Several other treatments have been studied or considered for the treatment of cancer-related fatigue but currently lack robust evidence for their clinical use. These include reversible cholinesterase inhibitors, nonsteroidal antiinflammatory agents, cyclooxygenase 2 inhibitors, cytokine antagonists, and bradykinin antagonists.[187]

13 Cancer survivors

There were approximately 17 million Americans living with a history of cancer in 2019, and this number is expected to grow to over 22 million people by 2030.[188] This population includes patients on current treatment, patients with no evidence of disease, and long-term survivors. Among male cancer survivors, the three leading types of cancer are prostate cancer, colorectal cancer, and melanoma. Among female cancer survivors, the three leading types of cancer are breast cancer, uterine cancer, and colorectal cancer. Most cancer survivors are older than 65.[6,188–190] Unfortunately, significant symptoms can persist after remission; among cancer survivors off of treatment, 27% report three or more moderate-to-severe symptoms. Poorly controlled symptoms can lead to impaired ability to work, nonadherence to follow-up care, and reduced quality of life.[191–194]

Data from an American cohort of 9535 long-term survivors of childhood cancer showed a 43.6% incidence of self-reported impairment in one or more of six domains: general health, mental health, functional impairment, activity limitations, pain resulting from cancer treatment, and anxiety/fears directly related to cancer and cancer treatment. Mental health concerns were the most frequently mentioned, and the risk factors included female gender, low educational attainment, and low household income.[189] Survivors were less likely to attend college and more likely to be unemployed as young

adults. Survivors of cancers and treatments affecting the central nervous system, particularly brain tumors and those with cognitive impairment, or impacting sensory function, such as hearing loss, were associated with greater risk for undesirable social outcomes.[195,196] A Dutch study found lower educational achievement, a lower rate of employment, a lower rate of marriage, and a higher incidence of men living with their family of origin compared with a healthy control group. A history of brain tumor or cranial irradiation was the strongest independent predictor but did not account for all variations. Another Danish study confirmed this; although long-term survivors had more psychological symptoms, they did not have an increased rate of psychiatric hospitalization unless they had had brain tumors or brain irradiation.[197] Psychosexual functioning has been found to be notably impaired, and fertility can be a painful concern. Many survivors have posttraumatic symptoms, as studies have shown that 10%–15% survivors of transplant meet criteria for PTSD and another 10%–15% experience subthreshold symptoms of PTSD.[6,190] Risk factors for the development of PTSD among patients with cancer include diagnosis of advanced disease, young age, history of prior trauma, poor social support, and low socioeconomic and educational status.[158,198,199] There are many survivor programs (e.g., The National Coalition of Cancer Survivors) that offer workshops, lectures, and groups for patients, couples, and families seeking support.

14 Families issues and bereavement

The families and caregivers of cancer patients are often overburdened, and their distress is frequently underrecognized or ignored. Primary caregivers have been noted to worsen over time, even after stabilization or improvement. Studies of families after bereavement or cancer survival show a significant incidence of impaired functioning that often worsens over time.[6,200] High-risk families can be identified early and can be significantly helped by mental health providers. A small number of families require formal family therapy, which must be done with a realistic understanding of the medical facts and the medical milieu. Most families manage with short-term crisis interventions. In family-focused grief therapy, patients and families meet to process events and anticipate upcoming loss.[6,201] This therapeutic modality continues for a few sessions after the patient's death, helping the family to consolidate the positive achievements made while the patient was alive. Children in the family also require attention, and it is important to provide guidance for the adult caregiver about answering children's questions in a developmentally appropriate manner.

15 Psychological issues for staff

Caregiver stress has been documented in studies across numerous settings. Communication problems between doctors, nurses, and patients lead not only to patient dissatisfaction, but also to lower job satisfaction and self-esteem for the physician.

Staff undergo a developmental process as they to work intensively with cancer patients for the first time. A high initial level of dysphoria, anxiety, sadness, and numbing recedes over the first few months, displaced by the need to demonstrate competence and the ability to survive. Over the next few months, existential issues surface more readily, and staff members shape a deeper adaptation that engages their whole personality. Informal support, peer group interactions, open communication, and adequate orientation are important to successful adaptation. A team-based approach encourages staff cohesion and the sharing of difficult tasks.[6]

References

1. Holland JC. History of psycho-oncology: overcoming attitudinal and conceptual barriers. *Psychosom Med.* 2002;64:206–221.
2. Holland JH, Friedlander MM. Oncology. In: Blumenfield M, Strain JJ, eds. *Psychosomatic Medicine*. Philadelphia, PA: Lippincott Williams & Wilkins; 2006:121–144.
3. Holland JC, Gooen-Piels J. Psycho-oncology. In: Holland JC, Frei E, eds. *Cancer Medicine*. 6th ed. Hamilton, Ontario: B.C. Decker Inc; 2003:1039–1053.
4. Derogatis LR, Morrow GR, Fetting D, et al. The prevalence of psychiatric disorders among cancer patients. *JAMA.* 1983;249:751–757.
5. Bylund CL, Brown RF, di Ciccone BL, et al. Training faculty to facilitate communication skills training: development and evaluation of a workshop. *Patient Educ Couns.* 2008;70(3):430–436.
6. Lederberg MS. Psycho-oncology. In: Sadock BJ, Sadock VA, eds. *Kaplan and Sadock's Synopsis of Psychiatry.* 9th ed. Philadelphia, PA: Lippincott Williams and Wilkins Press; 2003:1351–1365.
7. Park C, Folkman S. Meaning in the context of stress and coping. *Rev Gen Psychol.* 1997;1:115–144.
8. Daniels J, Kissane DW. Psychosocial interventions for cancer patients. *Curr Opin Oncol.* 2008;20(4):367–371. https://doi.org/10.1097/CCO.0b013e3283021658.
9. Spencer S, Carver C, Price A. Psychological and social factors in adaptation. In: Holland JC, ed. *Psycho-Oncology*. New York, NY: Oxford University Press; 1998:211–222.
10. Rowland JH. Interpersonal resources: developmental stage of adaptation: adult model. In: Holland JC, Rowland JH, eds. *Handbook of Psycho-Oncology: Psychological Care of the Patient with cancer.* New York, NY: Oxford University Press; 1989:25–43.
11. Russak SM, Lederberg M, Fitchett G. Spirituality and coping with cancer. *Psychooncology.* 1999;8:375–466.
12. Marmot M. Social determinants of health inequalities. *Lancet.* 2005;365:1099–1104.
13. Chochinov HM, Wilson KG, Enns M, Lander S. Prevalence of depression in the terminally ill: effects of diagnostic criteria and symptom threshold judgments. *Am J Psychiatry.* 1994;151:537–540.
14. Massie MJ. Prevalence of depression in patients with cancer. *J Natl Cancer Inst Monogr.* 2004;32:57–71.
15. Singer S, Das-Munshi J, Brähler E. Prevalence of mental health conditions in cancer patients in acute care—a meta-analysis. *Ann Oncol.* 2010;21(5):925–930. https://doi.org/10.1093/annonc/mdp515.
16. Mitchell AJ, Chan M, Bhatti H, et al. Prevalence of depression, anxiety, and adjustment disorder in oncological, haematological, and palliative-care settings: a meta-analysis of 94 interview-based studies. *Lancet Oncol.* 2011;12(2):160–174. https://doi.org/10.1016/S1470-2045(11)70002-X.
17. Mehnert A, Brähler E, Faller H, et al. Four-week prevalence of mental disorders in patients with cancer across major tumor entities. *J Clin Oncol.* 2014;32(31):3540–3546. https://doi.org/10.1200/JCO.2014.56.0086.
18. Gopalan MR, Karunakaran V, Prabhakaran A, Jayakumar KL. Prevalence of psychiatric morbidity among cancer patients—hospital-based, cross-sectional survey. *Indian J Psychiatry.* 2016;58(3):275–280. https://doi.org/10.4103/0019-5545.191995.
19. Anuk D, Özkan M, Kizir A, Özkan S. The characteristics and risk factors for common psychiatric disorders in patients with cancer seeking help for mental health. *BMC Psychiatry.* 2019;19(1):269. Published 2019 Sep 3 https://doi.org/10.1186/s12888-019-2251-z.
20. Walker J, Holm Hansen C, Martin P, et al. Prevalence of depression in adults with cancer: a systematic review. *Ann Oncol.* 2013;24(4):895–900. https://doi.org/10.1093/annonc/mds575.
21. Walker J, Hansen CH, Martin P, et al. Prevalence, associations, and adequacy of treatment of major depression in patients with cancer: a cross-sectional analysis of routinely collected clinical data. *Lancet Psychiatry.* 2014;1(5):343–350. https://doi.org/10.1016/S2215-0366(14)70313-X.
22. Casarett DJ, Inouye SK. Diagnosis and management of delirium near the end of life. *Ann Intern Med.* 2001;135(1):32–40.
23. Breitbart W, Friedlander M. Confusion/delirium. In: Bruera E, Higginson I, Ripamonti C, von Gunten C, eds. *Palliative Medicine.* London, UK: London Hodder Press; 2006:688–700.
24. Holland JC, Andersen B, Breitbart WS, et al. Distress management. *J Natl Compr Canc Netw.* 2007;5(1):66–98.
25. The National Comprehensive Cancer Network. *Distress Management Clinical Practice Guidelines in Oncology, version 3*; 2019.
26. Riba MB, Donovan KA, Andersen B, et al. Distress management, version 3.2019, NCCN clinical practice guidelines in oncology. *J Natl Compr Canc Netw.* 2019;17(10):1229–1249. https://doi.org/10.6004/jnccn.2019.0048.
27. Carlson LE, Groff SL, Maciejewski O, Bultz BD. Screening for distress in lung and breast cancer outpatients: a randomized controlled trial. *J Clin Oncol.* 2010;28(33):4884–4891. https://doi.org/10.1200/JCO.2009.27.3698.
28. Lin C, Clark R, Tu P, Bosworth HB, Zullig LL. Breast cancer oral anti-cancer medication adherence: a systematic review of psychosocial motivators and barriers. *Breast Cancer Res Treat.* 2017;165(2):247–260. https://doi.org/10.1007/s10549-017-4317-2.
29. Nipp RD, El-Jawahri A, Moran SM, et al. The relationship between physical and psychological symptoms and health care utilization in hospitalized patients with advanced cancer. *Cancer.* 2017;123(23):4720–4727. https://doi.org/10.1002/cncr.30912.
30. Greer S, Moorey S, Baruch JD, et al. Adjuvant psychological therapy for patients with cancer: a prospective randomised trial. *BMJ.* 1992;304(6828):675–680. https://doi.org/10.1136/bmj.304.6828.675.
31. Gielissen MF, Verhagen CA, Bleijenberg G. Cognitive behaviour therapy for fatigued cancer survivors: long-term follow-up. *Br J Cancer.* 2007;97(5):612–618. https://doi.org/10.1038/sj.bjc.6603899.

32. Jacobsen PB. Promoting evidence-based psychosocial care for cancer patients. *Psychooncology.* 2009;18(1):6–13. https://doi.org/10.1002/pon.1468.

33. Henke Yarbro C, Hansen Frogge M, Goodman M. *Cancer Symptom Management.* 3rd ed. Sudbury, MA: Jones and Bartless; 2004:24. 52, 89, 133, 466–467, 470.

34. Wolin KY, Schwartz AL, Matthews CE, Courneya KS, Schmitz KH. Implementing the exercise guidelines for cancer survivors. *J Support Oncol.* 2012;10(5):171–177. https://doi.org/10.1016/j.suponc.2012.02.001.

35. Segal R, Zwaal C, Green E, et al. Exercise for people with cancer: a systematic review. *Curr Oncol.* 2017;24(4):e290–e315. https://doi.org/10.3747/co.24.3619.

36. Spiegel D, Bloom JR, Yalom I. Group support for patients with metastatic cancer. A randomized outcome study. *Arch Gen Psychiatry.* 1981;38(5):527–533. https://doi.org/10.1001/archpsyc.1980.017803000390.

37. Fawzy FI, Canada AL, Fawzy NW. Malignant melanoma: effects of a brief, structured psychiatric intervention on survival and recurrence at 10-year follow-up. *Arch Gen Psychiatry.* 2003;60(1):100–103. https://doi.org/10.1001/archpsyc.60.1.100.

38. Andersen BL, Thornton LM, Shapiro CL, et al. Biobehavioral, immune, and health benefits following recurrence for psychological intervention participants [published correction appears in Clin Cancer Res. 2010 Sep 1;16(17):4490]. *Clin Cancer Res.* 2010;16(12):3270–3278. https://doi.org/10.1158/1078-0432.CCR-10-0278.

39. Lieberman MA, Golant M, Giese-Davis J, et al. Electronic support groups for breast carcinoma: a clinical trial of effectiveness. *Cancer.* 2003;97(4):920–925. https://doi.org/10.1002/cncr.11145.

40. Breitbart W, Gibson C, Poppito SR, Berg A. Psychotherapeutic interventions at the end of life: a focus on meaning and spirituality. *Can J Psychiatry.* 2004;49(6):366–372.

41. McMillan SC, Small BJ. Using the COPE intervention for family caregivers to improve symptoms of hospice homecare patients: a clinical trial. *Oncol Nurs Forum.* 2007;34(2):313–321. https://doi.org/10.1188/07.ONF.313-321.

42. Houts PS, Nezu AM, Nezu CM, Bucher JA. The prepared family caregiver: a problem-solving approach to family caregiver education. *Patient Educ Couns.* 1996;27(1):63–73. https://doi.org/10.1016/0738-3991(95)00790-3.

43. *Diagnostic and Statistical Manual of Mental Disorders: DSM-5.* 5th ed. American Psychiatric Association; 2013. DSM-V, doi-org.db29.linccweb.org/10.1176/appi.

44. Fainsinger R, Young C. Cognitive failure in a terminally ill patient. *J Pain Symptom Manage.* 1991;6:492–494.

45. Gagnon P, Charbonneau C, Allard P, et al. Delirium in terminal cancer: a prospective study using daily screening, early diagnosis, and continuous monitoring. *J Pain Symptom Manage.* 2000;19(6):412–426.

46. Coyle N, Breitbart W, Weaver S, et al. Delirium as a contributing factor to "crescendo" pain: three case reports. *J Pain Symptom Manage.* 1994;9(1):44–47.

47. Caraceni A, Nanni O, Maltoni M, et al. Impact of delirium on the short term prognosis of advanced cancer patients. Italian Multicenter Study Group on Palliative Care. *Cancer.* 2000;89(5):1145–1149.

48. Maltoni M, Caraceni A, Brunelli C, et al. Prognostic factors in advanced cancer patients: evidence-based clinical recommendations—a study by the Steering Committee of the European Association for Palliative Care. *J Clin Oncol.* 2005;23(25):6240–6248.

49. Inouye SK. Delirium in older persons. *N Engl J Med.* 2006;354(11):1157–1165.

50. Trzepacz PT, Breitbart W, Franklin J, et al. Practice guideline for the treatment of patients with delirium. American Psychiatric Association. *Am J Psychiatry.* 1999;156(5 suppl):1–20.

51. Williams ST, Dhesi JK, Partridge JSL. Distress in delirium: causes, assessment and management. *Eur Geriatr Med.* 2020;11:63–70. https://doi.org/10.1007/s41999-019-00276-z.

52. Breitbart W, Gibson C, Tremblay A. The delirium experience: delirium recall and delirium related distress in hospitalized patients with cancer, their spouses/caregivers, and their nurses. *Psychosomatics.* 2002;43(3):183–194.

53. Morita T, Hirai K, Sakaguchi Y, et al. Family-perceived distress from delirium-related symptoms of terminally ill cancer patients. *Psychosomatics.* 2004;45(2):107–113.

54. Lawlor PG, Gagnon B, Mancini IL, et al. Occurrence, causes, and outcome of delirium in patients with advanced cancer: a prospective study. *Arch Intern Med.* 2000;160(6):786–794.

55. Lawlor PG, Davis DHJ, Ansari M, et al. An analytical framework for delirium research in palliative care settings: integrated epidemiologic, clinician-researcher, and knowledge user perspectives. *J Pain Symptom Manage.* 2014;48(2):159–175. https://doi.org/10.1016/j.jpainsymman.2013.12.245.

56. Fong T, Racine A, Schmitt E, et al. The distress of delirium in patients with dementia. *Innov Aging.* 2018;2(suppl 1):141. Published 2018 Nov 11 https://doi.org/10.1093/geroni/igy023.512.

57. Titov A, Petukhov A, Staliarova A, et al. The biological basis and clinical symptoms of CAR-T therapy-associated toxicities. *Cell Death Dis.* 2018;9(9):897. https://doi.org/10.1038/s41419-018-0918-x. 30181581. PMC6123453.

58. Ruark J, Mullane E, Cleary N, et al. Patient-reported neuropsychiatric outcomes of long-term survivors after chimeric antigen receptor T cell therapy. *Biol Blood Marrow Transplant.* 2020;26(1):34–43. https://doi.org/10.1016/j.bbmt.2019.09.037.

59. Gaudreau JD, Gagnon P, Harel F, et al. Psychoactive medications and risk of delirium in hospitalized cancer patients. *J Clin Oncol.* 2005;23(27):6712–6718.

60. Ross CA, Peyser CE, Shapiro I, et al. Delirium: phenomenologic and etiologic subtypes. *Int Psychogeriatr.* 1991;3(2):135–147.

61. Stagno D, Gibson C, Breitbart W. The delirium subtypes: a review of prevalence, phenomenology, pathophysiology, and treatment response. *Palliat Support Care.* 2004;2(2):171–179.

62. Krewulak KD, Stelfox HT, Leigh JP, Ely EW, Fiest KM. Incidence and prevalence of delirium subtypes in an adult ICU: a systematic review and meta-analysis. *Crit Care Med.* 2018;46(12):2029–2035. https://doi.org/10.1097/CCM.0000000000003402.

63. Smith M, Breitbart W, Platt M. A critique of instruments and methods to detect, diagnose, and rate delirium. *J Pain Symptom Manage.* 1994;10:35–77.

64. Trzepacz P, Baker R, Greenhouse J. A symptom rating scale of delirium. *Psychiatry Res.* 1988;23:89–97.

65. Trzepacz PT. The delirium rating scale: its use in consultation-liaison research. *Psychosomatics.* 1999;40:193–204.

66. Breitbart W, Rosenfeld B, Roth A. The Memorial Delirium Assessment Scale. *J Pain Symptom Manage.* 1997;13:128–137.

67. Lawlor P, Nekolaichuck C, Gagnon B, Mancini I, Pereira J, Bruera E. Clinical utility, factor analysis and further validation of the Memorial Delirium Assessment Scale (MDAS). *Cancer.* 2000;88:2859–2867.

68. Ely EW, Margolin R, Francis J, et al. Evaluation of delirium in critically ill patients: validation of the Confusion Assessment Method for the Intensive Care Unit (CAM-ICU). *Crit Care Med.* 2001;29(7):1370–1379. https://doi.org/10.1097/00003246-200107000-00012.

69. Neufeld KJ, Hayat MJ, Coughlin JM, et al. Evaluation of two intensive care delirium screening tools for non-critically ill hospitalised patients. *Psychosomatics.* 2011;52:133–140. https://doi.org/10.1016/j.psym.2010.12.018.

70. Bush SH, Lawlor PG, Ryan K, et al. Delirium in adult cancer patients: ESMO clinical practice guidelines. *Ann Oncol.* 2018;29(suppl 4):iv143–iv165. https://doi.org/10.1093/annonc/mdy147.

71. Grassi L, Caraceni A, Beltrami E, et al. Assessing delirium in cancer patients: the Italian versions of the Delirium Rating Scale and the Memorial Delirium Assessment Scale. *J Pain Symptom Manage.* 2001;21(1):59–68. https://doi.org/10.1016/s0885-3924(00)00241-4.

72. Trzepacz PT, Mittal D, Torres R, Kanary K, Norton J, Jimerson N. Validation of the Delirium Rating Scale-revised-98: comparison with the delirium rating scale and the cognitive test for delirium [published correction appears in J Neuropsychiatry Clin Neurosci 2001 Summer;13(3):433]. *J Neuropsychiatry Clin Neurosci.* 2001;13(2):229–242. https://doi.org/10.1176/jnp.13.2.229.

73. Spiller JA, Keen JC. Hypoactive delirium: assessing the extent of the problem for inpatient specialist palliative care. *Palliat Med.* 2006;20(1):17–23.

74. Morita T, Tei Y, Inoue S. Impaired communication capacity and agitated delirium in the final week of terminally ill cancer patients: prevalence and identification of research focus. *J Pain Symptom Manage.* 2003;26:827–834.

75. Fainsinger R, Bruera E. Treatment of delirium in a terminally ill patient. *J Pain Symptom Manage.* 1992;7:54–56.

76. Morita T, Tsunoda J, Inoue S, et al. Survival prediction of terminally ill cancer patients by clinical symptoms: development of a simple indicator. *Jpn J Clin Oncol.* 1999;29(3):156–159.

77. Barr J, Fraser GL, Puntillo K, et al. Clinical practice guidelines for the management of pain, agitation, and delirium in adult patients in the intensive care unit. *Crit Care Med.* 2013;41(1):263–306. https://doi.org/10.1097/CCM.0b013e3182783b72.

78. Davis D, Searle SD, Tsui A. The Scottish Intercollegiate Guidelines Network: risk reduction and management of delirium. *Age Ageing.* 2019;48(4):485–488. https://doi.org/10.1093/ageing/afz036.

79. Fok MC, Sepehry AA, Frisch L, et al. Do antipsychotics prevent postoperative delirium? A systematic review and meta-analysis. *Int J Geriatr Psychiatry.* 2015;30(4):333–344. https://doi.org/10.1002/gps.4240.

80. Hirota T, Kishi T. Prophylactic antipsychotic use for postoperative delirium: a systematic review and meta-analysis. *J Clin Psychiatry.* 2013;74(12):e1136–e1144. https://doi.org/10.4088/JCP.13r08512.

81. Skrobik Y, Duprey MS, Hill NS, Devlin JW. Low-dose nocturnal dexmedetomidine prevents ICU delirium. A randomized, placebo-controlled trial. *Am J Respir Crit Care Med.* 2018;197(9):1147–1156. https://doi.org/10.1164/rccm.201710-1995OC.

82. Hatta K, Kishi Y, Wada K, et al. Preventive effects of ramelteon on delirium: a randomized placebo-controlled trial. *JAMA Psychiatry.* 2014;71(4):397–403. https://doi.org/10.1001/jamapsychiatry.2013.3320.

83. Al-Aama T, Brymer C, Gutmanis I, Woolmore-Goodwin SM, Esbaugh J, Dasgupta M. Melatonin decreases delirium in elderly patients: a randomized, placebo-controlled trial [published correction appears in Int J Geriatr Psychiatry. 2014 May;29(5):550]. *Int J Geriatr Psychiatry.* 2011;26(7):687–694. https://doi.org/10.1002/gps.2582.

84. Pitkala KH, Laurila JV, Strandberg TE, Tilvis RS. Multicomponent geriatric intervention for elderly inpatients with delirium: a randomized, controlled trial. *J Gerontol A Biol Sci Med Sci.* 2006;61(2):176–181.

85. Pitkala KH, Laurila JV, Strandberg TE, Kautiainen H, Sintonen H, Tilvis RS. Multicomponent geriatric intervention for elderly inpatients with delirium: effects on costs and health-related quality of life. *J Gerontol A Biol Sci Med Sci.* 2008;63(1):56–61.

86. Milisen K, Lemiengre J, Braes T, Foreman MD. Multicomponent intervention strategies for managing delirium in hospitalized older people: systematic review. *J Adv Nurs.* 2005;52(1):79–90.

87. Cole MG, McCusker J, Bellavance F, et al. Systematic detection and multidisciplinary care of delirium in older medical inpatients: a randomized trial. *CMAJ.* 2002;167(7):753–759.

88. Laitinen H. Patients' experience of confusion in the intensive care unit following cardiac surgery. *Intensive Crit Care Nurs.* 1996;12(2):79–83. https://doi.org/10.1016/s0964-3397(96)80994-3. 8845628.

89. Boettger S, Breitbart W. Atypical antipsychotics in the management of delirium: a review of the empirical literature. *Palliat Support Care.* 2005;3(3):227–237.

90. Lacasse H, Perreault MM, Williamson DR. Systematic review of antipsychotics for the treatment of hospital-associated delirium in medically or surgically ill patients. *Ann Pharmacother.* 2006;40(11):1966–1973.

91. Michaud L, Bula C, Berney A, et al. Delirium: guidelines for general hospitals. *J Psychosom Res.* 2007;62(3):371–383.

92. Schwartz TL, Masand P. The role of atypical antipsychotics in the treatment of delirium. *Psychosomatics.* 2002;43:171–174.

93. Seitz DP, Gill SS, van Zyl LT. Antipsychotics in the treatment of delirium: a systematic review. *J Clin Psychiatry.* 2007;68(1):11–21.

94. Lonergan E, Britton AM, Luxenberg J, Wyller T. Antipsychotics for delirium. *Cochrane Database Syst Rev.* 2007;(2):CD005594.

95. Jackson KC, Lipman AG. Drug therapy for delirium in terminally ill patients. *Cochrane Database Syst Rev.* 2004;2:CD004770.

96. Neufeld KJ, Yue J, Robinson TN, Inouye SK, Needham DM. Antipsychotic medication for prevention and treatment of delirium in hospitalized adults: a systematic review and meta-analysis [published correction appears in J Am Geriatr Soc. 2016 Oct;64(10):2171–2173]. *J Am Geriatr Soc.* 2016;64(4):705–714. https://doi.org/10.1111/jgs.14076.

97. Oh ES, Needham DM, Nikooie R, et al. Antipsychotics for preventing delirium in hospitalized adults: a systematic review. *Ann Intern Med.* 2019;171(7):474–484. https://doi.org/10.7326/M19-1859.

98. Agar MR, Lawlor PG, Quinn S, et al. Efficacy of oral risperidone, haloperidol, or placebo for symptoms of delirium among patients in palliative care: a randomized clinical trial [published correction appears in JAMA Intern Med. 2017 Feb 1;177(2):293]. *JAMA Intern Med.* 2017;177(1):34–42. https://doi.org/10.1001/jamainternmed.2016.7491.

99. Hocking CM, Kichenadasse G. Olanzapine for chemotherapy-induced nausea and vomiting: a systematic review. *Support Care Cancer.* 2014;22(4):1143–1151. https://doi.org/10.1007/s00520-014-2138-y.

100. Breitbart W, Marotta R, Platt MM, et al. A double-blind trial of haloperidol, chlorpromazine, and lorazepam in the treatment of delirium in hospitalized AIDS patients. *Am J Psychiatry.* 1996;153(2):231–237.

101. Meagher D, Agar MR, Teodorczuk A. Debate article: antipsychotic medications are clinically useful for the treatment of delirium. *Int J Geriatr Psychiatry.* 2018;33(11):1420–1427. https://doi.org/10.1002/gps.4759.

102. Rummel-Kluge C, Komossa K, Schwarz S, et al. Second-generation antipsychotic drugs and extrapyramidal side effects: a systematic review and meta-analysis of head-to-head comparisons. *Schizophr Bull.* 2012;38(1):167–177. https://doi.org/10.1093/schbul/sbq042.

103. Ralph SJ, Espinet AJ. Increased all-cause mortality by antipsychotic drugs: updated review and meta-analysis in dementia and general mental health care. *J Alzheimers Dis Rep.* 2018;2(1):1–26. Published 2018 Feb 2 https://doi.org/10.3233/ADR-170042.

104. Hu H, Deng W, Yang H. A prospective random control study comparison of olanzapine and haloperidol in senile delirium. *Chongqing Med J.* 2004;8:1234–1237.

105. Schneider LS, Dagerman KS, Insel P. Risk of death with atypical antipsychotic drug treatment for dementia, meta-analysis of randomized placebo-controlled trials. *JAMA.* 2005;294:1934–1943.

106. Wang PS, Schneeweiss S, Avorn J, et al. Risk of death in elderly users of conventional vs. atypical antipsychotic medications. *N Engl J Med.* 2005;353(22):2335–2341.

107. Zareba W, Lin DA. Antipsychotic drugs and QT interval prolongation. *Psychiatry Q*. 2003;74:291–306.

108. Sher Y, Miller Cramer AC, Ament A, Lolak S, Maldonado JR. Valproic acid for treatment of hyperactive or mixed delirium: rationale and literature review. *Psychosomatics*. 2015;56(6):615–625. https://doi.org/10.1016/j.psym.2015.09.008.

109. Gagnon DJ, Fontaine GV, Smith KE, et al. Valproate for agitation in critically ill patients: a retrospective study. *J Crit Care*. 2017;37:119–125. https://doi.org/10.1016/j.jcrc.2016.09.006.

110. Crowley KE, Urben L, Hacobian G, Geiger KL. Valproic acid for the management of agitation and delirium in the intensive care setting: a retrospective analysis. *Clin Ther*. 2020;42(4):e65–e73. https://doi.org/10.1016/j.clinthera.2020.02.007.

111. Gidal BE, Sheth R, Parnell J, Maloney K, Sale M. Evaluation of VPA dose and concentration effects on lamotrigine pharmacokinetics: implications for conversion to lamotrigine monotherapy. *Epilepsy Res*. 2003;57(2–3):85–93. https://doi.org/10.1016/j.eplepsyres.2003.09.008.

112. Pursche S, Schleyer E, von Bonin M, et al. Influence of enzyme-inducing antiepileptic drugs on trough level of imatinib in glioblastoma patients. *Curr Clin Pharmacol*. 2008;3(3):198–203. https://doi.org/10.2174/157488408785747656.

113. Coulter DW, Walko C, Patel J, et al. Valproic acid reduces the tolerability of temsirolimus in children and adolescents with solid tumors. *Anticancer Drugs*. 2013;24(4):415–421. https://doi.org/10.1097/CAD.0b013e32835dc7c5.

114. Ng KT, Shubash CJ, Chong JS. The effect of dexmedetomidine on delirium and agitation in patients in intensive care: systematic review and meta-analysis with trial sequential analysis. *Anaesthesia*. 2019;74(3):380–392. https://doi.org/10.1111/anae.14472.

115. Joseph SG. Melatonin supplementation for the prevention of hospital-associated delirium. *Ment Health Clin*. 2018;7(4):143–146. Published 2018 Mar 26 https://doi.org/10.9740/mhc.2017.07.143.

116. Campbell AM, Axon DR, Martin JR, et al. Melatonin for the prevention of postoperative delirium in older adults: a systematic review and meta-analysis. *BMC Geriatr*. 2019;19:272. https://doi.org/10.1186/s12877-019-1297-6.

117. Traube C, Silver G, Reeder RW, et al. Delirium in critically ill children: an international point prevalence study. *Crit Care Med*. 2017;45(4):584–590. https://doi.org/10.1097/CCM.0000000000002250.

118. Strain J. Adjustment disorders. In: Holland JC, ed. *Psycho-Oncology*. New York, NY: Oxford University Press; 1998:509–517.

119. van Beek FE, Wijnhoven LMA, Jansen F, et al. Prevalence of adjustment disorder among cancer patients, and the reach, effectiveness, cost-utility and budget impact of tailored psychological treatment: study protocol of a randomized controlled trial. *BMC Psychol*. 2019;7(1):89. Published 2019 Dec 23 https://doi.org/10.1186/s40359-019-0368-y.

120. Stark D, Kiely M, Smith A, Velikova G, House A, Selby P. Anxiety disorders in cancer patients: their nature, associations, and relation to quality of life. *J Clin Oncol*. 2002;20:3137–3148.

121. Smith MY, Redd WH, Peyser C, Vogl D. Posttraumatic stress disorder in cancer: a review. *Psychooncology*. 1999;8:521–537.

122. Kangas M, Henry JL. Bryant RA correlates of acute stress disorder in cancer patients. *J Trauma Stress*. 2007;20(3):325–334.

123. Kangas M, Henry JL, Bryant RA. Posttraumatic stress disorder following cancer. A conceptual and empirical review. *Clin Psychol Rev*. 2002;22:499–524.

124. Palmer SC, Kagee A, Coyne JC, DeMichelle A. Experience of trauma, distress, and posttraumatic stress disorder among breast cancer patients. *Psychosom Med*. 2004;66:258–264.

125. Kroll J. Posttraumatic symptoms and the complexity of responses to trauma. *JAMA*. 2003;290:267–270.

126. Stark DP, House A. Anxiety in cancer patients. *Br J Cancer*. 2000;83(10):1261–1267. https://doi.org/10.1054/bjoc.2000.1405.

127. Caruso R, Giulia Nanni M, Riba MB, Sabato S, Grassi L. Depressive spectrum disorders in cancer: diagnostic issues and intervention. A critical review. *Curr Psychiatry Rep*. 2017;19(6):33. https://doi.org/10.1007/s11920-017-0785-7.

128. Dalton SO, Mellemkjaer L, Olsen JH, Mortensen PB, Johansen C. Depression and cancer: a register-based study of patients hospitalized with affective disorders, Denmark 1969–1993. *Am J Epidemiol*. 2002;155:1088–1095.

129. Kessler RC, McGonagle KA, Swartz M, Blazer DG, Nelson CB. Sex and depression in the national comorbidity survey. I: lifetime prevalence, chronicity and recurrence. *J Affect Disord*. 1993;29:85–96.

130. DeFlorio ML, Massie MJ. Review of depression in cancer: gender differences. *Depression*. 1995;3:66–80.

131. Badger T, Segrin C, Dorros SM, Meek P, Lopez AM. Depression and anxiety in women with breast cancer and their partners. *Nurs Res*. 2007;56(1):44–53.

132. Coups EJ, Winell J, Holland JC. Depression in the context of cancer. In: Licinio J, Wong M-L, eds. Weinheim, Germany: Wiley; 2005:365–385. Biology of Depression: From Novel Insights to Therapeutic Strategies; vol. 1.

133. Akechi T, Okuyama T, Sugawara Y, Nakano T, Shima Y, Uchitomi Y. Major depression, adjustment disorders, and posttraumatic stress disorder in terminally ill cancer patients: associated and predictive factors. *J Clin Oncol*. 2004;22(10):1957–1965.

134. Lander M, Wilson K, Chochinov HM. Depression and the dying older patient. *Clin Geriatr Med*. 2000;16(2):335–356.

135. Musselman DL, Miller AH, Porter MR, et al. Higher than normal plasma interleukin-6 concentrations in cancer patients with depression: preliminary findings. *Am J Psychiatry*. 2001;158:1252–1257.

136. Ebrahimi B, Tucker SL, Li D, et al. Cytokines in pancreatic carcinoma. *Cancer*. 2004;101:2727–2736.

137. van Wilgen CP, Dijkstra PU, Stewart RE, Ranchor AV, Roodenburg JL. Measuring somatic symptoms with the CES-D to assess depression in cancer patients after treatment: comparison among patients with oral/oropharyngeal, gynecological, colorectal, and breast cancer. *Psychosomatics*. 2006;47(6):465–470.

138. Schroevers MJ, Ranchor AV, Sanderman R. The role of social support and self-esteem in the presence and course of depressive symptoms: a comparison of cancer patients and individuals from the general population. *Soc Sci Med*. 2003;57:375–385.

139. Spiegel D, Sand S, Koopman C. Pain and depression in patients with cancer. *Cancer*. 1994;74:2570–2578.

140. American Psychiatric Association. *Practice Guidelines for the Treatment of Patients with Major Depressive Disorder*. 2nd ed. Arlington, VA: American Psychiatric Publishing, Inc; 2000.

141. Kathol RG, Mutgi A, Williams J, Clamon G, Noyes Jr R. Diagnosis of major depression in cancer patients according to four sets of criteria. *Am J Psychiatry*. 1990;147:1021–1024.

142. Potash M, Breitbart W. Affective disorders in advanced cancer. *Hematol Oncol Clin North Am*. 2002;16(3):671–700.

143. Fava GA, Freyberger HJ, Bech P, et al. Diagnostic criteria for use in psychosomatic research. *Psychother Psychosom*. 1995;63(1):1–8. https://doi.org/10.1159/000288931.

144. de Figueiredo JM. Depression and demoralization: phenomenologic differences and research perspectives. *Compr Psychiatry*. 1993;34(5):308–311. https://doi.org/10.1016/0010-440x(93)90016-w.

145. Grassi L, Rossi E, Sabato S, Cruciani G, Zambelli M. Diagnostic criteria for psychosomatic research and psychosocial variables in breast cancer patients. *Psychosomatics*. 2004;45(6):483–491. https://doi.org/10.1176/appi.psy.45.6.483.

146. Mangelli L, Fava GA, Grandi S, et al. Assessing demoralization and depression in the setting of medical disease. *J Clin Psychiatry*. 2005;66(3):391–394. https://doi.org/10.4088/jcp.v66n0317.

147. Vehling S, Mehnert A. Symptom burden, loss of dignity, and demoralization in patients with cancer: a mediation model. *Psychooncology*. 2014;23(3):283–290. https://doi.org/10.1002/pon.3417.

148. Clarke DM, Kissane DW, Trauer T, Smith GC. Demoralization, anhedonia and grief in patients with severe physical illness. *World Psychiatry*. 2005;4(2):96–105.

149. Tecuta L, Tomba E, Grandi S, Fava GA. Demoralization: a systematic review on its clinical characterization. *Psychol Med*. 2015;45(4):673–691. https://doi.org/10.1017/S0033291714001597.

150. Robinson S, Kissane DW, Brooker J, Burney S. A systematic review of the demoralization syndrome in individuals with progressive disease and cancer: a decade of research. *J Pain Symptom Manage*. 2015;49(3):595–610. https://doi.org/10.1016/j.jpainsymman.2014.07.008.

151. DeVane CL. Differential pharmacology of newer antidepressants. *J Clin Psychiatry*. 1998;59(suppl 20):85–93.

152. Dugan SE, Fuller MA. Duloxetine: a dual reuptake inhibitor. *Ann Pharmacother*. 2004;38:2078–2085.

153. Wisor JP, Eriksson KS. Dopaminergic-adrenergic interactions in the wake promoting mechanism of modafinil. *Neuroscience*. 2005;132:1027–1034.

154. Goodwin PJ, Leszcz M, Ennis M, et al. The effect of group psychosocial support on survival in metastatic breast cancer. *N Engl J Med*. 2001;345:1719–1726.

155. Classen C, Butler LD, Koopman C, et al. Supportive-expressive group therapy and distress in patient with metastatic cancer: a randomized clinical intervention trial. *Arch Gen Psychiatry*. 2001;58(5):494–501.

156. Kissane DW, Bloch S, Smith GC, et al. Cognitive-existential group psychotherapy for women with primary breast cancer: a randomized controlled trial. *Psychooncology*. 2003;12:532–546.

157. Barnes EA, Bruera E. Fatigue in patients with advanced cancer: a review. *Int J Gynecol Cancer*. 2002;12(5):424–428.

158. Wachen JS, Patidar SM, Mulligan EA, Naik AD, Moye J. Cancer-related PTSD symptoms in a veteran sample: association with age, combat PTSD, and quality of life. *Psychooncology*. 2014;23(8):921–927. https://doi.org/10.1002/pon.3494.

159. Bolund C. Suicide and cancer: II. Medical and care factors in suicide by cancer patients in Sweden. 1973-1976. *J Psychosoc Oncol*. 1985;3:17–30.

160. Weisman AD. Coping behavior and suicide in cancer. In: Cullen JW, Fox BH, Ison RN, eds. *Cancer: The Behavioral Dimensions*. New York: Raven press; 1976.

161. Breitbart W. Suicide risk and pain in cancer and AIDS patients. In: Chapman CR, Foley KM, eds. *Current and Emerging Issues in Cancer Pain: Research and Practice*. New York, NY: Raven Press; 1993:49–65.

162. Zaorsky NG, Zhang Y, Tuanquin L, Bluethmann SM, Park HS, Chinchilli VM. Suicide among cancer patients [published correction appears in Nat Commun. 2020 Jan 31;11(1):718]. *Nat Commun*. 2019;10(1):207. Published 2019 Jan 14 https://doi.org/10.1038/s41467-018-08170-1.

163. Chochinov HM, Wilson KG, Ennis M, et al. Desire for death in the terminally ill. *Am J Psychiatry*. 1995;152:1185–1191.

164. Chochinov HM, Wilson KG, Ennis M, Lander S. Depression, hopelessness, and suicidal ideation in the terminally ill. *Psychosomatics*. 1998;39(4):336–370.

165. Louhivuori KA, Hakama J. Risk of suicide among cancer patients. *Am J Epidemiol*. 1979;109:59–65.

166. Labisi O. Assessing for suicide risk in depressed geriatric cancer patients. *J Psychosoc Oncol*. 2006;24(1):43–50.

167. Schairer C, Brown LM, Chen BE, et al. Suicide after breast cancer: an international population-based study of 723 810 women. *J Natl Cancer Inst*. 2006;98(19):1416–1419.

168. Recklitis CJ, Lockwood RA, Rothwell MA, Diller LR. Suicidal ideation and attempts in adult survivors of childhood cancer. *J Clin Oncol*. 2006;24(24):3852–3857.

169. Mock V, Atkinson A, et al. NCCN practice guidelines for cancer-related fatigue. *Oncology (Williston Park)*. 2000;14(11A):151–161.

170. Cella D, Davis K, et al. Cancer-related fatigue: prevalence of proposed diagnostic criteria in a United States sample of cancer survivors. *J Clin Oncol*. 2001;19(14):3385–3391.

171. Cella D, Peterman A, Passik S, Jacobsen P, Breitbart W. Progress toward guidelines for the management of fatigue. *Oncology*. 1998;12:369–377.

172. Curt GA, Breitbart W, Cella D, et al. Impact of cancer-related fatigue on the lives of patients: new findings from the fatigue coalition. *Oncologist*. 2000;5:353–360.

173. Flechtner H, Bottomley A. Fatigue and quality of life: lessons from the real world. *Oncologist*. 2003;8(Suppl 1):5–9.

174. Lawrence DP, Kupelnick B, et al. Evidence report on the occurrence, assessment, and treatment of fatigue in cancer patients. *J Natl Cancer Inst Monogr*. 2004;32:40–50.

175. Ahlberg K, Ekman T, et al. Assessment and management of cancer-related fatigue in adults. *Lancet*. 2003;362(9384):640–650. *ft*, Ros.

176. Mock V. Evidence-based treatment for cancer-related fatigue. *J Natl Cancer Inst Monogr*. 2004;32:112–118.

177. Morrow GR, Shelke AR, Roscoe JA, et al. Management of cancer related fatigue. *Cancer Invest*. 2005;23:229–239.

178. Breitbart W, Rosenfeld B, Kaim M, et al. A randomized, double-blind, placebo-controlled trial of psychostimulants for the treatment of fatigue in ambulatory patients with human immunodeficiency virus disease. *Arch Intern Med*. 2001;161:411–420.

179. Bruera E, Driver L, et al. Patient-controlled methylphenidate for the management of fatigue in patients with advanced cancer: a preliminary report. *J Clin Oncol*. 2003;21(23):4439–4443.

180. Bruera E, Valero V, et al. Patient-controlled methylphenidate for cancer fatigue: a double-blind, randomized, placebo-controlled trial. *J Clin Oncol*. 2006;24(13):2073–2078.

181. Cullum JL, Wojciechowski AE, Pelletier G, Simpson JS. Bupropion sustained release treatment reduces fatigue in cancer patients. *Can J Psychiatry*. 2004;49(2):139–144.

182. Morrow GR, Hickok JT, et al. Differential effects of paroxetine on fatigue and depression: a randomized, double-blind trial from the University of Rochester Cancer Center Community Clinical Oncology Program. *J Clin Oncol*. 2003;21(24):4635–4641.

183. Musselman DL, Lawson DH, Gumnick JF, et al. Paroxetine for the prevention of depression induced by high-dose interferon alfa. *N Engl J Med*. 2001;344:961–966.

184. Roscoe JA, Morrow GR, et al. Effect of paroxetine hydrochloride on fatigue and depression in breast cancer patients receiving chemotherapy. *Breast Cancer Res Treat*. 2005;89(3):243–249.

185. Jean-Pierre P, Morrow GR, Roscoe JA, et al. A phase 3 randomized, placebo-controlled, double-blind, clinical trial of the effect of modafinil on cancer-related fatigue among 631 patients receiving chemotherapy: a University of Rochester Cancer Center Community Clinical Oncology Program Research base study. *Cancer*. 2010;116(14):3513–3520. https://doi.org/10.1002/cncr.25083.

186. Debeljak N, Solár P, Sytkowski AJ. Erythropoietin and cancer: the unintended consequences of anemia correction. *Front Immunol*. 2014;5:563. Published 2014 Nov 11 https://doi.org/10.3389/fimmu.2014.00563.

187. McFarland D, Bjerre-Real C, Alici Y, Breitbart W. Cancer-related fatigue. In: *Psycho-Oncology*. 4th ed. Oxford University Press; 2021. https://doi.org/10.1093/med/9780190097653.003.0035.

188. Miller KD, Nogueira L, Mariotto AB, et al. Cancer treatment and survivorship statistics, 2019. *CA Cancer J Clin.* 2019;69(5):363–385. https://doi.org/10.3322/caac.21565.

189. Hudson MM, Mertens AC, Yasui Y, et al. Health status of adult long-term survivors of childhood cancer: a report from the Childhood Cancer Survivor Study. *JAMA.* 2003;290:1583–1592.

190. Bloom JR. Special issue on survivorship. *Psychooncology.* 2002;11:89–180.

191. Esther Kim JE, Dodd MJ, Aouizerat BE, Jahan T, Miaskowski C. A review of the prevalence and impact of multiple symptoms in oncology patients. *J Pain Symptom Manage.* 2009;37(4):715–736. https://doi.org/10.1016/j.jpainsymman.2008.04.018.

192. Murphy CC, Bartholomew LK, Carpentier MY, Bluethmann SM, Vernon SW. Adherence to adjuvant hormonal therapy among breast cancer survivors in clinical practice: a systematic review. *Breast Cancer Res Treat.* 2012;134(2):459–478. https://doi.org/10.1007/s10549-012-2114-5.

193. Cleeland CS, Zhao F, Chang VT, et al. The symptom burden of cancer: evidence for a core set of cancer-related and treatment-related symptoms from the Eastern Cooperative Oncology Group Symptom Outcomes and Practice Patterns study. *Cancer.* 2013;119(24):4333–4340. https://doi.org/10.1002/cncr.28376.

194. Sun Y, Shigaki CL, Armer JM. Return to work among breast cancer survivors: a literature review. *Support Care Cancer.* 2017;25(3):709–718. https://doi.org/10.1007/s00520-016-3446-1.

195. Gurney JG, Krull KR, Kadan-Lottick N, et al. Social outcomes in the Childhood Cancer Survivor Study cohort. *J Clin Oncol.* 2009;27(14):2390–2395. https://doi.org/10.1200/JCO.2008.21.1458.

196. Schulte F, Brinkman TM, Li C, et al. Social adjustment in adolescent survivors of pediatric central nervous system tumors: a report from the Childhood Cancer Survivor Study. *Cancer.* 2018;124(17):3596–3608. https://doi.org/10.1002/cncr.31593.

197. Ross L, Johanson C, Dalton SO, et al. Psychiatric hospitalizations among survivors of cancer in childhood or adolescence. *N Engl J Med.* 2003;349:650–657.

198. Abbey G, Thompson SB, Hickish T, Heathcote D. A meta-analysis of prevalence rates and moderating factors for cancer-related post-traumatic stress disorder. *Psychooncology.* 2015;24(4):371–381. https://doi.org/10.1002/pon.3654.

199. Swartzman S, Booth JN, Munro A, Sani F. Posttraumatic stress disorder after cancer diagnosis in adults: a meta-analysis. *Depress Anxiety.* 2017;34(4):327–339. https://doi.org/10.1002/da.22542.

200. Baider L, Cooper CL, De-Nour AK, eds. *Cancer and the Family.* 2nd ed. Chichester: John Wiley and Sons; 2000.

201. Kissane D, Lichtenthal WG, Zaider T. Family care before and after bereavement. *Omega.* 2007–2008;56(1):21–32.

Chronic cancer pain syndromes and their treatment

Nathan Cherney[a], Alan Carver[b], and Herbert B. Newton[c,d,e,f]

[a]Department of Medical Oncology, Shaare Zedek Medical Center, Jerusalem, Israel, [b]Department of Neuro-Oncology, Memorial Sloan-Kettering Cancer Center, New York, NY, United States, [c]Neuro-Oncology Center, Orlando, FL, United States, [d]CNS Oncology Program, Advent Health Cancer Institute, Advent Health Orlando Campus & Advent Health Medical Group, Orlando, FL, United States, [e]Neurology, UCF School of Medicine, Orlando, FL, United States, [f]Neurology & Neurosurgery (Retired), Division of Neuro-Oncology, Esther Dardinger Endowed Chair in Neuro-Oncology, James Cancer Hospital & Solove Research Institute, Wexner Medical Center at the Ohio State University, Columbus, OH, United States

Surveys indicate that pain is experienced by 30%–60% of cancer patients during active therapy and more than two-thirds of those with advanced disease.[1] Unrelieved pain is incapacitating and precludes a satisfying quality of life; it interferes with physical functioning and social interaction and is strongly associated with heightened psychological distress. It can provoke or exacerbate existential distress, disturb normal processes of coping and adjustment, and augment a sense of vulnerability, contributing to a preoccupation with the potential for catastrophic outcomes.[2] Persistent pain interferes with the ability to eat, sleep, think, interact with others and is correlated with fatigue in cancer patients.

Cancer pain syndromes are defined by the association of particular pain characteristics and physical signs with specific consequences of the underlying disease or its treatment. Syndromes are associated with distinct etiologies and pathophysiologies, and they have important prognostic and therapeutic implications. Pain syndromes associated with cancer can be either acute or chronic. Whereas acute pains experienced by cancer patients are usually related to diagnostic and therapeutic interventions, chronic pains are most commonly caused by direct tumor infiltration. Adverse consequences of cancer therapy, including surgery, chemotherapy, and radiation therapy, account for 15%–25% of chronic cancer pain problems, and a small proportion of the chronic pains experienced by cancer patients are caused by pathology unrelated to either the cancer or the cancer therapy.

Cancer-related acute pain syndromes are most commonly due to diagnostic or therapeutic interventions, and they generally pose little diagnostic difficulty.[3] Although some tumor-related pains have an acute onset (such as pain from a pathological fracture), most of these will persist unless effective treatment for the underlying lesion is provided. In contrast, most chronic cancer-related pains are caused directly by the tumor. Data from the largest prospective survey of cancer pain syndromes revealed that almost one-quarter of the patients experienced two or more pains. Over 90% of the patients had one or more tumor-related pains and 21% had one or more pains caused by cancer therapies. Somatic pains (71%) were more common than neuropathic (39%) or visceral pains (34%).[4] Bone pain and compression of neural structures are the two most common causes.[5–9]

1 Bone pain

Bone metastases are the most common cause of chronic pain in cancer patients.[10] Cancers of the lung, breast, and prostate most often metastasize to bone, but any tumor type may be complicated by painful bony lesions. Although bone pain is usually associated with direct tumor invasion of bony structures, more than 25% of patients with bone metastases are pain-free, and patients with multiple bony metastases typically report pain in only a few sites.

1.1 Differential diagnosis

Bone pain due to metastatic tumor needs to be differentiated from less common causes.[10] Nonneoplastic causes in this population include osteoporotic fractures (including those associated with multiple myeloma), focal osteonecrosis, which may be idiopathic or related to chemotherapy, corticosteroids or radiotherapy (see below), and osteomalacia.[11,12] Rarely, a paraneoplastic osteomalacia, which is associated with elevated levels of fibroblast growth factor 23, can mimic multiple metastases.[13,14]

1.2 Multifocal or generalized bone pain

Bone pain may be focal, multifocal, or generalized.[10] Multifocal bone pains are most commonly experienced by patients with multiple bony metastases. A generalized pain syndrome is occasionally produced by the replacement of bone marrow.[15–18] This bone marrow replacement syndrome has been observed in hematogenous malignancies and, less commonly, in solid tumors and brain tumors.[16,19–23] This syndrome can occur in the absence of abnormalities on bone scintigraphy or radiography, increasing the difficulty of diagnosis. It is best demonstrated on MRI imaging.[24]

1.3 Vertebral syndromes

The vertebrae are the most common sites of bony metastases.[10] More than two-thirds of vertebral metastases are located in the thoracic spine; lumbosacral and cervical metastases account for approximately 20% and 10%, respectively. Multiple-level involvement is common, occurring in greater than 85% of patients.[25] The early recognition of pain syndromes due to tumor invasion of vertebral bodies is essential, since pain usually precedes compression of adjacent neural structures and prompt treatment of the lesion may prevent the subsequent development of neurologic deficits.

Atlantoaxial destruction and odontoid fracture: Nuchal or occipital pain is the typical presentation of destruction of the atlas or fracture of the odontoid process. Pain often radiates over the posterior aspect of the skull to the vertex and is exacerbated by movement of the neck, particularly flexion.[26] Pathologic fracture may result in secondary subluxation with compression of the spinal cord at the cervicomedullary junction.

C7-T1 syndrome: Invasion of the C7 or T1 vertebra can result in pain referred to the interscapular region. These lesions may be missed if radiographic evaluation is mistakenly targeted to the painful area caudal to the site of damage.

T12-L1 syndrome: A T12 or L1 vertebral lesion can refer pain to the ipsilateral iliac crest or the sacroiliac joint.

Imaging procedures directed at pelvic bones can miss the source of the pain.

Sacral syndrome: Severe focal pain radiating to the buttocks, perineum, or posterior thighs may accompany destruction of the sacrum.[27] The pain is often exacerbated by sitting or lying and is relieved by standing or walking.[28] The neoplasm can spread laterally to involve muscles that rotate the hip (e.g., the pyriformis muscle). This may produce severe incident pain induced by motion of the hip, or a malignant "pyriformis syndrome," characterized by buttock or posterior leg pain that is exacerbated by internal rotation of the hip. Local extension of the tumor mass may also involve the sacral plexus (see below).

1.4 Back pain and epidural compression (see Chapter 7—Spinal Cord Compression)

Epidural compression of the spinal cord or cauda equina is the second most common neurologic complication of cancer, occurring in up to 10% of patients.[29] Back pain is the initial symptom in almost all patients with epidural compression, and in 10%, it is the only symptom at the time of diagnosis.[29–31] Since pain usually precedes neurologic signs by a prolonged period, it should be viewed as a potential indicator of epidural compression, which can lead to treatment at a time that a favorable response is most likely. Back pain, however, is a nonspecific symptom that can result from bony or paraspinal metastases without epidural encroachment, from retroperitoneal or leptomeningeal tumor, epidural lipomatosis due to steroid administration or from a large variety of other benign conditions.[32] Since it is not feasible to pursue an extensive evaluation in every cancer patient who develops back pain, the complaint should impel an evaluation that determines the likelihood of epidural compression and thereby selects patients appropriate for definitive imaging of the epidural space. The selection process is based on symptoms and signs and the results of simple imaging techniques.

1.5 Pain syndromes of the bony pelvis and hip

The pelvis and hip are common sites of metastatic involvement. Lesions may involve any of the three anatomic regions of the pelvis (ischiopubic, iliosacral, or periacetabular), the hip joint itself, or the proximal femur.[29,33] The weight-bearing function of these structures, essential for normal ambulation, contributes to the propensity of disease at these sites to cause incident pain with walking and weight-bearing.

Hip joint syndrome: Tumor involvement of the acetabulum or head of femur typically produces localized hip pain that is aggravated by weight-bearing and movement of the hip.[29,34] The pain may radiate to the knee or medial thigh, and occasionally, pain is limited to these structures.[33] Medial extension of an acetabular tumor can involve the lumbosacral plexus as it traverses the pelvic sidewall. Evaluation of this region is best accomplished with CT or MRI, both of which can demonstrate the extent of bony destruction and adjacent soft tissue involvement more sensitively than other imaging techniques.[35] Important differential diagnoses include avascular necrosis, radicular pain (usually L1), or, occasionally, occult infections.[36]

1.6 Arthritidis

1.6.1 Hypertrophic pulmonary osteoarthropathy

Hypertrophic pulmonary osteoarthropathy (HPOA) is a paraneoplastic syndrome that incorporates clubbing of the fingers, periostitis of long bones, and occasionally a rheumatoid-like polyarthritis.[37] Periostitis and arthritis can produce pain, tenderness, and swelling in the knees, wrists, and ankles. The onset of symptoms is usually subacute, and it may proceed the discovery of the underlying neoplasm by several months. It is most commonly associated with nonsmall-cell lung cancer. Less commonly, it may be associated with benign mesothelioma, pulmonary metastases from other sites, smooth muscle tumors of the esophagus, breast cancer, and metastatic nasopharyngeal cancer. HPOA is diagnosed on the basis of physical findings, radiological appearance, and radionuclide bone scan .[37–39] Effective antitumor therapy is sometimes associated with symptom regression, and bisphosphonate therapy may help relieve symptoms.[40–43]

2 Pain syndromes of the viscera and miscellaneous tumor-related syndromes

Pain may be caused by pathology involving the luminal organs of the gastrointestinal or genitourinary tracts, the parenchymal organs, the peritoneum, or the retroperitoneal soft tissues. Obstruction of hollow viscus, including intestine, biliary tract, and ureter, produces visceral nociceptive syndromes that are well described in the surgical literature.[44] Pain arising from retroperitoneal and pelvic lesions may involve mixed nociceptive and neuropathic mechanisms if both somatic structures and nerves are involved.

2.1 Hepatic distention syndrome

Pain-sensitive structures in the region of the liver include the liver capsule, blood vessels, and biliary tract.[45] Nociceptive afferents that innervate these structures travel via the celiac plexus, the phrenic nerve, and the lower-right intercostal nerves. Extensive intrahepatic metastases, or gross hepatomegaly associated with cholestasis, may produce discomfort in the right subcostal region, and less commonly in the right mid-back or flank.[45–48] Referred pain may be experienced in the right neck or shoulder, or in the region of the right scapula.[46]

Occasional patients who experience chronic pain due to hepatic distension develop an acute intercurrent subcostal pain that may be exacerbated by respiration. Physical examination may demonstrate a palpable or audible rub. These findings suggest the development of an overlying peritonitis, which can develop in response to some acute event, such as a hemorrhage into a metastasis.[49]

2.2 Midline retroperitoneal syndrome

Retroperitoneal pathology involving the upper abdomen may produce pain by injury to deep somatic structures of the posterior abdominal wall, distortion of pain-sensitive connective tissue, vascular and ductal structures, local inflammation, and direct infiltration of the celiac plexus. The most common causes are pancreatic cancer and retroperitoneal lymphadenopathy, particularly celiac lymphadenopathy.[50–56] The reasons for the high frequency of perineural invasion and the presence of pain in pancreatic cancer may be related to locoregional secretion and activation of nerve growth factor (NGF) and its high-affinity receptor TrkA. These factors are involved in stimulating epithelial cancer cell growth and perineural invasion.[57] In contrast, tumors with overexpression of a low-affinity receptor, p75NGFR, were associated with less pain.[58]

In some instances of pancreatic cancer, obstruction of the main pancreatic duct with subsequent ductal hypertension generates pain, which can be relieved by stenting of the pancreatic duct.[59]

The pain is experienced in the epigastrium, in the low thoracic region of the back, or in both locations. It is often diffuse and poorly localized. It is usually dull and boring in character, exacerbated with recumbency, and improved by sitting. The lesion can usually be demonstrated by CT, MRI, or ultrasound scanning of the upper abdomen.

2.3 Intestinal obstruction

Abdominal pain is an almost invariable manifestation of intestinal obstruction, which may occur in patients with abdominal or pelvic cancers.[60] The factors that contribute to this pain include smooth muscle contractions, mesenteric tension, and mural ischemia. Obstructive symptoms may be due primarily to the tumor, or more likely, to a combination of mechanical obstruction and other processes, such as autonomic neuropathy and ileus from metabolic derangements or drugs. Both continuous and colicky pains occur, which may be referred to the dermatomes represented by the spinal segments supplying the affected viscera. Vomiting, anorexia, and constipation are important associated symptoms.

2.4 Peritoneal carcinomatosis

Peritoneal carcinomatosis occurs most often by transcoelomic spread of abdominal or pelvic tumor; except breast cancer, hematogenous spread of an extra-abdominal neoplasm in this pattern is rare. Carcinomatosis can cause peritoneal inflammation, mesenteric tethering, malignant adhesions, and ascites, all of which can cause pain. Pain and abdominal distension are the most commonly presenting symptoms. Adhesions can also cause obstruction of hollow viscus, with intermittent colicky pain.[61] CT scanning may demonstrate the evidence of ascites, omental infiltration, and peritoneal nodules.[62]

2.5 Malignant perineal pain

Tumors of the colon or rectum, female reproductive tract, and distal genitourinary system are most commonly responsible for perineal pain.[63–67] Severe perineal pain following resection of pelvic tumors often precede evidence of detectable disease and should be viewed as a potential harbinger of progressive or recurrent cancer.[63,64,67] There is evidence to suggest that this phenomenon is caused by microscopic perineural invasion by recurrent disease.[68] The pain, which is typically described as constant and aching, is often aggravated by sitting or standing and may be associated with tenesmus or bladder spasms.[63]

Tumor invasion of the musculature of the deep pelvis can also result in a syndrome that appears similar to the so-called tension myalgia of the pelvic floor.[69] The pain is typically described as a constant ache or heaviness that exacerbates with upright posture. When due to tumor, the pain may be concurrent with other types of perineal pain. Digital examination of the pelvic floor may reveal local tenderness or palpable tumor.

2.6 Adrenal pain syndrome

Large adrenal metastases, common in lung cancer, may produce unilateral flank pain, and less commonly, abdominal pain. Pain is of variable severity, and it can be severe.[70] Adrenal metastases can be complicated by hemorrhage, which may cause severe abdominal pain.[71]

2.7 Ureteric obstruction

Ureteric obstruction is most frequently caused by tumor compression or infiltration within the true pelvis.[72,73] Less commonly, obstruction can be more proximal, associated with retroperitoneal lymphadenopathy, an isolated retroperitoneal metastasis, mural metastases, or intraluminal metastases. Cancers of the cervix, ovary, prostate, and rectum are most commonly associated with this complication. Pain may or may not accompany ureteric obstruction. When present, it is typically a dull chronic discomfort in the flank, with radiation into the inguinal region or genitalia.[74] If pain does not occur, ureteric obstruction may be discovered when hydronephrosis is discerned on abdominal imaging procedures or renal failure develops. Ureteric obstruction can be complicated by pyelonephritis or pyonephrosis, which often presents with features of sepsis, loin pain, and dysuria. Diagnosis of ureteric obstruction can usually be confirmed by the demonstration of hydronephrosis on renal sonography. The level of obstruction can be identified by pyelography, and CT scanning techniques will usually demonstrate the cause.[73]

2.8 Ovarian cancer pain

Moderate-to-severe chronic abdominopelvic pain is the most common symptom of ovarian cancer; it is reported by almost two-thirds of patients in the 2 weeks prior to the onset or recurrence of the disease.[75] Pain is experienced in the low back or abdomen.[76,77] In patients who have been previously treated, it is an important symptom of potential recurrence.[75]

2.9 Lung cancer pain

Even in the absence of involvement of the chest wall or parietal pleura, lung tumors can produce a visceral pain syndrome. In a large case series of lung cancer patients, pain was unilateral in 80% of the cases and bilateral in 20%. Among patients with hilar tumors, the pain was reported to the sternum or the scapula. Upper- and lower-lobe tumors referred to the shoulder and to the lower chest, respectively.[78,79] As previously mentioned, early lung cancers can generate ipsilateral facial pain.[80] It is postulated that this pain syndrome is generated via vagal afferent neurons.

2.10 Other uncommon visceral pain syndromes

Sudden-onset severe abdominal or loin pain may be caused by nontraumatic rupture of a visceral tumor. This has been most frequently reported with hepatocellular cancer but also with other liver metastases.[81,82] Kidney rupture due to a renal metastasis from an adenocarcinoma of the colon, splenic rupture in acute leukemia, rupture of adrenocortical cancers, and metastasis-induced perforated appendicitis have been reported.[83–86] Torsion of pedunculated visceral tumors can produce a cramping abdominal pain.

3 Headache and facial pain

Headache in the cancer patient results from traction, inflammation, or infiltration of pain-sensitive structures in the head or neck. Early evaluation with appropriate imaging techniques may identify the lesion and allow a prompt treatment, which may reduce pain and prevent the development of neurological deficits.[87] The most common causes of headache and facial pain in patients with cancer are intracerebral tumor, leptomeningeal metastases, base of skull metastases, and painful cranial neuralgias—some of which are discussed in other sections of this volume.

3.1 Ear and eye pain syndromes

OTALGIA: Otalgia is the sensation of pain in the ear, while referred otalgia is pain felt in the ear but originating from a nonotologic source. The rich sensory innervation of the ear derives from four cranial nerves and two cervical nerves which also supply other areas in the head, neck, thorax, and abdomen. Pain referred to the ear may originate in areas far removed from the ear itself. Otalgia may be caused by acoustic neuroma and metastases to the temporal bone or infratemporal fossa.[87–91] Referred otalgia is reported among patients with tumors involving the oropharynx or hypopharynx.[92]

EYE PAIN: Blurring of vision and eye pain are the two most common symptoms of choroidal metastases.[93] More commonly, chronic eye pain is related to metastases to the bony orbit, intraorbital structures such as the rectus muscles or optic nerve.[94–96]

3.2 Uncommon causes of headache and facial pain

Headache and facial pain in cancer patients may have many other causes.[87] Unilateral facial pain can be the initial symptom of an ipsilateral lung tumor.[80] Presumably, this referred pain is mediated by vagal afferents. Facial squamous cell carcinoma of the skin may present with facial pain due to extensive perineural invasion.[97] Patients with Hodgkin's disease may have transient episodes of neurological dysfunction that has been likened to migraine.[98] In some cases, this may be a reversible posterior leukoencephalopathy syndrome (RPLS), which is characterized by headache, conscious disturbance, seizure, and cortical visual loss with neuroimaging finding of edema in the posterior regions of the brain.[99]

Headache may occur with cerebral infarction or hemorrhage, which may be due to nonbacterial thrombotic endocarditis or disseminated intravascular coagulation. Headache is also the usual presentation of sagittal sinus occlusion, which may be due to tumor infiltration, hypercoagulable state, or treatment with L-asparaginase therapy.[100] Headache due to pseudotumor cerebri has also been reported to be the presentation of superior vena caval obstruction in a patient with lung cancer.[101] Tumors of the sinonasal tract may present with deep facial or nasal pain.[102]

4 Neuropathic pain involving the peripheral nervous system

Neuropathic pain involving the peripheral nervous system is common. The syndromes include painful radiculopathy, plexopathy, mononeuropathy, or peripheral neuropathy.

4.1 Painful radiculopathy

Radiculopathy or polyradiculopathy may be caused by any process that compresses, distorts, or inflames nerve roots. Painful radiculopathy is an important presentation of epidural tumor and leptomeningeal metastases (see above).

Postherpetic neuralgia (PHN): The precise clinical definition of PHN has been a matter of dispute. There is agreement that acute herpetic neuralgia refers to pain preceding or accompanying the eruption of rash that persists for up to 30 days from its onset. Some authorities refer to all persistent pain as PHN; others describe a subacute herpetic neuralgia that persists beyond healing of the rash but which resolves within 4 months of onset, reserving the term "PHN" for pain persisting beyond 4 months from the initial onset of the rash.[103] One study suggests that postherpetic neuralgia is two to three times more frequent in the cancer population than in the general population.[104]

4.2 Plexopathies

Among cancer patients, injuries to the cervical, brachial, or lumbosacral plexi are frequently due to tumor infiltration or treatment (including surgery or radiotherapy) to neoplasms in these regions.[105] Since specific plexopathies are described in greater detail elsewhere in this text, they will not be described in this chapter. Suffice it to say that they must be distinguished from other causes of painful plexopathy, including trauma during surgery or anesthesia, radiation-induced secondary neoplasms, plexus ischemia, and paraneoplastic neuritis.

5 Paraneoplastic nociceptive pain syndromes

Tumor-related gynecomastia: Tumors that secrete chorionic gonadotrophin (HCG), including malignant and benign tumors of the testis and rarely cancers from other sites, may be associated with chronic breast tenderness or gynecomastia.[106–111] Approximately 10% of patients with testis cancer have gynecomastia or breast tenderness at presentation, and the likelihood of gynecomastia is greater with increasing HCG level.[112] Breast pain can be the first presentation of an occult tumor.[113–115]

PARANEOPLASTIC PEMPHIGUS: Paraneoplastic pemphigus is a rare mucocutaneous disorder associated with non-Hodgkin's lymphoma and chronic lymphocytic leukemia. The condition is characterized by widespread shallow ulcers with hemorrhagic crusting of the lips, conjunctival bullae, and, uncommonly, pulmonary lesions. Characteristically, histopathology reveals intraepithelial and subepithelial clefting, and immunoprecipitation studies reveal autoantibodies directed against desmoplakins and desmogleins.[116,117]

PARANEOPLASTIC RAYNAUD'S SYNDROME: Paraneoplastic Raynaud's syndrome is a rare manifestation of solid tumors. It has been reported with lung cancer, ovarian cancer, testicular cancer, and melanoma.[118,119]

6 Chronic pain syndromes associated with cancer therapy

Most treatment-related pains are caused by tissue-damaging procedures. These pains are acute, predictable, and self-limited. Chronic treatment-related pain syndromes are associated with either a persistent nociceptive complication of an invasive treatment (such as a postsurgical abscess), or more commonly, neural injury.[120] In some cases, these syndromes occur long after the therapy is completed, resulting in a difficult differential diagnosis between recurrent disease and a complication of therapy.

6.1 Postchemotherapy pain syndromes

Toxic peripheral neuropathy: Chemotherapy-induced peripheral neuropathy is a common problem, which is typically manifested by painful paresthesias in the hands and/or feet, and signs consistent with an axonopathy, including "stocking-glove" sensory loss, weakness, hyporeflexia, and autonomic dysfunction.[120,121] The pain is usually characterized by continuous burning or lancinating pains, either of which may be increased by contact. The drugs most commonly associated with a peripheral neuropathy are the vinca alkaloids (especially vincristine), *cis*-platinum, oxaliplatin, and paclitaxel. Procarbazine, carboplatinum, misonidazole, and hexamethylmelamine are less common causes. Data from several studies indicate that the risk of neuropathy associated with *cis*-platinum and oxaliplatinum can be diminished by amifostine, glutathione, and calcium and magnesium infusion at the time of treatment.[122–126]

Recent data indicate that prophylactic vitamin E may reduce paclitaxel neuropathy.[127]

Avascular (aseptic) necrosis of femoral or humeral head: Avascular necrosis of the femoral or humeral head may occur either spontaneously or as a complication of intermittent or continuous corticosteroid therapy or high-dose chemotherapy with bone marrow transplantation.[128–130] Osteonecrosis may be unilateral or bilateral. Involvement of the femoral head is most common and typically causes pain in the hip, thigh, or knee. Involvement of the humeral head usually presents as pain in the shoulder, upper arm, or elbow. Pain is exacerbated by movement and relieved by rest. There may be local tenderness over the joint, but this is not universal. Pain usually precedes radiological changes by weeks to months; bone scintigraphy and MRI are sensitive and complementary diagnostic procedures for the detection of radiographically occult AVN. Radionuclide bone scanning and MRI are both sensitive methods, but MRI is preferred because it has greater sensitivity and a greater specificity than bone scanning.[131] Early treatment consists of analgesics, decrease or discontinuation of steroids, and sometimes surgery. With progressive bone destruction, joint replacement may be necessary.

Plexopathy: Lumbosacral or brachial plexopathy may follow *cis*-platinum infusion into the iliac artery or axillary artery, respectively.[132,133] Affected patients develop pain, weakness, and paresthesias within 48h of the infusion. The mechanism for this syndrome is thought to be due to small vessel damage and infarction of the plexus or nerve. The prognosis for neurologic recovery is not known.

Raynaud's phenomenon: Among patients with germ cell tumors treated with cisplatin, vinblastine, and bleomycin, persistent Raynaud's phenomenon is observed in 20%–30%.[134] This effect has also been observed in patients with carcinoma of the head and neck treated with a combination of cisplatin, vincristine, and bleomycin.[135] Pathophysiological studies have demonstrated that hyperreactivity in the central sympathetic nervous system results in a reduced function of the smooth muscle cells in the terminal arterioles.[136]

6.2 Chronic pain associated with hormonal therapy

Gynecomastia with hormonal therapy for prostate cancer: Chronic gynecomastia and breast tenderness are common complications of antiandrogen therapies for prostate cancer.[137] The incidence of this syndrome varies among drugs; it is frequently associated with diethyl stilbesterol and bicalutamide, is less common with flutamide and cyproterone, and is uncommon among patients receiving LHRH agonist therapy. Gynecomastia in the elderly must be distinguished from primary breast cancer or a secondary cancer in the breast.[138,139]

6.3 Chronic postsurgical pain syndromes

Surgical incision at virtually any location may result in chronic pain. Although persistent pain is occasionally encountered after nephrectomy, sternotomy, craniotomy, inguinal dissection, and other procedures, these pain syndromes are not well described in the cancer population. In contrast, several syndromes are now clearly recognized as sequelae of specific surgical procedures. The predominant underlying pain mechanism in these syndromes is neuropathic, resulting from injury to peripheral nerves or plexus.

Breast surgery pain syndromes: Chronic pain of variable severity is a common sequel of surgery for breast cancer. Although chronic pain has been reported to occur after almost any surgical procedure on the breast (from lumpectomy to radical mastectomy), it is most common after procedures involving axillary dissection.[140–143] This is a common pain syndrome after axillary lymph node dissection, occurring in 30%–70% of patients.[140,144–151]

The pain is usually characterized as a constricting and burning discomfort that is localized to the medial arm, axilla, and anterior chest wall.[143,152–155] Pain may begin immediately or as late as many months following surgery. The natural history of this condition appears to be variable, and both subacute and chronic courses are possible.[156,157] The onset of pain later than 18 months following surgery is unusual, and a careful evaluation to exclude recurrent chest wall disease is recommended in this setting. On examination, there is often an area of sensory loss within the region of the pain.[154] Chronicity of pain is related to the intensity of the immediate postoperative pain, postoperative complications, and subsequent treatment with chemotherapy and radiotherapy.[148,149,158] In many cases, pain is chronic and persists over many years.[159]

It is most commonly associated with neuropraxia of the intercostobrachial nerve during the process of axillary lymph node dissection.[143,154,160] There is a marked anatomic variation in the size and distribution of the intercostobrachial nerve, and this may account for some of the variability in the distribution of pain observed in patients with this condition.[161] In some cases, pain may be caused by hematoma in the axilla.[162]

The risk for, and severity of, pain is correlated positively with the number of lymph nodes removed[163,164] and is inversely correlated with age.[147,150,163] There are conflicting data as to whether preservation of the intercostobrachial nerve during axillary lymph node dissection can reduce the incidence of this phenomenon.[165–167] The incidence is reduced when axillary dissection is avoided either by sentinel node excision without full dissection or when nodes are irradiated without dissection.

This syndrome must be differentiated from postmastectomy frozen shoulder, axillary web syndrome,[168] and breast cellulitis.[169] In some cases of pain after breast surgery, a trigger point can be palpated in the axilla or chest wall.

Postradical neck dissection pain: Chronic neck and shoulder pain after radical neck dissection is common.[170] Shoulder pain is most often caused by damage to the spinal accessory nerve (CN XI).[171] In other cases, it can result from musculoskeletal imbalance in the shoulder girdle following surgical removal of neck muscles.[172] Similar to the droopy shoulder syndrome, this syndrome can be complicated by the development of a thoracic outlet syndrome or suprascapular nerve entrapment, with selective weakness and wasting of the supraspinatus and infraspinatus muscles.[173,174]

Escalating pain in patients who have undergone radical neck dissection may signify recurrent tumor or soft tissue infection. These lesions may be difficult to diagnose in tissues damaged by radiation and surgery. Repeated CT or MRI scanning may be needed to exclude tumor recurrence. Empiric treatment with antibiotics should be considered.[175,176]

Postthoracotomy pain: There have been two major studies of postthoracotomy pain.[177,178] In the first, three groups were identified: the largest (63%) had prolonged postoperative pain that abated within 2 months after surgery.[178] Recurrent pain, following resolution of the postoperative pain, was usually due to neoplasm. A second group (16%) experienced pain that persisted following the thoracotomy, then increased in intensity during the follow-up period. Local recurrence of disease and infection were the most common causes of the increasing pain. A final group had a prolonged period of stable or decreasing pain that gradually resolved over a maximum 8-month period. This pain was not associated with tumor recurrence. Overall, the development of late or increasing postthoracotomy pain was due to recurrent or persistent tumor in greater than 95% of patients. This finding was corroborated in the more recent study that evaluated the records of 238 consecutive patients who underwent thoracotomy which identified recurrent pain in 20 patients, all were found to have tumor regrowth.[177]

Patients with recurrent or increasing postthoracotomy pain should be carefully evaluated, preferably with a chest CT scan or MRI. Chest radiographs are insufficient to evaluate recurrent chest disease. In some patients, postthoracotomy pain appears to be caused by a taut muscular band within the scapular region. In such cases, pain may be amenable to trigger point injection of local anesthetic.[179]

Postoperatve frozen shoulder: Patients with postthoracotomy or postmastectomy pain are at risk for the development of a frozen shoulder.[140] This lesion may become an independent focus of pain, particularly if complicated by reflex sympathetic dystrophy. Adequate postoperative analgesia and active mobilization of the joint soon after surgery are necessary to prevent these problems.

Phantom pain syndromes: Phantom limb pain is perceived to arise from an amputated limb, as if the limb were still contiguous with the body. Phantom pain is experienced by 60%–80% of patients following limb amputation, but is only severe in about 5%–10% of cases.[180–182] The incidence of phantom pain is significantly higher in patients with a long duration of pre-amputation pain and those with pain on the day before amputation.[183,184] Patients who had pain prior to the amputation may experience phantom pain that replicates the earlier one.[185]

Phantom pain is more prevalent after tumor related than traumatic amputations, and postoperative chemotherapy is an additional risk factor.[180,182,186] The pain may be continuous or paroxysmal and is often associated with bothersome paresthesias. The phantom limb may assume painful and unusual postures and may gradually telescope and approach the stump. Phantom pain may initially magnify and then slowly fade over time. There is growing evidence that preoperative or postoperative neural blockade reduces the incidence of phantom limb pain during the first year after amputation.[180,187–190]

Some patients have spontaneous partial remission of the pain. The recurrence of pain after such a remission, or the late onset of pain in a previously painless phantom limb, suggests the appearance of a more proximal lesion, including recurrent neoplasm.[191]

Phantom pain syndromes have also been described after other surgical procedures.[180] Phantom breast pain after mastectomy, which occurs in 15%–30% of patients, also appears to be related to the presence of preoperative pain.[148,192–194] The pain tends to start in the region of the nipple and then spread to the entire breast. The character of the pain is variable and may be lancinating, continuous, or intermittent.[193,194] A phantom rectum pain syndrome occurs in approximately 15% of patients who undergo abdominoperineal resection of the rectum.[64,195] Phantom rectal pain may develop either in the early postoperative period or after a latency of months to years. Late-onset pain is almost always associated with tumor recurrence. Rare cases of phantom bladder pain after cystectomy and phantom eye pain after enucleation have also been reported.

Stump pain: Stump pain occurs at the site of the surgical scar several months to years following amputation.[196] It is usually the result of neuroma development at a site of nerve transection. This pain is characterized by burning or lancinating dysesthesias, which are often exacerbated by movement or pressure and blocked by an injection of a local anesthetic.

Postsurgical pelvic floor myalgia: Surgical trauma to the pelvic floor can cause a residual pelvic floor myalgia, which like the neoplastic syndrome described previously, mimics the so-called tension myalgia.[69] The risk of disease recurrence associated with this condition is not known, and its natural history has not been defined. In patients who have undergone anorectal resection, this condition must be differentiated from the phantom anus syndrome (see above).

6.4 Chronic postradiation pain syndromes

Chronic pain complicating radiation therapy tends to occur late in the course of a patient's illness. These syndromes must always be differentiated from recurrent tumor. Radiation-induced brachial and lumbosacral plexopathies and chronic radiation myelopathy are discussed elsewhere in this volume.

Chronic radiation enteritis and proctitis: Chronic enteritis and proctocolitis occur as a delayed complication in 2%–10% of patients who undergo abdominal or pelvic radiation therapy.[197,198]

Radiation cystitis: Radiation therapy used in the treatment of tumors of the pelvic organs (prostate, bladder, colon/rectum, uterus, ovary, and vagina/vulva) may produce chronic radiation cystitis.[199–201]

Lymphedema pain: One-third of patients with lymphedema as a complication of breast cancer or its treatment experience pain and tightness in the arm, and pain is a major part of the morbidity among affected patients.[202,203]

Burning perineum syndrome: Persistent perineal discomfort is an uncommon delayed complication of pelvic radiotherapy. After a latency of 6–18 months, burning pain can develop in the perianal region; the pain may extend anteriorly to involve the vagina or scrotum.[204,205]

Postprostate brachytherapy pelvic pain: Brachytherapy patients with prostate cancer may produce a chronic radiation-related pelvic pain syndrome that is exacerbated by urination or perineal pressure. Data suggest that it may be partly related to higher central prostatic radiation doses.[206]

Osteoradionecrosis: Osteoradionecrosis is another late complication of radiotherapy. Bone necrosis, which occurs as a result of endarteritis obliterans, may produce focal pain. Overlying tissue breakdown can occur spontaneously or as a result of trauma, such as dental extraction or denture trauma.[207,208] Delayed development of a painful ulcer must be differentiated from tumor recurrence.

7 Clinical evaluation of the patients with difficult pain

Clinical evaluation of the patient, the pain syndrome, and therapeutic options is critical. Since knowledge in the assessment and management of cancer pain is frequently deficient among physicians who lack specific training in this field, expert evaluation or consultation should be considered.[209] At Memorial Sloan-Kettering Cancer Center, a recent survey observed that a Pain Service consultation identified a previously undiagnosed etiology for the pain in 64% of 376 referred patients, including new neurological diagnoses in 36% of patients and an unsuspected infection in 4%; a substantial proportion of these patients received radiotherapy, surgery, or chemotherapy based on the findings of the pain consultant.[210] Similarly, Coyle et al. reported a series of cases of crescendo pain in which patient evaluation identified an unsuspected delirium, treatment of which restored adequate pain control.[211]

7.1 Therapeutic strategy

Table 1 presents a stepwise approach to the management of difficult pain problems. By addressing these questions sequentially, the clinician is able to explore therapeutic options in a rational and ordered manner that emphasizes the use of noninvasive approaches with low-potential morbidity and treatment burden.

In addressing a difficult cancer pain problem, a case conference approach is often useful, particularly since individual clinician bias can influence decision making.[212,213] This conference may involve the participation of oncologists, palliative care physicians, anesthesiologists, neurosurgeons, psychiatrists, nurses, social workers, and others. The discussion attempts to clarify the therapeutic options and the goals of care. When local expertise is limited, telephone consultation with physicians who are expert in the management of cancer pain is strongly encouraged.

TABLE 1   A stepwise approach to the management of difficult pain problems.

1	Are primary therapies (chemotherapy, radiation therapy, surgery or antibiotic therapy) feasible, and if so, are they likely to improve patient outcome?
2	Have opioids been titrated up to maximal tolerated doses?
3	Have side effects been addressed through appropriate drug therapy or by trials of alternative opioids?
4	Have appropriate adjuvant analgesics been considered or tried?
5	Have spinal opioids been considered?
6	Have other anesthetic or neurosurgical options been considered?
7	Is this a refractory pain state that may require sedation to achieve adequate relief?

7.1.1 Strategy 1: Are primary therapies (chemotherapy, radiation therapy, surgery, or antibiotic therapy) feasible, and if so, are they likely to improve patient outcome?

The assessment process may reveal a cause for the pain that is amenable to primary therapy (i.e., therapy that is directed at the etiology of the pain). This therapy may improve comfort, function, or duration of survival. For example, pain generated by tumor infiltration may respond to antineoplastic treatment with surgery, radiotherapy, or chemotherapy; and pain caused by infections may be relieved with antibiotic therapy or drainage procedures. Specific analgesic treatments are usually required as an adjunct to the primary therapy.

Radiotherapy: The analgesic effectiveness of radiotherapy is documented by abundant data and a favorable clinical experience in the treatment of painful bone metastases, epidural neoplasm, and headache due to cerebral metastases.[214–219] In other settings, however, there is a lack of data, and the use of radiotherapy is largely anecdotal. For example, the results with perineal pain due to low sacral plexopathy appear to be encouraging, and hepatic radiotherapy (e.g., 2000–3000 cGy) can be well tolerated and effective for the pain of hepatic capsular distention in 50%–90% of patients.[220–226]

Chemotherapy: Despite a paucity of data concerning the specific analgesic benefits of chemotherapy, there is a strong clinical impression that tumor shrinkage is generally associated with relief of pain. Although there are some reports of analgesic value even in the absence of significant tumor shrinkage, the likelihood of a favorable effect on pain is generally related to the likelihood of tumor response.[227–230] In all situations, the decision to administer chemotherapy solely for the treatment of symptoms should be promptly reconsidered unless the patient demonstrates a clearly favorable balance between relief and adverse effects.

Surgery: Surgery may have a role in the relief of symptoms caused by specific problems, such as obstruction of a hollow viscus, unstable bony structures, and compression of neural tissues.[231–237] In cases of otherwise uncontrollable extremity pain due to unresponsive locally advanced disease, amputation of the effected limb may have a dramatic impact on pain control.[238,239] The potential benefits must be weighed against the risks of surgery, the anticipated length of hospitalization and convalescence, and the predicted duration of benefit.

Antibiotic therapy: Antibiotics may be analgesic when the source of the pain involves infection. Illustrative examples include cellulitis, chronic sinus infections, pelvic abscess, pyonephrosis, and osteitis pubis.[169,240] In some cases, infection may be occult and confirmed only by the symptomatic relief provided by empiric treatment with these drugs.[175,176,210]

Radiofrequency tumor ablation: Radiofrequency ablation involves percutaneous or intraoperative insertion
of an electrode into a lesion under ultrasonic guidance.
Radiofrequency energy is emitted through the electrode
and generates heat, leading to coagulative necrosis.
Radiofrequency tumor ablation may result in substantial
relief from cancer pain.[241] A growing anecdotal literature
describes and supports the use of this approach in presacral and pelvic tumor recurrences, osteoid osteoma,
painful pancreatic renal and adrenal tumors, and painful
bony metastases, including vertebral metastases.[242–250]

Methylmethacrylate vertebroplasty and acetabuloplasty:
These approaches involve the radiographically guided
percutaneous injection of polymethyl methacrylate to
stabilize painful bony metastases. "Vertebroplasty" so-
lidifies and stabilizes the osteolytic lesion and can result
in rapid (1–3 days) disappearance of pain, with restoration of spinal stability.[251,252] Patients most suitable for
vertebroplasty are those suffering from an osteolytic lesion of the vertebral body without disruption of the posterior wall, with or without vertebral collapse, and with
severe pain. It is relatively contraindicated when there
is epidural invasion by tumor. Rare but serious complications can occur, including the extravasation of cement
into adjacent nerve foramen, venous emboli, and spinal
cord compression requiring decompression. A similar
technique, acetabuloplasty, has been described in a few
small series. Rare major complications include intraar-
ticular extension of cement into the hip joint causing
chondrolysis and extravasation around the sciatic nerve,
worsening the pretreatment level of pain.

7.1.2 Strategy 2: Titrated opioids up to maximal tolerated dose

Efficacy evidence suggests that the symptomatic
therapy of strong pain should start with an opioid, irrespective of the mechanism of the pain. The factors that
influence opioid selection in chronic pain states include
pain intensity, pharmacokinetic and formulatory considerations, previous adverse effects, and the presence of
co-existing disease.

Traditionally, patients with moderate pain have been
conventionally treated with a combination product
containing acetaminophen or aspirin plus codeine, dihydrocodeine, hydrocodone, oxycodone, and propoxy-
phene. The doses of these combination products can
be increased until the maximum dose of the nonopioid
co-analgesic is attained (e.g., 4000 mg acetaminophen).
Recent years have witnessed the proliferation of new
opioid formulations that may improve the convenience
of drug administration for patients with moderate pain.
These include controlled-release formulations of codeine, dihydrocodeine, oxycodone, morphine, and tramadol in dosages appropriate for moderate pain and,
most recently, patches of buprenorphine.

Patients who present with strong pain are usually
treated with morphine, hydromorphone, oxycodone,
oxymorphone, fentanyl, or methadone. Of these, the
short half-life opioid agonists (morphine, hydromorphone, fentanyl, oxycodone, or oxymorphone) are generally favored because they are easier to titrate than the
long half-life drugs, which require a longer period to approach steady-state plasma concentrations.

If the patient is currently using an opioid that is well
tolerated, it is usually continued unless difficulties in
dose titration occur or the required dose cannot be administered conveniently. A switch to an alternative opioid is considered if the patient develops dose-limiting
toxicity, which precludes adequate relief of pain without excessive side effects, or if a specific formulation,
not available with the current drug, is either needed or
may substantially improve the convenience of opioid
administration.

Because of enormous variability in opioid responsiveness and other interindividual factors, inadequate relief
should be addressed through gradual escalation of dose
until adequate analgesia is reported or excessive side effects supervene. Because opioid response increases lin-
early with the log of the dose, a dose increment of less
than 30%–50% is not likely to significantly improve analgesia. Patients vary greatly in the opioid dose required
to manage pain, and some patients have been reported
to require very high doses of systemic opioids to control pain.[253–262] The absolute dose is immaterial as long
as administration is not compromised by excessive side
effects, inconvenience, discomfort, or cost.

The need for escalating doses is a complex phenomenon. Most patients reach a dose that remains constant
for prolonged periods.[254,263–265] When the need for dose
escalation arises, any of a variety of distinct processes
may be involved.[266] Clinical experience suggests that
true analgesic tolerance is a much less common reason
than disease progression or increasing psychological
distress.[211,265–268] Indeed, most patients who require
an escalation in dose to manage increasing pain have
demonstrable progression of disease, and worsening
pain in a patient receiving a stable dose of opioids
should not be attributed to tolerance, but should be
assessed as presumptive evidence of disease progression or, less commonly, increasing psychological
distress.[266–268]

7.1.3 Strategy 3: Treat side effects through appropriate drug therapy or by trials of alternative opioids?

For any opioids, poor pain control is often caused by
dose-limiting side effects that preclude dose escalation
to a potentially efficacious dose. In this setting, it is essential to re-establish a better balance between opioid
analgesia and side effects. In general, four different ap-

proaches to the management of opioid adverse effects have been described[269]:

Dose reduction of systemic opioid: Reducing the dose of administered opioid usually results in a reduction in dose-related adverse effects. When patients have well-controlled pain, a gradual reduction in the opioid dose will often result in the resolution of dose-related adverse effects, while preserving adequate pain relief.[270] When opioid doses cannot be reduced without the loss of pain control, reduction in dose must be accompanied by the addition of an accompanying synergist approach. Two common approaches include:

(1) The addition of a nonopioid co-analgesic: The analgesia achieved from nonopioid co-analgesics from the nonsteroidal antiinflammatory class of agents is additive and often synergistic with that achieved by opioids.
(2) The addition of an adjuvant analgesic that is appropriate to the pain syndrome and mechanism; this is discussed further below:
(3) The application of a regional anesthetic or neuroablative intervention: The results of the WHO "analgesic ladder" validation studies suggest that 10%–30% of patients with cancer pain do not achieve a satisfactory balance between relief and side effects using systemic pharmacotherapy alone without unacceptable drug toxicity. Anesthetic and neurosurgical techniques may reduce or eliminate the requirement for systemically administered opioids to achieve adequate analgesia (see below).

Symptomatic management of the adverse effect: Symptomatic drugs used to prevent or control opioid adverse effects are commonly employed (see Table 2). Most of these approaches are based on cumulative anecdotal experience. With few exceptions, the literature describing these approaches is anecdotal or "expert opinion." Very few studies have prospectively evaluated efficacy and no studies have evaluated the toxicity

TABLE 2 Medication used in the symptomatic management of opioid side effects.

Sedation	Methylphenidate Pemoline Modafinil
Confusion	Haloperidol Risperdal
Myoclonus	Clonazepam
Constipation	Osmotic laxative Stimulant laxatives
Nausea/vomiting	Metoclopramide 5HT3 blockers Scopolamine

of these approaches over long term. In general, this approach involves the addition of a new medication, adding to medication burden and with the associated risks of adverse effects or drug interaction.

Opioid rotation: Over the past 10 years, numerous clinicians and cancer pain services have reported a successful reduction in opioid side effects by switching to an alternative opioid.[260,271–284] Improvements in cognitive impairment, sedation, hallucinations, nausea, vomiting, and myoclonus have been commonly reported. This approach requires familiarity with a range of opioid agonists and with the use of equianalgesic tables to convert doses when switching between opioids. While this approach has the practical advantage of minimizing polypharmacy, outcomes are variable and unpredictable. When switching between opioids, even with prudent use of equianalgesic tables, patients are at risk for under- or over-dosing by virtue of individual sensitivities.

7.1.4 Strategy 4: Consider the role of adjuvant analgesics

The term "adjuvant analgesic" describes a drug that has a primary indication other than pain but is analgesic in some conditions. Adjuvant analgesics may be combined with primary analgesics to improve the outcome for patients who cannot otherwise attain an acceptable balance between relief and side effects.[285] There is great interindividual variability in the response to all adjuvant analgesics and, for most, the likelihood of benefit is limited. Furthermore, many of the adjuvant analgesics have the potential to cause side effects, which may be additive to the opioid-induced adverse effects that are already problematic. In evaluating the utility of an adjuvant agent in a particular patient setting, one must consider the likelihood of benefit, the risk of adverse effects, the ease of administration, and patient convenience. In the management of cancer pain, adjuvant analgesics can be broadly classified based on conventional use. Four groups are distinguished:

(1) Multipurpose adjuvant analgesics.
(2) Adjuvant analgesics used for neuropathic pain.
(3) Adjuvant analgesics used for bone pain.
(4) Adjuvant analgesics used for visceral pain.

Multipurpose adjuvant analgesics

CORTICOSTEROIDS: Corticosteroids are among the most widely used adjuvant analgesics.[286,287] Painful conditions that commonly respond to corticosteroids include raised intracranial pressure headache, acute spinal cord compression, superior vena cava syndrome, metastatic bone pain, neuropathic pain due to infiltration or compression by tumor, symptomatic lymphedema, and hepatic capsular distension. The mechanism of analgesia produced by these drugs may involve antiedema effects, antiinflammatory effects, and a direct influence on the electrical activity in damaged nerves.[288] The most commonly used drug is dexamethasone, a choice that gains theoretical support from the relatively low mineralocorticoid effect of this agent. Dexamethasone also has been conventionally used for raised intracranial pressure and spinal cord compression.

Patients with advanced cancer who experience pain and other symptoms may respond favorably to a relatively small dose of corticosteroid (e.g., dexamethasone 1–2 mg twice daily). In some settings, however, a high-dose regimen may be appropriate. For example, in patients with spinal cord compression, an acute episode of very severe bone pain or neuropathic pain that cannot be promptly reduced with opioids may respond dramatically to a short course of relatively high doses (e.g., dexamethasone 100 mg, followed initially by 96 mg per day in divided doses).[287,289] This dose can be tapered over weeks, concurrent with the initiation of other analgesic approaches, such as radiotherapy. Although the effects produced by corticosteroids in patients with advanced cancer are often very gratifying, side effects are potentially serious and increase with prolonged usage.[290]

Topical local anesthetics: Topical local anesthetics can be used in the management of painful cutaneous and mucosal lesions. Viscous lidocaine is frequently used in the management of oropharyngeal ulceration. Although the risk of aspiration appears to be very small, caution with eating is required after oropharyngeal anesthesia.

Adjuvant medications used for neuropathic pain

Neuropathic pains are generally less responsive to opioid therapy than nociceptive pain, and in many cases, the outcome of pharmacotherapy may be improved by the addition of an adjuvant analgesic.

Antidepressant drugs are commonly used to manage continuous neuropathic pains, and the evidence for analgesic efficacy is greatest for the tertiary amine tricyclic drugs, such as amitriptyline, doxepin, and imipramine.[291] The secondary amine tricyclic antidepressants (such as desipramine, clomipramine, and nortriptyline) have fewer side effects and are preferred when concern about sedation, anticholinergic effects, or cardiovascular toxicity is high. There is less evidence for the analgesic effectiveness of selective serotonin-uptake inhibitor antidepressants, but given their reduced tendency to adverse effects, they may be considered in the management of neuropathic pain.[292]

Selected anticonvulsant drugs may be effective for diverse types of neuropathic pain. Although the evidence is best for carbamazepine, the potential for adverse hematologic effects associated with this drug has limited its use in the medically ill patients. Due to its proven analgesic effect in several neuropathic pains, its good tolerability, and a paucity of drug-drug interactions, gabapentin has been recommended as a first-line agent for the treatment of neuropathic pain of diverse etiologies.[293] A number of the newer anticonvulsants such as lamotrigine, topiramate, felbamate, and oxcarbazepine are also promising.

Occasionally, systemically administered local anesthetic drugs may be useful in the management of neuropathic pain characterized by either continuous or lancinating dysesthesias. It is reasonable to undertake a trial with an oral local anesthetic in patients with continuous dysesthesias who fail to respond adequately to, or who cannot tolerate, the tricyclic antidepressants, and for patients with lancinating pains refractory to trials of anticonvulsant drugs and baclofen. Long-term systemic local anesthetic therapy now is usually accomplished using an oral formulation such as flecainide, tocainide, or mexiletine. Analgesic response to a trial of intravenous lidocaine (5 mg/kg, over 45 min) may predict for likelihood of response to oral mexiletine.[294] Less compelling data support the use of clonidine, baclofen, calcitonin, and subcutaneously administered ketamine.[295]

Adjuvant medications used for bone pain

The management of bone pain frequently requires the integration of opioid therapy with multiple ancillary approaches. Although a meta-analysis of NSAID therapy in cancer pain that reviewed data from 3084 patients in 42 trials found no specific efficacy in bone pain and analgesic effects equivalent only to "weak" opioids, some patients appear to benefit greatly from the addition of such a drug. Corticosteroids are often advocated in difficult cases.[286,296]

Bisphosphonates are analogs of inorganic pyrophosphate that inhibit osteoclast activity and reduce bone resorption in a variety of illnesses. Controlled trials of intravenous pamidronate and clodronate in patients with advanced cancer have demonstrated a reduction of bone pain.[297] The analgesic effect of pamidronate appears to be dose and schedule dependent, a dose response is evident at doses between 15 and 30 mg/week, and it has been noted that 30 mg every 2 weeks is less effective than 60 mg every 4 weeks.[298] Similar effects have been observed with orally administered intravenous zoledronate and oral ibandronate.[299–301]

Radiolabeled agents that are absorbed into areas of high bone turnover have been evaluated as potential therapies for metastatic bone disease. It has the advantages of addressing all sites of involvement and relatively selective absorption, thus limiting radiation exposure to normal tissues. Excellent clinical responses with acceptable hematological toxicity have been observed with a range of radiopharmaceuticals. The best studied and most commonly used radionuclide is strontium-89.[302,303] This approach is contraindicated with patients who have a platelet count less than 60,000 or a WCC < 2.4, and is not advised for patients with very poor performance status.[304] Using another approach, bone-seeking radiopharmaceuticals that link a radioisotope with a bisphosphonate compound have been synthesized. Positive experience has been reported with samarium-153-ethylenediaminetetramethylene phosphonic acid, samarium-153-lexidronam, and rhenium-186-hydroxyethylidene diphosphonate.[305–307]

Adjuvant analgesics for visceral pain

There are limited data that support the potential efficacy of a range of adjuvant agents for the management of bladder spasm, tenesmoid pain, and colicky intestinal pain. Oxybutynin chloride, a tertiary amine with anticholinergic and papaverine-like, direct muscular antispasmodic effects, is often helpful for bladder spasm pain, as is flavoxate.[308–310] Based on limited clinical experience and in vitro evidence that prostaglandins play a role in bladder smooth muscle contraction, a trial of NSAIDs may be justified for patients with painful bladder spasms.[311] Limited data support a trial of intravesical capsaicin.[312,313]

There is no well-established pharmacotherapy for painful rectal spasms. A recent double-blinded study demonstrated that nebulized salbutamol can reduce the duration and severity of attacks.[314] There is anecdotal support for trials of diltiazem, clonidine, chlorpromazine, and benzodiazepines.[315–319]

Colicky pain due to inoperable bowel obstruction has been treated empirically with intravenous scopolamine (hyoscine) butylbromide and sublingual scopolamine (hyoscine) hydrobromide.[320–323] There are also data supporting the use of octreotide for this indication.[324,325]

Other more recent considerations for nonopioid adjuvant cancer pain management would include medical cannabis and the use of acupuncture and acupressure. Medical cannabis has been under evaluation for cancer-related pain since 1975, including double-blind and placebo-controlled trials.[326] Cannabinoid-based products containing tetrahydrocannabinol (THC) and cannabidiol (CBD) have been tested in various types of cancer patients, with some mild-to-moderate evidence for efficacy against chronic cancer pain and cancer-related neuropathic pain. Additional phase III trials are necessary to establish the optimal dosages and efficacy of different types of cannabis-based therapies. Evidence has also been accumulating for more than a decade on the use of acupuncture and acupressure for the treatment of chronic cancer pain.[327] A recent meta-analysis of 17 randomized clinical trials concluded that acupuncture and acupressure were associated with reduced pain intensity. In addition, these interventions were also associated with reduced opioid utilization.

7.1.5 Strategy 5: Consider the role of spinal opioids

In general, regional analgesic techniques such as intraspinal opioid and local anesthetic administration or intrapleural local anesthetic administration are usually considered first because they can achieve this end without compromising neurological integrity. Neurodestructive procedures, however, are valuable in a small subset of patients; some of these procedures, such as celiac plexus blockade in patients with pancreatic cancer, may have a favorable enough risk: benefit ratio that early treatment is warranted.

Epidural and intrathecal opioids: The delivery of low opioid doses near the sites of action in the spinal cord may decrease supraspinally mediated adverse effects. In the one randomized trial comparing intraspinal opioid therapy to routine systemic therapy, the intraspinal route was found to have advantages in terms of better analgesia and fewer adverse effects.[328] In general, intrathecal administration is preferred to epidural administration.

Opioid selection for intraspinal delivery is influenced by several factors. Hydrophilic drugs, such as morphine and hydromorphone, have a prolonged half-life in cerebrospinal fluid and significant rostral redistribution.[329] Lipophilic opioids, such as fentanyl and sufentanil, have less rostral redistribution and may be preferable for segmental analgesia at the level of spinal infusion. The addition of a low concentration of a local anesthetic, such as 0.125%–0.25% bupivacaine, has been demonstrated to increase analgesic effect without increasing toxicity.[330–332] Other agents have also been co-administered with intraspinal opioids, including clonidine, octreotide, ketamine, and calcitonin, but additional studies are required to assess their potential utility.[333–337]

Intraventricular opioids. A growing international experience suggests that the administration of low doses of an opioid (particularly morphine) into the cerebral ventricles can provide long-term analgesia in selected patients.[338,339] This technique has been used for patients with upper body or head pain, or severe diffuse pain, and has been generally very well tolerated. Schedules have included both intermittent injection via an Ommaya reservoir and continual infusion using an implanted pump.[338–340]

Regional local anesthetic. Several authors have described the use of intrapleural local anesthetics in the management of chronic postthoracotomy pain and cancer-related pains involving the head, neck, chest, arms, and upper abdominal viscera.[341–343] For patients with localized upper limb pain, intermittent infusion of bupivacaine through an interscalene brachial plexus catheter may be of benefit.[344]

7.1.6 Strategy 6: Consider other invasive neuroablative interventions

Consideration of invasive approaches requires a word of caution. Interpretation of data regarding the use of alternative analgesic approaches and extrapolation to the presenting clinical problem requires care. The literature is characterized by the lack of uniformity in patient selection, inadequate reporting of previous analgesic therapies, inconsistencies in outcome evaluation, and paucity of long-term follow-up. Furthermore, reported outcomes in the literature may not predict the outcomes of a procedure performed on a medically ill patient by a physician who has more limited experience with the techniques involved.

For most pain syndromes, there exists a range of techniques that may theoretically be applied. In choosing between a range of procedures, the following principles are salient:

1	Ablative procedures are generally deferred as long as pain relief is obtainable by nonablative modalities.
2	The procedure most likely to be effective should be selected. If there is a choice, however, the one with the fewest and least serious adverse effects is preferred.
3	In progressive stages of cancer, pain is likely to be multifocal and a procedure aimed at a single locus of pain, even if completed flawlessly, is unlikely to yield complete relief of pain until death. A realistic and sound goal is a lasting decrease in pain to a level that is manageable by pharmacotherapy with minimal side effects.
4	Whenever possible, somatic neurolysis should be proceeded by the demonstration of effective analgesia with a local anesthetic prognostic block.
5	Since there is a learning curve with all of the procedures, performance by a physician who is experienced in the specific intervention may improve the likelihood of a successful outcome.

7.2 Neuroablative sympathetic blocks for visceral pain

Celiac plexus block: Neurolytic celiac plexus blockade may be helpful in the management of pain caused by neoplastic infiltration of the upper abdominal viscera, including the pancreas, upper retroperitoneum, liver, gall bladder, and proximal small bowel.[345,346] Reported analgesic response rates in patients with pancreatic cancer are 50%–90%, and the reported duration of effect is generally 1–12 months.[345,347] Given the generally favorable response to this approach, and supportive data from two small studies, some clinicians recommend this intervention at an early stage; other experts differ and recommend celiac plexus block only for patients who do not maintain an adequate balance between analgesia and side effects from an oral opioid.[345–349] Common transient complications include postural hypotension and diarrhea. Traditionally, celiac plexus block has been performed under fluoroscopic or CT control; more recently, a transgastric approach has been developed using accurate anatomic localization with endoscopic ultrasound.[350,351]

Sympathetic blocks for pelvic visceral pain: Limited anecdotal experience has been reported with two techniques. Phenol ablation of the superior hypogastric nerve plexus has been reported to relieve chronic cancer pain arising from the descending colon, rectum, and the lower genitourinary structures.[352,353] Similarly, neurolysis of the ganglion impar (ganglion of Walther) (a solitary retroperitoneal structure at the sacrococcygeal junction) has been reported to relieve visceral sensations referred to the rectum, perineum, or vagina.[354–356]

Sympathetic blockade of somatic structures: Sympathetically maintained pain syndromes may be relieved by the interruption of sympathetic outflow to the affected region of the body. Lumbar sympathetic blockade should be considered for sympathetically maintained pain involving the legs, and stellate ganglion blockade may be useful for sympathetically maintained pain involving the face or arms.[357]

7.3 Neuroablative techniques for somatic and neuropathic pain

Rhizotomy: Segmental or multisegmental destruction of the dorsal sensory roots (rhizotomy), achieved by surgical section, chemical neurolysis, or radiofrequency lesioning, can be an effective method of pain control for patients with otherwise refractory localized pain syndromes. These techniques are most commonly used in the management of chest wall pain due to tumor invasion of somatic and neural structures.[358] Other indications include refractory upper limb, lower limb, pelvic, or perineal pain.[359] Satisfactory analgesia is achieved in about 50% of patients and the average duration of relief is 3–4 months, but with a wide range of distribution. Specific complications of the procedure depend on the site of neurolysis. For example, the complications of lumbosacral neurolysis include paresis (5%–20%), sphincter dysfunction (5%–60%), impairment of touch and proprioception, and dysesthesias. Although neurological deficits are usually transient, the risk of increased disability through weakness, sphincter incompetence, and loss of positional sense suggests that these techniques should be reserved for patients with limited function and preexistent urinary diversion. Patient counseling regarding the risks involved is essential.

Neurolysis of primary afferent nerves or their ganglia: Neurolysis of primary afferent nerves may also provide a significant relief for selected patients with localized pain. The utility of these approaches is limited by the potential for concurrent motor or sphincteric dysfunction. Refractory unilateral facial or pharyngeal pain may be amenable to trigeminal neurolysis (gasserian gangliolysis) or glossopharyngeal neurolysis.[360,361] Unilateral pain involving the tongue or floor of the mouth may be amenable to blockade of the sphenopalatine ganglion.[362]

Intercostal or paravertebral neurolysis is an alternative to rhizotomy for patients with chest wall pain. Unilateral shoulder pain may be amenable to suprascapular neurolysis.[363] Arm pain that is more extensive may be effectively relieved by brachial plexus neurolysis, but this approach will result in extreme motor weakness.[364]

Cordotomy: During cordotomy, the anterolateral spinothalamic tract is ablated to produce contralateral loss of pain and temperature sensibility.[365,366] This approach is generally reserved for patients with severe unilateral pain arising in the torso or lower extremities. The percutaneous technique is generally preferred; open cordotomy is usually reserved for patients who are unable to lie in the supine position or are not cooperative enough to undergo a percutaneous procedure. Significant pain relief is achieved in more than 90% of patients during the period immediately following cordotomy, but 50% of patients have recurrent pain after one year.[365–367] Repeat cordotomy can sometimes be effective. The neurological complications of cordotomy include paresis, ataxia, and bladder and "mirror-image" pain.[366] The complications are usually transient, but are protracted and disabling in approximately 5% of cases. Rarely, patients with a long duration of survival (> 12 months) develop a delayed-onset dysesthetic pain. The most serious potential complication is respiratory dysfunction, which may occur in the form of phrenic nerve paralysis or as sleep-induced.[368,369] Because of the latter concern, bilateral high cervical cordotomies or a unilateral cervical cordotomy ipsilateral to the site of the only functioning lung is not recommended.

7.3.1 Strategy 7: If pain is refractory, consider the role of sedation to achieve adequate relief

Through the vigilant application of analgesic care, pain is often relieved adequately without compromising the sentience or function of the patient beyond that caused by the natural disease process itself. Occasionally, however, this cannot be achieved and pain is perceived to be "refractory".[370] In deciding that a pain is refractory, the clinician must perceive that the further applications of standard interventions are either (1) incapable of providing adequate relief, (2) associated with excessive and intolerable acute or chronic morbidity, or (3) unlikely to provide relief within a tolerable time frame. In this situation, sedation may be the only therapeutic option capable of providing adequate relief. This approach is described as "sedation in the management of refractory symptoms at the end of life".[371]

The justification of sedation in this setting is that it is goal appropriate and proportionate. At the end of life, when the overwhelming goal of care is the preservation of patient comfort, the provision of adequate relief of symptoms must be pursued even in the setting of a narrow therapeutic index for the necessary palliative treatments.[372,373] In this context, sedation is a medically indicated and proportionate therapeutic response to refractory symptoms, which cannot be otherwise relieved. Appeal to patients' rights also underwrites the moral legitimacy of sedation in the management of otherwise intolerable pain at the end of life. Patients have a right, affirmed by the Supreme Court, to palliative care in response to unrelieved suffering.

Once a clinical consensus exists that pain is refractory, it is appropriate to present this option to the patient or their surrogate. When presented to a patient with refractory symptoms, the offer of sedation can demonstrate the clinician's commitment to the relief of suffering. This can enhance trust in the doctor-patient relationship and influence the patient's appraisal of their capacity to cope. Indeed, patients commonly decline sedation, acknowledging that pain will be incompletely relieved but secure in the knowledge that if the situation becomes intolerable to them, this option remains available. Other patients reaffirm comfort as the predominating consideration and request the initiation of sedation.

The published literature describing the use of sedation in the management of refractory pain at the end of life is anecdotal and refers to the use of opioids, neuroleptics, benzodiazepines, barbiturates, and propofol.[370] In the absence of relative efficacy data, guidelines for drug selection are empirical. Irrespective of the agent or agents selected, administration initially requires dose titration to achieve adequate relief, followed subsequently by provision of ongoing therapy to ensure maintenance of effect.

References

1. Neufeld NJ, Elnahal SM, Alvarez RH. Cancer pain: a review of epidemiology, clinical quality and value impact. *Future Oncol.* 2017;13:833–841.
2. Strang P. Existential consequences of unrelieved cancer pain. *Palliat Med.* 1997;11(4):299–305.
3. Stull DM, Hollis LS, Gregory RE, Sheidler VR, Grossman SA. Pain in a comprehensive cancer center: more frequently due to treatment than underlying tumor (meeting abstract). *Proc Annu Meet Am Soc Clin Oncol.* 1996;15:1717. abstract.
4. Caraceni A, Portenoy RK. An international survey of cancer pain characteristics and syndromes. IASP task force on cancer pain. International Association for the Study of Pain. *Pain.* 1999;82(3):263–274.
5. Foley KM. Pain syndromes in patients with cancer. *Med Clin North Am.* 1987;71(2):169–184.
6. Banning A, Sjogren P, Henriksen H. Pain causes in 200 patients referred to a multidisciplinary cancer pain clinic. *Pain.* 1991;45(1):45–48.
7. Twycross R, Harcourt J, Bergl S. A survey of pain in patients with advanced cancer. *J Pain Symptom Manage.* 1996;12(5):273–282.
8. Daut RL, Cleeland CS. The prevalence and severity of pain in cancer. *Cancer.* 1982;50(9):1913–1918.
9. Grond S, Zech D, Diefenbach C, Radbruch L, Lehmann KA. Assessment of cancer pain: a prospective evaluation in 2266 cancer patients referred to a pain service. *Pain.* 1996;64(1):107–114.

10. Jimenez-Andrade JM, Mantyh WG, Bloom AP, et al. Bone cancer pain. *Ann N Y Acad Sci.* 2010;1198:173–181.

11. Gray MR, Leinster SJ, al-Janabi M, Critchley M, Pearce CJ. Osteomalacia mimicking metastases: a treatable cause of bone pain. *J R Coll Surg Edinb.* 1992;37(5):344–345.

12. Shane E, Parisien M, Henderson JE, et al. Tumor-induced osteomalacia: clinical and basic studies. *J Bone Miner Res.* 1997;12(9):1502–1511.

13. Jonsson KB, Zahradnik R, Larsson T, et al. Fibroblast growth factor 23 in oncogenic osteomalacia and X-linked hypophosphatemia. *N Engl J Med.* 2003;348(17):1656–1663.

14. Edmister KA, Sundaram M. Oncogenic osteomalacia. *Semin Musculoskelet Radiol.* 2002;6(3):191–196.

15. Jonsson OG, Sartain P, Ducore JM, Buchanan GR. Bone pain as an initial symptom of childhood acute lymphoblastic leukemia: association with nearly normal hematologic indexes. *J Pediatr.* 1990;117(2 Pt 1):233–237.

16. Wong KF, Chan JK, Ma SK. Solid tumour with initial presentation in the bone marrow—a clinicopathologic study of 25 adult cases. *Hematol Oncol.* 1993;11(1):35–42.

17. Hesselmann S, Micke O, Schaefer U, Willich N. Systemic mast cell disease (SMCD) and bone pain. A case treated with radiotherapy. *Strahlenther Onkol.* 2002;178(5):275–279.

18. Lin JT, Lachmann E, Nagler W. Low back pain and myalgias in acute and relapsed mast cell leukemia: a case report. *Arch Phys Med Rehabil.* 2002;83(6):860–863.

19. Golembe B, Ramsay NK, McKenna R, Nesbit ME, Krivit W. Localized bone marrow relapse in acute lymphoblastic leukemia. *Med Pediatr Oncol.* 1979;6(3):229–234.

20. Lembersky BC, Ratain MJ, Golomb HM. Skeletal complications in hairy cell leukemia: diagnosis and therapy. *J Clin Oncol.* 1988;6(8):1280–1284.

21. Beckers R, Uyttebroeck A, Demaerel P. Acute lymphoblastic leukaemia presenting with low back pain. *Eur J Paediatr Neurol.* 2002;6(5):285–287.

22. Cohen Y, Zidan J, McShan D. Bone marrow biopsy in solid cancer. *Acta Haematol.* 1982;68(1):14–19.

23. Kleinschmidt-Demasters BK. Diffuse bone marrow metastases from glioblastoma multiforme: the role of dural invasion. *Hum Pathol.* 1996;27(2):197–201.

24. Ollivier L, Gerber S, Vanel D, Brisse H, Leclere J. Improving the interpretation of bone marrow imaging in cancer patients. *Cancer Imaging.* 2006;6:194–198.

25. Constans JP, de Divitiis E, Donzelli R, Spaziante R, Meder JF, Haye C. Spinal metastases with neurological manifestations. Review of 600 cases. *J Neurosurg.* 1983;59(1):111–118.

26. Bilsky MH, Shannon FJ, Sheppard S, Prabhu V, Boland PJ. Diagnosis and management of a metastatic tumor in the atlantoaxial spine. *Spine.* 2002;27(10):1062–1069.

27. Nader R, Rhines LD, Mendel E. Metastatic sacral tumors. *Neurosurg Clin N Am.* 2004;15(4):453–457.

28. Payer M. Neurological manifestation of sacral tumors. *Neurosurg Focus.* 2003;15(2):E1.

29. Taylor JW, Schiff D. Metastatic epidural spinal cord compression. *Semin Neurol.* 2010;30:245–253.

30. Ruckdeschel JC. Early detection and treatment of spinal cord compression. *Oncology (Huntingt).* 2005;19(1):81–86. discussion 86, 89–92.

31. Greenberg HS, Kim JH, Posner JB. Epidural spinal cord compression from metastatic tumor: results with a new treatment protocol. *Ann Neurol.* 1980;8(4):361–366.

32. Stranjalis G, Jamjoom A, Torrens M. Epidural lipomatosis in steroid-treated patients. *Spine.* 1992;17(10):1268.

33. Sim FH. Metastatic bone disease of the pelvis and femur. *Instr Course Lect.* 1992;41:317–327.

34. Singh PC, Patel DV, Chang VT. Metastatic acetabular fractures: evaluation and approach to management. *J Pain Symptom Manage.* 2006;32(5):502–507.

35. Beatrous TE, Choyke PL, Frank JA. Diagnostic evaluation of cancer patients with pelvic pain: comparison of scintigraphy, CT, and MR imaging [see comments]. *AJR Am J Roentgenol.* 1990;155(1):85–88.

36. Mackey JR, Birchall I, Mac DN. Occult infection as a cause of hip pain in a patient with metastatic breast cancer. *J Pain Symptom Manage.* 1995;10(7):569–572.

37. Martinez-Lavin M. Hypertrophic osteoarthropathy. *Curr Opin Rheumatol.* 1997;9(1):83–86.

38. Greenfield GB, Schorsch HA, Shkolnik A. The various roentgen appearances of pulmonary hypertrophic osteoarthropathy. *Am J Roentgenol Radium Ther Nucl Med.* 1967;101(4):927–931.

39. Sharma OP. Symptoms and signs in pulmonary medicine: old observations and new interpretations. *Dis Mon.* 1995;41(9):577–638.

40. Hung GU, Kao CH, Lin WY, Wang SJ. Rapid resolution of hypertrophic pulmonary osteoarthropathy after resection of a lung mass caused by xanthogranulomatous inflammation.[in process citation]. *Clin Nucl Med.* 2000;25(12):1029–1030.

41. Kishi K, Nakamura H, Sudo A, et al. Tumor debulking by radiofrequency ablation in hypertrophic pulmonary osteoarthropathy associated with pulmonary carcinoma. *Lung Cancer.* 2002;38(3):317–320.

42. Amital H, Applbaum YH, Vasiliev L, Rubinow A. Hypertrophic pulmonary osteoarthropathy: control of pain and symptoms with pamidronate. *Clin Rheumatol.* 2004;23(4):330–332.

43. Suzuma T, Sakurai T, Yoshimura G, et al. Pamidronate-induced remission of pain associated with hypertrophic pulmonary osteoarthropathy in chemoendocrine therapy-refractory inoperable metastatic breast carcinoma. *Anticancer Drugs.* 2001;12(9):731–734.

44. Cope's SW. *Early Diagnosis of the Acute Abdomen.* 16th ed. New York: Oxford; 1983.

45. Coombs DW. Pain due to liver capsular distention. In: Ferrer-Brechner T, ed. *Common Problems in Pain Management.* Chicago: Year Book Medical Publishers; 1990:247–253.

46. Mulholland MW, Debas H, Bonica JJ. Diseases of the liver, biliary system and pancreas. In: Bonica JJ, ed. *The Management of Pain.* Philadelphia: Lea & Febiger; 1990:1214–1231.

47. De Conno F, Polastri D. Clinical features and symptomatic treatment of liver metastasis in the terminally ill patient. *Ann Ital Chir.* 1996;67(6):819–826.

48. Harris JN, Robinson P, Lawrance J, et al. Symptoms of colorectal liver metastases: correlation with CT findings. *Clin Oncol (R Coll Radiol).* 2003;15(2):78–82.

49. La Fianza A, Alberici E, Biasina AM, Preda L, Tateo S, Campani R. Spontaneous hemorrhage of a liver metastasis from squamous cell cervical carcinoma: case report and review of the literature. *Tumori.* 1999;85(4):290–293.

50. Grahm AL, Andren-Sandberg A. Prospective evaluation of pain in exocrine pancreatic cancer. *Digestion.* 1997;58(6):542–549.

51. Kelsen DP, Portenoy R, Thaler H, Tao Y, Brennan M. Pain as a predictor of outcome in patients with operable pancreatic carcinoma. *Surgery.* 1997;122(1):53–59.

52. Kelsen DP, Portenoy RK, Thaler HT, et al. Pain and depression in patients with newly diagnosed pancreas cancer. *J Clin Oncol.* 1995;13(3):748–755.

53. Sponseller PD. Evaluating the child with back pain. *Am Fam Physician.* 1996;54(6):1933–1941.

54. Neer RM, Ferrucci JT, Wang CA, Brennan M, Buttrick WF, Vickery AL. A 77-year-old man with epigastric pain, hypercalcemia, and a retroperitoneal mass. *N Engl J Med.* 1981;305(15):874–883.

55. Krane RJ, Perrone TL. A young man with testicular and abdominal pain. *N Engl J Med.* 1981;305(6):331–336.

56. Schonenberg P, Bastid C, Guedes J, Sahel J. Percutaneous echography-guided alcohol block of the celiac plexus as treatment of painful syndromes of the upper abdomen: study of 21 cases. *Schweiz Med Wochenschr.* 1991;121(15):528–531.

57. Zhu Z, Friess H, diMola FF, et al. Nerve growth factor expression correlates with Perineural invasion and pain in human pancreatic cancer. *J Clin Oncol.* 1999;17(8):2419.

58. Dang C, Zhang Y, Ma Q, Shimahara Y. Expression of nerve growth factor receptors is correlated with progression and prognosis of human pancreatic cancer. *J Gastroenterol Hepatol.* 2006;21(5):850–858.

59. Tham TC, Lichtenstein DR, Vandervoort J, et al. Pancreatic duct stents for "obstructive type" pain in pancreatic malignancy. *Am J Gastroenterol.* 2000;95(4):956–960.

60. Ripamonti C. Management of bowel obstruction in advanced cancer. *Curr Opin Oncol.* 1994;6(4):351–357.

61. Averbach AM, Sugarbaker PH. Recurrent intraabdominal cancer with intestinal obstruction. *Int Surg.* 1995;80(2):141–146.

62. Archer AG, Sugarbaker PH, Jelinek JS. Radiology of peritoneal carcinomatosis. *Cancer Treat Res.* 1996;82:263–288.

63. Stillman M. Perineal pain: diagnosis and management, with particular attention to perineal pain of cancer. In: Foley KM, Bonica JJ, Ventafrida V, eds. *Second International Congress on Cancer Pain.* New York: Raven Press; 1990:359–377.

64. Boas RA, Schug SA, Acland RH. Perineal pain after rectal amputation: a 5-year follow-up. *Pain.* 1993;52(1):67–70.

65. Miaskowski C. Special needs related to the pain and discomfort of patients with gynecologic cancer. *J Obstet Gynecol Neonatal Nurs.* 1996;25(2):181–188.

66. Hagen NA. Sharp, shooting neuropathic pain in the rectum or genitals: pudendal neuralgia. *J Pain Symptom Manage.* 1993;8(7):496–501.

67. Rigor BM. Pelvic cancer pain. *J Surg Oncol.* 2000;75(4):280–300.

68. Seefeld PH, Bargen JA. The spread of carcinoma of the rectum: invasion of lymphatics, veins and nerves. *Ann Surg.* 1943;118:76–90.

69. Sinaki M, Merritt JL, Stillwell GK. Tension myalgia of the pelvic floor. *Mayo Clin Proc.* 1977;52(11):717–722.

70. Berger MS, Cooley ME, Abrahm JL. A pain syndrome associated with large adrenal metastases in patients with lung cancer. *J Pain Symptom Manage.* 1995;10(2):161–166.

71. Karanikiotis C, Tentes AA, Markakidis S, Vafiadis K. Large bilateral adrenal metastases in non-small cell lung cancer. *World J Surg Oncol.* 2004;2(1):37.

72. Harrington KJ, Pandha HS, Kelly SA, Lambert HE, Jackson JE, Waxman J. Palliation of obstructive nephropathy due to malignancy. *Br J Urol.* 1995;76(1):101–107.

73. Russo P. Urologic emergencies in the cancer patient. *Semin Oncol.* 2000;27(3):284–298.

74. Little B, Ho KJ, Gawley S, Young M. Use of nephrostomy tubes in ureteric obstruction from incurable malignancy. *Int J Clin Pract.* 2003;57(3):180–181.

75. Portenoy RK, Kornblith AB, Wong G, et al. Pain in ovarian cancer patients. Prevalence, characteristics, and associated symptoms. *Cancer.* 1994;74(3):907–915.

76. Goff BA, Mandel LS, Melancon CH, Muntz HG. Frequency of symptoms of ovarian cancer in women presenting to primary care clinics. *JAMA.* 2004;291(22):2705–2712.

77. Webb PM, Purdie DM, Grover S, Jordan S, Dick ML, Green AC. Symptoms and diagnosis of borderline, early and advanced epithelial ovarian cancer. *Gynecol Oncol.* 2004;92(1):232–239.

78. Marino C, Zoppi M, Morelli F, Buoncristiano U, Pagni E. Pain in early cancer of the lungs. *Pain.* 1986;27(1):57–62.

79. Marangoni C, Lacerenza M, Formaglio F, Smirne S, Marchettini P. Sensory disorder of the chest as presenting symptom of lung cancer. *J Neurol Neurosurg Psychiatry.* 1993;56(9):1033–1034.

80. Sarlani E, Schwartz AH, Greenspan JD, Grace EG. Facial pain as first manifestation of lung cancer: a case of lung cancer-related cluster headache and a review of the literature. *J Orofac Pain.* 2003;17(3):262–267.

81. Miyamoto M, Sudo T, Kuyama T. Spontaneous rupture of hepatocellular carcinoma: a review of 172 Japanese cases. *Am J Gastroenterol.* 1991;86(1):67–71.

82. Marini P, Vilgrain V, Belghiti J. Management of spontaneous rupture of liver tumours. *Dig Surg.* 2002;19(2):109–113.

83. Wolff JM, Boeckmann W, Jakse G. Spontaneous kidney rupture due to a metastatic renal tumour. Case report Scand. *J Urol Nephrol.* 1994;28(4):415–417.

84. Rajagopal A, Ramasamy R, Martin J, Kumar P. Acute myeloid leukemia presenting as splenic rupture. *J Assoc Physicians India.* 2002;50:1435–1437.

85. Stamoulis JS, Antonopoulou Z, Safioleas M. Haemorrhagic shock from the spontaneous rupture of an adrenal cortical carcinoma. A case report. *Acta Chir Belg.* 2004;104(2):226–228.

86. Ende DA, Robinson G, Moulton J. Metastasis-induced perforated appendicitis: an acute abdomen of rare aetiology. *Aust N Z J Surg.* 1995;65(1):62–63.

87. Bossi P, Giusti R, Tarsitano A, et al. The point of pian in head and neck cancer. *Crit Rev Oncol Hematol.* 2019;138:51–59.

88. Morrison GA, Sterkers JM. Unusual presentations of acoustic tumours. *Clin Otolaryngol.* 1996;21(1):80–83.

89. Shapshay SM, Elber E, Strong MS. Occult tumors of the infratemporal fossa: report of seven cases appearing as preauricular facial pain. *Arch Otolaryngol.* 1976;102(9):535–538.

90. Hill BA, Kohut RI. Metastatic adenocarcinoma of the temporal bone. *Arch Otolaryngol.* 1976;102(9):568–571.

91. Leonetti JP, Li J, Smith PG. Otalgia. An isolated symptom of malignant infratemporal tumors. *Am J Otol.* 1998;19(4):496–498.

92. Scarbrough TJ, Day TA, Williams TE, Hardin JH, Aguero EG, Thomas Jr CR. Referred otalgia in head and neck cancer: a unifying schema. *Am J Clin Oncol.* 2003;26(5):E157–E162.

93. De Potter P. Ocular manifestations of cancer. *Curr Opin Ophthalmol.* 1998;9(6):100–104.

94. Weiss R, Grisold W, Jellinger K, Muhlbauer J, Scheiner W, Vesely M. Metastasis of solid tumors in extraocular muscles. *Acta Neuropathol (Berl).* 1984;65(2):168–171.

95. Friedman J, Karesh J, Rodrigues M, Sun CC. Thyroid carcinoma metastatic to the medial rectus muscle. *Ophthal Plast Reconstr Surg.* 1990;6(2):122–125.

96. Laitt RD, Kumar B, Leatherbarrow B, Bonshek RE, Jackson A. Cystic optic nerve meningioma presenting with acute proptosis. *Eye.* 1996;10(Pt 6):744–746.

97. Schroeder TL, Farlane DF, Goldberg LH. Pain as an atypical presentation of squamous cell carcinoma [in process citation]. *Dermatol Surg.* 1998;24(2):263–266.

98. Dulli DA, Levine RL, Chun RW, Dinndorf P. Migrainous neurologic dysfunction in Hodgkin's disease [letter]. *Arch Neurol.* 1987;44(7):689.

99. Miyazaki Y, Tajima Y, Sudo K, et al. Hodgkin's disease-related central nervous system angiopathy presenting as reversible posterior leukoencephalopathy. *Intern Med.* 2004;43(10):1005–1007.

100. Sigsbee B, Deck MD, Posner JB. Nonmetastatic superior sagittal sinus thrombosis complicating systemic cancer. *Neurology.* 1979;29(2):139–146.

101. Portenoy RK, Abissi CJ, Robbins JB. Increased intracranial pressure with normal ventricular size due to superior vena cava obstruction [letter]. *Arch Neurol.* 1983;40(9):598.

102. Marshall JA, Mahanna GK. Cancer in the differential diagnosis of orofacial pain. *Dent Clin N Am.* 1997;41(2):355–365.

103. Dworkin RH, Portenoy RK. Pain and its persistence in herpes zoster. *Pain.* 1996;67(2–3):241–251.

104. Rusthoven JJ, Ahlgren P, Elhakim T, et al. Varicella-zoster infection in adult cancer patients. A population study. *Arch Intern Med.* 1988;148(7):1561–1566.

105. Jaeckle KA. Nerve plexus metastases. *Neurol Clin.* 1991;9(4):857–866.

106. Daniels IR, Layer GT. Testicular tumours presenting as gynaecomastia. *Eur J Surg Oncol.* 2003;29(5):437–439.

107. Duparc C, Boissiere-Veverka G, Lefebvre H, et al. An oestrogen-producing seminoma responsible for gynaecomastia. *Horm Metab Res.* 2003;35(5):324–329.

108. Foppiani L, Bernasconi D, Del Monte P, Marugo A, Toncini C, Marugo M. Leydig cell tumour-induced bilateral gynaecomastia in a young man: endocrine abnormalities. *Andrologia.* 2005;37(1):36–39.

109. Forst T, Beyer J, Cordes U, et al. Gynaecomastia in a patient with a hCG producing giant cell carcinoma of the lung. Case report. *Exp Clin Endocrinol Diabetes.* 1995;103(1):28–32.

110. Wurzel RS, Yamase HT, Nieh PT. Ectopic production of human chorionic gonadotropin by poorly differentiated transitional cell tumors of the urinary tract. *J Urol.* 1987;137(3):502–504.

111. Liu G, Rosenfield Darling ML, Chan J, Jaklitsch MT, Skarin AT. Gynecomastia in a patient with lung cancer. *J Clin Oncol.* 1999;17(6):1956.

112. Tseng Jr A, Horning SJ, Freiha FS, Resser KJ, Hannigan Jr JF, Torti FM. Gynecomastia in testicular cancer patients. Prognostic and therapeutic implications. *Cancer.* 1985;56(10):2534–2538.

113. Cantwell BM, Richardson PG, Campbell SJ. Gynaecomastia and extragonadal symptoms leading to diagnosis delay of germ cell tumours in young men [see comments]. *Postgrad Med J.* 1991;67(789):675–677.

114. Mellor SG, McCutchan JD. Gynaecomastia and occult Leydig cell tumour of the testis. *Br J Urol.* 1989;63(4):420–422.

115. Haas GP, Pittaluga S, Gomella L, et al. Clinically occult Leydig cell tumor presenting with gynecomastia. *J Urol.* 1989;142(5):1325–1327.

116. Allen CM, Camisa C. Paraneoplastic pemphigus: a review of the literature. *Oral Dis.* 2000;6(4):208–214.

117. Camisa C, Helm TN, Liu YC, et al. Paraneoplastic pemphigus: a report of three cases including one long-term survivor. *J Am Acad Dermatol.* 1992;27(4):547–553.

118. DeCross AJ, Sahasrabudhe DM. Paraneoplastic Raynaud's phenomenon. *Am J Med.* 1992;92(5):571–572.

119. Borenstein A, Seidman DS, Ben-Ari GY. Raynaud's phenomenon as a presenting sign of metastatic melanoma. *Am J Med Sci.* 1990;300(1):41–42.

120. Portenoy RK, Ahmed E. Cancer pain syndromes. *Hematol Oncol Clin North Am.* 2018;32:371–386.

121. Ocean AJ, Vahdat LT. Chemotherapy-induced peripheral neuropathy: pathogenesis and emerging therapies. *Support Care Cancer.* 2004.

122. Spencer CM, Goa KL. Amifostine. A review of its pharmacodynamic and pharmacokinetic properties, and therapeutic potential as a radioprotector and cytotoxic chemoprotector. *Drugs.* 1995;50(6):1001–1031.

123. Penz M, Kornek GV, Raderer M, Ulrich-Pur H, Fiebiger W, Scheithauer W. Subcutaneous administration of amifostine: a promising therapeutic option in patients with oxaliplatin-related peripheral sensitive neuropathy. *Ann Oncol.* 2001;12(3):421–422.

124. Cascinu S, Catalano V, Cordella L, et al. Neuroprotective effect of reduced glutathione on oxaliplatin-based chemotherapy in advanced colorectal cancer: a randomized, double-blind, placebo-controlled trial. *J Clin Oncol.* 2002;20(16):3478–3483.

125. Cascinu S, Cordella L, Del Ferro E, Fronzoni M, Catalano G. Neuroprotective effect of reduced glutathione on cisplatin-based chemotherapy in advanced gastric cancer: a

126. Gamelin L, Boisdron-Celle M, Delva R, et al. Prevention of oxaliplatin-related neurotoxicity by calcium and magnesium infusions: a retrospective study of 161 patients receiving oxaliplatin combined with 5-fluorouracil and leucovorin for advanced colorectal cancer. *Clin Cancer Res.* 2004;10(12 Pt 1):4055–4061.

127. Argyriou AA, Chroni E, Koutras A, et al. Preventing paclitaxel-induced peripheral neuropathy: a phase II trial of vitamin E supplementation. *J Pain Symptom Manage.* 2006;32(3):237–244.

128. Virik K, Karapetis C, Droufakou S, Harper P. Avascular necrosis of bone: the hidden risk of glucocorticoids used as antiemetics in cancer chemotherapy. *Int J Clin Pract.* 2001;55(5):344–345.

129. Cook AM, Dzik-Jurasz AS, Padhani AR, Norman A, Huddart RA. The prevalence of avascular necrosis in patients treated with chemotherapy for testicular tumours. *Br J Cancer.* 2001;85(11):1624–1626.

130. Fink JC, Leisenring WM, Sullivan KM, Sherrard DJ, Weiss NS. Avascular necrosis following bone marrow transplantation: a case-control study. *Bone.* 1998;22(1):67–71.

131. DeSmet AA, Dalinka MK, Alazraki N, et al. Diagnostic imaging of avascular necrosis of the hip. American College of Radiology. ACR appropriateness criteria. *Radiology.* 2000;215(Suppl):247–254.

132. Castellanos AM, Glass JP, Yung WK. Regional nerve injury after intra-arterial chemotherapy. *Neurology.* 1987;37(5):834–837.

133. Kahn Jr CE, Messersmith RN, Samuels BL. Brachial plexopathy as a complication of intraarterial cisplatin chemotherapy. *Cardiovasc Intervent Radiol.* 1989;12(1):47–49.

134. Berger CC, Bokemeyer C, Schneider M, Kuczyk MA, Schmoll HJ. Secondary Raynaud's phenomenon and other late vascular complications following chemotherapy for testicular cancer. *Eur J Cancer.* 1995;31A(13–14):2229–2238.

135. Kukla LJ, McGuire WP, Lad T, Saltiel M. Acute vascular episodes associated with therapy for carcinomas of the upper aerodigestive tract with bleomycin, vincristine, and cisplatin. *Cancer Treat Rep.* 1982;66(2):369–370.

136. Hansen SW, Olsen N, Rossing N, Rorth M. Vascular toxicity and the mechanism underlying Raynaud's phenomenon in patients treated with cisplatin, vinblastine and bleomycin [see comments]. *Ann Oncol.* 1990;1(4):289–292.

137. McLeod DG, Iversen P. Gynecomastia in patients with prostate cancer: a review of treatment options. *Urology.* 2000;56(5):713–720.

138. Ramamurthy L, Cooper RA. Metastatic carcinoma to the male breast. *Br J Radiol.* 1991;64(759):277–278.

139. Olsson H, Alm P, Kristoffersson U, Landin-Olsson M. Hypophyseal tumor and gynecomastia preceding bilateral breast cancer development in a man. *Cancer.* 1984;53(9):1974–1977.

140. Maunsell E, Brisson J, Deschenes L. Arm problems and psychological distress after surgery for breast cancer. *Can J Surg.* 1993;36(4):315–320.

141. Hladiuk M, Huchcroft S, Temple W, Schnurr BE. Arm function after axillary dissection for breast cancer: a pilot study to provide parameter estimates. *J Surg Oncol.* 1992;50(1):47–52.

142. Vecht CJ. Arm pain in the patient with breast cancer. *J Pain Symptom Manage.* 1990;5(2):109–117.

143. Vecht CJ, Van de Brand HJ, Wajer OJ. Post-axillary dissection pain in breast cancer due to a lesion of the intercostobrachial nerve. *Pain.* 1989;38(2):171–176.

144. Kakuda JT, Stuntz M, Trivedi V, Klein SR, Vargas HI. Objective assessment of axillary morbidity in breast cancer treatment. *Am Surg.* 1999;65(10):995–998.

145. Keramopoulos A, Tsionou C, Minaretzis D, Michalas S, Aravantinos D. Arm morbidity following treatment of breast cancer with total axillary dissection: a multivariated approach. *Oncology.* 1993;50(6):445–449.

146. Kuehn T, Klauss W, Darsow M, et al. Long-term morbidity following axillary dissection in breast cancer patients—clinical assessment, significance for life quality and the impact of demographic, oncologic and therapeutic factors. *Breast Cancer Res Treat.* 2000;64(3):275–286.

147. Warmuth MA, Bowen G, Prosnitz LR, et al. Complications of axillary lymph node dissection for carcinoma of the breast: a report based on a patient survey. *Cancer.* 1998;83(7):1362–1368.

148. Tasmuth T, von Smitten K, Kalso E. Pain and other symptoms during the first year after radical and conservative surgery for breast cancer. *Br J Cancer.* 1996;74(12):2024–2031.

149. Tasmuth T, von SK, Hietanen P, Kataja M, Kalso E. Pain and other symptoms after different treatment modalities of breast cancer. *Ann Oncol.* 1995;6(5):453–459.

150. Smith WC, Bourne D, Squair J, Phillips DO, Alastair Chambers W. A retrospective cohort study of post mastectomy pain syndrome. *Pain.* 1999;83(1):91–95.

151. Carpenter JS, Andrykowski MA, Sloan P, et al. Postmastectomy/postlumpectomy pain in breast cancer survivors. *J Clin Epidemiol.* 1998;51(12):1285–1292.

152. Wood KM. Intercostobrachial nerve entrapment syndrome. *South Med J.* 1978;71(6):662–663.

153. Paredes JP, Puente JL, Potel J. Variations in sensitivity after sectioning the intercostobrachial nerve. *Am J Surg.* 1990;160(5):525–528.

154. van Dam MS, Hennipman A, de Kruif JT, van der Tweel I, de Graaf PW. Complications following axillary dissection for breast carcinoma (see comments). *Ned Tijdschr Geneeskd.* 1993;137(46):2395–2398.

155. Granek I, Ashikari R, Foley KM. Postmastectomy pain syndrome: clinical and anatomic correlates. *Proc Am Soc Clin Oncol.* 1983;3:122. Abstract.

156. International Association for the Study of Pain. Subcommittee on taxonomy. Classification of chronic pain. *Pain.* 1986;3(suppl):135–138.

157. Ernst MF, Voogd AC, Balder W, Klinkenbijl JH, Roukema JA. Early and late morbidity associated with axillary levels I-III dissection in breast cancer. *J Surg Oncol.* 2002;79(3):151–155 [discussion 156].

158. Stevens PE, Dibble SL, Miaskowski C. Prevalence, characteristics, and impact of postmastectomy pain syndrome: an investigation of women's experiences. *Pain.* 1995;61(1):61–68.

159. Macdonald L, Bruce J, Scott NW, Smith WC, Chambers WA. Long-term follow-up of breast cancer survivors with postmastectomy pain syndrome. *Br J Cancer.* 2005;92(2):225–230.

160. Bratschi HU, Haller U. Significance of the intercostobrachial nerve in axillary lymph node excision. *Geburtshilfe Frauenheilkd.* 1990;50(9):689–693.

161. Assa J. The intercostobrachial nerve in radical mastectomy. *J Surg Oncol.* 1974;6(2):123–126.

162. Blunt C, Schmiedel A. Some cases of severe post-mastectomy pain syndrome may be caused by an axillary haematoma. *Pain.* 2004;108(3):294–296.

163. Hack TF, Cohen L, Katz J, Robson LS, Goss P. Physical and psychological morbidity after axillary lymph node dissection for breast cancer. *J Clin Oncol.* 1999;17(1):143–149.

164. Johansson S, Svensson H, Larsson LG, Denekamp J. Brachial plexopathy after postoperative radiotherapy of breast cancer patients—a long-term follow-up. *Acta Oncol.* 2000;39(3):373–382.

165. Torresan RZ, Cabello C, Conde DM, Brenelli HB. Impact of the preservation of the intercostobrachial nerve in axillary lymphadenectomy due to breast cancer. *Breast J.* 2003;9(5):389–392.

166. Temple WJ, Ketcham AS. Preservation of the intercostobrachial nerve during axillary dissection for breast cancer. *Am J Surg.* 1985;150(5):585–588.

167. Salmon RJ, Ansquer Y, Asselain B. Preservation versus section of intercostal-brachial nerve (IBN) in axillary dissection for breast cancer—a prospective randomized trial. *Eur J Surg Oncol.* 1998;24(3):158–161.

168. Moskovitz AH, Anderson BO, Yeung RS, Byrd DR, Lawton TJ, Moe RE. Axillary web syndrome after axillary dissection. *Am J Surg.* 2001;181(5):434–439.

169. Hughes LL, Styblo TM, Thoms WW, et al. Cellulitis of the breast as a complication of breast-conserving surgery and irradiation. *Am J Clin Oncol.* 1997;20(4):338–341.

170. Dijkstra PU, van Wilgen PC, Buijs RP, et al. Incidence of shoulder pain after neck dissection: a clinical explorative study for risk factors. *Head Neck.* 2001;23(11):947–953.

171. van Wilgen CP, Dijkstra PU, van der Laan BF, Plukker JT, Roodenburg JL. Shoulder complaints after neck dissection; is the spinal accessory nerve involved? *Br J Oral Maxillofac Surg.* 2003;41(1):7–11.

172. Talmi YP, Horowitz Z, Pfeffer MR, et al. Pain in the neck after neck dissection. *Otolaryngol Head Neck Surg.* 2000;123(3):302–306.

173. Swift TR, Nichols FT. The droopy shoulder syndrome. *Neurology.* 1984;34(2):212–215.

174. Brown H, Burns S, Kaiser CW. The spinal accessory nerve plexus, the trapezius muscle, and shoulder stabilization after radical neck cancer surgery. *Ann Surg.* 1988;208(5):654–661.

175. Bruera E, Mac Donald N. Intractable pain in patients with advanced head and neck tumors: a possible role of local infection. *Cancer Treat Rep.* 1986;70(5):691–692.

176. Coyle N, Portenoy RK. Infection as a cause of rapidly increasing pain in cancer patients. *J Pain Symptom Manage.* 1991;6(4):266–269.

177. Keller SM, Carp NZ, Levy MN, Rosen SM. Chronic post thoracotomy pain. *J Cardiovasc Surg (Torino).* 1994;35(6 Suppl 1):161–164.

178. Kanner R, Martini N, Foley KM. Nature and incidence of post-thoracotomy pain. *Proc Am Soc Clin Oncol.* 1982;1:590. Abstract.

179. Hamada H, Moriwaki K, Shiroyama K, Tanaka H, Kawamoto M, Yuge O. Myofascial pain in patients with postthoracotomy pain syndrome. *Reg Anesth Pain Med.* 2000;25(3):302–305.

180. Collins KL, Russell HG, Schumacher PJ, et al. A review of current theories and treatments for phantom limb pain. *J Clin Investig.* 2018;128:2168–2176.

181. Ehde DM, Czerniecki JM, Smith DG, et al. Chronic phantom sensations, phantom pain, residual limb pain, and other regional pain after lower limb amputation. *Arch Phys Med Rehabil.* 2000;81(8):1039–1044.

182. Flor H. Phantom-limb pain: characteristics, causes, and treatment. *Lancet Neurol.* 2002;1(3):182–189.

183. Nikolajsen L, Ilkjaer S, Kroner K, Christensen JH, Jensen TS. The influence of preamputation pain on postamputation stump and phantom pain. *Pain.* 1997;72(3):393–405.

184. Weinstein SM. Phantom pain. *Oncology (Huntingt).* 1994;8(3):65–70. discussion 70, 73–4.

185. Katz J, Melzack R. Pain 'memories' in phantom limbs: review and clinical observations. *Pain.* 1990;43(3):319–336.

186. Smith J, Thompson JM. Phantom limb pain and chemotherapy in pediatric amputees. *Mayo Clin Proc.* 1995;70(4):357–364.

187. Enneking FK, Morey TE. Continuous postoperative infusion of a regional anesthetic after an amputation of the lower extremity. A randomized clinical trial [letter]. *J Bone Joint Surg Am.* 1997;79(11):1752–1753.

188. Katz J. Prevention of phantom limb pain by regional anaesthesia. *Lancet.* 1997;349(9051):519–520.

189. Nikolajsen L, Ilkjaer S, Christensen JH, Kroner K, Jensen TS. Randomised trial of epidural bupivacaine and morphine in prevention of stump and phantom pain in lower-limb amputation [see comments]. *Lancet.* 1997;350(9088):1353–1357.

190. Pavy TJ, Doyle DL. Prevention of phantom limb pain by infusion of local anaesthetic into the sciatic nerve [see comments]. *Anaesth Intensive Care.* 1996;24(5):599–600.

191. Chang VT, Tunkel RS, Pattillo BA, Lachmann EA. Increased phantom limb pain as an initial symptom of spinal-neoplasia [published erratum appears in J pain symptom manage 1997 Sep; 14(3):135]. *J Pain Symptom Manage.* 1997;13(6):362–364.

192. Kwekkeboom K. Postmastectomy pain syndromes. *Cancer Nurs.* 1996;19(1):37–43.

193. Kroner K, Krebs B, Skov J, Jorgensen HS. Immediate and long-term phantom breast syndrome after mastectomy: incidence, clinical characteristics and relationship to pre-mastectomy breast pain. *Pain.* 1989;36(3):327–334.

194. Rothemund Y, Grusser SM, Liebeskind U, Schlag PM, Flor H. Phantom phenomena in mastectomized patients and their relation to chronic and acute pre-mastectomy pain. *Pain.* 2004;107(1–2):140–146.

195. Ovesen P, Kroner K, Ornsholt J, Bach K. Phantom-related phenomena after rectal amputation: prevalence and clinical characteristics. *Pain.* 1991;44(3):289–291.

196. Davis RW. Phantom sensation, phantom pain, and stump pain. *Arch Phys Med Rehabil.* 1993;74(1):79–91.

197. Yeoh EK, Horowitz M. Radiation enteritis. *Surg Gynecol Obstet.* 1987;165(4):373–379.

198. Nussbaum ML, Campana TJ, Weese JL. Radiation-induced intestinal injury. *Clin Plast Surg.* 1993;20(3):573–580.

199. Joly F, Brune D, Couette JE, et al. Health-related quality of life and sequelae in patients treated with brachytherapy and external beam irradiation for localized prostate cancer. *Ann Oncol.* 1998;9(7):751–757.

200. Pillay PK, Teh M, Chua EJ, Tan EC, Tung KH, Foo KT. Haemorrhagic chronic radiation cystitis—following treatment of pelvic malignancies. *Ann Acad Med Singapore.* 1984;13(4):634–638.

201. Perez CA, Grigsby PW, Lockett MA, Chao KS, Williamson J. Radiation therapy morbidity in carcinoma of the uterine cervix: dosimetric and clinical correlation. *Int J Radiat Oncol Biol Phys.* 1999;44(4):855–866.

202. Newman ML, Brennan M, Passik S. Lymphedema complicated by pain and psychological distress: a case with complex treatment needs. *J Pain Symptom Manage.* 1996;12(6):376–379.

203. McWayne J, Heiney SP. Psychologic and social sequelae of secondary lymphedema. *Cancer.* 2005.

204. Minsky BD, Cohen AM. Minimizing the toxicity of pelvic radiation therapy in rectal cancer. *Oncology (Huntingt).* 1988;2(8):21–25. 28–9.

205. Mannaerts GH, Rutten HJ, Martijn H, Hanssens PE, Wiggers T. Effects on functional outcome after IORT-containing multimodality treatment for locally advanced primary and locally recurrent rectal cancer. *Int J Radiat Oncol Biol Phys.* 2002;54(4):1082–1088.

206. Wallner K, Elliott K, Merrick G, Ghaly M, Maki J. Chronic pelvic pain following prostate brachytherapy: a case report. *Brachytherapy.* 2004;3(3):153–158.

207. Epstein JB, Rea G, Wong FL, Spinelli J, Stevenson-Moore P. Osteonecrosis: study of the relationship of dental extractions in patients receiving radiotherapy. *Head Neck Surg.* 1987;10(1):48–54.

208. Epstein J, van der Meij E, McKenzie M, Wong F, Lepawsky M, Stevenson-Moore P. Postradiation osteonecrosis of the mandible: a long-term follow-up study. *Oral Surg Oral Med Oral Pathol Oral Radiol Endod.* 1997;83(6):657–662.

209. Von Roenn JH, Cleeland CS, Gonin R, Hatfield AK, Pandya KJ. Physician attitudes and practice in cancer pain management. A survey from the eastern cooperative oncology group. *Ann Intern Med.* 1993;119(2):121–126.

210. Gonzales GR, Elliott KJ, Portenoy RK, Foley KM. The impact of a comprehensive evaluation in the management of cancer pain. *Pain.* 1991;47(2):141–144.

211. Coyle N, Breitbart W, Weaver S, Portenoy R. Delirium as a contributing factor to "crescendo" pain: three case reports. *J Pain Symptom Manage.* 1994;9(1):44–47.

212. Feldman HA, McKinlay JB, Potter DA, et al. Nonmedical influences on medical decision making: an experimental technique using videotapes, factorial design, and survey sampling. *Health Serv Res.* 1997;32(3):343–366.

213. Christakis NA, Asch DA. Biases in how physicians choose to withdraw life support. *Lancet.* 1993;342(8872):642–646.

214. Hoskin PJ. Radiotherapy for bone pain. *Pain.* 1995;63(2):137–139.

215. Bates T. A review of local radiotherapy in the treatment of bone metastases and cord compression. *Int J Radiat Oncol Biol Phys.* 1992;23(1):217–221.

216. Janjan NA. Radiation for bone metastases: conventional techniques and the role of systemic radiopharmaceuticals. *Cancer.* 1997;80(8 Suppl):1628–1645.

217. Vermeulen SS. Whole brain radiotherapy in the treatment of metastatic brain tumors. *Semin Surg Oncol.* 1998;14(1):64–69.

218. Thiagarajan A, Yamada Y. Radiobiology and radiotherapy of brain metastases. *Clin Exp Metastasis.* 2017;34:411–419.

219. Wang TJC, Brown PD. Brain metastases: fractionated whole-brain radiotherapy. *Handb Clin Neurol.* 2018;149:123–127.

220. Dobrowsky W, Schmid AP. Radiotherapy of presacral recurrence following radical surgery for rectal carcinoma. *Dis Colon Rectum.* 1985;28(12):917–919.

221. Bosch A, Caldwell WL. Palliative radiotherapy in the patient with metastatic and advanced incurable cancer. *Wis Med J.* 1980;79(4):19–21.

222. Leibel SA, Pajak TF, Massullo V, et al. A comparison of misonidazole sensitized radiation therapy to radiation therapy alone for the palliation of hepatic metastases: results of a radiation therapy oncology group randomized prospective trial. *Int J Radiat Oncol Biol Phys.* 1987;13(7):1057–1064.

223. Mohiuddin M, Chen E, Ahmad N. Combined liver radiation and chemotherapy for palliation of hepatic metastases from colorectal cancer. *J Clin Oncol.* 1996;14(3):722–728.

224. Sherman DM, Weichselbaum R, Order SE, Cloud L, Trey C, Piro AJ. Palliation of hepatic metastasis. *Cancer.* 1978;41(5):2013–2017.

225. Turek-Maischeider M, Kazem I. Palliative irradiation for liver metastases. *JAMA.* 1975;232(6):625–628.

226. Borgelt BB, Gelber R, Brady LW, Griffin T, Hendrickson FR. The palliation of hepatic metastases: results of the radiation therapy oncology group pilot study. *Int J Radiat Oncol Biol Phys.* 1981;7(5):587–591.

227. Patt YZ, Peters RE, Chuang VP, Wallace S, Claghorn L, Mavligit G. Palliation of pelvic recurrence of colorectal cancer with intraarterial 5-fluorouracil and mitomycin. *Cancer.* 1985;56(9):2175–2180.

228. Rothenberg ML. New developments in chemotherapy for patients with advanced pancreatic cancer. *Oncology (Huntingt).* 1996;10(9 Suppl):18–22.

229. Thatcher N, Anderson H, Betticher DC, Ranson M. Symptomatic benefit from gemcitabine and other chemotherapy in advanced non-small cell lung cancer: changes in performance status and tumour-related symptoms. *Anticancer Drugs.* 1995;6(Suppl 6):39–48.

230. Cattell E, Arance A, Middleton M. Assessing outcomes in palliative chemotherapy. *Expert Opin Pharmacother.* 2002;3(6):693–700.

231. Pothuri B, Vaidya A, Aghajanian C, Venkatraman E, Barakat RR, Chi DS. Palliative surgery for bowel obstruction in recurrent ovarian cancer: an updated series. *Gynecol Oncol.* 2003;89(2):306–313.

232. Miner TJ, Jaques DP, Shriver CD. A prospective evaluation of patients undergoing surgery for the palliation of an advanced malignancy. *Ann Surg Oncol.* 2002;9(7):696–703.

233. Ward WG, Holsenbeck S, Dorey FJ, Spang J, Howe D. Metastatic disease of the femur: surgical treatment. *Clin Orthop Relat Res.* 2003;415 Suppl:S230–S244.

234. Frassica FJ, Frassica DA. Evaluation and treatment of metastases to the humerus. *Clin Orthop.* 2003;415(Suppl):S212–S218.

235. Katzer A, Meenen NM, Grabbe F, Rueger JM. Surgery of skeletal metastases. *Arch Orthop Trauma Surg.* 2002;122(5):251–258.

236. Abrahm JL. Assessment and treatment of patients with malignant spinal cord compression. *J Support Oncol.* 2004;2(5):377–388. 391; discussion 391–3, 398, 401.

237. Yen D, Kuriachan V, Yach J, Howard A. Long-term outcome of anterior decompression and spinal fixation after placement of the Wellesley Wedge for thoracic and lumbar spinal metastasis. *J Neurosurg.* 2002;96(1 Suppl):6–9.

238. Paz IB. Major palliative amputations. *Surg Oncol Clin N Am.* 2004;13(3):543–547. x.

239. Wittig JC, Bickels J, Kollender Y, Kellar-Graney KL, Meller I, Malawer MM. Palliative forequarter amputation for metastatic carcinoma to the shoulder girdle region: indications, preoperative evaluation, surgical technique, and results. *J Surg Oncol.* 2001;77(2):105–113 [discussion 114].

240. Lopez MR, Stock JA, Gump FE, Rosen JS. Carcinoma of the breast metastatic to the ureter presenting with flank pain and recurrent urinary tract infection. *Am Surg.* 1996;62(9):748–752.

241. Neeman Z, Wood BJ. Radiofrequency ablation beyond the liver. *Tech Vasc Interv Radiol.* 2002;5(3):156–163.

242. Campos FG, Habr-Gama A, Kiss DR, Leite AF, Seid V, Gama-Rodrigues J. Management of the pelvic recurrence of rectal cancer with radiofrequency thermoablation: a case report and review of the literature. *Int J Colorectal Dis.* 2005;20(1):62–66.

243. Ohhigashi S, Nishio T, Watanabe F, Matsusako M. Experience with radiofrequency ablation in the treatment of pelvic recurrence in rectal cancer: report of two cases. *Dis Colon Rectum.* 2001;44(5):741–745.

244. Cioni R, Armillotta N, Bargellini I, et al. CT-guided radiofrequency ablation of osteoid osteoma: long-term results. *Eur Radiol.* 2004;14(7):1203–1208.

245. Varshney S, Sharma S, Pamecha V, et al. Radiofrequency tissue ablation: an early Indian experience. *Indian J Gastroenterol.* 2003;22(3):91–93.

246. Mirza AN, Fornage BD, Sneige N, et al. Radiofrequency ablation of solid tumors. *Cancer J.* 2001;7(2):95–102.

247. Rohde D, Albers C, Mahnken A, Tacke J. Regional thermoablation of local or metastatic renal cell carcinoma. *Oncol Rep.* 2003;10(3):753–757.

248. Goetz MP, Callstrom MR, Charboneau JW, et al. Percutaneous image-guided radiofrequency ablation of painful metastases involving bone: a multicenter study. *J Clin Oncol.* 2004;22(2):300–306.

249. Posteraro AF, Dupuy DE, Mayo-Smith WW. Radiofrequency ablation of bony metastatic disease. *Clin Radiol.* 2004;59(9):803–811.

250. Gronemeyer DH, Schirp S, Gevargez A. Image-guided radiofrequency ablation of spinal tumors: preliminary experience with an expandable array electrode. *Cancer J.* 2002;8(1):33–39.

251. Fourney DR, Schomer DF, Nader R, et al. Percutaneous vertebroplasty and kyphoplasty for painful vertebral body fractures in cancer patients. *J Neurosurg.* 2003;98(1 Suppl):21–30.

252. Masala S, Lunardi P, Fiori R, et al. Vertebroplasty and kyphoplasty in the treatment of malignant vertebral fractures. *J Chemother.* 2004;16(Suppl 5):30–33.

253. Bercovitch M, Adunsky A. Patterns of high-dose morphine use in a home-care hospice service: should we be afraid of it? *Cancer.* 2004;101(6):1473–1477.

254. Scarborough B, Smith CB. Optimal pain management for patients with cancer in the modern era. *CA Cancer J Clin.* 2018;68:182–196.

255. Wood H, Dickman A, Star A, Boland JW. Updates in palliative care—overview and recent advancements in the pharmacological management of cancer pain. *Clin Med.* 2018;18:17–22.

256. Boisvert M, Cohen SR. Opioid use in advanced malignant disease: why do different centers use vastly different doses? A plea for standardized reporting. *J Pain Symptom Manage.* 1995;10(8):632–638.

257. Subramanian AV, Yehya AHS, Oon CE. Molecular basis of cancer pain management: an updated review. *Medicina.* 2019;55:584. https://doi.org/10.3390/medicina55090584.

258. Collins JJ, Grier HE, Kinney HC, Berde CB. Control of severe pain in children with terminal malignancy. *J Pediatr.* 1995;126(4):653–657.

259. Fulton JS, Johnson GB. Using high-dose morphine to relieve cancer pain. *Nursing.* 1993;23(2):34–39.

260. Fitzgibbon DR, Ready LB. Intravenous high-dose methadone administered by patient controlled analgesia and continuous infusion for the treatment of cancer pain refractory to high-dose morphine. *Pain.* 1997;73(2):259–261.

261. Lilley LL, Guanci R. Using high-dose fentanyl patches. *Am J Nurs.* 1996;96(7):18–20. 22.

262. Radbruch L, Grond S, Zech DJ, Bischoff A. High-dose oral morphine in cancer pain management: a report of twelve cases. *J Clin Anesth.* 1996;8(2):144–150.

263. Brescia FJ, Portenoy RK, Ryan M, Krasnoff L, Gray G. Pain, opioid use, and survival in hospitalized patients with advanced cancer. *J Clin Oncol.* 1992;10(1):149–155.

264. Kanner RM, Foley KM. Patterns of narcotic drug use in a cancer pain clinic. *Ann N Y Acad Sci.* 1981;362:161–172.

265. Schug SA, Zech D, Grond S, Jung H, Meuser T, Stobbe B. A long-term survey of morphine in cancer pain patients. *J Pain Symptom Manage.* 1992;7(5):259–266.

266. Portenoy RK. Tolerance to opioid analgesics: clinical aspects. *Cancer Surv.* 1994;21:49–65.

267. Collin E, Poulain P, Gauvain-Piquard A, Petit G, Pichard-Leandri E. Is disease progression the major factor in morphine 'tolerance' in cancer pain treatment? *Pain.* 1993;55(3):319–326.

268. Paice JA. The phenomenon of analgesic tolerance in cancer pain management. *Oncol Nurs Forum.* 1988;15(4):455–460.

269. Cherny N, Ripamonti C, Pereira J, et al. Strategies to manage the adverse effects of oral morphine: an evidence-based report. *J Clin Oncol.* 2001;19(9):2542–2554.

270. Fallon MT, B ON. Substitution of another opioid for morphine. Opioid toxicity should be managed initially by decreasing the opioid dose [letter; comment]. *BMJ.* 1998;317(7150):81.

271. Lawlor P, Turner K, Hanson J, Bruera E. Dose ratio between morphine and hydromorphone in patients with cancer pain: a retrospective study. *Pain.* 1997;72(1–2):79–85.

272. Bruera E, Pereira J, Watanabe S, Belzile M, Kuehn N, Hanson J. Opioid rotation in patients with cancer pain. A retrospective comparison of dose ratios between methadone, hydromorphone, and morphine. *Cancer.* 1996;78(4):852–857.

273. Bruera E, Franco JJ, Maltoni M, Watanabe S, Suarez-Almazor M. Changing pattern of agitated impaired mental status in patients with advanced cancer: association with cognitive monitoring, hydration, and opioid rotation. *J Pain Symptom Manage.* 1995;10(4):287–291.

274. Cherny NJ, Chang V, Frager G, et al. Opioid pharmacotherapy in the management of cancer pain: a survey of strategies used by pain physicians for the selection of analgesic drugs and routes of administration. *Cancer.* 1995;76(7):1283–1293.

275. de Stoutz ND, Bruera E, Suarez-Almazor M. Opioid rotation for toxicity reduction in terminal cancer patients. *J Pain Symptom Manage.* 1995;10(5):378–384.

276. Thomas Z, Bruera E. Use of methadone in a highly tolerant patient receiving parenteral hydromorphone. *J Pain Symptom Manage.* 1995;10(4):315–317.

277. Galer BS, Coyle N, Pasternak GW, Portenoy RK. Individual variability in the response to different opioids: report of five cases. *Pain.* 1992;49(1):87–91.

278. Ripamonti C, Groff L, Brunelli C, Polastri D, Stavrakis A, De Conno F. Switching from morphine to oral methadone in treating cancer pain: what is the equianalgesic dose ratio? [see comments]. *J Clin Oncol.* 1998;16(10):3216–3221.

279. Maddocks I, Somogyi A, Abbott F, Hayball P, Parker D. Attenuation of morphine-induced delirium in palliative care by substitution with infusion of oxycodone. *J Pain Symptom Manage.* 1996;12(3):182–189.

280. Vigano A, Fan D, Bruera E. Individualized use of methadone and opioid rotation in the comprehensive management of cancer pain associated with poor prognostic indicators. *Pain.* 1996;67(1):115–119.

281. Paix A, Coleman A, Lees J, et al. Subcutaneous fentanyl and sufentanil infusion substitution for morphine intolerance in cancer pain management. *Pain*. 1995;63(2):263–269.

282. Ashby MA, Martin P, Jackson KA. Opioid substitution to reduce adverse effects in cancer pain management. *Med J Aust*. 1999;170(2):68–71.

283. Hagen N, Swanson R. Strychnine-like multifocal myoclonus and seizures in extremely high- dose opioid administration: treatment strategies [see comments]. *J Pain Symptom Manage*. 1997;14(1):51–58.

284. Makin MK, Ellershaw JE. Substitution of another opioid for morphine. Methadone can be used to manage neuropathic pain related to cancer [letter; comment]. *BMJ*. 1998;317(7150):81.

285. Portenoy RK. Adjuvant analgesic agents. *Hematol Oncol Clin North Am*. 1996;10(1):103–119.

286. Watanabe S, Bruera E. Corticosteroids as adjuvant analgesics. *J Pain Symptom Manage*. 1994;9(7):442–445.

287. Rousseau P. The palliative use of high-dose corticosteroids in three terminally ill patients with pain. *Am J Hosp Palliat Care*. 2001;18(5):343–346.

288. Devor M, Govrin-Lippmann R, Raber P. Corticosteroids suppress ectopic neural discharge originating in experimental neuromas. *Pain*. 1985;22(2):127–137.

289. Sorensen S, Helweg-Larsen S, Mouridsen H, Hansen HH. Effect of high-dose dexamethasone in carcinomatous metastatic spinal cord compression treated with radiotherapy: a randomised trial. *Eur J Cancer*. 1994;30A(1):22–27.

290. Twycross R. The risks and benefits of corticosteroids in advanced cancer. *Drug Saf*. 1994;11(3):163–178.

291. McQuay HJ, Tramer M, Nye BA, Carroll D, Wiffen PJ, Moore RA. A systematic review of antidepressants in neuropathic pain. *Pain*. 1996;68(2–3):217–227.

292. Mattia C, Paoletti F, Coluzzi F, Boanelli A. New antidepressants in the treatment of neuropathic pain. A review. *Minerva Anestesiol*. 2002;68(3):105–114.

293. Bennett MI, Simpson KH. Gabapentin in the treatment of neuropathic pain. *Palliat Med*. 2004;18(1):5–11.

294. Galer BS, Harle J, Rowbotham MC. Response to intravenous lidocaine infusion predicts subsequent response to oral mexiletine: a prospective study. *J Pain Symptom Manage*. 1996;12(3):161–167.

295. Lipman AG. Analgesic drugs for neuropathic and sympathetically maintained pain. *Clin Geriatr Med*. 1996;12(3):501–515.

296. McNicol E, Strassels S, Goudas L, Lau J, Carr D. Nonsteroidal anti-inflammatory drugs, alone or combined with opioids, for cancer pain: a systematic review. *J Clin Oncol*. 2004;22(10):1975–1992.

297. Wong R, Wiffen PJ. Bisphosphonates for the relief of pain secondary to bone metastases (Cochrane review). *Cochrane Database Syst Rev*. 2002;(2), CD002068.

298. Strang P. Analgesic effect of bisphosphonates on bone pain in breast cancer patients: a review article. *Acta Oncol*. 1996;5(50):50–54.

299. Li EC, Davis LE. Zoledronic acid: a new parenteral bisphosphonate. *Clin Ther*. 2003;25(11):2669–2708.

300. Menssen HD, Sakalova A, Fontana A, et al. Effects of long-term intravenous ibandronate therapy on skeletal-related events, survival, and bone resorption markers in patients with advanced multiple myeloma. *J Clin Oncol*. 2002;20(9):2353–2359.

301. Tripathy D, Lichinitzer M, Lazarev A, et al. Oral ibandronate for the treatment of metastatic bone disease in breast cancer: efficacy and safety results from a randomized, double-blind, placebo-controlled trial. *Ann Oncol*. 2004;15(5):743–750.

302. Robinson RG, Preston DF, Schiefelbein M, Baxter KG. Strontium 89 therapy for the palliation of pain due to osseous metastases. *JAMA*. 1995;274(5):420–424.

303. Gunawardana DH, Lichtenstein M, Better N, Rosenthal M. Results of strontium-89 therapy in patients with prostate cancer resistant to chemotherapy. *Clin Nucl Med*. 2004;29(2):81–85.

304. Schmeler K, Bastin K. Strontium-89 for symptomatic metastatic prostate cancer to bone: recommendations for hospice patients. *Hosp J*. 1996;11(2):1–10.

305. Wang RF, Zhang CL, Zhu SL, Zhu M. A comparative study of samarium-153-ethylenediaminetetramethylene phosphonic acid with pamidronate disodium in the treatment of patients with painful metastatic bone cancer. *Med Princ Pract*. 2003;12(2):97–101.

306. Sartor O, Reid RH, Hoskin PJ, et al. Samarium-153-lexidronam complex for treatment of painful bone metastases in hormone-refractory prostate cancer. *Urology*. 2004;63(5):940–945.

307. Palmedo H, Manka-Waluch A, Albers P, et al. Repeated bone-targeted therapy for hormone-refractory prostate carcinoma: randomized phase II trial with the new, high-energy radiopharmaceutical rhenium-188 hydroxyethylidenediphosphonate. *J Clin Oncol*. 2003;21(15):2869–2875.

308. Paulson DF. Oxybutynin chloride in control of post-transurethral vesical pain and spasm. *Urology*. 1978;11(3):237–238.

309. Baert L. Controlled double-blind trail of flavoxate in painful conditions of the lower urinary tract. *Curr Med Res Opin*. 1974;2(10):631–635.

310. Milani R, Scalambrino S, Carrera S, Pezzoli P, Ruffmann R. Flavoxate hydrochloride for urinary urgency after pelvic radiotherapy: comparison of 600 mg versus 1200 mg daily dosages. *J Int Med Res*. 1988;16(1):71–74.

311. Abrams P, Fenely R. The action of prostaglandins on smooth muscle of the human urinary tract in vitro. *Br J Urol*. 1976;47:909–915.

312. Lazzeri M, Beneforti P, Benaim G, Maggi CA, Lecci A, Turini D. Intravesical capsaicin for treatment of severe bladder pain: a randomized placebo controlled study. *J Urol*. 1996;156(3):947–952.

313. Barbanti G, Maggi CA, Beneforti P, Baroldi P, Turini D. Relief of pain following intravesical capsaicin in patients with hypersensitive disorders of the lower urinary tract. *Br J Urol*. 1993;71(6):686–691.

314. Eckardt VF, Dodt O, Kanzler G, Bernhard G. Treatment of proctalgia fugax with salbutamol inhalation. *Am J Gastroenterol*. 1996;91(4):686–689.

315. Boquet J, Moore N, Lhuintre JP, Boismare F. Diltiazem for proctalgia fugax [letter]. *Lancet*. 1986;1(8496):1493.

316. Castell DO. Calcium-channel blocking agents for gastrointestinal disorders. *Am J Cardiol*. 1985;55(3):210B–213B.

317. Swain R. Oral clonidine for proctalgia fugax. *Gut*. 1987;28(8):1039–1040.

318. Patt RB, Proper G, Reddy S. The neuroleptics as adjuvant analgesics. *J Pain Symptom Manage*. 1994;9(7):446–453.

319. Hanks GW. Psychotropic drugs. *Postgrad Med J*. 1984;60(710):881–885.

320. Ventafridda V, Ripamonti C, Caraceni A, Spoldi E, Messina L, De Conno F. The management of inoperable gastrointestinal obstruction in terminal cancer patients. *Tumori*. 1990;76(4):389–393.

321. De Conno F, Caraceni A, Zecca E, Spoldi E, Ventafridda V. Continuous subcutaneous infusion of hyoscine butylbromide reduces secretions in patients with gastrointestinal obstruction. *J Pain Symptom Manage*. 1991;6(8):484–486.

322. Baines MJ. ABC of palliative care. Nausea, vomiting, and intestinal obstruction. *BMJ*. 1997;315(7116):1148–1150.

323. Baines MJ. Management of intestinal obstruction in patients with advanced cancer. *Ann Acad Med Singapore*. 1994;23(2):178–182.

324. Dean A. The palliative effects of octreotide in cancer patients. *Chemotherapy*. 2001;47(Suppl 2):54–61.

325. Ripamonti C, Panzeri C, Groff L, Galeazzi G, Boffi R. The role of somatostatin and octreotide in bowel obstruction: pre-clinical and clinical results. *Tumori*. 2001;87(1):1–9.

326. Blake A, Wan BA, Malek L, et al. A selective review of medical cannabis in cancer pain management. *Ann Palliat Med*. 2017;6(Suppl 2):S215–S222.

327. He Y, Guo X, May BH, et al. Clinical evidence for association of acupuncture and acupressure with improved cancer pain. A systematic review and meta-analysis. JAMA. *Oncologia.* 2020;6:271–278.

328. Smith TJ, Staats PS, Deer T, et al. Randomized clinical trial of an implantable drug delivery system compared with comprehensive medical management for refractory cancer pain: impact on pain, drug-related toxicity, and survival. *J Clin Oncol.* 2002;20(19):4040–4049.

329. Brose WG, Tanelian DL, Brodsky JB, Mark JB, Cousins MJ. CSF and blood pharmacokinetics of hydromorphone and morphine following lumbar epidural administration. *Pain.* 1991;45(1):11–15.

330. Nitescu P, Sjoberg M, Appelgren L, Curelaru I. Complications of intrathecal opioids and bupivacaine in the treatment of "refractory" cancer pain. *Clin J Pain.* 1995;11(1):45–62.

331. Sjoberg M, Nitescu P, Appelgren L, Curelaru I. Long-term intrathecal morphine and bupivacaine in patients with refractory cancer pain. Results from a morphine: bupivacaine dose regimen of 0.5:4.75 mg/ml. *Anesthesiology.* 1994;80(2):284–297.

332. Mercadante S. Intrathecal morphine and bupivacaine in advanced cancer pain patients implanted at home. *J Pain Symptom Manage.* 1994;9(3):201–207.

333. Eisenach JC, DuPen S, Dubois M, Miguel R, Allin D. Epidural clonidine analgesia for intractable cancer pain. The Epidural Clonidine Study Group. *Pain.* 1995;61(3):391–399.

334. Penn RD, Paice JA, Kroin JS. Octreotide: a potent new nonopiate analgesic for intrathecal infusion [see comments]. *Pain.* 1992;49(1):13–19.

335. Yang CY, Wong CS, Chang JY, Ho ST. Intrathecal ketamine reduces morphine requirements in patients with terminal cancer pain. *Can J Anaesth.* 1996;43(4):379–383.

336. Yaksh TL. Epidural ketamine: a useful, mechanistically novel adjuvant for epidural morphine? *Reg Anesth.* 1996;21(6):508–513.

337. Blanchard J, Menk E, Ramamurthy S, Hoffman J. Subarachnoid and epidural calcitonin in patients with pain due to metastatic cancer. *J Pain Symptom Manage.* 1990;5(1):42–45.

338. Karavelis A, Foroglou G, Selviaridis P, Fountzilas G. Intraventricular administration of morphine for control of intractable cancer pain in 90 patients. *Neurosurgery.* 1996;39(1):57–61.

339. Cramond T, Stuart G. Intraventricular morphine for intractable pain of advanced cancer. *J Pain Symptom Manage.* 1993;8(7):465–473.

340. Dennis GC, DeWitty RL. Long-term intraventricular infusion of morphine for intractable pain in cancer of the head and neck. *Neurosurgery.* 1990;26(3):404–407. discussion 407–8.

341. Symreng T, Gomez MN, Rossi N. Intrapleural bupivacaine v saline after thoracotomy—effects on pain and lung function—a double-blind study [see comments]. *J Cardiothorac Anesth.* 1989;3(2):144–149.

342. Dionne C. Tumour invasion of the brachial plexus: management of pain with intrapleural analgesia [letter]. *Can J Anaesth.* 1992;39(5 Pt 1):520–521.

343. Lema MJ, Myers DP, De Leon-Casasola O, Penetrante R. Pleural phenol therapy for the treatment of chronic esophageal cancer pain. *Reg Anesth.* 1992;17(3):166–170.

344. Cooper MG, Keneally JP, Kinchington D. Continuous brachial plexus neural blockade in a child with intractable cancer pain. *J Pain Symptom Manage.* 1994;9(4):277–281.

345. Caraceni A, Portenoy RK. Pain management in patients with pancreatic carcinoma. *Cancer.* 1996;78(3):639–653.

346. Eisenberg E, Carr DB, Chalmers TC. Neurolytic celiac plexus block for treatment of cancer pain: a meta-analysis [published erratum appears in Anesth Analg 1995 Jul;(81)1:213]. *Anesth Analg.* 1995;80(2):290–295.

347. Wong GY, Schroeder DR, Carns PE, et al. Effect of neurolytic celiac plexus block on pain relief, quality of life, and survival in patients with unresectable pancreatic cancer: a randomized controlled trial. *JAMA.* 2004;291(9):1092–1099.

348. Kawamata M, Ishitani K, Ishikawa K, et al. Comparison between celiac plexus block and morphine treatment on quality of life in patients with pancreatic cancer pain. *Pain.* 1996;64(3):597–602.

349. Mercadante S. Celiac plexus block versus analgesics in pancreatic cancer pain. *Pain.* 1993;52(2):187–192.

350. Gress F, Schmitt C, Sherman S, Ciaccia D, Ikenberry S, Lehman G. Endoscopic ultrasound-guided celiac plexus block for managing abdominal pain associated with chronic pancreatitis: a prospective single center experience. *Am J Gastroenterol.* 2001;96(2):409–416.

351. Wiersema MJ, Wong GY, Croghan GA. Endoscopic technique with ultrasound imaging for neurolytic celiac plexus block. *Reg Anesth Pain Med.* 2001;26(2):159–163.

352. Plancarte R, de Leon-Casasola OA, El-Helaly M, Allende S, Lema MJ. Neurolytic superior hypogastric plexus block for chronic pelvic pain associated with cancer. *Reg Anesth.* 1997;22(6):562–568.

353. Plancarte R, Amescua C, Patt RB, Aldrete JA. Superior hypogastric plexus block for pelvic cancer pain. *Anesthesiology.* 1990;73(2):236–239.

354. Plancarte R, Velazquez R, Patt RB. Neurolytic block of the sympathetic axis. In: Patt RB, ed. *Cancer Pain.* Philadelphia: Lippincott; 1993:377–425.

355. Wemm Jr K, Saberski L. Modified approach to block the ganglion impar (ganglion of Walther) [letter]. *Reg Anesth.* 1995;20(6):544–545.

356. Nebab EG, Florence IM. An alternative needle geometry for interruption of the ganglion impar [letter]. *Anesthesiology.* 1997;86(5):1213–1214.

357. Lamacraft G, Cousins MJ. Neural blockade in chronic and cancer pain. *Int Anesthesiol Clin.* 1997;35(2):131–153.

358. Patt RB, Reddy S. Spinal neurolysis for cancer pain: indications and recent results. *Ann Acad Med Singapore.* 1994;23(2):216–220.

359. Saris SC, Silver JM, Vieira JF, Nashold Jr BS. Sacrococcygeal rhizotomy for perineal pain. *Neurosurgery.* 1986;19(5):789–793.

360. Ischia S, Luzzani A, Polati E. Retrogasserian glycerol injection: a retrospective study of 112 patients. *Clin J Pain.* 1990;6(4):291–296.

361. Rizzi R, Terrevoli A, Visentin M. Long-term results of alcoholization and thermocoagulation of trigeminal nerve for cancer pain. In: Erdmann W, Oyama T, Pernak MJ, eds. *The Pain Clinic I. Proceedings of the First International Symposium.* Utrecht: VNU Science Press; 1985:360.

362. Prasanna A, Murthy PS. Sphenopalatine ganglion block and pain of cancer [letter]. *J Pain Symptom Manage.* 1993;8(3):125.

363. Meyer-Witting M, Foster JM. Suprascapular nerve block in the management of cancer pain. *Anaesthesia.* 1992;47(7):626.

364. Neill RS. Ablation of the brachial plexus. Control of intractable pain, due to a pathological fracture of the humerus. *Anaesthesia.* 1979;34(10):1024–1027.

365. Stuart G, Cramond T. Role of percutaneous cervical cordotomy for pain of malignant origin. *Med J Aust.* 1993;158(10):667–670.

366. Sanders M, Zuurmond W. Safety of unilateral and bilateral percutaneous cervical cordotomy in 80 terminally ill cancer patients. *J Clin Oncol.* 1995;13(6):1509–1512.

367. Cowie RA, Hitchcock ER. The late results of antero-lateral cordotomy for pain relief. *Acta Neurochir.* 1982;1(2):39–50.

368. Chevrolet JC, Reverdin A, Suter PM, Tschopp JM, Junod AF. Ventilatory dysfunction resulting from bilateral anterolateral high cervical cordotomy. Dual beneficial effect of aminophylline. *Chest.* 1983;84(1):112–115.

369. Polatty RC, Cooper KR. Respiratory failure after percutaneous cordotomy. *South Med J.* 1986;79(7):897–899.

370. Cherny NI, Portenoy RK. Sedation in the management of refractory symptoms: guidelines for evaluation and treatment. *J Palliat Care.* 1994;10(2):31–38.

371. Burt RA. The Supreme Court speaks—not assisted suicide but a constitutional right to palliative care. *N Engl J Med.* 1997;337(17):1234–1236.
372. *President's Commission for the Study of Ethical Problems in Medical and Biomedical and Behavioral Research. Deciding to Forgo Life Sustaining Treatment: Ethical and Legal Issues in Treatment Decisions.* Washington: U.S. Government Printing Office; 1983.
373. American Medical Association. Good care of the dying patient. Council on scientific affairs, American Medical Association. *JAMA.* 1996;275(6):474–478.

Further reading

Sawe J. High-dose morphine and methadone in cancer patients. Clinical pharmacokinetic considerations of oral treatment. *Clin Pharmacokinet.* 1986;11(2):87–106.

36

Psychosocial issues in cancer patients with neurological complications

Ashlee R. Loughan[a], Kelcie Willis[b], Autumn Lanoye[c], Deborah Allen[d], Morgan Reid[b], Scott Ravyts[b], Rachel Boutte[b], and Julia Brechbeil[b]

[a]Department of Neurology, Virginia Commonwealth University, Richmond, VA, United States, [b]Department of Psychology, Virginia Commonwealth University, Richmond, VA, United States, [c]Department of Health Behavior, Virginia Commonwealth University, Richmond, VA, United States, [d]Nursing Research, Duke University, Durham, NC, United States

1 Introduction

Providing care to patients with cancer is complex and requires a multidisciplinary team to recognize and monitor the dynamic interplay between the psychological and behavioral manifestations that may occur throughout the cancer trajectory. Psychosocial issues range in severity and duration, and they may occur at any time after diagnosis through treatment and recovery. The consequences of many psychosocial issues can be distressing and result in life-long disability.

2 Neurocognitive concerns

While the reported incidence of cognitive change within the cancer population as a whole is approximately 75%,[1–3] this number rises substantially to 91% when focusing on neuro-oncology.[4] Individuals with a brain tumor consistently identify cognitive impairment as their greatest concern following treatment,[5] and research supports this perception, finding cognitive dysfunction across a variety of domains in comparison with their non-cancer peers.[6] However, variations in the assessment of cognitive dysfunction have contributed to a fluctuating definition, which has ultimately impacted the reported incidence across the larger literature base. For example, cognitive functioning can be measured objectively (i.e., via performance-based tasks) or subjectively (i.e., via self-report measures), and the standards for classifying "impairment" include a statistically sig-nificant difference compared to a healthy control group, 1.5 standard deviations below established normative data, or 1.5 standard deviations below an individual's estimated premorbid level of functioning. Impairment severity may be subtle, where only patients and close family notice, to severe enough to interfere with daily functioning, employment capability, and social interactions.[7–10] For many individuals, the resultant disability can be life-long.[11] Neurocognitive assessments can identify deficits and prompt referrals for planned rehabilitation, therapeutic recommendations, focused strategies, pharmacological endorsements, and supportive care.[12] Deficits demonstrated by brain tumor patients are most consistent with subcortical abnormalities: slowed mental processing; reduced attention, encoding, and retrieval of information; and psychomotor slowing. Obtaining accurate and valid neurocognitive performance data is paramount in providing quality of life (QoL) care.

2.1 Neuroanatomy

While directly linking cognitive deficits to tumor size and location makes sense, the clinical manifestations of treatment-related neurotoxicity are more complex and multivariate (dose, length of exposure time, mode of administration, and genetic background of the patient). Due to compression, displacement effects, and treatment neurotoxicity, posttreatment impairments have been shown to be more diffuse, and, at times, completely independent of tumor location.[13,14] This can make assessing brain tumor patients challenging and unlike other brain injuries which tend to be more site specific. What does appear to be consistent across neuro-oncological subgroups is the dose–response relationship between treatment and cognitive impairment, as do the predictive factors of age and cognitive reserve (i.e., premorbid level of cognitive functioning). Regardless, tumor size and location continue to play a crucial role in the neurocognitive symptom interplay.[15] Presence of cognitive impairment in patients with brain tumors that precedes surgery, radiotherapy, or chemotherapy is generally directly related to lesion location.[15–17]

2.1.1 Frontal lobe

Attention and executive function disorders frequently result from disruptions of frontal cortical areas or disturbances of the subcortical pathways connecting the frontal lobes to other parts of the brain. Individuals with frontal lobe brain tumors often report increased distractibility, slowed processing, difficulty planning or scheduling and managing one's own activities, acting inappropriately for the situation, and displaying flat or inappropriate affect.[18–21] Depending on the severity of the deficit, these individuals may need to be closely supervised in their everyday lives, particularly in cases where they demonstrate anosognosia, as they do not realize the extent of their deficits.[22,23] Searching for words, difficulty building phrases and whole passages of speech, trouble initiating speech, and/or echolalia are also characteristic of patients who have frontal lesions.[18,19,24,25] Additionally, the tumor itself and associated edema may damage the nerve tracts that the frontal region shares with other brain regions, which can interrupt modulatory frontal influences on brain function.[4,26] Thus, even if a patient does not have a frontal brain lesion, many patients exhibit executive dysfunction manifested by apathy, lack of motivation, lack of spontaneity, impaired attention, reduced working memory, and difficulty shifting mental sets.[27] While most deficits are reported during and shortly following a patient's chemotherapy regimen, some patients report persisting impairments.[28]

2.1.2 Temporal lobe

Memory deficits are common following temporal lobe disruptions, including both daily forgetfulness and difficulty retaining new information. An individual's ability to learn new information relies upon acquiring, encoding, and storing this material. An interruption in any of these processes can interfere with successful retrieval. Disruptions can be caused by treatment regimens (surgery, radiation, chemotherapy), hormones, antiepileptic agents, corticosteroids, and even extreme psychological distress. In most cases, a combination of these factors negatively impacts the patient's memory. Neurocognitive testing often reveals reduced learning efficiency and memory retrieval problems following chemotherapy treatment, even in the context of relatively preserved memory consolidation. A fear from oncology patients is that chemotherapy can lead to extensive memory loss in the form of dementia. However, epidemiological studies have demonstrated that if there is no sign of cognitive impairment prior to cancer treatment, chemotherapy has no bearing on long-term risk of dementia.[29] However, following radiation therapy, most brain tumor patients experience some degree of cerebral atrophy and hemispheric white matter changes that correlate with memory deficits.[30] Late-delayed (several months to years) encephalopathy—an irreversible and extremely serious complication that can result following radiation therapy—may also occur in approximately 30% of cases involving brain radiation.[13] Although severity can vary, severe neurocognitive deterioration can result in radiation-induced dementia as a late effect. This tends to be a greater issue within the low-grade glioma population given the extended length of overall survival. As the prefrontal cortex performs a pivotal role in remote memory recall, a thorough investigation of both temporal and frontal lobe capability, including attention and information processing, should be completed when patients report memory deficits.

Patients with temporal lobe lesions may also demonstrate a variety of language problems. Tumors within or close to the language centers are often found to be inoperable or only partially resected. As such, these tumors can cause detrimental language effects, which, in turn, significantly impacts patient QoL. Those with dominant hemisphere temporal lesions usually manifest disturbances of verbal speech components, such as difficulties with phonemic analysis, naming objects or word finding, and comprehending ideas from speech or written text, especially when complicated grammatical constructions and long sentences are used. Temporal lobe tumors located in the nondominant hemisphere may cause some intonation and prosodic speech component changes, difficulties in perceiving emotions in the speech of others, and an inability to express emotions in one's own speech. They also have difficulties with visual–perceptual skills and left-sided motor dexterity.[18,20,31]

2.1.3 Parietal/occipital lobe

Parietal lobe tumors may cause spatial orientation problems and difficulties expressing the relationships between objects and/or events.[32] The impact of treatment on visuospatial functioning is varied, ranging from no discernable impact to severe dysfunction.[1,33–35] There may be an observed indirect impact of other neurocognitive domains on visuospatial functioning, such as the effect of reduced planning from frontal lobe deficits, which impedes nonverbal, visual problem solving. Tumors in the superior parietal lobe, intraparietal sulcus, or frontal eye fields may produce visuospatial attention concerns[36,37]; likewise, tumors located in the putamen, pulvinar, caudate nucleus, insula, superior temporal gyrus, or periventricular white matter can cause visuo-spatial neglect.[38] Patients with right parietal lobe tumors may develop anosognosia. Patients with occipital lobe tumors may have difficulties with visual perception of objects by either misinterpreting the images or requiring more time and effort to perceive them correctly.[18,20,21]

2.1.4 Cerebellar region

Cerebellar posterior-lateral regions seem to be relevant for cognition, while vermal lesions seem to be associated with changes in affect.[39] The impact of cerebellar lesions has been attributed to a variety of disorders ranging in severity and extent. Schmahmann and Sherman postulated a "cerebellar cognitive affective syndrome" (now eponymously named "Schmahmann's syndrome"[40]), which depicts impairment of executive functions and disturbances in spatial cognition, language, and personality.[41] The deficits have been attributed to the disruption of neural circuits linking prefrontal, temporal, posterior parietal, and limbic cortices with the cerebellum and have now been identified with a validated brief battery of cognitive tests.[42] Right cerebellar lesions lead to verbal deficits because of the crossed pathways, whereas in left cerebellar lesions, spatial deficits are most prominent. This hypothesis has been confirmed for children with cerebellar lesions arising from tumor resection.[43,44] Other studies have shown deficits resulting from cerebellar lesions in the areas of metalinguistics, verbal fluency, error detection, planning, effortful memory, non-motor associative learning, spatial attention, social cognition, and shifting attention.[45–54] However, other researchers have failed to replicate these results or doubt the cognitive function of the cerebellum, explaining that the possible deficits are due to motor impairment or methodological problems instead.[55–57] Patients with third ventricle tumors are at risk for developing impairments in memory, executive function, and fine motor speed and dexterity, which are domains associated with frontal subcortical functions.[58]

2.2 Effects of treatment

Cognitive impairment may result temporarily or permanently from damage to neural tissues, cells and neurons, and the blood–brain barrier, either directly or indirectly. This damage may be due to compression from the tumor or treatment-related damage from surgery, chemotherapy, or radiation therapy.[59] Current treatment paradigms utilize multimodalities, including surgical resection, chemotherapy, and radiation therapy.

2.2.1 Surgery

Despite gross-total resection favoring survival rates,[60,61] subtotal resection for a brain tumor is the most common surgical procedure due to location and preservation of brain tissue for vital everyday function.

However, microsurgical technological advancements have substantially improved resection outcomes. Increased intracranial pressure is generally controlled preoperatively with steroid therapy, which is continued and tapered postoperatively. Seizure prophylaxis may be initiated intraoperatively and may be continued through radiation therapy, depending on seizure risk from tumor location and preoperative deficit presentation. Anesthesia, steroids, and antiepileptic medications can negatively impact cognitive functioning.[62] In addition, stress associated with the need for surgical intervention after an initial brain tumor diagnosis or disease recurrence can exacerbate cognitive outcomes.

2.2.2 Chemotherapy

Most research regarding the impact of chemotherapy on the development of cognitive impairment has been performed in patients with noncentral nervous system cancers,[63] and demonstrates long-term effects of chemotherapy reducing cortical connectivity in the frontal, striatal, and temporal regions as compared to age-related healthy controls.[64] Commonly used chemotherapeutic agents for the treatment of brain tumors include anthracyclines, alkylating agents, platinum-based agents, and antimetabolites—all of which may impact cognitive functioning through initiating or exacerbating inflammation, oxidative stress, increased free radical formation, and increased levels of cytokines.[65] These agents modulate neurogenesis, thereby potentially reducing the capacity for neuroplasticity and repair of damaged cortical pathways involved in cognitive function.[59,66] The cognitive impact may be observed through difficulties in new learning, memory, attention, executive function, spatial learning, processing speed, as well as inciting depression or other mood disturbances.[67] Studies of immunotherapies and targeted biotherapies have indicated cognitive impairment as reversible side effects, but long-term impact has yet to be determined.[68]

2.2.3 Radiation therapy

Radiation impact on subsequent cognitive impairment has been observed in 90% of patients with brain tumors who received whole-brain radiation therapy.[69] Consequently, technological advances have significantly improved the delivery of radiation therapy from global applications to focal-directed sites, sparing healthy surrounding tissues, particularly hippocampal tissue.[70,71] A variety of mechanisms underpin the dynamic interplay of actions between cortical tissues, cells, neurons, and endothelium that ensue during radiation treatments.[66,69] Radiation can induce prolonged inflammatory responses and oxidative stress, blood–brain barrier disruptions, and impaired neurogenesis and neuroplasticity.[69,72] Deficits in memory, attention, and problem solving have been observed following radiation therapy.[69,70]

2.2.4 Tumor-specific genetic mutations

Recent investigations examining the association between tumor-specific genetic mutations and their impact on cognitive functioning are underway.[73,74] As this line of research advances, differential neurocognitive functioning profiles associated with specific tumor genotypes may emerge.

2.3 Neurocognitive assessment

Cognitive impairment is described as the decline of function in one or more cognitive domains. A neuropsychologist will integrate the patient's current complaints and history (developmental, medical, psychiatric, educational), with available imaging, laboratory data, treatment planning, knowledge of the brain-behavioral relationships, functional neuroanatomy, neuropathology, cognitive psychology, psychometrics and test theory, psychopathology, and neurodevelopment. As stated previously, a decline in function can be measured *objectively* (neuropsychological battery) or *subjectively* (self-report). Objectively, identifying cognitive impairment is often dependent on multiple variables: (a) the neurocognitive instruments or battery selected, (b) the decision to compare scores with a normed healthy population versus the patient's baseline, premorbid functioning, and (c) use of scoring cut-points that may or may not define normal and impaired. Subjective assessments can be gathered either during the clinical interview where the neuropsychologist gathers relevant history and current symptomology from patients and/or their informants' perspective, or are performed via self-reported instruments in the form of questionnaires, which can also be completed by the patient themselves or their loved ones (e.g., informant reports). To simplify, objective assessments measure patients' current capability, whereas subjective assessment relies on patients' perceptions of change. Often, self-reported instruments ask the participant to reflect on their recent functioning over a period of time (e.g., last 7 days or 1 month), whereas objective assessments reflect a single snapshot of time.

2.3.1 Objective assessment

Most comprehensive neurocognitive evaluations include extensive sets of assessment tools measuring premorbid functioning, simple and sustained attention, speed of processing, executive function, confrontational naming and verbal fluency, learning and recall, motor capabilities, and emotional adjustment—all the while being mindful of examinee burden and adjusting to the functional level of each patient. Commonly utilized screening tools within the healthcare field such as the Mini Mental Status Examination (MMSE) or the Montreal Cognitive Assessment (MoCA) have been shown to be rather insensitive in the cancer population, specifically

in those undergoing brain tumor treatments. In fact, when measured, the sensitivity of the MMSE within a brain tumor population was dismal, demonstrating only 50% sensitivity: no more accurate than a coin toss.[75] The Repeatable Battery and Neuropsychological Status (RBANS), a more comprehensive screen than the MMSE or MoCA, has demonstrated greater ability to identify cognitive concern[76]; however, the feasibility of incorporating this 20-min battery into a busy neuro-oncology clinic has not been established to date. The International Cognitive and Cancer Task Force (ICCTF)[77] continues to stand behind their statement that objective (i.e., neuropsychological) tests remain the gold standard for measuring cognitive function when assessing change following cancer diagnosis and treatment. The ICCTF has provided recommendations for cognitive research batteries to be included in clinical trials (See Table 1), yet there remains no consensus regarding clinical evaluations. Ongoing discussions are occurring among experts with a hope for standards in the future. Unfortunately, to date there continues to be debate as to the specificity and sensitivity of cognitive assessments commonly used in clinical practice. Additionally, many patients do not undergo neuropsychological testing prior to diagnosis, thereby limiting objective determination of change in cognitive function that occurs from cancer or its treatment.

2.3.2 Subjective assessment

One method of obtaining subjective data is in the form of a clinical interview. The clinical interview, often described as an indispensable source of information regarding functioning, is conducted at the onset of a neuropsychological evaluation. This open-ended approach of communication frequently elicits specific problem areas, the patient's perspective regarding their difficulties, and other dimensions of historical or current functioning not revealed through formal, objective testing. Another method of obtaining subjective assessment is through questionnaires. Although standardly used in clinical care and during cognitive evaluations, there has been variability and limited correlations found between patient-reported concerns and measured impairments,[81] with patients' perception of impairment generally being worse than that detected by objective testing.[82,83] As such, patient questionnaires remain a controversial topic. In

TABLE 1 International cognition and cancer task force recommended cognitive battery.[77]

Measure
Hopkins Verbal Learning Test-Revised (HVLT-R)[78]
Controlled Oral Word Association (COWAT)[79]
Trail Making Test (TMT)[80]

TABLE 2 An example of a self-report scale for neuropsychological functioning: The Webexec Scale.

1. Do you find it difficult to keep your attention on a particular task?

2. Do you find yourself having problems concentrating on a task?

3. Do you have difficulty carrying out more than one task at a time?

4. Do you tend to "lose" your train of thoughts?

5. Do you have difficulty seeing through something that you have started?

6. Do you find yourself acting on "impulse"?

addition, patient report has been more strongly related to affective problems than objective impairment.[84,85] However, these self-report measures remain central to clinical practice and may often directly speak to the impact on an individual's QoL rather than the deficits statistically uncovered by standardized cognitive testing. Further, the use of subjective assessments—even brief screening measures—can cue physicians to refer the patient for a neuropsychological assessment. See Table 2 for an example of a short executive function subjective questionnaire, the Webexec Scale.

2.4 Neurocognitive treatment

There remain limited accepted and validated evidence-based interventions available to improve cancer-related cognitive impairment. Over the past decade, investigations have explored both pharmacological and non-pharmacological treatments, yet it is safe to say these explorations are in their infancy and many more are warranted. Additionally, the exclusion of brain tumor patients from many randomized controlled trials makes it difficult to gauge whether the same treatment strategies proven safe and effective in other cancers or neurological conditions should be pursued in patients with neuro-oncological cancers.

2.4.1 *Pharmacological treatments*

Clinical trials using psychostimulants have yielded promising cognitive function improvements in patients with brain tumors: methylphenidate or dexmethylphenidate, modafinil, armodafinil, memantine, or donepezil. Most agents require monitoring and titration over time to yield optimal cognitive benefit. Methylphenidate/dexmethylphenidate was observed to improve patient processing speed, attention, memory, and executive function.[86,87] Likewise, donepezil improved attention, concentration, memory, and processing speed.[88,89] Memantine was found to delay cognitive decline over time, specifically in domains of memory, executive function, and processing speed; this benefit is similar to that observed in patients with dementias.[90] Although modafinil/armodafinil studies did not yield significant changes in cognitive function between placebo and intervention arms,[91,92] patients overall reported improved physical health, information processing, memory, and less fatigue.[91]

2.4.2 *Non-pharmacological treatments*

Non-pharmacological approaches pertain to a range of complementary and alternative interventions, physical/behavioral rehabilitation, and cognitive training programs, including vitamin E, *Ginkgo biloba*, melatonin, natural restorative environmental, exercise, and mindfulness-based stress reduction, EEG biofeedback/neurofeedback, imagery, meditation, physical rehabilitation programs, and cognitive training programs. Frequently founded and borrowed from a broader base of neuro-rehabilitation treatment paradigms (e.g., traumatic brain injury and stroke), several non-pharmacological programs have been developed and demonstrate preliminary promise; however, the general lack of rigorous high-quality studies calls for caution in fully touting their efficacy.[93] Further, these approaches can be implemented in isolation or in combination with one another, and the improvements attributable to each component are yet to be disentangled. These approaches are the subject of ongoing research in order to provide an alternative to improving cognitive function for cancer survivors who prefer non-pharmacological solutions.[94] Importantly, such treatments may provide benefits beyond cognitive improvement, such as enhanced mood, fatigue, QoL, and general health.

Cognitive rehabilitation

Cognitive rehabilitation targeting cancer-related neurological complications consists of guided computer-based and/or strategy-oriented training; however, a standardized evidence-based program has yet to be developed. Such approaches focus on "cognitive retraining," the goal of which is to restore or circumvent damaged neural pathways.[93,95] Two common approaches to cognitive rehabilitation are *restorative* and *compensatory*.[96] The restorative approach is based on the assumption that repetitive stimulation of an area of impairment will promote neurogenesis and ultimately restore lost function. In contrast, the compensatory approach holds the assumption that lost neurological function cannot be restored; rather, teaching alternative strategies utilizing residual strengths is necessary to accommodate areas of weakness and promote cognitive improvement. Cognitive training and rehabilitation therapy in brain tumor patients has shown feasibility and acceptability—with promising results improving domains such as attention and memory[97–103] (see Table 3). In a recent systematic review, Fernandes and colleagues[95] found that cognitive

TABLE 3 Cognitive training/rehabilitation interventions for cognitive impairment specific to glioma patients.

Cognitive training/rehabilitation			
Study	N	Intervention	Findings
Gehring et al.[97]	140	2 h/wk for 6 wks of computer and compensatory retraining	Improvement in perceived cognitive function immediate and 6-mo post-intervention, improvement in attention and memory 6-mo post only
Hassler, et al.[98]	11	90 min/wk for 10 wks of perception, concentration, attention, memory, retentiveness, and verbal memory training	Improvement in attention and learning
Sherer et al.[99]	13	Cognitive and vocational outpatient rehabilitation (avg 2.6 ± 1.9 mo)	Improvement in independence (46%); improvement in productivity (62%)
Zucchella et al.[100]	58	16–1 h individual cognitive training (including computer exercises and metacognitive training) over 4 wks	Improvement in visual attention and verbal memory
Yang et al.[101]	38	30 min/day, 3 days/wk for 4 wks of virtual reality training	Improvement in visual and auditory working memory, sustain attention, and processing speed
Maschio et al.[102]	16	10–1 h/wk cognitive training with multimedia software	Improvement in verbal and episodic memory, fluency and long-term visuospatial immediately and 6-mo post
Richard et al.[103]	25	8–2 h/wk goal management training sessions + homework	Improvement in executive functioning and functional goal attainment at both immediate and 4-mo post

Note: *avg*, average; *h*, hours; *min*, minutes; *mo*, months; *wk*, week.

rehabilitation interventions were generally effective in improving at least one domain of cognition. In other reports, brain tumor patients have demonstrated superior benefits when compared to inpatient treatment for stroke and TBI.[97,104,105] Remaining issues include how best to achieve sustainable cognitive benefits and the extent to which these improvements transfer from task performance and generalize to real-world performance. Although potentially viewed as a lengthy and time-consuming intervention, rehabilitation addresses functional goals and may also facilitate coping. It should be viewed as an integral part of care following brain tumor treatment.

Cognitive behavioral therapy (CBT)

CBT extends beyond the strategy-oriented training implemented in cognitive rehabilitation by contextualizing cognitive difficulty within a wider framework of patients' thoughts and emotions.[95,106] Thus, in addition to teaching strategies specific to addressing cognitive deficits, CBT typically also includes components such as relaxation training, combatting dysfunctional beliefs about cognitive abilities, activity scheduling, and emotion regulation strategies.[106] To date, a manualized Memory and Attention Adaption Training (MAAT) is the most well-researched CBT approach to cancer-related cognitive dysfunction and comprises 8 weekly sessions. Compared to a no-treatment control group, MAAT participants demonstrated improvements in verbal memory in addition to improvements in QoL.[107] Compared to an active control group (supportive therapy), MAAT participants significantly improved with respect to self-reported cognitive impairment and processing speed as measured by the Symbol Digit Modalities Test.[108] To date, MAAT has not been examined among individuals with primary brain tumor; thus, it is unknown how this population would fare with such an approach.

Physical activity

Increased blood flow to the brain in combination with other known benefits of exercise (e.g., stress reduction, inflammation reduction, decreased fatigue, improved sleep, mood enhancement) may help to improve cognitive functioning following cancer-related injury.[93,109] In a review of 26 studies investigating the association between exercise and cancer-related cognitive dysfunction, 21 studies were found to demonstrate positive results spanning multiple cognitive domains, including memory, attention, and executive function.[110] In brain tumor patients, it appears that walking is the preferred form of exercise during treatment, with the majority of patients preferring in-home routines and wanting information about physical activity after treatment (70%) as compared to during treatment (45%).[111] Molassiotis and colleagues[112] found that those who added walking to their daily routine had improvements in their mood, coping skills, and QoL. While the variation in treatment type, duration, and intensity across these studies dilutes the strength of the evidence, exercise as a treatment approach for cognitive dysfunction following systemic cancer treatment is a nascent field of study and future research is warranted. Brain tumor patients who have neurological limitations with focal motor or sensory deficits should have physical therapy evaluations prior to initiating exercise during and after their cancer treatment.[113]

Meditation

Meditative practices such as mindfulness, qi gong, yoga, and other natural restorative environmental interventions have been shown to improve mood and reduce stress[114,115]; however, they may also directly improve cancer-related cognitive dysfunction by reducing chronic inflammation and improving distress. This is also an area of recent attention; as such, though promising, only preliminary data are available to support its potential efficacy.[93]

3 Psychological concerns

The diagnosis of a brain tumor marks a traumatic and stressful event. Individuals can react with a range of emotional responses, which include disbelief, fear, excessive crying, guilt, sadness, or rage given the uncertainty and uncontrollability of the disease. At times, these emotional reactions may even manifest as clinical depression, clinical anxiety, and existential distress. A comprehensive biopsychosocial model must be employed when evaluating the emotional and psychological effects of a brain tumor. Biological factors include tumor and treatment effects; psychological factors include premorbid personality, coping mechanisms, and pre-existing psychological diagnoses; and social factors include an individual's support network, financial stability, and access to medical care. A recent review found that more than half of patients with brain tumors had psychiatric symptoms.[116] Further, when investigating distress in brain tumor patients, Loughan and colleagues[117] identified four different distress profiles: (1) global distress, (2) emotional distress, (3) resilience, and (4) existential distress. These cluster profiles suggest brain tumor patients psychologically respond differently to diagnosis, treatment, and prognosis. Some exhibit no distress (resilience); others display typical psychological profiles of depression and anxiety (emotional distress); and a small number of patients show high levels of distress on measures across the board (global distress). However, most patients endorsed death-related distress, a type of existential distress. One notable difference found between these profiles was time since diagnosis. Patients in the resilience cluster were significantly further from their initial diagnosis compared to the existential distress cluster. Longitudinal data have shown similar conclusions, in that anxiety often decreases significantly over time, while depressive symptoms remain constant.[118] Together, this finding suggests that some patients may learn to adapt and accept the diagnosis as time passes, becoming more resilient. Still, depression and anxiety appear to be the most commonly assessed and thus studied in brain tumor patients.[119,120] Both have been shown to be associated with poorer prognosis.[121]

3.1 Depression

For brain tumor patients, symptoms of depression may present at various stages across the disease trajectory (before, during, or after diagnosis and treatment). Symptoms of depression may be psychological (e.g., anhedonia, sadness), behavioral (e.g., increased time in bed), or cognitive (e.g., low self-esteem) in nature. Nearly all patients with brain tumors (93%–95%) report symptoms of depression,[122,123] with rates of clinical diagnosis being greater than both other oncology populations and the general population (22%–41%, 13%, and 7%, respectively).[123–128] Yet, the exact cause of depression is unknown. Among explanations, the most common is that depression is a normative, psychological reaction to a poor prognosis. Upon diagnosis, patients are forced to face not only the medically induced symptoms of the tumor, but also the psychological meaning of their diagnosis and the effect of this will have on both the quality and the longevity of their life. Neuro-chemical side effects from chemotherapy, adjuvant therapy, and radiation therapy may also contribute to depression.[119] Regardless of etiology, symptoms of depression appear to be significantly associated with functional impairment, cognitive dysfunction, and reduced QoL.[116,129,130] In fact, depression has been found to be the main predictor of reduced QoL in brain tumor patients across various measurements and may even serve as an indicator for poor medical prognosis.[121] At a 5-year follow-up with low-grade brain tumor patients, Mainio and colleagues[129] found that those with depression had a significantly shortened survival time when compared to their nondepressed peers; a corresponding difference was not found in patients with high-grade or benign tumors. Similar longitudinal findings suggest that the severity of depressive symptoms from diagnosis through treatment is significantly predicted by high depression at baseline.[118] Moreover, emotional, cognitive, and social role factors appear to be increasingly related to depression over physical function—which was confirmed in Mainio's study.[121] It should be noted that depressive symptoms such as poor appetite and low energy are commonly seen in brain tumor patients and may result from medical treatment instead of depression. Thus, symptoms of depression should be thoroughly teased apart by a medical professional to ensure appropriate treatment planning.

3.2 Anxiety

Like depression, anxiety is a frequent and normative reaction to a brain cancer diagnosis. Approximately 17%–31% of patients endorsed moderate to severe symptoms of anxiety.[117,118] Anxiety is more often reported in female patients,[118] younger patients, and is slightly more frequent pre-radiotherapy.[131] Anxiety disorders in brain tumor patients appear to be related to tumor localization and may also be connected with tumor laterality,[132,133] but the type or grade of tumor does not seem to play a role.[132] Anxiety in brain tumor patients may be either a transient experience of elevated stress or worry or reach clinical significance (e.g., diagnosable as generalized anxiety disorder, specific phobia, panic disorder).[134] Many brain tumor patients find that their symptoms of anxiety wax and wane: often, patients report feeling the most anxious prior to receiving surgery or before regular MRIs, given the high rate of recurrence for this disease.[135] Indeed, a patient's worries may manifest as a fear of cancer recurrence or a fear of death, though research has found existential distress to be a separate, yet related type of distress from generalized anxiety.[117] Nevertheless, increased symptoms of anxiety symptoms are predictive of a worse QoL and may exacerbate cognitive dysfunction.[136,137] When distress lasts for longer than a week, worsens, or interferes with the patient's ability to cooperate with planned treatment, intervention is indicated.[138]

3.3 Existential distress

Existential distress can be defined as concern about one's existence, such as the purpose and meaning of life or beliefs around suffering, and fear of death.[139] Although existential distress is not unique to brain tumor patients, this population may be faced with a novel experience of distress, due to factors such as the ambiguity of prognosis and rapid physical and cognitive decline.[140,141] Though to date a thorough understanding of existential distress in neuro-oncology is lacking due to the common exclusion of brain tumor patients in research studies,[142] a recent study found that death-related distress was more frequently endorsed than general psychological distress in brain tumor patients.[117] This finding hold true when comparing brain tumor patients to other advanced cancers. Moreover, in longitudinal interviews conducted with brain tumor patients and their caregivers, Cavers et al.[140] identified a theme called the "dynamic existential trajectory," providing evidence for existential distress that evolves over time: from managing the diagnosis, to the ups and downs of intensive treatment and, lastly, to finding meaning toward the end of life.[140] Patients and caregivers expressed a deep desire to find meaning in their diagnosis and some found comfort in their religious faith. Related, the acceptance of death was met with a pragmatic view by some in an attempt to savor every remaining moment while alive.[140] These findings highlight the myriad responses across and within individuals attempting to understand their diagnosis and cope with the existential distress it brings. These concerns are important considerations, as the severity of existential distress impacts other important outcomes such as QoL, depression, and fatigue.[124,141,143]

3.4 Tumor-related psychiatric effects

While certain psychiatric symptoms can be a direct effect of any cancer diagnosis, some mood and personality changes are directly attributable to the disease itself in brain tumor patients. In fact, in a subset of those with neuro-oncological cancers, psychiatric symptoms are the first clinical indications of the disease.[144] As stated previously, the prevalence of distress among brain tumor patients appears to be more elevated than that of the general public[117,145]; however, this is not due solely to the previously discussed emotional response and psychosocial stressors. Proinflammatory cytokines associated with cancer development also produce classical symptoms of depression such as anhedonia, sleep and appetite changes, and psychomotor slowing.[146] Patients with glioblastomas may be at more risk for personality and behavioral changes because healthy brain tissue can be more easily destroyed from the increased cerebral pressure and swelling associated with these fast-growing tumors. Slow-growing tumors, such as pilocytic astrocytomas, may not cause as much impairment because their slow growth enables the brain to compensate for the tumor. Psychiatric symptoms may also depend on the location of the tumor.[116,147–158] See Table 4 for a summary.

TABLE 4 Psychiatric symptoms associated with various tumor locations.[116,153–155,158–160]

Tumor location	Associated psychiatric symptom
Right hemisphere	Paranoid ideation, mania, hallucinations, and agitation
Left hemisphere	Depressive symptoms
Frontal lobe	Confused states or disorientation, dementia, disinhibition, apathy, poor judgment, euphoria
Temporal lobe	Personality changes and mood (depression or anxiety)
Occipital lobe	Visual hallucinations
Thalamus and hypothalamus	Anxiety, depression, emotional lability, belligerence, recalcitrance, garrulousness, hypersexuality, dysnomy, dyspraxia, reduced attention, memory loss, decreased reasoning ability, eating disorders, and hypersomnia
Corpus callosum	Personality changes, psychosis, and, most commonly, affective symptoms
Basal ganglia	Personality changes and depression
Pituitary gland	Emotional lability, depression, and psychotic symptoms, typically due to endocrine changes
Cerebellum and brain stem	Affective disorders, including emotional blunting and disinhibition, paranoid delusions, personality changes, cerebellar cognitive affective syndrome

3.5 Psychological treatment

Despite the high incidence of distress among brain tumor patients, psychological concerns often go undertreated. If any of the above-mentioned conditions are recognized throughout the disease trajectory, different therapeutic strategies should be offered for both patients and their loved ones (see Table 5). A better understanding of the emotional distress experienced by brain tumor patients could lead to more effective interventions to enhance their ability to cope with the illness and enjoy a better QoL.[124] When treatment is indicated, a two-pronged approach is typically suggested: *pharmacological* and *behavioral*.

3.5.1 *Pharmacological*

Commonly prescribed psychotropic medications in brain tumor patients include antidepressants, benzodiazepines and, at times, antipsychotics. It has been found that nearly half of brain tumor patients diagnosed with depression postoperatively go on to receive antidepres-

TABLE 5 Practical suggestions for clinicians caring for patients with glioma and their caregivers.[140]

Disease timepoint	Practical caring suggestions
Referral	• Acknowledge the distress of uncertainty while waiting for a diagnosis
Initial diagnosis	• Offer more detailed information about the likely process of what will happen and the expected diagnosis
During treatment	• Provide an opportunity for patients to voice their problems or questions • Give supportive advice to caregivers and allow them to ask questions • Ensure immediate communication between the hospital and primary care facility
Follow-up	• Identify a named person (possibly a specialist nurse or family physician) who will maintain contact and offer psychological support and practical advice • Discuss what the future may hold and possible symptoms • Involve caregivers in discussions about practical help, where appropriate
Disease progression	• Sensitively give patients the opportunity to plan their future care, and to consider options such as their resuscitation preferences, living wills, and their preferred place of death • Involve caregivers in discussions about end-of-life care (with patients' consent) • Communicate the patient's care plan with all other services, including after-hours services • Maintain contact and support the caregiver
Bereavement	• Provide an initial bereavement visit from a known and key professional • Offer further contact

sant medications,[123] though there are few high-quality studies examining their effectiveness. Moreover, antidepressants may lower seizure threshold and cause additional memory impairment and fatigue; therefore, future controlled studies are necessary to examine whether the benefits of antidepressants outweigh potential side effects in this population.[161] At present, there is no "gold standard" antidepressant for use in oncology. The choice of antidepressants is often based on the side-effect profile. Antidepressants with more activating characteristics (i.e., fluoxetine, bupropion) may be good choices for depressed patients with significant fatigue. If insomnia is a prominent symptom of the mood disorder and is contributing to daytime fatigue, then a more sedating antidepressant (e.g., some tricyclic antidepressants, mirtazapine) may be a better choice for nighttime use. Benzodiazepines might be effective for mitigating symptoms of anxiety[162] or agitation caused by the brain tumor[163] but have the potential for negative drug interactions or side effects, including possible drug dependence.[164] Hallucinations and other psychotic symptoms in brain tumor patients may be treated with antipsychotics (e.g., haloperidol), though it is important that physicians ensure these symptoms are not better explained by other medications such as steroids.[165] Drug choice and dosage may require modification because psychotropics can induce seizures, delirium, and other side effects in brain tumor patients.[166] It is recommended that medications are only prescribed for a short duration and that brain tumor patients are highly monitored; referral to psychiatry may be indicated.

3.5.2 *Behavioral*

Various forms of psychotherapy have been used to treat emotional concerns in brain tumor patients. One of the most promising brain-tumor specific treatments—Making Sense of Brain Tumor (MSoBT)[143]—is a 10-week manualized program that includes psychoeducation, neuropsychological feedback, cognitive rehabilitation, psychotherapy, and increasing couple/family support. Both face-to-face and telehealth versions of MSoBT have been found to be both feasible and acceptable for glioma patients and produce significant decreases in depression and improvements in well-being.[167] Other psychosocial interventions, including relaxation training and cognitive-behavioral therapy, have successfully shown to reduce emotional distress in mixed-cancer samples that include BT patients.[168] Interventions to manage existential distress such as Managing Cancer and Living Meaningfully (CALM),[169–171] Meaning-Centered Psychotherapy,[172] and Dignity Therapy[173] have proven beneficial in other cancer populations, yet, brain tumor patients continue to be excluded from these trials. Lastly, regular exercise has also been found to decrease depression and anxiety and improve QoL in BT patients.[174]

4 Behavioral concerns

In addition to the aforementioned emotional challenges, the diagnosis and treatment of a brain tumor might consequently affect patients' everyday behaviors, such as appetite, sleep and fatigue, movement/gait, and sexual desire. These marked changes might be a result of cognitive or emotional symptoms or direct treatment effects. For example, while fluctuations in appetite, sleep, and sexual desire are common behavioral changes associated with a diagnosis of depression, the use of corticosteroids also may explain and alter these behaviors. Thus, a medical team of both physicians and psychologists might be best suited for assigning etiology and recommending treatment. As changes in behavior have a profound impact on patient QoL, it is important that these are thoroughly assessed. Additionally, other medical specialists (e.g., occupational therapists, physicians in sleep medicine) should be utilized when indicated to address symptomology.

4.1 Appetite

Patients with brain tumors can experience changes in appetite depending on a variety of factors. Treatment-specific changes in appetite are well documented. For example, among patients with brain tumors prescribed corticosteroids, 37% endorse significant *increases* in their appetite and 23% report weight gain.[175] Moreover, both steroid dose and duration contribute to increases in appetite.[175] Patients who endorse significant increases in appetite might benefit from an adjustment of steroid dose or a closer examination of other medications that might contribute to weight gain, such as carbamazepine or other agents used for behavioral or psychiatric disturbance.[176] On the other hand, the prevalence of *decreased* appetite is approximately 10%–14% and is consistent among patients with both high- and low-grade tumors, throughout the course of treatment.[177] Radiation and chemotherapy are both associated with decreased appetite.[178,179] McCall and team[180] found a very low prevalence of malnutrition in patients with brain tumors, < 20%, as compared to other cancer populations. Because malignant brain tumor patients are more likely to be at risk for malnourishment, this study recommends nutritional screening parameters of 5% weight loss or more, persistent nausea and vomiting, presence of dysphagia, and/or headache. Finally, depressed mood, nausea, and dysphagia—three common symptoms among patients with brain tumors—may also adversely affect appetite.[181] Patients who endorse decreases in appetite or significant nausea and vomiting might benefit from medications such as antiemetics, synthetic dronabinol, or steroids.[182] Additionally, psychotherapy might also secondarily address and regulate appetite.[183]

4.2 Fatigue

Fatigue is a common complaint present in oncology patients throughout their illness trajectory, with a prevalence rate exceeding 80% in brain tumor patients specifically. The National Comprehensive Cancer Network (NCCN) recognizes the contribution of fatigue to overall QoL and has developed guidelines[184] for oncology providers to consider when assessing and measuring fatigue. The NCCN defines cancer-related fatigue as "a distressing persistent, subjective sense of physical, emotional, and/or cognitive tiredness or exhaustion related to cancer or cancer treatment that is not proportional to recent activity and interferes with usual functioning" (p. FT-1). Fatigue is typically described by patients as tiredness, weakness, or exhaustion and may have additional, complementary components such as difficulty thinking or concentrating.[185] Fatigue can cause significant changes to patients' daily routines, including changes in employment status, social activities, and sleep patterns.[186] Patients who suffer from mental fatigue often report that they are easily overwhelmed and have difficulty organizing, meeting deadlines, and working efficiently in their daily activities. In addition, patients with fatigue report that activities once automatic require more effort.[187] Fatigue is also associated with poor survival[188] and is negatively associated with patients' perception of hope for their future.[189] The pathophysiology of fatigue in brain tumor patients is multifactorial, involving biological, medical, social, and behavioral factors.[190] Women, patients with active disease, and those with greater cognitive and motor impairments are at an increased risk of fatigue.[191] Fatigue has been found to increase from time of diagnosis to 2 weeks after radiotherapy.[192] Greater fatigue is associated with increased caregiver support[193] and decreased independence in activities of daily living.[194] The management of fatigue can include both pharmacologic and behavioral intervention. Neurostimulants have been used to treat fatigue in this population; however, controlled studies on this treatment are lacking.[87] Despite patients reporting lower fatigue severity and better motivation, modafinil did not exceed the effect of placebo with respect to symptom management.[91] Exercise and physical activity demonstrate strong evidence as effective treatments for cancer-related fatigue; however, there is a paucity of research on exercise specific to the brain tumor population.[195] Lastly, treatments for sleep (e.g., CBT-I) might improve symptoms of fatigue if the fatigue is a direct result of sleep disturbance.

4.3 Sleep

Sleep disturbance is among the most common and severe consequences experienced by patients with brain tumors.[196,197] In fact, in a recent investigation, 53% of brain tumor patients reported sleep disturbance, 15% endorsed clinically significant symptoms of insomnia, and 27% reported moderate-to-severe daytime drowsiness.[198] Such symptoms are particularly noteworthy given that they are unlikely to remit throughout the disease course.[199] Specifically, increased psychological distress, difficulty breathing, and pain each predicted greater sleep disturbance at the end of life in those with CNS cancers.[199] In fact, for those nearing end of life, sleep disruptions are observed in up to 95% of patients, primarily due to increasing intracranial pressure displacing cerebral structures—including the reticular activating system—eventually causing altered levels of consciousness and coma.[200] Multiple factors have been posited to contribute to sleep disturbance in this patient population, including the direct effect of the tumor; the indirect effects of medication and radiation; and co-occurring psychological symptoms, such as depression or anxiety.[197] Regardless of etiology, sleep disturbance is associated with a range of adverse consequences, including higher levels of anxiety, depression, fatigue, pain, and neurocognitive symptoms.[198] Sleep disturbance may be particularly distressing when coupled with daytime somnolence, which, by one estimate, occurs in approximately 90% of patients with brain tumors receiving cranial radiotherapy.[201] Unfortunately, few neuro-oncology providers routinely screen for sleep disturbance,[202] and if treated, most opt for pharmacological interventions (e.g., zolpidem), which may further exacerbate memory impairment in an already-vulnerable population or lead to falls, fractures, or motor vehicle accidents.[203] Providers should assess patients' current medications and determine if adjustments should be made, such as changing the time of the patient's last dose of steroids. Non-pharmacological interventions for sleep among this cancer population are lacking,[196] despite nearly half (47%) expressing a preference for behavioral treatment options.[204] A recent meta-analysis provided support for cognitive behavioral therapy for insomnia (CBT-I) in oncology patients,[205] and initial evidence also suggests that mindfulness-based stress reduction interventions may also improve sleep[206]; however, these treatments have yet to focus specifically on brain tumor patients.

4.4 Motor skills

Motor deficits occur in approximately 21% of brain tumor patients; however, this estimate significantly varies depending on whether the patient is receiving treatment (10%) or in the end-of-life phase (44%).[181] Difficulties with physical functioning is significantly predictive of decreased QoL,[207] including the loss of functional independence and the increased risk of falls and other complications due to immobility.[207] Motor deficits may be related to tumor location, as tumors located in or adjacent to the primary motor cortex of the frontal lobe, the parietal lobe, and the cerebellum have been associated with gait impairment, incoordination, and ataxia.[208] Treatment can also impact motor function; motor dysfunction, including increased tremor, gait instability, coordination difficulties, and decreased facial rigidity can be an indirect manifestation of encephalopathy.[209] Depending on dose, location, and severity, radiation treatment predicts encephalopathy in 7%–24% of brain tumor patients.[210] Chemotherapy-induced neuropathy refers to damage to the peripheral nerves that increases in severity with duration of treatment and manifests as numbness, tingling, or weakness, primarily of the extremities.[211] Medications, like corticosteroids, that are used to alleviate the side effects of treatment, such as edema, may increase risk for myopathy and other motor deficits.[212] Pharmacologic interventions, such as neuroprotective and anti-convulsant medication, can reduce tremors and neuropathy.[211] The neuro-oncology team should be actively aware of any treatment-induced motor deficits and attempt to minimize dose toxicity when possible.[211] Physical therapy that focuses on balance and coordination may be an important resource; the patient may also need assistive mobility devices.[207] Multidisciplinary rehabilitation interventions, including neurology, rehabilitative, and palliative care services, have been shown to significantly increase self-care abilities, sphincter control, locomotion, and mobility in a sample of posttreatment brain tumor patients.[213]

4.5 Sexual behavior

A brain tumor and/or its treatments may impact sexual behavior and sexuality. In a qualitative study of adult brain tumor patients' life experiences, patients reported loss of physical and emotional intimacy within their relationship.[214] Additionally, if a brain tumor patient is prescribed an antidepressant, they may also report a decreased sexual motivation.[215] In some cases, these changes led to distancing or divorce. Although research on sexual behavior has been conducted with other cancers,[216] little is known about the factors that underlie a sexual desire change in brain tumor patients. Moreover, adolescent PBT patients are often classified as pediatric, and health professionals may not be attentive to the physical and sexual changes that are occurring in these patients.[217] Patients who endorse changes in sexual desire and behavior may benefit from being referred to clinicians who specialize in sex counseling and relationships.[214]

5 Social concerns

The diagnosis and treatment of a brain tumor not only affects various aspects of the individual—such as the patient's cognitive functioning, affect and mood, and/or behaviors—but the experience of a brain tumor also, inevitably, extends beyond the individual and impacts the lives of close others, especially that of the primary caregiver. We summarize the effects on the patient's family, friends, and co-workers. Some relationships will encounter strain and others will grow; nevertheless, social support is one of the strongest predictors for QoL in brain tumor patients and their caregivers.

5.1 Relationships with others

Individuals with brain tumor may notice changes in their social relationships over the course of the disease.[218,219] Extant research has found evidence for both social decline and social development upon receiving a brain tumor diagnosis. Many patients report feeling like a "burden" or find their social lives limited by myriad medical appointments and various treatment side effects.[220,221] Others find that the stress of the disease brings them closer to the ones they love most, including primary caregivers, and some patients may even forge new relationships through advocacy or peer support.[222–224] Regardless of whether old bonds are strengthened or new relationships acquired, social support has important implications for QoL in brain tumor patients and may mitigate the negative psychological effects of functional impartment.[218]

5.2 Caring for brain tumor patients

Caregivers are one group particularly affected by the diagnosis and treatment of brain tumors. Given the aforementioned cognitive, psychological, and behavioral challenges associated with the presence of a brain tumor, it is perhaps no surprise that the caregiving experience might be both emotionally and physically taxing, depending on the level of patient impairment—which is subject to change rapidly.[225,226] Studies show that caregivers of patients with brain tumors experience a clinically significant reduction in overall QoL as compared to the general population.[227] Family caregivers must take on more responsibility for monitoring the illness; arranging treatments; communicating with healthcare professionals; and providing physical, emotional, and social support to the patient. They are challenged to make decisions as care needs change, yet often feel untrained and unprepared as they adjust to new roles and responsibilities. They may have difficulty understanding complicated medical directions, be unskilled in medical duties, or find this type of care too emotionally burdensome.[228]

Family caregivers of brain tumor patients report high levels of stress and poor physical and emotional health, as well as career sacrifices, monetary losses, and workplace discrimination. Because the focus is on the patient, the needs of the caregivers are often neglected.[229]

In Sherwood and colleagues'[230] conceptual model, a caregiver's stress response is a function of the caregiver's primary appraisal of their loved one's diagnosis (e.g., tumor status, extent of neuropsychological impairment) in relation to the caregiver's secondary appraisal of available resources (e.g., education, financial support, sense of self-efficacy; see Fig. 1). Caregiver burden increases when the patient's needs are greater than the available caregiving resources. A seminal qualitative analysis found that caregiver burden frequently centers around four main types of challenges: (1) *family issues*, including changing roles and family dynamics; (2) *managing challenging patient behaviors*, such as neurocognitive and personality changes; (3) *personal feelings*, such as feeling overwhelmed, depressed, or angry; and (4) *navigating the medical system*, including communication with the medical team and understanding the illness.[229] Moreover, because caregivers do not live in a vacuum, the stress of the caregiver may impact the well-being of the patient and vice versa,[231,232] though the type of stress experienced by caregivers and patients often differs.[233] There is even evidence to suggest that caregivers experience more symptoms of anxiety and depression than the patients themselves.[233,234]

To address caregiver burden, healthcare providers might intervene by identifying families at risk and providing the caregiver with additional resources, such as medical education or appropriate supportive referrals.[230,235] Additionally, social support—whether that be formal or informal—has been found to alleviate caregiver burden. Caregivers may find brain tumor support groups, individual or family therapy, spiritual networks, or simply the support of close friends and family members especially valuable during this stressful period.[219,220,236–239] As suggested previously, a caregiver's use of social support may, in turn, positively impact the patient's overall QoL.[240] Finally, while most of the literature has focused on deficits, there is evidence to suggest that the experience of caring for a brain tumor patient might ultimately lead to positive changes within the caregiver, including stronger relationships between family members or the discovery of inner strength and resilience.[241,242]

5.3 Employment

The ability to return to work is an important patient-centered outcome, which may contribute to increased QoL, sense of individual purpose, and family support among brain tumor patients. The likelihood of returning to work may vary by diagnosis and time since surgery or treatment. For patients with diffuse low-grade

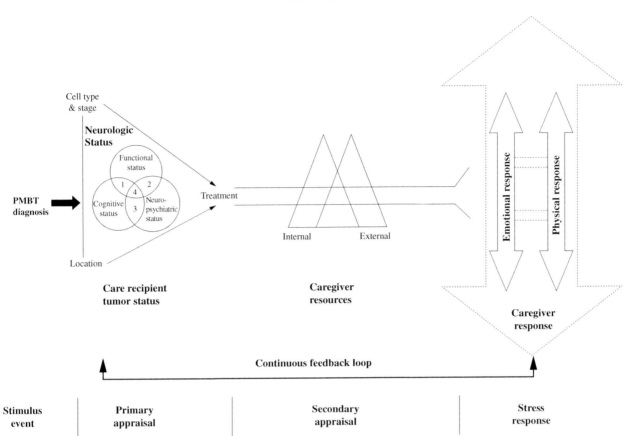

FIG. 1   Conceptual model of caring for persons with a brain tumor.[230]

gliomas, 97.1% were found to return to work within 1-year post-surgery.[243] For those with predominately grade II and III gliomas, between 57% and 70% returned to work 1-year post-surgery.[244] By contrast, rates of employment were significantly lower among patients diagnosed with glioblastomas, where only 13% of patients returned to work 1-year post-diagnosis.[245] Factors associated with a higher likelihood of returning to work include smaller tumor volume, less fatigue, fewer comorbidities, younger age, and the patient's identification as the primary bread-winner.[245–247] However, even those who are able to return to work may experience significant challenges as a result of their medical condition and its treatment. For example, specific neuropsychological deficits are associated with work-related performance, including poor time management, difficulty meeting physical work demands, and difficulty meeting mental/interpersonal work demands.[248] Specifically, the most common reported workplace difficulties include following the flow of events, remembering one's train of thought while speaking, putting together materials for a task, shifting between tasks, and following written instructions.[249] While vocational rehabilitation programs may increase patients' ability to return to work,[250] further research is needed to determine whether such programs can mitigate work-related impairments within neuro-oncological populations.

6 Disparity concerns

There are very limited data available on health disparities within individuals with brain tumors.[251] However, findings from wider bodies of literature highlight potential inequities that may be applicable to this population. Curry and Barker[251] explain that racial and ethnic minorities as well as individuals with low socioeconomic resources generally suffer from poorer health and less access to care—these disparities may be even more insidious in the case of brain cancer and its treatment.

6.1 Cognitive considerations

Ideally, neuropsychological assessments include a "baseline" or "premorbid" timepoint to evaluate an individual's cognitive functioning prior to some event (e.g., an intervention). With respect to neuropsychological testing in the context of medical diagnoses, however, it is more often the case that providers only have access to data following diagnosis—or even following treatment. In these cases, individual scores are compared to normative data to evaluate the presence and magnitude of cognitive impairment. There are also mechanisms available to estimate ones functioning prior to brain insult (e.g., education or occupational level, reading capability). To date, no comprehensive normative data exist for racial/ethnic brain tumor minorities in the United States. In the context of identifying and diagnosing neurological concerns, the consideration of race-based norms serves as a proxy for social constructs such as acculturation, education quality, and early life experiences.[252] Thus, the lack of availability of such data may result in diminished ability to accurately detect—and in turn, refer and treat—minorities with cognitive impairment. However, adjusting for demographics is not always warranted; thus, neuropsychologists and clinicians must be sure to stay up to date with best practices regarding this issue.[252,253] Other concerns that may produce or exacerbate disparities are the extent to which racial and ethnic minorities have access to neuropsychological services, trust in the medical community, and/or are aware of the potential benefits of neuropsychological testing.[252,254]

6.2 Psychological considerations

There is also a relative paucity of data examining disparities in the psychosocial context of brain tumor patients. The broader neuro-oncology literature indicates that Black and Hispanic patients have less access to high-quality neuro-oncological services, which likely extends to multidisciplinary teams which would typically provide supportive care.[255] In a recent study examining supportive care in older patients with brain metastases, the authors found that compared to non-Hispanic Whites, racial minority individuals received far fewer supportive medications to reduce symptoms that dramatically affect QoL, including antipsychotic/antidelirium and antidepressant medications among others.[256] Likewise, in a study examining palliative care in brain tumor patients, the authors state that providers have an obligation to investigate and address the psychosocial and palliative needs of elderly and racially and ethnically diverse patients.[257] Further, they also go on to report that although there are systemic barriers evident, it is important and necessary to engage these patients at the individual level in clinical encounters to ensure that they are provided access to the same level of support services and care as patients from more privileged groups.[257]

7 End-of-life concerns

While the end-of-life phase is an important aspect of the cancer care continuum generally,[258] this is especially true for brain tumor patients. The diagnosis and treatment of a brain tumor is associated with unique challenges, such as limited and noncurative treatment options, high likelihood of disease recurrence, rapid cognitive and physical decline, and overall poor prognosis.

Each of these challenges may prompt "double awareness," where brain tumor patients must simultaneously balance their hope for survival with death acceptance.[259] Indeed, as discussed earlier, many patients encounter existential distress, such as a fear of death or trouble-making sense of this disease, throughout this process. Yet only a small percentage of brain tumor patients will receive a consultation for palliative care,[260] create an advanced care directive with their family or medical team,[261] or receive therapeutic services from a chaplain or counselor for existential distress.[262] Given the current prognosis of the disease and the ubiquity of existential distress in this population, it is important that the medical team treats every stage of the disease equally, including patient and familial concerns at the end of life.

7.1 Palliative care

The goal of palliative care is to manage symptoms associated with serious medical illness in order to reduce patient suffering.[263] Palliative care can be divided into two domains with regard to delivery of services: (1) *primary palliative care*, which may be provided by any medical practitioner, and (2) *specialty palliative care*, which is provided by a team of providers specifically trained to work with palliative issues.[264] Although palliative care originated in the hospice movement and was historically reserved for end of life, recent data indicate that the introduction of palliative care services at earlier stages in disease progression likely elicits better QoL.[263]

The timing of the introduction of palliative care services is especially pertinent for patients with brain tumors, as they often experience a swift decline in cognitive and physical functioning, creating the potential for significant turmoil regarding end-of-life care and decisions. Extant data indicate that the median time of referral for palliative care services in brain tumor patients is 28–70 days prior to death.[263] At the same time, as many as 64% of patients lose their ability to communicate effectively in the final weeks of life, a time when palliative care is often introduced for many patients.[260,263] As such, there is a strong case to be made for introducing the services of palliative care and involving patients and their caregivers as soon as possible, in order to maximize the potential benefit that patients could gain from palliation. As the primary function of palliative care is to minimize suffering and maximize QoL, the high symptom burden experienced in this population in many ways makes neuro-oncological patients the prime candidates for palliative care; however, data indicate that this population utilizes these services less frequently. In their retrospective study, Gofton et al.[260] found that only 12% of brain tumor patients received a consultation for palliative care. A recent review of palliative care in patients with high-grade glioma explored some of the barriers such as conflation of hospice and palliative care, lack of a clear model for integrating palliative care into neuro-oncology, and limited education on neurologic disease among palliative care specialists.[264] Fig. 2 provides a summary of the complex challenges involved in integrating palliative care into neuro-oncology.

7.2 Advanced care planning

Advanced care planning (ACP) is the ongoing process that involves decisions made by patients, in consultation with surrogate decision makers, family, and healthcare providers regarding values, beliefs, life-sustaining treatment preferences, goals of care, and palliative care options, should they later become incapable of expressing such wishes.[265] ACP may further include the patient completing an advance directive that documents their wishes and/or appointment of a substitute decision maker. Palliative care literature emphasizes that ACP increases the likelihood of a person dying in their preferred place, increases hospice use, and reduces hospitalization,

producing lower stress levels and depression in surviving relatives.[265] However, a recent review of ACP in patients with brain tumors demonstrated that 12%–40% of patients did not have end-of-life discussions; when these discussions did occur, the majority occurred within 1 month of death.[261] A recent qualitative study of brain tumor stakeholders (patients, caregiver, and providers) identified three key themes toward improving ACP: (1) attitudes as a barrier, (2) amplify patient's voice, and (3) optimal timing of ACP.[266] Patients shared previous negative experiences of medical providers and caregivers avoiding or discouraging these conversations due to societal taboo, cultural barriers, lack of experience, anxiety or fear, or the perception that ACP implies diminished hope (e.g., "not fighting" or "giving up"). Rather, following ACP discussions, patients expressed an enhanced sense of autonomy and control, while caregivers and clinicians similarly reported that ACP improved communication, decreased confusion, and reduced emotional distress during decision making in emergency situations or at the end of life. The timing of ACP elicited conflicting

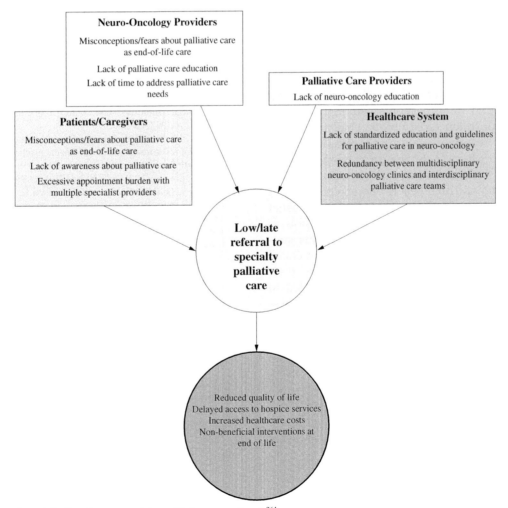

FIG. 2   Conceptual model of barriers to specialty palliative care referral.[264]

preferences among patients, caregivers, and providers. Medical providers expressed difficulty having ACP conversations too early in diagnosis due to patients and loved ones still processing the diagnosis; further, providers feel these conversations project message of "giving up" or "not offering hope" to patients and their loved ones. However, caregivers shared that they often delayed ACP discussions only to realize that it was too late as the patient had already cognitively declined beyond their ability to discuss their preferences. As outlined earlier in this chapter, patients with brain tumors typically experience a high symptom burden from tumor location and treatment that negatively impacts QoL, cognition, and long-term ability to make treatment decisions independently.[267] Therefore, involving medical providers, patients, and caregivers in ACP discussions earlier in the disease course is recommended with a focus on QoL through end-of-life care (aka Quality of Dying).

7.3 Religion and spirituality

End-of-life care for cancer patients often includes the assessment and treatment of a patient's spiritual needs.[268] Indeed, in the larger oncology literature, one's religion and/or spirituality has been found to be a source of strength and resilience, an aspect of personal development, and even, at times, a component of existential distress or inquiry for advanced cancer patients.[171,269,270] Limited research has considered the role of religion and spirituality in brain tumor patients, although this oncology population often faces unique challenges such as non-curative treatments, poor prognoses, and rapid cognitive and/or physical decline. In previous qualitative studies,[220,271,272] brain tumor patients expressed that their religious and spiritual beliefs offered comfort in a time of uncertainty and isolation; specifically, the belief in an afterlife and omnipotent God allowed patients to feel "calm."[220] Additional evidence from patients suggests that religious and spiritual beliefs were beneficial in making sense of stressful events or reappraising their lives.[220,271,272] Nevertheless, other patients struggled to reconcile their beliefs with their diagnosis and questioned why God had sent them such illness.[272] Though not all patients will encounter religious or spiritual challenges, brain tumor patients may benefit from the availability of hospital chaplains[273] or existential psychotherapies (e.g., CALM therapy)[171] that address religious and spiritual concerns as part of treatment.[271] In a retrospective study of services utilized, the authors found that only 31% of brain tumor patients accessed chaplains or had the ongoing support from their local religious community; furthermore, only 8% of patients utilized a counseling service.[262] In the absence of chaplains or existential psychotherapy services, patients have expressed

a desire to pray or discuss their end-of-life concerns with medical staff, including nurses.[271] Given the aforementioned rate of existential distress in this population, further attention to brain tumor patients' religious and spiritual needs is warranted.

8 Conclusion

Receiving a cancer diagnosis has a profound impact on patients and their caregivers. Treatment advances continue to improve survival for patients with brain tumors, necessitating attention to the array of psychosocial needs that may present during the neuro-oncology trajectory. This requires an interdisciplinary team to collaboratively offer interventions aimed to improve QoL, maximize everyday function, minimize symptom burden, and adhere to patient and family care priorities and preferences. Further research is warranted to better our understanding of underlying mechanisms of changes in, and the dynamic interactions between, neurocognitive and psychological functions. The field of oncology has finally realized that quality of life, along with quantity, matters.

References

1. Wefel JS, Lenzi R, Theriault RL, Davis RN, Meyers CA. The cognitive sequelae of standard-dose adjuvant chemotherapy in women with breast carcinoma: results of a prospective, randomized, longitudinal trial. *Cancer*. 2004;100(11):2292–2299. https://doi.org/10.1002/cncr.20272.
2. Moleski M. Neuropsychological, neuroanatomical, and neurophysiological consequences of CNS chemotherapy for acute lymphoblastic leukemia. *Arch Clin Neuropsychol*. 2000;15(7):603–630. https://doi.org/10.1016/S0887-6177(99)00050-5.
3. Shilling V, Jenkins V, Morris R, Deutsch G, Bloomfield D. The effects of adjuvant chemotherapy on cognition in women with breast cancer—preliminary results of an observational longitudinal study. *Breast*. 2005;14(2):142–150. https://doi.org/10.1016/j.breast.2004.10.004.
4. Tucha O, Smely C, Preier M, Lange KW. Cognitive deficits before treatment among patients with brain tumors. *Neurosurgery*. 2000;47:324–333 [discussion 333–334] https://doi.org/10.1097/00006123-200008000-00011.
5. Locke DEC, Cerhan JH, Wu W, et al. Cognitive rehabilitation and problem-solving to improve quality of life of patients with primary brain tumors: a pilot study. *J Support Oncol*. 2008;6(8):383–391. http://www.ncbi.nlm.nih.gov/pubmed/19149323%5Cnhttp://www.scopus.com/inward/record.url?eid=2-s2.0-59449097658&partnerID=40&md5=f0f909f7f954f-47216cec2758e57c203.
6. Gehrke AK, Baisley MC, Sonck ALB, Wronski SL, Feuerstein M. Neurocognitive deficits following primary brain tumor treatment: systematic review of a decade of comparative studies. *J Neuro-Oncol*. 2013;115(2):135–142. https://doi.org/10.1007/s11060-013-1215-2.
7. Mahalakshmi P, Vanisree AJ. Quality of life measures in glioma patients with different grades: a preliminary study. *Indian J Cancer*. 2015;52(4):580–585. https://doi.org/10.4103/0019-509X.178395.

8. de Boer AGEM, Taskila T, Ojajärvi A, van Dijk FJH, Verbeek JHAM. Cancer survivors and unemployment: a meta-analysis and meta-regression. *JAMA*. 2009;301(7):753–762. https://doi.org/10.1001/jama.2009.187.

9. Feuerstein M, Hansen JA, Calvio LC, Johnson L, Ronquillo JG. Work productivity in brain tumor survivors. *J Occup Environ Med*. 2007;49(7):803–811. https://doi.org/10.1097/JOM.0b013e318095a458.

10. Salander P, Bergenheim AT, Henriksson R. How was life after treatment of a malignant brain tumour? *Soc Sci Med*. 2000;51(4):589–598. http://ovidsp.ovid.com/ovidweb.cgi?T=JS&PAGE=referen ce&D=emed5&NEWS=N&AN=2000165064.

11. Burg MA, Adorno G, Lopez EDS, et al. Current unmet needs of cancer survivors: analysis of open-ended responses to the American Cancer Society study of cancer survivors II. *Cancer*. 2015;121(4):623–630. https://doi.org/10.1002/cncr.28951.

12. Ownsworth T, Hawkes A, Steginga S, Walker D, Shum D. A biopsychosocial perspective on adjustment and quality of life following brain tumor: a systematic evaluation of the literature. *Disabil Rehabil*. 2009;31(13):1038–1055. https://doi.org/10.1080/09638280802509538.

13. Crossen JR, Garwood D, Glatstein E, Neuwelt EA. Neurobehavioral sequelae of cranial irradiation in adults: a review of radiation-induced encephalopathy. *J Clin Oncol*. 1994;12(3):627–642. https://doi.org/10.1200/jco.1994.12.3.627.

14. Taphoorn MJ, Klein M. Cognitive deficits in adult patients with brain tumours. *Lancet Neurol*. 2004;3(3):159–168. https://doi.org/10.1016/S1474-4422(04)00680-5.

15. Meyers C, Kayl AE. Neurocognitive function. In: Levin V, ed. *Cancer in the Nervous System*. 2nd ed. New York: Oxford University Press; 2002.

16. Kayl A, Meyers C. Neuropsychological impact of brain metastases and its treatment. In: Sawaya R, ed. *Intracranial Metastases*. Malden: Blackwell Publishing; 2004:430–459.

17. Kayl A, Meyers C. Neuropsychological complications in patients with brain tumors. In: *Palliative Care Consultations in Primary and Metastatic Brain Tumors*. New York: Oxford University Press; 2004:83–92.

18. Heilman K, Valenstein E. *Clinical Neuropsychology*. 4th ed. New York: Oxford University Press; 2003.

19. Halligan P, Kischka U, Marshall J. *Handbook of Clinical Neuropsychology*. New York: Oxford University Press; 2003.

20. Martin G. *Human Neuropsychology*. 2nd ed. Upper Saddle River: Prentice-Hall; 2006.

21. Lezak M. *Neuropsychological Assessment*. 3rd ed. New York: Oxford University Press; 1995.

22. Weitzner M, Meyers C. Quality of life and neurobehavioral functioning in patients with malignant gliomas. In: Yung W, ed. *Bailliere's Clinical Neurology: Cerebral Gliomas*. London: Bailliere Tindall; 1996:425–439.

23. Meyers C. Quality of life of brain tumor patients. In: *Neuro-Oncology: The Essentials*. 1st ed. New York: Thieme Medical Publishers; 2000.

24. Coltheart M, Sartori G, Job R. *The Cognitive Neuropsychology of Language*. London: Lawrence Erlbaum Associates Ltd; 1987.

25. Hillis A. *The Handbook of Adult Language Disorders: Integrating Cognitive Neuropsychology, Neurology, and Rehabilitation*. New York: Psychology Press; 2002.

26. Packer R, Miller D, Shaffrey M. Intracranial neoplasms. In: Rosenberg R, Pleasure D, eds. *Comprehensive Neurology*. New York: John Wiley; 1998:187–243.

27. Lilja A, Smith GJW, Salford LG. Microprocesses in perception and personality. *J Nerv Ment Dis*. 1992. https://doi.org/10.1097/00005053-199202000-00003.

28. Janelsins MC, Kohli S, Mohile SG, Usuki K, Ahles TA, Morrow GR. An update on cancer- and chemotherapy-related cognitive dysfunction: current status. *Semin Oncol*. 2011. https://doi.org/10.1053/j.seminoncol.2011.03.014.

29. Koppelmans V, Breteler MMB, Boogerd W, Seynaeve C, Schagen SB. Late effects of adjuvant chemotherapy for adult onset non-CNS cancer; cognitive impairment, brain structure and risk of dementia. *Crit Rev Oncol Hematol*. 2013;88(1):87–101. https://doi.org/10.1016/j.critrevonc.2013.04.002.

30. Douw L, Klein M, Fagel SS, et al. Cognitive and radiological effects of radiotherapy in patients with low-grade glioma: long-term follow-up. *Lancet Neurol*. 2009;8(9):810–818. https://doi.org/10.1016/S1474-4422(09)70204-2.

31. Scheibel RS, Meyers CA, Levin VA. Cognitive dysfunction following surgery for intracerebral glioma: influence of histopathology, lesion location, and treatment. *J Neuro-Oncol*. 1996;30(1):61–69. https://doi.org/10.1007/BF00177444.

32. Spreen O, Strauss E. *A Compendium of Neuropsychological Tests: Administration, Norms, and Commentary*. Oxford University Press; 1998:1216.

33. Wieneke MH. Neuropsychological assessment of cognitive functioning following chemotherapy for breast cancer. *Psychooncology*. 1995;4(1):61–66.

34. Freeman JR, Broshek DK. Assessing cognitive dysfunction in breast cancer: what are the tools? *Clin Breast Cancer*. 2002;3(Suppl. 3):S91–S99. http://www.embase.com/search/results?subaction=viewrecord&from=export&id=L36495688%5Cnhttp://sfx.umd.edu/hs?sid=EMBASE&issn=15268209&id=doi:&title=Assessing+cognitive+dysfunction+in+breast+cancer:+what+are+the+tools?&stitle=Clin.+Breast+Cancer&title=Cli.

35. Ahles TA, Saykin AJ, Furstenberg CT, et al. Neuropsychologic impact of standard-dose systemic chemotherapy in long-term survivors of breast cancer and lymphoma. *J Clin Oncol*. 2002;20(2):485–493. https://doi.org/10.1200/JCO.20.2.485.

36. Corbetta M, Patel G, Shulman GL. *Review the Reorienting System of the Human Brain: From Environment to Theory of Mind*; 2008:306–324. https://doi.org/10.1016/j.neuron.2008.04.017.

37. Corbetta M, Kincade JM, Shulman GL. Neural systems for visual orienting and their relationships to spatial working memory. *J Cogn Neurosci*. 2002;14(3):508–523. https://doi.org/10.1162/089892902317362029.

38 Karnath H-O, Milner D, Vallar G. *The Cognitive and Neural Bases of Spatial Neglect*. Oxford University Press; 2002.

39. Rønning C, Sundet K, Due-Tønnessen B, Lundar T, Helseth E. Persistent cognitive dysfunction secondary to cerebellar injury in patients treated for posterior fossa tumors in childhood. *Pediatr Neurosurg*. 2005. https://doi.org/10.1159/000084860.

40. Manto M, Mariën P. Schmahmann's syndrome—identification of the third cornerstone of clinical ataxiology. *Cerebellum Ataxias*. 2015;2(2). https://doi.org/10.1186/s40673-015-0023-1.

41. Schmahmann JD, Sherman JC. The cerebellar cognitive affective syndrome. *Brain*. 1998. https://doi.org/10.1093/brain/121.4.561.

42. Hoche F, Guell X, Vangel MG, Sherman JC, Schmahmann JD. The cerebellar cognitive affective/Schmahmann syndrome scale. *Brain*. 2018;141:248–270. https://doi.org/10.1093/brain/awx317.

43. Riva D, Giorgi C. The cerebellum contributes to higher functions during development. Evidence from a series of children surgically treated for posterior fossa tumours. *Brain*. 2000. https://doi.org/10.1093/brain/123.5.1051.

44. Levisohn L, Cronin-Golomb A, Schmahmann JD. Neuropsychological consequences of cerebellar tumour resection in children. Cerebellar cognitive affective syndrome in a paediatric population. *Brain*. 2000. https://doi.org/10.1093/brain/123.5.1041.

45. Leggio MG, Silveri MC, Petrosini L, Molinari M. Phonological grouping is specifically affected in cerebellar patients: a verbal

fluency study. *J Neurol Neurosurg Psychiatry*. 2000. https://doi.org/10.1136/jnnp.69.1.102.

46. Fiez JA, Petersen SE, Cheney MK, Raichle ME. Impaired non-motor learning and error detection associated with cerebellar damage: a single case study. *Brain*. 1992. https://doi.org/10.1093/brain/115.1.155.

47. Grafman J, Litvan I, Massaquoi S, Stewart M, Sirigu A, Hallett M. Cognitive planning deficit in patients with cerebellar atrophy. *Neurology*. 1992. https://doi.org/10.1212/wnl.42.8.1493.

48. Appollonio IM, Grafman J, Schwartz V, Massaquoi S, Hallett M. Memory in patients with cerebellar degeneration. *Neurology*. 1993. https://doi.org/10.1212/wnl.43.8.1536.

49. Drepper J, Timmann D, Kolb FP, Diener HC. Non-motor associative learning in patients with isolated degenerative cerebellar disease. *Brain*. 1999. https://doi.org/10.1093/brain/122.1.87.

50. Townsend J, Courchesne E, Covington J, et al. Spatial attention deficits in patients with acquired or developmental cerebellar abnormality. *J Neurosci*. 1999. https://doi.org/10.1523/jneurosci.19-13-05632.1999.

51. Courchesne E, Townsend J, Akshoomoff NA, et al. Impairment in shifting attention in autistic and cerebellar patients. *Behav Neurosci*. 1994. https://doi.org/10.1037/0735-7044.108.5.848.

52. Molinari M, Leggio MG, Silveri MC. Verbal fluency and agrammatism. *Int Rev Neurobiol*. 1997. https://doi.org/10.1016/s0074-7742(08)60358-x.

53. Hoche F, Guell X, Sherman JC, Vangel MG, Schmahmann JD. Cerebellar contribution to social cognition. *Cerebellum*. 2016;15:732–743. https://doi.org/10.1007/s12311-015-0746-9.

54. Guell X, Hoche F, Schmahmann JD. Metalinguistic deficits in patients with cerebellar dysfunction: empirical support for the dysmetria of thought theory. *Cerebellum*. 2015;14:50–58. https://doi.org/10.1007/s12311-014-0630-z.

55. Beldarrain G, Garcia-Moncó C, Quintana M, Llorens V, Rodeño E. Diaschisis and neuropsychological performance after cerebellar stroke. *Eur Neurol*. 1997. https://doi.org/10.1159/000117415.

56. Helmuth LL, Ivry RB, Shimizu N. Preserved performance by cerebellar patients on tests of word generation, discrimination learning, and attention. *Learn Mem*. 1997. https://doi.org/10.1101/lm.3.6.456.

57. Daum I, Ackermann H. Neuropsychological abnormalities in cerebellar syndromes-fact or fiction? *Int Rev Neurobiol*. 1997. https://doi.org/10.1016/s0074-7742(08)60365-7.

58. Friedman MA, Meyers CA, Sawaya R, Bruce JN, Hodge CJ, Piepmeier JM. Neuropsychological effects of third ventricle tumor surgery. *Neurosurgery*. 2003. https://doi.org/10.1227/01.NEU.0000053367.94965.6B.

59. Dietrich J, Prust M, Kaiser J. Chemotherapy, cognitive impairment and hippocampal toxicity. *Neuroscience*. 2015;309:224–232. https://doi.org/10.1016/j.neuroscience.2015.06.016.

60. McGirt MJ, Chaichana KL, Gathinji M, et al. Independent association of extent of resection with survival in patients with malignant brain astrocytoma: clinical article. *J Neurosurg*. 2009;110(1):156–162. https://doi.org/10.3171/2008.4.17536.

61. Chang EF, Clark A, Smith JS, et al. Functional mapping-guided resection of low-grade gliomas in eloquent areas of the brain: improvement of long-term survival—clinical article. *J Neurosurg*. 2011;114(3):566–573. https://doi.org/10.3171/2010.6.JNS091246.

62. Allen DH, Loughan AR. Impact of cognitive impairment in patients with gliomas. *Semin Oncol Nurs*. 2018;34(5). https://doi.org/10.1016/j.soncn.2018.10.010.

63. Hodgson KD, Hutchinson AD, Wilson CJ, Nettelbeck T. A meta-analysis of the effects of chemotherapy on cognition in patients with cancer. *Cancer Treat Rev*. 2013. https://doi.org/10.1016/j.ctrv.2012.11.001.

64. Hosseini SMH, Koovakkattu D, Kesler SR. Altered small-world properties of gray matter networks in breast cancer. *BMC Neurol*. 2012. https://doi.org/10.1186/1471-2377-12-28.

65. Pendergrass JC, Targum SD, Harrison JE. Cognitive impairment associated with cancer: a brief review. *Innov Clin Neurosci*. 2018;15(1–2):36–44.

66. Seigers R, Schagen SB, Van Tellingen O, Dietrich J. Chemotherapy-related cognitive dysfunction: current animal studies and future directions. *Brain Imaging Behav*. 2013;7(4):453–459. https://doi.org/10.1007/s11682-013-9250-3.

67. Nokia MS, Anderson ML, Shors TJ. Chemotherapy disrupts learning, neurogenesis and theta activity in the adult brain. *Eur J Neurosci*. 2012;36(11):3521–3530. https://doi.org/10.1111/ejn.12007.

68. Gilbert MR, Dignam JJ, Armstrong TS, et al. A randomized trial of bevacizumab for newly diagnosed glioblastoma. *N Engl J Med*. 2014;370(8):699–708. https://doi.org/10.1056/NEJMoa1308573.

69. Greene-Schloesser D, Robbins ME. Radiation-induced cognitive impairment-from bench to bedside. *Neuro-Oncology*. 2012;14(Suppl. 4). https://doi.org/10.1093/neuonc/nos196.

70. Wilke C, Grosshans D, Duman J, Brown P, Li J. Radiation-induced cognitive toxicity: pathophysiology and interventions to reduce toxicity in adults. *Neuro-Oncology*. 2018. https://doi.org/10.1093/neuonc/nox195.

71. Tomé WA, Gökhan Ş, Brodin NP, et al. A mouse model replicating hippocampal sparing cranial irradiation in humans: a tool for identifying new strategies to limit neurocognitive decline. *Sci Rep*. 2015;5(August):14384. https://doi.org/10.1038/srep14384.

72. Monje ML, Vogel H, Masek M, Ligon KL, Fisher PG, Palmer TD. Impaired human hippocampal neurogenesis after treatment for central nervous system malignancies. *Ann Neurol*. 2007;62(5):515–520. https://doi.org/10.1002/ana.21214.

73. Wefel JS, Noll KR, Scheurer ME. Neurocognitive functioning and genetic variation in patients with primary brain tumours. *Lancet Oncol*. 2016;17(3):e97–e108. https://doi.org/10.1016/S1470-2045(15)00380-0.

74. Correa DD, Satagopan J, Martin A, et al. Genetic variants and cognitive functions in patients with brain tumors. *Neuro-Oncology*. 2019;21(10):1297–1309. https://doi.org/10.1093/neuonc/noz094.

75. Meyers CA, Wefel JS. The use of the mini-mental state examination to assess cognitive functioning in cancer trials: no ifs, ands, buts, or sensitivity. *J Clin Oncol*. 2003;21(19):3557–3558. https://doi.org/10.1200/JCO.2003.07.080.

76. Aslanzadeh F, Braun SE, Brechbiel J, et al. *Screening for Cognitive Impairment in Primary Brain Tumor Patients: A Preliminary Investigation With the MMSE-2 and RBANS*. International Neuropsychological Society; 2020.

77. Wefel JS, Vardy J, Ahles T, Schagen SB. International Cognition and Cancer Task Force recommendations to harmonise studies of cognitive function in patients with cancer. *Lancet Oncol*. 2011;12(7):703–708. https://doi.org/10.1016/S1470-2045(10)70294-1.

78. Benedict RHB, Schretlen D, Groninger L, Brandt J. Hopkins verbal learning test-revised: normative data and analysis of inter-form and test-retest reliability. *Clin Neuropsychol*. 1998;12(1):43–55. https://doi.org/10.1076/clin.12.1.43.1726.

79. Patterson J. Multilingual aphasia examination. In: *Encyclopedia of Clinical Neuropsychology*; 2011:1674–1676. https://doi.org/10.1007/978-0-387-79948-3_900.

80. Tombaugh TN. Trail Making Test A and B: normative data stratified by age and education. *Arch Clin Neuropsychol*. 2004;19(2):203–214. https://doi.org/10.1016/S0887-6177(03)00039-8.

81. Rugo HS, Ahles T. The impact of adjuvant therapy for breast cancer on cognitive function: current evidence and directions for research. *Semin Oncol*. 2003;30(6). https://doi.org/10.1053/j.seminoncol.2003.09.008.

82. Cull A, Hay C, Love SB, Mackie M, Smets E, Stewart M. What do cancer patients mean when they complain of concentration and memory problems? *Br J Cancer.* 1996;74(10):1674–1679. https://doi.org/10.1038/bjc.1996.608.

83. Poppelreuter M, Weis J, Külz AK, Tucha O, Lange KW, Bartsch HH. Cognitive dysfunction and subjective complaints of cancer patients: a cross-sectional study in a cancer rehabilitation centre. *Eur J Cancer.* 2004;40(1):43–49. https://doi.org/10.1016/j.ejca.2003.08.001.

84. Gehring K, Taphoorn MJB, Sitskoorn MM, Aaronson NK. Predictors of subjective versus objective cognitive functioning in patients with stable grades II and III glioma. *Neurooncol Pract.* 2015;2(1):20–31. https://doi.org/10.1093/nop/npu035.

85. van der Linden SD, Gehring K, De Baene W, Emons WHM, Rutten GJM, Sitskoorn MM. Assessment of executive functioning in patients with meningioma and low-grade glioma: a comparison of self-report, proxy-report, and test performance. *J Int Neuropsychol Soc.* 2020;26:187–196. https://doi.org/10.1017/S1355617719001164.

86. Meyers CA, Weitzner MA, Valentine AD, Levin VA. Methylphenidate therapy improves cognition, mood, and function of brain tumor patients. *J Clin Oncol.* 1998;16(7):2522–2527. http://www.ncbi.nlm.nih.gov/pubmed/9667273.

87. Gehring K, Patwardhan SY, Collins R, et al. A randomized trial on the efficacy of methylphenidate and modafinil for improving cognitive functioning and symptoms in patients with a primary brain tumor. *J Neuro-Oncol.* 2012;107(1):165–174. https://doi.org/10.1007/s11060-011-0723-1.

88. Shaw EG, Rosdhal R, D'Agostino RB, et al. Phase II study of donepezil in irradiated brain tumor patients: effect on cognitive function, mood, and quality of life. *J Clin Oncol.* 2006. https://doi.org/10.1200/JCO.2005.03.3001.

89. Rapp SR, Case LD, Peiffer A, et al. Donepezil for irradiated brain tumor survivors: a phase III randomized placebo-controlled clinical trial. *J Clin Oncol.* 2015;33(15):1653–1659. https://doi.org/10.1200/JCO.2014.58.4508.

90. Brown PD, Pugh S, Laack NN, et al. Memantine for the prevention of cognitive dysfunction in patients receiving whole-brain radiotherapy: a randomized, double-blind, placebo-controlled trial. *Neuro-Oncology.* 2013;15(10):1429–1437. https://doi.org/10.1093/neuonc/not114.

91. Boele FW, Douw L, De Groot M, et al. The effect of modafinil on fatigue, cognitive functioning, and mood in primary brain tumor patients: a multicenter randomized controlled trial. *Neuro-Oncology.* 2013;15(10):1420–1428. https://doi.org/10.1093/neuonc/not102.

92. Page BR, Shaw EG, Lu L, et al. Phase II double-blind placebo-controlled randomized study of armodafinil for brain radiation-induced fatigue. *Neuro-Oncology.* 2015;17(10):1393–1401. https://doi.org/10.1093/neuonc/nov084.

93. Treanor CJ, Mcmenamin UC, O'Neill RF, et al. Non-pharmacological interventions for cognitive impairment due to systemic cancer treatment. *Cochrane Database Syst Rev.* 2016. https://doi.org/10.1002/14651858.CD011325.pub2.

94. Mar Fan HG, Clemons M, Xu W, et al. A randomised, placebo-controlled, double-blind trial of the effects of d-methylphenidate on fatigue and cognitive dysfunction in women undergoing adjuvant chemotherapy for breast cancer. *Support Care Cancer.* 2008. https://doi.org/10.1007/s00520-007-0341-9.

95. Fernandes H, Richard N, Edelstein K. Cognitive rehabilitation for cancer-related cognitive dysfunction: a systematic review. *Support Care Cancer.* 2019;27:3253–3279. https://doi.org/10.1007/s00520-019-04866-2.

96. Walsh S, Primeau M. Neuropsychological rehabilitation and habituation. In: Noggle C, Dean R, Barisa M, eds. *Neuropsychological Rehabilitation.* Springer; 2013.

97. Gehring K, Sitskoorn MM, Gundy CM, et al. Cognitive rehabilitation in patients with gliomas: a randomized, controlled trial. *J Clin Oncol.* 2009;27(22):3712–3722. https://doi.org/10.1200/JCO.2008.20.5765.

98. Hassler MR, Elandt K, Preusser M, et al. Neurocognitive training in patients with high-grade glioma: a pilot study. *J Neuro-Oncol.* 2010;97(1):109–115. https://doi.org/10.1007/s11060-009-0006-2.

99. Sherer M, Meyers CA, Bergloff P. Efficacy of postacute brain injury rehabilitation for patients with primary malignant brain tumors. *Cancer.* 1997;80(2):250–257. https://doi.org/10.1002/(SICI)1097-0142(19970715)80.

100. Zucchella C, Capone A, Codella V, et al. Cognitive rehabilitation for early post-surgery inpatients affected by primary brain tumor: a randomized, controlled trial. *J Neuro-Oncol.* 2013;114(1):93–100. https://doi.org/10.1007/s11060-013-1153-z.

101. Yang S, Chun MH, Son YR. Effect of virtual reality on cognitive dysfunction in patients with brain tumor. *Ann Rehabil Med.* 2014;38(6):726–733. https://doi.org/10.5535/arm.2014.38.6.726.

102. Maschio M, Dinapoli L, Fabi A, Giannarelli D, Cantelmi T. Cognitive rehabilitation training in patients with brain tumor-related epilepsy and cognitive deficits: a pilot study. *J Neuro-Oncol.* 2015;125:419–426. https://doi.org/10.1007/s11060-015-1933-8.

103. Richard NM, Bernstein LJ, Mason WP, et al. Cognitive rehabilitation for executive dysfunction in brain tumor patients: a pilot randomized controlled trial. *J Neuro-Oncol.* 2019;142:565–575. https://doi.org/10.1007/s11060-019-03130-1.

104. Huang ME, Cifu DX, Keyser-Marcus L. Functional outcome after brain tumor and acute stroke: a comparative analysis. *Arch Phys Med Rehabil.* 1998;79(11):1386–1390.

105. O'Dell MW, Barr K, Spanier D, Warnick RE. Functional outcome of inpatient rehabilitation in persons with brain tumors. *Arch Phys Med Rehabil.* 1998;79(12):1530–1534.

106. Kucherer S, Ferguson RJ. Cognitive behavioral therapy for cancer-related cognitive dysfunction. *Curr Opin Support Palliat Care.* 2017. https://doi.org/10.1097/SPC.0000000000000247.

107. Ferguson RJ, McDonald BC, Rocque MA, et al. Development of CBT for chemotherapy-related cognitive change: results of a waitlist control trial. *Psychooncology.* 2012. https://doi.org/10.1002/pon.1878.

108. Ferguson RJ, Sigmon ST, Pritchard AJ, et al. A randomized trial of videoconference-delivered cognitive behavioral therapy for survivors of breast cancer with self-reported cognitive dysfunction. *Cancer.* 2016. https://doi.org/10.1002/cncr.29891.

109. Schmitz KH, Courneya KS, Matthews C, et al. American college of sports medicine roundtable on exercise guidelines for cancer survivors. *Med Sci Sports Exerc.* 2010;42(7):1409–1426. https://doi.org/10.1249/MSS.0b013e3181e0c112.

110. Myers JS, Erickson KI, Sereika SM, Bender CM. Exercise as an intervention to mitigate decreased cognitive function from cancer and cancer treatment: an integrative review. *Cancer Nurs.* 2018. https://doi.org/10.1097/NCC.0000000000000549.

111. Jones LW, Guill B, Keir ST, et al. Exercise interest and preferences among patients diagnosed with primary brain cancer. *Support Care Cancer.* 2007;15(1):47–55. https://doi.org/10.1007/s00520-006-0096-8.

112. Molassiotis A, Zheng Y, Denton-Cardew L, Swindell R, Brunton L. Symptoms experienced by cancer patients during the first year from diagnosis: patient and informal caregiver ratings and agreement. *Palliat Support Care.* 2010;8(03):313–324. https://doi.org/10.1017/S1478951510000118.

113. Lovely MP. Symptom management of brain tumor patients. *Semin Oncol Nurs.* 2004;20(4):273–283.

114. Saeed SA, Cunningham K, Bloch RM. Depression and anxiety disorders: benefits of exercise, yoga, and meditation. *Am Fam Physician.* 2019;99(10):620–627.

115. Wang CW, Chan CHY, Ho RTH, Chan JSM, Ng SM, Chan CLW. Managing stress and anxiety through qigong exercise in healthy adults: a systematic review and meta-analysis of randomized controlled trials. *BMC Complement Altern Med*. 2014. https://doi.org/10.1186/1472-6882-14-8.

116. Madhusoodanan S, Ting MB, Farah T, Ugur U. Psychiatric aspects of brain tumors: a review. *World J Psychiatry*. 2015;5(3):273–285. https://doi.org/10.5498/wjp.v5.i3.273.

117. Loughan AR, Aslanzadeh FJ, Brechbiel J, et al. Death-related distress in adult primary brain tumor patients. *Neurooncol Pract*. 2020;7(5):498–506. https://doi.org/10.1093/nop/npaa015.

118. Tibbs MD, Huynh-Le MP, Reyes A, et al. Longitudinal analysis of depression and anxiety symptoms as independent predictors of neurocognitive function in primary brain tumor patients. *Int J Radiat Oncol Biol Phys*. 2020;108(5):1229–1239. https://doi.org/10.1016/j.ijrobp.2020.07.002.

119. Litofsky NS, Resnick AG. The relationships between depression and brain tumors. *J Neuro-Oncol*. 2009;94:153–161. https://doi.org/10.1007/s11060-009-9825-4.

120. Bunevicius A, Deltuva V, Tamasauskas S, Tamasauskas A, Bunevicius R. Screening for psychological distress in neurosurgical brain tumor patients using the Patient Health Questionnaire-2. *Psychooncology*. 2013;22(8):1895–1900. https://doi.org/10.1002/pon.3237.

121. Mainio A, Hakko H, Niemelä A, Koivukangas J, Räsänen P. Depression in relation to anxiety, obsessionality and phobia among neurosurgical patients with a primary brain tumor: a 1-year follow-up study. *Clin Neurol Neurosurg*. 2011;113(8):649–653. https://doi.org/10.1016/j.clineuro.2011.05.006.

122. Fox SW, Lyon D, Farace E. Symptom clusters in patients with high-grade glioma. *J Nurs Scholarsh*. 2007;39(1):61–67.

123. Litofsky NS, Farace E, Anderson F, et al. Depression in patients with high-grade glioma: results of the glioma outcomes project. *Neurosurgery*. 2004;54:358–367. https://doi.org/10.1227/01.NEU.0000103450.94724.A2.

124. Pelletier G, Verhoef MJ, Khatri N, Hagen N. Quality of life in brain tumor patients: the relative contributions of depression, fatigue, emotional distress, and existential issues. *J Neuro-Oncol*. 2002;57(1):41–49.

125. Wellisch DK, Kaleita TA, Freeman D, Cloughesy T, Goldman J. Predicting major depression in brain tumor patients. *Psychooncology*. 2002;11(3):230–238. https://doi.org/10.1002/pon.562.

126. Arnold SD, Forman LM, Brigidi BD, et al. Evaluation and characterization of generalized anxiety and depression in patients with primary brain tumors. *Neuro-Oncology*. 2008;10(2):171–181. https://doi.org/10.1215/15228517-2007-057.

127. Linden W, Vodermaier A, MacKenzie R, Greig D. Anxiety and depression after cancer diagnosis: prevalence rates by cancer type, gender, and age. *J Affect Disord*. 2012;121(2–3):343–351. https://doi.org/10.1016/j.jad.2012.03.025.

128. American Psychiatric Association. *Diagnostic and Statistical Manual of Mental Disorders: DSM-5*. American Psychiatric Association; 2013:991. https://doi.org/10.1176/appi.books.9780890425596.744053.

129. Mainio A, Hakko H, Timonen M, Niemelä A, Koivukangas J, Räsänen P. Depression in relation to survival among neurosurgical patients with a primary brain tumor: a 5-year follow-up study. *Neurosurgery*. 2005;56(6):1234–1241. https://doi.org/10.1227/01.NEU.0000159648.44507.7F.

130. Huang J, Zeng C, Xiao J, et al. Association between depression and brain tumor: a systematic review and meta-analysis. *Oncotarget*. 2017;8(55):94932–94943. https://doi.org/10.18632/oncotarget.19843.

131. Kilbride L, Smith G, Grant R. The frequency and cause of anxiety and depression amongst patients with malignant brain tumours between surgery and radiotherapy. *J Neuro-Oncol*. 2007;84(3):297–304. https://doi.org/10.1007/s11060-007-9374-7.

132. Pringle AM, Taylor R, Whittle IR. Anxiety and depression in patients with an intracranial neoplasm before and after tumour surgery. *Br J Neurosurg*. 1999;13(1):46–51. https://doi.org/10.1080/02688699944177.

133. Mainio A, Hakko H, Niemelä A, Koivukangas J, Räsänen P. Gender difference in relation to depression and quality of life among patients with a primary brain tumor. *Eur Psychiatry*. 2006;21(3):194–199. https://doi.org/10.1016/j.eurpsy.2005.05.008.

134. Mirijello A, Leggio L, Ferrulli A, Miceli A. State and trait anxiety and depression in patients with primary brain tumors before and after surgery: 1-year longitudinal study celiac disease view project genetics of hemangioblastoma of the central nervous system view project. *J Neurosurg*. 2008. https://doi.org/10.3171/JNS/2008/108/2/0281.

135. Rosenblum ML, Kalkanis S, Goldberg W, et al. Odyssey of hope: a physician's guide to communicating with brain tumor patients across the continuum of care. *J Neuro-Oncol*. 2009;92(3 SPEC. ISS):241–251. https://doi.org/10.1007/s11060-009-9828-1.

136. Bunevicius A, Tamasauskas S, Deltuva V, Tamasauskas A, Radziunas A, Bunevicius R. Predictors of health-related quality of life in neurosurgical brain tumor patients: focus on patient-centered perspective. *Acta Neurochir*. 2014;156(2):367–374. https://doi.org/10.1007/s00701-013-1930-7.

137. Goebel S, Mehdorn HM. Development of anxiety and depression in patients with benign intracranial meningiomas: a prospective long-term study. *Support Care Cancer*. 2013;21(5):1365–1372. https://doi.org/10.1007/s00520-012-1675-5.

138. Morantz RA, Walsh WJ. *Brain Tumors: A Comprehensive Text*. 1st ed. New York, NY: Informa HealthCare; 1993.

139. Cohen SR, Mount BM, Tomas JJN, Mount LF. Existential well-being is an important determinant of quality of life: evidence from the McGill Quality of Life Questionnaire. *Cancer*. 1996;77(3):576–586. https://doi.org/10.1002/(SICI)1097-0142(19960201)77:3<576::AID-CNCR22>3.0.CO;2-0.

140. Cavers D, Hacking B, Erridge SE, Kendall M, Morris PG, Murray SA. Social, psychological and existential well-being in patients with glioma and their caregivers: a qualitative study. *Can Med Assoc J*. 2012;184(7):373–382. https://doi.org/10.1503/cmaj.111622.

141. Ownsworth T, Nash K. Existential well-being and meaning making in the context of primary brain tumor: conceptualization and implications for intervention. *Front Oncol*. 2015;5(APR):1–6. https://doi.org/10.3389/fonc.2015.00096.

142. Loughan AR, Lanoye A, Aslanzadeh F, et al. Fear of Cancer Recurrence and Death Anxiety: unaddressed concerns for adult neuro-oncology patients. *J Clin Psychol Med Settings*. 2019;1–15.

143. Ownsworth T, Chambers S, Damborg E, Casey L, Walker DG, Shum DHK. Evaluation of the making sense of brain tumor program: a randomized controlled trial of a home-based psychosocial intervention. *Psychooncology*. 2014. https://doi.org/10.1002/pon.3687.

144. Keschner M, Bender M, Strauss I. Mental symptoms associated with brain tumor: a study of 530 verified cases. *JAMA*. 1938;110:714–718.

145. Massie MJ. Prevalence of depression in patients with cancer. *J Natl Cancer Inst Monogr*. 2004. https://doi.org/10.1093/jncimonographs/lgh014.

146. Sotelo JL, Musselman D, Nemeroff C. The biology of depression in cancer and the relationship between depression and cancer progression. *Int Rev Psychiatry*. 2014. https://doi.org/10.3109/09540261.2013.875891.

147. Irle E, Peper M, Wowra B, Kunze S. Mood changes after surgery for tumors of the cerebral cortex. *Arch Neurol*. 1994;51(2):164–174. https://doi.org/10.1001/archneur.1994.00540140070017.

148. Ismail MF, Lavelle C, Cassidy EM. Steroid-induced mental disorders in cancer patients: a systematic review. *Future Oncol.* 2017. https://doi.org/10.2217/fon-2017-0306.

149. Filley CM, Kleinschmidt-DeMasters BK. Neurobehavioral presentations of brain neoplasms. *West J Med.* 1995.

150. Madhusoodanan S, Opler MG, Moise D, et al. Brain tumor location and psychiatric symptoms: is there any association? A meta-analysis of published case studies. *Expert Rev Neurother.* 2010. https://doi.org/10.1586/ern.10.94.

151. Lezak MD, Howieson DB, David WL. *Neuropsychological Assessment.* Oxford University Press; 2004.

152. Nachev P, Husain M. Disorders of visual attention and the posterior parietal cortex. *Cortex.* 2006;42:766–773.

153. Price BH, Mesulam M. Psychiatric manifestations of right hemisphere infarctions. *J Nerv Ment Dis.* 1985;173(10):610–614. https://doi.org/10.1097/00005053-198510000-00006.

154. Mechanick JI, Hochberg FH, LaRocque A. Hypothalamic dysfunction following whole-brain irradiation. *J Neurosurg.* 1986;65(4):490–494. https://doi.org/10.3171/jns.1986.65.4.0490.

155. Scharre DW. Neuropsychiatric aspects of neoplastic, demyelinating, infectious, and inflammatory brain disorders. In: Coffey CE, Jeffrey C, eds. *Textbook of Geriatric Neuropsychiatry. Washington.* Washington, DC: American Psychiatric Press, Inc; 2000:523–547.

156. Price TRP, Goetz KL, Lovell MR. Neuropsychiatric aspects of brain tumors. In: Yodofsky SC, Hales R, eds. *The American Psychiatric Publishing Textbook of Neuropsychiatry and Clinical Neurosciences.* 4th ed. Washington, DC: American Psychiatric Publishing, Inc; 2002:735–764.

157. Nasrallah HA, McChesney CM. Psychopathology of corpus callosum tumors. *Biol Psychiatry.* 1981;16:663–669.

158. Schmahmann JD. Disorders of the cerebellum: ataxia, dysmetria of thought, and the cerebellar cognitive affective syndrome. *J Neuropsychiatr Clin Neurosci.* 2004;16(3):367–378. https://doi.org/10.1176/jnp.16.3.367.

159. Heilman KM, Bowers D, Valenstein E, Watson RT. Disorders of visual attention. *Baillieres Clin Neurol.* 1993;2(2):389–413.

160. Madhusoodanan S, Danan D, Moise D. Psychiatric manifestations of brain tumors: diagnostic implications. *Expert Rev Neurother.* 2007;7(4):343–349. https://doi.org/10.1586/14737175.7.4.343.

161. Rooney A, Grant R. Pharmacological treatment of depression in patients with a primary brain tumour. *Cochrane Database Syst Rev.* 2013;5:1–19. https://doi.org/10.1002/14651858.CD006932.pub3.

162. Vaidya R, Sood R, Karlin N, Jatoi A. Benzodiazepine use in breast cancer survivors: findings from a consecutive series of 1,000 patients. *Oncology.* 2011;81(1):9–11. https://doi.org/10.1159/000330814.

163. Pace A, Di Lorenzo C, Guariglia L, et al. End of life issues in brain tumor patients. *J Neuro-Oncol.* 2009;91(1):39–43. https://doi.org/10.1007/s11060-008-9670-x.

164. Platt LM, Whitburn AI, Platt-Koch AG, Koch RL. Nonpharmacological alternatives to benzodiazepine drugs for the treatment of anxiety in outpatient populations: a literature review. *J Psychosoc Nurs Ment Health Serv.* 2016;54(8):35–42. https://doi.org/10.3928/02793695-20160725-07.

165. Boele FW, Rooney AG, Grant R, Klein M. Psychiatric symptoms in glioma patients: from diagnosis to management. *Neuropsychiatr Dis Treat.* 2015;11:1413–1420. https://doi.org/10.2147/NDT.S65874.

166. Yudofsky SC, Hales R, eds. *The American Psychiatric Publishing Textbook of Neuropsychiatry and Behavioral Neurosciences.* 5th ed. Arlington, VA: American Psychiatric Publishing Co; 2007.

167. Ownsworth T, Cubis L, Prasad T, et al. Feasibility and acceptability of a telehealth platform for delivering the Making Sense of Brain Tumour programme: a mixed-methods pilot study. *Neuropsychol Rehabil.* 2020;1–29. https://doi.org/10.1080/09602011.2020.1826331.

168. Clark MM, Rummans TA, Atherton PJ, et al. Randomized controlled trial of maintaining quality of life during radiotherapy for advanced cancer. *Cancer.* 2013;119(4):880–887. https://doi.org/10.1002/cncr.27776.

169. Rodin G, Lo C, Mikulincer M, Donner A, Gagliese L, Zimmermann C. Pathways to distress: the multiple determinants of depression, hopelessness, and the desire for hastened death in metastatic cancer patients. *Soc Sci Med.* 2009;68(3):562–569. https://doi.org/10.1016/j.socscimed.2008.10.037.

170. Lo C, Zimmermann C, Rydall A, et al. Longitudinal study of depressive symptoms in patients with metastatic gastrointestinal and lung cancer. *J Clin Oncol.* 2010;28(18):3084–3089. https://doi.org/10.1200/JCO.2009.26.9712.

171. Rodin G, Lo C, Rydall A, et al. Managing Cancer and Living Meaningfully (CALM): a randomized controlled trial of a psychological intervention for patients with advanced cancer. *J Clin Oncol.* 2018;36(23):2422–2432. https://doi.org/10.1200/JCO.2017.77.1097.

172. Breitbart W, Poppito S, Rosenfeld B, et al. Pilot randomized controlled trial of individual meaning-centered psychotherapy for patients with advanced cancer. *J Clin Oncol.* 2012;30(12):1304–1309. https://doi.org/10.1200/JCO.2011.36.2517.

173. Houmann LJ, Chochinov HM, Kristjanson LJ, Petersen MA, Groenvold M. A prospective evaluation of Dignity Therapy in advanced cancer patients admitted to palliative care. *Palliat Med.* 2014;28(5):448–458. https://doi.org/10.1177/0269216313514883.

174. Levin GT, Greenwood KM, Singh F, Tsoi D, Newton RU. Exercise improves physical function and mental health of brain cancer survivors: two exploratory case studies. *Integr Cancer Ther.* 2015;15(2):190–196. https://doi.org/10.1177/1534735415600068.

175. Armstrong TS, Ying Y, Wu J, et al. The relationship between corticosteroids and symptoms in patients with primary brain tumors: utility of the Dexamethasone Symptom Questionnaire-Chronic. *Neuro-Oncology.* 2015;17(8):1114–1120. https://doi.org/10.1093/neuonc/nov054.

176. Ness-Abramof R, Apovian CM. Drug-induced weight gain. *Drugs Today (Barc).* 2005;41(8):547–555. https://doi.org/10.1358/dot.2005.41.8.893630.

177. Armstrong TS, Vera-Bolanos E, Acquaye AA, Gilbert MR, Ladha H, Mendoza T. The symptom burden of primary brain tumors: evidence for a core set of tumor- and treatment-related symptoms. *Neuro-Oncology.* 2016;18(2):252–260. https://doi.org/10.1093/neuonc/nov166.

178. Bitterlich C, Vordermark D. Analysis of health-related quality of life in patients with brain tumors prior and subsequent to radiotherapy. *Oncol Lett.* 2017;14(2):1841–1846. https://doi.org/10.3892/ol.2017.6310.

179. Erharter A, Giesinger J, Kemmler G, et al. Implementation of computer-based quality-of-life monitoring in brain tumor outpatients in routine clinical practice. *J Pain Symptom Manag.* 2010;39(2):219–229. https://doi.org/10.1016/j.jpainsymman.2009.06.015.

180. McCall M, Leone A, Cusimano MD. Nutritional status and body composition of adult patients with brain tumours awaiting surgical resection. *Can J Diet Pract Res.* 2014;75(3):148–151. https://doi.org/10.3148/cjdpr-2014-007.

181. Ijzerman-Korevaar M, Snijders TJ, de Graeff A, Teunissen SCCM, de Vos FYF. Prevalence of symptoms in glioma patients throughout the disease trajectory: a systematic review. *J Neuro-Oncol.* 2018;140(3):485–496. https://doi.org/10.1007/s11060-018-03015-9.

182. Allen D. Dronabinol therapy: central nervous system adverse events in adults with primary brain tumors. *Clin J Oncol Nurs.* 2019;23(1):1–4. https://doi.org/10.1188/19.CJON.23-26.

183. Jacobsen PB, Jim HS. Psychosocial interventions for anxiety and depression in adult cancer patients: achievements and challenges. *CA Cancer J Clin.* 2008;58(4):214–230. https://doi.org/10.3322/ca.2008.0003.

184. National Comprehensive Cancer Network. Cancer-related fatigue. Clinical practice guidelines in oncology. *J Natl Compr Cancer Netw.* 2003;1(3):308–331.

185. Lovely M, Stewart-Amidei C, Arzbaecher J, et al. *Care of the Adult Patient With a Brain Tumor.* Chicago: American Association of Neuroscience Nurses; 2014.

186. Curt GA, Breitbart W, Cella D, et al. Impact of cancer-related fatigue on the lives of patients: new findings from the fatigue coalition. *Oncologist.* 2000;5(5):353–360. https://doi.org/10.1634/theoncologist.5-5-353.

187. Zandvoort V, Kappelle LJ, Algra A, De Haan F. Decreased capacity for mental eVort after single supratentorial lacunar infarct may aVect performance in everyday life. *J Neurol Neurosurg Psychiatry.* 1998. https://doi.org/10.1136/jnnp.65.5.697.

188. Portenoy RK, Itri LM. Cancer-related fatigue: guidelines for evaluation and management. *Oncologist.* 1999;4.

189. Lai Y-H, Chang JT-C, Keefe FJ, et al. Symptom distress, catastrophic thinking, and hope in nasopharyngeal carcinoma patients. *Cancer Nurs.* 2003;26:485–493.

190. Asher A, Fu JB, Bailey C, Hughes JK. Fatigue among patients with brain tumors. *CNS Oncol.* 2016;5(2):91–100. https://doi.org/10.2217/cns-2015-0008.

191. Armstrong TS, Cron SG, Vera Bolanos E, Gilbert MR, Kang DH. Risk factors for fatigue severity in primary brain tumor patients. *Cancer.* 2010;116(11):2707–2715. https://doi.org/10.1002/cncr.25018.

192. Lovely MP, Miaskowski C, Dodd M. Relationship between fatigue and quality of life in patients with glioblastoma multiformae. *Oncol Nurs Forum.* 1999;26(5):921–925.

193. Janda M, Steginga S, Dunn J, Langbecker D, Walker D, Eakin E. Unmet supportive care needs and interest in services among patients with a brain tumour and their carers. *Patient Educ Couns.* 2008;71(2):251–258. https://doi.org/10.1016/j.pec.2008.01.020.

194. Maqbool T, Agarwal A, Sium A, Trang A, Chung C, Papadakos J. Informational and supportive care needs of brain metastases patients and caregivers: a systematic review. *J Cancer Educ.* 2016. https://doi.org/10.1007/s13187-016-1030-5.

195. Puetz TW, Herring MP. Differential effects of exercise on cancer-related fatigue during and following treatment: a meta-analysis. *Am J Prev Med.* 2012;43(2). https://doi.org/10.1016/j.amepre.2012.04.027, e1.

196. Jeon MS, Dhillon HM, Agar MR. Sleep disturbance of adults with a brain tumor and their family caregivers: a systematic review. *Neuro-Oncology.* 2017;19(8):1035–1046. https://doi.org/10.1093/neuonc/nox019.

197. Armstrong TS, Shade MY, Breton G, et al. Sleep-wake disturbance in patients with brain tumors. *Neuro-Oncology.* 2017;19(3):323–335. https://doi.org/10.1093/neuonc/now119.

198. Jeon MS, Dhillon HM, Koh E-S, et al. Exploring sleep disturbance among adults with primary or secondary malignant brain tumors and their caregivers. *Neurooncol Pract.* 2020. https://doi.org/10.1093/nop/npaa057.

199. Jeon MS, Dhillon HM, Descallar J, et al. Prevalence and severity of sleep difficulty in patients with a CNS cancer receiving palliative care in Australia. *Neurooncol Pract.* 2019;6(6):499–507. https://doi.org/10.1093/nop/npz005.

200. Thier K, Calabek B, Tinchon A, Grisold W, Oberndorfer S. The Last 10 days of patients with glioblastoma: assessment of clinical signs and symptoms as well as treatment. *Am J Hosp Palliat Care.* 2016;33(10):985–988. https://doi.org/10.1177/1049909115609295.

201. Powell C, Guerrero D, Sardell S, et al. Somnolence syndrome in patients receiving radical radiotherapy for primary brain tumours: a prospective study. *Radiother Oncol.* 2011;100(1):131–136. https://doi.org/10.1016/j.radonc.2011.06.028.

202. Jeon MS, Dhillon HM, Koh ES, Nowak AK, Hovey E, Agar MR. Sleep disturbance in people with brain tumours and caregivers: a survey of healthcare professionals' views and current practice. *Support Care Cancer.* 2020;1–12. https://doi.org/10.1007/s00520-020-05635-2.

203. Proctor A, Bianchi MT. Clinical pharmacology in sleep medicine. *ISRN Pharmacol.* 2012;1–14. https://doi.org/10.5402/2012/914168.

204. Willis K, Ratvys S, Lanoye A, Loughan A. *Insomnia in Brain Tumor Patients: Assessing Needs in Neuro-Oncology.* Austin, TX: Society for NeuroOncology; 2020.

205. Campbell T, Garland S, Johnson J, et al. Sleeping well with cancer: a systematic review of cognitive behavioral therapy for insomnia in cancer patients. *Neuropsychiatr Dis Treat.* 2014;10:1113. https://doi.org/10.2147/NDT.S47790.

206. Garland SN, Carlson LE, Stephens AJ, Antle MC, Samuels C, Campbell TS. Mindfulness-based stress reduction compared with cognitive behavioral therapy for the treatment of insomnia comorbid with cancer: a randomized, partially blinded, noninferiority trial. *J Clin Oncol.* 2014;32(5):449–457. https://doi.org/10.1200/JCO.2012.47.7265.

207. Kushner DS, Amidei C. Rehabilitation of motor dysfunction in primary brain tumor patients†. *Neurooncol Pract.* 2015;2(4):185–191. https://doi.org/10.1093/nop/npv019.

208. Amidei C, Kushner DS. Clinical implications of motor deficits related to brain tumors†. *Neurooncol Pract.* 2015;2(4):179–184. https://doi.org/10.1093/nop/npv017.

209. Mez J, Stern RA, McKee AC. Chronic traumatic encephalopathy: where are we and where are we going? Topical collection on dementia. *Curr Neurol Neurosci Rep.* 2013;13(12):1–12. https://doi.org/10.1007/s11910-013-0407-7.

210. Chao ST, Ahluwalia MS, Barnett GH, et al. Challenges with the diagnosis and treatment of cerebral radiation necrosis. *Int J Radiat Oncol Biol Phys.* 2013;87(3):449–457. https://doi.org/10.1016/j.ijrobp.2013.05.015.

211. Windebank AJ, Grisold W. Chemotherapy-induced neuropathy. *J Peripher Nerv Syst.* 2008;13:27–46.

212. Dietrich J, Rao K, Pastorino S, Kesari S. Corticosteroids in brain cancer patients: benefits and pitfalls. *Expert Rev Clin Pharmacol.* 2014. https://doi.org/10.1586/ecp.11.1.

213. Khan F, Amatya B, Drummond K, Galea M. Effectiveness of integrated multidisciplinary rehabilitation in primary brain cancer survivors in an Australian community cohort: a controlled clinical trial. *J Rehabil Med.* 2014;46(8):754–760. https://doi.org/10.2340/16501977-1840.

214. Lovely MP, Stewart-Amidei C, Page M, et al. A new reality: long-term survivorship with a malignant brain tumor. *Oncol Nurs Forum.* 2013;40(3):267–274. https://doi.org/10.1188/13.ONF.267-274.

215. Zaini S, Guan NC, Sulaiman AH, Zainal NZ, Huri HZ, Shamsudin SH. The use of antidepressants for physical and psychological symptoms in cancer. *Curr Drug Targets.* 2018;19(12):1431–1455. https://doi.org/10.2174/1389450119666180226125026.

216. Andersen BL. Surviving cancer: the importance of sexual self-concept. *Med Pediatr Oncol.* 1999;33:15–23. https://doi.org/10.1002/(SICI)1096-911X(199907)33:1<15::AID-MPO4>3.0.CO;2-L.

217. Kieran MW, Walker D, Frappaz D, Prados M. Brain tumors: from childhood through adolescence into adulthood. *J Clin Oncol.* 2010;28(32):4783–4789. https://doi.org/10.1200/JCO.2010.28.3481.

218. Cubis L, Ownsworth T, Pinkham MB, Foote M, Legg M, Chambers S. The importance of staying connected: mediating and moderating effects of social group memberships on psychological well-being after brain tumor. *Psychooncology.* 2019;28(7):1537–1543. https://doi.org/10.1002/pon.5125.

219. Janda M, Eakin EG, Bailey L, Walker D, Troy K. Supportive care needs of people with brain tumours and their carers. *Support Care Cancer*. 2006;14(11):1094–1103. https://doi.org/10.1007/s00520-006-0074-1.

220. Cavers D, Hacking B, Erridge S, Kendall M, Morris PG, Murray SA. Social, psychological and existential well-being in patients with glioma and their caregivers: a qualitative study. *CMAJ*. 2012. https://doi.org/10.1503/cmaj.111622.

221. Sterckx W, Coolbrandt A, Clement P, et al. Living with a high-grade glioma: a qualitative study of patients' experiences and care needs. *Eur J Oncol Nurs*. 2015;19(4):383–390. https://doi.org/10.1016/j.ejon.2015.01.003.

222. Adelbratt S, Strang P. Death anxiety in brain tumour patients and their spouses. *Palliat Med*. 2000;14(6):499–507.

223. Ownsworth T, Chambers S, Hawkes A, Walker DG, Shum D. Making sense of brain tumour: a qualitative investigation of personal and social processes of adjustment. *Neuropsychol Rehabil*. 2010;21(1):117–137. https://doi.org/10.1080/09602011.2010.537073.

224. Ownsworth T. *Self-Identity After Brain Injury*. Psychology Press; 2014.

225. Mcconigley R, Halkett G. Caring for someone with high-grade glioma: a time of rapid change for caregivers. *Palliat Med*. 2010. https://doi.org/10.1177/0269216309360118.

226. Ownsworth T, Henderson L, Chambers SK. Social support buffers the impact of functional impairments on caregiver psychological well-being in the context of brain tumor and other cancers. *Psychooncology*. 2009;19(10):1116–1122. https://doi.org/10.1002/pon.1663.

227. Janda M, Steginga S, Langbecker D, Dunn J, Walker D, Eakin E. Quality of life among patients with a brain tumor and their carers. *J Psychosom Res*. 2007;63(6):617–623. https://doi.org/10.1016/j.jpsychores.2007.06.018.

228. Nezu AM, Nezu CM, Friedman SH, Faddis S, Houts PS. *Helping Cancer Patients Cope: A Problem-Solving Approach*. American Psychological Association; 2004. https://doi.org/10.1037/10283-000.

229. Schubart JR, Kinzie MB, Farace E. Caring for the brain tumor patient: family caregiver burden and unmet needs. *Neuro-Oncology*. 2008;10(1):61–72. https://doi.org/10.1215/15228517-2007-040.

230. Sherwood P, Given B, Given C, Schiffman R, Murman D, Lovely M. Caregivers of persons with a brain tumor: a conceptual model. *Nurs Inq*. 2004. https://doi.org/10.1111/j.1440-1800.2004.00200.x.

231. Boele FW, Heimans JJ, Aaronson NK, et al. Health-related quality of life of significant others of patients with malignant CNS versus non-CNS tumors: a comparative study. *J Neuro-Oncol*. 2013;115(1):87–94. https://doi.org/10.1007/s11060-013-1198-z.

232. Li Q, Lin Y, Xu Y, Zhou H. The impact of depression and anxiety on quality of life in Chinese cancer patient-family caregiver dyads, a cross-sectional study. *Health Qual Life Outcomes*. 2018;16(1):1–15. https://doi.org/10.1186/s12955-018-1051-3.

233. Baumstarck K, Leroy T, Hamidou Z, et al. Coping with a newly diagnosed high-grade glioma: patient-caregiver dyad effects on quality of life. *J Neuro-Oncol*. 2016;129(1):155–164. https://doi.org/10.1007/s11060-016-2161-6.

234. Petruzzi A, Finocchiaro CY, Lamperti E, Salmaggi A. Living with a brain tumor: reaction profiles in patients and their caregivers. *Support Care Cancer*. 2013;21(4):1105–1111. https://doi.org/10.1007/s00520-012-1632-3.

235. Coolbrandt A, Sterckx W, Clement P, et al. Family caregivers of patients with a high-grade glioma. *Cancer Nurs*. 2015;38(5):406–413. https://doi.org/10.1097/NCC.0000000000000216.

236. Lo C, Hales S, Chiu A, et al. Managing Cancer And Living Meaningfully (CALM): randomised feasibility trial in patients with advanced cancer. *BMJ Support Palliat Care*. 2016;9:209–218. https://doi.org/10.1136/bmjspcare-2015-000866.

237. Milbury K, Weathers SP, Durrani S, et al. Online couple-based meditation intervention for patients with primary or metastatic brain tumors and their partners: results of a pilot randomized controlled trial. *J Pain Symptom Manag*. 2020;59(6):1260–1267. https://doi.org/10.1016/j.jpainsymman.2020.02.004.

238. Northouse LL, Katapodi MC, Song L, Zhang L, Mood DW. Interventions with family caregivers of cancer patients: meta-analysis of randomized trials. *CA Cancer J Clin*. 2010;60(5). https://doi.org/10.3322/caac.20081.

239. Parvataneni R, Polley MY, Freeman T, et al. Identifying the needs of brain tumor patients and their caregivers. *J Neuro-Oncol*. 2011;104(3):737–744. https://doi.org/10.1007/s11060-011-0534-4.

240. Baumstarck K, Chinot O, Tabouret E, et al. Coping strategies and quality of life: a longitudinal study of high-grade glioma patient-caregiver dyads. *Health Qual Life Outcomes*. 2018;16(1):157. https://doi.org/10.1186/s12955-018-0983-y.

241. Lipsman N, Skanda A, Kimmelman J, Bernstein M. The attitudes of brain cancer patients and their caregivers towards death and dying: a qualitative study. *BMC Palliat Care*. 2007;6(1):7. https://doi.org/10.1186/1472-684X-6-7.

242. Ownsworth T, Goadby E, Chambers SK. Support after brain tumor means different things: family caregivers' experiences of support and relationship changes. *Front Oncol*. 2015;5(Feb):33. https://doi.org/10.3389/fonc.2015.00033.

243. Ng S, Herbet G, Moritz-Gasser S, Duffau H. Return to work following surgery for incidental diffuse low-grade glioma: a prospective series with 74 patients. *Neurosurgery*. 2020;87(4):720–729. https://doi.org/10.1093/neuros/nyz513.

244. Yoshida A, Motomura K, Natsume A, et al. Preoperative predictive factors affecting return to work in patients with gliomas undergoing awake brain mapping. *J Neuro-Oncol*. 2020;146(1):195–205. https://doi.org/10.1007/s11060-019-03371-0.

245. Starnoni D, Berthiller J, Idriceanu TM, et al. Returning to work after multimodal treatment in glioblastoma patients. *Neurosurg Focus*. 2018;44(6). https://doi.org/10.3171/2018.3.FOCUS1819, E17.

246. Yoshida M, Sato Y, Akagawa Y, Hiasa K. Correlation between quality of life and denture satisfaction in elderly complete denture wearers. *Int J Prosthodont*. 2001;14(1):77–80.

247. Senft C, Behrens M, Lortz I, et al. The ability to return to work: a patient-centered outcome parameter following glioma surgery. *J Neuro-Oncol*. 2020;149(3):403–411. https://doi.org/10.1007/s11060-020-03609-2.

248. Nugent BD, Weimer J, Choi CJ, et al. Work productivity and neuropsychological function in persons with skull base tumors. *Neurooncol Pract*. 2014;1(3):106–113. https://doi.org/10.1093/nop/npu015.

249. Collins C, Gehrke A, Feuerstein M. Cognitive tasks challenging brain tumor survivors at work. *J Occup Environ Med*. 2013;55(12):1426–1430. https://doi.org/10.1097/JOM.0b013e3182a64206.

250. Rusbridge SL, Walmsley NC, Griffiths SB, Wilford PA, Rees JH. Predicting outcomes of vocational rehabilitation in patients with brain tumours. *Psychooncology*. 2013;22(8):1907–1911. https://doi.org/10.1002/pon.3241.

251. Curry WT, Barker FG. Racial, ethnic and socioeconomic disparities in the treatment of brain tumors. *J Neuro-Oncol*. 2009;93(1):25–39. https://doi.org/10.1007/s11060-009-9840-5.

252. Romero HR, Lageman SK, Kamath V, et al. Challenges in the neuropsychological assessment of ethnic minorities: summit proceedings. *Clin Neuropsychol*. 2009. https://doi.org/10.1080/13854040902881958.

253. Lageman SK. Cultural diversity in neuropsychology. In: *Encyclopedia of Clinical Neuropsychology*. Springer; 2018.

254. Rivera Mindt M, Byrd D, Saez P, Manly J. Increasing culturally competent neuropsychological services for ethnic minority populations: a call to action. *Clin Neuropsychol*. 2010. https://doi.org/10.1080/13854040903058960.

255. Mukherjee D, Zaidi HA, Kosztowski T, et al. Disparities in access to neuro-oncologic care in the United States. *Arch Surg*. 2010;145(3):247–253. https://doi.org/10.1001/archsurg.2009.288.

256. Lamba N, Mehanna E, Kearney RB, et al. Racial disparities in supportive medication use among older patients with brain metastases: a population-based analysis. *Neuro-Oncology*. 2020;22(9):1339–1347. https://doi.org/10.1093/neuonc/noaa054.

257. Wilcox JA, Boire AA. Palliation for all people: alleviating racial disparities in supportive care for brain metastases. *Neuro-Oncology*. 2020;22(9):1239–1240. https://doi.org/10.1093/neuonc/noaa174.

258. Taplin SH, Anhang Price R, Edwards HM, et al. Introduction: understanding and influencing multilevel factors across the cancer care continuum. *J Natl Cancer Inst Monogr*. 2012;2012(44):2–10. https://doi.org/10.1093/jncimonographs/lgs008.

259. Colosimo K, Nissim R, Pos AE, Hales S, Zimmermann C, Rodin G. "Double awareness" in psychotherapy for patients living with advanced cancer. *J Psychother Integr*. 2018;28(2):125–140. https://doi.org/10.1037/int0000078.

260. Gofton TE, Graber J, Carver A. Identifying the palliative care needs of patients living with cerebral tumors and metastases: a retrospective analysis. *J Neuro-Oncol*. 2012;108(3):527–534. https://doi.org/10.1007/s11060-012-0855-y.

261. Song K, Amatya B, Voutier C, Khan F. Advance care planning in patients with primary malignant brain tumors: a systematic review. *Front Oncol*. 2016;6. https://doi.org/10.3389/fonc.2016.00223.

262. Faithfull S, Cook K, Lucas C. Palliative care of patients with a primary malignant brain tumour: case review of service use and support provided. *Palliat Med*. 2005. https://doi.org/10.1191/0269216305pm1068oa.

263. Walbert T. Integration of palliative care into the neuro-oncology practice: patterns in the United States. *Neurooncol Pract*. 2014;1(1):3–7. https://doi.org/10.1093/nop/npt004.

264. Crooms RC, Goldstein NE, Diamond EL, Vickrey BG. Palliative care in high-grade glioma: a review. *Brain Sci*. 2020;10(10):1–24. https://doi.org/10.3390/brainsci10100723.

265. Walbert T. Palliative care, end-of-life care, and advance care planning in neuro-oncology. *Continuum (Minneap Minn)*. 2017;23(6, Neuro-oncology):1709–1726. https://doi.org/10.1212/CON.0000000000000538.

266. Cutshall NR, Kwan BM, Salmi L, Lum HD. "It makes people uneasy, but it's necessary. #BTSM": using twitter to explore advance care planning among brain tumor stakeholders. *J Palliat Med*. 2020;23(1):121–124. https://doi.org/10.1089/jpm.2019.0077.

267. Fritz L, Dirven L, Reijneveld JC, et al. Advance care planning in glioblastoma patients. *Cancers (Basel)*. 2016;8(102):1–9. https://doi.org/10.3390/cancers8110102.

268 World Health Organization. *National Cancer Control Programmes: Policies and Managerial Guidelines*. 2nd. World Health Organization; 2002.

269. Salsman JM, Pustejovsky JE, Jim HSL, et al. A meta-analytic approach to examining the correlation between religion/spirituality and mental health in cancer. *Cancer*. 2015;121(21):3769–3778. https://doi.org/10.1002/cncr.29350.

270. Mystakidou K, Tsilika E, Parpa E, Galanos A, Vlahos L. Post-traumatic growth in advanced cancer patients receiving palliative care. *Br J Health Psychol*. 2008;13(4):633–646. https://doi.org/10.1348/135910707X246177.

271. Nixon A, Narayanasamy A. The spiritual needs of neuro-oncology patients from patients' perspective. *J Clin Nurs*. 2010;19(15–16). https://doi.org/10.1111/j.1365-2702.2009.03112.x.

272. Strang S, Strang P. Spiritual thoughts, coping and "sense of coherence" in brain tumour patients and their spouses. *Palliat Med*. 2001;15:127–134.

273. Piderman KM, Radecki Breitkopf C, Jenkins SM, et al. A chaplain-led spiritual life review pilot study for patients with brain cancers and other degenerative neurologic diseases. *Rambam Maimonides Med J*. 2015;6(2). https://doi.org/10.5041/rmmj.10199, e0015.

37

Supportive care

Alicia M. Zukas[a], *Mark G. Malkin*[b,c,d], *and Herbert B. Newton*[e,f,g,h]

[a]Department of Neurosurgery, Division of Neuro-Oncology, Hollings Cancer Center, Medical University of South Carolina, Charleston, SC, United States, [b]Neuro-Oncology Program at Massey Cancer Center, Virginia Commonwealth University School of Medicine, Richmond, VA, United States, [c]Neuro-Oncology Division, Department of Neurology, Virginia Commonwealth University School of Medicine, Richmond, VA, United States, [d]Neurology and Neurosurgery, William G. Reynolds, Jr. Chair in Neuro-Oncology, Virginia Commonwealth University School of Medicine, Richmond, VA, United States, [e]Neuro-Oncology Center, Orlando, FL, United States, [f]CNS Oncology Program, Advent Health Cancer Institute, Advent Health Orlando Campus & Advent Health Medical Group, Orlando, FL, United States, [g]Neurology & Neurosurgery (Retired), Division of Neuro-Oncology, Esther Dardinger Endowed Chair in Neuro-Oncology, James Cancer Hospital & Solove Research Institute, Wexner Medical Center at the Ohio State University, Columbus, OH, United States, [h]Neurology, UCF School of Medicine, Orlando, FL, United States

1 Introduction

The modern treatment of neuro-oncology patients usually involves a multidisciplinary team, including physicians, nurses, and support staff that specialize in various aspects of neuro-oncology, along with a Tumor Board specific for neuro-oncology patients.[1] In addition to the focus on therapeutic strategies to control tumor growth (e.g., surgical resection, radiotherapy, chemotherapy, tumor-treating fields), supportive care is fundamental for improving quality of life. Supportive care includes palliative care, which aims to help provide relief for physical, mental, social, and spiritual needs. The World Health Organization has supported that early initiation of palliative care is appropriate with the International Palliative Care Initiative.[2]

The challenge for the treatment team begins at the moment of diagnosis, when the bad news must be communicated to the patient and the family. Poor provider-patient communication may lead to distress, misunderstanding, and poor adherence with treatment. There are several important factors to be considered when breaking bad news, and it is important to note that neuro-oncology patients can receive numerous pieces of bad news. This can include the diagnosis, discussion about a new disability, and advanced care planning. The patient may also ask about the implications on their family, employment, ability to drive, and even inheritable risk. Breaking bad news includes the verbal component of the news, as well as addressing the patient's emotional response. Evidence-based delivery can help guide the challenging conversation. The ABCDE model for delivering bad news advises a stepwise approach of advance preparation, building a therapeutic relationship, communicating well, dealing with a patient's emotions and reactions, and encouraging/validating emotions.[3] The SPIKES protocol is specifically designed for disclosing bad news to a patient with cancer.[4] The six-step protocol first suggests establishing an appropriate setting for the conversation, including privacy, involvement of caregivers, and managing interruptions. Then, open the conversation by understanding the patient's perception of events. This helps determine the patient's readiness and expectations. Some patients and families may be medically literate and already realize a negative prognosis.[5] In contrast, other patients and families may be ambivalent or actively disinterested in learning about the disease or prognosis.[6] Invite the patient to learn more information and proceed accordingly. Confer the knowledge or diagnosis, assess the patient's emotions, provide empathy, and conclude with strategizing/summarizing information by including the patient in the decision making.[4] It is critical that the physician is empathic and uses simple, nontechnical language in a nonpatronizing manner. The physician should sit close to the patient and maintain good eye contact.

The role of support is crucial for neuro-oncology patients and their families; it continues for the caregivers even after the patient succumbs to disease.[7] Primary brain tumor (PBT) patients have a high prevalence of clinically significant death-related distress, and one study identified an 81% prevalence of death anxiety.[8] At the moment the patient and family hear the words "brain cancer," they enter into a crisis mode and often feel a loss of control, fear of the unknown, and sense of helplessness. To regain some aspect of control of their lives, they can form a partnership with their care team, establish a support system, and take an active role in the plan for treatment and recovery. Informational brochures and other written materials are helpful, as are the websites of organizations that provide services and resources for neuro-oncology patients and families. These organizations include the American Brain Tumor Association, the Pediatric Brain Tumor Foundation, the Brain Tumor Foundation of Canada, the National Brain Tumor Society, and The Childhood Brain Tumor Foundation. Although these groups are more focused on primary and metastatic brain tumors, additional organizations have resources that are more dedicated to patients with other neurological complications of systemic cancer, such as the American Cancer Society, Canadian Cancer Society, European Cancer Association, and the Union for International Cancer Control. During the course of a devastating illness such as an aggressive brain cancer, the patient's family will usually be the greatest source of support and comfort, as well as active caregivers in the home setting.[9,10] In this context, family members often take on the role of information seekers and patient advocates. It is important to note that family caregivers are also at risk for depression and other signs of stress, and require a strong support network to function effectively in this role.[10] Other sources of support for the patient and caregivers include the nurses of the treatment team, oncological social workers, Chaplains affiliated with the hospital or from the private sector, hospital-based support groups, and even online social networks.

The remaining sections of this chapter will review the various aspects of supportive care that may be necessary in the management of neuro-oncology patients.

1.1 Seizures and anticonvulsant therapy

Seizure activity is a frequent complication in neuro-oncology patients and compromises quality of life with seizure-related injuries, restriction of driving privileges, and anxiety related to subsequent ictal events.[11] Quality of life can be further affected by the side effects of seizure medications, drug interactions, and the economic burden of epilepsy and epilepsy drugs. Seizures occur at presentation in 20%–50% of patients with PBT and metastatic (MBT) brain tumors, as well as in patients with leptomeningeal disease and cerebrovascular complications.[11–14] In a series of 1028 patients with primary brain tumor, the prevalence of epilepsy was 85%, 69%, and 49% in patients with low-grade glioma, anaplastic glioma, and glioblastoma, respectively.[15] For patients with low-grade glioma, gross total resection is the biggest predictor of seizure freedom.[16] There may a tumor-genetic explanation for epileptogenicity, as the by-product of the mutant IDH1 enzyme, D-2-hydroxyglutarate, has been found to be epileptogenic.[17] Patients with tumors harboring this IDH mutation are twice as likely to have seizures and are more likely to be of a younger population.[17] It is important to note that more than 25% of adults between 25 and 64 years of age with newly diagnosed seizures will have an underlying brain tumor.[11] At the time of neurological disease progression, seizure activity often becomes more frequent and severe, affecting another 10%–20% of patients. Younger patients (e.g., children and young adults) tend to have a higher incidence of seizure activity when tumor involvement is supratentorial in location. In general, supratentorial tumors are most likely to cause seizures, especially when located within or near the cortex, and with proximity to the rolandic fissure and central sulcus.[16] Less common PBTs like ganglioglioma and dysembryoplastic neuroepithelial tumors have a higher incidence of epilepsy and tend to be located in the temporal lobe and paralimbic structures. Patients with low-grade glioneuronal tumors can achieve seizure freedom 80% of the time after gross total resection, regardless of the tumor's location relative to the temporal lobe.[18] Multifocal or bi-hemispheric tumors are also known to cause frequent ictal events, with metastases more likely to cause seizures if both brain and leptomeninges have disease.[19] Seizures are less common with tumors that are deep-seated or confined to the white matter. Studies indicate that tumor-associated seizures tend to originate from intact noninfiltrated neural tissue adjacent to the tumors and not from within the tumor mass.[11,14,20,21]

Patients with less infiltrative and irritative brain involvement typically manifest seizures that are equally divided between focal aware, focal impaired awareness, and focal to bilateral tonic-clonic varieties.[11–14,22] For patients with MBT, focal aware seizures are the predominant variety, with less common focal to bilateral tonic–clonic and focal impaired awareness. The neurological examination can be relatively normal in some neuro-oncology patients with seizures. However, patients with MBT are more likely to have seizure activity associated with focal neurological deficits on examination.[11,14]

The diagnosis of a seizure in a neuro-oncology patient is usually made clinically based on the history and description of the event from the patient and family members.[11–14] Testing with routine electroencephalography (EEG) may not be helpful, since only 25%–33% will demonstrate any focal interictal epileptiform activity.

Prolonged EEG monitoring (with or without a video component) may be more helpful in diagnosing seizures in subtle cases, or in order to rule out nonconvulsive status epilepticus or behavior arrest.

The presence of seizure activity does not necessarily impact the overall survival of brain tumor patients.[11] Seizures can convey both favorable and unfavorable factors. For example, a new-onset seizure may lead to a prompter workup and earlier diagnosis of a brain tumor, when the tumor burden is smaller and more amenable to therapy. An increase in seizure frequency may suggest tumor progression, reflect a change in tumor activity or hemorrhage, or be a mimic of tumor progression. It is also possible to have an increase in seizure activity at the onset of certain therapies that may cause irritation to the surrounding brain, such as at the initiation of external beam radiotherapy.

There is general consensus that any neuro-oncology patient with a well-documented, unequivocal seizure should be placed on an antiepileptic drug (AED).[11–14,20] Monotherapy tends to start with a drug that will not interact with steroids or chemotherapy, typically a nonenzyme-inducing agent. For example, phenytoin can interact with corticosteroids, as it reduces its half-life and bioavailability. Dexamethasone can induce cytochrome P450 enzymes, leading to a reduction in serum phenytoin levels. Nonenzyme-inducing drugs include levetiracetam, brivaracetam, lacosamide, topiramate, lamotrigine, gabapentin, and pregabalin. If there is concern about hematologic toxicity, which might lead to treatment delays, carbamazepine and zonisamide may be avoided. Thrombocytopenia can be seen with phenytoin or valproate. Phenytoin, carbamazepine, and valproate have relatively equivalent efficacy for reducing seizure activity.[23,24] There are no studies to suggest that one drug is superior to another for reducing seizure frequency in brain tumor patients. Monotherapy of the lowest effective dose is recommended in order to minimize drug side effects, and seizure freedom will be achieved about 50% of the time using monotherapy.[25] The use of monotherapy has also been found to improve compliance and be more cost-effective for the patient. In some patients, a second drug must be added if high therapeutic concentrations of the initial monotherapy drug are unable to control seizure activity or lead to intolerable side effects. Serum drug concentrations can be monitored for compliance purposes, adjusting for volume of distribution changes such as pregnancy, or when following a therapeutic window is appropriate, such as when using phenytoin or valproate.[26]

In the brain tumor population, seizures can remain difficult to control despite the use of AEDs. Patients who present with seizures tend to be more refractory to therapy than those who develop seizures later in the course of their disease.[11,14] Low-grade gliomas can be treatment-refractory in 15%–50% of cases.[16,27] Prognostic factors for improved seizure outcomes include a lower presurgical duration of epilepsy, gross total resection of tumor, and presence of a solitary lesion. In general, recurrent seizure activity is common, despite aggressive anticonvulsant therapy. The location of tumors near eloquent cortex reduces the chance of gross total resection and therefore seizure freedom. There is the current thought of supratotal resection for patients with refractory seizures, removing the entirety of tumor as well as nearby noneloquent cortex using functional boundaries.[28]

Reported side effects of AEDs are higher in patients with brain tumors as compared to patients without brain tumors.[29] Common side effects include slowed mentation, drowsiness, dizziness, and fatigue. These side effects can be especially challenging for patients with brain tumors, as these symptoms may be preexisting from underlying brain disease and tumor treatments. This further stresses the importance of establishing the lowest effective dose for the patient. Drug rashes can also be more common with brain tumor patients. They can arise with a reduction of steroid dose or the concomitant use of external beam radiation causing a hypersensitivity reaction. Because of severe skin reaction, the USA FDA has made a labeling change to carbamazepine's drug information. The HLA-B*1502 allele should be tested in patients with Asian ancestry prior to prescribing carbamazepine in order to reduce the risk for developing Stevens-Johnson syndrome and toxic epidermal necrolysis.[30] Oxcarbazepine is also not advised for patients intolerant to carbamazepine, with some cross-reactivity for skin reactions. Oxcarbazepine, phenytoin, and phenobarbital have also been implicated with HLA-B*1502 rash susceptibility.[31]

In general, there is no role for primary prophylaxis for seizures in a patient with a brain tumor without a history of seizures. This approach was first supported by a meta-analysis by Glantz and associates for the American Academy of Neurology.[29] Zimmerman et al. also found no role for primary seizure prophylaxis for gliomas, metastases, and meningiomas.[32] Additionally, a large Cochrane Review in 2008 found that the use of seizure prophylaxis should be made on an individual basis.[33] Data have also suggested that seizure prophylaxis could be considered in patients specifically with brain metastases from melanoma.[34] A second-generation AED provided a statistically significant reduction in seizure rate within 3 months of diagnosis in patients with no prior history of seizures.

Prophylactic anticonvulsant drugs are given at the time of craniotomy and can be discontinued typically 1–2 weeks after surgery.[29] This is based on data from several trials, one of which included 147 patients undergoing craniotomy for a supratentorial tumor.[35] Patients were randomized to either levetiracetam or phenytoin, and those who received the levetiracetam had significantly lower incidence of postoperative seizures as compared to the phenytoin group. Incidentally, the phenytoin group also experienced more drug side effects such as hepatic dysfunction or subtherapeutic levels, while patients on levetiracetam had more mood disturbances.

1.2 Corticosteroids

The use of corticosteroids is often necessary in both PMT and MBT to control symptoms caused by increased intracranial pressure (e.g., headache, nausea and emesis, confusion, weakness).[12,21,36] Peritumoral edema is the principal cause of elevated intracranial pressure and is mediated through numerous mechanisms, including the leaky neovasculature associated with tumor angiogenesis, as well as increased permeability induced by factors secreted by the tumor and surrounding tissues, such as oxygen free radicals, arachidonic acid, glutamate, histamine, bradykinin, atrial natriuretic peptide, and vascular endothelial growth factor (VEGF).[37–39] Dexamethasone is the high-potency steroid used most often to treat the edema associated with brain tumors.[14,36] It has several advantages over other synthetic glucocorticoids, including a longer half-life, reduced mineralocorticoid effect, lower incidence of cognitive and behavioral complications, and diminished inhibition of leukocyte migration.[40] The mechanisms by which dexamethasone and other glucocorticoids reduce peritumoral edema remain unclear. It is known that MBT have high concentrations of glucocorticoid receptors. The effects of these drugs on tumor-induced edema are most likely mediated through binding to these receptors, with subsequent transfer to the nucleus and the expression of novel genes.[39] In a MRI study, dexamethasone was able to induce a dramatic reduction in blood-tumor barrier permeability and regional cerebral blood volume, without significant alteration of cerebral blood flow or the degree of edema.[41] The inhibition of production and/or release of vasoactive factors secreted by tumor cells and endothelial cells, such as VEGF and prostacyclin, appears to be involved in this process.[38,39] In addition, glucocorticoids appear to inhibit the reactivity of endothelial cells to several substances that induce capillary permeability. Steroids can also induce a lymphomatous neoplasm into cell cycle arrest and cell death, which can be useful when rapid effects are desired.[42]

The exact dose of steroids necessary for each patient will vary depending on the neuro-oncological process, size and location of the tumor, if present, and amount of peritumoral edema. They can also be used to minimize side effects while undergoing radiation therapy or central nervous system symptoms during checkpoint inhibitor therapy.[43] In general, dexamethasone has become the drug of choice due to its long half-life, low mineralocorticoid

effect, and low delirium potential. Most patients with MBT will require between 4 and 12 mg of dexamethasone per day to remain clinically stable.[44] The lowest dose of steroid that can control the patient's pressure-related symptoms should be used.[14,36] This approach will minimize some of the toxicity and complications that can arise from long-term corticosteroid usage, which includes hyperglycemia, peripheral edema, proximal myopathy, gastritis, infection, osteopenia, weight gain, bowel perforation, cataracts, avascular necrosis, and psychiatric or behavioral changes (e.g., euphoria, hypomania, depression, psychosis, sleep disturbance).[14,45–50] Patients with dexamethasone-induced proximal myopathy will often improve when the dosage is reduced, but this may take months of recovery.[49,50] In addition, the proximal leg muscles can usually be strengthened if the patient is placed on a lower-extremity exercise regimen. Some authors have also reported an improvement in the myopathy when dexamethasone is replaced by an equivalent dosage of prednisone or hydrocortisone.[49,50]

The neuropsychiatric complications of steroids can affect up to 60% of patients on chronic dosing and can often be improved by dosage reduction or discontinuation of the drug.[42] When continued steroid usage is necessary, symptomatic pharmacological intervention is appropriate. For example, patients experiencing steroid-induced delirium or psychosis will often improve with a low-dose atypical antipsychotic such as olanzapine. Typical antipsychotics should be avoided in order to reduce risk for dystonic reactions. One study found that, at baseline, 50% of PBT patients have sleep dysfunction manifesting as insomnia.[51] This is likely multifactorial and related to altered brain parenchyma, radiation effects, anxiety, and concurrent psychoactive medications or substances. Steroid-induced sleep disturbances often respond to dosage reduction or by eliminating any doses after dinner, behavioral therapy, and a strict presleep routine. In refractory cases, the use of a sedative-hypnotic medication at bedtime will often be of benefit.

Corticosteroid-induced osteoporosis is a common problem, affecting 30%–50% of patients receiving treatment for a year or more.[45,52,53] Patients on long-term dexamethasone require a preventive program to minimize osteoporosis, including calcium and vitamin D supplements, and weight-bearing exercises. These measures should be started early, since bone loss is greatest in the first 2–4 months of chronic steroid treatment. For patients on long-term steroid therapy (i.e., ≥ 3 months), or in those with established osteoporosis or evidence of an osteoporotic fracture, bisphosphonate therapy (e.g., risedronate, 2.5–5.0 mg/day; alendronate, 5–10 mg/day) should be added to the regimen of calcium and vitamin D supplements.[53]

Neuro-oncology patients can be immunosuppressed for a variety of reasons, including long-term steroid use, immunomodulatory factors secreted by the tumor, and the effects of chemotherapy.[40,45] Chronic steroid usage can lead to lymphopenia, mainly through a reduction in the concentration of CD4 + T cells, and an associated increased risk of systemic infection. Studies have had mixed data for the use of pneumocystis carinii pneumonia (PCP) prophylaxis, a serious infection with a 50% case fatality rate.[54] Most PCP studies have been completed in HIV-infected patients, who have a different immunodeficiency profile than neurooncology patients. A prior study indicated that the rate of PCP in brain tumor patients was less than 1.0%.[55,56] The authors did not recommend PCP prophylaxis for every brain tumor patient on long-term steroids or chemotherapy. Rather, they suggested careful monitoring of all patients for the onset of lymphopenia, including an assessment in high-risk cases of the concentration of CD4 + T cells. A different single-institution retrospective review was completed with 240 brain tumor patients on temozolomide who did not receive PCP prophylaxis.[54] One patient of 240 (0.4%) developed PCP, but also had other signs of additional underlying immunodeficiency, and a total of 89% of the patients also had high steroid exposure. With mixed data, it can be said that for especially high-risk patients with lymphopenia and CD4 counts below 200 cells/mL, a prophylactic anti-PCP regimen should be instituted.[56] The most commonly used prophylactic antibiotic is trimethoprim-sulfamethoxazole (TMP-SMX, 160 + 800 mg), at a dose of one double-strength tablet three times per week. For patients with a sulfa allergy or deleterious interactions between TMP-SMX and other drugs (e.g., methotrexate), alternative prophylactic medications include pentamidine (300 mg/month by nebulizer) and dapsone (100 mg/day by mouth).

2 Novel antiangiogenic agents

Corticosteroids typically manage the vascular permeability and subsequent vasogenic edema associated with brain tumors. Antibodies to vascular endothelial growth factor (VEGF) and VEGF tyrosine kinase inhibitors are being increasingly used in cancer patients. They are approved for use in multiple cancer types including nonsmall-cell lung cancer, renal cell carcinoma, cervical cancer, ovarian cancer, colorectal cancer, hepatocellular carcinoma, and gastrointestinal stromal tumors. They can also be used alone or in combination with steroids in the treatment of radiation necrosis of the brain.[57] Antiangiogenic agents have steroid-sparing characteristics, which lead to a reduction of tumor-associated edema demonstrated on MRI.[58] Bevacizumab has received approval for the treatment of recurrent glioblastoma, so the use of this at disease recurrence can prompt a reduction

in the overall steroid requirement. Side effects of agents against VEGF can include gastrointestinal perforation, hemorrhage, venous thromboembolism, hypertension, impaired wound healing, and proteinuria.

2.1 Gastric acid inhibitors

The common perception is that steroid use increases the risk for peptic ulcer disease; however, a large meta-analysis and controlled trials have not shown a significant association between steroid use and peptic ulcer formation.[59] The National Institute for Health and Clinical Excellence has reported that patients with advanced cancer or poor functional status may be at a higher risk for serious gastrointestinal complications, and recommends primary prophylaxis.[59] Patients at risk should be treated prophylactically with a gastric acid inhibitor such as famotidine (20 mg po bid) or omeprazole (20–40 mg po qd).[60,61] These medications can be discontinued after the patient has been completely tapered off dexamethasone or health status improves.

2.2 Thromboembolic complications and anticoagulation

The risk of thromboembolism (i.e., deep venous thrombosis [DVT], pulmonary embolism [PE]) is high in cancer patients, with an antemortem incidence of symptomatic events approaching 15%.[62–65] However, at autopsy, the incidence rates are much higher, between 45% and 50% in some series. For patients with brain tumors, the risk for DVT and PE appears to be even higher than the general cancer population.[63,65] In the perioperative period, the overall incidence of thrombosis after brain tumor resection was 45%, as detected by ^{125}I-labeled fibrinogen scans.[64] The incidence varied depending on the tumor type and was 20% for MBT patients. Venous thromboembolic (VTE) risk continues to remain high in brain tumor patients after the perioperative period. For example, a meta-analysis of glioma patients noted a DVT incidence rate that ranged from 0.013 to 0.023 per patient-month of follow-up, corresponding to overall rates of 7%–24%.[66] The only prospective study included in the analysis followed 75 patients until death and had a DVT incidence rate of 24% (0.015 DVT/patient-month).[67] In addition to biological factors related to individual tumor histology, several clinical factors are also associated with an increased risk of DVT and PE, including arm paresis, leg paresis, history of prior DVT or PE before tumor diagnosis, concurrent medications such as VEGF inhibitors or erythropoietin-stimulating agents, and longer operative time.[65,66,68] Other factors that may also be relevant are older age, larger tumor size, and the use of other chemotherapies.

The preferred choice anticoagulation treatment is a common question for the neuro-oncology treatment team and continues to be studied in the literature. While vitamin K antagonists are still used globally, these drugs can be difficult to manage in patients with malnutrition and liver dysfunction. Two randomized clinical trials, the CANTHANOX and ONCENOX trials, found that the enoxaparin groups had lower risk of fatal bleeding as compared to warfarin.[69] In 2006, low molecular weight heparin (LMWH) was identified as standard of care for the treatment of VTE in patients with cancer. Quality of life measures were also established and identified an overall improved quality of life using LMWH over vitamin K antagonists. In comparison with unfractionated heparin, LMWHs have a more predictable anticoagulant response due to better bioavailability, a longer half-life, and more dose-dependent clearance.[70] In addition, the LMWHs can be administered subcutaneously in the outpatient setting and do not require monitoring of coagulation status. The LMWH treatment duration is for a period of 3–6 months or even indefinitely if risk factors indicate.[69] Novel oral anticoagulants have also been approved for VTE treatment; however, there are limited data available pertaining to cancer or brain tumor patients. These have a fixed oral dose, minimal food and drug interactions, low systemic side effects, and do not require routine laboratory monitoring.[69]

In general, the risk for symptomatic hemorrhage into a brain tumor is low during conservative anticoagulation, even with heparin and coumadin.[65,67,68,71] Most authors report a hemorrhage rate of 5%–7% for MBT. In order to minimize the potential for intratumoral hemorrhage, the parameters for heparin and coumadin therapy need to be very conservative if used, with PTT and PT values in the 1.5–2.0 times control range.[68] Once the patient has shifted completely over to coumadin, the INR should be maintained between 1.5 and 2.5.[65,72]

The utility of inferior vena cava (IVC) filters in brain tumor and other cancer patients remains controversial. The indications in this population are if the patient has a contraindication for anticoagulation, inability to maintain therapeutic anticoagulation, or recurrent VTE while on anticoagulation.[73] They have also failed to show a survival benefit in patients with late-stage cancer.[73] Several studies have demonstrated a significant complication rate for IVC filters in primary and metastatic brain tumor patients (range 40%–62%) and suggest that biological factors related to the tumor may be involved.[71] Complications after placement include filter thrombosis, recurrent DVT, recurrent PE, and thrombosis of the inferior vena cava. Although most authors now suggest that acute and long-term anticoagulation is superior to filter placement for brain tumor patients with DVT and/or PE, selected patients should still be considered for this approach. Patients with large regions of intratumoral

hemorrhage, impaired neurological function and excessive risk for falling episodes, and gastrointestinal bleeding could be evaluated for IVC filter placement instead of anticoagulation.

2.3 Dysphagia and swallowing disorders

Dysphagia and disorders of swallowing are common in patients with neurological disease and can be associated with stroke, multiple sclerosis, motor neuron disease, neurodegenerative disorders, and structural lesions such as a MBT or leptomeningeal disease.[74–80] Swallowing dysfunction can lead to serious morbidity from malnutrition, dehydration, and aspiration pneumonia. It is reported that dysphagia affects 63% of brain tumor patients, with a significant difference found according to tumor location.[81] Patients with infratentorial lesions have a higher proportion of dysphagia and oropharyngeal residue. The most well-described presentation involves dysfunction of the brainstem, either from compression to, or growth within, this region.[80,82,83] Tumors that can induce dysphagia in this manner include brainstem glioma, brainstem metastases, ependymoma, choroid plexus papilloma, large pineal region tumors (i.e., pinealoma, astrocytoma), and neoplasms of the cerebellopontine angle such as acoustic schwannoma and meningioma. Direct tumor compression causes impairment of the brainstem circuitry that underlies swallowing, including the nucleus tractus solitarius, ventromedial reticular formation, and cranial nerve motor efferents (V_3, VII, IX, X, XII, and ansa cervicalis).[84–86] Other reports contend that unilateral, supratentorial tumors can also cause dysphagia.[87,88] In a prospective analysis of dysphagia in PBT patients and a set of nonbrain tumor neurological controls, Newton and colleagues noted that 17 of 117 (14.5%) tumor patients complained of swallowing problems.[88] Formal swallowing assessment of the symptomatic cohort revealed that most patients significantly underestimated their degree of dysfunction. It was also noted that symptomatic patients with decreased level of alertness (LOA) were more likely to have abnormalities during bedside and videofluoroscopic testing. Twelve of the 17 symptomatic patients (70.5%) had large and diffuse, unilateral, supratentorial lesions with surrounding edema and mass effect, often associated with decreased LOA. The neuroanatomical basis for dysphagia from a unilateral lesion remains unclear. However, it is probably due to a combination of several factors, including reduced awareness of oral sensory feedback cues during mastication in patients with reduced LOA, contralateral weakness of the face and tongue, and oral apraxia with impaired motor programming ability for oral-lingual feeding behavior.

Based on the available literature, it would seem prudent to routinely screen all advanced brain tumor patients for dysphagic symptoms, especially in the latter stages of their disease, with or without reduced LOA. Another single-institution study found that 72.5% of their brain tumor patients admitted to their rehab unit had significant dysphagia.[81] Therefore, all symptomatic patients should undergo a formal swallowing evaluation, even when the complaint seems trivial.[88] The initial bedside screening examination can assess oral and laryngeal function and identify patients at risk for aspiration.[89] In addition, bedside testing can allow the modification of eating behavior to diminish the risk of aspiration. Further examination is often needed after the initial bedside evaluation to allow more detailed assessment of the swallowing mechanism, such as delays during the pharyngeal swallow, the degree of laryngeal elevation, pharyngeal symmetry, pooling or coating of pharyngeal recesses, and silent aspiration. The modified barium swallow is used for this assessment and can accurately reveal the abnormalities of the swallowing mechanism, the degree of aspiration, and how best to modify the diet.[89–91]

Management of dysphagic brain tumor patients can often be a complex issue. Patients must be able to demonstrate the necessary cognitive and communication skills to actively participate in a swallowing management program.[89,90] Tumor patients with diminished LOA or significant cognitive alterations may be unable to pursue complex rehabilitation strategies similar to those used for patients with other neurologic disorders (e.g., stroke). In those patients with adequate LOA, swallowing rehabilitation should be pursued. If compensatory techniques do not improve oral efficiency, an alternate route of nutrition may be required, such as a gastric feeding tube.

2.4 Hospice care and end-of-life issues

The Hospice movement originated in England in the 1960s when Dr. Cicely Saunders founded the first multidisciplinary hospice to care for terminally ill patients.[92–94] The movement expanded and eventually spread to the United States in the 1970s. In England, most of the Hospice care was administered within inpatient facilities, while in the United States, the care was shifted, whenever possible, to the home setting. Data would suggest that the quality of Hospice care is superior to the care available in the non-Hospice setting for patients in the end stages of life.[95] The improved quality of care was equivalent whether the patient was in a hospital-based unit, inpatient nonhospital-based unit, or home Hospice setting. Because of the success of the early Hospice programs, the Congress passed legislation resulting in the establishment of the Medicare Hospice Benefit in 1982.[93,94] The Medicare Hospice Benefit subsidizes care for terminally ill patients with a life expectancy

of 6 months or less, as certified by their attending physician and the Hospice medical director. In addition to the life expectancy criteria, other qualifications for Medicare Hospice Benefit include eligibility for Medicare (i.e., at least 65 years of age or certified as disabled), foregoing further aggressive or "curative" therapy, being able to receive most care in the home, and having a primary caregiver present at home. The Medicare Benefit will continue to pay for patients who live longer than 6 months, as long as the attending physician continues to certify that the patient is terminally ill. Most patients admitted to Hospice do not live longer than the typical 6 months allowed by the Medicare Benefit. In fact, the majority of them die within 4–6 weeks, suggesting that physicians are not referring their patients to Hospice care early enough.[94] Patients are dying pursuing aggressive treatments in the hospital and 56% of patients die within 10 days of entering hospice.[96] Still, more than one half of all terminally ill cancer patients in the United States are not being offered Hospice services at all or are entering Hospice care too late to achieve maximal benefit.[93,96]

For MBT and other neuro-oncology patients for whom continued treatment is futile, the most important and critical step will be to broach the subject of hospice care and palliative symptom control.[97–99] Similar to when the physician first tells the patient their diagnosis, this must be done with the utmost compassion and sensitivity. Patients should be referred to Hospice as soon as curative or stabilizing therapy has been discontinued and the primary goal becomes comfort. Hospice care works best when there is time for members of the Hospice team to develop meaningful relationships with the patient and family (i.e., over weeks to months).

The purpose of Hospice care is to provide medical, psychosocial, and spiritual support for terminally ill patients and their families.[92–97] This is an especially critical time when hope for cure is lost and anxiety about the future is overwhelming. Hospice care attempts to alleviate the physical and emotional suffering of the neuro-oncology patient. Hope for cure is shifted to hope for maximizing dignity, comfort, quality of life (QOL), and the process of enjoying each remaining day to its fullest. In addition, support is provided for the family members, who are also suffering and attempting to cope with the imminent loss of their loved one.[98] One of the most common fears about advanced incurable cancer is isolation from family and loved ones. The presence of the Hospice care team, especially in the home setting, alleviates this fear and ensures that isolation and loneliness are minimized.

The use of Hospice is an important aspect of the complete care of virtually all patients with neuro-oncologic disease. It should be offered to all patients in the terminal stages of their disease once efforts to halt the progression of the malignancy have ceased. In addition to expertise at treating many aspects of cancer-related pain, Hospice physicians and nurses can also effectively manage some of the problems specific to MBT patients, such as swallowing difficulties and seizures.[99] During the terminal phases of disease, many patients cannot take oral anticonvulsants properly, either due to neurological compromise of the swallowing mechanism or due to a reduced level of consciousness. It is often helpful for the caregiver to provide intranasal, buccal, sublingual, or intramuscular administration—these have been found to be as effective as intravenous or rectal routes.[100] Lorazepam liquid is stable for nearly a month at room temperature and midazolam is stable for 60 days. These can be kept at the patient's bedside and used quickly as needed. Patients with respiratory compromise and dyspnea may benefit from the use of opioids, with the dose depending on the level of prior exposure to this class of medications.[101]

2.5 Ethical issues

The care of neuro-oncology patients is often complicated by numerous ethical dilemmas and discussions.[102–105] In no other subspecialty of medicine are there such large numbers of seriously ill patients in which the day-to-day care involves life and death decisions. These patients are not only adversely affected by their disease, but frequently suffer deleterious side effects and complications from treatment, which is often very intense and rigorous. Physicians caring for neuro-oncologic patients should be well versed in ethical principles and theory. This foundation will better prepare the physician for the many complex ethical predicaments that inevitably develop during the course of therapy and, in many cases, subsequent palliative care.

There are several basic ethical principles that require definition. The most important ethical principles are respect for autonomy, justice, beneficence, and nonmaleficence.[102–104] *Respect for Autonomy* refers to recognition by the physician of the patient's right and ability to make his own decisions. These decisions are unique, are influenced by the patient's value system, and may differ from what is advised by the physician. *Justice* relates to fairness and what people are legitimately entitled to once they enter the medical system. In this context, justice demands that patients with neuro-oncologic disease have access to care (e.g., treatment, pain control, nutritional support) equal to patients with other diseases that may have a less grave prognosis. *Beneficence* refers to actions by the physician toward the patient that will maximize positive outcomes and avoid unnecessary pain, injury, and suffering. These activities can include treatment of the cancer and extension of quality survival, control of pain and other disease-related symptoms, and interpersonal support. *Nonmaleficence* means that the physician

should "do no harm" while providing care to the patient. This principle has a broad scope and can refer to many issues, including withholding relevant diagnostic or prognostic information, improper treatment of pain, inappropriate undertreatment, and persistent overtreatment.

Physicians usually have an ethical position or frame of reference that incorporates these basic ethical principles. The most common ethical stance is that in which the physician makes a decision based on an assessment of the good or bad consequences of each course of action. This ethical position, called *consequentialism* or *utilitarianism*, justifies a given decision by comparing probable good or benefit with potential harm or pain.[106] The second most common ethical position, *respect for persons*, relies heavily on the ethical principles of autonomy and respect.[102] This approach emphasizes the importance of allowing patients to be involved in all decisions about their care and treatment. An alternative to *respect for persons* is *paternalism*, in which the physician assumes that all decisions should be made for the good of the patient, without regard to his or her specific wishes or needs.[102–104,106]

It is often difficult to be honest with a patient when discussing a new diagnosis as devastating as cancer, especially when it is a MBT or similar neuro-oncologic problem.[104,107,108] In fact, a survey of ethical issues in the oncology literature determined that *truth-telling* was the most commonly debated subject.[106] Between 1961 and 1979, most physicians took the paternalistic approach and withheld information regarding diagnosis and prognosis in order to maintain hope and minimize psychological damage to their patients. Since 1979, attitudes have changed so that many physicians now prefer to reveal accurate information about their patient's diagnosis and prognosis.[107–109] This trend away from paternalism, toward a more "patient-oriented" or "respect for persons" approach when discussing diagnosis and prognosis, is important since the vast majority of patients want to know as much as possible about their disease, treatment options, and chances of survival.[108]

The physician caring for a neuro-oncology patient needs to balance the ethically appropriate duty to convey accurate information about diagnosis and prognosis with the equally important responsibility to nurture and maintain hope. However, it is now clear that a more honest and accurate diagnostic interview does not remove hope and is more likely to strengthen the physician-patient relationship.[108,110] The physician should explore what hope means to each patient, since it can represent many different things, some of which will be separate from the hope for cure or lengthy survival.

Ethical issues arise frequently during the design and administration of clinical trials for oncology patients. The moral cornerstone of any clinical trial is the concept and practice of valid informed consent.[108] Valid consent has three features: the provision of adequate information, the absence of coercion, and the competence of the patient. Adequate information must be provided about tests, procedures, and treatments inherent in any clinical trial. Significant risks and benefits, if any, must be outlined. All serious risks that are likely to occur should be included. Any risk of death beyond a trivial risk should also be included, because death is such a serious evil that the patient must be made aware of any chance that it may occur.[111] Rational alternative treatment to the clinical trial in question must be presented by the physician, in an open, objective, and unbiased manner. Alternative treatments should include those offered at other medical centers. One of the duties of the physician is to inform the patient of the consequences and probable outcome with no treatment at all. The patient should know that the final decision concerning any clinical trial is his or hers alone to make. Competency, in this setting, implies that the patient understands the information provided during the consent process and appreciates that it applies to him/her at that particular point in time.

The protection of the patient's best interests falls squarely and heavily on the shoulders of the physician seeking participation in the clinical trial.[105] The physician must take into account the influence of personal beliefs, biases, and academic ambitions before embarking on such endeavors. The focus of the physician who designs and performs clinical trials must always be on the need for conclusive proof of efficacy. A study designed according to rigorous scientific and ethical criteria can accrue patients with confidence and good faith.

The decision to stop therapy is often very difficult for patients, family members, and the treatment team.[102–104,112,113] It signals the "beginning of the end," when all reasonable hope for cure or prolonged stabilization is gone and the patient's death is imminent. These decisions usually arise when the patient has just progressed through the latest protocol and has often suffered further neurological deterioration. In many cases, the neurological status is quite poor, to a degree that functional ambulation, cognition, and verbal interaction are severely compromised. Although there are often other treatment options that could be offered, the physician must state clearly and honestly that further therapy will not significantly affect outcome. In this situation, it is critical to weigh the adverse effects of further therapy on QOL against the potential benefits for improvement of QOL and prolongation of survival, which would be extremely limited. The physician must reassure the patient and family that the termination of active treatment does not mean the physician will abandon them. Even though the focus of subsequent care will shift to comfort, pain relief, and symptom control, the physician will remain actively involved in the patient's care. In addition to questions about the potential for

extension of survival, many patients and family members want to know if further treatment might improve neurological function. In other words, could the patient's current neurological status be reversed somewhat to enhance QOL for the time they have left? Neurological function is rarely restored or improved at these late stages of disease; it would be optimistic even to expect further therapy to stabilize the patient's condition.

Because QOL is so subjective and the behavior of neuro-oncologic disease can be so variable, the proper time to stop treatment will differ from patient to patient. Some patients accustomed to a high level of function cannot tolerate living their life in a severely compromised fashion while suffering the rigors of treatment. For others, the alterations of function and lifestyle are more easily accepted, so that simple survival is adequate, with less regard for the quality of existence.

Is it ethically appropriate to terminate therapy? If the physician has explained the situation properly and is acting in accordance with the wishes of the patient or family, the decision would be consistent with the principles of respect for autonomy, beneficence, and nonmaleficence.[102–104,112,113] The physician would be acting to allow a more dignified, peaceful death without the rigors of active therapy. Active treatment is terminated to "promote the good," which is to let the patient die on their own terms. It would be ethically improper and contrary to the principle of nonmaleficence for the physician to coerce or force the patient into undergoing further therapy.

3 Conclusion

Although the focus of the Treatment Team will be on curative or stabilizing therapy for most patients, it will still be very important for the treating physician to be aware of the many aspects of supportive care outlined above. Common problems related to seizure control, toxicity of anticonvulsants, prophylaxis and treatment of thromboembolic complications, and corticosteroids must be assiduously monitored in every patient. As each patient enters the final stages of their disease, the physician must also be aware of end-of-life issues and the appropriate utilization of Hospice resources. All of these aspects of care should be performed in the context of proper ethical conduct and under the principles of Respect for Autonomy and Beneficence.

References

1. Ruhstaller T, Roe H, Thurlimann B, Nicoll JJ. The multidisciplinary meeting: an indispensable aid to communication between different specialties. *Eur J Cancer.* 2006;42(15):2459–2462.
2. Callaway MV, Connor SR, Foley KM. World Health Organization public health model: a roadmap for palliative care development. *J Pain Symptom Manage.* 2018;55(2S):S6–S13.

3. VandeKieft GK. Breaking bad news. *Am Fam Physician*. 2001;64(12):1975–1978.

4. Baile WF, Buckman R, Lenzi R, Glober G, Beale EA, Kudelka AP. SPIKES—a six-step protocol for delivering bad news: application to the patient with cancer. *Oncologist*. 2000;5(4):302–311.

5. Back AL, Arnold RM. Discussing prognosis: "how much do you want to know?" Talking to patients who are prepared for explicit information. *J Clin Oncol*. 2006;24:4209–4213.

6. Back AL, Arnold RM. Discussing prognosis: "how much do you want to know?" Talking to patients who do not want information or who are ambivalent. *J Clin Oncol*. 2006;24:4214–4217.

7. Feldman GB. The role of support in treating the brain tumor patient. In: Black PM, Loeffler JS, eds. *Cancer of the Nervous System*. vol. 17. Cambridge: Blackwell Science; 1997:335–345.

8. Loughan AR, Aslanzadeh FJ, Brechbiel J, et al. Death-related distress in adult primary brain tumor patients. *Neurooncol Pract*. 2020;7(5):498–506.

9. Given BA, Given CW, Kozachik S. Family support in advanced cancer. *CA Cancer J Clin*. 2001;51:213–231.

10. Nijboer C, Tempelaar R, Triemstra M, van den Bos GAM, Sanderman R. The role of social and psychologic resources in caregiving of cancer patients. *Cancer*. 2001;91:1029–1103.

11. Glantz M, Recht LD. Epilepsy in the cancer patient. In: Vecht CJ, ed. *Handbook of Clinical Neurology*. Amsterdam: Elsevier Science; 1997:9–18. Neuro-Oncology, Part III; vol. 25. 69. 2.

12. Chamberlain MC. Neoplastic meningitis. *J Clin Oncol*. 2005;23:3605–3613.

13. Rogers LR. Cerebrovascular complications in cancer patients. *Neurol Clin North Am*. 2003;21:167–192.

14. Newton HB. Neurological complications of systemic cancer. *Am Fam Physician*. 1999;59:878–886.

15. Lote K, Stenwig AE, Skullerud K, Hirschberg H. Prevalence and prognostic significance of epilepsy in patients with gliomas. *Eur J Cancer*. 1998;34(1):98.

16. Ruda R, Trevisan E, Soffietti R. Epilepsy and brain tumors. *Curr Opin Oncol*. 2010;22:611.

17. Chen H, Judkins J, Thomas C, et al. Mutant IDH1 and seizures in patients with glioma. *Neurology*. 2017;88(19):1805–1813.

18. Englot DJ, Berger MS, Barbaro NM, Chang EF. Factors associated with seizure freedom in the surgical resection of glioneuronal tumors. *Epilepsia*. 2012;53(1):51–57.

19. Oberndorfer S, Schmal T, Lahrmann H, et al. The frequency of seizures in patients with primary brain tumors or cerebral metastases. An evaluation from the Ludwig Boltzmann Institute of Neuro-Oncology and Department of Neurology. *Wien Klin Wochenschr*. 2002;114(21–22):911–916.

20. Schaller B, Rüegg SJ. Brain tumor and seizures: pathophysiology and its implications for treatment revisited. *Epilepsia*. 2003;44:1223–1232.

21. Ettinger AB. Structural causes of epilepsy: tumors, cysts, stroke, and vascular malformation. *Neurol Clin*. 1994;12:41–56.

22. Fisher RS, Cross JH, French JA, et al. Operational classification of seizure types by the International League Against Epilepsy: Position Paper of the ILAE Commission for Classification and Terminology. *Epilepsia*. 2017;58(4):522–530.

23. Brodie MJ, Dichter MA. Antiepileptic drugs. *N Engl J Med*. 1996;334:168–175.

24. Britton JW, So EL. Selection of antiepileptic drugs: a practical approach. *Mayo Clin Proc*. 1996;71:778–786.

25. Van Breemen MS, Rijsman RM, Taphoorn MJ, et al. Efficacy of anti-epileptic drugs in patients with gliomas and seizures. *J Neurol*. 2009;256(9):1519–1526.

26. St Louis EK. Monitoring antiepileptic drugs: a level-headed approach. *Curr Neuropharmacol*. 2009;7(2):115–119.

27. Vecht CJ, Kerkhof M, Duran-Pena A. Seizure prognosis in brain tumors: new insights and evidence-based management. *Oncologist*. 2014;19(7):751–759.

28. Duffau H. Surgery of low-grade gliomas: towards a 'functional neurooncology'. *Curr Opin Oncol*. 2009;21(6):543–549.

29. Glantz MJ, Cole BF, Forsyth PA, et al. Practice parameter: anticonvulsant prophylaxis in patients with newly diagnosed brain tumors. Report of the Quality Standards Subcommittee of the American Academy of Neurology. *Neurology*. 2000;54(10):1886–1893.

30. Ferrell Jr PB, McLeod HL. Carbamazepine, HLA-B*1502 and risk of Stevens-Johnson syndrome and toxic epidermal necrolysis: US FDA recommendations. *Pharmacogenomics*. 2008;9(10):1543–1546.

31. Sun D, Yu CH, Liu ZS, et al. Association of HLA-B*1502 and *1511 allele with antiepileptic drug-induced Stevens-Johnson syndrome in central China. *J Huazhong Univ Sci Technolog Med Sci*. 2014;34(1):146–150.

32. Sirven JI, Wingerchuk DM, Drazkowski JF, Lyons MK, Zimmerman RS. Seizure prophylaxis in patients with brain tumors: a meta-analysis. *Mayo Clin Proc*. 2004;79(12):1489–1494.

33. Tremont-Lukats IW, Ratilil BO, Armstrong T, Gilbert MR. Antiepileptic drugs for preventing seizures in people with brain tumors. *Cochrane Database Syst Rev*. 2008;2, CD004424.

34. Goldlust SA, Hsu M, Lassman AB, Panageas KS, Avila EK. Seizure prophylaxis and melanoma brain metastases. *J Neurooncol*. 2012;108(1):109–114.

35. Wu AS, Trinh VT, Suki D, et al. A prospective randomized trial of perioperative seizure prophylaxis in patients with intraparenchymal brain tumors. *J Neurosurg*. 2013;118(4):873–883.

36. Newton HB, Turowski RC, Stroup TJ, McCoy LK. Clinical presentation, diagnosis, and pharmacotherapy of patients with primary brain tumors. *Ann Pharmacother*. 1999;33:816–832.

37. Ohnishi T, Sher PB, Posner JB, Shapiro WB. Capillary permeability factor secreted by malignant brain tumor. Role in peritumoral edema and possible mechanism for anti-edema effect of glucocorticoids. *J Neurosurg*. 1990;72:245–251.

38. Del Maestro RF, Megyesi JF, Farrell CL. Mechanisms of tumor-associated edema: a review. *Can J Neurol Sci*. 1990;17:177–183.

39. Samdani AF, Tamargo RJ, Long DM. Brain tumor edema and the role of the blood-brain barrier. In: Vecht CJ, ed. *Handbook of Clinical Neurology*. Amsterdam: Elsevier Science; 1997:71–102. Neuro-Oncology, Part I; vol. 23. 67. 4.

40. Mukwaya G. Immunosuppressive effects and infections associated with corticosteroid therapy. *Pediatr Infect Dis J*. 1988;7:499–504.

41. Østergaard L, Hochberg FH, Rabinov JD, et al. Early changes measured by magnetic resonance imaging in cerebral blood flow, blood volume, and blood-brain barrier permeability following dexamethasone treatment in patients with brain tumors. *J Neurosurg*. 1999;90:300–305.

42. Roth P, Happold C, Weller M. Corticosteroid use in neuro-oncology: an update. *Neurooncol Pract*. 2015;2(1):6–12.

43. Petrelli F, Signorelli D, Ghidini M, et al. Association of steroids use with survival in patients treated with immune checkpoint inhibitors: a systematic review and meta-analysis. *Cancer*. 2020;12(3):546.

44. Dietrich J, Rao K, Pastorino S, Kesari S. Corticosteroids in brain cancer patients: benefits and pitfalls. *Expert Rev Clin Pharmacol*. 2011;4(2):233–242.

45. Lester RS, Knowles SR, Shear NH. The risks of systemic corticosteroid use. *Dermatol Clin*. 1998;16:277–286.

46. Weissman DE, Dufer D, Vogel V, Abeloff DD. Corticosteroid toxicity in neuro-oncology patients. *J Neurooncol*. 1987;5:125–128.

47. Fadul CE, Lemann W, Thaler HT, Posner JB. Perforation of the gastrointestinal tract in patients receiving steroids for neurologic disease. *Neurology*. 1988;38:348–352.

48. Stiefel FC, Breitbart WS, Holland JC. Corticosteroids in cancer: neuropsychiatric complications. *Cancer Invest*. 1989;7:479–491.

49. Dropcho EJ, Soong SJ. Steroid-induced weakness in patients with primary brain tumors. *Neurology*. 1991;41:1235–1239.

50. Batchelor TT, Taylor LT, Thaler HT, Posner JB, DeAngelis LM. Steroid myopathy in cancer patients. *Neurology.* 1997;48:1234–1238.

51. Wellisch DK, Kaleita TA, Freeman D, Cloughesy T, Goldman J. Predicting major depression in brain tumor patients. *Psychooncology.* 2002;11(3):230–238.

52. Joseph JC. Corticosteroid-induced osteoporosis. *Am J Hosp Pharm.* 1994;51:188–197.

53. McIlwain HH. Glucocorticoid-induced osteoporosis: pathogenesis, diagnosis, and management. *Prev Med.* 2003;36:243–249.

54. Neuwelt AJ, Nguyen TM, Fu R, et al. Incidence of *Pneumocystis jirovecii* pneumonia after temozolomide for CNS malignancies without prophylaxis. *CNS Oncol.* 2014;3(4):267–273.

55. Mahindra AK, Grossman SA. *Pneumocystis carinii* pneumonia in HIV negative patients with primary brain tumors. *J Neurooncol.* 2003;63:263–270.

56. Mathew BS, Grossman SA. Pneumocystis carinii pneumonia prophylaxis in HIV negative patients with primary CNS lymphoma. *Cancer Treat Rev.* 2003;29:105–119.

57. Zhuang H, Shi S, Yuan Z, Chang JY. Bevacizumab treatment for radiation brain necrosis: mechanism, efficacy and issues. *Mol Cancer.* 2019;18(1):21.

58. Brastianos PK, Batchelor TT. VEGF inhibitors in brain tumors. *Clin Adv Hematol Oncol.* 2009;7(11):753–760.

59. Liu D, Ahmet A, Ward L, et al. A practical guide to the monitoring and management of the complications of systemic corticosteroid therapy. *Allergy Asthma Clin Immunol.* 2013;9(1):30.

60. Garnett WR, Garabedian-Ruffalo SM. Identification, diagnosis, and treatment of acid-related diseases in the elderly: implications for long-term care. *Pharmacotherapy.* 1997;17:938–958.

61. Sachs G. Proton pump inhibitors and acid-related diseases. *Pharmacotherapy.* 1997;17:22–37.

62. Lee AYY, Levine MN. Management of venous thromboembolism in cancer patients. *Oncologia.* 2000;14:409–421.

63. Gomes MPV, Deitcher SR. Diagnosis of venous thromboembolic disease in cancer patients. *Oncologia.* 2003;17:126–139.

64. Sawaya R, Zuccarello M, Elkalliny M, Nighiyama H. Postoperative venous thromboembolism and brain tumors: part I. Clinical profile. *J Neuro-Oncol.* 1992;14:119–125.

65. Hamilton MG, Hull RD, Pineo GF. Venous thromboembolism in neurosurgery and neurology patients: a review. *Neurosurgery.* 1994;34:280–296.

66. Marras LC, Geerts WH, Perry JR. The risk of venous thromboembolism is increased throughout the course of malignant glioma. An evidence-based review. *Cancer.* 2000;89:640–646.

67. Brandes AA, Scelzi E, Salmistraro E, et al. Incidence and risk of thromboembolism during treatment of high-grade gliomas: a prospective study. *Eur J Cancer.* 1997;33:1592–1596.

68. Quevedo JF, Buckner JC, Schmidt JL, Dinapoli RP, O'Fallon JR. Thromboembolism in patients with high-grade glioma. *Mayo Clin Proc.* 1994;69:329–332.

69. Barbosa M. What is the best treatment for a cancer patient with thrombosis? *Clin Med Insights Oncol.* 2014;8:49–55.

70. Weitz JI. Low-molecular-weight heparins. *N Engl J Med.* 1997;337:688–698.

71. Schiff D, DeAngelis LM. Therapy of venous thromboembolism in patients with brain metastases. *Cancer.* 1994;73:493–498.

72. Altshuler E, Moosa H, Selker RG, Vertosick FT. The risk and efficacy of anticoagulant therapy in the treatment of thromboembolic complications in patients with primary brain tumors. *Neurosurgery.* 1990;27:74–77.

73. Pandhi MB, Desai KR, Ryu RK, Lewandowski RJ. The role of inferior vena cava filters in cancer patients. *Semin Interv Radiol.* 2016;33(2):71–74.

74. Buchholz D. Neurologic causes of dysphagia. *Dysphagia.* 1987;1:152–156.

75. Kirshner HS. Causes of neurogenic dysphagia. *Dysphagia.* 1989;3:184–188.

76. Barer DH. The natural history and functional consequences of dysphagia after hemispheric stroke. *J Neurol Neurosurg Psychiatry.* 1989;52:236–241.

77. Lieberman AN, Horowitz L, Redmond P, Pachter L, Lieberman I, Leibowitz M. Dysphagia in Parkinson's disease. *Am J Gastroenterol.* 1980;74:157–160.

78. Daly DD, Code CF, Anderson HA. Disturbances of swallowing and esophageal motility in patients with multiple sclerosis. *Neurology.* 1962;59:250–256.

79. Robbins J. Swallowing in ALS and motor neuron disease. *Neurol Clin.* 1987;5:213–229.

80. Buchholz D. Neurologic evaluation of dysphagia. *Dysphagia.* 1987;1:187–192.

81. Park DH, Chun MH, Lee SJ, Song YM. Comparison of swallowing functions between brain tumor and stroke patients. *Ann Rehabil Med.* 2013;37(5):633–641.

82. Frank Y, Schwartz SB, Epstein NE, Beresford HR. Chronic dysphagia, vomiting and gastroesophageal reflux as manifestations of a brain stem glioma: a case report. *Pediatr Neurosci.* 1989;15:265–268.

83. Straube A, Witt TN. Oculo-bulbar myasthenic symptoms as the sole sign of tumour involving or compressing the brain stem. *J Neurol.* 1990;237:369–371.

84. Cunningham ET, Donner MW, Jones B, Point SM. Anatomical and physiological overview. In: Jones B, Donner MW, eds. *Normal and Abnormal Swallowing. Imaging in Diagnosis and Therapy.* vol. 2. New York: Springer-Verlag; 1991:7–32.

85. Dodds WJ, Stewart ET, Logemann JA. Physiology and radiology of the normal oral and pharyngeal phases of swallowing. *Am J Roentgenol.* 1989;154:953–963.

86. Sessle BJ, Henry JL. Neural mechanisms of swallowing: neurophysiological and neurochemical studies on brain stem neurons in the solitary tract region. *Dysphagia.* 1989;4:61–75.

87. Meadows JC. Dysphagia in unilateral cerebral lesions. *J Neurol Neurosurg Psychiatry.* 1973;36:853–860.

88. Newton HB, Newton C, Pearl D, Davidson T. Swallowing assessment in primary brain tumor patients with dysphagia. *Neurology.* 1994;44:1927–1932.

89. Emick-Herring B, Wood P. A team approach to neurologically based swallowing disorders. *Rehabil Nurs.* 1990;15:126–132.

90. Dodds WJ, Logemann JA, Stewart ET. Radiologic assessment of abnormal oral and pharyngeal phases of swallowing. *Am J Roentgenol.* 1990;154:965–974.

91. Bloch AS. Nutritional management of patients with dysphagia. *Oncologia.* 1993;7:127–137.

92. Rhymes J. Hospice care in America. *JAMA.* 1990;264:369–372.

93. Kinzbrunner BM. Ethical dilemmas in hospice and palliative care. *Support Care Cancer.* 1995;3:28–36.

94. Von Gunten CF, Neely KJ, Martinez J. Hospice and palliative care: program needs and academic issues. *Oncologia.* 1996;10:1070–1074.

95. Finlay IG, Higginson IJ, Goodwin DM, et al. Palliative care in hospital, hospice, at home: results from a systematic review. *Ann Oncol.* 2002;13:257–264.

96. Mulville AK, Widick NN, Srivastava Makani N. Timely referral to hospice care for oncology patients: a retrospective review. *Am J Hosp Palliat Care.* 2019;36(6):466–471.

97. Kahn MJ, Lazarus CJ, Owens DP. Allowing patients to die: practical, ethical, and religious concerns. *J Clin Oncol.* 2003;21:3000–3002.

98. Brenner PR. Managing patients and families at the ending of life: hospice assumptions, structures, and practice in response to staff stress. *Cancer Invest.* 1997;15:257–264.

99. D'Olimpio J. Contemporary drug therapy in palliative care: new directions. *Cancer Invest.* 2001;19:413–423.

100. Gronheit W, Popkirov S, Wehner T, Schlegel U, Wellmer J. Practical management of epileptic seizures and status epilepticus in adult palliative care patients. *Front Neurol.* 2018;9:595.

101. Thomas JR, Von Gunten CF. Treatment of dyspnea in cancer patients. *Oncologia.* 2002;16:745–750.

102. Latimer E. Ethical challenges in cancer care. *J Palliat Care.* 1992;8:65–70.

103. Smith TJ, Bodurtha JN. Ethical considerations in oncology: balancing the interests of patients, oncologists, and society. *J Clin Oncol.* 1995;13:2464–2470.

104. Newton HB, Malkin MG. Ethical issues in neuro-oncology. *Semin Neurol.* 1997;17:219–226.

105. Vick NA, Wilson CB. Total care of the patient with a brain tumor. With considerations of some ethical issues. *Neurol Clin.* 1985;3:705–710.

106. Vanderpool HY, Weiss GB. Ethics and cancer: a survey of the literature. *South Med J.* 1987;80:500–506.

107. Gert B, Culver CM. Moral theory in neurologic practice. *Semin Neurol.* 1984;4:9–14.

108. Butow PN, Kazemi JN, Beeney LJ, Griffin AM, Dunn SM, Tattersall MHN. When the diagnosis is cancer. Patient communication experiences and preferences. *Cancer.* 1996;77:2630–2637.

109. Gert B, Nelson WA, Culver CM. Moral theory and neurology. *Neurol Clin.* 1989;7:681–696.

110. Clayton JM, Butow PN, Arnold RM, Tattersall MHN. Fostering coping and nurturing hope when discussing the future with terminally ill cancer patients and their caregivers. *Cancer.* 2005;103:1965–1975.

111. Culver CM, Gert B. Basic ethical concepts in neurologic practice. *Semin Neurol.* 1984;4:1–8.

112. Thomasma DC. The ethics of caring for the older patient with cancer: defining the issues. *Oncologia.* 1992;6:124–130.

113. Nelson WA, Bernat JL. Decisions to withhold or terminate treatment. *Neurol Clin.* 1989;7:759–774.

Index

Note: Page numbers followed by *f* indicate figures and *t* indicate tables.